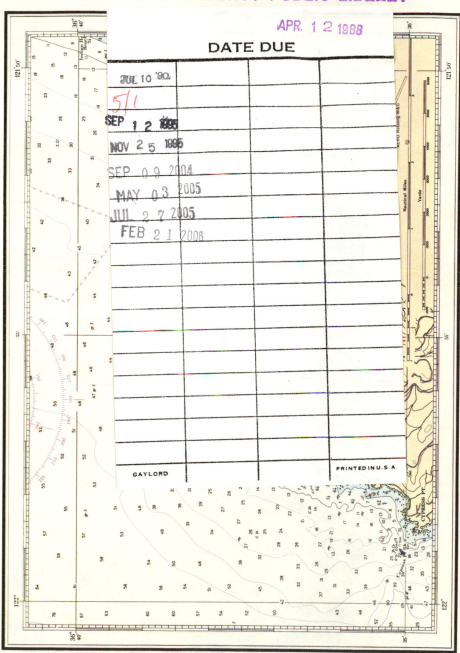

Hydrographic chart. (*Courtesy National Ocean Survey.*)

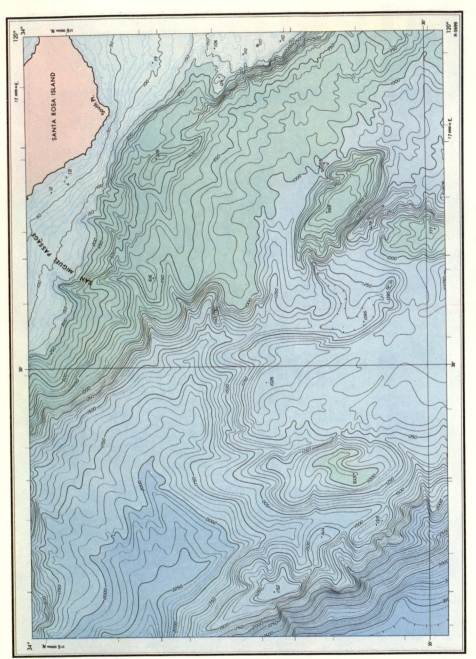

Bathymetric map. (*Courtesy National Ocean Survey.*)

# SURVEYING
# Theory and Practice
## SIXTH EDITION

**Raymond E. Davis**, M.S., C.E., D.Eng.
Late Professor of Civil Engineering
University of California

**Francis S. Foote**, E.M.
Late Professor of Railroad Engineering
University of California

**James M. Anderson**, Ph.D.
Professor of Civil Engineering
University of California, Berkeley

**Edward M. Mikhail**, Ph.D.
Professor of Civil Engineering
Purdue University

**McGraw-Hill Book Company**

New York    St. Louis    San Francisco
Auckland    Bogotá    Hamburg    Johannesburg
London    Madrid    Mexico    Montreal    New Delhi
Panama    Paris    São Paulo    Singapore
Sydney    Tokyo    Toronto

**Library of Congress Cataloging in Publication Data**
Main entry under title:

Surveying, theory and practice.

Fifth ed. published in 1966 entered under R. E. Davis
Includes bibliographies and index.
1. Surveying. I. Davis, Raymond Earl, dates
II. Davis, Raymond Earl, dates
Surveying, theory and practice.
TA545.D45  1981      526.9      80-15878
ISBN 0-07-015790-1

## SURVEYING: THEORY AND PRACTICE

90DODO8987

This book was set in Times Roman by Science Typographers. The editors
were Julienne V. Brown and Madelaine Eichberg; the designer was
Ben Kann; the production supervisor was Phil Galea. New drawings
were done by Fine Line Illustrations, Inc.
R. R. Donnelley & Sons Company was printer and binder.

# Contents

## PART III   SURVEY OPERATIONS

## PART V   TYPES OF SURVEYS

# Contents of field and office problems

# Preface

This edition, while maintaining the thoroughness and extensive coverage of material, offers significant improvements over the preceding edition which will make the book beneficial both as a text and as a reference for the practicing professional. Because of the substantial advances in the techniques and particularly the equipment used in surveying in the last decade and a half, careful examination of the preceding edition was undertaken. The philosophy underlying this edition is based on careful rearrangement and consolidation of material and minimizing the scattering of related parts, and a clear distinction between surveying and mapping. This consolidation, while preserving the depth and breadth characteristics of the book, provided room for several timely additions made necessary by recent significant developments in the technologies influencing surveying and mapping.

One such new technology involves the wide availability of electronic computing aids—from hand-held calculators, to desktop computers, to large central processors (it is assumed that all students using this text own and know how to use a hand-held calculator with trigonometric functions). The ready availability of these aids requires a fundamental change in the philosophy for survey computations. Thus, slide rules, log tables, and tabular entries for various calculations become unnecessary. Instead, direct treatment of the algebraic and geometric relationships involved in surveying is not only shorter and easier to understand, but can also be readily evaluated numerically. Furthermore, consolidated and rigorous concepts of the theory of observations and errors are now not only desirable but quite essential when the student recognizes that many desktop computing systems contain software that is based on these concepts. Consequently, an entire chapter (Chapter 2) is devoted to survey measurements and computations. Another consequence of the ready availability and power of modern computing systems is that the method of least-squares adjustment, frequently avoided in the past, has now become a standard technique. Since its application is essentially unlimited with respect to survey problems, an extensive appendix (Appendix B) covering the method of least squares was added. In this format, Chapter 2 and Appendix B can be consulted by the reader at various junctures in the book when a rigorous adjustment is appropriate.

An additional feature to this edition is the introduction of matrix algebra as a means of concisely describing linear relationships in surveying. Appendix A is an introduction to the topic for those who are unfamiliar with matrix notation. Reference to this appendix will reveal that matrix algebra, although relatively uncomplicated, is very powerful when applied to least-squares adjustments.

It should also be noted that throughout the book an abundance of worked numerical examples are included to demonstrate the various technical and reduction operations.

The book is divided into five distinct parts. Part I addresses general concepts and includes three chapters. Chapter 1 is introductory, containing basic definitions and differences between surveying and mapping. Chapter 2 describes the basic concepts of the theory of observations and errors as related to survey measurements and calculations. This first part is concluded with Chapter 3 on general aspects of field and office work.

Part II, containing Chapters 4 through 7, deals with basic survey measurements: Chapter 4 on distance measurement; Chapter 5 on vertical distance measurement or leveling; Chapter 6 on direction and angle measurement; and finally Chapter 7 on stadia and tacheometry. Note the logical order of these chapters and the wisdom on having one part devoted to fundamental measurements without encumbering the reader with techniques, methodology, and operations all at once.

Part III, which is composed of Chapters 8 through 12, follows naturally after Part II by covering basic surveying operations. This section relies with minimum repetition on the basic methods for survey measurement covered in Part II. Traverse is discussed in Chapter 8; the operations of resection and intersection are developed in Chapter 9; triangulation and trilateration are addressed in Chapter 10; and an introduction to astronomy is given in Chapter 11. These chapters are written to reflect the newest techniques and introduce the students to the most recent equipment. The modern concepts of errors and measurements introduced in Chapter 2 are utilized in each chapter for error propagation, with references being given where necessary to appropriate least-squares adjustments in Appendix B. In this fashion, the depth of coverage is essentially an option left to the discretion of the instructor, who can select material applicable to the course level.

Part III is concluded with another new chapter on modern positioning systems. Chapter 12 is a direct result of the tremendous advances in optical-electronic and space technologies. It covers three highly promising positioning systems, electronic, inertial, and doppler, which will certainly have considerable impact on the future surveyor's activities.

Whereas Parts II and III are on surveying, Part IV is devoted to mapping and is composed of two chapters. Chapter 13 is an introduction to mapping and map drafting. In this chapter, the student is apprised of the allied modern techniques available in automated cartography and computer graphics. Chapter 14 could also be considered as a new chapter, since it is as complete a coverage of map projections as is likely to be found in any general textbook on surveying and mapping. It covers not only the basic theory of mapping the curved earth on a map sheet, but also the practical aspects accompanied by substantial descriptions of the Lambert conformal, transverse Mercator, and universal transverse Mercator plane coordinate systems. Numerical examples demonstrate the application of not only the published map projection tables but also the rigorous formulation. The latter is included to facilitate the task for those who wish to program the transformations on an electronic computer.

The last section, Part V, covers seven distinct types of surveys in as many chapters. It begins logically with control and topographic surveys in Chapter 15, the material that is used in subsequent chapters. Chapter 16, on photogrammetric surveying and mapping, is extensively revised to reflect the substantial advances that have occurred in this important field. Chapters 17 and 18, on route and construction surveying, are similarly revised and brought up to date, including all the new methodology and particularly instruments of modern design. In addition to land surveys, covered in Chapter 19, Part V is concluded with two contributed chapters: Chapter 20 on mine surveys; and Chapter 21 on hydrographic surveys. The gentlemen contributing these chapters are experts in their areas and bring the latest information in their respective fields to these chapters.

In essence, then, this edition preserves the main contents of the book from previous editions. However, the more logical rearrangement of material makes it more in tune with modern times and in the process provides a text that is timely and up to date. As in the past, each chapter is followed by a list of problems and references for additional reading. Throughout, error and measurement analyses based on modern statistical concepts are introduced in each chapter. This is done in such a manner as to prepare the future surveyor for the task of properly handling the measured data that will be acquired. After all, the reader should recognize that surveying is essentially a science of metrology or measurement.

## ACKNOWLEDGMENTS

The authors would like to acknowledge the assistance offered by their colleagues and users of the text and authors of contributed chapters. Professors Warren Marks of Penn State University, Arthur J. McNair of Cornell University, Dean Merchant of Ohio State University, Francis H. Moffitt of the University of California, and Fareed Nader of California State University, Joseph Dracup of the U.S. National Ocean Survey, George Katibah of the California Department of Transportation, J. C. McGlone of Purdue University, and James E. Thompson of Spectra Physics, Inc., all provided useful information and rendered valuable suggestions and advice. Dr. Adam Chrzanowski, University of New Brunswick in Canada, and Dr. A. J. Robinson, University of New South Wales, Australia, wrote Chapter 20, Mining Surveys and Commander James Collins, Ph.D., NOAA National Ocean Survey, prepared Chapter 21, Hydrographic Surveying. Their efforts are very much appreciated.

Many of the illustrations and tables have been taken or adapted from publications or furnished directly by federal, state, or county agencies, specifically the U.S. Bureau of Land Management, U.S. National Ocean Survey, U.S. Geological Survey, California Department of Transportation, and Riverside County of California Flood Control and Conservation District. Use has also been made of photographs and tabular information from the American Society of Civil Engineers, U.S. Air Force, U.S. Army, U.S. Naval Observatory, Institute for Photogrammetry at the University of Stuttgart (West Germany), and the Department of Mining and Metallurgy at the University of Krakow (Poland).

Illustrative material was furnished by the following manufacturers of surveying, photogrammetric, and electronic equipment and consulting services; Abrams Aerial Survey Corporation; AGA Corporation; Bausch & Lomb, Inc.; California Computer Products, Inc.; CONCAP Computing Systems; The Cooper Group (Lufkin); DBA Systems, Inc.; Geodetic Services, Inc.; Gerber Scientific Instrument Co.; Hewlett-Packard Co.; JMR Instruments, Inc.; Kern Instruments, Inc.; Kelsh Instrument Division of Danko Arlington, Inc.; Keuffel & Esser Co.; The Lietz Co.; Litton Guidance and Control Systems; Raytheon Co., Raymond Vail and Associates; SPAN International, Inc.; Spectra Physics, Inc.; Tellurometer, U.S.A.; Wang Laboratories, Inc.; Wild Heerbrugg Instruments, Inc.; Carl Zeiss, Inc.; and Zena Co.

During the checking of galley and page proofs, substantial editing assistance was rendered by Dr. L. Holderly, Miss D. Schwartz, and Messrs. J. Thurgood and F. Paderes, all of Purdue University. The authors are grateful to these individuals for their contributions of time and expertise.

Finally, a very special note of gratitude is expressed to our wives, Ruth Anderson and LaVerne Mikhail and our children, who persevered with us through the period of preparation of the manuscript.

*James M. Anderson*
*Edward M. Mikhail*

# PART I
# CONCEPTS

# CHAPTER 1
# Surveying and mapping

**1.1. Surveying** Surveying has to do with the determination of the relative spatial location of points on or near the surface of the earth. It is the art of measuring horizontal and vertical distances between objects, of measuring angles between lines, of determining the direction of lines, and of establishing points by predetermined angular and linear measurements.

Concomitant with the actual measurements of surveying are mathematical calculations. Distances, angles, directions, locations, elevations, areas, and volumes are thus determined from data of the survey. Also, much of the information of the survey is portrayed graphically by the construction of maps, profiles, cross sections, and diagrams.

The equipment available and methods applicable for measurement and calculation have changed tremendously in the past decade. Aerial photogrammetry, satellite observations, remote sensing, inertial surveying, and electronic distance measurement techniques are examples of modern systems utilized to collect data usable in the surveying process. The relatively easy access to electronic computers of all sizes facilitates the rigorous processing and storage of large volumes of data.

With the development of these modern data acquisition and processing systems, the duties of the surveyor have expanded beyond the traditional tasks of the field work of taking measurements and the office work of computing and drawing. Surveying is required not only for conventional construction engineering projects, mapping, and property surveys, but is also used increasingly by other physical sciences, such as geology and geophysics; biology, including agriculture, forestry, grasslands, and wildlife; hydrology and oceanography; and geography, including human and cultural resources. The tasks in these physical science operations need to be redefined to include design of the surveying procedure and selection of equipment appropriate for the project; acquisition of data in the field or by way of remote stations; reduction or analysis of data in the office or in the field; storage of data in a form compatible with future retrieval; preparation of maps or other displays in the graphical (including photographic) or numerical form needed for the purpose of the survey; and setting of monuments and boundaries in the field as well as control for construction layout. Performance of these tasks requires a familiarity with the uses of surveying, knowledge of the fundamentals of the surveying process, and a knowledge of the various means by which data can be prepared for presentation.

**1.2. Uses of surveys** The earliest surveys known were for the purpose of establishing the boundaries of land, and such surveys are still the important work of many surveyors.

Every construction project of any magnitude is based to a greater or less degree upon measurements taken during the progress of a survey and is constructed about lines and points established by the surveyor. Aside from land surveys, practically all surveys of a private nature and most of those conducted by public agencies are of assistance in the conception, design, and execution of engineering works.

For many years the government, and in some instances the individual states, have conducted surveys over large areas for a variety of purposes. The principal work so far accomplished consists of the fixing of national and state boundaries; the charting of coastlines and navigable streams and lakes; the precise location of definite reference points throughout the country; the collection of valuable facts concerning the earth's magnetism at widely scattered stations; the establishment and observation of a greater network of gravity stations throughout the world; the establishment and operation of tidal and water level stations; the extension of hydrographic and oceanographic charting and mapping into the approximately three-fourths of the world which is essentially unmapped; and the extension of topographic mapping of the land surfaces of the earth, for which the United States has achieved coverage of nearly one-half of its surface area to map scales as large as 1 : 24,000 or 1 in to 2000 ft.

Observations of a worldwide net of satellite triangulation stations were made during the decade 1964–1974. Results of the computations are expected to be completed by 1983. They will allow determination of the shape of the earth from one to two orders of magnitude better than has heretofore been known. Consequently, surveys of global extent have been performed and will become common in the future.

Thus surveys are divided into three classes: (1) those for the primary purpose of establishing the boundaries of land, (2) those providing information necessary for the construction of public or private works, and (3) those of large extent and high precision conducted by the government and to some extent by the states. There is no hard and fast line of demarcation between surveys of one class and those of another as regards the methods employed, results obtained, or use of the data of the survey.

**1.3. The earth as a spheroid**   The earth has the approximate shape of an oblate spheroid of revolution, the length of its polar axis being somewhat less than that of its equatorial axis. The lengths of these axes are variously computed, as follows:

| Reference | Polar (minor) axis, m | Equatorial (major) axis, m |
|-----------|----------------------|---------------------------|
| Clarke (1866) | 12,713,168 | 12,756,602 |
| Hayford (1909) | 12,713,824 | 12,756,776 |
| Fischer (1960) | 12,713,546 | 12,756,310 |

The lengths computed by Clarke have been generally accepted in the United States and have been used in government land surveys. Hayford's dimensions were the first to be used for an international figure of the earth. The values calculated by Fischer have been adopted by the National Aeronautics and Space Administration (NASA) for extraterrestrial spatial computations.

It is seen that the polar axis is shorter than the equatorial axis by about 27 mi, or about 43 km. Relative to the diameter of the earth this is a very small quantity, less than 0.34 percent. Imagine the earth as shrunk to the size of a billiard ball, still retaining the same shape. In this condition, it would appear to the eye as a smooth sphere, and only by precise measurements could its lack of true sphericity be detected.

Let us consider that the irregularities of the earth have been removed. The surface of this imaginary spheroid is a curved surface every element of which is normal to the plumb line. Such a surface is termed a *level surface*. The particular surface at the average sea level is termed *mean sea level*.

The student's understanding of the various concepts regarding the earth is greatly facilitated by assuming it to be a sphere. Imagine a plane as passing through the center of the earth, which is assumed to be spherical, as in Fig. 1.1. Its intersection with the level

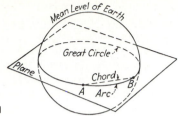

**Fig. 1.1**

surface forms a continuous line around the earth. Any portion of such a line is termed a *level line*, and the circle defined by the intersection of such a plane with the mean level of the earth is termed a *great circle* of the earth. The distance between two points on the earth, as *A* and *B* (Fig. 1.1), is the length of the arc of the great circle passing through the points, and is always more than the chord intercepted by this arc. The arc is a level line; the chord is a mathematically straight line.

If a plane is passed through the poles of the earth and any other point on the earth's surface, as *A* (Fig. 1.2), the line defined by the intersection of the level surface and plane is called a *meridian*. Imagine two such planes as passing through two points as *A* and *B* (Fig. 1.2) on the earth, and the section between the two planes removed like the slice of an orange, as in Fig. 1.3. At the equator the two meridians are parallel; above and below the equator they converge, and the angle of convergency increases as the poles are approached. No two meridians are parallel except at the equator.

Imagine lines, normal to the meridians, drawn on the two cut surfaces of the slice. If the earth is as a perfect sphere, these lines converge at a point at the center of the earth. Considering the lines on either or both of the cut surfaces, no two are parallel. The radial lines may be considered as vertical or plumb lines, and hence we arrive at the deductions that all plumb lines converge at the earth's center and that no two are parallel. (Strictly speaking, this is not quite true, owing to the unequal distribution of the earth mass and to the fact that normals to an oblate spheroid do not all meet at a common point.)

Consider three points on the mean surface of the earth. Let us make these three points the vertices of a triangle, as in Fig. 1.4. The surface within the triangle *ABC* is a curved surface, and the lines forming its sides are arcs of great circles. The figure is a spherical triangle. In the figure the dashed lines represent the plane triangle whose vertices are points *A*, *B*, and *C*. (Actually, the "auxiliary plane triangle" has sides equal in length to the *arcs* of the corresponding spherical triangle.) Lines drawn tangent to the sides of the spherical triangle at its vertices are shown. The angles $\alpha$, $\beta$ and $\gamma$ of the spherical triangle are seen to be greater than the corresponding angles $\alpha'$, $\beta'$, and $\gamma'$ of the plane triangle. The amount of this excess would be small if the points were close together, and the surface forming the triangle would not depart far from a plane passing through the three points. If the points were far apart, the difference would be considerable. Evidently, the same conditions would exist for a figure of any number of sides. Hence, we see that angles on the surface of the earth are spherical angles.

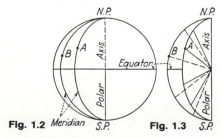

**Fig. 1.2**  **Fig. 1.3**

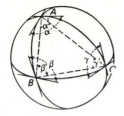

**Fig. 1.4**

In everyday life we are not concerned with these facts, principally because we are dealing with only a small portion of the earth's surface. We think of a line passing along the surface of the earth directly between two points as being a straight line, we think of plumb lines as being parallel, we think of a level surface as a flat surface, and we think of angles between lines in such a surface as being plane angles.

As to whether the surveyor must regard the earth's surface as curved or may regard it as plane (a much simpler premise) depends upon the character and magnitude of the survey and upon the precision required.

In either case, to provide a suitable framework to which all types of surveys are referenced, it is necessary to establish a *horizontal datum* and a *vertical datum*. A horizontal datum is the surface to which horizontal distances are referred and consists of (1) an initial point or origin, (2) the direction of a line from this origin, and (3) the polar and equatorial axes of the figure of the earth that best fits the area to be surveyed. The horizontal datum now being used in the United States is the North American Datum of 1927. A vertical datum is the surface to which all vertical distances are referred. The vertical datum currently being used in the United States is the National Geodetic Vertical Datum (NGVD) of 1929, formerly called the Sea Level Datum of 1929. Additional details concerning the datum can be found in Art. 13.2.

It was mentioned earlier (Art. 1.2) that a global survey and adjustment which would provide an improved figure of the earth was in progress. In conjunction with this program, the United States National Geodetic Survey is working on a recomputation of the North American Datum. This project is scheduled for completion in 1983. At that time new values for the major and minor axes of the spheroid will be published. Accompanying the development of these new parameters defining the figure of the earth will be slight modifications to the horizontal datum and vertical datum for the United States, Canada, and Mexico. The nature of these modifications is already being publicized in the technical literature for surveyors, to prepare for a smooth transition from the old to the new datum. Final results are scheduled to be published in 1983.

**1.4. Geodetic surveying**   That type of surveying which takes into account the true shape of the earth is defined as *geodetic surveying*. Surveys employing the principles of geodesy are of high precision and generally extend over large areas. Where the area involved is not great, as for a state, the required precision may be obtained by assuming that the earth is a perfect sphere. Where the area is large, as for a country, the true spheroidal shape of the earth is considered. Surveys of the latter character have been conducted primarily through the agencies of governments. In the United States such surveys have been conducted principally by the U.S. National Geodetic Survey and the U.S. Geological Survey. Geodetic surveys have also been conducted by the Great Lakes Survey, the Mississippi River Commission, several boundary commissions, and others. Surveys conducted under the assumption that the earth is a perfect sphere have been made by such large cities as Washington, Baltimore, Cincinnati, and Chicago.

Though relatively few engineers and surveyors are employed full time in geodetic work, the data of the various geodetic surveys are of great importance in that they furnish precise points of reference to which the multitude of surveys of lower precision may be tied. For each state, a system of plane coordinates has been devised, to which all points in the state can be referred without an error of more than one part in 10,000 in distance or direction arising from the difference between the reference surface and the actual mean surface of the earth.

**1.5. Plane surveying**   That type of surveying in which the mean surface of the earth is considered as a plane, or in which its spheroidal shape is neglected, is generally defined as *plane surveying*. With regard to horizontal distances and directions, a level line is considered as mathematically straight, the direction of the plumb line is considered to be the same at all points within the limits of the survey, and all angles are considered to be plane angles.

By far the greater number of all surveys are of this type. When it is considered that the length of an arc 11.5 mi or 18.5 km long lying in the earth's surface is only 0.02 ft or 0.007 m greater than the subtended chord, and further that the difference between the sum of the angles in a plane triangle and the sum of those in a spherical triangle is only 1 second for a triangle at the earth's surface having an area of 75.5 mi$^2$ or 196 km$^2$, it will be appreciated that the shape of the earth must be taken into consideration only in surveys of precision covering large areas. However, with the increasing size and sophistication of engineering and other scientific projects, surveyors who restrict their practice to plane surveying alone are severely limited in the types of surveys in which they can be engaged.

Surveys for the location and construction of highways, railroads, canals, and, in general, the surveys necessary for the works of human beings are plane surveys, as are the surveys made for the purpose of establishing boundaries, except state and national. The United States system of subdividing the public lands employs the methods of plane surveying but takes into account the shape of the earth in the location of certain of the primary lines of division.

The operation of determining *elevation* is usually considered as a division of plane surveying. Elevations are referred to a spheroidal surface, a tangent at any point in the surface being normal to the plumb line at that point. The curved surface of reference, usually mean sea level, is called a "datum" or, curiously and incorrectly, a "datum plane." The procedure ordinarily used in determining elevations automatically takes into account the curvature of the earth, and elevations referred to the curved surface of reference are secured without extra effort on the part of the surveyor. In fact, it would be more difficult for the surveyor to refer elevations to a true plane than to the imaginary spheroidal surface which was selected. Imagine a true plane, tangent to the surface of mean sea level at a given point. At a horizontal distance of 10 mi from the point of tangency, the vertical distance (or elevation) of the plane above the surface represented by mean sea level is 67 ft, and at a distance of 100 mi from the point of tangency the elevation of the plane is 6670 ft above mean sea level. In metric units at a horizontal distance of 1 km from the point of tangency, the vertical distance above mean sea level is 78.5 mm; at 10 km the vertical distance is 7.85 m; and at 100 km it is 785 m. Evidently, the curvature of the earth's surface is a factor that cannot be neglected in obtaining even very rough values of elevations.

This book deals chiefly with the methods of plane surveying.

**1.6. Operations in surveying**   A *control survey* consists of establishing the horizontal and vertical positions of arbitrary points. A *land, boundary, or property survey* is performed to determine the length and direction of land lines and to establish the position of these lines on the ground. A *topographic survey* is made to secure data from which may be made a *topographic map* indicating the configuration of the terrain and the location of natural and human-made objects. *Hydrographic surveying* refers to surveying bodies of water for the purposes of navigation, water supply, or subaqueous construction. *Mine surveying* utilizes the principles for control, land, geologic, and topographic surveying to control, locate, and map underground and surface works related to mining operations. *Construction surveys* are performed to lay out, locate, and monitor public and private engineering works. *Route surveying* refers to those control, topographic, and construction surveys necessary for the location and construction of lines of transportation or communication, such as highways, railroads, canals, transmission lines, and pipelines. *Photogrammetric surveys* utilize the principles of aerial and terrestrial photogrammetry, in which measurements made on photographs are used to determine the positions of photographed objects. Photogrammetric surveys are applicable in practically all the operations of surveying and in a great number of other sciences.

**1.7. Summary of definitions** A *level surface* is a curved surface every element of which is normal to a plumb line. Disregarding local deviations of the plumb line from the vertical, it is parallel with the mean spheroidal surface of the earth. A body of still water provides the best example.

A *horizontal plane* is a plane tangent to a level surface at a particular point.

A *horizontal line* is a line tangent to a level surface. In surveying, it is commonly understood that a horizontal line is straight.

A *horizontal angle* is an angle formed by the intersection of two lines in a horizontal plane.

A *vertical line* is a line perpendicular to the horizontal plane. A plumb line is an example.

A vertical line in the direction toward the center of the earth is said to be in the direction of the nadir. A vertical line in the direction away from the center of the earth and above the observer's head is said to be directed toward the zenith.

A *vertical plane* is a plane in which a vertical line is an element.

A *vertical angle* is an angle between two intersecting lines in a vertical plane. In surveying, it is commonly understood that one of these lines is horizontal, and a vertical angle to a point is understood to be the angle in a vertical plane between a line to that point and the horizontal plane.

A *zenith angle* is an angle between two lines in a vertical plane where it is understood that one of the lines is directed toward the zenith. A *nadir angle* is an angle between two lines in a vertical plane where it is understood that one of the lines is directed toward the nadir.

In plane surveying, distances measured along a level line are termed *horizontal distances*. The distance between two points is commonly understood to be the horizontal distance from the plumb line through one point to the plumb line through the other. Measured distances may be either horizontal or inclined, but in practically all cases the inclined distances are reduced to equivalent horizontal lengths.

The *elevation* of a point is its vertical distance above (or below) some arbitrarily assumed level surface, or datum.

A *contour* is an imaginary line of constant elevation on the ground surface. The corresponding line on the map is called a *contour line*.

The vertical distance between two points is termed the *difference in elevation*. It is the distance between an imaginary level surface containing the high point and a similar surface containing the low point. The operation of measuring difference in elevation is called *leveling*.

The *grade*, or *gradient*, of a line is its slope, or rate of ascent or descent.

**1.8. Units of measurement** The operations of surveying entail both angular and linear measurements.

The sexagesimal units of angular measurement are the *degree*, *minute*, and *second*. A plane angle extending completely around a point equals 360 degrees; 1 degree = 60 minutes; 1 minute = 60 seconds. In Europe, the centesimal unit, the *grad*, or *grade*, is the angular unit;

400 grads$^{(g)}$ equals 360°; 1 grad$^{(g)}$ = 100 centesimal minutes$^{(c)}$ = 0.9°; 1 centesimal minute$^{(c)}$ = 100 centesimal seconds$^{(cc)}$ = 0°00′32.4″. Grads are usually expressed in decimals. For example, 100$^{(g)}$42$^{(c)}$88$^{(cc)}$ is expressed as 100.4288 grads.

The international unit of linear measure is the metre. Originally, the metre was defined as 1/10,000,000 of the earth's meridional quadrant. The metre as used in the United States, where it has always been employed for geodetic work, was initially based on an iron metre bar standardized at Paris in 1799. In 1866, by act of Congress, the use of metric weights and measures was legalized in the United States and English equivalents were specified. Among these equivalents, the metre was given as 39.37 in, or the ratio of the foot (12 in) to the metre was specified as 1200/3937. This ratio yields a yard (3 ft) equal to 0.914401829 m.

In 1875, a treaty was signed in Paris by representatives of seventeen nations (the United States included) to establish a permanent International Bureau of Weights and Measures. As a direct result of this treaty, the standard for linear measure was established as the International Metre, defined by two marks on a bar composed of 90 percent platinum and 10 percent iridium. The original International Metre Bar was deposited at Sèvres, near Paris, France. Two copies of this metre bar (called prototype metres) were sent to the United States in 1889. These bars were first acquired by the United States Coast and Geodetic Survey (now called the National Ocean Survey or NOS) and later were transferred to the National Bureau of Standards in Washington, D.C. By executive order in 1893, these new standards were established as *fundamental standards* for the nation. In this executive order the foot-to-metre ratio was defined to be 1200/3937, the same as specified by Congressional action in 1866.

Also established by the Treaty of 1875 was a General Conference on Weights and Measures (Conference General de Poids et Mesures, abbreviated to CGPM) that was to meet every 6 years. In 1960, the CGPM modernized the metric system and established the *Système International d'Unités* (SI). The standard unit for linear measure, as fixed by this conference, was the metre, defined as a length equal to 1,650,763.73 wavelengths of the orange-red light produced by burning the element krypton at a specified energy level. Such a standard has the advantage of being reproducible and is immune to inadvertent or purposeful damage.

The metre is subdivided into the following units:

$$1 \text{ decimetre (dm)} = 0.1 \text{ metre (m)}$$
$$1 \text{ centimetre (cm)} = 0.01 \text{ metre}$$
$$1 \text{ millimetre (mm)} = 0.001 \text{ metre}$$
$$1 \text{ micrometre } (\mu\text{m}) = 0.001 \text{ mm} = 10^{-6} \text{ m}$$
$$1 \text{ nanometre (nm)} = 0.001 \text{ } \mu\text{m} = 10^{-9} \text{ m}$$

In the United States by executive order on July 1, 1959, the yard (3 ft) was redefined to be 0.9144 m. This order changed the foot-to-metre ratio as previously established so as to shorten the foot by 1 part per 500,000. The relationships between the English system of linear units of measure (the foot and the inch) as agreed upon in 1959 are:

$$1 \text{ U.S. inch} = 1 \text{ British inch} = 2.54 \text{ cm}$$
$$1 \text{ U.S. foot} = 30.48 \text{ cm}$$

The English system was used by the United States and the United Kingdom and is now employed officially only in the United States, where conversion to the International system is slowly but surely occurring.

The difference between the U.S. *survey foot* (1200/3937) and the International foot is very small and amounts to 0.2 m/100,000 m or 0.66 ft/60 mi. When six digits or fewer are used in surveying computations, no significant difference will result. However, note that many hand-held calculators contain automatic conversion factors based on the International

System. The surveyor must be aware of this possibility and recognize that when six to ten digits are used in computations, the proper conversion factor of 1200/3937 must be calculated or significant differences will occur.

In the United States, the *rod* and *Gunter's chain* were units used in land surveying. Gunter's chain is still employed in the subdivision of U.S. public lands. The Gunter's chain is 66 ft long and is divided into 100 *links* each 7.92 in in length. One mile = 80 chains = 320 *rods, perches,* or *poles* = 5280 ft.

The *vara* is a Spanish unit of measure used in Mexico and several other countries that were under early Spanish influence. In portions of the United States formerly belonging to Spain or Mexico, the surveyor will frequently have occasion to rerun property lines in which lengths are given in terms of the vara. Commonly, 1 vara equals 32.993 in (Mexico), 33 in (California), or $33\frac{1}{3}$ in (Texas); but other somewhat different values have been used for many surveys.

The units of area, as used in the United States, are the *square foot, acre,* and *square mile.* Formerly, the *square rod* and *square Gunter's chain* were also used. One acre = 10 square Gunter's chains = 160 square rods = 43,560 square feet. One square mile equals 640 acres.

In the metric system the unit of area is the square metre. A European metric unit of area is 1 hectare = 10,000 $m^2$ = 107,639.10 $ft^2$ = 2.471 acres. One acre equals 0.4047 hectares. The unit of volume is the *cubic metre,* where 1 cubic metre = 35.315 $ft^3$ = 1.308 $yd^3$.

The units of volume measurement in the United States have been the *cubic foot* and the *cubic yard.* One cubic foot = 0.0283 $m^3$ and 1 cubic yard = 0.7646 $m^3$.

**1.9. The drawings of surveying**   The drawings of surveying consist of maps, profiles, cross sections, and to a certain extent graphical calculations. The usefulness of these drawings is largely dependent upon the accuracy with which the points and lines are projected on paper. For the most part, few dimensions are shown, and the person who makes use of the drawings must rely on distances measured with a scale or angles measured with a protractor. Moreover, the drawings of surveying are so irregular and the data upon which the drawings are based are of such a nature that the use of conventional drafting tools for the construction of these drawings is the exception rather than the rule. The drawings may be constructed manually by the application of traditional drafting methods or may be compiled from numerical data stored in the central memory of an electronic computer, on magnetic tape, or in some other peripheral storage unit and plotted by a computer-controlled automatic flat-bed plotter.

An important concept involved in the construction of these drawings is that of *scale.* Scale is defined as the ratio of a distance on the drawing to the corresponding distance on the object. Scale can be expressed as a ratio: for example, 1:1000, where the units of measure are the same or by stating that 1 cm on the drawing represents 1000 cm on the object. Another way is to use different units; for example, 1 in on the drawing representing 100 ft on the object, which is equivalent to the scale ratio of 1:1200.

**1.10. Map projections**   A map shows graphically the location of certain features on or near the surface of the earth. Since the surface of the earth is curved and the surface of the map is flat, no map can be made to represent a given territory without some distortion. If the area considered is small, the earth's surface may be regarded as a plane, and a map constructed by orthographic projection, as in mechanical drawing, will represent the relative locations of points without measurable distortion. The maps of plane surveying are constructed in this manner, points being plotted by rectangular coordinates or by horizontal angles and distances.

As the size of the territory increases, this method becomes inadequate, and various forms of projection are employed to minimize the effect of map distortion. Control points are plotted by spherical coordinates, or the latitude and longitude of the points. *Latitude* is the angular distance above or below the plane of the equator and *longitude* is the angular distance in the plane of the equator east or west of the Greenwich Meridian (see Fig. 11.3, Art. 11.3). It is also customary to show meridians and parallels of latitude on the finished map. The maps of states and countries, as well as those of some smaller areas, such as the U.S. Geological Survey topographic quadrangle maps, are constructed in this manner. Map projection, or the projection of the curved surface of the earth onto the plane of the map sheet, is usually *conformal*, which means that angles measured on the earth's surface are preserved on the map. There are several different conformal map projection methods, each suited for a particular purpose or area of specific configuration. Each method is based on minimizing one or more of the distortions inherent in map production, which include distortions in length, distortions in azimuth and angle, and distortions in area. The various methods of map projection are discussed in Chap. 14.

State plane coordinate systems have been designed for each state in the United States, whereby, even over large areas, points can be mapped accurately using a simple orthogonal coordinate system without the direct use of spherical coordinates.

On a global basis, the Universal Transverse Mercator projection system (Chap. 14) is defined according to parameters of an international ellipsoid. This system can be used throughout the world for plane coordinate computations.

**1.11. Maps**  Maps may be divided into two classes: those that become a part of public records of land division, and those that form the basis of studies for private and public works. The best examples of the former are the plans filed as parts of deeds in the county registry of deeds in most states of the United States. Good examples of the latter are the preliminary maps along the proposed route of a highway. It is evident that any dividing line between these two classes is indistinct, since many maps might serve both purposes.

Maps that are the basis for studies may be divided into two types: (1) historical *line-drawn maps* conventionally produced by a draftsman, although at the present time an increasing proportion of such maps is being drawn on a flat-bed plotter under the control of an electronic computer; and (2) *orthophoto maps* photogrammetrically produced from a stereo pair of vertical aerial photographs (Chap. 16).

Maps are further divided into the following two general categories: (1) *planimetric maps*, which graphically represent in plan such natural and artificial features as streams, lakes, boundaries, condition and culture of the land, and public and private works; and (2) *topographic maps*, which include not only some or all of the preceding features but also represent the relief or contour of the ground.

Maps of large areas, such as a state or country, which show the locations of cities, towns, streams, lakes, and the boundary lines of principal civil divisions are called *geographic maps*. Maps of this character, which show the general location of some kind of the works of human beings, are designated by the name of the works represented. Thus we have a *railroad map of the United States* or an *irrigation map of California*. Maps of this type, which emphasize a single topic and where the entire map is devoted to showing the geographic distribution or concentration of a specific subject, are called *thematic maps*.

*Topographic maps* indicate the relief of the ground in such a way that elevations may be determined by inspection. Relief is usually shown by irregular lines, called contour lines, drawn through points of equal elevation (Chaps. 13 and 15). General topographic maps represent topographic and geographic features, public and private works, and are usually drawn to a small scale. The quadrangle maps of the U.S. Geological Survey are good examples.

*Hydrographic maps* show shorelines, locations and depths of soundings or lines of equal depth, bottom conditions, and sufficient planimetric or topographic features of lands adjacent to the shores to interrelate the positions of the surface with the underwater features (Chap. 21). Charts of the U.S. National Ocean Survey are good examples of hydrographic maps. Such maps form an important part of the basic information required for the production of environmental impact statements.

Maps may be constructed using data acquired in the field by traditional field survey methods. Under these conditions, the map is usually drawn manually using traditional drafting techniques. Maps are more frequently compiled photogrammetrically using data from aerial photography or other remote sensors supported by either aircraft or satellite. Maps of this type are constructed using a combination of manual, photographic, and automatic compilation techniques. Note that for a map compiled by photogrammetric methods, a sufficient number of ground control points of known positions must be identifiable in the photographic or sensed record to allow scaling the map and selecting the proper datum.

An increasing number of maps is also being produced from map data obtained from a wide range of sources. Thus, data from old maps, from aerial photography, and data obtained by traditional field methods can be digitized and stored in the memory of an electronic computer or on magnetic tape. Information stored electronically in this way is frequently referred to as a *data bank*. The data bank so formed is called a *digital map*. When desired, these data can be retrieved and plotted automatically on a flat-bed plotter (see Art. 13.13) under the control of an electronic computer. The overall procedure involved in acquiring and processing data in this way is called *automatic mapping*, which is described in more detail in Art. 13.36.

**1.12. Precision of measurements**  In dealing with abstract quantities, we have become accustomed to thinking in terms of exact values. The student of surveying should appreciate that the physical measurements acquired in the process of surveying are correct only within certain limits, because of errors that cannot be eliminated. The degree of precision of a given measurement depends on the methods and instruments employed and upon other conditions surrounding the survey. It is desirable that all measurements be made with high precision, but unfortunately a given increase in precision is usually accompanied by more than a directly proportionate increase in the time and effort of the surveyor. It therefore becomes the duty of the surveyor to maintain a degree of precision as high as can be justified by the purpose of the survey, but not higher. It is important, then, that the surveyor have a thorough knowledge of (1) the sources and types of errors, (2) the effect of errors upon field measurements, (3) the instruments and methods to be employed to keep the magnitude of the errors within allowable limits, and (4) the intended use of the survey data.

A complete discussion of error analysis as related to surveying procedures is presented in Chap 2.

**1.13. Principles involved in surveying**  Plane surveying requires a knowledge of geometry and trigonometry, to a lesser degree of physics and astronomy, and of the behavior of random variables and the adjustment of survey data. Some knowledge of mathematical statistics is invaluable for an understanding of error propagation, which, in turn, is needed for design of survey procedures and selection of equipment. In this respect, a knowledge of differential calculus is also necessary. Applicable portions of physics, astronomy, and statistics are developed in succeeding chapters when the need arises.

Data processing with high-speed electronic computers, minicomputers, and desktop electronic calculators is so much a part of the recording, reduction, storage, and retrieval of survey data that familiarity with these systems is absolutely essential to the surveyor. No attempt is made to cover these subjects in this book because such tools are being changed continuously and newer models introduced. It is assumed that the student will have acquired these skills in other courses and from other textbooks.

**1.14. The practice of surveying** The practice of surveying is complex. No amount of theory will make a good surveyor unless the requisite skill in the art and science of measuring and in field and office procedures is obtained. The importance of the practical phases of the subject cannot be overemphasized.

Surveying is frequently one of the first professional subjects studied by civil engineering students. Even though these students may not expect to major in surveying, they should understand that the education received in the art and science of measuring and computing and in the practice of mapping will contribute directly to success in other subjects, regardless of the branch of engineering ultimately chosen. The same principles are equally important in the other geosciences.

## References

1. Bomford, G., *Geodesy*, 2d ed., Oxford at the Clarendon Press, London, 1962.
2. Ewing, C. E. and Mitchell, M. M., *Introduction to Geodesy*, American Elsevier Publishing Company, Inc., New York, 1970.
3. Hamilton, A. C., "Surveying: 1976–2176," *Surveying and Mapping*, Vol. 37, No. 2, June 1977, pp. 119–126.
4. Moffitt, F. H., and Bouchard, H., *Surveying*, 6th ed., Harper & Row, Publishers, New York, 1975.
5. National Bureau of Standards Publication, *The International System of Units (SI)*, U.S. Government Printing Office, Washington, D.C., 1974 ed.
6. Strasser, Georg, "The Toise, the Yard, and the Metre—The Struggle for a Universal Unit of Length," *Surveying and Mapping*, Vol. 35, No. 1, March 1975, pp. 25–46.
7. Whitten, C. A., "Surveying 1776 to 1976," *Surveying and Mapping*, Vol. 37, No. 2, June 1977, pp. 109–118.
8. Whitten, C. A., "Metric Corner," *ACSM Bulletin*, No. 54, August 1976.

# CHAPTER 2
# Survey measurements and adjustments

**2.1. Introduction** Measurements are essential to the functions of the surveyor. The surveyor's task is to design the survey; plan its field operations; designate the amount, type, and acquisition techniques of the measurements; and then adjust and analyze these measurements to arrive at the required survey results. It is important, then, that the individual studying surveying understand the basic idea of a measurement or an observation (the two terms are used alternatively as having the same meaning).

Except for counting, measuring entails performing a physical operation which usually consists of several more elementary operations, such as preparations (either instrument setup or calibration or both), pointing, matching, and comparing. Yet, it is the result of these physical operations that is assigned a numerical value and called "measurement." Therefore, it is important to note that a measurement is really an indirect thing, even though in some simple instances it may appear to be direct. Consider, for example, the simple task of determining the length of a line using a tape. This operation involves several steps: setting up the tape and stretching it, aligning the zero mark to the left end of the line, and observing the reading on the tape opposite to the right end. The value of the distance as obtained by subtracting zero from the second reading (40.3 in Fig. 2.1) is what we call the measurement, although actually two alignments have been made. To make this point clear, visualize the case when the tape is simply aligned next to the line and the tape is read opposite to its ends. In this case, the reading on the left end may be 11.9 and the one on the right end 52.2, with the net length measurement of 40.3.

Another more common situation regarding observations involves the determination of an angle. This raises the question of distinguishing between directions and angles. Directions are more fundamental than angles, since an angle can be derived from two directions. In the cases where both directions and angles may be used, care must be exercised in treating each group, since an angle is the difference between two directions. Similar to the case of measuring a line, an angle is directly obtained if the first pointing (or direction) is taken specifically at zero; then the reading at the second direction will be the value of the angle

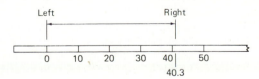

**Fig. 2.1** Length of line using a tape.

and may be considered as the observation. However, if the surveying process is formulated to be fundamentally in terms of directions, the values read for these directions should then be the observations. In this case, any angle determination will be a simple linear function of the observed directions and its properties can be evaluated from the properties of the directions and the known function, as will be shown later.

**2.2. Observations and the model** In surveying, the measurements or observations are rarely used directly as the required information. In general, they are utilized in some other subsequent operations, to derive, often computationally, other quantities, such as directions, lengths, relative positions, areas, shapes, volumes, and so on. The relationships applied in the computational effort are the mathematical representation of the geometric and/or physical conditions of the problem. Such a representation, in the context of present activities, is referred to as the *mathematical model*. The moment you have observations that must be reduced in some fashion to yield required useful information, you have a model to use in such a reduction process.

Suppose that we are interested in the area of a rectangular tract of land (assumed reasonably small to use computations in a plane). To compute that area, we need the length $a$ and the width $b$. The area will naturally be computed from $A = ab$, which is the "functional" model. Now, it appears that all we have to do is go out with a tape, measure the length and width, each once, and apply the two values into the formula to compute the area. This sounds simple enough until we begin taking a closer look at the problem and analyzing the factors involved in its solution.

Begin by assuming that a standardized tape (Art 4.16) is available (which may or may not be true!). To make a length measurement, we have to align the zero mark and note the reading at the other end of the line. Being human beings, working with a manufactured tool (the tape), we have no assurance that our determination of length is the best value we can obtain. We are at least *uncertain* about whether this *one measurement* will in fact be the best we can do. Perhaps if we repeat the measurement twice we would feel more confident. And if we do, we will probably end up with two different values for the length. Obviously, the same applies to the measurement of the width. We have no reason to accept one measurement over the other; both appear equally reasonable.

Return to the assumption that the tape is standardized. What if the tape is too short by a certain amount and we do not know it? Would not all our measurements with that tape be too large by factors of the amount by which the tape is short? When the measured length is used directly in the model, the computed area will obviously be incorrect.

Finally, examine the adequacy of the model $A = ab$. This model is formulated on the basic assumption that the parcel of land is exactly rectangular. What if one, or more, of the four angles are checked and differences from the right-angle assumption are discovered? The simple model given could no longer be used; instead, another which more properly reflects the *geometric* conditions should be utilized.

What began as a simple straightforward proposition has somehow managed to become very complicated. The point is that in the real world, where information is gathered by human beings and/or manufactured tools, there will always be an element of uncertainty, however small. A primary objective of this chapter is to dwell on this question of uncertainty in observations, together with the errors that might occur due to inadequate modeling. How to reconcile these uncertainties and errors and how to make the best utilization of the data will be shown.

One last point in our preliminary discussion of the model merits mentioning. Suppose that we are interested in the shape of a plane triangle. All that is required for this operation is to measure two of its angles, and the shape of the triangle will be uniquely determined. However, if we were to decide, for safety's sake, to measure all three angles, any attempt to construct such a triangle will immediately show *inconsistencies* among the three observed angles. In this case the model is simply that the sum of the three angles must equal 180°. If three observations are used in this model, it is highly unlikely that the sum will equal 180°. Therefore, when redundant observations or more observations than are absolutely necessary are acquired, these observations will rarely fit the model exactly. Intuitively, and relying on our previous discussion, this is due to something that is characteristic of the observations and makes them inconsistent in the case of redundancy. In the remainder of this chapter we discuss the causes of and treatment for inconsistencies in redundant observational data.

**2.3. Functional and stochastic models**   We said in the previous section that survey measurements are planned with a mathematical model in mind which describes the physical situation or set of events for which the survey is designed. This mathematical model is composed of two parts: a functional model and a stochastic model. The *functional model* is the more obvious part, since it usually describes the geometric or physical characteristics of the survey problem. Thus the functional model for the example concerning the triangle discussed at the end of the preceding section *involves the determination of the shape* of a plane triangle through the measurement of interior angles. Thus, if three angles in a triangle are available, redundant measurements are present with respect to the functional model. All this does not say anything about the properties of the measured angles. For example, in one case, each angle may be measured by the same observer, using the same instrument, applying the same measuring technique, and performing the measurements under very similar environmental conditions. In such a case, the three measured angles are said to be equally "reliable." But there are certainly a variety of other cases in which the resulting measurements are not of equal quality. In fact, as most practicing surveyors know from experience, measurements are always subject to unaccountable influences which result in variability when observations are repeated. Such statistical variations in the observations are important and must be taken into consideration when using the survey measurements to derive the required information. The *stochastic model* is the part of the mathematical model that describes the statistical properties of all the elements involved in the functional model. For example, in the case of the plane triangle, to say that each interior angle was measured and to give its value is insufficient. Additional information should also be included as to how well each angle was measured and if there is reason to believe that there is statistical "correlation" (or interaction) among the angles, and if so, how much. The outcome of having a unique shape for the triangle from the redundant measurements depends both on knowing that the sum of its internal angles is 180° (the functional model) and knowing the statistical properties of the three observed angles (the stochastic model).

**2.4. Observations and errors**   As indicated, variability in repeated measurements (under similar conditions) is an inherent quality of physical processes and must be accepted as a basic property of observations. Therefore, observations or measurements are numerical values for random variables which are subject to statistical fluctuations. Once this is recognized, we may proceed to treat the observations, employing established statistical techniques, to derive estimates or make appropriate inferences. In the past, these statistical variations in the observations were said to be due to observational errors and the *theory* of *errors* was developed. At the present time we speak instead of the *theory of observations*. The term "error" can, in general, be considered as referring to the difference between a given measurement and the "true" or "exact" value of the measured quantity. Since such a true value is almost never known, the term "error" is actually somewhat misleading and currently is used less frequently. Because most of the foundation of present knowledge was formed on the basis of the idea of errors, and since many in the profession still use the concept, it will be defined and utilized occasionally to relate it to the modern concept of observations.

Classically, errors are considered to be of three types: blunders or mistakes, systematic errors, and random errors. Each type of error will be discussed in turn.

**2.5. Blunders or mistakes** Blunders or mistakes are actually not errors, because they are usually so gross in magnitude compared to the other two types of errors. One of the most common reasons for mistakes is simple carelessness on the part of the observer, who may take the wrong reading of a scale or a dial, or if the proper value was read, may record the wrong one by, for example, transposing numbers. If the operation of collecting the observations is performed through an automatic recording technique, mistakes may still occur through failure of the equipment, although they may be less frequent in this case. Another cause of blunders is failure in technique, as in the case of reading the fraction on a tape on the wrong side of the zero mark, or by selecting the wrong whole degree in the measurement of angles that are very close to an integer of degrees (e.g., 71°59′58″ instead of 72°00′02″). Finally, a mistake may also occur due to misinterpretation, such as sighting to the wrong target.

Once the possibility that mistakes and blunders will occur is recognized, observational procedures and methodology must be designed to allow for their detection and elimination. Here, a variety of ways can be employed, such as taking multiple independent readings (not mere replications) and checking for reasonable consistency; careful checking of both sighting on targets and recording; using simple and quick techniques for verification, applying logic and common sense; checking and verifying the performance of equipment, particularly that with automatic readout; repeating the experiment with perhaps slightly different techniques or adopting a different datum or indices; increasing the redundancy of the observations used in a model; in the case of relatively complex models, applying simplified geometric or algebraic checks to detect the mistakes; and simply noting that most mistakes have a large magnitude, which may lead directly to their detection. For example:

1. Measuring an angle or a distance several times and computing the average. Any single measurement deviating from that average by an amount that is larger than a preset value can be assumed to contain a blunder.
2. Taking two readings on the horizontal circle of a transit differing by 180°.
3. Using two types of units, such as metres and feet, on a level rod and converting one into the other for comparison.
4. Realizing simple facts such as that a tape is usually 100 ft long; vertical angles must lie between ±90°; and so on.
5. Making a quick check on a spherical triangle using plane trigonometry.
6. Using small portable calculators or even slide rules to check surveying computations to a low number of significant digits.

Modern statistical concepts designate observations as samples from probability distributions, and their variability is therefore governed by the rules of probability. Consequently, observations containing mistakes or blunders are considered not to belong to the same distributions from which the observational samples are drawn. In other words, an observation with a mistake is not useful unless the mistake is removed; otherwise, that observation must be discarded.

**2.6. Systematic errors** Systematic errors or effects occur according to a *system* which, if known, can always be expressed by mathematical formulation. They follow a defined pattern, and if the experiment is repeated while maintaining the same conditions, the same pattern will be duplicated and the systematic errors will reoccur. The system causing the pattern can be dependent on the observer, the instrument used, or the physical environmental conditions of the observational experiment. Any change in one or more of the elements of the system will cause a change in the character of the systematic effect if the observational process is repeated. It must be emphasized, however, that repetition of the measurement under the same conditions will not result in the elimination of the systematic error. One must find the system causing the systematic errors before they can be eliminated from the observations.

Systematic effects take on different forms depending on the value and sign of each of the effects. If the value and sign remain the same throughout the measuring process, so-called constant error is present. Making distance measurements with a tape that is either too short or too long by a constant value is an example of a constant error. All lengths measured by that tape will undergo the same systematic effect due to the tape alone. If the sign of the systematic effect changes, perhaps due to personal bias of an observer, the resulting systematic errors are often called *counteracting*.

In surveying, systematic errors occur due to natural causes, instrumental factors, and the observer's human limitations. Temperature, humidity, and barometric pressure are examples of natural sources that will affect angle measurements and distance measurements either by tapes or electronic distance measuring equipment. Instrumental factors are caused by either imperfections in construction or lack of adequate adjustment of equipment before their use in data acquisition. Examples include unequal graduations on linear and circular scales; lack of centering of different components of the instrument; compromise in optical design, thus leaving certain amounts of distortions and aberrations; and physical limitations in machining parts, such as straight ways and pitch of screws.

Although automation has been considered and in some cases introduced (with its own sources of systematic effects!) to several tasks, the human observer remains an important element in the activities of surveying. The observer relies mostly on the natural senses of vision and hearing, both of which have limitations and vary due to circumstances and from one individual to another. Although some personal systematic errors are constant and some are counteracting, many others may be erratic. In a way, the set of errors committed by an observer will depend on the precise physical, psychological, and environmental conditions that exist during the particular observational experiment.

Although the sources of systematic effects discussed above pertain to the experiment as such, other systematic errors may occur as a result of the choice of geometric or mathematical model used to treat the measurements. For example, three options exist in treating a triangle on the earth's surface—plane, spherical, or ellipsoidal—and the choice of one rather than another may result in systematic error.

**2.7. Systematic errors, corrections, and residuals**   Classically, if an "error" is removed from a measurement, the value of that measurement should improve. This idea is applicable to *systematic errors*. Thus, if $x$ is an observation and $e_s$ is a systematic error in that observation,

$$x_c = x - e_s \qquad (2.1)$$

is the value of the observation "corrected" for the systematic error. Again, in the classical error theory, a *correction* $c_s$ for systematic error was taken equal in magnitude but opposite in sign to the systematic error; thus

$$x_c = x + c_s \qquad (2.2)$$

In fact, the concept of error and correction was not limited to the systematic type but was regarded as general in nature, with $x_c$ termed as the "true value," which is never known. In the theory of observations, there is no need for true values or true errors; instead, the observations are sample values from random probability distribution with mean $\mu$ and some specified variation (this concept will be explained in detail in Art. 2.10). Like the true value, the distribution mean $\mu$ is usually unknown, and we can only make an improved estimate for the observation. The improved estimate for an observation $x$ is designated $\hat{x}$, and instead of error or correction we have the so-called "residual" $v$, such that

$$\hat{x} = x + v \tag{2.3}$$

The residual, which will be used throughout this development, although having the same sign sense as the "correction," is not a hypothetical concept, since it does have value in the reduction of redundant observational data.

**2.8. Examples of compensation for systematic errors**   Examples of a variety of systematic errors abound in geodetic and surveying operations. Consider the operation of taping for the determination of distances between points on the earth's surface. The length of a given tape may be physically different from the values indicated by the numbers written on its graduations, owing to some or all of the following factors:

1. Change in temperature between that used for tape standardization (calibration) and the temperature actually recorded in the field during observation [Eq. (4.13), Art. 4.18].
2. The tension or pull applied to the tape during measurement is different from that used during calibration [Eq. (4.14), Art. 4.19].
3. The method of tape support is different during measurement from that used during calibration [Eq. (4.15), Art. 4.20].
4. The end points of the distance to be measured are at different elevations and the horizontal distance is desired. In this case a correction is needed due to the fact that one would be measuring the slope distance instead of the horizontal distance [Eq. (4.4), Art. 4.13].

Electronic (and electro-optical) distance measuring techniques are also subject to a number of systematic effects whose sources and characteristics must be determined and alleviated. Of these sources the following are mentioned here: the change in the density of air through which the signal travels, as it causes a change in the signal frequency (due to variations in wave propagation velocity); the instrument (and sometimes the reflector or remote unit) may not be properly centered on the ends of the line to be measured; and the path of propagation of the signal may not conform to the straight-line assumption and may be deviated due to environmental and other factors.

Another surveying operation for which systematic effects must be determined is leveling, or the observation of differences in elevation between points on the earth's surface. The "level" is the instrument used for acquisition of data by placing it at one point and sighting on a graduated rod on another point. In the instrument there is a spirit bubble which must be centered to ensure that the line of sight is horizontal. But if the axis of the telescope is not parallel to the axis of the bubble, the line of sight will be inclined to the horizontal even when the bubble is centered. This error, however, may be compensated for by a relatively simple observational procedure of selecting equal backsight and foresight distances. In this fashion, the systematic effects will be counteracting and would cancel each other.

In addition to the level itself, the rod may also cause systematic errors. For example, the rod may not be held vertical, or it may change in dimensions due to thermal expansion (being in the sun as opposed to being in the shade). The environment also contributes its share as a source of systematic effects. The geodetic concept of the deflection of the vertical due to local gravity anomalies is one source. Another is the fact that the earth is not a plane and the line of sight is not a straight line, because of atmospheric refraction.

Another instrument that is used extensively in surveying and geodesy to measure horizontal and vertical angles is the transit. The following are some of the sources of systematic effects that may be associated with such instruments:

1. Horizontal circle may be off center.
2. Graduations on either or both horizontal and vertical circles may not be uniform.
3. The horizontal axis of the telescope (about which it rotates) may not be perpendicular to the vertical axis of the instrument.

4. The longitudinal (or optical) axis of the telescope may not be normal to the horizontal axis of rotation. (In this case, the axis of the telescope would describe a cone instead of a plane as it rotates through a complete revolution.)
5. When the optical axis of the telescope is horizontal, the reading on the vertical circle is different from zero.
6. The telescope axis and the axis of the leveling bubble may not be parallel.
7. This source does not pertain to the transit but to the target on which sightings are taken. If the natural illuminating conditions are such that part of the target is in shadow, the observer will tend to center the transit's cross hair so as to bisect the illuminated portion of the target. This type of error is often referred to as *phase error*.

These are only some examples of systematic sources in surveying and geodesy. There are numerous others, such as spherical and spheroidal excess, gravimeter and other instrument errors, and timing and other errors in astrogeodetic work. Once the systematic errors are recognized, they can often be accounted for, by:

1. Actual formulation and computation of corrections, which are then applied to the raw observations.
2. Careful calibration and adjustment of equipment and measuring under the same conditions specified by calibration results.
3. Devising observational procedures which will result in the elimination of systematic errors that would otherwise occur.
4. Extension of the functional model to include the effect of the systematic errors.

**2.9. Random errors**    After mistakes are detected and eliminated and all sources of systematic errors are identified and corrected, the values of the observations will be free of any biases and are regarded as sample values for random variables. A *random variable* may be defined as a variable that takes on *several possible values*, and with each value is associated a probability. Probability may be defined as the number of chances for success divided by the total number of chances, or as the limit value to which the relative frequency of occurrence tends as the number of repetitions is increased indefinitely. As a simple illustration, consider the experiment of throwing a die and noting *the number of dots on the top face*. That number is a random variable, because there are six possible values, from 1 to 6. For example, the probability that a three-dot will occur tends to be $\frac{1}{6}$ as the number of throws gets to be very large. This probability value of $\frac{1}{6}$ or 0.166 is the limit of the relative frequency, or the ratio between the number of times a three-dot shows and the total number of throws.

A survey measurement, such as a distance or an angle, after mistakes are eliminated and systematic errors are corrected, is a random variable such as the number of dots in the die example. If the nominal value of an angle is $41°13'36''.0$, and the angle is measured 20 times, it is not unusual to get values for each of the measurements which differ slightly from the nominal angle. Each one of these values has a probability that it will occur. The closer the value approaches $41°13'36''$, the higher the probability, and the farther away it is, the lower the probability. This is, in very simple terms, the basic concept of the theory of observations, or the treatment of observational variables as random variables.

*In the past*, when the theory of errors was more prevalent, the value $41°13'36''$ was designated the "true value," which was never known. Then, when an observation was given which, owing to random variability, was different from the true value, an "error" was defined as

$$\text{error} = \text{measured value} - \text{true value} \qquad (2.4)$$

and a correction, which is the negative of the error, was defined as

$$\text{correction} = \text{true value} - \text{measured value} \qquad (2.5)$$

Whereas for systematic effects, the concept of error and correction is reasonable, in the case of random variation it is not, since there is no reason to say that "errors" have been committed. The so-called "random error" is actually nothing but a random variable, since it represents the difference between a random variable, the measurement, and a constant, the "true value." The ideal value of the error is zero (which in statistics is called the expected value); that for the observation is the true value (or the expectation or distribution mean, as will be explained later). The variation of the random errors around zero is identical to the variation of the observations around the expectation $\mu$—or the true value! Thus, it is better to talk about the observations themselves and seek better estimates for these observations rather than to discuss errors, since, strictly speaking, the values being analyzed are not errors.

Other classically used terms include *discrepancy*, which is the difference between two measurements of the same variable and obviously has no particular significance in the theory of observations. *Best value*, *most probable value*, and *corrected value* are all terms that refer to a new estimate of a random variable in the presence of redundancy. Such an estimate is usually obtained by some adjustment technique, such as least squares. If the random variable is $x$, the estimate from the adjustment is called the "least-squares estimate" and is denoted by $\hat{x}$. The "residual" has been defined by Eq. (2.3) and will be used in the same sense as it has been in the past. The "deviation" is simply the negative of the residual, and may be used on occasion in the course of statistical computation.

Next, consider some of the basic statistical concepts which are pertinent to survey computations and adjustment.

**2.10. Probability functions**  The concepts of random variable and probability were introduced and defined in the preceding section. It was said that a random variable can assume many different values, each having a specific probability. The *population* is defined as the totality of *all* possible values of a random variable. Because of the large size of the population, it is either impossible or totally impractical to seek all of its elements in order to evaluate its characteristics. Instead, only a certain number of observations, called a *sample*, is selected and studied. From the results of the sample study, inferences and statistical statements regarding the population can be made. As an example, consider the case of a horizontal angle. The population in this case may be composed of essentially an infinite number of measurements. In practice, a few measurements of the angle are usually considered sufficient. Specifically, assume a sample of size 10, and use the ten measurements to estimate parameters that belong to the population. One such parameter is the population mean, $\mu$ (the classical "true value"), for which the sample mean, or the mean of the ten observations, is an estimate. Since there is a probability associated with each value of the random variable, the relation between the values and their corresponding probabilities is expressed by the *probability functions*. There are two types: the cumulative function and the density function. The *density function* is easier to understand in the discrete case such as that of throwing a pair of dice. The probability of obtaining a particular number of dots (from 2 to 12) is plotted as the ordinate against the number of dots. In this case one can determine the probability of a certain outcome directly from the density function. In the continuous case, however, the probability is the area under the curve representing the density function $f(x)$, as shown in Fig. 2.2a. Thus, the probability that the random variable $\tilde{x}$ takes values in the interval between $x_1$ and $x_2$ is equal to the shaded area in Fig. 2.2a and is given by

$$P(x_1 < \tilde{x} < x_2) = \int_{x_1}^{x_2} f(x)\,dx \tag{2.6}$$

It should be obvious that the probability that $\tilde{x}$ takes either a $-\infty$ or $+\infty$ value is zero. Furthermore, the probability that the value of $\tilde{x}$ falls between $-\infty$ and $+\infty$ is 1. Thus, the total area under the density function curve is equal to $+1$ (there is no such thing as negative probability).

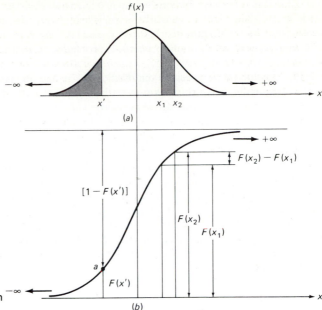

**Fig. 2.2** (a) Density distribution curve. (b) Cumulative distribution curve.

The *cumulative function* is defined by

$$F(x) = P(\tilde{x} < x) = \int_{-\infty}^{x} f(r)\, dr \tag{2.7}$$

where $f(r)$ is the density function, which leads to

$$f(x) = \frac{dF(x)}{dx} \tag{2.8}$$

The cumulative function is shown in Fig. 2.2b directly beneath the density function. This function goes to zero as $x$ approaches $-\infty$, and to $+1$ as $x$ approaches $+\infty$. At the abscissa $x = x'$, the probability that $\tilde{x} < x'$ is represented by the shaded area under the density function to the left of $x'$. In the cumulative function it is represented by the ordinate at $x = x'$. Thus,

$$P(\tilde{x} < x') = P(-\infty < \tilde{x} < x') = F(x') \tag{2.9}$$

Also, the probability that $\tilde{x} > x'$ would be represented by the area under the density function to the right of $x'$. Since the total area under the density function equals 1, then $P(\tilde{x} > x') = 1 - P(\tilde{x} < x')$. Therefore, on the cumulative curve, the probability that $\tilde{x} > x'$ would be represented by the vertical distance from the point on the curve $a$ to the maximum value possible (i.e., $+1$), or $1 - F(x')$. Hence,

$$P(x' < \tilde{x}) = P(x' < \tilde{x} < +\infty) = 1 - F(x') \tag{2.10}$$

The probability that $\tilde{x}$ lies between $x_1$ and $x_2$ is represented by the shaded area under the density function between $x = x_1$ and $x = x_2$. Correspondingly, on the cumulative function, it is the difference between the two ordinates, or $F(x_2) - F(x_1)$. Therefore,

$$P(x_1 < \tilde{x} < x_2) = F(x_2) - F(x_1) \tag{2.11}$$

The *density function* expresses the whole *population* and is characterized by a number of variables called *parameters*. Knowing the parameters totally specifies the density function. There are a number of density functions which are useful in the reduction of survey data. The most pertinent and commonly used density function is the normal or gaussian distribution density function, which is discussed in the following section.

There are other distribution functions (e.g., Student $t$, Chi square $\chi^2$, and $F$ distributions) which are also useful particularly in performing statistical tests on survey data. Tabulated values for these different distributions are usually given in terms of the cumulative function to simplify their application to practical problems, as will be shown in subsequent sections.

**2.11. The normal or gaussian distribution**  The one-dimensional normal distribution is the most frequently used distribution in statistical analysis of all types of data, including survey measurements. The normal density function is given by[†]

$$f(x) = \frac{1}{\sigma_x \sqrt{2\pi}} \exp\left[ -\frac{(x-\mu_x)^2}{2\sigma_x^2} \right] \qquad (2.12)$$

where $\mu_x$ and $\sigma_x$ are the only two parameters necessary to specify the function; $\mu_x$ is called the *mean* or *expectation* of the random variable $\tilde{x}$, and $\sigma_x$ is called its *standard deviation*, the square of which ($\sigma_x^2$) is called the *variance*. Said another way, $\mu_x$ and $\sigma_x^2$ are the mean and variance of the normal population of values which the random variable $\tilde{x}$ may take. The significance of $\mu_x$ and $\sigma_x$ is graphically depicted in Fig. 2.3. The variance is called a parameter of dispersion or spread. The value of $f(x)$ is a maximum at $x = \mu_x$, about which the normal curve is symmetric.

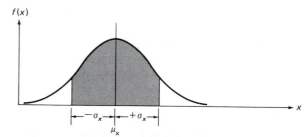

**Fig. 2.3** Points of inflection on the density distribution curve.

The points of inflection of the curve lie at a distance $\sigma_x$ on either side of $\mu_x$, as shown in Fig. 2.3. The *mean* $\mu_x$ is called a parameter of location; others include the median and mode. The *median* $x_m$ is defined by

$$\int_{-\infty}^{x_m} f(x)\,dx = \tfrac{1}{2} \qquad (2.13)$$

The *mode* is the value of the random variable at which the density function has a relative maximum. Because of the symmetry of the normal distribution, the median and mode are equal to the mean $\mu_x$.

Some of the useful features of the normal distribution are:

1. The normal density function approaches zero as $x$ goes to $\pm\infty$.
2. The probability that $x$ falls in the interval between $x_1$ and $x_2$ is the area under the curve bounded by $x = x_1$ and $x = x_2$. The shaded area in Fig. 2.3 represents the probability that $\tilde{x}$ falls within $\pm\sigma$ of the mean $\mu_x$. The probabilities for deviations from the mean of the first three integral multiples of $\sigma$ are

$$P(-\sigma_x < \tilde{x} - \mu_x < \sigma_x) = 0.6827$$
$$P(-2\sigma_x < \tilde{x} - \mu_x < +2\sigma_x) = 0.9545 \qquad (2.14)$$
$$P(-3\sigma_x < \tilde{x} - \mu_x < +3\sigma_x) = 0.9973$$

3. Sometimes it is useful to fix the probability and find the multiplier of $\sigma_x$ for the interval about $\mu_x$:

$$P(-1.645\sigma_x < \tilde{x} - \mu_x < +1.645\sigma_x) = 0.90$$
$$P(-1.960\sigma_x < \tilde{x} - \mu_x < +1.960\sigma_x) = 0.95 \qquad (2.15)$$
$$P(-2.576\sigma_x < \tilde{x} - \mu_x < +2.576\sigma_x) = 0.99$$

[†]The symbol "exp," pronounced exponential, is used such that $\exp x$ means $e$ to the power $x$, or $e^x$, where $e$ is the base of natural logarithm.

4. The probability that $\tilde{x}$ takes on values on either side of $\mu_x$ (i.e., either larger or smaller than $\mu_x$) is equal to 0.5.

Frequently, instead of working with the random variable $\tilde{x}$ it is much easier to use another random variable $\tilde{z}$ derived from $\tilde{x}$ by

$$\tilde{z} = \frac{\tilde{x} - \mu_x}{\sigma_x} \qquad (2.16)$$

This new random variable $\tilde{z}$, called the standardized normal random variable, has a mean $\mu_z = 0$ and variance $\sigma_z^2 = 1$.

The normal distribution is used extensively in surveying and geodesy, and in many other engineering fields. Although many of the random physical phenomena involved may not precisely obey the normal law, the normal distribution often provides a good approximation for other distributions. It also has many properties that make it easy to manipulate and apply.

**2.12. Expectation**  There are a number of properties which relate a random variable and its probability density function that are useful in our understanding of its behavior. The first of these expresses our intuitive concept of the mean as being the most probable value of the random variable. It is referred to by any of several terms—expectation, expected value, mean, or average—and will be denoted by $E(\tilde{x})$.

The *expected value* $E(\tilde{x})$ of a random variable $\tilde{x}$ is defined as the weighted sum $\mu_x$ of all possible values, where the weights are the corresponding probabilities. In mathematical terms

$$E(\tilde{x}) = \mu_x = \sum_{i=1}^{n} x_i P(x_i) \qquad \text{for a discrete distribution} \qquad (2.17)$$

and

$$E(\tilde{x}) = \mu_x = \int_{-\infty}^{\infty} x f(x)\, dx \qquad \text{for a continuous distribution} \qquad (2.18)$$

where $f(x)$ is the density function. Expectation may also be defined for a function $g(\tilde{x})$ of a random variable $\tilde{x}$ having a density function $f(x)$ as (assuming continuous function)

$$E(g(\tilde{x})) = \int_{-\infty}^{\infty} g(x) f(x)\, dx \qquad (2.19)$$

In the discrete case (2.17), if there are $n$ values of equal probability, then $P(x_i) = 1/n$ and $E(\tilde{x}) = \sum_{i=1}^{n} x_i/n = \mu_x$ or the mean, as previously stated.

If the probabilities are not equal, the weighted mean would result. As an example, consider a random variable $\tilde{y}$ equal to the sum of two fair dice yielding the following data:

| $y_i$ | $P(y_i)$ | $y_i P(y_i)$ |
|---|---|---|
| 2 | 1/36 | 2/36 |
| 3 | 2/36 | 6/36 |
| 4 | 3/36 | 12/36 |
| 5 | 4/36 | 20/36 |
| 6 | 5/36 | 30/36 |
| 7 | 6/36 | 42/36 |
| 8 | 5/36 | 40/36 |
| 9 | 4/36 | 36/36 |
| 10 | 3/36 | 30/36 |
| 11 | 2/36 | 22/36 |
| 12 | 1/36 | 12/36 |
| | | $\Sigma = 252/36$ |

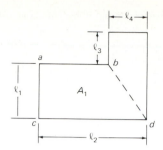

**Fig. 2.4** Expected value of an area.

so that

$$E(\tilde{y}) = \sum_{i=2}^{12} [y_i P(y_i)] = 7 = \mu_y$$

Several useful properties of expectation include:

$$E(c) = c \tag{2.20}$$
$$E(c\tilde{x}) = cE(\tilde{x}) \qquad (c = \text{a constant}) \tag{2.21}$$
$$E(E(\tilde{x})) = E(\mu_x) = \mu_x \tag{2.22}$$
$$E(\tilde{x} + \tilde{y}) = E(\tilde{x}) + E(\tilde{y}) \tag{2.23}$$
$$E(\tilde{x}\tilde{y}) \neq E(\tilde{x})E(\tilde{y}) \tag{2.24}$$

The relation in (2.24) is in general true unless $\tilde{x}$ and $\tilde{y}$ are independent random variables, in which case $E(\tilde{x}\tilde{y}) = E(\tilde{x})E(\tilde{y})$.

As an example of the application of the properties of expectation, consider the expected value of the area $A_1$ of the trapezoid $abcd$ in Fig. 2.4 assuming that $\ell_1$, $\ell_2$, $\ell_3$, and $\ell_4$ are not correlated.

$$E(A_1) = E\left[\ell_1\left(\frac{2\ell_2 - \ell_4}{2}\right)\right]$$

$E(A_1) = \frac{1}{2}\{E(\ell_1)[2E(\ell_2) - E(\ell_4)]\}$ and assuming that each of the values $\ell_1, \ell_2, \ell_4$ has respective means of $\mu_1, \mu_2, \mu_4$, then

$$E(A) = \frac{1}{2}(2\mu_1\mu_2 - \mu_1\mu_4)$$

**2.13. Variance, covariance, and correlation**   If the function $g(\tilde{x})$ in 2.19 is specifically $(\tilde{x} - \mu_x)^2$, then the expectation is called the *variance* and is designated by either var$(x)$ or $\sigma_x^2$; therefore,

$$\text{var}(x) = \sigma_x^2 = E\left[(\tilde{x} - \mu_x)^2\right] = \int_{-\infty}^{\infty} (x - \mu_x)^2 f(x)\, dx \tag{2.25}$$

The *positive square root* of the variance, called the *standard deviation*, is designated $\sigma_x$ and is a measure of the dispersion or spread of the random variable.

Similarly, if in Eq. (2.19) the specific function $(\tilde{x} - \mu_x)(\tilde{y} - \mu_y)$ relating to two random variables $\tilde{x}$ and $\tilde{y}$ is used, the expectation is called the *covariance* and is referred to by cov$(x,y)$ or $\sigma_{xy}$; thus,

$$\text{cov}(x,y) = \sigma_{xy} = E\left[(\tilde{x} - \mu_x)(\tilde{y} - \mu_y)\right] \tag{2.26}$$

The covariance expresses the mutual interrelation between the two random variables. (Unlike the standard deviation, the square root of the covariance, if $\sigma_{xy}$ is positive, has no meaning and is therefore not used.) Another term is the *correlation coefficient* $\rho_{xy}$, which is defined as

$$\rho_{xy} = \frac{\sigma_{xy}}{\sigma_x \sigma_y} = E\left[\left(\frac{\tilde{x} - \mu_x}{\sigma_x}\right)\left(\frac{\tilde{y} - \mu_y}{\sigma_y}\right)\right] \tag{2.27}$$

where $\sigma_x$ and $\sigma_y$ are the standard deviations of $x$ and $y$, respectively.

**2.14. Covariance, cofactor, and weight matrices**   In the one-dimensional case, there is one random variable $\tilde{x}$ with mean or expectation $\mu_x$ and a variance $\sigma_x^2$. In the two-dimensional case, there are two random variables $\tilde{x}$ and $\tilde{y}$, with means $\mu_x$ and $\mu_y$, and variances $\sigma_x^2$ and $\sigma_y^2$, respectively, and covariance $\sigma_{xy}$. The latter three parameters can be collected in a *square symmetric* matrix, $\Sigma$, of order 2 and called the *variance-covariance matrix* or simply the *covariance matrix*. It is constructed as

$$\Sigma = \begin{bmatrix} \sigma_x^2 & \sigma_{xy} \\ \sigma_{xy} & \sigma_y^2 \end{bmatrix} \tag{2.28}$$

where the variances are along the main diagonal and the covariance off the diagonal. The concept of the covariance matrix can be extended to the multidimensional case by considering $n$ random variables $\tilde{x}_1, \tilde{x}_2, \ldots, \tilde{x}_n$ and writing

$$\Sigma_{xx} = \begin{bmatrix} \sigma_1^2 & \sigma_{12} & \cdots & \sigma_{1n} \\ \sigma_{12} & \sigma_2^2 & \cdots & \sigma_{2n} \\ \cdots & \cdots & \cdots & \cdots \\ \sigma_{1n} & \sigma_{2n} & \cdots & \sigma_n^2 \end{bmatrix} \tag{2.29}$$

which is an $n \times n$ square symmetric matrix.

Often in practice, the variances and covariances are not known in absolute terms but only to a scale factor. The scale factor is given the symbol $\sigma_0^2$ and is termed the *reference variance*, although other names, such as "variance factor" and "variance associated with weight unity," have also been used. The square root $\sigma_0$, of $\sigma_0^2$, is called the "reference standard deviation" and was classically known as the "standard error of unit weight." The relative variances and covariances are called *cofactors* and are given by

$$q_{ii} = \frac{\sigma_i^2}{\sigma_0^2} \quad \text{and} \quad q_{ij} = \frac{\sigma_{ij}}{\sigma_0^2} \tag{2.30}$$

Collecting the cofactors in a square symmetric matrix produces the *cofactor matrix*, $\mathbf{Q}$, with the obvious relationship with covariance matrix

$$\mathbf{Q} = \frac{1}{\sigma_0^2} \Sigma \tag{2.31}$$

When $\mathbf{Q}$ is nonsingular, its inverse is called the *weight matrix* and designated by $\mathbf{W}$; thus,

$$\mathbf{W} = \mathbf{Q}^{-1} = \sigma_0^2 \Sigma^{-1} \tag{2.32}$$

If $\sigma_0^2$ is equal to 1, or in other words if the covariance matrix is known, the weight matrix becomes its inverse.

Equation (2.32) should be carefully understood, particularly in view of the *classical* definition of weights as being inversely proportional to variances. This is clearly not true *unless all covariances are equal to zero*, which means that all the random variables are mutually uncorrelated. Only then would $\Sigma$ (and $\mathbf{Q}$) become a diagonal matrix, and the weight of the random variable becomes equal to $\sigma_0^2$ divided by its variance.

**2.15. Sampling**   So far our discussion has been directed toward the total population as represented by the distribution functions and their parameters. Also, attention has been focused on the impracticality of considering the entire population. The logical alternative is to draw *samples* from the population. Using these samples, *estimates* are derived for the parameters of the probability distribution. These estimates are usually called *sample statistics*.

From the sample data some or all of the following can usually be computed:

1. Frequency diagrams (histograms and stereograms).
2. Sample statistics for location (mean, median, mode, midrange).
3. Sample statistics for dispersion (variances and covariances).

***Histograms and stereograms***   When a sample of repeated measurements for the same random variable is given, a *histogram* may be constructed to represent the probability density function. The sample values are divided into equal intervals of appropriate size and the numbers of observations in each interval are determined. For example, Table 2.1 contains 127 micrometer readings observed using a Wild T-2 theodolite. If an interval range of 1″ is selected, the number of micrometer observations falling into each interval can be determined. Table 2.2 shows the interval ranges, number of observations in each interval, and relative frequencies for each interval. The first interval is from 26.7 to 27.7″ inclusive, the second from 27.8 to 28.7″ inclusive, and so on. The number of measurements falling in each interval is determined and divided by 127 (the total number of observations) to give the corresponding relative frequencies which represent probabilities. Rectangles are then constructed over the intervals, with each having an area equal to the relative frequency resulting in the histogram depicted in Fig. 2.5. For this example, the number of observations in the interval 33.8 to 34.7″ inclusive is 22, yielding a relative frequency of $22/127 = 0.17$. Since the interval is 1″, the heights of rectangles in the histogram for respective intervals equal the relative frequencies. Note that the sum of the relative frequencies equals 1.

In the case of a single random variable a histogram is constructed, but for two random variables the frequency diagram is called a *stereogram*. The base would be a plane with square units when equal intervals are used for both variables, and square columns erected on them whose volumes correspond to the relative frequencies. Beyond two random variables, frequency diagrams become impractical.

In practice, histograms are frequently not used because their construction requires large sample sizes. Furthermore, the type of distribution (such as the normal distribution) is often known or assumed and estimates for its parameters (as, for example, the mean and variance) are sought using the sample data. In the following two sections the most commonly used sample statistics are discussed.

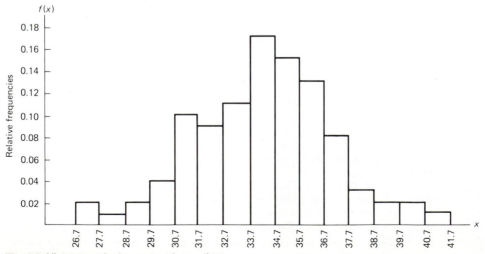

**Fig. 2.5** Histogram of micrometer observations.

**Table 2.1   127 Micrometer Readings Taken with the Wild T-2 Theodolite**

| n | Value, " of arc | n | Value, " of arc | n | Value, " of arc |
|---|---|---|---|---|---|
| 1 | 27.2 | 43 | 31.9 | 85 | 34.1 |
| 2 | 29.0 | 44 | 30.5 | 86 | 32.9 |
| 3 | 37.4 | 45 | 35.7 | 87 | 36.2 |
| 4 | 37.0 | 46 | 30.0 | 88 | 33.9 |
| 5 | 32.0 | 47 | 36.7 | 89 | 28.2 |
| 6 | 32.5 | 48 | 35.6 | 90 | 33.0 |
| 7 | 36.0 | 49 | 33.8 | 91 | 31.2 |
| 8 | 35.0 | 50 | 30.8 | 92 | 32.1 |
| 9 | 33.8 | 51 | 37.1 | 93 | 34.0 |
| 10 | 34.0 | 52 | 33.9 | 94 | 33.0 |
| 11 | 35.0 | 53 | 32.4 | 95 | 33.6 |
| 12 | 30.0 | 54 | 33.9 | 96 | 33.5 |
| 13 | 34.7 | 55 | 33.9 | 97 | 33.8 |
| 14 | 35.5 | 56 | 34.5 | 98 | 36.2 |
| 15 | 36.0 | 57 | 36.7 | 99 | 37.3 |
| 16 | 37.0 | 58 | 32.9 | 100 | 33.9 |
| 17 | 35.0 | 59 | 32.1 | 101 | 35.8 |
| 18 | 36.0 | 60 | 36.3 | 102 | 36.2 |
| 19 | 34.8 | 61 | 30.1 | 103 | 38.3 |
| 20 | 33.5 | 62 | 32.8 | 104 | 39.9 |
| 21 | 35.0 | 63 | 31.8 | 105 | 37.4 |
| 22 | 35.7 | 64 | 33.1 | 106 | 37.1 |
| 23 | 38.0 | 65 | 30.8 | 107 | 37.9 |
| 24 | 31.3 | 66 | 35.7 | 108 | 33.2 |
| 25 | 33.0 | 67 | 34.5 | 109 | 35.8 |
| 26 | 31.0 | 68 | 30.6 | 110 | 39.1 |
| 27 | 31.2 | 69 | 35.6 | 111 | 31.0 |
| 28 | 31.5 | 70 | 37.2 | 112 | 34.0 |
| 29 | 32.0 | 71 | 31.7 | 113 | 34.0 |
| 30 | 36.1 | 72 | 32.0 | 114 | 31.0 |
| 31 | 37.5 | 73 | 35.7 | 115 | 33.1 |
| 32 | 36.1 | 74 | 35.5 | 116 | 32.0 |
| 33 | 33.0 | 75 | 32.1 | 117 | 34.1 |
| 34 | 36.0 | 76 | 35.2 | 118 | 36.5 |
| 35 | 28.8 | 77 | 34.5 | 119 | 39.1 |
| 36 | 28.8 | 78 | 34.9 | 120 | 36.0 |
| 37 | 37.4 | 79 | 33.9 | 121 | 41.1 |
| 38 | 27.0 | 80 | 32.7 | 122 | 35.6 |
| 39 | 31.2 | 81 | 35.6 | 123 | 38.7 |
| 40 | 31.1 | 82 | 34.3 | 124 | 40.7 |
| 41 | 31.2 | 83 | 34.2 | 125 | 35.2 |
| 42 | 34.0 | 84 | 35.0 | 126 | 40.6 |
|  |  |  |  | 127 | 33.1 |

First interval, 26.7–27.7 ".
Last interval, 40.8–41.7 ".

$n = 127$
Mean value = 34.2 "

**Table 2.2**

| Interval range, " of arc | 26.7–27.7 | 27.8–28.7 | 28.8–29.7 | 29.8–30.7 | 30.8–31.7 | 31.8–32.7 | 32.8–33.7 | 33.8–34.7 | 34.8–35.7 | 35.8–36.7 | 36.8–37.7 | 37.8–38.7 | 38.8–39.7 | 39.8–40.7 | 40.8–41.7 | Σ |
|---|---|---|---|---|---|---|---|---|---|---|---|---|---|---|---|---|
| Number of observations in each interval | 2 | 1 | 3 | 5 | 13 | 12 | 14 | 22 | 19 | 16 | 10 | 4 | 2 | 3 | 1 | 127 |
| Relative frequency | 0.02 | 0.01 | 0.02 | 0.04 | 0.10 | 0.09 | 0.11 | 0.17 | 0.15 | 0.13 | 0.08 | 0.03 | 0.02 | 0.02 | 0.01 | 1.0 |

## 2.16. Sample statistics for location

**The sample mean**  The first and most commonly used measure of location is the *mean* of a sample, which is defined as

$$\bar{x} = \frac{1}{n} \sum_{i=1}^{n} x_i \qquad (2.33)$$

where $\bar{x}$ is the mean, $x_i$ are the observations, and $n$ is the total number of observations in the sample. It can be shown that

$$E(\bar{x}) = \mu = \text{population mean} \qquad (2.34)$$

Thus, the arithmetic mean of a set of independent observations is an unbiased estimate of the mean of the population describing the random variable for which observations are made. The sample mean of the 127 micrometer readings in Table 2.1 is $\bar{x} = 4339.9/127 = 34.2''$.

**The sample median**  Another measure of location is the median, $x_m$, which is obtained by arranging the values in the sample in their order of magnitude. The median is the value in the middle if the number of observations $n$ is odd, and is the mean of the middle two values if $n$ is even. Thus, the number of observations larger than the median equals the number smaller than the median.

The sample median from Table 2.1 is 34.0″.

**The sample mode**  The sample mode is the value that occurs most often in the sample. Therefore, if a histogram is developed from the sample data, it is the value at which the highest rectangle is constructed.

The sample mode from Table 2.1 is 33.9″.

**Midrange**  If the observation of smallest magnitude is subtracted from that of the largest magnitude, a value called the range is obtained. The value of the observation that is midway along the range is called the *midrange* and may be used, though infrequently, as a measure of position for a given sample. It is simply the arithmetic mean of the largest and smallest observations.

The midrange of the 127 micrometer readings is $27.0 + (41.1 - 27.0)/2 = 27.0 + 14.1/2 = 34.05''$.

## 2.17. Sample statistics for dispersion

**The range**  The simplest of dispersion (scatter or spread) measures is the range as defined in the preceding section. It is not, however, as useful a measure as others.

**The mean (or average) deviation**  This is a more useful measure of dispersion and has been conventionally called the average error. It is the arithmetic mean of the absolute values of the deviations from any measure of position (usually the mean). Thus, the mean deviation from the mean for a sample of $n$ observations would be given by

$$\text{mean deviation} = \frac{1}{n} \sum_{i=1}^{n} |(x_i - \bar{x})| \qquad (2.35)$$

**Sample variance and standard deviation**  The mean deviation, although useful in certain cases, does not reflect the dispersion or scatter of the measured values as effectively as the standard deviation, which was defined previously as the positive square root of the variance. The variance of a sample is

$$\hat{\sigma}_x^2 = \frac{1}{n-1} \sum_{i=1}^{n} (x_i - \bar{x})^2 \tag{2.36}$$

where $\bar{x}$ is the sample mean and $n$ the sample size. The reason for using $(n-1)$ instead of $n$ in Eq. (2.36) is that in this case the expectation of $\hat{\sigma}_x^2$ is equal to the population variance, $\sigma^2$. Thus, $\hat{\sigma}_x^2$ is an unbiased estimate of $\sigma^2$. It is often called the estimate of the variance of any one observation. On the other hand, the estimate of the variance of the mean $\bar{x}$ is given by

$$\hat{\sigma}_{\bar{x}}^2 = \frac{\hat{\sigma}_x^2}{n} \tag{2.37}$$

for $n$ independent observations. Equation (2.37) can be derived using the least-squares method as shown in Appendix B. For two-dimensional and higher cases the sample covariance between, for example, $x$ and $y$ is computed from

$$\hat{\sigma}_{xy} = \frac{1}{n-1} \sum_{i=1}^{n} [(x_i - \bar{x})(y_i - \bar{y})] \tag{2.38}$$

where $\bar{x}$ and $\bar{y}$ are the sample means for the random variables $\tilde{x}$ and $\tilde{y}$ computed from the application of Eq. (2.33).

**Example 2.1**   Twenty measurements of an angle are given in Table 2.3. Compute the mean of the angle $\bar{x}$, the standard deviation $\sigma_x$, and the standard deviation of the mean $\sigma_{\bar{x}}$.

**Table 2.3**

| Observation number | Value of angle | Observation number | Value of angle |
|---|---|---|---|
| 1 | 31°02′  29.3″ | 11 | 31°02′  24.1″ |
| 2 | 31°02′  24.0″ | 12 | 31°02′  26.2″ |
| 3 | 31°02′  27.9″ | 13 | 31°02′  30.1″ |
| 4 | 31°02′  26.8″ | 14 | 31°02′  29.7″ |
| 5 | 31°02′  26.1″ | 15 | 31°02′  24.1″ |
| 6 | 31°02′  25.9″ | 16 | 31°02′  26.2″ |
| 7 | 31°02′  26.1″ | 17 | 31°02′  27.1″ |
| 8 | 31°02′  27.8″ | 18 | 31°02′  24.9″ |
| 9 | 31°02′  27.2″ | 19 | 31°02′  25.7″ |
| 10 | 31°02′  28.0″ | 20 | 31°02′  25.2″ |

SOLUTION   Applying Eq. (2.33), the sample mean is

$$\bar{x} = 31°02′ \frac{532.4″}{20} = 31°02′26.6″$$

The deviations, $v_i^2$, from $\bar{x}$ are calculated and the variance $\hat{\sigma}_x^2$ is equal to the sum of $v_i^2$ divided by $(n-1)$ [Eq. (2.36)]. Thus,

$$\hat{\sigma}_x^2 = \frac{61.12}{19} = 3.22 \quad \text{or} \quad \hat{\sigma}_x = 1.8″$$

Also, the standard deviation of the mean may be obtained from Eq. (2.37) as

$$\hat{\sigma}_{\bar{x}} = \frac{\hat{\sigma}_x}{\sqrt{n}} = \frac{1.8}{\sqrt{20}} = 0.4″$$

**Example 2.2**   Refer to the sample of 127 micrometer observations illustrated in Table 2.1. Compute the mean of observations, $\bar{x}$, the estimated standard deviation $\hat{\sigma}_x$, the estimated standard deviation of the mean $\hat{\sigma}_{\bar{x}}$, and ordinates for the density distribution curve and plot this curve superimposed on the histogram.

SOLUTION   By Eq. (2.33) the sample mean is

$$\bar{x} = \frac{4339.9}{127} = 34.2''$$

The variance can be obtained using Eq. (2.36).

$$\hat{\sigma}_x^2 = \frac{1}{(127-1)} \sum_1^{127} (x_i - 34.2)^2 = 7.552$$

$$\hat{\sigma}_x = 2.75''$$

The standard deviation of the mean is calculated using Eq. (2.37) and is

$$\hat{\sigma}_{\bar{x}} = \frac{\hat{\sigma}_x}{\sqrt{n}} = \frac{2.75}{\sqrt{127}} = 0.24''$$

Since the mean $\bar{x}$ and standard deviation $\hat{\sigma}_x$ are estimated from the sample, if one assumes a normal distribution, then by Eq. (2.12),

$$y = f(x) = \frac{1}{\hat{\sigma}_x \sqrt{2\pi}} \exp\left[ -\frac{(x-\bar{x})^2}{2\hat{\sigma}_x^2} \right]$$

or

$$y = A_1 e^{-A_2(x-\bar{x})^2} = \frac{A_1}{e^{A_2(x-\bar{x})^2}}$$

in which

$$A_1 = \frac{1}{\hat{\sigma}_x \sqrt{2\pi}} \qquad \text{and} \qquad A_2 = \frac{1}{2\hat{\sigma}_x^2}$$

In this way the density distribution curve can be plotted to the scale of the histogram of the relative frequencies illustrated in Fig. 2.5. For this example,

$$A_1 = \frac{1}{(2.75)\sqrt{2\pi}} = 0.145 \qquad \text{and} \qquad A_2 = \frac{1}{(2)(7.552)} = 0.066$$

The remainder of the calculations are best performed in tabular form as follows:

| $x$, " of arc | $(x-\bar{x})^2$, (" of arc)$^2$ | $A_2(x-\bar{x})^2$ | $e^{A_2(x-\bar{x})^2}$ | $y = \dfrac{A_1}{e^{A_2(x-\bar{x})^2}}$ |
|---|---|---|---|---|
| 34.2 | 0 | 0 | 1.000 | 0.145 |
| 35.2 | 1 | 0.0662 | 1.068 | 0.136 |
| 36.2 | 4 | 0.2648 | 1.303 | 0.111 |
| 37.2 | 9 | 0.5959 | 1.815 | 0.080 |
| 38.2 | 16 | 1.059 | 2.88 | 0.050 |
| 39.2 | 25 | 1.655 | 5.23 | 0.028 |
| 40.2 | 36 | 2.384 | 10.84 | 0.013 |
| 41.2 | 49 | 3.244 | 25.63 | 0.006 |

Figure 2.6 shows the ordinates $y$ listed above plotted over the histogram. The density distribution curve is the continuous smooth curve formed by connecting these plotted points. Note that this curve has the characteristic bell shape of the normal density function previously discussed and that the area under the curve is the same as that bounded by the histogram and is equal to 1.

The values of standard deviation obtained in Examples 2.1 and 2.2 are actually a measure of the "quality" or precision of the estimate to which they refer. This concept is explained further in the next section.

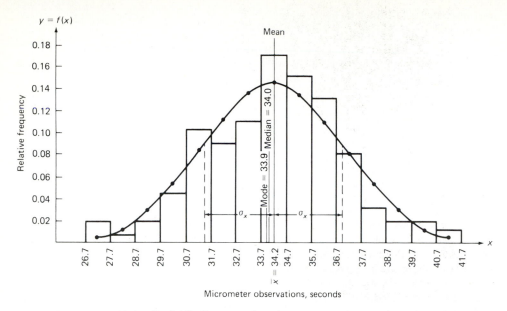

**Fig. 2.6** Histogram and density distribution curve for micrometer readings.

**2.18. Accuracy and precision**   The term *accuracy* refers to the closeness between measurements and their expectations (or, in conventional terms, to their true values). The farther a measurement is from its expected value, the less accurate it is. *Precision*, on the other hand, pertains to the closeness to one another of a set of repeated observations of a random variable. Thus, if such observations are closely clustered together, then the observations are said to have been obtained with high precision. It should be apparent, then, that observations may be precise but not accurate if they are closely grouped together but about a value that is different from the expectation (or true value) by a significant amount. Also, observations may be accurate but not precise if they are well distributed about the expected value but are significantly disbursed from one another. Finally, observations will be both precise and accurate if they are closely grouped around the expected value (or the distribution mean).

One of the examples used most often to demonstrate the difference between the two concepts of accuracy and precision is that of rifle shot groupings. Figure 2.7 shows three different types of groupings that it is possible to obtain. From the discussion above, group (*a*) is both accurate and precise, group (*b*) is precise but not accurate, and group (*c*) is accurate but not precise. One of the harder notions to accept is that case (*c*) is in fact *accurate*, even though the scatter between the different shots is rather large. A justification which may help is that we can visualize that the center of mass (which is equivalent to the expected value of the different shots) turns out to be very close to the target center (which is the true value).

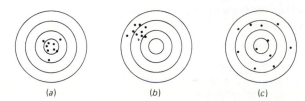

**Fig. 2.7** Rifle shot groupings.   (*a*)          (*b*)          (*c*)

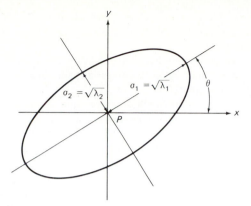

**Fig. 2.8** Error ellipse.

**2.19. Error ellipses**   The variance or standard deviation are measures of precision for the one-dimensional case of an angle or a distance, for example. In the case of two-dimensional problems, such as the horizontal position of a point, error ellipses may be established around the point to designate precision regions of different probabilities. The orientation of the ellipse relative to the $x,y$ axes system (Fig. 2.8) depends on the correlation between $x$ and $y$. If they are uncorrelated, the ellipse axes will be parallel to $x$ and $y$. If the two coordinates are of equal precision, or $\sigma_x = \sigma_y$, the ellipse becomes a circle.

Considering the general case where the covariance matrix for the position of point $P$ is given as

$$\Sigma = \begin{bmatrix} \sigma_x^2 & \sigma_{xy} \\ \sigma_{xy} & \sigma_y^2 \end{bmatrix} \tag{2.39}$$

The semimajor and semiminor axes of the corresponding ellipse are computed in the following manner. First, a second-degree polynomial (called the characteristic polynomial) is set up using the elements of $\Sigma$ as

$$\lambda^2 - (\sigma_x^2 + \sigma_y^2)\lambda + (\sigma_x^2 \sigma_y^2 - \sigma_{xy}^2) = 0 \tag{2.40}$$

The two roots $\lambda_1, \lambda_2$ of Eq. (2.40) (which are called the eigenvalues of $\Sigma$) are computed and their square roots are the semimajor and semiminor axes of the *standard error ellipse*, as shown in Fig. 2.8. The orientation of the ellipse is determined by computing $\theta$ between the $x$ axis and the semimajor axis from

$$\tan 2\theta = \frac{2\sigma_{xy}}{\sigma_x^2 - \sigma_y^2} \tag{2.41}$$

The quadrant of $2\theta$ is determined from the fact that the sign of $\sin 2\theta$ is the same as the sign of $\sigma_{xy}$, and $\cos 2\theta$ has the same sign as $(\sigma_x^2 - \sigma_y^2)$. Whereas in the one-dimensional case, the probability of falling within $+\sigma$ and $-\sigma$ is 0.6827 [see Eq. (2.14)], the probability of falling on or inside the standard error ellipse is 0.3935. In a manner similar to constructing intervals with given probabilities as in Eq. (2.15) for the one-dimensional case, different-size ellipses may be established, each with a given probability. It should be obvious that the larger the size of the error ellipse, the larger is the probability. Using the standard ellipse as a base, Table 2.4 gives the scale multiplier $k$ to enlarge the ellipse and the corresponding probability (see Mikhail, Ref. 8).

**Table 2.4**

| $k$ | 1.000 | 1.177 | 2.146 | 2.447 | 3.035 |
|-----|-------|-------|-------|-------|-------|
| $p$ | 0.394 | 0.500 | 0.900 | 0.950 | 0.990 |

As an example, for an ellipse with axes $a = 2.447a_s$ and $b = 2.447b_s$ where $a_s, b_s$ are the semimajor and semiminor axes, respectively, of the standard ellipse, the probability that the point falls inside the ellipse is 0.95.

In the three-dimensional case, where the horizontal position as well as the elevation of the point is involved, the precision region becomes an ellipsoid. For discussion of error ellipsoids, the reader may consult Ref. 8.

The concepts of error ellipse and error ellipsoid are quite useful in establishing confidence regions about points determined by surveying techniques. These regions are measures of the reliability of the positional determination of such points. They could also be specified in advance as a means of establishing specifications. For an indication of the numerical calculations involved in evaluating an error ellipse, refer to Example 2.9.

**2.20. Error propagation**    Our discussion has so far concentrated on having directly the observations or measurements of the random variables of interest and evaluating their means and measures of their precision. It is not always true in practice that direct measurements of the quantities needed will be available. Therefore, it is an important matter to show how to obtain the precision of a quantity or quantities computed from measurements through known mathematical relationships. For example, in trigonometric leveling one measures the slope distance $s$ and the inclination angle $\beta$ and computes the height $h$ from $h = s \sin \beta$. The question now is: If $\sigma_s$ and $\sigma_\beta$ are known, how do we compute $\sigma_h$? The answer to this question involves the application of what is conventionally known as the technique of "error propagation." Actually, it may be better termed the propagation of variances and covariances. The general case of propagation can be expressed as follows. Let **y** be a set (vector) of $m$ quantities each of which is a function of another set (vector) **x** of $n$ random variables. Given the covariance matrix $\Sigma_{xx}$ (or the cofactor matrix $\mathbf{Q}_{xx}$) for the variables **x**, the covariance matrix $\Sigma_{yy}$ (or cofactor matrix $\mathbf{Q}_{yy}$) for the new quantities **y** may be evaluated from

$$\Sigma_{yy} = \mathbf{J}_{yx} \Sigma_{xx} \mathbf{J}_{yx}^t \tag{2.42}$$

or

$$\mathbf{Q}_{yy} = \mathbf{J}_{yx} \mathbf{Q}_{xx} \mathbf{J}_{yx}^t \tag{2.43}$$

where $\mathbf{J}_{yx}$ is $m \times n$ and is called the jacobian matrix, or the partial derivative of **y** with respect to **x**, with the following elements:

$$\mathbf{J}_{yx} = \begin{bmatrix} \dfrac{\partial y_1}{\partial x_1} & \dfrac{\partial y_1}{\partial x_2} & \cdots & \dfrac{\partial y_1}{\partial x_n} \\[2mm] \dfrac{\partial y_2}{\partial x_1} & \dfrac{\partial y_2}{\partial x_2} & \cdots & \dfrac{\partial y_2}{\partial x_n} \\[2mm] \cdots & \cdots & \cdots & \cdots \\[2mm] \dfrac{\partial y_m}{\partial x_1} & \dfrac{\partial y_m}{\partial x_2} & \cdots & \dfrac{\partial y_m}{\partial x_n} \end{bmatrix} \tag{2.44}$$

Equation (2.42) or (2.43) is quite general inasmuch as multiple functions in terms of several variables are considered and, more important, no restrictions are imposed on the structure of the given covariance matrix $\Sigma_{xx}$. Therefore, the given random variables could in general be of unequal precision and correlated, so that $\Sigma_{xx}$ would be a full matrix. From the general propagation relationships of Eq. (2.42) several relationships could be obtained. First consider the case of a single function $y$ of several $(n)$ variables $x_1, x_2, \ldots, x_n$, which are *uncorrelated* and with variances $\sigma_1^2, \sigma_2^2, \ldots, \sigma_n^2$, respectively. Equation (2.42) becomes

$$\sigma_y^2 = \begin{bmatrix} \dfrac{\partial y}{\partial x_1} & \dfrac{\partial y}{\partial x_2} & \cdots & \dfrac{\partial y}{\partial x_n} \end{bmatrix} \begin{bmatrix} \sigma_1^2 & & & 0 \\ & \sigma_2^2 & & \\ & & \ddots & \\ 0 & & & \sigma_n^2 \end{bmatrix} \begin{bmatrix} \partial y/\partial x_1 \\ \partial y/\partial x_2 \\ \vdots \\ \partial y/\partial x_n \end{bmatrix} \tag{2.45}$$

or

$$\sigma_y^2 = \left( \frac{\partial y}{\partial x_1} \right)^2 \sigma_1^2 + \left( \frac{\partial y}{\partial x_2} \right)^2 \sigma_2^2 + \cdots + \left( \frac{\partial y}{\partial x_n} \right)^2 \sigma_n^2 \tag{2.46}$$

Of course, if the variables $x_i$ were correlated, $\Sigma_{xx}$ in Eq. (2.45) would not be a diagonal matrix, and Eq. (2.46) would include cross-product terms in all combinations. In such a case it would be unwise to write the expanded form of Eq. (2.46), but instead work with the matrix form directly.

Note that Eqs. (2.42) through (2.46) are given in terms of variances and covariances of the distributions. However, since such parameters are rarely known in practice, the equations apply equally using sample variances and covariances.

**Example 2.3** In trigonometric leveling, the slope distance is $s = 50.00$ m with $\sigma_s = 0.05$ m, and $\beta = 30°00'$ with $\sigma_\beta = 00°30'$. Compute $h$ and $\sigma_h$. (Assume $s$ and $\beta$ to be uncorrelated.)

SOLUTION

$h = s \sin \beta = (50.00)(0.5) = 25.00$ m

$$\sigma_h^2 = \left( \frac{\partial h}{\partial s} \right)^2 \sigma_s^2 + \left( \frac{\partial h}{\partial \beta} \right)^2 \sigma_\beta^2$$

$\quad = (\sin \theta)^2 (0.5)^2 + (s \cos \beta)^2 (0.0087)^2 = 0.2053$ m$^2$

$\sigma_h = 0.45$ m

In this example note that $\sigma_\beta$ was converted to radians to balance the dimensions in the relationship.

**Example 2.4** The area of a rectangular parcel of land is required together with its standard deviation. The length is $a = 100$ m with $\sigma_a = 0.50$ m, and the width is $b = 40$ m with $\sigma_b = 0.30$ m. (Assume $a$ and $b$ to be uncorrelated.)

SOLUTION

$A = ab = 4000$ m$^2$

$\sigma_A^2 = b^2 \sigma_a^2 + a^2 \sigma_b^2$

$\quad = (40)^2 (0.5)^2 + (100)^2 (0.3)^2 = 1300$ m$^4$

$\sigma_A = 36.06$ m$^2$

A further simplification for the case given by Eq. (2.46) is to consider $y$ as a *linear* function of $x_i$, or

$$y = a_1 x_1 + a_2 x_2 + \cdots + a_n x_n \tag{2.47}$$

in which case Eq. (2.46) becomes

$$\sigma_y^2 = a_1^2 \sigma_1^2 + a_2^2 \sigma_2^2 + \cdots + a_n^2 \sigma_n^2 \tag{2.48}$$

If all $a_i$ in Eq. (2.47) are equal to either $+1$ or $-1$ (addition and/or subtraction of the random variables), that is,

$$y = x_1 \pm x_2 \pm \cdots \pm x_n \tag{2.49}$$

then the variance of $y$ is simply the *sum* of all the variances of $x_i$, or

$$\sigma_y^2 = \sigma_1^2 + \sigma_2^2 + \cdots \sigma_n^2 \tag{2.50}$$

Finally, if in addition to being uncorrelated, all $x_i$ have the same precision with a variance $\sigma_x^2$ for each, then $\sigma_y^2$ for the function in Eq. (2.49) will be

$$\sigma_y^2 = n\sigma_x^2 \tag{2.51}$$

**Example 2.5**  Three adjacent distances along the same line were independently measured, with the following results: $x_1 = 51.00$ m with $\sigma_1 = 0.05$ m; $x_2 = 36.50$ m with $\sigma_2 = 0.04$ m; and $x_3 = 26.75$ m with $\sigma_3 = 0.03$ m. Compute the total distance and its standard deviation.

SOLUTION

$y = x_1 + x_2 + x_3 = 114.25$ m

and from Eq. (2.50),

$\sigma_y^2 = \sigma_1^2 + \sigma_2^2 + \sigma_3^2 = (0.05)^2 + (0.04)^2 + (0.03)^2 = 0.005$ m$^2$

$\sigma_y = 0.07$ m

**Example 2.6**  If in Example 2.5 all three distances were measured with a standard deviation $\sigma_x = 0.05$ m, what would $\sigma_y$ be?

SOLUTION  From Eq. (2.51),

$\sigma_y^2 = 3(0.05)^2 = 0.0075$ m$^2$

$\sigma_y = 0.09$ m

**Example 2.7**  A distance is independently measured by two observers to be $x_1 = 110.00$ m and $x_2 = 110.8$ m. If the weights of these two measurements are $w_1 = 2$ and $w_2 = 3$, respectively, and the best estimate is the weighted mean given by $\hat{x} = (w_1 x_1 + w_2 x_2)/(w_1 + w_2)$, compute the standard deviation of $\hat{x}$ assuming a reference variance of 1.

SOLUTION  Referring to Art. 2.14, since the two values $x_1$ and $x_2$ are uncorrelated and $\sigma_0^2 = 1$, then from Eq. (2.32),

$$\mathbf{W} = \begin{bmatrix} w_1 & 0 \\ 0 & w_2 \end{bmatrix} = \Sigma^{-1} = \begin{bmatrix} \dfrac{1}{\sigma_1^2} & 0 \\ 0 & \dfrac{1}{\sigma_2^2} \end{bmatrix}$$

Next,

$$\sigma_{\hat{x}}^2 = \left(\frac{\partial \hat{x}}{\partial x_1}\right)^2 \sigma_1^2 + \left(\frac{\partial \hat{x}}{\partial x_2}\right)^2 \sigma_2^2$$

$$= \left(\frac{w_1}{w_1 + w_2}\right)^2 \left(\frac{1}{w_1}\right) + \left(\frac{w_2}{w_1 + w_2}\right)^2 \frac{1}{w_2}$$

$$= \frac{w_1 + w_2}{(w_1 + w_2)^2} = \frac{1}{w_1 + w_2}$$

which means that the weight $w_{\hat{x}}$ of the weighted mean of two observations is equal to the sum of their weights, or $w_{\hat{x}} = w_1 + w_2$.

Finally,

$$\sigma_{\hat{x}}^2 = \frac{1}{w_1 + w_2}$$

When more than one function $y_j$ are involved, it is best to use the general relationship (2.42) directly rather than try to write an expanded form that will turn out to be unduly involved. One statement is important, though; even if the original variables $x_i$ are uncorrelated, the computed quantities $y_j$ are not necessarily uncorrelated, as demonstrated by the following example.

**Example 2.8**   The position of point $A$, in Fig. 2.9 is determined by the radial distance $r = 100$ m, with $\sigma_r = 0.5$ m, and the azimuth angle $\alpha = 60°$, with $\sigma_\alpha = 0°30'$. Compute rectangular coordinates $x$, $y$ and the associated covariance matrix for point $A$. (Assume $r$ and $\alpha$ to be uncorrelated.)

SOLUTION   From Fig. 2.9,

$$x = r\sin\alpha = 86.60$$
$$y = r\cos\alpha = 50.00$$

$$J = \begin{bmatrix} \dfrac{\partial x}{\partial \alpha} & \dfrac{\partial x}{\partial r} \\ \dfrac{\partial y}{\partial \alpha} & \dfrac{\partial y}{\partial r} \end{bmatrix} = \begin{bmatrix} r\cos\alpha & \sin\alpha \\ -r\sin\alpha & \cos\alpha \end{bmatrix} = \begin{bmatrix} 50 & 0.866 \\ -86.6 & 0.5 \end{bmatrix}$$

The covariance matrix of the known variables $r, \alpha$ is a diagonal matrix because they are uncorrelated and is equal to

$$\Sigma = \begin{bmatrix} \sigma_\alpha^2 & 0 \\ 0 & \sigma_r^2 \end{bmatrix} = \begin{bmatrix} (0.0087)^2 & 0 \\ 0 & (0.5)^2 \end{bmatrix}$$

Applying Eq. (2.42), the covariance matrix of the cartesian coordinates is

$$\Sigma_{coord} = \begin{bmatrix} 50 & 0.866 \\ -86.6 & 0.5 \end{bmatrix} \begin{bmatrix} 0.7569 \times 10^{-4} & 0 \\ 0 & 0.25 \end{bmatrix} \begin{bmatrix} 50 & -86.6 \\ 0.866 & 0.5 \end{bmatrix}$$

$$= \begin{bmatrix} 0.3767 & -0.2195 \\ -0.2195 & 0.630 \end{bmatrix}$$

This matrix expresses the reliability of the cartesian coordinates of point $A$.

**Example 2.9**   Using the covariance matrix $\Sigma_c$ for the cartesian coordinates computed in Example 2.8, compute the semimajor and semiminor axes of the standard ellipse and its orientation.

SOLUTION   According to Eq. (2.40), the characteristic polynomial is

$$\lambda^2 - (0.3767 + 0.6300)\lambda + (0.3767)(0.6300) - (-0.2195)^2 = 0$$
$$\lambda^2 - 1.0067\lambda + 0.189141 = 0$$

from which the two roots (eigenvalues) are

$$\lambda_1 = 0.7567 \qquad \lambda_2 = 0.25$$

The semimajor axis is $a = \sqrt{\lambda_1} = 0.87$ m and the semiminor axis is $b = \sqrt{\lambda_2} = 0.50$ m. To get the

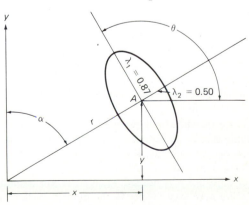

**Fig. 2.9** Error ellipse for rectangular coordinates.

orientation of the semimajor axis, use Eq. (2.41).

$$\tan 2\theta = \frac{2\sigma_{xy}}{\sigma_x^2 - \sigma_y^2} = \frac{-2(0.2195)}{0.3767 - 0.6300} = \frac{-0.4390}{-0.2533} = 1.732$$

Because both $\sin 2\theta$ and $\cos 2\theta$ are negative, then $2\theta$ is in the third quadrant, or $2\theta = 240°$, and $\theta = 120°$.

The results obtained here are rather important. They show, as demonstrated in Fig. 2.9, that the semiminor axis is along the extension of $r$ and is equal to $\sigma_r = 0.5$ m. Similarly, the semimajor axis is oriented normal to $r$ and is equal to $r\sigma_\theta = (100)(0.0087) = 0.87$ m. Thus, the error ellipse axes are always oriented in the directions where the variables are uncorrelated, as in this case $r$ and $\alpha$. Along any other pair of directions the variables are correlated, as in the case of $x$ and $y$, for which the correlation coefficient $\rho_{xy}$ may be computed from Eq. (2.27) as

$$\rho_{xy} = \frac{-0.2195}{(0.3767)(0.6300)} = -0.45$$

**2.21. Estimation**   In surveying, the values of certain quantities are obtained through the use of observations which are considered as random variables. Such random variables follow some type of distribution, which may be specified by a certain number of parameters. Determination of the actual values of these parameters requires an infinitely large amount of observational data, which is obviously impractical. Instead, a finite sample of data is selected and used to make inferences regarding the population parameters. Making inferences concerning the parameters of probability distributions on the basis of sample data is called *estimation*. From the sample data, sample *statistics* are computed. Each statistic corresponds to a parameter and is called an *estimator* for the corresponding parameter. For example, the sample mean $\bar{x}$ [Eq. (2.33)] is an estimator for the population mean $\mu$, and the sample variance $\hat{\sigma}_x^2$ and covariance $\hat{\sigma}_{xy}$ [Eqs. (2.36) and (2.38)] are estimators for the population variance $\sigma_x^2$ and covariance $\sigma_{xy}$, respectively. These estimators are called *unbiased* because the expectation of each is equal to its corresponding parameter. The numerical value of each statistic computed from the sample data is called an *estimate* or, more precisely, a *point estimate*. The reason for the use of the term "point" is that each parameter will have a single estimate value as opposed to a range of values in the case of interval estimation to be explained shortly. Methods of point estimation include the *maximum likelihood*, which requires knowledge of the distribution, and *least squares*, which does not. Least-squares estimation, by far the most common technique used in surveying, is discussed in the following section and presented in detail in Appendix B.

In *interval estimation* an attempt is made to specify "how good" an estimate is by establishing an interval such that the probability that it includes the parameter in question is high. In other words, if the measurements are repeated many times, then each time an interval is established, a high percentage of the intervals would be expected to contain the parameter. Such intervals are also called *confidence intervals*, in which the term confidence is implied by the probability associated with the interval. For a detailed discussion of confidence intervals, see Refs. 8 and 9.

**2.22. Least-squares adjustment**   As mentioned in the preceding section, least squares is the one method of estimation or adjustment most commonly used in surveying and geodesy. Most people refer to least squares as an *adjustment* technique which is equivalent to estimation in statistics. Although adjustment is not the most precise term, it is appropriate since adjustment is needed when there are redundant observations (i.e., more observations than are necessary to specify the model). In this case the total set of observations given are not consistent with the model and are replaced by another set of estimates, classically called *adjusted observations* (which is also not a precise term), which satisfy the model. As an example, consider the determination of a distance between two points. The distance may be considered as a random variable, and if measured once would have one estimate, so that no adjustment is needed. On the other hand, if the distance were measured three times, there would likely be three slightly different values, $x_1$, $x_2$, and $x_3$. Since the model concerns a single distance that would be *uniquely* specified by one measurement, it is obvious that there are two *redundant* measurements. It is also clear that a situation exists which requires adjustment in order to have a unique solution. Otherwise, there are several different possibilities for the required distance: we can take any one of $x_1, x_2, x_3$, or a combination of $x_1$ and $x_2$; or $x_2$ and $x_3$; or $x_1$ and $x_3$; or $x_1$, $x_2$, and $x_3$; and so on. Therefore, having redundant observations makes it possible to have numerous ways of computing the desired values. The multiplicity of possibilities and arbitrariness of choice in obtaining the required information is obviously undesirable. Instead, a process or technique must be found such that one would always get one unique answer, which is derived from the data and is the "best" that can be obtained. This is why the relative confidence in, or merit of, the different observations should be taken into account when computing the best estimate, which is defined as the estimate that deviates least from all the observations while considering their relative reliability. This is basically the role of least squares adjustment. As another example, consider the simple case of a plane triangle in which the three angles must add up to 180°. The fact that the three angles must add to 180° represents a functional relationship which reflects the geometrical system involved in the problem. If the shape and not the size of the triangle is of interest, it is not necessary to observe the magnitudes of three angles, since two angles will be sufficient to determine the third from the functional relationship just mentioned. However, in practice the three angles $\alpha$, $\beta$, and $\gamma$ are measured whenever possible, and their sum will likely be different from 180°. Suppose that the sum of the angles exceeds 180° by 3″ of arc. Any two of the three measured angles would give the shape of the triangle, but all three possibilities will be, in general, different. Therefore, in order to satisfy the condition that the sum of the angles must be 180°, the values of the observed angles must be altered. Here, there are numerous possibilities: 3″ may be subtracted from any one of the three angles; perhaps 2″ could be subtracted from the largest angle, 1″ from the second largest, and nothing from the third angle; or it may appear to be more satisfactory if 1″ were subtracted from each angle assuming that they are equally reliable, taken by the same instrument and observer under quite similar conditions. An alternative criterion to those given above may be to apply alterations which are proportionate to the relative magnitudes of the angles, or the magnitudes of their complements or supplements; or even inversely proportionate to such magnitudes. It is clear, then, that while adjustment is necessary, the large number of possibilities given are quite arbitrary and a *criterion* is required in addition to the satisfaction of the functional model of summing the angle to 180°. Such is the *least-squares* criterion.

Let $\ell$ designate the vector of given observation and $\mathbf{v}$ the vector of *residuals* (or alterations) which when added to $\ell$ yields a set of new estimates $\hat{\ell}$ which is consistent with the model. Thus,

$$\hat{\ell} = \ell + \mathbf{v} \tag{2.52}$$

The statistical or stochastic properties of the observations are expressed by either the covariance or cofactor matrix $\Sigma$ or $\mathbf{Q}$, or by the weight matrix $\mathbf{W}$. (Note that $\mathbf{W} = \mathbf{Q}^{-1}$, and $\mathbf{W} = \Sigma^{-1}$ if the reference variance $\sigma_0^2$ is equal to unity.) With these variables, the general form of the least-squares criterion is given by

$$\phi = \mathbf{v}^t \mathbf{W} \mathbf{v} \rightarrow \text{minimum} \tag{2.53}$$

Note that $\phi$ is a scalar, for which a minimum is obtained by equating to zero its partial derivative with respect to **v**, as elaborated upon in Appendix B. In Eq. (2.53) the weight matrix of the observations **W** may be full, implying that the observations are correlated. If the observations are uncorrelated, **W** will be a diagonal matrix and the criterion simplifies to

$$\phi = \sum_{i=1}^{n} w_i v_i^2 = w_1 v_1^2 + w_2 v_2^2 + \cdots + w_n v_n^2 \rightarrow \text{minimum} \tag{2.54}$$

which says that the sum of the weighted squares of the residuals is a minimum. Another and simpler case involves observations which are uncorrelated and of equal weight (precision), for which **W** = **I** and $\phi$ becomes

$$\phi = \sum_{i=1}^{n} v_i^2 = v_1^2 + v_2^2 + \cdots + v_n^2 \rightarrow \text{minimum} \tag{2.55}$$

The case covered by Eq. (2.55) is the oldest and may have accounted for the name "least squares," since it seeks the "least" sum of the squares of the residuals.

If we refer back to the example of measuring a distance three times, and assume that $x_1$, $x_2$, and $x_3$ are of equal precision (weight) and uncorrelated, it can be shown that the $\Sigma v_i^2$ is a minimum if the best estimate $\hat{x}$ is taken as the arithmetic mean of the three observations. Similarly, if the three interior angles $\alpha$, $\beta$, and $\gamma$ in the plane triangle example have a unit weight matrix, the method of least squares will yield all three residuals equal to $-1''$. Thus, when each angle is reduced by $1''$, their sum will be $180°$ and the functional model will be satisfied. These two examples, as well as several others, are worked out in detail in Appendix B.

**2.23. Concluding remarks**   In this chapter the basic concepts involved in the acquisition and reduction of survey measurements have been introduced to the surveying student. In succeeding chapters many of the techniques developed in this chapter will be used. For instance, in distance measurement (Chap. 4), leveling (Chap. 5), and direction and angle measurements (Chap. 6), the various systematic errors (discussed in Arts. 2.6 to 2.8) are identified and relations for their compensation developed.

When repeated survey measurements are obtained, sample statistics for location (Art. 2.16) and for dispersion (Art. 2.17) are calculated. The most common of these are the sample mean [Eq. (2.33)] and sample variance [Eq. (2.36)]. When two or more interrelated variables (e.g., the $X$ and $Y$ coordinates of a survey point) are involved, the sample covariance [Eq. (2.38)] becomes important. In the two-dimensional case, the confidence with which a survey point is located is expressed by an *error ellipse* centered at the point. The size of that ellipse reflects the degree of this confidence: the larger the size, the more confident we are that the true point will be inside the ellipse (Art. 2.19). Error ellipses are often used in the reverse case of survey design, where their sizes are specified beforehand and techniques and equipment selected to meet the specifications. Discussions concerning the relation of error ellipses to surveying problems can be found in Example 8.4, Art. 8.23, and Examples 20.1 and 20.2, Art. 20.8.

Another very useful concept is that of error propagation covered in Art. 2.20. Wherever a quantity is to be calculated as a function of other variables, error propagation rules [Eqs. (2.42), (2.43), or (2.46)] are used to obtain its precision from the given precisions of the variables in the function. The reader will encounter the application of error propagation in almost every chapter.

This chapter concludes with an introduction to the method of least-squares adjustment (Art. 2.22), where it is explained that whenever more survey measurements are obtained than are necessary for a unique solution, adjustment is required. It is shown that least squares as a procedure of adjustment has certain optimum qualities and is therefore commonly used in surveying and mapping. Derivation of the techniques of least squares is deferred, however, to Appendix B, so that its application to various survey nets (yet to be covered in succeeding chapters) can be explained.

## 2.24. Problems

**2.1**  Explain what is meant by mathematical model and the basic difference between its two components: the functional and stochastic models.

**2.2**  A survey measurement, corrected for blunders and systematic errors, is considered as a random variable; so is its associated random error. Discuss the similarities and differences between the two.

**2.3**  Describe the three types of errors and their influence on survey measurements.

**2.4**  Explain several methods by which one can compensate for recognized systematic errors.

**2.5**  A linear distance is assumed to be normally distributed with a mean $\mu = 113.15$ m and variance $\sigma^2 = 0.20$ m$^2$. Sketch the distribution showing the 0.9545 probability region. Give the range in distance corresponding to this probability.

**2.6**  Define, in terms of the concept of expectation, the mean, the standard deviation, the covariance, and the correlation coefficient.

**2.7**  What is meant by reference variance and what is its significance in the treatment of survey measurements?

**2.8**  Three independent distance measurements have the following standard deviations: 2 cm, 4 cm, and 6 cm. Construct the corresponding covariance matrix. What is the correlation coefficient between the first and third measurements?

**2.9**  For Prob. 2.8 calculate the weight matrix $\mathbf{W}$ (see Art. 2.14), if a reference variance is selected as $\sigma_{01}^2 = 16$. Using another value, $\sigma_{02}^2 = 8$, what is the new weight matrix?

**2.10**  Calculate the ratio $w_{11} : w_{22} : w_{33}$ for both weight matrices in Prob. 2.9 such that the number corresponding to $w_{11}$ is 1. What conclusion can you draw?

**2.11**  Two adjacent angles are measured such that the standard deviations are $\sigma_1 = 1'$ of arc and $\sigma_2 = 1.5'$ of arc. If the correlation coefficient between the two angles is $\rho_{12} = 0.5$, calculate the covariance matrix using radians as the angular measure. Selecting a suitable value for the reference variance, calculate the corresponding cofactor and weight matrices.

**2.12**  The following are 25 independent measurements of a distance:

| | | | | |
|---|---|---|---|---|
| 731.68 (m) | 731.64 | 731.60 | 731.58 | 731.65 |
| 731.59 | 731.56 | 731.61 | 731.63 | 731.59 |
| 731.61 | 731.61 | 731.69 | 731.60 | 731.61 |
| 731.64 | 731.62 | 731.60 | 731.55 | 731.67 |
| 731.62 | 731.60 | 731.63 | 731.64 | 731.62 |

(a) Construct a histogram from these sample data.

(b) Calculate the sample mean, sample median, sample mode, and sample midrange.

(c) Calculate the sample variance and standard deviation, and the sample mean deviation.

**2.13**  Using the sample data in Prob. 2.12, calculate the parameters of a normal density curve and plot this curve superimposed over the histogram.

**2.14**  Assuming that the calculated standard deviation for the curve in Prob. 2.13 is in fact the population parameter, evaluate three ranges within which the population mean $\mu$ falls, corresponding to the three probabilities 0.6827, 0.95, and 0.99.

**2.15**  The random errors in the planimetric position of a survey point have standard deviations of $\sigma_x = 0.2$ m and $\sigma_y = 0.2$ m in the $x$ and $y$ directions and a correlation coefficient of $\rho_1 = 0.5$. Evaluate the standard error ellipse and the 95 percent ellipse. Sketch both ellipses.

**2.16**  For the same $\sigma_x = \sigma_y = 0.2$ m in Prob. 2.15, evaluate and sketch the error ellipses for the cases with $\rho_2 = -0.5$, $\rho_3 = 0$, and $\rho_4 = 1$.

**2.17**  Three adjacent angles have the following mean values, standard deviations, and correlation coefficients.

| Angle | Mean value | $\sigma$ | $\rho$ |
|-------|-----------|----------|--------|
| 1 | 20°07'30" | 20" | |
| 2 | 15°43'00" | 15" | 0.5 |
| 3 | 19°00'45" | 10" | 0.5 |
| 1 | | | 0.0 |

Calculate the mean value and standard deviation of the total angle (sum of all three angles).

**2.18** A tract of land has a trapezoidal shape with two parallel sides $a, b$ and the distance between them (i.e., height of trapezoid) $h$. You are given the following data:

$a = 125.12$ m    $\sigma_a = 0.10$ m

$b = 150.08$ m    $\sigma_b = 0.12$ m

$h = 20.00$ m    $\sigma_h = 0.08$ m

The three measurements are uncorrelated. Calculate the area of the tract and its standard deviation.

**2.19** A side and two internal angles of a plane triangle are measured as follows:

side $a = 84.22$ m    with $\sigma_a = 3$ cm

angle $B = 90°00'$    with $\sigma_B = 1'$

angle $C = 32°17'$    with $\sigma_C = 1'$

The three measurements are uncorrelated. In keeping with the general notation of geometry, the two angles $B$ and $C$ are at the extremities of the side $a$.

(a) Calculate the two sides $b$ and $c$ and their covariance matrix.

(b) Calculate the third angle $A$ and its standard deviation.

(c) Calculate the area of the triangle and its standard deviation.

**2.20** When does adjustment become necessary in a survey net?

**2.21** What is the basic objective of a survey adjustment?

**2.22** What is the least-squares criterion and why is it needed?

**2.23** Why do we need to add residuals to the observations?

**2.24** Is the covariance matrix of the survey measurements used in least squares? Show where it is used and why.

## References

1. Aguilar, A. M., "Principles of Survey Error Analysis and Adjustment," *Surveying and Mapping*, September 1973.
2. Allman, J. S., et al., "Angular Measurement in Traverses," *Australian Surveyor*, December 1973.
3. Barry, A., *Engineering Measurements*, John Wiley & Sons, Inc., New York, 1964.
4. Cooper, M. A. R., *Fundamentals of Survey Measurements and Analysis*, Crosby Lookwood, London, 1974.
5. Cross, P. A., "The Effect of Errors in Weights," *Survey Review*, July 1972.
6. Cross, P. A., "Ellipse Problem—More Solutions," *Survey Review*, January 1972.
7. Cross, P. A., "Error Ellipse: A Simple Example," *Survey Review*, October 1971.
8. Mikhail, E. M., *Observations and Least Squares*, Harper & Row, Publishers, 1976.
9. Mikhail, E. M., and Gracie, G., *Analysis and Adjustment of Survey Measurements*, Van Nostrand Reinhold Company, New York, 1980.
10. Richardus, P., *Project Surveying*, North-Holland Publishing Company, Amsterdam, 1966.
11. Thompson, E. H., "A Note on Systematic Errors," *Photogrammetria*, Vol. 10, 1953–1954.
12. Uotila, U. A., "Useful Statistics for Land Surveyors," *Surveying and Mapping*, March 1973.

# CHAPTER 3
# Field and office work

**3.1. General**  The nature of surveying measurements has already been indicated. Much of the field and office work involved in the acquisition and processing of measurement is performed concurrently. Field and office work for a complete survey consists of:

1. Planning and design of the survey; adoption of specifications; adoption of a map projection and coordinate system, and of a proper datum; selection of equipment and procedures.
2. Care, handling, and adjustment of the instruments.
3. Fixing the horizontal location of objects or points by horizontal angles and distances.
4. Determining the elevations of objects or points by one of the methods of leveling.
5. Recording field measurements.
6. Field computations for the purpose of verifying the data.
7. Office computations in which data are reduced, adjusted, and filed or stored for current utilization or for use in the near or distant future.
8. The setting of points in the field to display land property location and to control construction layout (as may be necessary).
9. Performing the final *as built survey*, in which all structures built as a part of the project are located with respect to the basic control network and/or established property lines.

Discussions in this chapter are restricted to general comments related to planning and design of the survey, care and adjustment of instruments, methods for recording data, computational methods, and computational aids.

**3.2. Planning and design of the survey**  There are many types of surveys related to an almost infinite variety of projects. Thus, the planning of the survey can be discussed only in very general terms at this point. Assuming that the nature of the project is established and the results desired from the survey are known, the steps involved in planning the survey are:

1. Establishing specifications for horizontal and vertical control accuracies.
2. Location and analysis of all existing control, maps, photographs, and other survey data.
3. Preliminary examination of the site in the office using existing maps and photographs and in the field to locate existing, and to set new, control points.
4. Selection of equipment and surveying procedures appropriate for the task.
5. Selection of computational procedures and the method for presentation of the data in final form.

To understand the procedure, consider an example. Assume that a private surveying firm has been engaged to prepare a topographic map of a 2000 acre (809.4 ha) tract at a scale of 1:1500 (see Art. 1.9 for a definition of scale) and with a contour interval of 2 m.[†] Control

---

[†] A contour is the line of intersection of a level surface with the terrain surface; a contour interval is the vertical distance between two successive such level surfaces (see Art. 13.8 for more details).

surveys are to be performed using traditional surveying methods while the major portion of the topography is to be compiled photogrammetrically. The tasks to be performed by the surveying firm are to select a proper coordinate grid and datum plane and to establish a horizontal and vertical control network throughout the area, including sufficient points to control aerial photographs for the photogrammetric mapping; and to map heavily wooded valleys not accessible to the photogrammetric methods. Another firm has been engaged to do the photogrammetric work.

The map is to be used by the client for presentation of plans of a proposed industrial park to the county planning commission. If plans are approved, earthwork quantities will be calculated using measurements from the map, and construction surveys will be made from control points set in the survey. Consider some of the problems involved in planning for this type of a survey.

**3.3. Specifications**  The accuracy of the measurements should be consistent with the purpose of the survey. Each survey is a problem in itself, for which the surveyor must establish the limits of error using a knowledge of the equipment involved, the procedure to be employed, error propagation, and judgment based on practical experience. The best survey is one that provides data of adequate accuracy without wasting time or money.

The National Geodetic Survey publishes specifications for first-, second-, and third-order horizontal and vertical control surveys (Chaps. 5 and 10). These specifications provide a starting point for establishing standards on most jobs that require basic control surveys.

For the example given, the work can be divided into three categories: establishing basic control, establishing supplementary control, and performing topographic surveys.

Basic control consists of a primary network of rather widely spaced horizontal and vertical control points which will coordinate with local, state, and national control and will be used to control all other surveys throughout the entire area to be studied. Second-order specifications are generally recommended for basic control networks. Since photogrammetric surveys will follow, particular attention should be paid to establishing basic horizontal control points around the perimeter of the area to be mapped, plus several vertical control points near the center of the project. Construction surveys are definitely a possibility, so that a sufficient number of basic control points should be established in the interior of the area to allow tying all subsequent construction surveys to second-order control points.

Supplementary control is set to provide additional control points for photogrammetric mapping and for topographic mapping by field methods. Third-order specifications or lower (down to a relative precision of 1 part in 3000) are usually considered adequate for supplementary horizontal control. On the example project, most of the supplementary control will be used for controlling the photogrammetric mapping, and a specification for error in position is appropriate. Thus, it could be specified that supplementary control points should be to within ± a tolerance based on the requirements of the photogrammetric mapping system. Vertical control is usually established to satisfy third-order specifications. Close cooperation is required between the surveyor and photogrammetrist in this phase of the work. Under certain circumstances, supplementary control can be established by photogrammetric methods. This could be a topic for discussion between the surveyor and the photogrammetrist.

Topographic measurements in the field are made from supplementary control points. Accuracies in these measurements are usually based on map accuracies. For the example, since the map scale is 1 : 1500, then 0.5 mm on the map is equivalent to 0.75 m on the ground or $\frac{1}{50}$ in represents 2.5 ft. Thus, the accuracy in position for mapping is ±0.8 m or ±2.5 ft. Elevations are governed by the contour interval. National map standards for accuracy require that 90 percent of all spot elevations determined from the map be within ± one-half contour interval of the correct elevation. Experience has shown that this accuracy of contour interpolation can normally be achieved for maps if spot elevations in the field are determined to within one-tenth of the contour interval. Heights of topographic points should be located to within $\pm 0.2$ m for the example project. Since it is possible to determine elevations well within this interval by both field and photogrammetric methods, this specification presents no problem for either the field or the photogrammetric survey method.

The specifications for photogrammetric mapping are largely a function of the map scale and contour interval. As the contour interval decreases, the cost of the photogrammetric map increases at an exponential rate. Thus, it is worth studying the desired contour interval to be sure that this interval is *really* necessary. For this example, earthwork will be calculated from measurements made on the map so that both contour interval and map scale represent optimum values that must be maintained. This being the case, aerial photography must be flown at an altitude and to specifications compatible with the stated scale and contour interval.

Unique situations on the project may require special treatment. If construction is to be concentrated in a particular area of the site, second-order control can be clustered in that region and mapping of that portion can be at a larger scale and with a smaller contour interval. For a relatively small area such as the example problem of 2000 acres, mapping at two scales would probably not be practical. Such an approach could be feasible under special circumstances in larger areas. The requirements for construction surveys may be the limiting factor. When structures for the transport of water or other liquids by gravity flow are involved, first-order levels are required. Bridges, complicated steel structures, tunnels, and city surveys are a few examples where first-order control surveys may be necessary.

**3.4. Existing maps and control points**  A thorough search should be performed to locate all survey data, existing maps, and aerial photographs of the area to be surveyed.

The National Geodetic Survey in Rockville, Maryland 20852, maintains records of all first-, second-, third-, and selected fourth-order and photogrammetrically derived control point locations in the horizontal and vertical control networks of the United States. Control point data plus descriptions will be provided for a fee when requested. The Bureau of Land Management keeps a file on property survey data related to public lands survey. State, county, city, and town engineering and surveyor's offices should also be consulted for useful survey data.

The U.S. Geological Survey is responsible for compiling maps at various scales of the United States. The U.S.G.S. *quadrangle* topographic maps at a scale of $1:24,000$ or $1:25,000$ in the metric series are the maps most commonly used for engineering studies. Information can be obtained from the National Cartographic Information Center (NCIC) in Reston, Virginia 22092. The National Ocean Survey compiles and sells nautical and aeronautical charts. State, county, city, and town engineering and survey offices should also be checked for maps at scales larger than the $1:24,000$.

Aerial photography is very useful for preliminary planning purposes and can be acquired from the following government agencies: U.S. Geological Survey; U.S. Department of Agriculture, Stabilization and Conservation Service, and Soil Conservation Service; U.S. Forest Service; Bureau of Land Management; National Ocean Survey; and the Tennessee Valley Authority (TVA). Many of these agencies have local or regional offices which maintain files of photography of that particular area. State highway departments generally have photography along the right of way for highways. Private photogrammetric firms usually keep files of photography for the areas that they have mapped.

Data from all sources should be gathered and studied so as to exploit all previous survey work in the area. Preliminary studies can be made in the office using existing maps and/or aerial photographs to choose possible locations for control points. These studies are followed by field reconnaissance to locate all existing control points and to set new horizontal and vertical control points in the basic and supplementary networks to be established. At this point, close cooperation between the surveyor and photogrammetrist is absolutely essential when photogrammetric mapping is involved.

Whenever possible, the basic control network should be tied to at least two existing government monuments so as to provide a basis for performing the entire survey in a proper state or national coordinate system.

**3.5. Selection of equipment and procedures**   The specifications for horizontal and vertical control provide an allowable tolerance for horizontal and vertical positions. Using estimates for the capabilities inherent in various types of instruments, errors in distance and direction can be propagated for a given procedure. These propagated errors are then compared with the allowable tolerance and modification in instrument or procedure can be made if necessary. Various instruments, procedures for use of these instruments, and error propagation methods are described in subsequent chapters. Considerable judgment and practical experience in surveying is required in this phase of the survey design.

**3.6. Selection of computational procedures and method for data presentation**   Systematic procedures for gathering, filing, processing, and dissemination of the survey data should be developed. Field notes and methods for recording data are described in Arts. 3.12 to 3.14.

The prevalence of modern, efficient electronic computing equipment virtually dictates that rigorous methods and adjustment techniques be utilized in all but the smallest of surveys. Computational aids are described in Art. 3.19.

Methods for filing or storing survey information should be standardized for all jobs and ought to allow current or future retrieval without complication.

The method for presenting the data must be carefully considered. For the example case, a conventional line map or an orthophoto map (see Chap. 16) with an overlay of contours would be appropriate for presentation to a planning commission.

**3.7. Relation between angles and distances**   In Art. 3.5 reference was made to the propagated errors in distance and in direction. These errors are used to determine the uncertainty in position for a point. Assume that estimated standard deviations in a distance $r$ and direction $\alpha$ are $\sigma_r$ and $\sigma_\alpha$, respectively. As developed in Arts. 2.19 and 2.20 and Example 2.8, the position of a point determined using $r$ and $\alpha$ has an uncertainty region defined by an ellipse centered about a point located by $r$ and $\alpha$ as shown in Fig. 3.1. In Example 2.8, $\sigma_r = 0.5$ m, $r = 100$ m, $\alpha = 60°$, and $\sigma_\alpha = 30'$, which produce an ellipse having a semiminor axis

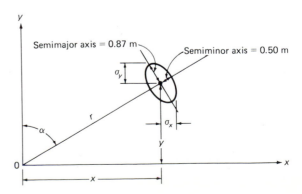

**Fig. 3.1** Error ellipse for $r = 100$ m, $\alpha = 60°$, $\sigma_r = 0.50$ m, and $\sigma_\alpha = 30'$.

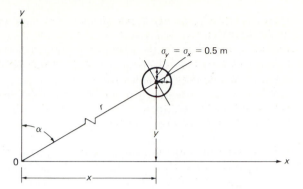

**Fig. 3.2** Error ellipse for $r = 100$ m, $\alpha = 30°$, and $\sigma_r = r\sigma_\alpha$.

of $\sigma_r = 0.5$ m that is parallel to $r$. The semimajor axis of the ellipse $= r\sigma_\alpha = 0.87$ m and is normal to line $r$. Note that $\sigma_x = 0.61$ m and $\sigma_y = 0.79$ m, both of which are less than the maximum uncertainty in the point $= r\sigma_\alpha = 0.87$ m.

Suppose that $\sigma_\alpha$ is chosen so that $r\sigma_\alpha = \sigma_r$, or $\sigma_\alpha = 0.5/100 = 0.00500$ rad. In this case the two axes of the region of uncertainty are equal and the ellipse becomes a circle, as illustrated in Fig. 3.2. Thus, to have the same contribution from distance and angle errors, $\sigma_\alpha$ should be about $0°17'$.

The preceding analysis illustrates the relationship between uncertainties in direction and distance and emphasizes the desirability of maintaining consistent accuracy in the two measurements. The error in distance is normally expressed as a relative precision or ratio of the error to the distance. Thus, in the example, the relative accuracy is $0.5/100$, or 1 part in 200. Similarly, the linear distance subtended by $\sigma_\alpha$ in a distance $r$ equals 0.5 and the tangent or sine of the error or its value in radians is 1 part in 200. Accordingly, a consistent relation between accuracies in angles and distances will be maintained if the estimated standard deviation in direction equals $\sigma_r/r$ radians or the relative precision in the distance.

It is impossible to maintain an exact equality between these two relative accuracies; but with some exceptions, to be considered presently, surveys should be conducted so that the difference between angular and distance accuracies is not great. Table 3.1 shows for various angular standard deviations the corresponding relative precision and the linear errors for

**Table 3.1   Corresponding Angular and Linear Errors**

| Standard deviation in angular measurement | Linear error in: | | Relative precision |
|:---:|:---:|:---:|:---:|
| | 1000 ft | 300 m | |
| 10′ | 2.9089 | 0.87267 | $\frac{1}{344}$ |
| 5′ | 1.4544 | 0.43633 | $\frac{1}{688}$ |
| 1′ | 0.2909 | 0.08727 | $\frac{1}{3440}$ |
| 30″ | 0.1454 | 0.04363 | $\frac{1}{6880}$ |
| 20″ | 0.0970 | 0.02909 | $\frac{1}{10,300}$ |
| 10″ | 0.0485 | 0.01454 | $\frac{1}{20,600}$ |
| 5″ | 0.0242 | 0.00727 | $\frac{1}{41,200}$ |
| 1″ | 0.004848 | 0.00145 | $\frac{1}{206,000}$ |

lengths of 1000 ft and 300 m. For a length other than 1000 ft or 300 m, the linear error is in direct proportion. A convenient relation to remember is that an angular error of 01′ corresponds to a linear error of about 0.3 ft in 1000 ft or 3 cm in 100 m.

To illustrate the use of the table, suppose that distances are to be taped with a relative precision of 1/10,000. From the table the corresponding permissible angular error is 20″. As another example, suppose that the distance from the instrument to a desired point is determined as 250 m with a standard deviation of 0.8 m. For an angular error of 10′ the corresponding linear error is (250/300)(0.87)=0.73 m. Therefore, the angle needs to be determined only to the nearest 10′.

The prevalence of EDM equipment for measuring distances creates a situation where an exception occurs. Distances can be measured using EDM with very good relative precision without additional effort. For example, suppose that a distance of 3000 m is observed with an estimated standard deviation of 0.015 m, producing a relative precision of 1/200,000. The corresponding angular standard deviation is 01″. For an ordinary survey this degree of angular accuracy would be entirely unnecessary. At the other end of the spectrum, distances may be roughly determined by taping or pacing and angles measured with more than the required accuracy. For example, in rough taping, the relative accuracy in distance might be 1/1000, corresponding to a standard deviation in angles of 03′. Even using an ordinary 01′ transit, angles could be observed as easily to the nearest 01′ as to the nearest 03′.

Often, field measurements are made on the basis of computations involving the trigonometric functions, and it is necessary that the computed results be of a required precision. If the values of these functions were exactly proportional to the size of the angles—in other words, if any increase in the size of an angle were accompanied by a proportional increase or decrease in the value of a function—the problem of determining the precision of angular measurements would resolve itself into that explained in the preceding section. However, since the rates of change of the sines of small angles, of the cosines of angles near 90°, and the tangents and cotangents of small and large angles are relatively large, it is evident that the degree of precision with which an angle is determined should be made to depend upon the size of the angle and upon the function to be used in the computations. It is not practical to measure each angle with exactly the precision necessary to ensure sufficiently accurate computed values, but at least the surveyor should have a sufficiently comprehensive knowledge of the purpose of the survey and of the properties of the trigonometric functions to keep the angles within the required precision.

The curves of Figs. 3.3 and 3.4 show the relative precision corresponding to various standard deviations in angles from 05″ to 01′ for sines, cosines, tangents, and cotangents. For the function under consideration these curves may be used as follows:

1. To determine the relative precision corresponding to a given angle and error.
2. To determine the maximum or minimum angle that for a given angular error will furnish the required relative accuracy.
3. To determine the precision with which angles of a given size must be measured to maintain a required relative precision in computations.

The following examples illustrate the use of the curves:

1. An angle measured with a 1′ transit is recorded as 32°00′. It is desired to know the relative precision of a computation involving the tangent of the angle if the error of the angle is 30″. In Fig. 3.4 it will be seen that the relative precision opposite the intersection of the curve $\sigma_\alpha = 30″$ and a line corresponding to 32° is 1/3000.
2. In a triangulation system the angles can be measured with $\sigma_\alpha$ not exceeding 05″. Computations involving the use of sines must maintain a precision not lower than 1/20,000. It is desired to determine the minimum allowable angle. In Fig. 3.3 (sines) the angle corresponding to a relative precision of 1/20,000 and $\sigma_\alpha$ of 05″ is about 26°.

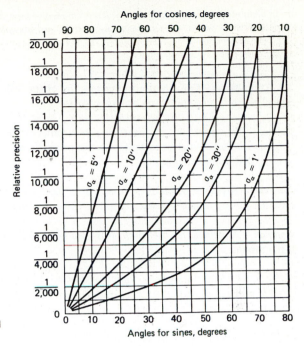

**Fig. 3.3** Relative precision for sines and cosines.

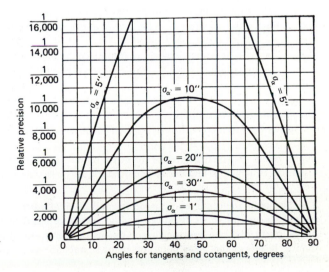

**Fig. 3.4** Relative precision for tangents and cotangents.

3. In computations involving the use of cosines a relative precision of 1/10,000 is to be maintained. It is desired to know with what precision angles must be measured. In Fig. 3.3 (cosines) opposite 1/10,000 it will be seen that for angles of about 76° the $\sigma_\alpha$ cannot exceed 05"; for angles of about 64° the $\sigma_\alpha$ cannot exceed 10"; and so on.

**3.8. Definitions**   For a better understanding of the following sections, brief definitions of a few of the terms of surveying are appropriate.

*Taping*   The operation of measuring horizontal or inclined distances with a tape. The persons who make such measurements are called "tapemen." Historically, this procedure has been called *chaining*.

*Flagman*   A person whose duty it is to hold the flagpole, or range pole, at selected points, as directed by the transit operator or other person in charge.

*Rodman*   A person whose duty it is to hold the rod and to assist level operator or topographer.

*Backsight*   (1) A sight taken with the level to a point of known elevation. (2) A sight or observation taken with the transit along a previous line of known direction to a reference point.

*Foresight*   (1) A sight taken with the level to a point the elevation of which is to be determined. (2) A sight taken with the transit to a point (usually in advance), along a line whose direction is to be determined.

*Grade or gradient*   The slope, or rate of regular ascent or descent, of a line. It is usually expressed in percent; for example, a 4 percent grade is one that rises or falls 4 ft in a horizontal distance of 100 ft or 4 m in 100 m. The term *grade* is also used to denote an established line on the profile of an existing or a proposed roadway. In such expressions as "at grade" or "to grade" it denotes the elevation of a point either on a grade line or at some established elevation as in construction work.

*Hub*   A transit station, or point over which the transit is set, in the form of a heavy stake set nearly flush with the ground, with a tack in the top marking the point.

*Line*   The path or route between points of control along which measurements are taken to determine distance or angle. To *give line* is to direct the placing of a flagpole, pin, or other object on line.

*Turning point*   A fixed point or object, often temporary in character, used in leveling where the rod is held first for a foresight, then for a backsight.

*Bench mark*   A fixed reference point or object, more or less permanent in character, the elevation of which is known. A bench mark may also be used as a turning point.

**3.9. Signals**   Field surveying is performed by *field parties* composed of from two to seven people. Except for short distances, a good system of hand signals between different members of the party makes a more efficient means of communication than is possible by word of mouth. A few of the more common hand signals are as follows.

*Right or left*  The corresponding arm is extended in the direction of the desired movement. A long, slow, sweeping motion of the hand indicates a long movement; a short, quick motion indicates a short movement. This signal may be given by the transit operator in directing the tapeman on line, by the level operator in directing the rodman for a turning point, by the chief of the party to any member, or by one tapeman to another tapeman.

*Up or down*  The arm is extended upward or downward, with wrist straight. When the desired movement is nearly completed, the arm is moved toward the horizontal. The signal is given by the level operator.

*All right*  Both arms are extended horizontally and the forearms waved vertically. The signal may be given by any member of any party.

*Plumb the flagpole or plumb the rod*  The arm is held vertically and moved in the direction in which the flagpole or rod is to be plumbed. The signal is given by the transit operator or level operator.

*Give a foresight*  The instrument operator holds one arm vertically above the head.

*Establish a turning point or set a hub*  The instrument operator holds one arm above the head and waves it in a circle.

*Turning point or bench mark*  In profile leveling the rodman holds the rod horizontally above the head and then brings it down on the point.

*Give line*  The flagman holds the flagpole horizontally in both hands above the head and then brings it down and turns it to a vertical position. If a hub is to be set, the flagpole is waved (with one end of it on the ground) from side to side.

*Wave the rod*  The level operator holds one arm vertically and moves it from side to side.

*Pick up the instrument*  Both arms are extended outward and downward, then inward and upward, as they would be in grasping the legs of the tripod and shouldering the instrument. The signal is given by the chief of the party or by the head tapeman when the transit is to be moved to another point.

When distances are sufficiently long to render hand signals inadequate, the use of a citizens' band radio provides an excellent method for communication among members of a field party.

**3.10.  Care and handling of instruments**  As the use of the various surveying instruments is discussed in the following chapters, suggestions for the care and manipulation of these instruments are given.

***Surveying Instruments***  The following suggestions apply to such instruments as the theodolite, transit, level, and plane table. More detailed information is given in the references at the end of the chapter.

1. Handle the instrument with care, especially when removing it from or replacing it in its case.
2. See that it is securely fastened to the tripod head.
3. Normally, carry the instrument mounted on the tripod and over one shoulder with the tripod legs forward and held together by the hand and forearm of that shoulder.

4. Avoid carrying the instrument on the shoulder while passing through doorways or beneath low-hanging branches; carry it under the arm, with the head of the instrument in front.

5. When walking along a sidehill always carry the instrument on the downhill shoulder to leave the uphill arm free to catch the body in case of tripping or stumbling and to prevent the instrument from slipping off the shoulder.

6. Before climbing over a fence or similar obstacle, place the instrument on the other side, with the tripod legs well spread.

7. Whenever the instrument is being carried or handled, the clamp-screws should be clamped very lightly so as to allow the parts to move if the instrument is struck.

8. Protect the instrument from impact and vibration.

9. If the instrument is to be shipped, pack paper, cloth, or styrofoam padding around it in the case; pack the case, well padded, in a larger box.

10. Never leave the instrument while it is set up in the street, on the sidewalk, near construction work, in fields where there are livestock, or in any other place where there is a possibility of accident.

11. Just before setting up the instrument, adjust the wing nuts controlling the friction between tripod legs and head so that each leg when placed horizontally will barely fall under its own weight.

12. Do not set the tripod legs too close together, and see that they are firmly planted. Push *along* the leg, not vertically downward. As far as possible, select solid ground for instrument stations. On soft or yielding ground, do not step near the feet of the tripod.

13. While an observation is being made, do not touch the instrument except as necessary to make a setting; and do not move about.

14. In tightening the various clamp-screws, adjusting screws, and leveling screws, bring them only to a firm bearing. *The general tendency is to tighten these screws far more than necessary.* Such a practice may strip the threads, twist off the screw, bend the connecting parts, or place undue stresses in the instrument, so that the setting may not be stable. Special care should be taken not to strain the small screws that hold the cross-hair ring.

15. For the plumb-bob string, learn to make a sliding bowknot that can be easily undone. Hard knots in the string indicate an inexperienced or careless instrument operator.

16. Before observations are begun, focus the eyepiece on the cross hairs and (by moving the eye slightly from side to side) see that no parallax is present.

17. *When the magnetic needle is not in use, see that it is raised off the pivot.* While the needle is resting on the pivot, impact is apt to blunt the point of the pivot or to chip the jewel, thus causing the needle to be sluggish.

18. Always use the sunshade. Attach or remove it by a clockwise motion, in order not to unscrew the objective.

19. If the instrument is to be returned to its box, put on the dust cap if one was supplied by the manufacturer, and wipe the instrument clean and dry.

20. Never rub the coated lenses of a telescope with the fingers or with a rough cloth. Use a camel's-hair brush to remove dust, or use clean chamois or lint-free soft cloth if the dust is caked or damp. Occasionally, the lenses may be cleaned with a mixture of equal parts of alcohol and water. Keep oil off the lenses.

21. Never touch the graduated circles, verniers, micrometer lenses, or prisms with the fingers. Do not wipe them more than necessary, and particularly do not rub the edges.

22. Do not touch the level vials or breathe on them while in use, as unequal heating of the level tube will cause the bubble to move out of its correct position.

23. In cold weather, the instrument should not be exposed to sudden changes in temperature (as by bringing it indoors); and the observer should be careful not to breathe on the eyepiece.

24. Return the instrument regularly to a manufacturer's representative or qualified instrument repair shop for cleaning, maintenance, and repair.

Suggestions regarding the adjustment of instruments are given in Art. 3.11.

**Taping equipment**   Keep the tape straight when in use; any tape will break when kinked and subjected to a strong pull. Steel tapes rust readily and for this reason should be wiped dry after being used.

Use special care when working near electric power lines. Fatal accidents have resulted from throwing a metallic tape over a power line.

Do not use the flagpole as a bar to loosen stakes or stones, such use bends the steel point and soon renders the point unfit for lining purposes.

To avoid losing pins, tie a piece of colored cloth (preferably bright red) through the ring of each.

**Leveling rod**   Do not allow the metal shoe on the foot of the rod to strike against hard objects, as this, if continued, will round off the foot of the rod and thus introduce a possible error in leveling. Keep the foot of the rod free of dirt. When not in use, long rods should be either placed upright or supported for their entire length; otherwise, they are likely to warp. When not in use, jointed rods should have all clamps loosened to allow for possible expansion of the wood.

**EDM equipment**   EDM instruments are precision surveying devices which require the same care and handling as a theodolite or level. These instruments should always be transported in the carrying case especially designed for that purpose. All EDM devices have a power unit. Excessive shock can damage this unit. Care must be exercised when transporting the power unit. Most EDM instruments require little service other than the normal cleaning of transmitter lens and other parts. The power unit does require periodic recharging and checking to ensure that power is not lost at a critical moment. EDM instruments should be calibrated and tested periodically to verify instrument constants and check for frequency drift (Art. 4.34).

**3.11. Adjustment of Instruments**   By "adjustment" of a surveying instrument is meant the bringing of the various fixed parts into proper relation with one another, as distinguished from the ordinary operations of leveling the instrument aligning the telescope, and so on.

The ability to perform the adjustments of the ordinary surveying instruments is an important qualification of the surveyor. Although it is a fact that the effect of instrumental errors may largely be eliminated by proper field methods, it is also true that instruments in good adjustment greatly expedite the field work. It is important that the surveyor:

1. Understand the principles upon which the adjustments are based.
2. Learn the method by which nonadjustment is discovered.
3. Know how to make the adjustments.
4. Appreciate the effect of one adjustment upon another.
5. Know the effect of each adjustment upon the use of the instrument.
6. Learn the order in which adjustments may most expeditiously be performed.

The frequency with which adjustments are required depends upon the particular adjustment, the instrument and its care, and the precision with which measurements are to be taken. Often in a good instrument, well cared for, the adjustments will be maintained with sufficient precision for ordinary surveys over a period of months or even years. On the other hand, blows that may pass unnoticed are likely to disarrange the adjustments at any time. On ordinary surveys, it is good practice to test the critical adjustments once each day, especially on long surveys where frequent checks on the accuracy of the field data are impossible. Failure to observe this simple practice sometimes results in the necessity of retracing lines, which may represent the work of several days. Testing the adjustments with reasonable frequency lends confidence to the work and is a practice to be strongly commended. The instrument operator should, if possible, make the necessary tests at a time that will not interfere with the general progress of the survey party. Some adjustments may be made with little or no loss of time during the regular progress of the work.

The adjustments are made by tightening or loosening certain screws. Usually, these screws have capstan heads which may be turned by a pin called an *adjusting pin*. Following are some general suggestions:

1. The adjusting pin should be carried in the pocket and not left in the instrument box. Disregard of this rule frequently leads to loss of valuable time.
2. The adjusting pin should fit the hole in the capstan head. If the pin is too small, the head of the screw is soon ruined.
3. Preferably make the adjustments with the instrument in the shade.
4. Before adjusting the instrument, see that no parts (including the objective) are loose. When an adjustment is completed, always check it before using the instrument.
5. When several interrelated adjustments are necessary, time will be saved by first making an approximate or rough series of adjustments and then by repeating the series to make finer adjustments. In this way, the several disarranged parts are gradually brought to their correct position. This practice does not refer to those adjustments which are in no way influenced by others.
6. Most of the more precise instruments, such as theodolites, precise levels, and EDM equipment, require more complicated tools, more stable or constant environmental conditions for the test and adjustment phase, and more specialized personnel to achieve the more refined adjustments. Consequently, such instruments are rarely adjusted in the field, but the operator does need to test the instrument periodically to determine if adjustment is necessary.

**3.12. Field notes**   No part of the operations of surveying is of greater importance than the field notes. The competency of a surveyor is reflected with great fidelity by the character of the notes recorded in the field. These notes should constitute a permanent record of the survey with data in such form as to be interpreted with ease by anyone having a knowledge of surveying. Unfortunately, this is often not the case. Many surveyors seem to think that their work is well done if the field record, reinforced by their own memories, is sufficiently comprehensive to make the field data of immediate use for whatever purpose the survey may have. On most surveys, however, it is impossible to predict to what extent the information gathered may become of value in the remote future. Often court proceedings involve surveys made long before. Often, it is desirable to rerun, extend, or otherwise make use of surveys made years previously. In such cases it is quite likely that the old field notes will be the only visible evidence, and their value will depend largely upon the clarity and completeness with which they are recorded.

The notes consist of numerical data, explanatory notes, and sketches. Also, the record of every survey or student problem should include a title indicating the location of the survey and its nature or purpose; the equipment used, including manufacturer and number for major instruments; the date, the weather conditions; and the names and duties of the members of the party.

All field notes should be recorded *in the field book* at the time the work is being done. Notes made later, from memory or copied from temporary notes, may be useful, but they are not field notes. Notes should be neat. They are generally recorded in pencil, but they should be regarded as a permanent record and not as memoranda to be used only in the immediate future.

It is not easy to take good notes. The recorder should realize that the notes may be used by persons not familiar with the locality, who must rely entirely upon what has been recorded. Not only should the notebook contain all necessary information, but the data should be recorded in a form which will allow only the correct interpretation. A good sketch will help to convey a correct impression, and sketches should be used freely. The use to be made of the notes will guide the recorder in deciding which data are necessary and which are not. To make the notes clear, recorders should put themselves in the place of one who is not in the field at the time the survey is made. Before any survey is made, the necessary data to be collected should be considered carefully, and in the field all such data should be obtained, but no more.

Although convenient forms of notes such as those shown in this book are in common use, it will generally be necessary to supplement these, and in many cases it will be necessary for surveyors to devise their own form of record. A code of symbols is desirable.

In some cases, as in locating details for mapping, the field notes may be supplemented by photographs taken with an ordinary camera. Frequently, in field completion surveys for photogrammetric mapping, an enlargement of the aerial photograph that covers the area to be mapped is annotated in the field, thus providing a record of additional information.

Other methods for recording field data are described in Art. 3.14.

**3.13. Notebook**   In practice the field notebook should be of good-quality rag paper, with a stiff board or leather cover, made to withstand hard usage, and of pocket size. Treated papers are available which will shed rain; some of these can be written on when wet.

Special field notebooks are sold by engineering supply companies for particular kinds of notes, such as cross sections or earthwork. For general surveying or for students in field work, where the problems to be done are general in character, an excellent form of notebook has the right-hand page divided into small rectangles with a red line running up the middle, and has the left-hand page divided into six columns; both pages have the same horizontal ruling. In general, tabulated numerical values are written on the left-hand page, sketches and explanatory notes on the right. This type is called a *field book* (see, e.g., Fig. 8.12). Another common form, used in leveling, has both pages ruled in columns and has wider horizontal spacing than the field book; this is called a *level book*.

The field notebook may be bound in any of three ways: conventional, spiral ring, or loose-leaf. The ring type, which consists of many metal rings passing through perforations in the pages, is not loose-leaf; it has the advantage over the conventional binding that the book opens quite flat and that the covers can be folded back against each other.

Loose-leaf notebooks are increasing in use because of the following advantages:

1. Only one book need be carried, as in it may be inserted blank pages of various rulings, together with notes and data relating to the current field work.
2. Sheets can be withdrawn for use in the field office while the survey is being continued.
3. Carbon copies can be made in the field, for use in the field or headquarters office. Carbon copies are also a protection against loss of data. Duplicating books are available.
4. Notes of a particular survey can be filed together. Files can be made consecutive and are less bulky than for bound books.
5. The cost of binders is less than that for bound books.

The disadvantages are:

1. Sheets may be lost or misplaced.
2. Sheets may be substituted for other sheets—an undesirable practice.
3. There may be difficulty in establishing the identity of the data in court, as compared with a bound book. (This feature is important in land surveying.) When loose-leaf books are used, *each sheet should be fully identified by date, serial number, and location.*

Loose leaves are furnished in either single or double sheets. Single sheets are ruled on both sides and are used consecutively. Double sheets comprising a left-hand and a right-hand page joined together are ruled on one side only. Sheets for carbon copies need not be ruled.

**3.14. Recording data**   A 4H pencil, well pointed, should be used. Lines made with a harder pencil are not as distinct; lines made with a softer pencil may become smeared. Reinhardt slope lettering (Fig. 13.20) is commonly considered to be the best form of lettering for taking notes rapidly and neatly. Office entries of reduced or corrected values should be made in red ink, to avoid confusion with the original data.

The figures used should be plain; one figure should never be written over another. In general, *numerical data should not be erased*; if a number is in error, a line should be drawn through it, and the corrected value written above.

In tabulating numbers, the recorder should place all figures of the tens column, etc., in the same vertical line. Where decimals are used, the decimal point should never be omitted. The number should always show with what degree of precision the measurement was taken; thus, a rod reading taken to the nearest 0.01 ft should be recorded not as 7.4 ft but as 7.40 ft. Notes should not be made to appear either more precise or less precise than they really are.

Sketches are rarely made to exact scale, but in most cases they are made approximately to scale. They are made freehand and of liberal size. The recorder should decide in advance just what the sketch is to show. A sketch crowded with unnecessary data is often confusing, even though all necessary features are included. Large detailed sketches may be made of portions having much detail. Many features may be most readily shown by conventional symbols (Art. 13.25); special symbols may be adopted for the particular organization or job.

Explanatory notes are employed to make clear what the numerical data and sketches fail to do. Usually, they are placed on the right-hand page in the same line with the numerical data that they explain. If sketches are used, the explanatory notes are placed where they will not interfere with other data and as close as possible to that which they explain.

If a page of notes is abandoned, either because it is illegible or because it contains erroneous or useless data, it should be retained and the word "void" written in large letters diagonally across the page. The page number of the continuation of the notes should be indicated.

**3.15. Other media for recording data**  The development of compact, portable, programmable electronic calculators, solid-state memory units, and magnetic tape cassette recorders, all of which have low power requirements, has caused the recording of field data to enter a new era.

Data observed in the field can be keyed into a portable, programmable calculator, partially processed to detect blunders, and stored on magnetic tape cassettes. At the end of the day's work, the tape cassette is taken to the field office, where the data are read into a desktop computer which produces a *hard copy* listing of the observed data. Then, using a remote terminal at the field office, the data are transmitted to a centrally located electronic computer for adjustment and analysis. This computer produces another hard copy listing that contains all the observed data, processed results, and pertinent statistics. The data stored in the central computer can then be filed on disk or tape for use in subsequent adjustment of the complete job of which this day's work was only a portion.

Another option is available in digitized EDM instruments and theodolites (Arts. 6.31 and 12.2). In these instruments, digitized distances and/or angles are partially processed by an on-board computer and stored in a portable solid-state memory unit or magnetic tape recorder fastened to the instrument tripod. These data are then transmitted to a central computer, by way of a remote terminal, for further processing. The computer makes the necessary calculations and furnishes a listing of the observations and processed data.

In each case, the computer listing constitutes the permanent record of the day's work and would be placed in the field notebook along with sketches and other information pertinent to the survey. The legality of such a process, particularly with respect to land surveys, is by no means certain at the present time. When the inevitable disputes arise, a court decision will no doubt set a precedent.

The examples cited represent only two of many similar possible schemes using available components. Obviously, a substantial investment must be made to initiate such methods and time will be required to standardize equipment and techniques for optimum efficiency. The National Ocean Survey is currently involved in utilizing the first option discussed, for first-order leveling (see Ref. 8).

Another potential method for recording data consists of entering the data into computer memory by voice using a microphone connected to a minicomputer which controls the measuring system. At the time of this writing, systems with this capability are strictly experimental (see Ref. 1). However, the potential exists and is certain to be exploited in more sophisticated surveying systems in the future.

When large amounts of survey information are stored systematically in the solid-state memory of an electronic computer or on other peripheral devices such as magnetic tapes and disk storage, this information is frequently said to be filed in a *data bank* (see also Art. 1.11).

**3.16. Computations** Calculations of one kind or another form a large part of the work of surveying, and the ability to compute with speed and accuracy is an important qualification for the surveyor. To become an expert computer, one must not only have practice but must also (1) possess a knowledge of the precision of measurements and the effects of errors in the given data upon the precision of values calculated therefrom, and (2) be familiar with the algebraic and graphical processes and computational aids which help reduce the labor of computing to a minimum.

Computations are made *algebraically* by the use of simple arithmetical procedures and trigonometric functions, *graphically* by accurately scaled drawings, or *electronically* by pocket or desk calculators or high-speed electronic computers.

Before making calculations of importance, the surveyor should plan a clear and orderly arrangement, using tabular forms wherever possible. Such steps save time, prevent mistakes, make the calculations legible to others, allow proper checks, and facilitate the work of the checker. The work should not be crowded.

Particular care should be exercised in preparing data for computation by a program on an electronic computer. The presence of undetected blunders or data out of order punched by an unsuspecting keypunch operator can lead to the wasting of inestimable amounts of valuable computer time.

All computations should be preserved in a notebook maintained for that purpose. Since a large portion of the computations will be output from an electronic computer, a loose-leaf notebook is advisable so that sheets from the computer printout can easily be included.

Office computations are generally a continuation of some field work. They should be easily accessible for future reference; for this reason, the pages should be numbered and a table of contents included. Parts of problems separated by other computations should be cross-referenced. Each problem should have a clear heading, which should include the name of the survey, the kind of computations, the field book number and page of original notes, the name of the computer, the name of the checker, and the dates of computing and checking. Usually, enough of the field notes should be transcribed to make computations possible without further reference to the field notebook. All transcripts should be checked.

**3.17. Checking** In practice, *no confidence is placed in results that have not been checked*, and important results are preferably checked by more than one method. Rarely is a lengthy computation made without a mistake. A good habit to develop is that of checking one's work until certain that the results are correct. These checks should be performed independently. Comparing each step with that of other students is not a true check and would not be countenanced in practice.

Many problems can be solved by more than one method. Since the use of the same method in checking may cause the same error to occur, results should be checked by a different method when this is feasible. Approximate checks to discover large arithmetical mistakes may be obtained by applying approximate solutions to the problem with a pocket calculator or checking the answer by scaling from a map or drawing with protractor and scale. Graphical methods may be used as an approximate check; they take less time than arithmetical solutions, and possible incorrect assumptions in the precise solution may be detected.

Each step in a long computation that cannot be verified otherwise should be checked by repeating the computation.

When work is being checked and a difference is found, the computation should be repeated before a correction is made, since the check itself may be incorrect.

In many cases, large mistakes such as faulty placing of a decimal point, can be located by inspection of the value to see if it looks reasonable.

The checking procedures outlined in the previous paragraphs are primarily directed toward manual computations performed with a pocket or desk calculator. When a computer program, known to be operating properly, is used for the solution, checking should be concentrated on the input and output listings for and from the computer. Exhaustive checking should be applied to input data for a computer program to eliminate tabulation and punching errors. It is also good practice to list the input data with the program output to provide another verification of the information used for the solution.

**3.18. Significant figures**  The expression *significant figures* is used to designate those digits in a number which have meaning. A significant figure can be any one of the digits $1, 2, 3, \ldots, 9$; and 0 is a significant figure except when used to fix a decimal point. Thus, the number 0.00456 has three significant figures and the number 45.601 has five significant figures.

In making computations you will have occasion to use numbers which are mathematically exact and numbers determined from observations (directly or indirectly) which inevitably contain observational errors. Mathematically exact numbers are absolute and have as many significant figures as are required. For example, if you multiply a given quantity by 2, the number 2 is absolute and is exact.

An observed value is *never* exact. A distance measured with a steel tape is an example of a directly observed quantity. If the distance is measured roughly, it may be recorded as 52 ft or as 16 m. If greater precision is required, the distance is observed as 52.3 ft (15.95 m) or if still more refinement is desired 52.33 ft (15.950 m) may be the value recorded. Note that none of these three numbers expresses the correct distance exactly, but that they contain two, three, four, and five significant figures, respectively. Thus, the number of significant figures in a directly observed quantity is related to the precision or refinement employed in the observation.

An indirectly measured quantity is obtained when the observed value is determined from several related, dependent observations. For example, a total distance determined by a summation of a number of directly measured quantities is an indirectly observed distance.

The number of significant figures in a directly observed quantity is evident. Thus, if you measure the width of a room using a 50-ft steel tape (graduated to hundredths of a foot), your answer can contain up to four significant figures. Assume that the width is recorded as 32.46 ft. This is an example of a directly observed measurement in which the number of significant figures is related to the precision of the equipment and the method of observation. Note that *precision* of the method of observation is a function of the *skill* and *experience of the observer*, as well as the *least count* of the instrument being used.

When numbers are determined indirectly from observed quantities, the number of significant figures is less easily determined. Suppose that the length of the same room is slightly over 50 ft and that two directly observed quantities 25.12 and 27.56 ft, yield a sum of 52.68 ft. This distance could also have been found by utilizing the full 50-ft length of tape and a short increment, resulting in measurements of 50.00 ft and 2.68 ft, to yield a total of 52.68 ft. The total distance is correctly expressed to four significant figures even though one distance contained only three significant digits. This example illustrates the fact that the number of significant figures in individual directly observed quantities does not control the precision of the quantity calculated by addition of these quantities.

   Consistency in computations requires that one distinguish between exact and observed quantities and that the rules of significant figures be followed. First consider the act of rounding numbers.

**Rounding of numbers**    Assume that 27 and 13.1 are exact numbers. The quotient of $27/13.1 = 2.061068702 \cdots$ never terminates. To use this quotient, superfluous digits are removed from the right and the quotient is retained as 2.06 or 2.0611. This process is called *rounding off*.

   To round a number, retain a certain number of digits counted from the left and drop the others. If it is desired to round $\pi$ to three, four, and five significant figures, the results are 3.14, 3.142, and 3.1416, respectively. Rounding is performed to cause the least possible error and should be done according to the following rule:

   To round a number to $n$ significant figures, discard all digits to the right of the $n$th place. If the discarded digit in the $(n+1)$st place is less than one-half a unit in the $n$th place, leave the $n$th digit unchanged; if the discarded digit is greater than one-half a unit in the $n$th place, add one to the $n$th digit. If the discarded digit is exactly one-half a unit in the $n$th place, leave the $n$th digit unaltered if it is even and increase the $n$th digit by one if it is odd.

   A number rounded according to this rule is said to be correct to $n$ significant figures. Examples of numbers rounded to four significant figures are:

| | | |
|---|---|---|
| 31.68234 | becomes | 31.68 |
| 45.6874 | becomes | 45.69 |
| 2.3453 | becomes | 2.345 |
| 10.475 | becomes | 10.48 |

**Rules of significant numbers applied to arithmetic operations**

ADDITION

Find the sum of approximate numbers 561.32, 491.6, 86.954, and 3.9462, where each number is correct to its last digit. Round all the numbers to one more decimal than the least significant number and add.

$$
\begin{array}{r}
561.32 \\
491.6 \\
86.95 \\
3.95 \\
\hline
1143.82
\end{array}
\quad \text{or} \quad 1143.8
$$

Retention of the extra significance in the more accurate numbers eliminates errors inherent in these numbers and reduces the total error in the sum. Round the final answer to tenths.

   The average of a set of measurements provides a more reliable result than any of the individual measurements from which the average is computed. Retention of one more figure in the average than in the numbers themselves is justified.

## SUBTRACTION

To subtract one approximate number from another, first round each number to the same decimal place before subtracting. Consider the difference between 821.8 and 10.464:

$$821.8 - 10.5 = 811.3$$

Errors resulting from subtraction of approximate numbers are most significant and serious when the numbers are very nearly equal and the leading significant figures are lost. Such errors in subtraction of nearly equal figures can cause computations to be worthless.

## MULTIPLICATION

Round the more accurate numbers to one more significant figure than the least accurate number. Answer should be given to the same number of significant figures as are found in the least accurate factor. For example:

$$(349.1)(863.4) = 301,412.94$$

which should be rounded to

$$301,400 = 3.014 \times 10^5$$

which has four significant figures.

## DIVISION

Same rules as for multiplication. For example, assume that 5 is an exact number. Then

$$\frac{56.5}{\sqrt{5}} = \frac{56.3}{2.236}$$

where $\sqrt{5} = 2.236$ is rounded to one more significant figure than the numerator and the quotient, $56.3/2.236 = 25.2$, is rounded to the same number of significant figures as are found in the least significant number in the calculation.

## POWERS (see Scarborough, Ref. 7)

If $k$ is the value of the first significant figure of a number having $n$ significant digits, its $p$th power is correct to

$$n-1 \quad \text{significant figures} \quad \text{if } p \leqslant k$$
$$n-2 \quad \text{significant figures} \quad \text{if } p \leqslant 10k$$

For example, raise 0.3862 to the fourth power. Thus, $k=3$, $n=4$, and $p=4$, so that $p<10k$ or $p<30$ and $(0.3862)^4 = 0.02225$ should be rounded to $n-2$ significant figures, or to 0.0223.

## ROOTS (see Scarborough, Ref. 7)

The $r$th root of a number having $n$ significant digits and in which $k$ is the first significant digit is correct to

$$n \quad \text{significant figures} \quad \text{if } rk \geqslant 10$$
$$n-1 \quad \text{significant figures} \quad \text{if } rk < 10$$

As examples, consider roots of the following numbers where the given numbers are correct to the last digit:

$$(25)^{1/2} = 5 \qquad \text{where } r=2,\ k=2,\ n=2, \text{ and } rk=4<10$$
$$(615)^{1/4} = 5.00 \qquad \text{where } r=4,\ k=6,\ n=3, \text{ and } rk=24>10$$
$$(32,768)^{1/5} = 8.0000 \qquad \text{where } r=5,\ k=3,\ n=5, \text{ and } rk=15>10$$

**3.19. Computational aids**   Electronic calculators and high-speed electronic computers are the major computational aids available to the surveyor and students of surveying. The various types of electronic devices listed according to increasing speed, capacity, and memory are: pocket calculators, including those which are programmable; programmable desktop calculators; minicomputers; and large-capacity high-speed computers.

Pocket calculators burst upon the market in 1973–1974 and have virtually eliminated mechanical desk calculators, slide rules, and logarithms from surveying schools, offices, and field operations. The simplest of these devices allows addition, subtraction, multiplication, and division. The more exotic models permit taking roots, powers, exponential functions, trigonometric functions, fixed- and floating-point arithmetic, and regression analysis. Many of these latter types have numerous storage registers, so that matrices of a limited size can be manipulated. Programmable pocket calculators are available in various models which can be programmed according to the following methods: internally by keying the program in from the keyboard, by means of programmed magnetic cards, and by plugging in preprogrammed solid-state modules for special routines. Most of these models have up to ten storage registers and several can be used with a portable printing and plotting machine so that hard copy output is available. All pocket calculators can be run on rechargeable batteries or may be plugged into ac current. These moderately priced calculators are extremely powerful tools for students and the practicing surveyor in the field and office.

Programmable desktop calculators have memories of moderate size of up to several thousand locations. Instructions are generally entered by way of a keyboard, magnetic cards, or magnetic tape. Output can be observed on a nixie tube display and is printed or plotted. Programs are available for many routine surveying calculations, such as for horizontal curves, earthwork computations, traverse, triangulation, and many others. Desktop calculators provide an intermediate step between manual calculations with the pocket calculators and the larger-capacity computers. With the proper type of batteries, desktop calculators could be used in the field.

Minicomputers are high-speed machines of moderate size which have central core memories of up to 32 thousand 16-bit words. These computers can accept input from punched cards, paper tape, and magnetic tape. Output is by line printer, magnetic tape, or punched cards. With the proper type of peripheral hardware, such as disk files and magnetic tape units, these machines can do jobs that formerly required large-capacity computers. These systems are permanently installed and are suitable for organizations of moderate size which do not need a large computer but want the independence of having their own machine.

Large-capacity high-speed computers have central core memories of perhaps several hundred thousand words and also additional disk and/or magnetic tape storage. For example, a typical system may have 500,000 words of central and extended core storage and 50,000,000 words of disk storage. These systems are designed for very large numerical computations and in particular are suitable for servicing a large number of remote terminals.

Remote terminals are input/output stations which allow communication with a large central computer by way of a special cable or a telephone line. These terminals are usually equipped with a teletypewriter, a punched-card reader, and a line printer. Instructions and data can be typed on the teletypewriter or may be read in the form of punched cards by the card reader. When suitable instructions are typed on the teletypewriter, the program and data are processed by the central computer. Answers are printed by the line printer in the remote terminal or, on request, may be output at the computer center on the line printer or on punched cards.

*Interactive* systems consist of remote terminals equipped with a keyboard for input and a cathode ray tube (CRT) for observing in real time the manipulation of the program and data. One way to operate such a system is to store an operating program on disk at the central computer center. Then when desired, data are input at the remote terminal using the keyboard. Also entered are the instructions requesting use of the desired program, which is stored on disk. Data and the program can be displayed on the CRT for checking and solution of the problem. When computations are completed, the answers may be displayed on the CRT or printed on the line printer. If changes are needed during the solution, they can be made immediately by entering the appropriate instructions at the keyboard. Data can be displayed on the CRT in numerical, alphanumeric, or graphical form. The potential for such systems in surveying and mapping is enormous.

The trend toward large centralized systems with many remote terminals was especially strong during the late 1960s. The emergence of high-speed minicomputers with very adequate core storage and well-designed peripheral hardware, all at a fairly reasonable cost, has reversed this trend at the time of this writing. In large government agencies for surveying and mapping, minicomputers are interfaced with a large central computer in a distributive array manner.

Technological advances in the computer industry have been so rapid during the past few years that it is very difficult for the practitioner to stay abreast of the most recent developments. Rational decisions as to what constitutes the best system for a particular organization are difficult even for those well acquainted with computer equipment and operations. Members of the surveying profession must utilize the best in computational aids, since computations are their business. To do this, they need to continually familiarize themselves with the various options and become well-enough acquainted with the technicalities of different systems to allow optimum use of the best facilities available.

An old-fashioned device still useful to the surveyor is the *polar planimeter*. It is of great value in finding areas of figures plotted to scale. The accuracy with which results may be obtained depends primarily on the skill of the operator in traversing the lines of the drawing with the tracing point. In general, results may be determined to three significant figures, which is consistent with many of the field data upon which calculations of area are based. It is a simple instrument to operate and furnishes a most efficient means of determining the area of figures with irregular or curved boundaries.

### 3.20. Problems

**3.1.** A point is to be established on the ground at a distance of about 200 m from a given point by means of one linear and one angular measurement. It is desired to establish the point within 3 cm of its true location. With what precision need the angle be measured?

**3.2.** An angle of 40° is measured with a transit having a vernier with a least count of 30″. The maximum error is half the least count of the vernier. What is the ratio of precision if the sine of the angle is to be used in computations? The cosine? The tangent?

**3.3.** If tangents or cotangents are involved, what is the highest precision corresponding to single measurements (of angles of any size) with the transit of the preceding problem?

**3.4.** What is the minimum allowable angle that for computations involving sines will permit a ratio of precision of 1/10,000 to be maintained if the angular error is 10″?

**3.5.** Round off the following numbers to four significant figures: 72.8432; 10.285; 0.0082495; 93,821; 521,242; 0.0800189; 6490.8.

**3.6.** The following distances are added: 167.8084 ft, 228.01 ft, 73.133 m, and 167.8084 ft. What is the sum of the distances? (1 m = 3.280833 ft.)

**3.7.** Perform the following subtractions: 779.9 − 36.543; 65.3956 − 65.214; 18,230 − 18,228.

**3.8.** The hypotenuse of a right triangle is 384.79 m long and makes an angle of 25°11′30″ with the long leg of the triangle. Calculate, to five significant figures, the lengths of both legs of the triangle.

**3.9.** In a 30°-60°-90° triangle, the long leg is 100.00 m in length. Calculate the length of the hypotenuse of this triangle.

**3.10.** In an oblique triangle, the two sides and the included angle are 100 m, 130 m, and 120°, respectively. Calculate the third side of the triangle.

**3.11.** Calculate the following values correct to the proper number of significant figures: $\sqrt{12} - 4$; $\sqrt{324}$ ; $\sqrt{1359}$ ; $(12.5)^2$; $(2.845)^3$; $(225)^2$; $(34.662)^{1/2}$; $(625)^{1/4}$.

**3.12.** How many significant figures are there in each of the following numbers?

(a) 0.208     (d) $30.0 \times 10^6$    (g) $28.46 \pm 0.08$

(b) 76248.0    (e) 0.006

(c) 1.4800    (f) $1.04 \pm 0.02$

**3.13.** Find the product of each of the following series of numbers, to the proper number of significant figures:

(a) $2.64 \times 79.18 \times 0.0767 \times 1.0028$

(b) $(3.415)^2 \times \sqrt[3]{72.849}$

(c) $(6.618)^{1/3} \times (68.627)^{1/2}$

**3.14.** Find the quotient for each of the following series of numbers, to the proper number of significant figures:

(a) $76.21 \div 49.20$

(b) $0.4092 \div 0.006314$

(c) $1.4823 \div 16.4917$

(d) $8.472 \div \tan 42°21'$

**3.15.** How many feet and how many metres are there in $3\frac{1}{2}$ rods? In 6 chains and 8 links?

**3.16.** How many acres are there in a rectangular field 18.25 rods wide and 920.0 ft long? A rectangular field 100.55 by 424.8 m? How many hectares are in each field?

**3.17.** If the angles of an approximately equilateral triangle are measured with a standard deviation of 30″ and the distances with corresponding precision, how many significant figures will there be in the area computed by sines and cosines?

## References

1. Beek, B., Broglie, J., and Vonusa, R., "Voice Data Entry for Cartographic Applications," *Proceedings, Fall Convention, ACSM*, 1977, p. 161.
2. Chapman, W. H., "Programmable Desk Calculators—Their Use in Topographic Mapping," *Proceedings, Fall Convention, ACSM*, 1972, p. 238.
3. Mayfield, G. M., "Engineering Surveying, Chapters 2, 5, 6, 8, 16, and 20," *Journal of the Surveying and Mapping Division, ASCE*, Vol. 102, No. SU1, Proceedings Paper 12648, December 7, 1976, pp. 59–94.
4. Meade, B. K., "Solution of Surveying Problems by Mini-Programmable Computers," *Proceedings, 38th Annual Meeting, ACSM*, 1978, p. 312.
5. Miller, M. L., "Perpetuation and Consolidation of Federal Survey Notes," *Proceedings, Fall Convention, ACSM* 1974, p. 287.
6. *National Geodetic Survey Data: Availability, Explanation, and Application*, NOAA Technical Memorandum NOS NGS-5, National Geodetic Survey, Rockville, Md., June 1976.
7. Scarborough, J. B., *Numerical Mathematical Analysis*, The Johns Hopkins Press, Baltimore, Md., 1962.
8. Whalen, C. T., and Balacz, E. I., "Test Results of First-Order Class III Leveling," *Surveying and Mapping*, Vol. 37, No. 1, March 1977, p. 45.

# PART II
# BASIC
# SURVEY MEASUREMENTS

# CHAPTER 4
# Distance measurement

**4.1. Distance**   The distance between any two random points in three-dimensional space is a spatial distance. This concept is particularly pertinent considering the advanced stage of electronic distance measurement (EDM) techniques, which automatically provide spatial or slope distances in the range of from 15 m or 50 ft to 60 km or 36 mi. Although slope distances are frequently observed in the surveying operation, these distances are then reduced to a horizontal projection for more convenient use in subsequent calculations and field layout. Horizontal distance is further reduced to its equivalent at sea level for geodetic surveys (Art. 10.17). In plane surveying, horizontal distances are reduced to sea level only when it is desired to convert them into equivalent distances at another elevation, such as that of the state plane coordinate system or the average elevation of a survey for which the variation in elevation over the area is large.

There are several methods of determining distance, the choice of which depends on the accuracy required, the cost, and other conditions. The methods in ascending order of accuracy are estimation, scaling from a map, pacing, odometer, tacheometry, taping, photogrammetry, inertial systems, and electronic distance measurement. For example, on rough reconnaissance, a relative precision of 1 part in 100 (i.e., a standard deviation of 1/100 of the distance) or less may be adequate for the survey; on the other hand, specifications for first-order surveys (refer to Table 10.1) call for base-line measurement to have a standard deviation of 1 part in 1 million of the distance measured. On certain surveys, a combination of methods may be warranted.

Table 4.1 classifies the principal methods of measuring distance according to the usual relative precisions obtained. The various methods are discussed further in the following sections.

## METHODS OF MEASUREMENT

**4.2. Pacing**   Pacing furnishes a rapid means of approximately checking more precise measurements of distance. It is used on reconnaissance surveys and, in small-scale mapping, for locating details and traversing with the plane table. Pacing over rough country has furnished a relative precision of 1/100; under average conditions, a person of experience will have little difficulty in pacing with a relative precision of 1/200.

Each two paces or double step is called a *stride*. Thus, for a pace the stride would be 5 ft, or there would be roughly 1000 strides per mile.

**Table 4.1  Methods of Measuring Distance**

| Method | Relative precision[a] | Use | Instruments required | Instruments for measuring angles with corresponding accuracy |
|---|---|---|---|---|
| Pacing odometer or mileage recorder | 1/100 | Reconnaissance, small-scale mapping, checking tape measurement, quantity surveys | Pedometer, odometer | Hand compass, peepsight alidade |
| Tacheometry Stadia | 1/300–1/1000 | Location of details for topographic mapping, rough traverse, checking more accurate measurements | Level rod or stadia board, calibrated optical line of sight with stand | Transit, telescopic alidade and plane table, surveyor's compass |
| Distance wedge | 1/5000–1/10,000 | Traverse for land surveys, control of route and topographic surveys and construction work | Horizontal graduated rod and support calibrated optical line of sight equipped with a distance wedge | Transit or theodolite |
| Subtense bar | 1/1000–1/9000 | Hydrographic surveys, traverse | Calibrated subtense bar and tripod; 1″ theodolite | Transit or theodolite |
| Ordinary taping | 1/3000–1/5000 | Traverse for land surveys and for control of route and topographic surveys and construction | Steel tape, chaining pins, plumb bobs | Transit |
| Precise taping | 1/10,000–1/30,000 | Traverses for city surveys, base lines for triangulation of low accuracy, and construction surveys requiring high accuracies | Calibrated steel tape, thermometer, tension handle, hand level, plumb bobs | Transit or theodolite |
| Photogrammetry | up to 1/50,000 | Location of detail for topographic mapping, second- and third-order ground control surveys | Stereoplotters, mono and stereo comparators, electronic computer | |
| Inertial systems | up to 1/50,000 | Rapid, reconnaissance surveying: large area surveys, second- and third-order ground control surveys | Inertial positioning system | |
| Base-line taping | 1/100,000–1/1,000,000 | First-, second-, and third-order triangulation for large areas, city surveys, long bridges, and tunnels | Calibrated steel tape, thermometers, tension handle, taping supports, level, level rod | Theodolite (1″) |
| EDM | ± 0.02 or 6 mm + 1 ppm | Traverse, triangulation, and trilateration for control surveys of all types and for construction surveys | EDM equipment | Theodolite (1″) which can be built into the EDM |

[a] Relative precision may be defined as the ratio of the allowed standard deviation to the distance measured.

Paces or strides are usually counted by means of a tally register operated by hand or by means of a pedometer attached to the leg. In hilly country, rough corrections for slope can be applied.

Beginners should standardize their pace by walking over known distances on level, sloping, and uneven ground.

**4.3. Mileage recorder, odometer, and other methods**   Distance may be measured by observing the number of revolutions of the wheel of a vehicle. The *mileage recorder* attached to the ordinary automobile speedometer registers distance to 0.1 mi and may be read by estimation to 0.01 mi. Special speedometers are available reading to 0.01 or 0.002 mi. The *odometer* is a simple device that can be attached to any vehicle and directly registers the number of revolutions of a wheel. With the circumference of the wheel known, the relation between revolutions and distance is fixed. The distance indicated by either the mileage recorder or the odometer is somewhat greater than the true horizontal distance, but in hilly country a rough correction based on the estimated average slope may be applied.

Distances are sometimes roughly estimated by *time interval of travel* for a person at walk, a saddle animal at walk, or a saddle animal at gallop.

By *mathematical* or *graphical* methods, unknown distances may be determined through their relation to one or more known distances. These methods are used in triangulation and plane-table work.

**4.4. Tacheometry**   Tacheometry includes stadia with transit and stadia rod; stadia with alidade, plane table, and rod; distance wedge and horizontal rod; and subtense bar and theodolite. The stadia method, described in detail in Chap. 7, offers a rapid indirect means of determining distances. It is used extensively and is particularly useful in topographic surveying. The telescope of the transit, level, or plane-table alidade is equipped with two horizontal hairs, one above and the other an equal distance below the horizontal cross hair. The distance from the instrument to a given point is indicated by the interval between these stadia hairs as shown on a graduated rod held vertically at the point. The precision of the stadia method depends upon the instrument, the observer, the atmospheric conditions, and the length of sights. Under average conditions the stadia method will yield a relative precision between 1/300 and 1/1000. The use of a self-reducing tacheometer or self-reducing alidade for the stadia method simplifies the operation substantially.

The distance wedge is used in conjunction with a horizontally supported rod placed perpendicular to the optical line of sight.

Another method for determining distance by tacheometry consists of using a subtense bar and precise transit or theodolite. The subtense bar is a 2-m calibrated bar mounted horizontally on a tripod and placed perpendicular to the line of sight. The subtended angle is then measured with a 1″ theodolite placed at the other end of the line to be measured so that the horizontal distance between the points can be calculated. Further details regarding this procedure can be found in Chap. 7.

**4.5. Taping**   Taping involves direct measurement of the distance with steel tapes varying in length from 3 ft (1 m) to 1000 ft (300 m). Graduations are in feet, tenths, and hundredths, or metres, decimetres, centimetres, and millimetres. Formerly, on surveys of ordinary precision it was the practice to measure the length of lines with the *engineer's chain* or *Gunter's chain*; for measurements of the highest precision, special bars were used.

The *engineer's chain* is 100 ft long and is composed of 100 links each 1 ft long. At every 10 links brass tags are fastened, notches on the tags indicating the number of 10-link segments between the tag and the end of the tape. Distances measured with the engineer's chain are recorded in feet and decimals.

The *surveyor's* or *Gunter's chain* is 66 ft long and is divided into 100 links each 0.66 ft = 7.92 in long; distances are recorded in chains and links. Thus,

$$1 \text{ (Gunter's) chain} = 66 \text{ ft} = 100 \text{ links} = 4 \text{ rods}$$
$$80 \text{ (Gunter's) chains} = 1 \text{ mi}$$
$$10 \text{ square (Gunter's) chains} = 1 \text{ acre} = 43{,}560 \text{ ft}^2$$

Measuring with chains was originally called "chaining." The term has survived and is often applied to the operation of measuring lines with tapes.

The precision of distance measured with tapes depends upon the degree of refinement with which measurements are taken. On the one hand, rough taping through broken country may be less accurate than the stadia. On the other hand, when extreme care is taken to eliminate all possible errors, measurements have been taken with a relative precision of less than 1/1,000,000. In ordinary taping over flat, smooth ground, the relative precision is about 1/3000 to 1/5000.

**4.6. Electronic distance measurement**  Recent scientific advances have led to the development of electro-optical and electromagnetic instruments which are of great value to the surveyor for accurate measurements of distances. Measurement of distance with electronic distance measuring (EDM) equipment is based on the invariant speed of light or electromagnetic waves in a vacuum. EDM equipment which can be used for traverse, triangulation, and trilateration as well as for construction layout is rapidly supplanting taping for modern surveying operations except for short distances and certain types of construction layout. A detailed explanation of EDM systems and procedures is given in Arts. 4.29 to 4.43.

**4.7. Choice of methods**  Most boundary, control, and construction surveys involving long lines and large areas can be performed most accurately and economically using modern EDM equipment. Where the distances involved are relatively short or specific construction layout requirements are present, taping the distances can be more practical. Stadia is still unsurpassed for small topographic surveys and preliminary surveys for projects of limited extent.

Each of the methods mentioned in the preceding sections has a field of usefulness. On the survey for a single enterprise, the surveyor may find occasion to employ a combination of methods to advantage.

**4.8. Tapes**  Tapes are made in a variety of materials, lengths, and weights. Those more commonly used by the surveyor and for engineering measurements are the steel tapes, sometimes called the engineer's or surveyor's tape, and woven nonmetallic and metallic tapes.

The woven *metallic tape* (Fig. 4.1) is a ribbon of waterproofed fabric into which are woven small brass or bronze wires to prevent its stretching. It is usually 50, 100, or 150 ft long; is graduated in feet, tenths, and half-tenths; and is $\frac{5}{8}$ in wide. The metric metallic tape is 10, 20, 30, or 50 m long; is graduated in metres, centimetres, and 2 mm; and is usually 13 mm wide. Metallic tapes are used principally in earthwork cross sectioning, in location of details, and in similar work where a light, flexible tape is desirable and where small errors in length are not of consequence. Nonmetallic glass fiber tapes which are nonconductors of electricity have been developed for use near power lines.

A tape of phosphor bronze is rustproof and is particularly useful for work in the vicinity of salt water.

For very precise measurements, such as those for base lines and in city work, the *invar tape* has come into general use. Invar is a composition of nickel and steel with a very low coefficient of thermal expansion, sometimes as small as one-thirtieth that of steel, and is

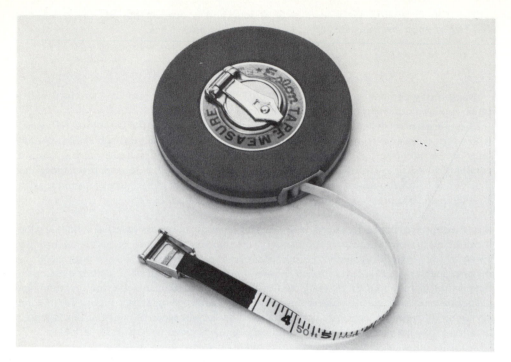

**Fig. 4.1** Metallic tapes graduated in feet and metric units. (*Courtesy Lietz Instruments, Inc.*)

affected little by temperature changes. Since the compositions having the lowest coefficients of expansion may not remain constant in length over a period of time, it is customary to use a composition having a larger coefficient; the thermal expansion of commercial invar tapes is about one-tenth that of steel tapes. Invar is a soft metal, and the tape must be handled very carefully to avoid bends and kinks. This property and its high cost make it impractical for ordinary use.

Steel tapes for which the foot is the unit of length are graduated as follows: lightweight tapes and some engineer's tapes are graduated to hundredths of feet throughout the length; normally, the heavier tapes have graduations with numbers every foot, with only the end feet graduated in tenths and/or hundredths of feet.

Metric tapes are 15, 25, 30, or 50 m long. The light box tapes are graduated throughout in metres, decimetres, and centimetres. Heavier tapes are graduated as follows: throughout to metres and decimetres; first and last metre in centimetres; and first and last decimetres in millimetres; or throughout to metres and half metres with end metres to decimetres. Various styles of steel tapes graduated in feet and metres are shown in Fig. 4.2*a* through *e*.

Fiberglass tapes are also available in units of feet and metres and may be obtained in lengths of 50, 100, and 150 ft or 15, 30, and 50 m.

Tapes for which the Gunter's chain is the unit of length are 1, 2, and 2.5 chains in length and are graduated in links throughout with end links in tenths.

Ordinarily, rawhide thongs serving as handles are fastened to the rings at each end of the chain tape. Wire handles are sometimes used, but they are objectionable when the tape must be dragged through grass or brush. Detachable clamp handles are available for grasping the tape at any point. Figure 4.3 illustrates the use of a tape clamp.

The tape should be kept straight when in use; any tape will break when kinked and subjected to a straight pull. Steel tapes rust readily and should be wiped dry after being used.

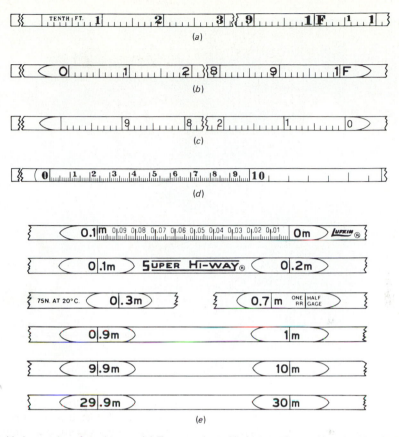

**Fig. 4.2** Various styles of steel tapes. (a) Tape graduated in feet, tenths, and hundredths throughout. (b) First foot graduated to hundredths, tape graduated throughout to feet. A "cut" tape. (c) Added foot graduated to hundredths of a foot, tape graduated to feet throughout. An "add" tape. (d) Metric tape. First decimetre graduated to millimetres, first metre graduated to centimetres, graduated throughout to decimetres. A "cut" tape. (e) Metric tape. Added decimetre graduated to millimetres. Tape graduated throughout to decimetres. An "add" tape. [*Courtesy the Cooper Group (Lufkin).*]

Special care is required when working near power lines. Fatal accidents have occurred from throwing a steel or metallic tape over a power line.

**4.9. Equipment for taping**   Additional equipment employed for determining the lengths of lines by direct measurement with a tape consists of plumb bobs, the hand level, the tension handle, chaining pins, and range poles.

The plumb bob is a pointed metal weight used to project the horizontal location of a point from one elevation to another.

The hand level is described in Art. 5.21 and can be used to keep the two ends of the tape at the same elevation when measuring over irregular terrain. A tension handle is a spring scale which can be attached to the end of the tape and allows applying the proper tension.

Steel taping pins, also called chaining pins, taping arrows, or surveyor's arrows, are commonly employed to mark the ends of the tape during the process of taping between two points more than a tape length apart. They are usually 10 to 14 in (25 to 35 cm) long. A set consists of 11 pins (see Fig. 4.4a).

**Fig. 4.3** Use of the tape clamp.

**Fig. 4.4a** Taping pins.

**Fig. 4.4b** Range pole.

For more precise taping or for future reference, nails may be driven into the earth. Marking tags, which can be stamped with identification as desired, may be threaded over the nails before they are driven. On paved surfaces, tape lengths and other points may be marked with keel (carpenter's chalk), pencil, or spray paint; for precise work a short piece of opaque adhesive tape may be stuck on the pavement and the point marked with a pencil or ball-point pen.

Wooden stakes or hubs are driven into the ground to mark the significant points of profile-level and transit-traverse lines. Taping tripods and wooden or concrete posts are used to mark tape ends in base-line taping (Chap. 10).

Metal, wooden, or fiberglass range poles, also called flags, flagpoles, or lining rods, are used as temporary signals to indicate the location of points or the direction of lines. They are of octagonal or circular cross section and are pointed at the lower end. Wooden and fiberglass range poles are shod with a steel point. The common length is 6 or 8 ft, or 2 or 3 m. Usually, the pole is painted with alternate bands of red and white 1 ft or $\frac{1}{2}$ m long (see Fig. 4.4$b$).

**4.10. Taping on smooth, level ground**   The procedure followed in measuring distances with the tape depends to some extent upon the required accuracy and the purpose of the survey. The following description represents the usual practice when the measurements are of ordinary relative precision of, say, 1/5000. Errors in taping are discussed in Art. 4.26 and mistakes in Art. 4.28.

The tape is supported throughout its length. If only the distance between two fixed points (as the corners of a parcel of land) is to be determined, the equipment will consist of one or more range poles, 11 chaining pins, and a 100-ft or 30-m heavy steel tape graduated as described in Art. 4.8. One range pole is placed behind the distant point to indicate its location.

The rear tapeman *with one pin* is stationed at the point of beginning. The head tapeman, with the zero (graduated) end of the tape and 10 pins, advances toward the distant point. When the head tapeman has gone nearly 100 ft, or 30 m, the rear tapeman calls "chain" or "tape," a signal for the head tapeman to halt. The rear tapeman holds the 100-ft or 30-m mark at the point of beginning and, by hand signals or by voice, lines in a taping pin (held by the head tapeman) with the range pole (or other signal) marking the distant point. During the lining-in process, the rear tapeman is in a kneeling position on the line and facing the distant point; the head tapeman is in a kneeling position to one side of and facing the line so that the tape can be held steady and so that the rear tapeman will have a clear view of the signal marking the distant point. The head tapeman with one hand sets the pin vertically on line and a short distance to the rear of the zero mark. With the other hand the tapeman then pulls the tape taut and, making sure that it is straight, brings it in contact with the pin. The rear tapeman, upon observing that the 100-ft mark is at the point of beginning, calls "good" or "mark." The head tapeman pulls the pin and sticks it at the zero mark of the tape, with the pin sloping away from the line (Fig. 4.5). As a check, the head tapeman again pulls the tape taut and notes that the zero point coincides with the pin at its intersection with the ground. The head tapeman then calls "good," the rear tapeman releases the tape, the head tapeman moves forward as before, and so the process is repeated.

As the rear tapeman leaves each intermediate point, the pin is pulled. Thus, there is always one pin in the ground, and the number of pins held by the rear tapeman at any time indicates the number of hundreds of feet, or *stations*, or multiples of 30 m, from the point of beginning to the pin in the ground.

At the end of each 1000 ft (10 full stations) or 300 m, the head tapeman has placed the last pin in the ground. The head tapeman signals for pins, the rear tapeman comes forward and gives the 10 pins to the head tapeman, and both check the total, which is recorded. The

**Fig. 4.5** Taping on smooth level ground.

procedure is then repeated. The count of pins is important, as the number of tape lengths is easily forgotten, owing to distractions.

When the end of the course is reached, the head tapeman halts and the rear tapeman comes forward to the last pin set. The head tapeman holds the zero mark at the terminal point. The rear tapeman pulls the tape taut and observes the number of *whole feet*, metres, or decimetres between the last pin and the end of the line (Fig. 4.6a). The rear tapeman then holds the next *larger* foot, metre, or decimetre mark at the pin, and the head tapeman pulls the tape taut and reads the decimal by means of the finer graduations at the end of the tape (Fig. 4.6b). For tapes graduated in feet and having fine graduations between the 0 and 1-ft

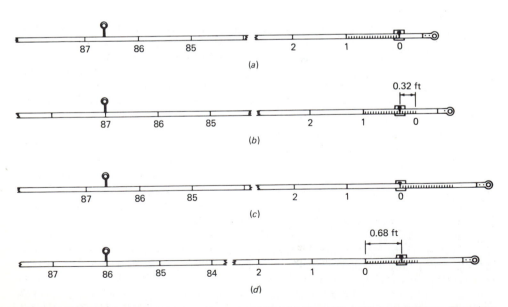

**Fig. 4.6** (a) and (b) Measuring with a "cut" tape. (c) and (d) Measuring with an "add" tape. Plan view of a tape graduated in feet.

points, (a "cut" tape), the distance is the whole number of feet held by the rear tapeman minus the fractional increment observed by the head tapeman. For example, if the rear tapeman holds 87 ft and the head tapeman observes 0.32 ft directly, the distance is $87 - 0.32 = 86.68$ ft as illustrated in Fig. 4.6*b*.

For tapes having an extra graduated foot beyond the zero point (an "add" tape), the rear tapeman holds the next *smaller* foot mark at the pin, and the head tapeman reads the decimal. Figure 4.6*c* and *d* illustrates this procedure for the same distance of 86.68 ft.

Metric tapes have the first decimetre graduated to millimetres and the first metre to centimetres (a "cut" tape; see Fig. 4.2*d*) or a decimetre graduated to millimetres is added to the zero end of the tape (an "add" tape; see Fig. 4.2*e*). Both styles are usually graduated throughout to decimetres.

Cut and add metric tapes are used in the same way as tapes graduated in feet. Suppose that when the zero end of a metric cut tape is at the terminal point, the rear tapeman observes that the rear pin falls just past 25.3 m. The rear tapeman "gives" a decimetre and holds 25.400 m at the rear taping pin. If the head tapeman then reads 0.075 m on the first graduated dm at the terminal point, the distance is $25.400 - 0.075 = 25.325$ m.

In the same situation, using a metric add tape, the rear tapeman would hold 25.300 m at the rear taping pin. The head tapeman would then read 0.025 m on the added decimetre at the terminal point, yielding a distance of $25.300 + 0.025 = 25.325$ m.

When the transit is set up on the line to be measured, the transitman usually directs the head tapeman in placing the pins on line.

On some surveys it is required that stakes be set on line at short intervals, usually 100 ft or 30 m. Sometimes stakes are driven by the rear tapeman after the pin is pulled. On surveys of low precision, no pins are used and the head tapeman sets the stakes in the manner previously described for pins, measuring the distance between centers of stakes at their junction with the ground. On more precise surveys, measurements are carried forward by setting a tack in the head of each stake. When the tack has been driven, it is tested for line and distance.

**4.11. Horizontal taping over sloping ground**    The process of taping over uneven or sloping ground or over grass and brush is much the same as that just described for smooth, level ground, except that the plumb bob is used. The tape is held horizontal, and a plumb bob is used by either, or at times by both, tapemen for projecting from tape to pin, or vice versa. For rough work, plumbing can be accomplished with the range pole.

To secure accuracy comparable to taping over level ground, considerable skill is required. Some experience is necessary to determine when the tape is nearly horizontal; the tendency is to hold the downhill end of the tape too low. A hand level is useful to estimate the proper height to hold the tape ends so as to have a horizontal tape. The tape is unsupported between its two ends, and either the pull must be increased to eliminate the effect of sag or a correction for sag must be applied. A firm stance is important; the tapemen should place the planes of their bodies parallel to the tape, with legs well apart. The forearms should be in line with the tape, and should be steadied or snubbed against the respective legs or bodies, as required by the height at which the tape must be held.

Where the slope is less than 5 ft in 100 ft, or 1.5 m in 30 m, the head tapeman advances a full tape length at a time. If the course is downhill, the head tapeman holds the plumb line at the zero point of the tape, with the tape horizontal and the plumb bob a few inches from the ground, then pulls the tape taut and is directed to line by the rear tapeman. When the plumb bob comes to rest, it is lowered carefully to the ground (see Fig. 4.7) and a pin is set in its place. As a check, the measurement is repeated. If the course is uphill, the rear tapeman holds a plumb line suspended from the 100-ft or 30-m mark and signals the head tapeman to

**Fig. 4.7** Plumbing at the downhill end of a horizontal tape.

give or take until the rear tapeman's plumb bob comes to rest over the pin. The head tapeman sets a pin, and the measurement is repeated.

Where the course is steeper and it is still desired to establish 100-ft or 30-m stations, as for a route survey, the following procedure is recommended. Assuming that the slope is downhill, the head tapeman advances a full tape length and then returns to an intermediate point from which the tape can be held horizontal. The head tapeman suspends the plumb line at a foot mark, is lined in by the rear tapeman, and sets a pin at the indicated point. The rear tapeman comes forward, *gives the head tapeman a pin*, and at the pin in the ground holds the tape at the foot mark, from which the plumb line was previously suspended. The head tapeman proceeds to another point from which the tape can be held horizontal, and so the process is repeated until the head tapeman reaches the zero mark on the tape. At each *intermediate* point of a tape length the rear tapeman gives the head tapeman a pin, but not at the point marking the full tape length. In this manner the tape is always advanced a full length at a time; the number of pins held by the rear tapeman at each 100-ft or 30-m point indicates the number of hundreds of feet or multiples of 30 m from the last tally, and the rear tapeman's count of pins is not confused. The process is called "breaking tape."

To illustrate, Fig. 4.8 represents the profile of a line to be measured in the direction of $A$ to $D$, and $A$ represents a pin marking the end of a 30-m interval from the point of beginning. The head tapeman goes forward until the 30-m mark is at $A$, where the rear tapeman is stationed. The head tapeman then returns to $B$, where the tape is held horizontal and is plumbed from the 25-m mark to set a pin at $B$. The rear tapeman gives the head tapeman a pin and holds the 25-m mark at $B$. The head tapeman plumbs from the 15-m mark and sets a pin at $C$. The rear tapeman gives the head tapeman a pin and holds the 15-m mark at $C$.

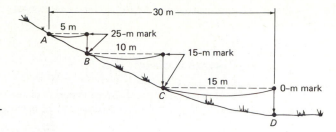

**Fig. 4.8** Horizontal measurements on a steep slope.

The head tapeman plumbs from the zero mark to set a pin at $D$ at the end of the full tape length. The rear tapeman goes forward but keeps the pin that was pulled at $C$.

Some surveyors prefer to measure distances less than the tape length individually and add these measurements. However, this practice requires recording and may lead to mistakes in addition.

Usually, the tape is estimated to be horizontal by eye. This practice commonly results in the downhill end's being too low, sometimes causing a significant error in horizontal measurement. The safe procedure in rough country is to use a hand level.

In horizontal measurements over uneven or sloping ground, the tape sags between supports and becomes effectively shorter. The effect of sag can be eliminated by standardizing the tape, by applying a computed correction, or by using the normal tension (Art. 4.21). In breaking tape or when the tape is supported for part of its length, the difference in effect of sag as between a full tape length and the unsupported length can be taken into account roughly by varying the pull on the tape.

In measurements over low obstructions, if the tape is partially supported between the tapemen, account should be taken of the effect this support makes on sag corrections.

**4.12. Slope taping**    Where the ground is fairly smooth, slope measurements are generally preferred because they are usually made more accurately and quickly than horizontal measurements. Some means of determining either the slope or the difference in elevation between successive tape ends and/or breaks in slope is required. For surveys of ordinary accuracy, either the clinometer or transit (for measuring vertical angles) or the hand level (for measuring difference in elevation) may be used to advantage. If only the distance between the ends of the line is required, the procedure of taping is the same as on level ground, but a record is kept either of the slope or of the difference in elevation of each tape length (or less at breaks in slope). The horizontal distances are then computed from the distances measured on the slope.

Where stakes are to be placed at regular intervals (e.g., 100 ft, 20 m, 30 m), corrections to the slope distances may be applied as the taping progresses.

**4.13. Correction for slope**    Where the slopes are measured with sufficient accuracy to warrant, the corresponding horizontal distances may be computed by exact trigonometric relations. In Fig. 4.9 let $s$ represent the slope distance between two points $A$ and $B$, $h$ the difference in elevation, and $H$ the horizontal distance, all in the same units. $\theta$ is the vertical angle observed with a clinometer or transit. If $\theta$ has been observed, then

$$H = s \cos \theta \qquad (4.1)$$

and the correction is

$$C_h = s - H$$

**Example 4.1** A slope measurement of 29.954 m was made between two points where $\theta$ is 4°30′. Determine the horizontal distance.

SOLUTION   Using Eq. (4.1) yields

$H = (29.954)(\cos 4°30′) = (29.954)(0.996917) = 29.862$ m

**Example 4.2** It is desired to set point $D$ a horizontal distance of 195.00 ft from point $E$ along a line which has a slope angle of 5°30′. What slope distance should be measured in the field?

SOLUTION   Using Eq. (4.1), we obtain

$$S = \frac{H}{\cos\theta} = \frac{195.00}{0.995396} = 195.90 \text{ ft}$$

Reduction of slope to horizontal distances can also be determined by using the difference in elevation between the two ends of the line. Referring to Fig. 4.9 yields

$$H^2 = s^2 - h^2$$

from which

$$s = (H^2 + h^2)^{1/2} \tag{4.2}$$

and

$$H = (s^2 - h^2)^{1/2} \tag{4.3}$$

which are the direct relations for the slope and horizontal distances, respectively.

Expanding the right side of Eq. (4.3) with the binomial theorem yields

$$H = s + \left( -\frac{h^2}{2s} - \frac{h^4}{8s^3} - \cdots \right) \tag{4.4}$$

Traditionally, the quantity enclosed by parentheses in Eq. (4.4) is designated the *slope correction*, labeled $C_h$ in Fig. 4.9. The same equation can be used to calculate the slope distance necessary to lay out a given horizontal distance.

For moderate slopes the first term within the parentheses of Eq. (4.4), $h^2/2s$, is usually adequate. The error introduced by Eq. (4.4) using only the first term is negligible for ordinary slopes. The degree of approximation is shown in Table 4.2.

Thus, for slopes not exceeding 15 percent, the second term is not significant unless relative precisions of 1 : 15,000 or better are required.

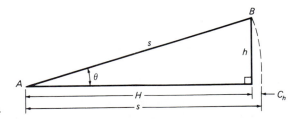

**Fig. 4.9** Slope correction.

**Example 4.3** A distance of 130.508 m was measured over terrain with a constant slope along a sloping line which has a difference in elevation between the two ends of 5.56 m. Calculate the horizontal distance between the two points.

SOLUTION   By Eq. (4.3),

$H = [(130.508)^2 - (5.56)^2]^{1/2} = 130.390$ m

**Table 4.2**

| Difference in elevation per | | Error caused by using Eq. (4.4) with only one term in | |
|---|---|---|---|
| 100 ft | 30 m | 100 ft | 30 m |
| of slope distance | | of slope distance | |
| ft | m | ft | m |
| 5 | 1.5 | 0.0001 | 0.00002 |
| 10 | 3 | 0.001 | 0.0004 |
| 15 | 4 | 0.006 | 0.0012 |
| 20 | 6 | 0.02 | 0.006 |
| 30 | 9 | 0.1 | 0.03 |
| 40 | 12 | 0.3 | 0.1 |
| 60 | 18 | 1.6 | 0.5 |

**Example 4.4**   Point $R$ is to be set at a horizontal distance of 98.25 ft from point $Q$ along a sloping line where the difference in elevation between $R$ and $Q$ is 4.35 ft. Calculate the slope distance to be measured in the field.

SOLUTION   Using Eq. (4.2) yields

$$s = [(98.25)^2 + (4.35)^2]^{1/2} = 98.35 \text{ ft}$$

**Example 4.5**   A distance was measured over irregularly sloping terrain. Slope distances and differences in elevation are tabulated in the two columns on the left of the following table. Calculate the horizontal distance.

| Slope distance, ft | Difference in elevation, ft | Horizontal distance, ft | $C_h = h^2/2s$, ft |
|---|---|---|---|
| 100.00 | 3.50 | 99.94 | 0.06 |
| 100.00 | 5.30 | 99.86 | 0.14 |
| 80.50 | 4.20 | 80.39 | 0.11 |
| 100.00 | 8.05 | 99.68 | 0.32 |
| 62.35 | 5.25 | 62.13 | 0.22 |
| 442.85 | | 442.00 | 0.85 |
| − 0.85 | | | |
| 442.00 | | | |

SOLUTION   $H$ is calculated using Eq. (4.3), individually for each length and tabulated in the third column. The total horizontal distance is the sum of the calculated $H$'s. An alternative is to compute the correction $C_h$ from Eq. (4.4) using only one term, since the slopes are less than 15 percent. These corrections are tabulated in the fourth column. In this procedure, the pocket calculator or the slide rule may be used. The horizontal distance is then computed by either subtracting the sum of the corrections, $C_h$, from the total slope distance; or applying individual corrections to corresponding slope distances and taking the sum of the horizontal distances.

**4.14. Precision required for $\theta$ and $h$**   The relative precision of a measured line is usually expressed as the ratio of the allowable discrepancy to the distance measured. Thus, a relative precision of 1 in 10,000 implies a discrepancy of 1 unit in 10,000 or 0.01 unit in 100 units. Required relative precisions in measured distances for various classes of work are given in Chaps. 8 and 10. Ordinary taping is generally said to have a relative precision of 1 part in 3000 to 1 part in 5000.

The final precision in $H$ will depend on the precision with which $s$, and either $\theta$ or $h$, are measured. In other words, the standard deviation $\sigma_H$ may be evaluated in terms of $\sigma_s$ and either $\sigma_\theta$ or $\sigma_h$, applying the principle of error propagation as developed in Chap. 2. The contribution of $\sigma_\theta$ alone is [see Eq. (2.56)]

$$\sigma_H^2 = \left(\frac{dH}{d\theta}\right)^2 \sigma_\theta^2 \tag{4.5}$$

in which $\sigma_H$ and $\sigma_\theta$ are estimated standard deviations in $H$ and $\theta$, respectively, and $dH/d\theta$ is found by differentiating Eq. (4.1) with respect to $\theta$ to yield

$$\frac{dH}{d\theta} = -s \sin\theta \tag{4.6}$$

so that

$$\sigma_H = s \sin\theta \sigma_\theta \tag{4.7}$$

The relative precision in $H$ due to $\sigma_\theta$ is thus

$$\frac{\sigma_H}{H} = \frac{s \sin\theta}{s \cos\theta} \sigma_\theta = \sigma_\theta \tan\theta \tag{4.8}$$

Thus, if the relative precision for the distance measurement is given, the precision necessary in obtaining $\theta$ can be calculated.

**Example 4.6**  In Example 4.1, assume that the contribution to $\sigma_H$ due to $\theta$ is to be within $\pm 0.005$ m. Determine the allowable tolerance in $\theta$ to meet this specification.

SOLUTION  Using Eq. (4.7), we obtain

$$\sigma_\theta = \frac{\sigma_H}{s \sin\theta} = \frac{(0.005)(206,265)}{(29.954)(0.07846)} = 439'' = 7'19''$$

Note that $\sigma_\theta$ in radians was converted to seconds of arc through multiplication by 206,265. Consequently, the vertical angle needs to be read to the nearest $\pm 5'$ to satisfy the stated specification.

**Example 4.7**  Assume that a relative precision of 1 part in 25,000 is due to the measurement of $\theta$ in Example 4.2. Compute the tolerance required in the measurement of $\theta$.

SOLUTION  Using Eq. (4.8), we obtain

$$\sigma_\theta \tan 5°30' = \frac{1}{25,000}$$

$$\sigma_\theta = \frac{206,265}{(25,000)(0.09629)} = 85.7'' = 1'26''$$

and $\theta$ should be read to the nearest minute to maintain the stated relative precision.

A similar analysis utilizing Eq. (4.4) yields the estimated precision required in $h$ to satisfy its contribution to horizontal distance measurement specifications. In this case

$$\sigma_H^2 = \left(\frac{dH}{dh}\right)^2 \sigma_h^2 \tag{4.9}$$

in which $\sigma_h$ is the estimated standard deviation in $h$, the difference in elevation, and $dH/dh$ is found by differentiating Eq. (4.3) with respect to $h$. Thus,

$$\frac{dH}{dh} = (1/2)(s^2 - h^2)^{-1/2}(-2h) = -\frac{h}{(s^2 - h^2)^{1/2}}$$

or

$$\frac{dH}{dh} = -\frac{h}{H} \approx -\frac{h}{s}$$

Thus, from Eq. (4.9),

$$\sigma_H = \frac{h}{s}\sigma_h \qquad (4.10)$$

and the relative precision relationship will be

$$\frac{\sigma_H}{s} = \frac{h}{s^2}\sigma_h \qquad (4.11)$$

**Example 4.8**  The tolerance in the distance in Example 4.3 due to error in $h$ is $\pm 0.005$ m. Compute the allowable tolerance in the determination of the elevation difference.

SOLUTION   Using Eq. (4.10) yields

$$\sigma_h = \frac{(130.508)(0.005)}{5.56} = 0.12 \text{ m}$$

so that in this case elevations to the nearest 3 cm would be adequate.

Where the total measurement consists of a series of full or partial tape lengths having different elevation differences, $\sigma_h$ may be evaluated with either Eq. (4.10) or (4.11) using the line having the maximum change in elevation.

The final precision $\sigma_H$ in the horizontal distance $H$ must take into account the precision $\sigma_s$ of the actual field taping in order to be theoretically correct. This point will be discussed in detail and illustrated with examples in Art. 4.27.

**4.15. Systematic errors in taping**   Systematic errors in taping linear distances are those attributable to the following causes: (1) tape not of standard length, (2) tape not horizontal, (3) variations in temperature, (4) variations in tension, (5) sag, (6) incorrect alignment of tape, and (7) tape not straight.

**4.16. Tape not of standard length**   The nominal length of a tape, as stated by the manufacturer, rarely corresponds exactly with the true length. The true length is taken as the length determined by comparison with a known standard length under given conditions of temperature, tension, and support. The National Bureau of Standards in Washington, D.C., will standardize tapes and provide a certificate of standardization upon payment of a fee. The absolute value for the tape correction $C_d$ is

$$C_d = \text{true length} - \text{nominal length} \qquad (4.12)$$

This discrepancy is normally assumed to be distributed uniformly throughout the tape and is directly proportional to the fractional portion of the tape used for a measurement.

Tapes are usually standardized at 20°C (68°F) and with tensions varying from 4 to 14 kg or 10 to 30 lb. Tapes may be supported throughout, at specified points, or at the two ends only. Desired conditions of standardization are specified when requesting the standardization.

The normal use and abuse of a tape in the field over a period of time can invalidate the standard length. Consequently, the standard tape should be kept in the office and used only for purposes of comparison. At regular intervals, field tapes should be compared with the standard tape or to a standard length measured with the standardized tape. Such comparisons are extremely important, if specified accuracies are to be maintained in the field.

**4.17. Tape not horizontal**   When horizontal taping is being performed over sloping ground, discrepancies frequently occur because the tape is not truly horizontal. Slopes are often deceptive, even to experienced surveyors; the tendency is to hold the downhill end of the tape too low. Inexperienced tapemen have been observed keeping very careful alignment, yet taping what they thought were horizontal distances on a slope of perhaps 10 percent. The resulting discrepancy is 0.5 ft/100 ft or 26 ft/mi (0.15 m/30 m or $\simeq$5 m/km). In ordinary taping, this is one of the largest of contributing errors. It will not be eliminated by repeated measurements, but it can be reduced to a negligible amount by leveling the tape with either a hand level or a clinometer.

**4.18. Variations in temperature**   The tape expands as the temperature rises and contracts as the temperature falls. Therefore, if the tape is standardized at a given temperature and measurements are taken at a higher temperature, the tape is too long. For a change in temperature of 15°F, a 100-ft steel tape will undergo a change in length of about 0.01 ft, introducing an error of about 0.5 ft/mi. Similarly, a change of 15°C produces a change in length of 5 mm/30 m or 0.17 m/km. Under a change of 50°F the error would be 1.5 ft/mi or 0.58 m/km. It is seen that, even for measurements of ordinary precision, the error due to thermal expansion becomes of consequence when the measurements are taken during cold or hot weather.

The coefficient of thermal expansion of steel is approximately 0.00000645/1°F or 0.0000116/1°C. If the tape is standard at a temperature of $T_0$ degrees and measurements are taken at a temperature of $T$ degrees, the correction $C_t$ for change in length is given by the formula

$$C_t = \alpha L(T - T_0) \tag{4.13}$$

in which $L$ is the measured length and $\alpha$ the coefficient of thermal expansion.

Errors due to variations in temperature are greatly reduced by using an invar tape. If a steel tape is used, one or more tape thermometers should be taped to it.

**4.19. Variations in tension**   If the tension or pull is greater or less than that for which the tape is verified, the tape is elongated or shortened accordingly. The correction for variation in tension in a steel tape is given by the formula

$$C_p = \frac{(P - P_0)L}{aE} \tag{4.14}$$

where $C_p$ = correction per distance $L$, ft or m
  $P$ = applied tension, lb or kg
  $P_0$ = tension for which the tape is standardized, lb or kg
  $L$ = length, ft or m
  $a$ = cross-sectional area, in$^2$ or cm$^2$
  $E$ = elastic modulus of the steel, lb/in$^2$ or kg/cm$^2$

The modulus of elasticity is taken as 28 to 30 million lb/in$^2$ or $2.1 \times 10^6$ kg/cm$^2$. The cross-sectional area of the tape can be computed from the weight and dimensions, since steel weighs approximately 490 lb/ft$^3$ or $7.85 \times 10^{-3}$ kg/cm$^3$. Light (1-lb) and heavy (3-lb) 100-ft tapes have cross-sectional areas of approximately 0.003 and 0.009 in$^2$, respectively. Light and heavy 30-m tapes have respective cross-sectional areas of about 0.019 and 0.058 cm$^2$.

Some idea of the effect of variation in tension can be obtained from the following examples.

**Example 4.9**   Assume that a light 100-ft tape is standard under a tension of 10 lb, $E$ = 30,000,000 lb/in$^2$, and the cross-sectional area of the tape is 0.003 in$^2$. Determine the elongation for an increase in tension from 10 to 30 lb.

SOLUTION   Using Eq. (4.14), we obtain

$$C_p = \frac{(30 - 10)(100)}{(30,000,000)(0.003)} = 0.0222 \text{ ft}$$

**Example 4.10**  A heavy 30-m tape having a cross-sectional area of 0.06 cm$^2$ has been standardized at a tension of 5 kg. If $E = 2.1 \times 10^6$ kg/cm$^2$, calculate the elongation of the tape for an increase in tension from 5 to 15 kg.

SOLUTION  Using Eq. (4.14) yields

$$C_p = \frac{(15 - 5)(30)}{(2,100,000)(0.06)} = 0.0024 \text{ m}$$

Corrections for variation in tension are seen to be significant for ordinary taping when light tapes are used and the differences in tension are substantial. For measurements of high accuracy such as base lines, tension variations are significant regardless of the weight of tape.

**4.20. Correction for sag**  When the tape sags between points of support, it takes the form of a catenary. The correction to be applied is the difference in length between the arc and the subtending chord. For the purpose of determining the correction, the arc may be assumed to be a parabola, and the correction is then given with sufficient precision for most purposes by the formula

$$C_s = \frac{w^2 L^3}{24 P^2} = \frac{W^2 L}{24 P^2} \qquad (4.15)$$

where $C_s$ = correction between points of support, ft or m
     $w$ = weight of tape, in lb/ft or kg/m
     $W$ = total weight of tape between supports, lb or kg
     $L$ = distance *between supports*, ft or m
     $P$ = applied tension, lb or kg

The correction is seen to vary directly as the cube of the unsupported length and inversely as the square of the pull. Although the equation is intended for use with a *level* tape, it may be applied without error of consequence to a tape held on a slope up to approximately 10°.

The changes in correction for sag due to variation in tension, weight of tape, and distance between supports are illustrated in Fig. 4.10 for 100-ft and 25-ft spans and in Fig. 4.11 for 10-m and 30-m tapes.

Analysis of Figs. 4.10 and 4.11 reveals the hazard of measuring with a heavy tape without using a tension handle. A variation of $-0.5$ kg, from a standard tension of 8 kg, with a very heavy tape over a 30-m span results in a discrepancy of 8 mm in the sag correction or a relative precision of 1 part in 3750. A similar variation in tension using a medium-weight 30-m tape causes discrepancies of 1.5 mm for a relative precision of 1 in 20,000. Estimation of tension to within $\pm0.5$ kg or about $\pm1$ lb is very difficult, especially with a heavy tape. Consequently, even for ordinary taping where a relative precision of 1 in part 5000 is sought, a tension handle should be used for measurements with heavy tapes.

The effect of reducing span length is also illustrated in Figs. 4.10 and 4.11, where discrepancies in $C_s$ are substantially reduced for the shorter 10-m and 25-ft spans. This factor should not be interpreted to indicate that shorter tape length should be used, as the error propagation due to setting and measuring additional points would probably offset any reduction in sag correction errors. Reduction of span length is most useful when using long heavy tapes for precise base-line measurement (Art. 10.16). Note that a 25-ft unsupported span of a very heavy 100-ft tape has the same sag corrections as a medium-weight 100-ft unsupported span.

**4.21. Normal tension**  By equating the right-hand members of Eqs. (4.14) and (4.15), the elongation due to increase in tension is made equal to the shortening due to sag; thus, the

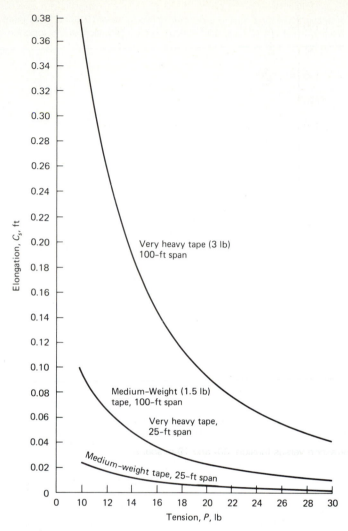

**Fig. 4.10** Sag correction versus tension, 100- and 25-ft spans.

effect of sag can be eliminated. The pull that will produce this condition, called *normal tension* $P_n$, is given by the formula

$$P_n = \frac{0.204\,W\sqrt{aE}}{\sqrt{P_n - P_0}} \tag{4.16}$$

This equation can be solved using successive approximations for $P_n$. Normal tensions for very heavy ($W = 3$ lb or 1.7 kg; $a = 0.01$ in² or 0.06 cm²) and medium-weight ($W = 1.5$ lb or 0.7 kg; $a = 0.005$ in² or 0.03 cm²) 100-ft and 30-m tapes assuming that $P_0 = (10$ lb, 5 kg) and $E = (30,000,000$ lb/in², 2,100,000 kg/cm²), respectively, are as follows:

|  | Very heavy tape | | Medium-weight tape | |
|---|---|---|---|---|
|  | 100 ft | 30 m | 100 ft | 30 m |
| Normal tension | 51.8 lb | 26.5 kg | 28.0 lb | 12.8 kg |

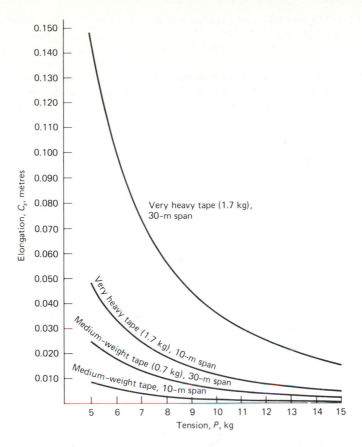

**Fig. 4.11** Sag correction versus tension, 30- and 10-m spans.

It can be seen that the normal tension for the very heavy tape over the long span is not practical.

**4.22. Imperfect alignment of tape**   The head tapeman is likely to set the pin sometimes on one side and sometimes on the other side of the correct line. This produces a variable systematic error, since the horizontal angle that the tape makes with the line is not the same for one tape length as for the next. The error cannot be eliminated, but it can be reduced to a negligible quantity by care in lining. Generally, it is the least important of the errors of taping, and extreme care in lining is not justified. The linear error when one end of the tape is off line a given amount can be computed by Eq. (4.3) in the same manner as the slope correction. For a 100-ft tape, the error amounts to 0.005 ft when one end with respect to the other is off line 1 ft and to only 0.001 ft when the error in alignment is 0.5 ft. Displacements at the ends of a 30-m tape of 3 dm and 1.5 dm produce discrepancies of 1.5 mm and 0.4 mm, respectively. Many surveyors use unnecessary care in securing good alignment without paying much attention to other more important sources of error. Errors in alignment tend to make the measured length between fixed points greater than the true length and hence are positive.

**4.23. Tape not straight**   In taping through grass and brush or when the wind is blowing, it is impossible to have all parts of the tape in perfect alignment with its ends. The error arising from this cause is systematic and variable and is of the same sign (positive) as that from measuring with a tape that is too short. Care must be exercised to stretch the tape taut and to observe that it is straight by sighting over it, so that the error is not of consequence.

**4.24. Summary of systematic errors in taping**   The various systematic errors discussed in Arts. 4.13 through 4.23 are summarized in Table 4.3. The apparently simple task of linear measurement is affected by a remarkable number of factors.

When corrections are applied to the observed length of a line measured between fixed points (designated *measurement*) with a tape that is too long, the correction is *added*. When layout of a distance is performed (i.e., a second point is established at a specified distance from a starting point) with a tape that is too long, the correction is subtracted from the specified distance to determine the distance to be laid out. For a tape that is too short, the corrections are opposite in direction to those just stated. Refer to the sample problems in Art. 4.25 for examples of how corrections are applied.

**4.25. Combined corrections**   Whenever corrections for several effects such as slope, tension, temperature, and sag are to be applied, for convenience they may be combined as a single net correction per tape length. Since the corrections are relatively small, the value of each is not appreciably affected by the others and each may be computed on the basis of the nominal tape length. For example, even if the verified length of a tape were 100.21 ft, the correction for temperature (within the required precision) would be found to be the same whether computed for the exact length or for a nominal length of 100 ft. The combined correction per tape length can be used as long as conditions remain constant; however, the temperature is likely to vary considerably during the day, and the ground may be such that it may be necessary to support the tape at various intervals of length and with different slopes so that corrections must be calculated for individual tape lengths (see Example 4.5). Several examples to illustrate applying corrections follow.

**Example 4.11**   A 30-m tape weighing 0.55 kg and with a cross-sectional area of 0.02 cm$^2$ was standardized and found to be 30.005 m at 20°C, with 5-kg tension and supported at the 0 and 30-m points. This tape was used to measure a distance of about 89 m over terrain of a uniform 5 percent slope. The temperature was constant at 30°C, the tape was fully supported throughout, and tension of 5 kg was applied to each tape length. The observed distances were 30.000 m, 30.000 m, and 29.500 m. Calculate the horizontal distance between the points.

SOLUTION   Slope, sag, temperature, and standardization are in direct proportion to the length and can be calculated for the total distance.

Slope correction [Eq. (4.4)]:

$$s - H = \frac{h^2}{2s} = \frac{[(0.05)(89.500)]^2}{(2)(89.500)} = 0.112 \text{ m} \qquad \text{tape is short; correction} = -0.112 \text{ m}$$

Sag correction [Eq. (4.15)] for two 30.000-m and one 29.500-m tape lengths is

$$C_s = (2)\frac{(0.55)^2(30.005)}{(24)(5)^2} + \frac{[0.55(29.5/30)]^2(29.500)(30.005/30.000)}{(24)(5)^2} = 0.045 \text{ m} \qquad \text{tape is long}$$

Temperature correction [Eq. (4.13)]:

$$C_t = (0.0000116)(89.50)(30 - 20) = 0.0104 \text{ m} \qquad \text{tape is long}$$

Standardization correction [Eq. (4.12)]:

$$C_d = (30.005 - 30.000)\left(\frac{89.50}{30}\right) = 0.0149 \text{ m} \qquad \text{tape is long}$$

The total correction is

$$C = -0.112 + 0.045 + 0.010 + 0.015 = -0.042 \text{ m}$$

and the corrected distance is 89.458 m.

## Table 4.3 Systematic Errors in Taping

| Source | Amount | Error of 0.01 ft/ 100 ft or 3 mm/ 30 m tape length caused by: | Makes tape too: | Importance in 1:5000 taping | Procedure to eliminate or reduce |
|---|---|---|---|---|---|
| Tape not of standard length | — | — | Long or short | Usually small but must be checked | Standardize tape and apply computed correction |
| Temperature | $C_t = \alpha L(T - T_0)$ | $15°F$ or $9°C$ | Long or short | Of consequence only in hot or cold weather | Measure temperature and apply computed correction; for precise work tape at favorable times and/or use invar tape |
| Change in pull or tension | $C_p = \dfrac{(P - P_0)L}{aE}$ | 15 lb or 4.2 kg | Long or short | Negligible | Apply computed correction; in precise work use a spring balance. Use normal tension $P_n = \dfrac{0.204W\sqrt{aE}}{\sqrt{P_n - P_0}}$ or use standard tension $P_0$ |
| Sag | $C_s = \dfrac{w^2 L^3}{24P^2} = \dfrac{W^2 L}{24P^2}$ | $\Delta P = 0.6$ lb or[a] 1 kg too small | Short | Large, especially with heavy tape | Apply computed correction; use tape fully supported |
| Slope | $H = (s^2 - h^2)^{1/2}$ $C_h = -\dfrac{h^2}{2s}$ (approx.) $H = s\cos\theta$ | 1.4 ft or[b] 0.42 m in $h$, $0°48'$ in $\theta$ (both from horizontal) | Short | — | At breaks in slope determine differences in elevation or slope angle; apply computed correction |
| Imperfect horizontal alignment | Same as slope | 1.4 ft or 0.42 m | Short | Not serious | Use reasonable care in aligning tape; keep tape taut and reasonably straight |

Note   In measuring a distance between fixed points with a tape that is too long, add the correction (assume a tape of medium weight).

[a] Error affected nonlinearly by the tension $P$.

[b] Nonlinear relationship, error increases nonlinearly as the slope increases.

**Example 4.12** A 100-ft tape has a standardized length of 100.000 ft at 68°F, with 20-lb tension and with the tape fully supported. The observed distance between two points, with the tape held horizontal, supported at the 0 and 100-ft points only, and 30-lb tension is 100.452 ft. The tape weighs 3.0 lb and has a cross-sectional area of 0.01 in$^2$. Calculate the true distance between the two points. Assume that $E = 3 \times 10^7$ lb/in$^2$.

SOLUTION First calculate the sag correction at the standard tension with Eq. (4.15):

$$C_s = \frac{(3)^2(100.452)}{(24)(30)^2} = -0.042 \text{ ft}$$

The tape is shorter, so the correction is negative when measuring between two fixed points. Next, compute the correction due to the increase in tension using Eq. (4.14):

$$C_p = \frac{(30-20)(100.452)}{(0.01)(3)(10^7)} = 0.003 \text{ ft}$$

The increased tension makes the tape long, so that the correction is positive. The corrected distance is

$$100.452 - 0.042 + 0.003 = 100.413 \text{ ft}$$

Note that the foregoing corrections can be calculated with adequate accuracy with a slide rule and using $L = 100$ ft without introducing significant errors.

$$C_s = \frac{(3)^2(100)}{(24)(20)^2} = -0.042 \text{ ft}$$

$$C_p = \frac{(30-20)(100)}{(0.01)(3)(10^7)} = 0.003 \text{ ft}$$

**4.26. Random errors in taping** Random errors in taping occur as a result of human limitations in observing measurements and in manipulating the equipment. Specific causes of random errors in taping are listed in Table 4.4 together with corresponding estimates of the errors per tape length, assuming experienced personnel, conditions as given in the table, and medium-weight (1.5 lb, 0.7 kg) 100-ft and 30-m tapes having cross-sectional areas of 0.006 in$^2$ and 0.040 cm$^2$, respectively.

Note that random errors can occur as a result of (1) direct observations, such as marking tape ends, reading graduations, and so on; or (2) indirect observation of the difference in elevation or slope between the tape ends, applying tension, and so on. In these latter cases the observational error must be propagated through the appropriate equations before arriving at an estimate for the error in taping. For example, the error in observing elevation differences is propagated through Eq. (4.3) (Art. 4.13) by using Eq. (4.10) in Art. 4.14.

Also note that $\sigma_v$, $\sigma_m$, $\sigma_p$, and $\sigma_h$ occur randomly as each tape length is measured and accumulate as the square root of the number of tape lengths $n$. On the other hand, $\sigma_d$ propagates systematically with each tape length and cumulates in direct proportion to the number of tape lengths $n$.

Comparison of Tables 4.4 and 4.3 reveals that systematic errors are of a much greater magnitude than are random errors. When every device is employed to eliminate systematic errors using the methods described in Arts. 4.13 through 4.23 and as detailed for precise base-line measurement in Art. 10.16, random errors of observation become more significant; for this reason long tapes (200 ft, 50 m) are frequently employed for base-line measurement.

**4.27. Error propagation in taping** It would be valuable if a definite outline of procedure could be established to produce any desired degree of precision in taping. Unfortunately, the conditions are so varied, and so much depends upon the skill of the individual, that the surveyor must be guided largely by experience and by knowledge of the errors involved. However, by making certain assumptions concerning the errors involved and procedures

**Table 4.4 Random Errors in Taping**

| Designation | Source | Governing conditions causes and manner of cumulation | Estimated value per tape length |
|---|---|---|---|
| $\sigma_v$ | Plumbing to mark tape ends | Rugged terrain, breaking tape frequently; cumulates randomly $\propto \sqrt{n}$ | 0.05–0.10 ft (15–30 mm) |
| $\sigma_m$ | Marking tape ends with tape fully supported | Tape graduated to hundredths of ft or mm; cumulates randomly $\propto \sqrt{n}$ | 0.01 ft (3 mm) |
| $\sigma_p$ | Applying tension | Change in sag correction due to variations in tension of $\pm 2$ lb or 0.9 kg from standard tension; cumulates $\propto \sqrt{n}$ | 0.01 ft (3 mm) |
| $\sigma_h$ | Determining elevation difference or slope angle (assume a maximum 6 percent slope) | In $h = \begin{array}{l}\pm 0.8 \text{ ft} \\ \pm 0.25 \text{ m}\end{array}$ <br> In $\theta = \pm 0°28'$; <br><br> cumulates $\propto n$ | 0.050 ft (15 mm) |
| $\sigma_d$ | Standardization | Field tapes compared to standardized tape kept in office; cumulates $\propto n$ | 0.005 ft (1.5 mm) |

employed a reasonably methodical approach can be developed which is very useful in establishing procedures related to measurements required for specific tasks.

The usual practice in rough taping through broken country (maximum slopes 6 percent) is to take measurements with the tape horizontal, plumbing at the downhill end, breaking tape where necessary, applying tension by estimation, and making no corrections for sag, temperature, or tension. The tape is usually 100 ft or 30 m long and weighs about 2 lb or 1 kg. For this case the most significant errors and assumed magnitudes are:

| Cause of error | Assigned error ($\sigma$) | | Effect on measured line |
|---|---|---|---|
| | ft | m | |
| $\sigma_h$, tape not level | 0.05 | 0.015 | Systematic |
| $\sigma_s$, sag in tape | 0.03 | 0.009 | Random |
| $\sigma_v$, plumbing | 0.05 | 0.015 | Random |

For a distance of 1000 ft (10 tape lengths) the total error would be $10 \times 0.05 = 0.50$ ft due to the tape not being level plus $(0.03 + 0.05)\sqrt{10} = 0.25$ ft due to sag and plumbing, for a total of 0.75 ft. Therefore, under these assumptions the relative accuracy is only about 1 part in 1300.

To achieve a relative accuracy of 1 part in 5000, more control must be exercised over the taping procedure. The following magnitudes for the most significant errors are:

| Cause of error | Assigned errors | | Effect on measured line |
| --- | --- | --- | --- |
| | ft | m | |
| Plumbing, $\sigma_v$ | 0.02 | 0.006 | Random |
| Sag, $\sigma_s$ | 0.01 | 0.003 | Random |
| Temperature, $\sigma_t$ | 0.004 | 0.001 | Systematic |
| Tape not level, $\sigma_h$ | 0.003 | 0.001 | Systematic |
| Standardization, $\sigma_d$ | 0.004 | 0.001 | Systematic |

Using the error propagation rules, the total error is 0.205 ft for the 1000-ft distance and 0.058 m for the 300-m distance, yielding the desired relative accuracy of 1 part in 5000.

To meet the stated conditions, the following specifications should govern the taping operation (assume a 100-ft medium-weight tape having a cross-sectional area of 0.006 in$^2$, and maximum slopes of 6 percent):

1. The tape shall be standardized to within $\pm 0.004$ ft ($\sigma_d = 0.004$ ft) with the tape supported at two ends at 68°F and with 12 lb of tension.
2. A tension handle shall be used with tension applied to within $\pm 2$ lb of the standardized tension.
3. End points of each tape length shall be marked to within $\pm 0.02$ ft ($\sigma_v + \sigma_m = 0.02$ ft).
4. A hand level shall be used to ensure that the two tape ends of a full tape length are within $\pm 0.5$ ft of horizontal, assuming the maximum 6 percent slope.
5. Temperatures at the site of the measurement shall be observed to within $\pm 7$°F.

Note that some rather stringent specifications are necessary to ensure a relative accuracy of 1 part in 5000. Note also that more than half the total error occurs in plumbing or marking the end points of each tape length. Results can be improved by using a longer tape with fewer tape lengths to be marked.

**4.28. Mistakes in taping**   Some of the mistakes commonly made by individuals inexperienced in taping are:

1. *Adding or dropping a full tape length*   This is not likely to occur if both tapemen count the pins, or when numbered stakes are used, if the rear tapeman calls out the station number of the rear stake in response to which the head tapeman calls out the number of the forward stake while marking it. A tape length may be added through failure of the rear tapeman to give the head tapeman a pin at breaks marking fractional tape lengths. A tape length may be dropped through failure of the rear tapeman to take a pin at the point of beginning.
2. *Adding a foot or a decimetre*   This usually happens in measuring the fractional part of a tape length at the end of the line. This distance should be checked by the head tapeman holding the zero mark on the tape at the terminal point and the rear tapeman noting the number of feet or decimetres and approximate fraction at the last pin set.
3. *Other points incorrectly taken as 0 or 100-ft (or 0 and 30-m) marks on tape*   The tapeman should note whether these marks are at the end of rings or on the tape itself, also whether there is an extra graduated unit at one end of the tape.
4. *Reading numbers incorrectly*   Frequently "68" is read 89" or "6" is read as "9". As a check, good practice is to observe the number of the foot or decimetre marks on each side of the one indicating the measurement, especially if the numbers are dirty or worn. Also, the tape should be read with the numbers right side up.
5. *Calling numbers incorrectly or not clearly*   For example, 50.3 might be called "fifty, three" and recorded as 53.0. If called as "fifty, point, three" or preferably "five, zero, point, three," the mistake would not be likely to occur. When a decimal point or a zero occurs in a number, it should be called. When numbers are called to a recorder, the notekeeper should repeat them as they are recorded.

Often, large mistakes will be prevented or discovered if those involved in the work form the habit of pacing distances or of estimating them by eye. If a transit is being used to give line, distances can be checked by reading the approximate stadia interval on a range pole.

**4.29. Electronic distance measurement**    Electronic distance measuring (EDM) equipment includes electro-optical (lightwaves) and electromagnetic (microwaves) instruments. The first generation of electro-optical instruments, typified by the Geodimeter, was originally developed in Sweden in the early 1950s. The basic principle of electro-optical devices is the indirect determination of the time required for a light beam to travel between two stations. The instrument, set up on one station, emits a modulated light beam to a passive reflector set up on the other end of the line to be measured. The reflector, acting as a mirror, returns the light pulse to the instrument, where a phase comparison is made between the projected and reflected pulses. The velocity of light is the basis for computation of the distance. A clear line of sight is required, and observations cannot be made if conditions do not permit intervisibility between the two stations. The maximum range for first-generation instruments was from 3 to 5 km (2 to 3 mi) in daylight and 25 to 30 km (15 to 20 mi) at night, depending on atmospheric conditions.

Electromagnetic EDM devices, first developed in South Africa during the late 1950s, utilize high-frequency microwave transmission. A typical electromagnetic device, such as the Tellurometer, consists of two interchangeable instruments, one being set up on each end of the line to be measured. The sending instrument transmits a series of microwaves which are run through the circuitry of the receiving unit and are retransmitted to the original sending unit, which measures the time required. Distances are computed on the basis of the velocity of the radio waves. An unobstructed measuring path between the two instruments is necessary. However, intervisibility is not required, and therefore observations can be made in fog and other unfavorable weather conditions. Distances up to 66 to 80 km (40 to 50 mi) can be measured under favorable conditions.

First-generation EDM equipment, although very precise for measuring long distances, was bulky, heavy, and expensive. The practicing surveyor needed a lighter, portable device useful for measuring distances of from a few metres or feet up to 1 km (0.6 mi).

During the mid-1960s miniaturization of electronics and development of small light-emitting diodes allowed the design of a second generation of electro-optical EDM instruments. These instruments were smaller, lighter, required less power, and were easier to operate but had substantially shorter range. Included in this group are instruments that have ranges of from 0.5 to 5 km (0.3 to 3 mi) with inherent standard deviations of $\pm 2$ to $\pm 10$ mm $+0.5$ to 5 ppm or $\pm 0.006$ to $\pm 0.032$ ft. $+0.5$ to 5 ppm.[†] These instruments are well adapted to the needs of the practicing surveyor, and those with ranges of up to 5 km (3 mi) are also applicable to high-order control surveys of fairly large extent, such as city or county surveys.

Incorporation of coherent laser light into the light-wave EDM devices gave rise to a third generation of electro-optical instruments. These instruments have fairly good portability, low power requirements, and ease of readout, coupled with longer range of from 15 m to 60 km (50 ft to 36 mi) and high accuracy of $\pm 5$ mm $+1$ ppm ($\pm 0.016$ ft $+1$ ppm).

Short- and long-range electro-optical EDM instruments of modern design are illustrated in Figs. 4.12 and 4.13, respectively. Concurrent with the rapid development of electro-optical instruments, microwave equipment was also improved, becoming more portable and being provided with direct readout. Figures 4.14 and 4.15, respectively, illustrate long-range and moderate-range microwave equipment. However, none of the aforementioned instruments was really adaptable to use for simultaneously measuring the length and direction of a line.

---

[†] ppm = parts per million. Thus, if a distance of 1000 m is measured, the standard deviation would be $\pm 2$ to $\pm 10$ mm $+0.5$ to 5 mm.

**Fig. 4.12** Hewlett-Packard Model 3805A distance meter. (*Courtesy Hewlett-Packard, Inc.*)

Only in the mid-1970s were EDM instruments reduced in size sufficiently to be mounted on theodolites. Continued development and refinement of EDM equipment can be expected in the future. As discussed in Chap. 12, EDM equipment has been combined with a theodolite equipped with horizontal and vertical axis encoders enabling real-time reduction of slope distance to positional data using an attached minicomputer. Another device under development is the laser range pole, which will allow proper orientation of a measuring path when the line of sight between two stations is obstructed.

**4.30. Basic principles of electro-optical EDM instruments** A continuous-wave carrier beam of light, generated in the transmitter, is modulated by an electronic shutter before entering the aiming optics and being transmitted to the other end of the line, where a reflector is placed. The original light-wave instruments, such as the Geodimeter, utilized incandescent or mercury light and a Kerr cell modulator to achieve the high-frequency modulation of the beam. More recent instruments use infrared or laser light as carrier beams and ultrasonic modulators to impose the intensity modulation on the continuous-wave beam. This intensity modulation is analogous to turning a light off and on using a switch. In the EDM device, the modulator chops the beam into wavelengths that are proportional to

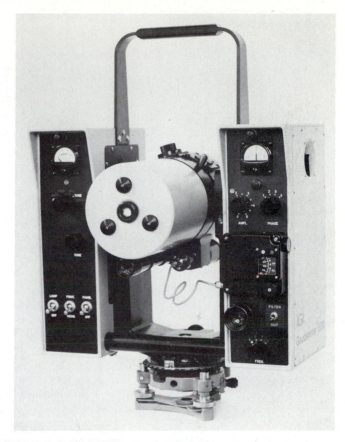

**Fig. 4.13** AGA Model 600 geodimeter. (*Courtesy AGA Corporation.*)

the modulating frequency, where the wavelength is given by

$$\lambda = V/f \tag{4.17}$$

where $\lambda$ = wave length,
$\quad$ $V$ = velocity of light through the atmosphere, m/s
$\quad$ $f$ = modulating frequency, Hz (hertz = cycles per second)

The velocity of light in the atmosphere varies with temperature, humidity, and the partial
pressure of water vapor. Corrections for these factors are described in Art. 4.33. The
intensity of the modulated light varies from zero or no light at 0° to maximum light at 90° to
a second zero at 180°, to a second maximum at 270° and back to zero at 360°, as illustrated
graphically in Fig. 4.16. Measurement of a distance is accomplished by placing the
electro-optical transmitter at one end and a reflector at the other end of the line to be
measured. The reflector is a corner cube of glass in which the sides are perpendicular to
within a very close tolerance. It has the characteristic that incident light is reflected parallel
to itself and is called a retrodirective prism or *retro-reflector*. The EDM instrument transmits
an intensity-modulated light beam to the reflector, which reflects the light beam back to the
transmitter, where the incoming light is converted to an electrical signal, allowing a phase
comparison between transmitted and received signals. The amount by which transmitted
and received wavelengths are out of phase can be measured electronically and registered on
a meter to within a few millimetres or hundredths of feet.

**Fig. 4.14** Long-range Tellurometer MRA5. (*Courtesy Tellurometer–U.S.A., Division of Plessey, Inc.*)

To visualize the operation, consider an electro-optical EDM instrument operating on one modulating frequency and with a null meter to record positive or negative maximum or zero intensity of modulation. This relationship is illustrated schematically in Fig. 4.17. Assume that the instrument centered over $A$ in Fig. 4.17 is an even number of wavelengths, $\lambda$, from the reflector at $C$ and the meter registers zero. Now, move the instrument slowly toward $C$ so that the meter will reach a maximum on the right at 90°, return to zero at 180°, a second maximum on the left at 270°, and a second zero at 360°. This procedure is repeated as the instrument is moved toward $C$, with the needle on the meter moving from zero to one maximum to zero to another maximum until the instrument is at a distance of $\lambda/2$ from the reflector. Note that the constant or uniform "tape length" between successive zero points is $\lambda/2$ for the double distance.

In practice, the instrument is rarely an even number of wavelengths from the reflector and is not moved along the line of sight toward the reflector. Assume that distance $AC$ is not an integral number of wavelengths. Figure 4.18 shows a section through $AC$ with the transmitter at $A$ and the reflector at $C$. The modulated beam transmitted at $A$ travels the double distance from $A$ to $C$ and back, where the received distance is out of phase with the

**Fig. 4.15** Moderate-range Tellurometer CA1000. (*Courtesy Tellurometer–U.S.A., Division of Plessey, Inc.*)

transmitted signal by an amount $d$. The incerment $d$ is measured electronically by adjusting some type of phase meter in the instrument, and the distance from $A$ to $C$ is

$$D = \tfrac{1}{2}(n\lambda + d) \tag{4.18}$$

in which $\lambda$ is the wavelength of the modulated beam and $n$ the integral number of wavelengths in the double path of the light. Note that if the reflector were moved $\lambda/2$ or any even number of $\lambda/2$'s, the increment $d$ does not change. Thus, if one knows the double distance to within $\lambda/2$ or the distance to within $\lambda/4$, the total distance can be resolved.

Knowledge of the distance to within $\pm\lambda/4$ is an impractical constraint on an EDM device. The ambiguity in $n$ is resolved by using some type of multiple-frequency technique. Light is transmitted on one frequency and $d_1$ is observed. Light is then transmitted on a second but slightly different frequency and $d_2$ is observed. Two equations (4.18) are formed

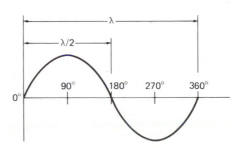

**Fig. 4.16** Wavelength of modulated light.

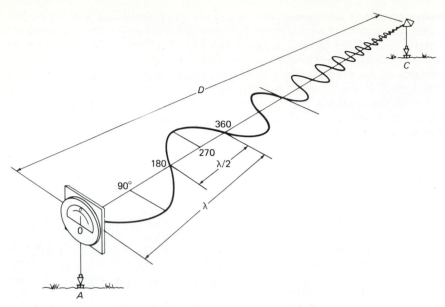

**Fig. 4.17** Modulation frequency of EDM. (*Courtesy AGA, Ref. 1.*)

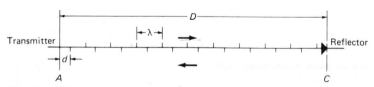

**Fig. 4.18** Measurement principle of EDM.

and solved for $n$. The multiple-frequency technique is built into the EDM device, permitting electronic determination of $n$ and a direct readout of the distance. Decade modulation is one of several techniques applicable and is used for the following example.

**Example 4.13** Assume that a modulation frequency of 15 MHz (megahertz = $10^6$ cps), giving $\lambda/2$ of 10 m, is used.[†] A full sweep of the phase meter represents this 10-m distance and allows reading the unit and decimal parts of a meter from 0 to 9.999 m. If a given distance is 5495.800 m, the phase meter records 5.800 m on this frequency. This operation is analogous to measuring the distance with a steel tape graduated at every 10-m point (but not labeled), and with one added 10-m section graduated to metres and millimetres, where the increment 5.800 is observed. Now switch to a second frequency with $f = 1.5$ MHz, so that $\lambda/2 = 100$ m. A full sweep of the phase meter allows reading 10 to 90 m and permits resolving the number of tens of metres, or 90 m for this example. Next, switch to 0.15 MHz, yielding $\lambda/2 = 1000$ m, permitting resolution of the number of hundreds of metres, in this case 400 m. Finally, turn to a frequency of 15 kHz (kilohertz = $10^3$ cps) with $\lambda/2 = 10,000$ m, which will allow resolving the number of 1000 m in the distance, 5000 m for the example. The total distance is $5.800 + 90 + 400 + 5000 = 5495.800$ m.

[†] Assume that $V \cong 300,000$ km/s.

**4.31. Electromagnetic or microwave EDM principles**    Microwave EDM transmitters generate electromagnetic carrier beams having frequencies of from 3 to 35 GHz (gigahertz $= 10^9$ cps) and modulated with frequencies varying from 10 to 75 MHz. The wavelength is given by Eq. (4.17), in which $\lambda$ is the wavelength of modulation in metres or feet, $V$ the velocity of microwaves through the atmosphere in metres m/s or ft/s, and $f$ the modulating frequency in Hz. The velocity $V$ is a function of temperature, atmospheric pressure, and the partial pressure of water vapor. Details of corrections for these factors are given in Art. 4.33.

The fundamental concepts of microwave instrument operations are similar to those involved in electro-optical EDM devices. The major difference is that two similar instruments and two operators, one for each end of the line, are required to measure a line using microwave equipment. The master unit transmits to the remote unit, which acts as a reflector for the transmitted microwaves. Voice communication is available for the operators at each unit. The operator selects a frequency at the master unit and by voice signal transmits this information to the operator at the other station, who sets the remote unit to correspond to the transmitted frequency. The signal is received at the remote unit and retransmitted to the master station, where the odd increment is measured with a phase meter.

The number of integral measuring units $n$ between the two ends of the line is determined by the method of multiple frequencies similar to that employed in the electro-optical EDM instrument. An example of a typical long-range microwave instrument is illustrated in Fig. 4.14. The multiple frequencies employed in this device are:

$A$ pattern:    10,000 MHz
$B$ pattern:     9.990 MHz
$C$ pattern:     9.900 MHz
$D$ pattern:     9.000 MHz

The $A$ pattern resolves the nearest 50 ft and decimal parts of 50 ft. Combinations of the $A$ and $D$, $A$ and $C$, and $A$ and $B$ patterns enable resolving the nearest 500, 5000, and 50,000 ft, respectively.

**4.32. Systematic errors in EDM equipment**    Effects of atmospheric conditions on wave velocity, uncertainties in the position of the electrical center of the transmitter, uncertainties in the effective center(s) of the reflectors(s), and transmitter nonlinearity all contribute to the systematic errors found in microwave instruments.

**4.33. Effects of atmospheric conditions on wave velocity**    The velocity $V$ of electromagnetic waves in air is a function of the speed of light in a vacuum $V_0$ and the refractive index of air $n$, yielding

$$V = \frac{V_0}{n} \qquad (4.19)$$

The constant $V_0$ is 299,792.5 km/s. The refractive indices of light waves and microwaves in air are functions of air temperature, atmospheric pressure, and the partial pressure of water vapor, which in turn depend on temperature and relative humidity. Thus, a knowledge of these atmospheric conditions is required to determine the refractive index and the consequent effects on the velocity of the propagated waves in air. Since light waves and microwaves react somewhat differently to varying atmospheric conditions, each is treated separately.

For light waves, it is first necessary to calculate the index of refraction $n_g$ of standard air given by the Barrell and Sears equation for an atmosphere at 0°C, 760 mmHg pressure, and 0.03 percent carbon dioxide.

$$n_g = 1 + \left( 287.604 + \frac{4.8864}{\lambda^2} + \frac{0.068}{\lambda^4} \right) 10^{-6} \qquad (4.20)$$

in which $\lambda$ is the wavelength of the carrier beam of light in micrometres. Values of $\lambda$ for various sources of light are:

| Carrier | $\lambda$, $\mu m$ |
|---|---|
| Mercury vapor | 0.5500 |
| Standard lamp | 0.5650 |
| Red laser | 0.6328 |
| Infrared | 0.900–0.9300 |

Owing to changes in temperature, pressure, and humidity, the refractive index of air becomes $n_a$, as given by

$$n_a = 1 + \frac{0.359474(n_g - 1)p}{273.2 + t} - \frac{1.5026e(10^{-5})}{273.2 + t} \tag{4.21}$$

where $p$ = atmospheric pressure, mmHg
   $t$ = air temperature, °C
   $e$ = vapor pressure, mmHg

The humidity represented by water vapor pressure in the second term of Eq. (4.21) has little effect on light waves and usually can be ignored. For measurements of high precision an average value of 0.5 ppm can be assumed for humidity correction and is substituted for the second term in Eq. (4.21).

**Example 4.14**   An electro-optical instrument utilizing infrared light with a wavelength of 0.9100 $\mu m$ has a modulation frequency at 24.5 MHz. At the time of measurement the temperature was 27.0°C and atmospheric pressure was 755.1 mmHg. Calculate the modulated wavelength of the light under the given atmospheric conditions.

SOLUTION   Using Eq. (4.20), the refractive index of air under standard conditions is

$$n_g = 1 + \left[ 287.604 + \frac{4.8864}{(0.91)^2} + \frac{0.068}{(0.91)^4} \right] 10^{-6} = 1.0002936$$

The refractive index of air under the given conditions by Eq. (4.21), neglecting the second term, is

$$n_a = 1 + \frac{(0.359474)(1.0002936 - 1)(755.1)}{273.2 + 27.0} = 1.00026547$$

The velocity of the infrared light through the atmosphere by Eq. (4.19), where $n = n_a$, is

$$V_a = \frac{299,792.5}{1.00026547} = 299,712.9 \text{ km/s}$$

and by Eq. 4.17 the modulated wavelength in the given atmosphere is

$$\lambda = \frac{299,712.9}{(24.5)(10^6)} = 0.01223318 \text{ km} = 12.23318 \text{ m}$$

Microwaves are more sensitive than light waves to humidity. The refractive index of the atmosphere for microwaves is

$$(n_r - 1)10^6 = \frac{103.49}{273.2 + t}(p - e) + \frac{86.26}{273.2 + t}\left(1 + \frac{5748}{273.2 + t}\right)e \tag{4.22}$$

where $n_r$ = refractive index
   $p$ = atmospheric pressure, mmHg
   $e$ = vapor pressure, mmHg
   $t$ = temperature, °C

The velocity of microwaves in air is given by Eq. (4.19), where $n = n_r$.

**Example 4.15**   The modulation frequency for microwave transmission is exactly 10 MHz at a temperature of 15.4°C, atmospheric pressure of 645 mmHg, and vapor pressure of 3.8 mmHg. Calculate the modulated wavelength under these conditions.

SOLUTION   Using Eq. (4.22), we obtain

$$(n_r - 1)(10^6) = \frac{103.49}{288.6}(645 - 3.8) + \frac{86.26}{288.6}\left(1 + \frac{5748}{288.6}\right)(3.8)$$

$$(n_r - 1)(10^6) = 253.69$$

$$n_r = 1.00025369$$

By Eq. (4.19),

$$V = \frac{299,792.5}{1.00025369} = 299,716.5 \text{ km/s}$$

and by Eq. (4.17) the modulated wavelength in the atmosphere is

$$\lambda = \frac{299,716.5}{(10)(10^6)} = 0.02997165 \text{ km} = 29.97165 \text{ m}$$

The examples indicate that as atmospheric conditions vary, the velocity of the modulated wave is altered, resulting in a corresponding change in the modulated wavelength and hence in the basic measuring unit of the EDM instrument.

The method of correcting the observed slope distance to account for varying atmospheric conditions is achieved differently for each type of EDM instrument. For light-wave instruments the temperature and atmospheric pressure are recorded at each end of the line. Corrections are then calculated using the average of the observed meteorological data, or the instrument circuitry is modified to permit automatic compensation as measurements progress. Most recent electro-optical EDM systems allow dialing in temperature and atmospheric pressure for this automatic compensation. In recording meteorological data, care should be exercised to obtain air temperatures. Estimating temperature to $\pm 10°F$ ($\simeq \pm 6°C$) and atmospheric pressure to $\simeq \pm 1$ in or 25 mmHg will introduce a relative error of 10 ppm in light-wave equipment. Thermometers should be placed above the ground and in the shade. Atmospheric pressure can be determined from readings on an aneroid barometer (Art. 5.6) converted to the appropriate pressure reading in mmHg or inHg using a chart provided by the instrument manufacturer.

The partial pressure of water vapor as determined by wet- and dry-bulb thermometer readings must be determined for microwave instruments. An error of 1°C between wet- and dry-bulb temperatures at normal conditions (temperature of 20°C, atmospheric pressure of 760 mmHg) produces a relative error of 7 ppm in the distance. This relative error at 45°C is 17 ppm. Consequently, wet- and dry-bulb temperatures must be carefully observed using calibrated equipment. Once the meteorological data are recorded, corrections to observed slope distances are made using charts and nomographs provided with the instrument.

**4.34. Systematic instrumental errors in EDM systems**   Systematic instrumental errors occurring in electro-optical systems include uncertainties in the position of the electrical center of the transmitter, uncertainties in the effective center of the reflectors, frequency drift, and instrument nonlinearity. The first two sources of error must be taken into account in all survey measurements, the third requires constant monitoring, and the fourth is critical only for measurements of high precision.

Microwave systems are affected by uncertainties in the electrical centers of the master and remote units and by a phenomenon called ground swing or reflection.

In EDM systems properly adjusted at the factory, the errors noted above will be very small and in a practical sense may be insignificant. However, it is important that users of these systems realize that periodic calibration of the instruments against a known distance is absolutely necessary to assure consistent results. First, consider calibration of an electro-optical system.

Uncertainties in the effective center of the reflector are illustrated by reference to Fig. 4.19, which shows a cross section through a corner-cube retro-reflector. The distance from the face of the cube to the back corner is $t$. The ray path within the cube is $a+b+c=2t$. Owing to the refractive properties of the glass in the cube, the equivalent travel in air of the ray path within the cube is $1.57\times2t$. Point $D$ represents the effective center of the corner cube and is literally the end of the line being measured. If it were possible to mount the reflector so that $D$ coincided with the plumb line, the reflector offset would be zero. However, $D$ is so far behind the face of the prism that such an arrangement would be unbalanced and difficult to manage in the field. Prisms are usually mounted so that $c_R$ is from 28 to 40 mm. Some manufacturers eliminate $c_R$ by making an adjustment in the transmitter to absorb the offset.

When slope distances are measured, light rays striking the reflector are not perpendicular to its front face of the reflector, as indicated in Fig. 4.19, thus altering the path of the rays within the reflector and changing the position of the effective center (Kivioja, Ref. 7). The amount of this change is a function of the degree of slope and varies from a few tenths of a millimetre at slopes of 4° to 6° to 7 to 14 mm at slopes from 30° to 40°. Given the angle of slope, corrections can be calculated to compensate for this error. Another option is to design the reflector with an adjustment allowing the front face of the reflector to be placed perpendicular to the incoming rays of light compensating for the error instrumentally. Details concerning the formulas for making corrections can be found in Ref. 7. Corrections of this type would be necessary on surveys of high precision where slope angles are consistently large.

At the other end of the line, if the plumb line does not coincide with the exact electrical center of the electro-optical transmitter, an instrument offset exists.

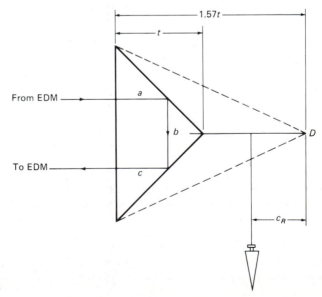

**Fig. 4.19** Reflector offset.

Similar errors exist in microwave systems where the instrument offsets in the master and remote transmitters are analogous to the instrument and reflector offsets of the light-wave system.

A precisely taped base line is required to calibrate an EDM system if high-precision results are to be obtained. In establishing this base line, all the procedures recommended for base-line taping (Arts. 10.15 and 10.16) should be followed. Preferably, the length of the base line should conform to the average length of line to be measured with the equipment.

The calibration procedure is the same for both the electro-optical and microwave systems. A series of EDM distances should be recorded (ten, for example). Meteorological data should be gathered with extreme care. The average of the EDM distances corrected for meteorological conditions and for slope of the line should agree with the taped distance for the base line. The difference between the two measurements represents the system constant, $\Delta_s$. In electro-optical systems a system constant should be determined for each reflector or group of reflectors to be used. Reflectors should be numbered and constants recorded for use in subsequent surveys.

Once a base line has been established for calibration purposes, the instrument should be recalibrated by remeasuring the base line at periodic intervals to guard against frequency drift. The time between check calibrations is a function of the age of the equipment and the user's confidence in it. With new equipment a check every week would not be unreasonable until confidence in the system stability is established. Thereafter, once per month would be sufficient. The same equipment should be used for each check calibration. The mean of the observations should be plotted versus the date of measurement and a record of meteorological conditions should be maintained. Ideally, the graph should be a horizontal straight line within the specification range for the system.

If a known base line is not available, calibration on a line of unknown length is possible. This type of calibration is accomplished by measuring a line of unknown length in several sections. Let the distance $D$ in Fig. 4.20 be divided into arbitrary increments $d_1, d_2, \ldots, d_n$. Measure the total length of line using the EDM system and then measure the $n$ sections separately, where $n = 2$ represents the minimum number of sections. For electro-optical systems the same reflector should be used for the entire set of measurements. All distances should be corrected for meteorological conditions and slope. Let the system constant be $\Delta_s$ so that the total distance is $D + \Delta_s$, each of the increments is $d_1 + \Delta_s, d_2 + \Delta_s, \ldots, d_n + \Delta_s$ and the following equation can be formed:

$$D + \Delta_s = (d_1 + \Delta_s) + (d_2 + \Delta_s) + \cdots + (d_n + \Delta_s)$$

from which

$$\Delta_s = \frac{-(d_1 + d_2 + \cdots + d_n) + D}{n - 1} \tag{4.23}$$

If the distance is divided into sections such that each section contains a distance with a different unit digit (e.g., for five sections $d_1 = 152$ m, $d_2 = 153$ m, $d_3 = 241$ m, $d_4 = 305$ m, $d_5 = 206$ m, and $D = 1057$ m), then errors due to nonlinearity (discussed in the next section) will be averaged and the net effect of this error on $\Delta_s$ will be small.

**4.35. Nonlinearity in electro-optical systems**  A nonlinear periodic bounded error occurs in all electro-optical instruments. As the distance between the transmitter and reflector is

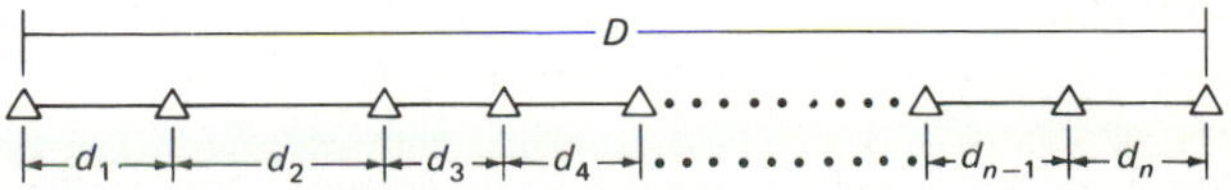

**Fig. 4.20** Base-line calibration using a line of unknown length.

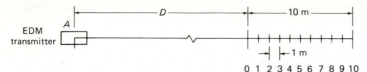

**Fig. 4.21** Set up for calculating nonlinear errors in EDM.

changed, the error will rise to a maximum, fall to a minimum, rise to a maximum, fall to a minimum, and so forth, repeating over and over again within the same boundary. To achieve maximum accuracy from a given electro-optical system, this cyclic error must be evaluated.

The nonlinearity of an instrument can be detected by taking a series of readings, from a single instrument setup, to a reflector set successively on a series of points located at intervals the sum of which should equal one full period of the nonlinearity. It is not necessary to know the total distance from the transmitter to this span of intervals, but the distances between the points must be determined with the highest possible precision.

For example, assume that the instrument to be calibrated has a basic measuring unit of 10 m. Figure 4.21 illustrates the configuration for the calibration base line where points are set at 1-m intervals over a span of 10 m, resulting in 11 stations. The series of measurements observed will be:

$$
\begin{array}{ll}
D + e_0 & D + 6 + e_6 \\
D + 1 + e_1 & D + 7 + e_7 \\
D + 2 + e_2 & D + 8 + e_8 \\
D + 3 + e_3 & D + 9 + e_9 \\
D + 4 + e_4 & D + 10 + e_{10} \\
D + 5 + e_5 &
\end{array}
$$

where $D$ is the distance from instrument to the first station of the 10-m span and $e_1, e_2, \ldots, e_{10}$ are the errors in each reading due to nonlinearity. Subtract $D, D+1, D+2, \ldots, D+10$ from each successive measurement to yield $e_1, e_2, \ldots, e_{10}$. Figure 4.22 shows the corrections or plot of $-e_0, -e_1, \ldots, -e_{10}$ versus distance for a short-range electro-optical system. Thus, if the

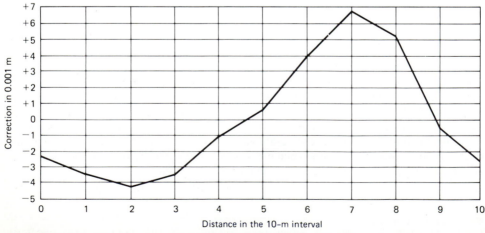

**Fig. 4.22** Typical calibration curve for nonlinearity. (*After Moffitt, Ref. 10.*)

system constant, as determined by one of the methods outlined in Art. 4.34, is $-0.005$ m and the observed slope distance is 342.541 m, the correct slope distance is $342.541 - 0.005 - 0.004 = 342.532$ m, where $-0.004$ m is taken at the intersection of the ordinate for a distance of 2.5 m with the calibration curve from the graph in Fig. 4.22. The correction is determined for the value 2.5 m because it represents the units and decimal parts of a metre which were measured.

In most electro-optical instruments factory calibration is performed so that the average nonlinearity and the true distance coincide. In other words, errors due to nonlinearity are uniformly distributed about the calibration point so that the total peak to peak error is within specified limits. From the practical standpoint, calibration for nonlinearity is rarely needed unless the instrument is to be used for projects in which very high accuracies are required, such as earth crustal movement and dam or structure deformation studies.

**4.36. Ground reflections of microwave EDM instruments**   The beam transmitted by microwave instruments is relatively wide. Thus, when transmitting over smooth level terrain or water, reflections can occur which yield erroneous distances. Occurrence of errors due to reflections, or "ground swing" as it is called, can be reduced by elevating the master and remote units above the surface over which the line is to be measured. Microwave instruments of recent design operate on higher frequencies with a more narrow beam width ($1\frac{1}{2}°$), so that errors due to ground swing have been reduced. It is recommended that a series of fine readings (the A pattern for the Tellurometer) be taken, each at a different frequency, from both ends of the line at the beginning and ending of each measurement. Strong reflections will cause these readings to vary in a cyclic manner and if the readings are plotted versus frequency, the resulting curve will ideally take the shape of a sine wave. When swing is present, considerable experience and judgment are required to interpret the results properly. Usually a straight-line average of the fine readings will be adequate. The National Ocean Survey recommends that the two measurements from the two ends of the line agree to within 1 part in 100,000 and that the spread of the fine readings not exceed 4 cm.

**4.37. Reduction of slope distances obtained by EDM**   For short lines less than 2 mi or 3.3 km in length and/or vertical angles of less than $5°$, EDM slope distances corrected for meteorological conditions and system constants can be reduced to horizontal with the usual slope correction equations as outlined in Art. 4.13. Thus, either the vertical angle or the difference in elevation between the two ends of every measured line must be obtained. When vertical angles are observed, using earlier EDM instruments separate from the theodolite, additional data required for the reduction include the heights of the EDM transmitter and the reflector or remote unit above the ground at respective stations, and the heights of the theodolite used to measure the angle and the target sighted in this measurement above the ground at their respective stations.

Figure 4.23 illustrates a typical situation for distance measurement by separate EDM equipment accompanied by observation of the vertical angle. The EDM transmitter is at $E$ while the reflector or remote unit is at $E'$, with respective heights above the ground of $EA$ and $E'B$. The theodolite and target, located at $T$ and $T'$, have heights above the ground (h.i.'s) of $AT$ and $BT'$, respectively. Angle $\alpha_T$ is the observed vertical angle and $\Delta\alpha$ is the correction to be calculated to determine the vertical angle $\alpha_E$ of the measured line $EE'$. The difference in heights of instruments required for this calculation is

$$\Delta\text{h.i.} = (\text{h.i. reflector} - \text{h.i. target}) - (\text{h.i. EDM} - \text{h.i. theodolite}) \tag{4.24}$$

and

$$\Delta\alpha = \frac{\Delta\text{h.i. } \cos\alpha}{EE' \text{ arc } 1''} \tag{4.25}$$

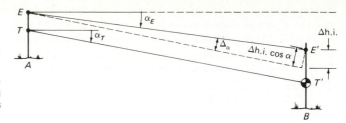

**Fig. 4.23** Difference between vertical angle and line along which EDM slope distance is measured.

leading to

$$\alpha_E = \alpha_T + \Delta\alpha \qquad (4.26)$$

The sign of $\Delta\alpha$ is a function of the sign of $\Delta$h.i. which can be negative or positive. Care should be exercised in calculating $\alpha_E$ so as to reflect the proper signs of $\alpha$, $\Delta$h.i., and $\Delta\alpha$.

**Example 4.16** The slope distance from $A$ to $B$ corrected for meteorological conditions and EDM system constants is 920.850 m. The EDM transmitter is 1.840 m above the ground and the reflector is 2.000 m above the ground. The observed vertical angle is $-4°30'00''$ with theodolite and target 1.740 m and 1.800 m above the ground, respectively. Calculate the horizontal distance.

SOLUTION  From Eq. (4.24), calculate $\Delta$h.i.

$\Delta$h.i. $= (2.000 - 1.800) - (1.840 - 1.740) = 0.100$

Using Eq. (4.25), determine $\Delta\alpha$.

$$\Delta\alpha = \frac{(0.100)(\cos 4°30'00'')}{(920.850)(4.848)(10^{-6})} = 22''$$

and the corrected vertical angle by Eq. (4.26) is

$\alpha_E = \alpha_T + \Delta\alpha = -4°30'22'' + 0°00'22'' = -4°29'38''$

so that the horizontal distance by Eq. (4.1) is

$H = s\cos\alpha_E = (920.850)(0.996926) = 918.019$ m

**Example 4.17** A slope distance from $C$ to $D$ corrected for EDM constants and meteorological conditions is 5005.45 ft with a vertical angle of $+3°30'00''$ observed from a theodolite having an h.i. of 4.40 ft to a target with h.i. of 4.60 ft. The h.i.'s of EDM and reflector are 5.00 and 4.80 ft, respectively. Compute the horizontal distance.

SOLUTION  By Eq. (4.24),

$\Delta$h.i. $= (4.80 - 4.60) - (5.00 - 4.40) = -0.40$ ft

Using Eq. (4.25), we obtain

$$\Delta\alpha = \frac{(-0.40)(\cos 3°30'00'')}{(5005.45)(4.848)(10^{-6})} = -16''$$

so the corrected angle using Eq. (4.26) is

$\alpha_E = \alpha_T + \Delta\alpha = 3°29'44''$

and the horizontal distance by Eq. (4.1) is

$H = s\cos\alpha_E = 4996.14$ ft

It is recommended that vertical angles be observed with the telescope direct and reversed using a 1″ instrument (see Art. 6.43), and that vertical angles be observed from both ends of every line if first-order horizontal distances are desired.

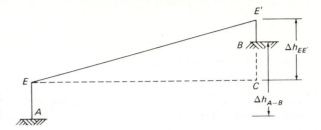

**Fig. 4.24** Reduction of short slope distances by EDM using differences in elevation.

Several types of electro-optical systems now have the EDM transmitter built into a theodolite, so that vertical angles can be observed simultaneously with the slope distance measurement. In these instruments there is a vertical offset between the electrical center of the transmitter and horizontal axis of the telescope. Also, the target sighted to observe the vertical angle may not be the same height above the ground as the reflector. These vertical offsets must be taken into account with such systems using the method outlined previously. For even more advanced systems with combined distance and direction measurement, refer to Chap. 12.

When slope distances are to be reduced using differences in elevation between the two ends of the line, the heights of the EDM instrument and reflector (or remote unit) above the ground at respective stations must be observed and recorded. In Fig. 4.24 the EDM instrument is at $E$ with a height above the ground of $AE$ and the reflector or remote unit is at $E'$ at a height of $BE'$ above the ground. Then the elevation difference between $E$ and $E'$ is

$$\Delta h_{EE'} = \Delta h_{A-B} - AE + BE'$$

or

$$\Delta h_{EE'} = \Delta h_{A-B} - \text{h.i. EDM} + \text{h.i. reflector or remote unit} \tag{4.27}$$

**4.38. Slope reductions for long lines**  When long distances are taped, the individual tape lengths or sections are very short in proportion to the total length of line. If these tape lengths or sections are sloping, Equation (4.1) or (4.2) can be used to reduce the increments individually to horizontal so that the sum of these distances follows the curvature of the earth, as illustrated in Fig. 4.25. When such a distance is reduced to sea level as described in Chap. 10, it is called a geodetic distance.

If this same distance is measured with an EDM system in a single observation, the assumption that the slope distance is the hypotenuse of a right triangle is no longer satisfied. In this case Eqs. (4.1) and (4.2) are not adequate for slope correction and a different procedure must be developed for slope reduction of lines longer than 2 mi or 3.3 km or which are 2 mi long and have vertical angles greater than 5°.

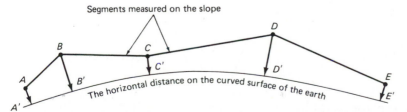

**Fig. 4.25** Tape lengths or sections (*ABCDE*) reduced to horizontal distance (*A'B'C'D'E'*) on the earth.

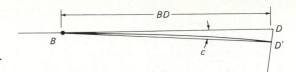

**Fig. 4.26** Curvature expressed as an angle.

Reduction of long slope lines involves use of vertical angles affected by curvature and refraction. Development of the formulation can be simplified by determining the effect of earth curvature as an angular value per unit of distance and using this value in the slope reduction. The fundamental equations for earth curvature and refraction are derived in Chap. 5. Assuming an average radius for the earth of 3959 mi or 6,371 km, curvature is 0.667 ft/mi or 0.0785 m/km. Thus, in Fig. 4.26 the curvature $DD'$ is 0.667 ft when $BD$ is 1 mi and 0.0785 m when $BD = 1$ km. The curvature $c$, expressed as an angle in seconds, is

$$c = \frac{0.667 \text{ ft/mi}}{5280 \text{ ft arc } 1''} = 26.06''/\text{mi}$$

or

$$c = \frac{26.06''/\text{mi}}{5.28} = 4.935''/1000 \text{ ft} \tag{4.28}$$

In the metric system

$$c = \frac{0.0785 \text{ m/km}}{1000 \text{ m arc } 1''} = 16.192''/1 \text{ km} \tag{4.29}$$

**4.39. Slope reduction with reciprocal vertical angles**   Figure 4.27 illustrates the geometry involved when a long slope distance $s$ is measured from $O'$, at elevation $O'L$ above mean

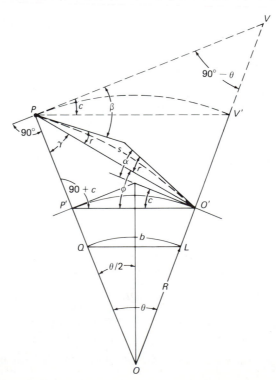

**Fig. 4.27** Reduction of long slope distances to horizontal sea level distance.

sea level, to $P$ at elevation $QP$ above mean sea level. The curved path of the light or electromagnetic waves from $O'$ to $P$ is approximated by the straight-line distance $O'P$. As illustrated in Art. 4.42, the difference is insignificant; for all practical purposes (in 90 mi or 150 km) the differences are less than 0.7 ft or 0.2 m for EDM instruments. The elevation vertical angle (Art. 6.46) from the horizontal at $O'$ to $P$ is angle $\alpha$. From $P$ at the other end of the observed line $PO'$, a horizontal line will intersect the plumb line through $O'$ at $V$, and point $O'$ itself will be observed with a depression angle $\beta$ (Art. 6.43). When both $\alpha$ and $\beta$ are measured, they are called *reciprocal vertical angles*.

In triangle $OPV$, $\theta$ is the angle subtended by $O'$ and $P'$ at the center of the earth and angle $PVO$ equals $90-\theta$.

$O'P'$ is the chord distance at elevation $O'L$ above mean sea level, $QL$ the sea level chord distance, $b$ the sea level arc distance, $r$ the angle of refraction, $c$ the angle of curvature, and by the geometry of the figure $c=\theta/2$.

Assume that reciprocal vertical angles $\alpha$ and $\beta$ have been observed at each end of the measured line $O'P$. In triangle $O'PV$

$$90-\theta+\beta+r+90-\alpha+r=180°$$

so that

$$r=\frac{\theta+\alpha-\beta}{2}$$

In triangle $PO'P'$

$$\text{angle } P'O'P=\phi=\alpha-r+c=\alpha-\left(\frac{\theta+\alpha-\beta}{2}\right)+c$$

or

$$\phi=\frac{\alpha+\beta}{2} \tag{4.30}$$

Thus, when reciprocal vertical angles are observed, curvature and refraction cancel.

Also in triangle $PO'P'$,

$$\gamma=180°-\left(\phi+90+\frac{\theta}{2}\right)=180°-(\phi+90+c) \tag{4.31}$$

Since

$$O'P'\simeq O'P\cos\phi \tag{4.32}$$

then

$$\sin\frac{\theta}{2}=\frac{O'P'}{2R} \tag{4.33}$$

is a sufficiently close approximation for $\theta$. $R$ is the radius of the earth for the average latitude of the observation in the direction of the line. An alternative approach is to calculate $c=\theta/2$ by either Eq. (4.28) or (4.29).

With the angles $\phi$, $\gamma$, and $c$ evaluated, solve triangle $PO'P'$ by the law of sines for $O'P'$.

$$O'P'=\frac{O'P\sin\gamma}{\sin(90°+c)} \tag{4.34}$$

to give the horizontal distance at elevation $O'L$ above sea level.

Reduce $O'P'$ to a sea level chord $QL$ using Eq. (10.24, Art. 10.17).

$$QL=\frac{(R)(O'P')}{R+h_{o'}} \tag{4.35}$$

in which $R$ is the radius of the earth for the average latitude of the area in the direction of the line measured and $h_{o'}=O'L$ is the elevation of $O'$ above mean sea level.

For the case where a single vertical angle is measured, an angle of refraction $r$ must be estimated. Assuming that refraction reduces the curvature correction by 14 percent, Eqs.

(4.28) and (4.29) become

$$(c \& r) = (4.935''/1000 \text{ ft})(0.86) = 4.244''/1000 \text{ ft} \qquad (4.36)$$

and

$$(c \& r) = (16.192''/\text{km})(0.86) = 13.925''/\text{km} \qquad (4.37)$$

If the elevation angle $\alpha$ is observed in triangle $PO'P'$,

$$\phi = \alpha + (c \& r) \qquad (4.38)$$

When $\beta$, the depression angle, is observed:

$$\phi = \beta + (c \& r) \qquad (4.39)$$

in which $\beta$ always has a negative sign. Thus, $\phi$ is calculated by Eq. (4.39) or (4.40), $\gamma$ with Eq. (4.31), and $O'P'$ is computed using Eq. (4.34).

**Example 4.18**   A slope distance of 23,457.500 m is measured with an electro-optical EDM system at station $O'$, 323.00 m above sea level, to station $P$. Reciprocal vertical angles $\alpha_{O'}$ and $\beta_P$ are $3°03'00''$ and $-3°13'10''$, respectively. Calculate the sea level distance for $O'P$. Refer to Fig. 4.27.

SOLUTION   Calculate $\phi$ by Eq. (4.30).

$$\phi = \frac{\alpha_{O'} + \beta_P}{2} = \frac{(3°03'00'') + (3°13'10'')}{2} = 3°08'05''$$

Compute $c$ by Eq. (4.29).

$c'' = (16.192''/\text{km})(23.457 \text{ km}) = 379.''8 = 6'19.8''$
$\phi + c = 3°14'24.8''$

Determine $\gamma$ by Eq. (4.31).

$\gamma = 180° - (\phi + 90 + c) = 180° - (93°14'24.8'') = 86°45'35.2''$

Solve the oblique triangle for the horizontal distance $P'O'$ with Eq. (4.34).

$$P'O' = \frac{O'P \sin \gamma}{\sin(90° + c)} = \frac{(23,457.500)(\sin 86°45'35.2'')}{\sin 90°06'19.8''}$$

$= 23,420.039$ m at elevation 323.00 m above mean sea level (msl)

Reduce to sea level by Eq. (4.35). Use an average radius for the earth of 6,370,000 m.

$$QL = \frac{(R)(O'P')}{R + h_{O'}} = \frac{(6,370,000)(23,420.039)}{6,370,000 + 323.00}$$

$=$ sea level chord distance $= 23,418.851$ m

**Example 4.19**   A slope distance of 58,050.30 ft is measured from station $O'$ in Fig. 4.27, which has an elevation of 2422.85 ft above mean sea level. A single vertical angle $\alpha = 2°30'00''$ is observed from $O'$ to $P$. Calculate the sea level chord distance. Assume $R = 20,906,000$ ft.

SOLUTION   Use Eq. (4.36) to evaluate $(c \& r)$.

$(c \& r) = (4.244'')(58.050) = 246.4'' = 0°04'06.4''$

Evaluate $c$ by Eq. (4.28).

$c = 4.935''/1000 \text{ ft}$
$c = (4.935)(58.050) = 286.5'' = 0°04'46.5''$

By Eq. (4.38),

$\phi = \alpha + (c \& r) = (2°30'00'') + (0°04'06.4'') = 2°34'06.4''$
$\phi + c = 2°38'52.9''$

and by Eq. (4.31)

$\gamma = 180° - (90 + \phi + c) = 87°21'07.1''$

Solve for $O'P'$ by Eq. (4.34).

$$O'P' = \frac{O'P\sin\gamma}{\sin(90+c)} = \frac{(58,050.30)(\sin 87°21'07.1'')}{\sin 90°04'46.5''}$$
$$= 57,988.37 \text{ ft}$$

and reduce to the sea level chord distance with Eq. (4.35).

$$QL = \frac{(20,906,000)(57,988.37)}{20,906,000 + 2423} = 57,981.65 \text{ ft}$$

**4.40. Reduction of long lines using elevation differences**   In Fig. 4.28 the slope distance $s$ is measured between points $O'$ and $P$ for which respective elevations equal to $LO'$ and $QP$ above sea level are known. The difference in elevation between $O'$ and $P = PP' = h_P - h_{O'} = \Delta h$. As in the previous example, $s$ is approximated by the straight-line distance $O'P$; $\theta$ is the angle subtended by $O'$ and $P'$; and the angle of curvature $c = \theta/2$.

In triangle $PFP'$, $\theta/2 = c$ can be calculated using Eq. (4.28) or (4.29), assuming that $O'P \approx O'P'$. Then the horizontal straight-line distance $O'P'$ is given by

$$O'P' = \left[(O'P)^2 - (PF)^2\right]^{1/2} - PP'\sin c \qquad (4.40)$$

where $PF = P'P\cos c = \Delta h\cos c$.

Since $c$ is calculated assuming that $O'P = O'P'$, the solution may need to be iterated once more using an improved value of $c$ to achieve a sufficiently accurate value of $P'O'$. The horizontal distance $P'O'$ should then be reduced to sea level by Eq. (4.35).

An exact solution for a direct reduction to sea level is also possible. In triangle $OQL$, by the law of cosines,

$$(QL)^2 = 2R^2 - 2R^2\cos\theta$$

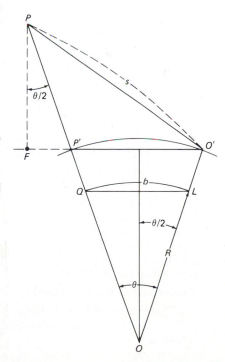

**Fig. 4.28** Reduction of long lines using elevations.

from which

$$\cos\theta = 1 - \frac{(QL)^2}{2R^2}$$

In triangle $OO'P$, by the law of cosines,

$$(O'P)^2 = (R+h_P)^2 + (R+h_{O'})^2 - 2(R+h_P)(R+h_{O'})\cos\theta$$

where $h_P = QP$, and $h_{O'} = LO'$. Substitution of the value found for $\cos\theta$ yields

$$(O'P)^2 = (h_P - h_{O'})^2 + \frac{(QL)^2(R+h_P)(R+h_{O'})}{R^2}$$

which can be further simplified by setting $\Delta h = h_P - h_{O'}$ and solving for $(QL)^2$.

$$(QL)^2 = \frac{R^2[(O'P)^2 - \Delta h^2]}{(R+h_P)(R+h_{O'})}$$

or

$$QL = \left\{ \frac{R^2[(O'P)^2 - \Delta h^2]}{(R+h_P)(R+h_{O'})} \right\}^{1/2} \tag{4.41}$$

$QL$ is the sea level chord distance and needs no further correction unless a chord to arc correction is warranted. Equation (4.41) is suitable for electronic computer computation. Extreme care should be exercised when using pocket or desk calculators, as significant figures may be lost.

**Example 4.20**   Stations $O'$ and $P$ (Fig. 4.28) have elevations above mean sea level of 414.52 m and 1517.91 m. The slope distance corrected for meteorological conditions and instrumental constants from $O'$ to $P$ is 28,545.600 m. Calculate the sea level chord distance. Assume $R = 6,371,390$ m.

SOLUTION   Calculate $c$ by Eq. (4.29) and the horizontal distance with Eq. (4.40).

$c = (16.192)(28.546) = 468.2'' = 0°07'42.2''$

The difference in elevation between $O'$ and $P$ is 1103.39 m.

$PF = P'P\cos c = \Delta h\cos c = (1103.39)\cos 0°07'42.2'' = 1103.387$ m

so that by Eq. (4.40),

$O'P = [(28,545.600)^2 - (1103.387)^2]^{1/2} - (1103.39)(\sin 0°07'42.2'')$
   $= (28,524.267) - 2.472 = 28,521.795$ m

Now recalculate $c$ with the improved estimate for the distance.

$c = (16.192)(28.522) = 461.8'' = 0°07'41.8''$

The second iteration involves evaluation of only the second term of Eq. (4.40), so that

$O'P = 28,524.267 - (1103.39)(\sin 0°07'41.8'')$
   $= 28,524.267 - 2.470 = 28,521.797$ m

Reduce to sea level chord distance by Eq. (4.35).

$QL = \dfrac{(28,521.797)(6,371,390)}{(6,371,390) + (414.52)} = 28,519.941$ m

As an alternative solution, the direct reduction to sea level is by Eq. (4.41), where

$QL = \left\{ \dfrac{(6,371,390)^2[(28,545.600)^2 - (1103.39)^2]}{(6,371,390 + 1517.91)(6,371,390 + 414.52)} \right\}^{1/2}$
   $= 28,519.942$ m

The difference of 1 mm can be attributed to round-off error in the second solution.

**4.41. Chord-to-arc correction**    In Fig. 4.28 the sea level distance $b$ exceeds the sea level chord $QL$ by a very small amount. In 100-mi and 200-km distances the respective differences are 11 ft and 3.5 m. The chord-to-arc correction is calculated as follows:

$$QL = 2R \sin\left(\frac{\theta}{2}\right)$$

Expanding $\sin \theta/2$ in series expansion yields

$$QL = 2R\left(\frac{\theta}{2} - \frac{\theta^3}{48} + \frac{\theta^5}{3840} - \cdots + \cdots\right)$$

Also,

$$b = R\theta = \frac{2R\theta}{2}$$

$$\theta = \frac{b}{R} \simeq \frac{QL}{R}$$

chord-to-arc correction $= b - QL = \dfrac{2R\theta}{2} - 2R\left(\dfrac{\theta}{2} - \dfrac{\theta^3}{48} + \dfrac{\theta^5}{3840}\right)$ which reduces to $b - QL = R\theta^3/24$, and since $\theta \simeq QL/R$,

$$b - QL = \frac{(QL)^3}{24R^2} = \text{arc} - \text{chord} \qquad (4.42)$$

The chord-to-arc correction is always positive.

In 20 mi or 33 km, this correction is 1 ppm, and for lines this length or less is insignificant.

**Example 4.21**    In Example 4.20, the sea level chord distance is 28,519.941 m. What is the chord-to-arc correction? As before $R = 6,371,390$ m and
$QL = 28,519.941$ m.

SOLUTION    By Eq. (4.42),

$$b - QL = \text{arc to Chord} = \frac{(28,519.941)^3}{24(6,371,390)^2} = 0.024 \text{ m}$$

Thus, the geodetic distance from $S$ to $L$ is
$b = 28,519.941 + 0.024 \text{ m} = 28,519.965$
Note that the correction for this length of line is 1 ppm and is insignificant except for very special cases.

**4.42. Correction for ray path**    Determination of the correction due to curvature of the ray path for propagated electronic waves is insignificant except for surveys with very long lines, where extraordinary precision is required. The radius of the ray path is a function of the coefficient of refraction in air $k$, which has values of approximately 0.13 and 0.20, respectively, for daytime and night observations with electro-optical systems. For microwave instruments, $k$ is approximately 0.25.

The equation for the correction is the same as for the chord-to-arc correction [Eq. (4.42)] but with the opposite sign. Referring to Fig. 4.29 and using $R' = R/k$,

$$PO' - s = -\frac{s^3}{24(R')^2} = -\frac{k^2 s^3}{24R^2} \qquad (4.43)$$

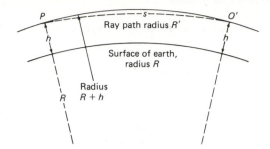

**Fig. 4.29** Curvature of ray path.

Assuming a radius of 6,370,000 m and the previously cited values for $k$, $k^2/24R^2$ takes the following values:

|  | $k$ | $k^2/24R^2$ | Correction in metres for a line of: | | |
|---|---|---|---|---|---|
|  |  |  | 100 km | 150 km | 200 km |
| Microwave, day | 0.13 | $1.74 \times 10^{-17}$ | 0.017 | 0.059 | 0.139 |
| Microwave, night | 0.20 | $4.11 \times 10^{-17}$ | 0.041 | 0.139 | 0.329 |
| Electro-optical | 0.25 | $6.42 \times 10^{-17}$ | 0.064 | 0.217 | 0.513 |

Thus, for lines less than 150 km, the correction is insignificant, so that the correction for ray path curvature is primarily of academic interest.

**4.43. Accuracies possible with EDM systems**  Short-range EDM systems (maximum range of 500 to 5000 m) have constant uncertainties which vary from 1.5 to 10 mm. Long-range systems (maximum range of 15 to 65 km) have constant uncertainties of from 0.5 to 1.5 cm. It is reasonable to assume that these figures represent legitimate estimates of the standard deviation of a measurement when corrections for meteorological conditions and instrumental systematic errors have been corrected as completely as possible. Then the worst of the short-range instruments would provide a relative precision of 1 part in 25,000 or better for lines equal to or in excess of 250 m. The worst of the long-range systems would provide similar relative precisions for lines equal to or in excess of 375 m. Achievement of relative precisions of 1 part in 25,000 over comparable distances by taping would require use of taping tripods in precise taping procedures. Hence, EDM equipment has a clear superiority over traditional taping for surveys with distances of these magnitudes. For longer distances relative precisions of 1 part in 100,000 to 1 part in 1,000,000 are attainable. It is important to recognize that relative precisions of these magnitudes are possible only when corrections for meteorological conditions are very carefully evaluated and the EDM system is in good adjustment and calibrated against a legal standard.

When very long lines and high precision are involved, acquisition and correction for meteorological conditions represent the limiting factors in relative precisions possible with EDM systems. For maximum accuracy meteorological conditions should always be recorded at both ends of the line. For best results temperatures should be recorded 30 ft or about 10 m above terminal stations. The next step would be to acquire meteorological data at specified intervals along the line so as to obtain a more realistic estimate for the index of refraction. An approach in this direction has been taken on long lines measured in conjunction with fault-line movements in California. In this case an aircraft was flown between the two terminal points of the line to acquire additional meteorological data along the line while observations were being performed.

The ultimate solution probably lies in the instrumental determination of the index of refraction simultaneously with the observations for measurement by utilizing a combination of light and microwaves. A system now under development uses modulated red and blue laser light in addition to simultaneous modulated microwave transmission. The difference in velocities of the two light-wave signals permits a real-time determination of the dry air density, leading to an average air density over the path. From the average air density, the average group index of refraction can be computed. The microwave index of refraction is much more sensitive to water vapor pressure than light waves are, so that simultaneous transmission of microwaves permits determination at the average water vapor pressure from the optical-microwave dispersion. A general-purpose minicomputer is built into the instrument to allow real-time processing of the data and output consisting of slope distances corrected for meteorological conditions.

## 4.44. Problems

**4.1**   The length of a line measured with a 30-m tape is 308.550 m. When the tape is compared to the standard it is found to be 0.009 m too long under the same conditions of support, tension, and temperature as existed during measurement of the line. Compute the length of the line.

**4.2**   A building 80.00 ft by 160.00 ft is to be laid out with a 50-ft tape which is 0.016 ft too long. What ground measurements should be made?

**4.3**   The slope measurement of a line is 243.840 m. The differences in elevation between successive 30-m points are 0.50, 0.46, 0.75, 1.20, 1.40, 1.50, 2.00, and 0.20 m. Calculate the horizontal distance along the line.

**4.4**   The slope measurement of a line is 1246.5 ft. Slope angles measured with a clinometer are as shown below. Determine the horizontal distance.

| Cumulated distance, ft | 0 | 300 | 800 | 1000 | 1246.5 |
|---|---|---|---|---|---|
| Slope angle, deg | $\frac{1}{2}$ | $1\frac{1}{4}$ | $2\frac{1}{2}$ | 4 | |

**4.5**   Two points at a slope distance of about 50 m apart have a difference in elevation of 4.00 m. Calculate the slope distance to be set off to establish a horizontal distance of 50 m.

**4.6**   Two points at a slope distance of about 25 m apart have a difference in elevation of 2.5 m. Determine the slope distance to be laid off to set the second point 25.000 m from the first.

**4.7**   The distance measured over smooth level ground between two monuments was recorded as 95.00 m. Measurements consisted of three full tape lengths and one partial tape length. If the first taping point was misaligned 0.4 dm left, the second 9.2 dm right, and the third 5.0 dm left, compute the correct distance between the points.

**4.8**   A tape that has a standardized length of 100.010 at 68°F with the tape fully supported is to be used to set the corner points of a foundation for a building that is 100.00 ft by 200.00 ft. What distances should be measured between the building corners with the tape fully supported over level terrain when the temperature is 80°F?

**4.9**   The tape in Prob. 4.8 was used to measure the distance between two monuments set at the corners of a city block. The tape was fully supported over level terrain and the temperature was 48°F. The distance recorded was 610.86 ft. What is the correct horizontal distance between the two monuments?

**4.10**   A standardized 30-m steel tape weighing 0.7 kg measures 29.9935 m supported at the two ends only, with 7.5 kg of tension and at 20°C.

(a) A horizontal distance of 3248.835 m is measured with the tape fully supported, with 7.5 kg of tension and at a temperature recorded at 38°C. Determine the correct horizontal distance;

(b) A horizontal distance of 200.843 m is to be measured from a monument to set an adjacent property corner. What horizontal distance must be measured under standardization conditions of support and tension to set this property corner where the temperature is 11°C?

**4.11**   A 100-ft tape weighing 2 lb is of standard length under a tension of 12 lb, supported for the full length of the tape. A line on smooth level ground is measured with the tape under a tension of 35 lb and found to be 4863.5 ft long. $E = 30,000,000$ lb/in$^2$ and 3.53 in$^3$ of steel weighs 1 lb. Make the correction for increase in tension.

**4.12**   A second line is measured with the tape of Problem 4.11, the tape being supported at intervals of 50 ft and the pull being 20 lb. The measured length is 1823.00 ft. Compute the corrections for sag and variation in tension and determine the corrected length of the line.

**4.13**   A 30-m steel tape weighing 0.7 kg has a cross-sectional area of 0.03 cm$^2$ and a standard length of 30.005 m when fully supported, at 20°C, and with 5 kg of tension.

   (a) A line on a smooth level ground is measured with the tape under a tension of 10 kg at a temperature of 20°C and is recorded as 1242.823 m long. Compute the correction for the increase in tension. $E = 2,100,000$ kg/cm$^2$.

   (b) This same tape is used to measure a distance recorded as 985.423 m over level terrain with the tape supported at the two ends and 5 kg of tension (temperature = 20°C). Calculate the correction due to sag for this distance.

**4.14**   Compute the normal tension for the tape in (a) Prob. 4.11; (b) Prob. 4.13.

**4.15**   For the purpose of establishing monuments in a city, a line along a paved street having a grade of 2.5 percent is measured on the slope. The applied tension is 7.5 kg, and observations of temperature are made at each application of the tape. The measured length on the slope is 402.351 m, and the mean of the observed temperatures is 29.3°C. The 30-m steel tape used for the measurements is standardized at 20°C, supported for its full length, and is found to be 0.001 m too short under a tension of 7.5 kg. Determine the horizontal length of the line.

**4.16**   The tolerance in measuring a tape length with a 30-m steel tape supported at the two ends and held horizontally is specified as ±0.010 m. If all other sources of error are neglected, how closely must the difference in elevation between the two ends of the tape be determined to satisfy this specification?

**4.17**   For rough taping with a 30-m steel tape over irregular terrain, the following standard deviations in taping operations are assumed: (1) tape not level, $\sigma_h = 0.02$ m; (2) sag in the tape caused by estimating tension, $\sigma_s = 0.01$ m; (3) plumbing the points, $\sigma_v = 0.02$ m.

   (a) Calculate the standard deviation in measuring a tape length due to these estimated errors.

   (b) A line 900 m long is to be measured with this tape using procedures that satisfy the estimates given above. Compute the standard deviation for the length of the measured line.

**4.18**   A line roughly 2 mi long along a railroad track is measured with a steel tape, and corrections are made for observed temperatures. What error will be introduced if the actual temperature of the tape is 2°F higher than the observed temperature? State the error in fractional form with 1 as the numerator.

**4.19**   Assume that an invar tape having a coefficient of thermal expansion of 0.00000083/1°F is used under the conditions of Prob. 4.18. Compute the error introduced.

**4.20**   A hedge along the line AB makes direct measurement impossible. A point C is established at an offset distance of 6.10 m from the line AB and roughly equidistant from A and B. The distances AC and CB are then chained; $AC = 392.34$ m and $CB = 412.39$ m. Compute the length of the line AB.

**4.21**   Describe and explain the fundamental differences between electro-optical and electromagnetic distance measuring instruments.

**4.22**   Describe and explain the major systematic errors that affect EDM systems.

**4.23**   An electro-optical EDM instrument using infrared light with a wavelength of 0.9050 μm has a modulation frequency of 30 MHz. Calculate the modulated wavelength of light at 26°C and an atmospheric pressure of 752.9 mmHg. Compare this wavelength with the wavelength of the modulation frequency in a standard atmosphere at 0°C and 760 mmHg.

**4.24**   Compute the velocity of red laser light at a temperature of 30°C and an atmospheric pressure of 758.2 mmHg

**4.25**   Compute the modulation wavelength of microwaves having a modulation frequency of 25 MHz at a temperature of 19.5°C, an atmospheric pressure of 758.2 mmHg, and a vapor pressure of 5.2 mmHg.

**4.26**   The vertical angle from point 1 to 2 is +2°27′01″ and the slope distance (corrected for atmospheric conditions and system constants) is 17,708.974 m, as measured by an EDM instrument. If the elevation of point 1 is 738.13 m above mean sea level, compute the horizontal sea level distance from 1 to 2.

**4.27** The vertical angles between stations $A$ and $B$ are 3°02′05″ from $A$ to $B$ and −3°12′55″ from $B$ to $A$. The slope distance $AB$ corrected for atmospheric conditions and system constants is 23,458.487 m and $A$ is 322.073 m above mean sea level. Compute the horizontal distance $AB$ reduced to sea level.

**4.28** The zenith angle from $C$ to $D$ is 87°32′59″ and the slope distance corrected for atmospheric conditions and EDM system constants is 58,100.31 ft. The elevation of $C$ is 2553.44 ft above mean sea level. Calculate the horizontal distance from $C$ to $D$ and reduce to sea level.

**4.29** The difference in elevation between points $A$ and $B$ is 778.540 m and the slope distance (corrected for atmospheric conditions and EDM system constants) is 16,421.825 m. Determine the horizontal distance reduced to sea level. The elevation of $A$ is 150.82 m above mean sea level.

**4.30** The standard deviations for measuring the zenith angle and corrected slope distance in Prob. 4.28 are $\sigma_{vert\ angle} = 05″$ and $\sigma_{slope\ distance} = 0.050$ m. Calculate the standard deviation in the horizontal distance from $C$ to $D$.

## 4.45. Field problems

### PROBLEM 1   TAPING OVER UNEVEN GROUND

**Object**   To standardize the 100-ft or 30-m steel tape; to find, by two methods, the horizontal length of an assigned course about 800 ft or 250 m long over uneven ground; and to correct for the error in length of tape as determined by the standardization tests.

**Procedure**   (1) Standardize the tape before and after the field work, by comparing it with the official standard of length. Maintain the required pull by means of a spring balance. Determine the error with a

| Sta. | Length in Feet | | | Correct-ion | Corr. Length |
|---|---|---|---|---|---|
| | Forward | B'kward | Mean | | |
| 351 | | | | | |
| | 311.73 | 311.65 | 311.69 | +0.07 | 311.76 |
| 352 | | | | | |
| | 198.60 | 198.62 | 198.61 | +0.04 | 198.65 |
| 353 | | | | | |
| | 213.84 | 213.78 | 213.81 | +0.05 | 213.86 |
| 354 | | | | | |

*TAPING*

*OVER UNEVEN GROUND*

Chaining Equipment — J. Pratt, H.C.
Locker No. 35 — S. Fulton, R.C.
Chicago 100 ft. Steel — Oct. 3, 1978 (2 hrs.)
Tape — Cold and Windy

Tk. in Stk. Ctr. Line M.C.R.R. at Haskell St.
East on Ctr. Line Haskell St. up Steep Hill
Nail in Pavement
North over Creek and R.R. Embankment
S.B.S.E. Corner Jones' Lot
North up Hill (Rough)
1″ Iron Pipe N.E. Corner Jones' Lot

| Standardization | |
|---|---|
| Actual length of standard | 100.038′ |
| Observed length of standard | |
| Before | 100.017′ |
| After | 100.015′ |
| Mean | 100.016′ |
| Actual length of tape | 100.022′ |
| Correction per tape length | +0.022′ |

Tape compared with Standard at
City Hall

**Fig. 4.30** Notes for taping over uneven ground with tape horizontal.

| MEASUREMENT OF A DISTANCE | | | | | BY TAPING | Oct. 7, 1976 |
|---|---|---|---|---|---|---|
| Station | Dist., m | Temp., C° | Temp. Corr., m | Calibr. Length, m | Corrected Dist., m | 30-m tape No. 6 / Clear, Warm |
| $A_7$ | 30.000 | 9.7 | -0.0036 | 29.9934 | 29.9898 | Hand Level / Morgan |
| | 30.000 | 9.7 | -0.0036 | 29.9934 | 29.9898 | Line Rod / Roberts |
| | 5.825 | 12.6 | -0.0005 | 5.8237 | 5.8232 | Thermometer No. 9 / Pental |
| | | | | | 65.8033 | Spring Balance No. 4 / Mennick |
| $B_7$ | 30.000 | 6.3 | -0.0048 | 29.9934 | 29.9886 | 2 Plumb Bobs |
| | 30.000 | 4.7 | -0.0053 | 29.9934 | 29.9881 | |
| | 5.822 | 3.8 | -0.0011 | 5.8207 | 5.8196 | Tape 6 measures 29.9934 at 20°C, |
| $A_7$ | | | | | 65.7963 | supported at the two ends, and with |
| | 30.000 | 6.9 | -0.0046 | 29.9934 | 29.9888 | 7-kg tension |
| | 30.000 | 4.1 | -0.0055 | 29.9934 | 29.9879 | |
| | 5.807 | 3.5 | -0.0011 | 5.8057 | 5.8046 | |
| $B_7$ | | | | | 65.7813 | |
| | 30.000 | 5.7 | -0.0050 | 29.9934 | 29.9884 | |
| | 30.000 | 8.1 | -0.0041 | 29.9934 | 29.9893 | |
| | 5.839 | 7.8 | -0.0008 | 5.8377 | 5.8369 | |
| $A_7$ | | | | | 65.8146 | |
| | 30.000 | 6.2 | -0.0048 | 29.9934 | 29.9886 | |
| | 30.000 | 7.1 | -0.0045 | 29.9934 | 29.9889 | |
| | 5.820 | 6.5 | -0.0009 | 5.8187 | 5.8178 | |
| $B_7$ | | | | | 65.7953 | |
| Average of 5 measurements | | | | = | 65.7982 | |

$$\hat{\sigma} = \left[ \frac{\sum_{i=1}^{5} \bar{x} - x_i}{n-1} \right]^{1/2} = \left[ \frac{0.0005926}{4} \right]^{1/2}$$

$$\hat{\sigma} = 0.012 \ m$$

**Fig. 4.31** Notes for precise taping over uneven ground.

finely divided scale. (2) Tape the course by horizontal measurement after the manner described in Art. 4.11. Maintain a pull estimated to be equal to that used during standardization. For measurements in which the line slopes more than 5 or 6 ft in 100, or 1.5 to 2 m in 30 m, "break tape." (3) Correct for the error in length of tape by adding to or subtracting from the observed distance the observed error multiplied by the number of tape lengths in the course. Record the data in a form similar to that of Fig. 4.30. (4) Tape the course by measurement on the slope, recording the difference in elevation (to the nearest foot or nearest 2 dm) of each tape length as determined by a hand level. If sharp breaks in slope occur between stations, treat each distance between breaks in the same manner as a full station. Correct the measurements for error in length of tape and for slope (Art. 4.13). Compare results obtained under (3) and (4).

## PROBLEM 2   PRECISE TAPING OVER UNEVEN GROUND

**Object**   To measure a distance of a line using a standardized steel tape, tension handle, and tape thermometer, and to make corrections for incorrect tape length and temperature.

**Procedure**   Obtain the standardized length of the steel tape and standardization conditions from the instructor or compare the steel tape with a standardized tape using a tension handle and tape thermometer attached to the tape. Set two hubs with tacks at the end points of the line to be measured. Measure the line five times as follows: (1) mark the ends of each tape length by pencil lines on the top of a hub; (2) apply the standard tension with a tension handle; (3) use a hand level to check the horizontal position of the tape; (4) with a tape thermometer attached to the tape, observe and record the temperature of the tape after each measurement; (5) estimate distances to thousandths of a foot or to millimetres.

Compute the (1) corrections for incorrect length of tape and temperature and apply to the measured distances; (2) average of the five measurements; (3) standard deviation of the measurements (Art. 2.17). Figure 4.31 illustrates a suggested form of notes.

## References

1. AGA Operating Manuals, AGA Corporation Geodimeter Division, S-181 20 Lidingö, Sweden.
2. Dracup, J. F., Kelley, C. F., Lesley, G. B., and Tomlinson, R. W., *National Geodetic Survey Lecture Notes for Use at Workshops on Surveying Instrumentation and Coordinate Computation*, National Ocean Survey, Rockville, Md.
3. Huggett, G. R., and Slater, S. L., "Precision Electromagnetic Distance Measuring Instrument for Determining Secular Strain and Fault Movement," Presented paper, International Symposium on Recent Crustal Movements, Eidgonössische Technische Hochschule, Zurich, Switzerland, August 26–31, 1974.
4. Huggett, G. R., "High Accuracy Techniques Using the HP 3800 Distance Meter," *Insight*, Hewlett-Packard Engineering Products, Vol. 2, No. 1, November 1973.
5. Huggett, G. R., "Instrument Non-linearity," *Insight*, Hewlett-Packard Engineering Products, Vol. 2, No. 4, August 1974.
6. Kester, J. M., "EDM Slope Reduction and Trigonometric Leveling," *Surveying and Mapping*, Vol. 33, No. 1, March 1973, p. 61.
7. Kivioja, L. A., "The EDM Corner Reflector Is Not Constant," *Surveying and Mapping*, Vol. 38, No. 2, June 1978, pp. 142–155.
8. Laurilla, S. H., *Electronic Surveying*, John Wiley & Sons, Inc., New York, 1976.
9. Mead, B. D., "Precision in Electronic Distance Measuring," *Surveying and Mapping*, Vol. 32, No. 1, March 1972, pp. 69–78.
10. Moffitt, F. H., "Field Evaluation of the Hewlett-Packard Model 3800 Distance Meter," *Surveying and Mapping*, Vol. 31, No. 1, March 1971.
11. Moffitt, F. H., "Calibration of EDM's for Precision Measurement," *Surveying and Mapping*, Vol. 35, No. 2, June 1975, pp. 147–154.
12. Moffitt, F. H., and Bouchard, H. B., *Surveying*, 6th ed., Harper & Row, Publishers, New York, 1975.
13. Romaniello, C. G., "EDM 1976," *Surveying and Mapping*, Vol. 37, No. 1, March 1977, p. 25.
14. Saxena, N. K., "Electro-optical Short Range Surveying Instruments," *Journal of the Surveying and Mapping Division, ASCE*, Vol. 101, No. SU1, October 1975, pp. 137–147.
15. Tomlinson, R. W., "Electronic Distance Measuring Instruments Calibration—Why, When, Where, and How," *Proceedings, Fall Convention, ACSM*, 1977, p. 56.

# CHAPTER 5
# Vertical distance measurement: leveling

**5.1. Definitions** The *elevation* of a point near the surface of the earth is its vertical distance above or below an arbitrarily assumed *level surface* or curved surface every element of which is normal to the plumb line. The level surface (real or imaginary) used for reference is called the *datum*. A *level line* is a line in a level surface.

The *difference in elevation* between two points is the vertical distance between the two level surfaces in which the points lie. *Leveling* is the operation of measuring vertical distances, either directly or indirectly, to determine differences in elevation.

A *horizontal line* is a line, in surveying taken as straight, tangent to a level surface.

A *vertical angle* is an angle between two intersecting lines in a vertical plane. In surveying, it is commonly understood that one of these lines is horizontal.

The datum most commonly used is mean sea level, particularly the Mean Sea Level Datum established by the United States government (Art. 13.2). If merely the *relative* elevations of points for a particular survey are desired, the datum may bear no known relation to sea level. For example, the initial point in a survey may be assumed to have given elevation and the elevations of all succeeding points computed accordingly. To avoid negative values of elevation, either the datum is taken below the lowest point of the survey or a positive value is assigned to it. However, the use of a local datum is to be discouraged and should not be considered good surveying practice. The presence of vertical control surveys, each related to its own unrelated datum, contributes to chaotic conditions and inevitably leads to complications and increased costs for subsequent surveys as well as for planning, construction, and operation of engineering and scientific projects.

**5.2. Curvature and refraction** In leveling, it is necessary to consider the effects of (1) the curvature of the earth and (2) atmospheric refraction, which affects the line of sight. Usually, these two effects are considered together.

In Fig. 5.1 is shown a horizontal line tangent at $A$ to a level line near the surface of the earth. The vertical distance between the horizontal line and the level line is a measure of the earth's curvature. It varies approximately as the square of the distance from the point of tangency. In Fig. 5.1, let $OA = r$, the average radius of the earth. Also, let $c = ED$, the correction for earth curvature. Then

$$r^2 + \overline{AE}^2 = (r+c)^2 = r^2 + 2rc + c^2$$
$$\overline{AE}^2 = c(2r + c)$$
$$c = \frac{\overline{AE}^2}{2r + c} \tag{5.1}$$

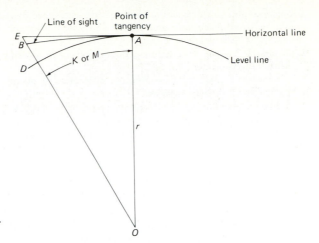

**Fig. 5.1** Earth curvature and refrac - tion.

Since $c$ is very small compared to $r$, a reasonable approximation for earth curvature is

$$c = \frac{\overline{AE}^2}{2r} \qquad (5.2)$$

Assuming a mean radius of the earth of 3959 mi or 6371 km, the curvature correction is

$$c = 0.667M^2 \text{ feet} \qquad (5.3)$$

in which $M$ is the distance from the point of tangency (station of the observer) in miles. In the metric system, the earth curvature correction $c_m$, in metres, is

$$c_m = 0.0785K^2 \qquad (5.4)$$

in which $K$ is the distance from the point of tangency in kilometres. Thus, the curvature correction is 0.67 ft/mi and 7.9 cm/km. For distances of 100 ft or 30 m the respective corrections would be 0.00024 ft and 0.07 mm.

Owing to the phenomenon of atmospheric refraction, rays of light are refracted, or bent downward slightly. This bending of the rays of light towards the center of the earth tends to diminish the effect of earth curvature by approximately 14 percent. In Fig. 5.1, $AB$ is the refracted line of sight and the distance $BD$ represents the combined effect of curvature and refraction. Let $(c \ \& \ r) = BD$ be computed by the following equations:

$$(c \ \& \ r) = 0.574M^2 \text{ feet} \qquad (5.5)$$
$$(c \ \& \ r) = (2.06)10^{-2}D^2 \text{ feet} \qquad (5.6)$$
$$(c \ \& \ r) = 0.0675K^2 \text{ metres} \qquad (5.7)$$

in which $M$ is in miles, $K$ is in kilometres as above, and $D$ is in thousands of feet.

In most ordinary spirit leveling operations, the line of sight is rarely more than 6 ft or approximately 2 m above the ground, where variations in temperature cause substantial uncertainties in the refractive index of air. Fortunately, most lines of sight in leveling are relatively short (about 100 ft or 30 m) and backsight and foresight distances are balanced (see Art. 5.3). Consequently, curvature and refraction corrections are rarely significant except for precise leveling (see Art. 5.46).

**5.3. Methods**  Difference in elevation may be measured by the following methods:

1. *Direct* or *spirit leveling*, by measuring vertical distances directly. Direct leveling is the most precise method of determining elevations and is the one commonly used (Art. 5.4).
2. *Indirect* or *trigonometric leveling*, by measuring vertical angles and horizontal or slope distances (Art. 5.5).

3. *Stadia leveling*, in which vertical distances are determined by tacheometry using the engineer's transit and level rod; plane table and alidade and level rod; or self-reducing tacheometer and level rod. Details for these procedures are given in Chap. 7.
4. *Barometric leveling*, by measuring the differences in atmospheric pressure at various stations by means of a barometer (Art. 5.6).
5. *Gravimetric leveling*, by measuring the differences in gravity at various stations by means of a gravimeter for geodetic purposes.
6. A ground "elevation meter," developed originally by the U.S. Corps of Engineers but now commercially available. This device consists of a vehicle on which is mounted an electromechanical device that integrates the vertical component of any longitudinal movement. Its accuracy is within 0.1 ft $(mi)^{1/2}$ and it can travel up to 30 mi/h. This is one of the inertial system instruments to appear in the mid-1970s.

*Differential leveling* is the operation of determining differences in elevation of points some distance apart or of establishing bench marks. Usually, differential leveling is accomplished by direct leveling. *Precise leveling* is a precise form of differential leveling.

*Profile leveling* is the operation—usually by direct leveling—of determining elevations of points at short measured intervals along a definitely located line, such as the center line for a highway or a sewer.

Direct leveling is also employed for determining elevations for cross sections, grades, and contours.

**5.4. Direct leveling**    In Fig. 5.2, *A* represents a point of known elevation and *B* represents a point the elevation of which is desired. In the method of direct or spirit leveling, the level is set up at some intermediate point as *L*, and the vertical distances *AC* and *BD* are observed by holding a leveling rod first at *A* and then at *B*, the line of sight of the instrument being horizontal. (Owing to refraction, the line of sight is slightly curved, as explained previously.)

If the difference in elevation between the points *A* and *E* is designated as $H_a$ and between *E* and *B* as $H_b$,

$$H_a = h_a - h_a' \quad \text{and} \quad H_b = h_b - h_b'$$

in which $h_a$ and $h_b$ are the vertical distances read at *A* and *B*, respectively, and $h_a' = (c \ \& \ r)_a$ and $h_b' = (c \ \& \ r)_b$ are the effects of curvature and refraction for the horizontal distances *LA* and *LB*, respectively, calculated using one of Eqs. (5.5), (5.6), or (5.7).

The difference in elevation *H* between *A* and *B* is then

$$H = H_a - H_b = (h_a - h_a') - (h_b - h_b')$$
$$= h_a - h_b - h_a' + h_b' \tag{5.8}$$

If the backsight distance *LA* is equal to the foresight distance *LB*, then $h_a' = h_b'$ and

$$H = h_a - h_b \tag{5.9}$$

Thus, if backsight and foresight distances are balanced, the difference in elevation between two points is equal to the difference between the rod readings taken to the two points, and no correction for curvature and refraction is necessary. In direct leveling, the work is usually

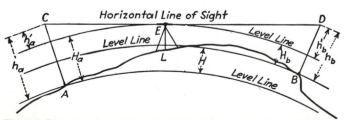

**Fig. 5.2** Direct leveling. (Owing to refraction, line of sight is slightly curved.)

conducted so that the effect of curvature and refraction is reduced to a negligible amount (Art. 5.39).

On lines of direct levels the usual procedure is as follows. Let $A$ and $M$ be two established points some distance apart whose difference in elevation is desired. With the level in some convenient location, not necessarily on a line joining $A$ and $M$, a backsight is taken to point $A$ and a foresight is taken to some convenient point $B$. The level is then moved ahead, a backsight is taken to $B$, and a foresight is taken to some accessible point as $C$. And so the process is repeated until the terminal point $M$ is reached.

**5.5.  Trigonometric leveling**    Trigonometric leveling involves observing the vertical angle and either the horizontal or slope distance between two points. The difference in elevation can then be calculated. In Fig. 5.3, $A$ represents a point of known elevation and $B$ a point the elevation of which is desired. Within the limits of ordinary practice, triangle $ECD$ can be assumed a right triangle and the distance $ED = EF$. If the horizontal distance $ED$ is known, then

$$DC = ED \tan \alpha$$
$$DF = (c \,\&\, r)_{ED}$$

in which the correction for earth curvature and refraction ($c \,\&\, r$) can be calculated using one of Eqs. (5.5), (5.6), or (5.7). The difference in elevation, $\Delta H_{AB}$, is

$$\Delta H_{AB} = AE + DF + DE \tan \alpha - BC \qquad (5.10)$$

where $AE$ and $BC$ are instrument height above $A$ and target height above $B$, respectively. These heights are measured using a level rod (Art. 5.23) or by means of the telescopic plumbing rod if the instrument tripod is so equipped.

A more common situation occurs when the slope distance $EC$ is measured very accurately using EDM equipment (Art. 4.34). In this case $DC = EC \sin \alpha$ and $\Delta H_{AB}$ becomes

$$\Delta H_{AB} = AE + DF + EC \sin \alpha - BC \qquad (5.11)$$

A major source of error in determining the difference in elevation by this method is the uncertainty in the curvature and refraction correction caused by variations in atmospheric conditions. To reduce the effects of this uncertainty, vertical angles are observed from both ends of the line (Art. 5.45). In Fig. 5.4, $\alpha$ is the *elevation* angle from instrument line of sight $E$ over point $A$ to target $C$ at $B$ and $\beta$ is the *depression* vertical angle from instrument line of sight $C'$ over $B$ to target $E'$ at $A$. The difference in elevation, $H_{AB}$ is calculated using Eq.

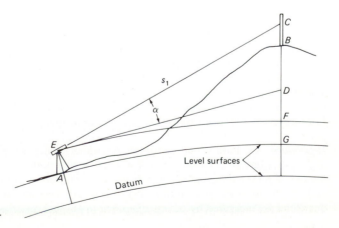

**Fig. 5.3**  Trigonometric leveling.

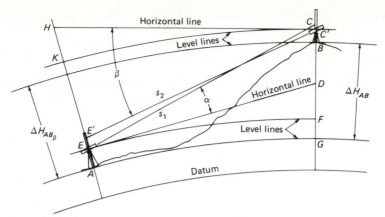

**Fig. 5.4** Trigonometric leveling with reciprocal vertical angles.

(5.11). The difference in elevation from $A$ to $B$, $\Delta H_{AB_\beta}$, using the absolute value of $\beta$ is

$$\Delta H_{AB_\beta} = C'E'\sin\beta - HK - C'B + E'A \tag{5.12}$$

in which $C'B$ is the instrument height above $B$ and $E'A$ is the target height above $A$.

If vertical angles $\alpha$ and $\beta$ are observed simultaneously, they are said to be *reciprocal* vertical angles. Since atmospheric conditions will be the same, $HK = DF$ and the addition of Eqs. (5.12) and (5.11) yields

$$\Delta H_{AB} + \Delta H_{AB_\beta} = EC\sin\alpha + C'E'\sin\beta + AE + E'A - BC - C'B$$

or

$$\Delta H = \frac{\Delta H_{AB} + \Delta H_{AB_\beta}}{2} = \frac{EC\sin\alpha + C'E'\sin\beta}{2} + \frac{(AE - BC) + (E'A - C'B)}{2} \tag{5.13}$$

Thus, when reciprocal vertical angles are observed, the difference in elevation is the average of the elevations obtained from the two ends of the line and the correction for earth curvature and refraction cancels.

When reciprocal vertical angles are employed with horizontal distances (as, for example, when using stadia), $\Delta H$ is

$$\Delta H = \frac{DE\tan\alpha + C'H\tan\beta}{2} + \frac{(AE - BC) + (E'A - C'B)}{2} \tag{5.14}$$

Normally, in stadia operations (see Chap. 7) $AE = BC$ and $E'A = C'B$, so that Eq. (5.14) becomes

$$\Delta H = \frac{DE\tan\alpha + C'H\tan\beta}{2} \tag{5.15}$$

**Uses**  In ordinary surveying, indirect trigonometric leveling furnishes a rapid means of determining the elevations of points in rolling or rough country. On reconnaissance surveys, angles may be measured with the clinometer and distances may be obtained by pacing. On more precise surveys, angles are measured with the transit and distances by the stadia. Indirect leveling is used extensively in plane-table work.

**Procedure**  On lines of indirect levels for which angles are measured with the transit, the usual procedure is as follows. Let $A$ and $D$ be two points whose difference in elevation is desired. With the transit at some intermediate location $T_1$, not necessarily on line between $A$ and $D$, the distance and vertical angle to $A$ are determined by a *backsight*, and similar quantities are measured by taking a *foresight* to another intermediate point $B$ at a convenient location not necessarily on line between $A$ and $D$. The instrument is then moved ahead to $T_2$, and similar observations are taken to $B$ and $C$. And so the process is repeated until the end of the line is reached. If the transit is *equidistant* from the points on either side of it to which sights are taken, the effect of curvature and refraction will be eliminated. Equation (5.10) is used for the reduction of data.

In practice, very little attention is paid to keeping the backsight and foresight distances balanced, since the effect of curvature and refraction is negligible (0.02 ft for a distance of 1000 ft and 6 mm in 300 m) as compared with the precision with which elevations can be determined by this method (generally not closer than tenths of feet). Generally, the transit is set at some convenient place where good sights can be obtained in both directions and which is about the same distance from adjacent points on which sights are to be taken.

In small-scale mapping, the method of indirect leveling is employed to determine the difference in elevation between the plane table and a point sometimes at a distance of several kilometres. In such cases the effect of curvature and refraction becomes large and the correction must be applied. For example, if the horizontal distance from the plane table to the point sighted were 10 kilometres, the correction would be 6.75 m or ~22 ft.

Using EDM equipment, slope distances are measured and vertical angles are observed with a 1″ theodolite. Equation (5.12) or (5.13) is employed to calculate differences in elevation. For this type of operation the limiting factor is the uncertainty in atmospheric conditions which causes uncertainty in the curvature and refraction correction. Consequently, observation of reciprocal vertical angles and the use of Eq. (5.13) is advisable. Assuming estimated standard deviations for slope distance, vertical angles, and height of instrument to be $\pm 0.020$ m, $\pm 4''$, and $\pm 0.005$ m, respectively, the propagated estimated standard deviation in determining $\Delta H$ for $s = 1000$ m and vertical angles of $5°$ is $\pm 0.016$ m. [Propagate the error through Eq. (5.13) using Eq. (2.51), Art. 2.20.] Thus, when using EDM equipment and observing simultaneous vertical angles with a 1″ theodolite, the error in trigonometric leveling is about 2 cm/km and will vary as the square root of the number of lines observed.

**Example 5.1** Referring to Fig. 5.3, the slope distance $s_1$ from $E$ to $C$ is 332.791 m with an average vertical angle of $9°46'29''$ from a height of instrument at $E$ of 1.558 m above $A$ to a target height $C$ of 1.372 m above $B$. If the elevation at $A$ is 21.935 m above mean sea level, calculate the difference in elevation from $A$ to $B$ and the elevation of $B$.

SOLUTION Use Eq. (5.11), where the correction for curvature and refraction is calculated with Eq. (5.7).

$\Delta H_{AB} = 1.558 + (0.067)(0.332791)^2 + (332.791)(\sin 9°46'29'') - 1.372$

$\qquad = 56.693$ m

elevation of $B = 21.935 + 56.693 = 78.628$ m

**Example 5.2** Referring to Fig. 5.4, the slope distance $s_1$ from $E$ to $C$ is 404.163 m with a vertical angle of $1°48'26''$ from an instrument height at $E$ of 1.558 m above $A$ to a target height at $C$ of 1.521 m above $B$. A reciprocal vertical angle of $-1°48'38''$ is observed from an instrument height of 1.560 m at $C'$ above $B$ to a target height of 1.587 m at $E'$ above $A$. If the observed slope distance $s_2$ from $C'$ to $E'$ is 404.161 m and the elevation of $A$ is 29.935 m above sea level, calculate the difference in elevation from $A$ to $B$ and the elevation of $B$.

SOLUTION Use Eq. (5.13) to compute $\Delta H_{AB}$.

$\Delta H_{AB} = \frac{1}{2}[(404.163)(\sin 1°48'26'') + (404.161)(\sin 1°48'38'')]$

$\qquad\quad + \frac{1}{2}[(1.558 - 1.521) + (1.587 - 1.560)]$

$\qquad = 12.758 + 0.032$

$\qquad = 12.790$ m

elevation of $B = 29.935 + 12.790 = 42.725$ m above sea level

***Error propagation for Examples 5.1 and 5.2***   Assume the following estimated standard deviations for Example 5.1: $\hat{\sigma}_{s_1} = 0.02$ m, $\hat{\sigma}_\alpha = 04''$, $\hat{\sigma}_{h.i.} = \hat{\sigma}_{\text{target height}} = 0.005$ m, and $\hat{\sigma}_{(c\&r)} = 0.002$ m. Referring to Eq. (5.11) and Fig. 5.3, the estimated standard deviation of $\Delta H$ from $A$ to $B$ is [assuming no correlation between the different variables in Eq. (5.11) and letting $s = EC$, $c$ & $r = DF$, h.i. $= AE$, h.i. $= BC = $ target height]

$$\hat{\sigma}^2_{\Delta H_{(A-B)}} = \left(\frac{\partial \Delta H}{\partial AE}\right)^2 \hat{\sigma}^2_{h.i.} + \left(\frac{\partial \Delta H}{\partial BC}\right)^2 \hat{\sigma}^2_{h.i.} + \left(\frac{\partial \Delta H}{\partial s}\right)^2 \hat{\sigma}^2_s + \left(\frac{\partial \Delta H}{\partial \alpha}\right)^2 \hat{\sigma}^2_\alpha + \left(\frac{\partial \Delta H}{\partial (c\ \&\ r)}\right) \hat{\sigma}^2_{(c\ \&\ r)}$$

$$= 2(0.005)^2 + (\sin \alpha)^2 (0.02)^2 + [(332.791)\cos \alpha]^2 [(4)(4.848)(10^{-6})]^2$$
$$+ (0.002)^2$$
$$= 0.000050 + 0.000012 + 0.000043 + 0.000004 = 0.000109$$
$$\hat{\sigma}_{\Delta H_{(A-B)}} = 0.010 \text{ m} = 10 \text{ mm}$$

The uncertainty in ($c$ & $r$) was estimated by neglecting refraction completely. Note that for short lines the effect is not significant. For long lines the uncertainty in ($c$ & $r$) is substantially larger and has a pronounced effect on the error propagation.

For Example 5.2, the error must be propagated using Eq. (5.13). Assume that $\hat{\sigma}_\beta = \hat{\sigma}_\alpha = 04''$, $\hat{\sigma}_{s_1} = \hat{\sigma}_{s_2} = 0.020$ m, $\sigma_{AE} = \sigma_{E'A} = \sigma_{BC} = \sigma_{C'B} = 0.005$ m, and let $s_1 = EC$, $s_2 = E'C'$. The estimated standard deviation for $\Delta H_{AB}$ is

$$\hat{\sigma}^2_{\Delta H_{(A-B)}} = \tfrac{1}{2}\left[4\hat{\sigma}^2_{H.I.} + (\sin^2 \alpha + \sin^2 \beta)\hat{\sigma}^2_s + (s_1^2 \cos^2 \alpha + s_2^2 \cos^2 \beta)\sigma^2_\alpha\right]$$

$$= \tfrac{1}{2}\Big[(4)(0.005)^2 + (\sin^2 1°48'26'' + \sin^2 1°48'38'')(0.02)^2$$
$$+ \{(404.163)^2(\cos 1°48'26'')^2 + (404.161)^2(\cos 1°48'38'')^2\}[(4)(4.848)(10^{-6})]^2\Big]$$

$$= \tfrac{1}{2}(0.0001 + 0.000000797 + 0.000122731)$$
$$= 0.0001118$$
$$\hat{\sigma}_{\Delta H_{(A-B)}} = 0.011 \text{ m} = 11 \text{ mm}$$

**5.6. Barometric leveling**   Since the pressure of the earth's atmosphere varies inversely with the elevation, the barometer may be employed for making observations of difference in elevation. Barometric leveling is employed principally on exploratory or reconnaissance surveys where differences in elevation are large, as in hilly or mountainous country. Since atmospheric pressure may vary over a considerable range in the course of a day or even an hour, elevations determined by one ordinary barometer carried from one elevation to another may be several feet in error. However, by means of sensitive barometers and special techniques, elevations can be determined within a foot or so.

Usually, barometric observations are taken at a fixed station during the same period that observations are made on a second barometer, which is carried from point to point in the field. This procedure makes it possible to correct the readings of the portable barometer for atmospheric disturbances.

***Instruments and methods***   The mercurial barometer is accurate, but it is cumbersome and is suitable only for observations at a fixed station. For field use, an aneroid barometer is commonly used because it is light and is easily transported. The usual type has a dial about 3 in in diameter, graduated both in inches of mercury and in feet of altitude (elevation); it is compensated for temperature. At a point of known altitude, the pointer can be set at the corresponding reading on the scale in order to index readings. The aneroid barometer can be calibrated against a mercurial barometer by comparing values at a given station over a range of temperatures.

In use, the barometer should be given time to reach the temperature of the air before an observation is made. In most instruments today this time is much less than 1 min.

A single aneroid barometer is sometimes used by topographers on small-scale surveys where the contour interval is large. Stops are made at frequent intervals during the day at a single point of assumed elevation or at various points of known elevation, instrument readings are made, and the rate of change in atmospheric conditions is calculated; suitable corrections are thus determined and are applied to the observed values. Where distances permit, it is preferable to return to the starting point and to correct the intermediate readings in proportion to the change in atmospheric pressure during the interval between observations.

**Sensitive barometers**   Extremely sensitive barometers known as altimeters have been developed, with which elevations can be determined within a foot or so. In one procedure used in topographic surveying, two of the instruments are employed at fixed bases and one or more instruments are carried from point to point over the area being surveyed. One fixed instrument is located at a point of known elevation near the highest elevation of the area, and one near the lowest elevation; these instrument stations are called the *upper base* and *lower base*, respectively. Other altimeters are carried to points whose elevations are desired, and readings are taken. Readings on the fixed instruments are taken either simultaneously (as determined by signaling) or at fixed intervals of time; in the latter case the readings at the desired instant are determined by proportion. The elevation of a portable instrument is then determined by interpolation. The horizontal location of each point at which a reading is taken is determined by conventional methods.

**Example 5.3**   Given elevation of upper base 275 ft, of lower base 56 ft; the difference in elevation between the bases is, therefore, $275 - 56 = 219$ ft. At a given instant, the three altimeter readings indicate that the difference in elevation of an intermediate point from the upper base is 209 ft and from the lower base is 25 ft; thus, the indicated total difference in elevation between bases is 234 ft. The corrected differences in elevation are proportionately $(219/234)(209) = 196$ ft (from upper base) and $(219/234)(25) = 23$ ft (from lower base); as a check, the total computed difference in elevation between bases is now $196 + 23 = 219$ ft. The elevation of the point is 79 ft, computed by difference from either base ($275 - 196 = 79$; or $56 + 23 = 79$).

Another procedure is to employ one barometer at a fixed base and one barometer which is carried to points whose elevations are desired, simultaneous readings being taken. The carried barometer is finally brought either back to the starting point or to another point of known elevation; the computation of the elevation of each point then takes account of the corresponding change in atmospheric pressure at the fixed base.

## INSTRUMENTS FOR DIRECT LEVELING

**5.7. Kinds of levels**   Any instrument commonly used for direct leveling has as its essential features a *line of sight* and *spirit level tube* or some other means of making the line of sight horizontal. The level tube is so mounted that its axis is parallel to the line of sight. The instrument used principally in the United States is the *engineer's level*. Figure 5.5*a* is a diagram of the principal parts of an engineer's level. The level consists of the telescope *A* mounted upon the level bar *B* which is rigidly fastened to the spindle *C*. Attached to the telescope or the level bar and parallel to the telescope is the level tube *D*. The spindle fits into a cone-shaped bearing of the leveling head *E*, so that the level is free to revolve about the spindle *C* as an axis. The leveling head is attached to a tripod *F*. In the tube of the telescope are cross hairs at *G*, which appear on the image viewed through the telescope, as illustrated by Fig. 5.5*b*. The bubble of the level is centered by means of the leveling screws *H*. The two distinct types of the engineer's level are the *dumpy level* for which the telescope tube is permanently fastened to the level bar and *self-leveling* or *automatic* levels. The

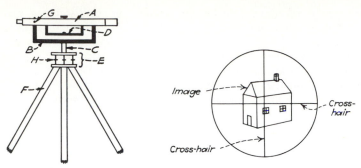

**Fig. 5.5** Principal parts of an engineer's level.

*architect's level* or *construction level*, a modified form of the engineer's level but with a telescope of lower magnifying power and with a less-sensitive level tube, is used in establishing grades for buildings. The *hand level* (Arts. 5.21 and 5.22) is a simple, hand held, useful device for determining differences in elevation roughly. Instruments which are frequently used for direct leveling but which are not primarily designed for this purpose are the *engineer's transit* (Chap. 6) and the telescopic alidade of the *plane table* (Chap. 7).

The measurements of difference in elevation are made by sighting on a graduated rod, called a *leveling rod* (Arts. 5.23 to 5.25). Other accessories sometimes used are a *rod level* (Art. 5.29) and a *turning point* (Art. 5.31).

Usually the leveling head of the engineer's level may be equipped with four leveling screws, but the three-screw type is favored by many engineers and surveyors, particularly for instruments of high precision. Reasons for this preference are that the three-screw type can be leveled rapidly, requires the use of only one hand, and is relatively stable as compared with the four-screw type when the latter is not perfectly set. With the three-screw leveling head, on the other hand, when one of the screws is turned, the elevation of the telescope may change slightly.

**5.8. Telescope**   Modern levels are equipped with internally focusing telescopes that have features as illustrated in Fig. 5.6a. Rays of light emanating from an object point are caught by the objective lens $A$ and are brought to a focus and form an image in the plane of cross hairs $B$. The lenses of the eyepiece $C$ form a microscope which is focused on the image at the cross hairs. The objective lens is fixed in the end of the telescope tube and a negative lens $L$ is attached to a slide $D$ within the telescope tube. This negative lens can be moved parallel to the line of sight by rotating the focusing knob $F$ and allows focusing the telescope on objects at different distances from the instrument. The eyepiece $C$ is held in position by

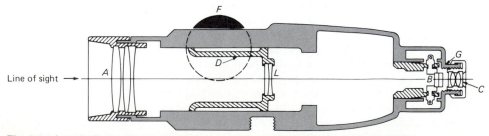

**Fig. 5.6a** Longitudinal section of internal-focusing telescope. (*Courtesy the Lietz Co.*)

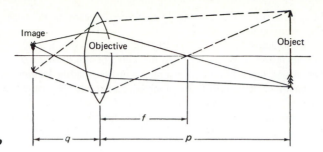

**Fig. 5.6***b*

threaded sleeve *G*, which may be moved in a longitudinal direction for focusing. By means of adjusting screws, the cross-hair reticule may be moved transversely so that the intersection of the cross hairs will appear in the center of the field of view.

The line of sight is defined by the intersection of the cross hairs and the optical center of the objective lens. The instrument is so constructed that the optical axis of the objective lens coincides (or practically coincides) with the axis of the negative or objective slide (*D*, Fig. 5.6*a*); in other words, a given ray of light passing through the optical center of the objective always occupies the same position in the telescope tube regardless of the longitudinal position of the negative lens. The cross hairs can be adjusted so that the line of sight and the optical axis coincide.

A few older levels still in use are equipped with externally focusing telescopes. The objective lenses on these telescopes are attached to the objective slide, which fits within the telescope and can be moved externally in a longitudinal direction to allow focusing on the object sighted. Since the objective end of the telescope cannot be sealed, dirt and moisture can enter, causing maintenance problems and excessive wear.

The internally focusing telescope has the following advantages:

1. Because both ends of the telescope are closed, the focusing slide is practically free of grit that would cause wear; and the entire interior of the telescope is practically free of dust and moisture.
2. Since the focusing slide is light in weight and is located near the middle of the telescope, the telescope tends to balance well.
3. In making measurements by the stadia method (Chap. 7), an instrumental constant is eliminated and the computations thus simplified.

Practically all modern surveying instruments are now manufactured with internally focusing telescopes equipped with coated lenses to reduce light reflections from each glass surface and thus provide better illumination for the images.

**5.9. Objective lens**   The principal function of the telescope objective lens is to form an image for sighting purposes. For accuracy of measurements, the objective should produce an image that is well lighted, accurate in form, distinct in outline, and free of discolorations. A single biconvex lens meets the first two of these requirements but is faulty in regard to the other two, for the following reasons:

1. Rays entering the lens near its edge come to a focus nearer the objective than do those entering near its center. The image does not lie in a plane, but in the surface of a sphere. Hence, as viewed through the telescope, portions of the object are blurred. This defect is called *spherical aberration*.
2. Rays of the various colors of the spectrum are deviated by different amounts as they pass through the lens; hence, the field of view appears discolored by lights of various hues. This is called *chromatic aberration*.

These two objectionable features of the single lens are nearly eliminated in most surveying instruments by providing an outer double-convex lens of crown glass and an inner concavoconvex lens of flint glass. The two lenses are usually cemented together with balsam but are sometimes separated by a thin spacer ring.

The *optical center* of the objective is that point in the lens through which a ray of light will pass without permanent deviation, regardless of the direction of the object from which the light emanates. In other words, the direction of the ray is the same after leaving the lens as before entering it. In a biconvex lens with faces of equal curvature, the optical center and geometrical center coincide.

The *optical axis* is the line taken by a light ray that experiences no deviation either on entering or on leaving the objective. It passes through the optical center and the centers of curvature of the lens.

The *principal focus* is a point on the optical axis back of the objective where rays entering the telescope parallel with the optical axis are brought to a focus; or it is a point in front of the objective from which diverging light rays entering the lens emerge from it parallel with the optical axis. Stated in another form, the image of a point on the optical axis and an infinite distance away is at the principal focus back of the objective. If a point is at the principal focus in front of the objective, however, it will have no image.

The *focal length* of the objective is the distance from its optical center to the principal focus. When the telescope is focused on a distant point, the focal length is very nearly the distance from the optical center of the objective to the plane of the cross hairs, for reasons which the preceding paragraph makes clear.

The basic principle (lens equation), which relates object distance $p$, image distance $q$, and focal length $f$ (for externally focusing telescopes), is

$$\frac{1}{p} + \frac{1}{q} = \frac{1}{f} \tag{5.16}$$

in which $f$ is a constant for the telescope. Figure 5.6$b$ illustrates the manner in which rays from an object are deviated by the objective and brought to a focus to form the image. Note that the image is inverted.

**5.10. Focusing**    When the telescope is to be used, the eyepiece is first moved in or out until the cross hairs appear sharp and distinct. This adjustment of the eyepiece should be tested frequently, as the observer's eye becomes tired.

When an object is sighted, the negative lens or objective slide is moved in or out until the image appears clear. At this point the image should be in the plane of the cross hairs. If a slight movement of the eye from side to side or up or down produces an apparent movement of the cross hairs over the image, the plane of the image and the plane of the cross hairs do not coincide, and *parallax* is said to exist. Since parallax is a source of error in observations, it should be eliminated by refocusing the objective, the eyepiece, or both until further trial shows no apparent movement. Figure 5.7$a$ through $c$ provides a graphical illustration of the concept of parallax. The negative or objective lens must be focused for each distance sighted.

The telescope cannot be focused on objects closer than about 6 ft or 2 m from the center of the instrument, unless special short-focus lenses are employed.

Most telescopes are equipped with peep sights on top or alongside the telescope. The observer should point the telescope approximately at the object by first sighting along the peep sights at the target. Then looking through the telescope, the observer should focus the objective lens so that there is a clear, distinct image of the target at which the telescope is pointed. A subsequent movement of the pointing of the telescope by a degree or two one way or another should point the line of collimation (cross hairs) precisely on the target.

The strain on the eyes will be reduced if the observer tries keeping both eyes open while sighting.

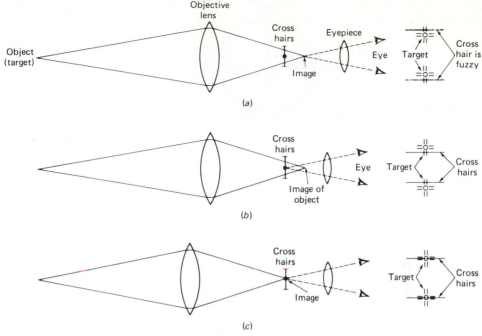

**Fig. 5.7** Presence of parallax. (*a*) Objective not focused, eyepiece not focused, parallax is present. (*b*) Objective not focused, eyepiece focused on cross hairs, parallax present. (*c*) Objective focused, eyepiece focused, no parallax.

Any lateral movement of the negative lens in an internally focusing telescope or the objective at the externally focusing telescope causes a deviation in the line of sight, introducing errors in measurements. The quality of work on any good instrument is sufficiently precise to ensure practical elimination of errors of this sort when the telescope is new, but in the course of long use, wear develops between the sliding parts and the mechanism becomes loose. This condition produces uncertainties in observation which no amount of adjustment can overcome. In this respect the internally focusing telescope, in which the focusing lens and mechanism are sealed within the telescope tube protected from dust, water, and other foreign matter, is definitely superior. Errors caused by changing focus are rarely significant for ordinary leveling but can be of consequence in precise leveling (Art. 5.46).

**5.11. Cross hairs**   The cross hairs, which define the line of sight, used to be made of threads from the cocoon of the brown spider and later of very fine platinum wire. In modern instruments the *cross-hair ring* or *reticule* consists of a glass plate on which are etched fine vertical and horizontal lines which serve as cross hairs. Special patterns of additional lines are used on some instruments: for example, stadia hairs, or double horizontal and vertical lines closely spaced for sighting between them precisely. The spacing between a double line can be pointed more precisely than can a single line.

As shown in Fig. 5.8*a* and *b*, the cross-hair ring is held in position by four capstan-headed screws which pass through the telescope tube and tap into the ring. The holes in the telescope tube are slotted so that when the screws are loosened, the ring may be rotated through a small angle about its own axis. To rotate the ring without disturbing its centering, two adjacent screws are loosened; and the same two screws are tightened after the ring has

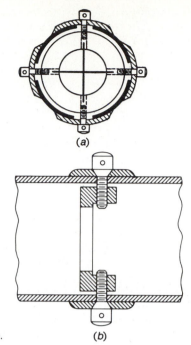

(a)

**Fig. 5.8** Cross-hair reticule.                    (b)

been rotated. The ring is smaller than the inside of the tube, and it may be moved either horizontally or vertically by means of the screws. Thus, to move it to the left, the right-hand screw is loosened and the left-hand screw is tightened. If the movement is to be large, first the top or bottom screw is loosened slightly; and after the movement to the left has been completed, the same (top or bottom) screw is tightened again.

**5.12. Eyepiece**  Attention has previously been drawn to the fact that the image formed by the objective is inverted. Eyepieces are of two general types:

The *erecting* or *terrestrial* eyepiece, the more common of the two types, reinverts the image so that the object appears to the eye in its normal position. Usually, it consists of four plano-convex lenses placed in a metal tube called the eyepiece slide (Fig. 5.9$a$). In the figure $A$ represents the object, $B$ the inverted image in the plane of the cross hairs, $C$ the image which is magnified by the lens nearest the eye, and $D$ the magnified image as it appears to the eye.

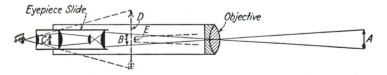

**Fig. 5.9$a$** Erecting eyepiece.

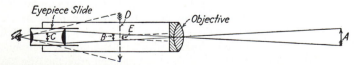

**Fig. 5.9$b$** Inverting eyepiece.

The *inverting* or *astronomical* eyepiece simply magnifies the image without reinverting it. It is composed of two plano-convex lenses generally arranged as shown in Fig. 5.9*b*. The arrangement is seen to be identical with that of the two lenses farthest apart in the erecting eyepiece. The magnified image *D* is seen to be inverted, and the object, as viewed through the telescope, is upside down.

For either eyepiece the ratio of the angle at the eye subtended by the magnified image, to that subtended by the object itself, is the magnifying power of the telescope. If, in either Fig. 5.9*a* or Fig. 5.9*b*, *D* is the apparent length of the magnified image and *E* is the apparent length of the object as seen by the naked eye, the ratio of *D* to *E* is the magnifying power.

Each lens which is interposed between the object and the eye absorbs some of the light that strikes it and each lens surface reflects some of the entering and departing light. Hence, other things being equal, the object is more brilliantly illuminated when viewed through the inverting eyepiece; this is a great advantage, particularly when observations are made during cloudy days or near nightfall. Another important advantage of the inverting eyepiece is that the telescope is shorter and the instrument lighter in weight. The beginner experiences some inconvenience on viewing things apparently upside down, but this difficulty is overcome with a little practice. The single advantage of the erecting eyepiece is that objects appear in their natural position, and this is the reason why its use is so common.

**5.13. Properties of the telescope**   The *illumination* of the image depends upon the effective size of the objective, the quality and number of lenses, and the magnifying power. Other conditions being the same, either the larger the objective or the smaller the magnifying power, the better the illumination; that is, the better lighted appears the object.

Distortion of the field of view so that it does not appear flat is mainly caused by what is termed the *spherical aberration* of the eyepiece. Although this introduces no appreciable error in ordinary measurements, it is not desirable when two points in the field are to be observed at the same time, as in stadia measurements.

The *definition* of a telescope is its power to produce a sharp image. It depends upon the quality of the glass, the accuracy with which the lenses are ground and polished, and the precision with which they are spaced and centered. Light rays passing through the lenses near their edges are particularly troublesome, and to improve the definition these rays are intercepted by diaphragms or screens placed between the lenses of the eyepiece and in the rear of the objective. The effect of these screens is to decrease the illumination somewhat.

The angular width of the field of view is the angle subtended by the arc, whose center is nearly at the eye and whose length is the distance between opposite points of the field viewed through the telescope. For a particular instrument this angle may be readily determined by observation. It is independent of the size of the objective. In general, the larger the telescope and the greater the magnifying power, the less the angle of the field of view. For most surveying of moderate precision, the work is greatly retarded if the instrument does not have a fairly large field of view, and this is one of the reasons why the telescopes are not usually made of high magnifying power. Usually, the angular width of the field ranges from about 1°30′ for a magnifying power of 20, to 45′ for a magnifying power of 40.

The magnifying power of the telescope for the better grade of engineer's level is about 30 diameters. The U.S. National Ocean Survey type of precise level has a magnification of about 40 diameters.

**5.14. Spirit level tube**   The level tube used in surveying instruments (Fig. 5.10) is a glass vial with the inside ground barrel-shaped, so that a longitudinal line on its inner surface is the arc of a circle (Fig. 5.11). The tube is nearly filled with sulfuric ether or with alcohol. The remaining space is occupied by a bubble of air which takes up a location at the high point in

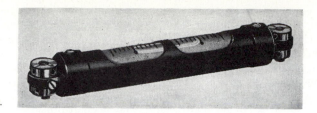

**Fig. 5.10** Level tube.

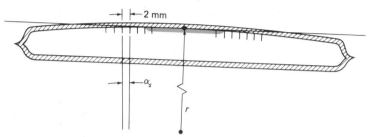

**Fig. 5.11** Cross section of spirit level tube.

the tube. The tube is usually graduated in both directions from the middle; thus, by observation of the ends of the bubble it may be "centered," or its center brought to the midpoint of the tube.

A longitudinal line tangent to the curved inside surface at its upper midpoint is called the *axis of the level tube* or *axis of the level*. When the bubble is centered, the axis of the level tube is horizontal.

The tube is set in a protective metal housing, usually with plaster of paris. The housing is attached to the instrument by means of screws which permit vertical adjustment at one end and lateral movement at the other end of the tube, as shown in Fig. 5.10.

Some leveling instruments are equipped with a prismatic viewing device by means of which one end of the bubble appears reversed in direction and alongside the other end. Figure 5.12*a* illustrates such a *coincidence* bubble when it is not centered. The bubble is centered by matching its ends as shown in Fig. 5.12*b*. Settings can be made more accurately by such coincidence than by matching the ends of the bubble with respect to the graduations on the level tube. Tilting levels (Art. 5.18) are usually equipped with coincidence viewing of the spirit level tube.

**5.15. Sensitivity of level tube**    If the radius of the circle to which the inner chamber level tube is ground is large, a small vertical movement of one end of the tube will cause a large displacement of the bubble; if the radius is small, the displacement will be small. Thus, the radius of the tube is a measure of its sensitivity. The sensitivity is generally expressed in

**Fig. 5.12** Coincidence bubble. (*a*) Bubble not centered. (*b*) Bubble centered. (*Courtesy Kern Instruments, Inc.*)

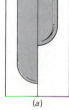

(a)

(b)

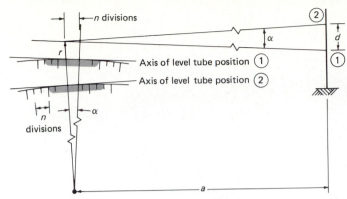

**Fig. 5.13** Determination of level-tube sensitivity.

seconds of the central angle, whose arc is one division of the tube. For most instruments the length of a division is 2 mm. The sensitivity expressed in seconds of arc is not a definite measure unless the spacing of graduations is known.

Sensitivities of bubble tubes, expressed in seconds per 2-mm division, vary from 1 to 2″ for precise levels up to 10 to 30″ for engineer's levels.

Should determination of bubble tube sensitivity be necessary, proceed as follows: (1) align the bubble tube axis with a pair of diagonally opposite level screws; (2) hold a rod in a vertical position at a measured distance from the level; (3) observe the rod reading; (4) tilt the telescope by manipulating the level screws, moving the bubble tube through $n$ divisions; and (5) observe the rod reading. Repeat steps (3) and (4) at least five times and calculate the average increment on the rod subtended by the angle caused by movement of the bubble through $n$ divisions. From Fig. 5.13, the sensitivity $\alpha_s$ in seconds per division is

$$\alpha_s = \frac{d}{(a)(n)\,\mathrm{arc}\ 1''} \qquad (5.17)$$

in which $d$ is the average increment on the rod, $a$ the distance from level to rod in compatible units, $n$ the number of divisions the bubble tube moved, and arc $1'' = 0.000004848$.

**5.16. Relation between magnifying power and sensitivity**   It is desirable that the sensitivity of the level tube be such that for the smallest noticeable movement of the bubble there is an apparent movement of the cross hairs on a level rod held at an average sighting distance from the instrument; and likewise for the smallest noticeable movement of the cross hairs there should be an observable movement of the bubble. The least noticeable movement of the cross hairs depends to some extent upon the definition and illumination of the image, but principally upon the magnification of the telescope.

If the level tube is more sensitive than is necessary, time is wasted in centering the bubble. If the magnifying power is higher than it need be, unnecessary labor is expended by reason of the more limited field of view and by reason of the increased difficulties of focusing the objective properly. A satisfactory test may be conducted by one person sighting at a rod while a second person bears down slightly on one end of the telescope and at the same time observes the level tube. If the first noticeable movement of the bubble is accompanied by an apparent movement of the cross hairs, there is a satisfactory balance between sensitivity and magnification. If the cross hairs move first, a level tube of greater radius might properly be employed.

**Fig. 5.14** Engineer's dumpy level. (*Courtesy Keuffel & Esser Co.*)

**5.17. Dumpy level**   Figure 5.14 shows the details of a conventional American dumpy level with erecting eyepiece. The telescope *A* is rigidly attached to the level bar *B*, and the instrument is so constructed that the optical axis of the telescope is perpendicular to the axis of the center spindle. The level tube *C* is permanently placed so that its axis lies in the same vertical plane as the optical axis, but it is adjustable in altitude by means of a capstan-headed screw at one end. The spindle revolves in the socket of the leveling head *D*, which is controlled in position by the four leveling screws *E*. At the lower end of the spindle is a ball-and-socket joint which makes a flexible connection between the instrument proper and the foot plate *F*. When the leveling screws are turned, the level is moved about this joint as a center. The sunshade *G* protects the objective from the direct rays of the sun. The adjusting screws *H* for the cross-hair ring are near the eyepiece end of the telescope.

The telescope of the dumpy level usually has a magnifying power of about 30 diameters, and the level tube usually has a sensitivity of 20″ of arc per graduation (2 mm).

The name "dumpy level" originated from the fact that formerly this level was usually equipped with an inverting eyepiece and therefore was shorter than other American manufactured levels of the same magnifying power. The current primary advantage of the dumpy-style level is that its weight lends stability under conditions where vibration or wind render the lighter automatic levels less stable.

**5.18. Tilting levels**   The distinctive feature of a *tilting level* (Fig. 5.15) is that the telescope is mounted on a transverse fulcrum at the vertical axis and on a micrometer screw at the eyepiece end of the telescope. After the instrument has been leveled in the usual manner, approximately, perhaps by use of a circular spirit level, the telescope is pointed in the direction desired and is then "tilted," or rotated slightly in the vertical plane of its axis by turning the micrometer screw, until the sensitive telescope-level bubble is centered. The line of sight is then horizontal, even though the instrument as a whole is not exactly level. On the micrometer screw is mounted a graduated drum called a gradienter. The relation between graduations, pitch of the tangent-screw, and length of the arm is usually such that one division on the drum corresponds to a movement of the line of sight of 0.01 ft at 100 ft from

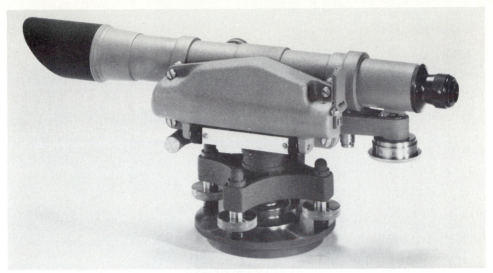

**Fig. 5.15** Tilting level. (*Courtesy Keuffel & Esser Co.*)

the instrument; for other distances the movement of the line of sight is in proportion to the distance. This relation is useful in setting grades and sometimes in measuring distances and differences in elevation where sights are nearly level. The gradienter can also be used to advantage in reciprocal leveling (Art. 5.45).

Modern tilting levels, although of the dumpy type, incorporate features which make them quite different from the dumpy-style tilting level. They are small, light in weight, and permit rapid setting up and observing. They are equipped with a circular spirit level and have three leveling screws. The telescope is internally focusing, usually has an inverted image, and has the cross hairs, including stadia lines, etched on a glass diaphragm.

The telescope-level bubble is centered by bringing the images of the ends of the bubble into coincidence as shown in Fig. 5.12. These coincidence bubbles are viewed through a system of prisms from the eyepiece end of the telescope either through a separate parallel eyepiece or off to one side of the field of view of the main telescope. Electric illumination can be provided for night observation. Figure 5.16 illustrates a modern tilting level of European design. This instrument has a level bubble with a sensitivity of 18″ per 2-mm division. The bubble can be viewed for coincidence leveling in the field of view of the telescope (see Fig. 5.16*b*). Horizontal angles can also be observed optically to 1′ of arc.

The coincidence bubble permits more accurate centering of the spirit bubble than is possible by observing the bubble directly. Consequently, tilting levels equipped with coincidence bubbles are used for first- and second-order precise leveling.

**5.19. Self-leveling levels**   A *self-leveling* or automatic level allows establishment of a horizontal line of sight by means of a system of prisms and mirrors supported by wires as in a pendulum. Figure 5.17 shows one form of a self-leveling level and Fig. 5.18 illustrates a cross-sectional view of the optics involved. Light enters the objective lens at $A$, passes through focusing lens $B$, is reflected by the optical compensator $C$, which is suspended by wires at $D$ and clamped magnetically at $E$. When the compensator $C$ is freely supported, the line of sight defined by the objective lens, compensator, and eyepiece is automatically horizontal. Most automatic levels are designed so that the line of sight is horizontal when the telescope barrel is within ± 10′ of being horizontal. At each setup, the instrument is leveled

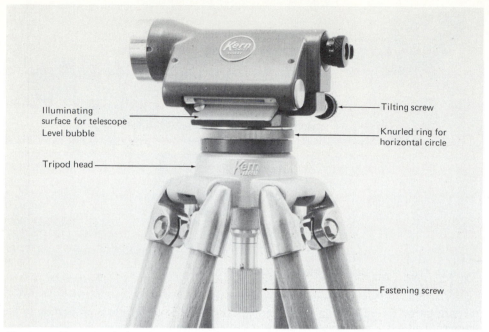

Illuminating
surface for telescope
Level bubble

Tilting screw

Knurled ring for
horizontal circle

Tripod head

Fastening screw

**Fig. 5.16a** European-designed tilting level. (*Courtesy Kern Instruments, Inc.*)

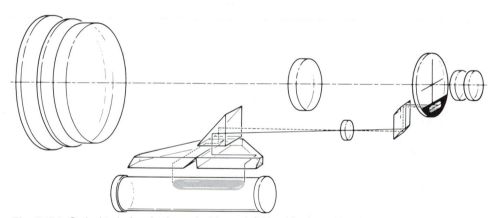

**Fig. 5.16b** Optical train for viewing coincidence bubble. (*Courtesy Lietz Instruments, Inc.*)

approximately by use of a circular spirit or bull's-eye level (Fig. 5.17) and the pendulum maintains a horizontal line of sight. The instrument is light, easy to handle, and its operation is quick and accurate. Note that the random error in centering the bubble is absent in this style of instrument.

**5.20. Geodetic levels**   The high order of accuracy required in geodetic leveling requires an extremely sensitive level of special design. The sensitivity of the spirit level is 2″ per 2-mm division on the tube. The bubble tube and middle portion of the telescope are protected from sudden and uneven changes in temperature by being completely encased by an outer shell. The telescope and adjacent parts are of material having a low coefficient of expansion.

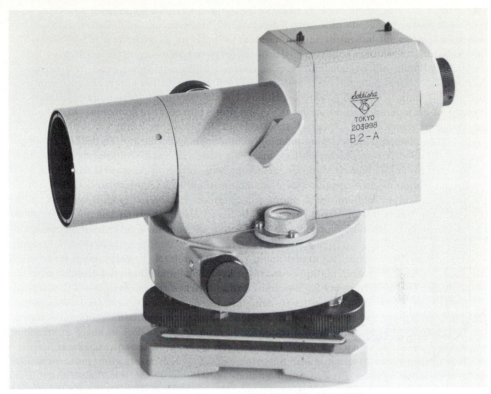

**Fig. 5.17** Lietz/Sokkisha B-2A self-leveling level. (*Courtesy The Lietz Co.*)

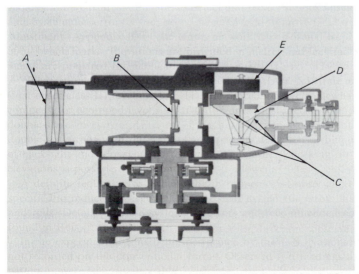

**Fig. 5.18** Sectional view of a Lietz/Sokkisha B-1 self-leveling level. (*Courtesy The Lietz Co.*)

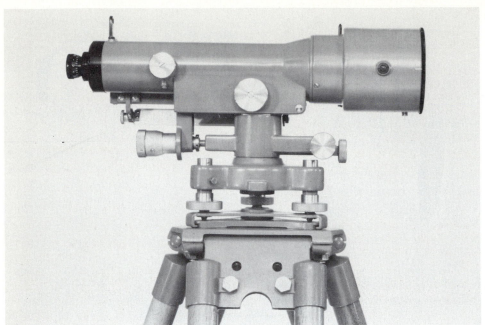

**Fig. 5.19** Tilting geodetic level equipped with optical micrometer. (*Courtesy Wild Heerbrugg Instruments, Inc.*)

The telescope is of high magnification (42 diameters) and sharpness of image, and the large objective lens provides a relatively large field of view. Modern geodetic levels for first-order class I and II leveling are of the tilting type, contain coincidence bubbles, and are equipped with an optical micrometer. Figure 5.19 shows a precise tilting level suitable for geodetic leveling.

The U.S. National Ocean Survey has also recently approved use of certain types of automatic-compensator-equipped levels for first-order class III work (Ref. 11). Automatic levels suitable for first-order leveling are illustrated in Fig. 5.20. These instruments are designed according to the principles described in Art. 5.19 for ordinary self-leveling automatic levels. However, they are heavier, more stable, have more powerful optics, and are equipped with optical micrometers. Reference 2 should be consulted for details concerning specifications and operational techniques for use of self-leveling levels for first-order leveling.

**5.21. Locke hand levels**   The Locke hand level is widely used for rough leveling. It consists of a metal sighting tube about 6 in long on which is mounted a level vial (Fig. 5.21*a*). In the tube beneath the vial is a prism which reflects the image of the bubble to the eye end of the level. Just beneath the level vial is a cross wire which is adjustable by means of a pair of screws, the heads of which protrude from the case; one screw is loosened and the other is tightened. The eyepiece consists of a peephole mounted in the end of a slide which fits inside the tube and is held in a given position by friction. Mounted on the right half of the inner end of the slide is a semicircular convex lens which magnifies the image of the bubble and cross wire as reflected by the prism. Both the object and the eye ends of the tube are closed by disks of plain glass so that dust will not collect on the prism and lens. The magnifying lens is focused by moving the eyepiece slide in or out.

In using the level the object is viewed directly through the left half of the sighting tube, without magnification, while with the same eye and at the same time the position of the

(a)                                                      (b)

**Fig. 5.20** Automatic level suitable for geodetic leveling. (*a*) NI 002 automatic level. (*b*) Zeiss NI1.
(*Courtesy Carl Zeiss Oberkochen.*)

bubble with respect to the cross wire is observed in the right half of the field of view. The
level is held with the level vial uppermost and is tipped up or down until the cross wire
bisects the bubble, when the line of sight is horizontal. After a little practice one may make
observations with greater facility by keeping both eyes open. Some observers steady the
hand level by holding it against, or fastening it to, a staff. Hand levels equipped with stadia
hairs are available.

**Fig. 5.21***a*  Locke hand level. (*Courtesy
Keuffel & Esser Co.*)

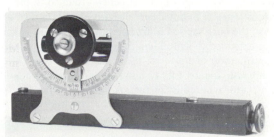

**Fig. 5.21***b*  Abney hand level and
clinometer. (*Courtesy The A. Lietz Co.*)

**5.22. Abney hand level and clinometer** As its name indicates, this level is suitable both for direct leveling and for measuring the angles of slopes. The instrument shown in Fig. 5.21*b* is graduated both in degrees and in percentage of slope, or grade. When it is used as a level, the index of the vernier is set at zero, and it is then used in the same way as the Locke hand level. When it is used as a clinometer, the object is sighted, and the level tube is caused to rotate about the axis of the vertical arc until the cross wire bisects the bubble as viewed through the eyepiece. Either the slope angle or the slope percentage is then read on the vertical arc.

**5.23. Leveling rods** These are graduated wooden rods of rectangular cross section by means of which difference in elevation is measured. The lower or ground end of the rod is shod with metal to protect it from wear and is usually the point of zero measurement from which the graduations are numbered. Aluminum alloy and fiberglass rods are also available in certain specific styles.

The rod is held vertically, and hence the reading of the rod as indicated by the horizontal cross hair of the level is a measure of the vertical distance between the point on which the rod is held and the line of sight.

Rods are obtainable in a variety of types, patterns, and graduations and are either in single pieces or in sections which are jointed together or slide past each other and are clamped together. Common lengths are 12 and 13 ft. In the United States, rods are graduated in hundredths of a foot and in metric units. On some government surveys the rods are graduated in decimals of the metre or the yard.

The two general classes of leveling rods are (1) *self-reading* rods, which may be read directly by the leveler while looking through the telescope of the level, and (2) *target* rods, for which a target sliding on the rod is set by the rodman as directed by the leveler. Under ordinary conditions, observations with the self-reading rod can be made with nearly the same precision and much more rapidly. The self-reading rod is the one commonly employed, even for precise leveling.

**5.24. Self-reading rods** The self-reading rod is held vertically; the leveler observes the graduation at which the line of sight intersects the rod and records the reading. Observations closer than the smallest division on the rod are made by estimation.

The self-reading rod should be so marked that the graduations appear sharp and distinct for any normal distance between level and rod. Normally, the background is white or yellow with graduations 0.01 ft or 0.010 m wide painted in black as shown in Fig. 5.22. The readings to 0.01 ft and to 0.010 m are made on the edges of the graduations. Usually, the rod is read to half-hundredths of a foot or to millimetres, by estimation. The numbers indicating feet or metres are in red, and those indicating tenths of feet or decimetres are in black. This style of graduation is satisfactory for self-reading when the length of sight is less than 400 to 500 ft or 120 to 150 m.

The *Philadelphia* rod is the most widely used rod. It is made in two sliding sections held in contact by two brass sleeves. A screw attached to the upper sleeve permits clamping the two sections together in any desired relative position. For readings of 7 ft (2 m on the metric rod) or less, the back section is clamped in its normal collapsed position. For greater readings, the rod is extended to its full length so that graduations on the front face of the back section are a continuation of those on the lower front strip. When thus extended, the rod is called a "high" rod. Figure 5.22*a* shows a partially extended Philadelphia rod graduated in feet, tenths, and hundredths of feet. A portion of the face of a European-style rod graduated in metric units is shown in Fig. 5.22*b*. This rod comes in 3- and 4-m lengths, is hinged for folding to simplify carrying, and contains a built in circular level bubble used to keep the rod vertical.

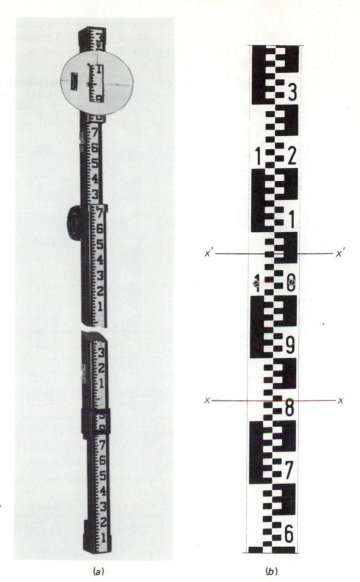

**Fig. 5.22** (a) Philadelphia rod graduated in ft, tenths, and hundredths of ft, partially extended. (*Courtesy Keuffel & Esser Co.*) (b) Portion of a rod graduated in metric units. (*Courtesy Wild Heerbrugg Instruments, Inc.*)

(a)                                    (b)

Direct observation of a self-reading rod involves estimation to obtain the third decimal place in metres and feet. In Fig. 5.22b, graduations are in metres numbered on the left, decimetres numbered on the right, and centimetres designated by alternating black and white squares. These squares are arranged in a checkerboard pattern so that the cross hair will always appear on a white square. Again referring to Fig. 5.22b, readings at lines xx and x'x' to the nearest centimetre yields 0.83 m and 1.07 m, respectively. Estimation to the nearest millimetre at these same lines gives readings of 0.831 m and 1.066 m, respectively.

Figure 5.23a and b shows the faces of Philadelphia-style rods graduated in feet and metres, respectively. In Fig. 5.23a the largest crosshatched (red) numerals indicate the whole numbers of feet, the black numerals tenths of feet, and the alternating black and white graduations represent hundredths of feet. Thus, the readings at lines aa', bb', and cc' to the

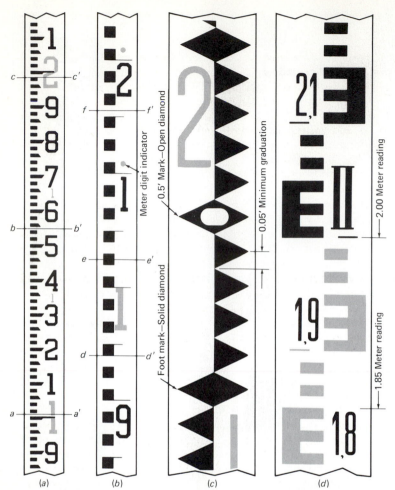

**Fig. 5.23** Direct reading of: (*a*) rod graduated in feet, tenths, and hundredths of feet; (*b*) rod graduated in metres, decimetres, and centimetres; (*c*) stadia rod graduated in feet, tenths, and 0.05 ft; (*d*) stadia rod graduated in metres, decimetres, and centimetres. (*Courtesy Lietz, Inc.*)

nearest hundredth are 1.01, 1.54, and 1.98 ft, respectively. Estimation to the nearest thousandth of a foot yields respective readings of 1.006, 1.545, and 1.980. In Fig. 5.23*b* the light gray numbers are metres, black numerals are decimetres, and alternating black and white graduations are centimetres. Readings to the nearest centimetre at lines *dd'*, *ee'*, and *ff'* are 0.94, 1.02, and 1.15 m, respectively. If millimetres are desired, they must be estimated yielding respective readings of 0.940, 1.024, and 1.155 m. Note that in Fig. 5.23*a* the whole number of feet is indicated by the small crosshatched (red) number between tenth foot numerals and the whole number of metres in Fig. 5.23*b* is given by the number of dots above each decimetre mark on the metric rod. Consequently, even though the field of view of the telescope may fall within full unit graduations, determination of the full unit is possible.

When using a Philadelphia rod, high rod readings are made with the rod fully extended. On these rods there is a metal stop against which the lower brass sleeve makes contact when the rod is fully extended. Occasionally, this metal stop becomes loose and out of adjustment, so that the rod may not be extended to the proper length, introducing errors in high rod readings. To guard against the occurrence of this type of error, examine the vernier on the back of the rod (Art. 5.26).

**5.25. Stadia rods**    This type of rod is used for stadia surveying as described in Chap. 7. Any type of self-reading rod may be used as a stadia rod, but the leveling rod graduated in hundredths of feet or in centimetres as in Fig. 5.23a and b is suitable only for sights of less than 400 ft or 120 m. For longer sights a rod with larger, heavier graduations arranged in a different pattern is necessary. Figure 5.23c and d illustrates two such patterns graduated in feet, tenths, and 0.05 ft; and metres, decimetres, and centimetres, respectively. These graduations are suitable for distances of up to about 700 ft or 210 m. Readings can be estimated to the nearest 0.01 ft and nearest millimetre on these two rods. These rods are equally suitable for use in stadia surveying and leveling.

**5.26. Verniers**    A vernier, or vernier scale, is a short auxiliary scale placed alongside the graduated scale of an instrument, by means of which fractional parts of the least division of the main scale can be measured precisely; the length of one space on the vernier scale differs from that on the main scale by the amount of one fractional part. The precision of the vernier depends on the fact that the eye can determine more closely when two lines coincide than it can estimate the distance between two parallel lines. The scale may be either straight (as on a leveling rod) or curved (as on the horizontal and vertical circles of a transit). The zero of the vernier scale is the index for the main scale.

Verniers are of two types: (1) the *direct* vernier, which has spaces slightly shorter than those of the main scale; and (2) the *retrograde* vernier, which has spaces slightly longer than those of the main scale. The use of the two types is identical, and they are equally sensitive and equally easy to read. Since they extend in opposite directions, however, one or the other may be preferred because it permits a more advantageous location of the vernier on the instrument. Both types are in common use.

*Direct vernier*    Figure 5.24a represents a scale graduated in hundredths of feet, and a direct vernier having each space 0.001 ft shorter than a 0.01-ft space on the main scale; thus, each vernier space is equal to 0.009 ft, and 10 spaces on the vernier are equal to 9 spaces on the scale. The index, or zero, of the vernier is set at 0.400 ft on the scale. If the vernier were moved upward 0.001 ft, its graduation numbered 1 would coincide with a graduation (0.41

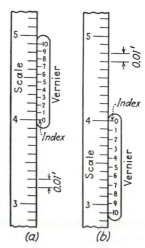

**Fig. 5.24**    (a) Direct vernier. (b) Retrograde vernier.        (a)                (b)

ft) on the scale, and the index would be at 0.401 ft; and so on. It is thus seen that the position of the index is determined to thousandths of feet without estimation, simply by noting which graduation on the vernier coincides with one on the scale. Note that the coinciding graduation on the main scale does *not* indicate the main-scale reading.

The *least count*, or fineness of reading of the vernier, is equal to the difference between a scale space and a vernier space. For a direct vernier, if $s$ is the length of a space on the scale and if $n$ is the number of vernier spaces of total length equal to that of $(n-1)$ spaces on the scale, then the least count is $s/n$.

**Retrograde vernier**   On the retrograde vernier shown in Fig. 5.24*b*, each space on the vernier is 0.001 ft *longer* than a 0.01-ft space on the main scale, and 10 spaces on the vernier are equal to 11 spaces on the scale. As before, the index is set at 0.400 ft on the scale. If the vernier were moved upward 0.001 ft, its graduation numbered 1 would coincide with a graduation (0.39 ft) on the scale; and so on. It is seen that, from the index, the retrograde vernier extends backward along the main scale, and that the vernier graduations are also numbered in reverse order; however, the retrograde vernier is read in the same manner as the direct vernier.

The least count of the retrograde vernier is equal to the difference between a scale space and a vernier space, as in the case of the direct vernier. For the retrograde vernier, if $s$ is the length of a space on the scale and if $n$ is the number of vernier spaces of total length equal to that of $(n+1)$ spaces on the scale, then the least count is $s/n$.

**Reading the vernier**   Figure 5.25 illustrates settings of direct verniers on the target (at the left) and on the back (at right) of a Philadelphia leveling rod (Art. 5.24). The rod reading indicated by the target [4.347 ft (Fig. 5.25*a*)] is determined by first observing the position of the vernier index on the scale to hundredths of feet (4.34 in the figure), next by observing the number of spaces *on the vernier* from the index to the coinciding graduations (7 spaces in the figure), and finally by adding the vernier reading (0.007 ft in the figure) to the scale reading (4.34 ft). On the back of the rod (Fig. 5.25*b*) both the main scale and the direct vernier read

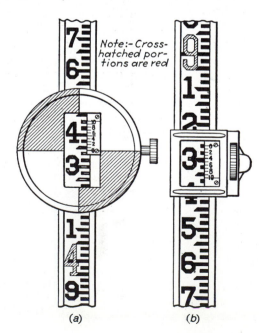

**Fig. 5.25** Direct vernier settings.         (*a*)              (*b*)

*down* the rod. The scale reading is 9.26 ft and the vernier reading is 0.004 ft, hence the rod reading is 9.264 ft.

A helpful check in reading the vernier is to note that the lines on either side of the coinciding line should depart from coincidence by the same amount, in opposite directions. As a check against possible mistakes, it is advisable to estimate the fractional part of the main-scale division by reading the index directly.

**5.27. Targets**  The usual target (Figs. 5.22*a* and 5.25*a*) is a circular or elliptical disk about 5 in in diameter, with horizontal and vertical lines formed by the junction of alternate quadrants of white and red. A rectangular opening in the front of the target exposes a portion of the rod to view so that readings can be taken. The attached vernier (Art. 5.26) fits closely to the rod, its zero point or index being at the horizontal line of the target. In Fig. 5.25*a* a direct vernier is shown, but both retrograde and direct verniers are in common use.

**5.28. Target rods**  With the target rod, the leveler signals the rodman to slide the target up or down until it is bisected by the line of sight. The target is then clamped, and the rodman, leveler, or both observe the indicated reading. Usually, the target is equipped with a vernier or other device by means of which fractional measurements of the rod graduations can be read without estimation.

*The Philadelphia rod* (Fig. 5.22*a*), previously described, is designed as a self-reading rod but may also be used as a target rod. Lugs on the target engage in a groove on either side of the front strip. For readings on the lower half of the rod, the reading is made to thousandths of feet by means of a vernier attached to the target. For example, the reading in Fig. 5.25*a* is 4.347 ft using the target.

Graduations on the back of the rear strip are a continuation of those on the front strip and read downward. On the back of the top sleeve is a vernier employed for observations with the rod extended. For readings greater than can be taken with a "short" rod, the target in Fig. 5.22*a* is clamped on the 7.000-ft graduation (the 2.000-m graduation on the metric rod shown in Fig. 5.23*b*) on the front face of the upper section. The vertically held rod is then extended until the target is bisected by the line of sight. The vertical distance from the foot of the rod is then indicated by the reading of the vernier on the back of the rod (Fig. 5.25*b*). Note that the back of the rod is a continuation of graduations on the front, with numbers increasing from top to bottom. For the example illustrated in Fig. 5.25*b* with the target set at 7.00 ft on the front of the rod, the vertical distance from the bottom of the rod to the target is 9.264 ft.

The principal advantage of the target rod is that mistakes are less likely to occur, particularly if both rodman and leveler read the rod. Under certain conditions its use materially facilitates the work: for example, where very long sights are taken, where the rod is partly obscured from view, or where it is necessary to establish a number of points all at the same elevation. However, under ordinary conditions its use retards progress without adding much, if anything, to the precision.

**5.29. Rod levels**  The rod level is an attachment for indicating the verticality of the leveling rod. One type (Fig. 5.26) consists of a circular or "bull's-eye" level vial mounted on a metal angle or bracket which is either attached by screws to the side of the rod or is held against the rod, as desired. As noted for the level rod shown in Fig. 5.22*b*, the rod level may be built into the rod.

Another type consists of a hinged casting on each wing of which is mounted a level tube that is held parallel to a face of the rod. When both of the bubbles are centered, the rod is plumb. The hinge makes it possible to fold the level compactly when it is not in use.

**Fig. 5.26** Rod level.

**5.30. Precise level rods** Precise or geodetic levels (Art. 5.46) require the use of a specially constructed self-reading rod which is a little over 3 m long and has a hardened steel foot 25 mm in diameter with a slightly convex bottom face. The rod is made of a single piece of well-seasoned hard wood or of extruded aluminum alloy. The front of the rod is graduated in metres or feet. Along a groove in the face of the rod runs a strip of invar metal graduated in centimetres or in feet with a least interval of 0.02 of a foot. The invar strip is fastened rigidly only at the bottom of the rod and is kept taut by means of a spring at the top; thus, it is free to expand or contract independently of the main rod. On older styles, the back of the rod may be graduated in feet and tenths of feet to provide a check reading. The rod may be equipped with a built-in rod level and thermometer. Precise level rods are usually manufactured in pairs, since the procedure for precise levels requires the use of two rods (Art. 5.46).

The most recent precise level rods have a double set of graduations with one scale offset from the other by a known amount. This arrangement permits two independent readings with a single observation using one cross hair. Figure 5.27a and b shows bottom and top views of a precise level rod having double scales with 1-cm intervals. On this rod, the constant difference between scales is 301.55 cm. The rod is shown supported on a special turning point (Fig. 5.27a) at the bottom and held rigidly in a vertical position by a pair of struts attached at the top of the rod and resting on the ground in back of the rod. Note that the graduations on this rod are upside down and reversed so as to provide an upright, right-reading image with an inverting telescope. Rods of this style having erect numerals for use with levels equipped with erecting telescopes are also available.

Another style of precise level rod, constructed of a light metal extruded T section and equipped with a graduated invar strip and built in circular level bubble, is illustrated in Fig. 5.28a. This rod has a double scale of the line graduation type. The least division is 0.005 m, with every tenth graduation numbered. There are two vertical rows of graduations. The main scale on the right (on the observer's left using an inverting telescope) has zero at the foot of the rod while the check scale on the left (observer's right) is offset exactly 0.025 m below the

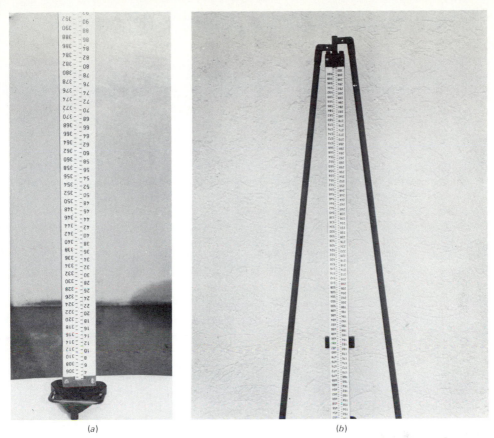

(a)                                                                                                    (b)

**Fig. 5.27** Precise level rod with double scales, special turning point, and supporting struts. (*Courtesy Wild Heerbrugg Instruments, Inc.*)

main scale and is numbered with a higher series of numbers. The first number on the check scale is 60 units placed so that the corresponding reading on the main scale is 59.250 units less. For example, in Fig. 5.28*b* the scale is shown as visible to an observer viewing through an inverting telescope. The reading on the main scale at cross hair *aa'* is 11.350 and on the check scale is 70.600, so that the difference is $70.600 - 11.350 = 59.250$. Thus, different readings are made on two different scales, providing a positive numerical check against blunders and two independent observations. This rod can be used with the special turning point shown in Fig. 5.28*a* or with any other type of turning point suitable for precise leveling (Art. 5.46).

**5.31. Turning points**   A metal plate or pin that will serve temporarily as a stable object on which the leveling rod may be held at turning points is a useful part of the leveling equipment for lines of differential levels. The iron pin shown in Fig. 5.29*a* is adapted for use in firm ground. Often a railroad spike or rivet pin is used.

In soft ground the steel plate of Fig. 5.29*b* makes a satisfactory turning point for ordinary leveling. The plate is also adapted for use where the ground is so solid as to make driving the pin impossible or at least impractical, as along highways. Under these conditions the plate with the dogs at its corners acts as a tripod, no special attempt being made to secure bearing between the lower surface of the plate and the ground. The turning points illustrated in Figs.

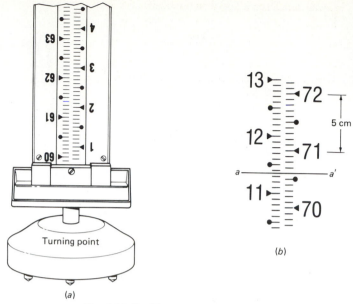

**Fig. 5.28** Double-scale precise level rod.

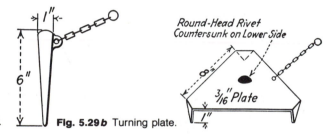

**Fig. 5.29a** Turning point.  **Fig. 5.29b** Turning plate.

5.27a and 5.28a are made of heavy cast iron, have three pointed spears on the base, and are suitable for use on pavement or soft ground.

**5.32. Optical micrometer**  Estimation to the nearest millimetre or to thousandths of a foot is inadequate in leveling operations where extremely high accuracy is required. In such cases, the engineer's precise level or tilting level is equipped with an optical micrometer mounted in front of the objective lens of the telescope. Figure 5.30a shows a cross-sectional view of a telescope equipped with an optical micrometer.

The basic element in an optical micrometer is a plane-parallel glass plate or optical flat which can be rotated about an axis perpendicular to the optical axis and in a horizontal plane. The principle of the optical flat is illustrated by Fig. 5.31. The light ray, obliquely incident to side $AB$ at an angle $\alpha$ with the normal, is refracted toward the normal (of the more dense medium) as it passes through the glass. This refracted ray is bent in the opposite direction by the same angle $\alpha$ on emerging in air (the less dense medium) at side $CD$, which is parallel to $AB$. Thus, the emergent ray is parallel to the incident ray and it can be shown that the displacement $d$ is

$$d = t \tan \alpha \left( \frac{n-1}{n} \right) \qquad \text{(approx.)} \qquad (5.18)$$

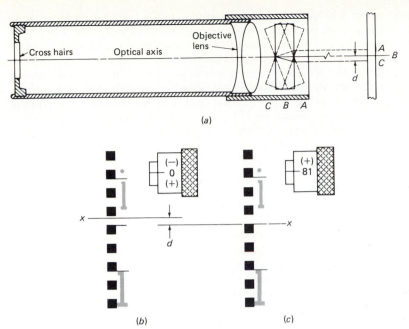

(a)

(b)                              (c)

**Fig. 5.30** Optical micrometer and micrometer settings.

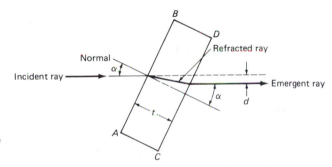

**Fig. 5.31** Principle of the
optical micrometer.

where $d$ = displacement between incident and emergent rays
  $t$ = thickness of the optical flat
  $\alpha$ = angle of incidence with the normal to the optical flat
  $n$ = refractive index of the glass in the optical flat

Rotation of the plane parallel plate in the optical micrometer produces an apparent vertical shift $d$ of the field of view (Fig. 5.30a), so that the image of the cross hair $x$ in Fig. 5.30b can be brought into coincidence with a graduation on the scale being observed (Fig. 5.30c).

From Eq. (5.18), the displacement $d$ varies as the tangent of the rotation angle $\alpha$. The knob that rotates the optical flat is linked to the plate, so that uniform rotation of the knob generates a tangent function of the tilting plate. A drum scale attached to the knob permits observing the amount of rotation in units of the vertical scale being observed.

The range on an optical micrometer must be sufficient to include one full least division plus one-tenth of this division on the scale to be observed. Figure 5.30 illustrates a cross section of the micrometer attachment and example readings on a metric scale with a least division of 10 mm. In this case the range of the micrometer should be 11 mm with a zero reading on the drum occurring with the optical flat in position $B$ or normal to the line of sight (Fig. 5.30$a$). Maximum upward (negative) displacement is produced by a counterclockwise rotation of the knob and occurs when the flat is in position $A$. Maximum downward (positive) displacement results from a clockwise rotation and occurs with the flat in position $C$. The estimated reading at cross hair $x$ in Fig. 5.30$b$ is 1.058 m. A clockwise rotation of the knob displaces the image of the graduations to yield coincidence between the cross hair $x$ and 1.050, where the increment is $+81 = +0.0081$, giving a reading of 1.0581 m. As a check, a counterclockwise rotation of the knob permits displacement of the image and coincidence of cross hair $x$ with 1.060 (not illustrated), yielding a micrometer reading of $-15 = -0.0015$. This gives a reading of $1.060 - 0.0015 = 1.0585$, so the final observation is the average of the two readings, or 1.0583 m. Extreme care must be exercised prior to each observation to ensure proper focusing and coincidence of the level bubble.

**5.33. Setting up engineer's levels**   The engineer's level is placed in a desired location, with the tripod legs well spread and firmly pressed into the ground, with the tripod head nearly level, and with the telescope at a convenient height for sighting. If the setup is on a slope, it is preferable to orient the tripod so that one of its legs extends up the slope.

When operating three screw tilting or automatic levels which have bull's-eye bubbles as illustrated in Fig. 5.32$a$, opposite rotation of level screws $A$ and $B$ moves the bubble along axis $xx'$, as shown in Fig. 5.32$b$, while rotation of screw $C$ moves the bubble in the direction of axis $yy'$ (Fig. 5.32$c$). When the bubble is centered within the circle, the instrument is sufficiently leveled to be within the range of the spirit level in a tilting instrument or the pendulum device in the automatic level.

If the leveling head has four screws, the telescope is brought over one pair of opposite leveling screws, and the bubble is centered approximately by turning the level screws in *opposite directions*; then the process is repeated with the telescope over the other pair. By repetition of this procedure the leveling screws are manipulated until the bubble remains centered, or nearly so, for any direction in which the telescope is pointed. If the instrument is in adjustment, the line of sight is then horizontal.

**5.34. Reading the rod**   For observations to millimetres and to hundredths or thousandths of feet, the rod is held on some well-defined point of a stable object. The rodman holds the rod vertical either by observing the rod level or by estimation. First, the cross hairs are carefully focused to suit the eye of the observer. This can be done by holding an open field

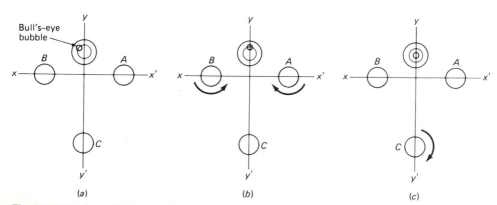

**Fig. 5.32** Leveling a three-screw instrument.

**Fig. 5.33** Waving the rod.

book in front of the objective lens of the telescope and rotating the eyepiece adjustment knob until the cross hairs are clear, sharp, and black. Next, the level operator revolves the telescope about the vertical axis until the rod is about in the middle of the field of view, focuses the objective for distinct vision, checks for presence of parallax (Arts. 5.10 and 5.42), and carefully centers the bubble. If the self-reading rod is used, the leveler observes and records the reading indicated by the line of sight, that is, the apparent position of the horizontal cross hair on the rod. Checking, the leveler again observes the bubble and the rod. If the target rod is used, the procedure is identical except that the target is set by the rodman as directed by the leveler.

For leveling of lower precision, as when rod readings for points on the ground are determined to the nearest 0.1 ft or nearest cm, the observations usually are not checked, and proportionally less care is exercised in keeping the rod vertical and the bubble centered, always bearing in mind the errors involved and the precision with which measurements are desired.

If no rod level is used, in calm air the rodman can plumb the rod accurately by balancing it upon the point on which it is held. By means of the vertical cross hair the leveler can determine when the rod is held in a vertical plane passing through the instrument, but cannot tell whether it is tipped forward or backward in this plane. If it is in either of these positions, the rod reading will be greater than the true vertical distance, as illustrated by Fig. 5.33. To eliminate this error, the rodman *waves the rod*, or tilts it forward and backward as indicated by the figure, and the leveler takes the least reading, which occurs when the rod is vertical. The larger the rod reading, the larger the error due to the rod's being held at a given inclination; hence, it is more important to wave the rod for large readings than for small readings. Further, whenever the rod is tipped backward about any support other than the front edge of its base, the graduated face rises and an error is introduced; for small readings this error is likely to be greater than that caused by not waving the rod.

**5.35. Differential leveling** Differential leveling is the operation of determining the elevations of points some distance apart. Usually, this is accomplished by direct leveling. Differential leveling requires a series of setups of the instrument along the general route and, for each setup, a rod reading back to a point of known elevation and forward to a point of unknown elevation.

**5.36. Bench marks** A *bench mark* (B.M.) is a definite point on an object, the elevation and location of which are known. It may be permanent (P.B.M.) or temporary (T.B.M.). Bench marks serve as points of reference for levels in a given locality. Their elevations are established by differential leveling, except that the elevation of the initial bench mark of a local project may be assumed.

Throughout the United States permanent bench marks are established by the U.S. Geological Survey and the U.S. National Ocean Survey. Similarly, bench marks have been established by various other federal, state, and municipal agencies and by such private interests as railroads and water companies, so that the surveyor has not far to go before finding some point of known elevation.

**Fig. 5.34** United States National Ocean Survey (formerly U.S. Coast & Geodetic Survey) bench mark.

The National Ocean Survey bench marks (Fig. 5.34) consist of bronze plates set in stone or concrete and marked with the elevation above mean sea level. Those of the other agencies are similar. Other objects frequently used as bench marks are stones, pegs or pipes driven in the ground, nails or spikes driven horizontally in trees or vertically in pavements, and marks painted or chiseled on street curbs.

For any survey or construction enterprise, levels are run from some initial bench mark of known or assumed elevation to scattered points in desirable locations for future reference as bench marks.

In some areas, the elevation of bench marks may be altered by earth movements such as those caused by earthquakes, slides, lowering of water tables, pumping from oil fields, mining, or construction.

**5.37. Definitions**   A *turning point* (T.P.) is an intervening point between two bench marks upon which point foresight and backsight rod readings are taken. It may be a pin or plate (see Art. 5.31) which is carried forward by the rodman after observations have been made, or it may be any stable object such as a street curb, railroad rail, or stone. The nature of the turning point is usually indicated in the notes, but no record is made of its location unless it is to be reused. A bench mark may be used as a turning point.

A *backsight* (B.S.) is a rod reading taken on a point of known elevation, as a bench mark or a turning point. Usually, it will be taken with the level sighting back along the line, hence the name. A backsight is sometimes called a *plus sight*. The horizontal distance from level to rod on a B.S. is called *backsight distance*.

A *foresight* (F.S.) is a rod reading taken on a point the elevation of which is to be determined, as on a turning point or on a bench mark that is to be established. A foresight is sometimes called a *minus sight*. The horizontal distance from the level to the rod on a F.S. is called *foresight distance*.

The *height of instrument* (H.I.) is the elevation of the line of sight of the telescope above the datum when the instrument is leveled.

In surveying with the transit, the terms backsight, foresight, and height of instrument have meanings different from those defined here.

**5.38. Procedure**   In Fig. 5.35, B.M.$_1$ represents a point of known elevation (bench mark), and B.M.$_2$ represents a bench mark to be established some distance away. It is desired to

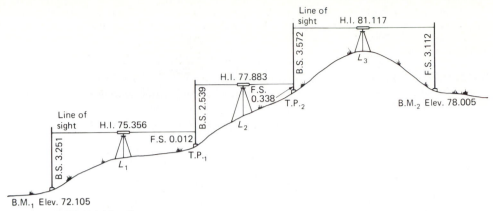

**Fig. 5.35** Differential leveling.

determine the elevation of B.M.$_2$. The rod is held at B.M.$_1$, and the level is set up in some convenient location, as $L_1$, along the general route B.M.$_1$ to B.M.$_2$. The level is placed in such a location that a clear rod reading is obtainable, but no attempt is made to keep on the direct line joining B.M.$_1$ and B.M.$_2$. A backsight is taken on B.M.$_1$. The rodman then goes forward and, as directed by the leveler, chooses a turning point T.P.$_1$ at some convenient spot within the range of the telescope along the general route B.M.$_1$ to B.M.$_2$. It is desirable, but not necessary, that each foresight distance, such as $L_1$-T.P.$_1$, be approximately equal to its corresponding backsight distance, such as B.M.$_1$-$L_1$. The chief requirement is that the turning point shall be a stable object at an elevation and in a location favorable to a rod reading of the required precision. The rod is held on the turning point, and a foresight is taken. The leveler then sets up the instrument at some favorable point, as $L_2$, and takes a backsight to the rod held on the turning point; the rodman goes forward to establish a second turning point T.P.$_2$; and so the process is repeated until finally a foresight is taken on the terminal point B.M.$_2$.

It is seen in Fig. 5.35 that a *backsight added to the elevation* of a point on which the backsight is taken gives the height of instrument, and that a *foresight subtracted from the height of instrument* determines the elevation of the point on which the foresight is taken. Thus if the elevation of B.M.$_1$ is 72.105 m and the B.S. is 3.251 m, the H.I. with the instrument set at $L_1$ is 72.105+3.251=75.356 m. If the following F.S. is 0.012 m, the elevation of T.P.$_1$, is 75.356−0.012=75.344 m. Also, the difference between the backsight taken on a given point and the foresight taken on the following point is equal to the difference in elevation between the two points. It follows that the difference between the sum of all backsights and the sum of all foresights gives the difference in elevation between the bench marks.

Since the level is normally at a higher elevation than that of the points on which rod readings are taken, the backsights are often called "plus" sights and the foresights "minus" sights and are so recorded in the field notes. Sometimes, however, in leveling for a tunnel or a building it is necessary to take rod readings on points which are at a higher elevation than that of the H.I. In such cases the rod is held inverted, and in the field notes each such backsight is indicated with a minus sign and each foresight with a plus sign.

When several bench marks are to be established along a given route, each intermediate bench mark is made a turning point in the line of levels. Elevations of bench marks are checked sometimes by rerunning levels over the same route but more often by "tying on" to a previously established bench mark near the end of the line or by returning to the initial bench mark. A line of levels that ends at the point of beginning is called a *level circuit*. The final observation in a level circuit is therefore a foresight on the initial bench mark. If each bench mark in a level circuit is also a turning point and the circuit checks within the prescribed limits of error, it is regarded as conclusive evidence that the elevations of all turning points in the circuit, including all bench marks used as turning points, are correct within prescribed limits.

**5.39. Balancing backsight and foresight distances**   In Art. 5.4 it has been shown that, if a foresight distance were equal to the corresponding backsight distance, any error in readings due to earth's curvature and to atmospheric refraction (under uniform conditions) would be eliminated. In ordinary leveling no special attempt is made to balance *each* foresight distance against the preceding backsight distance. Whether or not such distances are approximately balanced between bench marks will depend upon the desired precision. The effect of the earth's curvature and atmospheric refraction is slight unless there is an abnormal difference between backsight and foresight distances. The effect of instrumental errors is likely to be of considerably greater consequence with regard to the balancing of these distances. The chances are that there is not absolute parallelism between the line of sight and the axis of the level tube, so that if the instrument were perfectly leveled, the line of sight would be inclined always slightly upward or always slightly downward. The error in a rod reading due to this imperfection of adjustment is proportional to the distance from the instrument to the rod and is of the same sign for a backsight as for a foresight. Since backsights are added and foresights are subtracted, this instrumental error is eliminated if, between bench marks, the *sum* of the foresight distances is made equal to the *sum* of the backsight distances.

In ordinary leveling no special attempt is made to equalize these distances *if there is assurance that the instrument is in good adjustment*. Normally, for levels run over flat or gently rolling ground, the line of sight will fall within the length of the rod regardless of the position of the instrument, and the distance between instrument and rod is governed by the optical qualities of the telescope. While moving forward, the leveler generally paces or estimates by eye a distance about the proper maximum length of sight; the rodman similarly estimates the proper distance from the instrument.

For leveling of moderately high precision it is necessary to equalize backsight and foresight distances between bench marks; hence these distances are recorded. In less refined leveling, distances are usually determined by pacing; in precise leveling, they are usually measured by stadia or the gradienter.

In leveling uphill or downhill the length of sight is usually governed by the slope of the ground. In order that maximum distances between turning points may be obtained and hence progress be most rapid, the leveler sets up the instrument in a position such that the line of sight will intersect the rod near its top if the route is uphill, or near its bottom if the route is downhill; the leveler directs the rodman to a similarly favorable location for the turning point. In leveling uphill or downhill, a balance between foresight and backsight distances can be obtained with a minimum number of setups by following a zigzag course.

When bench marks are at roughly the same elevation, the backsight and foresight distances will tend to balance in the long run, regardless of the character of the terrain. However, for levels between two points having a large difference in elevation, a very small inclination of the line of sight will produce a marked error unless some attempt is made to equalize backsight and foresight distances.

**5.40. Differential-level notes**   For ordinary differential leveling when no special effort is made to equalize backsight and foresight distances between bench marks, usually the record of field work is kept in the form indicated by Fig. 5.36, in which the levels from B.M.$_1$ to B.M.$_2$ are the same as shown by Fig. 5.35. The left page is divided into columns for numerical data, and the right page is reserved for descriptive notes concerning bench marks

| Sta. | B.S. | H.I. | F.S. | Elev. |
|------|------|------|------|-------|
| LEVELS FOR BENCH MARKS ALONG RIDGE ROAD | | | | |
| B.M.₁ | 3.251 | 75.356 | | 72.105 |
| T.P.₁ | 2.539 | 77.883 | 0.012 | 75.344 |
| T.P.₂ | 3.572 | 81.117 | 0.338 | 77.545 |
| B.M.₂ | 0.933 | 78.938 | 3.112 | 78.005 |
| T.P.₃ | 0.317 | 75.949 | 3.306 | 75.632 |
| T.P.₄ | 0.835 | 74.068 | 2.716 | 73.233 |
| T.P.₅ | 0.247 | 70.773 | 3.542 | 70.526 |
| B.M.₃ | | | 3.786 | 66.987 |
| Σ B.S.= | 11.694 | Σ F.S.= | −16.812 | |
| | | | 11.694 | |
| | B.M.₁ | 72.105 − | 5.118 = | 66.987 |

June 20, 1977        J.G. Sutter
Fair  70°F           W.R. Knowles Rod

Level #42, Rod #12

Top of Hydrant Corner of Oak & Ridge
Elev. in metres above sea level.

Spike in Pole North of Williams House
No. 260 Oak St. Marked B.M. 75.902 m
Stone

Top of Brass Plate in Concrete Monument
4 m North of N. Edge Pvt. Oak St. at
County Line

**Fig. 5.36** Differential-level notes.

and turning points. In the same horizontal line with each turning point or bench mark shown in the first column are all data concerning that point. The heights of instrument and the elevations are computed as the work progresses. Thus, when the backsight (3.251) has been taken on B.M.₁ it is added to the elevation (72.105) to determine the H.I. (75.356). The height of instrument is recorded on the same line with the backsight by means of which it is determined. When the first foresight (0.012) is observed, it is recorded on the line below and is subtracted from the preceding H.I. (75.356) to determine the elevation of T.P.₁ (75.344). And so the notes are continued. Usually, at the foot of each page of level notes the *computations* are checked by comparing the difference between the sum of the backsights and the sum of the foresights with the difference between the initial and the final elevation, as illustrated at the bottom of Fig. 5.36. Agreement between these two differences signifies that the additions and subtractions are correct but does not check against mistakes in observing or recording.

Bench marks should be briefly but definitely described and should be so marked in the field that they can be readily identified. They are usually marked with paint or with crayon that will withstand the effects of the weather. When the bench mark is on stone or concrete the position is often indicated by a cross cut with a chisel. A bench mark may or may not be marked with its elevation. Whenever there might arise any question as to the exact location of the point on which the rod was held, its nature should be clearly indicated in the notes. A description of turning points is of no particular importance unless the points are on objects that can be identified and might therefore become of some value in future leveling operations. Such points are usually marked with crayon and briefly described in the notes.

**5.41. Mistakes in leveling**    Some of the mistakes commonly made in leveling are:

1. Confusion of numbers in reading the rod, as, for example, reading and recording 4.92 when it should be 3.92. The mistake is not likely to occur if the observer notices the numbers on both sides of the observed reading.
2. Recording backsights in foresight column and vice versa.
3. Faulty additions and subtractions; adding foresights and subtracting backsights. As a check, the difference between the sum of the backsights and the sum of the foresights should be computed for each page or between bench marks.
4. Rod not held on same point for both foresight and backsight. This is not likely to occur if the turning points are marked or otherwise clearly defined.
5. Not having the Philadelphia rod fully extended when reading the long rod. Before a reading on a turning point is taken, the clamp should be inspected to see that it has not slipped.
6. Wrong reading of vernier when the target rod is used.
7. When the long target rod is used, not having the vernier on the target set to read exactly the same as the vernier on the back of the rod when the rod is short.

**5.42. Errors in leveling**    In leveling, errors are due to some or all of the following causes (see also Table 5.1):

*1. Imperfect adjustment of the instrument*    Insofar as results are concerned, the only essential relation is that the line of sight should be parallel to the axis of the level tube. Any inclination between these lines causes a systematic error, for then if the bubble were perfectly centered, the line of sight would be inclined always slightly upward or downward. Evidently, the error in a rod reading due to this imperfection would be proportional to the distance from the instrument to the rod, and for a given distance would be of the same magnitude and sign for a backsight as for a foresight. Since backsights are added and foresights are subtracted, it is clear that the error in elevations will be eliminated to the extent that, between bench marks, the sum of the backsight distances is made equal to the sum of the foresight distances. Conversely, a systematic error will result to the extent that these distances are not equalized between any two bench marks. Often, these distances will be sufficiently balanced in the long run, regardless of the terrain, to yield a satisfactory final result; but that fact does not ensure a corresponding accuracy for the bench marks established along the line.

The effect of imperfect adjustment of the instrument is minimized by adjusting the instrument and by balancing backsight and foresight distances. In precise leveling this error is also further reduced by computing a collimation correction as shown in Art. 5.47.

*2. Parallax*    This condition is present when either or both of the following occur: (*a*) the objective lens is not focused on the object; (*b*) the observer's eye is not focused on the plane of the cross hairs (see Fig. 5.7). The effect of parallax is to cause relative movement between the image of the cross hairs and image of the object when the eye is moved up and down. Parallax causes a random error and can be practically eliminated by careful focusing as described in Art. 5.10.

*3. Earth's curvature*    This produces an error only when backsight and foresight distances are not balanced. Under ordinary conditions these distances do not tend to vary greatly, and whatever resultant error arises from this source in ordinary leveling is so small as to be of no consequence. When backsight distances are consistently made greater than foresight distances, or *vice versa*, a systematic error of considerable magnitude is produced, particularly when the sights are long. The effect is the same as that due to the line of sight being inclined. The error varies as the square of the distance from instrument to rod and hence will be eliminated not merely by equalizing the sum total of backsight and foresight distances between bench marks but rather by balancing *each* length of backsight by a corresponding length of foresight.

**4. Atmospheric refraction** This varies as the square of the distance, but under normal conditions is only about one-seventh of that due to the Earth's curvature, and its effect is opposite in sign. It is usually considered together with the earth's curvature, but though the effect of the latter will be entirely eliminated if each backsight distance is made equal to the following foresight distance, the atmospheric refraction often changes rapidly and greatly in a short distance. It is particularly uncertain when the line of sight passes close to the ground. Hence, it is impossible to eliminate entirely the effect of refraction even though the backsight and foresight distances are balanced. In ordinary leveling its effect is negligible. In leveling of greater precision the change in refraction can be minimized by keeping the line of sight well above the ground (at least 2 ft or ─0.7 m) and by taking the backsight and foresight readings in quick succession. In the long run the error is random, but over a short period, as a day, it may be systematic. So-called heat waves are evidence of rapidly fluctuating refraction. Errors from this source can be reduced by shortening the length of sight until the rod appears steady. In precise geodetic leveling, corrections for Earth curvature and refraction are calculated using the equations given in Art. 5.2.

**5. Variations in temperature** The sun's rays falling on top of the telescope, or on one end and not on the other, will produce a warping or twisting of its parts and hence may influence rod readings through temporarily disturbing the adjustments. Although this effect is not of much consequence in leveling of ordinary precision, it may produce an appreciable error in more refined work. The error is usually random, but under certain conditions it may become systematic. It is practically eliminated by shielding the instrument from the rays of the sun.

**6. Rod not standard length** If the error is distributed over the length of the rod, a systematic error is produced which varies directly as the difference in elevation and bears no relation to the length of the line over which levels are run. The error can be eliminated by comparing the rod with a standard length and applying the necessary corrections. The case is analogous to measurement of distance with a tape that is too long or too short. If the rod is too long, the correction is added to a measured difference in elevation; if the rod is too short, the correction is subtracted.

Most manufactured rods are nearly of standard length, but where large differences in elevation are to be determined, few rods are near enough to the standard that corrections can be ignored in precise work.

If the rod is worn uniformly at the bottom, an erroneous height of instrument is shown at each setup, but the error in backsight is balanced by that in the following foresight, and no error results in the elevation of the foresight point.

**7. Expansion or contraction of the rod** Owing to change in moisture content or change in temperature, the leveling rod may expand or contract. The resultant error is systematic. Wood when well-seasoned and painted will shrink or swell but little in the direction of the grain. Also, its coefficient of thermal expansion is small. The error is of no particular consequence in ordinary leveling. For precise leveling, gage points may be established by inserting metal plugs in the rod, and corrections for shrinkage may be determined by observing any change in distance between the gage points. Corrections for thermal expansion may be based upon observed temperatures of the rod, as indicated by an attached thermometer, the temperature being recorded in the notes. An invar tape may be used on the rod.

**8. Rod not held plumb**  This condition produces rod readings which are too large. In running a line of levels uphill or downhill it becomes a systematic error, inasmuch as the backsights are larger than the foresights, or *vice versa*. Over rolling or level ground the resultant error is random since the backsights are, on the average, about equal to the foresights. The error varies directly with the first power of the rod reading and directly as the square of the inclination. Thus, if a 3 m rod is 0.06 m out of plumb, the error amounts to 0.0006 m for a 3 m reading and 0.0002 m for a 1 m reading; but if the rod is 0.12 m out of plumb, the corresponding errors are 0.002 and 0.0008 m, respectively. It is therefore evident that appreciable inclinations of the rod must be avoided, particularly for high rod readings. The error can be eliminated by swinging the rod, or by using a rod level.

**9. Faulty turning points**  A random error results when turning points are not well defined. Even a flat, rough stone, for example, does not make a good turning point for precise leveling for the reason that no definite point exists on which to hold the rod, which is not likely to be held in the same position for both backsight and foresight.

**10. Settlement of tripod or turning points**  If the tripod settles in the interval that elapses between taking a backsight and the following foresight, the foresight will be too small and the observed elevation of the forward turning point will be too large. Similarly, if a turning point settles in the interval between foresight and backsight readings, the height of instrument as computed from the backsight reading will be too great. It is thus seen that by the normal leveling procedure, if either the level tripod or the turning point settles, as may occur to some extent when leveling over soft ground, the error will be systematic and the resulting elevations will always be too high.

   Few occasions arise when turning points cannot be so selected or established as to eliminate the possibility of settlement, but care should be taken not to strike the bottom of the rod against the turning point between sights or to exert any pressure whatsoever on the turning point.

   On the other hand, some settlement of the instrument is nearly certain to occur when leveling over muddy, swampy, or thawing ground or over melting snow. The errors due to such settlement can be greatly reduced by employing two rods and two rodmen, one rodman setting the turning point ahead while the other remains at the turning point in the rear. Backsight and foresight readings can then be made in quick succession. Small errors remaining from this source can be minimized by reversing the order of sights at alternate setups, as described in Art. 5.46.

**11. Bubble not exactly centered at instant of sighting**  This condition produces a random error which tends to vary as the distance from instrument to rod. Hence, the longer the sight, the greater is the care that should be observed in leveling the instrument.

**12. Inability of observer to read the rod exactly or to set the target exactly on the line of sight**  This inability causes a random error of a magnitude depending upon the instrument, weather conditions, length of sight, and observer. It can be confined within reasonable limits through proper choice of length of sight.

**5.43. Error propagation in leveling**  The sources of errors discussed in Art. 5.42 are summarized in Table 5.1. Obviously, a great many factors affect the differential leveling operation. Although the final accuracy achieved is influenced by the instrument used, a great deal depends on the skill of the leveler and the degree of refinement with which the work is executed. Assuming that proper leveling procedures are employed and care is given to detail, systematic errors can be nearly eliminated. Thus, the remaining errors are random

**Table 5.1    Errors in leveling**

| Source | Type | Cause | Remarks | Procedure to eliminate or reduce |
|---|---|---|---|---|
| Instrumental | Systematic | Line of sight not parallel to axis of level tube | Error of each sight proportional to distance[a] | Adjust instrument; balance sum of backsight and foresight distances |
| | | Rod not standard length (throughout length)[b] | May be due to manufacture, moisture, or temperature; error usually small | Standardize rod and apply corrections, same as for tape |
| Personal | Random | Parallax | ... | Focus carefully |
| | | Bubble not centered at instant of sighting | Error varies as length of sight | Check bubble before making each sight |
| | | Rod not held plumb | Readings are too large; error of each sight proportional to square of inclination[a] | Wave the rod, or use rod level |
| | | Faulty reading of rod or setting of target | ... | Check each reading before recording; for self-reading rod, use fairly short sights |
| | | Faulty turning points | ... | Choose definite and stable points |
| Natural | Random | Temperature | May disturb adjustment of level | Shield level from sun |
| | Systematic | Earth's curvature | Error of each sight proportional to square of distance[a] | Balance *each* backsight and foresight distance; or apply computed correction |
| | Random | Variations in atmospheric refraction | Error of each sight proportional to square of distance[a] | Same as for Earth's curvature; also take short sights, well above ground, and take backsight and foresight readings in quick succession |
| | Systematic | Settlement of tripod or turning points | Observed elevations are too high | Choose stable locations; take backsight and foresight readings in quick succession preferably alternating order of sights |

[a] The error of *each sight* is systematic, but the resultant error is the difference between the systematic error for foresights and that for backsights; hence, the resultant error tends to be random.
[b] Uniform wear of the bottom of the rod causes no error.

**Table 5.2  Classification, Standards of Accuracy, and General Specifications for Vertical Control**

| Classification | First-order: Class I, class II | Second-order Class I | Second-order Class II | Third-order |
|---|---|---|---|---|
| Principal uses | Basic framework of the National Network and of metropolitan area control. Extensive engineering projects. Regional crustal movement investigations. Determining geopotential values | Secondary control of the National Network and of metropolitan area control. Large engineering projects. Local crustal movement and subsidence investigations. Support for lower-order control | Control densification, usually adjusted to the National Net. Local engineering projects. Topographic mapping. Studies of rapid subsidence. Support for local surveys | Miscellaneous local control may not be adjusted to the National Network. Small engineering projects. Small-scale topographic mapping. Drainage studies and gradient establishment in mountainous areas |
| Minimum standards; higher accuracies may be used for special purposes | | | | |
| *Recommended spacing of lines* National Network | Net A; 100–300 km *class I* Net B; 50–100 km *class II* | Secondary net; 20–50 km | Area control; 10–25 km | As needed |
| Metropolitan control; | 2–8 km | 0.5–1 km | As needed | As needed |
| other purposes | As needed | As needed | As needed | As needed |
| *Spacing of marks along lines* | 1–3 km | 1–3 km | Not more than 3 km | Not more than 3 km |
| *Gravity requirement* | $0.20 \times 10^{-3}$ gpu | — | — | — |
| *Instrument standards* | Automatic or tilting levels with parallel-plate micrometers; invar scale rods | Automatic or tilting levels with optical micrometers or three-wire levels; invar scale rods | Geodetic levels and invar scale rods | Geodetic levels and rods |

| *Field procedures* | Double-run; forward and backward, each section | Double-run; forward and backward, each section | Double- or single-run | Double- or single-run |
|---|---|---|---|---|
| Section length | 1–2 km | 1–2 km | 1–3 km for double-run | 1–3 km for double-run |
| Maximum length of sight | 50 m *class I*; 60 m *class II* | 60 m | 70 m | 90 m |
| *Field procedures*[a] | | | | |
| Max. difference in lengths Forward and backward sights per setup | 2 m *class I*; 5 m *class II* | 5 m | 10 m | 10 m |
| per section (cumulative) | 4 m *class I*; 10 m *class II* | 10 m | 10 m | 10 m |
| Maximum length of line between connections | Net A; 300 km Net B; 100 km | 50 km | 50 km double-run 25 km single-run | 25 km double-run 10 km single-run |
| *Maximum closures*[b] | | | | |
| Section; forward and backward | 3 mm $\sqrt{K}$ *class I*; 4 mm $\sqrt{K}$ *class II* | 6 mm $\sqrt{K}$ | 8 mm $\sqrt{K}$ | 12 mm $\sqrt{K}$ |
| Loop or line | 4 mm $\sqrt{K}$ *class I*; 5 mm $\sqrt{K}$ *class II* | 6 mm $\sqrt{K}$ | 8 mm $\sqrt{K}$ 0.035 ft $\sqrt{M}$ | 12 mm $\sqrt{K}$ 0.05 ft $\sqrt{M}$ [c] |

[a] The maximum length of line between connections may be increased to 100 km for double-run for second-order class II, and to 50 km for double-run for third-order in those areas where the first-order control has not been fully established.

[b] Check between forward and backward runnings, where $K$ is the distance in kilometres.

[c] $M$ is distance leveled in miles.

and can be attributed to centering the spirit level, reading the rod, and variations in refraction in air. The first two errors are proportional to the length of sight. Although refraction varies as the square of the distance, the exact relationship cannot be determined, since these variations may occur anywhere between the instrument and the rod. Consequently, refraction is also assumed proportional to the length of sight, and these three sources of random error act together.

To evaluate the propagated error in a line of levels, the random errors in a sight from an instrument setup are combined into a single value and the backsights are assumed equal to the foresights.

Let

$\hat{\sigma}_s$ = estimated standard deviation of the combined random errors in a single sight from one instrument setup expressed in units/unit of length of sight

$\ell$ = length of a single sight

Since $\Delta$ elevation = B.S. − F.S. [by Eq. (2.46), Art. 2.20],

$$\hat{\sigma}_\Delta = \ell \hat{\sigma}_s \sqrt{2} \tag{5.19}$$

in which $\hat{\sigma}_\Delta$ is the estimated standard deviation for a complete instrument setup.

Next let,

$\hat{\sigma}_d$ = estimated standard deviation in a line of levels

$d$ = length of the line of levels

$n$ = number of instrument setups required for closure

Then, also by Eq. (2.46),

$$\hat{\sigma}_d = \hat{\sigma}_\Delta \sqrt{n} \tag{5.20}$$

Since $n = d/2\ell$,

$$\hat{\sigma}_d = \hat{\sigma}_\Delta \sqrt{\frac{d}{2\ell}} = \hat{\sigma}_s \sqrt{d\ell} \tag{5.21}$$

Equations (5.20) and (5.21) show that the propagated error in a line of levels varies as the square root of the number of instrument setups or as the square root of the distance leveled and the square root of the lengths of sights. Thus, most specifications governing closures in lines or circuits of levels are expressed as a function of the square of the distance leveled and the maximum length of sights is specified.

Levels to establish vertical control are classified as first, second, and third order, depending on the equipment and procedures employed. Table 5.2 shows the general specifications and standards of accuracy for first-, second-, and third-order vertical control. These standards were prepared by the Federal Geodetic Control Committee and are published by the National Ocean Survey.

**5.44. Levels with two rods and two sets of turning points**  The advantage of this method lies primarily in the inherent self-checking aspects of the procedure as the work progresses. It is particularly useful in running levels that do not close on points of known elevation.

Two sets of turning points are established so that at each setup of the level, two independent backsights and two independent foresights are taken. The turning points on one line are usually a foot or more higher than corresponding points on the other line so as to eliminate the possibility of making the same mistake in reading the foot marks on both rods. When two rodmen are employed, one gives readings for points along the "high" line and the other for points along the "low" line.

An appropriate form of notes is illustrated by Fig. 5.37. The observations are seen to give two independent determinations for the height of instrument at each set up. Were it not for errors of

| | | | | *LEVELS,* | | DIXFIELD TO PERU | | | | |
|---|---|---|---|---|---|---|---|---|---|---|
| | | *Double-Rodded Line* | | | | Lietz Level No. 52 | | F.K. Burton, | | |
| | | *Along P. & R.F. Ry.* | | | | 2-Phila. Rods | | J.J. Hamel, Recorder | | |
| | | | | | | | | Lowe & Smith, Rods | | |
| Sta. | B.S. | H.I. | F.S. | Elev. | | | | July 9, 1977 | | |
| B.M.$_1$ | 5.241 | 532.871 | | 527.630 | | U.S.G.S. in Culvert | | Fair & Warm | | |
| B.M.$_1$ | 5.239 | 532.869 | | | | 800 ft. S. of Mile Post 13 | | | | |
| | | | | | | | | | | |
| T.P.$_1$ H | 6.943 | 535.898 | 3.916 | 528.955 | | | | | | |
| T.P.$_1$ L | 7.897 | 535.893 | 4.873 | 527.996 | | | | | | |
| | | | | | | | | 531.043 | 531.039 | |
| | | | | | | | | 527.630 | 527.630 | |
| T.P.$_2$ H | 8.337 | 541.804 | 2.431 | 533.467 | | | 87.597 | 3.413 | | 3.409 |
| T.P.$_2$ L | 9.746 | 541.797 | 3.842 | 532.051 | | | 80.775 | 3.409 | | |
| | | | | | | | 6.822 = | 6.822 ck. | | |
| T.P.$_3$ H | 5.173 | 541.508 | 5.469 | 536.335 | | | | | | |
| T.P.$_3$ L | 7.549 | 541.504 | 7.842 | 533.955 | | | | | | |
| | | | | | | | | | | |
| T.P.$_4$ H | 3.411 | 536.731 | 8.188 | 533.320 | | | | | | |
| B.M.$_2$ L | 4.963 | 536.725 | 9.742 | 531.762 | | Spk. in Tel. Pole at Road to Abbotts Mills | | | | |
| | | | | | | | | | | |
| T.P.$_5$ H | 2.344 | 531.837 | 7.238 | 529.493 | | | | | | |
| T.P.$_5$ L | 5.729 | 531.830 | 10.624 | 526.101 | | | | | | |
| | | | | | | | | | | |
| B.M.$_3$ H | 7.004 | 531.043 | 7.798 | 524.039 | | d.h. in Culvert at Alder Brook | | | | |
| T.P.$_6$ L | 8.021 | 531.039 | 8.812 | 523.018 | | | | | | |
| | 87.597 | | 80.775 | | | | | | | |

**Fig. 5.37** Notes for levels with two rods and two sets of turning points.

observation these H.I.'s should exactly agree. If at any set up the discrepancy between the two H.I.'s shows a material variation from the discrepancy between H.I.'s for the preceding setup, observations are repeated. In careful leveling the maximum allowable variation between the discrepancies at two successive setups is usually two or three thousandths of a foot or 0.6 to 0.9 mm. Normally, the difference between H.I.'s may be expected to increase as the length of the line increases, and hence the two independent determinations of the elevation of a bench mark along the route may be expected to show a difference which in general will increase with the distance from the point of beginning. Thus, in the notes, the discrepancies between the two H.I.'s are seen to be successively 0.002, 0.005, 0.007, 0.004, and 0.006. The variations between the discrepancies for succeeding setups are therefore 0.003, 0.002, 0.003, and 0.002. On the right-hand page of the notes are computations for checking the additions and subtractions. The difference between the total of all backsights and the total of all foresights is double the difference in elevation between the initial and final bench marks.

It is desirable that both rods be read on the terminal bench marks. Intermediate bench marks are employed as turning points on either of the two lines. When the discrepancy between H.I.'s becomes sufficiently large to be of consequence, the elevation of each bench mark is adjusted to conform to the mean of the two heights of instrument for the setup from which a foresight to the bench mark was taken.

**5.45. Reciprocal leveling**  Occasionally, it becomes necessary to determine the relative elevations of two widely separated intervisible points between which levels cannot be run in the ordinary manner. For example, it may be desired to transfer levels from one side to the other of a deep canyon, or from bank to bank of a wide stream.

**Fig. 5.38** Reciprocal leveling. (*a*) Plan view. (*b*) Section.

If *A* and *B* are two such points, then the level is set up near *A* and one or more rod readings are taken on both *A* and *B* (Fig. 5.38*a*). Then the level is set up in a similar location near *B*, and rod readings to near and distant points are taken as before. The mean of the two differences in elevation thus determined is taken to be the difference between the two points. Usually, the distance between points is large (often a half mile or a kilometre) so that it is necessary to use a target on the distant rod. If more accurate results are desired, a series of foresights is taken on the distant rod and sometimes also a series of backsights on the near rod, the bubble being recentered and the target reset after each observation. The difference in elevation is then computed by using the mean of the backsights and the mean of the foresights. The three-wire method described in Art. 5.46 can be used here.

More refined results can be obtained if a tilting level equipped with a gradienter is available (Art. 5.18). In this case, two targets are placed on the distant rod at *B*, one above and one below the approximate point where the horizontal line of sight strikes the rod. The optical line of sight is tilted using the micrometer screw and a series of gradienter readings are observed and recorded to each of the targets on the distant rod and also when the spirit level bubble is centered. Readings for each target position on the rod are recorded and the position of the line of sight is calculated by proportion using the gradienter readings. The entire operation is repeated with a setup near *B*.

Two factors that may appreciably affect the results are variations in temperature, causing unequal expansion of parts of the instrument, and variations in atmospheric refraction. To minimize these effects, the instrument should be shaded and corrections for curvature and refraction applied (Art. 5.2). A more satisfactory approach is to use two levels, each having about the same magnifying power and sensitiveness of level bubble. With one level set up near *A* and the other near *B*, sets of observations are obtained simultaneously from both locations. Thus, the effects of curvature and refraction cancel. The two instruments are interchanged and the operation is repeated. The mean of the differences in elevations is taken as the best estimate of actual difference in elevation.

**5.46. Precise, three-wire leveling**   The subject of precise leveling as practiced on government surveys is not to be considered in detail, as it is thoroughly covered in several U.S. National Ocean Survey manuals (see the references at the end of the chapter). However, it is appropriate to call attention to certain refinements by means of which a relatively high degree of accuracy may be obtained with the ordinary dumpy, self-leveling or tilting levels, and the self-reading rod.

The rod should be calibrated at frequent intervals by comparison with a standard length. Rods constructed with the graduations on a strip of invar metal are preferable. The rod should have an attached rod level for plumbing. Turning points *must* be on solid objects with rounded tops so that the base of the rod can be held in the same position for backsights and foresights.

The level must be equipped with stadia hairs in addition to the regular cross hairs (Art. 5.11). Engineers' levels in the United States have stadia cross hairs spaced so that a reading interval between the upper and lower cross hairs of 1 ft or 1 m is equivalent to 100 ft or 100 m, respectively, of horizontal distance. A tilting level with a coincidence bubble is preferable. To prevent unequal thermal expansion, the level should be protected from the sun by an umbrella. The level tripod should be set very firmly to prevent settlement. To eliminate, as nearly as possible, the effects of variations in atmospheric refraction, settlement of the tripod, or warping of the level, the shortest possible time elapse between backsight and succeeding foresight is desirable. For each sight, the three cross hairs are read by estimation to thousandths of feet or millimetres and recorded. The mean of the readings is taken as the correct rod reading for each sight. The interval between the reading on the upper cross hair and the reading on the lower cross hair is a measure of the distance from level to rod.

A suitable form for notes is shown in Fig. 5.39. Note that the entire two-page spread is used for the data, with backsights on the left and foresights on the right page. The terms "thread" and "wire" are used interchangeably to designate cross hair. The station refers to the instrument setup and not to the turning point number. In the example of Fig. 5.39, levels are carried in the forward direction from B.M.$_{25}$ to B.M.$_{26}$.

At the beginning of each day the collimation correction or $C$ factor for the level should be determined, as outlined in Art. 5.47. When this has been done, rod $A$ is held on B.M.$_{25}$ and a turning point is set in a forward position for rod $B$. The level is set up so as to balance backsight and foresight distances. A backsight is taken on rod $A$ held at B.M.$_{25}$, and the three thread readings 2037, 1843, and 1648 are recorded in the second column. The level operator must exercise care to be sure the bubble is centered at the instant of reading the

**THREE WIRE LEVELING**    Date: Oct. 10, 1977    Sun: 8    Forward

**U.S. 24, LAKE CO., COLO. (NORTH)**    From: BM 25 Level No. 1623    Wind: 1    To: BM 26    Time: 0900 MST

| No. of Station | Thread Reading, mm | Mean, mm | Middle Thread, ft | Thread Interval | Σ Intervals | Rod No. € Temp. | Thread Reading, mm | Mean, mm | Middle Thread, ft | Thread Interval mm | Σ Intervals, mm |
|---|---|---|---|---|---|---|---|---|---|---|---|
| (1) | (2) | (3) | (4) | (5) | (6) | (7) | (8) | (9) | (10) | (11) | (12) |
|   | 2037 |   |   | 194 |   | A | 0850 |   |   | 0209 |   |
| 1 | 1843 | 18427 | 6.04 | 195 |   | 22°C | 0641 | 06403 | 2.09 | 0211 |   |
|   | 1648 |   |   | 389 | 389 |   | 0430 |   |   | 0420 | 0420 |
|   | 2446 |   |   | 222 |   | B | 0712 |   |   | 0245 |   |
| 2 | 2224 | 22240 | 7.30 | 222 |   | 21°C | 0467 | 04677 | 1.53 | 0243 |   |
|   | 2002 |   |   | 444 | 833 |   | 0224 |   |   | 0488 | 0908 |
|   | 1725 |   |   | 114 |   | A | 1324 |   |   | 055 |   |
| 3 | 1611 | 16110 | 5.29 | 114 |   | 22°C | 1269 | 12687 | 4.16 | 056 |   |
|   | 1497 |   |   | 228 | 1061 |   | 1213 |   |   | 111 | 1019 |
|   | 2089 |   |   | 198 |   | B | 0658 |   |   | 167 |   |
| 4 | 1891 | 18913 | 6.21 | 197 |   | 22°C | 0491 | 04917 | 1.61 | 165 |   |
|   | 1694 |   |   | 395 | 1456 |   | 0326 |   |   | 332 | 1351 |
|   | 2240 |   |   | 089 |   | A | 1116 |   |   | 171 |   |
| 5 | 2151 | 21527 | 7.07 | 084 |   | 20°C | 0945 | 09453 | 3.10 | 170 |   |
|   | 2067 |   |   | 173 | 1629 |   | 0775 |   |   | 341 | 1692 |
|   | 29165 | 97217 | 31.91 ft |   |   |   | 11441 | 38137 | 12.49 ft |   | 1629 |
|   | −11441 | −38137 | (9.73m) |   |   |   |   |   | 3.81 m |   | 63 |
|   | 3)17724 | 59080 |   |   |   |   |   |   |   |   |   |
|   | 5908 |   |   |   |   |   |   |   |   |   |   |

**Fig. 5.39** Three-wire level notes.

rod. If precise level rods graduated on the back side in feet (Art. 5.70) are being used, rod $A$ is turned and a reading is taken in feet and recorded in the fourth column. The notekeeper immediately calculates and records the half intervals 194, 195 mm in column 5. If these intervals do not agree to within $\pm 2$ mm, the observations should be repeated. The mean thread reading 1842.7 mm is calculated and recorded in column 3. This mean should compare closely with the middle thread reading. The sum of half intervals, 389, are recorded in column 6. The temperature at rod $A$ should be recorded.

The telescope is now sighted on rod $B$ at the forward turning point and the foresight thread readings 850, 641, and 430 mm are observed and recorded in column 8 of the notes. The middle thread is read on the back side of rod $B$ as 2.09 ft and is recorded in column 10. The half intervals 209, 211 mm agree satisfactorily and are recorded in column 11. The mean thread reading 640.3 mm is calculated and recorded in column 9. At this point, instrument setup at station 1 is complete.

Rod $A$ is moved ahead and a turning point is established in the direction of B.M.$_{26}$. The level is set at station 2 and the first sight is a foresight on rod $A$, with the data being recorded on the right-hand page as before. This step is followed by a backsight on rod $B$ on the previous turning point, with data being recorded on the left page. This completes instrument setup 2. Rod $B$ is moved ahead to a forward turning point and the level is moved to station 3. At station 3, the backsight is taken first and the foresight second. This routine of observing backsight-foresight, foresight-backsight, and so on, is alternated at each successive setup so as to minimize the effects of instrumental settlement (Art. 5.42). The backsight and foresight distances should balance to within the specifications given in Table 5.2. For third-order levels, the difference between backsight and foresight thread intervals should not exceed 100 mm or 10 m horizontal distance. Similarly, the difference between cumulated backsight and foresight thread intervals should not exceed 100 mm or 10 m on the ground. For example, the first specification is exceeded at station 3, where the difference is $228 - 111 = 117$ mm, which is equal to a horizontal distance of 11.7 m. However, the difference between the cumulated B.S. and F.S. distances between B.M.$_{25}$ and B.M.$_{26}$ is 63 mm (or 6.3 m), well within the latter specifications. Levels are run forward and backward in sections the lengths of which are not to exceed the specification given in Table 5.2. Thus, even on long lines, each day's work provides a closed loop.

When the leveling is completed, thread readings, mean thread readings, and middle thread readings (on back of rod) are cumulated and listed in their respective columns at the bottom of the note sheet. The sum of the thread readings divided by 3 should equal the sum of the mean readings, and the sum of the middle thread readings converted to metres should check closely with the sum of the mean values, as shown in Fig. 5.39. The difference in elevation between B.M.$_{25}$ and B.M.$_{26}$, uncorrected for collimation error or curvature and refraction, equals the difference between the sum of the mean backsights and the sum of the mean foresights.

**5.47. Collimation correction** In precise three-wire and geodetic leveling when the sum of backsights and sum of foresights are unbalanced, a correction can be applied when the slope of the line of sight is known. This slope of the optical line of sight occurs when the axis of the spirit level tube and optical axis are not parallel. The amount of slope in units per unit of length of sight is the collimation correction and is referred to as the $C$ factor.

Effects of an inclined line of sight were discussed in Art. 5.42 and are further illustrated in Fig. 5.40, where the dashed line of sight is inclined with an angle $\alpha$ with the true horizontal line. In Fig. 5.40$a$ the line of sight deviates downward from true horizontal, causing the backsight and foresight rod readings $N_{s1}$ and $N_{f1}$ to be too small. Thus, for this condition the sign of $\alpha$ (and $C$) is taken as positive. Conversely, when the line of sight deviates upward from true horizontal, the sign of $C$ is negative.

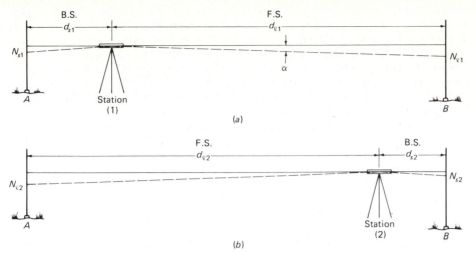

**Fig. 5.40** Determination of the $C$ factor.

Determination of $C$ involves a procedure similar to the *peg test* described in Art. 5.58 for adjustment of the level. Set two points $A$ and $B$ about 200 ft or 60 m apart (Fig. 5.40a). Set up the level to be checked at station 1 about 20 ft or 6 m from $A$. Read and record the backsight thread readings, half-intervals, sum of intervals, and backsight mean in columns 2, 4, 5, and 3, as indicated in Fig. 5.41b. The backsight mean 1356.0 mm $= N_{s1}$ and the sum of intervals 92 mm $= d_{s1}$ in Fig. 5.40a. Next, read and record foresight thread readings, half-intervals, sum of intervals, and foresight mean in columns 6, 8, 9, and 7 of Fig. 5.41b. The foresight mean 1070.7 mm $= N_{\ell 1}$ and the sum of intervals 441 mm $= d_{\ell 1}$ in Fig. 5.40a. The level is then moved to station 2 (Fig. 5.40b) and the entire procedure as outlined for station 1 is repeated. From station 2, $N_{s2} = 1281.7$ mm, $d_{s2} = 119$ mm, $N_{\ell 2} = 1574.0$ mm, and $d_{\ell 2} = 426$ mm.

The true differences in elevation between $A$ and $B$ from stations 1 and 2 are

$$\Delta H_{AB} = (N_{s1} + Cd_{s1}) - (N_{\ell 1} + Cd_{\ell 1}) \qquad (5.22a)$$
$$\Delta H_{AB} = (N_{\ell 2} + Cd_{\ell 2}) - (N_{s2} + Cd_{s2}) \qquad (5.22b)$$

where $C =$ correction per unit of stadia interval and is positive when the line of sight is below horizontal

$N_{s1}$, $N_{s2}$, $d_{s1}$, and $d_{s2}$ are defined in Fig. 5.40.

Equating Eq. (5.22a) with Eq. (5.22b) yields

$$(N_{s1} + Cd_{s1}) - (N_{\ell 1} + Cd_{\ell 1}) = (N_{\ell 2} + Cd_{\ell 2}) - (N_{s2} + Cd_{s2}) \qquad (5.23)$$

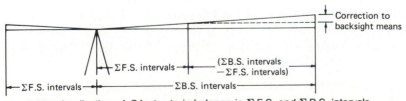

**Fig. 5.41a** Application of $C$-factor to imbalance in $\Sigma$ F.S. and $\Sigma$ B.S. intervals.

| | C FACTOR DETERMINATION | | | | | Level No. 438 | | | Temp. 80° Clear |
|---|---|---|---|---|---|---|---|---|---|
| Sta. (1) | Thread Reading (2) | B.S. Mean (3) | Thread Interval (4) | Sum Intervals (5) | | Thread Reading (6) | F.S. Mean (7) | Thread Interval (8) | Sum Intervals (9) |
| | 1402 | | 46 | | | 1291 | | 220 | |
| 1 | 1356 | 1356.0 | 46 | | | 1071 | 1070.7 | 221 | |
| | 1310 | $=N_{s1}$ | 92 | $92=d_{s1}$ | | 850 | $=N_{\ell 1}$ | 441 | $441=d_{\ell 1}$ |
| | 1341 | | 59 | | | 1787 | | 213 | |
| 2 | 1282 | 1281.7 | 60 | | | 1574 | 1574.0 | 213 | |
| | 1222 | $=N_{s2}$ | 119 | $211=d_{s1}+d_{s2}$ | | 1361 | $=N_{\ell 2}$ | $426=d_{\ell 2}$ | $867=d_{\ell 1}+d_{\ell 2}$ |
| | | | | $=d_{s2}$ | | | | | |
| | 7913 | 2637.7 | $=N_{s1}+N_{s2}$ | | | 7934 | $2644.7=N_{\ell 1}+N_{\ell 2}$ | $-211=d_{s1}+d_{s2}$ | |
| | | $-2644.7$ | $=N_{\ell 1}+N_{\ell 2}$ | | | | | | |
| | | $-7.0$ | $=(N_{s1}+N_{s2})-(N_{\ell 1}+N_{\ell 2})$ | | | | $(d_{\ell 1}+d_{\ell 2})-(d_{s1}+d_{s2})=656$ | | |

$$C=\frac{-7.0}{656}=-0.011 \text{ mm} \Big/ \text{unit of stadia interval}$$

**Fig. 5.41b** C-factor determination.

Solving Eq. (5.23) for $C$ gives

$$C=\frac{(N_{s1}+N_{s2})-(N_{t1}+N_{t2})}{(d_{t1}+d_{t2})-(d_{s1}+d_{s2})} \tag{5.24}$$

Thus, the $C$ factor equals the difference between the sum of the short backsights minus the sum of the long foresights, the quantity divided by the difference between the sum of the long foresight distances and short backsight distances. For the example given in Fig. 5.41, $C$ is $-0.011$ mm/mm of stadia interval. If rods graduated in feet were used, the units would be feet per foot of stadia interval. The minus sign indicates that the line of sight is above the horizontal.

The value for $C$ is applied to the lack of balance between cumulated foresight and backsight distances. The difference in elevation corrected for inclination of the line of sight is

$$\Delta H \text{ corrected}=\Delta H \text{ observed}+C(\Sigma \text{ B.S. intervals}-\Sigma \text{ F.S. intervals}) \tag{5.25}$$

This correction is shown graphically in Fig. 5.41a. If the $C$ factor calculated in Fig. 5.41b is applied to the change in elevation between bench marks 25 and 26 in Fig. 5.39, the elevation corrected for inclination of the line of sight is

$$\Delta H_{25-26}=5.9080 \text{ m}+(-0.011 \text{ mm/mm})(1629 \text{ mm}-1692 \text{ mm})$$

$$=5.9080 \text{ m}+0.0007 \text{ m}=5.9087 \text{ m}$$

**5.48. Geodetic leveling** Geodetic leveling is direct leveling of a high order of accuracy and precision. It is based on the irregular surfaces associated with the geoid rather than the regular surface of a spheroid or an ellipsoid representing the surface of the earth. The geoid is the figure of the earth considered as a sea level surface extended continuously through the continents. At every point, the geoid's surface is perpendicular to the plumb line; it is determined by observing deflections of the vertical.

Geodetic leveling is of first-order class I and II accuracy (Table 5.2) and is normally conducted in connection with triangulation and/or trilateration over large areas to furnish vertical control for the framework of the National Network, for extensive engineering surveys, and regional crustal movement studies, to mention a few application areas.

The methods employed are generally the same as described in Art. 5.46 for precise three-wire levels. However, procedure and equipment are more refined and more attention is paid to details. For example, a tilting or automatic level equipped with an optical micrometer is specified. Figure 5.19 is an example of a tilting level suitable for first-order work. Calibrated, matched pairs of precise level rods similar to those illustrated in Figs. 5.27 and 5.28 must be used. The level must be protected from the sun's rays by an umbrella at all times.

For first-order, class I leveling, the maximum length of sight is 50 m, the maximum unbalance between cumulated backsight and foresight distances is 2 m, and the allowable closure error forward and backward per 2-km section is 3 mm times $\sqrt{\text{distance leveled in kilometres}}$. These and other specifications are listed in Table 5.2. Details concerning instruments and procedures can be found in National Ocean Survey Publications given in Refs. 8 and 9. Notes are kept in a form similar to Fig. 5.39.

Bench marks are spaced at intervals not to exceed 1 to 2 km. The level routes are almost invariably along railways or highways; therefore, the bench marks of the system of vertical control are usually not near triangulation stations of horizontal control, which are elevated for reasons of visibility. The usual form of bench mark is an inscribed bronze disk (Fig. 5.34) set solidly into a concrete post, a masonry structure, or rock.

Field computations are checked in the office. Corrections for calibrated length and temperature of rod are calculated and applied. An *orthometric* correction (Art. 5.49), which is a function of latitude and elevation, is made to account for the irregular shape of the geoid. Circuit closures are made to detect and rectify serious misclosures. When all mistakes and systematic errors have been detected and removed, a simultaneous least squares adjustment (Appendix B) of the entire network is performed to provide adjusted elevations of all bench marks. Elevations are provided in metres and feet and descriptions of the bench marks are prepared for publication.

When automatic compensator equipped levels (Art. 5.19) are used for geodetic leveling (first-order class III leveling, Ref. 11), the procedures are somewhat modified. An automatic compensator equipped level similar to the ones illustrated in Fig. 5.20 is used with a pair of double-scale level rods (Fig. 5.27), each rod having a different constant offset. Use of double-scale rods with different offsets permits detection of blunders such as transposition of backsight and foresight rod readings. Thus, when checking for blunders against existing elevations is possible, single-run leveling is allowed.

A definite sequence for leveling the instrument and observing the low and high scales is specified to reduce the accumulation of small systematic errors that occur in the instrument and rods. Details of these systematic errors and recommended compensatory steps can be found in Refs. 2 and 11.

In the experimental runs being performed by the U.S. National Ocean Survey, notes are not recorded on the conventional forms. Observed rod readings are entered directly into a battery-powered, programmable calculator connected to a cassette tape recorder. The calculator is programmed to provide a warning signal when discrepancies occur in the observations. For example, if there is a lack of balance between backsight and foresight distances or lack of agreement between differences in elevations determined by left and right scale readings, a warning signal is displayed. This feature permits an immediate check of the quality in the measurements and correction of irregularities before the level is moved to the next setup. Measurements from approximately 28 km of leveling can be stored on one cassette. At the end of the day's work, data recorded on the cassette tape are read into a desktop calculator in the field office. This calculator provides a hard copy listing of the observations and stores the data on another cassette tape unit. Using a terminal in the field office, data stored on this cassette can be transmitted to a large central computer, which contains programs for further analysis of the data, correction for all systematic errors, and eventual adjustment of the level net when the field operations are completed.

The procedure outlined above represents a genuine departure from the traditional geo-detic leveling methods. Detailed procedures and specifications for first-order class III leveling using compensator-type levels are given in Ref. 11.

**5.49. Orthometric correction** As noted in Art. 5.1, the datum for leveling is a *level surface* everywhere perpendicular to gravity. The most commonly used datum is the mean sea level surface. The force of gravity is less at the equator than at the poles, owing primarily to the action of centrifugal force. Since the level surface is a function of gravity, a given level surface is farther from sea level at the equator than at a point nearer the pole. What this means in terms of leveling is that level lines at different elevations which progress along northerly or southerly directions are referenced to different level surfaces. Figure 5.42 illustrates the divergence of a level surface from the mean sea level surface.

For example, assume that levels referred to sea level are run along an inland route from San Diego, California, to Seattle, Washington, where closure in both cases is on a bench mark at or near mean sea level. Since the level surface at San Diego is farther from mean sea level than the level surface at Seattle, the error due to this convergence of level surfaces is about 1.3 m. This amount in that distance is significant for first-order levels.

The orthometric correction is applied to account for this convergence of level surfaces at different elevations and can be calculated using the equation

$$\text{correction} = -0.005288 \sin 2\phi \, h \, \Delta\phi \, \text{arc } 1' \qquad (5.26)$$

where   $\phi$ = latitude at the starting point
 $h$ = mean sea level elevation, ft or m, at the starting point
 $\Delta\phi$ = change in latitude in minutes between the two points (+ in the direction of increasing latitude or toward the pole)

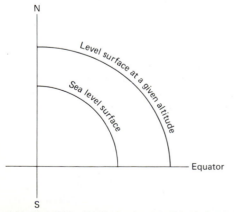

**Fig. 5.42** Divergence of a level surface from sea level.

**Example 5.4**  The elevation of a lake running in a north-south direction is 200 m above mean sea level at the northern end, where $\phi = 42°00'$N. A line of levels is run along the lake to the south end, where $\phi = 40°40'$N. Calculate the orthometric correction.

SOLUTION  By Eq. (5.26),

orthometric correction $= (-0.005288)(\sin 84°)(200 \text{ m})(-80')(0.000290)$
$$= 0.024 \text{ m}$$

Thus, the elevation at the south end of the lake $= 200.000 \text{ m} + 0.024 = 200.0244 \text{ m}$.

**Example 5.5**  A level line is run from latitude $38°00'$N along a south to north line to latitude $38°05'$N at an average elevation of 7000 ft above mean sea level. Compute the orthometric correction.

SOLUTION  By Eq. (5.26),

orthometric correction $= (-0.005288)(\sin 76°)(7000)(05')(0.000290)$
$$= -0.052 \text{ ft}$$

The elevation at $38°05'$N latitude $= 7000.00 - 0.052 = 6999.948$ ft above mean sea level.

**5.50. Adjustment of intermediate bench marks**  When a line of levels makes a complete circuit, almost invariably the final elevation of the initial bench mark as computed from the level notes will not agree with the initial elevation of this point. The difference is the error of running the circuit and is called the *error of closure*. It is evident that elevations of intermediate bench marks established while running the circuit will also be in error, and there arises the problem of determining the errors for these intermediate points and of adjusting their elevations accordingly.

It has been shown that the principal errors of leveling are random, hence the error tends to propagate as the square root of the number of opportunities for error, or as the square root of the number of setups (Art. 5.43). In the adjustment of elevations it will usually be sufficiently exact to assume that the number of setups per unit of distance leveled is the same for one portion of the circuit as for any other, and that therefore the error propagates as the square root of the distance. Since corrections to such related quantities are proportional to the variance or the square of the standard deviations, it follows that the appropriate correction to the observed elevation of a given bench mark in the circuit is directly proportional to the distance to the bench mark from the point of beginning. Thus, if $E_c$ is the error of closure of a level circuit of length $L$, and if $C_a, C_b, ..., C_n$ are the respective corrections to be applied to observed elevations of bench marks $A, B, ..., N$ whose respective distances from the point of beginning are $a, b, ..., n$, then

$$C_a = -\frac{a}{L} E_c; \quad C_b = -\frac{b}{L} E_c; \quad \cdots; \quad C_n = -\frac{n}{L} E_c \qquad (5.27)$$

**Example 5.6**  The accepted elevation of the initial bench mark B.M.$_i$ of a level circuit is 470.46 ft. The length of the circuit is 10 mi. The final elevation of the initial bench mark as calculated from the level notes is 470.76. The observed elevations of bench marks established along the route and the distances to the bench marks from B.M.$_i$ are as shown in the third and second columns of the accompanying tabulation. The elevations of these intermediate points are required.

$E_c = 470.76 - 470.46 = +0.30$ ft

By Eq. (5.27),

$C_a = -\frac{2}{10} \times 0.30 = -0.06$ ft
$C_b = -\frac{5}{10} \times 0.30 = -0.15$ ft
$C_c = -\frac{7}{10} \times 0.30 = -0.21$ ft

| Point | Distance from B.M.$_j$, miles | Observed elevation, ft | Correction, ft | Adjusted elevation, ft |
|---|---|---|---|---|
| B.M.$_j$ | 0 | 470.46 | 0.0 | |
| B.M.$_a$ | 2 | 780.09 | $-0.06$ | 780.03 |
| B.M.$_b$ | 5 | 667.41 | $-0.15$ | 667.26 |
| B.M.$_c$ | 7 | 544.32 | $-0.21$ | 544.11 |
| B.M.$_i$ | 10 | 470.76 | $-0.30$ | 470.46 |

These corrections subtracted from the corresponding observed elevations give the adjusted elevations as tabulated. It is to be noted that, if the error of closure is positive, all corrections are to be subtracted, and *vice versa*.

If desired, corrections may be made directly proportional to the number of setups instead of the distances.

The principles described in this section apply also to the adjustment of elevations of bench marks on a line of levels run between two points whose difference in elevation has previously been determined by more accurate methods and is assumed to be correct.

**5.51. Adjustment of levels over different routes**   When differential levels are run over several different routes from a fixed bench mark to establish a bench mark, there will be as many observed elevations for the bench mark as there are lines terminating at that point. In other words, redundant data are available which must be resolved. As indicated in Chap. 2, the best answer in this case is the *least-squares estimate*. In order to make such an adjustment, weights must be established for each line.

Assume that the standard deviation for each individual observed difference in elevation, $\sigma_\ell$ varies as the square root of the distance leveled, $d_i$ or the number of instrument setups, $N$ (Art. 5.43). Consequently, $\sigma_{\ell i} = \sqrt{d_i}\,\sigma_s$, where $\sigma_s$ is the estimated standard deviation in units per unit of distance leveled. If $i = 1,2,\ldots,n$ lines are involved and noncorrelation is assumed between observed differences in elevation, then according to Eq. (2.28), Art. 2.14, the covariance matrix for observations is

$$\Sigma_{\ell i} = \begin{bmatrix} \sigma_s^2 d_1 & & & \mathbf{0} \\ & \sigma_s^2 d_2 & & \\ & & \cdot & \\ \mathbf{0} & & & \sigma_s^2 d_n \end{bmatrix} \tag{5.28}$$

and according to Eqs. (2.30) and (2.31), the cofactor matrix for the observations is

$$Q_{\ell i} = \frac{1}{\sigma_0^2} \Sigma_{\ell i} \tag{5.29}$$

By definition [Eq. (2.32)] the matrix of weights is

$$W_{\ell i} = Q_{\ell i}^{-1} = \sigma_0^2 \Sigma_{\ell i}^{-1} = \sigma_0^2 \begin{bmatrix} 1/\sigma_s^2 d_1 & & & \mathbf{0} \\ & 1/\sigma_s^2 d_2 & & \\ & & \cdot & \\ \mathbf{0} & & & 1/\sigma_s^2 d_n \end{bmatrix} \tag{5.30}$$

Now, $\sigma_0$ is the standard deviation of unit weight, an arbitrary constant which can be taken equal to $\sigma_s$ so that the weight matrix becomes

$$
\mathbf{W}_{ti} = \begin{bmatrix} 1/d_1 & & & \mathbf{0} \\ & 1/d_2 & & \\ & & \ddots & \\ \mathbf{0} & & & 1/d_n \end{bmatrix}
\tag{5.31}
$$

Thus, weights of individual lines are inversely proportional to the distance leveled or the number of instrument setups. Consider an example to illustrate the use of such a weight matrix in the least-squares adjustment of a leveling problem with redundant data.

**Example 5.7**   Lines of levels between B. M.$_A$ and B.M.$_B$ are run over four different routes. The fixed elevation of B.M.$_A$ is 640.000 m. The lengths of lines, corresponding weights [according to Eq. (5.31)], and observed differences in elevation are shown in the accompanying tabulation. It is required to determine the least-squares estimate for B.M.$_B$.

| Route | Length $d_i$, km | Weight $w_i$ | Observed D.E., m | Observed elevation, m |
|-------|-----------------|--------------|-------------------|------------------------|
| 1 | 2 | $\frac{1}{2}$ | 0.720 | 640.720 |
| 2 | 4 | $\frac{1}{4}$ | 0.560 | 640.560 |
| 3 | 10 | $\frac{1}{10}$ | 1.080 | 641.080 |
| 4 | 20 | $\frac{1}{20}$ | 0.260 | 640.260 |

SOLUTION   To perform the adjustment, a mathematical model must be formed. In this case the model consists of condition equations of the following general form for each line [see Eq. (B.72), Art. B.11].

(observed difference in elevation + residual)$_i$

$$= \text{(adjusted elevation of B.M.}_B - \text{elevation of B.M.}_A)$$
$$\tag{5.32}$$

For the example problem, the equations (5.33) can be written for the four lines

$$
\begin{aligned}
0.720 + v_1 &= \Delta - 640.000 \\
0.560 + v_2 &= \Delta - 640.000 \\
1.080 + v_3 &= \Delta - 640.000 \\
0.260 + v_4 &= \Delta - 640.000
\end{aligned}
\tag{5.33}
$$

in which $\Delta$ is the adjusted elevation of B.M.$_B$. Equations (5.33) become

$$
\begin{aligned}
v_1 - \Delta &= -640.720 \\
v_2 - \Delta &= -640.560 \\
v_3 - \Delta &= -641.080 \\
v_4 - \Delta &= -640.260
\end{aligned}
\tag{5.34}
$$

which can be stated in matrix form as [see Eq. (B.11)]

$$
\underset{4,1}{\mathbf{V}} + \underset{(4,1)}{\mathbf{B}} \underset{(1,1)}{\Delta} = \underset{4,1}{\mathbf{f}}
\tag{5.35}
$$

The criterion for the least-squares adjustment is (Art. B.7)

$$w_1 v_1^2 + w_2 v_2^2 + w_3 v_3^2 + w_4 v_4^2 = \text{a minimum}
\tag{5.36}$$

from which the normal equations are [see Eq. (B.33)]

$$
(\underset{1,4}{\mathbf{B}^T} \underset{4,4}{\mathbf{W}} \underset{4,1}{\mathbf{B}}) \underset{1,1}{\Delta} = \underset{1,4}{\mathbf{B}^T} \underset{4,4}{\mathbf{W}} \underset{4,1}{\mathbf{f}}
\tag{5.37}
$$

so that

$$\Delta = (\mathbf{B}^T \mathbf{W} \mathbf{B})^{-1} \mathbf{B}^T \mathbf{W} \mathbf{f}
\tag{5.38}$$

in which

$$B = \begin{bmatrix} -1 \\ -1 \\ -1 \\ -1 \end{bmatrix}$$

$\Delta$ is the adjusted elevation of point $B$, $\mathbf{W}$ is defined by Eq. (5.31) and the values $d_i$ in the table, and $\mathbf{f}$ is defined in Eqs. (5.34) and (5.35).

Substitution of these values into (5.38) yields

$$\Delta = \left\{ \begin{bmatrix} -1 & -1 & -1 & -1 \end{bmatrix} \begin{bmatrix} \tfrac{1}{2} & & & 0 \\ & \tfrac{1}{4} & & \\ & & \tfrac{1}{10} & \\ 0 & & & \tfrac{1}{20} \end{bmatrix} \begin{bmatrix} -1 \\ -1 \\ -1 \\ -1 \end{bmatrix} \right\}^{-1}$$

$$\begin{bmatrix} -1 & -1 & -1 & -1 \end{bmatrix} \begin{bmatrix} \tfrac{1}{2} & & & 0 \\ & \tfrac{1}{4} & & \\ & & \tfrac{1}{10} & \\ 0 & & & \tfrac{1}{20} \end{bmatrix} \begin{bmatrix} -640.720 \\ -640.560 \\ -641.080 \\ -640.260 \end{bmatrix}$$

which gives the solution for $\Delta$. Thus,

$$\Delta = \frac{(640.720)(\tfrac{1}{2}) + (640.560)(\tfrac{1}{4}) + (641.080)(\tfrac{1}{10}) + (640.260)(\tfrac{1}{20})}{\tfrac{1}{2} + \tfrac{1}{4} + \tfrac{1}{10} + \tfrac{1}{20}}$$

$$= 640.690 \text{ m} \tag{5.39}$$

is the least-squares estimate for the elevation of B.M.$_B$.

The astute reader would already have noted that the solution for $\Delta$ is actually nothing more than the weighted mean of the given four observed values, which in this case provides the least-squares estimate. In fact, for the example provided, the least-squares solution is trivial. The logical approach would be to simply form the weights and calculate the *weighted mean* with, of course, the knowledge that such a procedure provides a least-squares estimate for this type of problem, since *no correlation* exists between the observed differences in elevation.

**Example 5.8**  Lines of levels are run from bench marks $A$, $B$, and $C$ to establish elevations at junction point $E$. The number of instrument setups per line, weights, and observed elevations are listed in the following table:

| Route | Number of setups $N$ | Weight $w_i$ | Observed elevation, ft |
|---|---|---|---|
| 1 | 9 | $\tfrac{1}{9}$ | 320.48 |
| 2 | 12 | $\tfrac{1}{12}$ | 320.32 |
| 3 | 4 | $\tfrac{1}{4}$ | 320.89 |

Calculate the least-squares estimate for the elevation of junction point $E$.

SOLUTION  Weights are taken as being inversely proportional to the number of setups, as shown in the third column of the table. As in the previous example, the least-squares adjustment is given by the

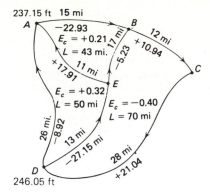

**Fig. 5.43** Simple level net.

weighted mean as follows:

$$\text{adjusted elevation of J.P.E.} = \frac{(320.48)(\frac{1}{9}) + (320.32)(\frac{1}{12}) + (320.89)(\frac{1}{4})}{\frac{1}{9} + \frac{1}{12} + \frac{1}{4}}$$

$$= 320.68 \text{ ft}$$

**5.52. Adjustment of level networks**  A level network consists of a number of intersecting lines of levels which are tied into known bench marks. Figure 5.43 shows a simple level net which contains known bench marks $A$ and $D$. The adjusted elevations of bench marks $B$, $C$, and $E$ are to be found. Lengths and differences in elevations for each line are indicated, with the arrow showing the direction in which levels were run. The errors of closure and total lengths of loops $ABEA$, $AEDA$, and $DEBCD$ are also given (Fig. 5.43). The least-squares adjustment provides the most rigorous and expedient solution to this problem. Using the method of indirect observations as outlined in Art. B.7, this problem would require forming seven condition equations, which are then solved by the method of least squares for adjusted elevations of bench marks $B$, $C$, and $E$. Solution of this type of problem using such a method is easily accomplished with a pocket calculator. More complex level nets involving many unknowns are also solvable by the method of least squares, using programmable pocket and desktop calculators or with computer programs designed for high-speed electronic computers. In any case, the fundamental principles involved are similar to those described in Art. 5.51 and described in detail in Example B.7, Art. B.11.

## PROFILE LEVELING; GRADES

**5.53. Profile leveling**  The process of determining the elevations of points at short measured intervals along a fixed line is called *profile leveling*. During the location and construction of highways, railroads, canals, and sewers, stakes or other marks are placed at regular intervals along an established line, usually the center line. Ordinarily, the interval between stakes is 100 ft, 50 ft, or 25 ft, with intervals of 100 m, 50 m, 20 m, and 10 m being utilized in the metric system. The 100-ft points or 100-m points, reckoned from the beginning of the line, are called *full stations*, and all other points are called *plus stations*. Each stake is marked with its station and plus. Thus, a stake set at 1600 ft from the point of beginning is numbered "16" or "16+00," and one set at 1625 ft from the point of beginning is numbered "16+25." Similarly, a point 450 m from the origin is station 4+50 and a point set 1410 m from the origin is number 14+10. Elevations by means of which the profile may be constructed are obtained by taking level-rod readings on the ground at each stake and at intermediate points where marked changes in slope occur.

Figure 5.44 illustrates in plan and elevation the steps in leveling for profile. In this case stakes are set every 100 ft, according to the common practice in highway and railway location. The instrument is set up in some convenient location not necessarily on the line (as at $L_1$), the rod is held on a bench mark (B.M. 28, elevation 564.31 ft above mean sea level), a backsight (1.56) is taken, and the height of instrument (565.87) is obtained as in differential leveling. Readings (in ft) are then taken with the rod held on the ground at successive stations along the line. These rod readings, being for points of unknown elevation, are foresights regardless of whether they are back or ahead of the level. They are frequently designated as *intermediate foresights* to distinguish them from foresights taken on turning points or bench marks. The intermediate foresights $(0.7, 2.9, \ldots, 11.9)$ subtracted from the H.I. (565.87) give ground elevations of stations. When the rod has been advanced to a point beyond which further readings to ground points cannot be observed, a turning point (T.P.$_1$) is selected, and a foresight (11.63) is taken to establish its elevation. The level is set up in an advanced position ($L_2$), and a backsight (0.41) is taken on the turning point (T.P.$_1$) just established. Rod readings on ground points are then continued as before. The rodman observes where changes of slope occur (as $609 + 50$, $610 + 40, \ldots, 610 + 65$), and readings are taken to these intermediate stations. The "plus," or distance from the preceding full station to the intermediate point, is measured by pacing or with a tape or the rod according to the precision required.

The care exercised in taking observations on turning points depends upon the distance between bench marks, the elevations of which have been determined previously, and upon the required precision of the profile. For a ground profile the backsights and foresights are usually read to hundredths of feet (mm) and no particular attention is paid to balancing backsight and foresight distances; the intermediate foresights to ground points are read to tenths of feet (cm) only. Occasions arise when it is desirable or necessary to determine intermediate foresights to hundredths of feet (~3 mm), for example, in securing the profile of railroad track or of the water grade in a canal; rod readings on turning points are then generally taken to thousandths of feet or to mm, and backsight and foresight distances are often balanced.

As the work of leveling for profile progresses, bench marks are generally established to facilitate later work; these are made turning points wherever possible. To check the elevation

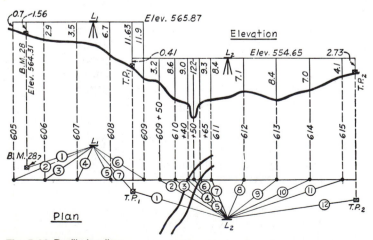

**Fig. 5.44** Profile leveling.

of turning points, it is necessary either to run levels back to the point of beginning or to run short lines of differential levels connecting with bench marks previously established by some other survey. The effect of an occasional error in the elevations of intermediate ground points on the profile is usually not of sufficient importance to justify the additional work which checking would make necessary, and if turning points are checked, it is regarded as sufficient.

**5.54. Profile-level notes**   The notes for profile leveling may be recorded as shown in Fig. 5.45, in which foresights to turning points and bench marks are in a separate column from intermediate foresights to ground points. The notes for turning points are kept in the same manner as for differential leveling.

The computations shown at the foot of the notes of Fig. 5.45 check all computations for H.I.'s and elevations of T.P.'s on the page and thus for the notes shown the difference between the sum of all backsights and the sum of all foresights is equal to the error of closure on B.M. 30. Elevations of ground points are recorded only to the *number of decimal places contained in the intermediate foresights*, regardless of the number of places in the H.I.

The right-hand page is reserved for concise descriptions of bench marks and for other pertinent items. Occasionally, as shown in Fig. 5.45, simple sketches are employed in conjunction with the explanatory notes.

**5.55. Plotting the profile**   To provide a useful product for the designer, the profile is plotted on profile paper for specified horizontal and vertical scales. Profile paper is a

| Sta. | B.S. | H.I. | F.S. | I.F.S. | Elev,m | | | |
|---|---|---|---|---|---|---|---|---|
| | | PROFILE LEVELS FOR STORM | | | | SEWER LOCATION STATE UNIV. | | |
| | | | | | | Lietz Level No. 21, Rod No. 12 | April 27, 1977 | |
| | | | | | | | Clear, windy 70°F | |
| B.M. 30 | 3.478 | 33.478 | | | 30.000 | Intersection West Rd., Gayley Ave. Top of fireplug N.W. corner | Jensen ⋆ | |
| 0+00 | | | | 3.617 | 29.861 | Top manhole | Laird, Notes | |
| 0+00 | | | | 5.141 | 28.337 | Flow line manhole | Lucas, Rod | |
| +10 | | | | 1.72 | 31.76 | | | |
| T.P.1 | 3.134 | 36.419 | 0.193 | | 32.285 | | 0+53.3 ◎ Exist. M.H. | |
| +20 | | | | 2.86 | 33.56 | | | |
| +29.5 | | | | 1.852 | 34.567 | Edge of asphalt sidewalk | +46.5 sidewalk | |
| +30 | | | | 1.805 | 34.614 | Asphalt walk | +43.8 | |
| +31.9 | | | | 1.738 | 34.681 | Edge of asphalt sidewalk | +38 | |
| +34.7 | | | | 1.250 | 35.169 | Edge of asphalt sidewalk | +34.7 Side walk | |
| +38.0 | | | | 0.951 | 35.468 | Edge of asphalt sidewalk | +31.9 Sidewalk | |
| +40 | | | | 0.65 | 35.77 | | +29.5 | |
| +43.8 | | | | 0.054 | 36.365 | Edge of asphalt sidewalk | Stakes | |
| T.P.2 | 3.551 | 39.844 | 0.126 | | 36.293 | | set at | |
| +46.5 | | | | 3.178 | 36.666 | Edge of asphalt sidewalk | 10-m | N |
| +48.1 | | | | 2.87 | 36.97 | | intervals | |
| +50 | | | | 2.74 | 37.10 | | on slope | |
| +53.3 | | | | 2.289 | 37.555 | Top of manhole | South of | |
| +53.3 | | | | 3.508 | 36.336 | Flow line manhole | Allison Hall | |
| T.P.3 | 0.081 | 37.636 | 2.289 | | 37.555 | | | |
| T.P.4 | 0.333 | 34.308 | 3.661 | | 33.975 | | | |
| T.P.5 | 0.515 | 31.972 | 2.851 | | 31.457 | | Exist M.H. ◎ | |
| B.M. 30 | | | 1.974 | | 29.998 | | | |
| | 11.092 | | −11.094 | | | | | |
| | | | 11.092 | | | ⎰check | | |
| | | 30.000 − 0.002 = 29.998 | | | | | | |

**Fig. 5.45** Profile notes in metric units.

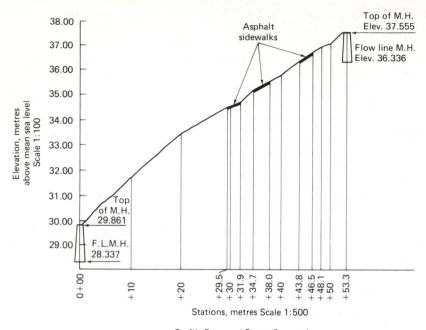

**Fig. 5.46** Plotted profile.

cross-sectioned paper of good quality with a specified number of lines per horizontal and vertical unit. Common scales are 1:600 (50 ft/in) horizontally and 1:120 (10 ft/in) vertically. Profile paper has 20 lines per inch vertically and 10 lines per inch horizontally. In the metric system comparable horizontal and vertical scales are 1:500 and 1:100 in the horizontal and vertical scales, respectively. Figure 5.46 illustrates a profile plotted using the elevations in Fig. 5.45. The horizontal and vertical scales in Fig. 5.46 are 1:500 and 1:100, respectively. Note that a certain balance between horizontal and vertical scale is necessary in order to portray the ground profile in a realistic manner. The plotted profile is utilized by the designer to establish grade lines for highways, roads, sewer lines, canals, and so on. Frequently, a plan view of the area through which the profile runs is placed above or below the profile on the same sheet. Further details concerning profiles and establishing grade lines can be found in Art. 17.29. A plan and profile sheet for a route alignment is shown in Fig. 17.37.

**5.56. Adjustment of levels** Regardless of the precision of manufacture, levels in process of use require certain field adjustments from time to time. It becomes an important duty of the surveyor to test his instrument at short intervals and to make with facility such adjustments as are found necessary.

In some instances one adjustment is likely to be altered by, or depends upon, some other adjustment made subsequently. For example, lateral movement of the cross-hair ring may likewise produce a small rotation, and the lateral adjustment of the level tube depends upon the vertical adjustment. Hence, if an instrument is badly out of adjustment, related adjustments must be repeated until they are gradually perfected.

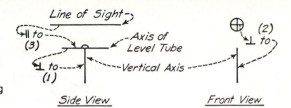

**Fig. 5.47** Desired relations among principal lines of dumpy level.

**5.57. Desired relations in the dumpy level**   For a dumpy level in perfect adjustment the following relations should exist (see Fig. 5.47):

1. The axis of the level tube should be perpendicular to the vertical axis.
2. The horizontal cross hair should lie in a plane perpendicular to the vertical axis, so that it will lie in a horizontal plane when the instrument is level.
3. The line of sight should be parallel to the axis of the level tube.

Also the optical axis, the axis of the objective slide, and the line of sight should coincide; but for the dumpy levels commonly manufactured in the United States the optical axis and the axis of the objective slide are fixed perpendicular to the vertical axis by the manufacturer, and no provision for further adjustment is made.

**5.58. Adjustment of the dumpy level**   The parts capable of and requiring adjustment are the cross hairs and the level tube. The basis for adjustments is the vertical axis. The adjustments for a level with four foot-screws are given in the following.

**1. To make the axis of the level tube perpendicular to the vertical axis**   Approximately center the bubble over each pair of opposite leveling screws; then carefully center the bubble over one pair. Rotate the level end for end about its vertical axis. If the level tube is in adjustment the bubble will retain its position. If the tube is not in adjustment, the displacement of the bubble indicates double the actual error, as shown by Fig. 5.48. If $(90° - \alpha)$ represents the angle between the vertical axis and axis of level tube, then when the bubble is centered the vertical axis makes an angle of $\alpha$ with the true vertical. When the level is reversed, the bubble is displaced through the arc whose angle is $2\alpha$. Hence, the correction is the arc whose angle is $\alpha$. Make the correction by bringing the bubble halfway back to the center by means of the capstan nuts at one end of the tube. Relevel the instrument with the leveling screws, and repeat the process until the adjustment is perfected. Usually, three or four trials are necessary. As a final check, the bubble should remain centered over each pair of opposite leveling screws. This adjustment involves the principle of *reversion*.

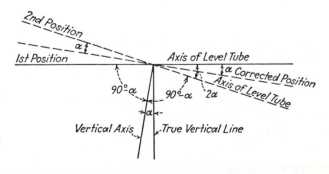

**Fig. 5.48** Adjustment of axis of level tube of dumpy level.

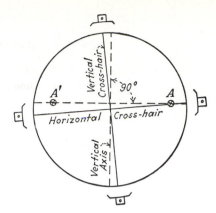

**Fig. 5.49** Adjustment of horizontal cross hair.

**2. To make the horizontal cross hair lie in a plane perpendicular to the vertical axis (and thus horizontal when the instrument is level)**   Sight the horizontal cross hair on some clearly defined point (as $A$, Fig. 5.49) and rotate the instrument slowly about its vertical axis. If the point appears to travel along the cross hair, no adjustment is needed.

If the point departs from the cross hair and takes some position as $A'$ on the opposite side of the field of view, loosen two adjacent capstan screws and rotate the cross-hair ring until by further trial the point appears to travel along the cross hair. Tighten the same two screws. The instrument need not be level when the test is made.

**3. To make the line of sight parallel to the axis of the level tube (two-peg test)**   Set two pegs 200 to 300 ft (60 to 90 m) apart on approximately level ground, and designate as $A$ the peg near which the second setup will be made (Fig. 5.50); call the other peg $B$. Set up and level the instrument at any point $M$ equally distant from $A$ and $B$, that is, in a vertical plane bisecting the line $AB$. Take rod readings $a$ on $A$ and $b$ on $B$; then $(a - b)$ will be the true difference in elevation, since any error would be the same for the two equal sight distances $L_m$. Due account must be taken of signs throughout the test.

Move the instrument to a point $P$ near $A$, preferably but not necessarily on line with the pegs; set up as before, and measure the distances $L_a$ to $A$ and $L_b$ to $B$. Take rod readings $c$ on $A$ and $d$ on $B$. Then $(c - d)$, taken in the same order as before, is the indicated difference in elevation; if $(c - d) = (a - b)$, the line of sight is parallel to the axis of the level tube, and the instrument is in adjustment. If not, $(c - d)$ is called the "false" difference in elevation, and the inclination (error) of the line of sight in the net distance $(L_b - L_a)$ is equal to

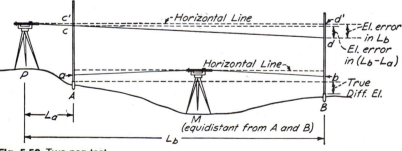

**Fig. 5.50** Two-peg test.

$(a-b)-(c-d)$. By proportion, the error in the reading on the far rod is

$$e_{fr} = \frac{L_b}{L_b - L_a}[(a-b)-(c-d)] \tag{5.40}$$

Subtract algebraically the amount of this error from the reading $d$ on the far rod to obtain the correct reading $d'$ at $B$ for a horizontal line of sight with the position of the instrument unchanged at $P$. Set the target at $d'$ and bring the line of sight on the target by moving the cross-hair ring vertically.

**Example 5.9**   With level at $M$, the rod reading $a$ is 0.970 and $b$ is 2.986; the true difference in elevation $(a-b)$ is then $0.970 - 2.986 = -2.016$ ft., with $B$ thus indicated as being lower than $A$. With level at $P$, the rod reading $c$ is 5.126 and $d$ is 7.018; the false difference in elevation $(c-d)$ is then $5.126 - 7.018 = -1.892$, with $B$ again indicated as being lower than $A$. The distance $L_a$ is observed to be 30 ft and $L_b$ to be 230 ft. The inclination of the line of sight in (230b-30 = 200) ft is $(-2.016)-(-1.892) = -0.124$ ft. The error in elevation of the line of sight at the far rod is $(230/200) \times (-0.124) = -0.143$ ft. The correct rod reading $d'$ for a horizontal line of sight is $7.018 - (-0.143) = 7.161$ ft.

As a partial check on the computations, the correct rod reading $c'$ at $A$ may be computed by proportion; the difference in elevation computed from the two corrected rod readings $c'$ and $d'$ should be equal to the true difference in elevation observed originally at $M$.

**Example 5.10**   In the preceding example, the error in elevation of the line of sight at the near rod is $(30/200) \times (-0.124) = -0.019$ ft. The correct rod reading $c'$ is $5.126 - (-0.019) = 5.145$ ft. The "false" difference in elevation is $5.145 - 7.161 = -2.016$ ft, which is equal to the true difference in elevation; hence, the computations are checked to this extent.

A sketch should always be drawn. Also, theoretically a correction for Earth's curvature and atmospheric refraction (see Art. 5.2) should be added numerically to the final rod reading $d'$, although in practice it is usually considered negligible.

Some surveyors prefer to set up at $P$ within 2 to 3 m or 6 to 10 ft of $A$ and to consider $[(a-b)-(c-d)]$ as being the *total* error in elevation, to be subtracted directly from $d$. This serves as a first approximation; the procedure for the setup at $P$ is then repeated. The amount of computation is thus reduced, but the amount of field work is increased.

**5.59. Adjustment of automatic levels**   The standard peg test described in Art. 5.58 can also be employed with automatic, self-leveling levels. If the slope of the optical line of sight is excessive, the cross hair can be adjusted. Reference should be made to the instrument manual of instructions for this operation, since most automatic levels do not have exposed adjusting screws for the cross-hair reticule. The cross-hair reticule for the level illustrated in Fig. 5.17 is exposed by unscrewing the metal cover around the eyepiece.

Prior to leveling, the circular bubble and activity of the compensator, which automatically makes the line of sight horizontal, should be checked. Center the circular bubble in the engraved circle (Fig. 5.51a) and rotate the instrument through 180° or to the direction of maximum bubble displacement (Fig. 5.51b). When the displacement $d$ exceeds one-half the bubble diameter, adjustment is necessary. Adjustment is performed in two steps by reducing the two components of the displacement, $\ell$ from left to right and $t$ from top to bottom. First, remove $\frac{1}{2}\ell$ using the appropriate level screws (Fig. 5.51c). Second, remove the remaining $\frac{1}{2}\ell$ by manipulating the adjusting screws (Fig. 5.51d). One-half the top to bottom displacement $t$ (Fig. 5.51e) is then removed by the level screws and the remainder $t/2$ is compensated using the adjusting screws to bring the bubble to the center of the engraved circle. Turn 180° and check the bubble. Repeat the entire procedure until the bubble remains centered in all positions.

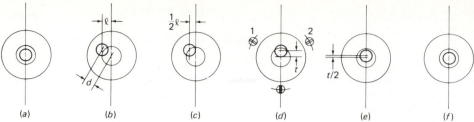

**Fig. 5.51** Adjustment of circular level bubble.

To check the compensator, sight a target about 30 m away. Center the bull's-eye bubble. Next, tap one tripod leg with a hand. The image of the target will appear to swing in the field of view but the target will return to its original position. Next, turn a level screw, causing the line of sight to slope slightly. Once again the target will appear to swing but will return to the original position.

### 5.60. Problems

**5.1** What is the combined effect of the Earth's curvature and mean atmospheric refraction in a distance of 300 ft? In a distance of 3000 ft? In a distance of 6 mi? In a distance of 60 mi?

**5.2** Calculate the combined effect of the Earth's curvature and refraction for the following distances: (a) 100 m; (b) 300 m; (c) 1 km; (d) 12 km; (e) 100 km.

**5.3** An observer standing on the shoreline of a lake can just see the top of a tower on an island. If the eye of the observer is 1.65 m above lake level and the top of the tower is 15 m above lake level, how far is the tower from the observer?

**5.4** The backsight to a bench mark is 2.125 m and the foresight from the same instrument setup is 4.587 m. If the distance from the level to the bench mark is 5 m and from the level to the turning point is 400 m, compute the difference in elevation between the bench mark and the turning point.

**5.5** Two points, A and B, are each distant 2000 ft from a third point C, from which the measured vertical angle to A is $+3°21'$ and that to B is $+0°32'$. What is the difference in elevation between A and B?

**5.6** Let A be a point of elevation 100.00 ft, and let B and C be points of unknown elevation. By means of an instrument set 4.00 ft above B, vertical angles are observed, that to A being $-1°55'$ and that to C being $+3°36'$. If the horizontal distance AB is 1500 ft and the horizontal distance BC is 5000 ft, what are the elevations of B and C, making due allowance for the Earth's curvature and atmospheric refraction?

**5.7** Two points, A and B, are 1000 ft apart. The elevation of A is 615.03 ft. A level is set up on the line between A and B and at a distance of 250 ft from A. The rod reading on A is 9.15 ft and that on B is 2.07 ft. Making the due allowance for curvature and refraction, what is the elevation of B? What would be the magnitude and sign of the error introduced if the correction for curvature and refraction were omitted?

**5.8** The zenith angles between stations A and B are $86°57'55''$ from A to B and $93°12'55''$ from B to A. The slope distance measured by EDM from A to B is 23,458.487 m and the elevation of station A above the datum is 322.073 m. The heights of theodolite and target above stations A and B are 1.55 m and 1.38 m, respectively. Similarly, the heights of theodolite and target above B and A for observing the zenith angle from B are 1.43 m and 1.51 m, respectively. The heights of EDM and reflector above stations A and B are 1.62 m and 1.41 m, respectively. Compute the elevation of station B.

**5.9** What is the radius of curvature of a level tube graduated to 0.1 in and having a sensitivity of $30''$ per division?

**5.10** What is the sensitivity of a level tube graduated to 2 mm having a radius of curvature of 10 m?

**5.11** A sight is taken with an engineer's level at a rod held 100 m away, and an initial reading of 1.927 m is observed. The bubble is then moved through five spaces on the level tube, when the rod reading is 2.008 m. What is the sensitivity of the level tube in seconds of arc? What is the radius of curvature of the level tube if one space is 2 mm?

**5.12**   Design (a) a direct vernier and (b) a retrograde vernier reading to thousandths of feet, each space on the rod being equal to 0.005 ft.

**5.13**   Design a direct vernier and retrograde vernier both reading to 1 mm for a rod graduated to centimetres. Draw a neat sketch of each vernier and a portion of the rod for a reading of 1.215 m with graduations shown and labeled.

**5.14**   On a rod graduated to 5 mm, a direct-reading vernier is to read to millimetres. State the following: (a) length of one vernier space; (b) number of spaces on the vernier; (c) number of spaces on the main scale corresponding to the full length of vernier scale; (d) least count of the vernier.

**5.15**   Design a direct vernier reading to 1 mm, applied to a scale graduated to 0.5 cm. For the same scale, design a retrograde vernier with least count of 0.2 mm. Sketch the verniers in relation to the scale.

**5.16**   A telescope is sighted on the 2.00-ft mark of a rod held on a distant point, and the corresponding reading of the gradienter is noted. The screw is turned until the 6.00-ft mark on the rod is sighted, when it is observed that 84 divisions have been turned off. How far is the rod from the instrument?

**5.17**   If the rod were inclined 0.4 ft forward in a length of 13 ft, what error would be introduced in a rod reading of 5.0 ft?

**5.18**   The rod placed on a pointed turning point is inclined backward 15 cm in a length of 3 m. What error is introduced in a rod reading of (a) 0.500 m; (b) 2.50 m?

**5.19**   A rod that is 50 mm square at the base is placed on a flat surface for a turning point. If the top of the 3-m rod is inclined backward 20 cm, what error is introduced in a reading of 0.300 m?

**5.20**   A line of differential levels was run between two bench marks 20 mi apart, and the measured difference in elevation was found to be 2163.4 ft. Later the rod whose nominal length was 13 ft was found to be 0.003 ft too short, the error being distributed over its full length. Correct the measured difference in elevation for erroneous length of rod.

**5.21**   Suppose that the levels of Prob. 5.20 had been run by using a rod which was 0.003 ft too short owing to wear on the lower end. What would have been the error?

**5.22**   Differential levels were run from B.M.$_1$ (el. 143.277 m) to B.M.$_2$, a distance of 60 km. On the average, the backsight distances were 100 m in length and the foresight distances were 50 m in length. The elevation of B.M.$_2$ as computed from the level notes was 1113.355 m. Compute the error due to Earth's curvature and atmospheric refraction, and correct the elevation of B.M.$_2$.

**5.23**   The levels of Prob. 5.22 were rerun using an average backsight distance of 50 m and an average foresight distance of 25 m. The elevation of B.M.$_2$ as deduced from the level notes was 1112.941 m. Compute the error due to curvature and refraction and correct the elevation of B.M.$_2$.

**5.24**   Suppose that the instrument used in running the levels of Probs. 5.22 and 5.23 was out of adjustment, so that when the bubble was centered the line of sight was inclined 0.001 m upward in a distance of 100 m. Correct the observed results of Probs. 5.22 and 5.23 for inclination of line of sight.

**5.25**   If in running levels between two points the rod were inclined 10 cm forward in a height of 3.5 m, what error would be introduced per setup when backsight readings averaged 3 m and foresight readings averaged 0.3 m?

**5.26**   If levels are run from B.M.$_1$ (el. 600 m) to B.M.$_2$ (observed el. 900 m) and the rod is on the average 6 cm out of plumb in a height of 3 m, what error is introduced due to the rod's not being plumb? What is the correct elevation of B.M.$_2$?

**5.27**   What would be the error if in Prob. 5.26 both bench marks were at the same elevation?

**5.28**   Complete the differential level notes shown below, putting the entire set of notes in proper field note form and making all the customary checks. All units are in metres.

| Station | B.S. | H.I. | F.S. | Elevation |
|---|---|---|---|---|
| B.M.$_A$ | 0.232 | | — | 45.272 |
| T.P.$_1$ | 0.503 | | 2.892 | |
| T.P.$_2$ | 0.212 | | 3.056 | |
| T.P.$_3$ | 1.246 | | 3.302 | |
| T.P.$_4$ | 2.169 | | 1.257 | |
| T.P.$_5$ | 2.895 | | 0.678 | |
| T.P.$_6$ | 3.004 | | 0.202 | |
| B.M.$_B$ | — | | 1.423 | |

**5.29** If sights average 200 ft in length and the standard deviation of a single observation is 0.004 ft/200 ft sight, what is the standard deviation of running a line of levels 25 mi long? 100 mi long?

**5.30** The error of closure of a level circuit 30 km long is 0.066 m. The average length of sight is 80 m. Assuming that all systematic errors have been eliminated, what is the estimated error per instrument setup? What is the estimated error for a single observation of the rod?

**5.31** If sights with a level average 100 m in length and the standard deviation per 100 m sight is 0.005 m, what is the standard deviation in running a line of levels 10 km long? 30 km?

**5.32** Complete the differential-level notes shown below, putting the entire set of notes in proper field note form. Determine the error of closure of the level circuit and adjust the elevations of B.M.$_2$ and B.M.$_3$, assuming that the error is a constant per setup.

| Station | B.S. | H.I. | F.S. | Elevation |
|---------|------|------|------|-----------|
| B.M.$_1$ | 4.127 | | — | 100.000 |
| T.P.$_1$ | 3.831 | | 9.346 | |
| T.P.$_2$ | 4.104 | | 10.725 | |
| T.P.$_3$ | 2.654 | | 12.008 | |
| B.M.$_2$ | 4.368 | | 7.208 | |
| T.P.$_4$ | 6.089 | | 6.534 | |
| T.P.$_5$ | 8.863 | | 4.736 | |
| B.M.$_3$ | 12.356 | | 2.100 | |
| T.P.$_6$ | 10.781 | | 3.662 | |
| T.P.$_7$ | 12.365 | | 4.111 | |
| B.M.$_1$ | — | | 9.059 | |

**5.33** The elevation of junction point $X$ has been determined by differential leveling from bench marks $A$, $B$ and $C$, which have elevations of 101.823 m, 95.342 m, and 93.243 m, respectively. The data are as follows:

| Line | Number of setups | Difference in elevation, m |
|------|------------------|----------------------------|
| AX | 4 | − 1.810 |
| BX | 6 | 4.660 |
| CX | 8 | 6.755 |

Compute the best estimate for the elevation of junction point 3.

**5.34** Lines of differential levels are run from B.M.$_1$ to B.M.$_2$ over three different routes. Following are the lengths of the routes and the observed elevations of B.M.$_2$. Determine the best estimate value of the elevation of B.M.$_2$.

| Route | Length, mi | Elevation of B.M.$_2$ |
|-------|-----------|------------------------|
| a | 10 | 742.81 |
| b | 16 | 742.58 |
| c | 40 | 743.27 |

**5.35** The following data are for a level net whose perimeter (reading clockwise) is *ABCDEFA*. Within the net, a line of levels extends from *B* to *F* and from *C* to *E*. The elevation of *A* is 100.00 ft. Adjust the elevations by the method of least squares (note, see Example C.7, Art. C.11).

| Circuit | From | To | Distance, mi | Difference in elevation, ft |
|---------|------|-----|------|------|
| ABFA | A | B | 40 | + 17.47 |
|  | B | F | 35 | − 10.87 |
|  | F | A | 52 | −  6.26 |
| BCEFB | B | C | 33 | + 11.88 |
|  | C | E | 16 | −  8.48 |
|  | E | F | 26 | − 14.01 |
|  | F | B | 35 | + 10.87 |
| CDEC | C | D | 27 | − 16.36 |
|  | D | E | 34 | +  7.59 |
|  | E | C | 16 | +  8.48 |

**5.36** The observations for determining the $C$ factor for level number 66564 are as follows: (1) from instrument position 1, the thread readings on rod $A$, the close backsight, are 1692, 1642, and 1592 mm; thread readings on rod $B$, the distant foresight, are 1703, 1479, and 1255 mm; (2) from position 2, the thread readings on rod $B$, the close backsight, are 1656, 1610, and 1564 mm; thread readings on distant rod $A$ are 2011, 1782, and 1555 mm. Compute the $C$ factor for this level.

**5.37** Thread readings in millimetres for precise levels run from B.M.$_K$ to B.M.$_L$ in the order taken (e.g., B.S., F.S.) are: (1) 3305, 3118, 2930; 0378, 0245, 0112; (2) 3266, 3122, 2978; 0472, 0355, 0238; (3) 2845, 2719, 2593; 1405, 1195, 0985; (4) 1371, 1158, 0945; 2803, 2681, 2560; (5) 0468, 0353, 0237; 3445, 3283, 3122; (6) 0500, 0377, 0252; 3265, 3085, 2903 mm. Put these observations in field note form, make all necessary checks, and determine the difference in elevation between B.M.$_K$ and B.M.$_L$. The $C$ factor for the level used was determined as $-0.0137$ mm/mm of interval.

**5.38** The data for profile levels in the order taken are as follows: B.S. on B.M.$_{10}$ = 3.845, F.S. on T.P.$_1$ = 1.234, B.S. on T.P.$_1$ = 3.005, I.F.S. on 0 + 00 = 2.54, I.F.S. on 0 + 10 = 1.52, I.F.S. on 0 + 20 = 0.80, I.F.S. on 0 + 30 = 0.12, F.S. on T.P.$_2$ = 0.064, B.S. on T.P.$_2$ = 3.945, I.F.S. on 0 + 35 = 3.80, I.F.S. on 0 + 40 = 3.00, I.F.S. on 0 + 50 = 2.35, I.F.S. on 0 + 60 = 2.00, I.F.S. on 0 + 70 = 1.50, I.F.S. on 0 + 80 = 1.00, I.F.S. on 0 + 90 = 2.42, I.F.S. on 1 + 00 = 2.68, I.F.S. on 1 + 08.50 = 2.785 (hub), F.S. on T.P.$_3$ = 3.002, B.S. on T.P.$_3$ = 0.105, F.S. on T.P.$_4$ = 3.421, B.S. on T.P.$_4$ = 0.062, F.S. on B.M.$_{10}$ = 3.231 m. Prepare profile-level notes for these data, make all necessary checks, adjust the H.I.'s in proportion to the number of setups, and calculate elevations for all stations. The elevation of B.M.$_{10}$ is 50.508 m above the datum.

**5.39** Plot the profile for the data of Prob. 5.38 at a horizontal scale of 1 : 500 and a vertical scale of 1 : 100.

**5.40** The data for profile levels in the order taken in the field are as follows: B.S. on B.M.$_{22}$ = 12.31, I.F.S. on 0 + 00 = 11.02 (hub), I.F.S. on 0 + 25 = 5.2, I.F.S. on 0 + 50 = 0.4, F.S. on T.P.$_1$ = 0.31, B.S. on T.P.$_1$ = 11.34, I.F.S. 0 + 75 = 7.9, I.F.S. on 0 + 85 = 8.60 (edge of sidewalk), I.F.S. on 0 + 97 = 6.42, I.F.S. 1 + 00 = 5.2, F.S. on T.P.$_2$ = 2.57, B.S. on T.P.$_2$ = 11.16, I.F.S. on 1 + 25 = 9.5, I.F.S. 1 + 44 = 6.8, I.F.S. 1 + 50 = 6.2, I.F.S. 1 + 52 = 6.1, I.F.S. 1 + 75 = 3.5, F.S. on T.P.$_3$ = 3.49, B.S. on T.P.$_3$ = 2.67, F.S. on T.P.$_3$ = 12.73, B.S. on T.P.$_3$ = 1.88, F.S. on T.P.$_4$ = 12.42, B.S. on T.P.$_4$ = 3.24, F.S. on B.M.$_{22}$ = 11.14, all units in feet. Prepare profile level notes for these data, make all necessary checks, adjust the H.I.'s in proportion to the number of instrument setups, and calculate elevations for all stations. The elevation of B.M.$_{22}$ is 155.25 ft above the datum.

**5.41** Plot the profile of the data in Prob. 5.40 using a horizontal scale of 1 : 240 and a vertical scale of 1 : 120. Include all the information considered necessary for a profile on the drawing.

**5.42** Reciprocal leveling across a canyon from station 1 to station 2, with simultaneous readings using two tilting levels and two rods on opposite sides of the canyon, yielded the following average readings:

| Instrument station | Average near reading, m | Average distant readings, m |
|---------------------|------|------|
| 1 | 2.685 | 3.241 |
| 2 | 1.521 | 0.973 |

The distance from station 1 to 2 is approximately 1 km. Compute the difference in elevation between stations 1 and 2.

**5.43**  In the two-peg test of a level, the following observations are taken:

|  | Instrument<br>at $M$ | Instrument<br>at $P$ |
|---|---|---|
| Rod reading on $A$ | 3.612 | 1.862 |
| Rod reading on $B$ | 3.248 | 0.946 |

$M$ is equidistant from $A$ and $B$; $P$ is 40 ft from $A$ and 240 ft from $B$. What is the true difference in elevation between the two points? With the level in the same position at $P$, to what rod reading on $B$ should the line of sight be adjusted? What is the corresponding rod reading on $A$ for a horizontal line of sight? Check these two rod readings against the true difference in elevation, previously determined.

## 5.61. Field problems

### PROBLEM 1   RADIUS OF CURVATURE AND SENSITIVITY OF LEVEL TUBE

**Object**  To determine, in the field without the use of special apparatus, the radius of curvature of the level tube of transit or level and the sensitivity of the bubble (Art. 5.15).

**Procedure**  (1) Hold the rod on a solid point 300 ft or 100 m from the instrument. With one end of the bubble at a division near the end of the level tube, take a careful rod reading to the nearest 0.001 ft or 0.001 m. Note the exact position of each end of the bubble. (2) Manipulate the leveling screws until the other end of the bubble falls near the other end of the tube. Take another rod reading, and measure the exact distance traversed by each end of the bubble. (3) Determine the bubble movement expressed as the number of divisions moved times the length of a single graduation on the bubble tube $= (n)(i)$. On most levels, $i = 2$ mm. Also determine the difference between the two rod readings $d$ (see Fig. 5.13), expressed in the same units as the bubble tube graduation $i$. (4) In this manner obtain a series of five bubble movements and their corresponding rod readings. (5) Compute the radius of curvature by the formula $r = (ni/d)a$, in which $r$ is the radius of curvature, $a$ is the distance from the instrument to the rod, $ni$ is the mean of the five bubble movements, and $d$ is the mean of the five differences in rod readings. (6) Compute the sensitivity of the level tube in seconds of arc per division using Eq. (5.17).

### PROBLEM 2   DIFFERENTIAL LEVELING WITH ENGINEER'S LEVEL AND SELF-READING ROD

**Object**  Given the elevation of an initial bench mark, to determine the elevations of points in an assigned level circuit.

**Procedure**  Follow the procedure outlined in Art. 5.38. Keep notes in the form of the sample notes shown in Fig. 5.36, but estimate rod readings to thousandths of feet or to mm. Check each rod reading by taking a second observation. Close the circuit, and compute the error of closure.

**Hints and precautions**  (1) When using *self-leveling instruments* (Art. 5.19): (a) Try to keep the base plate of the tripod as close to a horizontal position as possible when setting the tripod, or it may be impossible to center the bull's-eye bubble, owing to insufficient run on the level screws. (b) Center the bull's-eye bubble as directed in Art. 5.33 and then turn the instrument through 180°. The bull's-eye bubble should stay within the circle (see Art. 5.59). (c) To check the compensator, sight the rod, note the reading at the middle cross hair, and then tap the leg of the tripod. If the compensator is working properly, the image will appear to move in the field of vision but the middle cross hair will return to its original position. (d) Always double-check to make sure the rod reading was taken at the middle and not at one of the stadia cross hairs. (2) When using *spirit level instruments* of the dumpy level type: (a) sight in the direction of the rod and focus before centering the bubble exactly; (b) center the bubble before each reading and check its position immediately afterward. For each type of instrument, (3) test for parallax (Art. 5.42). (4) Choose the turning points with an eye to simplicity of field operations, but roughly balance the backsight and foresight distances between bench marks; keep no record of these distances. (5) Use signals (Art. 3.9). (6) Keep the foot of the rod free from dirt. (7) Be sure that the rod is held vertical while a sight is being taken. When the "long rod" is used, wave the rod, and before each reading check the position of the back vernier. (8) Describe each bench mark in the field notes. (9) Check the computations.

## References

1. Berry, R. M., "History of Geodetic Leveling in the United States," *Surveying and Mapping*, Vol. 36, No. 2, June 1976, pp. 137–154.
2. Berry, R. M., "Observational Techniques for Use with Compensator Leveling Instruments for First-Order Levels," *Surveying and Mapping*, Vol. 37, No. 1, March 1977, p. 17.
3. Federal Geodetic Control Committee, *Specifications to Support Classification, Standards of Accuracy, and General Specifications of Geodetic Control Surveys*, U.S. Department of Commerce, National Oceanic and Atmospheric Administration, Rockville, Md., 1975, 1976.
4. Hou, G.-Y., Veress, S. A., and Colcord, J. E., "Refraction in Precise Leveling," *Surveying and Mapping*, Vol. 32, No. 2, June 1972, p. 231.
5. Hradilek, L., "Refraction in Trigonometric and Three-Dimensional Terrestrial Networks," *The Canadian Surveyor*, Vol. 26, No. 1, March 1972, pp. 59–70.
6. Huether, G., "New Automatic Level NI 002 of the Jena Optical Works for First and Second Order Leveling," *Proceedings, Fall Convention, ACSM*, 1974, p. 365.
7. Moffitt, F. H., and Bouchard, H., *Surveying*, 6th ed., Harper & Row Publishers, New York.
8. Rappleye, H. S., *Manual of Leveling Computation and Adjustment*, Spec. Publ. No. 240, U.S. Department of Commerce, Coast and Geodetic Survey (now National Ocean Survey, NOAA), Washington, D.C., 1948; reprinted 1975.
9. Rappleye, H.S., *Manual of Geodetic Leveling*, Spec. Publ. No. 239, U.S. Department of Commerce, Coast and Geodetic Survey (now National Ocean Survey, NOAA), Washington, D.C., 1948; reprinted 1963/1976.
10. Selley, A. D., "A Trigonometric Level Crossing of the Strait of Belle Isle," *The Canadian Surveyor*, Vol. 31, No. 3, September 1977.
11. Whalen, C. T., and Balacz, E. I., "Test Results of First-Order Class III Leveling," *Surveying and Mapping*, Vol. 37, No. 1, March 1977, pp. 45–58.

# CHAPTER 6
# Angle and direction measurement

**6.1. Location of points**  As previously stated, the purpose of a survey is to determine the relative locations of points below, on, or above the surface of the earth. Since the earth is three-dimensional, the most logical reference framework with which to locate points is a three-dimensional rectangular coordinate system. Figure 6.1 shows an earth-centered, three-dimensional rectangular coordinate system $X'Y'Z'$ called the geocentric coordinate system. It is a right-handed system, with $X'$ passing through Greenwich in England and $Z'$ passing through the North Pole ($Y'$ is uniquely fixed because $X'Y'Z'$ is specified as right-handed). In addition to the $X'Y'Z'$ coordinates, any point may also be located by a set of spherical coordinates consisting of latitude $\phi$, longitude $\lambda$, and the distance $\rho+h$ along the normal to the earth ellipsoid. The latter system is used mainly for geodetic surveys and is not discussed in further detail in this text.

A reference more suitable for plane surveying is a local $XYZ$ right-handed rectangular coordinate system similar to the one illustrated in Fig. 6.2. The origin of such a system is usually chosen near the center of the area to be surveyed. Its $XY$ plane is tangent to the reference ellipsoid at the point of origin, and the $Y$ axis is generally directed toward the North Pole. Any point $P$ may be located either by $XYZ$ or by the two angles $\alpha$ (in the $XY$ plane) and $\beta$, and the distance $r$ from $O$ to $P$, as shown in Fig. 6.2.

Visualization of the local $XYZ$ system is more convenient when the $Z$ axis coincides with the reader's local vertical or the plumb line, as illustrated in Fig. 6.3. In this figure, point $A$, which is assumed to be of known position, is also introduced. With respect to Fig. 6.3, the position of point $P$ in space may be determined in one of the following measurement procedures:

1. The direction of line $OP$ (angles $\alpha,\beta$) and the distance $r=OP$.
2. The directions of lines $OP$ (angles $\alpha,\beta$) and $AP$ (angles $\alpha'$ in the $XY$ plane and, $\beta'$ in plane $APP'A'$) from two known points $O$ and $A$.
3. The distances $r$ (from $O$ to $P$), $r'$ (from $A$ to $P$), and $h'$ (height of $P$ above the $XY$ plane).

For most surveys of small extent, the positions of $P$ and $A$ and the slope distances $r$ and $r'$ are projected onto the horizontal $XY$ plane. Still referring to Fig. 6.3, the position of $P'$ in the horizontal $XY$ plane is defined when measurements in one of the following cases are given:

1. Distance and direction from one known point, such as $r_h$ and $\alpha$ from known point $O$.
2. Directions from two known points such as $\alpha$ and $\alpha'$ from known points $O$ and $A$, respectively.
3. Distances from two known points such as $r_h$ and $r'_h$ from points $O$ and $A'$.

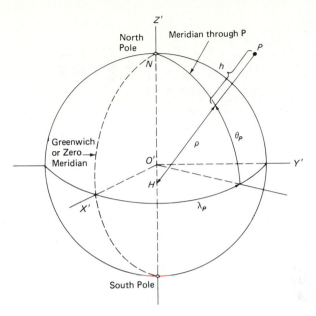

**Fig. 6.1** Position of $P$ in geocentric system.

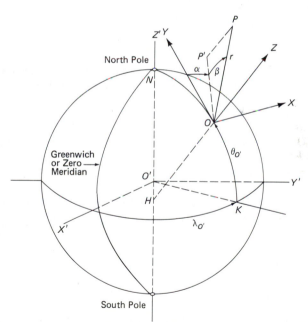

**Fig. 6.2** Local coordinate system.

Note that the direction of spatial line $OP$ is specified by angles $\alpha$ in the horizontal $XY$ plane and $\beta$ in the vertical plane $OPP'$. Similarly, the direction of line $AP$ is determined by $\alpha'$ and $\beta'$ in the horizontal $XY$ plane and the vertical plane $APP'A'$, respectively. The directions of horizontal lines $OP'$ and $A'P'$ are specified only by horizontal angles $\alpha$ and $\alpha'$, respectively. Thus, in general the angular measurements of surveying are either horizontal or vertical.

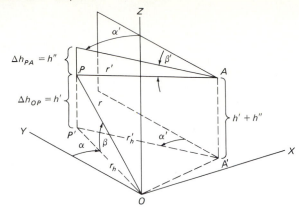

**Fig. 6.3** Position of a point
in local coordinate system.

The angle between two points is understood to mean the horizontal angle or angle between projections in the horizontal plane of two lines passing through the two points and converging at a third point. Thus, at $P$ in Fig. 6.3, the angle between $A$ and $O$ is the horizontal angle $OP'A'$.

The vertical angle to a point is the angle of elevation or depression measured from the horizontal. In Fig. 6.3, $\beta$ is an elevation vertical angle from $O$ to $P$ and $\beta'$ is a depression vertical angle from $A$ to $P$. Vertical angles can also be measured from the zenith (the $OZ$ axis). For example, in Fig. 6.3 the zenith angle from $O$ to $P$ is angle $ZOP$.

**6.2. Meridians**  The relative directions of lines connecting survey points may be obtained in a variety of ways. Figure 6.4 shows lines intersecting at a point. The direction of any line (as $OB$) with respect to an adjacent line (as $OA$) is given by the horizontal angle between the two lines (as $\alpha_2$) and the direction of rotation (as clockwise). The direction of any line (as $OC$) with respect to a line not adjacent (as $OA$) is not given by any of the measured angles but may be computed by adding the intervening angles (as $\alpha_2 + \alpha_3$).

Figure 6.5 shows the same system of lines but with all angles measured from a line of reference $OM$. The direction of any line (as $OA$) with respect to the line of reference (as $OM$) is given by the angle between the lines (as $\beta_1$) and its direction of rotation (as clockwise). The angle between any two lines (as $AOC$) is not given directly, but may be computed by taking the difference between the direction angles of the two lines (as $\angle \beta_3 - \angle \beta_1 = \angle AOC$).

The fixed line of reference may be any line in the survey, or it may be purely imaginary. It is termed a *meridian*. If it is arbitrarily chosen, it is called an *assumed meridian*; if it is a north-and-south line passing through the geographical poles of the earth, it is called a *true* or *astronomic meridian*; if it is a line parallel to a central true meridian, it is called a *grid meridian*; or if it lies parallel with the magnetic lines of force of the earth as indicated by the direction of a magnetized needle, it is called a *magnetic meridian*.

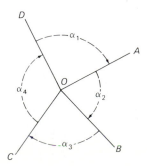

**Fig. 6.4** Directions by angles.

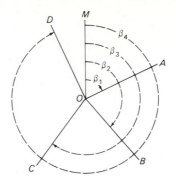

**Fig. 6.5** Directions referred to meridian.

**6.3. True meridian**   The *true meridian*, or *astronomic meridian*, is determined by astronomical observations as described in Chap. 11. In Fig. 6.2, arc *NOK* is a true meridian. For any given point on the earth its direction is always the same, and hence directions referred to the true meridian remain unchanged regardless of time. The lines of most extensive surveys and usually the lines marking the boundaries of landed property are referred to the true meridian.

At many triangulation stations established throughout the United States by the National Ocean Survey, reference lines of known true direction have been established for use by surveyors.

**6.4. Grid meridian**   Examination of Fig. 6.1 shows that true meridians converge at the poles. In plane surveys of limited extent, it is convenient to perform the work in a rectangular *XY* coordinate system in which one central meridian coincides with a true meridian. All remaining meridians are parallel to this central true meridian, eliminating the need to calculate the convergence of meridians when determining positions of points in the system. These parallel meridians are called *grid meridians*.

Each state in the United States has plane coordinate projections established by the National Ocean Survey. These projections are designed so that surveys can be performed as on a horizontal surface within tolerable levels of accuracy. The two projections used for this purpose in the United States are the *Lambert conformal projection* and the *Universal Transverse Mercator projection*, which are described in Chap. 14.

**6.5. Magnetic meridian**   The direction of the *magnetic meridian* is that taken by a freely suspended magnetic needle. The magnetic poles are at some distance from the true geographic poles; hence, in general the magnetic meridian is not parallel to the true meridian. The location of the magnetic poles is constantly changing; therefore, the direction of the magnetic meridian is not constant. However, the magnetic meridian is employed as a line of reference on rough surveys where a magnetic compass is used and often is employed in connection with more precise surveys in which angular measurements are checked approximately by means of the compass. It was formerly used extensively for land surveys.

Details concerning magnetic compasses and meridians can be found in Arts. 6.13 to 6.20.

**6.6. Angles and directions**   Angles and directions may be defined by means of *bearings*, *azimuths, deflection angles, angles to the right*, or *interior angles*, as described in the following sections. These quantities are said to be *observed* when obtained directly in the field, and *calculated* when obtained indirectly by computation. Conversion from one means of expressing angles and directions to another is a simple matter if a sketch is drawn to show the existing relations.

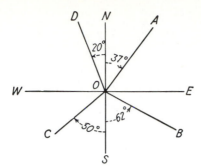

**Fig. 6.6** Bearings.

**6.7. Bearings**  The direction of any line with respect to a given meridian may be defined by the *bearing*. Bearings are called *true (astronomic) bearings, magnetic bearings,* or *assumed bearings* depending on whether the meridian is true, magnetic, or assumed. The bearing of a line is indicated by the quadrant in which the line falls and the acute angle which the line makes with the meridian in that quadrant. Thus, in Fig. 6.6 the bearing of the line *OA* is read north 37° east and is written N37°E. The bearings of *OB*, *OC*, and *OD* are, respectively, S62°E, S50°W, and N20°W. In all cases, values of bearing angles lie between 0° and 90°. If the direction of the line is parallel to the meridian and north, it is written as N0° or *due North*; if perpendicular to the meridian and east, it is written as N90°E or *due East*.

In Fig. 6.7, if the observed bearing of *OA* is N37°E and the angle *AOB*=81°, the calculated bearing of *OB* is S62°E.

**6.8. Azimuths**  The *azimuth* of a line is its direction as given by the angle between the meridian and the line measured in a clockwise direction usually from the north branch of the meridian. In astronomical observations azimuths are generally reckoned from the true south; in surveying, some surveyors reckon azimuths from the south and some from the north branch of whatever meridian is chosen as a reference, but on any given survey the direction of zero azimuth is either always south or always north. Thus, it is necessary to designate whether the azimuth is from the north or from the south. Azimuths are called *true (astronomic) azimuths, magnetic azimuths,* or *assumed azimuths* depending on whether the meridian is true, magnetic, or assumed. Azimuths may have values between 0 and 360°.

In Fig. 6.8, azimuths measured from the south point are $A_{OA}=217°$, $A_{OB}=298°$, $A_{OC}=50°$, and $A_{OD}=160°$; or in Fig. 6.9, in which are shown the same lines with azimuths measured from the north point, $A_{OA}=37°$, $A_{OB}=118°$, $A_{OC}=230°$, and $A_{OD}=340°$. In Fig. 6.8 if the observed azimuth of *OA* as reckoned from the south is 217° and the observed angle *AOB* is 81°, the calculated azimuth of *OB* is 298°.

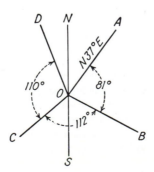

**Fig. 6.7** Angles and bearings.

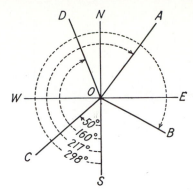

**Fig. 6.8** Azimuths from the south.

Azimuths may be calculated from bearings, or vice versa, preferably with the aid of a sketch. For example, if the bearing of a line is N16°E, its azimuth (from south) is $180 + 16 = 196°$; and if the azimuth (from south) of a line is 285°, its bearing is $360 - 285 =$ S75°E.

In some special cases, the term "azimuth" is used in the sense of a bearing and therefore may be taken either clockwise or counterclockwise, as in "azimuth of Polaris" (Art. 11.39).

It is assumed that when an azimuth is given as being from $O$ to $A$, as in Fig. 6.9, where the azimuth from north $A_{OA} = 37°$, this specifies the direction of the line from an origin at $O$ to a terminal point $A$ and the azimuth is called a *forward azimuth*. Conversely, the azimuth from $A$ to $O$ is called the *back azimuth* of $OA$. When the azimuth of a line is less than 180°, the back azimuth of the line is the forward azimuth plus 180°. For example, the back azimuth of $OA$ in Fig. 6.9 is $37° + 180° = 217°$. When the forward azimuth of a line is greater than 180°, the back azimuth equals the forward azimuth minus 180°. In Fig. 6.9, the back azimuth of line $OC = 230° - 180° = 50°$. The concept of forward and back azimuth is further illustrated in Fig. 6.10, where the azimuth from the north of line $CD = 60°$. Now, suppose that the initial point was $D$. Then the azimuth of $DC$ is 240° and the back azimuth of $DC$ is $240° - 180° = 60°$. Note that the idea of forward and back azimuth as developed here is valid only for plane surveys of limited extent where grid north is used as the reference meridian.

**6.9. Interior angles**   In a closed polygon, angles inside the figure between adjacent lines are called *interior angles*. Figure 6.11 illustrates interior angles $\alpha_1, \alpha_2, \ldots, \alpha_6$ in polygon $ABCDEF$. If $n$ equals the number of sides in a closed polygon, the sum of the interior angles is $(n-2)(180°)$. In Fig. 6.11, $\Sigma_{i=1}^{6} \alpha_i = (4)(180°) = 720°$.

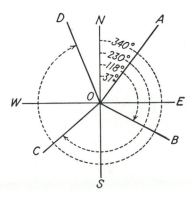

**Fig. 6.9** Azimuths from the north.

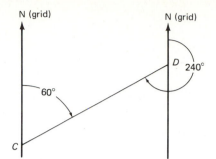

**Fig. 6.10** Forward and
back azimuths.

**6.10. Deflection angles**    The angle between a line and the prolongation of the preceding
line is called a *deflection angle*. Deflection angles are recorded as *right* or *left* depending on
whether the line to which measurement is taken lies to the right (clockwise) or left
(counterclockwise) of the prolongation of the preceding line. Thus, in Fig. 6.12 the deflection
angle at $B$ is 22°R, and at $C$ is 33°L. Deflection angles may have values between 0° and
180°, but usually they are not employed for angles greater than 90°. In any closed polygon
the algebraic sum of the deflection angles (considering right deflections as of sign opposite to
left deflections) is 360°. Figure 6.13 shows polygon *ABCDE* with deflection angles
$\alpha_1, \alpha_2, \ldots, \alpha_5$. In this example $\alpha_1 + \alpha_2 - \alpha_3 + \alpha_4 + \alpha_5 = 360°$.

**6.11. Angles to the right**    Angles may be determined by clockwise measurements from the
preceding to the following line, as illustrated by Fig. 6.14. Such angles are called *angles to
right* or *azimuths from back line*.

**6.12. Methods of determining angles and directions**    Angles are normally measured with
a transit but can also be determined by means of a tape, plane-table alidade, sextant, or
compass. Directions are observed with a direction theodolite or with a magnetic compass.

**1. Transit**    The engineer's transit is designed to enable observing horizontal and vertical
angles. Most transits allow measuring the horizontal angle to the nearest minute of arc but
on some instruments the horizontal angle can be observed to the nearest 10″ of arc. Vertical
angles can be read to the nearest minute of arc. A detailed description of the engineer's
transit and its use can be found in Arts. 6.22 to 6.29.

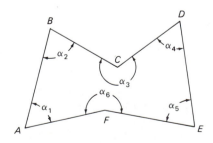

**Fig. 6.11** Interior angles in
polygon.

**Fig. 6.12** Deflection angles.

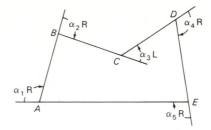

**Fig. 6.13** Deflection angles in polygon.

**2. Tape**  If the sides of a triangle are measured, sufficient data are obtained for computing the angles in the triangle. The error in the computed values of the angles depends on the care with which points are established and the accuracy with which measurements are taken. For acute angles on level ground the error need not exceed 05′ to 10′. For angles greater than 90°, the corresponding acute angle should be observed. The method is slow and is generally used only as a check.

**3. Plane table and alidade**  Using a plane table, which is essentially a drawing board mounted on a tripod, and an alidade, which is a straight edge with parallel line of sight, angles can be determined graphically. The plane table and alidade and its use are described in Art. 7.19.

**4. Sextant**  Although the sextant is used primarily by the navigator, it is also employed by the surveyor, principally on hydrographic surveys. Using the sextant, an angle may be measured while the observer is moving; hence, angles may be read from a boat from which soundings are taken. The angle measured by the sextant is in the plane defined by the two points sighted and the telescope and therefore is not in general a horizontal angle. The sextant is not an instrument of high accuracy for small angles of less than 15° or for short distances of less than 300 m or 1000 ft. The sextant and its use are described in Arts. 6.59 and 6.60.

**5. Direction theodolite**  The terms "transit" and "theodolite" are frequently used interchangeably. In the United States, *direction theodolite* refers to an extremely precise transit which has only one horizontal motion. With this type of instrument, directions are observed and the angle is then computed as the difference between the two directions. Direction theodolites usually have a nominal accuracy of ±1″ of arc. The characteristics and use of these instruments are given in Arts. 6.30, 6.40, and 6.41.

**6. Magnetic compass**  The use of the magnetic compass is described in the following sections. By itself the compass is useful in making rough surveys and retracing early land surveys. Mounted on the transit, the compass is useful as a means of approximately checking horizontal angles measured by more precise methods.

**6.13. Magnetic compass**  Any slender symmetrical bar of magnetized iron when freely suspended at its center of gravity takes up a position parallel with the lines of magnetic force

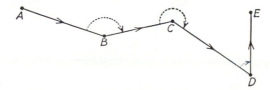

**Fig. 6.14** Angles to right.

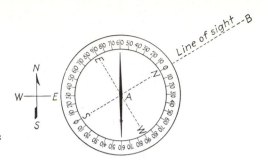

**Fig. 6.15** Features of magnetic compass used in surveying.

of the earth. In horizontal projection these lines define the magnetic meridians. In elevation, the lines are inclined downward toward the north in the northern hemisphere, and downward toward the south in the southern hemisphere. Since the bar takes a position parallel with the lines of force, it becomes inclined with the horizontal. This phenomenon is called the *magnetic dip*. The angle of dip varies from 0° at or near the equator to 90° at the magnetic poles. The needle of the magnetic compass rests on a pivot. To counteract the effect of dip, so that the needle will take a horizontal position when directions are observed, a counterweight is attached to one end (the south end in the Northern Hemisphere). The counterweight usually consists of a short piece of fine brass wire wound around the needle and held in place by spring action. As long as the needle is used in a given locality and loses

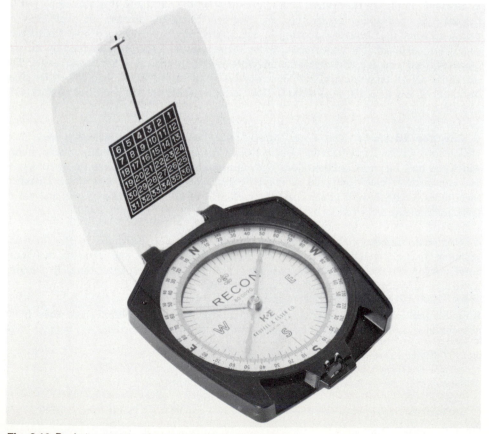

**Fig. 6.16** Pocket compass. (*Courtesy Keuffel & Esser Co.*)

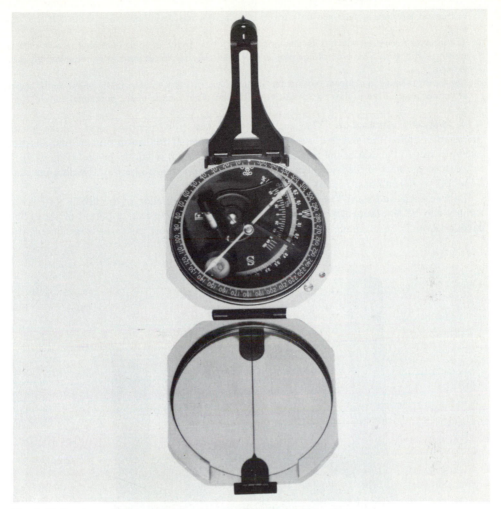

**Fig. 6.17** Brunton pocket transit. (*Courtesy Keuffel & Esser Co.*)

none of its magnetism, it will remain balanced. When for any reason it becomes unbalanced, it is adjusted to the horizontal by sliding the counterweight along the needle. At the midpoint of the needle is a jewel which forms a nearly frictionless bearing for the pivot.

The essential features of the magnetic compass used by the surveyor are (1) a compass box with a circle graduated from 0° to 90° in both directions from the N and S points and usually having the E and W points interchanged, as illustrated in Fig. 6.15; (2) a line of sight in the direction of the SN points of the compass box; and (3) a magnetic needle. When the line of sight is pointed in a given direction, the compass needle (when pivoted and brought to rest) gives the magnetic bearing. Thus, in the figure the bearing of *AB* is N60°E. If the N point of the compass box is nearest the object sighted, the bearing is read by observing the north end of the needle.

The varieties of compasses exhibiting the features just mentioned are:

1. Various *pocket compasses*, which are generally held in the hand when bearings are observed and are used in reconnaisance or other rough surveys. Figure 6.16 illustrates one type of pocket compass. Another style, the Brunton pocket transit, is shown in Fig. 6.17. It is designed primarily as a hand

instrument but may be mounted on a tripod or a Jacob's staff [a pointed stick about 5 ft (1.52 m) long].

2. The *surveyor's compass*, which is usually mounted on a light tripod or sometimes on a Jacob's staff, is shown in Fig. 6.18. This type of compass is now used only for forest surveys or retracing old land surveys.

3. The *transit compass*, a compass box similar to that of the surveyor's compass, mounted on the upper or vernier plate of the engineer's transit (Art. 6.22) and often used to check horizontal angles.

**6.14. Magnetic declination**   The angle between the true meridian and the magnetic meridian is called the *magnetic declination*, or *variation*. If the north end of the compass needle points to the east of the true meridian, the declination is said to be east (Fig. 6.19); if it points to the west of the true meridian, the declination is said to be west. Declination may be set off on the compass as shown in Fig. 6.20.

If a true north-and-south line is established, the mean declination of the needle for a given locality can be determined by compass observations extending over a period of time. The declination may be estimated with sufficient precision for most purposes from an *isogonic*

**Fig. 6.18** Surveyor's compass. (*Courtesy Keuffel & Esser Co.*)

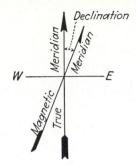

**Fig. 6.19** Declination east.

*chart* published by the U.S. National Ocean Survey; specific values for a particular locality can be obtained from the Survey.

**6.15. Isogonic chart**   The isogonic chart of the continental United States shown in Fig. 6.21 applies to January 1, 1975. It is based upon observations made by the U.S. National Ocean Survey at stations widely scattered throughout the country. The solid lines are lines of equal magnetic declination, or *isogonic lines*. East of the heavy solid line of zero declination, or *agonic line*, the north end of the compass needle points west of north; west of that line it points east of north. The north end of the compass needle is moving eastward over the area of eastward annual change and westward elsewhere over the chart at an annual rate indicated by the lines of equal annual change.

**6.16. Variations in magnetic declination**   The magnetic declination changes more or less systematically in cycles over periods of (1) approximately 300 years, (2) 1 year, and (3) 1 day, as follows:

**1. Secular variation**   Like a pendulum, the magnetic meridian swings in one direction for perhaps a century and a half until it gradually comes to rest and then swings in the other direction, and as with a pendulum the velocity of movement is greatest at the middle of the swing. The rate of change per year, however, varies irregularly. The causes of this secular variation are not well understood. In the United States, it amounts to several degrees in a half-cycle of approximately 150 years. In Fig. 6.21 the annual rates of change in the secular variation for the year 1975 are shown by dashed lines. Because of its magnitude, the secular variation is of considerable importance to the surveyor, particularly in retracing lines the directions of which are referred to the magnetic meridian as it existed years previously.

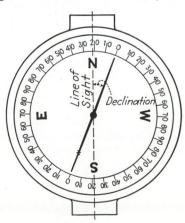

**Fig. 6.20** Declination set off on compass circle.

**Fig. 6.21** Isogonic chart for the United States, showing magnetic declination in the United States for 1975. (*Courtesy U.S. Geological Survey.*)

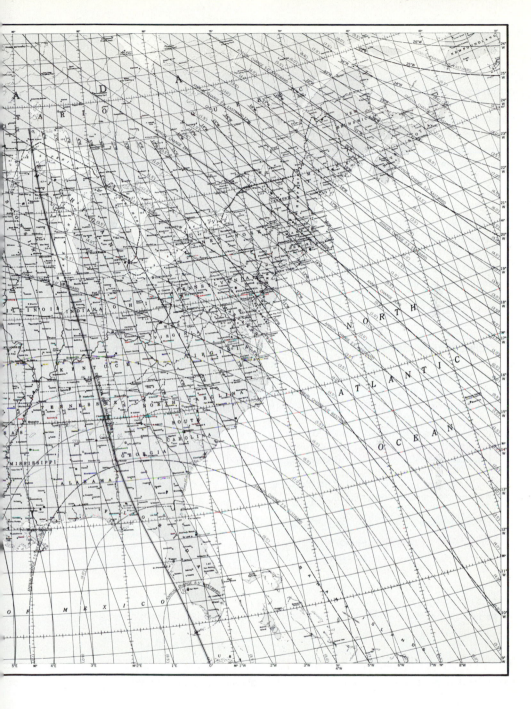

When *variation* is mentioned without further qualification, it is taken to mean the secular variation.

**2. Annual variation**   This is a small annual swing distinct from the secular variation. For most places in the United States, it amounts to less than 01′.

**3. Daily variation**   This variation, also called *solar-diurnal* variation, is a periodic swing of the magnetic needle occurring each day. For points in the United States the north end of the needle reaches its extreme easterly swing at about 8 or 9 A.M. and its extreme westerly swing at about 1 or 2 P.M. The needle usually reaches its mean position between 10 and 11 A.M. and between 7 and 11 P.M. In general, the higher the latitude, the greater the range in the daily variation. The average range for points in the United States is less than 08′, a quantity so small as to need no consideration for most of the work for which the compass needle is employed. However, in the United States in summer, a line run 1000 ft by compass at 8 A.M. would end as much as 3 ft to the right of the point where it would end if run at 1 P.M.

**4. Irregular variations**   Irregular variations are due to magnetic disturbances usually associated with sunspots. They cannot be predicted but are most likely to occur during magnetic storms, when auroral displays occur and radio transmission is disturbed. They may amount to a degree or more, particularly at high latitudes.

**6.17. Correction for declination**   When magnetic directions are used to obtain coarse estimates for bearings or when an old survey must be retraced, it is necessary to reduce the magnetic directions to true bearings or azimuths. Conversion from magnetic to true azimuths, or vice versa, is most easily accomplished by using azimuths. Consider the following examples.

**Example 6.1**   A magnetic azimuth of 54°30′ was observed along line *AD* in June 1977. The declination for the area surveyed is found by interpolation from an isogonic chart dated 1970 to be 17°30′E with an annual change of 1′ westward. Compute the true azimuth of line *AD*.

SOLUTION   First draw a careful sketch of the relationship among true north, magnetic north, and the direction of the line as illustrated by Fig. 6.22a.

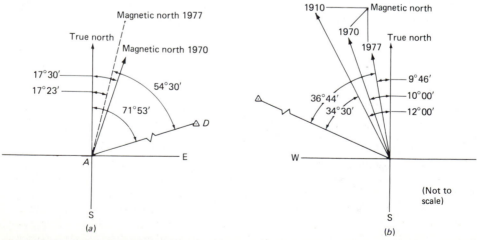

**Fig. 6.22** Corrections for declination.

| | |
|---|---|
| Magnetic azimuth *AD* | 54°30′ |
| Declination 1970 | 17°30′ |
| Change = (7 yr)(1′/yr) | − 0°07′ |
| True azimuth *AD* | 71°53′ |

**Example 6.2**   A magnetic bearing of N34°30′W is recorded on an old survey plan dated August 20, 1910. It is desired to reestablish this direction on the site in 1977. The 1970 isogonic chart shows a declination of 10°W for the area, with an annual change of 2′ eastward. Determine the magnetic bearing that must be used to relocate the direction of the line in the field.

SOLUTION   As before, draw a careful sketch of the lines involved, as shown in Fig. 6.22*b*.

| | |
|---|---|
| Declination in 1970 | 10°00′W |
| Change in 60 yr = 60 × 2′ | 2°00′ |
| Declination 1910 | 12°00′W |
| Magnetic bearing 1910 | N34°30′W |
| True bearing of line | N46°30′W |
| Declination 1977 = 10°W − (7 yr)(2′E) | 9°46′W |
| Magnetic bearing, 1977 | N36°44′W |

**6.18. Local attraction**   Objects of iron or steel, some kinds of iron ore, and currents of direct electricity alter the direction of the lines of magnetic force in their vicinity and hence are likely to cause the compass needle to deviate from the magnetic meridian. The deviation arising from such local sources is called *local attraction* or *local disturbance*. In certain localities, particularly in cities, its effect is so pronounced as to render the magnetic needle of no value for determining directions. It is not likely to be the same at one point as at another, even though the points may be but a short distance apart. It is even affected by such objects as the steel tape, chaining pins, axe, and small objects of iron or steel that are on the person. Usually, its magnitude can be determined, and directions observed with the compass can be corrected accordingly. Local attraction can usually be detected by observing the compass bearing of a line at two or more points on the line.

**6.19. Use of the compass**   In order that *true* bearings may be read directly, some compasses, such as the one shown in Fig. 6.20, are designed so that the compass circle may be rotated with respect to the box in which it is mounted. When the circle is in its normal position, the line of sight as defined by the vertical slits in the sight vanes is in line with the N and S points of the compass circle, and the observed bearings are magnetic. If the magnetic declination is set off by means of the circle, the observed bearings will be true, as is evident from Fig. 6.20. If the declination is east, as in the figure, the circle is rotated clockwise with respect to the plate; if the declination is west, counterclockwise.

When the direction of a line is to be determined, the compass is set up on line and is leveled. The needle is released, and the compass is rotated about its vertical axis until a range pole or other object on line is viewed through the slits in the two sight vanes. When the needle comes to rest, the bearing is read. Ordinarily, the sight vane at the end of the compass box marked "S" is held next to the eye; in this case the bearing is given by the north end of the needle.

The following suggestions apply to compass observations: At each observation the compass box should be tapped lightly as the needle comes to rest, so that the needle may swing freely. In order not to confuse the north and south ends of the needle when taking bearings, the observer should always note the position of the counterbalancing wire (which is on the south end in the northern hemisphere). Since the *precision* with which angles may be read depends on the delicacy of the needle, special care should be taken to avoid jarring between the jewel bearing of the needle and the pivot point. *Never move the instrument without making certain that the needle is lifted and clamped.*

Sources of magnetic disturbance such as chaining pins and axe should be kept away from the compass while a reading is being taken. Care should be taken not to produce static charges of electricity by rubbing the glass; a moistened finger pressed against the glass will remove such charges. Ordinarily the amount of metal about the person of the instrument-man is not large enough to deflect the needle appreciably, but a change of position between two readings should be avoided.

Surveying with the compass is usually by traversing. Only alternate stations need be occupied, but a check is secured and local attraction is detected if both a backsight and a foresight are taken from each station. Unlike a transit traverse, in which an error in any angle affects the observed or computed directions of all following lines, an error in the observed bearing of one line in a compass traverse has no effect upon the observed *directions* of any of the other lines. This is an important advantage, especially in the case of a traverse having many angles. Another advantage of the compass is that obstacles such as trees can be passed readily by offsetting the instrument a short measured distance from the line. The procedure for performing and adjusting a compass traverse is outlined in Art. 8.11.

### 6.20. Sources of error; adjustment of compass

*1. Needle bent*  If the needle is not perfectly straight, a constant error is introduced in all observed bearings. As shown by Fig. 6.23*a*, one end of the needle will read higher than the correct value whereas the other end will read lower. For each observation the error can be eliminated by reading both ends of the needle and averaging the two values. The needle can be straightened with pliers.

*2. Pivot bent*  If the point of the pivot supporting the needle is not at the center of the graduated circle, there is introduced a variable systematic error, the magnitude of which

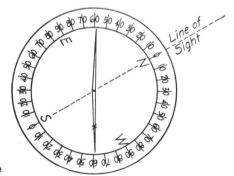

**Fig. 6.23*a***  Bent needle

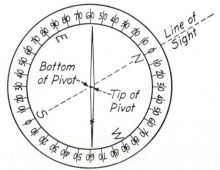

**Fig. 6.23*b***  Bent pivot.

depends upon the direction in which the compass is sighted. For one direction, the error is zero; for the normal to this direction, it is a maximum. In this case also, one end of the needle will read higher than the correct value whereas the other end will read lower (Fig. 6.23b); for each observation the error can be eliminated by reading both ends of the needle and averaging the two values. The instrument can be corrected by bending the pivot until the end readings of the needle are 180° apart for any direction of pointing.

**3. Plane of sight not vertical, or graduated circle not horizontal**   This misalignment introduces a systematic error, but it is usually so small as to be of no consequence. However, the sight vanes may become bent so that, even though the instrument is leveled, an appreciable error is introduced, particularly if the line of sight is steeply inclined when taking a bearing. The vanes may be tested by leveling the compass and sighting at a plumb line. The adjustment of the level tubes may be tested by reversal, as described for the transit in Art. 6.51.

**4. Sluggish needle**   The needle is not likely to come to rest exactly on the magnetic meridian. This lag produces random error. As the needle comes nearly to rest, tapping the glass lightly will tend to prevent the needle from sticking to the pivot. If the needle is "weak," it may be remagnetized by drawing its ends over a bar magnet, from the center to the ends of the magnet. The south-seeking end of the compass needle is drawn over the north-seeking half of the bar magnet, and vice versa. On each return stroke the needle should be lifted well above the magnet. If the pivot point is blunt, it may be sharpened by rubbing it on a fine-grained oilstone.

**5. Reading needle**   The inability of the observer to determine exactly the point on the graduated circle at which the needle comes to rest is generally the source of the most important and largest random error in compass work. The needle should be level, and the eye of the observer should be above the coinciding graduation and in line with the needle. If the needle dips perceptibly, its counterweight should be adjusted. Other conditions being equal, the longer the needle, the smaller the error of observing. With the 6-in needle used on many surveyor's compasses, the estimated standard deviation need not exceed $\pm 10'$; with the $3\frac{1}{2}$- or 4-in needle on the engineer's transit, the standard deviation is likely to be as much as 15'.

**6. Magnetic variations**   Undetected deviations of the magnetic needle from whatever cause are the source of the largest and most important systematic errors in compass work. Largely because of such variations, the compass, no matter how finely constructed, is not a suitable instrument for any except rough surveys. Deviations due to local attraction can be detected and corrections can be applied as described earlier.

**6.21. Theodolite and transit**   A theodolite or transit is an instrument designed to measure horizontal and vertical angles. It consists of a telescope mounted so as to rotate vertically on a horizontal axis supported by a pair of vertical standards attached to a revolvable circular plate containing a graduated circle for reading horizontal angles. Another graduated arc is attached to one standard so that vertical angles can be observed. Strictly speaking, "theodolite" is the correct word to be used in describing such an instrument. However, through common usage in the United States, the term "transit" has come to be applied to the conventional, double-center instrument.

Emphasis in the American design has been concentrated on retaining the traditional configuration of the instrument (with four-screw leveling) equipped with metal graduated scales and verniers for direct reading of horizontal and vertical angles using a pocket magnifier. European designers concentrated on developing instruments equipped with graduated scales etched on glass which are viewed by way of optical trains that contain optical micrometers for measuring horizontal and vertical angles. These instruments have three-screw leveling and optical plummets. Regardless of the differences cited, both groups of instruments are designed according to the same fundamental principles.

Instruments of the first group are normally called *engineer's transits* and have been used in the United States for property and construction surveying. In general, instruments of the second group possess higher nominal angular accuracies approaching 1″ of arc. These instruments have been used primarily in control and construction surveys. They are usually designated as *repeating* or *direction theodolites*.

In this chapter all angle-measuring instruments in which the telescope can be reversed will be classified as follows: (1) Double-center instruments with direct-reading verniers. These instruments will be designated *transits* or engineer's transits. (2) Double-center instruments with optical scales. These instruments will be referred to as *repeating theodolites*. (3) Single-center instruments with optical scales. These instruments will be called *direction theodolites*.

**6.22. Engineer's transit**   The engineer's transit is sometimes called the "universal surveying instrument" by reason of the wide variety of uses for which it is adapted. It may be employed for measuring and laying off horizontal angles, directions, vertical angles, differences in elevation, and distances and for prolonging lines.

Some of the modern types of transit differ considerably in design and construction from those long in use, but their essential features do not differ greatly, and their use not at all. Some differences do exist and are typified by the transit illustrated in Fig. 6.24, which has

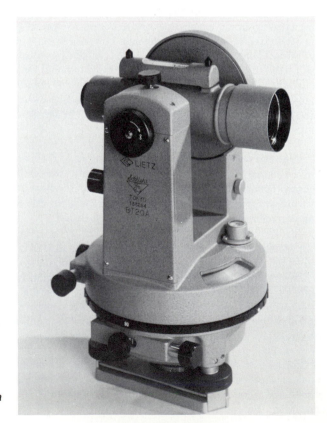

**Fig. 6.24** Engineer's transit with three-screw leveling and optical plummet. (*Courtesy Lietz/Sokkisha Co.*)

three-screw leveling and an optical plummet. The optical plummet can be found on other engineer's transits, but three-screw leveling is not typical of this style of instrument. In this discussion, the more traditional engineer's transit with four-screw leveling will be discussed first. Figure 6.25 shows such an instrument and Fig. 6.26 illustrates a vertical section of the transit. It is seen to consist of an *upper*, or *vernier, plate* to which are attached standards supporting the telescope and a *lower plate* to which is fixed a horizontal graduated circle. The upper and lower plates are fastened, respectively, to vertical inner and outer spindles, the two axes of rotation being coincident with and at the geometric center of the graduated circle. The outer spindle is seated in the tapered socket of the leveling head. Near the bottom of the leveling head is a ball-and-socket joint which secures the instrument to the foot plate yet permits rotation of the instrument about the joint as a center.

The outer spindle carrying the lower plate may be clamped in any position by means of the lower *clamp-screw*. Similarly, the inner spindle carrying the upper plate may be clamped to the outer spindle by means of the upper clamp-screw. After either clamp has been tightened, small movements of the spindle may be made by turning the corresponding *tangent-screw*. The axis about which the spindles revolve is called the *vertical axis* of the instrument.

Level tubes, called *plate levels*, are mounted at right angles to each other on the upper plate. Four *leveling screws*, or foot screws, are threaded into the leveling head and bear against the foot plate; when the screws are turned, the instrument is tilted about the ball-and-socket joint. When all four screws are loosened, pressure between the sliding plate and the foot plate is relieved, and the transit may then be shifted laterally with respect to the foot plate. From the end of the spindle and at the center of curvature of the ball-and-socket

**Fig. 6.25** Traditional-style engineer's transit. (*Courtesy Keuffel & Esser Co.*)

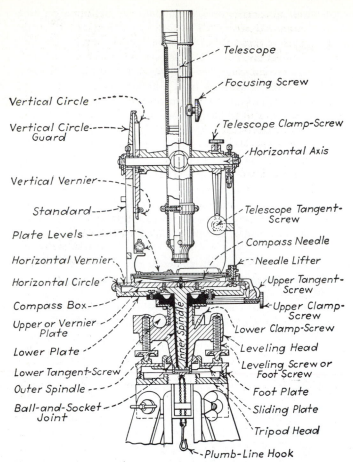

Telescope

Focusing Screw

Vertical Circle

Vertical Circle Guard

Telescope Clamp-Screw

Horizontal Axis

Vertical Vernier

Standard

Telescope Tangent-Screw

Plate Levels

Compass Needle

Horizontal Vernier

Needle Lifter

Horizontal Circle

Upper Tangent-Screw

Compass Box

Upper Clamp-Screw

Upper or Vernier Plate

Lower Clamp-Screw

Lower Plate

Leveling Head

Lower Tangent-Screw

Leveling Screw or Foot Screw

Outer Spindle

Foot Plate

Ball-and-Socket Joint

Sliding Plate

Tripod Head

Plumb-Line Hook

Inner Spindle

**Fig. 6.26** Section of engineer's transit.

joint is suspended a chain with hook for the plumb line. The instrument is mounted on a tripod by screwing the foot plate onto the tripod head.

The spindle design is modified in the transit equipped with an optical plummet. Figure 6.27 shows a vertical section of an instrument similar to the transit shown in Fig. 6.25 but with an optical plummet. Figure 6.28 illustrates the cross-slide arrangement on the tripod designed to accommodate a transit with the optical centering.

The *telescope* is fixed to a transverse *horizontal axis* which rests in bearings on the standards. The telescope may be rotated about this horizontal axis and may be fixed in any position in a vertical plane by means of the telescope clamp-screw; small movements about the horizontal axis may then be secured by turning the telescope tangent-screw. Fixed to the horizontal axis is the *vertical circle*, and attached to one of the standards is the vertical vernier. Beneath the telescope is the *telescope level tube*.

Attached to the upper plate is the *compass box*. If the compass circle is fixed, its N and S points are in the same vertical plane as the line of sight of the telescope. The compass boxes of some transits are so designed that the compass circle may be rotated with respect to the upper plate, so that the magnetic declination may be laid off and true bearings may be read. At the side of the compass box is a screw, or needle lifter, by means of which the magnetic needle may be lifted from its pivot and clamped.

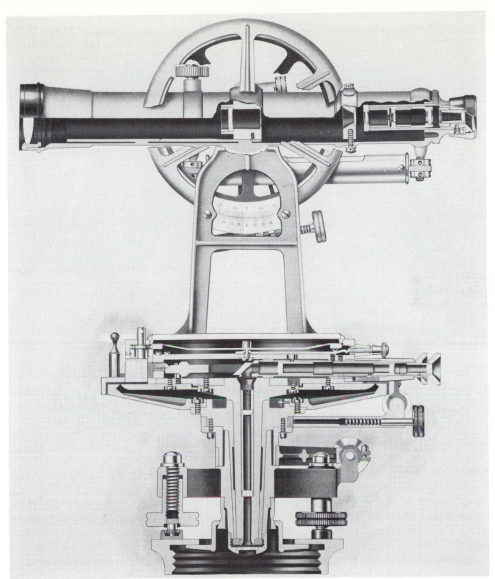

**Fig. 6.27** Vertical section, transit with an optical plummet. (*Courtesy Keuffel & Esser Co.*)

Summing up the main features: (1) the center of the transit can be brought over a given point by loosening the leveling screws and shifting the transit laterally; (2) the instrument can be leveled by means of the plate levels and the leveling screws; (3) the telescope can be rotated about either the horizontal or the vertical axis; (4) when the upper clamp-screw is tightened and the telescope is rotated about the vertical axis, there is no relative movement between the verniers and the horizontal circle; (5) when the lower clamp-screw is tightened and the upper one is loose, a rotation of the telescope about the vertical axis causes the vernier plate to revolve but leaves the horizontal circle fixed in position hence the designation *double-center* instrument; (6) when both upper and lower clamps are tightened, the telescope cannot be rotated about the vertical axis; (7) the telescope can be rotated about the horizontal axis and can be fixed in any direction in a vertical plane by means of the

**Fig. 6.28** Parallel-shift tripod head for transit with optical plummet. (*Courtesy Keuffel & Esser Co.*)

telescope clamp-screws and tangent-screws; (8) the telescope can be leveled by means of the telescope level tube, and hence the transit can be employed for direct leveling; (9) by means of direct observation of the vertical circle and vernier, vertical angles can be determined, and hence the transit is suitable for indirect leveling; (10) by means of direct observation of the horizontal circle and vernier, horizontal angles can be measured; and (11) by means of the compass, magnetic bearings can be determined.

**6.23. Level tubes**   The sensitivity of the several spirit levels of the transit should be such as to produce a well-balanced instrument and hence should correspond to the fineness of graduation of the circles and the optical properties of the telescope. If the levels are more sensitive than necessary to maintain this balance, time is wasted in centering the bubbles; if less sensitive than necessary, the precision of measurements is less than it should be for the transit as otherwise designed.

The plate levels of the ordinary transit reading to 01′ usually are alike in sensitivity and have a value of about 60″ per 0.1-in graduation, or about 75″ per 2-mm graduation, of the level tube. When horizontal angles are measured between points nearly in the same horizontal plane, it can be shown that no appreciable error is introduced even if the bubbles are some distance off center. On the other hand, where there is a large difference in vertical angle between the points sighted, a small displacement of the bubble in the tube that is parallel to the horizontal axis causes a relatively large error in the horizontal angle. Some instruments are equipped with a 20″ or 30″ striding level which is employed for leveling the horizontal axis whenever sights are sharply inclined.

The telescope level has a sensitivity of about 30″ per 2-mm graduation, depending upon the magnifying power of the telescope. The sensitivity of the vertical-vernier control bubble should depend upon the least reading of the vernier; for a vertical circle reading to 01′, a level tube having a sensitivity of 40″ per 2-mm graduation is commonly employed. For further details concerning level tubes, see Arts. 5.14 and 5.15.

**6.24. Telescope**  The telescope of the transit is similar to that of the engineer's level (Art. 5.8). When the transit is used as an instrument for either direct or trigonometric leveling, any point on the horizontal cross hair may be used in sighting; when the transit is used for establishing lines, measuring angles, or taking bearings, any point on the vertical cross hair may be used. Most instruments are equipped with stadia hairs (Chap. 7), which are usually mounted in the same plane with the cross hairs. The magnifying power is usually 20 to 28 diameters. For the transit, as for the level, usually the erecting eyepiece is employed; but the superior optical properties of the inverting eyepiece make it the favorite of some surveyors, and it is the type used in instruments of precision. The instruments illustrated in Fig. 6.24 and 6.25 have erecting eyepieces. For the relative merits of inverting and erecting eyepieces, see Art. 5.8. Telescopes on modern transits are usually of the internal focusing type (Art. 5.8).

**6.25. Graduated circles**  The vertical circle has two opposite zero points and is graduated usually in half-degrees, the numbers increasing to 90° in both directions from the zero points, as illustrated in Fig. 6.29a. When the telescope is level, the index of the vernier is at 0°.

The horizontal circle is likewise usually graduated in half-degrees, but may be graduated in third-degrees or quarter-degrees. It may be numbered from 0° to 360° clockwise (Fig. 6.29b), 0° to 360° clockwise and 0° to 90° in quadrants (Fig. 6.29c), or 0° to 360° in both

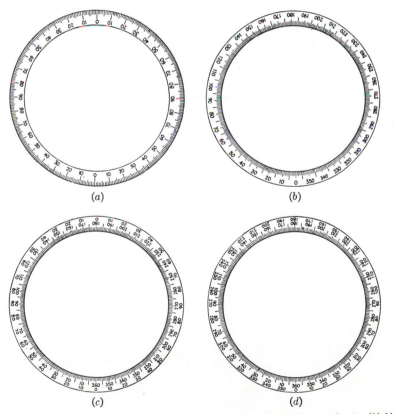

(a)  (b)

(c)  (d)

**Fig. 6.29**  Numbering of circles. (a) Vertical circle numbered in quadrants. (b) Horizontal circle numbered 0 to 360. (c) Horizontal circle numbered 0 to 360 and in quadrants. (d) Horizontal circle numbered 0 to 360 and 360 to 0.

directions (Fig. 6.29d). Most surveyors prefer the numbering system illustrated in Fig. 6.29d. Usually, the numbers slope in the direction of reading.

The horizontal circles of transits designed for work of moderately high precision are graduated to 20′ or to 15′. Those for repeating theodolites are often graduated to 10′. In special cases, circles may be graduated in divisions based on the *mil* or the *grad*.

**6.26. Verniers** The verniers employed for reading the horizontal and vertical circles of the transit are identical in principle with those for the target rod, described in Art. 5.26. Practically all transit verniers are of the direct type.

Figure 6.30a shows the usual type of double direct vernier reading to minutes. The circle is graduated to half-degrees and the vernier contains 30 divisions on either side of the index. One division on the vernier is $\frac{1}{30}$ less than the least division (half-degree interval) on the main scale, so that 30 divisions on the vernier cover 29 divisions on the main scale. Thus, the least count is 30′/30, or 1′. In Fig. 6.30a, the clockwise angle is 342°30′ + 05′ (on the vernier reading to the left of the index) = 342°35′. Reading the counterclockwise circle and the vernier to the right of the index yields an angle of 17°25′.

Figure 6.30b shows a double direct vernier on a circle (Fig. 6.29c) with a least division of 20′. The vernier contains 40 divisions which cover 39 divisions on the circle. The least count

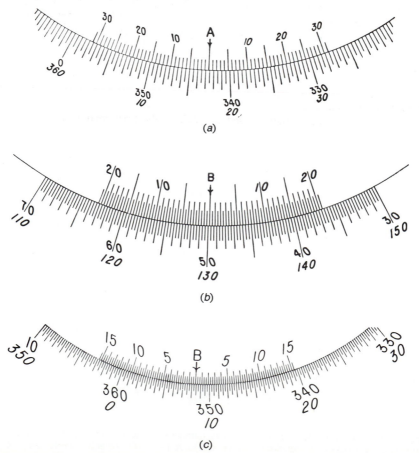

(a)

(b)

(c)

**Fig. 6.30** Vernier styles. (a) One-minute double direct vernier. (b) Thirty-second double direct vernier. (c) Twenty-second double direct vernier. (*Courtesy Keuffel & Esser Co.*)

is $20'/40 = \frac{1}{2}'$ or $30''$. The clockwise angle is $49°40' + 10'30'' = 49°50'30''$. The counterclockwise angle is $130° + 9'30'' = 130°09'30''$. The reader should attempt to determine the least count and the clockwise and counterclockwise angles illustrated by Fig. 6.30c.

**6.27. Eccentricity of verniers and centers**    Transits have two verniers for reading the horizontal circle, their indexes being 180° apart. The one nearest the upper clamp and tangent-screw is known as the *A* vernier, and the one opposite is known as the *B* vernier. The verniers are attached to the upper plate and are adjusted by the instrument maker so that they are much nearer to being truly 180° apart than their least count. Failure of the two verniers to register readings exactly 180° apart on the circle may be due to either or both of the two following causes:

*1. Eccentricity of verniers*    The verniers may have become displaced, so that a line joining their indexes does not pass through the center of rotation of the upper plate. The error will be the same for all parts of the graduated circle.

*2. Eccentricity of centers*    The spindles may have become worn or otherwise damaged, so that the center of rotation of the upper plate does not coincide with the geometrical center of the graduated horizontal circle. The error varies according to the position of the verniers with respect to the horizontal plate. As indicated by Fig. 6.31, there will be one setting on the graduated circle for which the indexes are exactly 180° apart (first position), and 90° therefrom there will be another setting for which the verniers fail to register 180° apart by a maximum amount (second position).

To correct either of these defects requires the services of an instrument maker, but neither defect limits the precision with which angles can be measured. Taking the mean of the two vernier readings (*A* and *B* verniers) eliminates errors due to either or both types of eccentricity. Furthermore, if the verniers only are eccentric, no error is introduced in an angle as long as the same vernier is used for making the final reading as for making the initial setting.

**6.28. Transit with three level screws**    The engineer's transit illustrated in Fig. 6.24 is fundamentally similar to the transit described in Arts. 6.22 to 6.27 with certain specific exceptions. There is no vertical spindle as such. Double centering is achieved by means of the black positioning ring attached to the horizontal circle, which permits the horizontal circle to be rotated with respect to the alidade that contains the verniers. A lower lock and lower tangent-screw serve the same function as on a conventional transit. A circular level bubble is provided for approximate leveling and a single-level tube is used for precise leveling. The three-screw leveling and lack of a vertical spindle mean that an optical

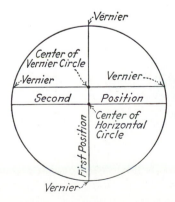

**Fig. 6.31** Eccentricity of verniers.

plummet is very easily adapted. Consequently, this instrument contains a combination of features found on both the traditional engineer's transit and the optical theodolites.

**6.29. Geometry of the transit**   A knowledge of the geometry of the transit is necessary for a thorough understanding of its operation. Figure 6.32b shows the geometry of the transit in isometric, schematic form. The intersecting lines labeled A, define the plane of the plate level tubes; the vertical axis is designated as B; the horizontal axis is C; the optical line of sight is D; and the axis of the spirit level attached to the telescope is E. In a perfectly constructed and adjusted instrument, the following relationships should exist among these axes: B is perpendicular to A; C is perpendicular to B; D is perpendicular to C; E is parallel to D; and B, C, and D pass through a single point. Achievement of these relationships by adjustment is detailed in Arts. 6.50 and 6.51.

**6.30. Theodolites**   Repeating and direction theodolites have features that are common to both types of instruments. Characteristics of modern optical theodolites will be discussed before pointing out the differences between repeating and direction models.

Although there are differences as among the various manufacturers, certain general features of theodolites are fairly common. The weight (without tripod) is about 10 lb or 5 kg as compared with about 15 lb or 7 kg for the conventional American transit; the difference is due both to compact size and to the use of light materials in manufacture. The finish is light in color to minimize temperature effects of sunlight. There is no vernier plate, and the center is cylindrical and rotates on ball bearings. A circular bubble of moderate sensitivity is used for approximate leveling, while a single plate level bubble of high sensitivity serves for precise leveling. The telescope level is tubular and of high sensitivity. There are three leveling

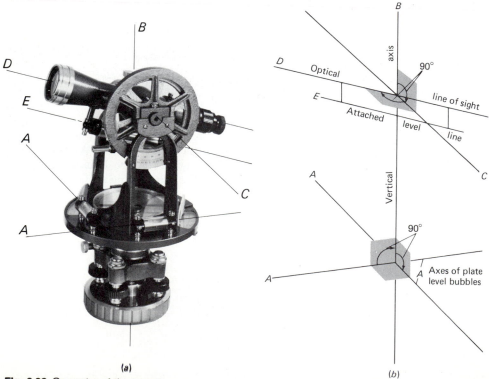

(a)

(b)

**Fig. 6.32** Geometry of the transit.

screws. An optical plummet is used instead of a plumb line. All motions are enclosed, so that the instrument is dustproof and moistureproof. Interior lighting is accomplished by means of mirrors and prisms, and provision is made for night lighting. The only field adjustments are those of the level bubble. These adjustments are similar to those for the conventional American transit.

The telescope of the theodolite is short and can be rotated completely about the horizontal axis. It is of the internal-focusing type, so that the stadia constant (Art. 7.5) is zero. The reticule is of glass with etched crosslines, and the stadia interval factor (Art. 7.6) is fixed at 100.0. For European use the eyepiece is of the inverting type; for American use it is usually of the erecting type.

Figure 6.33 shows a repeating theodolite; Figs. 6.34 and 6.35 illustrate direction theodolites.

The graduated horizontal and vertical circles are of glass and are relatively small; they are viewed simultaneously from the eye end of the telescope by means of a system of microscopes, prisms, and mirrors. Figure 6.36 shows the optical train of a direction theodolite. The horizontal circle is read by means of an optical micrometer, in accordance with systems devised by the various manufacturers. Each reading is obtained by one observation which is the average of two readings at opposite sides of the circle and which is therefore free of errors due to eccentricity. Direction theodolites are usually read directly to 01″. Repeating theodolites are read directly to 20″ or 01′ and by estimation to one-tenth the corresponding direct reading.

Most manufacturers produce instruments on which the theodolite, a sighting target, an EDM instrument, or subtense bar may be mounted interchangeably.

The repeating theodolite has all the features described above plus a lower lock and tangent-screw which allow angles to be set and repeated. Thus, it is a double-center instrument with the same capabilities as the engineer's transit. The repeating theodolite

(a)

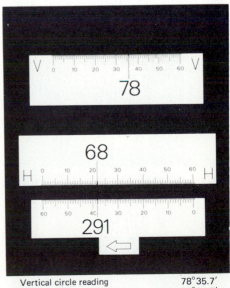

| | |
|---|---|
| Vertical circle reading | 78°35.7′ |
| Horizontal clockwise circle | 68°21.8′ |
| Horizontal counterclockwise circle | 291°38.2′ |

(b)

**Fig. 6.33** (a) Repeating theodolite. (b) Vertical and horizontal circle readings. (*Courtesy Kern Instruments, Inc.*)

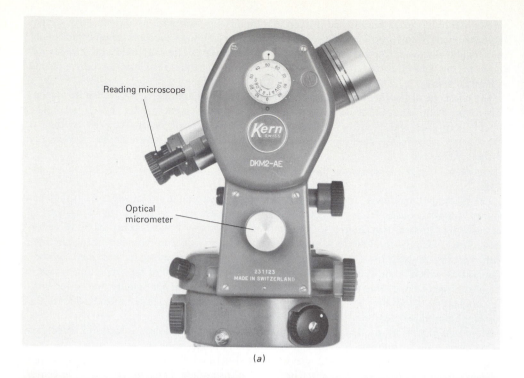

(a)

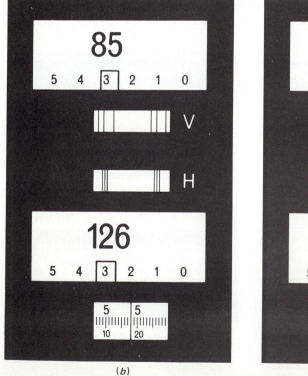

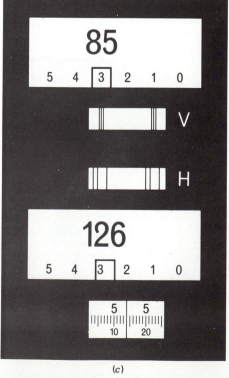

(b)                                             (c)

**Fig. 6.34** (a) Direction theodolite, Kern DKM2-AE. (b) Horizontal circle reading. (c) Vertical circle reading yielding a zenith angle in degrees. (*Courtesy Kern Instruments, Inc.*)

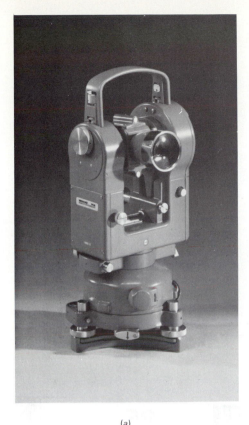

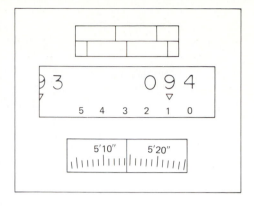

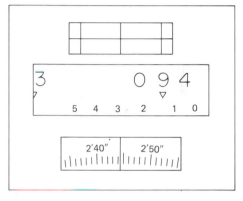

(a)                                                                 (b)

**Fig. 6.35** (*a*) Wild theodolite, T2. (*b*) Horizontal circle reading. (*Courtesy Wild Heerbrugg Instruments, Inc.*)

illustrated in Fig. 6.33*a* allows reading horizontal and zenith angles directly to minutes and estimating to tenths of minutes.

Direction theodolites have only a single horizontal motion in contrast to the double-center instruments. *Directions* are observed with these theodolites and the angle is the difference between the two directions (Art. 6.40). In general, direction instruments are more precise than are repeating theodolites. The plate level bubble has a sensitivity of 20″ per 2-mm division, and the vertical circle, on instruments of recent design, has an automatic compensator setting accuracy of from 0.3′ to 0.5′. The direction theodolite illustrated in Fig. 6.34*a* has a glass horizontal circle divided into 20′ intervals. The micrometer permits reading both horizontal and zenith angles directly to 1″ and estimating to 0.2″.

The method of observing horizontal and vertical angles on both repeating and direction theodolites involves viewing through a reading microscope mounted parallel to the telescope (Fig. 6.34*a*). By means of an optical train (Fig. 6.36), the observer's eye can be focused on the horizontal or vertical circle. The optical train is illuminated by adjustable mirrors (Fig. 6.36) for daylight operations. These mirrors can be replaced by attachments with light bulbs for night observations.

The procedure for reading angles varies with the make of instrument. First, the eyepiece of the reading microscope should be focused to the observer's eye so that a sharp image free of parallax is presented. The field of view visible when looking through the reading microscope on the repeating theodolite in Fig. 6.33*a* is illustrated in Fig. 6.33*b*. Three illuminated scales in separate frames are visible: the scale at the top, labeled *V*, shows the

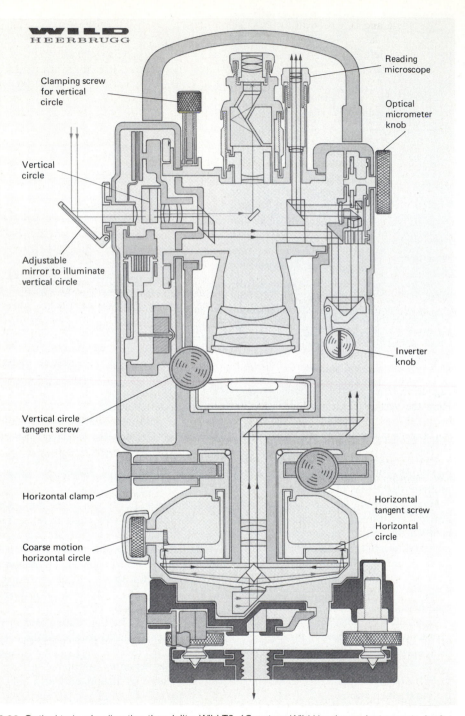

**Fig. 6.36** Optical train of a direction theodolite, Wild T2. (*Courtesy Wild Heerbrugg Instruments, Inc.*)

zenith angle and can be read directly to minutes and estimated to tenths of minutes; the second scale, labeled $H$, is the horizontal clockwise circle; and the bottom scale, with the arrow to the left, is the horizontal counterclockwise circle reading. Each of the three scales can be read directly to degrees and minutes, and estimated to tenths of minutes. The circle readings for this example are indicated on Fig. 6.33b. On theodolites of older design, the control spirit level attached to the vertical circle must be centered before observing the vertical or zenith angle. The instrument illustrated in Fig. 6.33a has automatic vertical circle indexing, so that this operation is not necessary.

A direction theodolite in which the horizontal and vertical circles can be read directly to seconds is illustrated in Fig. 6.34. The scales visible when viewing through the reading microscope are shown in Fig. 6.34b and c. The uppermost frame contains the vertical circle reading in degrees and tens of minutes. The large numbers are degrees and the small numerals are the number of tens of minutes. In Fig. 6.34b, the approximate vertical circle reading is 85°30′. Immediately below is a much smaller frame, labeled $V$, which contains a pair of split graduations and two index marks which are controlled by the optical micrometer. These index marks must be centered between respective split graduations in order to observe vertical circle readings to seconds. Below this $V$ coincidence scale is a similar coincidence scale, labeled $H$, for horizontal circle observations. Note that index marks are centered between their respective split graduations. Thus, Fig. 6.34b illustrates a horizontal circle reading. The large frame at the bottom contains horizontal circle readings in degrees and tens of minutes. The smaller scale at the very bottom contains horizontal or vertical circle readings to minutes and seconds. Readings can be estimated to tenths of seconds. First, coincidence of the index mark and the appropriate split graduations is obtained by rotating the optical micrometer. The horizontal circle reading is 126°30′ plus 05′18″, or 126°35′18.0″. The 05′18.0″ is read directly from the lowest scale. A similar array is shown in Fig. 6.34c, where the index marks are centered on respective split graduations in the $V$ scale. Here the vertical circle reading is 85°30′ from the topmost frame plus 05′14.0″ from the lowest scale, or 85°35′14.0″. This reading is a zenith angle. In each case the reading to minutes and seconds read from the lowest scale is obtained by rotating the optical micrometer and centering the index marks between split graduations on either the $V$ or $H$ scale, depending on which circle observation is desired.

Another style of direction theodolite with a different scheme for obtaining circle observations is shown in Fig. 6.35. The field of view visible when looking through the reading microscope is illustrated before and after coincidence in Fig. 6.35b. In the top diagram, before coincidence, the top frame contains a scale that shows degrees and minutes to the nearest 10′, while the bottom frame has a scale graduated in minutes and seconds. The horizontal and vertical circles are divided into 5′ intervals and the optical micrometer is employed to measure directly to seconds. Before coincidence, the horizontal circle reading is estimated as 94°10′. Coincidence is obtained by rotating the micrometer knob until the vertical marks in the top scale coincide. This rotation causes the pointer in the middle scale to indicate the nearest 10′, while the bottom scale provides minutes and seconds. The setting shown on the bottom of Fig. 6.35b yields a horizontal circle reading of 94°10′ plus 02′44.4″, or 94°12′44.4″. Horizontal circle readings are made with the indentation on the inverter knob (Fig. 6.36) in a horizontal position. Vertical circle readings are obtained in exactly the same way but with the inverter knob rotated 90° so that the indentation is in a vertical position. Turning this knob rotates a reflective prism so that the optical path is directed toward either the horizontal or vertical circle. In Fig. 6.36, illustrating the optical train, this knob is set to permit viewing the vertical circle.

When making vertical circle observations, due regard must be given to taking into account instrumental vertical circle index error. On engineer's transits, where zero corresponds to the horizontal, the index error is noted and applied to observed vertical angles (see Art. 6.45). The procedure is different in theodolites in which vertical circles are oriented with zero at

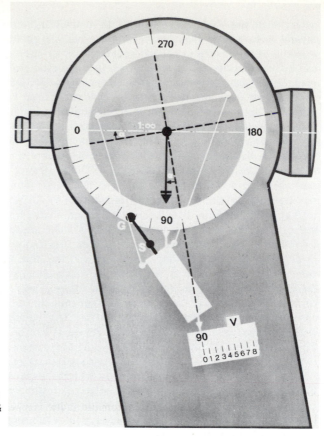

**Fig. 6.37** Pendulum prism V-type compensator. (*Courtesy Keuffel & Esser Co.*)

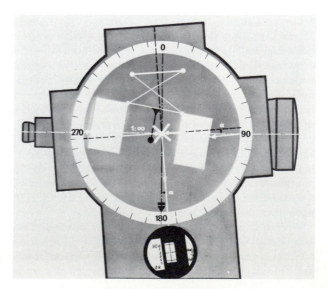

**Fig. 6.38** Pendulum prism X-type compensator. (*Courtesy Keuffel & Esser Co.*)

the zenith. In order that vertical circle readings be referred to the true zenith, theodolites of the most recent design have automatic vertical circle indexing. One method of achieving automatic indexing is to have a prism or pairs of prisms suspended by a pendulum apparatus incorporated into the ray path of the optical-reading microscope for vertical circle observations. Figures 6.37 and 6.38 show in schematic form, how typical V types and X types of pendulum compensators operate.

Another style of theodolite utilizes a special colorless liquid in a sealed glass container as a compensator. This unit is mounted at the top of the telescope standard, where it reflects light in the optical path for reading the vertical circle. When the theodolite is perfectly leveled, and zero on the vertical circle coincides with the zenith, the optical path of the reading rays forms an isosceles triangle as shown in Fig. 6.39a. In this case the vertical circle reading yields the correct angle referred to the zenith. When the instrumental vertical axis does not coincide with the zenith, the free surface of the liquid in the compensator remains horizontal and the rays in the optical path are deviated, as illustrated in Fig. 6.39b. The optics of the reading system are designed so that this deviation is observed as an equal amount acting in the opposite direction, as indicated by Fig. 6.39b. Thus, the vertical circle reading is referred to the true zenith even though the vertical axis is not truly vertical.

All of the several types of compensators have the same function, to refer vertical circle readings to the direction of gravity regardless of small inclinations of the vertical axis of the instrument.

Theodolites of earlier design do not contain automatic vertical circle indexing but do have sensitive spirit levels attached to the vertical circle so that zero on this circle can be made to correspond with the zenith. This bubble usually has a sensitivity of 20″ per 2-mm division and is viewed by a prism arrangement which allows split bubble coincidence. Assuming that the bubble is in adjustment, no index error results if it is centered each time a vertical circle observation is made.

Accuracies possible using either automatic index compensation or the level bubble are comparable assuming an instrument in good adjustment and careful centering of the spirit level in the latter case. Automatic index compensation does reduce the time required for observation and also relieves the strain on the operator.

All optical theodolites have optical plummets, which consist of a vertical collimation line that coincides with the instrumental vertical axis. The eyepiece, which allows viewing along this collimation line, is usually located near the base of the instrument. To center the

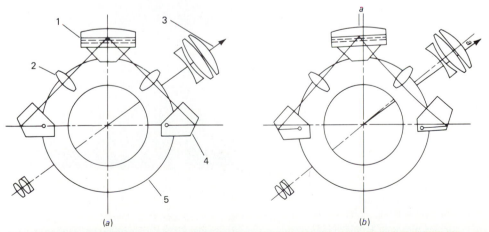

(a)                                                    (b)

**Fig. 6.39** Optical path of automatic index compensator, Kern DKM2A. (*a*) Vertical axis plumb: (1) liquid reservoir; (2) objective; (3) optical line of sight; (4) deviating prism; (5) vertical circle. (*b*) Vertical axis not plumb. (*Courtesy Kern Instruments, Inc.*)

theodolite over a station with the optical plummet, the instrument must first be leveled, then centered and again releveled. This process is repeated until the plate level bubble is centered, making the instrumental vertical axis vertical so that the cross in the vertical collimation line coincides with the point over which the theodolite is being centered. The optical plummet permits centering to within ±0.5 mm and is immune to the effects of wind.

Another method for centering which utilizes a telescopic centering rod attached directly to the leveling head is shown in Fig. 6.40. Note that the centering tripod does not support the instrument directly but is attached to a shifting tripod head, the upper portion of which has a spherical surface (Fig. 6.40a). This spherical surface supports a plate on which the instrument is clamped. The plate is perpendicular to and is rigidly connected to the centering rod. An adjustable circular level bubble is attached to the centering rod. A clamping grip holds the centering rod against the lower surface of the tripod head. In order to make a setup, the tripod is set over the station and with the clamping grip loose, the lower pointed end of the rod is placed on the station point. Next, the legs of the tripod are adjusted or moved to roughly center the circular level bubble on the centering rod. Force the legs of the tripod firmly into the ground to complete this rough centering. Now, shift the head on the tripod plate until the circular bubble is accurately centered. The centering rod is vertical and the tripod plate is horizontal. Tighten the clamping grip and place the theodolite on the support plate, fastening it securely with the lever provided for that purpose. The instrument is then leveled accurately using the level screws to center the plate level bubble

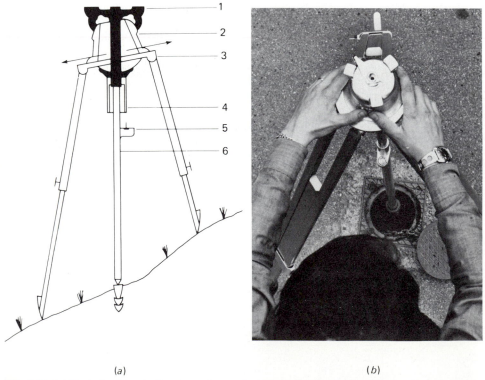

(a)                                             (b)

**Fig. 6.40** (a) Centering tripod with telescopic plumbing rod: (1) supporting plate; (2) tripod head with spherical surface; (3) tripod plate; (4) clamping grip; (5) circular level; (6) centering rod. (b) Centering with the telescopic plumbing rod. (*Courtesy Kern Instruments, Inc.*)

and the setup is completed. The theodolite can be released from the support plate simply by releasing the clamping lever. Thus, it is very convenient to replace the theodolite with a target (see Fig. 6.67) or EDM reflector without disturbing the setup. This procedure is called *forced centering*.

**6.31. Digitized theodolites**   Theodolites containing digitized horizontal and vertical scales are now on the market. Figure 6.41 illustrates a digitized direction theodolite. Horizontal and vertical angles are output from this instrument with a least count of 3″.

One method of digitizing angular output is to use a glass encoder disk such as the one shown in Fig. 6.42. This disk contains a metallic film pattern deposited on the surface. Illuminators are placed on one side of the disk, which are opposite to photodiodes on the other side of the disk.

Several sets of illuminators and diodes are placed at specific positions about the circle. Each diode senses the collimated illumination transmitted through the pattern. The amount of current generated by each photodiode is a function of the amount of light passing through the circle and hence depends on the position of the circle. The impulses generated by the photodiodes are converted by an on-board electronic computer to angular values. In the most elementary form of such an instrument, angles in degrees, minutes, and seconds; grads, centesimal minutes, and centesimal seconds; or mils are output on small nixie tubes. The raw data can also be stored in a memory unit connected to the on-board computer or on magnetic tape. On the more sophisticated models, repeated angles can be stored and

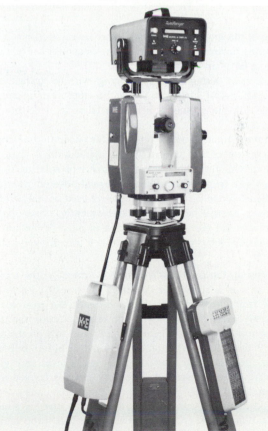

**Fig. 6.41** Digitized theodolite.
(*Courtesy Keuffel & Esser Co.*)

**Fig. 6.42** Glass encoder disk. (*Courtesy Hewlett Packard, Inc.*)

processed by the on-board computer so that output consists of the average angle. Further discussions concerning electronic surveying systems incorporating digitized theodolites can be found in Art. 12.2. Technical details about these devices are outlined in Refs. 3 and 4.

**6.32. Use of the engineer's transit**    Succeeding sections describe operations with a transit for measuring horizontal and vertical angles as well as for running lines. Transit surveys are described in detail in Chap. 8.

The process of taking magnetic bearings with the transit is the same as with the magnetic compass (Art. 6.13). The transit may be employed for running direct levels in the same manner as with the engineer's level, the telescope level bubble being centered each time a reading is taken.

The primary function of the transit is for observing horizontal and vertical angles. One of the major assets of the transit in this application is that the telescope can be reversed. *Reversing* or plunging the telescope consists of rotating it about the horizontal axis. The telescope is said to be in the *direct* or *normal* position when the focusing knob is upright or on top; the telescope is said to be in the *reversed* or *inverted* position when the focus knob is upside down. As will be explained subsequently, observations repeated with the telescope direct and reversed permit compensation of most of the systematic instrumental errors.

First, consider the procedure for setting a transit over a station point.

**6.33. Setting up the engineer's transit**    Ordinarily, the transit is set over a definite point, such as a tack in a stake. For centering the transit with four level screws, a plumb line is suspended from the hook and chain beneath the instrument. First the transit is placed approximately over the point, with the telescope near the observer's eye level. On sidehill setups, one tripod leg should be up the hill, with two on the downhill side of the point. Each tripod leg is then moved as required to bring the plumb bob within about 0.02 ft or 6 mm of being over the tack, with the foot plate nearly level and with the shoe of each tripod leg pressed firmly into the ground. The instrument is leveled approximately by means of the leveling screws.

The procedure for manipulating the level screws is similar to that described for the four-screw dumpy level (Art. 5.33). Each plate level tube is aligned with a pair of diagonally opposite level screws, as illustrated in Fig. 6.43*a*, where the axis of tube *A* is in line with level

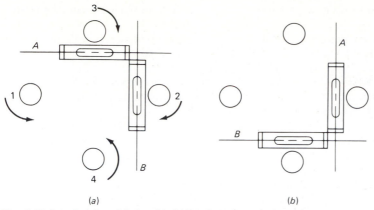

**Fig. 6.43** Leveling the plate level bubbles, four-screw instrument.

screws 1 and 2. The position of the bubble in tube $A$ is controlled by screws 1 and 2 and the bubble in tube $B$ is controlled by screws 3 and 4. Level bubble $A$ is brought to center by turning screws 1 and 2, keeping both screws lightly in contact with the foot plate. It is convenient to remember that the bubble travels in the *same direction as the left thumb*. The motions indicated in Fig. 6.43$a$ will cause the bubble in tube $A$ to move from left to right. Next, the bubble in tube $B$ is centered in similar fashion using screws 3 and 4. With the two bubbles approximately centered in this position, two adjacent screws (2 and 4, for example) are loosened by turning them in the same direction. This action releases the pressure of the footscrews on the foot plate and permits lateral movement of the leveling head so that the plumb bob can be centered accurately over the tack. The length of plumb bob line is changed as necessary to make the plumb bob just clear the tack. Tighten the two adjacent screws to a firm but not tight bearing and recenter the bubbles in tubes $A$ and $B$ very carefully. Rotate the instrument through 90°, as shown in Fig. 6.43$b$, and center the bubbles in tubes $A$ and $B$ as before, using the foot screws. The instrument is now rotated back to the original position (Fig. 6.43$a$), where the positions of the level bubbles are checked. The bubbles are recentered if necessary. When the plate level bubbles remain centered in both positions, the transit is rotated 180° from the original position and the centering of the level bubbles is checked again. If the bubbles remain centered and the plumb bob is over the tack, the setup is complete. If the bubbles move off center a small amount (about one 2-mm division), bring each bubble halfway to center using the appropriate pair of diagonally opposite level screws. If the deviation is large, adjustment of the bubbles is necessary (Art. 6.54). The telescope should be tested for parallax before observations begin.

The traditional-style engineer's transit equipped with a vertical optical collimator or optical plummet (Figs. 6.27 and 6.28) is set up in a similar fashion except for centering over the point. Initially, a plumb bob is attached to the vertical spindle and the transit is approximately leveled and centered as described in the preceding paragraph. Then the plumb bob is removed and the instrument is centered by loosening the knurled screw which secures the cross-slide apparatus that supports the transit on the tripod (Fig. 6.28). Centering with the optical collimator is then possible. Releveling is necessary, followed by recentering, since adjustment of the level screws alters the vertical collimation line. When the optical collimation line is centered over the point and the plate level bubbles are centered, the instrument is turned through 90° and the entire procedure is repeated. The process of alternately leveling and recentering is repeated in the original position and with the instrument rotated through 90° until the bubbles remain centered and the optical plumb coincides with the point. As before, the instrument should be rotated through 180° from the original position as a check.

Just before the transit is moved, the instrument is centered on the foot plate, the leveling screws are roughly equalized, the upper motion is clamped, the lower motion is either unclamped or clamped lightly, and the telescope is pointed vertically with the vertical motion lightly clamped.

**6.34. Setting up three-screw Instruments**  These instruments are equipped with a circular level bubble, only one plate level bubble, and an optical plummet. Consequently, certain significant differences in the setting up procedure exist.

The instrument is set approximately over the point using a plumb bob attached to a hook which can be inserted into the screw that secures the instrument to the top of the tripod. Care should be exercised to ensure that the foot plate is nearly level, so as to simplify succeeding operations. Next, center the circular level bubble by manipulating the three level screws as described in Art. 5.33. Remove the plumb bob and focus the optical plummet on the point, so as to be free of parallax. After slightly loosening the central screw, the instrument is moved laterally until the cross mark in the optical plummet coincides with the station point. The circular level bubble is checked and if recentering is required, centering of the optical plummet should be checked. When the optical plummet is centered over the station point and the circular level bubble remains centered, the instrument is rotated so that the axis of the plate level bubble is aligned with any two level screws, as shown in Fig. 6.44$a$. The bubble is centered by rotating screws 1 and 2 in opposite directions. The rotations indicated in Fig. 6.44$a$ will cause the bubble to move from left to right. Next, rotate the instrument so that one end of the bubble tube is aligned with the remaining screw (Fig. 6.44$b$). Center the bubble in this position by rotating this remaining screw. The rotation indicated in Fig. 6.44$b$ will cause the bubble to move away from level screw 3. Return to the original position (Fig. 6.44$a$) and check centering of the bubble. Rotate through 180° so that end $A$ of the bubble tube is on line with level screw 1. If the bubble moves off center, bring the bubble halfway to center using level screws 1 and 2. Rotate 90° so as to be in the position of Fig. 6.44$b$ and check the level bubble. Rotate through 180° so that end $B$ of the bubble tube coincides with level screw 3 (Fig. 6.44$b$). If the bubble moves off center, bring halfway to center with level screw 3. If this deviation from center is large, the bubble tube must be adjusted (see Art. 6.56). Finally, check the centering of the optical plummet.

The plate level bubble on most direction theodolites is quite sensitive (about 20″ per 2-mm division) and centering the bubble throughout a 360° revolution is a challenging task unless the theodolite is protected from the rays of the sun.

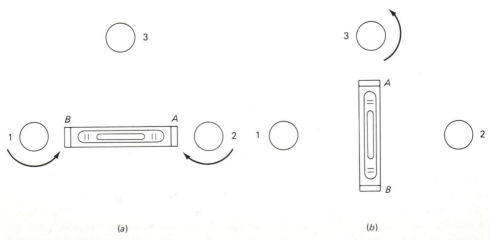

(a)  (b)

**Fig. 6.44**  Leveling the plate level bubble, three-screw instrument.

The process of sequentially leveling and centering is made much easier when the top of the tripod is close to being level initially. Care should be exercised to ensure that this condition is satisfied. Some tripods are equipped with level bubbles built into the supporting plate to facilitate this operation.

Setup of an engineer's transit or theodolite equipped with a tripod having a telescopic centering rod was described in the last paragraph of Art. 6.30. When using the centering tripod, the support plate on which the instrument rests is automatically level after the rod is plumbed. Thus, leveling the instrument requires only a few turns of the level screws. One precaution should be observed. After the circular level bubble on the centering rod has been centered, it should be rotated 180° and checked to see if the bubble remains in the center. If the bubble deviates from center, loosen the clamping grip (Fig. 6.40a), bring the bubble halfway to center, and then tighten the clamping grip. If the discrepancy is excessive, the circular level bubble must be adjusted as described in Art. 5.59.

**6.35. Horizontal angles with the engineer's transit**  If a horizontal angle such as $AOB$ in Fig. 6.45 is to be measured, the transit is set over $O$. The upper motion is clamped with one of the verniers set near zero, and by means of the upper tangent screw one vernier is set to zero. The telescope is sighted approximately to $A$ (telescope in the direct position) by first sighting over the top of the telescope to get near the point and then sighting through the telescope. The lower clamp screw is locked, the telescope is raised or lowered so that the cross-hair intersection is near the point, and the telescope is focused on point $A$. Care should be exercised to detect and remove parallax. By turning the lower tangent-screw, the vertical cross hair is set exactly on the plumbed line, range pole, or other target which marks point $A$. At this stage the optical line of sight is on line between $O$ and $A$ and the vernier is set to zero on the horizontal circle. Next, the upper clamp is loosened and the telescope is turned until the line of sight is approximately on $B$. The upper clamp is tightened and the line of sight is set exactly on $B$ by turning the upper tangent screw. The reading of the vernier that was set to zero on point $A$ gives the value of the angle. If point $B$ falls to the right of point $A$, as shown in Fig. 6.45, the objective lens of the telescope has moved to the right and the angle $AOB$ is said to be *turned to the right*. Assume that the horizontal circle setting when sighted on $B$ is as shown in Fig. 6.30b. The horizontal angle $\alpha = AOB$ turned to the right is read on the clockwise circle as being 49°40′ plus 10′30″ = 49°50′30″, where the 10′30″ is observed on the vernier to the left of the index mark. From this same setup, the counterclockwise angle from $A$ to $B$ or $\beta$ (Fig. 6.45) is observed on the counterclockwise horizontal circle and the

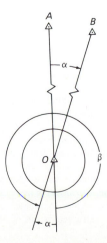

**Fig. 6.45** Horizontal angle to the right.

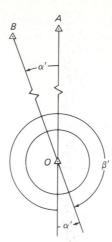

**Fig. 6.46** Horizontal angle to the left.

vernier is read on the right side of the index. This counterclockwise angle $\beta = AOB$ is $130°09'30'' + 180° = 310°09'30''$.

If $B$ falls to the left of $A$, the objective lens of the telescope is turned to the left and angle $\alpha' = AOB$ is said to be turned to the left, as shown in Fig. 6.46. Assume that the horizontal circle settings are as illustrated in Fig. 6.30a. An angle turned to the left is observed on the counterclockwise circle. Observation of the counterclockwise circle and the right side of the vernier yields $\alpha' = 17°25'00''$. The clockwise angle from $A$ to $B$ or $\beta'$, observed on the clockwise circle, is $342°35'00''$.

The following is a list of suggestions for turning horizontal angles with an engineer's transit:

1. Make reasonably close settings by hand so that the tangent-screws will not need to be turned through more than one or two revolutions.
2. Make the last movement of the tangent-screw clockwise, thus compressing the opposing spring.
3. When reading the vernier, have the eye directly over the coinciding graduation, to avoid parallax. It is also helpful to observe that the graduations on both sides of those coinciding fail to concur by the same amount.
4. As a check on the reading of one vernier, the other vernier may be read also. Or, check readings may be taken at each end of the vernier scale; these differ from the vernier reading by a value which is constant for the given type of vernier.
5. The plate bubbles should be centered before measuring an angle, but between initial and final settings of the line of sight the leveling screws should not be disturbed. When an angle is being measured by repetition (Art. 6.39), the plate may be releveled after each turning of the angle before again sighting on the initial point.
6. The flagman should stand directly behind the range pole, holding it lightly with the fingers of both hands, and balancing it on the tack or other mark indicating the point.
7. In sighting at a range pole the bottom of which is not visible, particular care should be taken to see that it is held vertical. When the view is obstructed for a considerable distance above the point to which the sight is taken, use a plumb line behind which a white card is held. For short sights a pencil or ruler held on the point makes a satisfactory target. Where the lighting is poor, the sight may be taken on a flashlight.
8. When a number of angles are to be observed from one point without moving the horizontal circle, the instrument operator should sight at some clearly defined object that will serve as a reference mark and should observe the angle. If occasionally the angle to the reference mark is read again, any accidental movement of the horizontal circle will be detected.
9. Whenever an angle is doubled, if the instrument is in adjustment, the two readings should not differ by more than the least count of the vernier. A greater discrepancy, if confirmed by repeating the measurement, will indicate that the instrument is out of adjustment.

**6.36. Horizontal angles with the repeating theodolite**  Figure 6.33 shows a repeating theodolite. Assume that this instrument is set over point $O$ in order to observe horizontal angle $AOB$. The initial setting of the horizontal circle is accomplished by viewing through the reading microscope and setting the horizontal circle reading and index mark to zero using the upper clamp and tangent-screw. A sight is then taken on $A$ using the lower clamp and tangent-screw. At this point the optical line of sight coincides with line $OA$ and the horizontal circle reading is zero. The upper clamp is loosened and the telescope is turned to point $B$, where a sight is taken with the upper clamp locked and using the upper tangent-screw. The angle is then observed by viewing through the reading microscope and following the instructions given in Art. 6.35. Figure 6.33*b* illustrates a horizontal circle reading of 68°21.8′. These scales are read directly. Note that both the clockwise and counterclockwise angles can be observed on this instrument.

**6.37. Laying off horizontal angles**  If an angle $AOB$ is to be laid off from line $OA$, the transit is set up at $O$, one vernier (or index and micrometer scale in optical-reading instruments) is set at zero, and the line of sight is set on $A$. The upper clamp is loosened, and the plate is turned until the index of the vernier or optical-reading system is approximately at the required angle. The upper clamp is tightened, and the vernier (or index and micrometer scale in an optical-reading instrument) is set exactly on the given angle by means of the upper tangent-screw. Point $B$ is then established on the line of sight.

**6.38. Common mistakes**  In measuring horizontal angles, mistakes often made are:

1. Turning the wrong tangent-screw.
2. Failing to tighten the clamp.
3. Reading numbers on the horizontal scale from the wrong row.
4. Reading angles in the wrong direction.
5. Dropping 30′ or 20′ by failure to take the full-scale reading before reading the vernier, for example, with a circle graduated to 30′ calling an angle 21°14′ when it is actually 21°44′, the vernier reading being 14′.
6. Reading the vernier in the wrong direction.
7. Reading the wrong vernier.

**6.39. Measuring horizontal angles by repetition**  By means of the engineer's transit or repeating theodolite, a horizontal angle may be mechanically cumulated and the sum can be read with the same precision as the single value. When this sum is divided by the number of repetitions, the resulting angle has a precision that exceeds the nominal least count of the instrument. Thus, with a transit having verniers reading to single minutes, an angle for which the true value is between the limits 30°00′30″ and 30°01′30″ will be read as 30°01′ and the limits of possible error will be ±30″. If the angle is cumulated six times on the horizontal circle, the sum, also read to the nearest minute, might be 180°04′, its true value being within the limits 180°03′30″ and 180°04′30″; the limits of possible error, as far as reading the horizontal circle is concerned, will also be ±30″. Dividing the observed sum 180°04′ by 6, the single value becomes 30°00′40″ for which the limits of possible "reading" error are ±30″/6 = ±05″. Similarly, a repeating theodolite (Fig. 6.33) with a least count of 30″ (estimating to 15″) would yield an angle having a possible reading error of about ±03″ in five repetitions. This method of determining an angle is called *measurement by repetition*. The precision with which an angle can be observed by this method increases directly with the number of times the angle is cumulated or repeated up to between six and twelve repetitions; beyond this number the precision is not appreciably increased by further repetition because of errors in graduations of the horizontal circle; eccentricity of instrument centers; play in the instrument; and random errors such as those due to setting the line of sight, setting the angle on the horizontal circle, and reading angles on the horizontal circle.

When angles are observed by repetition, the principle of *double centering* is also employed. As noted in Art. 6.29, on geometry of the transit, the lines and planes defined by vertical axis, horizontal axis, optical line of sight, and so on, theoretically have exact relationships with respect to one another. In practice, these relationships are not exact, even in the best-adjusted transit or theodolite, so that systematic instrumental errors are present (see Art. 6.55). One of the excellent features of the transit or theodolite is that the telescope can be transited about the horizontal axis so that observations can be made with the telescope in the direct and reversed positions (Art. 6.32). When operations on the instrument include a direct and reversed observation or setting, systematic instrumental errors will occur in opposite directions. Thus, if the average of the observations or settings is used, the effect of the systematic errors cancels. Implementation of this process is called *double centering* or *double sighting*.

Another reason for repeating an angle is to decrease the possibility of mistakes in observations. Thus, the primary reasons for observing an angle by repetition with the telescope direct and reversed are to increase the accuracy of reading the angle, compensate for systematic errors, and eliminate mistakes. Consider some examples.

To repeat an angle once direct and once reversed, as for angle *AOB* in Fig. 6.45, the transit is set up at *O*, where it is very carefully centered and leveled. A sight is taken on point *A* with the telescope direct and a single value of the angle is observed as described in Art. 6.35 or 6.36. Next, with this angle still set on the plate, the lower motion is unlocked, the telescope is reversed, the instrument is turned on its lower motion, and a second sight is taken using the lower clamp and lower tangent-screw. Note that on this sighting from *O* to *A* the value of the angle *AOB* is on the horizontal circle. The upper clamp is loosened and the telescope, now in the reversed position, is again sighted on point *B*. At this stage, approximately twice the original value of the first angle appears at the index of the horizontal circle, and the angle has been repeated with the telescope direct and reversed. The accompanying notes show how data can be recorded for observation of angle *AOB* by repetition once direct and once reversed using an engineer's transit with a least count of 30″.

| Station | From/to | | Rep. | Tel. | Circle reading | Angle |
|---------|---------|---|------|------|----------------|-------|
| *O* | *A* | | 0 | Direct | 0°00′00″ | |
| | | *B* | 1 | Direct | 34°20′15″ | 34°20′15″ |
| | | *B* | 2 | Reversed | 68°41′00″ | 34°20′45″ |

The first angle is obtained by subtracting the initial circle reading on *A* (0°00′00″ in this example) from the circle reading on *B*, yielding a first angle of 34°20′15″. The second angle equals the second circle reading minus the first reading on point *B*, or $(68°41′00″) - (34°20′15″) = 34°20′45″$. This second angle should agree with the first angle to within ± the least count of the transit, ±30″ in this case. Should this difference be greater, the set should be repeated. This set is acceptable. The final angle is then the average of angle 1 and angle 2: $[(34°20′15″) + (34°20′45″)]/2 = 34°20′30″$. The final angle can also be calculated by dividing the cumulated angle minus the initial observation by the number of repetitions; in this case, $[(68°41′00″) - (0°00′00″)]/2$ equals 34°20′30″.

The preceding example dealt with an interior angle. Next, consider the procedure for turning a deflection angle by repetition (Art. 6.10). Assume that the deflection angle at *B* in Fig. 6.12 is to be observed. The transit is set at point *B* and with the *A* vernier set on zero, a backsight is taken on point *A* with the telescope in the direct position. Next, the telescope is plunged so that the line of sight falls along the extension of line *AB* and the *A* vernier is still on zero. The upper clamp is loosened and a sight is taken on point *C* using the upper clamp and tangent-screw. The deflection angle is observed on the clockwise circle using the *A* vernier. Assume that the first deflection angle at *B* from *A* to *C* is 22°35′00″ to the right. To *double* the angle, the lower clamp is loosened and a second backsight is taken on point *A* using the lower clamp and lower tangent-screw. The telescope is reversed along the backsight and the angle 22°35′00″ is on the horizontal circle at the $A$ vernier. Again the telescope is plunged, the upper clamp loosened, and a sight is taken on point $C$ with the upper clamp and tangent-screw. The telescope is now in the direct position and the angle observed on the horizontal circle at the $A$ vernier is the doubled deflection angle. Notes for the example are shown below.

| Station | From/to | Rep. | Tel. | Circle reading | Deflection angle |
|---------|---------|------|------|----------------|------------------|
| B | A | 0 | Direct | 0°00′00″ | |
| | C | 1 | Reversed | 22°35′00″ | 22°35′00″ R |
| | C | 2 | Direct | 45°10′30″ | 22°35′30″ R |

*Note:* Engineer's transit number 37, least count 30″.

The procedure for calculating the first and second deflection angle is similar to that described for interior angles in the preceding example. Since the first angle agrees with the second to within the least count of the transit, the set is acceptable. The final average angle is 22°35′15″, the average of the two angles. The procedure just outlined is called observing angles by *double deflection*. Note that when deflection angles are observed, the direction of the deflection must be recorded.

Whenever possible, angles should always be turned by repetition the minimum number of times, once direct and once reversed. In this way systematic instrumental errors are compensated and blunders are eliminated, even though a high order of accuracy is not achieved.

When an angle is turned more than twice, the number of repetitions is always some multiple of two. Thus, an angle repeated twice direct and twice reversed involves four repetitions. Similarly, angles can be turned three, four, five, and six times direct and reversed involving six, eight, ten, and twelve repetitions, respectively. There should always be an equal number of direct and reversed observations.

When observing an angle by repetition using an engineer's transit with direct-reading verniers, it is customary to observe both the $A$ and $B$ verniers. In this way the systematic error between the two verniers is averaged and the accuracies possible by repetition are more nearly achieved.

The note form for turning an angle with six repetitions using a 30″ engineer's transit is shown below:

| Station | From/to | Rep. | Tel. | Circle reading | Vernier | |
|---------|---------|------|------|----------------|---------|---|
| | | | | | A | B |
| O | M | 0 | Direct | 0° 0′ | 00″ | 00″ |
| | N | 1 | Direct | 80°20′ | 00″ | 00″ |
| | N | 6 | Reversed | 122°01′ | 00″ | 00″ |

Note that the circle reading for the first repetition on point $N$ is recorded and then the angles are cumulated without recording until the final repetition is completed. The first angle is the difference between circle observation one on point $N$ and circle observation zero on point $M$. For this set, the initial angle, used *only* as a check value, is 80°20′00″. To calculate the average angle, it is necessary to know the number of times the horizontal circle index has passed zero. The total number of degrees passed in six repetitions is the initial approximate angle times the number of repetitions or (6)(80°20′)=482°. Since zero was passed once, 360° must be added to the cumulated angle recorded for the sixth repetition before calculating the average angle. Thus, the average angle for this set is [(122°01′00″)+ 360°]/6=80°20′10″. This average angle agrees with the initial check angle to within ± the least count of the transit, so that the set of observations is acceptable. Figure 6.47a shows a field note form for angles by repetition.

The circle reading for the first sight does not have to be zero, as illustrated by the following notes for six repetitions of angle *EFG* using a transit having a least count of 20″:

| Station | From/to | Rep. | Tel. | Circle reading | Vernier | |
|---------|---------|------|------|----------------|---------|-----|
| | | | | | **A** | **B** |
| F | E | 0 | Direct | 136°02′ | 00″ | $\overline{40″}$ |
| | G | 1 | Direct | 216°21′ | 00″ | 00″ |
| | G | 6 | Reversed | 257°58′ | 00″ | $\overline{40″}$ |

The reading on the *B* vernier for repetition zero on point *E* is $\overline{40″}$. The bar means that the next lower minute was read, so the average circle reading is $\frac{1}{2}[(136°02'00″)+(136°01'40″)]=136°01'50″$. The approximate angle is therefore $(216°21'00″)-(136°01'50″)=80°19'10″$. Six repetitions of this angle or $6(80°19'10″)=481°55'00″$. This value added to $136°01'50″$ equals $617°56'50″$, indicating that zero was passed once and 360° must be added to the final circle reading to calculate the final angle. Thus, $[(257°57'50″+360°)-(136°01'50″)]/6=80°19'20″$, the final average angle.

When angles are repeated with a repeating theodolite, readings 180° apart on the horizontal circle are automatically averaged in the optical reading process. Notes for angle

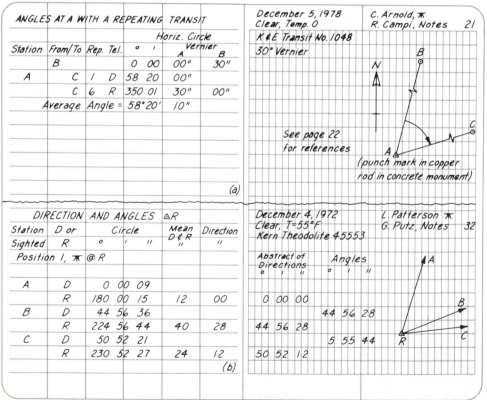

**Fig. 6.47** (a) Field notes for angles by repetition using a repeating instrument. (b) Field notes for directions with a theodolite.

SRT observed 12 times with a repeating theodolite having a least count of 10″ are given below:

| Station | From/to | Rep. | Tel. | Circle reading |
|---------|---------|------|------|----------------|
| R | S | 0 | Direct | 70°10′30″ |
| | T | 1 | Direct | 170°30′30″ |
| | T | 12 | Reversed | 194°11′30″ |

The approximate angle is 100°20′00″. Twelve times this approximate angle plus the first circle reading equals 1274°10′30″. Thus, the circle index passed zero three times. The average angle equals $[(194°11′30″)+(1080°)-(70°10′30″)]/12 = 100°20′05″$, which is within the least count or ±10″ of the approximate angle.

Angles by repetition should not be attempted unless the instrument operator is experienced both in the instrument setup and in observing angles. Speed and accuracy in observation are essential if the repetition procedure is to be successful. Otherwise, considerable time can be wasted without obtaining the desired results.

**6.40. Directions and angles with the direction theodolite** As noted in Art. 6.30, the direction theodolite has only a single horizontal motion equipped with one clamp and tangent-screw for azimuth. Thus, it is possible to observe *directions only*, and angles are computed by subtracting one direction from another. For example, assume that the angle $AOB$ (Fig. 6.45) is to be measured with a direction instrument. The theodolite is set up over point $O$, leveled and centered, and a sight is taken on point $A$ using the horizontal clamp and tangent-screw for azimuth (Fig. 6.36). The horizontal circle is then viewed through the optical-viewing system and the circle reading is observed and recorded (Art. 6.30). Assume that this reading is 89°34′22″. The horizontal clamp is then released and the telescope is sighted on point $B$. Once again the sight is taken with the horizontal clamp and the tangent-screw for azimuth. The horizontal circle is observed through the optical-viewing system and this reading is recorded as being 102°10′52″. These two observations constitute directions which have a common reference direction that is completely arbitrary. The clockwise angle = $(102°10′52″)-(89°34′22″)=12°36′30″$. Note that on a direction theodolite, there is only one horizontal circle graduated in a clockwise direction so that only clockwise angles can be derived from observations of directions. In Fig. 6.46, point $B$ is to the left of point $A$. Suppose that the theodolite is set up on point $O$, the initial sight is on point $A$, and the horizontal circle observation is 10°15′12″. The horizontal clamp is loosened and the telescope is rotated to sight on point $B$, where the circle reading is 351°50′32″. The clockwise angle from $A$ to $B$ is therefore $(351°50′32″)-(10°15′12″)=341°35′20″$. Again, the angle is clockwise from the initial pointing on $A$. Note that in neither case was any attempt made to set the horizontal circle on zero or on an integral value. There is a control which allows the horizontal circle to be advanced to a desired approximate value (Fig. 6.36). This control is a coarse motion only and does not permit fine settings.

**6.41. Angles by repetition using the direction theodolite** Repeating angles with the direction theodolite involves observing directions with the telescope direct and reversed, calculating mean directions to the observed points, and determining mean angles from the mean directions. Figure 6.47*b* shows notes for directions observed from a theodolite set at point $R$ to points $A$, $B$, and $C$. Referring to these notes, the procedure for observing the directions is as follows:

1. Make an initial pointing and record the horizontal circle observation. Normally, the telescope is in the direct position and the left most point is selected as the initial sighting. The value of the initial circle reading is governed by the number of repetitions to be observed (Art. 10.18).

2. Release the horizontal clamp and sight each point in succession, recording the circle reading for each sight. These observations constitute the circle readings to $A$, $B$, and $C$ with the telescope direct.
3. Reverse the telescope, sight point $C$, record the circle reading, and then sight on points $B$ and $A$ in reverse directions, recording the circle observation with the telescope in the reversed position in each case. These reversed readings are different by exactly 180° from the direct observation plus or minus a few seconds which represent the collimation error of the instrument (Art. 6.51). The observations taken as described constitute *one position*.

The means of the direct and reversed readings are calculated and recorded in column 4 of the notes (Fig. 6.47b). Then an *abstract* of *directions* is calculated and recorded in column 5 of the notes. This abstract is determined by assuming a direction of 0°00′00″ along the initial sight from $R$ to $A$. Thus, the direction from $R$ to $B$ is $(44°56′40″)-(0°00′12″)=44°56′28″$ and the direction from $R$ to $C=(50°52′24″)-(0°00′12″)=50°52′12″$. Finally, angles $ARB$ and $RBC$ are computed by subtracting successive directions as given in the abstract.

Depending on the accuracy desired, anywhere from 1 to 16 positions of directions may be observed. In such cases the horizontal circle would be advanced for each successive position so as to use the entire circle. The final abstract of directions and angles would be the mean of all the positions. Details concerning the procedures involved can be found in Art. 10.18.

**6.42. Laying off angles by repetition**   If it is desired to establish an interior angle with a precision greater than that possible by a single observation, the methods of the preceding section may be employed in the following manner: In Fig. 6.48a, $OA$ represents a fixed line and $AOB$ the angle which is to be laid off to establish the line $OB$. The engineer's transit is set up at $O$, the vernier is set at 0°, and a sight is taken to $A$. The vernier is set as closely as possible to the given angle and a trial point $B'$ is established with the line of sight in its new position. The angle $AOB'$ is then measured by repetition, and the line $OB'$ is measured. The angle $AOB'$ must be corrected by an angular amount $B'OB$ to establish the correct angle $AOB$. The correction, which is too small to be laid off accurately by angular measurement, is applied by offsetting the distance $B'B=OB'\tan$ (or sin) $B'OB$, thus establishing the point $B$ beside $B'$. It is convenient to remember that the tangent or sine of $1'=0.0003$ and the tan or sine of $1''=4.848\times10^{-6}$. As a check, the angle $AOB$ is measured by repetition.

**Example**   Suppose that an angle of 30°00′ correct to the nearest 05″ is to be laid off and that the transit to be employed reads to the nearest 01′. Let the total value of $AOB'$ after six repetitions be 180°02′, correct to the nearest 30″. Then the measured value of $AOB'$ is 180°02′/6=30°00′20″ correct to the nearest 05″, and the correction to be applied to $AOB'$ is 20″. Suppose that $OB'=400$ ft. Then the length of the offset $B'B$ equals $\tan 20'' \times 400$ ft $=(0.0001)(400)=0.04$ ft.

In route surveying, highway center lines are frequently set by deflection angles. Referring to Fig. 6.48b assume that it is desired to establish the direction of the line from 16+10.450 to station 18+84.900 by turning the deflection angle 45°10′00″ to the right from a backsight on 14+55.250. The instrument is set on station 16+10.450, the $A$ vernier is set on zero, and with the telescope in the direct position a backsight is taken on 14+55.250 using the lower clamp and tangent-screw. The telescope is plunged, the upper clamp is loosened, the angle 45°10′00″ is set off on the clockwise circle with the $A$ vernier, and point $A$ is set on line along the line from 16+10.450 to 18+84.900. With 45°10′00″ still set on the horizontal circle, release the lower clamp and take a second backsight on 14+55.250 using the lower clamp and tangent-screw. At this point the telescope is in the reversed position. Next, the telescope is plunged a second time, the upper clamp is loosened, and twice the deflection angle, or 90°20′00″, is set off on the horizontal circle using the upper clamp and upper tangent-screw. This procedure is called setting the line by *double deflection angle*. If the instrument is in excellent adjustment, the last direction will coincide with the first point set at $A$. However, this rarely occurs, so that a second point is set at $B$ on the top of the same

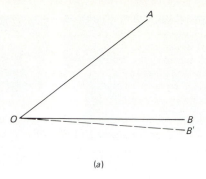

(a)

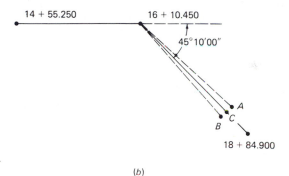

**Fig. 6.48** Laying off horizontal angles.
(a) Interior angle. (b) Deflection angle.

(b)

stake set for $A$. Point $C$ is set such that $BC = AC$ and the line from $16 + 10.450$ to $C$ represents the best estimate of the line that deflects $45°10'00''$ to the right.

**6.43. Measurement of vertical angles** The method for measuring vertical angles varies with the type of instrument used. The engineer's transit has a fixed vertical vernier and spirit level attached to the telescope. Repeating and direction theodolites have vertical circles on which zero corresponds to the zenith when a control spirit level bubble is centered. Theodolites of a more recent design are equipped with some form of vertical circle index compensator actuated by force of gravity, which automatically ensures that zero on the vertical circle coincides with the zenith (Art. 6.30).

Using the engineer's transit, the vertical angle to a point is its angle of elevation (+) or depression (−) from the horizontal. The transit is set up and leveled as when measuring horizontal angles. The plate bubbles should be centered carefully. The telescope is sighted approximately at the point, and the horizontal axis is clamped. The horizontal cross hair is set exactly on the point by turning the telescope tangent-screw, and the angle is read by means of the vertical vernier. Note that the vernier is on the outside or below the vertical circle which moves with respect to the fixed vernier when the telescope is rotated about the horizontal axis (Fig. 6.49). For a positive or elevation vertical angle with the telescope in the direct position, the vertical circle is read to the right of zero and the vernier is read on the right side of the index. A negative or depression vertical angle (telescope direct) is read to the left of zero and on the left side of the vernier index. Figure 6.49 illustrates positive and negative vertical angles of $+10°53'$ and $-9°43'$, respectively.

 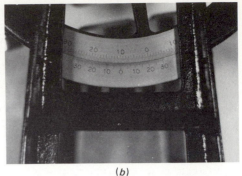

(a) (b)

**Fig. 6.49** Vertical circle readings with engineer's transit. (a) Positive vertical angle. (b) Negative vertical angle.

On theodolites, zenith angles from 0° to 90° are above horizontal (+ vertical angles, Fig. 6.50a), while zenith angles from 90° to 180° are below horizontal (− vertical angles, Fig. 6.50b). Before reading the vertical circle, the vertical circle control bubble is carefully centered. This ensures that zero on the vertical circle coincides with the zenith. The vertical circle is then viewed through the reading microscope (Fig. 6.34a). Figures 6.33b and 6.34c provide example vertical circle readings on different styles of theodolites. Zenith angles are observed in a similar manner on theodolites equipped with vertical circle index compensators. On these instruments, there is no need to center a vertical circle bubble, since the compensator automatically makes the vertical circle index coincide with the zenith. Details concerning the compensators can be found in Art. 6.30. Theodolites shown in Figs. 6.33a, 6.34a, and 6.35a are equipped with compensators.

Given the zenith angle, observed on a theodolite, the vertical angle can be computed. For example, a zenith angle of $85°10'45''$ yields a vertical angle of $(90°00'00'') - (85°10'45'') = + 4°49'15''$. A zenith angle of $94°12'44''$ corresponds to a vertical angle of $(90°00'00'') - (94°12'44'') = -4°12'44''$. Both of these angles are observed with the telescope in the direct position. With the telescope reversed and assuming zero index error (Art. 6.45), corresponding zenith angles are $360° - (85°10'45'') = 274°49'15''$ and $360° - (94°12'44'') = 265°47'16''$. These zenith angles then yield respective vertical angles of $(274°49'15'') - 270° = +4°49'15''$ and $(265°47'16'') - 270° = -4°12'44''$. Note that the zenith angles are **unambiguous**, since they are positive in all cases.

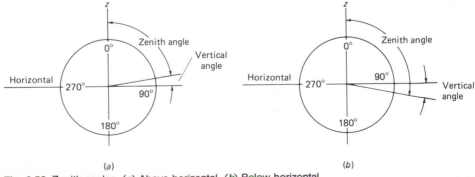

(a) (b)

**Fig. 6.50** Zenith angles. (a) Above horizontal. (b) Below horizontal.

**6.44. Vertical angles by double centering**   The method of observing vertical or zenith angles by double centering consists of reading once with the telescope direct and once with it reversed and taking the mean of the two values thus obtained.

Double centering is used in astronomical observations and in similar measurements of vertical angles to distant objects. In traversing, a similar result is obtained by measuring the vertical angle of each traverse line from each end, with the telescope the same side up for the two observations, and taking the mean of the two values.

**6.45. Index error**   *Index error* is the error in an observed angle due to (1) lack of parallelism between the line of sight and the axis of the telescope level bubble or between the line of sight and axis of the vertical circle level bubble, (2) displacement (lack of adjustment) of the vertical vernier, and/or (3) for a transit having a fixed vertical vernier, inclination of the vertical axis. If the instrument were in perfect adjustment and were leveled perfectly for each observation, there would be no index error; however, in practice these conditions seldom exist.

The effect of index errors due to lack of adjustment of the instrument can be eliminated either by double centering for each observation or by applying to each observation a correction determined (by double centering) for the instrument in its given condition of adjustment. For the common type of transit with a fixed vertical vernier, the effect of imperfect leveling cannot be eliminated by double centering, but—provided that the line of sight is in adjustment—for each direction of pointing a correction can be determined (as described later) and applied. Often, it is more convenient to apply the correction than to ensure that the instrument is perfectly adjusted and leveled.

The index correction is equal in amount but opposite in sign to the index error. Thus, if the observed vertical angle is $+12°14'$ and if the index error is determined to be $+02'$, the correct value of the angle is

$$+12°14' - 02' = +12°12'$$

Methods of determining the index error (and, therefore, the correction) are given in the following paragraphs.

***1. Lack of parallelism between line of sight and axis of telescope level***   If the axis of the telescope level is not parallel to the line of sight and if the vertical vernier reads zero when the bubble is centered (Fig. 6.51$a$), an error in vertical angle results. This error can be rendered negligible for ordinary work by careful adjustment of the instrument (Art. 6.51, adjustment number 5). The combined error due to this cause and to displacement of the vertical vernier can be eliminated by double centering. The index error due to the two causes can be determined by comparing a single reading on any given point with the mean of the two readings obtained by double centering to the same point. Thus, if the observed vertical angle to a point is $+2°58'30''$ with telescope normal and is $+2°55'30''$ with telescope

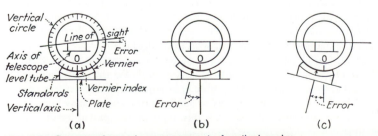

**Fig. 6.51** Sources of error in measurement of vertical angles.

reversed, the index error for readings with telescope normal is
$$(+2°58'30'' - 2°55'30'')/2 = +1'30''$$
Repeating and direction theodolites are not equipped with telescope level bubbles. However, there can be index error in these instruments due to lack of parallelism between the line of sight and the level bubble used to orient the vertical circle. As with the engineer's transit, this type of error can be detected by observing vertical circle readings on a given point with the telescope direct and reversed. Thus, if direct and reversed zenith angles are 94°12'44'' and 265°47'24'', respectively, the index error is
$$\frac{94°12'44'' - 360° + 265°47'24''}{2} = 0°00'04''$$
This type of error can be eliminated by double centering or can be made nearly negligible for most ordinary surveys by adjustment of the vertical circle level bubble (Art. 6.53, adjustment number 4).

**2. Displacement of vertical vernier** Displacement of the vertical vernier on an engineer's transit (Fig. 6.51*b*) introduces a constant index error. The error can be rendered negligible by careful adjustment (Art. 6.51, adjustments 6 and 6*a*). The combined index error due to this cause and to lack of parallelism between the line of sight and the axis of the telescope level can be eliminated by double centering; or the combined error can be determined as described in the preceding paragraph. For a transit having a fixed vertical vernier, the error due to displacement of the vertical vernier alone can be determined—provided that the line of sight is in adjustment—by leveling the transit carefully, leveling the telescope, and reading the vertical vernier. For a transit having a movable vertical vernier with control level, the error due to displacement of the vertical vernier alone can be determined by leveling both the telescope level and the vernier level, and reading the vertical vernier.

**3. Inclination of vertical axis** For an engineer's transit having a fixed vertical vernier, any inclination of the vertical axis (Fig. 6.51*c*) due to erroneous leveling of the instrument introduces an index error which varies with the direction in which the telescope is pointed and which is equal in amount to the angle through which the fixed vertical vernier is displaced about the horizontal axis while the instrument is directed toward the point. This index error can be rendered negligible by careful leveling of the transit before each observation, making sure that the plate level bubbles remain in position for any direction of pointing. It is not eliminated by double centering, since the condition causing the error is not changed by reversal (and plunging) of the instrument (see Fig. 6.51*c*). If the line of sight and the vertical vernier are in adjustment, the index error due to inclination of the vertical axis alone can be determined for each direction of pointing by leveling the telescope and reading the vertical vernier.

When a series of horizontal and vertical angles is to be measured from a given station, recentering the plate bubbles necessitates taking a new backsight (and thus additional work) before additional horizontal angles can be measured correctly, yet the plate bubbles may be considerably displaced without appreciably affecting the correctness of *horizontal* angles. Often, it involves less work to make the index correction for vertical angles than to relevel the instrument each time the plate level bubbles are seen to be displaced.

For theodolites having vertical circle spirit levels any moderate inclination of the vertical axis does not introduce an appreciable error in vertical angles, provided the instrument is in adjustment and provided the vertical circle level bubble is in adjustment and centered each time an observation is made. On topographic surveys or similar work where many horizontal and vertical angles are to be observed, the use of this type of equipment results in a considerable saving of time.

**6.46. Precise leveling of the transit or theodolite**   When using the engineer's transit for astronomical observations or for measurement of horizontal angles requiring steeply inclined sights, it is usually desired to level the transit with greater precision than that which is possible through the use of the plate levels, and in such cases the vertical axis is made truly vertical by means of the telescope level as follows. First the transit is leveled by means of the plate levels in the usual manner. With the telescope over one pair of opposite leveling screws, the bubble of the telescope level is centered using the telescope tangent-screw. The telescope is rotated end for end about the vertical axis; then the bubble is brought halfway back to center by means of the leveling screws, the plate levels being disregarded. The process is repeated alternately for both pairs of opposite leveling screws until the bubble of the telescope level remains centered for any direction of pointing.

For astronomic observations, repeating and direction theodolites can be equipped with a *striding level*. A striding level is a spirit level with a sensitivity of from 05″ to 15″ per 2-mm division and is constructed so that it can be placed on the horizontal or trunion axis of the instrument and removed when the need for it is completed. Using a striding level, the vertical axis of a theodolite can be made truly vertical by ensuring that the bubble is centered throughout a horizontal rotation of the instrument about the vertical axis. This condition is achieved completely by manipulation of the level screws. For high-precision work, the striding level should be calibrated.

**6.47. Operations with the transit or theodolite: prolonging a straight line**   If a straight line such as $AB$ (Fig. 6.52) is to be prolonged to $P$ (not already defined on the ground), which is beyond the limit of sighting distance or is invisible from $A$ and $B$, the line is extended by establishing a succession of stations $C$, $D$, etc., each of which is occupied by the transit. Three methods may be employed. The third method is the most reliable and is the recommended procedure.

**Method 1**   The transit is set up at $A$, a sight is taken to $B$, and a point $C$ is established on a line beyond $B$. The transit is moved to $B$, a sight is taken to $C$, and point $D$ is set on line beyond $C$. The process is continued until the desired distance is traversed and point $P$ is set.

**Method 2**   The transit is set up at $B$, and a backsight is taken to $A$. With both upper and lower motions clamped, the telescope is plunged, and a point $C$ is set on line. If the line of sight is perpendicular to the horizontal axis, as it will be if the instrument is in perfect adjustment, it will generate a vertical plane as the telescope is revolved, and point $C$ will lie on the prolongation of $AB$. The transit is moved to $C$, a backsight is taken to $B$, a point $D$ is set in similar manner, and the process is repeated until point $P$ is set.

If the line of sight is not perpendicular to the horizontal axis of the transit, as the telescope is plunged (say from the reversed to the direct position), the line of sight will generate a portion of a cone whose vertex is at the center of the instrument and two of whose elements are $AB$ and $BC'$, and $C'$ will not lie on the true prolongation of $AB$. If the instrument is set up at $C'$, a backsight taken to $B$ with the telescope reversed as before, and the telescope plunged to the direct position, a second and similar cone is generated and $D'$ will not lie on the prolongation of $BC'$. Thus, if the line is extended by the method outlined and all backsights are taken with the telescope in one position (either direct or reversed), the points established will lie along a curve instead of a straight line, and each segment of the line will be deflected in the same direction (to the right or to the left) by double the error of

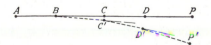

**Fig. 6.52** Prolonging a straight line.

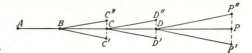

**Fig. 6.53** Double sighting to prolong line.

adjustment of the line of sight. On the other hand if, say, at the even-numbered stations $B$, $D$, $F$, etc., backsights were taken with the telescope reversed and at odd-numbered stations, $C$, $E$, etc., backsights were taken with the telescope direct, a zigzag line would be established with some points on one side of the line joining the terminals and some perhaps on the other. This method should not be used where the adjustment of the instrument is question-able and should never be employed where more than a low order of accuracy is required.

**Method 3**    This procedure, known as *double centering*, should be employed for all surveys requiring from moderate to high accuracy. If the line $AB$ (Fig. 6.53) is to be prolonged to some point $P$, the transit is set up at $B$ and a backsight is taken to $A$ with the telescope in its *direct* position. The telescope is plunged, and a point $C'$ is set on line. The transit is then revolved about its vertical axis, and a second backsight is taken to $A$ with telescope *reversed*. The telescope is plunged, and a point $C''$ is established on line beside $C'$. It is evident that $C'$ will be as far on one side of the true prolongation of $AB$ as $C''$ is on the other. Midway between $C'$ and $C''$, a point $C$ is set defining a point on the correct prolongation of $AB$. In a similar manner the next point $D$ is established by setting up at $C$, double centering to $B$, and setting points at $D'$, $D''$, and $D$. The process is repeated until the desired distance is traversed.

**6.48. Prolonging a line past an obstacle**    Figure 6.54 illustrates one method of prolonging a line $AB$ past an obstacle where the offset space is limited. The transit is set up at $A$, a right angle is turned, and a point $C$ is established at a convenient distance from $A$. Similarly the point $D$ is established, the distance $BD$ being made equal to $AC$. The line $CD$, which is parallel to $AB$, is prolonged; and points $E$ and $F$ are established in convenient locations beyond the obstacle. From $E$ and $F$ right-angle offsets are made, and $G$ and $H$ are set as were $C$ and $D$; $GH$ then defines the prolongation of $AB$. The distance $AH$ is determined by measuring the length of the lines $AB$, $DE$, and $GH$. If measurement of distance is to be carried forward with high accuracy, it is necessary to erect the perpendiculars $AC$, $BD$, etc. with greater than ordinary care; and if the line is to be prolonged with high accuracy, it is essential not only that the offset distances be measured carefully but also that $AB$ and $EF$, the distances between offsets, be as long as practicable.

Another method of prolonging a line $AB$ past an obstacle is illustrated by Fig. 6.55. A small angle $\alpha$ is turned off at $B$, and the line is prolonged to some convenient point $C$ which will enable the obstacle to be cleared. At $C$, the angle $2\alpha$ is turned off in the reverse direction, and the line is prolonged to $D$, with $CD$ made equal to $BC$. The point $D$ is then on the prolongation of $AB$; and $DE$, the further prolongation of $AB$, is established by turning off the angle $\alpha$ at $D$. If there were another obstacle between $D$ and $E$, as there might often be in wooded country, the line $CD$ is prolonged to some point, as $F$, from which the obstacle

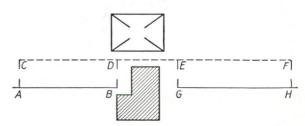

**Fig. 6.54** Prolonging line past obstacle by perpendicular offsets.

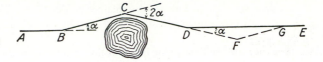

**Fig. 6.55** Prolonging line past obstacle by angles.

can be cleared; and so a zigzag course is followed until it is possible to resume traversing on the direct prolongation of the main line $AB$. As compared with the method of perpendicular offsets, this method is more convenient in the field, but it requires computation to determine the length $BD$. However, if the angle $\alpha$ is small, say not greater than a degree or so, often it will be sufficiently precise to take the distance along the main line as equal to that along the auxiliary lines.

**6.49. Running a straight line between two points**   If the terminal points $A$ and $B$ of a line are fixed and it is desired to establish intervening points on the straight line joining the terminals, the method to be employed depends upon the length of the line and the character of the terrain. Two common cases are considered below:

*Case 1. Terminals intervisible*   The transit is set up at $A$, a sight is taken to $B$, and intervening points are established on line (Fig. 6.56).

If the intervening points thus established were to lie in the same plane with the center of the instrument and the terminal point $B$, they would define a truly straight line regardless of whether or not the horizontal axis of the transit were truly horizontal. If the horizontal axis is inclined with the horizontal, the line of sight will not generate a vertical plane as the telescope is revolved; and thus if it is necessary to rotate the telescope about the horizontal axis in order to set the intervening points, the points so established will not lie on a truly straight line (as seen in plan view) joining the terminals.

Ordinarily, the vertical angles through which it is necessary to rotate the telescope will be small, and if the horizontal axis is in fair adjustment and the plate bubbles are centered, the error arising from this source is negligible. Occasionally, however, when the intervening points are to be set with high precision or when the adjustment of the instrument is uncertain and the vertical angles are large, the intervening stations are set by double sighting.

*Case 2. Terminals not intervisible, but visible from an intervening point on line*   The location of the line at the intervening point $C$ is determined by trial, as follows: In Fig. 6.56, $A$ and $B$ represent the terminals both of which can be seen from the vicinity of $C$. The transit is set up on the estimated location of the line near $C$, a backsight is taken to $A$, the telescope is plunged, and the location of the line of sight at $B$ is noted. The amount that the transit must be shifted laterally is estimated; and the process is repeated until, when the telescope is plunged, the line of sight falls on the point at $B$. This process is known as "balancing in." The location of the transit should then be tested by double centering; the test will also disclose whether or not the line of sight and the horizontal axis are in adjustment.

If the instrument were in perfect adjustment, its center would be on the true line joining $AB$. On the other hand, if the line of sight were not perpendicular to the horizontal axis, a cone would be generated by the line of sight when the telescope was plunged, as explained in Art. 6.47; also if the horizontal axis were not truly horizontal, the line of sight in its rotation

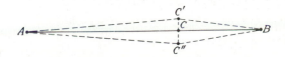

**Fig. 6.56** Balancing in.

would not generate a vertical plane, as explained under case 1. Hence, if the transit were not in adjustment, its center might be at $C'$ and still the line of sight would bisect $B$ when the telescope is plunged.

To locate the intermediate point truly on line, trials are made first with the telescope in, say, its direct position for backsights to $A$, until an intermediate point, as $C'$, is determined. Then a second series of trials is made with the telescope reversed for backsights to $A$, until the corresponding point $C''$ is located. Then, for reasons previously explained, the true line is at $C$, halfway between $C'$ and $C''$. Other intermediate points may then be established by setting up the transit at $C$ and proceeding as in case 1.

**6.50. Adjustment of the engineer's transit: desired relations**   For a transit in perfect adjustment the relations stated in the following paragraphs should exist. The number of each paragraph is the same as that of the corresponding adjustment described in the following section. For adjustments 1 to 5, Fig. 6.57 shows the desired relations among the principal lines of the transit.

1. The vertical cross hair should lie in a plane perpendicular to the horizontal axis so that any point on the hair may be employed when measuring horizontal angles or when running lines.
2. The axis of each plate level should lie in a plane perpendicular to the vertical axis so that when the instrument is leveled the vertical axis will be truly vertical; thus horizontal angles will be measured in a horizontal plane, and vertical angles will be measured without index error due to inclination of the vertical axis.
3. The line of sight should be perpendicular to the horizontal axis at its intersection with the vertical axis. Also, the optical axis, the axis of the objective slide, and the line of sight should coincide. If these conditions exist, when the telescope is rotated about the horizontal axis the line of sight will generate a plane when the objective is focused for either a near sight or a far sight, and that plane will pass through the vertical axis.
4. The horizontal axis should be perpendicular to the vertical axis so that when the telescope is plunged the line of sight will generate a vertical plane.
5. The axis of the telescope level should be parallel to the line of sight so that the transit may be employed in direct leveling and so that vertical angles may be measured without index error due to lack of parallelism.
6. If the transit has a fixed vernier for the vertical circle, the vernier should read zero when the plate bubbles and telescope bubble are centered, in order that vertical angles may be measured without index error due to displacement of the vernier.

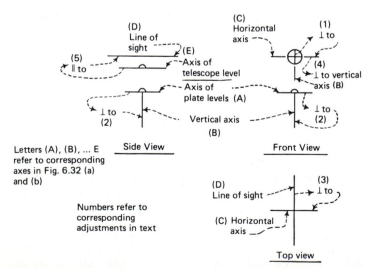

**Fig. 6.57** Desired relations among principal lines of transit.

6*a*. If the vertical vernier is movable and has a control level, the axis of the control level should be parallel to that of the telescope level when the vernier reads zero.
  7. The optical axis and the line of sight should coincide (see 3, above).
  8. The axis of the objective slide should be perpendicular to the horizontal axis (see 3, above).
  9. The intersection of the cross hairs should appear in the center of the field of view of the eyepiece.
 10. If the transit is equipped with a striding level for the horizontal axis, the axis of the striding level should be parallel to the horizontal axis. Thus, when the bubble of the striding level is centered and the instrument is plunged, the line of sight (if in adjustment) will generate a vertical plane.
 11. On transits and theodolites equipped with an optical plummet, the vertical collimation line should coincide with the plumb line.

**6.51. Adjustments**    In the description of the following adjustments (except 7 and 9) it is assumed that the objective slide does not permit adjustment, but that it is permanently fixed in the telescope tube so far as lateral motion is concerned; and that the maker has so constructed the instrument that the optical axis and the axis of the objective slide coincide and are perpendicular to the horizontal axis. This ideal construction is never exactly attained; but in most instruments the departure is so slight that it need not be considered in ordinary transit work, and in precise surveying the resulting errors are eliminated by methods of procedure.

For those adjustments which involve sighting through the telescope, particular attention should be given to proper focusing of both the eyepiece and the objective prior to testing the adjustments.

The adjustments of the transit are more or less dependent upon one another. For this reason, if the instrument is badly out of adjustment, time will be saved by first making corrections roughly for related adjustments until all the tests have been tried and then repeating the tests and corrections in the same order. The plate levels will not be disturbed by other adjustments and should be corrected exactly before other adjustments are attempted.

Any movement of the screws controlling the cross-hair ring is likely to produce both lateral displacement and rotation of the ring; hence any considerable adjustment of the line of sight is likely to disturb the vertical hair so that it will no longer remain on a point when the telescope is rotated about the horizontal axis. The adjustment of the telescope level depends upon the unaltered position of the horizontal cross hair and hence should not be tested until the line of sight and horizontal axis have been corrected.

The transit adjustments commonly made are 1 to 6*a*. Adjustments 7 to 11 may be required occasionally for some instruments.

### 1. To make the vertical cross hair lie in a plane perpendicular to the horizontal axis

*Test*. Sight the vertical cross hair on a well-defined point not less than 200 ft or 60 m away. With both horizontal motions of the instrument clamped, swing the telescope through a small vertical angle, so that the point traverses the length of the vertical cross hair. If the point appears to move continuously on the hair, the cross hair lies in a plane perpendicular to the horizontal axis (see Fig. 6.58).

*Correction*. If the point appears to depart from the cross hair, loosen two adjacent capstan screws and rotate the cross-hair ring in the telescope tube until the point traverses the entire length of the hair. Tighten the same two screws. This adjustment is similar to adjustment 2 of the dumpy level (Art. 5.58), with the terms *vertical* and *horizontal* interchanged.

### 2. To make the axis of each plate level lie in a plane perpendicular to the vertical axis

*Test*. Rotate the instrument about the vertical axis until each level tube is parallel to a pair of opposite leveling screws. Center the bubbles by means of the leveling screws. Rotate the instrument end for end about the vertical axis. If the bubbles remain centered, the axis of each level tube is in a plane perpendicular to the vertical axis (see Fig. 6.59).

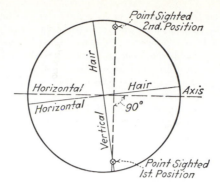

**Fig. 6.58** Adjustment of vertical cross hair of transit.

*Correction*. If the bubbles become displaced, bring them *halfway* back by means of the adjusting screws. Level the instrument again and repeat the test to verify the results. This is the method of reversion.

The steps involved in adjustment are shown in Fig. 6.59. In view (*a*) is shown the level tube out of adjustment by the amount of the angle α, but with the bubble centered; the support is therefore not level, and the vertical axis is not vertical. In view (*b*) the level tube has been lifted and reversed end for end, which is effected by rotating the instrument 180° about the vertical axis. The axis of the level tube now departs from the horizontal by 2α, or *double* the error of the setting. In view (*c*) the bubble has been brought back *halfway* to the middle of the tube by means of the adjusting screw *C*, without moving the support; the tube is now in adjustment. Finally, in view (*d*) the bubble is again centered by raising the low end (and/or lowering the high end) of the support; the support is now level and the adjustment may be checked by reversing the tube again.

If it were desired to level the support in the direction of the tube without taking time to adjust the tube, this could be accomplished by centering the bubble as in view (*a*), Fig. 6.59; reversing the tube as in view (*b*); and raising the low end (and/or lowering the high end) of the support until the bubble is brought halfway back to the center of the tube. This position of the bubble corresponds to the error of setting of the tube; and whenever the bubble is in this position the support will be level.

### 3. To make the line of sight perpendicular to the horizontal axis

*Test*. Level the instrument. Sight on a point *A* (see Fig. 6.60) about 500 ft or 150 m away, with telescope in the direct position. With both horizontal motions of the instrument clamped, plunge the telescope and set another point *B* on the line of sight and about the

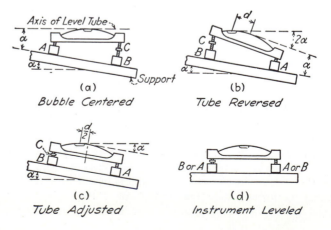

**Fig. 6.59** Adjustment of level tube by reversion.

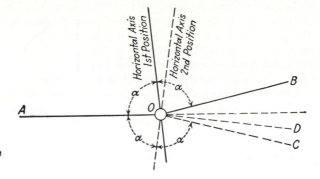

**Fig. 6.60** Adjustment of line of sight of transit.

same distance away on the opposite side of the transit. Unclamp the upper motion, rotate the instrument end for end about the vertical axis, and again sight at *A* (with telescope reversed). Clamp the upper motion. Plunge the telescope as before; if *B* is on the line of sight, the desired relation exists.

*Correction.* If the line of sight does not fall on *B*, set a point *C* on the line of sight beside *B*. Mark a point *D*, *one-fourth* of the distance from *C* to *B*, and adjust the cross-hair ring (by means of the two opposite horizontal screws) until the line of sight passes through *D*. The points sighted should be at about the same elevation as the transit.

### 4. To make the horizontal axis perpendicular to the vertical axis

*Test.* Set up the transit near a building or other object on which is some well-defined point *A* at a considerable vertical angle. Level the instrument very carefully, thus making the vertical axis truly vertical. Sight at the high point *A* (see Fig. 6.61) and, with the horizontal

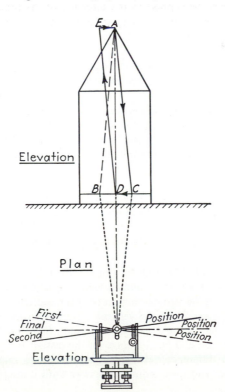

**Fig. 6.61** Adjustment of horizontal axis of transit.

motions clamped, depress the telescope and set a point $B$ on or near the ground. If the horizontal axis is perpendicular to the vertical axis, $A$ and $B$ will be in the same vertical plane. Plunge the telescope, rotate the instrument end for end about the vertical axis, and again sight on $A$. Depress the telescope as before; if the line of sight falls on $B$, the horizontal axis is perpendicular to the vertical axis.

*Correction.* If the line of sight does not fall on $B$, set a point $C$ on the line of sight beside $B$. A point $D$, halfway between $B$ and $C$, will lie in the same vertical plane with $A$. Sight on $D$; elevate the telescope until the line of sight is beside $A$; loosen the screws of the bearing cap, and raise or lower the adjustable end of the horizontal axis until the line of sight is in the same vertical plane with $A$.

The high end of the horizontal axis is always on the same side of the vertical plane through the high point as the point last set.

In readjusting the bearing cap, care should be taken not to bind the horizontal axis, but it should not be left so loose as to allow the objective end of the telescope to drop of its own weight when not clamped.

### 5.  *To make the axis of the telescope level parallel to the line of sight*

*Test and correction.* Proceed the same as for the two-peg adjustment of the dumpy level (Art. 5.58, adjustment 3), except as follows: With the line of sight set on the rod reading established for a horizontal line, the correction is made by raising or lowering one end of the telescope level tube until the bubble is centered.

### 6.  (For Transit Having a Fixed Vertical Vernier) *To make the vertical circle read zero when the telescope bubble is centered*

*Test.* With the plate bubbles centered, center the telescope bubble and read the vertical vernier.

*Correction.* If the vernier does not read zero, loosen it and move it until it reads zero. Care should be taken that the vernier will not bind on the vertical circle as the telescope is rotated about the horizontal axis.

### 6a.  (For Transit Having a Movable Vertical Vernier with Control Level) *To make the axis of the auxiliary level parallel to the axis of the telescope level when the vertical vernier reads zero*

*Test.* Center the telescope bubble, and by means of the vernier tangent-screw move the vertical vernier until it reads zero.

*Correction.* If the bubble of the control level which is attached to the vertical vernier is not at the center of the tube, bring it to the center by means of the capstan screws at one end of the tube.

**6.52. Special adjustments**   In addition to the usual adjustments just described, the following adjustments may be made as required by the type or the condition of the transit.

### 7.  *To make the line of sight, insofar as defined by the horizontal cross hair, coincide with the optical axis*

*Test.* Set two pegs, one about 25 ft or about 8 m and the other 300 to 400 ft or about 90 to 100 m from the transit. With the vertical motion clamped, take a rod reading on the distant point, and without disturbing the vertical motion read the rod on the near point. Plunge the telescope, rotate the instrument about the vertical axis, and set the horizontal cross hair at the last rod reading with the rod held on the near point. Sight to the distant point. If the desired relation exists, the first and last readings on the distant rod will be the same.

*Correction.* If there is a difference between the rod readings, move the horizontal cross hair by means of the upper and lower adjusting screws until it intercepts the distant rod at the mean of the two readings. Repeat the process, until by successive approximations the error is reduced to zero.

**8.** (For Transit Equipped with Striding Level) *To make the axis of the striding level parallel to the horizontal axis*

*Test*. By means of the leveling screws center the striding level bubble. Lift the level from its supports and turn it end for end. If it is in adjustment, the bubble will again be centered.

*Correction*. If the bubble is displaced, bring it halfway back to the center by means of the capstan screw at one end of the level tube (Fig. 6.59). Relevel the instrument by means of the leveling screws and repeat the test until the adjustment is perfected.

**9. *Adjustment of the optical plummet***

Field adjustment is performed as follows. The instrument is leveled and centered over a clearly marked point using a plumb bob suspended from the screw that secures the instrument to the foot plate. The plumb line is removed and the cross in the optical plummet is very carefully focused on the point. Any deviations between the point and cross are due to maladjustment of the optical plummet and can be corrected by means of adjusting screws usually located near the optical plummet's eyepiece. For best results, this test should be conducted in an area protected from the wind. A heavy plumb bob should be used.

A more accurate procedure for checking and adjusting an optical plummet is to apply the reversion principle. After centering the instrument over a carefully marked point, it is necessary to rotate the entire assembly of instrument, leveling head, foot plate, and tripod (or trivet) 180° about the vertical axis. This procedure requires auxiliary equipment and is best performed in a laboratory.

**6.53. Adjustment of theodolites**    Adjustment of repeating and direction theodolites requires that the same fundamental relationships must be satisfied in the same sequence as detailed for the engineer's transit. However, there are certain differences which are outlined in the following paragraphs.

**1. *Adjustment of the plate level bubble***

The theodolite is set up and the circular level bubble is centered by means of the three level screws (Fig. 6.44). Next, the axis of the plate level bubble tube is aligned with a pair of level screws and the bubble is centered. Rotate the theodolite 180° about the vertical axis; if the bubble leaves its central position, correct one-half the error with the level screws and one-half the error using the adjusting screw at one end of the bubble tube. If the error is only one or two graduations, do not make the adjustment. Bring the bubble halfway to center with the level screws and note the position of the bubble for future operations. Next, rotate the instrument 90° about the vertical axis from its original position and center the plate level bubble with the remaining screw. Then rotate the instrument 180° and note the position of the bubble. If the deviation is small (one or two divisions), note the position for future reference and bring the bubble halfway to center with the level screws but do not make an adjustment. If the deviation is large, bring the bubble halfway to center with the level screws and remove the remaining half with the adjusting screw usually located at one end of the level bubble tube.

**2. *Adjustment of the circular level bubble***

Level the theodolite using the plate level bubble (Art. 6.34). The circular bubble in the tribach of the leveling head should be centered. If the circular bubble is not centered, it should be adjusted by using the adjusting screws located around the bubble case. The adjustment procedure is the same as described for automatic levels in Art. 5.59.

### 3. Optical plummet

Adjustment of the optical plummet in a theodolite is the same as described for the engineer's transit equipped with an optical collimator (Art. 6.52, adjustment number 9).

### 4. Vertical circle level bubble

The instrument is leveled and the vertical circle level bubble is centered using the altitude tangent screw. Using the vertical lock and tangent screw, the horizontal cross hair is then sighted on a well-defined object point about 400 ft or 120 m away and the vertical circle reading is observed. Assume that this observation is 92°01′30.0″. The telescope is reversed, the vertical circle level bubble is centered, and the horizontal cross hair is sighted on the same object point with the vertical tangent motion. If the zenith angle from this second sight is 360° − (267°58′40.0″) = 92°01′20.0″, the index error is 05″ and the true value of the zenith angle is 92°01′25.0″. To make the adjustment, 267°58′35.0″ is set on the vertical circle using the altitude tangent-screw and observing the optical micrometer while the horizontal cross hair remains on the object point. This manipulation causes the vertical circle level bubble to move off center. The bubble is centered by loosening one of the adjusting screws at one end of the level bubble and tightening the other. When the correction has been made, the entire procedure should be repeated to test the goodness of the adjustment.

Locations of adjusting screws are not standardized on theodolites. Reference should be made to the manufacturer's instruction manual prior to attempting any adjustment.

**6.54. Errors in determining angles and directions: general** Except in field astronomy, a measured angle is always closely related to a measured distance; and in general there should be a consistent relation between the precision of measured angles and that of measured distances. From the standpoint of both precision achieved and expediting the work, it is important that the surveyor be able to: (1) visualize the effect of errors in terms of both angle and distance, (2) appreciate what degree of care must be exercised to keep certain errors within specified limits, and (3) know under what conditions various instrumental errors can be eliminated.

On surveys of ordinary accuracy it usually requires much more care to keep *linear* errors within prescribed limits than to maintain a corresponding degree of *angular* precision. Often undue attention is paid to securing precision in angular measurements, and at the same time large and important errors in the measurement of distances are overlooked.

Errors in transit work may be instrumental (Art. 6.55), personal (Art. 6.56), or natural (Art. 6.57).

**6.55. Instrumental errors** Errors due to instrumental imperfections and/or nonadjustment are all systematic, and they can either be eliminated or reduced to a negligible amount by proper methods of procedure.

### 1. Errors in horizontal angles caused by nonadjustment of plate levels

When the bubbles of plate levels in nonadjustment are centered, the vertical axis is inclined, and hence measured angles are not truly horizontal angles. Also the horizontal axis is inclined to a varying degree depending upon the direction in which the telescope is sighted. There will be one vertical plane which will include the vertical axis in its inclined position; this is illustrated by Fig. 6.62, in which the horizontal axis and the vertical axis are in the plane of the paper. When the horizontal axis is rotated about the vertical axis until it is normal to the plane of the paper, it becomes truly horizontal and the line of sight will generate a vertical plane when the telescope is plunged; hence no error in direction is introduced regardless of the angle of elevation to the point sighted. As the transit is rotated about its vertical axis, the horizontal axis becomes inclined, making a maximum angle with the horizontal when it reaches the plane of the paper. With the horizontal axis in this position, the line of sight

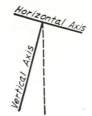

**Fig. 6.62** Inclination of vertical axis due to maladjustment of plate level bubbles.

generates a plane making an angle with the vertical equal to the error in the position of the vertical axis; and with the line of sight inclined at a given angle, the maximum error in determining the direction of a line is introduced. The larger the vertical angle, the greater the error in direction. The error cannot be eliminated by double centering.

The diagram of Fig. 6.63 shows for various vertical angles (values of $\alpha$) the errors introduced in horizontal angles due to an inclination of 01′ in the vertical axis or one space on the plate levels of the ordinary transit. The values of $H$ are the horizontal angles which the line of sight makes with the vertical plane in which lies the vertical axis in its inclined position (i.e., with the plane of the paper, Fig. 6.62). Within reasonable limits the error in horizontal angle varies directly as the inclination of the vertical axis, hence a similar diagram for an inclination of 02′ would show ordinates twice as great as those of Fig. 6.63.

Although the diagram may not be of much practical value, it serves to illustrate some noteworthy facts.

(*a*) For observations of ordinary accuracy taken in flat country where the vertical angles are rarely greater than 3° and usually much less, the plate bubble may be out several spaces without appreciably affecting the precision of horizontal angular measurements. For example, if an angle were measured between $H=0°$ and $H=90°$, with bubble out two spaces and $\alpha=3°$, the error would be about 06″; or if in prolonging a straight line the telescope were plunged from the position $H=90°$ to $H=270°$ with both backsight and foresight taken at a vertical angle of $+3°$, the error would be doubled and for the bubbles out two spaces (vertical axis in error 02′), the angular error introduced in the direction of the line would be 12″.

(*b*) For angular measurements of higher precision, such as when measuring an angle by repetition, the plate levels must be in good adjustment and the bubbles must be centered with reasonable care even though the survey is conducted over fairly smooth ground. For example, if a horizontal angle were measured between the positions $H=0°$ and $H=90°$, $\alpha=5°$, and the vertical axis were inclined 30″, the error in horizontal angle would amount to 02.2″.

(*c*) In rough country where the vertical angles are large, even for surveys of ordinary precision the plate levels must be in good adjustment and the bubbles must be carefully centered if errors in horizontal angles or in the prolongation of lines are to be kept within negligible limits. For example, if a line were prolonged by plunging the telescope from the position $H=90°$ to $H=270°$, $\alpha$ for both backsight and foresight being $+30°$ and the vertical axis being inclined 01′, the diagram (Fig. 6.63) shows that the error introduced is $2\times34.6″=01′09.2″$. In other words, the angle at the station at which

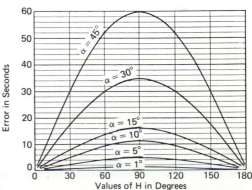

**Fig. 6.63** Errors in horizontal angles for a 1′ inclination of the vertical axis ($\alpha$ = vertical angle).

the instrument was set instead of being a true 180° would be 180°01′09.2″, and beyond the station the established line would depart from the true prolongation about 0.1 ft in each 300 ft or 33 mm in 100 m.

**2. Errors In vertical angles due to nonadjustment of plate levels**   These errors obviously vary with the direction in which the instrument is pointed. With the fixed vertical vernier they are eliminated by observing (for each sighting) the index error of the corresponding observed vertical angle (Art. 6.43).

Certain styles of repeating and direction theodolites have vertical circle level bubbles which are centered prior to each vertical angle eliminating the error caused by nonadjustment of plate level bubbles (Art. 6.51, test for adjustment 2). Theodolites equipped with automatic vertical circle indexing are also immune to this type of error.

It may be noted further that nonadjustment of the plate levels causes an inclination of the plane of the vertical arc. This source of error may be considered negligible.

**3. Line of sight not perpendicular to horizontal axis**   If the telescope is not reversed between backsight and foresight, if the sights are of the same length so that it is not required to refocus the objective, and if both points sighted are at the same angle of inclination of the line of sight, no error is introduced in the measurement of horizontal angles even though the error due to this lack of adjustment is large. If the instrument is plunged between backsight and foresight, the telescope rotating about the horizontal axis generates a cone, the apex of which falls at the intersection of the line of sight and the horizontal axis. The resultant error in the observed angle is double the error of adjustment.

With the line of sight out of adjustment by a given amount, the effect of the error depends on the vertical angle to the point sighted. In Fig. 6.64, $OA$ and $DB$ are horizontal and are perpendicular to the horizontal axis $OH$ of the instrument; $e$ is the angle between the nonadjusted line of sight and a vertical plane normal to $OH$ (that is, $e$ is the error in direction for a horizontal sight $OB$); $E$ is the error in direction for an inclined sight $OC$; $h$ is the actual vertical angle to $C$, the point sighted; and $OB$ is made equal to $OC$. Then

$$\sin e = \frac{AB}{OB}$$
$$= \frac{AB}{OB}\frac{OD}{OD} = \frac{AB}{OD}\cdot\frac{OD}{OB} = \frac{AB}{OD}\frac{OD}{OC}$$
$$= \sin E \cos h$$

or

$$\sin E = \sin e \sec h \qquad\qquad (6.1)$$

If $\alpha$ is the observed vertical angle, it can similarly be shown that

$$\tan E = \tan e \sec \alpha \qquad\qquad (6.2)$$

For all ordinary cases Eq. (6.1) or Eq. (6.2) may be taken as

$$E = e \sec h = e \sec \alpha \qquad \text{(approx.)} \qquad\qquad (6.3)$$

For example, assume that the optical line of sight makes an angle of 89°58′00″ with the horizontal axis of the instrument. In turning a horizontal angle, if the line of sight on the backsight is horizontal and is inclined with a vertical angle of 40°00′ on the foresight, the

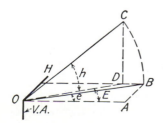

**Fig. 6.64**

error in the observed horizontal angle due to the lack of adjustment is

$$E = (02')\sec 40°00' = (02')(1.30541) = 0°02'37''$$

For two direct pointings (the backsight and the foresight) there will be a value of $E$ for each, and the error in the angle is the difference between them. In the measurement of a deflection angle by the method in which the telescope is inverted between backsight and foresight, the error in angle is the *sum* of the two values of $E$.

The error may be eliminated by taking the mean of two angular observations, one with the telescope in the direct position and the other with the telescope reversed. In prolonging a line, errors are avoided by the method of double centering described in Art. 6.47.

**4. Horizontal axis not perpendicular to vertical axis**  No error is introduced in horizontal angles so long as the points sighted are at the same angle of inclination of the line of sight. The angular error in the observed direction of any line depends both on the angle by which the horizontal axis departs from the perpendicular to the vertical axis and on the vertical angle to the point sighted. In Fig. 6.65, $OH$ is perpendicular to the vertical axis; $OH'$ is the horizontal axis in nonadjustment with the vertical axis by the angular amount $e'$; $OA$ is horizontal and is perpendicular to $OH$ and $OH'$; $\alpha$ is the observed vertical angle to $C$, the point sighted; $B$ is directly beneath $C$ and in the same horizontal plane with $OA$; angles $OAB$, $OAC$, and $ABC$ are right angles; and $\theta$ is the angular error in direction. From the figure,

$$
\begin{aligned}
\tan\theta &= \frac{BA}{OA} \\
&= \frac{BA}{OA}\frac{CA}{CA} = \frac{BA}{CA}\frac{CA}{OA} \\
&= \sin e' \tan\alpha
\end{aligned}
$$
(6.4)

or with sufficient precision,

$$\theta = e' \tan\alpha \text{ (approx.)}$$
(6.5)

Thus, for example, if a horizontal angle were measured between $D$, to which the vertical angle is $-30°$, and $E$, to which the vertical angle is $+15°$, the horizontal axis being inclined $02'$, then the error in horizontal angle is

$$\theta = 02'[\tan 15° - (-\tan 30°)] = 32'' + 01'09'' = 01'41''$$

The preceding example is sufficient to show that the error in an observed horizontal angle may become large. Obviously, the sign of the error in a horizontal angle depends upon the direction of displacement of the horizontal axis from its correct position. Hence, if any angle is measured with the telescope first in the direct and then in the reversed position, one value will be too great by the amount of the error and the other will be correspondingly too small; thus, the error is eliminated by taking the mean of the two values.

**5. Effect of lack of coincidence between line of sight and optical axis**  Under these conditions if the line of sight is perpendicular to the horizontal axis for one position of the objective, it will not be perpendicular for other positions, but will swing through an angle as the objective is moved in or out. If an angle is measured without disturbing the position of

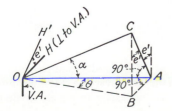

**Fig. 6.65**

the objective, no error is introduced. For most instruments the error from this source is not sufficiently large to be of consequence in ordinary transit or theodolite work. It is eliminated by taking the mean of two angles, one observed with the telescope direct and the other with it reversed.

**6. Errors due to eccentricity**  With the transit in good condition, errors due to eccentricity of verniers and/or eccentricity of centers are of no consequence in the ordinary measurement of angles. In any case, such errors are eliminated by taking the mean of readings indicated by opposite verniers.

**7. Imperfect graduations**  Errors from this source are of consequence only in work of high precision. They can be reduced to a negligible amount by taking the mean of several observations for which the readings are distributed over the circle and over the vernier.

**8. Lack of parallelism between axis of telescope level and line of sight**  This introduces an error in leveling which is compensated by equalizing backsight and foresight distances (see Art. 5.42). An error is also introduced into vertical angles where compensation is achieved by observing two vertical angles, one with the telescope direct and one with the telescope reversed (Art. 6.43). The mean of the two observations is the best estimate of the vertical angle.

**9. Nonadjustment of the vertical vernier**  This produces a constant error in the measurement of vertical angles. If the transit is equipped with a full vertical circle, the error can be eliminated by taking the mean of two values, one observed with the telescope direct and the other with the telescope reversed.

**Summary**  Summing up, it is seen that with regard to instrumental errors:

1. Errors in horizontal angles due to nonadjustment of plate levels or of horizontal axis become large as the inclination of the sights increases.

2. The maximum error in horizontal angles due to nonadjustment of the line of sight is introduced when the telescope is plunged between backsight and foresight readings. If the telescope is not plunged between backsight and foresight readings when observing horizontal angles, no error is introduced, as these distances are equal and the inclination of the line of sight is the same for both backsight and foresight.

3. Errors due to instrumental imperfections and/or nonadjustment are all systematic, and without exception they can be either eliminated or reduced to a negligible amount by proper procedure. In general, this procedure consists in obtaining the mean of two values—one observed before and one after a reversal of the horizontal plate by plunging the telescope and rotating it about the vertical axis. One of these values is as much too large as the other is too small. An exception is the error in either horizontal or vertical angle due to inclination of the vertical axis, which cannot be so eliminated but which can be eliminated, so far as its *systematic* character is concerned, by releveling the plate bubbles in addition to the reversal of the plate. However, for precise work the usual practice would be to make the vertical axis truly vertical by means of the telescope level, and then to proceed in the ordinary manner.

**6.56. Personal errors**  Personal errors arise from the limitations of the human eye in setting up and leveling the transit and in making observations.

**1. Effect of not setting up exactly over the station**  This produces an error in all angles measured at a given station, the magnitude of the error varying with the direction of pointing and inversely with the length of sight. It is convenient to remember that 0.1 ft in approximately 300 ft or 3 cm in about 100 m yields an angle of 01′. Thus, if the transit were offset 0.05 ft or 1.5 cm from the ends of lines 150 ft or 50 m long, respectively, the error in the observed direction would be 01′, but if the lines were 600 ft or 200 m long, the error would be only 15″. In general, the error may be kept within negligible limits by reasonable care. Many instrument operators waste time by exercising needless care in setting up when the sights to be taken are long.

When using a plumb bob, centering to within 0.02 ft or 6 to 9 mm is possible without difficulty. Centering errors of less than 0.003 ft or 1 mm are possible with the optical plummet or telescopic centering rod.

**2. Effect of not centering the plate bubbles exactly**   This produces an error in horizontal angles after the manner described in the preceding section for plate levels out of adjustment. The error from this source is small when the sights are nearly level, but may be large for steeply inclined sights (see Fig. 6.63). The average transit operator does not appreciate the importance of careful leveling for steeply inclined sights; on the other hand, unnecessary care in leveling is often used when sights are nearly horizontal. Since the error in horizontal angle is caused largely by inclination of the horizontal axis, the striding level is a necessity on precise work.

**3. Errors in setting and reading the vernier**   These are functions of the least count of the vernier and of the legibility of scale and vernier lines. For the usual 01′ transit the readings and settings can be made to the nearest 30″; for the 30″ transit these values are to the nearest 15″. The use of a reading glass enables closer reading, particularly for finely graduated circles. Also, in reading the vernier it is helpful to observe the position of the graduations on both sides of the ones that appear to coincide, and to note that the unmatched graduations appear to lack coincidence by the same amount.

Horizontal and vertical circle readings with an optical theodolite having a least count of 1″ can be observed with a standard deviation of from 2″ to 3.″

**4. Not sighting exactly on the point**   This is likely to be a source of rather large error on ordinary surveys where sights are taken on the range pole of which often only the upper portion is visible from the transit. The effect upon a direction is, of course, the same as the effect of not setting up exactly over the station. For short sights greater care should be taken than for long sights, and the plumb line should be employed instead of the range pole.

**5. Imperfect focusing (parallax)**   The error due to imperfect focusing is always present to a greater or less degree, but with reasonable care it can be reduced to a negligible quantity. The manner of detecting parallax is described in Art. 5.10.

**Summary**   All the personal errors are random and hence cannot be eliminated. They form a large part of the resultant error in transit and theodolite work. Of the personal errors, those due to inaccuracies in reading and setting the vernier or reading and setting the optical micrometer and to not sighting exactly on the point are likely to be of greater magnitude.

**6.57. Natural errors**   Sources of natural errors are (1) settlement of the tripod, (2) unequal atmospheric refraction, (3) unequal expansion of parts of the telescope due to temperature changes, and (4) wind, producing vibration of the transit or making it difficult to plumb correctly.

In general, the errors resulting from natural causes are not large enough to affect appreciably the measurements of ordinary precision. However, large errors are likely to arise from settlement of the tripod when the transit or theodolite is set up on boggy or thawing ground; in such cases the instrument may be kept relatively stable through the use of extra-long tripods or of small platforms for the tripod legs. Settlement is usually accompanied by an angular movement about the vertical axis as well as linear movements both vertically and horizontally. When horizontal angles are being measured, usually a larger error is produced by the angular displacement of the circle between backsight and corresponding foresight than by the movement of the transit laterally from the point over which it is set. Errors due to adverse atmospheric conditions can usually be rendered negligible by choosing appropriate times for observing.

For measurements of high precision the methods of observing are such that instrumental and personal errors are kept within very small limits, and natural errors become of relatively great importance. Natural errors are generally random, but under certain conditions systematic errors may arise from natural causes. On surveys of very high precision, special attempt is made to establish a procedure which will as nearly as possible eliminate natural systematic errors. Thus the instrument may be set up on a masonry pier and protected from sun and wind; also certain readings may be made at night when temperature and atmospheric conditions are nearly constant.

**6.58. Error propagation in angles**   Estimation of attainable precision in observing angular measurements is essential for proper planning and design of a survey. Realistic estimation of the standard deviations achievable in angles observed with the transit or theodolite is possible by analyzing the observational procedure and assigning procedural specifications to eliminate known systematic errors (Art. 6.55). Then, estimates can be assigned to major sources of personal and natural random errors (Arts. 6.56 and 6.57); the error propagated in the final angle can be evaluated using error propagation procedures detailed in Art. 2.20.

As an example, consider the error propagation in a horizontal angle determined by repetition using an engineer's transit (Art. 6.39). As mentioned in Art. 6.56, the major sources of random error in angles are those due to reading and setting the vernier, and pointing on signals which mark the backsight and foresight. The final averaged angle for an angle measured by repetition with an engineer's transit or repeating theodolite is

$$\alpha = \frac{R_2 - R_1}{n} \tag{6.6}$$

where   $\alpha$ = average angle
   $n$ = number of repetitions
   $R_1$ = initial circle reading on the backsight, the average of vernier readings $A_1$ and $B_1$
   $R_2$ = final circle reading on the foresight after $n$ repetitions of the angle, the average of vernier readings $A_2$ and $B_2$, and includes errors due to $2n$ pointings

The initial circle reading $R_1 = (A_1 + B_1)/2$. By Eq. (2.51), the estimated variance in $R_1$ is

$$\sigma_{R_1}^2 = \frac{\sigma_s^2}{2} \tag{6.7}$$

where $\sigma_s$ is the estimated standard deviation in reading or setting the horizontal circle.

The final circle reading is the average of vernier readings $A_2$ and $B_2$ but is also affected by $2n$ pointings. Thus, the estimated variance in $R_2$ is

$$\sigma_{R_2}^2 = \frac{\sigma_s^2}{2} + 2n\sigma_p^2 \tag{6.8}$$

where $\sigma_p$ is the estimated standard deviation in pointing on a signal.

The final average angle is calculated by Eq. (6.6). Propagating the error through Eq. (6.6) by Eq. (2.46) and using Eqs. (6.7) and (6.8) yields the estimated variance in the average angle:

$$\sigma_\alpha^2 = \frac{\sigma_s^2}{n^2} + \frac{2\sigma_p^2}{n} \tag{6.9}$$

Assuming that $\sigma_s = 15''$ and $\sigma_p = 02''$ for a $30''$ engineer's transit and using Eq. (6.9), estimated standard deviations have been calculated according to several different numbers of repetitions and are listed in Table 6.1. The table offers ample evidence of the value of repeating an angle.

Angles determined from direction theodolite observations are calculated by taking the difference between final and initial observed *directions* (Art. 6.41). Thus, if $D_1$ and $D_2$ are the averaged directions of $n$ backsights and foresights, respectively, the final angle is

$$\alpha = D_2 - D_1 \tag{6.10}$$

in which

$$D_1 = \frac{\sum\limits_{i=1}^{n} D_{1i}}{n} \tag{6.11a}$$

$$D_2 = \frac{\sum\limits_{i=1}^{n} D_{2i}}{n} \tag{6.11b}$$

where $D_{1i} = i$th circle observation on the backsight
$D_{2i} = i$th circle observation on the foresight
$n =$ number of repetitions

Assume that

$\sigma_m^2 =$ estimated variance in optical micrometer observations on the horizontal circle

$\sigma_p^2 =$ estimated variance in pointing on a signal

The error in a single backsight or foresight is a combination of reading and pointing errors. Thus, the propagated estimated variance for a single backsight or foresight is

$$\sigma_{D_{1i}}^2 = \sigma_{D_{2i}}^2 = \sigma_m^2 + \sigma_p^2 \tag{6.12}$$

The estimated variance in an average of $n$ backsights or foresights can be found by propagating through Eqs. (6.11a) or (6.11b) by Eq. (2.46) and using Eq. (6.12) to yield

$$\sigma_{D_1}^2 = \sigma_{D_2}^2 = \frac{\sigma_m^2 + \sigma_p^2}{n} \tag{6.13}$$

Then the estimated variance in the angle calculated from average directions $D_1$ and $D_2$, propagated through Eq. (6.7) by Eq. (2.46) and using Eq. (6.13), is

$$\sigma_\alpha^2 = \frac{2(\sigma_m^2 + \sigma_p^2)}{n} \tag{6.14}$$

**Table 6.1    Propagated Errors in Angles by Repetition Using an Engineer's Transit**

| Number of repetitions | Propagated variance $\sigma_\alpha^2$ ($''$ of arc)$^2$ | Standard deviation $\sigma_\alpha$ ($''$ of arc) |
|:---:|:---:|:---:|
| 2 | 60.25 | 7.8 |
| 4 | 16.06 | 4.0 |
| 6 | 7.58 | 2.8 |
| 8 | 4.52 | 2.1 |
| 10 | 3.05 | 1.8 |
| 12 | 2.23 | 1.5 |

*Note:* $30''$ engineer's transit, $\sigma_s = 15''$, $\sigma_p = 02''$

**Table 6.2   Propagated Errors in Angles by Repetition Using a Direction Theodolite**

| Number of repetitions | $\sigma_\alpha^2$ (" or arc)$^2$ | $\sigma_\alpha$ (" of arc) |
|---|---|---|
| 2 | 8.00 | 2.8 |
| 4 | 4.00 | 2.0 |
| 6 | 2.67 | 1.6 |
| 8 | 2.00 | 1.4 |

*Note:* 1" direction theodolite, $\sigma_m = 02"$, $\sigma_p = 02"$.

Estimated standard deviations for different numbers of repetitions of an angle using a direction theodolite for which $\sigma_m = 02"$ and $\sigma_p = 02"$ are listed in Table 6.2.

Note that an important difference exists between angles by repetition with an engineer's transit and angles obtained from directions observed with a direction theodolite. When the angle is turned by repetition using a repeating instrument, the angle is the observed quantity. If more than one angle is observed about a point, as shown in Fig. 6.66, each angle can be considered as an independent observed value with an estimated standard deviation propagated by Eq. (6.9). When this procedure is followed, no correlation exists between $\alpha_1$ and $\alpha_2$.

On the other hand, if a direction theodolite is used to observe directions $D_1$, $D_2$, and $D_3$ (Fig. 6.66), these directions are the independent, noncorrelated observed values for which respective variances can be calculated by Eq. (6.12). The covariance matrix for these directions [Eq. (2.28)] is

$$\Sigma_{DD} = \begin{bmatrix} \sigma_{D_1}^2 & 0 & 0 \\ 0 & \sigma_{D_2}^2 & 0 \\ 0 & 0 & \sigma_{D_3}^2 \end{bmatrix} \tag{6.15}$$

The angles $\alpha_1$ and $\alpha_2$ are functions of these observations [Eq. (6.10)] as follows:

$$\begin{bmatrix} \alpha_1 \\ \alpha_2 \end{bmatrix} = \begin{bmatrix} D_2 - D_1 \\ D_3 - D_2 \end{bmatrix} = \begin{bmatrix} -1 & 1 & 0 \\ 0 & -1 & 1 \end{bmatrix} \begin{bmatrix} D_1 \\ D_2 \\ D_3 \end{bmatrix} \tag{6.16}$$

Thus, the covariance matrix for angles derived from directions must be propagated through Eq. (6.16) by Eq. (2.42) to give

$$\Sigma_{\alpha\alpha} = J_{\alpha D} \Sigma_{DD} J_{\alpha D}^T \tag{6.17}$$

where $\Sigma_{DD}$ is defined by Eq. (6.15) and

$$J_{\alpha D} = \begin{bmatrix} \partial\alpha_1/\partial D_1 & \partial\alpha_1/\partial D_2 & \partial\alpha_1/\partial D_3 \\ \partial\alpha_2/\partial D_1 & \partial\alpha_2/\partial D_2 & \partial\alpha_2/\partial D_3 \end{bmatrix} = \begin{bmatrix} -1 & 1 & 0 \\ 0 & -1 & 1 \end{bmatrix} \tag{6.18}$$

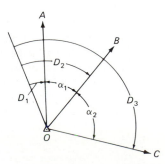

**Fig. 6.66** Angles and directions about a point.

is the matrix of partial derivatives of $\alpha_1, \alpha_2$ with respect to the observed directions. Substitution of Eq. (6.15) into Eq. (6.17) yields

$$\Sigma_{\alpha\alpha} = \begin{bmatrix} (\sigma_{D_1}^2 + \sigma_{D_2}^2) & -\sigma_{D_2}^2 \\ -\sigma_{D_2}^2 & \sigma_{D_2}^2 + \sigma_{D_3}^2 \end{bmatrix} \qquad (6.19)$$

in which the diagonal terms represent the propagated variances and the off-diagonal elements are covariances of angles $\alpha_1$ and $\alpha_2$, respectively. The presence of these covariance elements in $\Sigma_{\alpha\alpha}$ means that the angles $\alpha_1$ and $\alpha_2$ are definitely correlated. Consequently, when a direction theodolite is employed to observe directions from which angles are calculated and these angles are then utilized in further adjustment computations, the entire covariance matrix must be used in these computations or erroneous results will be obtained.

**Example 6.4**   Consider the observations of Fig. 6.66 taken with a 1″ theodolite using four repetitions of each direction. Assume that estimated standard deviations in reading the horizontal circle and pointing on the signals are $\sigma_m = \sigma_p = 02$″, and signals A, B, and C are of equal quality. Determine covariance and weight matrices for angles $\alpha_1$ and $\alpha_2$. Assume that the reference variance $\sigma_0^2 = 1$.

SOLUTION   By Eq. (6.13), the variance in the direction $D_1$ is

$$\hat{\sigma}_{D_1}^2 = \frac{\sigma_m^2 + \sigma_p^2}{n} = \frac{(02)^2 + (02)^2}{4} = 02\,''^2$$

Since signals are of equal quality, $\sigma_{D_1}^2 = \sigma_{D_2}^2 = \sigma_{D_3}^2$. Propagation of the covariance matrix by Eq. (6.19) yields

$$\Sigma_{\alpha_1 \alpha_2} = \begin{bmatrix} 2\sigma_{D_1}^2 & -\sigma_{D_1}^2 \\ -\sigma_{D_1}^2 & 2\sigma_{D_1}^2 \end{bmatrix} = \begin{bmatrix} 4 & -2 \\ -2 & 4 \end{bmatrix}$$

By definition the weight matrix is [Eq. (2.32), Art. 2.14]

$$\mathbf{W}_{\alpha_1 \alpha_2} = \sigma_0^2 \Sigma_{\alpha_1 \alpha_2}^{-1} = \begin{bmatrix} \sigma_{\alpha_1}^2 & \sigma_{\alpha_1 \alpha_2}^2 \\ \sigma_{\alpha_2 \alpha_1}^2 & \sigma_{\alpha_2}^2 \end{bmatrix}^{-1} = \begin{bmatrix} 4 & -2 \\ -2 & 4 \end{bmatrix}^{-1}$$

which reduces to

$$\mathbf{W}_{\alpha_1 \alpha_2} = \begin{bmatrix} \frac{1}{3} & \frac{1}{6} \\ \frac{1}{6} & \frac{1}{3} \end{bmatrix}$$

This is the weight matrix that must be used in all subsequent calculations involving $\alpha_1$ and $\alpha_2$.

To ensure the results cited in Tables 6.1 and 6.2, all angles must be observed with an equal number of direct and reversed repetitions so as to eliminate systematic instrumental errors. The instrument should be in good general adjustment and in particular, the plate level bubbles must be in adjustment (Art. 6.51). Targets to mark backsights and foresights must be selected so as to ensure the pointing accuracy specification. For example, a range pole is 0.10 ft or 30 mm in diameter. A plumbed range pole can be sighted to within 0.02 ft or 6 mm. Given a length of 2000 ft or about 600 m, a pointing accuracy of ±02″ could be achieved. However, for a sight of 500 ft or about 150 m in length on the same range pole, pointing accuracy becomes ±8″ to ±10″. Consequently, for shorter sights a plumbed line or other well-defined target, compatible with the desired pointing accuracy, must be employed. Figure 6.67 shows a target that can be used for daylight operation and may also be illuminated for night operation. This target is mounted on the same tripod as is used for the transit or theodolite and is adaptable to forced centering where instrument and target may be interchanged from one tripod to another.

**Example 6.5**   A horizontal control survey, where the angles are to be observed with a standard deviation of 01′, is to be run for a route alignment. Sights are to have a minimum length of 500 ft or

**Fig. 6.67** Target adaptable to tripod. (*Courtesy The Lietz Co.*)

about 152 m, and range poles plumbed by eye are to be used to give line. Owing to intervening irregularities in the terrain and brush, many sights will have to be taken well above ground level. A 01′ engineer's transit is available for observations. An experienced instrument operator is available, but inexperienced personnel must be utilized to give line. The site is wooded and has relatively steep slopes. Evaluate the procedures necessary to provide 01′ accuracy in the angles.

SOLUTION   A range pole held by an inexperienced worker over traverse points on steep terrain in the woods may be out of plumb by ± the diameter of the pole a large percentage of the time. Thus, a 0.10-ft or 30-mm deviation can be expected in a distance of 500 ft or about 152 m. This deviation yields an angular error of $\sim \pm 40''$, so that $\sigma_p$ can be estimated as $40''$. A 01′ transit permits observation of the horizontal circle to within $\pm 30''$ by an experienced operator.

If angles are turned once [Eq. (6.9)],

$$\sigma_\alpha^2 = \sigma_s^2 + 2\sigma_p^2$$
$$= (30)^2 + 2(40)^2 = 4100\,''^2$$
$$\sigma_\alpha = 64''$$

which exceeds the allowable limit assuming a transit in near perfect adjustment. Good surveying practice would dictate turning the angle at least twice, once direct and once reversed (Art. 6.39). Under these conditions the estimated standard deviation in the angles by Eq. (6.9) is

$$\sigma_\alpha^2 = \frac{\sigma_s^2}{n^2} + \frac{2\sigma_p^2}{n}$$
$$= \frac{(30)^2}{4} + \frac{(2)(40)^2}{2} = 1825\,''^2$$
$$\sigma_\alpha = 42.7''\quad\text{or}\quad 43''$$

If $\sigma_s = 01'$ and by Eq. (6.9),

$$\sigma_\alpha^2 = \frac{(60)^2}{4} + \frac{(2)(40^2)}{2} = 2500\,''^2$$

$$\sigma_\alpha = 50''$$

The specifications necessary to guarantee $\sigma_\alpha = 50''$ are as follows:

1. A 01' engineer's transit in fair adjustment should be used. Plate level bubbles must be in good adjustment.
2. Angles must be turned once direct and once reversed. Vernier readings should be observed to the nearest $\pm 01'$.
3. Sights must be no less than 500 ft or 150 m in length.
4. Sights may be taken on range poles plumbed by eye.
5. An experienced transit operator must be available.
6. Inexperienced personnel may be utilized to give line.

Consider the effect of random errors in centering the instrument over the point as applied to the preceding example. For a given random error of centering equal to $r$, the point may occur anywhere on a circle of radius $r$ centered about the point over which the transit is to be set. In Fig. 6.68a, the angle $ACB = \theta$ is to be observed from a set up at $C$. Assume that the random error in centering is $CC' = r$. The angle actually turned is $AC'B = \theta + e_2 - e_1$, where $\tan e_1 = CC'/AC$ and $\tan e_2 \cong r\sin(90° - \theta)/BC$. Assuming that $CB$ and $CA$ are approximately equal, $\theta$ is less than 90°, and the error is approximately normal to line $CA$, the total angular error introduced is $e_2 - e_1$. For $CC' = 0.05$ ft or 15 mm, $CA = CB = 500$ ft or 150 m, and $\theta = 45°$, $e_1 = 21''$, and $e_2 = 15''$, yielding a total angular error of 06''. When the centering error falls on the bisector of the angle $\theta$ as shown in Fig. 6.68b, the observed angle is $AC''B = \theta + e_1' + e_2'$ and the total angular error introduced is $e_1' + e_2'$. In this case, $\tan e_1' \cong (r\sin\theta/2)/AC$ and $\tan e_2' \cong (r\sin\theta/2)/CB$. Under the previous assumptions and given values $e_1' = e_2' = 08''$, the total angular error is 16''. Thus, the maximum angular error will occur when the centering error falls on the bisector of the observed angle.

In Example 6.5, the propagated angular error due to reading ($\sigma_s = 01'$) and pointing is 50''. According to the analysis just completed, the uncertainty in angular observations due to a centering error of 0.05 ft or 15 mm is $\pm 16''$. Thus, the total estimated standard deviation in the angle due to reading, pointing, and centering errors is

$$\sigma_{\alpha,\,total}^2 = (50)^2 + (16)^2 = 2756\,''^2$$

$$\sigma_{\alpha,\,total} = 53''$$

and the effect of centering error is seen to be negligible.

Centering error can be significant if the lengths of sights are short and high accuracy is desired.

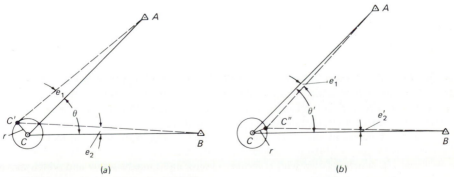

**Fig. 6.68**  Errors due to improper centering.

**Example 6.6** Relative positions of targets in a camera calibration test array are to be determined by triangulation from a base line 6 m in length using a 01 ″ theodolite. Lengths of sight, $D$, are from 6 to 8 m and angles vary from 45 to 90°. Angular measurements are specified to be within ± 30 ″ of arc. Angles are to be observed twice direct and twice reversed, for a total of four repetitions.

SOLUTION Assuming that the standard deviation in reading, $\sigma_m$ = standard deviation in pointing $\sigma_p$ = 02 ″, then $\sigma_\alpha$ = 02 ″ (Table 6.2). A centering accuracy of 0.001 m or 0.003 ft will lead to

$$e_1' + e_2' = \frac{2r\sin(\theta/2)}{D\sin 01''} = \frac{(0.001)(\sin 45°)(2)}{(6)(4.848)(10^{-6})} = 49''$$

so that $\sigma_{\alpha,\text{total}} = [(02)^2 + (49)^2]^{1/2} = 49''$, which exceeds the specified value. Using an optical plummet, centering to within 0.0005 m is possible. This value leads to an angular error of

$$e_1' + e_2' = \frac{(0.0005)(\sin 45°)(2)}{(6)(4.848)(10^{-6})} = 24.3''$$

so that

$$\sigma_{\alpha,\text{total}} = [(02)^2 + (24.3)^2]^{1/2} = 24.4''$$

Note that in this problem, centering is the critical factor.

**Example 6.7** A river is to be triangulated for a preliminary bridge survey. The shortest length of sight is about 1000 ft or 305 m. The average triangle closure is not to exceed ± 05 ″ with maximum closure not to exceed ± 10 ″. Observation of each angle with an estimated standard deviation of 02 ″ will more than satisfy this requirement. Determine equipment and methods required for this accuracy.

SOLUTION As shown in Table 6.2, Art. 6.57, two positions or four repetitions of an angle with a 01 ″ theodolite will yield an estimated standard deviation of $\sigma_{\alpha D}$ = 02 ″ (based on $\sigma_m = \sigma_p$ = 02 ″). Ten repetitions using a 30 ″ engineer's transit would be required to achieve comparable accuracy (based on $\sigma_s$ = 15 ″, $\sigma_p$ = 02 ″, Table 6.1). From the standpoint of time consumed, use of the 01 ″ theodolite would be most economical. The following specifications should be written to ensure attaining $\sigma_{\alpha D}$ = 02 ″.

1. A 01″ optical-reading direction theodolite in good adjustment and equipped with an optical plummet or centering tripod shall be used for observing directions. An experienced operator should be available.
2. Two positions of directions shall be observed at each triangulation station. One direct and one reversed direction to each point sighted constitutes one position. The initial circle readings for each position should be distributed about the horizontal circle so as to eliminate the effects of graduation errors (Art. 10.18).
3. Targets must be chosen, plumbed, and maintained so as to provide the 0.01-ft or 3-mm centering accuracy required in a 1000-ft or 305-m sight to ensure a pointing accuracy of ±02″. The use of theodolites and targets compatible (Fig. 6.67) with the same tripods so as to permit forced centering would simplify this task (Art. 6.30).
4. Observations should be scheduled to take advantage of atmospheric conditions. Observations should not be made looking directly into the sun or when the atmosphere is excessively hazy.

Angles determined from directions observed about a single station according to these specifications will be correlated. This correlation must be taken into account in subsequent calculations with angles in triangulation computations.

**6.59. Sextant** The sextant (Fig. 6.69) is well suited to hydrographic work and has the added advantage of measuring angles in any plane (see Art. 21.3). It is used principally by navigators and surveyors for measuring angles from a boat, but it is also employed on exploratory, reconnaissance, and preliminary surveys on land.

The essential features of the sextant are illustrated in Fig. 6.70, with the instrument in position for measuring a horizontal angle *FEL*. An index mirror *I* is rigidly attached to a movable arm *ID*, which is fitted with a vernier, clamp, and tangent-screw and which moves along the graduated arc *AB*. A second mirror *H*, called the *horizon glass*, having the lower

**Fig. 6.69** Sextant.

half of the glass silvered and the upper half clear, is rigidly attached to the frame. A telescope $E$, also rigidly attached to the frame, points into the mirror $H$.

With signals at $L$ and $F$ and the eye at $E$, it is desired to measure the angle $FEL$. The ray of light from signal $L$ passes through the clear portion of glass $H$ on through the telescope to the eye at $E$. The ray of light from signal $F$ strikes the index mirror at $I$ and is reflected to $h$ and then through the telescope to $E$. Each set of rays forms its own image on its respective half of the objective. When the arm $ID$ is moved, these images can be made to move over each other and there will be one position in which they coincide. An observation with the sextant consists in bringing the two images into exact coincidence and reading the angle on the vernier scale on limb $AB$.

To prove that the angle $FEL$ equals two times angle $IDh$, that is, that the angle between the signals is equal to twice the angle between the mirrors: Draw $IP$ and $hp$ normal to the two mirrors; then the angles of incidence and reflection of the two mirrors are $i$ and $i'$, respectively. By trigonometry,

$$FEL = FIh - IhE$$
$$= 2i - 2i'$$
$$= 2(i - i')$$

Also,

$$IDh = HhI - hID$$
$$= (90° - i') - (90° - i)$$
$$= i - i'$$

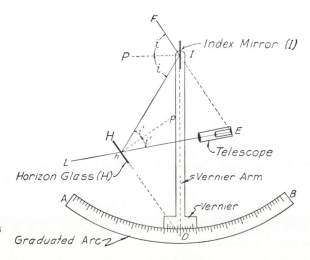

**Fig. 6.70** Essential features of sextant.

Therefore, *FEL* (the angle between the objects) equals twice the angle *IDh* (the angle between the mirrors).

**6.60. Measuring angles with the sextant**   The handle of the sextant is held in the right hand, and the plane of the arc is made to coincide with the plane of the two objects between which the angle is to be measured. The sextant is turned in the plane of the objects until the left-hand object can be viewed through the telescope and the clear portion of the horizon glass. With the instrument held in this position, the index arm is moved with the left hand until the images of the two objects coincide. The final setting is made with the tangent-screw, and a test for coincidence is made by twisting the sextant slightly in the hand to make the reflected image move back and forth across the position of coincidence. When the setting is thus verified, the vernier arm is clamped and the angle is read.

The precision of horizontal angle measurement with the sextant depends upon the size of the angle and upon the length of sight. It is evident from Fig. 6.70 that the angle *FEL* actually measured has its vertex *E* not at the eye but at the intersection of the sight rays *FE* and *LE* from the points sighted. The distance to this intersection will increase as the angle decreases, and for small angles the vertex may be at a considerable distance back of the observer. The effect of this factor on the horizontal angle is essentially the same as the error caused by improper centering of the transit (Fig. 6.68, Art. 6.58). Hence the sextant is not an instrument of precision for small angles (say, less than 15°) and short distances (say, less than 1,000 ft or about 300 m). If objects sighted are at a great distance away, the angular error is usually small (about 1′ of arc).

Vertical angles can be measured with the sextant in a manner similar to that just described for horizontal angles. Most commonly, the altitude of a celestial body is observed, as in navigation. The vertical angle between the body and the sea horizon is measured and is corrected for dip, or angular distance of the observer above the horizon. On land, if the sea horizon is not visible, an artificial horizon is used; this consists of a horizontal reflecting surface, such as that of a small vessel of mercury, near the sextant. The vertical angle between the celestial body and its reflection in the artificial horizon is measured; this angle is twice the altitude of the body.

**6.61. Problems**

**6.1**   Describe the methods by which points on or near the surface of the earth may be located.

**6.2**   Describe the differences among true, grid, and magnetic meridians.

**6.3**   Given the following bearings: N35°40′E, S46°32′E, N60°30′W, S32°59′W, N89°40′W, N79°59′E, N65°20′W, S78°42′W, N10°13′W, S15°16′W. Express these directions as (*a*) azimuths from the north; (*b*) azimuths from the south.

**6.4**   Express the following azimuths from the north as bearings: 329°15′, 45°00′, 181°01′, 172°31′, 146°40′, 242°13′, 10°38′, 92°59′, 349°21′, 319°40′.

**6.5**   Given the following forward azimuths: $A_{AB} = 273°30′$, $A_{AC} = 42°15′$, $A_{DC} = 195°25′$, and $A_{DE} = 160°45′$. Convert these directions to back azimuths.

**6.6**   In Prob. 6.5, determine the clockwise horizontal angles between lines *AB* and *AC*, and *DC* and *DE*.

**6.7**   The following azimuths are reckoned from the north: *AB*, 187°12′; *BC*, 273°47′; *CD*, 318°48′; *DE*, 0°48′; *EF*, 73°00′. What are the corresponding bearings? What are the deflection angles between consecutive lines?

**6.8**   The interior angles of a five-sided closed polygon *ABCDE* are as follows: *A*, 117°36′; *B*, 96°32′; *C*, 142°54′; *D*, 132°18′. The angle at *E* is not measured. Compute the angle at *E*, assuming the given values to be correct.

**6.9**   The magnetic bearing of a line is S47°30′W and the magnetic declination is 12°10′W. What is the true bearing of the line?

**6.10**   The true bearing of a line is N18°17′W and the magnetic declination is 7°12′E. What is the magnetic bearing of the line?

**6.11** In an old survey made when the declination was 2°10′W, the magnetic bearing of a given line was N35°15′E. The declination in the same locality is now 3°15′E. What are the true bearing and the present magnetic bearing that would be used in retracing the line?

**6.12** Thirty spaces on a transit vernier are equal to 29 spaces on the graduated circle, and 1 space on the circle is 15′. What is the least count of the vernier?

**6.13** Sixty spaces on a transit vernier are equal to 59 on the graduated circle, and 1 space on the circle is 15′. What is the least count of the vernier?

**6.14** A transit for which the circle is graduated 0° to 360° clockwise is used to measure an angle by 10 clockwise "repetitions," 5 with telescope direct and 5 with the telescope reversed. Compute the value of the angle from the following data:

| | | Vernier | |
|---|---|---|---|
| Telescope | Reading | A | B |
| Direct | Initial | 48°46′ | 228°46′ |
| Direct | After first turning | 161°09′ | — |
| Reversed | After tenth turning | 92°41′ | 272°42′ |

**6.15** An angle is repeated three times direct and three times reversed with a repeating theodolite having a least count of 0.1′. The circle reading for the initial backsight is 330°28.1′ and after the first repetition is 129°33.0′. The circle reading after the sixth repetition is 204°52.0′. Calculate the average angle.

**6.16** A direction theodolite with a least count of 1″ is set over station $D$ to measure directions to stations $C$, $B$, and $A$. The observed directions for one position are as follows:

| Station | Tel. | Circle |
|---|---|---|
| C | Direct | 0°10′16″ |
| | Reversed | 180°10′26″ |
| B | Direct | 48°52′06″ |
| | Reversed | 228°52′40″ |
| A | Direct | 83°06′48″ |
| | Reversed | 263°06′48″ |

Compile an abstract of average directions and compute the average angles.

**6.17** In laying out the lines for a building, a 90° angle was laid off as precisely as possible with a 01′ transit. The angle was then measured by repetition and found to be 89°59′40″. What offset should be made at a distance of 250 ft from the transit to establish the true line?

**6.18** The following observations were made to determine an index correction: Vertical angle to point $A = +7°16′$ with telescope direct and $+7°14′$ with telescope inverted. Compute the index correction for observations with telescope direct.

**6.19** A vertical angle measured by a single observation is $-12°02′$, and the index error is determined to be $+06′$. What is the correct value of the angle?

**6.20** A theodolite equipped with a vertical circle level bubble has zero on the vertical circle oriented toward the zenith. The following direct and reversed zenith angles were observed to determine the index correction: 80°05′20″ and 279°54′16″. Compute the index correction for the instrument. The same instrument is used to measure a zenith angle of 91°16′35″. Calculate the (a) corrected zenith angle; (b) corrected vertical angle.

**6.21** Two points $A$ and $B$, 5280 ft apart, are to be connected by a straight line. A random line run from $A$ in the general direction of $B$ is found by computation to deviate 03′18″ from the true line. On the random line at a distance 1250.6 ft from $A$ an intermediate point $C$ is established. What must be the offset from $C$ to locate a corresponding point $D$ on the true line?

**6.22** What error would be introduced in the measurement of a horizontal angle, with sights taken to points at the same elevation as the transit if, through nonadjustment, the horizontal axis was inclined (a) 03′? (b) 3°? (c) If the horizontal axis was inclined 03′, what error would be introduced if both sights were inclined at angles of $+30°$? (d) If one sight was inclined at $+30°$ and the other at $-30°$?

**6.23** The estimated standard deviation in reading and setting a vernier on a 01′ repeating transit is $\sigma_s = 30″$. The standard deviation in pointing on a specified target is $\sigma_p = 40″$. Compute the standard deviation for an angle turned: (*a*) once, with the telescope in the direct position; (*b*) twice, once with the telescope direct and once with the telescope reversed; (*c*) six times, three direct and three reversed.

**6.24** In Prob. 6.16, estimated standard deviations in observing the horizontal circle and pointing on the target are $\sigma_m = 05''$ and $\sigma_p = 10''$, respectively. Propagate the covariance matrix for the angles *CDB* and *BDA* determined from the observed directions. Assume that the reference variance $\sigma_0^2 = 1.0$.

**6.25** The angles in a simple triangulation figure are to be determined with a standard deviation of 10". A 01' engineer's transit is to be used. The standard deviation in setting and reading the vernier is $\sigma_s = 30''$. The standard deviation in pointing on the targets is estimated to be $\sigma_p = 05''$. Determine the number of repetitions of the angle that are required to satisfy the specified accuracy.

**6.26** Write specifications to ensure that the standard deviation in measured angles as specified in Prob. 6.25 is obtained. Assume an average length of sight of 300 m.

**6.27** The random error in centering a theodolite over a point is estimated as being 0.002 m. What is the maximum angular error introduced by this centering error to an angle of 30° assuming lengths of sights of *a*) 50 m? (*b*) 300 m?

**6.28** Determine the estimated standard deviation in measuring the angle under the conditions of Prob. 6.23 if the random error in centering the instrument over the point is estimated to be 0.003 m, the lengths of sights are 150 m, and the angle is 45°.

## 6.62. Field problems

### PROBLEM 1   MEASUREMENT OF ANGLES BY REPETITION

**Object**   To obtain a more precise determination of the horizontal angles between various stations about a point than would be possible by a single measurement (see Art. 6.39).

**Procedure**   (1) Set up the repeating theodolite or engineer's transit very carefully over the point labeled *O*. (2) Set four chaining pins about 150 ft or 50 m from the instrument, forming four angles about the occupied station *O*. (3) When using a repeating theodolite, set the horizontal circle to zero. If a transit with direct-reading verniers is being used, set the *A* vernier to zero and read the *B* vernier and record the readings. (4) Keep notes in a form similar to those employed for repeated angles in Art. 6.39 and illustrated in Fig. 6.47a or b. (5) With the telescope direct, measure one of the angles clockwise. Record the horizontal circle reading to the least count of the circle or both vernier readings to the least count of the verniers. (6) Leaving the upper motion clamped, again set on the first point and again measure the first angle clockwise with the telescope direct (thus doubling the angle). (7) Continue until three repetitions with the telescope direct have been obtained. (8) In like manner, without resetting the horizontal circle, measure the horizontal angle three repetitions with the telescope reversed, always measuring clockwise. (9) Record the total angle observing to the least count of the instrument used. With instruments having direct-reading verniers, observe both the *A* and *B* verniers and take the average. (10) Compute the average angle by dividing the total angle by the number of repetitions or six in this case. Compare this average angle with the first angle turned. (11) If the average angle fails to agree with the first single observation to within the least count of the instrument, repeat the entire set. (12) Go through the same process (steps 5 through 11) for all other angles about the point. For an instrument with a least count of 1', the error of horizon closure should not exceed $10''\sqrt{\text{number of angles}}$ . (13) Adjust the angles so their sum will equal 360° by distributing the error equally among the mean values.

**Hints and precautions**   (1) Level the instrument very carefully before each repetition but do not disturb the leveling screws while a measurement is being made. (2) Be careful not to loosen the wrong clamp-screw. (3) Do not become confused when computing the total angle turned. Observe how the horizontal circle is graduated and do not omit a full turn. (4) The instrument should be handled very carefully. When the lower motion is being turned, the hands should be in contact with the *lower* plate, not the upper motion. When making an exact setting on a point, the last movement of the tangent-screw should be *clockwise* or against the opposing spring. (5) After each repetition the instrument should be turned on its lower motion in the same direction as that of the measurement. (6) Do not walk around the transit to read the second vernier; rotate it to you, always turning the instrument clockwise in order to avoid possible errors due to slackness and twist. (7) The single measurement is taken as a check on the number of repetitions; it should agree closely but not necessarily exactly with the mean value.

## PROBLEM 2   LAYING OFF AN ANGLE BY REPETITION

**Object**   To lay off a given horizontal angle more precisely than is possible with a single setting of the horizontal circle (see Art. 6.42).

**Procedure**   (1) Drive and set tacks in two stakes about 500 ft or 150 m apart. (2) Carefully set up the transit over one end of the line. Sight at the point at the other end, and lay off the given angle. (3) Set a stake on the line of sight about 500 ft or 150 m from the instrument (distance by pacing), and carefully set a tack. (4) By repetition measure the angle laid off, as in the previous problem, making three "repetitions" with telescope direct and three with it reversed. (5) Find the difference between the angle laid off and the required angle, and by trigonometry compute the linear distance that the tack must be moved perpendicular to the line of sight. (6) Set the tack accordingly.

## PROBLEM 3   OBSERVING DIRECTIONS WITH A DIRECTION THEODOLITE

**Object**   To observe four positions of directions about a station using a direction theodolite (see Art. 6.40).

**Procedure**   (1) Set a hub with a tack in it as the station to be occupied (station *O*). Set three chaining pins about 100 m from the station in an array such that the angles subtended at the station are less than 100° and more than 30°. Existing points, such as lightning rods, flagpoles, and so on, may be used if they are suitably located and provide well-defined targets. (2) Set the theodolite over the central station leveling and centering according to the instructions in Art. 6.34. (3) Designate the points to be sighted as *A*, *B*, and *C* in a clockwise direction, with *A* to be the initial point sighted. (4) Following the instructions in Arts. 6.40 and 6.41, observe directions to points *A*, *B*, and *C* with the telescope direct and then backward from *C* to *B* to *A* with the telescope reversed. Use the note form shown in Fig. 6.47*b*. (5) Take the mean of the direct and reversed readings and reduce to an abstract of directions assuming a direction of 0°00′00″ from station *O* to *A* (Art. 6.41). This set of measurements constitutes *one position*. (6) Repeat steps (4) and (5) three more times, advancing the horizontal circle about 45° for each initial pointing on *A*. (7) When the four positions have been completed, calculate the averages of the directions in the four positions to yield a final abstract of directions to stations *A*, *B*, and *C* from *O*. Compute the angles *AOB* and *BOC* from these directions.

## References

1. Allan, A. L., "A Note on Centering the Instrument," *Survey Review*, October 1977.
2. Berthon-Jones, P., "A Note on Run Error and Error of Half-interval in Mean Reading Theodolites," *Australian Surveyor*, December 1972.
3. Ericson, K. E., "Electronic Surveying Systems," *Proceedings, 37th Annual Meeting, ACSM*, 1977, p. 209.
4. Gort, A. E., "The Hewlett-Packard 3829A Electronic Total Station," *Proceedings, Fall Technical Meeting, ACSM*, October 1977, pp. 308–326.
5. Jackson, F. W., "A Method of Adjusting the Optical Plummet of the Wild T-2," *Survey Review*, July, 1971.
6. Kissam, P., *Surveying for Civil Engineers*, McGraw-Hill Book Company, New York, 1956.
7. Moffitt, F. H., and Bouchard, H., *Surveying*, 6th ed., Harper & Row, Publishers, New York, 1975.
8. Navyasky, M., "Adjusting Observed Directions by Angles Using Equitable Weights," *Surveying and Mapping*, December 1974.
9. Webber, J. T., "A Source of Error in Optical Plummets," *Australian Surveyor*, March 1971.

# CHAPTER 7
# Stadia and tacheometry

**7.1. General**  Tacheometry is the procedure by which horizontal distances and differences in elevation are determined indirectly using subtended intervals and angles observed with a transit or theodolite on a graduated rod or scale. The distances and elevations thus obtained are usually of a lower order of accuracy than is possible by taping, EDM, or differential leveling. However, the results are very adequate for many purposes. Horizontal distances can be determined with a relative accuracy of 1 part in 300 to 400, while differences in elevation can be obtained to within ±0.1 ft or ±0.03 m.

Applications of tacheometry include traverse and leveling for topographic surveys, location of detail for topographic surveys, leveling as well as field completion surveys for photogrammetric mapping, and hydrographic surveys.

The most common tacheometric method practiced in the United States is stadia surveying, in which an engineer's transit or telescopic alidade and graduated leveling rod are used. Stadia surveying by this procedure is discussed first. Sections on self-reducing theodolites and subtense bars follow.

**7.2. Stadia method**  Equipment for stadia measurements consists of a telescope with two horizontal cross hairs called *stadia hairs* and a graduated rod called a *stadia rod*.

The process of taking stadia measurements consists of observing, through the telescope, the apparent locations of the two stadia hairs on the rod, which is held in a vertical position. The interval thus determined, called the *stadia interval* or *stadia reading*, is a direct function of the distance from the instrument to rod as developed in Art. 7.4. The ratio of distance to stadia interval is 100 for most instruments.

**7.3. Stadia hairs and stadia rods**  The telescopes of transits, theodolites, plane-table alidades, and many levels are furnished with stadia hairs in addition to the regular cross hairs. One stadia hair is above and the other an equal distance below the horizontal cross hair. Stadia hairs are usually mounted on the same reticule and in the same plane as the horizontal and vertical cross hairs. Under these conditions, the stadia hairs are not adjustable and the distance between the hairs remains unchanged. Figure 7.1 shows two styles of stadia cross hairs: the stub cross hair (Fig. 7.1a), and the stadia hairs, which go completely across the cross-hair ring (Fig. 7.1b).

A conventional rod such as the Philadelphia rod is adequate for stadia work where the sights do not exceed 200 ft or 60 m. For longer sights stadia rods are advisable and provide more accurate results. Stadia rods are described in Art. 5.25 and illustrated in Fig. 5.23.

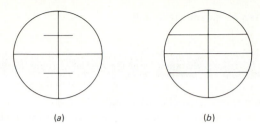

**Fig. 7.1** Two styles of stadia hairs.          (a)                                        (b)

**7.4. Principle of stadia**   Figure 7.2a illustrates the principle upon which the stadia method is based. In the figure, the line of sight of the telescope is horizontal and the stadia rod is vertical. The stadia hairs are indicated by points $a$ and $b$. The distance between the stadia hairs is $i$. The apparent locations of the stadia hairs on the rod are points $A$ and $B$ and the stadia interval is $s$.

Rays from $a$ passing through the optical center of the lens $O$ and the focal point of the lens $F$ are brought to a focus at $A$. Similarly, the reverse is true, so that rays from $A$ which pass through $F$ and $O$ are brought to a focus at $a$.

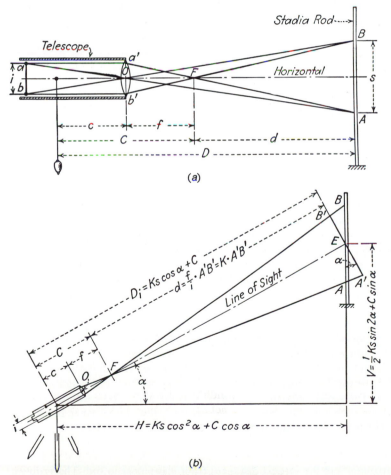

**Fig. 7.2** (a) Horizontal stadia sight. (b) Inclined stadia sight.

Since $ab = a'b'$, by similar triangles

$$\frac{f}{i} = \frac{d}{s}$$

Hence, the horizontal distance from the principal focus to the rod is

$$d = (f/i)s = Ks$$

in which $K = f/i$ is a coefficient called the *stadia interval factor*, which for a particular instrument is a constant as long as conditions remain unchanged. Thus, for a horizontal sight the distance from principal focus to rod is obtained by multiplying the stadia interval factor by the stadia interval. The horizontal distance from center of instrument to rod is then

$$D = Ks + (f + c) = Ks + C \tag{7.1}$$

in which $C$ is the distance from center of instrument to principal focus. This formula is employed in computing horizontal distances from stadia intervals when sights are horizontal.

**7.5. Stadia constants**   The focal distance $f$ is a constant for a given instrument. It can be determined with all necessary accuracy by focusing the objective on a distant point and then measuring the distance from the cross-hair ring to the objective. The distance $c$, though a variable depending upon the position of the objective, may for all practical purposes be considered a constant. Its mean value can be determined by measuring the distance from the vertical axis to the objective when the objective is focused for an average length of sight.

Usually, the value of $C = f + c$ is determined by the manufacturer and is stated on the inside of the instrument box. For external-focusing telescopes, under ordinary conditions $C$ may be considered as 1 ft without error of consequence. Internal-focusing telescopes (Art. 5.8) are so constructed that $C$ is zero or nearly so; this is an important advantage of internal-focusing telescopes for stadia work.

**7.6. Stadia interval factor**   The nominal value of the stadia factor $K = f/i$ is usually 100. The interval factor can be determined by observation. The usual procedure is to set up the instrument in a location where a horizontal sight can be obtained. With a tape lay off, from a point $C = f + c$ in front of the center of the instrument, distances of 100 ft, 200 ft, etc., up to perhaps 1000 ft, and set stakes at the points established. The stadia rod is then held on each of the stakes, and the stadia interval is read. The stadia interval factor is computed for each sight by dividing the distance from the principal focus to the stake by the corresponding stadia interval, and the mean is taken as the best estimate. Owing to errors in observation and perhaps to errors from natural sources, the values of $K$ for the several distances are not likely to agree exactly.

To overcome any prejudicial tendencies on the part of the instrument operator, observations may be made on the rod held on stakes at random distances from the instrument; these distances being measured later with the tape.

For use on long sights, where the full stadia interval would exceed the length of the rod, the stadia interval factor may be determined separately for the upper stadia hair and horizontal cross hair and for the lower stadia hair and horizontal cross hair.

With adjustable stadia hairs, the interval factor is made 100 by moving the hairs until their rod interval is $\frac{1}{100}$ of the distance from the principal focus to the rod, this distance being measured with a tape. The stadia hairs are so adjusted that the horizontal cross hair bisects the space between them, each in its turn being moved vertically until the distance between it and the horizontal hair is the proper half-interval, as indicated by the rod interval.

**7.7. Inclined sights**   In stadia surveying, most sights are inclined, and usually it is desired to find both the horizontal and the vertical distances from instrument to rod. The problem therefore resolves itself into finding the horizontal and vertical projections of an inclined line of sight. For convenience in field operations the rod is always held vertical.

Figure 7.2*b* illustrates an inclined line of sight, $AB$ being the stadia interval on the vertical rod and $A'B'$ being the corresponding projection normal to the line of sight. The length of the inclined line of sight from center of instrument is

$$D_i = \frac{f}{i}(A'B') + C \tag{7.2}$$

For all practical purposes the angles at $A'$ and $B'$ may be assumed to be 90°. Let $AB = s$; then $A'B' = s\cos\alpha$. Making this substitution in Eq. (7.2), and letting $K = f/i$, the inclined distance is

$$D_i = Ks\cos\alpha + C \tag{7.3}$$

The horizontal component of this inclined distance is

$$H = Ks\cos^2\alpha + C\cos\alpha \tag{7.4}$$

which is the general equation for determining the horizontal distance from center of instrument to rod, when the line of sight is inclined.

The vertical component of the inclined distance is

$$V = Ks\cos\alpha\sin\alpha + C\sin\alpha \tag{7.4a}$$

The equivalent of $\cos\alpha\sin\alpha$ is conveniently expressed in terms of double the angle $\alpha$, or

$$V = \tfrac{1}{2}Ks\sin2\alpha + C\sin\alpha \tag{7.5}$$

which is the general equation for determining the difference in elevation between the center of the instrument and the point where the line of sight cuts the rod. To determine the difference in ground elevations, the height of instrument and the rod reading of the line of sight must be considered.

Equations (7.4) and (7.5) are known as the *stadia formulas for inclined sights.*

**7.8. Permissible approximations**  More approximate forms of the stadia formulas are sufficiently precise for most stadia work. Usually, distances are computed only to feet or decimetres and elevations to tenths of feet or centimetres. Under these conditions, for side shots where vertical angles are less than 3°, Eq. (7.4) for horizontal distances may properly be reduced to the form

$$H = Ks + C \tag{7.6}$$

which is the same as for horizontal sights (Art. 7.4). But for traverses of considerable length, owing to the systematic error introduced, this approximation should not be made for vertical angles greater than perhaps 2°.

Owing to unequal refraction and to accidental inclination of the rod, observed stadia intervals are in general slightly too large. To offset the systematic errors from these sources, frequently on surveys of ordinary precision the constant $C$ is neglected. Hence, in any ordinary case Eq. (7.4) may, with sufficient precision, be expressed in the form

$$H = Ks\cos^2\alpha \text{ (approx.)} \tag{7.7}$$

Also, Eq. (7.5) may often be expressed with sufficient precision for ordinary work in the form

$$V = \tfrac{1}{2}Ks\sin2\alpha \text{ (approx.)} \tag{7.8}$$

However, the error in elevation introduced through using Eq. (7.8) may not be negligible, as for large vertical angles it amounts to several tenths of a foot or up to a decimetre.

Equations (7.7) and (7.8) are simple in form and are most generally employed.

When $K$ is 100, the common practice is to multiply mentally the stadia interval by 100 at the time of observation, and to record this value in the field notebook. This distance $Ks$ is often called the *stadia distance*. Thus, if the stadia interval were 7.37 ft, the stadia distance recorded would be 737 ft.

For external-focusing telescopes, the degree of approximation in using Eqs. (7.7) and (7.8) may be greatly reduced either by adding 0.01 ft or 3 mm to the observed stadia interval $s$ or —when $K$ is 100—by adding 1 ft or 0.3 m to the observed stadia distance $Ks$. It is convenient to add the correction mentally and to record the corrected value in the field notebook. The notes should state that corrected values are recorded.

**7.9. Stadia reductions** Horizontal distance and difference in elevation can be computed from stadia observations by using (1) an electronic calculator by the exact Eqs. (7.4) and (7.5) or approximate Eqs. (7.7) and (7.8), (2) stadia tables, and (3) stadia slide rules or diagrams.

**Example 7.1** The following data were obtained by stadia observation: vertical angle $= +8°10'$, $s = 2.50$ ft. The stadia interval factor is known to be 100 and $C = 0.75$ ft. Calculate $H$ and $V$.

SOLUTION By exact equations (7.4) and (7.5),

$H = Ks\cos^2\alpha + C\cos\alpha$
$= (100)(2.50)(\cos^2 8°10') + (0.75)(\cos 8°10')$
$= (100)(2.50)(0.9798) + (0.75)(0.9899)$
$= 245.0 + 0.7$
$= 245.7 = 246$ ft
$V = \frac{1}{2}Ks\sin 2\alpha + C\sin\alpha$
$= \frac{1}{2}(100)(2.50)(0.2812) + (0.75)(0.1421)$
$= 35.2 + 0.11$
$= 35.3$ ft

If approximate equations (7.7) and (7.8) are used and $C = 1.0$ is assumed, we may add 0.01 ft to the stadia interval. Thus, $s = 2.50 + 0.01 = 2.51$:

$H = (100)(2.51)(\cos^2 8°10') = (251)(0.9798)$
$= 246$ ft
$V = \frac{1}{2}(100)(2.51)\sin 2\alpha = (125.5)(0.2812)$
$= 35.3$ ft

and the approximate equations are seen to yield adequate accuracy for these data.

**Example 7.2** The stadia interval was 1.372 m at a vertical angle of 20°32'. $C = 0.30$ m and the stadia interval factor is 100. Calculate $H$ and $V$.

SOLUTION By Eq. (7.4),

$H = (100)(1.372)(\cos^2 20°32') + (0.30)(\cos 20°32')$
$= 120.6$ m

By Eq. (7.5),

$V = \frac{1}{2}(100)(1.372)(\sin 41°04') + (0.30)(\sin 20°32')$
$= 45.17$ m

If approximate equations (7.7) and (7.8) are used for this problem, add 0.003 to the stadia interval so that

$$s = 1.372 + 0.003 = 1.375$$

and

$$H = 100(1.375)\cos^2\alpha = 120.6 \text{ m}$$
$$V = \frac{1}{2}(100)(1.375)\sin 41°04' = 45.16 \text{ m}$$

Use of the approximate equations for solution with the large vertical angle resulted in an error of 1 cm in the difference in elevation. Had the telescope of the instrument been internally focusing, as is the case with most modern transits and theodolites, $C$ is equal to zero and approximate equations (7.7) and (7.8) can be used without danger of errors.

**7.10. Observation of stadia interval**   On transit or plane-table surveys the stadia interval is usually determined by setting the lower stadia hair on a foot or decimetre mark and then reading the location of the upper stadia hair. Thus, the stadia interval is mentally computed more easily and with less chance of mistake than if the lower hair were allowed to take a random position on the rod. When the vertical angle is taken to a given mark on the rod, the corresponding stadia interval is observed with the lower hair on the foot or decimetre mark that renders a minimum displacement of the horizontal cross hair from the mark to which the vertical angle is referred.

Thus, if a vertical angle were taken with the line of sight cutting the rod at 4.9 ft and the lower stadia hair fell at 2.3 ft, the telescope would be rotated about the horizontal axis until the lower hair was at 2.0 ft. The center horizontal cross hair would then fall at 4.6 ft.

On metric rods it is more convenient to set the bottom cross hair on the nearest even decimetre. For example, if the vertical angle were taken when the middle cross hair is on 1.495 m, and the lower stadia hair was at 1.23, rotate the telescope in a vertical plane about the horizontal axis until the lower stadia hair coincides with 1.200 and read the top stadia hair as 1.730 m. The interval is then $1.730 - 1.200 = 0.530$ m.

Whenever the stadia interval is in excess of the length of the rod, the separate half-intervals are observed and their sum is taken.

For precise stadia work, the readings may be made by means of two targets on the rod, one target on, say, the 2-ft or 0.500-m mark and the other set by the rodman as directed by the instrumentman. To avoid excessive effects of atmospheric refraction, the intercept of the lower cross hair with the rod should not fall nearer the ground than necessary.

Note forms that can be used for recording stadia data in stadia traverses (Art. 8.12) and topographic surveys by the stadia method (Art. 15.10) are shown in Figs. 8.17 and 15.6, respectively.

**7.11. Beaman stadia arc**   The Beaman stadia arc, in modified form known also as the *stadia circle*, is a specially graduated arc on the vertical circle of the transit or the plane-table alidade. It is used to determine distances and differences in elevation by stadia without reading vertical angles and without the use of tables, diagrams, or stadia slide rules. The stadia arc has no vernier, but settings are read by an index mark.

*Horizontal distance*   In the type of stadia arc shown in Fig. 7.3, the graduations for determining distances are at the left, inside the vertical circle. When the telescope is level (vertical vernier reading zero), the reading of the arc is 100, indicating that the horizontal distance is 100 percent of the observed stadia distance. When an inclined sight is taken, the observed stadia interval is multiplied by the reading of the "Hor." stadia arc, to obtain the horizontal distance from principal focus to rod. Thus, the "Hor." scale is a non-uniform logarithmic scale graduated according to $(100)\cos^2\alpha$, where $\alpha$ is the vertical angle the telescope makes with the horizontal. For example, if the stadia interval is 4.11 ft, $K = 100$, $C = 0.0$, and the "Hor." reading on the stadia arc is 99, the stadia distance is $(100)(4.11) = 411$ and the horizontal distance $= (4.11)(99) = 407$ ft. In Fig. 7.3, the "Hor." reading is 98. If $s = 1.325$ m, the horizontal distance is $(1.325)(98) = 129.9$ m.

Another type of stadia arc (Fig. 7.4) is graduated to give the *correction* in percent to be subtracted from the observed stadia interval times the stadia interval factor. In Fig. 7.4, the "$H$" reading opposite the index is estimated to be 2.6. If $Ks$ is 208.4 m, the horizontal

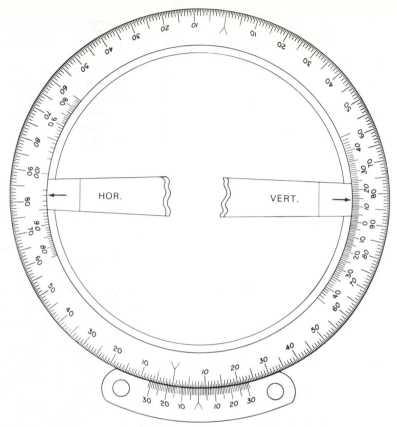

**Fig. 7.3** Stadia arc (multiplier type).

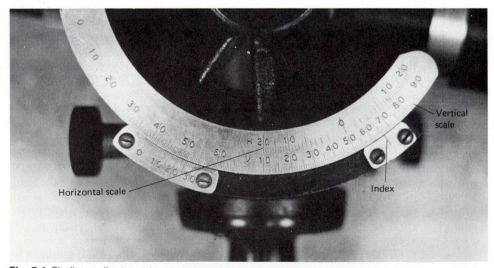

**Fig. 7.4** Stadia arc (horizontal scale subtraction type: vertical scale index = 50).

distance is $208.4 - (208.4)(0.026) = 203.0$ m. Thus, the $H$ scale is graduated according to $100(1 - \cos^2 \alpha)$, where $\alpha$ is the vertical angle of the telescope with the horizontal.

**Difference in elevation**    In Fig. 7.3, the graduations for determining differences in elevation are at the right, inside the vertical circle. When the telescope is level (vertical vernier reading zero), the reading of the arc is zero. When an inclined sight is taken, first the stadia distance is observed in the usual manner, that is, with the lower stadia hair on a foot mark of the rod. The telescope is then either elevated or depressed slightly until the nearest graduation on the "Vert." scale of the stadia arc coincides with the index of the arc (in order to avoid interpolation), and a rod reading is taken at the point where the line of sight strikes the rod. The observed stadia interval is multiplied by the reading of the "Vert." stadia arc, to obtain the vertical distance from center of the instrument to point last sighted on the rod. This difference in elevation, combined with the height of instrument and the rod reading, gives the difference in elevation between the instrument station and the point on which the rod is held (see Art. 7.12).

In Fig. 7.3, the "Vert." reading is $+14$ ($+$ indicates that the line of sight is above the horizon). Assuming that $s = 4.11$, the difference in elevation between the center of the telescope and the point sighted on the rod is $(14)(4.11) = 57.5$ ft. This value is analogous to $V$ in Eq. (7.5) or (7.8) and the "Vert." scale is graduated according to $\frac{1}{2} K \sin 2\alpha$.

The other type of stadia arc (Fig. 7.4) is so graduated that, when the telescope is level, the reading of the "Vert." stadia arc is 50 instead of 0; when the telescope is elevated, the reading is greater than 50, and when the telescope is depressed, the reading is less than 50. In all cases, 50 is subtracted from the reading of the stadia arc, and the remainder (positive or negative as determined by the subtraction) is multiplied by the observed stadia interval to obtain the vertical distance from center of the telescope to the point sighted on the rod. This arrangement of the scale on the stadia arc avoids mistakes of reading a positive value for a negative value, or vice versa, but it introduces an additional step (subtraction) in the computations. In Fig. 7.4, the $V$ reading has been set to 66. If the stadia interval is 1.431 m, the difference in elevation between the center of the telescope and the point sighted is $(1.431)(66 - 50) = 22.9$ m.

Observations with the Beaman stadia arc do not include the effect of the instrumental constant $C = f + c$. If, for an external-focusing telescope, more precise results are desired than are yielded by the approximate formulas [Eqs. (7.7) and (7.8)], the observed values should be corrected, particularly if the vertical angles are large. The simplest method is to add 0.01 ft or 0.003 m to the observed stadia interval.

**7.12. Difference in elevation**    In Fig. 7.5 the instrument is at station $A$ with a height of instrument or h.i. above $A$ equal to $AB$ and the rod is held in a vertical position at station $D$. It is desired to determine the difference in elevation between points $A$ and $D$ given stadia observations and the vertical angle $\alpha$. First consider the problem when the difference in elevation between points $A$ and $D$ on the ground is desired. This situation occurs in stadia traversing and in topographic surveying from a known point. The h.i. $= AB$, the stadia interval $s$, the vertical angle $\alpha$ with the middle cross hair set on $E$, and the rod reading $DE$ are observed and recorded. The difference in elevation between the telescope and $E$, which is equal to $V$, can be calculated by using Eq. (7.5) or (7.8) or if the Beaman arc is used, by the methods described in Art. 7.11. The difference in elevation is $\Delta$el. $= AB + V - DE$ or $\Delta$el. $=$ h.i. $+ V -$ rod reading at $D$. Thus, the elevation at $D =$ the elevation at $A + (AB + V - DE)$. If the middle cross hair is sighted on the rod such that $DE = AB =$ h.i., then $\Delta$el. $= V$ and the elevation at $D =$ the elevation at $A + V$. Obviously, setting the horizontal cross hair

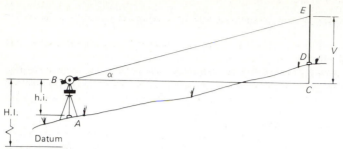

**Fig. 7.5** Difference in elevation between instrument and rod stations.

on the rod at a value equal to the h.i. prior to reading the vertical angle simplifies calculating the change in elevation between the transit station and the point on which the rod is held.

In another situation also common in topographic surveying, it may be desired to determine the difference in elevation between the center of the telescope and the ground point on which the rod is held. Referring to Fig. 7.5, assume that the elevation of station $A$ and the height of instrument above $A$ or the h.i. are known. Thus, the height of instrument above the datum or H.I. can be calculated and the elevation of $D$ can be computed directly from the instrument H.I. Given the stadia interval and $\alpha$, $V$ can be computed so that the elevation of $D = (\text{H.I. at } B) + V - DE$, or elevation of $D = (\text{H.I. at } B) + V - (\text{rod reading})$. Note that it is important that the rod reading always be recorded.

Figure 7.6 illustrates the situation where the difference in elevation is desired between points $B$ and $F$ utilizing an intermediate instrument set up at $G$. Assume that a backsight is taken on a rod held at $B$. The stadia interval, rod reading $BC$, and $\alpha_B$ are recorded. Using these data, $V_B$ can be calculated. Next, a foresight is taken on the rod held at $F$. The stadia interval, rod reading $FE$, and vertical angle are observed and recorded so that $V_F$ can be computed. A general expression for difference in elevation between two points $B$ (the backsight) and $F$ (the foresight) is

$$\Delta \text{el.}_{BF} = \text{rod}_B \pm V_B \pm V_F - \text{rod}_F \tag{7.9}$$

in which the sign on $V_B$ is opposite to the sign of $\alpha_B$ and the sign of $V_F$ corresponds to the sign on $\alpha_E$. Thus, in Fig. 7.6, $\Delta \text{el.}_{BF} = BC - V_B - V_F - EF$.

In the reverse direction, if the backsight is on $F$ and the foresight on $B$, then $\Delta \text{el.} = EF + V_F + V_B - BC$.

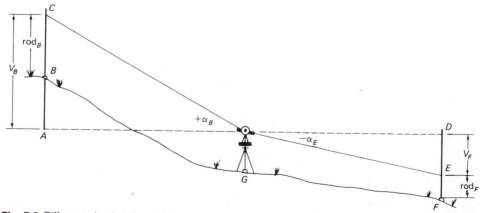

**Fig. 7.6** Difference in elevation between two points.

### 7.13. Uses of stadia    Uses of stadia are as follows:

1. In differential leveling, the backsight and foresight distances are balanced conveniently if the level is equipped with stadia hairs.
2. In profile leveling and cross sectioning, stadia is a convenient means of finding distances from level to points on which rod readings are taken.
3. In rough trigonometric, or indirect, leveling with the transit, the stadia method is more rapid than any other. The line of trigonometric levels is run as described in Art. 5.35, except that stadia intervals are observed and that differences in elevation are computed by the stadia formula. The method is described in Art. 7.14.
4. On traverses of low relative accuracy, where only horizontal angles and distances are required, the stadia method is more rapid than taping. It may be used either in running traverse lines or in locating details from traverse lines. Stadia intervals are observed as each point is sighted. Horizontal angles are measured, but vertical angles are observed only when of sufficient magnitude to make the horizontal distance appreciably different from the stadia distance (say, when greater than 3°) and then are estimated without reading the vernier. The stadia traverse method of running such surveys is described in Art. 8.12.
5. On surveys of low relative accuracy—particularly topographic surveys—where both the relative location of points in a horizontal plane and the elevation of these points are desired, stadia is useful. Both horizontal and vertical angles are measured, and the stadia interval is observed, as each point is sighted; these three observations define the location of the point sighted. The transit-stadia method of making observations when both the horizontal location and the elevation are desired is described in Art. 8.12.
6. Where the plane table is used (Art. 7.19), stadia observations are made with the telescopic alidade (Art. 7.19) in the same manner as with the transit, but horizontal distances and differences in elevation are computed in the field and are plotted immediately instead of being recorded in the form of notes. Methods for plane-table traverse are described in Art. 8.13. Uses of the plane table and alidade in topographic surveying are given in Art. 15.11.
7. Transit or plane-table stadia are useful for field completion surveys required for photogrammetric topographic mapping (Art. 16.14).

### 7.14. Indirect leveling by stadia    Where the required precision is low and the country is rolling or rough, the stadia method of indirect leveling is rapid. The transit should preferably be provided with a sensitive control level for the vertical vernier in order that index error may be readily eliminated.

Theodolites and self-reducing tacheometers with automatic vertical indexing are ideal for stadia leveling. With the ordinary transit having a vertical circle reading to single minutes, differences in elevation are usually computed only to the nearest 0.1 ft or to the nearest 0.03 m. The average length of sight in stadia surveying is usually considerably greater than in differential leveling.

In running a line of levels by this method, the transit is set up in a convenient location. A backsight is taken on the rod held at the initial bench mark, first by observing the stadia interval and then by measuring the vertical angle to some arbitrarily chosen mark on the rod. A turning point is then established in advance of the transit, and similar observations are taken, the vertical angle being measured with the middle cross hair set on the same mark as before. The transit is moved to a new location in advance of the turning point, and the process is repeated. The stadia distances and vertical angles are recorded, as is the rod reading, which is used as an index when vertical angles are measured. If it is impractical to sight at this chosen index reading, the vertical angle is measured to some other graduation and this rod reading is recorded in the notes.

Figure 7.7 shows a form of notes for stadia leveling. The bench mark sighted for a backsight is recorded in column 1 and its known elevation is shown on the same line in column 11. The stadia interval for the backsight is recorded in column 2, the vertical angle in column 3, and the rod reading, $\text{rod}_B$, where the middle cross hair cuts the rod is given in

| Station (1) | Backsight | | | | | Foresight | | | | | |
|---|---|---|---|---|---|---|---|---|---|---|---|
| | (2) Interval | (3) Vert Angle | (4) (rod)$_B$ | (5) $V_B$ | (6) Interval | (7) Vert Angle | (8) (rod)$_F$ | (9) $V_F$ | (10) ΔElev | (11) Elev. ft | Transit |
| BM 42 | 4.22 | -0°06' | 3.2 | +0.7 | | | | | | 5972.4 | No.132 |
| TP1 | 3.16 | 0°00' | 9.7 | 0.0 | 2.64 | +4°16' | 4.4 | +19.6 | +19.1 | 5991.5 | C = 0 |
| TP2 | 4.05 | -2°15' | 6.0 | +15.9 | 4.41 | +1°02' | 7.7 | + 8.0 | +10.0 | 6001.5 | K = 100 |
| TP3 | | | | | 3.94 | +2°20' | 6.0 | +16.0 | +31.9 | 6033.4 | |

*STADIA LEVEL FOR INDIAN* — *TRAIL RECONNAISANCE* — *July 20, 1976 Fair 70°F J.L.Black π V.A. Alden, Rod*

**Fig. 7.7** Notes for stadia leveling.

column 4. Similar readings for the foresight are recorded one line down in columns 6, 7, and 8. The values $V_B$ and $V_F$ (Fig. 7.6) for backsight and foresight are calculated by Eq. (7.8) and are recorded in columns 5 and 9, respectively. The sign of $V_B$ is opposite the sign of the vertical angle to the backsight, while the sign of $V_F$ corresponds to the sign of the vertical angle to the foresight. The values of rod$_B$, $V_B$, $V_F$, and rod$_F$ are then used in Eq. (7.9) to calculate Δel. between the point on the ground where the rod was held for the backsight and the point on the ground where the rod was held for the foresight. This value is recorded in column 10. Thus, for the notes in Fig. 7.7 for the first instrument setup between B.M.$_{42}$ and T.P.$_{-1}$, Δel. = 3.2 + 0.7 + 19.6 − 4.4 = 19.1 ft. This Δel. (recorded in column 10) is applied to the elevation on the ground at the backsight (B.M.$_{42}$) to yield the elevation on the ground at the foresight (T.P.$_1$). This method of recording the data is then continued for subsequent setups, as indicated in Fig. 7.7.

**7.15. Errors in stadia**  Many of the errors of stadia are those common to all similar operations of measuring horizontal angles (Arts. 6.54 to 6.58) and differences in elevation (Art. 5.42). Sources of error in horizontal and vertical distances computed from observed stadia intervals are as follows:

**1. Stadia interval factor not that assumed**  This condition produces a systematic error in distances, the error being proportional to that in the stadia interval factor. The case is parallel to that of the tape which is too long or too short. When the value of the interval factor is closely determined by observations as described in Art. 7.6 and the stadia measurements are taken under similar conditions, the error from this source may be made negligible.

**2. Rod not standard length**  If the spaces on the rod are uniformly too long or too short, a systematic error proportional to the stadia interval is produced in each distance. Errors from this source may be kept within narrow limits if the rod is standardized and corrections for erroneous length are applied to observed stadia intervals. Except for stadia surveys of more than ordinary precision, errors from this source are usually of no consequence.

**3. Incorrect stadia interval**  The stadia interval varies randomly owing to the inability of the instrument operator to observe the stadia interval exactly. In a series of connected observations (as a traverse) the error may be expected to vary as the square root of the number of sights. This is the principal error affecting the precision of distances. It can be kept to a minimum by proper focusing to eliminate parallax, by taking observations at favorable times, and by care in observing. Where high precision is required, stadia measurements may be taken by sighting on a rod with two targets, one fixed and the other movable.

**4. Rod not plumb**   This condition produces a small error in the vertical angle. It also produces an appreciable error in the observed stadia interval and hence in computed distances, this error being greater for large vertical angles than for small angles. It can be eliminated by using a rod level.

**5. Unequal refraction**   Unequal refraction of light rays in layers of air close to the Earth's surface affects the sight on the lower stadia hair more than the sight on the upper stadia hair and thus introduces systematic positive errors in stadia measurements. Although errors from this source are of no consequence in ordinary stadia surveying, they may be important on the more precise surveys. The periods most favorable for equal refraction are at times when it is cloudy or, if the sun is shining, during the early morning or late afternoon. On precise stadia surveys where it is necessary to work under a variety of atmospheric conditions, it is appropriate to determine the stadia interval factor for each condition and to apply the factor to all observations taken under that condition. Whenever atmospheric conditions are unfavorable, the sights should not be taken near the bottom of the rod.

**6. Errors in vertical angles**   Errors in vertical angles are relatively unimportant in their effect upon *horizontal distances*. For example, analysis of Eq. (7.7) shows that an uncertainty of 01′ in a vertical angle of 5° yields a discrepancy of 0.02 ft in a 300-ft sight and 0.005 m in a 100-m sight. Under the same conditions, an error of 01′ in a 15° vertical angle produces discrepancies of 0.05 ft in the 300-ft sight and 0.02 m in the 100-m sight.

With respect to differences in elevation, analysis of Eq. (7.8) reveals that an uncertainty of 01′ in the magnitude of the vertical angle results in a discrepancy of 0.1 ft in a 300-ft sight and 0.03 m in a 100-m sight. Uncertainty in the stadia interval, especially at higher vertical angles, will have a more pronounced effect on elevation differences than will errors in vertical angles.

**7.16. Error propagation in stadia**   Estimation of accuracies possible in horizontal distances and differences in elevation calculated from stadia observations is necessary for the planning of stadia surveys. Evaluation of these accuracies is best achieved by error propagation using the basic equations developed in Arts. 7.4 and 7.7.

If the stadia rod is standardized and proper corrections applied for erroneous length, if the stadia interval factor $K$ is carefully determined, if the instrument constant $C=0.0$ or is carefully determined, and if the rod is plumbed using a rod level, the major source of error affecting both horizontal and vertical distances is that of observing the stadia interval. Errors in observing the vertical angle have an effect but are of secondary importance.

**1. Horizontal distance error propagation**   Propagate the error through Eq. (7.4) using Eq. (2.46) in Art. 2.20. Thus, the estimated variance in the horizontal distance $H$ is

$$\sigma_H^2 = \left(\frac{\partial H}{\partial s}\right)^2 \sigma_s^2 + \left(\frac{\partial H}{\partial \alpha}\right)^2 \sigma_\alpha^2 \tag{7.10}$$

where

$$\frac{\partial H}{\partial s} = K\cos^2\alpha \qquad \text{and} \qquad \frac{\partial H}{\partial \alpha} = -(Ks\sin 2\alpha + C\sin\alpha)$$

are obtained by partial differentiation of Eq. (7.4). Assume that $\sigma_s = 0.01$ ft or 0.003 m and $\sigma_\alpha = 01′$ or $\sigma_\alpha = (60″)(\text{arc}\,01″) = (60″)(0.000004848) = 0.0003$ rad. Further assuming that $s=$

3.00 ft, $\alpha = 5°$, $C = 0.0$, and $K = 100$, the estimated variance and standard deviation in $H$ are

$$\sigma_H^2 = [(100)\cos^2 5°]^2 (0.01)^2 + [(100)(3)\sin 10°]^2 (0.0003)^2$$
$$\sigma_H^2 = 0.9849 + 0.0002 = 0.9851 \text{ ft}^2$$
$$\sigma_H = 0.99 \text{ ft} = 1.0 \text{ ft in a 300-ft sight} \qquad \text{or}$$
$$\sigma_H = 0.3 \text{ m in a 100-m sight}$$

A similar analysis using vertical angles of 10° and 15° yields estimated values for $\sigma_H$ which are virtually the same. The contribution of observational error in the vertical angle (the second term) does become somewhat larger with the higher vertical angles but is still insignificant. Even if $\sigma_\alpha = 05'$, the second term in Eq. (7.10) is only 0.0476 ft² and contributes little to the propagated standard deviation of 0.97 ft, assuming a vertical angle of 15°. Thus, the error in horizontal distance determined from stadia is primarily dependent upon the magnitude of the error in the stadia interval. A value of $\sigma_s$ of 0.01 ft or 0.003 m is quite reasonable for sights up to 400 ft or 120 m. For this reason, the maximum relative accuracy that can be expected in horizontal stadia distances is 1 part in 300 to 1 part in 400.

**2. Differences in elevation** The difference in elevation between the center of the telescope and the point on which the rod is held is found by Eq. (7.5), Art. 7.7. Under similar assumptions as made for horizontal distance, the estimated variance in $V$ by propagating the errors through Eq. (7.5) using Eq. (2.46) in Art. 2.20 is

$$\sigma_V^2 = \left(\frac{\partial V}{\partial s}\right)^2 \sigma_s^2 + \left(\frac{\partial V}{\partial \alpha}\right)^2 \sigma_\alpha^2 \tag{7.11}$$

where

$$\left(\frac{\partial V}{\partial s}\right) = \tfrac{1}{2} K \sin 2\alpha \qquad \text{and} \qquad \left(\frac{\partial V}{\partial \alpha}\right) = Ks \cos 2\alpha + C \cos \alpha$$

Assuming that $s = 3.00$ ft, $\alpha = 5°$, $K = 100$, $C = 0.0$, $\sigma_s = 0.01$ ft or 0.003 m, and $\sigma_\alpha = 01'$, the estimated variance and standard deviation by Eq. (7.11) are

$$\sigma_V^2 = \left[\tfrac{1}{2}(100)\sin 10°\right]^2 (0.01)^2 + [(100)(3)(\cos 10°)]^2 (0.0003)^2$$
$$= 0.007538 + 0.007856 = 0.015394 \text{ ft}^2$$
$$\sigma_V = 0.12 \text{ ft for a 300-ft sight} \qquad \text{or}$$
$$\sigma_V = 0.04 \text{ m in a 100-m sight}$$

If $\alpha = 15°$ is assumed,

$$\sigma_V^2 = [(50)(\sin 30°)]^2 (0.01)^2 + [(300)(\cos 30°)]^2 (0.0003)^2$$
$$= 0.0625 + 0.0061 = 0.0686 \text{ ft}^2$$
$$\sigma_V = 0.26 = 0.3 \text{ ft in a 300-ft sight} \qquad \text{or}$$
$$\sigma_V = 0.08 = 0.1 \text{ m in a 100-m sight}$$

Thus, for vertical angles of about 5°, which is normal for most stadia operations, the effects of errors in stadia intervals and in the vertical angle observations on the calculated value of $V$ are about equal. For very steep angles, which do occur in rugged terrain, the effects of errors in the stadia interval on $V$ increase substantially while effects of errors in vertical angles on $V$ remain about the same.

Consequently, if care is exercised in handling the equipment and observing the stadia interval, differences in elevation between the center of the telescope and the point where the rod is held can be determined by stadia for sights of up to 300 ft or 100 m to within $\pm 0.1$ ft or $\pm 0.030$ m, respectively. However, when the vertical angles are steep, the estimated standard deviation is approximately doubled and to maintain this accuracy, extreme care must be exercised in observing the stadia interval.

Differences in elevation by stadia leveling are determined with Eq. (7.9) using values of $V$ and rod readings from backsights and foresights. The estimated standard deviation in elevation change between two points is found by propagating the errors through Eq. (7.9) to give

$$\sigma_{\Delta\text{el.}}^2 = \left(\frac{\partial \Delta\text{el.}}{\partial V_B}\right)^2 \sigma_{V_B}^2 + \left(\frac{\partial \Delta\text{el.}}{\partial V_F}\right)^2 \sigma_{V_F}^2 + \left(\frac{\partial \Delta\text{el.}}{\partial \text{rod}_B}\right)^2 \sigma_{\text{rod}_B}^2 + \left(\frac{\partial \Delta\text{el.}}{\partial \text{rod}_F}\right)^2 \sigma_{\text{rod}_F}^2$$

$$= \sigma_{V_B}^2 + \sigma_{V_F}^2 + \sigma_{\text{rod}_B}^2 + \sigma_{\text{rod}_F}^2$$

Assuming that $\sigma_{\text{rod}_B} = \sigma_{\text{rod}_F} = 0.01$ ft or 0.003 m, sights of 300 ft or 100 m, and using values of $\sigma_{V_B}$ and $\sigma_{V_F}$ from previous examples, yields

$$\sigma_{\Delta\text{el.}}^2 = 2\sigma_V^2 + 2\sigma_{\text{rod}}^2$$
$$\sigma_{\Delta\text{el.}}^2 = 2(0.01539) \text{ ft}^2 + 2(0.01)^2 = 0.03098 \text{ ft}^2$$
$$\sigma_{\Delta\text{el.}} = 0.18 \text{ ft assuming backsights and foresights of 300 ft} \qquad \text{or}$$
$$= 0.05 \text{ m with backsights and foresights of 100 m}$$

**7.17. Self-reducing tacheometers**   The Beaman arc provided an example of how an analog device could be incorporated into the engineer's transit so as to simplify the stadia reduction procedure. A similar concept is used for optical theodolites where the major use of the instrument is for tacheometry. In self-reducing, optical-reading tacheometers, the fixed stadia cross hairs are replaced by three variable slightly curved lines as illustrated in Fig. 7.8 for one style of tacheometer. These curves are etched on the vertical circle and are projected onto the plane of the reticule so as to be visible when viewing through the telescope. The bottom curve is a base-line or zero curve, the middle curve is for elevation differences, and the top curve is for the horizontal distances. The interval between the zero curve and middle curve varies according to the $\sin\alpha\cos\alpha$ and when multiplied by the constant above the curve, yields the difference in elevation between the center of the telescope and the point sighted on the rod directly. In Fig. 7.8, the bottom line is on 1.000 m and the middle line is on 1.215 m, yielding an interval of 0.215 m. The factor of 0.1, labeled above the middle line

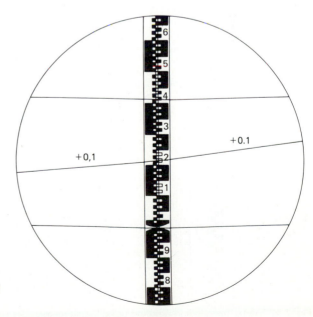

**Fig. 7.8** Curved stadia intercepts in a self-reducing tacheometer. (*Wild Heerbrugg Instruments, Inc.*)

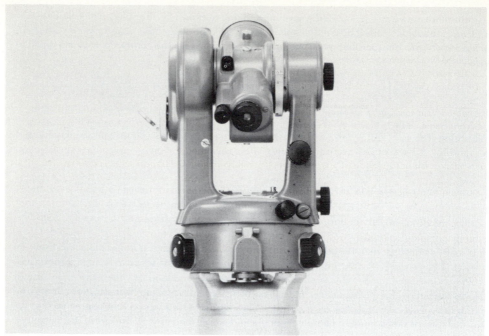

**Fig. 7.9** Self-reducing tacheometer, Kern K1-RA. (*Kern Instruments, Inc.*).

in the field of view, must be multiplied times the interval in centimetres. Thus, the elevation difference between the telescope and the point sighted is (215)(0.1)=21.5 m. The interval between the bottom and top lines varies according to $\cos^2\alpha$ and when multiplied by 100 yields horizontal distance. In Fig. 7.8 this interval is 0.410 m, so that the horizontal distance is (100)(0.410)=41.0 m. As the inclination of the telescope varies, the shapes of the lines change and the multiplier on the middle curve changes. The constant for the horizontal interval is always 100.

Another style of self-reducing tacheometer equipped with a pendulum compensator for automatic vertical circle index compensation is illustrated in Fig. 7.9. In this instrument, two

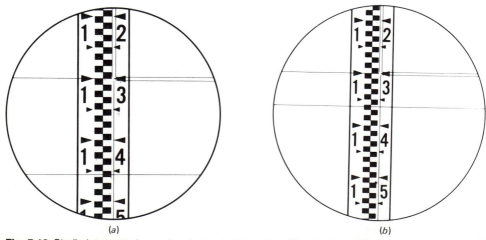

(a)                                                                 (b)

**Fig. 7.10** Stadia intercepts in a self-reducing tacheometer. (*Kern Instruments, Inc.*)

horizontal lines are projected into the field of view of the telescope (Fig. 7.10a). The upper line is a fixed base line. The lower line is on a movable reticule which is raised or lowered mechanically by a cam that changes position as the inclination of the telescope varies. The distance between the two lines changes according to the stadia equation, so that the interval is the distance directly. In Fig. 7.10a, the top line is on 1.300 m and the bottom (movable) line is on 1.456 m, yielding an interval of 0.156 m. The horizontal distance is (100)(0.156) = 15.6 m. The same horizontal base line is used to obtain an interval for difference in elevation. A control on the tacheometer permits rotating the cam through 180°. In this position, the cam moves the lower line in such a way that the interval gives the elevation difference between the telescope and the point sighted with the fixed top line. In Fig. 7.10b, the upper fixed line is on 1.300 m and the bottom line is on 1.364 m, giving an interval of 0.064 m. Thus, the difference in elevation between the telescope and the point sighted is (100)(0.064) = 6.4 m.

**7.18. Subtense bar**   Another method for indirect distance determination is to observe the angle subtended by the two ends of a horizontal rod of fixed length called a *subtense bar* (Fig. 7.11). The subtense bar has targets at the two ends which are connected by invar wires that are under a slight but constant tension. In this way, effects of temperature changes are minimized. The subtense bar is mounted on a tripod and can be centered over the station to which it is desired to measure. The bar is leveled by centering a circular level bubble and three level screws. The bar can then be oriented perpendicular to the line to be measured, by sighting a small telescope attached to the midpoint of the bar.

Referring to Fig. 7.12, the subtense bar is set over point $S$ and a 1″ direction theodolite is set over point $R$. Then the *horizontal angle* $\beta$ between the two targets at the ends of the bar is measured by repetition. The *horizontal* distance $RS$ is

$$RS = \frac{BB'}{2} \cot \frac{\beta}{2}$$

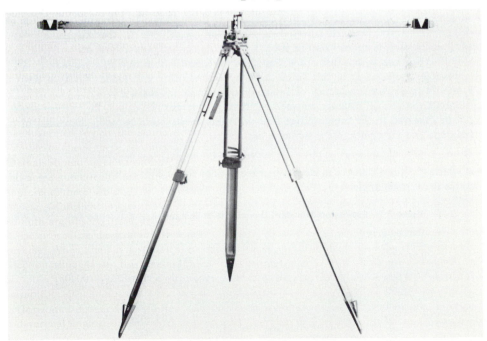

**Fig. 7.11** Wild invar subtense bar. (*Wild Heerbrugg Instruments, Inc.*)

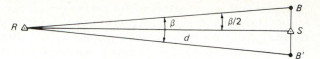

**Fig. 7.12** Geometry of subtense-bar reduction.

or

$$d = \frac{b}{2}\cot\frac{\beta}{2} \tag{7.12}$$

in which $d$ is the horizontal distance between the two points, $b$ the calibrated length of the subtense bar, and $\beta$ the observed angle subtended by the two targets at the ends of the subtense bar. Normally, subtense bars are 2 m in length, so that $b = 2$ m.

The angle $\beta$ must be measured a sufficient number of times to yield accurate values for the distance $d$. Assuming that $\sigma_\beta = 01''$ and $\sigma_b = 0.0002$ m, the propagated estimated standard deviation in the distance $d$ is calculated as follows

$$\sigma_d^2 = \left(\frac{\partial d}{\partial b}\right)^2 \sigma_b^2 + \left(\frac{\partial d}{\partial \beta}\right)^2 \sigma_\beta^2 \tag{7.13}$$

in which

$$\left(\frac{\partial d}{\partial b}\right) = \frac{d}{b} \quad \text{and} \quad \left(\frac{\partial d}{\partial \beta}\right) = -\frac{b}{4\sin^2(\beta/2)} \tag{7.13a}$$

so that for a distance of 50 m for which $\beta/2 = \arctan(1/50) = 1°08'45''$,

$$\sigma_d^2 = \left(\frac{50}{2}\right)^2 (0.0002)^2 + \left[\frac{2}{(4)\sin^2 1°08'45''}\right]^2 (4.84 \times 10^{-6})^2$$
$$= 0.000025 + 0.00003662 = 0.00006$$
$$\sigma_d = 0.008 \text{ m}$$

Listed in Table 7.1 are estimated standard deviations for various distances measured by subtense bar. Note that the relative accuracy in the distance $d$ decreases rapidly as the distance increases. With distances over 400 m or about 1300 ft, the maximum relative accuracy possible is about 1 part in 1000.

If 12 repetitions with a first-order theodolite were made to observe $\beta$, a value of $\sigma_\beta = 0.6''$ is possible. Propagation of this value through Eq. (7.13) for a distance of 150 m leads to $\sigma_d = 0.036$ or a relative accuracy of 1 part in 4000 can be maintained.

Should the vertical distance between occupied points also be desired, the vertical angle $\alpha$ can be observed to the subtense bar. The difference in elevation between the center of the telescope and the center of the subtense bar is

$$V = d\tan\alpha \tag{7.14}$$

in which $V$ is the difference in elevation, $d$ the horizontal distance computed from Eq. (7.12), and $\alpha$ the vertical angle.

**Table 7.1  Estimated Standard Deviations in Distances by Subtense Bar**

| Distance, m | $\sigma_d$, m | Relative accuracy, $\sigma_d/d$ |
|:---:|:---:|:---:|
| 50 | 0.008 | 1:6250 |
| 100 | 0.026 | 1:3846 |
| 150 | 0.057 | 1:2630 |
| 200 | 0.099 | 1:2020 |
| 300 | 0.220 | 1:1364 |
| 400 | 0.390 | 1:1026 |
| 500 | 0.608 | 1:822 |
| 600 | 0.875 | 1:686 |

**7.19. Plane table and alidade**  The surveying system made up of the *plane table* and *alidade* consists of (1) a drawing board mounted on a tripod, and (2) an alidade having the vertical plane of the line of sight fixed parallel to a straightedge that rests upon but is not attached to the board (Fig. 7.13). A sheet of drawing paper, called a *plane-table sheet*, is fastened to the board.

The location of any object is determined as follows (Fig. 7.13*a*). With the straightedge through the plotted point *o* representing the station *O* occupied by the instrument, the line of sight is directed to the object *A*, and a line *oa* of indefinite length is drawn along the straightedge on the plane-table sheet; this line represents the direction from station to object. The measured distance *OA* between station and object is then plotted to scale from *o*, thus locating *A* on the map at *a*.

The term "plane table" is somewhat ambiguous, being used sometimes to designate only the board with its supporting tripod and sometimes (more generally) both the table proper and its accompanying alidade.

By means of the plane table, points on the ground to which observations are made can be plotted immediately in their correct relative positions on the drawing, all angles being plotted graphically. The plane-table method is especially adapted to securing the details of the map. On extensive surveys the primary points of horizontal and vertical control are generally established by other more precise methods. It is a valuable and commonly used means of completing the compilation of maps from aerial photographs (Chap. 16), especially where the ground is hidden by overgrowth.

It is helpful to consider the similarity between angular measurements made with the transit, together with the office procedure of plotting the notes, and the corresponding operations with the plane table.

The plane-table board may be said to correspond to the graduated horizontal circle of the transit. After the board is oriented, the direction of any line passing through the instrument station is observed by turning the alidade about the plotted location of the instrument station until the line of sight coincides with the line (just as with the transit, the upper motion is rotated); the direction of the line is given graphically by the position of the straightedge on the paper (just as with the transit, the numerical value of the azimuth is given by the vernier reading). The corresponding line on the paper is then established by drawing a line along the straightedge and by laying off to scale the measured length of the line on the ground. Thus, with the plane table, a combination of transit and drafting-room

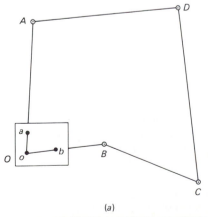

(*a*)    (*b*)

**Fig. 7.13**  (*a*) Plotting positions of points on a plane table. (*b*) Plane table and alidade.

(a)

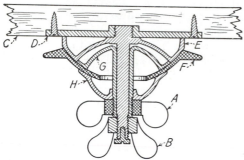

**Fig. 7.14** (a) Johnson head for the plane table. (*Keuffel & Esser Co.*). (b) Cross section of the Johnson head.

methods is employed, but no record of numerical values is secured. The plane table is therefore useful for mapping only.

The *plane table* is mounted on the tripod by means of a Johnson head (Fig. 7.14), which permits the table to be rotated in azimuth and leveled without disturbing the tripod. The drawing board is 18 by 18 in (45.7×45.7 cm), 18 by 24 in (45.7×61.0 cm), or 24 by 31 in (61.0×78.7 cm). Into the underside of the board is fastened a circular brass plate, shown as *D* in Fig. 7.14*b*. By means of the threaded opening in the plate, the board can be screwed to the upper casting *E* of the tripod head. The head comprises a ball-and-socket joint and a vertical spindle; the cup *F* is supported by the tripod (not shown). When clamp *A* is loosened, the grip of parts *G* and *H* on the cup is released, and thus the board can be leveled. The clamp is then tightened to fix the board in a horizontal plane. When clamp *B* is loosened, the board can be rotated about the vertical axis and thus be oriented. The plane-table sheet is held in position by countersunk screws in the top of the board or by drafting tape placed along the edges of the sheet.

There are other styles of plane-table mounts. One is illustrated in Fig. 7.15, where the plane table is mounted on a heavy metal casting, so arranged that the board can be leveled by means of three leveling screws. A clamp and tangent-screw permit the board to be rotated to any position in azimuth.

**Telescopic alidade**   The telescopic alidade is designed to provide precision in the control of the table, especially to make possible the stadia method of measuring distances. The base, or plate, of the alidade consists of a brass ruler or straightedge beveled on one edge and

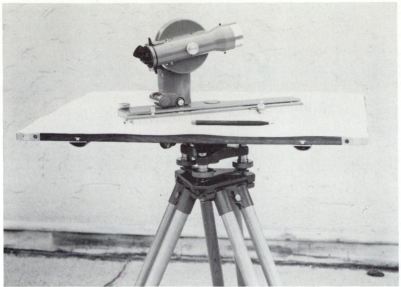

**Fig. 7.15** Plane table with European-style Wild RK1 self-reducing alidade. (*Wild Heerbrugg Instruments, Inc.*)

chrome-plated on the bottom. Upon one end of the plate is mounted either a circular level or a pair of level tubes at right angles to each other. Upon the other end of the plate is mounted a trough compass consisting of a magnetic needle mounted in a narrow box with a short graduated arc at the end. In the center of the plate is mounted a column which supports a telescope similar to that of the transit.

Most alidades of modern design are equipped with an enclosed glass vertical circle which is observed by means of a microscope. The glass circle is graduated according to the stadia equations so that $H$ and $V$ readings (similar to the Beaman arc, Art. 7.11) are read directly and can be easily converted to horizontal distance and difference in elevation. A vertical circle that can be read directly to 30″ is also visible. Figure 7.16 shows a *self-indexing* alidade (of the American style) in which a damped pendulum automatically brings the index of the vertical arc to the correct scale reading even though the board is not quite level. The telescope is internal focusing; therefore, the stadia constant $f + c$ is zero. The reticule is fixed in the telescope, and the telescope is fixed to its axis; thus, neither the line of sight nor the axis of the telescope needs adjustment. No striding level or collars are required. The design is simple, and most parts are enclosed. A telltale circle on the bull's-eye level indicates when the blade is tilted beyond the range of the pendulum. A detachable elbow eyepiece is provided. The only field adjustment is to zero the index of the vertical arc when the line of sight is horizontal, as follows. With the line of sight on the rod reading established for a horizontal line by means of the two-peg test (Art. 5.58, adjustment 3), the index is brought to the zero mark (or equivalent scale graduation) by means of a capstan-head screw which controls its setting.

Figure 7.17 shows the field of view visible when looking through the reading microscope of the alidade shown in Fig. 7.16. Three sets of graduations are seen: the zenith angle of 103°42′; the $V$ reading [an optical version of the Beaman stadia arc (Art. 7.11)] which is 27.0, from which 50 (the $V$ reading when the telescope is horizontal) must be subtracted to yield −23.0 that must be multiplied by the stadia interval to get the difference in elevation between the center of the telescope and the point sighted on the rod; and the $H$-scale

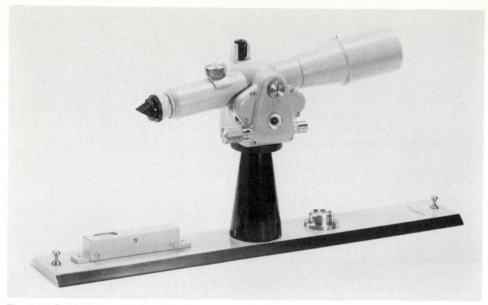

**Fig. 7.16** Self-indexing alidade of the American style. (*Keuffel & Esser Co.*)

multiplier, which is 94.4, multiplied by the stadia interval to give horizontal distance. If the stadia interval is 1.431 m, $V = (-23.0)(1.431) = -32.9$ m and $H = (94.4)(1.431) = 135.1$ m.

When using this style of self-indexing alidade, the instrument operator sights on the rod; observes $s$, the stadia interval; sets to the nearest even $V$ scale reading; reads $H$; and observes where the center cross hair cuts the rod. The values of $s$, $V$, and $H$ and the rod reading are recorded and the horizontal distance and difference in elevation are calculated.

Another style of alidade with optical reading of the $H$, $V$ scales and vertical angle but without the automatic self-indexing is shown in Fig. 7.18. This alidade is equipped with a vertical clamp and tangent-screw to control movement of the telescope and another tangent-screw to control a spirit level bubble attached to the vertical circle. When this bubble is centered, the vertical circle index coincides with the horizontal. The telescope contains stadia hairs with $K = 100$ and is internally focusing so that $f + c$ equals zero. The operation of this instrument is similar to that of the self-indexing instrument with one exception: prior to reading or setting the $V$ scale and observing the rod, the vertical circle

Vertical scale reads 27
Vertical multiplier
$= 27 - 50 = -23$
Zenith angle $= 103°42'$
Horizontal multiplier
$= 94.4$

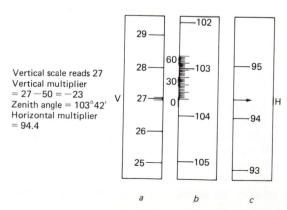

**Fig. 7.17** View through scale-reading eyepiece of an American-style self-index-ing alidade. (*Keuffel & Esser Co.*)

**Fig. 7.18** Telescopic microptic alidade.

index bubble must be centered. Figure 7.19 illustrates the circle readings visible when looking through the microscope of this style of alidade.

Note that for the vertical angle and $V$ scale, there is a sign visible in the field of view. These signs appear for angular values of up to $\pm 3°$ and up to the $V$ graduation of $\pm 5$. Beyond these graduations, the tilt of the telescope is sufficient so that the operator can easily tell whether the inclination is plus or minus.

For this example, $V = +6$, the vertical circle $= +3°26'$, and $H$ is estimated as 0.4. With a stadia interval of 0.348 m, the difference in elevation between the center of the telescope and the point sighted is $(6)(0.348) = +2.09$ m and the horizontal distance is $[100 - (0.4)](0.348) = 34.7$ m. Assuming the H. I. of the alidade to be 31.22 m, the middle horizontal cross hair on 1.25 m with $V$ on $+6$, and the vertical circle index bubble centered, the elevation of the point $= 31.22 + 2.09 - 1.25 = 32.06$ m.

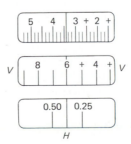

**Fig. 7.19** Circle readings on a microptic alidade.

A European-style self-reducing alidade is shown in Fig. 7.15. This alidade contains cross hairs with intervals which vary with the inclination of the telescope. The stadia intervals, when multiplied by a constant that appears in the field of view, provide horizontal distance and difference in elevation directly. The principle is similar to that described in Art. 7.17 for self-reducing tacheometers.

**7.20. Plane-table sheet** Since the plane-table sheet is exposed to outdoor conditions, specially prepared papers are required to avoid undue expansion or shrinkage. Only the best drawing papers should be used. The paper can be seasoned, that is, rendered more resistant to changes in humidity of the air, by exposing it alternately to very moist and very dry atmospheres for a number of cycles.

The drawing paper can be mounted on muslin or on each side of a sheet of muslin with the grain of the paper of one sheet laid transverse to the grain of the other. These forms of plane-table sheet are excellent but are not sufficiently flexible to be rolled under the edge of the plane-table board if a sheet larger than the board is desired. For accurate work, such as graphical triangulation, the drawing paper may be mounted on a thin aluminum sheet. A sheet of celluloid with roughened surface is sometimes used for work in light rains; the details thus plotted are later transferred to the regular sheet. Cellulose-acetate or drafting film sheets are used in compilation of data from and in connection with aerial photographs.

If the plotting is to be continued for several days, the map sheet is protected by a cover sheet of tough paper which is torn away piece by piece to expose the map sheet as the work progresses.

Sharply pointed, hard (6H to 9H) pencils are used for drawing lines and plotting details, and a fine needle is used for plotting control stations. Special care should be taken not to smear the drawing; the alidade should be lifted instead of slid into position.

**7.21. Setting up and orienting the plane table** The plane table is set up approximately waist-high, so that the topographer can bend over the board without resting against it. The tripod legs are spread well apart and planted firmly in the ground. The board is leveled by centering the circular bubble on the blade of the alidade. When the board is level, the upper locking clamp is tightened. Since few tables are sufficiently rigid to remain level as the alidade is shifted about, no special attempt is made to see that the board is perfectly level each time an observation is made.

For plotted angles to be theoretically correct, the plotted location of the station at which the plane table is set should be exactly over the corresponding point on the ground. Practically, the degree of care exercised in bringing the plotted point over the ground point depends upon the scale of the map. For map scales smaller than perhaps 1:600, the plane table is set up over the station without any attempt to place the plotted point vertically above the station point. For maps of larger scale, the table is set up roughly and oriented approximately, and then it is shifted bodily until the point on the paper is practically over the station point, as indicated by plumbing. Either a hook-shaped plumbing arm may be used to support the plumb line under the plotted point, or the plumb line and point may be sighted from two directions approximately at right angles to each other. In any case, the aim is to set up with sufficient care so that the plotted position of lines drawn from the station will be shown correctly within the scale of the map.

The table may be oriented (1) by use of the *magnetic compass*, (2) by *backsighting*, or (3) by solving the *three-point problem* (Arts. 9.9 and 9.10). As soon as the table is oriented, it is clamped in position and all mapping at the station is carried on without disturbing the board.

**1. Orientation by compass**  For rough mapping at small scale, orientation by the magnetic compass is sufficiently accurate. This method is susceptible to the same errors as those encountered when using the surveyor's compass, but an error in the plotted direction of one line introduces no systematic errors in the lines plotted from succeeding stations.

If the compass is fixed to the drawing board, the board is oriented by rotating it about the vertical axis until the fixed bearing (usually magnetic north) is observed. If the compass is attached to the alidade or to a movable plate, the edge of the ruler or the plate is aligned with a meridian previously established on the plane-table sheet, the board is turned until the needle reads north and the lower locking clamp $B$ (Fig. 7.14$b$) is tightened.

**2. Orientation by backsighting**  For mapping at intermediate or large scale, the board is oriented by backsighting along an established line, the direction of which has been plotted previously but the length of which need not be known. The method is equivalent to that employed in azimuth traversing with the transit. Greater precision is obtainable than with the compass, but an error in direction of one line is transferred to succeeding lines.

The plane table is set up as at $B$ (Fig. 7.20) on the line $AB$, which has previously been plotted as $ab$, the board is leveled, the straightedge of the alidade is placed along the line $ba$, and the board is oriented by rotating it until the line of sight falls at $A$. Finally, the lower clamp $B$ (Fig. 7.14$b$) is tightened. The length of $AB$ or $ab$ need not be known. Preferably the longest line available should be used for precision in orienting.

When the straightedge of the alidade is placed along a previously plotted line passing through the plotted location of the station occupied and that of another station or object (such as line $ab$, Fig. 7.20), and then the board is turned until the line of sight cuts the station or object, the sight is called a *backsight*. When a station or object is sighted and a ray is drawn through the plotted location of the station occupied toward the station or object sighted, the sight is called a *foresight*. Line $bc$, the plotted location of line $BC$ in Fig. 7.20, is an example of a foresight.

When a known station is sighted and a line is drawn through the plotted location of that station toward the station occupied, such as line $be$ (Fig. 7.20), the sight is called a *resection*. Resection is also a general term applied to the process of determining the location of the station occupied (Art. 9.9).

When the table has been oriented, the direction to any object in the landscape may be drawn on the map by pivoting the alidade about the plotted location of the plane-table station, pointing the alidade toward the distant object, and drawing a line along the straightedge.

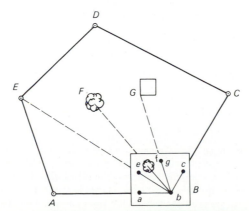

**Fig. 7.20** Plane-table setup.

Thus, in Fig. 7.20 the plane table is shown in position over station $B$. The plotted location of the plane-table station is indicated at $b$ on the plane-table sheet. The alidade is pivoted about this point; and as sights are taken to points as $A$, $E$, and $C$, rays are drawn along the edge of the ruler.

Stadia data are recorded for each sight and are reduced to horizontal distance and difference in elevation to each point. As the distance to each point is calculated, the position to the point is plotted to scale along the corresponding ray, thus locating points such as $F$ and $G$ on the plane-table sheet at $f$ and $g$. This procedure is called *radiation*.

**7.22. Sources of error**    The major sources of error in plane-table work are the same as those which affect transit work and plotting, and the discussion relating to those subjects need not be repeated here. However, the following three sources of error should be considered:

**1. *Setting over a point***    Because plotted results only are required, it is not necessary to set the plotted location of the plane table over the corresponding ground point with any greater precision than is required by the scale of the map (see Art. 7.21).

**2. *Drawing rays***    The accuracy of plane-table mapping depends largely upon the precision with which the rays are drawn; consequently, the rays should be of considerable length. To avoid confusion, however, only enough of each ray is drawn to ensure that the plotted point will fall upon it, with one or two additional dashes drawn near the end of the alidade straightedge to mark its direction. Fine lines are desirable both for precision and for legibility.

**3. *Instability of the table***    If it is manipulated with care, the plane table can be oriented with considerable precision; however, a principal source of error in its use arises from the fact that its position is subject to continual disturbance by the topographer while work is progressing. Errors from this source can be kept within reasonable limits ($a$) by planting the tripod firmly in the ground, ($b$) by setting the table approximately waist-high so that the topographer can bend over it without leaning against it, ($c$) by avoiding undue pressure upon or against the table, and ($d$) by testing the orientation of the board occasionally and correcting its position if necessary. This test is always applied before a new instrument station is plotted.

**7.23. Applications for plane table and alidade**    The plane table and alidade are especially adapted for securing details for mapping. The use of the plane table for topographic mapping is discussed in Art. 15.11. It is also a valuable and commonly used method for field completion and checking of photogrammetrically compiled maps (Art. 16.17), especially where the ground is concealed by vegetation. The plane table can also be used for indirect leveling (Art. 7.14), graphical triangulation (Art. 10.24), graphical resection and intersection (Arts. 9.8 and 9.9), and for traversing (Art. 8.13).

**7.24. Problems**

**7.1**    To determine the stadia interval factor, a transit is set up at a distance $C$ back of the zero end of a level base line 800.0 ft long, the base line being marked by stakes set every 100.0 ft. A rod is then held at successive stations along the base line. The stadia interval and each half-interval observed at each location of the rod are tabulated below. Compute the lower, upper, and full interval factor for each distance, and find the average value for the lower interval, the upper interval, and the full interval.

| Distance | Lower interval | | Upper interval | | Full interval | |
|---|---|---|---|---|---|---|
| − C, ft | Feet | Factor | Feet | Factor | Feet | Factor |
| 100 | 0.49 | | 0.50 | | 0.99 | |
| 200 | 0.98 | | 0.99 | | 1.97 | |
| 300 | 1.47 | | 1.48 | | 2.95 | |
| 400 | 1.97 | | 1.98 | | 3.96 | |
| 500 | 2.46 | | 2.47 | | 4.94 | |
| 600 | 2.95 | | 2.97 | | 5.92 | |
| 700 | 3.45 | | 3.47 | | 6.91 | |
| 800 | 3.94 | | 3.96 | | 7.89 | |
| Average | | | | | | |

**7.2**   A stadia interval of 2.415 m is observed with a theodolite for which the stadia interval factor is 100.0 and $C$ is 0.305 m. The zenith angle is 98°10′30″ with the middle cross hair set on 2.000 m. If the instrument has a height of instrument h.i. of 1.38 m above the point over which it is set and the point has an elevation of 100.05 m, calculate the horizontal distance and elevation of the point sighted by the (a) exact stadia equations; (b) approximate stadia equations.

**7.3**   The following observations are taken with a transit for which the interval factor is 100.0 and $C$ is 1.00 ft:

| Observation | Stadia interval | Vertical angle |
|---|---|---|
| | (i) | |
| (a) | 10.00 ft | +0°30′ |
| (b) | 10.00 ft | +10°00′ |
| (c) | 10.00 ft | +25°00′ |
| | (ii) | |
| (d) | 3.000 m | +0°40′ |
| (e) | 3.000 m | +10°10′ |
| (f) | 3.000 m | +25°10′ |

By means of Eqs. (7.4) and (7.5), Art. 7.7, compute the horizontal distance and differences in elevation. By means of the approximate equations (7.7) and (7.8), Art. 7.8, determine the same quantities and note the errors introduced by the approximations.

**7.4**   What would be the amount and sign of error introduced in each computed horizontal distance and difference in elevation if the observations of Prob. 7.3(i) were taken (a) with a 12-ft rod that was unknowingly 0.5 ft out of plumb with top leaning toward the transit? (b) With the top of the rod leaning 0.5 ft away from the transit? (c) What conclusions may be drawn from these results?

**7.5**   What error will be introduced in each computed horizontal distance and difference in elevation if in the observations of Prob. 7.3, parts (i) or (ii), (a) each vertical angle contains an error of 01′?; (b) each stadia interval is in error by 1/1000 of the interval? (c) What conclusions may be drawn from these results?

**7.6**   In determining the elevation of point B and the distance between two points A and B, a transit equipped with a stadia arc is set up at A and the following data are obtained: V = 38, H = 3.0, stadia interval = 1.311 m, h.i. = 1.28 m, and the line of sight at 2.62 m on rod. The instrument constants are K = 100.0 and C = 0.305 m. The stadia arc has index marks of H = 0 and V = 50 for a horizontal line of sight. The elevation of point A is 38.28 m. Compute the distance AB and the elevation of point B.

**7.7**   Following are the notes for a line of stadia levels. The elevation of B.M.$_1$ is 637.05 ft. The stadia interval factor is 100.0 and C = 1.25 ft. Rod readings are taken at height of instrument. Determine the elevations of remaining points. Record notes and elevations in a proper note form.

| Station | Backsight | | Foresight | |
| --- | --- | --- | --- | --- |
| | Stadia interval, ft | Vertical angle | Stadia interval, ft | Vertical angle |
| B.M.$_1$ | 4.26 | $-3°38'$ | | |
| T.P.$_1$ | 2.85 | $-1°41'$ | 3.18 | $+2°26'$ |
| T.P.$_2$ | 3.30 | $+0°56'$ | 2.71 | $-4°04'$ |
| T.P.$_3$ | 2.26 | $+2°09'$ | 4.45 | $-0°38'$ |
| B.M.$_2$ | | | 3.09 | $+7°27'$ |

**7.8**   Notes for a line of stadia levels are shown below.

| Station | Backsight | | | Foresight | | |
| --- | --- | --- | --- | --- | --- | --- |
| | Stadia interval, m | Vertical angle | Rod, m | Stadia interval, m | Vertical angle | Rod, m |
| B.M.$_{10}$ | 1.241 | $-4°24'$ | 2.41 | | | |
| T.P.$_1$ | 2.041 | $+3°23'$ | 1.05 | 1.506 | $+8°10'$ | 1.52 |
| T.P.$_2$ | 1.721 | $-6°30'$ | 1.45 | 1.432 | $-3°05'$ | 3.05 |
| T.P.$_3$ | 2.421 | $-10°05'$ | 2.04 | 0.901 | $+10°08'$ | 0.24 |
| B.M.$_{11}$ | | | | 1.321 | $+4°20'$ | 1.43 |

The elevation of B.M.$_{10}$ is 42.852 m. The stadia interval factor is 100.0 and $C=0.31$ m. Determine elevations for the turning points and B.M.$_{11}$. Record notes and elevations in a proper note form.

**7.9**   Following are stadia intervals and vertical angles for a transit-stadia traverse. The elevation of station $A$ is 418.6 ft. The stadia interval factor is 100.0 and $C=1.00$ ft. Rod readings are taken at height of instrument. Compute the horizontal lengths of the courses and the elevations of the transit stations.

| Station | Object | Stadia interval, ft | Vertical angle |
| --- | --- | --- | --- |
| B | A | 8.50 | $+0°48'$ |
| | C | 4.37 | $+8°13'$ |
| C | B | 4.34 | $-8°14'$ |
| | D | 12.45 | $-2°22'$ |
| D | C | 12.41 | $+2°21'$ |
| | E | 7.18 | $-1°30'$ |

**7.10**   Following are stadia intervals and vertical angles taken to locate points from a transit station the elevation of which is 415.7 ft. The height of instrument above the transit station is 4.6 ft, and rod readings are taken at 4.6 ft except as noted. The stadia interval factor is 100.0 and $C=1.00$ ft. Compute the horizontal distances and the elevations.

| Object | Stadia interval, ft | Vertical angle |
| --- | --- | --- |
| 43 | 7.04 | $-0°58'$ |
| 44 | 8.25 | $+2°30'$ on 2.1 |
| 45 | 7.56 | $-0°44'$ on 9.2 |
| 46 | (7.25) | $+1°20'$ on 6.0 |
| 47 | 3.72 | $-5°36'$ |

**7.11**   A transit equipped with a stadia arc is used in locating points from a transit station the elevation of which is 233.39 m. The stadia arc has index marks of $H=100$ and $V=50$ for a horizontal line of sight. The instrument constants are $K = 100.0$ and $C = 0$ (internal-focusing telescope). The height of instrument above the transit station is 1.37 m. Compute the horizontal distances and the elevations.

| Object | Stadia interval, m | Rod reading, m | Stadia arc readings | |
|---|---|---|---|---|
| | | | V | H |
| 114 | 0.994 | 1.10 | 18 | 88.3 |
| 115 | 2.390 | 1.77 | 35 | 97.7 |
| 116 | 0.664 | 1.43 | 39 | 98.8 |
| 117 | 0.506 | 1.31 | 76 | 92.6 |
| 118 | 2.481 | 1.95 | 69 | 96.2 |

**7.12** A stadia interval of 2.341 m and vertical angle of $12°45'$ are observed between points $C$ and $D$. The stadia interval factor is 100.0 and $C = 0.31$ m. Calculate the horizontal distance and difference in elevation between points $C$ and $D$ using the exact stadia equations. If $\sigma_s = 0.005$ m, $\sigma_\alpha = 01'$, $\sigma_K = 0.1$, and $\sigma_C = 0.01$ m, determine the estimated standard deviation in the calculated distance and difference in elevation. Comment on the relative importance of uncertainties in $s$, $\alpha$, $K$, and $C$.

**7.13** A transit is set over station 10 and stadia observations are taken on point $B$. The h.i. $= 4.50$ ft, the stadia interval, $s = 4.24$ ft, $\alpha = +10°00'$, $K = 100$, $C = 0.0$, $\sigma_s = 0.02$ ft, $\sigma_{h.i.} = 0.02$ ft, and $\sigma_\alpha = 05'$. The elevation of 10 is 532.45 ft above mean sea level and $K = 100$, assumed errorless. Compute the horizontal distance from 10 to $B$, the elevation of $B$, and their respective standard deviations.

**7.14** The following data are used to determine the elevation of a photo-control point from a bench mark by stadia: the backsight on B.M.$_{10}$ yields $s = 1.234$ m, $\alpha = -5°10'$, and rod$_B = 1.829$ m; the foresight on point $P1$ yields $s = 1.201$ m, $\alpha = -2°20'$, and rod$_F = 1.524$ m. The constants $K = 100$ and $C = 0.0$. (a) Compute the elevation of $P1$. (b) If $\sigma_s = 0.005$ m, $\sigma_\alpha = 01'$, $\sigma_{rod} = 0.005$ m, determine the standard deviation for the elevation of $P1$ calculated in (a).

**7.15** For the data given in Fig. 7.7 (stadia leveling), the elevation of B.M.$_{42}$ and the constant $K$ are assumed errorless. If $\sigma_s = 0.01$ ft, $\sigma_\alpha = 01'$, and $\sigma_{rod} = 0.01$ ft, compute the standard deviation for the elevation of T.P.$_3$.

## References

1. Allan, A. L., "An Alidade for Use with Electro-Optical Distance Measurement," *Survey Review*, July 1977.
2. Asplund, E. C., "A Review of Self Reducing, Double Image Tacheometers," *Survey Review*, January 1971.
3. Moffitt, F. H., and Bouchard, H., *Surveying*, 6th ed., Harper & Row, Publishers, New York, 1975.
4. Instruction Manual, *Self-reducing Engineer's Tachymeter, K1-RA*, Kern & Co. Ltd., Aaru, Switzerland.
5. Wolf, P. R., Wilder B. P., and Mahun, G., "An Evaluation of Accuracies and Applications of Tacheometry," *Surveying and Mapping*, Vol. 38, No. 3, September 1978.

# PART III
# SURVEY OPERATIONS

# CHAPTER 8
# Traverse

**8.1. General** A *traverse* consists of a series of straight lines connecting successive established points along the route of a survey. The points defining the ends of the traverse lines are called *traverse stations* or *traverse points*. Distances along the line between successive traverse points are determined either by direct measurement using a tape or electronic distance measuring (EDM) equipment, or by indirect measurement using tacheometric methods. At each point where the traverse changes direction, an angular measurement is taken using a transit or theodolite. With minor modifications, these practices are common to land, city, topographic, and hydrographic surveys and to location surveys such as those for highways and railways.

**8.2. Traverse party** The traverse party is usually composed of an instrument operator, a head tapeman, and a rear tapeman. The transit operator directs the party, operates and cares for the instrument, and takes notes. The head tapeman performs the duties of that position as described in Arts. 4.10 to 4.12, gives line as directed by the transit operator, is responsible for the accuracy and speed of taping operations, and attends to the setting and marking of stakes. The rear tapeman carries out the duties of that position as described in Arts. 4.10 to 4.12, often carries and drives stakes, and assists in removing obstructions to the vision of the transit operator. In heavily wooded terrain, additional employees are engaged to clear the transit line of trees and brush. The head tapeman keeps in close communication with those clearing the line and assists in their direction. Where sights are long, a flagman may be employed to give sights to the transit operator. Where many observations are taken, the notes may be taken by a recorder, who may also act as the chief of party.

**8.3. Equipment for the traverse party** The equipment of the party usually consists of a transit or theodolite, steel tape of desired length, or EDM device plus supporting equipment: range poles and plumb bobs, stakes and hubs, tacks, axe or hammer, field notebook, taping pins, and marking crayon. Where *forced centering* (Art. 6.30) is to be employed, three tripods compatible with theodolite, sighting targets, and/or reflectors are required. Also included are devices for marking stations, such as ordinary or special nails, a cold chisel, spray paint, and colored plastic flagging or tape. A light tripod with plumb bob is useful for sighting on traverse stations where vision is obscured near the ground. Portable sending and receiving radios (such as citizens' band radios) may be used to expedite communication.

For stadia or tacheometric traverses (Art. 8.12), distance measuring equipment is omitted and a level or stadia rod is included in addition to the other items listed above.

**8.4. Traverse stations** Any temporary or permanent point of reference over which the transit is set up is called a *traverse station*. On most surveys the traverse station is a peg, called a *hub*, driven flush with the ground and having a tack driven in its top to mark the exact point of reference for measurements. On pavements the traverse station may be a driven nail, a cross cut in the pavement or curb, or a tack set in a hole drilled with a star drill and filled with lead wool. In land surveying, the stations are often iron pipes, stones, or other more or less permanent monuments set at the corners. In mountainous country, often the station marks are cut in the natural rock.

The location of a hub is usually indicated by a flat *guard stake* extending above the ground and driven sloping so that its top is over the hub. This guard stake carries the number or letter of the traverse station over which it stands. Usually, the number is marked with keel or lumber crayon on the underside of the guard stake and reads down the stake. The hubs are usually square, say 2 by 2 in (5 by 5 cm) and the guard stakes are usually flat, perhaps $\frac{3}{4}$ by 3 in (2 by 8 cm).

To avoid the necessity for using a rear flagman to give backsights, as far as practical each traverse station is marked by a temporary signal, such as a lath or a stick set on line, with a piece of paper, cloth, or tape attached. When forced centering is employed, sighting targets mounted on tripods are used instead.

Lines connecting traverse stations are called *traverse lines*. Both in the field notes and as part of the identification mark left in the field, a traverse station number is preceded by the symbol $\odot$.

**8.5. Purpose of the traverse** Traversing is a convenient, rapid method for establishing horizontal control. It is particularly useful in densely built up areas and in heavily forested regions where lengths of sight are short so that neither triangulation nor trilateration is suitable. Traverses are made for numerous purposes, including:

1. Property surveys to locate or establish boundaries.
2. Supplementary horizontal control for topographic mapping surveys.
3. Location and construction layout surveys for highways, railways, and other private and public works.
4. Ground control surveys for photogrammetric mapping.

Frequently, the traverse is employed to densify the number of coordinated control points in a triangulated or trilaterated horizontal control network.

**8.6. Types of traverse** The two general classes of traverse are the open traverse and the closed traverse.

The *open traverse* originates at a point of known position and terminates at a point of unknown position. No computational check is possible to detect errors or blunders in distances or directions in this type of traverse. To minimize errors, distances can be measured twice, angles turned by repetition, magnetic bearings observed on all lines, and astronomic observations made periodically. In spite of these precautions, an open traverse is a risky proposition and should not be used for any of the applications discussed in Art. 8.5, since the results are uncertain and will always be subject to question. Open traverses are necessary in certain types of mine surveying (Chap. 20). Hence, there are applications of an open traverse when circumstances dictate their use.

*Closed traverses* originate at a point of known position and close on another point of known horizontal position. A point of known horizontal position is given either by geographic latitude and longitude or by $X$ and $Y$ coordinates on a rectangular coordinate grid system. In Fig. 8.1, traverse $MABCDQ$ originates at known point $M$ with a backsight along line $MN$ of known azimuth and closes on known point $Q$, with a foresight along line $QR$ also of known azimuth. This type of traverse is preferable to all others since computational checks are possible which allow detection of systematic errors in both distance and direction.

A traverse that originates and terminates on a single point of known horizontal position is called a *closed-loop traverse*. In Fig. 8.2, traverse 1-2-3-4-5-6-1, which originates and

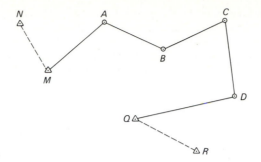

**Fig. 8.1** Closed traverse.

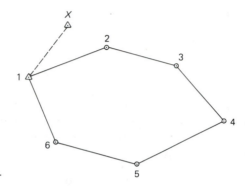

**Fig. 8.2** Closed-loop traverse.

closes on point 1, is an example of a closed-loop traverse. This type of traverse permits an internal check on the angles, but there is no way to detect systematic errors in distance or errors in the orientation of the traverse. A closed-loop traverse should not be used for major projects. To minimize the chance of errors in a closed-loop traverse, all distance measuring equipment must be carefully calibrated and astronomic observations must be made periodically.

A closed-loop traverse that originates and terminates on a point of assumed horizontal position provides an internal check on angles but no check on systematic errors in distance. An additional disadvantage is that the points are not located on any datum and have no relationship with points in any other survey, although the relative positions of the points are determined. The use of such a method of running a traverse is to be discouraged except under extenuating circumstances, where no feasible alternative is possible.

Each of the types of traverse discussed above can also be described according to the method of turning the angles in the traverse. Thus, there are traverses by deflection angles, interior angles, angles to the right, and the azimuth method. Each of these methods for running traverse is discussed in the following articles.

**8.7. Deflection-angle traverse**  This method of running traverses is probably more commonly employed than any other, especially on open traverses where only a few details are located as the traverse is run. It is used almost entirely for the location surveys for roads, railroads, canals, and pipelines. It is employed to a less extent in land surveying and in establishing control traverses for topographic and hydrographic surveys.

Successive transit stations are occupied, and at each station a backsight is taken with the *A* vernier set at zero and the telescope reversed. The telescope is then plunged, the foresight is taken by turning the instrument about the vertical axis on its upper motion, and the

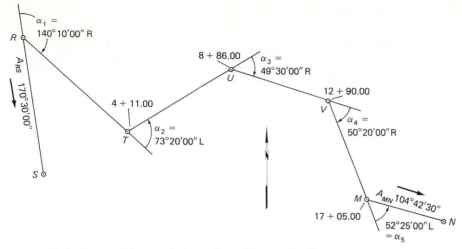

**Fig. 8.3** Deflection-angle traverse between lines of known direction.

deflection angle is observed. The angle is recorded as right R or left L, according to whether the upper motion is turned clockwise or counterclockwise. Usually, it is considered good practice to observe the deflection angle at least twice, once with the telescope direct and once reversed as described in Art. 6.42. This process is called turning the angle by double deflection.

Figure 8.3 shows a closed deflection angle traverse that originates at point $R$ at one end of line $RS$ having a known azimuth of $170°30'00''$ as determined from a previous survey. The traverse closes on point $M$ at one end of line $MN$ which has an azimuth of $104°42'30''$ also determined from a previous survey.

A portion of the field notes for the traverse is illustrated in Fig. 8.4. Angles are recorded on the left with the necessary details sketched on the right. All data are recorded from the bottom to the top of the page, a procedure that is also followed in preliminary route surveys. Angles are turned by double deflection and magnetic bearings are observed forward and backward from each traverse station as a rough check on the angles. To check the angular closure, azimuths are calculated from known line $RS$ to known line $MN$.

To compute the azimuth of an unknown line, a deflection angle to the *right* is added to the *forward azimuth* of the previous line. A deflection angle to the *left* is subtracted from the forward azimuth of the previous line. Thus, the azimuth of line $RT$ is $350°30'00'' + 140°10'00'' - 360°00'00'' = 130°40'00''$. Similarly, the azimuth of line $TU$ is $130°40'00'' - 73°20'00'' = 57°20'00''$. Figure 8.5 illustrates computation of azimuths using deflection angles to the left and right, respectively. The results of these computations for the example traverse in Fig. 8.3 are shown in Table 8.1. The process of computing azimuths for unknown lines is continued through lines $UV$, $VM$, to $MN$, where the calculated azimuth fails to agree with the fixed azimuth of $104°42'30''$ by $02'30''$, which is the angular closure for the traverse. The equation by which this error of closure may be computed for the example traverse is

$$A_{SR} + \alpha_1 - \alpha_2 + \alpha_3 + \alpha_4 - \alpha_5 - 360° = A_{MN} \qquad (8.1)$$

The general relationship for a traverse that contains $n$ deflection angles is

$$A_1 + \sum_{i=1}^{n} \alpha_{R_i} - \sum_{i=1}^{n} \alpha_{L_i} = A_2 + 360° \qquad (8.2)$$

in which $A_1$ is the forward azimuth at station 1, the origin of the traverse, and $A_2$ is the azimuth of station 2, the closing point for the traverse.

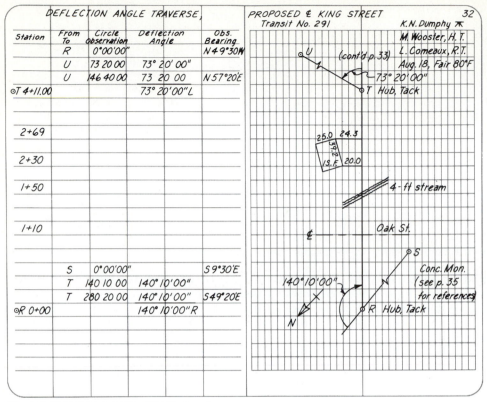

| Station | From To | Circle observation | Deflection Angle | Obs. Bearing |
|---|---|---|---|---|
| | R | 0°00'00" | | N 49°30'W |
| | U | 73 20 00 | 73° 20' 00" | |
| | U | 146 40 00 | 73  20  00 | N 57°20'E |
| ⊙T 4+11.00 | | | 73° 20'00"L | |
| | | | | |
| 2+69 | | | | |
| | | | | |
| 2+30 | | | | |
| | | | | |
| 1+50 | | | | |
| | | | | |
| 1+10 | | | | |
| | | | | |
| | S | 0°00'00" | | S 9°30'E |
| | T | 140 10 00 | 140° 10'00" | |
| | T | 280 20 00 | 140° 10'00" | S 49°20'E |
| ⊙R 0+00 | | | 140° 10'00"R | |

DEFLECTION ANGLE TRAVERSE,

PROPOSED ℄ KING STREET
Transit No. 291                                  32
                                    K.N.Dumphy ✗
                                    M.Wooster, H.T.
                                    L.Comeaux, R.T.
                                    Aug. 18, Fair 80°F

(cont'd p. 33)

73° 20'00"
⊙T Hub, Tack

25.0  24.3

15.F  20.0

4-ft stream

℄ ———— Oak St.

⊙S

Conc. Mon.
140°10'00"  (see p. 35
for references)
⊙R Hub, Tack

N

**Fig. 8.4** Closed traverse by deflection angles.

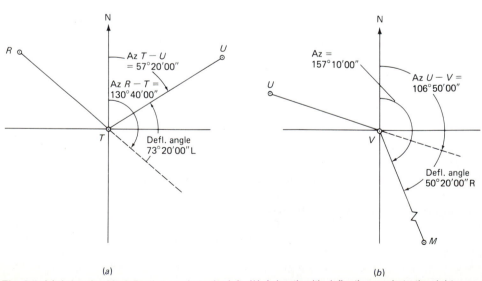

(a)                                              (b)
**Fig. 8.5** (a) Azimuth with deflection angle to the left. (b) Azimuth with deflection angle to the right.

**Table 8.1  Calculation and Adjustment of Azimuths Closed Traverse by Deflection Angles**

| From | To | Azimuth | Correction | Adjusted azimuth | Adjusted bearing |
|---|---|---|---|---|---|
| S | R | 350°30′00″ | | 350°30′00″ | N9°30′00″W |
| | + ∠R | 140°10′00″R | | | |
| | | 490°40′00″ | | | |
| | − | 360°00′00″ | | | |
| R | T | 130°40′00″ | −30″ | 130°39′30″ | S49°20′30″E |
| | − ∠T | 73°20′00″L | | | |
| T | U | 57°20′00″ | −1′00″ | 57°19′00″ | N57°19′00″E |
| | + ∠U | 49°30′00″R | | | |
| U | V | 106°50′00″ | −1′30″ | 106°48′30″ | S73°11′30″E |
| | + ∠V | 50°20′00″R | | | |
| V | M | 157°10′00″ | −2′00″ | 157°08′00″ | S22°52′00″E |
| | − ∠M | 52°25′00″L | | | |
| M | N | 104°45′00″ | −2′30″ | 104°42′30″ | S75°17′30″E |
| M | N | 104°42′30″ | | | |
| Angular error of closure = + 2′30″ | | | | | |

If adjustment of the observed angles to fit the known directions is desired (assuming equal weights for all angles), the closure error may be distributed equally among the five deflection angles. For this example 30″ will be subtracted from each deflection angle where angles to the right are positive and angles to the left are negative. When azimuths have already been calculated, as in this case, it is simpler to apply the cumulative correction directly to the unadjusted azimuth. Thus, the correction to the first azimuth from R to T is −30″, to the second from T to U is −01′00″, etc., as shown in Table 8.1. The same result is achieved using adjusted deflection angles to compute the azimuths.

Figure 8.6 illustrates a closed-loop traverse with deflection angles. This traverse originates and closes on traverse point A. The notes are recorded down the page in the order of occupying the points and a sketch is drawn on the right-hand page. Magnetic bearings have been observed forward and backward from each station. For any closed-loop traverse the initial azimuth, $A_1$ equals the closing azimuth $A_2$. Then from Eq. (8.2), the sum of the deflection angles should equal 360°. The sum of the average values of the angles is recorded at the bottom of the right-hand page, where a closure error of 02′30″ is indicated. If an adjustment is desired and assuming that angles are of equal weight, this error is distributed equally among the five angles. Since the left deflection angles have a sum that is too large, 30″ is subtracted from each left deflection angle and 30″ is added to the right deflection angle.

The next step is to calculate directions for each of the traverse lines. Since this is a closed-loop traverse and no angle was observed to a line of known direction, line AB will be assumed to have the observed bearing of S30°00′W which with 18°E declination (Art. 6.19) gives a true bearing of S48°00′W. This bearing is then converted to an azimuth of 228°00′00″ from the north for determination of azimuths for the balance of the lines. The procedure is similar to that employed for the previous example except that the closure check is on the original line AB, as shown in Table 8.2, which contains the calculation of the directions. Note that the closure is exact, as it should be, since the angles had been balanced previously. The directions, as determined for this example, are not absolute since they are based on a magnetic bearing that is only approximate. In practice, an angle should be turned to a line of known azimuth or an astronomic direction should be determined.

The arithmetical check of the sum of the angles should be performed immediately on completing the traverse in order to detect blunders or excessively large errors before leaving

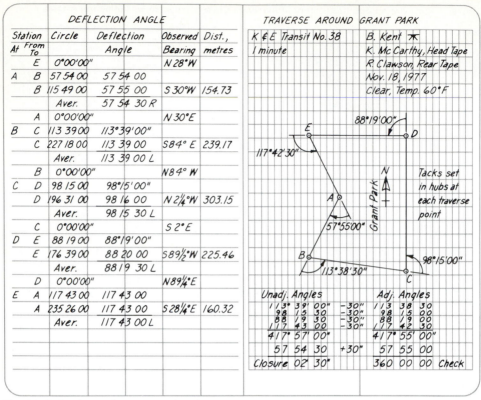

**Fig. 8.6** Deflection-angle traverse, closed loop.

**Table 8.2  Azimuths and Bearings in a Closed-Loop Traverse**

| From | To | Azimuth | Bearing |
|------|------|------|------|
| A | B | 228°00′00″ | S48°00′00″W |
|  | − ∠B | 113°38′30″L |  |
| B | C | 114°21′30″ | S65°38′30″E |
|  | − ∠C | 98°15′00″L |  |
| C | D | 16°06′30″ | N16°06′30″E |
|  | − ∠D | 88°19′00″L |  |
|  | + | 360°00′00″ |  |
| D | E | 287°47′30″ | N72°12′30″W |
|  | − ∠E | 117°42′30″L |  |
| E | A | 170°05′00″ | S9°55′00″E |
|  | + ∠B | 57°55′00″R |  |
| A | B | 228°00′00″  checks | S48°00′00″W |

the field. The angular error of closure should not exceed the estimated standard deviation for observing an angle from a single setup times the square root of the number of instrument stations. In practice, this estimate is usually taken as the least count of the instrument used to turn the angles.

**8.8. Interior-angle traverse**  Field operations for this method of traversing are not materially different from those used for the deflection-angle traverse. At each station the vernier is

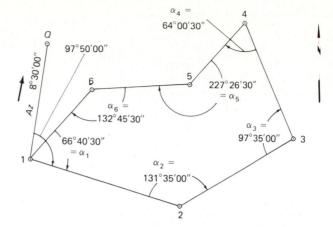

**Fig. 8.7** Closed-loop traverse
with interior angles observed.

set at zero, and a backsight to the preceding station is taken. The instrument is then turned
on its upper motion until the advance station is sighted and the interior angle is observed.
Except for surveys of a very low order of accuracy, all interior angles should be turned at
least twice, once with the telescope direct and once with the telescope reversed. Notes may
be kept in a form similar to that shown for deflection angles in Fig. 8.6.

A closed-loop traverse run by the method of interior angles is illustrated in Fig. 8.7. In this
traverse, point 1 was occupied first and an angle was observed between line $1Q$, which has a
known azimuth, and traverse line 12. After the clockwise interior angle at 1 was observed,
points 2, 3, 4, 5, and 6 were occupied with a clockwise interior angle being observed by
repetition (at least twice) at each traverse station.

Directions for the traverse are reckoned using the angle $Q12$ turned from the line of
known azimuth $1Q$ to line 12. Thus, the azimuth of line 12 is $8°30'00'' + 97°50'00'' =$
$106°20'00''$. The azimuth of each succeeding traverse line is then calculated by adding the
clockwise interior angle to the *back azimuth* of the preceding line. For example, the forward
azimuth of line $23 = 106°20'00'' + 180°00'00'' + 131°35'00'' - 360°00'00'' = A_{12} + \alpha_2 - 180° =$
$106°20'00'' + 131°35'00'' - 180° = 57°55'00''$, as illustrated in Fig. 8.8. Similarly, the azimuth
of line $34 = A_{23} + 180° + \alpha_3 = 57°55'00'' + 180° + 97°35'00'' = 335°30'00''$, etc., as illustrated in
Table 8.3.

For the example traverse, the angular error of closure can be checked by taking the sum
of the angles $\alpha_i$ in Table 8.3 or

$$\alpha_1 + \alpha_2 + \alpha_3 + \alpha_4 + \alpha_5 + \alpha_6 = 4(180°)$$

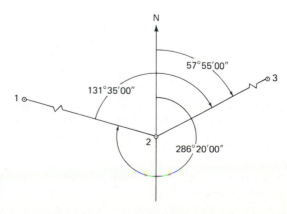

**Fig. 8.8** Calculation of an azimuth using
an interior angle.

**Table 8.3 Azimuths Using Interior Angles**

| Line | Angle | Unadjusted azimuth | Correction | Adjusted azimuth |
|------|-------|--------------------|------------|------------------|
| 12 | | 106°20′00″ | | 106°20′00″ |
| 21 | | 286°20′00″ | | |
| | $+\alpha_2$ | 131°35′00″ | | |
| 23 | | 57°55′00″ $=A_{12}-180°+\alpha_2$ | −30″ | 57°54′30″ |
| 32 | | 237°55′00″ | | |
| | $+\alpha_3$ | 97°35′00″ | | |
| 34 | | 335°30′00″ $=A_{23}+180°+\alpha_3$ | −01′00″ | 335°29′00″ |
| 43 | | 155°30′00″ | | |
| | $+\alpha_4$ | 64°00′30″ | | |
| 45 | | 219°30′30″ $=A_{34}-180°+\alpha_4$ | −01′30″ | 219°29′00″ |
| 54 | | 39°30′30″ | | |
| | $+\alpha_5$ | 227°26′30″ | | |
| 56 | | 266°57′00″ $=A_{45}-180°+\alpha_5$ | −02′00″ | 266°55′00″ |
| 65 | | 86°57′00″ | | |
| | $+\alpha_6$ | 132°45′30″ | | |
| 61 | | 219°42′30″ $=A_{56}-180°+\alpha_6$ | −02′30″ | 219°40′00″ |
| 16 | | 39°42′30″ | | |
| | $+\alpha_1$ | 66°40′30″ | | |
| 12 | | 106°23′00″ $=A_{61}-180°+\alpha_1$ | −03′00″ | 106°20′00″ |
| 12 | | 106°20′00″ | | |
| | Angular error of closure = 03′00″ | | | |

In general, for any polygon with $n$ sides, the sum of the interior angles is given by

$$\alpha_1 + \alpha_2 + \cdots + \alpha_n = (n-2)(180°) \tag{8.3}$$

Thus, the most rapid and simple field check of observed interior angles is to calculate the sum of the observations and compare with $(n-2)180°$.

Table 8.4 shows the sum of the observed interior angles for the example traverse. The angular error of closure is 03′, the same value as determined previously.

If adjusted angles are desired at this point, then assuming that all angles were observed with equal precision, the error is distributed equally among the angles and $180″/6=30″$ is subtracted from each observed angle, as shown in Table 8.4.

The simplest way to compute adjusted azimuths is to apply a cumulative correction to the unadjusted azimuths in Table 8.3. Thus, corrections of $30″$, $01′00″$, $01′30″$, $02′00″$, $02′30″$, and $03′00″$ are applied respectively to $A_{23}, A_{34}, \ldots, A_{61}$. These corrections and adjusted

**Table 8.4 Checking and Adjustment of Interior Angles**

| Station | Observed angles | Correction | Adjusted angles |
|---------|-----------------|------------|-----------------|
| 1 | 66°40′30″ | −30″ | 66°40′00″ |
| 2 | 131°35′00″ | −30″ | 131°34′30″ |
| 3 | 97°35′00″ | −30″ | 97°34′30″ |
| 4 | 64°00′30″ | −30″ | 64°00′00″ |
| 5 | 227°26′30″ | −30″ | 227°26′00″ |
| 6 | 132°45′30″ | −30″ | 132°45′00″ |
| Sum = | 720°03′00″ | | 720°00′00″   checks |
| $(n-2)180°=$ | 720°00′00″ | | |
| Closure = | 03′00″ | | |

**Table 8.5   Azimuth Calculation Using Interior Angles**

| Line | Angle | Azimuth | Bearing |
|------|-------|---------|---------|
| 12 | | 106°20′00″ | S73°40′00″E |
| 21 | | 286°20′00″ | |
| | $+\alpha_2$ | 131°34′30″ | |
| 23 | | 57°54′30″ | N57°54′30″E |
| 32 | | 237°54′30″ | |
| | $+\alpha_3$ | 97°34′30″ | |
| 34 | | 335°29′00″ | N24°31′00″W |
| 43 | | 155°29′00″ | |
| | $+\alpha_4$ | 64°00′00″ | |
| 45 | | 219°29′00″ | S39°29′00″W |
| 54 | | 39°29′00″ | |
| | $+\alpha_5$ | 227°26′00″ | |
| 56 | | 266°55′00″ | S86°55′00″W |
| 65 | | 86°55′00″ | |
| | $+\alpha_6$ | 132°45′00″ | |
| 61 | | 219°40′00″ | S39°40′00″W |
| 16 | | 39°40′00″ | |
| | $+\alpha_7$ | 66°40′00″ | |
| 12 | | 106°20′00″   check | |

azimuths are listed in Table 8.3. An alternative procedure is to use adjusted angles to compute the adjusted azimuths and bearings, as shown in Table 8.5.

Note that in this closed-loop traverse, although the calculated azimuths are internally consistent, the absolute orientation is based entirely on one angle observed between lines 1Q and 12. This aspect of a closed-loop traverse is a weakness in this procedure which was mentioned in Art. 8.6. To eliminate the weakness, another angle should be observed from some other traverse point to another independent line of known azimuth. If points are not available to permit this second tie, an astronomic observation for azimuth should be made for line 34 or 45 so as to provide a redundancy in orientation of the traverse. Short traverses, such as the one illustrated in Fig. 8.7, may not require such a tie, depending on the nature of the survey. Long closed-loop traverses should always be tied to at least two lines of known direction.

Allowable closures for interior traverses are similar to those for deflection angle traverses. The sum of the interior angles should not deviate from $(n-2)180°$ by more than the square root of the number of instrument setups times the estimated standard deviation in observing the angles where this estimate is usually taken as the least count of the instrument.

**8.9. Traverse by angles to the right** This method can be used in open, closed, or closed-loop traverses. A repeating transit or theodolite or a direction theodolite may be employed to turn the angles. When a repeating instrument is used, the backsight to the preceding station is taken with the $A$ vernier set on zero. Upper clamp loosened, the instrument is turned on the upper motion, a foresight is taken on the next station, and the angle turned to the right is observed on the clockwise circle using the $A$ vernier. The characteristics of a direction theodolite are such that angles derived from directions observed between backsight and foresight are always angles to the right (Art. 6.40). As in the other methods for observing traverse angles, it is advisable to observe the angles twice, once with the telescope direct and once with the telescope reversed.

A traverse by angles to the right is shown in Fig. 8.9. This traverse originates at line $YZ$ for which the azimuth is 250°00′00″ and closes on line $TU$ with a given azimuth of

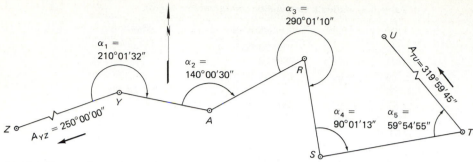

**Fig. 8.9** Traverse by angles to the right with a direction instrument.

319°59′45″ from $T$ to $U$. Angles for this traverse were derived from directions observed with a direction theodolite. Figure 8.10 shows the notes from which the angles were determined. Note that directions were observed with the telescope direct and reversed and angles are calculated from the average directions.

Azimuths are calculated by adding the angle to the right to the back azimuth of the previous line. Computation of these azimuths is shown in Table 8.6. The angular closure is 25″. This condition of closure can be expressed by the following equation for this example:

$$A_{YZ} + \alpha_1 + \alpha_2 + \alpha_3 + \alpha_4 + \alpha_5 - (4)(180°) - A_{TU} = 0 \qquad (8.4)$$

| Station | From/To | Tel. | Circle | Average Direction | | Angle | | R. Hoyt ☀ |
|---------|---------|------|--------|-------------------|---|-------|---|-----------|
| | | | GRANT FARM TRAVERSE | | | | | Dec. 1, 1977 |
| | | | | | | | | Clear 50°F      11 |
| | Z | D | 10° 01′ 13″ | 10° 01′ 15″ | | | | C. Roup R.T. |
| Y | | R | 190 01 17 | | | 210° 01′ 32″ | | K. Krone H.T. |
| | A | D | 220 02 45 | 220 02 47 | | | | |
| | | R | 40 02 49 | | | | | Wild T-2 |
| | Y | D | 90 05 22 | 90 05 25 | | | | Direction Theodolite |
| A | | R | 270 05 28 | | | 140 00 30 | | No. 5238 |
| | R | D | 230 05 52 | 230 05 55 | | | | |
| | | R | 50 05 58 | | | | | |
| | A | D | 5 10 02 | 5 10 05 | | | | |
| R | | R | 185 10 08 | | | 290 01 10 | | |
| | S | D | 295 11 13 | 295 11 15 | | | | Note: Traverse alignment |
| | | R | 115 11 17 | | | | | and references |
| | R | D | 170 04 11 | 170 04 14 | | | | are on Page 10 |
| S | | R | 350 04 17 | | | 90 01 13 | | |
| | T | D | 260 05 26 | 260 05 27 | | | | |
| | | R | 80 05 28 | | | | | |
| | S | D | 270 00 00 | 270 00 01 | | | | |
| T | | R | 90 00 02 | | | 59 54 55 | | |
| | U | D | 329 54 53 | 329 54 56 | | | | |
| | | R | 149 54 59 | | | | | |

**Fig. 8.10** Notes for angles to the right traverse.

**Table 8.6    Azimuths Calculated by Angles to the Right**

| Line | Azimuth | Correction | Corrected azimuth |
|------|---------|------------|-------------------|
| YZ | 250°00′00″ | | |
| | 210°01′32″ | | |
| | 460°01′32″ | | |
| | − 180°00′00″ | | |
| AY | 280°01′32″ | 05″ | 280°01′37″ |
| | 140°00′30″ | | |
| | 420°02′02″ | | |
| | − 180°00′00″ | | |
| RA | 240°02′02″ | 10″ | 240°02′12″ |
| | 290°01′10″ | | |
| | 530°03′12″ | | |
| | − 180°00′00″ | | |
| SR | 350°03′12″ | 15″ | 350°03′27″ |
| | 90°01′13″ | | |
| | 440°04′25″ | | |
| | − 180°00′00″ | | |
| TS | 260°04′25″ | 20″ | 260°04′45″ |
| | 59°54′55″ | | |
| TU | 319°59′20″ | 25″ | 319°59′45″ |
| TU | 319°59′45″  fixed | | |
| | Angular closure = 25″ | | |

which in general becomes

$$A_1 + \alpha_1 + \alpha_2 + \cdots + \alpha_n - (n-1)(180°) - A_2 = 0 \qquad (8.5)$$

in which $A_1$ and $A_2$ are fixed azimuths from the north at lines of origin and closing, respectively, and there are $n$ traverse stations (not counting fixed stations $Z$ and $U$).

If adjusted azimuths are desired at this point, assuming observations of equal precision, the error of closure is distributed equally among the five angles, or $25″/5 = 05″$ per angle. The adjustment is made by applying a cumulative correction to each unadjusted azimuth. The corrections and adjusted azimuths are also listed in Table 8.6.

**8.10. Azimuth traverse**    An advantage of the azimuth method is that the simple statement of one angular value, the azimuth, fixes the direction of the line to which it refers. The method is used extensively on topographic and other surveys where a large number of details are located by angular and linear measurements from the traverse stations. Any angular error of closure of a traverse becomes evident by the difference between initial and final observations taken along the first line. The reference meridian may be either true or assumed.

Successive stations are occupied, beginning with the line of known or assumed azimuth. At each station the transit is "oriented" by setting the $A$ vernier or horizontal circle index to read the back azimuth (forward azimuth ± 180°) of the preceding line and then backsighting to the preceding traverse station. The instrument is then turned on the upper motion, and a foresight on the following traverse station is secured. The reading indicated by the $A$ vernier on the clockwise circle is the azimuth of the forward line.

Figure 8.11 shows a closed-loop traverse run by the azimuth method. The traverse is begun at station 1 with a backsight along line 15 for which the azimuth is known to be 270°28′ by a solar observation. With the instrument set over station 1, the $A$ vernier is set to 270°28′ on the clockwise circle and a backsight is made on station 5 using the lower motion.

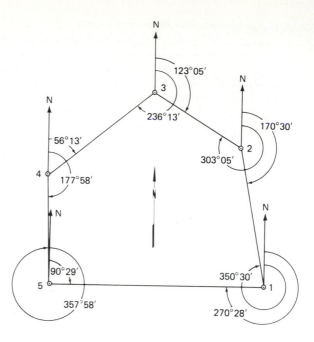

**Fig. 8.11** Azimuth traverse.

### AZIMUTH TRAVERSE AT HIGH–WATER LINE

Proposed Mill Pond, El. 741.36
Silver Creek, Penn.
(For Land Damage Est.)

| Sta. | Obj. | Dist. | Azimuth | Mag.B. | Cal.Bear. |
|------|------|-------|---------|--------|-----------|
| 1 | 5 | | 270°28' | N80½°W | N89°32'W |
|   | 2 | 689.32 | 350°30' | N | N9°30'W |
| 2 | 1 | | 170°30' | S | S9°30'E |
|   | 3 | 509.66 | 303°05' | N48°W | N56°55'W |
| 3 | 2 | | 123°05' | S48¼°E | S56°55'E |
|   | 4 | 678.68 | 236°13' | S65½°W | S56°13'W |
| 4 | 3 | | 56°13' | N65¼°E | N56°13'E |
|   | 5 | 572.50 | 177°58' | S 7°W | S2°02'E |
| 5 | 4 | | 357°58' | N 7¼°E | N2°02'W |
|   | 1 | 1082.71 | 90°29' | S80°E | S89°31'E |
|   |   |   | Error = 01' | | |

Gurley transit    J. Stanbois
No. 191           F. Lowe
                  June 15, 1977
                  Cloudy, Warm

True azimuth of line 1-5 found
by solar observation.
Mag. declination = 9°10'W

**Fig. 8.12** Notes for a short closed azimuth traverse.

The instrument is then turned on the upper motion and a foresight is taken on station 2. Since zero is oriented toward true north (Fig. 8.11), the clockwise circle read at the $A$ vernier (350°30′) is the azimuth from the north of line 12. Note that any other angle observed on the clockwise circle to another point from this setup is also an azimuth from the north. When the instrument is moved to point 2, the back azimuth of line 12 is computed by subtracting 180° from the forward azimuth (350°30′ − 180° = 170°30′), and this value is set on the clockwise circle before a backsight, using the lower motion, is taken on station 1. Once again, the clockwise circle is oriented with zero toward true north and when the instrument is turned on the upper motion for a foresight on station 3, the angle observed on the clockwise circle at the $A$ vernier (303°05′) is the azimuth of line 23 from true north. This operation is repeated for each successive traverse station. When the final setup is made over station 5, the azimuth observed on the foresight along line 51 should equal the azimuth of line 15 minus 180°, or in this case 270°28′ − 180° = 90°28′. Any deviation between this last observation and the fixed direction of the closing line is the angular closure. Figure 8.12 illustrates the notes recorded for the traverse shown in Fig. 8.11. The closure error is 01′. Since angles have been observed only to the nearest minute, this error could be applied to one angle to provide adjusted directions. Magnetic bearings are observed and a check against blunders is made by noting that computed bearings vary from corresponding magnetic bearings by about 9°10′, the amount of the magnetic declination.

**8.11. Compass traverse**  Use of the compass was described in Art. 6.19. When a compass traverse is performed, forward and back bearings are observed from each traverse station and distances are taped. If local attraction exists at any traverse station, both the forward and back bearings are affected equally. Thus, interior angles computed from forward and back bearings are independent of local attraction. Field notes for a closed-loop compass traverse are shown in Fig. 8.13. The declination of 20°15′E was set off on the compass (Fig. 6.20, Art. 6.14) so that bearings are referred to the true meridian.

Checking the compass traverse is accomplished by computing the interior angle from the observed bearings. Since these angles are independent of local attraction, the sum of these interior angles provides a legitimate indication of the angular error in the traverse. The method for computing interior angles from bearings is illustrated in Fig. 8.14. Thus, at station $A$ the unadjusted interior angle is 180° + 28°00′ + 30°40′ = 238°40′ and the unadjusted angle at $B$ is 180° − (30°40′ + 83°50′) = 65°30′. Computed interior angles are listed in column four of the notes shown in Fig. 8.13. The sum of these angles reveals a closure error of 25′. Assuming that all bearings are of equal precision and noncorrelated, this error is distributed equally among the five interior angles so that 05′ is added to each unadjusted interior angle to provide an adjusted angle. Adjusted interior angles are listed in column 5 of the notes in Fig. 8.13.

Since none of the traverse lines has an absolute direction that is known to be correct, it is necessary to select a line affected least by local attraction. Examination of column 4 of the notes in Fig. 8.13 indicates that points $A$ and $B$ are most likely to be free of local attraction since the forward and back bearings of $AB$ are numerically equal and opposite in direction. Hence, the correct forward bearing from $A$ to $B$ can be taken as S30°40′W. Holding the direction of $AB$ fixed, the bearing of $BC$ is 180° − (30°40′ + 65°35′) = S83°45′E, where 65°35′ is the adjusted interior angle at $B$. The remainder of the adjusted bearings are computed similarly, as illustrated graphically in Fig. 8.15. These adjusted bearings are listed in column 6 of the notes shown in Fig. 8.13.

If the error in the sum of interior angles exceeds 10′ to 15′ times the square root of the number of angles, it is likely that a blunder in reading the compass has occurred, and the field measurements should be repeated. If the error is within permissible limits but cannot be divided equally among the angles in amounts of 05′ or 10′, the greater corrections (in

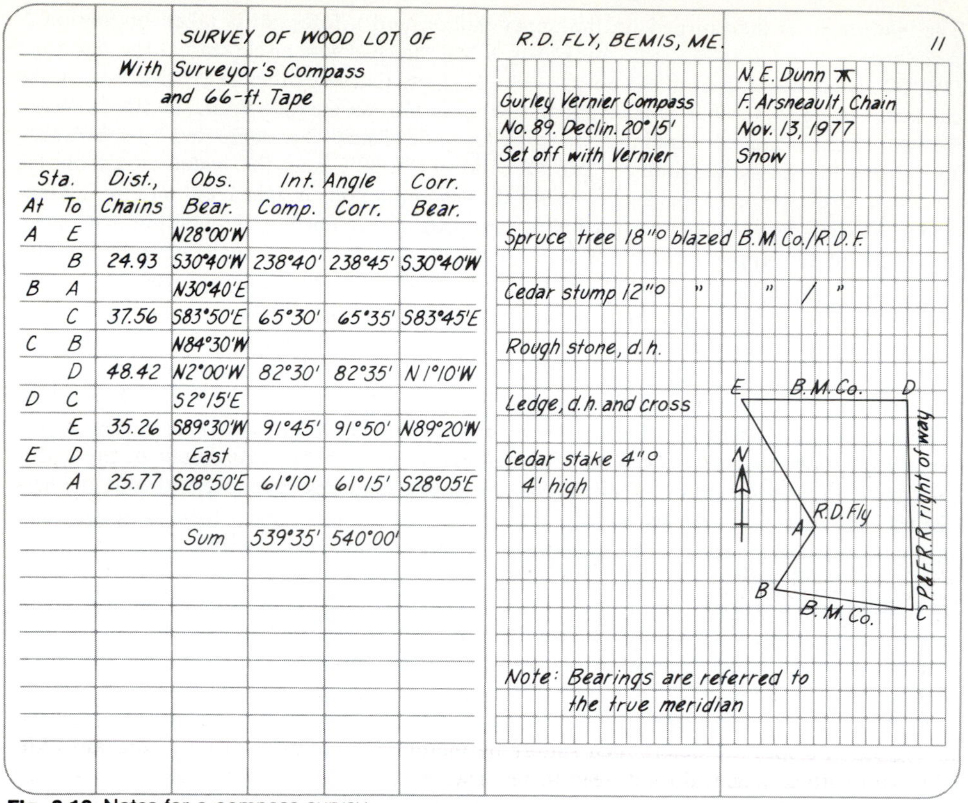

| Sta. At | To | Dist., Chains | Obs. Bear. | Int. Angle Comp. | Corr. | Corr. Bear. | |
|---|---|---|---|---|---|---|---|
| A | E | | N28°00'W | | | | Spruce tree 18"⌀ blazed B.M.Co./R.D.F. |
| | B | 24.93 | S30°40'W | 238°40' | 238°45' | S30°40'W | |
| B | A | | N30°40'E | | | | Cedar stump 12"⌀  "   "  /  " |
| | C | 37.56 | S83°50'E | 65°30' | 65°35' | S83°45'E | |
| C | B | | N84°30'W | | | | Rough stone, d.h. |
| | D | 48.42 | N2°00'W | 82°30' | 82°35' | N1°10'W | |
| D | C | | 52°15'E | | | | Ledge, d.h. and cross |
| | E | 35.26 | S89°30'W | 91°45' | 91°50' | N89°20'W | |
| E | D | | East | | | | Cedar stake 4"⌀ 4' high |
| | A | 25.77 | S28°50'E | 61°10' | 61°15' | S28°05'E | |
| | | | Sum | 539°35' | 540°00' | | |

*SURVEY OF WOOD LOT OF — With Surveyor's Compass and 66-ft. Tape*

*R.D. FLY, BEMIS, ME.   11*

*Gurley Vernier Compass No. 89. Declin. 20°15' Set off with Vernier — N.E. Dunn, F. Arsneault, Chain, Nov. 13, 1977, Snow*

Note: Bearings are referred to the true meridian

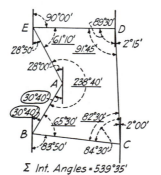

**Fig. 8.13** Notes for a compass survey.

**Fig. 8.14** Observed bearings and computed interior angles.

Σ Int. Angles = 539°35'

multiples of 05') should be applied arbitrarily to those angles for which the conditions of observing were estimated to be least favorable. The precision of most compass measurements does not justify computations with a precision closer than multiples of 05'.

**8.12. Stadia traverse**   When distances between traverse stations are determined by the stadia method (Chap. 7), the resulting survey is called a stadia traverse. This type of traverse is appropriate for certain types of preliminary and reconnaissance surveys, rough surveys for boundary locations, and topographic mapping surveys. Horizontal position can be de-

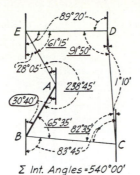

**Fig. 8.15** Corrected bearings and corrected interior angles.

Σ *Int. Angles* = 540°00'

termined with relative accuracies of 1 part in 300 to 400 by stadia traverses. The stadia traverse is sufficiently accurate and considerably more rapid and economical than corresponding surveys made with a transit or theodolite and a tape. One advantage of the stadia traverse is that, if desired, elevations may be determined concurrently with horizontal position.

**Elevations not required**   Where only horizontal locations for a control survey are required, the field party consists of a transit operator, rodman, and recorder. Stadia intervals and horizontal angles (or directions) are observed as each point is sighted. Vertical angles are observed only if large enough to make the horizontal distance appreciably different from stadia distance (when greater than 3°) and then are read to the nearest degree without reading the vernier. When a self-reducing tacheometer is employed, horizontal distances for all degrees of slopes are determined simply by multiplying the interval by a constant (Art. 7.17), and vertical angles do not need to be observed at all for this type of stadia traverse. Using any type of instrument, reduced horizontal distances are expressed to the nearest foot or to the nearest decimetre.

When performing a stadia traverse, it is customary to observe the stadia interval both backward and forward from each setup of the instrument. In this way two independent observations are made for each distance in the traverse. The closeness of agreement between the two values for each line is a check against blunders, and the mean is taken as the best estimate for the distance.

Figure 8.16 is a sample page of notes for a short closed-loop stadia traverse. The recorded value of the interval factor is 100.2. The directions of lines are azimuths and are checked by magnetic bearings. The stadia intervals are given (column 4 in notes) rather than stadia distances, since $K$ is not 100. Vertical angles are observed only to the nearest 10'. The vertical angle is recorded only for lines $AB$ and $DE$, for which the angles are large enough to make a horizontal correction necessary. If a self-reducing instrument is used (Art. 7.17), the vertical angle is not recorded and it is necessary to record only the rod interval and horizontal distance.

**Elevations required**   On topographic and similar surveys, the elevation and horizontal position of each traverse station are desired. The procedure consists of observing directions by the azimuth method and distances by stadia, as described in the preceding paragraph. In addition, differences in elevation are determined by observing vertical angles or stadia arc readings and stadia intervals. As in the preceding case, if a self-reducing instrument is being used only the rod interval, horizontal distance (rod interval times a given constant), and difference in elevation (rod interval times a given constant) are recorded. The party consists of an observer, a rodman, and a recorder.

STADIA TRAVERSE

For Horizontal Control

| Sta. | Obj. | Az. | Mag. B. | Rod Int. | Vert. Ang. | Aver. Hor. Dist. ft. |
|---|---|---|---|---|---|---|
| B | A | 279°00′ | N81°W | 9.09 | +2°40′ | 908 |
| | C | 356°14′ | N4°W | 8.94 | | |
| C | B | 176°14′ | S3°30′E | 8.98 | | 899 |
| | D | 296°56′ | N63°W | 13.45 | | |
| D | C | 116°56′ | S63°E | 13.50 | | 1351 |
| | E | 221°49′ | S42°W | 8.49 | +4°40′ | |
| E | D | 41°49′ | N41°E | 8.47 | −4°40′ | 845 |
| | A | 127°57′ | S52°E | 11.90 | | |
| A | E | 307°57′ | N52°W | 12.00 | | 1199 |
| | B | 98°58′ | S81°E | 9.05 | −2°40′ | |

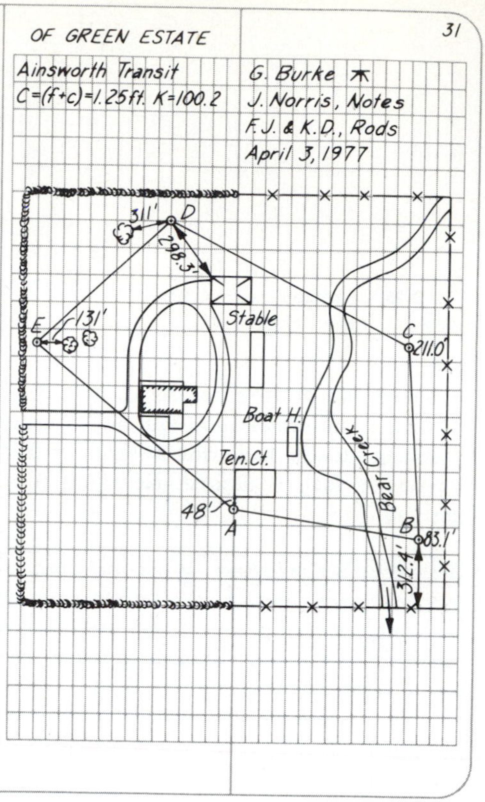
OF GREEN ESTATE
31
Ainsworth Transit
C=(f+c)=1.25ft. K=100.2
G. Burke
J. Norris, Notes
F.J. & K.D., Rods
April 3, 1977
311′
D
298.3′
Stable
E
131′
C
211.0′
Boat H.
Bear Creek
Ten. Ct.
48′
A
B
83.1′
312.4′

**Fig. 8.16** Stadia traverse for horizontal control.

Field notes for a small closed-loop traverse run with a transit or theodolite are shown in Fig. 8.17. The instrument is set up over traverse station 1, the elevation and location of which are known. The index error of the vertical circle is observed and recorded and the height of instrument (h.i.) above the station over which the instrument is set is measured with a rod or tape to the nearest centimetre or nearest tenth of a foot. If the direction of a line in the traverse is not known, a magnetic direction along one line can be observed. In the example traverse, a magnetic bearing is first observed along line 14. This value is converted to an azimuth and corrected for declination. These calculations are shown at the top left of the sample notes in Fig. 8.17. This azimuth is then set off on the clockwise horizontal circle using the *A* vernier and station 4 is resighted with the lower motion. The stadia interval and vertical angle for the sight on station 4 are observed and recorded in columns 2 and 5, respectively, in the notes shown in Fig. 8.17. In this example, the middle cross hair was set on the value of the h.i. = 1.49 m before reading the vertical angle, so that it was not necessary to record a rod reading. Should it not be possible to set the middle cross hair on the h.i., due to some obstruction on the line of sight, the rod reading sighted is recorded directly above the vertical angle in column 5 of the notes (Fig. 8.17 from point 2 to 3). When observations along the backsight to station 4 are completed, the upper clamp is loosened and a foresight is taken on station 2, where the stadia interval, vertical angle, and azimuth are observed and recorded. The azimuth is then observed on the clockwise horizontal circle. Thus, this and all subsequent azimuths in the traverse are observed according to the *azimuth method*, as outlined in Art. 8.10. If a self-reducing instrument is being used for the observations, only the rod interval and rod reading at the middle or fixed cross hair are observed and recorded

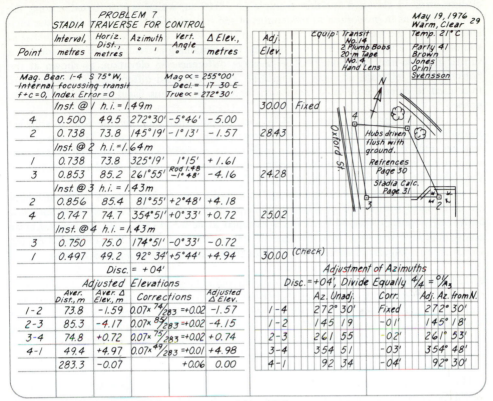

PROBLEM 7
STADIA TRAVERSE FOR CONTROL

May 19, 1976  29
Warm, Clear
Temp. 21° C

| Point | Interval, metres | Horiz. Dist., metres | Azimuth ° ' | Vert. Angle ° ' | Δ Elev., metres | Adj. Elev. |
|---|---|---|---|---|---|---|
| Mag. Bear. 1-4 S 75° W, | | | Mag ∝ = 255°00' | | | |
| Internal focussing transit | | | Decl. = 17-30 E | | | |
| f+c = 0, Index Error = 0 | | | True ∝ = 272°30' | | | |
| Inst. @ 1 h.i. = 1.49m | | | | | | 30.00  Fixed |
| 4 | 0.500 | 49.5 | 272°30' | -5°46' | -5.00 | |
| 2 | 0.738 | 73.8 | 145°19' | -1°13' | -1.57 | 28.43 |
| Inst. @ 2 h.i.=1.64m | | | | | | |
| 1 | 0.738 | 73.8 | 325°19' | 1°15' | +1.61 | |
| 3 | 0.853 | 85.2 | 261°55' | Rod 1.48 -1°48' | -4.16 | 24.28 |
| Inst. @ 3 h.i. = 1.43m | | | | | | |
| 2 | 0.856 | 85.4 | 81°55' | +2°48' | +4.18 | |
| 4 | 0.747 | 74.7 | 354°51' | +0°33' | +0.72 | 25.02 |
| Inst. @ 4 h.i. = 1.43m | | | | | | |
| 3 | 0.750 | 75.0 | 174°51' | -0°33' | -0.72 | |
| 1 | 0.497 | 49.2 | 92° 34' | +5°44' | +4.94 | 30.00  (Check) |
| | | Disc. = +04' | | | | |

Adjusted Elevations

| | Aver. Dist., m | Aver. Δ Elev., m | Corrections | Adjusted Δ Elev. |
|---|---|---|---|---|
| 1-2 | 73.8 | -1.59 | 0.07× 74/283 =+0.02 | -1.57 |
| 2-3 | 85.3 | -4.17 | 0.07× 85/283 =+0.02 | -4.15 |
| 3-4 | 74.8 | +0.72 | 0.07× 75/283 =+0.02 | +0.74 |
| 4-1 | 49.4 | +4.97 | 0.07× 49/283 =+0.01 | +4.98 |
| | 283.3 | -0.07 | +0.06 | 0.00 |

Equip: Transit No.14
2 Plumb Bobs
20-m Tape
No. 4
Hand Lens

Party 41
Brown
Jones
Orini
Svensson

N

Oxío St.

Hubs driven flush with ground.

Refrences Page 30

Stadia Calc. Page 31

Adjustment of Azimuths

Disc. = +04', Divide Equally 4/4 = 0/4 /4s

| | Az. Unadj. | Corr. | Adj. Az. from N. |
|---|---|---|---|
| 1-4 | 272 30 | Fixed | 272° 30' |
| 1-2 | 145 19 | -01' | 145° 18' |
| 2-3 | 261 55 | -02' | 261° 53' |
| 3-4 | 354 51 | -03' | 354° 48' |
| 4-1 | 92 34 | -04' | 92° 30' |

**Fig. 8.17** Stadia traverse for horizontal and vertical control.

in columns 2 and 5, respectively. If the middle cross hair is set on a rod reading equal to the value of the h.i., it need not be recorded and the difference in elevation is the rod interval multiplied by the observed constant (Art. 7.17).

When all observations are completed at the first station, the instrument is moved to station 2, where the h.i. is measured (1.64 m) and a backsight is taken on station 1 with the back azimuth of line 12 (145°19' + 180° = 325°19') set on the clockwise horizontal circle at the $A$ vernier. When the stadia interval and vertical angle from 2 to 1 are observed, it is possible to check these readings against the corresponding readings from 1 to 2. The stadia intervals should agree to within 2 or 3 mm and the vertical angles should have opposite signs and agree numerically to within a few minutes. Note that if index error in the vertical angle is present, there will be a systematic difference between forward and backward vertical angles. A foresight is then taken on station 3, where the stadia interval, vertical angle, and azimuth are observed and recorded.

The procedure described in the previous paragraphs is repeated for each station in the traverse. When the final station is occupied, the azimuth along the initial line of known direction (line 41 in this example) should correspond to 180° ± the known azimuth or 272°30' - 180°00' = 92°30'. The difference is the angular error of closure, which is 04' in this example.

Horizontal distances and differences in elevation are computed using the stadia intervals and vertical angles. These values are recorded in columns 3 and 6 in the notes. Horizontal distances are averaged and average differences in elevation are used to check the closure error in elevation. These values are tabulated at the bottom of the left page in the notes. The closure error in elevation for this example is −0.07 m, which is distributed among the traverse stations in proportion to the lengths of the traverse sides (Art. 5.50). Since the total length around the traverse is 283.3 m, the correction to the difference in elevation between stations 1 and 2 is $(74/283)(0.07) = +0.02$ m. The remaining corrections are computed in a similar manner and are applied to the average differences in elevation, as shown at the bottom of the left page of notes in Fig. 8.17. These corrected differences in elevation are then used to calculate adjusted elevations for traverse stations listed in column 7 of the notes.

The error in angular closure of 04′ is distributed equally among the four unadjusted azimuths by applying a cumulative correction to each observed direction. Thus, the correction per interior angle is $04'/4 = 01'$ per angle or the corrections are 01′, 02′, 03′, and 04′ per azimuth for lines 12, 23, 34, and 41, respectively. These values are tabulated at the bottom of the right page of the sample notes in Fig. 8.17.

Larger angular closures can be tolerated in a stadia traverse than one in which distances are determined with a tape or EDM equipment. If the estimated standard deviation of a horizontal distance measured by stadia is assumed to be 0.3 m (Art. 7.16), an angular error compatible with this error in a distance of 100 m (Art. 3.7) is $\tan^{-1} 0.3/100 \simeq 10'$ of arc. Consequently, angular closure errors of up to 10′ are reasonable in a stadia traverse.

**8.13. Plane-table traverse**    Traversing with the plane table involves the same principles as running a traverse with a transit or theodolite. As each successive station is occupied, the table is oriented, sometimes with the compass but usually by taking a backsight on the preceding station. A foresight is then taken to the next station and its location is plotted on the plane-table sheet. Distances and differences in elevations are determined by stadia using a conventional or self-reducing alidade. The plane table and alidade are described in Art. 7.19. Use of the plane table is outlined in Art. 7.21.

The plane-table traverse is used for the control of small surveys and extension of control for larger surveys in order to locate points from which map details can be plotted advantageously.

In Fig. 8.18, a series of traverse stations $A$, $B$, $C$, $D$, and $E$ are represented. The initial setup is on station $A$, from which $E$ and $B$ are visible. The plane table is set up at $A$, leveled, and $a$, representing station $A$ on the ground, is plotted on the plane-table sheet in such a location that the other stations will fall within the limits of the sheet. Using the plotted position of $a$ as a pivot, the blade of the alidade is directed toward station $E$, stadia readings are observed and reduced to a horizontal distance, and the position of $E$ is plotted as $e$ on the plane-table sheet to the scale of the traverse. Next, the blade of the alidade is rotated about $a$ and a foresight is taken on station $B$. Stadia observations are made on station $B$ and are reduced to the horizontal distance $AB$. A line in the direction of $B$ is drawn from $a$ on the plane table and point $b$ is plotted to scale.

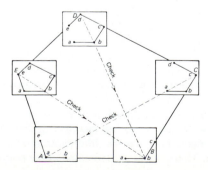

**Fig. 8.18** Traverse with a plane table.

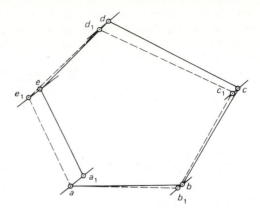

**Fig. 8.19** Adjustment of the plane-table traverse.

The plane table is then set over station $B$, is leveled, the blade of the alidade is aligned with $ba$, and the plane table is rotated until the vertical cross hair of the alidade is on line with station $A$. Stadia readings are made on $A$ and reduced to a second determination of the distance $BA$. The average of the two distances is plotted from $a$ to establish an improved position for $b$ on the plane table. The blade of the alidade is rotated about $b$ and a foresight is taken on $C$ along with the stadia observations necessary to determine distance $BC$. A line in the direction of $C$ is drawn from $b$ and the position of $c$ is plotted.

The procedure outlined is repeated for each station in the traverse. When the last station $E$ is occupied, the board is oriented by a backsight on station $D$, where the necessary stadia observations are made to permit a second determination of the horizontal distance $DE$, and an improved position for $e$ is plotted using the average distance $DE$. Finally, a foresight is made on station $A$, permitting a second determination for the horizontal distance $EA$. The average distance $EA$ is plotted to scale along the line from $e$ in the direction of $A$ so as to establish a final position for $a$ which does not necessarily coincide with its original plotted position. The difference between the initial and final positions for $a$ represents the *closure* for the stadia traverse. In Fig. 8.19, the closure is represented by distance $aa_1$, where $a$ and $a_1$ are the initial and final locations of the first station. To adjust the traverse graphically, draw lines parallel to $aa_1$ through each plotted traverse station. The increments $bb_1$, $cc_1$, $dd_1$, and $ee_1$ are in proportion to the distance of the station from the fixed initial station $a$. The adjusted traverse lines are dashed.

**8.14. Referencing traverse stations**    Many of the hubs marking the location of highways, railroads, and other public and private works are bound to be uprooted or covered during the progress of construction and must be replaced, often more than once, before construction is completed. Such hubs marking traverse stations are usually tied by angular and/or linear measurements to temporary wooden hubs called *reference hubs*, or to other objects that are not likely to be disturbed. A traverse station is said to be *referenced* when it is so tied to nearby objects that it can be replaced readily. The manner of referencing a traverse station is indicated in the notes by an appropriate sketch. The precision of referencing should be comparable to that of the measurements between traverse stations. Particularly on land surveys, the corners should be tied to nearby objects which can be found readily, which are not likely to be moved or obliterated, and which are of a more-or-less permanent character.

Often in land surveying, a corner is incorrectly said to be "witnessed" when angular and linear measurements are taken to nearby objects of the character just mentioned. This unfortunate designation leads to the confusion of reference marks for a corner which has

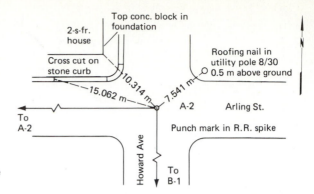

**Fig. 8.20** Reference to a traverse station in an urban area.

been established with *witness corners* (Art. 19.42), which are markers set on one or more of the land lines leading to a corner when the corner falls in a place where it would be either impossible or impractical to establish or to maintain a monument.

Figures 8.20 through 8.22 illustrate several methods of referencing a traverse station. The station shown in Fig. 8.20 is tied by linear measurements to three specific points located on semipermanent structures or objects located around the traverse station. Location or relocation of the traverse station is accomplished by intersecting the reference distances. Where possible, reference points should be chosen so as to be easily identifiable and ought to be located on objects or structures of a permanent nature. Tie lines should intersect at favorable angles (greater than 30° and less than 150°), so that the station can be relocated with certainty.

Figure 8.21 shows a station referenced by setting reference hubs on a line passing through the station and taking linear measurements from these hubs to the station. In this example, four reference hubs are set at the ends of lines *AB* and *CD*, which intersect at traverse station *Y*-1. With the measurements as shown, any two reference hubs may be destroyed and the station may still be relocated.

A typical method for referencing property corners in land surveying is illustrated in Fig. 8.22. Both angular and linear ties are employed.

**8.15. Traverse computations**   Traverse operations in the field yield observed angles or directions and measured distances for a set of lines connecting a series of traverse stations.

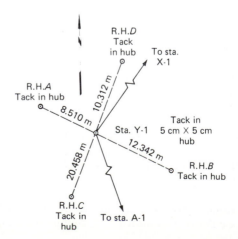

**Fig. 8.21** Referencing by reference hubs.

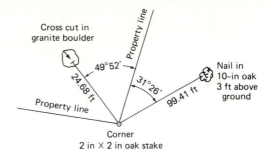

**Fig. 8.22** References to a property corner.

Angles, or directions converted to angles, can be checked to obtain the angular error of closure that may be distributed among the angles to provide preliminary adjusted values from which preliminary azimuths or bearings are computed. When the observed distances have been corrected for all systematic errors (Arts. 4.15 to 4.24), the preliminary directions and reduced distances are suitable for use in traverse computations which are performed in a plane rectangular coordinate system.

Computations with plane rectangular coordinates are extremely adaptable for use with pocket calculators, desktop calculators, and high-speed electronic computers. Coordinated points have an unambiguous designation that is ideal for storage and retrieval from data banks (see Arts. 1.11 and 3.15) and is compatible with many methods for plotting the locations of points, including automatic plotting techniques.

The system within which the coordinates are computed may have a national, state, regional, local, or completely arbitrary origin. Plane coordinate computations in the United States are usually performed in one of the state plane coordinate systems established by the National Ocean Survey for each state. Characteristics of these systems are described in Chap. 14.

Computations with plane coordinates are illustrated by reference to Fig. 8.23, where traverse line $ij$ has a reduced horizontal distance of $d_{ij}$ and azimuth of $A_i$. Let $x_{ij}$ and $y_{ij}$ be designated as the departure and latitude for line $ij$, so that

$$x_{ij} = d_{ij} \sin A_i$$
$$y_{ij} = d_{ij} \cos A_i$$

(8.6)

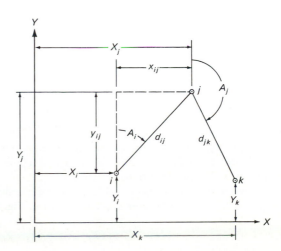

**Fig. 8.23** Departure and latitude of lines *ij*.

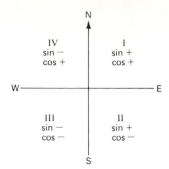

**Fig. 8.24** Signs of azimuth functions.

The algebraic signs of the departure and latitudes for a traverse line depend on the signs of the sine and cosine of the azimuth of that line. Figure 8.24 shows the algebraic signs of the sine and cosine functions for each of the four quadrants. Pocket calculators, desktop calculators, and electronic computers with internal routines for trigonometric functions yield the proper sign automatically given the azimuth. When bearings are used to describe the direction of the line, the sine and cosine are always considered as positive and the algebraic sign of the latitude and departure is taken from the quadrant. Consequently, the azimuth is the logical means of defining direction in traverse calculations when modern calculators are used.

Please observe that the terms "departure" and "latitude," as well as $X$ and $Y$, are specified in that order throughout this text. Note that several hand-held calculators with direct rectangular to polar or polar to rectangular keys use the reverse notation of latitude and departure, $Y$ and $X$.

For example, in Fig. 8.23, let $A_i = 43°30'00''$ and $d_{ij} = 432.182$ m. Applying Eq. (8.6), the departure $= x_{ij} = 432.182 \sin 43°30'00'' = 432.182(0.6883546) = 297.494$ m and the latitude $= y_{ij}$ $= 432.182 \cos 43°30'00'' = 432.182(0.7253744) = 313.494$ m. Continuing, if $A_j = 153°25'40''$, $d_{jk}$ $= 385.151$ m, then the departure $= x_{jk} = 385.151(0.4473255) = 172.288$ m and the latitude $= y_{jk}$ $= 385.151(-0.8943712) = -344.468$ m.

Given that the coordinates of station $i$ are $X_i$ and $Y_i$ (Fig. 8.23), the coordinates of station $j$ are

$$X_j = X_i + x_{ij}$$
$$Y_j = Y_i + y_{ij} \tag{8.7}$$

If the coordinates of station $i$ in the example are $X_i = 142,482.352$ m, $Y_i = 43,560.805$ m, then the coordinates of station $j$ are

$$X_j = 142,482.352 + 297.494 = 142,779.846 \text{ m}$$
$$Y_j = \phantom{1}43,560.805 + 313.494 = \phantom{1}43,874.299 \text{ m}$$

and the coordinates of station $k$ are

$$X_k = 142,779.846 + 172.288 = 142,952.134 \text{ m}$$
$$Y_k = \phantom{1}43,874.299 - 344.468 = \phantom{1}43,529.831 \text{ m}$$

Consequently, a major portion of traverse computations consists of calculating departures and latitudes for successive traverse lines and cumulating the values to determine the coordinates for consecutive traverse stations.

The inverse solution is also possible, given the coordinates for the two ends of a traverse line. Thus, the distance between stations $i$ and $j$ in Fig. 8.23 is

$$d_{ij} = \left[ (X_j - X_i)^2 + (Y_j - Y_i)^2 \right]^{1/2} \tag{8.8}$$

and the azimuth of line $ij$ from the north is

$$A_{Nij} = \arctan \frac{X_j - X_i}{Y_j - Y_i} \tag{8.9}$$

The azimuth of line $ij$ from the south is

$$A_{Sij} = \arctan \frac{X_i - X_j}{Y_i - Y_j} \tag{8.10}$$

In Eqs. (8.6) through (8.10), due regard should be paid to algebraic signs of the trigonometric functions (Fig. 8.24) and the differences between coordinates at the two ends of a line.

When coordinates for all the traverse points (or all of the departures and latitudes) for all lines have been computed, a check is necessary on the accuracy of the observations and the validity of the calculations. In a closed traverse, the algebraic sum of the departures should equal the difference between the $X$ coordinates at the beginning and ending stations of the traverse. Similarly, the algebraic sum of the latitudes should equal the difference between the $Y$ coordinates at the beginning and ending stations. If Eq. (8.7) is used to calculate coordinates directly, the calculated coordinates should agree with the given values for the final or closing station. It follows that in a closed-loop traverse, the algebraic sum of the departures and the algebraic sum of the latitudes each must equal zero. For a traverse containing $i = 1, 2, \ldots, n$ stations, starting at station $i = 1$ and terminating at station $i = n$, the foregoing conditions can be expressed as follows:

$$X_n - X_1 = \sum_{i=1}^{n-1} x_{i,i+1} = \sum_{i=1}^{n-1} \text{departures}$$
$$Y_n - Y_1 = \sum_{i=1}^{n-1} y_{i,i+1} = \sum_{i=1}^{n-1} \text{latitudes} \tag{8.10a}$$

Equations (8.10a) are rarely satisfied exactly in practice, owing to random observational errors, uncorrected systematic errors in observations, and inaccuracies in the given coordinates for a closed traverse. The amounts by which Eqs. (8.10a) fail to be satisfied are called the *errors in closure in position* or simply the *closures* for a traverse. The closure *corrections* $dX$ and $dY$, which are of opposite signs to errors, for a traverse defined as for Eqs. (8.10a) are

$$dX = (X_n - X_1) - \sum_{i=1}^{n-1} x_{i,i+1}$$
$$dY = (Y_n - Y_1) - \sum_{i=1}^{n-1} y_{i,i+1} \tag{8.10b}$$

When preliminary traverse computations are completed, closure corrections are evaluated and an adjustment is performed. The subject of traverse adjustment is covered in Art. 8.18. First consider the complete traverse computations for a closed traverse.

**8.16. Computations for a closed traverse**   Figure 8.25 shows a traverse that originates at station 1, and closes on station 5. Observed distances, corrected for systematic errors, and angles to the right observed once direct and once reversed with a 30″ repeating theodolite, are given in Table 8.7.

The angular error of closure is 01′ [Eq. (8.5) and Table 8.8]. Adjusted azimuths are calculated according to the procedure in Art. 8.9. Distribution of the angular error equally among the five angles leads to a correction of $-12″$ per angle. Observation of the angles twice with a 30″ instrument does not warrant this precision. Thus, the corrections to angles are rounded arbitrarily to $-10″$ each to angles 1, 2, and 3 and $-15″$ to angles 4 and 5. Table 8.8 shows azimuth calculations and adjusted azimuths.

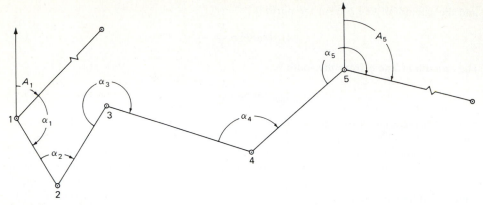

**Fig. 8.25** Closed traverse.

**Table 8.7  Data for Closed Traverse**

| Line | Distance, m | Angle to the right | |
|---|---|---|---|
| 12 | 305.41 | $\alpha_1 = 104°00'30''$ | $A_{N_1} = \;\;43°31'30''$ |
| 23 | 359.61 | $\alpha_2 = \;\;63°20'00''$ | $A_{N_5} = 106°17'30''$ |
| 34 | 612.15 | $\alpha_3 = 258°11'00''$ | |
| 45 | 485.12 | $\alpha_4 = 120°45'00''$ | $\sigma_d = 0.01$ m |
| | | $\alpha_5 = 236°30'30''$ | $\sigma_\alpha = 30''$ |

**Table 8.8  Azimuth Calculations**

| Line | Unadjusted azimuth | | Correction | Adjusted azimuth |
|---|---|---|---|---|
| $A_1$ | | 43°31'30'' | | |
| | $\alpha_1$ | 104°00'30'' | | |
| 12 | | 147°32'00'' | − 10'' | 147°31'50'' |
| | $\alpha_2$ | 63°20'00'' | | |
| 32 | | 210°52'00'' | | |
| 23 | | 30°52'00'' | − 20'' | 30°51'40'' |
| | $\alpha_3$ | 258°11'00'' | | |
| 43 | | 289°03'00'' | | |
| 34 | | 109°03'00'' | − 30'' | 109°02'30'' |
| | $\alpha_4$ | 120°45'00'' | | |
| 54 | | 229°48'00'' | | |
| 45 | | 49°48'00'' | − 45'' | 49°47'15'' |
| | $\alpha_5$ | 236°30'30'' | | |
| | | 286°18'30'' | | |
| | − | 180°00'00'' | | |
| $A_5$ | | 106°18'30'' | − 01'00'' | 106°17'30'' |
| | | 106°17'30'' | fixed | |
| Angular error | | 01'00'' | | |

**Table 8.9   Traverse Computations for Closed Traverse of Fig. 8.25**

| Sta-tion | Distance, m | Azimuth | Departures $x$ | Latitudes $y$ | Coordinates, m $X$ | $Y$ |
|---|---|---|---|---|---|---|
| 1 | | | | | 4321.404 | 6240.562 |
| | 305.41 | 147°31′50″ | 163.959 | − 257.668 | | |
| 2 | | | | | 4485.363 | 5982.894 |
| | 359.61 | 30°51′40″ | 184.465 | + 308.694 | | |
| 3 | | | | | 4669.828 | 6291.588 |
| | 612.15 | 109°02′30″ | 578.654 | − 199.717 | | |
| 4 | | | | | 5248.482 | 6091.871 |
| | 485.12 | 49°47′15″ | 370.464 | + 313.205 | | |
| 5 | | | | | 5618.946 | 6405.076 |
| | 1762.29 | | $\Sigma = 1297.542$ | $\Sigma = 164.514$ | [5619.243] | [6405.272] |
| | | | $X_5 - X_1 = 1297.839$ | $Y_5 - Y_1 = 164.710$ | | |
| Closure corrections: | | | $dX = +0.297$ | $dY = +0.196$ | $dX = 0.297$ | $dY = 0.196$ |

$$d_c = (dX^2 + dY^2)^{1/2} = [(0.297)^2 + (0.196)^2]^{1/2} = 0.36 \text{ m}$$

Distances, adjusted azimuths, fixed coordinates for the beginning and ending stations, calculated azimuths, and calculated coordinates are listed in Table 8.9. Note that coordinates can be calculated directly using Eq. (8.7) and tabulation of departures and latitudes is not necessary unless these values are to be used for some specific purpose. They are included here for illustrative purposes. The closure corrections in $X$ and $Y$ coordinates at station 5 are 0.297 m and 0.196 m, respectively. These values may also be calculated by applying Eqs. (8.10b) to yield $dX = (X_5 - X_1) - \Sigma \text{dep.} = 1297.839 - 1297.542 = 0.297$ and $dY = (Y_5 - Y_1) - \Sigma \text{lat.} = 164.710 - 164.514 = 0.196$ m. The values for closure are here shown graphically in Fig. 8.26; the fixed and calculated positions for station 5 are shown to a larger scale. Using Eq. (8.8), the resultant closure, $d_c = (dX^2 + dY^2)^{1/2}$, is the distance from the calculated position at 5′ to the fixed location 5. In a closed traverse of this type, the resultant closure is caused by random variations in the observations and uncorrected systematic errors in distance and direction. For this example, $d_c = [(0.196)^2 + (0.297)^2]^{1/2} = 0.36$ m.

The ratio of the closure $d_c$ to the distance traversed provides an indication of the goodness of the survey and is often referred to as the relative accuracy ratio for the traverse. Surveys are frequently classified according to relative accuracy ratios such as 1/5000, 1/10,000, etc. When blunders or uncorrected systematic errors in distance or directions are present, the closure and the consequent relative accuracy ratio will be too large. The relative accuracy ratio provides a measure of the relative merits of various traverses, but should not be misconstrued as an indication of the absolute accuracy in position for each station in the traverse. For the example traverse, the relative accuracy ratio is 0.36/1762 = 1/4894.

Specifications for various classifications of traverse and corresponding relative accuracies are discussed in Art. 8.23.

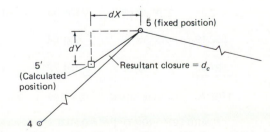

**Fig. 8.26** Closure in a closed traverse.

**8.17. Computations for a closed-loop traverse**   When a traverse originates and closes on the same station, the algebraic sum of the departures and the algebraic sum of the latitudes each must equal zero. As an example, consider the traverse illustrated in Fig. 8.7 for which the adjusted azimuths are listed in Table 8.5, and station 1 is held fixed with given coordinates.

Distances, azimuths, calculated departures and latitudes, given coordinates for station 1, and calculated coordinates for the rest of the traverse stations are listed in Table 8.10. The sum of the departures and the sum of the latitudes are 0.22 and $-0.17$ ft, respectively. These algebraic sums represent the errors of closures for the loop traverse. According to Eqs. (8.10$b$), the closure corrections are $dX = 0.0 - 0.22 = -0.22$ ft, $dY = 0.0 - (-0.17) = +0.17$ ft. Figure 8.27 shows that the resultant closure $d_c$, calculated by Eq. (8.8), is 0.28 ft. For a closed-loop traverse of this type, the resultant closure is a function of random variations in the observations and uncorrected systematic errors in angles or directions. Systematic errors in distance measuring equipment will cancel and are not revealed by the mathematical closure of the traverse. Note also that although the directions are consistent within the traverse, the entire polygon could be rotated with respect to the grid system with no indication of this rotation in the results from the traverse computations. A second tie to a line of known direction is necessary to reveal this type of error.

**8.18. Traverse adjustment**   Traverse adjustment is for the purpose of providing a mathematically closed figure and at the same time yielding the best estimates for horizontal positions of all the traverse stations. Methods of adjustment may be classified as approximate and rigorous. Consider the approximate adjustments first.

Traditional methods of approximate traverse adjustment have been developed to accommodate prevailing conditions in certain combinations of angular and linear precision in the observations. In this respect there are three combinations which are still common.

1. The precision in angles or directions exceeds its equivalent in linear distance observation (Art. 3.7). This combination exists in the stadia survey, where azimuths can be determined to within $\pm 30''$ with little difficulty, while the relative precision in distance measurements will never exceed 1 part in 300 to 400.

**Table 8.10   Computations for a Closed-Loop Traverse of Fig. 8.7**

| Station | Distance, ft | Azimuth | Departures $x$ | Latitudes $y$ | Coordinates, ft $X$ | Coordinates, ft $Y$ |
|---|---|---|---|---|---|---|
| 1 | | | | | 4382.09 | 6150.82 |
| | 405.24 | 106°20′00″ | + 388.89 | − 113.96 | | |
| 2 | | | | | 4770.98 | 6036.86 |
| | 336.60 | 57°54′30″ | + 285.17 | + 178.83 | | |
| 3 | | | | | 5056.15 | 6215.69 |
| | 325.13 | 335°29′00″ | − 134.92 | + 295.82 | | |
| 4 | | | | | 4921.23 | 6511.51 |
| | 212.91 | 219°29′00″ | − 135.38 | − 164.33 | | |
| 5 | | | | | 4785.85 | 6347.18 |
| | 252.19 | 266°55′00″ | − 251.82 | − 13.56 | | |
| 6 | | | | | 4534.03 | 6333.62 |
| | 237.69 | 219°40′00″ | − 151.72 | − 182.97 | | |
| 1 | | | | | 4382.31 | 6150.65 |
| | 1769.76 | Closure errors | Σ = 0.22 | Σ = − 0.17 | (4382.09) | (6150.82) |
| | | | | | $dX = -0.22$ | $dY = +0.17$ |
| | $d_c = [(0.22)^2 + (0.17)^2]^{1/2} = 0.28$ ft | | | | Closure corrections | |

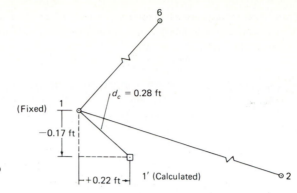

**Fig. 8.27** Closure in a closed-loop traverse.

2. Precision in angles or directions is essentially equal to its equivalent in the precision of distances. This combination is typified by a traverse with an engineer's transit and a steel tape, where the terrain is fairly regular and the distances are corrected for all possible systematic errors.
3. Precision in distances exceeds that in angles and directions. This combination is prevalent in traverses containing long lines measured with calibrated EDM equipment.

Approximate adjustments include the *transit rule*, the *compass rule*, and the *Crandall method*. The transit rule was developed for the first combination of precisions in observations. Unfortunately, the transit rule is valid only when the traverse lines are parallel with the grid system used for the traverse computations and therefore is not considered further in this text. The compass rule was developed for the second combination and can be shown to be rigorous when the condition that the angular precision equals the precision in linear distances is rigidly enforced. The compass rule is developed further in Art. 8.19. The Crandall method is a rather complicated procedure which is more rigorous than either the compass or transit rules but requires substantially more computations. This method is not developed further in this book. The interested reader is referred to Refs. 3, 6, and 12 at the end of the chapter for detailed developments of these and other approximate methods.

The method of least squares provides the most rigorous adjustment, which allows for variation in precision in the observations, minimizes random variations in the observations, provides the best estimates for positions of all traverse stations, and yields statistics relative to the accuracies of adjusted observations and positions. This method does require more of a computational effort than the approximate adjustments. However, it is well within the capabilities of desktop and even some of the more sophisticated hand-held calculators. Theoretical development and examples of traverse adjustment by the method of least squares can be found in Art. B.14. The student of surveying and the practicing surveyor should study these examples and become familiar with adjustment of traverse by the method of least squares. The results are well worth the effort.

**8.19. Adjustment of a traverse by the compass rule**    For any traverse station $i$, let

$\delta X_i = $ correction to $X_i$

$\delta Y_i = $ correction to $Y_i$

$dX_t = $ total closure correction of the traverse in the $X$ coordinates [by Eq. (8.10b)]

$dY_t = $ total closure correction of the traverse in the $Y$ coordinates [by Eq. (8.10b)]

   $L_i = $ distance from station $i$ to the initial station

    $L = $ total length of traverse

then the corrections are

$$\delta X_i = \left(\frac{L_i}{L}\right) dX_t \quad \text{and} \quad \delta Y_i = \left(\frac{L_i}{L}\right) dY_t \qquad (8.11a)$$

In an alternative procedure, corrections may be applied to the departures and latitudes prior to calculating coordinates. In this case,

$$\delta x_{ij} = \left(\frac{d_{ij}}{L}\right)dX_t \qquad \text{and} \qquad \delta y_{ij} = \left(\frac{d_{ij}}{L}\right)dY_t \qquad (8.11b)$$

where $\delta x_{ij}$ and $\delta y_{ij}$ are respective corrections to the departure and latitude of line $ij$, which has a length of $d_{ij}$, and $dX_t$, $dY_t$, and $L$ are as defined previously.

As an example, consider adjustment, by the compass rule, of the coordinates for the traverse shown in Fig. 8.25 and calculated in Table 8.9. The closure corrections in departures and latitudes are 0.297 and 0.196 m, respectively. According to Eqs. (8.11a), $\delta X_2 = (305.41/1762)(0.297) = 0.051$ m, so that $X_2$ adjusted $= 4485.363 + 0.051 = 4485.41$ m rounded to two decimal places. Similarly, $\delta Y_2 = (305.41/1762)(0.196) = 0.034$ m and $Y_2$ adjusted $= 5982.894 + 0.034 = 5982.93$ m. To calculate corrections to $X_3$ and $Y_3$, the cumulated distance to station 3 is used in Eqs. (8.11a). Thus, $\delta X_3 = (665/1762)(0.297) = 0.112$ m and $\delta Y_3 = (665/1762)(0.196) = 0.074$ m, yielding $X_3$ adjusted $= 4669.828 + 0.112 = 4669.94$ m and $Y_3$ adjusted $= 6291.588 + 0.074 = 6291.66$ m. Distances, corrections, and adjusted coordinates are listed in Table 8.11.

Traverse adjustment by the compass rule can also be accomplished by balancing (adjusting) the departures and latitudes using Eqs. (8.11b) prior to computing the coordinates.

To illustrate this procedure of applying the compass rule adjustment, consider the traverse shown in Fig. 8.7 for which calculations are given in Table 8.10. For this traverse, $dX$ and $dY$, calculated by substituting the sums of departures and latitudes into Eqs. (8.10b), are $-0.22$ ft and $+0.17$ ft, respectively. Thus, applying Eqs. (8.11b), the correction to the departure of line 1-2, $\delta x_{12} = (405.24/1770)(-0.22) = -0.05$ ft and the correction to the latitude of line 1-2, $\delta y_{12} = (405.24/1770)(+0.17) = +0.04$.

Corrections and balanced (adjusted) departures and latitudes are listed in Table 8.12. The sums of corrections to latitudes and departures should be computed and checked against their respective closures. Note that the sum of the latitudes was different from the closure in latitudes by 0.01 ft due to roundoff error. This discrepancy was corrected by examining the third place of the corrections and noting that the correction to the latitude of line 5-6 lay closest to the next highest hundredth, so that 0.01 ft was added to this correction. The sums of the adjusted departures and latitudes should also be checked to ensure against any error in applying the corrections.

Adjusted coordinates computed using the balanced departures and latitudes are given in Table 8.13.

When calculating coordinates with balanced departures and latitudes, the computations should always be checked by cumulating the balanced departures and latitudes of the final line so as to provide a check on the coordinates of the closing station. Regardless of whether coordinates are adjusted directly or departures and latitudes are balanced first, the final adjusted coordinates must be identical.

**Table 8.11    Adjustment of Coordinates by the Compass Rule (Traverse of Fig. 8.25, Calculated in Table 8.9)**

| Station | Distance, m | Corrections | | Adjusted Coordinates, m | |
|---|---|---|---|---|---|
| | | $\delta X_i$ | $\delta Y_i$ | $X$ | $Y$ |
| 1 | | | | 4321.404 | 6240.562 |
| 2 | 305.41 | 0.051 | 0.034 | 4485.41 | 5982.93 |
| 3 | 665.02 | 0.112 | 0.074 | 4669.94 | 6291.66 |
| 4 | 1277.17 | 0.215 | 0.142 | 5248.70 | 6092.01 |
| 5 | 1762.29 | 0.297 | 0.196 | 5619.243 | 6405.272 |

**Table 8.12   Corrections to Latitudes and Departures (Traverse of Fig. 8.7, Calculated in Table 8.10)**

| Station | Distance ft | Departures, ft | Latitudes, ft | Corrections Dep. | Corrections Lat. | Adjusted Departures | Adjusted Latitudes |
|---|---|---|---|---|---|---|---|
| 1 | | | | | | | |
| | 405.24 | + 388.89 | − 113.96 | − 0.05 | 0.04 | + 388.84 | − 113.92 |
| 2 | | | | | | | |
| | 336.60 | + 285.17 | + 178.83 | − 0.04 | 0.03 | + 285.13 | + 178.86 |
| 3 | | | | | | | |
| | 325.13 | − 134.92 | + 295.82 | − 0.04 | 0.03 | − 134.96 | + 295.85 |
| 4 | | | | | | | |
| | 212.91 | − 135.38 | − 164.33 | − 0.03 | 0.02 | − 135.41 | − 164.31 |
| 5 | | | | | | | |
| | 252.19 | − 251.82 | − 13.56 | − 0.03 | 0.03 | − 251.85 | − 13.53 |
| 6 | | | | | | | |
| | 237.69 | − 151.72 | − 182.97 | − 0.03 | 0.02 | − 151.75 | − 182.95 |
| 1 | | | | | | | |
| | 1769.76 | 0.22 ft | − 0.17 ft | − 0.22 | 0.17 | Σ = 0.00 | Σ = 0.00 |

**8.20. Adjusted distances and azimuths**   When the traverse has been adjusted, the final step in traverse computations is to calculate adjusted (corrected) distances and directions using the adjusted coordinates or adjusted departures and latitudes in Eqs. (8.8) and (8.9).

For the example traverse (Fig. 8.7) calculated in Table 8.10, and using adjusted coordinates from Table 8.13, the corrected distance from 2 to 3 using Eq. (8.8) is

$$d_{23} = \left[ (5056.06 - 4770.93)^2 + (6215.76 - 6036.90)^2 \right]^{1/2}$$
$$= 336.59 \text{ m}$$

and the direction of line 23 (from the north) by Eq. (8.9) is

$$A_{N23} = \arctan \frac{X_3 - X_2}{Y_3 - Y_2}$$
$$= \arctan \frac{285.13}{178.86} = 57°54'01''$$

**Table 8.13   Adjusted Coordinates (Traverse of Fig. 8.7, Calculated in Table 8.10)**

| Station | Adjusted Departures, ft | Adjusted Latitudes, ft | Adjusted Coordinates, ft X | Adjusted Coordinates, ft Y |
|---|---|---|---|---|
| 1 | | | 4382.09 | 6150.82 |
| | + 388.84 | − 113.92 | | |
| 2 | | | 4770.93 | 6036.90 |
| | + 285.13 | + 178.86 | | |
| 3 | | | 5056.06 | 6215.76 |
| | − 134.96 | + 295.85 | | |
| 4 | | | 4921.10 | 6511.61 |
| | − 135.41 | − 164.31 | | |
| 5 | | | 4785.69 | 6347.30 |
| | − 251.85 | − 13.53 | | |
| 6 | | | 4533.84 | 6333.77 |
| | − 151.75 | − 182.95 | | |
| 1 | | | 4382.09 | 6150.82 |

**Table 8.14 Adjusted Distances and Azimuths (Closed traverse in Fig. 8.7)**

| Station | Adjusted Distance, ft | Azimuth |
|---|---|---|
| 1 | | |
| | 405.18 | 106°19'45" |
| 2 | | |
| | 336.59 | 57°54'01" |
| 3 | | |
| | 325.18 | 335°28'43" |
| 4 | | |
| | 212.92 | 219°29'33" |
| 5 | | |
| | 252.21 | 266°55'30" |
| 6 | | |
| | 237.69 | 219°40'28" |
| 1 | | |

Corrected distances and azimuths are listed for all the lines in this example traverse in Table 8.14. These are the values that would be used on the plan or map for which the survey was intended.

**8.21. Computations with rectangular coordinates**  The use of rectangular coordinates permits application of the principles of analytical geometry to solving surveying problems. Some of the more useful relationships are outlined in this section.

Equations (8.8) to (8.10), developed in Art. 8.15, allow calculation of distance and azimuth of a line given the coordinates for the two end points. Two additional equations for the distance of line $ij$ are

$$d_{ij} = \frac{X_j - X_i}{\sin A_{ij}} \tag{8.12a}$$

$$d_{ij} = \frac{Y_j - Y_i}{\cos A_{ij}} \tag{8.12b}$$

When $|X_j - X_i|$ exceeds $|Y_j - Y_i|$, use Eq. (8.12a). Conversely, when $|Y_j - Y_i|$ exceeds $|X_j - X_i|$, use Eq. (8.12b).

Equations (8.8) to (8.12b) are the best known and most frequently used relationships from analytical geometry involving coordinates that are employed in surveying calculations. Next, consider some equally useful but less used relationships.

If the coordinates of point $i$ (Fig. 8.28) and the azimuth of line $iP$ are given, the *point slope equation* for the straight line $iP$ is

$$Y - Y_i = (X - X_i) \cot A_{iP} \tag{8.13}$$

where the $\cot A_{iP}$ represents the slope of the line measured from the positive $X$ axis.

When the azimuth and $Y$ intercept are known, the *slope intercept form* for the equation of a line is

$$Y - b = X \cot A_{iP} \tag{8.14}$$

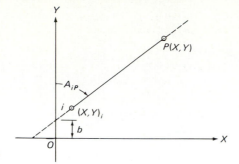

**Fig. 8.28**

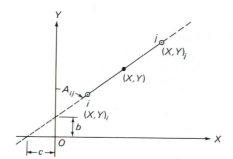

**Fig. 8.29**

If the coordinates for two ends of a line are given, as for line $ij$ in Fig. 8.29, then the *two-point form* for the equation of the straight line $ij$ becomes

$$\frac{Y - Y_i}{Y_j - Y_i} = \frac{X - X_i}{X_j - X_i} \tag{8.15}$$

Expansion and rearrangement of Eq. (8.15) leads to the general equation for a straight line,

$$AX + BY + C = 0 \tag{8.16}$$

in which

$$A = Y_j - Y_i \qquad B = X_i - X_j \qquad C = X_j Y_i - X_i Y_j \tag{8.16a}$$

which then leads to

$$X \text{ intercept} = -\frac{C}{A} = \frac{X_j Y_i - X_i Y_j}{Y_i - Y_j} \tag{8.17}$$

$$Y \text{ intercept} = -\frac{C}{B} = \frac{X_j Y_i - X_i Y_j}{X_j - X_i} \tag{8.18}$$

$$\tan A_{ij} = -\frac{B}{A} = \frac{X_j - X_i}{Y_j - Y_i} \tag{8.19}$$

Note that the symbol $A$ in Eq. (8.16) should not be confused with the notation $A_{ij}$ used to designate azimuth.

Referring to Fig. 8.30, the slope of line $ij$ is $m = \cot A_{ij}$, and the slope of the normal $ON = -1/m = \tan \omega$. Therefore, $m = -\cot \omega$. If the length of the normal is $p$, the $Y$ intercept is $b = p/\sin \omega$ and the *normal form* of the equation of the line $ij$ according to Eq. (8.14) is

$$Y - \frac{p}{\sin \omega} = -X \cot \omega$$

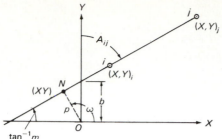

**Fig. 8.30**   $\tan^{-1} m$

or

$$Y \sin \omega - p = -X \cos \omega$$

which becomes

$$X \cos \omega + Y \sin \omega - p = 0 \qquad (8.20)$$

Since $\cot A_{ij} = -\cot \omega = -A/B$ from Eq. (8.19), then

$$\sin \omega = \frac{B}{\pm (A^2 + B^2)^{1/2}} \qquad \cos \omega = \frac{A}{\pm (A^2 + B^2)^{1/2}} \qquad (8.21)$$

Because the slope of the line is less than 180°, $\sin \omega$ is positive and the sign of the radical must equal the sign of $B$. Dividing Eq. (8.16) by $\pm (A^2 + B^2)^{1/2}$ yields

$$\frac{A}{\pm \sqrt{A^2 + B^2}} X + \frac{B}{\pm \sqrt{A^2 + B^2}} Y + \frac{C}{\pm \sqrt{A^2 + B^2}} = 0 \qquad (8.22)$$

the normal form of the equation for straight line $ij$ in which $A$, $B$, and $C$ are defined by Eq. (8.16).

The normal form of the equation for a line [Eq. (8.22)] can be applied so as to obtain the perpendicular distance from any straight line $ij$ to any coordinated point $P$, as illustrated in Fig. 8.31. In this case, the distance $d$ is given by

$$d = \frac{A}{\pm \sqrt{A^2 + B^2}} X_P + \frac{B}{\pm \sqrt{A^2 + B^2}} Y_P + \frac{C}{\pm \sqrt{A^2 + B^2}} \qquad (8.23)$$

where $X_P$, $Y_P$ are coordinates of point $P$. The coordinates $(XY)_i$ and $(XY)_j$ are used to evaluate $A$, $B$, and $C$ with Eqs. (8.16) to (8.19).

The preceding equations are related to straight lines and are of the first degree. An example of an equation of the second degree, applicable in surveying calculations is the equation for a circle.

Let the coordinates for the center of a circle be $(H, K)$, the length of the radius be $r$, and the coordinates of the point $P$ tracing the circle be $(X, Y)$ (Fig. 8.32). By Eq. (8.8),

$$r = \left[ (X - H)^2 + (Y - K)^2 \right]^{1/2}$$

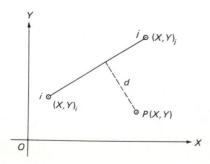

**Fig. 8.31**   $O$

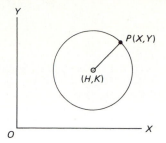

**Fig. 8.32**

or

$$r^2 = (X - H)^2 + (Y - K)^2 \tag{8.24}$$

is the standard equation for a circle. If the center of the circle is taken as the origin of the coordinate system, Eq. (8.24) becomes

$$r^2 = X^2 + Y^2 \tag{8.25}$$

Consider application of the foregoing equations to some surveying problems.

**Example 8.1**   Determine the coordinates for the intersection of a line passing from station 1 to 3 with a line going from station 2 to 5 in the closed-loop traverse calculated in Art. 8.17 and shown in Fig. 8.7.

SOLUTION   The coordinates of stations 1, 3, 2, and 5 are as follows (Table 8.13):

| Station | $X$, ft | $Y$, ft | $X' = X - 4300$ ft | $Y' - Y - 6000$ ft |
|---------|---------|---------|--------------------|--------------------|
| 1 | 4382.09 | 6150.82 | 82.09 | 150.82 |
| 3 | 5056.06 | 6215.76 | 756.06 | 215.76 |
| 2 | 4770.93 | 6036.90 | 470.93 | 36.90 |
| 5 | 4785.69 | 6347.30 | 485.69 | 347.30 |

Let $X$ and $Y$ be the unknown coordinates for the point of intersection. Using the translated $X'$ and $Y'$ coordinates for stations 1 $\cdots$ 5 so as to reduce the size of the numbers, and thus possible roundoff error, the equation for line 13 by Eq. (8.15) is

$$\frac{Y' - 150.82}{215.76 - 150.82} = \frac{X' - 82.09}{756.06 - 82.09} \tag{1}$$

and the equation for line 25 is

$$\frac{Y' - 36.90}{347.30 - 36.90} = \frac{X' - 470.93}{485.69 - 470.93} \tag{2}$$

Expansion and rearrangement of Eqs. (1) and (2) yield

$$-673.97Y' + 64.94X' = \qquad 96,317.23 \tag{3a}$$
$$14.76Y' - 310.40X' = -145,632.028 \tag{3b}$$

Equations (3a) and (3b) can be solved simultaneously for $Y'$ by multiplying Eq. (3a) by $310.40/64.94$ and adding it to Eq. (3b).

$$14.76Y' - 310.40X' \quad = -145,632.0280$$
$$-3221.4396Y' + 310.40X' = -460,376.7815$$
$$-3206.6796Y' \qquad = -606,008.8095$$
$$Y' \qquad = 188.9833 \text{ ft}$$

Substitution of this value for $Y'$ into Eqs. (3a) and (3b) as a check gives
$X' = 478.1618$ ft

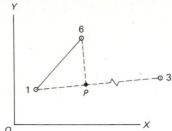

**Fig. 8.33**

The final coordinates of the intersection point are

$X = 4300.00 + 478.16 = 4778.16$ ft
$Y = 6000.00 + 188.98 + 6188.98$ ft

**Example 8.2** Using data from Example 8.1 and the traverse calculated in Art. 8.17, compute the shortest distance from station 6 to the straight line connecting stations 1 and 3.

SOLUTION   In Fig. 8.33, let the unknown coordinates of $P$ be $(X, Y)$. The coordinates of 6 (Table 8.13) are $X_6 = 4533.84$ ft and $Y_6 = 6333.77$ ft, which may be reduced to $X_6' = 233.84$ ft and $Y_6' = 333.77$ ft as was done in Example 8.1. The distance $6P$ can be found by evaluating the normal form [Eq. (8.22)] of Eq. (1) for line 13 as developed in Example 8.1, with the coordinates of station 6. Thus,

$A = +64.94 \qquad B = -673.97 \qquad C = 96{,}317.23$

and $\sqrt{A^2 + B^2} = 677.09$, so that the normal form of Eq. (1) is

$$\frac{-64.94}{677.09} X + \frac{673.97}{677.09} Y + \frac{96{,}317.23}{677.09} = 0$$

Since $B$ is negative, $\sqrt{A^2 + B^2}$ is also negative. According to Eq. (8.23), substitution of $X_6'$ and $Y_6'$ into the normal form of the equation yields the normal distance $d$, or

$d = (-0.095910)(233.84) + (0.995390)(333.77) - 142.25 = 167.55$ ft

that is, the distance $6P$.

   An alternative solution is possible by forming the point-slope equation for the line passing through 6 and perpendicular to line 13. The equation for line 13 is [ from Eq. (8.13)]

$(Y_3 - Y_1) = (X_3 - X_1)\cot_{13}$

Since line $6P$ is perpendicular to line 1-3,

$\tan A_{6P} = -\dfrac{6215.76 - 6150.82}{5056.06 - 4382.09}$

$\qquad = -0.096354$

or

$\cot A_{6P} = -10.378349$

so the point-slope equation for line $6P$ is

$\qquad Y - 333.77 = (-10.378349)(X - 233.84)$

$Y + 10.378349X = 2760.64$

This equation and the equation for line 13 (Example 8.1) are

$\qquad Y + 10.378349X = 2760.64 \qquad$ eq. line $6P$

$673.97Y - 64.94X = 96{,}317.23 \qquad$ eq. line 13

Multiplication of the first equation by 673.97 and subtracting the second from the first yields

$\qquad 673.97Y + 6994.70X = 1{,}860{,}588.54$

$\underline{-673.97Y + \qquad 64.94X = - \quad 96{,}317.23}$

$\qquad\qquad 7059.64X = 1{,}764{,}271.31$

$\qquad\qquad\qquad X = \qquad 249.91 \text{ ft} = X_P'$

Substitution of $X$ into both equations, as a check, gives

$Y = 166.99 \text{ ft} = Y'_P$

The distance $6P$ is evaluated using Eq. (8.8):

$d_{6P} = [(249.91 - 233.84)^2 + (166.99 - 333.77)^2]^{1/2}$

$\quad = 167.55 \text{ ft}$

the same as calculated previously. The coordinates for $P$ in the nontranslated system are $X = 4300.00 + 249.91 = 4549.91$ ft and $Y_P = 6000.00 + 166.99 = 6166.99$ ft.

**Example 8.3**   In route and property surveying, it is often necessary to determine coordinates for the intersection of a straight property line and a curved center line or property line. Figure 8.34 illustrates a curved property line with a radius of $r = 750.000$ m that is intersected by line $QR$, the center line of a sewer right of way. The azimuth of $QR$ is $30°50'00''$. Calculate the coordinates of point $P$ if the coordinates of $Q$ and $O$ (the center of the circle) are as follows:

| Station | X<br>m | Y<br>m |
|---|---|---|
| Q | 2310.638 | 2560.921 |
| O | 2561.051 | 2110.452 |

For convenience, these values are translated $-2100$ m in $Y$ and $-2300$ m in $X$ to reduce the size of the numbers thus:

| Station | X'<br>m | Y'<br>m |
|---|---|---|
| Q | 10.638 | 460.921 |
| O | 261.051 | 10.452 |

SOLUTION   Write the point-slope equation [Eq. (8.13)] for line $QR$ and the equation for the circle of radius $r = 750.000$ m. Both equations are formed in terms of the unknown coordinates for $P[X, Y]_P$. Solve these two equations simultaneously for $[X, Y]_P$.

The point-slope equation for line $QR$ is

$Y'_P - Y'_Q = (X'_P - X'_Q)\cot A_{QR}$

$Y'_P - 460.921 = (X'_P - 10.638)(1.675299)$

$Y'_P = 1.675299 X'_P + 443.0992$ \hfill (1)

The equation for the circle $r = 750.00$ using Eq. (8.24) is

$(X'_P - 261.051)^2 + (Y'_P - 10.452)^2 - (750.00)^2 = 0$ \hfill (2)

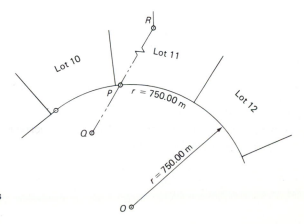

**Fig. 8.34** Intersection of straight lines and curved.

Substitution of $Y'_p$ from (1) into (2) yields

$$3.806627X_p'^2 + 927.5247X_p' - 307,168.7757 = 0 \qquad (3)$$

Equation (3) is a quadratic equation the roots of which can be computed from

$$x = \frac{-b \pm (b^2 - 4ac)^{1/2}}{2a}$$

where $a = 3.806627$, $b = 927.5247$, and $c = -307,168.7757$, from Eq. (3). Thus, the two roots of Eq. (3) and corresponding values for $Y'_p$ are

$X'_p = 187.258$ m   or   $-430.919$ m
$Y'_p = 756.813$ m   or   $-278.820$ m

Examination of Fig. 8.34 reveals that the first set of answers is correct, since the second set yields the intersection point along the extended line $RQ$ to the other side of the circle. To check the solution, $X'_p = 187.258$ and $Y'_p = 756.813$ are substituted into Eq. (2). The final coordinates for $P$ are

$X_p = 187.258 + 2300.000 = 2487.258$ m
$Y_p = 756.813 + 2100.000 = 2856.813$ m

**8.22. Coordinate transformations**   When surveyed points are coordinated, a number of survey problems can be solved conveniently by *coordinate transformations*. Coordinates in a local system may be required in the state plane coordinate system; or it may be desired to convert coordinates in several local systems into one regional framework. Any time that it becomes necessary to convert from one coordinate system to another, a transformation is needed.

Coordinate transformations in two- and three-dimensional space can be found in Ref. 9.

**8.23. Traverse accuracy and error propagation**   In general, the accuracy of traverses is judged on the basis of the resultant closure (Art. 8.16) of the traverse. This resultant closure is a function of the accuracies in the measurement of lengths and directions and hence varies with the length of the traverse. Consequently, a *relative accuracy* based on the resultant closure and length of traverse is computed. Thus, for the example traverse computed in Table 8.9, the resultant closure is 0.36 m for a traverse having a total length of 1762 m. The relative accuracy for this traverse is 0.36/1762, or 1 part in 4894.

To achieve a desired relative accuracy for a given traverse, specifications are written to govern the traverse field operations and the types of equipment that should be employed. Table 8.15 shows specifications recommended by the U.S. Federal Geodetic Control Committee for first-, second-, and third-order traverses. Note that second- and third-order traverses are each divided into class I and class II surveys. The relative accuracy or *positional closure* for each order and class of survey is expressed as a ratio (e.g., 1 : 10,000 for a third-order class I traverse) or as a constant times the square root of the distance traversed, which is designated by $K$ [e.g., $0.4(K)^{1/2}$ for a third-order class I traverse]. The latter criterion is to be used for longer traverses. It is recommended that the relationship yielding the smallest possible positional closure be utilized.

Although the positional closure is an indication of the overall quality of the traverse and is used for traverse classification, it does not yield information on the precision of point locations determined in a traverse. Therefore, in the following paragraphs, the techniques of error propagation (see Art. 2.20) are employed to determine the covariance matrix for each point in the traverse.

For simplicity, consider the very first point in a traverse, such as point 2, which follows control point 1 in Fig. 8.35. The azimuth of line 12, $A_{12}$, is given by

$$A_{12} = A_r + \alpha_1 - 180°$$

in which $A_r$ is the beginning reference azimuth. If the reasonable assumption is made that $A_r$

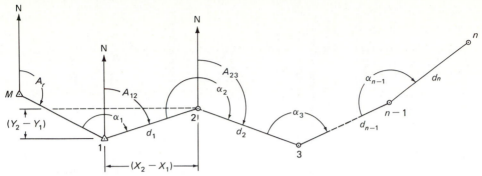

**Fig. 8.35** Open-end traverse.

is relatively error-free and has a standard deviation of zero, then $\sigma_{A_{12}}=\sigma_{\alpha_1}$. Further, let $\sigma_{d_1}$ represent the standard deviation of the measured distance $d_1$, and assume that $\alpha_1$ and $d_1$ are uncorrelated (i.e., $\sigma_{\alpha_1 d_1}=0$), which is a very logical assumption. Then

$$\Sigma_{m1}=\begin{bmatrix} \sigma_{\alpha_1}^2 & 0 \\ 0 & \sigma_{d_1}^2 \end{bmatrix}$$

represents the covariance matrix (actually a variance matrix in this case) for the measurements in the first leg of the traverse. Next, the coordinates of point 1 are given by

$$X_2 = X_1 + d_1 \sin A_{12} = X_1 + d_1 \sin(A_r + \alpha_1 - 180°)$$
$$Y_2 = Y_1 + d_1 \cos A_{12} = Y_1 + d_1 \cos(A_r + \alpha_1 - 180°) \tag{8.26}$$

in which $X_1, Y_1$, the coordinates of the control point 1, are assumed to be errorless. Applying error propagation equation (2.42) to Eq. (8.26),

$$\Sigma_{C2}=\begin{bmatrix} \sigma_{X_2}^2 & \sigma_{X_2 Y_2} \\ \sigma_{X_2 Y_2} & \sigma_{Y_2}^2 \end{bmatrix}=J\Sigma_{m1}J^t \tag{8.27}$$

in which $\Sigma_{C2}$ is the covariance matrix of the *coordinates* of point 2 and $J$ is the jacobian matrix of partial derivatives given by

$$J=\begin{bmatrix} \dfrac{\partial X_2}{\partial \alpha_1} & \dfrac{\partial X_2}{\partial d_1} \\ \dfrac{\partial Y_2}{\partial \alpha_1} & \dfrac{\partial Y_2}{\partial d_1} \end{bmatrix}=\begin{bmatrix} d_1 \cos A_{12} & \sin A_{12} \\ -d_1 \sin A_{12} & \cos A_{12} \end{bmatrix}$$

or

$$J=\begin{bmatrix} Y_2 - Y_1 & \dfrac{X_2 - X_1}{d_1} \\ -(X_2 - X_1) & \dfrac{Y_2 - Y_1}{d_1} \end{bmatrix} \tag{8.28}$$

where we used the substitution $(Y_2 - Y_1)/d_1 = \cos A_{12}$ and $(X_2 - X_1)/d_1 = \sin A_{12}$.

**Table 8.15 Classification, Standards of Accuracy, and General Specifications for Horizontal Control**

| Classification | First-order | Second-order Class I | Second-order Class II | Third-order Class I | Third-order Class II |
|---|---|---|---|---|---|
| | | Traverse | | | |
| *Recommended spacing of principal stations* | Network stations 10–15 km; other surveys seldom less than 3 km | Principal stations seldom less than 4 km except in metropolitan area surveys where the limitation is 0.3 km | Principal stations seldom less than 2 km except in metropolitan area surveys where the limitation is 0.2 km | Seldom less than 0.1 km in tertiary surveys in metropolitan area surveys; as required for other surveys | |
| *Horizontal directions or angles* | | | | | |
| Instrument | 0."2 | 0."2 } or { 1."0 | 0."2 } or { 1."0 | 1."0 | 1."0 |
| Number of observations | 16 | 8 / 12$^b$ | 6 / 8$^b$ | 4 | 2 |
| Rejection limit from mean | 4" | 4" / 5" | 4" / 5" | 5" | 5" |
| *Length measurements* | | | | | |
| Standard error$^c$ | 1 part in 600,000 | 1 part in 300,000 | 1 part in 120,000 | 1 part in 60,000 | 1 part in 30,000 |
| *Reciprocal vertical angle observations* | | | | | |
| Number of and spread between observations | 3 D/R—10" | 3 D/R—10" | 2 D/R—10" | 2 D/R—10" | 2 D/R—20" |
| Number of stations between known elevations | 4–6 | 6–8 | 8–10 | 10–15 | 15–20 |

| | 5–6 | 10–12 | 15–20 | 20–25 | 30–40 |
|---|---|---|---|---|---|
| **Astro azimuths** | | | | | |
| Number of courses between azimuth checks | | | | | |
| Number of observations/ night | 16 | 16 | 12 | 8 | 4 |
| Number of nights | 2 | 2 | 1 | 1 | 1 |
| Standard error | 0".45 | 0".45 | 1".5 | 3".0 | 8".0 |
| Azimuth closure at azimuth checkpoint not to exceed [a] | 1".0 per station or 2"$\sqrt{N}$ | 1".5 per station or 3"$\sqrt{N}$ Metropolitan area surveys seldom to exceed 2".0 per station or 3"$\sqrt{N}$ | 2".0 per station or 6"$\sqrt{N}$ Metropolitan area surveys seldom to exceed 4".0 per station or 8"$\sqrt{N}$ | 3".0 per station or 10"$\sqrt{N}$ Metropolitan area surveys seldom to exceed 6".0 per station or 15"$\sqrt{N}$ | 8" per station or 30"$\sqrt{N}$ |
| **Position closure [d]** | | | | | |
| after azimuth adjustment | 0.04 m $\sqrt{K}$ or 1:100,000 | 0.08 m $\sqrt{K}$ 1:50,000 | 0.2 m $\sqrt{K}$ 1:20,000 | 0.4 m $\sqrt{K}$ 1:10,000 | 0.8 m $\sqrt{K}$ 1:5,000 |

[a] $N$ is the number of stations for carrying an azimuth.
[b] May be reduced to 8 and 4, respectively, in metropolitan areas.
[c] Use Eq. 2.37, Art. 2.17.
[d] $K$ is the distance in kilometres.

By substitution of Eq. (8.28) into Eq. (8.27) and matrix multiplication (see Art. A.3.4), the covariance matrix for the coordinates of point 2 becomes

$$\Sigma_{C2} = \left[\begin{array}{c:c}
(d_1^2\sigma_{\alpha_1}^2\cos^2 A_{12} + \sigma_{d_1}^2\sin^2 A_{12}) & (\sigma_{d_1}^2 - d_1^2\sigma_{\alpha_1}^2)\sin A_{12}\cos A_{12} \\ \hdashline
(\sigma_{d_1}^2 - d_1^2\sigma_{\alpha_1}^2)\sin A_{12}\cos A_{12} & (d_1^2\sigma_{\alpha_1}^2\sin^2 A_{12} + \sigma_{d_1}^2\cos^2 A_{12})
\end{array}\right]$$

$$= \left[\begin{array}{c:c}
(Y_2 - Y_1)\sigma_{\alpha_1}^2 + \dfrac{(X_2 - X_1)^2}{d_1^2}\sigma_{d_1}^2 & -(X_2 - X_1)(Y_2 - Y_1)\sigma_{\alpha_1}^2 + \dfrac{(X_2 - X_1)(Y_2 - Y_1)}{d_1^2}\sigma_{d_1}^2 \\ \hdashline
\text{symmetric} & (X_2 - X_1)^2\sigma_{\alpha_1}^2 + \dfrac{(Y_2 - Y_1)^2}{d_1^2}\sigma_{d_1}^2
\end{array}\right]$$

$$(8.29)$$

When performing calculations, Eq. (8.27) is usually applied without going through the expanded form. In this case, the matrix multiplication is carried out to illustrate the following two points:

1. Although the original measurements $\alpha_1, d_1$ may be uncorrelated, the resulting quantities are in general correlated, as shown in Eq. (8.29).
2. In order for the covariance term in Eq. (8.29) to be zero,

$$\sigma_{d_1}^2 = d_1^2\sigma_{\alpha_1}^2 \qquad \text{or} \qquad \sigma_{d_1} = d_1\sigma_{\alpha_1}$$

Note that the propagation of error to point 2 in the example is somewhat of a special case since the beginning point 1 is a control point assumed to be error-free. Now, consider the precision of the coordinates for succeeding point 3 in Fig. 8.35. The coordinates of point 3 are given by

$$X_3 = X_2 + d_2\sin A_{23}$$
$$Y_3 = Y_2 + d_2\cos A_{23} \qquad\qquad (8.30)$$

The four variables $X_2, Y_2, \alpha_2, d_2$ (which are implicit in $A_{23}$, see Fig. 8.35) are the random variables. Their covariance matrix is given symbolically by

$$\left[\begin{array}{cccc}
\sigma_{X_2}^2 & \sigma_{X_2 Y_2} & \sigma_{X_2 \alpha_2} & \sigma_{X_2 d_2} \\
\sigma_{X_2 Y_2} & \sigma_{Y_2}^2 & \sigma_{Y_2 \alpha_2} & \sigma_{Y_2 d_2} \\
\sigma_{X_2 \alpha_2} & \sigma_{Y_2 \alpha_2} & \sigma_{\alpha_2}^2 & \sigma_{\alpha_2 d_2} \\
\sigma_{X_2 d_2} & \sigma_{Y_2 d_2} & \sigma_{\alpha_2 d_2} & \sigma_{d_2}^2
\end{array}\right] = \left[\begin{array}{cc}
\Sigma_{C2} & \Sigma_{C2m2} \\
\Sigma_{C2m2}^t & \Sigma_{m2}
\end{array}\right]$$

in which $\Sigma_{C2}$ is the covariance matrix for the coordinates of point 2 as derived in Eq. (8.29) and $\Sigma_{m2}$ is the covariance matrix for the two measurements $d_2, \alpha_2$ that is usually diagonal with elements $\sigma_{d_2}^2, \sigma_{\alpha_2}^2$. The matrix $\Sigma_{C2m2}$ represents the correlation between the coordinates of point 2 and the new measurements $d_2$ and $\alpha_2$. At first glance, it may appear that no such correlation exists. However, except for an azimuth traverse (Art. 8.10) or a traverse run using a gyroscopic equipped theodolite (Arts. 11.46 and 20.15), the azimuth $A_{23}$ is computed not only from $\alpha_2$ but also from $A_{12}$, which in turn is a function of $\alpha_1$. Since $X_2, Y_2$ were computed in terms of $\alpha_1$, then in general they will be correlated with $\alpha_2$ and $\Sigma_{C2m2}$ will not necessarily be zero. To evaluate $\Sigma_{C2m2}$ requires *cross-covariance* propagation, which is beyond the scope of this textbook (for those interested, see Ref. 9 at the end of the chapter). Instead, Eq. (8.30) is extended so that it is expressed in terms of the original observations.

Thus,

$$X_3 = X_1 + d_1 \sin A_{12} + d_2 \sin A_{23}$$
$$Y_3 = Y_1 + d_1 \cos A_{12} + d_2 \cos A_{23}$$

or

$$X_3 = X_1 + d_1 \sin(A_r + \alpha_1 - 180°) + d_2 \sin[A_r + \alpha_1 + \alpha_2 - (2)(180°)]$$
$$Y_3 = Y_1 + d_1 \cos(A_r + \alpha_1 - 180°) + d_2 \cos[A_r + \alpha_1 + \alpha_2 - (2)(180°)]$$

The jacobian matrix for these equations is (see Art. A.10)

$$\mathbf{J} = \begin{bmatrix} \dfrac{\partial X_3}{\partial \alpha_1} & \dfrac{\partial X_3}{\partial d_1} & \dfrac{\partial X_3}{\partial \alpha_2} & \dfrac{\partial X_3}{\partial d_2} \\[3mm] \dfrac{\partial Y_3}{\partial \alpha_1} & \dfrac{\partial Y_3}{\partial d_1} & \dfrac{\partial Y_3}{\partial \alpha_2} & \dfrac{\partial Y_3}{\partial d_2} \end{bmatrix}$$

$$= \begin{bmatrix} (d_1 \cos A_{12} + d_2 \cos A_{23}) & \vdots & \sin A_{12} & \vdots & d_2 \cos A_{23} & \vdots & \sin A_{23} \\ -(d_1 \sin A_{12} + d_2 \sin A_{23}) & \vdots & \cos A_{12} & \vdots & -d_2 \sin A_{23} & \vdots & \cos A_{23} \end{bmatrix}$$

$$= \begin{bmatrix} Y_3 - Y_1 & \vdots & \dfrac{X_2 - X_1}{d_1} & \vdots & Y_3 - Y_2 & \vdots & \dfrac{X_3 - X_2}{d_2} \\[3mm] -(X_3 - X_1) & \vdots & \dfrac{Y_2 - Y_1}{d_1} & \vdots & -(X_3 - X_2) & \vdots & \dfrac{Y_3 - Y_2}{d_2} \end{bmatrix}$$

The four measurements $\alpha_1, d_1, \alpha_2, d_2$ may be assumed to be uncorrelated, and therefore their covariance matrix is diagonal or $\Sigma_{m_{12}} = \text{diag}(\sigma_{\alpha_1}^2 \sigma_{d_1}^2 \sigma_{\alpha_2}^2 \sigma_{d_2}^2)$. Using $\mathbf{J}$ just evaluated and this diagonal covariance matrix for the measurements in the error propagation Eq. (2.42), the covariance matrix for the coordinates of point 3 is

$$\Sigma_{C3} = \underset{2,4}{\mathbf{J}} \ \underset{4,4}{\Sigma_{m_{12}}} \ \underset{4,2}{\mathbf{J}^t}$$

$$\Sigma_{C3} = \begin{bmatrix} \left[(Y_3 - Y_1)^2 \sigma_{\alpha_1}^2 + (Y_3 - Y_2)^2 \sigma_{\alpha_2}^2\right] + \left[\dfrac{(X_2 - X_1)^2}{d_1^2} \sigma_{d_1}^2 + \dfrac{(X_3 - X_2)^2}{d_2^2} \sigma_{d_2}^2\right] \\[4mm] \hline \\ -\left[(Y_3 - Y_1)(X_3 - X_1)\sigma_{\alpha_1}^2 + (Y_3 - Y_2)(X_3 - X_2)\sigma_{\alpha_2}^2\right] + \left[\dfrac{(X_2 - X_1)(Y_2 - Y_1)}{d_1^2}\sigma_{d_1}^2 + \dfrac{(X_3 - X_2)(Y_3 - Y_2)}{d_2^2}\sigma_{d_2}^2\right] \\[4mm] \text{symmetric} \\[4mm] \left[(X_3 - X_1)^2 \sigma_{\alpha_1}^2 + (X_3 - X_2)^2 \sigma_{\alpha_2}^2 + \dfrac{(Y_2 - Y_1)^2}{d_1^2}\sigma_{d_1}^2 + \dfrac{(Y_3 - Y_2)^2}{d_2^2}\sigma_{d_2}^2\right] \end{bmatrix}$$

which becomes

$$\Sigma_{C3} =$$

$$\begin{bmatrix} \left[\sum_{i=1}^{2}(Y_3 - Y_i)^2\sigma_{\alpha_i}^2 + \sum_{i=1}^{2}\left(\dfrac{X_{i+1} - X_i}{d_i}\right)^2\sigma_{d_i}^2\right] & \left[-\sum_{i=1}^{2}(Y_3 - Y_i)(X_3 - X_i)\sigma_{\alpha_i}^2 + \sum_{i=1}^{2}\dfrac{(X_{i+1} - X_i)(Y_{i+1} - Y_i)}{d_i^2}\sigma_{d_i}^2\right] \\[4mm] \hline \\ \text{symmetric} & \left[\sum_{i=1}^{2}(X_3 - X_i)^2\sigma_{\alpha_i}^2 + \sum_{i=1}^{2}\left(\dfrac{Y_{i+1} - Y_i}{d_i^2}\right)^2\sigma_{d_i}^2\right] \end{bmatrix}$$

$$(8.31)$$

Equation (8.31) illustrates the error propagation for the special case of two measured lines. For the general case of $n$ traverse stations in an open traverse, starting with point 1 and terminating at point $n$, the coordinates for point $n$ can be expressed as

$$X_n = X_1 + d_1 \sin(A_r + \alpha_1 - 180°) + d_2 \sin[A_r + \alpha_1 + \alpha_2 - (2)180°] + \cdots$$
$$+ d_n \sin[A_r + \alpha_1 + \alpha_2 + \cdots + \alpha_{n-1} - (n-1)180°]$$
$$Y_n = Y_1 + d_1 \cos(A_r + d_1 - 180°) + d_2 \cos[A_r + \alpha_1 + \alpha_2 - (2)180°] + \cdots$$
$$+ d_n \cos[A_r + \alpha_1 + \alpha_2 + \cdots + \alpha_{n-1} - (n-1)180°]$$

Assuming uncorrelated measurements of angles and distances and errorless coordinates for station 1, application of error propagation Eq. (2.42) yields the covariance matrix for point $n$ as follows:

$$\Sigma_{Cn} = \underset{2n,2}{J} \ \underset{2n,2n}{\Sigma_{m(n-1)}} \ \underset{2n,2}{J^t} = \begin{bmatrix} \sigma_{Xn}^2 & \sigma_{XnYn} \\ \sigma_{XnYn} & \sigma_{Yn}^2 \end{bmatrix} \tag{8.32}$$

in which the elements $\sigma_{Xn}^2, \sigma_{Yn}^2, \sigma_{XnYn}$ are a direct extension to those in Eq. (8.31), or

$$\sigma_{Xn}^2 = \sum_{i=1}^{n-1} (Y_n - Y_i)^2 \sigma_{\alpha_i}^2 + \sum_{i=1}^{n-1} \left( \frac{X_{i+1} - X_i}{d_i} \right)^2 \sigma_{d_i}^2$$

$$\sigma_{Yn}^2 = \sum_{i=1}^{n-1} (X_n - X_i)^2 \sigma_{\alpha_i}^2 + \sum_{i=1}^{n-1} \left( \frac{Y_{i+1} - Y_i}{d_i} \right)^2 \sigma_{d_i}^2 \tag{8.33}$$

$$\sigma_{Y_n X_n} = -\sum_{i=1}^{n-1} (Y_n - Y_i)(X_n - X_i)\sigma_{\alpha_i}^2 + \sum_{i=1}^{n-1} \left[ \frac{(X_{i+1} - X_i)(Y_{i+1} - Y_i)}{d_i^2} \right] \sigma_{d_i}^2$$

The practical value of this error propagation as developed lies in the application of Eqs. (8.33) to the data for proposed traverses in order to estimate accuracies possible at specific traverse stations using assumed values for $\sigma_{\alpha_i}$ and $\sigma_{d_i}$.

**Example 8.4** Consider the traverse used in Examples B.10 and B.11 (Art. B.14, Appendix B). Figure B.10a shows a closed traverse starting at point $B$ and closing on point $E$, with two intermediate stations $C$ and $D$. For convenience, the data are repeated here.

| Angle | Value | $\sigma$ | Distance | Value, m | $\sigma$, m |
|---|---|---|---|---|---|
| $\alpha_1$ | 172°53′34″ | 2″ | $d_1$ | 281.832 | 0.016 |
| $\alpha_2$ | 185°22′14″ | 2″ | $d_2$ | 271.300 | 0.016 |
| $\alpha_3$ | 208°26′19″ | 2″ | $d_3$ | 274.100 | 0.016 |
| $\alpha_4$ | 205°13′51″ | 2″ | Point | X, m | Y, m |
| $A_{z_B}$ | 68°15′20.7″ | 0 | B | 8478.139 | 2483.826 |
| $A_{z_E}$ | 300°11′30.5″ | 0 | E | 7709.336 | 2263.411 |

The observations are assumed to be uncorrelated and with the precisions indicated in the tabulation.

It is desired to determine the estimated covariance matrix for point $D$ treating stations $B$, $C$, and $D$ as an open traverse and to calculate the parameters of the resulting standard error ellipse at $D$.

SOLUTION  Distances, azimuths, and unadjusted coordinates for lines $BC$, $CD$, and points $C$ and $D$ are as follows:

| Sta-tion | Distance, m | Azimuth | Departures x, m | Latitudes y, m | Coordinates, m X | Coordinates, m Y |
|---|---|---|---|---|---|---|
| B | | | | | 8478.139 | 2483.826 |
| | 281.832 | 241°08′57.7″ | −246.851 | −135.993 | | |
| C | | | | | 8231.288 | 2347.833 |
| | 271.300 | 246°31′14.6″ | −248.838 | −108.091 | | |
| D | | | | | 7982.450 | 2239.742 |

The standard deviation in angles is $\sigma_{\alpha_i} = (2'')(4.8481368)(10^{-6}) = (9.6962736)(10^{-6})$ rad. Thus, the variance, $\sigma_{\alpha_i}^2 = (9.401772)10^{-11}$ rad$^2$. The variance in each measured distance is $\sigma_{d_i}^2 = (0.016)^2 = (2.56)10^{-4}$ m$^2$. Using these estimates, the measured distances and directions, and calculated coordinates in Eq. (8.33), the propagated variances and covariances are

$$\sigma_{X_n}^2 = \sum_{i=1}^{2}(Y_3 - Y_i)^2\sigma_{\alpha_i}^2 + \sum_{i=1}^{2}\left(\frac{X_{i+1}-X_i}{d_i}\right)^2\sigma_{d_i}^2 = (4.18457)10^{-4}\,m^2$$

$$\sigma_{Y_n}^2 = \sum_{i=1}^{2}(X_3 - X_i)^2\sigma_{\alpha_i}^2 + \sum_{i=1}^{2}\left(\frac{Y_{i+1}-Y_i}{d_i}\right)^2\sigma_{d_i}^2 = (1.29166)10^{-4}\,m^2$$

$$\sigma_{X_n Y_n} = -\sum_{i=1}^{2}(Y_3 - Y_i)(X_3 - X_i)\sigma_{\alpha_i}^2 + \sum_{i=1}^{2}\left[\frac{(X_{i+1}-X_i)(Y_{i+1}-Y_i)}{d_i^2}\right]\sigma_{d_i}^2 = (1.87843)10^{-4}\,m^2$$

so that the covariance matrix for point $D$ is

$$\Sigma_{DD} = \begin{bmatrix} (4.18457)10^{-4} & (1.87843)10^{-4} \\ (1.87843)10^{-4} & (1.29166)10^{-4} \end{bmatrix} m^2$$

Analysis of this covariance matrix using the methods developed in Art. 2.19 and illustrated by Examples 2.8 and 2.9 yields the following error ellipse data:

semimajor axis, $a = 0.118$ m
semiminor axis, $b = 0.110$ m
orientation angle $\theta = 26°12'$ ( = angle between semi-major and x-axes)

The procedure outlined in Example 8.4 is particularly useful in the design and planning of traverses. Thus, given approximate distances and directions for a proposed traverse, estimated values for $\sigma_{\alpha_i}$ and $\sigma_{d_i}$ are chosen and the covariance matrices for specified traverse points are calculated using Eqs. (8.32) and (8.33). The values of $\sigma_{\alpha_i}$ and $\sigma_{d_i}$ selected are those which produce covariance matrices yielding error ellipses that satisfy specifications for the proposed survey.

It must be emphasized that a covariance matrix propagated for a point using unadjusted data and estimated standard deviations as illustrated in Ex. (8.4) does not provide a correct measure for the accuracy of the *adjusted* coordinates for a point. For example, if the method were used for error propagation with coordinates adjusted by the compass rule, the resulting covariance matrix would be incorrect. Consequently, the procedure is of primary value as a design tool for open traverses.

A legitimate measure of the accuracy of an adjusted point can be obtained only by analysis of the covariance matrices propagated as a by-product of a least-squares adjustment of the traverse. Adjustment of traverses by the method of least squares is outlined in Art. B.14 and illustrated by Examples B.10 and B.11.

**8.24. Omitted measurements** When it is impossible or impractical to determine by field observations the length and bearing of every side of a closed traverse, the missing data may generally be calculated, provided not more than two quantities (lengths and/or azimuths) are omitted. (If only one measurement is omitted, a partial check is obtained on the work.) It must be assumed that the observed values are without error, and hence all errors of measurement are thrown into the computed lengths or directions. Measurements which may be supplied in this manner are:

1. Length and direction of one side.
2. Length of one side and direction of another.
3. Lengths of two sides for which the directions have been observed.
4. Directions of two sides for which the lengths have been observed.

There are three general cases: (1) length and direction of one side unknown, (2) omitted measurements in *adjoining* sides of the traverse, and (3) omitted measurements in *nonadjoin-*

*ing* sides. In case 3, the solution involves changing the order of sides in the traverse in such a way as to make the two partly unknown sides adjoin.

When one of the sides is known in direction but unknown in length, the solution can be facilitated by assuming that side to lie on the reference meridian.

Methods of parting land, which involve the calculation of lengths and directions of unknown sides of a traverse, are described in Arts. 8.34 to 8.37.

**Length and direction of one side unknown**   In a closed-loop traverse containing $k$ stations,

$$x_{12}+x_{23}+\cdots+x_{ij}+\cdots+x_{k,1}=0$$
$$y_{12}+y_{23}+\cdots+y_{ij}+\cdots+y_{k,1}=0$$

where $x$ and $y$ denote departures and latitudes calculated by Eq. (8.6). If the line $ij$ is of unknown distance and/or direction, then

$$x_{ij}=-(x_{12}+x_{23}+\cdots+x_{k,1})=-\sum D$$
$$y_{ij}=-(y_{12}+y_{23}+\cdots+y_{k,1})=-\sum L \tag{8.34}$$

and the distance of the unknown line is

$$d_{ij}=\left[\left(\sum D\right)^{2}+\left(\sum L\right)^{2}\right]^{1/2} \tag{8.35}$$

where $\sum D$, $\sum L$ are the sums of departures and latitudes of the known sides of the traverse, respectively.

The direction of an unknown line is

$$\tan A_{ij}=\frac{-\sum D}{-\sum L} \tag{8.36}$$

with due regard to signs. Note that Eqs. (8.35) and (8.36) are similar in form to Eqs. (8.8) and (8.9).

**Length of one side and direction of another side unknown**   Figure 8.36 represents a closed traverse for which the direction of the line $DE=d$ and the length of the line $EA=e$ are not determined by field measurements. Let an imaginary line extend from $D$ to $A$, cutting off the unknown sides from the remainder of the traverse. Then $ABCDA$ forms a closed traverse for which the side $DA=f$ is unknown in both direction and length. By Eq. (8.36),

$$\tan A_{DA}=\frac{x_{DA}}{y_{DA}}=\frac{-(x_a+x_b+x_c)}{-(y_a+y_b+y_c)}$$

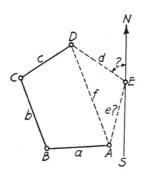

**Fig. 8.36**

and the distance $DA$ is [Eqs. (8.12$a$) and (8.12$b$)]

$$d_{DA} = \frac{x_{DA}}{\sin A_{DA}} = \frac{y_{DA}}{\cos A_{DA}}$$

In computing the length of $DA$ it is desirable to use the larger of the two quantities, departure or latitude.

The angle between the lines $e$ and $f$ in triangle $ADE$ is

$$\angle DAE = \text{azimuth of } AE - \text{azimuth of } AD$$

In the triangle $ADE$ the length of the two sides $d$ and $f$ and one angle $DAE$ are known. By the sine law in a triangle,

$$\sin DEA = \frac{f}{d} \sin DAE$$

With angle $DEA$ known, angle $ADE$ can be computed, and the remaining unknown length is given by the expression

$$e = f \frac{\sin ADE}{\sin DEA} = d \frac{\sin ADE}{\sin DAE}$$

Also,

$$\text{azimuth of } DE = \text{azimuth of } DA - \angle ADE$$

**Example 8.5** In the accompanying tabulation are given the measured lengths and bearings for the courses of a closed traverse $a$ to $f$ (Fig. 8.37), together with the departures and latitudes of the known sides. The length of $b$ and the bearing of $e$ are not observed. The general direction of $e$ is southwest. It is desired to compute the unknown length and direction. Quantities in parentheses are derived from following calculations.

| Line | Length, ft | Bearing | Departure E (+) | Departure W (−) | Latitude N (+) | Latitude S (−) |
|------|-----------|---------|-----------------|-----------------|----------------|----------------|
| a | 500.0 | N0°00′E | 0.0 | | 500.0 | |
| c | 854.4 | S69°27′E | 800.0 | | | 299.9 |
| d | 1019.8 | S11°19′E | 200.1 | | | 1000.0 |
| f | 656.8 | N54°06′W | | 532.0 | 385.1 | |
| e | 1118.0 | Unknown | | | | |
|   |   | (S78°56′30″W) | | (1097.2) | | (214.4) |
| b | Unknown | N45°00′E | | | | |
|   | (889.8) |   | (629.2) | | (629.2) | |
| g | (625.4) | (N48°27′20″W) | | (468.1) | (414.8) | |

In Fig. 8.37 the lines $a$, $c$, $d$, and $f$ are the courses for which the length and bearing are known. The line $g$ is the closing side of the figure formed by the known courses. From the tabulated quantities, and using Eq. (8.34), the departure for line $g$, $x_g = -(800.0 + 200.1 - 532.0) = -468.1$ and the latitude

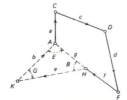

**Fig. 8.37**

$y_g = -(500.0 + 385.1 - 299.9 - 1000.0) = 414.8$. Thus, by Eq. (8.36),

$$\tan A_{HA} = \frac{-468.1}{414.8} = -1.12850$$

so that the azimuth of $HA$ or line $g$ is $311°32'40''$, yielding a bearing of N48°27'20''W. The length of $g$ is

$$d_g = \frac{x_g}{\sin A_{HA}} = \frac{-468.1}{-0.74851} = 625.38 = 625.4 \text{ ft}$$

Since the direction of line $b$ from $K$ to $A$ is N45°00'E,

$$\angle E = 45°00' + 48°27'20'' = 93°27'20''$$

$$\sin G = \frac{g}{e} \sin E = \frac{625.38}{1118.0} \times 0.99818 = 0.55836$$

$$\angle G = 33°56'30''$$

$$\angle B = 180°00' - 93°27'20'' - 33°56'30'' = 52°36'10''$$

$$\text{length } b = \text{length } e \cdot \frac{\sin B}{\sin E} = 1{,}118 \times \frac{0.79444}{0.99818} = 889.8 \text{ ft}$$

bearing of $e = 180° - 48°27'20'' - 52°36'10'' = 78°56'30''$

As a check on the calculations, the departures and latitudes of the lines $b$ and $e$ are computed and the values are shown in parentheses. The sum of the latitudes and the sum of the departures for courses $a$, $b$, $c$, $d$, $e$, and $f$ are found to be approximately zero, and hence the computations for determining the unknown length and bearing are correct.

**Unknown courses not adjoining**   The preceding method of solution is generally applicable even though two partly unknown courses are not adjoining. Obviously, the departure and the latitude of any line of fixed direction and length are the same for one location of the line as for any other. In other words, a line may be moved from one location to a second location parallel with the first, and its departure and latitude will remain unchanged. Since this is the case, then it must also be true that the algebraic sum of the latitudes and the algebraic sum of the departures of any system of lines forming a closed figure must be zero, regardless of the order in which the lines are placed. Thus, the courses which are shown in the order $a$, $b$, $c$, $d$, $e$, in Fig. 8.36 are given in the order $a$, $e$, $b$, $c$, $d$ in Fig. 8.38. If now it is assumed that the direction of $d$ and the length of $a$ (Fig. 8.36) are unknown, the problem of determining these unknown quantities is seen to be identical with that explained in the preceding section for the case where the partly unknown sides were adjoining.

**Length of two sides unknown**   This problem commonly occurs where angular observations are taken from two or more points in the main traverse to some landmark, the measurements being introduced as a check. It occasionally occurs on main traverse lines where there are obstacles to the direct measurement of length but where angles are observed. The solution is nearly identical with that for the case where the direction of one side and the length of another are unknown.

In Fig. 8.39, $ABCD$ represents the portion of a closed traverse for which the courses are known in both direction and length, and the lines $DE$ and $EA$ represent courses for which

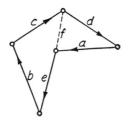

**Fig. 8.38**

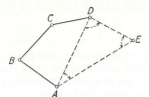

**Fig. 8.39**

the direction is known but the length is unknown. From the departures and latitudes of the known sides, the length and bearing of the closing line *DA* are computed; and in the triangle *ADE* the angles *A*, *D*, and *E* are computed from the known directions of the sides. The lengths *DE* and *EA* are determined through the relation

$$\frac{DE}{\sin A} = \frac{EA}{\sin D} = \frac{DA}{\sin E}$$

If the two lines are not adjoining, the problem may be solved as though they were. As the angle between the partly unknown lines approaches 90°, the solution becomes strong, and as the angle approaches 0° or 180°, the solution becomes weak, the problem being indeterminate when the lines are parallel.

**Direction of two sides unknown**   In Fig. 8.39, if *DA* is the closing side of the known portion of the traverse, its direction and length are computed; then the lengths of the three sides of the triangle *ADE* are known, and the angles *A*, *D*, and *E* can be computed.

The general direction of at least one of the partly unknown lines must be observed, as the values of the trigonometric functions merely determine the shape of the triangle but do not fix its position. Thus in Fig. 8.40, if *DA* is the closing line of the known portion of the traverse forming the base of the triangle of which the courses of unknown direction but of known length are the legs, then it is evident that the vertex may fall at either *E* or *E'*.

**8.25. Calculation of areas of land: general**   One of the primary objects of most land surveys is to determine the area of the tract. A closed traverse is run, in which the lines of the traverse are made to coincide with property lines where possible. Where the boundaries are irregular or curved or where they are occupied by objects which make direct measurement impossible, they are located with respect to the traverse line by appropriate angular and linear measurements. The lengths and bearings of all straight boundary lines are determined either directly or by computation, the irregular boundaries are located with respect to traverse lines by perpendicular offsets taken at appropriate intervals, and the radii and central angles of circular boundaries are obtained. The following sections explain the several common methods by means of which these data are employed in calculating areas.

In ordinary land surveying, as discussed herein, the area of a tract of land is taken as its projection upon a horizontal plane, and it is not the actual area of the surface of the land. For precise determinations of the area of a large tract, such as state or nation, the area is taken as the projection of the tract upon the earth's spheroidal surface at mean sea level.

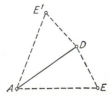

**Fig. 8.40**

**8.26. Methods of determining area** The area of a tract may be determined by any of the following methods:

1. By plotting the boundaries to scale as described in Chap. 13; the area of the tract may then be found by use of the planimeter or it may be calculated by dividing the tract into triangles and rectangles, scaling the dimensions of these figures, and computing their areas mathematically. This method is useful in roughly determining areas or in checking those that have been calculated by more exact methods. Its advantage lies in the rapidity with which calculations can be made.
2. By mathematically computing the areas of individual triangles into which the tract may be divided (Art. 8.27). This method is employed when it is not expedient to compute the departures and latitudes of the sides.
3. By calculating the area from the coordinates of the *corners* of the tract (Art. 8.28).
4. By calculating the area from the double meridian distances and the latitudes of the *sides* of the tract (Art. 8.29).
5. For tracts having irregular or curved boundaries, the methods of Arts. 8.30 to 8.33 are employed.

For computation of the areas of cross sections, see Art. 17.31.

**8.27. Area by triangles** When the lengths of two sides and the included angle of any triangle are known, its area is given by the expression

$$\text{area} = \tfrac{1}{2}ab\sin C \tag{8.37}$$

When the lengths of the three sides of any triangle are given, its area is determined by the equation

$$\text{area} = \sqrt{s(s-a)(s-b)(s-c)} \tag{8.38}$$

in which $s = \tfrac{1}{2}(a+b+c)$.

In surveying small lots as for a city subdivision, it is common practice to omit the determination of the error of closure of each lot (a practice not condoned in this book) and hence the computation of departures and latitudes is unnecessary. Under such circumstances the area may be calculated by dividing the lot, usually quadrangular in shape, into triangles, as illustrated by Fig. 8.41, for each of which two sides and the included angle have been measured. By Eq. (8.37), the areas of $ABD$ and $BCD$ are computed; the sum of these two areas is the area of the lot. The area thus found can be checked independently by computing the areas of the two triangles $ABC$ and $CDA$ formed by dividing the quadrilateral by a line from $A$ to $C$.

The accuracy of the field work may be investigated by determining the lengths of the diagonals. $BD$ can be determined by solving either triangle $ABD$ or triangle $BCD$. The field measurements are without error if the length of $BD$, computed by solving one of the triangles, is the same as that computed by solving the other.

Figure 8.42 illustrates a survey made by a single setup of the transit at $O$, such as might be the case for a small lot where the property lines $ABCD$ are obstructed or where the transit can not be set up at the corners. Under these circumstances the angles about $O$ and the

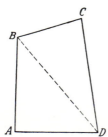

**Fig. 8.41**

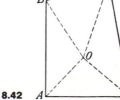

**Fig. 8.42**

distances $OA$, $OB$, $OC$, and $OD$ are measured in the field. Since in each triangle two sides and the included angle are known, the area can be determined by use of Eq. (8.37). If in addition to the foregoing measurements, the lengths of the sides $AB$, $BC$, etc., are measured, the area of the lot can be checked independently by solving each triangle by Eq. (8.38). In general, the area of a triangle computed from two different sets of measurements will not be identical, owing to random measuring errors. Therefore, an alternative would be to use the method of least squares (Appendix B) to compute the area using all redundant measurements.

**8.28. Area by coordinates** When the points defining the corners of a tract of land are coordinated with respect to some arbitrarily chosen coordinate axes or are given in a regional system (such as a state plane coordinate system), these coordinates are useful not only in finding the lengths and bearings of the boundaries but also in calculating the area of the tract. Essentially the calculation involves finding the areas of trapezoids formed by projecting the lines upon one of a pair of coordinate axes, usually a true meridian and a parallel at right angles thereto.

In Fig. 8.43, 12345 represents a tract the area of which is to be determined, where each point $1, 2, \ldots, 5$ has coordinates $X_1, Y_1, \ldots, X_5, Y_5$, as shown in the figure.

The area of the tract can be computed by summing algebraically the areas of the trapezoids formed by projecting the lines upon the reference meridian; thus,

$$\text{area } 12345 = \text{area } 23cb + \text{area } 34dc - \text{area } 45fd$$
$$- \text{area } 15fa - \text{area } 21ab$$

or

$$\text{area} = \tfrac{1}{2}(X_2 + X_3)(Y_2 - Y_3) + \tfrac{1}{2}(X_3 + X_4)(Y_3 - Y_4)$$
$$- \tfrac{1}{2}(X_4 + X_5)(Y_5 - Y_4) - \tfrac{1}{2}(X_5 + X_1)(Y_1 - Y_5)$$
$$- \tfrac{1}{2}(X_1 + X_2)(Y_2 - Y_1) \tag{8.39}$$

By multiplication and a rearrangement of terms in Eq. (8.39), there is obtained

$$2 \times \text{area} = -[Y_1(X_5 - X_2) + Y_2(X_1 - X_3) + Y_3(X_2 - X_4)$$
$$+ Y_4(X_3 - X_5) + Y_5(X_4 - X_1)] \tag{8.40}$$

and in general for any polygon having $n$ stations

$$2 \times \text{area} = Y_1(X_2 - X_n) + Y_2(X_3 - X_1) + \cdots + Y_{n-1}(X_n - X_{n-2})$$
$$+ Y_n(X_1 - X_{n-1}) \tag{8.41}$$

or, equivalently,

$$2 \times \text{area} = X_1(Y_2 - Y_n) + X_2(Y_3 - Y_1) + \cdots$$
$$+ X_{n-1}(Y_n - Y_{n-2}) + X_n(Y_1 - Y_{n-1}) \tag{8.42}$$

Note that a negative sign for the final value is of no concern and that the total area will always be positive.

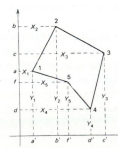

**Fig. 8.43** Area by coordinates.

**Example 8.6**  Given the following data, find the required area by applying Eq. (8.41).

| Corner | 1 | 2 | 3 | 4 | 5 |
|---|---|---|---|---|---|
| X coordinate, m | 300 | 400 | 600 | 1,000 | 1,200 |
| Y coordinate, m | 300 | 800 | 1,200 | 1,000 | 400 |

SOLUTION

$$2 \times \text{area} = 300(400 - 1200) + 800(600 - 300) + 1200(1000 - 400) + 1000(1200 - 600)$$
$$+ 400(300 - 1000)$$
$$= -240,000 + 240,000 + 720,000 + 600,000 - 280,000$$
$$= 1,040,000 \text{ m}^2$$
$$= \frac{1,040,000}{2} = 520,000 \text{ m}^2$$

Equation (8.41) can also be expressed in the form

$$2 \times \text{area} = X_2 Y_1 + X_3 Y_2 + X_4 Y_3 + \cdots + X_n Y_{n-1} + X_1 Y_n$$
$$- X_1 Y_2 - X_2 Y_3 - \cdots - X_n Y_1 \tag{8.43}$$

When this form is employed, computations can be made conveniently by tabulating each $X$ coordinate below the corresponding $Y$ coordinate as follows:

$$\frac{Y_1}{X_1} \times \frac{Y_2}{X_2} \times \frac{Y_3}{X_3} \times \cdots \frac{Y_n}{X_n} \times \frac{Y_1}{X_1} \tag{8.44}$$

Then in expression (8.44), the difference between the sum of the products of the coordinates joined by full lines and the sum of the products of the coordinates joined by dotted lines is equal to twice the area of the tract.

**Example 8.7**  Compute the area enclosed by the polygon formed by the closed-loop traverse calculated in Art. 8.18 and adjusted in Art. 8.19. The adjusted coordinates are listed in Table 8.13. Use Eq. (8.42).

SOLUTION

$$2 \times \text{area} = 4382.09(6036.90 - 6333.77) + 4770.93(6215.76 - 6150.82)$$
$$+ 5056.06(6511.61 - 6036.90) + 4921.10(6347.30 - 6215.76)$$
$$+ 4785.69(6333.77 - 6511.61) + 4533.84(6150.82 - 6347.30)$$

Observing the rules of significant figures the double area is

$$2 \times \text{area} = 314,501 \text{ ft}^2$$
and
$$\text{area} = 157,250 \text{ ft}^2 = 3.6100 \text{ acres}$$

**8.29. Area by double meridian distances and latitudes**  Certain closed-loop traverses, such as those performed to locate property corners, are not coordinated. The closure for this type of survey should always be checked by calculating departures and latitudes for the traverse. When the area enclosed by the traverse is also desired, it can be calculated with the adjusted departures and latitudes using the *double meridian distance* method as described in this section.

The departures and latitudes of all the courses are determined as described in Art. 8.17, and the survey is adjusted. A reference meridian is then assumed to pass through some corner of the tract, usually for convenience the most westerly point of the survey; the double meridian distances of the lines are computed as described herein; and double the areas of the trapezoids or triangles formed by orthographically projecting the several traverse lines upon the meridian are computed. The algebraic sum of these double areas is double the area within the traverse.

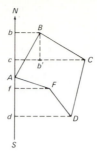

**Fig. 8.44** Area by double meridian distances.

The meridian distance of a point is the total departure or perpendicular distance from the reference meridian; thus in Fig. 8.44 the meridian distance of $B$ is $Bb$ and is positive. The meridian distance of a straight line is the meridian distance of its midpoint. The *double meridian distance* of a straight line is the sum of the meridian distances of the two extremities; thus, the double meridian distance of $BC$ is $Bb + Cc$. It is clear that if the meridian passes through the most westerly corner of the traverse, the double meridian distance of all lines will be positive, which is a convenience (although not a necessity) in computing.

The length of the orthographic projection of a line upon the meridian is the latitude of the line; thus, in Fig. 8.44 the latitude of $BC$ is $bc$ and is negative, and that of $DF$ is $df$ and is positive.

From the figure it is seen that each projection trapezoid or triangle, for which a course in the traverse forms one side, is bounded on the north and south by meridian distances and on the west by the latitude of that course. Thus, the projection trapezoid for $BC$ is $BCcb$. Therefore, the double area of any triangle or trapezoid formed by projecting a given course upon the meridian is the product of the double meridian distance (D.M.D.) of the course and the latitude of the course, or

$$\text{double area} = \text{D.M.D.} \times \text{latitude} \tag{8.45}$$

In computing double areas, account is taken of signs. If the meridian extends through the most westerly point, all double meridian distances are positive; hence, the sign of a double area is the same as that of the corresponding latitude. Thus, in the figure the double areas of $AbB$, $DdfF$, and $FfA$ are positive, the latitudes $Ab$, $df$, and $fA$ being positive; while the double areas of $CcbB$ and $DdcC$ are negative, the latitudes $bc$ and $cd$ being negative. Since the projected areas *outside* the traverse are considered once as positive and once as negative, the algebraic sum of their double areas is zero. Therefore, the algebraic sum of all double areas is equal to twice the area of the tract *within* the traverse.

Whether this algebraic sum of the double areas is a positive or negative quantity is determined solely by the order in which the lines of the traverse are considered. If the reference meridian passes through the most westerly corner, then a clockwise order of lines, as in the figure, results in a negative double area, and a counterclockwise order results in a positive double area. The sign of the area is not significant.

***Computation of D.M.D.***   When the reference meridian passes through a traverse point, there is an intimate relation between the departures and double meridian distances. In Fig. 8.44 it is seen that the D.M.D of $AB$ is $bB$, which is equal to the departure of the course in both magnitude and sign. The D.M.D. of $BC$ is equal to $bB + cb' + b'C$, which is equal to the D.M.D. of $AB$, plus the departure of $AB$, plus the departure of $BC$. Similar quantities make up the D.M.D.'s of $CD$ and $DF$. For the last line $FA$ the D.M.D. is $fF$, which is equal in magnitude but opposite in sign to the departure of $FA$.

Following are three convenient rules for determining D.M.D.'s, which are deduced from the relations just illustrated:

1. The D.M.D. of the first course (reckoned from the point through which the reference meridian passes) is equal to the departure of that course.
2. The D.M.D. of any other course is equal to the D.M.D. of the preceding course, plus the departure of the preceding course, plus the departure of the course itself.
3. The D.M.D. of the last course is numerically equal to the departure of the course but with opposite sign.

The first two rules are employed in computing values. The third rule is useful as a check on the correctness of the computations. Assuming the departures as balanced, the D.M.D. of the last line, as found by computing the D.M.D.'s in succession around the traverse from the first line, should be numerically equal to the departure of the last line if no mistake in addition or subtraction has been made.

***Area within a closed polygon by the D.M.D. method***   The steps employed in calculating the area within a closed-loop traverse by the D.M.D. method are as follows:

1. Compute and balance latitudes and departures for all lines in the traverse (Arts. 8.17 to 8.19).
2. Assume that the reference meridian passes through the most westerly point in the survey and compute the D.M.D.'s by the rules of the preceding section, using the corrected departures.
3. Compute the double areas by multiplying each D.M.D. by the corresponding corrected latitude.
4. Find the algebraic sum of the double areas, and determine the area by dividing this sum by 2.

**Example 8.8**   Determine the area enclosed by the closed-loop traverse of Fig. 8.7 computed in Art. 8.19. Use the adjusted departures and latitudes listed in Table 8.13. Establish the reference meridian through station 1 and compute D.M.D.s in a counterclockwise direction. The D.M.D. of the first line equals the departure of $12 = 388.84$ ft. The D.M.D. of the second line 23 equals the departure of the previous course, $388.84 +$ the D.M.D. of the previous course, $388.84 +$ the departure of the course itself, $285.13 = 1062.81$. The double area for line $12 = (388.84)(-113.92) = -44,296.7$ ft$^2$. There are five significant figures in the data, so that six digits are carried in the calculations. D.M.D.'s for each line, latitudes, double areas, and the final area are listed in Table 8.16.

**8.30. Area of tract with irregular or curved boundaries**   If the boundary of a tract of land follows some irregular or curved line, such as a stream or road, it is customary to run a traverse in some convenient location near the boundary and to locate the boundary by offsets from the traverse line. Figure 8.45 represents a typical case, $AB$ being one of the traverse lines. The determination of area of the entire tract involves computing the area within the closed traverse, by methods which have already been described, and adding to this the area of the irregular figure between the traverse line $AB$ and the curved boundary. The offset distances are $aa'$, $bb'$, etc., and the corresponding distances along the traverse line are $Aa$, $Ab$, etc. Where the boundary is irregular, as from $a'$ to $f'$, it is necessary to take offsets at points of change and hence generally at irregular intervals. Where a segment of the boundary is straight, as from $f'$ to $g'$, offsets are taken only at the ends. Where the boundary is a gradual curve, as from $g'$ to $m'$, ordinarily the offsets are taken at regular intervals.

If the offsets are taken sufficiently close together, the error involved in considering the boundary as straight between offsets is small as compared with the inaccuracies of the measured offsets. When this assumption is made, the assumed boundary takes some such form as that illustrated by the dotted lines $g'h'$, $h'k'$, etc., in Fig. 8.45, and the areas between offsets are of trapezoidal shape. Under such an assumption, irregular areas are said to be calculated by the *trapezoidal rule* (Art. 8.31).

Where the curved  boundaries are of such definite character as to make it justifiable, the area may be calculated somewhat more accurately by assuming that the boundary is made

**Table 8.16   Computations for Area by Double Meridian Distance**

| Line | | | D.M.D. ft | Latitude ft | Double area + | Double area − |
|------|---|---|-----------|-------------|-----|-----|
| 12 | | | 388.84 | − 113.92 | | 44,296.7 |
| | Dep. 12 | | 388.84 | | | |
| | Dep. 23 | | 285.13 | | | |
| 23 | D.M.D. | = | 1062.81 | + 178.86 | 190,094 | |
| | Dep. 23 | | 285.13 | | | |
| | Dep. 34 | | − 134.96 | | | |
| 34 | D.M.D. | = | 1212.98 | + 295.85 | 358.860 | |
| | Dep. 34 | | − 134.96 | | | |
| | Dep. 45 | | − 135.41 | | | |
| 45 | D.M.D. | = | 942.61 | − 164.31 | | 154,880 |
| | Dep. 45 | | − 135.41 | | | |
| | Dep. 56 | | − 251.85 | | | |
| 56 | D.M.D. | = | 555.35 | − 13.53 | | 7,513.89 |
| | Dep. 56 | | − 251.85 | | | |
| | Dep. 61 | | − 151.75 | | | |
| 61 | D.M.D. | = | 151.75 | − 182.95 | | 27,762.7 |

$$548,954 \qquad -234,453$$
$$-234,453$$
$$2 | 314,501 = 2 \cdot \text{area}$$
$$\text{area} = 157,250 \text{ ft}^2$$

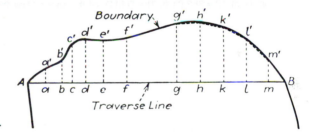

**Fig. 8.45** Irregular boundary.

up of segments of parabolas as first suggested by Simpson. Under this assumption, irregular areas are said to be computed by *Simpson's one-third rule* (Art. 8.32).

**8.31. Offsets at regular intervals: trapezoidal rule**   Let Fig. 8.46 represent a portion of a tract lying between a traverse line $AB$ and an irregular boundary $CD$, offsets $h_1, h_2, \ldots, h_n$ having been taken at the regular intervals $d$. The summation of the areas of the trapezoids comprising the total area is

$$\text{area} = \frac{h_1 + h_2}{2} d + \frac{h_2 + h_3}{2} d + \cdots + \frac{h_{n-1} + h_n}{2} d \qquad (8.46)$$

$$= d\left( \frac{h_1 + h_n}{2} + h_2 + h_3 + \cdots + h_{n-1} \right) \qquad (8.47)$$

Equation (8.46) may be expressed conveniently in the form of the following rule:

**Trapezoidal rule**   *Add the average of the end offsets to the sum of the intermediate offsets. The product of the quantity thus determined and the common interval between offsets is the required area.*

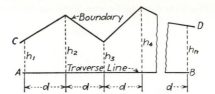

**Fig. 8.46** Area by trapezoidal rule.

**Example 8.9**   By the trapezoidal rule find the area between a traverse line and a curved boundary, rectangular offsets being taken at intervals of 5 m, and the values of the offsets in metres being $h_1 = 3.2$, $h_2 = 10.4$, $h_3 = 12.8$, $h_4 = 11.2$, and $h_5 = 4.4$. By the foregoing rule,

$$\text{area} = 5\left( \frac{3.2 + 4.4}{2} + 10.4 + 12.8 + 11.2 \right) = 191 \text{ m}^2$$

**8.32. Offsets at regular intervals: Simpson's one-third rule**   In Fig. 8.47 let $AB$ be a portion of a traverse line, $DFC$ a portion of the curved boundary assumed to be the arc of a parabola, and $h_1$, $h_2$, and $h_3$ any three consecutive rectangular offsets from traverse line to boundary taken at the regular interval $d$.

The area between traverse line and curve may be considered as composed of the trapezoid $ABCD$ plus the area of the segment between the parabolic arc $DFC$ and the corresponding chord $DC$. One property of a parabola is that the area of a segment (as $DFC$) is equal to two-thirds the area of the enclosing parallelogram (as $CDEFG$). Then the area between the traverse line and curved boundary within the length of $2d$ is

$$\text{area}_{1,2} = \frac{h_1 + h_3}{2} 2d + \left( h_2 - \frac{h_1 + h_3}{2} \right) 2d \left( \frac{2}{3} \right)$$
$$= \frac{d}{3} (h_1 + 4h_2 + h_3)$$

Similarly for the next two intervals,

$$\text{area}_{3,4} = \frac{d}{3} (h_3 + 4h_4 + h_5)$$

The summation of these partial areas for $(n-1)$ intervals, $n$ being an odd number and representing the number of offsets, is

$$\text{area} = \frac{d}{3} [h_1 + h_n + 2(h_3 + h_5 + \cdots + h_{(n-2)}) + 4(h_2 + h_4 + \cdots + h_{(n-1)})] \qquad (8.48)$$

Equation (8.48) may be expressed conveniently in the form of the following rule, which is applicable if the number of offsets is odd.

**Simpson's one-third rule**   *Find the sum of the end offsets, plus twice the sum of the odd intermediate offsets, plus four times the sum of the even intermediate offsets. Multiply the quantity thus determined by one third of the common interval between offsets, and the result is the required area.*

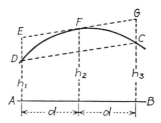

**Fig. 8.47** Area by Simpson's rule.

**Example 8.10**   By Simpson's one-third rule find the area between the traverse line and the curved boundary of Example 8.9.

SOLUTION

$$\text{area} = \tfrac{5}{3}[3.2 + 4.4 + 2(12.8) + 4(10.4 + 11.2)] = 199 \text{ m}^2$$

If the total number of offsets is *even*, the partial area at either end of the series of offsets is computed separately, in order to make $n$ for the remaining area an odd number and thus make Simpson's rule applicable.

Simpson's rule is also useful in other applications, such as finding centers of areas. The prismoidal formula for computing volumes of earthwork (Art. 17.34) embodies Simpson's rule for area and a factor for the third dimension.

Results obtained by using Simpson's rule are greater or smaller than those obtained by using the trapezoidal rule, depending upon whether the boundary curve is concave or convex toward the traverse line. Some appreciation of the variations between the two methods will be gained by studying the foregoing examples. It will be seen that the two results differ by more than 4 percent. Under average conditions the difference will be much less than this, but in an extreme case it may be much larger.

In general, Simpson's rule is always more accurate than the trapezoidal rule. The latter approaches the former in accuracy to the extent that the irregular boundary has curves of contrary flexure thereby producing the compensative effects mentioned above.

When the interval along the traverse line is irregular, the trapezoidal rule may be applied by using Eq. (8.46) and using the appropriate interval $d_1, d_2, \ldots, d_n$. Simpson's one-third rule is based on an equal interval for each pair of adjacent segments. Thus, if it is desired to utilize Simpson's one-third rule, a regular interval should be used in the field measurements.

**8.33. Area of segments of circles**   A problem of frequent occurrence in the surveying of city lots and of rural lands adjacent to the curves of highways and railways is that of finding the area where one or more of the lines of the boundary is the arc of a circle.

In Fig. 8.48, $ABCDEQF$ may be taken as a boundary of this character, for which it is convenient to run a traverse along the straight portions of the boundary and to make the chord $EF$ the closing side of the traverse, the length of the chord $EF = L$ and the middle ordinate $PQ = M$ being measured in the field.

In calculating the area, it is convenient to divide the tract into two parts: (1) that within the polygon formed by the traverse $ABCDEF$, for which the area is found by the coordinate

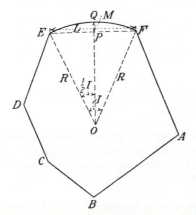

**Fig. 8.48**

method or the double-meridian-distance method, and (2) that between the chord *EPF* and the arc *EQF*, which is the segment of a circle. The area of this segment is found exactly by subtracting the area of the triangle *OEPF* from the area of the circular sector *OEQF*. If *I* is the angle and *R* is the radius whose arc is *EQF*, then by Art. 17.4,

$$\tan\frac{1}{4}I=\frac{2M}{L} \tag{8.49}$$

and

$$R=\frac{L}{2\sin\frac{1}{2}I} \tag{8.50}$$

The area of the circular sector *OEQF* is $A_s=\pi R^2 I°/360$, in which $I°$ is expressed in degrees. The area of triangle *OEF* is

$$A_t=\frac{LR\cos(I/2)}{2}=\frac{R^2}{2}\sin I$$

The area of the segment is exactly

$$\text{area}=A_s-A_t=R^2\left(\frac{\pi I°}{360}-\frac{\sin I}{2}\right) \tag{8.51}$$

**Example 8.11** Find the area of a circular segment when the chord length is 80.00 m and the middle ordinate is 10.00 m.

SOLUTION By Eq. (8.49),

$\tan\frac{1}{4}I=0.2500$

$\frac{1}{4}I=14°036$    $I=56°145$

By Eq. (8.50),

$$R=\frac{\dfrac{L}{2}}{(\sin\frac{I}{2})}=\frac{80}{(2)(0.47059)}=85.00 \text{ m}$$

By Eq. (8.51),

$$\text{area}=(85.000)^2\left[\frac{(56°145)\pi}{360}-\frac{\sin 56°145}{2}\right]$$

$$=539.9 \text{ m}^2$$

An alternative method of finding the area of the tract *ABCDEQF* (Fig. 8.48) is to divide the area into a rectilinear polygon *ABCDEOF* and the circular sector *OEQF*, and to add the two areas. The polygon has one more side than the one used above, but there is no need to compute the area of a circular segment.

**Approximation by parabolic segment** The area of a parabolic segment is

$$(\text{area})_p=\frac{2}{3}LM \tag{8.52}$$

in which the letters have the same significance as before. This expression may be employed for finding the approximate areas of circular segments, the precision decreasing as the size of the central angle *I* increases. The following example illustrates the error involved in applying this expression to the conditions of Example 8.11.

**Example 8.12** By Eq. (8.52) find the approximate area of the circular arc of Example 8.11, and determine the percentage of error introduced through using the approximate expression Eq. (8.52).

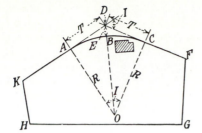

**Fig. 8.49**

SOLUTION    The area by Eq. (8.52) is

$(\text{area})_P = \frac{2}{3} \times 80 \times 10 = 533.3$ m$^2$. This value is

$\dfrac{539.9 - 533.3}{539.9}\, 100 = 1.2$ percent too low

When the central angle is small, the error involved in using Eq. (8.52) for circular arcs is often negligible; thus, when $I = 30°$, the error is less than 0.2 percent. But for large values of $I$, the error introduced is so great as to render the approximate expression of little use; thus, when $I = 90°$, the error is about 3 percent, and when $I = 180°$, the error is about 15 percent.

**Alternative method**    When tangents to the curve are property lines, it is sometimes more convenient to establish the traverse, as illustrated by Fig. 8.49. Here $KA$ and $FC$, which are tangent to the curve $ABC$, are run to an intersection at $D$, and the distances $AD$ and $CD$ and the angle $I$ are measured. Also, $E$ is usually measured as a check.

The work of finding the area is conveniently divided into two parts: (1) that of calculating the area within the polygon $ADCFGHK$ by the coordinate method or the double-meridian-distance method, and (2) that of calculating the external area between the arc $ABC$ and the tangents $AD$ and $CD$. The latter area subtracted from the former is the required area.

The external area may be found by subtracting the area of the circular sector $OABC = A_s$ $= \pi R^2 I° / 360$ from the area $OADC = TR$, in which $T$ is the tangent distance $AD = CD$, and $R$ is the radius of the curve. If $R$ is unknown, it may be found by the relation

$$R = \frac{T}{\tan\frac{1}{2}I} \qquad (\text{see Art. 17.4}) \qquad (8.53)$$

**8.34. Partition of land**    Four of the simpler cases frequently encountered in the subdivision of irregular tracts of land will be described in the succeeding articles. Methods of subdividing the U.S. public lands are given in Chap. 19.

Where a given tract is to be divided into two or more parts, a resurvey is run, the departures and latitudes are computed, the survey is adjusted, and the area of the entire tract is determined. The corrected departures and latitudes are further employed in the computations of subdivision.

**Area cut off by a line between two points**    In Fig. 8.50 let $ABCDEFG$ represent a tract of land to be divided into two parts by a line extending from $A$ to $D$. A survey of the tract has been made, the adjusted coordinates have been computed for the corners, and the area has been computed.

It is desired to determine the length and direction of the cutoff line $AD$ without additional field measurements, and to calculate the area of each of the two parts into which the tract is divided.

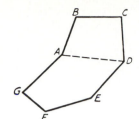

**Fig. 8.50**

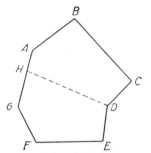

**Fig. 8.51**

When coordinates are available for $A$ and $D$, the simplest approach is to use Eqs. (8.8) and (8.9) developed in Art. 8.15 to calculate the distance and azimuth of $AD$. Then the areas of the traverse $ABCD$ and $ADEFG$ can be computed by the coordinate method.

**8.35. Area cut off by a line running in a given direction**   In Fig. 8.51 $ABCDEFG$ represents a tract of known dimensions, for which the corrected departures, latitudes, and adjusted coordinates are given; and $DH$ represents a line running in a given direction which passes through the point $D$ and divides the tract into two parts.

It is desired to calculate from the given data the lengths $DH$ and $HA$ and the area of each of the two parts into which the tract is divided.

The direction of $DH$ is given and the coordinates of $D$ are known, so that the point slope Eq. (8.13), Art. 8.21, can be formed. Since coordinates are known for $G$ and $A$, Eq. (8.15), Art. 8.21, can be written for line $AG$. These two equations, one for line $DH$ and one for line $GA$, can be solved simultaneously to obtain the coordinates of $H$ (see Example 8.1). The distance $HA$ can be calculated with Eq. (8.8), using the coordinates of $H$ just computed and the known coordinates for $A$. Areas for $ABCDHA$ and $HDEFGH$ can be computed by the coordinate method [Eq. (8.41) or (8.42)].

In the field the length and direction of the side $DH$ are laid off from $D$, and a check on field work and computations is obtained if the point $H$ thus established lies on the line $GA$ and if the computed distance $HA$ agrees with the observed distance. The area computations may be checked by observing that the sum of the areas of the two parts, each computed independently, is equal to the area of the entire tract.

**8.36 To cut off a required area by a line through a given point**   In Fig. 8.52, $ABCDEF$ represents a tract of land of known dimensions, for which the corrected departures, latitudes, and adjusted coordinates for each point are given; and $G$ represents a point on the boundary through which a line is to pass cutting off a required area from the tract. The area for the total tract can be calculated by the coordinate method and a sketch of the tract has been prepared.

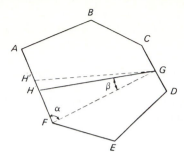

**Fig. 8.52**

To find the length and direction of the dividing line, the procedure is as follows. A line $GF$ is drawn to that corner of the traverse which, from inspection of the sketch, will come nearest to being on the required line of division. Since distance $CG$ is specified, the coordinates of $G$ can be calculated. Then the distance and azimuth of $GF$ are calculated using Eqs. (8.8) and (8.9). The area enclosed by the traverse and the cutoff line $GF$, $ABCGFA$ is computed by the coordinate method. The difference between this area and the amount specified is found.

In the figure it is assumed that $FABCG$ has an area greater than the desired amount, $GH$ being the correct position of the dividing line. Then, the triangle $GFH$ represents this excess area; and as the angle $\alpha$ may be computed from known directions, there are given in this triangle one side $FG$, one angle $\alpha$, and the area. The length $HF$ is computed from the equation for area, area $= \frac{1}{2}(ab \sin C)$, so that from Fig. 8.52,

$$ HF = \frac{2 \cdot \text{area } GFH}{FG \sin \alpha} \tag{8.54} $$

The triangle is then solved for angle $\beta$ and length $GH$. From the known direction of $GF$ and the angle $\beta$, the azimuth of $GH$ is computed. The departures and latitudes of the lines $FH$, $GH$, and $HA$ are computed and coordinates are calculated for point $H$.

In the field the length $GH$ is laid off in the required direction, and a check on field work and computations is obtained if the point $H$ thus established falls on the line $FA$ and if the computed distance $HF$ or $HA$ agrees with the measured distance.

**8.37. To cut off a required area by a line running in a given direction** In Fig. 8.53, $ABCDEF$ represents a tract of land of known dimensions and area, which is to be divided into two parts, each of a required area, by a line running in a given direction. The figure is assumed to be drawn at least roughly to scale, and the corrected departures and latitudes, and adjusted coordinates for the corners are known.

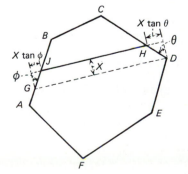

**Fig. 8.53**

Through the corner that seems likely to be nearest the line cutting off the required area, a trial line *DG* is drawn in the given direction. Then in the closed traverse *GBCDG* the departures and latitudes of *BC* and *CD* and the directions of *DG* and *GB* are known, and the lengths of two sides *DG* and *GB* are unknown.

Using these data, the coordinates of *G* can be found by forming the point-slope equation, Eq. (8.13), for line *DG* and Eq. (8.15) for line *AB* and solving these two equations simultaneously for $X_G$ and $Y_G$. It would also be possible to solve for *DG* and *GB* using the methods outlined in Art. 8.24. By either procedure the distance for *DG* and the distance for *AG* are calculated.

Next, the area cut off by trial line *DG* is calculated. The difference between this area and that required is represented in the figure by the trapezoid *DGJH* in which the side *DG* is known. The angles at *D* and *G* can be computed from the known directions of adjacent sides, and in this way $\theta$ and $\phi$ are determined. Tan $\theta$ and tan $\phi$ are positive if they fall inside the trapezoid and negative if they fall outside. With this convention, the

$$\text{area of trapezoid} = (DG)x + \frac{x^2}{2}(\tan\theta + \tan\phi) \tag{8.55}$$

in which $x$ is the altitude of the trapezoid. (In the figure both angles lie outside the trapezoid so that both tangents are negative.) When the known values for *DG*, $\theta$, and $\phi$ are substituted into Eq. (8.55), it is a quadratic equation in the form $ax^2 + bx + c = 0$ which may be solved for $x$ using $x = [-b \pm (b^2 - 4ac)^{1/2}]/2a$.

In the field points *H* and *J* are established on the lines *CD* and *AB*, at the calculated distances from the adjacent corners. The side *JH* is then measured. If this measured value agrees with the computed value, the field work and portions of the computations are verified. A further check on the computations is introduced by calculating the area *BCHJ* and comparing it with the required area of this figure.

**Example 8.13** Determine the length and direction of the line passing through point 5, which divides the area of the tract enclosed by the traverse of Fig. 8.7 into two equal areas. Adjusted coordinates are listed in Table 8.13 and the total area within the tract is 157,250 ft$^2$ (see Example 8.7).

SOLUTION Use of coordinates and analytical geometry provides the most direct solution. To simplify the computations, all coordinates in Table 8.14 are translated so that the $Y'$ axis passes through station 1 and the $X'$ axis through station 2 (Fig. 8.54). The $X'$ and $Y'$ translated values are given in Table 8.17.

Assume line 5-2 as a first approximation to the division line. The distance and direction of 5-2 by Eqs. (8.8) and (8.9) are

$$d_{5\text{-}2} = [(403.60 - 388.84)^2 + (310.40 - 0.0)^2]^{1/2} = 310.751 \text{ ft}$$

$$\tan A_{52} = \frac{388.84 - 403.60}{0 - 310.40}$$

$$A_{52} = 182°43'21''$$

angle $521 = A_{52} - A_{12} = 76°23'36''$

The total area of the tract, calculated in Example 8.7 by the coordinate method, is 157,250 ft$^2$, so that area/2 = 78,625 ft$^2$.

The area enclosed by stations 1256, using expression (8.44), is

$$2(\text{area})_{1256} = \frac{113.92}{0.00} \diagdown\diagup \frac{0.00}{388.84} \diagdown\diagup \frac{310.40}{403.60} \diagdown\diagup \frac{296.87}{151.75} \diagdown\diagup \frac{113.92}{0.00}$$

$$(\text{area})_{1256} = 83,200 \text{ ft}^2$$

$$-\frac{\text{total area}}{2} = 78,625$$

$$\text{error} = 4,575 \text{ ft}^2$$

so that the area enclosed by 1256 is too large by 4575 ft$^2$. By Eq. (8.54) and referring to Fig. 8.54, the

**Table 8.17   Data for Computing Parting-off Land from Traverse of Fig. 8.7 (See Tables 8.13 and 8.14)**

| Station | Corrected | | Translated Coordinate | |
| --- | --- | --- | --- | --- |
| | Distances, ft | Azimuths | $X'$, ft | $Y'$, ft |
| 1 | | | 0.00 | 113.92 |
| | 405.18 | 106°19′45″ | | |
| 2 | | | 388.84 | 0.00 |
| | 336.59 | 57°54′01″ | | |
| 3 | | | 673.97 | 178.86 |
| | 325.18 | 335°28′43″ | | |
| 4 | | | 539.01 | 474.71 |
| | 212.92 | 219°29′33″ | | |
| 5 | | | 403.60 | 310.40 |
| | 252.21 | 266°55′30″ | | |
| 6 | | | 151.75 | 296.87 |
| | 237.69 | 219°40′28″ | | |
| 1 | | | | |

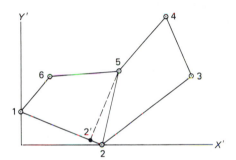

**Fig. 8.54**

distance 22′ from station 2 to the corrected division line 52′ is

$$22' = \frac{(2)(4575)}{(310.751)\sin 76°23'36''} = 30.2951 \text{ ft}$$

The coordinates of 2′ by Eqs. (8.6) and (8.7) are

$X'_{2'} = 388.84 + (30.2951)(\sin 286°19'45'') = 388.84 + (30.2951)(-0.959662) = 359.767$ ft
$Y'_{2'} = 0.00 + (30.2951)(\cos 286°19'45'') = 0.00 + (30.2951)(0.281155) = 8.5176$ ft

The area enclosed by 12′561 is

$$2(\text{area})_{12'561} = \frac{0.00}{113.92} \diagdown \frac{359.767}{8.5176} \diagdown \frac{403.60}{310.40} \diagdown \frac{151.75}{296.87} \diagdown \frac{0.00}{113.92}$$

$(\text{area})_{12'561} = 78{,}625 \text{ ft}^2$

that is equal to total area/2. The area enclosed by 2′23452′ is computed as a check

$$2(\text{area})_{2'23452'} = \frac{359.767}{8.5176} \diagdown \frac{388.84}{0.00} \diagdown \frac{673.97}{178.86} \diagdown \frac{539.01}{474.71} \diagdown \frac{403.60}{310.40} \diagdown \frac{359.767}{8.5176}$$

$(\text{area})_{2'23452'} = 78{,}625 \text{ ft}^2$

By Eqs. (8.8) and (8.9), $d_{52'} = 305.05$ ft and $A_{52'} = 188°15'39''.$

**Example 8.14**   The tract enclosed by the traverse of Fig. 8.7 and also shown in Fig. 8.55 is to be divided so that one-fourth of the enclosed area lies south of a line parallel to the X and X′ coordinate axes. Compute coordinates for the intersection of this dividing line with the traverse sides. Data listed in Table 8.17 may be used for the computations.

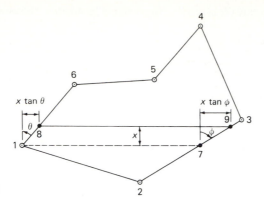

**Fig. 8.55**

SOLUTION  The initial position for the dividing line is taken passing through station 1, as illustrated in Fig. 8.55. An approximate area for triangle 127 made using distances scaled from the sketch reveals that this line results in an area that is ~20 percent too small, but it does provide a suitable starting line. As in Example 8.13, analytical geometry and coordinates provide the best method of solving the problem. Since the line 17 is parallel to the $X'$ axis,

$$Y_7' = Y_1' = 113.92 \tag{1}$$

Equation (8.15) for line 23 is

$$\frac{Y_7' - Y_2'}{Y_3' - Y_2'} = \frac{X_7' - X_2'}{X_3' - X_2'}$$

$$\frac{Y_7' - 0.00}{178.86 - 0.00} = \frac{X_7' - 388.84}{673.97 - 388.84}$$

which yields

$$285.13\,Y_7' - 178.86X_7' = -69{,}547.92 \tag{2}$$

Substitution of $Y_7'$ from (1) into (2) gives

$$-178.86X_7' = -102{,}029.93$$
$$X_7' = 570.446 \text{ ft} = d_{17}$$

The area of 127 by area $= \frac{1}{2}(ab)$, where $a = 570.446$ and $b = 113.92$, is

$$\text{area} = \tfrac{1}{2}(570.446)(113.92) = 32{,}493 \text{ ft}^2 \tag{3}$$

The total area, calculated in Example 8.7 is 157,250 ft$^2$, so that (area)/4 = 39,313 ft$^2$. Thus, the area enclosed by 127 is too small by $39{,}313 - 32{,}493 = 6820$ ft$^2$. Accordingly, the dividing line is moved parallel to itself to position 89, a distance $x$ from the original line. Using Eq. (8.55), the trapezoidal area enclosed by 1897 is

$$\text{area}_{1897} = \frac{x}{2}(d_{17} + d_{17} - x\tan\theta + x\tan\phi) = xd_{17} + x^2\left(\frac{\tan\phi - \tan\theta}{2}\right) \tag{4}$$

in which $\theta = 39°40'28''$ and $\phi = 57°54'01''$ are the azimuths of lines 16 and 23 as found in Table 8.17 and shown on Fig. 8.55. Substitution of known values into Eq. (4) gives

$$6820 = 570.446x + \left(\frac{1.59415 - 0.829463}{2}\right)x^2 = 570.446x + 0.382345x^2$$

which is of the form $ax^2 + bx + c = 0$ and can be solved by

$$x = \frac{-b \pm (b^2 - 4ac)^{1/2}}{2a} = \frac{-570.446 \pm [(570.446)^2 - (4)(0.382345)(-6820)]^{1/2}}{(2)(0.382345)}$$

$$= \frac{-570.446 \pm 579.516}{0.764691} = \frac{9.070}{0.764691} = 11.8613 \text{ ft}$$

Distances $d_{18}$ and $d_{79}$ are calculated using $x$ and the cosines of angles $\theta$ and $\phi$

$$d_{18} = \frac{x}{\cos\theta} = \frac{11.8613}{0.769684} = 15.411 \text{ ft}$$

$$d_{79} = \frac{x}{\cos\phi} = \frac{11.8613}{0.531394} = 22.321 \text{ ft}$$

Coordinates for 8 and 9 are computed by Eqs. (8.6) and (8.7) as follows:

$$X_8'' = X_1' + d_{18}\sin39°40'28'' = 0.00 + (15.411)(0.638425) = 9.84 \text{ ft}$$
$$Y_8'' = Y_1' + d_{18}\cos39°40'28'' = 113.92 + 11.86 = 125.78$$
$$X_9'' = X_7' + d_{79}\sin57°54'01'' = 570.446 + (22.321)(0.847124) = 589.35 \text{ ft}$$
$$Y_9'' = Y_7' + d_{79}\cos57°54'01'' = 113.92 + 11.86 = 125.78 \text{ ft}$$

As a final check, the area of the segment enclosed by points 12981 is calculated using the computed coordinates and expression (8.44).

$$2(\text{area}) = \frac{0.00}{113.92} \diagdown\diagup \frac{388.84}{0.00} \diagdown\diagup \frac{589.35}{125.7813} \diagdown\diagup \frac{9.84}{125.7813} \diagdown\diagup \frac{0.00}{113.92}$$

$$= 78,625 \text{ ft}^2$$

area = 39,313 ft$^2$ that is equal to one-fourth of the total area. Final coordinates for 8 and 9 are:
$X_8 = 4391.93$, $Y_8 = 6162.68$, $X_9 = 4971.44$, $Y_9 = 6162.68$ ft.

**8.38. Radial surveys**   A traverse performed from a single station is called a *radial traverse*. The station utilized should be a coordinated point in the primary control network for the project. A second station is necessary for a backsight and a third known point is useful for check purposes. A theodolite and EDM device or an electronic tacheometer (Chap. 12) are required since the procedure involves observing directions and distances.

The method is best used where the terrain is open with few restrictions to the line of sight. It may be employed to determine locations for traverse stations or to set off the positions of precalculated points.

Figure 8.56 shows a radial traverse executed from known control point $A$ with a backsight on known point $H$. It is desired to locate the positions of traverse points $B$, $C$, $D$, $E$, and $F$. If a theodolite and separate EDM device are being used, the theodolite is set over station $A$, a backsight taken on station $H$, and directions are observed to all traverse stations $B$ through $E$. In addition, a sight is taken on known control point $K$ and auxiliary point $A'$ is set along the line $AK$. When the directions to these points have been observed a sufficient number of times to ensure the desired accuracy, the theodolite is removed from $A$ and is replaced by the EDM device. Distances are observed from $A$ to stations $B$, $C$, $D$, $E$, $A'$, and $F$. If an electronic surveying system or electronic tacheometer is employed (see Art. 12.2), directions and distances can be obtained simultaneously. Since lines $AH$ and $AK$ are of known directions, the azimuths of lines $AB$, $AC$, $AD$, $AE$, $AA'$ and $AF$ can be determined.

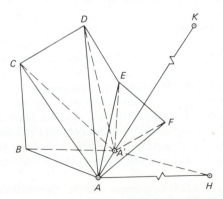

**Fig. 8.56** Radial traverse.

Using EDM distances (corrected for all systematic errors) and azimuths, departures and latitudes are calculated [Eq. (8.6)] for lines *AB*, *AC*, *AD*, *AE*, *AA'*, and *AF*, permitting computation of coordinates for *B*, *C*, *D*, *E*, *A'*, and *F* [Eq. (8.7)].

To provide a check on the work, the entire operation is repeated from auxiliary station *A'* with a backsight on station *K* yielding a second set of azimuths and horizontal distances from *A'* to all of the traverse stations *A* through *F*. These extra measures provide a second set of coordinates for all traverse points. In this way, blunders and undetected systematic errors can be found and corrected. If the values for the two sets of coordinates for each point are in reasonable agreement, the mean of each pair of coordinates is used as the final position for the station. Alternatively, the data from both points *A* and *A'* can be combined in a simultaneous least-squares solution. Distances and directions between these coordinated stations can then be calculated for lines *AB*, *BC*, *CD*, *DE*, *EF*, and *FA* using Eqs. (8.8) and (8.9). If distance *A'H* is observable, the triangle *AA'H* should also be calculated as a check.

Note that the entire procedure could be performed with a backsight on a single line such as *AH* or *AK*. However, the extra known point provides an additional check on direction and if the distance can be observed from *A'* to *K* a redundancy on the position of *A'* is furnished. In either case, an auxiliary station such as *A'* is absolutely necessary to provide a check on each point located.

In order to improve the final positions adopted for the stations, rigorous adjustment methods can be performed to reconcile the redundant data. The linearized form of the distance and direction Eqs. (8.8) and (8.9) can be formed for each ray from *A* and *A'* to the traverse points. The system of equations so formed could then be solved by the method of least squares in a procedure similar to that outlined and developed in Arts. B.13 and B.15 for directions and in Art. B.16 for distances. Adjustment of combined triangulation and trilateration may also be applicable. This procedure is illustrated in Art. B.17.

Procedures for setting off the precalculated locations of points such as in construction layout are developed in Chap. 18.

It should be emphasized that there is no standard solution for the application of radial methods to surveying problems. Each problem is unique and the basic rules as outlined must be used to suit the situation in the field.

**8.39. Traverse in three dimensions** When elevations are also determined in the process of running a stadia or plane-table traverse (Arts. 8.12 and 8.13), these methods constitute traverses in three dimensions. Such three-dimensional surveys are suitable when horizontal and vertical positional accuracy requirements are low (horizontal, 1 part in 300 to 400 and vertical position to tenths of feet or to a few centimetres). When EDM equipment and direction theodolites or electronic tacheometers or surveying systems (Chap. 12) are employed for establishing horizontal and vertical control, relatively high accuracies in both horizontal and vertical positions are possible. In other words, trigonometric leveling as described in Art. 5.5 is performed in conjunction with any one of the several methods of running a traverse described in Arts. 8.7 to 8.10. Since the direction theodolite with its 01″ nominal least count in the vertical circle is necessary for trigonometric leveling, the angles-to-the-right traverse (Art. 8.9) is the most appropriate.

The minimum equipment (also the least expensive) required for a three-dimensional traverse consists of a direction theodolite with a nominal least count in horizontal and vertical circles of 01″ (Art. 6.30); an EDM instrument and reflectors or remote unit; two sighting targets adaptable to a tripod (Fig. 6.67); and three tripods compatible with theodolite, EDM instrument, reflectors, and target so that forced centering can be employed (Art. 6.30).

Assume that a light-wave EDM instrument and 01″ theodolite are to be used to run a three-dimensional traverse from station *A* to station *E* as shown in Fig. 8.57. Line *AK* is of

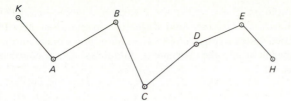

**Fig. 8.57**

known direction and stations $A$ and $E$ have known $XY$ coordinates and elevations. Line $EH$ is also of known direction. The procedure for running the traverse is as follows. The theodolite is set over station $A$ and tripods are centered over stations $K$ and $B$. Sighting targets are placed on the tripods at $K$ and $B$. All heights of instrument and targets above the traverse stations (h.i.'s) are observed and recorded. The angle to the right from $K$ to $B$ and the zenith angle from $A$ to $B$ are observed. The theodolite is removed from the tripod at $A$ and is replaced by the EDM device. A reflector replaces the sighting target on the tripod at $B$. The h.i. for the EDM at $A$ and reflector at $B$ are recorded. While the slope distance from $A$ to $B$ is being observed and recorded, the tripod at $K$ is moved ahead and centered over station $C$ where the sighting target is set in place and the h.i. observed and recorded. The theodolite is moved to the tripod at $B$, the EDM is moved toward station $B$, and a sighting target is placed on the tripod at $A$. After the necessary h.i.'s are recorded, the zenith angle from $B$ to $A$, the angle to the right from $A$ to $C$, and the zenith angle from $B$ to $C$ are observed. The sighting targets at $A$ and $C$ are replaced by reflectors and the theodolite at $B$ is replaced by the EDM instrument. All h.i.'s must be measured and recorded. The slope distances $BA$ and $BC$ are observed and recorded. Next, the tripod at $A$ is moved ahead to $D$ with a sighting target, the theodolite is set on the tripod at $C$, the EDM device is moved toward $C$, and a sighting target is placed on the tripod at $B$. The necessary h.i.'s are recorded, and the zenith angles from $C$ to $B$ and from $C$ to $D$ and the horizontal angle to the right from $B$ to $D$ are measured and recorded. The EDM instrument replaces the theodolite at $C$, reflectors are set on the tripods at $B$ and $D$, and slope distances are measured from $C$ to $B$ and $C$ to $D$. This sequence of moving theodolite, EDM device, sighting targets, and reflectors ahead is continued until station $E$ is occupied by the theodolite and the EDM instrument in that order, yielding the zenith angle from $E$ to $D$, the closing horizontal angle from $D$ to known line $EH$, and the slope distance $ED$. Care must be observed throughout to measure and record all h.i.'s to reflectors, targets, theodolite, and EDM instrument. It is also necessary to observe and record the necessary meteorological data to permit correcting the distances measured by EDM for systematic errors caused by atmospheric conditions (Art. 4.33).

Zenith angles should be measured a minimum of once with the telescope direct and once with the telescope reversed (Art. 6.44). Directions ought to be measured with a sufficient number of repetitions (Art. 6.41) to provide the accuracies specified for horizontal position (see Table 8.15). Azimuths are then computed from the angles determined from the averaged repeated directions. Measured slope distances should be corrected for all EDM systematic errors (Arts. 4.32, 4.34, and 4.35) and reduced to horizontal distance using the average zenith angles (Arts. 4.37 to 4.42). Distances and directions so determined are then used to perform traverse computations as described in Art. 8.16.

Differences in elevations can also be calculated using average zenith angles and corrected slope or horizontal distances according to the methods outlined in Art. 5.5.

Electronic tacheometers or electronic surveying systems (Art. 12.2) used with combined sighting targets and reflectors (Fig. 12.2) expedite the running of three-dimensional traverse considerably. In this case, distance and directions are measured from one instrument setup and there is no need to replace sighting targets with reflectors, and vice versa. Also, the data-reduction process is simplified. The simplest of the electronic tacheometers (Fig. 12.1*b*) provides direct output of horizontal distance and difference in elevation between points sighted. The more sophisticated systems (Figs. 12.4 to 12.6) have data storage and processing capacity using a minicomputer built into the instrument. Additional details concerning these systems can be found in Art. 12.2.

## 8.40. Problems

**8.1**  Describe the two general classes of traverses, pointing out advantages, disadvantages, and applications for each class.

**8.2**  A traverse is initiated and closed on station *George*, which has known *X* and *Y* coordinates. What type of traverse is this? Should a traverse of this type be employed for a major project? Explain.

**8.3**  The interior angles of a five-sided closed traverse are as follows: *A*, 114°34′; *B*, 94°30′; *C*, 140°50′; *D*, 130°20′. The angle *E* is not measured. Compute the angle *E*, assuming the given values to be correct.

**8.4**  (*a*) What are the deflection angles of the traverse of Prob. 8.3? (*b*) What are the computed bearings if the bearing of *AB* is due north?

**8.5**  Following are the deflection angles of a closed traverse: *A*, 85°20′L; *B*, 10°11′R; *C*, 83°32′L; *D*, 63°27′L; *E* 34°18′L; *F*, 72°56′L; *G*, 30°45′L. Compute the error of closure. Adjust the angular values on the assumption that the error is the same for each angle.

**8.6**  In Prob. 8.5, the azimuth of line *AB* from the north is 72°30′. Compute azimuths for the rest of the lines in the traverse and show a check on the calculations.

**8.7**  The clockwise interior angles observed with a 01′ transit in a traverse that starts and closes on point *J* are as follows: *J* = 127°30′; *A* = 83°30′; *B* = 251°50′; *C* = 101°20′; *D* = 85°50′; *E* = 91°40′; *F* = 251°00′; and *G* = 89°27′. Compute the error of closure and distribute this error assuming that all angles are of equal precision. Calculate azimuths for the lines in the traverse assuming that line *AJ* has a known azimuth from the north of $A_{AJ} = 206°30′$.

**8.8**  Comment on the relative and absolute angular accuracy of the data in Prob. 8.7.

**8.9**  The following deflection angle traverse was run with a 30″ repeating theodolite from a line of known direction *LM* to another independent line of known direction *TU*. Compute the angular error of closure and adjusted azimuths for each line. Comment on the relative and absolute angular accuracy of this traverse.

| Station | Deflection angle | Azimuth from north |
|---------|------------------|--------------------|
| *L*     |                  | 212°45′            |
| *M*     | 39°47′L          |                    |
| *N*     | 17°28′L          |                    |
| *O*     | 14°08′L          |                    |
| *P*     | 3°11′L           |                    |
| *Q*     | 49°59′L          |                    |
| *R*     | 32°18′R          |                    |
| *S*     | 18°44′R          |                    |
| *T*     | 7°31′L           |                    |
| *U*     |                  | 131°43′            |

**8.10**  The following are bearings taken for an open compass traverse. Correct for local attraction.

| Line | Forward bearing | Back bearing |
|------|-----------------|--------------|
| *AB* | N37°15′E        | S36°40′W     |
| *BC* | S65°30′E        | N66°15′W     |
| *CD* | S31°00′E        | N31°00′W     |
| *DE* | S89°15′W        | N89°45′E     |
| *EF* | N46°30′W        | S46°45′E     |
| *FG* | N15°00′W        | S14°45′E     |

**8.11**   The following are bearings taken on a closed compass traverse. Compute the interior angles and correct them for observational errors. Assuming the observed bearing of the line *AB* to be correct, adjust the bearings of the remaining sides.

| Line | Forward bearing | Back bearing |
|------|-----------------|--------------|
| AB | S37°30′E | N37°30′W |
| BC | S43°15′W | N44°15′E |
| CD | N73°00′W | S72°15′E |
| DE | N12°45′E | S13°15′W |
| EA | N60°00′E | S59°00′W |

**8.12**   Following are stadia intervals and vertical angles for a transit-stadia traverse. The elevation of station *A* is 418.6 ft; the stadia interval factor, 100.0; *C* = 1.00 ft. Rod readings are taken at height of instrument. Compute the horizontal lengths of the courses and the elevations of the traverse stations.

| Station | Object | Stadia interval, ft | Vertical angle |
|---------|--------|---------------------|----------------|
| B | A | 8.50 | +0°48″ |
|   | C | 4.37 | +8°13′ |
| C | B | 4.34 | −8°14′ |
|   | D | 12.45 | −2°22′ |
| D | C | 12.41 | +2°21′ |
|   | E | 7.18 | −1°30′ |

**8.13**   Following are stadia intervals and vertical angles for a transit-stadia traverse. The elevation of station *A* is 163.434 m. The stadia interval factor is 100 and *C* = 0. Rod readings are taken at the height of instrument.

| Station | Azimuth | Interval, m | Vertical angle | Horizontal distance | Differences in elevation | Elevation, m |
|---------|---------|-------------|----------------|---------------------|--------------------------|--------------|
| A |  |  |  |  |  | 160.434 |
|   | 85°06′ | 0.997 | +4°32′ |  |  |  |
| B |  |  |  |  |  |  |
|   | 10°18′ | 1.896 | +3°51′ |  |  |  |
| C |  |  |  |  |  |  |
|   | 265°00′ | 1.551 | −5°04′ |  |  |  |
| D |  |  |  |  |  |  |
|   | 173°04′ | 1.859 | −2°10′ |  |  |  |
| A |  |  |  |  |  |  |

(*a*) Calculate the horizontal distance for each line to the nearest decimetre. Determine differences in elevation between stations to the nearest centimetre. (*b*) Determine elevations of transit stations and distribute the error of closure in proportion to the distance.

**8.14**   Given the following notes for a closed-loop traverse. The coordinates for station *A* are $X_A$ = 457,200.054 m, $Y_A$ = 167,640.842 m. Compute the error of closure and coordinates for each traverse station adjusted according to the compass rule.

| Course | Azimuth | Distance, m |
|--------|---------|-------------|
| AB | 0°42′ | 372.222 |
| BC | 94°03′ | 164.988 |
| CD | 183°04′ | 242.438 |
| DA | 232°51′ | 197.145 |

**8.15** Given the following adjusted azimuths and distances for a closed traverse that starts at station $B$ ($X_B = 450,237.821$ m and $Y_B = 152,321.065$ m) and closes on station $J$ ($X_J = 449,032.330$ m and $Y_J = 150,140.510$ m). Compute the error in closure for the traverse and coordinates for each traverse station adjusted by the compass rule. Determine the relative or positional accuracy for this traverse. Discuss the value of this measure of accuracy in terms of absolute position. Describe a better way to describe the absolute positional accuracy of a traverse station for which coordinates are available.

| Course | Azimuth | Distance, m |
|--------|---------|-------------|
| AB | 142°08′ | |
| BC | 181°37′ | 349.301 |
| CD | 296°13′ | 158.740 |
| DE | 323°46′ | 248.869 |
| EF | 249°51′ | 221.407 |
| FG | 214°03′ | 567.538 |
| GH | 195°45′ | 852.068 |
| HJ | 191°28′ | 750.936 |
| JK | 138°42′ | |

**8.16** Given the following plane rectangular coordinates for traverse stations 3 and 4. Compute the distance and azimuth of the line from 3 to 4.

| Station | X, m | Y, m |
|---------|------|------|
| 3 | 525,341.821 | 148,241.305 |
| 4 | 524,210.613 | 148,000.006 |

**8.17** Compute adjusted distances and azimuths between all traverse points using adjusted coordinates from Prob. 8.14.

**8.18** Using the data for the traverse in Table 8.7 and shown in Fig. 8.25 (Art. 8.16), propagate $\sigma_x$, $\sigma_y$, and $\sigma_{xy}$ for point 4, treating lines 12, 23, and 34 as an open traverse and assuming that $\sigma_\alpha = 30''$, $\sigma_d = 0.01$ m (see Art. 8.23).

**8.19** Using adjusted coordinates in Table 8.13 for the traverse of Fig. 8.7, compute the coordinates of the intersection of line 62 with the line from 1 to 3.

**8.20** Using adjusted coordinates in Table 8.13 for the traverse of Fig. 8.7, determine the shortest distance from station 6 to a straight line from 1 to 5.

**8.21** It is necessary to determine the coordinates for the intersection of a straight property line and the curved line having a radius of 750.000 m, which defines a road right of way. The azimuth of the property line from $M$ to $N$ is 210°50′00″. The coordinates for $M$ and $O$ (the center of the circle) are as follows:

| Station | X, m | Y, m |
|---------|------|------|
| M | 5679.533 | 6072.099 |
| O | 5571.166 | 5120.567 |

**8.22** Given the following data for a closed traverse, for which the lengths of $BC$ and $DE$ have not been measured in the field. Compute the unknown lengths.

| Course | Bearing | Distance, ft |
|--------|---------|--------------|
| AB | N9°30′W | 689.32 |
| BC | N56°55′W | unknown |
| CD | S56°13′W | 678.68 |
| DE | S2°02′E | unknown |
| EA | S89°31′E | 1,082.71 |

**8.23**   Given the following data for a closed traverse. Compute the length and bearing of the unknown side.

| Course | Bearing | Distance, m |
|--------|---------|-------------|
| AB | N82°00′W | 140.5 |
| BC | unknown | unknown |
| CD | N68°15′E | 252.7 |
| DA | N80°45′E | 134.4 |

**8.24**   A square field contains 40 acres. What are its dimensions in chains, in rods, in feet, and in metres?

**8.25**   How many acres are there in a rectangular tract $50 \times 100$ ft? In a tract $400 \times 400$ ft? In a tract $2640 \times 2640$ ft? Express these areas in square metres and in hectares (Art. 1.8).

**8.26**   A triangle has sides 150.30, 300.85, and 245.62 m in length. Calculate the area of the triangle expressed in square metres, hectares, and acres.

**8.27**   What is the area of a triangle having sides of length 219.0, 317.2, and 301.6 ft? Of a triangle having two sides of length 1167.1 and 392.7 ft and an included angle of 39°46′?

**8.28**   The mutually bisecting diagonals of a four-sided field are 480 and 360 ft. The angle of intersection between the diagonals is 100°. Find the interior angles and the lengths of the sides.

**8.29**   In the following tabulation are given $X$ and $Y$ coordinates (in metres) for stations in a closed traverse. Calculate the area enclosed by the traverse using the coordinate method.

| Coordinate | Station | | | |
|------------|---------|---------|---------|---------|
| | A | B | C | D |
| X | 5000.0 | 5102.5 | 5202.8 | 5102.5 |
| Y | 5000.0 | 5153.4 | 4900.0 | 4874.5 |

**8.30**   In the following tabulation are given the departures and latitudes of an adjusted closed traverse. Calculate the area (a) by the D.M.D. method; (b) by the coordinate method.

| Course | Departure, ft | Latitude, ft |
|--------|---------------|--------------|
| AB | W213.6 | S198.7 |
| BC | W174.4 | N181.1 |
| CD | E 89.2 | N334.1 |
| DE | E110.7 | N224.9 |
| EA | E188.1 | S541.4 |

**8.31**   (a) Find the error of closure of the following traverse. Adjust the survey by the compass rule, and calculate the area in square metres by the D.M.D. method.

| Course | Bearing | Length, m |
|--------|---------|-----------|
| AB | S45°45′E | 89.733 |
| BC | N65°30′E | 80.284 |
| CD | N35°15′E | 95.585 |
| DE | N64°15′W | 119.482 |
| EF | S59°00′W | 60.107 |
| FA | S25°30′W | 73.152 |

(b) The coordinates for A are $X_A = 520,484.183$ m and $Y_A = 424,323.640$ m. Calculate coordinates for the traverse stations and determine the area in hectares by the coordinate method.

**8.32**   A traverse ABCD is established inside a four-sided field, and the corners of the field are located by angular and linear measurements from the traverse stations, all as indicated by the following data.

Compute the departures and latitudes, and adjust the traverse by the compass rule. Compute the coordinates of each traverse point and of each property corner, using $D$ as an origin of coordinates. Compute the length and bearing of each side of the field *EFGH*, and tabulate results. Calculate the area of the field by the coordinate method.

| Course | Bearing | Length, ft |
|---|---|---|
| AB | S89°38'E | 296.4 |
| AE | N20°00'W | 34.2 |
| BC | S43°20'W | 333.9 |
| BF | N35°20'E | 16.9 |
| CD | S80°21'W | 215.6 |
| CG | S73°00'E | 27.6 |
| DA | N27°24'E | 314.2 |
| DH | S36°30'W | 15.7 |

**8.33** Given the following offsets from traverse line to irregular boundary, measured at points 10 m apart. By the trapezoidal rule (Art. 8.31) calculate the area between traverse line and boundary.

| Distance, m | Offset, m | Distance, m | Offset, m |
|---|---|---|---|
| 0 | 0.0 | 50 | 8.60 |
| 10 | 5.06 | 60 | 3.63 |
| 20 | 10.70 | 70 | 9.36 |
| 30 | 11.98 | 80 | 13.23 |
| 40 | 12.80 | 90 | 6.86 |

**8.34** Given the data of Prob. 8.33, calculate the required area by Simpson's one-third rule. Note that the number of offsets is even.

**8.35** Following are offsets from a traverse line to an irregular boundary, taken at irregular intervals. Calculate the area between traverse line and boundary by means of the trapezoidal rule.

| Distance, ft | Offset, ft | Distance, ft | Offset, ft |
|---|---|---|---|
| 0 | 18.5 | 100 | 44.1 |
| 25 | 37.7 | 170 | 53.9 |
| 60 | 58.2 | 200 | 46.0 |
| 70 | 40.5 | 220 | 34.2 |

**8.36** Following are offsets from a traverse line to an irregular boundary taken at irregular intervals. Calculate the area between the traverse line and the boundary using the trapezoidal rule.

| Distance, m | Offset, m | Distance, m | Offset, m |
|---|---|---|---|
| 0 | 5.15 | 30 | 13.40 |
| 8 | 10.49 | 55 | 17.54 |
| 20 | 16.74 | 60 | 14.05 |
| 22 | 13.54 | 70 | 9.52 |

**8.37** In Fig. 8.48, what is the area of the circular segment *EQF* if the length of the chord $L$ is 249.08 m and the middle ordinate $M$ is 27.179 m?

**8.38** In Fig. 8.48, what is the area of the circular segment *EQF* if the chord length $L$ is 600 ft and the middle ordinate $M$ is 7.85 ft?

**8.39** Solve Probs. 8.37 and 8.38 using the approximate expression [Eq. (8.52)] of Art. 8.33.

Compare the results with those of Probs. 8.37 and 8.38, and for each case compute the percentage of error introduced through use of the approximate expression.

**8.40** A curved corner lot is similar in shape to that shown in Fig. 8.49. The tangent distances *T* are each 50.0 ft and the intersection angle *I* is 40°. What is the area between the circular curve *ABC* and the tangents *AD* and *CD*? What is the external distance *E*?

**8.41** Given the data of Table 8.13, Art. 8.19, and illustrated in Fig. 8.7. Find the area of each of the two parts into which the tract is divided by a meridian line through the point 5.

**8.42** Given the data of Prob. 8.41. Find the length and direction of a line that runs through 3 and divides the tract into two equal parts.

**8.43** Given the data of Prob. 8.41. The tract is to be divided into two equal parts by a north-south line. Compute the length of the dividing line, and compute the distances from the ends of the line to adjacent traverse stations.

## 8.41. Field problems

### PROBLEM 1   DEFLECTION ANGLE TRAVERSE

**Object**   To measure deflection angles at each station of a traverse having *n* sides using the method of double deflection angles as described in Art. 6.39. A repeating transit or theodolite having a least count of 01′ or 30″ should be used. It is desirable that the instrument be equipped with a magnetic compass.

**Procedure**

1. Set *n* traverse stations each consisting of a hub with a tack at locations which provide maximum length of sight and optimum geometric strength for the area designated. If the traverse is to be used again, reference each station (Art. 8.14).
2. Occupy each station with the repeating transit or theodolite moving either direction around the traverse.
3. Level and center the instrument over the first station and loosen the compass needle.
4. Set the horizontal circle index or the *A* vernier to zero and take a backsight on the previous station with the telescope direct. Observe and record the magnetic bearing along that line.
5. Measure the deflection angle twice, once with the telescope direct and once with it reversed according to the procedure outlined in Art. 6.39. Be sure to observe and record the magnetic bearing along the line of the foresight when the telescope is in the direct position. The second angle should agree with the first angle to within the least count of the instrument. If it does not, repeat the measurement.
6. Repeat steps 3 to 5 for each station, recording the angles in a form similar to that illustrated in Fig. 8.6. Be sure to indicate the direction of the deflection angle as right (R) or left (L).
7. Calculate the sum of the average deflection angles. This sum should check the theoretical value [by Eq. (8.2)] to within $\sqrt{n}$ (least count of the instrument).
8. Calculate adjusted angles by distributing the angular error of closure equally among the angles.

**Hints and precautions**

1. When lengths of sights are short, best results can be obtained by sighting directly on a plumb bob or pencil held on the traverse station being sighted.
2. When sighting into the sun, the notekeeper should shade the objective lens and eyes of the instrument operator.
3. Be sure to tighten the clamp before using the tangent-screw when using either upper or lower motions.
4. Concentrate on each sight to be sure that the lower clamp and tangent-screw are used on the backsight and upper clamp and tangent-screw are used on the foresight.

### PROBLEM 2   INTERIOR ANGLE TRAVERSE

**Object**   To measure interior angles at each station of a closed traverse having *n* sides, using the method for measuring angle by repetition described in Art. 6.39. Use a repeating transit or theodolite having a least count of 30″ or 01′.

**Procedure**

1. Set *n* traverse stations as described for Field Problem 1.
2. Occupy each station, moving around the traverse in either a clockwise or a counterclockwise direction.
3. Level and center the instrument over the first station and set the horizontal circle index or *A* vernier to zero.
4. Take a backsight on the previous station with the telescope direct and using the lower clamp and tangent-screw. Record 0°00′00″ for this sight in a form similar to that used for deflection angles (Fig. 8.6).
5. Loosen the upper clamp and turn the alidade toward the next traverse station with the telescope direct, taking a foresight with the upper clamp and upper tangent-screw.
6. Read and record the *interior* angle.
7. Loosen the lower clamp and resight the first station with the telescope reversed using the lower clamp and tangent-screw.
8. Repeat step 5, doubling the angle with the telescope reversed. The difference between the first and second angles should not exceed the least count of the instrument. If it does, repeat the set.
9. Repeat steps 3 to 8 for each traverse station. When all angles have been measured, check the sum of the average angles. This sum should not deviate from $(n-2)(180°)$ by more than $(\sqrt{n})$ (least count of the instrument). The difference is the angular error of closure.
10. Distribute the error of closure equally among the *n* angles.

**Hints and precautions**

1. Turn all interior angles in the same direction.
2. Be sure to observe the correct horizontal circle; that is, when turning an interior angle to the right, read the clockwise circle; when turning an angle to the left, read the counterclockwise circle. Observe all precautions given for Field Problem 1.

PROBLEM 3   STADIA TRAVERSE WITH TRANSIT

**Object**   To establish horizontal and vertical control for a topographic survey. The azimuth method (Art. 8.10) is used to establish directions and stadia (Chap. 7 and Art. 8.12) is employed to get distances and differences in elevations.

**Procedure**

1. Set stakes around the perimeter of the area to be surveyed, locating stations so as to provide maximum length of sight and strong geometry.
2. Establish the elevation of one station by differential leveling from the nearest benchmark (Arts. 5.35–5.38).
3. Occupy the point of known elevation (the initial station), measure the height of instrument above the station (h.i.) with a rod, and take a sight on the station that is to be the backsight.
4. Release the magnetic compass and observe the magnetic bearing along the line of the backsight. Convert this value to a magnetic azimuth and apply the declination to give a true azimuth. Release both motions of the repeating transit, set the true azimuth of the line on the clockwise horizontal circle, and resight the backsight station using the lower motion.
5. Have a rod placed on the station used for a backsight. Sight the rod observing the stadia interval and vertical angle according to the procedure described in Art. 8.12 under "Elevations required."
6. Record stadia interval, vertical angle, and true azimuth using the note form of Fig. 8.17.
7. Release the upper motion and take a foresight on the next station using the upper clamp and tangent-screw.
8. Observe and record the stadia interval, vertical angle, and azimuth (on the clockwise horizontal circle) to this station.
9. Compute and record the back azimuth of this line by adding 180° to the forward azimuth.
10. Set the instrument up on the station just sighted and with the back azimuth just computed set on the clockwise horizontal circle, sight the station previously occupied using the lower clamp and tangent-screw. The instrument is now oriented.

11. Release the upper motion and sight the next station forward, observing and recording the stadia interval, vertical angle, and azimuth. Proceed in this manner until all stations have been occupied.

12. The forward azimuth from the last station to the initial station should coincide with $180° \pm$ the azimuth previously recorded for this line (step 4). Any discrepancy is the angular error of closure which should not exceed ($\sqrt{n}$) (least count of the instrument) where $n =$ number of instrument setups. Distribute the discrepancy to each forward azimuth in proportion to the number of angles involved in determining that azimuth, leaving the first observed side fixed.

## References

1. Angus-Leppan, P., "Practical Application of Accuracy Standards in Traversing," *Australian Surveyor*, March 1973.
2. Adler, R. K., et al., "Precise Traverse in Major Geodetic Networks," *Canadian Surveyor*, September 1971.
3. Baarda, W., and Albercla, J., "The Connection of Geodetic Adjustment Procedures with Methods of Mathematical Statistics," *Bulletin Géodésique*, Vol. 66, 1962.
4. Berton-Jones, P., "Simplified Traverse Design," *Australian Surveyor*, September 1971.
5. Berton-Jones, P., "The Notion of Permissible Misclosure in Traversing," *Australian Surveyor*, September 1970.
6. Crandall, C. L., "The Adjustment of a Transit Survey as Compared with that of a Compass Survey," *Transactions of the American Society of Civil Engineers*, Vol. 45, pp. 453–464, 1901.
7. Chrzanowski, A., *Design and Error Analysis of Surveying Projects*, Lecture Notes No. 47, Department of Surveying Engineering, University of New Brunswick, Fredericton, New Brunswick, Canada, 1977.
8. Matson, D. F., "Angular Reference Ties," *Surveying and Mapping*, Vol. 33, No. 1, March 1973, p. 33.
9. Mikhail, E. M., *Observations and Least Squares*, Harper & Row, Publishers, New York, 1976.
10. Moffitt, F. H., and Bouchard, H., *Surveying*, Harper & Row, Publishers, New York, 1975.
11. Obenson, G., "Absolute Traverse Distortions of Some Adjustment Methods," *Survey Review*, July 1975.
12. Richardus, P., *Project Surveying*, John Wiley & Sons, Inc., New York, 1966, p. 277.
13. Tárczy-Hornoch, A., "Remarks on Computations for Missing Elements of Closed Traverses," *Surveying and Mapping*, Vol. 32, No. 4, p. 523.

# CHAPTER 9
# Intersection and resection

**9.1. Location of points by intersection**  When coordinates of a point are given and the azimuth and distance to a second point are also known, it is possible to compute the coordinates of the second point (Art. 8.15). Similarly, if the coordinates are given for the two ends of a line and directions are observed from each end of this line to a third point not on the line, then coordinates of that third point can be calculated. This procedure is called *location by intersection*.

In Fig. 9.1, $B$ and $D$ are points of known coordinates from which angles $\alpha$ and $\beta$ have been observed so as to locate point $C$ by intersection. The distance $c$ and azimuth $A_{BD}$ from $B$ to $D$ can be found using Eqs. (8.8) and (8.9):

$$c = \left[ (X_D - X_B)^2 + (Y_D - Y_B)^2 \right]^{1/2}$$

$$\tan A_{BD} = \frac{X_D - X_B}{Y_D - Y_B}$$

Also in triangle $DBC$, $\gamma = 180° - (\alpha + \beta)$, so that

$$d = \frac{c \sin \beta}{\sin \gamma} \quad \text{and} \quad b = \frac{c \sin \alpha}{\sin \gamma} \tag{9.1}$$

where $d$ is the distance $B$ to $C$ and $b$ the distance $D$ to $C$. The azimuths of lines $BC$ and $DC$ can be calculated and the coordinates for $C$ may be computed by Eqs. (8.6) and (8.7) using data from line $BC$ as a check on calculations with data from line $DC$.

**9.2. Intersection by the base solution**  A more direct solution is possible using the observed angles, $\alpha$, $\beta$, and the base line $c$ computed from the coordinates of $B$ and $D$. In triangle $BCD$ from Eqs. (8.6), (8.7), and (9.1),

$$Y_C = Y_D + b \cos A_{DC} = Y_D + \frac{c \sin \alpha}{\sin \gamma} \cos(A_{DB} + \beta)$$

$$Y_C = Y_D + \frac{c \sin \alpha}{\sin(\alpha + \beta)} \cos(A_{DB} + \beta) \tag{9.2}$$

in which

$$\cos(A_{DB} + \beta) = \cos A_{DB} \cos \beta - \sin A_{DB} \sin \beta$$

and the sign of $\beta$ is determined by careful inspection of the figure. By Eq. (8.11) and (8.12) (Art. 8.21),

$$\sin A_{DB} = \frac{X_B - X_D}{c} \quad \text{and} \quad \cos A_{DB} = \frac{Y_B - Y_D}{c}$$

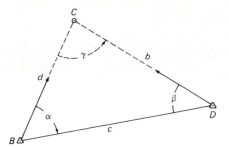

**Fig. 9.1** Location of a point by intersection.

so that

$$\cos(A_{DB}+\beta)=\frac{Y_B-Y_D}{c}\cos\beta-\frac{X_B-X_D}{c}\sin\beta$$

and by substitution into Eq. (9.2),

$$Y_C=Y_D+(Y_B-Y_D)\frac{\sin\alpha\cos\beta}{\sin(\alpha+\beta)}-(X_B-X_D)\frac{\sin\beta\sin\alpha}{\sin(\alpha+\beta)}$$

$$=Y_D+\frac{(Y_B-Y_D)\cot\beta}{\cot\alpha+\cot\beta}-\frac{(X_B-X_D)}{\cot\alpha+\cot\beta}$$

$$=\frac{(X_D-X_B)+Y_D\cot\alpha+Y_B\cot\beta}{\cot\alpha+\cot\beta} \tag{9.3}$$

and in a similar manner,

$$X_C=\frac{(Y_B-Y_D)+X_D\cot\alpha+X_B\cot\beta}{\cot\alpha+\cot\beta} \tag{9.4}$$

Using Eqs. (9.3) and (9.4), azimuths for the lines to the intersected point do not have to be determined, and the coordinates can be calculated using adjusted base angles $\alpha$ and $\beta$. This method is advantageous when computing through a chain of triangles where adjusted angles may be available but no azimuths have been calculated.

**9.3. Intersection when azimuths are given**   When azimuths are given for lines $DC$ and $BC$, in addition to the coordinates of points $D$ and $B$, the point-slope equations [Eq. (8.13), Art. 8.21] for lines $DC$ and $BC$ are

$$Y_C-Y_D=(X_C-X_D)\cot A_{DC}$$
$$Y_C-X_C\cot A_{DC}+X_D\cot A_{DC}-Y_D=0 \tag{9.5a}$$

and

$$Y_C-Y_B=(X_C-X_B)\cot A_{BC}$$
$$Y_C-X_C\cot A_{BC}+X_B\cot A_{BC}-Y_B=0 \tag{9.5b}$$

Subtraction of Eq. (9.5b) from Eq. (9.5a) yields

$$X_C=\frac{(Y_D-Y_B)-X_D\cot A_{DC}+X_B\cot A_{BC}}{\cot A_{BC}-\cot A_{DC}} \tag{9.6}$$

and solution for $Y_C$ in Eqs. (9.5a) and (9.5b) gives

$$Y_C=Y_D+(X_C-X_D)\cot A_{DC} \tag{9.7}$$
$$Y_C=Y_B+(X_C-X_B)\cot A_{CA} \tag{9.8}$$

Some examples serve to illustrate applications of the different methods.

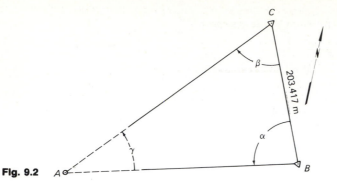

**Fig. 9.2**

**Example 9.1**   In Fig. 9.2, the coordinates in metres for triangulation stations $B$ and $C$ are

| Station | $X$, m | $Y$, m |
|---|---|---|
| $B$ | 3369.287 | 2890.836 |
| $C$ | 3300.259 | 3082.183 |

The observed angles are

$\beta = 64°32'28''$    $\alpha = 81°17'38''$

Determine $X$ and $Y$ coordinates using the first method (Art. 9.1) of solving the intersection problem.

SOLUTION   The distance from $C$ to $B$ is

$CB = [(3300.259 - 3369.287)^2 + (3082.183 - 2890.836)^2]^{1/2} = 203.41716$ m

The azimuth of $BC$ is

$$\tan A_{BC} = \frac{3300.259 - 3369.287}{3082.183 - 2890.836} = -0.36074775$$

$A_{BC} = 340°09'47''$

Compute azimuths for $CA$ and $BA$:

$A_{BC} = 340°09'47''$

$-\alpha = -81°17'38''$

$A_{BA} = 258°52'09''$

$A_{CB} = 160°09'47''$

$+\beta = 64°32'28''$                                    $\alpha = 81°17'38''$

$A_{CA} = 224°42'15''$                                 $\beta = 64°32'28''$

$\gamma = \angle CAB = 258°52'09'' - 224°42'15'' = 34°09'54''$

Check sum of angles $\alpha + \beta + \gamma$        $= 180°00'00''$

In triangle $BCA$, solve for sides $CA$ and $BA$ using the law of sines:

$$CA = \frac{(203.4172)\sin\alpha}{\sin\gamma} = \frac{(203.4172)(0.9884777)}{0.5615780} = 358.051 \text{ m}$$

$$BA = \frac{203.4172\sin\beta}{\sin\gamma} = \frac{(203.4172)(0.9028940)}{0.5615780} = 327.050 \text{ m}$$

The coordinates of $A$ using line $CA$ and Eq. (8.7) are

$X_A = 3300.259 + (358.051)(\sin 224°42'15'') = 3048.389$ m

$Y_A = 3082.183 + (358.051)(\cos 224°42'15'') = 2827.699$ m

As a check, calculate the coordinates of $A$ using line $BA$:

$X_A = 3369.287 + (327.050)(\sin 258°52'09'') = 3048.389$ m

$Y_A = 2890.836 + (327.050)(\cos 258°52'09'') = 2827.699$ m

For purposes of comparison, determine the coordinates from Example 9.1 by Eqs. (9.3) and (9.4). First solve for $Y_A$ using Eq. (9.3), which for this example is

$$Y_A = \frac{(X_C - X_B) + Y_C \cot\alpha + Y_B \cot\beta}{\cot\alpha + \cot\beta} = \frac{U_Y}{V_Y}$$

$X_C = 3300.259$        $\cot\alpha = \cot 81°17'38'' = 0.15313066$

$X_B = 3369.287$        $\cot\beta = \cot 64°32'28'' = 0.47609507$

$X_C - X_B = -69.028$        $\cot\alpha + \cot\beta = V_Y = 0.62922573$

$Y_C \cot\alpha = 471.9767$        $Y_C = 3082.183$

$Y_B \cot\beta = 1376.3128$        $Y_B = 2890.836$

$U_Y = 1779.2615$

$$Y_A = \frac{U_Y}{V_Y} = 2827.700 \text{ m}$$

Equation (9.3) for $X_A$ is

$$X_A = \frac{(Y_B - Y_C) + X_C \cot\alpha + X_B \cot\beta}{\cot\alpha + \cot\beta} = \frac{U_X}{V_X}$$

$Y_B = 2890.836$        $X_C \cot\alpha = (3300.259)(0.15313066)$

$Y_C = 3082.183$                 $= 505.3708$

$Y_B - Y_C = -191.347$        $X_B \cot\beta = (3369.287)(0.47609507)$

$X_C \cot\alpha = 505.3708$            $= 1604.1009$

$X_B \cot\beta = 1604.1009$

$U_X = 1918.1247$        $V_X = V_Y = 0.62922573$

$$X_A = \frac{U_X}{V_X} = 3048.389 \text{ m}$$

This method is more compact and efficient computationally but does not have the inherent mathematical check found in the preceding procedure.

**Example 9.2**   In Fig. 9.3, point $A$ is intersected from stations $C$ and $D$ having the following coordinates:

| Station | X, m | Y, m |
|---------|----------|----------|
| C | 3300.259 | 3082.183 |
| D | 3047.954 | 3048.344 |

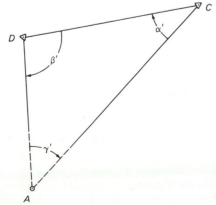

**Fig. 9.3**

with observed angles

$\alpha' = $ angle $DCA = 37°39'28''$
$\beta' = $ angle $ADC = 97°31'31''$

SOLUTION   Use the method developed in Art. 9.3, which requires the azimuths to be calculated for lines $DA$ and $CA$. First compute the azimuth of $DC$.

$$\text{azimuth of } DC = A_{DC} = \tan^{-1}\frac{X_C - X_D}{Y_C - Y_D} = \frac{3300.259 - 3047.954}{3082.183 - 3048.344}$$

$\tan A_{DC} = 7.4560418 \qquad A_{DC} = 82°21'40''$

Next, compute the azimuths of $DA$ and $CA$.

| Line | Azimuth |
|------|---------|
| DC | 82°21′40″ |
| +β′ | 97°31′31″ |
| DA | 179°53′11″ |
| CD | 262°21′40″ |
| −α′ | 37°39′28″ |
| CA | 224°42′12″ |
| −DA | −179°53′11″ |
| ∠DAC=γ′ | =44°49′01″ |

| | |
|---|---|
| β′ | 97°31′31″ |
| α′ | 37°39′28″ |
| γ′ | 44°49′01″ |
| | 180°00′00″  check |

The equation for line $DA$ is

$$Y_A - Y_D = (X_A - X_D)\cot A_{DA} \tag{1}$$

and for line $CA$

$$Y_A - Y_C = (X_A - X_C)\cot A_{CA} \tag{2}$$

Subtraction of Eq. (2) from Eq. (1) gives

$$X_A = \frac{(Y_D - Y_C) - X_D\cot A_{DA} + X_C\cot A_{CA}}{\cot A_{CA} - \cot A_{DA}} \tag{3}$$

Substitution of the given values into Eq. (3) can be tabulated as follows.

$Y_D - Y_C = 3048.344 - 3082.183 \qquad = \qquad -33.839$

$-X_D\cot A_{DA} = -(3047.954)(-504.31421) = 1,537,126.5$

$X_C\cot A_{CA} = (3300.259)(1.0104096) \qquad = \qquad 3,334.6134$

$(Y_D - Y_C) - X_D\cot A_{DA} + X_C\cot A_{CA} = \qquad 1,540,427.3$

$\cot A_{CA} - \cot A_{DA} \quad = \qquad\qquad\qquad 505.32462$

$X_A = 3048.3916$ m

Calculate $Y_A$ by Eq. (9.8).

$$Y_A = Y_C + (X_A - X_C)\cot A_{CA}$$
$$= (3082.183) + (3048.3916 - 3300.259)(1.0104096)$$
$$= 2827.694 \text{ m}$$

As in the method of base-line angles, there is no built-in check on the arithmetic in this solution. A solution using the first method of Art. 9.1 reveals that these answers are correct.

Note that in all three solutions contained in Examples 9.1 and 9.2, one more significant figure was carried than the number of significant digits in the original coordinates.

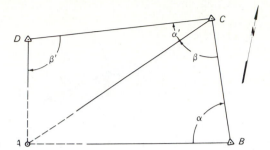

**Fig. 9.4** Intersection with redundant data.

**9.4. Intersection using redundant data**   Frequently, a point can be located by intersection from more than two known points. In Fig. 9.4, points $B$, $C$, and $D$ are triangulation stations of known positions, from which the angles $\alpha$, $\beta$, $\alpha'$, $\beta'$ have been observed so as to locate point $A$. Thus, the position of $A$ can be determined using triangle $ABC$ and another separate solution is possible using the data in triangle $CDA$. Note that there are two extra observations in this example or there is a redundancy of two. Using base line $BC$ and angles $\alpha$ and $\beta$, one set of coordinates may be obtained for $A$. With base line $DC$ and angles $\alpha'$ and $\beta'$, another set of coordinates can be calculated. The likelihood that these two sets will agree exactly is very small.

The observant reader will have noticed that triangle $ABC$ in Fig. 9.4 was used for Example 9.1 and triangle $ADC$ was used in Example 9.2. If the data were perfect, the coordinates in Example 9.1 would have agreed exactly with those computed in Example 9.2. The data are real observations of high quality but cannot be perfect, since they contain random errors. Note that the values obtained for the coordinates in the two examples differ by 3 mm in $X$ and 5 mm in $Y$. These observations were made in triangulation for bridge construction, and exceptional care was exercised in the execution of the survey. In an ordinary survey the difference between the two might be several centimetres or even decimetres and the discrepancy would have to be resolved. One method of resolving the discrepancy is to take the average of the two sets as the final values, particularly if the four angles are of equal precision.

A more rigorous adjustment procedure consists of applying the method of least squares. Such a procedure requires that an equation be formed for each line that intersects at the unknown point. These equations are linearized and solved by the method of least squares so as to provide a best estimate for the coordinates of the intersected point. Approximate values of coordinates for the unknown point, which are necessary for linearization, can be computed by any one of the methods discussed above using any pair of angles. Details and an example of this procedure can be found in Example B.9, Art. B.13, Appendix B.

**9.5. Resection**   When angles between lines to three points of known position are observed from a point of unknown position, the coordinates of the unknown point can be calculated. This procedure is called location by *resection*.

The resection problem is illustrated graphically in Fig. 9.5. In the figure, $B$ and $A$ are of known position and $O$ is of unknown position. If the angle $\alpha$ is observed between $OB$ and $OA$, the position of $O$ is indeterminate because $O$ can be anywhere on the circle circumscribing the triangle $OAB$. Additional information is needed to make the problem determinate. When the direction of $OA$ or $OB$ is known, the problem is determinate. However, this information is usually not available. It is more convenient to observe angle $\beta$, subtended by $AC$, where $C$ is a third known point. Then $O$, $A$, and $C$ lie on the circle that circumscribes triangle $OAC$. Since $O$ is on the circles that circumscribe both of the triangles $OAB$ and $OAC$, it must lie at one of the two intersection points of the two circles. As $A$ is one of these

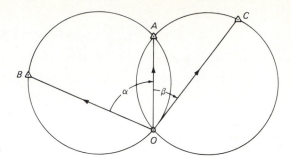

**Fig. 9.5** Three-point resection.

two intersections, $O$ is uniquely determined. This solution presumes that the three known points $A$, $B$, $C$ and $O$ do not fall on the circumference of one circle.

Should the two circumscribing circles tend to merge into one circle, the problem will be less stable and finally becomes indeterminate again when the two circles coincide. Points should be selected in the field so as to avoid this situation. It can be shown that this condition is present when $\angle BAC + \alpha + \beta = 180°$.

**9.6. Resection calculation**  The method presented here is somewhat similar to the procedure given by the U.S. National Ocean Survey in Ref. 5 at the end of the chapter.

In Fig. 9.6, stations $B$, $A$, and $C$ are coordinated control points and $O$ is a point of unknown position from which angles $\alpha$ and $\beta$ are observed. The problem is to determine angles $\theta$ and $\gamma$ so that the distances and directions of $AO$ and $BO$ or $CO$ can be calculated in order to compute and check the position of point $O$.

In polygon $OBAC$ (Fig. 9.6)

$$\theta + \gamma = 360 - (\phi + \alpha + \beta) = R \tag{9.9a}$$

for the cases represented by Fig. 9.6a and b, and

$$R = \phi - \alpha - \beta \tag{9.9b}$$

for the case shown in Fig. 9.6c.

By the law of sines in triangle $OAC$,

$$AO = \frac{b \sin \gamma}{\sin \beta} = \frac{c \sin \theta}{\sin \alpha} \tag{9.9c}$$

so that

$$\sin \theta = \frac{b \sin \gamma \sin \alpha}{c \sin \beta} \tag{9.9d}$$

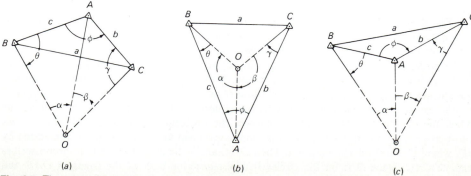

(a)  (b)  (c)

**Fig. 9.6** Three possible cases for three-point resection.

Since $\theta = R - \gamma$ from Eq. (9.9a), Eq. (9.9d) becomes

$$\sin(R - \gamma) = \frac{b \sin \gamma \sin \alpha}{c \sin \beta} = \sin R \cos \gamma - \cos R \sin \gamma \qquad (9.9e)$$

Division of Eq. (9.9e) by $\sin R \sin \gamma$ yields

$$\cot \gamma = \cot R + \frac{b \sin \alpha}{c \sin \beta \sin R} \qquad (9.10)$$

The computational procedure is as follows:

1. Obtain $R$ from Eq. (9.9a) or Eq. (9.9b).
2. Solve for $\gamma$ with Eq. (9.10).
3. Compute $\theta = R - \gamma$.
4. Solve for the distance $AO$ using $b$, $\gamma$, and $\beta$ in Eq. (9.9c). Check using $c$, $\theta$, and $\alpha$.
5. Compute angle $CAO = 180 - (\gamma + \beta)$ so as to determine the direction of $AO$. Check by calculating angle $BAO$.
6. Solve for $OC$ and/or $OB$ using law of sines.
7. Calculate and check the coordinates of $O$.

**Example 9.3**   Given the following data for a three-point resection:

| Point | X, ft | Y, ft |
|-------|-----------|-----------|
| B | 10,000.00 | 20,000.00 |
| A | 16,672.50 | 20,000.00 |
| C | 27,732.76 | 14,215.24 |

$c = $ 6672.5 ft   $\alpha = 20°05'53''$

$b = 12,481.7$ ft   $\beta = 35°06'08''$

$\phi = 152°23'22''$

$\alpha = 20°05'53''$   $R = 360° - (\alpha + \beta + \phi)$

$\beta = 35°06'08''$   $= (360°) - 207°35'23''$

$\phi = 152°23'22''$   $R = 152°24'37''$

$\alpha + \beta + \phi = 207°35'23''$

$\cot R$ $= -1.91365958$

$b \sin \alpha = (12,481.7)(0.34362782)$ $= 4289.0594$

$c \sin \beta \sin R = (6672.5)(0.57503698)(0.46313706) = 1777.0265$

$\cot \gamma = \cot R + \dfrac{b \sin \alpha}{c \sin \beta \sin R}$ $= 0.49995638$

$\gamma$ $= 63°26'13''$

$\theta = R - \gamma = (152°24'37'') - (63°26'13'')$ $= 88°58'24''$

Solve for $AO$ using Eq. (9.9c).

$$AO = \frac{(12,481.7)(0.89444280)}{0.57503698} = 19,414.693 \text{ ft}$$

$$AO = \frac{(6672.5)(0.99983946)}{0.34362782} = 19,414.693 \text{ ft} \quad \text{checks}$$

$\angle CAO = 180° - (\gamma + \beta) = 81°27'39''$

$\angle BAO = 180° - (\theta + \alpha) = \dfrac{70°55'43''}{152°23'22''} = \phi \quad \text{checks}$

Calculate the length of $OC$.

$$OC = \frac{b \sin CAO}{\sin \beta} = \frac{(12,481.7)(0.98891459)}{0.57503698} = 21,465.289 \text{ ft}$$

Determine the azimuths of lines *AO* and *OC*.

| Line | Azimuth | |
|------|---------|---|
| *AC* | 117°36'38" | computed from given coordinates |
| + ∠*CAO* | 81°27'39" | |
| *AO* | 199°04'17" | |
| *OA* | 19°04'17" | |
| +β | 35°06'08" | |
| *OC* | 54°10'25" | |
| *CO* | 234°10'25" | |
| +γ | 63°26'13" | |
| *CA* | 297°36'38" | checks |

Finally, compute coordinates for *O*.

| Point | Distance, ft | Direction | X, ft | Y, ft |
|-------|-------------|-----------|-------|-------|
| A | | | 16,672.50 | 20,000.00 |
| | 19,414.693 | 199°04'17" | | |
| O | | | 10,328.83 | 1,650.94 |
| | 21,465.289 | 54°10'25" | | |
| C | | | 27,732.76 | 14,215.24 |

$X_O = 10{,}328.8$ ft    $Y_O = 1{,}650.9$ ft

Many other methods exist for computing the three-point resection problem. References 2, 4, and 6 provide a sample of the literature on this subject.

**9.7. Resection with redundant data**  As in the case of intersection, resections are frequently performed on more than the minimum number of three known control points. A resection from point *O* with directions observed to known control points *B*, *A*, *C*, and *D* so that angles $\alpha_1$, $\alpha_2$, $\alpha_3$, can be measured is illustrated in Fig. 9.7. Obviously, more than one solution exists for this problem. A three-point resection can be computed to locate *O* using *B*, *A*, and *C*; *A*, *C*, and *D*; *B*, *A*, and *D*; or *B*, *C*, and *D*. Each of these resections would yield a slightly different result. The most rigorous method of reconciling these differences is to perform a least squares adjustment to obtain the best estimate for the coordinates of the resected point. Details concerning the least squares adjustment of the resection problem and an example solution can be found in Example B.8, Art. B.12, Appendix B. The coordinates

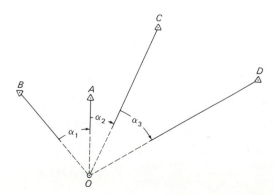

**Fig. 9.7** Resection on four points.

of the point from any one of the four possibilities above would serve as excellent approximations for the linearization of the three condition equations.

**9.8. Intersection with a plane table**   A graphical solution to the intersection problem can be performed on the plane table. It is useful for locating objects when distances to them are not otherwise conveniently obtainable. The location of an object is determined by sighting at the object from each of two plane-table stations (previously plotted) and by drawing rays as in the method of radial traverse, Art. 7.21; the intersection of the two rays thus drawn marks the plotted location of the object. No linear measurements are required except to determine the length of the line joining the two plane-table stations.

Thus, the locations of the objects $A$, $B$, $C$, etc. (Fig. 9.8) may be plotted as follows. The plane table is set up at station $M$, a foresight to station $N$ is taken as in the method of traversing, and the line $mn$ representing the line $MN$ is drawn to scale. Rays of indefinite length are drawn from $m$ toward the objects $A$, $B$, $C$, etc. The plane table is then set up at station $N$ and is oriented by backsighting to station $M$. Rays are drawn from $n$ toward the same objects. The intersections of these rays with the corresponding rays drawn from $m$ mark the plotted locations of the objects at $a$, $b$, $c$, etc. Distances to the objects are not measured but may be scaled from the map. If the angle between the intersecting rays is small, the location will be indefinite.

**9.9. Resection**   Resection with a plane table and alidade is the graphical process of determining the plotted location of a station occupied by the instrument by means of sights taken toward known points the locations of which have been plotted. Resection enables the topographer to select advantageous plane-table stations not previously plotted.

With the plane table oriented over the desired station of unknown map location, two or more objects of known locations are sighted; as each object is sighted, a line of indefinite length is drawn through the plotted location of that object on the map. The intersection of these lines marks the plotted location of the station occupied by the plane table. For precision, it is desirable to resect from nearby stations rather than distant stations.

The table may be oriented by any of the methods stated in Art. 7.21. It is emphasized that, for the methods of orientation by magnetic compass and by backsighting, resection can be accomplished only *after the board has been oriented*. If resection is by the three-point problem or the two-point problem, orientation and resection are accomplished in the same operation.

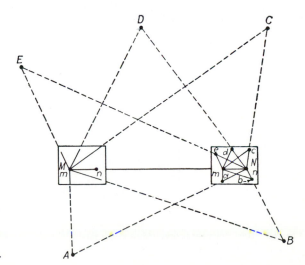

**Fig. 9.8** Intersection with plane table.

**9.10. Resection and orientation with a plane table**  Frequently, the topographer wishes to occupy an advantageous station which has not been located on the map and toward which no ray from located stations has been drawn, and at the same time orientation by use of the compass is not sufficiently accurate. If three located stations are visible, three-point resection offers a convenient method of orienting and resecting in the same operation. There are several solutions of the three-point problem. In the United States, experienced topographers commonly employ a method of direct trial, guided by rules (Art. 9.11). The mechanical or tracing-cloth solution (Art. 9.12) is simpler to understand but is not as satisfactory or as expeditious under the usual field conditions.

**9.11. Trial method**  The plane table is set up over the station of unknown location and is oriented approximately either by compass or by estimation. Resection lines from the three stations of known location are drawn through the corresponding plotted points. These lines will not intersect at a common point unless the trial orientation happens to be correct. (An exception to this statement occurs when the station of unknown location happens to fall on the circumference of a circle passing through the three stations of known location, as discussed later in this section.) Usually, a small triangle called the *triangle of error* is formed by the three lines.

Thus, in Fig. 9.9, suppose that the plane table has been set up over a ground point $P$ and oriented approximately. Resection lines are drawn from $A$, $B$, and $C$ through the corresponding plotted points $a$, $b$, and $c$ respectively, forming a triangle of error. The correct plotted location $p$ of the plane-table station, called the *point sought*, is then determined more closely.

One method is to draw arcs of circles which circumscribe points $a$, $b$, and $ab$; $b$, $c$, and $bc$; and $a$, $c$, and $ac$. The circles will intersect at $p$, the point sought.

Usually, the correct location of the point sought is estimated more conveniently by means of rules 1 and 2, given below.

First, the board is reoriented by backsighting through the estimated location of $p$ toward one of the known stations (preferably the most distant); and the orientation is checked by resecting from the other two known stations. If the three lines still do not meet at a point, the process is repeated until they do; the orientation is then correct, and the common intersection of the three lines is the correct plotted location of the plane-table station.

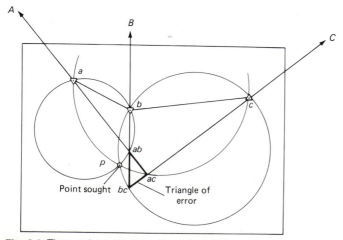

**Fig. 9.9** Three-point resection with a plane table.

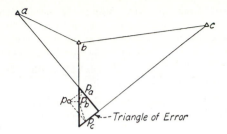

**Fig. 9.10** Triangle of error.

**Rule 1**   *The point sought is on the same side of all resection lines.* That is, it lies either to the right of each line (as the observer faces the corresponding station) or to the left of each line.

**Rule 2**   *The distance from each resection line to the point sought is proportional to the length of that line.* By "length" is meant either the actual distance from plane-table station to known station or the corresponding plotted distance.

Rule 2 can be proved by reference to Fig. 9.10, which shows the triangle of error as in Fig. 9.9. The triangle of error is actually small, and distances from $a$, $b$, and $c$ to any point of it may be taken as the distance to $p$ without error of consequence. Lines $pp_a$, $pp_b$, and $pp_c$ are perpendicular to the resection rays from $a$, $b$, and $c$, respectively. Then, by similar triangles,

$$\frac{pp_a}{pa} = \frac{pp_b}{pb} = \frac{pp_c}{pc}$$

Rules 1 and 2 are general and apply to any location of the plane-table station except on the *great circle* passing through the three known stations (Fig. 9.11). In this case, regardless of the orientation of the table, the lines will meet in a common point (see points 2 in the figure) which will not necessarily be the point sought. If it is suspected but not known that the plane-table station is on the great circle, either the great circle should be plotted on the plane-table sheet or the orientation of the board should be changed slightly and a second trial made. If it is found that the station is on the great circle, one (or more) of the three known stations must be replaced by a known station (or stations) suitably located.

**Auxiliary rules**   Rules 1 and 2 are supplemented by the following auxiliary rules which apply to particular locations of the plane-table station:

1. (*a*) If the new station is inside the great triangle, the point sought is within the triangle of error and is in the same position relative to the triangle of error that the triangle of error occupies in the great triangle (Fig. 9.11, point 1).
2. If the new station is on one of the three segments of the great circle formed by the sides of the great triangle, the point sought is on the *opposite* side of the resection line through the middle known point from the intersection of the other two lines (Fig. 9.11, point 3).
3. If the new station is outside the great circle, the point sought is always on the *same* side of the

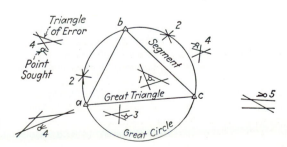

**Fig. 9.11** Three-point problem, solution by trial.

resection line from the most distant point as the point of intersection of the other two lines (Fig. 9.11, points 4).

4. If the new station is outside the great triangle, of the six sectors formed by the resection lines there are only two in which the point sought can be on the same side (right or left) of all lines.
5. If the new station is so located that the triangle of error is not formed within the limits of the plane-table sheet, that is, if two of the resection lines are almost parallel (Fig. 9.11, point 5), the foregoing rules will apply.
6. If the new station is on line between two of the known stations, the resection lines drawn from those two stations will be parallel. The foregoing rules still apply.

**Strength of determination**   The strength of the determination varies with the location of the plane-table station, as described below. The strength of determination should be considered not only in selecting the most favorable of the available known stations but also in deciding whether the three-point problem can satisfactorily be used with the plane table at a given location.

1. When the new station is inside the great circle, the nearer the new station to the center of gravity of the great triangle, the stronger the determination.
2. When the new station is on the great circle, its location is indeterminate.
3. When the new station is near the great circle, the determination is weak.
4. When the new station is outside the great circle, for given angles the nearer the new station to the middle known station, the stronger the determination.
5. When either one angle is small or the new station is on line with two known stations, the larger the angle to the third known station (up to 90°), the stronger the determination. The two known stations near or on line should not be near each other.

## 9.12. Tracing-cloth method

A simple solution of the three-point problem, known as the *tracing-cloth method*, is as follows. A piece of tracing cloth or tracing paper is fastened on the plane table over the map. Any convenient point on the tracing cloth is chosen to represent the unknown station over which the plane table is set, and from it rays are drawn toward the three known stations or objects. Then the cloth is loosened and is shifted over the map until the three rays pass through the corresponding plotted points. The intersection of the rays marks the plotted location of the plane-table station. It is pricked through onto the map, and the table is oriented by backsighting on one of the known stations (preferably the most distant). (This procedure is similar to graphical three-point resection in photogrammetry, Chap. 16.)

## 9.13. Problems

**9.1**   The coordinates for stations $C$ and $D$ are as follows:

| Station | $X$, m | $Y$, m |
|---------|--------|--------|
| $C$ | 323,484.123 | 124,231.305 |
| $D$ | 323,888.059 | 124,832.170 |

Horizontal angles measured with a $01''$ theodolite from stations $C$ and $D$ to station $B$ are angle $BCD = 65°37'24.8''$ and angle $CDB = 80°49'41.9''$. The azimuth from the north of line $CD$ is $A_{CD} = 30°00'00.0''$. (i) Compute the coordinates for station $B$ by intersection using the base solution. (ii) Compute coordinates for station $B$ by intersection using the azimuths calculated for lines $DB$ and $CB$.

**9.2**   Station $C$ is located by intersection from triangulation points $B$ and $A$ for which the coordinates are $X_A = 1,421,231.304$ ft and $Y_A = 521,304.009$ ft. The distance and azimuth from north of the line from $A$ to $B$ are 897.859 ft and $235°20'32''$, respectively. The measured horizontal angles are: angle $BAC = 80°27'35.8''$ and angle $CBA = 54°14'37.8''$. Compute the coordinates of point $C$ by intersection.

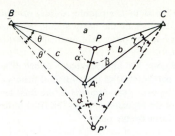

**Fig. 9.12**

**9.3**   Horizontal angles were observed from unknown station Campbell to three stations of known position having the following coordinates:

| Station | X, m | Y, m |
|---------|------|------|
| A | 26,984.819 | 24,424.243 |
| B | 25,078.670 | 29,693.183 |
| C | 24,933.356 | 30,082.605 |

The measured angles are:

| Station | From | To | Angle |
|---------|------|-----|-------|
| | A | B | 44°49′01.4″ |
| Campbell | | | |
| | B | C | 25°50′52.4″ |

Calculate the coordinates of station Campbell by resection.

**9.4**   Given the data listed below. Assume the instrument at station $P$, within the triangle $ABC$ (Fig. 9.12), and solve the three-point problem (case $b$, Fig. 9.6) for the angles $\gamma$ and $\theta$.

$\angle BAC = 102°45′20″$    $\alpha = 89°15′30″$
$b = 6,883.4$ ft    $\beta = 128°20′10″$
$c = 6,605.3$ ft

**9.5**   Given the data of Prob. 9.4 and the additional data shown below. Assume the instrument at station $P'$ (Fig. 9.12), outside the triangle $ABC$, and solve the three-point problem (case c, Fig. 9.6) for the angles $\gamma$ and $\theta$.

$\alpha' = 26°34′50″$
$\beta' = 44°15′15″$

## 9.14. Field problems

### PROBLEM 1   LOCATION OF AN INACCESSIBLE POINT BY INTERSECTION

**Object**   To determine the horizontal position of an inaccessible object such as a church steeple, top of a tower, top of a flag pole, etc., by intersection from the two ends of a measured base line.

**Procedure**

1. Select an object to locate which is visible from the ground in an area suitable for measuring a base line with moderate precision.
2. Set hubs containing tacks at the two ends of the base line which should be of sufficient length to permit a fairly strong determination of the position of the object to be located.
3. If the base line is measured with a tape, follow the procedure given in Field Problem 2, Art. 4.45.
4. When the base line is measured by EDM equipment, observe and record the meteorological data necessary (temperature and atmospheric pressure) to correct for atmospheric conditions. The correction can be made either by dialing the necessary information into the EDM instrument or analytically (Art. 4.33). Record instrument and reflector constants (Art. 4.34). Measure twice, once in each direction.

5. Occupy the stations at the two ends of the base line with a repeating transit or theodolite and measure the horizontal angles by repetition. Each member of the party should measure the angle three times direct and three times reversed according to the procedure outlined in Art. 6.39. Notes should be recorded in the form shown in Fig. 6.47a.

6. Reduce the base-line measurements making corrections for all systematic errors. The difference between the two measurements should not exceed 1 part in 5000 of the measured distance. Assuming this requirement to be satisfied, use the average of the two measurements as the final length.

7. Compute the average horizontal angle for each set of six repeated angles. The final horizontal angle from each station is the average of the several sets taken by the party.

8. Compute the position of the inaccessible point by intersection using the base solution (Art. 9.2).

9. As a check on the computations, assume coordinates of $X = 0$, $Y = 0$ for one end of the base line and calculate the position of the unknown station using Eqs. (9.7) and (9.8).

## PROBLEM 2   LOCATION OF A POINT BY RESECTION

**Object**   To determine the horizontal position of a point by measuring the horizontal angles subtended by three points (visible from the unknown station) of known $X$ and $Y$ coordinates.

### Procedure

1. Occupy the unknown station with a repeating transit or theodolite.

2. Locate the stations to be sighted and make a careful sketch of relative locations of all involved points and nearby details in the field book. If the station to be located is for subsequent surveys, it should be thoroughly referenced (Art. 8.14). Use the note form shown in Fig. 6.47a.

3. Measure each horizontal angle three times direct and three times reversed (one set or position) using the procedure outlined in Art. 6.39. When a set of angles is completed, calculate the average angle and compare with the first angle observed. If the two angles do not agree within the least count of the instrument, repeat the set.

4. If time permits, measure two sets of angles. The final angles are the averages of the two sets.

5. Using these average angles and the known $X$ and $Y$ positions for the stations sighted, calculate the $X$ and $Y$ coordinates for the occupied station by the procedure given in Art. 9.6 or using the equations outlined in Art. 21.5.

**Hint**   If more than three stations of known positions are visible, measure angles between all stations by repetition as described in step 3 and solve the resection problem using the redundant data according to the method of least squares as described in Example B.8, Art. B.12.

### References

1. Allan, A. L., "A Proof of the Barycentric Resection Formulae," *Survey Review*, July 1975.

2. Klinkenberger, H., "Coordinates Systems and the Three-Point Problem," *The Canadian Surveyor*, July 1953.

3. Mattson, D. F., "Determination of Position with Intersecting Circles," *Surveying and Mapping*, September 1973.

4. Rainsford, H. F., *Survey Adjustments and Least Squares*, Constable and Company Ltd., London, 1957.

5. Reynolds, W. F., *Manual of Triangulation Computation and Adjustment*, Spec. Publ. No. 138, U.S. Department of Commerce, Coast and Geodetic Survey (now National Ocean Survey, NOAA), Washington, D.C., 1965.

6. Richardus, P., *Project Surveying*, John Wiley & Sons, Inc., New York, 1966.

7. Vincenty, T., "Desk Top Computers: Resection," *Survey Review*, April 1973.

# CHAPTER 10
# Triangulation and trilateration

**10.1. Introduction**  Triangulation and trilateration are employed extensively to establish horizontal control for topographic mapping; charting lakes, rivers, and ocean coast lines; and for the surveys required for the design and construction of public and private works of large extent.

A *triangulation system* consists of a series of joined or overlapping triangles in which an occasional line is measured and the balance of the sides are calculated from angles measured at the vertices of the triangles. The lines of a triangulation system form a network that ties together all the *triangulation stations* at the vertices of the triangles.

A *trilateration system* also consists of a series of joined or overlapping triangles. However, for trilateration all of the lengths of the triangle's sides are measured and the few directions or angles observed are only those required to establish azimuth. Trilateration has become feasible with the development of EDM equipment (Chap. 4) which makes practical the measurement of all lengths with a high order of accuracy under almost all field conditions. Field procedures for the establishment of trilateration stations are similar to those discussed in this chapter for triangulation.

A combined triangulation and trilateration system consists of a network of triangles in which all the angles and all the distances are observed. Such a combined system represents the strongest network for creating horizontal control that can be established by conventional terrestrial methods.

Triangulation and/or trilateration may be used for a simple topographic survey covering a few acres or hectares; for design and construction of bridges and pipeline crossings; for topographic maps of counties or regions; or for extending first-, second-, and third-order horizontal control throughout an entire continent.

The most notable example of control establishment by triangulation and trilateration in the United States is the transcontinental horizontal system established by the U.S. National Ocean Survey. This system was designed to provide horizontal control for the entire continent. A permanent reference point for datum, called the North American Datum of 1927, has been established at Meades Ranch in Osborne County, Kansas. All of the horizontal control surveys of the United States, Canada, and Mexico are referred to this point. Originally, this control network was extended purely by triangulation. With the advent of EDM devices, trilateration and traverse were used to supplement and strengthen the original network. Recently, data from all of the modern positioning systems such as analytical photogrammetric triangulation (Art. 16.18), inertial systems (Art. 12.3), and satellite doppler measurements (Art. 12.5) are being employed with traditional triangulation and trilateration data in a readjustment of the entire system to a new datum that is expected to be available in 1983 (see Ref. 7 at the end of the chapter).

**10.2. Accuracy of horizontal control systems**  Accuracies required for horizontal control depend on the type of survey and the ultimate use of the control points. In the United States, standards of accuracy for geodetic control surveys are prepared by the Federal Geodetic Control Committee (FGCC) and have been reviewed by the American Society of Civil Engineers, the American Congress on Surveying and Mapping, and the American Geophysical Union.

These standards provide for three orders of accuracy: first, second, and third; the latter two of which are subdivided into classes I and II. First-order or primary horizontal control provides the principal framework for the national control network. It is used for earth crustal movement studies in areas of seismic and tectonic activity, for testing defense and scientific equipment, for studying the performance of space vehicles, for engineering projects of high precision and extending over long distances, and for surveys used in metropolitan expansion.

Second-order class I or secondary horizontal control consists of the networks between first-order arcs and detailed surveys in areas where the land values are high. Surveys of this class include the basic framework for densification of control. Secondary horizontal control strengthens the entire network and is adjusted as part of the national network.

Second-order class II surveys are utilized to establish control for inland waterways, the interstate highway system, and for extensive land subdivision and construction. This class of control contributes to and is published as part of the National Network.

Third-order class I and class II or supplementary surveys are used to establish control for local improvements and developments, topographic and hydrographic surveys, or for other such projects for which they provide sufficient accuracy. Third-order control may or may not be adjusted to the national network. The surveying engineer should know that third-order class I surveys constitute the lowest order of accuracy permissible for specifying points in the state plane coordinate systems (Chap. 14).

Table 10.1 contains a tabulation of standards and principal uses of geodetic control taken from the U.S. National Ocean Survey (NOS) publication *Classification, Standards of Accuracy and General Specifications of Geodetic Control Surveys*. Standards of accuracy and general specifications for horizontal control established by triangulation and trilateration are shown in Table 10.2. Terms such as strength of figure (Art. 10.11), side checks (Art. 10.9), and closure in length are explained in subsequent sections.

**10.3. Triangulation figures**  In a narrow triangulation system a chain of figures is employed, consisting of *single triangles*, *polygons*, *quadrilaterals*, or combinations of these figures. A triangulation system extending over a wide area is likewise divided into figures irregularly overlapping and intermingling, as illustrated by the system of Fig. 10.1. The computations for such a system can be arranged to afford checks on the computed values of most of the sides. As many sides as possible are included in the routes through which the computations are carried from one base line to the next.

**1. Chain of triangles**  In the chain of single triangles (Fig. 10.2) there is but one route by which distances can be computed through the chain. If *AC* is the base line whose length *b* is measured, and if all the angles of the triangles are observed, the length of the triangle sides in the chain (as *AB*, *BC*, *CE*, etc.) may be calculated progressively along the chain from the measured base line to the triangle side farthest removed from the base line. If two lines are measured as base lines, one at each end of the system, the calculations may be carried from each toward the other to a triangle side somewhere between them.

**2. Chain of polygons**  In triangulation, a polygon, or "central-point figure," is composed of a group of triangles, the figure being bounded by three or more sides and having within it a station which is at a vertex common to all the triangles. A chain of such composite figures is

**Table 10.1 Standards for the Classification of Geodetic Control and Principal Recommended Uses**

**Horizontal Control**

| Classification | First-order | Second-order Class I | Second-order Class II | Third-order Class I | Third-order Class II |
|---|---|---|---|---|---|
| Relative accuracy between directly connected adjacent points (at least) | 1 part in 100,000 | 1 part in 50,000 | 1 part in 20,000 | 1 part in 10,000 | 1 part in 5000 |
| Recommended uses | Primary National Network, metropolitan area surveys, scientific studies | Area control that strengthens the National Network; subsidiary metropolitan control | Area control that contributes to, but is supplemental to, the National Network | General control surveys referenced to the National Network; local control surveys | |

**Vertical Control**

| Classification | First-order Class I | First-order Class II | Second-order Class I | Second-order Class II | Third-order |
|---|---|---|---|---|---|
| Relative accuracy between directly connected points or bench marks (standard error) | 0.5 mm $\sqrt{K}$ | 0.7 mm $\sqrt{K}$ | 1.0 mm $\sqrt{K}$ | 1.3 mm $\sqrt{K}$ | 2.0 mm $\sqrt{K}$ |
| | | | ($K$ is the distance in kilometres between points) | | |
| Recommended uses | Basic framework of the National Network and metropolitan area control; regional crustal movement studies; extensive engineering projects; support for subsidiary surveys | | Secondary framework of the National Network and metropolitan area control; local crustal movement studies; large engineering projects; tidal boundary reference; support for lower order surveys | Densification within the National Network; rapid subsidence studies; local engineering projects; topographic mapping | Small-scale topographic mapping; establishing gradients in mountainous areas; small engineering projects; may or may not be adjusted to the National Network |

388 Triangulation and trilateration  Chapter 10

**Table 10.2  Classification, Standards of Accuracy, and General Specifications for Horizontal Control**

| Classification | First-order | Second-order Class I | Second-order Class II | Third-order Class I | Third-order Class II |
|---|---|---|---|---|---|
| | | **Triangulation** | | | |
| *Recommended spacing of principal stations* | Network stations seldom less than 15 km; metropolitan surveys 3 to 8 km and others as required | Principal stations seldom less than 10 km; other surveys 1 to 3 km or as required | Principal stations seldom less than 5 km or as required | As required | As required |
| *Strength of figure* | | | | | |
| $R_1$ between bases | | | | | |
| Desirable limit | 20 | 60 | 80 | 100 | 12b |
| Maximum limit | 25 | 80 | 120 | 130 | 175 |
| Single figure | | | | | |
| Desirable limit | | | | | |
| $R_1$ | 5 | 10 | 15 | 25 | 25 |
| $R_2$ | 10 | 30 | 70 | 80 | 120 |
| Maximum limit | | | | | |
| $R_1$ | 10 | 25 | 25 | 40 | 50 |
| $R_2$ | 15 | 60 | 100 | 120 | 170 |
| *Base measurement* | | | | | |
| Standard error[a] | 1 part in 1,000,000 | 1 part in 900,000 | 1 part in 800,000 | 1 part in 500,000 | 1 part in 250,000 |
| *Horizontal directions* | | | | | |
| Instrument | 0."2 | 0."2 | 0."2 or 1."0 | 1."0 | 1."0 |
| Number of positions | 16 | 16 | 8 or 12 | 4 | 2 |
| Rejection limit from mean | 4" | 4" | 5" or 5" | 5" | 5" |
| *Triangle closure* | | | | | |
| Average not to exceed | 1."0 | 1."2 | 2."0 | 3."0 | 5."0 |
| Maximum seldom to exceed | 3."0 | 3."0 | 5."0 | 5."0 | 10."0 |

| | | | | |
|---|---|---|---|---|
| **Side checks** | | | | |
| In side equation test, average correction to direction not to exceed | | | | |
| 0".3 | 0".4 | 0".6 | 0".8 | 2" |
| **Astro azimuths** | | | | |
| Spacing figures | | | | |
| 6–8 | 6–10 | 8–10 | 10–12 | 12–15 |
| Number of observations/night | | | | |
| 16 | 16 | 16 | 8 | 4 |
| Number of nights | | | | |
| 2 | 2 | 1 | 1 | 1 |
| Standard error | | | | |
| 0".45 | 0".45 | 0".6 | 0".8 | 3".0 |
| **Vertical angle observations** | | | | |
| Number of and spread between observations | | | | |
| 3 D/R—10" | 3 D/R—10" | 2 D/R—10" | 2 D/R—10" | 2 D/R—20" |
| Number of figures between known elevations | | | | |
| 4–6 | 6–8 | 8–10 | 10–15 | 15–20 |
| **Closure in length** (also position when applicable) after angle and side conditions have been satisfied should not exceed | | | | |
| 1 part in 100,000 | 1 part in 50,000 | 1 part in 20,000 | 1 part in 10,000 | 1 part in 5000 |

*(Continued)*

**Table 10.2** *(Continued)*

| | Trilateration | | | | |
|---|---|---|---|---|---|
| *Recommended spacing of principal stations* | Network stations seldom less than 10 km; other surveys seldom less than 3 km | Principal stations seldom less than 10 km; other surveys seldom less than 1 km | Principal stations seldom less than 5 km; for some surveys a spacing of 0.5 km between stations may be satisfactory | Principal stations seldom less than 0.5 km | Principal stations seldom less than 0.25 km |
| *Geometric configuration* Minimum angle contained within, not less than | 25° | 25° | 25° | 20° | 20°15° |
| *Length measurement* Standard error[a] | 1 part in 1,000,000 | 1 part in 750,000 | 1 part in 450,000 | 1 part in 250,000 | 1 part in 150,000 |
| *Vertical angle observations* Number of and spread between observations | 3 D/R—10" | 3 D/R—10" | 2 D/R—10" | 2 D/R—10" | 2 D/R—20" |
| Number of figures between known elevations | 4–6 | 6–8 | 8–10 | 10–15 | 15–20 |
| *Astro azimuths* Spacing figures | 6–8 | 6–10 | 8–10 | 10–12 | 12–15 |
| Number of observations / night | 16 | 16 | 16 | 8 | 4 |
| Number of nights | 2 | 2 | 1 | 1 | 1 |
| Standard error | 0."45 | 0."45 | 0."6 | 0."8 | 3."0 |
| *Closure in position* after geometric conditions have been satisfied should not exceed | 1 part in 100,000 | 1 part in 50,000 | 1 part in 20,000 | 1 part in 10,000 | 1 part in 5000 |

[a] Computed from Eq. (2.37), Art. 2.17.

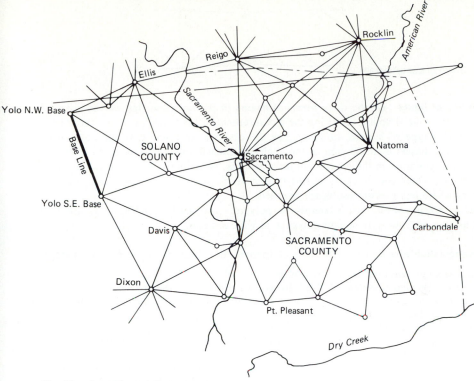

**Fig. 10.1** Area triangulation.

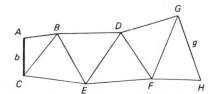

**Fig. 10.2** Chain of single triangles.

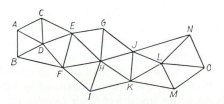

**Fig. 10.3** Chain of polygons.

illustrated in Fig. 10.3, in which *BACEF* is a five-sided polygon with *D* as the central point, and *FEGJKI* is a six-sided polygon with *H* as the central point.

**3. Chain of quadrilaterals**   Figure 10.4 illustrates another type of triangulation figure, in which the individual triangles more or less overlap one another. This type usually occurs in the form of the quadrilateral, of which the figures *ABDC*, *DCEF*, etc., are examples. In the individual quadrilateral there is no triangulation station at the intersection of the diagonals.

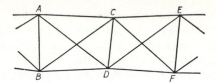

**Fig. 10.4** Chain of quadrilaterals.

**10.4. Choice of figure**  Of the three forms of chains of triangulation figures, the chain of single triangles is the simplest, requiring the measurement of fewer angles than either of the other two. This type of system, however, has the obvious weakness that the only check (except between bases) is in the sum of the angles of each triangle considered by itself. To reach the same precision in the determination of lengths, base lines would need to be placed closer together. This type of chain is not employed in work of high precision, but it is satisfactory where less precise results are required.

For more precise work, quadrilaterals or polygons are used; quadrilaterals are best adapted to long, narrow systems and polygons to wide systems.

**10.5. Triangulation procedure**  The work of triangulation consists of the following steps:

1. Reconnaissance, to select the location of stations.
2. Error propagation and/or evaluation of the strength of figure for the proposed network.
3. Erection of signals and, in some cases, tripods or towers for elevating the signals and/or instruments.
4. Observations of directions or angles.
5. Measurement of the base lines.
6. Astronomic observation at one or more stations in order to determine the true meridian to which azimuths are referred.
7. Computations including: reduction to sea level, reduction to center (where necessary), calculation of spherical excess (when necessary), calculation of all lengths of triangle sides and coordinates for all triangulation stations, and adjustment of the triangulation network to provide the best estimates of coordinates for all points.

**10.6. Reconnaissance**  The initial and one of the most important stages of a triangulation project is the preliminary reconnaissance. Reconnaissance consists of selection of stations, determination of the size and shape of the resulting triangles, the number of stations to be occupied, and the number of angles or directions to be observed. The intervisibility and accessibility of stations, the usefulness of stations in later work, the cost of the necessary signals, and the convenience of base-line measurements are considered. Information acquired during reconnaissance is utilized for error propagation and/or strength of figure determination (Arts. 10.10 and 10.11).

After a preliminary study of all available maps, survey information, and aerial photographs of the area, the person in charge makes an on-site inspection, choosing the most favorable locations for stations. If the information required cannot be obtained from existing maps, photographs, etc., angles and distances to other stations are estimated or measured roughly en route, so that the suitability of the system as a whole can be evaluated before detailed work is begun. If the base line is to be measured by EDM equipment, it should be located so as to provide maximum strength of figure (Art. 10.11). When the base line is to be taped, it should be located so as to allow high-precision measurement and also to permit accurate expansion to other lines in the triangulation network (Art. 10.14). If towers are required at any of the stations, the necessary heights and number of towers should be evaluated.

Reconnaissance for triangulation of low precision (lower than third order) is very limited in extent or is omitted entirely, the stations being selected as work progresses.

**10.7. Angle and side conditions in triangulation**  The computations for triangulation involve calculation of the lengths of the sides in successive triangles, polygons, or quadrilaterals using the initial measured base line and the observed directions or angles. To ensure homogeneous results from these computations, the network must have adequate geometric strength. One important factor that affects geometric strength of a network is the magnitude of the angles observed. This aspect is discussed in detail in Art. 10.11. Another factor that influences the geometric strength of a configuration is the number of angle and side conditions in the network. Angle and side conditions are discussed in Arts. 10.8 and 10.9, respectively.

**10.8. Angle conditions**  The angle-condition equations in a figure express the following: (1) the sum of the interior angles in a polygon must equal some multiple of $180°$; (2) if one or more directly observed angles $\alpha_i$ at a station can be expressed as a function of other angles $\beta_j$ also observed at that station, there is a station equation; and (3) if all angles about a point are observed (i.e., the horizon is closed), then a center-point equation which states that the sum of these angles is equal to $360°$ is required. First consider the number of angle conditions in a polygon.

Start with one line $a$ as illustrated in Fig. 10.5$a$, where there are no angles. Add one line $b$ to yield one angle $\alpha_1$ (Fig. 10.5$b$), so that there are two lines and one angle. Next, add a third line $c$ (Fig. 10.5$c$) to yield $\alpha_2$, so that there are three lines and two angles. This process continues until line $n$ is added, yielding angle $(n-1)$, so that in general, $n$ lines result in $(n-1)$ angles to produce a determinant figure. Any angles measured in excess of those required for the determinant case are redundant and need a condition equation. For example, in a plane triangle, if three angles are measured, there is one redundancy, for which the angle condition is

$$\alpha_1 + \alpha_2 + \alpha_3 = 180°$$

It is necessary to be able to evaluate the number of angle conditions in a given figure. Let

$C_A$ = total number of angle conditions (including
   center-point equations) in a polygon
$L$ = number of lines in the polygon
$A$ = number of angles measured in the polygon

Then

$$C_A = A - (L-1) = A - L + 1 \qquad (10.1)$$

Consider as an example the triangle with a center point illustrated in Fig. 10.6. The number of angles $A = 10$ and the number of lines $L = 6$. Thus, by Eq. (10.1),

$$C_A = 10 - 6 + 1 = 5$$

in which $C_A$ is the total number of angle conditions, including a center-point equation and a

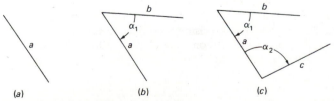

(a)  (b)  (c)

**Fig. 10.5** Angle conditions in a polygon.

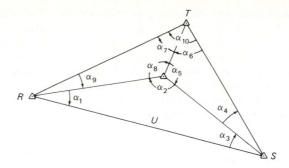

**Fig. 10.6** Center-point triangle.

station equation. The angle-condition equations that can be written for the figure are

$$
\begin{aligned}
\alpha_1 + \alpha_2 + \alpha_3 &= 180° & (a)\\
\alpha_4 + \alpha_5 + \alpha_6 &= 180° & (b)\\
\alpha_7 + \alpha_8 + \alpha_9 &= 180° & (c)\\
\alpha_1 + \alpha_3 + \alpha_4 + \alpha_6 + \alpha_7 + \alpha_9 &= 180° & (d)\\
\alpha_2 + \alpha_5 + \alpha_8 &= 360° & (e)\\
\alpha_6 + \alpha_7 - \alpha_{10} &= 0° & (f)
\end{aligned}
$$

in which Eqs. $(a)$, $(b)$, $(c)$, $(e)$, and $(f)$; $(b)$, $(c)$, $(d)$, $(e)$, and $(f)$; or $(a)$, $(b)$, $(d)$, $(e)$, and $(f)$ are permissible sets of independent angle-condition equations for the example.

**10.9. Side conditions** The angle conditions in a figure can be satisfied without having consistent lengths in the sides. Equation (10.1) gives the number of angle conditions necessary for adjusting a triangulation figure. The *total* number of conditions will always be larger than $C_A$ except for a simple triangle in which case $C_A$ will be the total number of conditions. Now, let $n$ be the total number of measured angles and $n_0$ be the minimum number of angles necessary to construct the designated figure. Then the total number of conditions is

$$
C = n - n_0 \tag{10.2}
$$

Note that this number is the same as the redundancy, $r$ (or the statistical degrees of freedom) discussed in Arts. 2.22 and B.4.

Figure 10.7 shows a braced quadrilateral with $n=8$ observed internal angles. It can be seen that a minimum of $n_0=4$ measured angles are necessary for fixing the shape of the figure. Consequently, the total number of condition equations according to Eq. (10.2) is $C=8-4=4$. If these four conditions are all angle conditions, it will not be possible to guarantee having a consistent quadrilateral. For instance, let line $CB$ be rotated a small amount $d\alpha$ about station $B$. Angle $\alpha_6$ will then become $\alpha_6 - d\alpha$. However, since $\alpha_7$ becomes $\alpha_7 + d\alpha$, the angle conditions involving $\alpha_7$ and $\alpha_6$ will still check in spite of the fact that $CB$ now has a different length. To avoid inconsistencies of this type, a *side condition* is required.

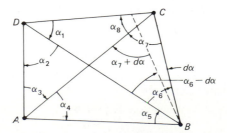

**Fig. 10.7** Braced quadrilateral.

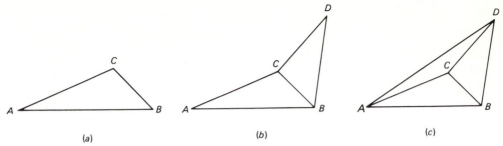

**Fig. 10.8** Side conditions in triangulation.

Side conditions are needed when lengths of a side in a triangle can be computed by more than one route using the law of sines. For example, in Fig. 10.8$a$ one triangle is formed by three points and three sides. Since there is only one route, no side condition is required. In Fig. 10.8$b$, two lines are added, the intersection of which creates point $D$. There is still only one route for calculating lengths, so no side condition is necessary. Next, add a line passing through points $A$ and $D$ as shown in Fig. 10.8$c$. Now there are two routes by which side $DA$ can be calculated using (1) triangles $ABC$ and $ACD$ or (2) triangles $BDC$ and $ACD$. Thus, line $AD$ is an extra or redundant line. For every one of these extra lines in a triangulation network, a side condition is required. Let

$$C_s = \text{number of side conditions}$$
$$n' = \text{number of sides in a figure}$$
$$s = \text{number of stations in the figure}$$

Note that three lines are needed to fix the location of the first point $C$ and two lines are required for each additional point such as $D$. Therefore, the number of extra lines or number of side conditions is

$$C_s = n' - 3 - 2(s - 3) = n' - 2s + 3 \tag{10.3}$$

For the quadrilateral in Fig. 10.7, the number of angle conditions according to Eq. (10.1) is $C_A = 8 - 6 + 1 = 3$, and the number of side conditions according to Eq. 10.3 is $C_s = 6 - 2(4) + 3 = 1$. Thus, the total number of conditions is 4, which is the same as obtained from Eq. (10.2), or $C = 8 - 4 = 4$.

As an example of the formulation of a side condition equation, consider the quadrilateral $ABCD$ in Fig. 10.7. In triangles $DBC$ and $ADC$, by the law of sines,

$$CB = \frac{DC \sin \alpha_1}{\sin \alpha_6} \quad \text{and} \quad DC = \frac{DA \sin \alpha_3}{\sin \alpha_8}$$

so that

$$CB = \frac{DA \sin \alpha_1 \sin \alpha_3}{\sin \alpha_6 \sin \alpha_8} \tag{10.4}$$

Similarly, in triangles $ABD$ and $ABC$,

$$AB = \frac{DA \sin \alpha_2}{\sin \alpha_5} \quad \text{and} \quad CB = \frac{AB \sin \alpha_4}{\sin \alpha_7}$$

so that

$$CB = \frac{DA \sin \alpha_2 \sin \alpha_4}{\sin \alpha_5 \sin \alpha_7} \tag{10.5}$$

Equating (10.4) with (10.5) yields

$$\frac{\sin \alpha_1 \sin \alpha_3 \sin \alpha_5 \sin \alpha_7}{\sin \alpha_2 \sin \alpha_4 \sin \alpha_6 \sin \alpha_8} = 1 \tag{10.6}$$

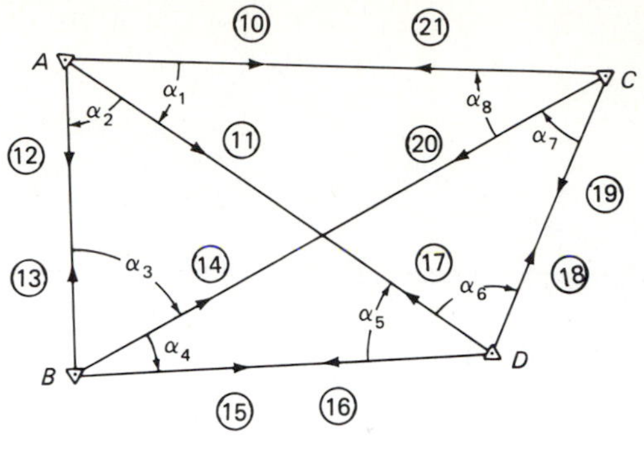

**Fig. 10.9** Angles and directions in a braced quadrilateral.

Other forms of Eq. (10.6) can be easily derived by using different combinations of triangles.

Equation (10.6) is valid for quadrilaterals in which the angles are directly measured. When directions are measured, the angles are functions of the directions [Eq. (6.16), Art. 6.61] and are expressed as follows:

$$\begin{aligned}
\alpha_1 &= d_{11} - d_{10} \\
\alpha_2 &= d_{12} - d_{11} \\
&\phantom{=}\cdots\cdots\cdots \\
\alpha_8 &= d_{21} - d_{20}
\end{aligned} \tag{10.7}$$

in which $d_{10}, d_{11}, d_{12}, \ldots, d_{20}, d_{21}$ are observed directions, as illustrated in Fig. 10.9. Equations (10.7) can be expressed in matrix form as

$$\begin{bmatrix} \alpha_1 \\ \alpha_2 \\ \alpha_3 \\ \vdots \\ \alpha_8 \end{bmatrix} = \begin{bmatrix} -1 & +1 & 0 & 0 & \cdots & & \cdots & & \cdots & 0 \\ 0 & -1 & +1 & 0 & & & & & & 0 \\ 0 & 0 & 0 & -1 & +1 & & & & & 0 \\ & & & & 0 & & -1 & +1 & 0 & \\ 0 & 0 & \cdots & & \cdots & & 0 & -1 & +1 \end{bmatrix} \begin{bmatrix} d_{10} \\ d_{11} \\ d_{12} \\ \vdots \\ d_{20} \\ d_{21} \end{bmatrix} \tag{10.8}$$

which can be written more compactly as

$$\underset{8,1}{\mathbf{a}} = \underset{8,12}{\mathbf{C}} \; \underset{12,1}{\mathbf{d}} \tag{10.9}$$

where $\mathbf{a}$, $\mathbf{C}$, and $\mathbf{d}$ are defined in Eq. (10.8).

In this case, the side condition as expressed in Eq. (10.6) becomes

$$\frac{\sin(d_{11} - d_{10})\sin(d_{14} - d_{13})\sin(d_{17} - d_{16})\sin(d_{20} - d_{19})}{\sin(d_{12} - d_{11})\sin(d_{15} - d_{14})\sin(d_{18} - d_{17})\sin(d_{21} - d_{20})} = 1 \tag{10.10}$$

Use of the angle- and side-condition equations is demonstrated in Art. 10.22 and Example B.12, Art. B.15, Appendix B.

**10.10. Error propagation in triangulation** A specification frequently applied to obtain the desired results in triangulation computations is that the length of a calculated side should be within a given tolerance specified by the variance (Art. 2.13) in the length. This variance is determined by error propagation through an equation that expresses the length in terms of the measured angles and distances. First consider this error propagation in a single triangle, as illustrated in Fig. 10.10. Assume that side $b$ is the measured base line and $A$ and $C$ are independent, uncorrelated measured angles with respective variances of $\sigma_A^2$ and $\sigma_C^2$. Then

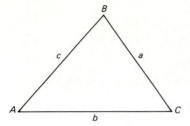

**Fig. 10.10**

sides $a$ and $c$ determined by the sine law are

$$a = \frac{b \sin A}{\sin B} = \frac{b \sin A}{\sin(A + C)} \tag{10.11a}$$

$$c = \frac{b \sin C}{\sin B} = \frac{b \sin C}{\sin(A + C)} \tag{10.11b}$$

Propagating the variances in sides $a$ and $c$ [use Eq. (2.46)] yields

$$\sigma_a^2 = \left( \frac{\partial a}{\partial A} \right)^2 \sigma_A^2 + \left( \frac{\partial a}{\partial C} \right)^2 \sigma_C^2 + \left( \frac{\partial a}{\partial b} \right)^2 \sigma_b^2$$

$$\sigma_c^2 = \left( \frac{\partial c}{\partial A} \right)^2 \sigma_A^2 + \left( \frac{\partial c}{\partial C} \right)^2 \sigma_C^2 + \left( \frac{\partial c}{\partial b} \right)^2 \sigma_b^2 \tag{10.12}$$

As an example, the partial derivatives of $a$ with respect to $A$, $C$, and $b$ are (the student should evaluate the partials of $c$ as an exercise)

$$\frac{\partial a}{\partial A} = \frac{b \sin(A + C)\cos A - b \sin A \cos(A + C)}{\sin^2(A + C)} = a \cot A + a \cot B$$

$$\frac{\partial a}{\partial C} = \frac{- b \sin A \cos(A + C)}{\sin^2(A + C)} = - a \cot(A + C) = a \cot B \tag{10.13}$$

$$\frac{\partial a}{\partial b} = \frac{\sin A}{\sin(A + C)} = \frac{a}{b}$$

If an errorless base line is assumed, the variances in calculated sides $a$ and $c$ become a minimum when the angles $A + C = 90°$ or $B = 90°$. It can be seen that for angles less than $45°$ (or more than $135°$), error propagation increases. In practice, angles less than $30°$ or greater than $150°$ are avoided.

If three angles $A$, $B$, and $C$ are measured and adjusted, then assuming uncorrelated observations of equal precision and a reference variance of 1.0, the cofactor matrix (Art. 2.14) for the adjusted angles is (see Example B.5a, Art. B.7, Appendix B)

$$\mathbf{Q}_{ABC} = \begin{bmatrix} \frac{2}{3} & -\frac{1}{3} & -\frac{1}{3} \\ -\frac{1}{3} & \frac{2}{3} & -\frac{1}{3} \\ -\frac{1}{3} & -\frac{1}{3} & \frac{2}{3} \end{bmatrix} \tag{10.14}$$

In this case (assuming no errors in the base line $b$), the propagated cofactor for side $a$ is

$$q_a = \mathbf{J Q}_{ABC} \mathbf{J}^T \tag{10.15}$$

in which $\mathbf{Q}_{ABC}$ is defined by Eq. (10.14) and

$$\mathbf{J} = \left[ \begin{array}{ccc} \dfrac{\partial a}{\partial A} & \dfrac{\partial a}{\partial B} & \dfrac{\partial a}{\partial C} \end{array} \right] \tag{10.16}$$

where $\partial a / \partial A$, $\partial a / \partial C$ are defined in Eq. (10.13), and $\partial a / \partial B = 0$. Expansion of Eq. (10.15) by substitution of Eqs. (10.14) and (10.16) yields

$$q_a = \frac{2}{3}\left[\left(\frac{\partial a}{\partial A}\right)^2 + \left(\frac{\partial a}{\partial C}\right)^2 - \left(\frac{\partial a}{\partial A}\right)\left(\frac{\partial a}{\partial C}\right)\right]$$

Using partial derivatives from Eq. (10.13), then

$$q_a = \frac{2}{3}\left[(a\cot A + a\cot B)^2 + (a\cot B)^2 - (a\cot A + a\cot B)(a\cot B)\right]$$

or

$$q_a = \frac{2}{3}a^2(\cot^2 B + \cot A \cot B + \cot^2 A)$$

and by Eq. (2.30) the variance in side $a$ is

$$\sigma_a^2 = \frac{2}{3}a^2(\cot^2 B + \cot A \cot B + \cot^2 A)\sigma^2 \tag{10.17}$$

in which $\sigma^2$ is the variance for the measured angles. Equation (10.17) implies an errorless base. Since this does not occur in practice, $(\partial a/\partial b)^2\sigma_b^2 = (a^2/b^2)\sigma_b^2$ is added to the right side of Eq. (10.17) to yield

$$\sigma_a^2 = \frac{a^2}{b^2}\sigma_b^2 + \frac{2}{3}a^2(\cot^2 A + \cot A \cot B + \cot^2 B)\sigma^2 \tag{10.18}$$

Equation (10.18) is for a single triangle. This expression may be propagated through a chain of triangles (see Fig. 10.2). Certain assumptions are made, which include (1) that the angles measured in each triangle are totally independent of those in any other triangles; and (2) that each triangle is adjusted independently with a resulting cofactor matrix of adjusted angles as given by Eq. (10.14); and (3) that there are no redundant lines in the triangulation figure such as occur in the chain of braced quadrilaterals shown in Fig. 10.4. Under these assumptions it can be shown that the variance in the final computed length $\ell$ is

$$\sigma_\ell^2 = \frac{\ell^2}{b^2}\sigma_b^2 + \frac{2}{3}\ell^2\sum(\cot^2 A + \cot A \cot B + \cot^2 B)\sigma^2 \tag{10.19}$$

The tolerance in the calculated side can also be expressed in terms of the variance in the logarithm of the side. In this case the equation is usually written as follows:

$$\sigma_{\log\ell}'^2 = \sigma_{\log b}^2 + \frac{2}{3}\sum\left(\delta_{A_i}^2 + \delta_{A_i}\delta_{B_i} + \delta_{B_i}^2\right)\sigma^2 \tag{10.20}$$

in which $\delta_{A_i}$ and $\delta_{B_i}$ are the differences for $01''$ in the sixth decimal place of the logarithmic sines for angles $A_i$ and $B_i$.

When directions are observed (Art. 6.40), the angles derived from directions are correlated (Art. 6.58) and the error propagation becomes considerably more complicated. In practice, Eqs. (10.19) and (10.20) are modified to account for the fact that the variance in an angle is twice the variance of a direction [Eq. (6.19), Art. 6.58)]. It does not accommodate correlation between angles in one triangle and those in adjacent triangles. Thus, Eq. (10.20) becomes

$$\sigma_{\log\ell}'^2 = \sigma_{\log b}^2 + \frac{4}{3}\sum\left(\delta_{A_i}^2 + \delta_{A_i}\delta_{B_i} + \delta_{B_i}^2\right)\sigma_d^2 \tag{10.21}$$

in which $\sigma_{\log\ell}'^2$ is the variance in the logarithm of the calculated side and $\sigma_d^2$ is the variance in the measured directions.

A more direct and rigorous criterion on which to base accuracies in triangulation is to specify the maximum allowable variance in terms of the propagated error ellipses (Art. 2.19) for the positions of adjusted triangulation points. This procedure is statistically valid and has the added advantage that propagated error ellipses are a direct by-product of the least squares adjustment of triangulation (Art. B.15).

**10.11. Strength of figure in triangulation**  When a triangulation project is being evaluated in the preliminary stages of the work, it is necessary to determine the *strength* of *figure* for the network. This step is required in order to ensure uniform accuracy throughout the network. The strength of figure is a function of:

1. The geometric strength of the triangles that make up the network. Ideally, the triangles should be equilateral.
2. The number of stations occupied for angle or direction measurements. Lines occupied at only one end should be avoided whenever possible.
3. The number of angle and side conditions used in adjusting the network. This number should be large in proportion to the number of observations.

The U.S. National Ocean Survey uses a method for evaluating strength of figure that is based on Eq. (10.21). This equation is modified to include a factor to account for the number of conditions and observations. The modified equation is

$$R = \frac{D-C}{D} \sum \left( \delta_{Ai}^2 + \delta_{Ai}\delta_{Bi} + \delta_{Bi}^2 \right) \qquad (10.22)$$

where    $R$ = strength of figure
$D$ = total number of directions in the network
$C$ = total number of angle and side conditions
$\delta_{A_i}$ = difference in the sixth decimal place of the log sine of the angle (labeled $A$) opposite the side to be calculated in the triangle.
$\delta_{B_i}$ = difference in the sixth decimal place in the log sine of the angle (labeled $B$) opposite the known side in the triangle

The number of directions $D$ refers to the directions observed with a theodolite. Thus, a line occupied at both ends would have two directions. The starting line in a network is assumed to be of known direction, so that the directions observed along this line are not included in figuring the value of $D$. For example, in Figs. 10.8$b$ and $c$ and 10.9, the respective values for $D$ are 8, 10, and 10.

The total number of conditions $C$ is

$$C = C_A + C_S = (A - L + 1) + (n' - 2s + 3) \qquad (10.23)$$

where $C_A$ and $C_S$ are defined in Eqs. (10.1) and (10.3).

Angles $A_i$ and $B_i$ are designated *distance angles* in triangle $i$. These angles may be scaled from a preliminary sketch of the triangulation on an existing map or may be unadjusted field measurements. Values to the nearest degree are adequate. In Fig. 10.11, angles $A_1$, $B_1$ are the distance angles in triangle $RST$; angles $A_2$, $B_2$ are distance angles in triangle $TSU$; etc. Note that these angles $A_i$ opposite the side to be computed and $B_i$ opposite the known side, correspond to the angles $A$ and $B$ used in Fig. 10.10, Art. 10.10, to propagate the variance in the calculated side, using the law of sines.

In Eq. (10.22), the two factors $(D - C)/D$ and $(\delta_{A_i}^2 + \delta_{A_i}\delta_{B_i} + \delta_{B_i}^2)$ are related only to the number of conditions and observations and the geometry of the triangles. Thus, the value for $R$ is independent of the precision of the measurements. Consequently, this procedure is useful primarily as a means of comparing various network configurations so as to obtain optimum geometric conditions and a desirable number of conditions versus number of

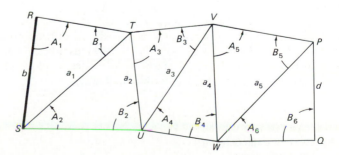

**Fig. 10.11** Strength of figure, single chain of triangles.

measurements. It is also used as a means for determining the most favorable route for calculating through a network of triangulation.

Note that as the number of conditions increases versus the number of measurements, the first term of Eq. (10.22) decreases. Also, as the distance angles $A_i$ and $B_i$ approach the ideal value of 45°, the change in the sixth decimal place of the log sines decreases and the second term, the summation within the parentheses becomes smaller. Consequently, the stronger the figure, the lower the value of $R$ becomes. In this respect, evaluation of strength of figure is analogous to the rigorous computation of the variance of the calculated side by Eq. (10.19); that is, the lower the variance, the stronger the figure. However, in the latter case the precision of the measurements is a factor, whereas in the former this has no effect.

To illustrate some of the problems involved in the preliminary evaluation of triangulation by determining strength of figure, consider an example.

**Example 10.1** It is desired to compute the strength of the quadrilateral *ACDB* in Fig. 10.12 for computation of side *CD* from known side *AB* when all lines are observed in both directions. Assume that line *AB* is of known distance and direction.

SOLUTION   Either Eq. (10.2) or Eq. (10.23) [using Eqs. (10.1) and (10.3)] may be used. Since the total number of measured angles $n$ is 8 and the minimum number required for a determinate solution of the quadrilateral $n_0$ is 4, then by Eq. (10.2),

$$C = n - n_0 = 8 - 4 = 4$$

Similarly, by Eq. (10.23),

$$C = (8 - 6 + 1) + (6 - 8 + 3) = 4$$

Since line *AB* is of fixed direction $D = 10$, then

$$R' = \frac{D - C}{D} = \frac{10 - 4}{10} = 0.60$$

The value of $R'$ will be a constant for any triangle in *ACDB*. The second term of Eq. (10.22), $(\delta_A^2 + \delta_A\delta_B + \delta_B^2)$, is evaluated from Table 10.3 using distance angles $A$ and $B$ for a specific triangle in quadrilateral *ACDB* as the arguments. Select triangles *ABC* and *ACD* as the first route. In triangle *ABC* the known side is *AB* and the side to be calculated is *AC*. Thus, $A = 43°$ and $B = 60°$. Using these values as arguments, $R$ from Table 10.3 is equal to 9.8. Next, in triangle *ACD* the known side is *AC* and the side to be calculated is *CD*, so that $A = 40°$ and $B = 36°$, yielding $R = 22.2$ from Table 10.3. Thus, $R = (0.60)(9.8 + 22.2) = 19$. The results of this computation and those for the other three possible routes in *ACDB* are tabulated in Table 10.4.

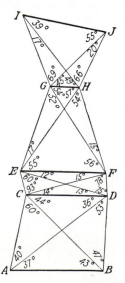

**Fig. 10.12** Strength of figure, chain of quadrilaterals.

The strongest chain of triangles is through triangles *BAC* and *BCD*, which yield the lowest value of *R* equal to 3. By similar computations for the remaining quadrilaterals, the remaining values of *R* are found to be *CEFD*, 0; *EGHF*, 29; *GIJH*, 20. Therefore, the strongest quadrilateral is *CEFD* and the weakest is *EGHF*. The strength of the entire figure (for *IJ* computed from *AB*) is represented by a value of $R_1 = 52$; that is, the sum of the lowest values for the four consecutive quadrilaterals in the chain. The next lowest value or $R_2 = 5 + 0 + 29 + 20 = 54$. Reference to Table 10.2 reveals that with respect to strength of figure, the triangulation network of Fig. 10.12 would be suitable for second-order Class I triangulation.

Some remarks concerning the triangulation figure of Fig. 10.12 are pertinent. Recall that the comment was made in Art. 10.10 that angles greater than 150° and less than 30° should be avoided. This conclusion was based on error propagation using the sine law in a single triangle. A visual inspection of Fig. 10.12 shows that three of the quadrilaterals, *CEFD*, *EGHF*, and *GIJH*, have angles that are less than 20°.

Consider quadrilateral *CEFD*. Calculations from *CD* determined in quadrilateral *ABCD* to *EF* can be made through triangles *CFD* and *ECF* or *DCE* and *EFD* without using one of the small angles individually in any of the calculations. In fact, quadrilateral *CEFD* is a stronger figure ($R = 0$) than *ABCD*, in which there are no angles less than 36°.

In a similar analysis of quadrilateral *EGHF*, there is no route that does not require calculations involving an angle of either 15° or 17°. Thus, by any possible route, the calculated length of side *GH* will be affected by the errors propagated by the use of small angles individually. Note that this quadrilateral has an *R* value of 29, the highest of the four braced figures.

Quadrilateral *GIJH* also has two small angles of 17° and 20°, but it is possible to follow a route through triangles *GHJ* and *GIJ* in which neither small angle is used individually. Note that the strength of figure value for this quadrilateral is 20.

Consequently, although small and large angles should be avoided, they can be tolerated provided these angles are not used individually. Hence, the advisability of using the quadrilateral as the basic figure in triangulation nets and the inadvisability of utilizing a chain of triangles as illustrated in Fig. 10.4.

**10.12. Signals and instrument supports**   Each triangulation station is marked by a signal visible from stations from which it is to be sighted. The form of the signal depends on the locality and the available materials. If the station is not to be used as an instrument station, but is merely to be sighted, a relatively simple structure is used. This signal may be one constructed for the purpose or it may be an object already in place such as a chimney or church steeple. A pole set vertically in the ground or held firmly vertical by a pile of stones, or by guys or bracing, makes an excellent signal on a bare summit or in open country. A white paint mark on a rock cliff is sometimes all that is required. To increase the visibility of a pole or tree signal, two rectangular targets are sometimes attached, being placed at right angles to each other.

Signals should be of such shape and orientation that they will be as nearly free as possible from phase, or asymmetry of sighting. If the signal is illuminated from the side, the observer tends to sight to one side of its center. This phenomenon is called *phase*.

The best time for observing is at late afternoon or at night. For night observations, an electric lamp can be used as a signal. The signal shown in Fig. 6.67 (Chap. 6) can be attached to a regular tripod and is illuminated for night observations.

At an instrument station, it is desirable to have a signal of a type that will permit centering the instrument directly over the station when measuring angles. If the station is to be used over a long period of time, the station may be marked by an iron pipe set vertically in the ground. A range pole can then be set in this pipe and used for a sight. The pole can be removed when it is necessary to set the instrument over the station. Where a tall mast is necessary for visibility, it may be supported in position by three guy wires attached to it near

**Table 10.3**  **Factors for Determining Strength of Figure (Courtesy U.S. National Ocean Survey)**
Values of $(\delta_A^2 + \delta_A\delta_B + \delta_B^2)$ for various combinations of distance angles $A$ and $B$ of a triangle

| | 10° | 12° | 14° | 16° | 18° | 20° | 22° | 24° | 26° | 28° | 30° | 35° | 40° | 45° | 50° | 55° | 60° | 65° | 70° | 75° | 80° | 85° | 90° |
|---|---|---|---|---|---|---|---|---|---|---|---|---|---|---|---|---|---|---|---|---|---|---|---|
| 10° | 428 | 359 | | | | | | | | | | | | | | | | | | | | | |
| 12° | 359 | 295 | 253 | | | | | | | | | | | | | | | | | | | | |
| 14° | 315 | 253 | 214 | 187 | | | | | | | | | | | | | | | | | | | |
| 16° | 284 | 225 | 187 | 162 | 143 | | | | | | | | | | | | | | | | | | |
| 18° | 262 | 204 | 168 | 143 | 126 | 113 | | | | | | | | | | | | | | | | | |
| 20° | 245 | 189 | 153 | 130 | 113 | 100 | 91 | | | | | | | | | | | | | | | | |
| 22° | 232 | 177 | 142 | 119 | 103 | 91 | 81 | 74 | | | | | | | | | | | | | | | |
| 24° | 221 | 167 | 134 | 111 | 95 | 83 | 74 | 67 | 61 | | | | | | | | | | | | | | |
| 26° | 213 | 160 | 126 | 104 | 89 | 77 | 68 | 61 | 56 | 51 | | | | | | | | | | | | | |
| 28° | 206 | 153 | 120 | 99 | 83 | 72 | 63 | 57 | 51 | 47 | 43 | | | | | | | | | | | | |
| 30° | 199 | 148 | 115 | 94 | 79 | 68 | 59 | 53 | 48 | 43 | 40 | 33 | | | | | | | | | | | |
| 35° | 188 | 137 | 106 | 85 | 71 | 60 | 52 | 46 | 41 | 37 | 33 | 27 | 23 | | | | | | | | | | |
| 40° | 179 | 129 | 99 | 79 | 65 | 54 | 47 | 41 | 36 | 32 | 29 | 23 | 19 | 16 | | | | | | | | | |
| 45° | 172 | 124 | 93 | 74 | 60 | 50 | 43 | 37 | 32 | 28 | 25 | 20 | 16 | 13 | 11 | | | | | | | | |
| 50° | 167 | 119 | 89 | 70 | 57 | 47 | 39 | 34 | 29 | 26 | 23 | 18 | 14 | 11 | 9 | 8 | | | | | | | |
| 55° | 162 | 115 | 86 | 67 | 54 | 44 | 37 | 32 | 27 | 24 | 21 | 16 | 12 | 10 | 8 | 7 | 5 | | | | | | |
| 60° | 159 | 112 | 83 | 64 | 51 | 42 | 35 | 30 | 25 | 22 | 19 | 14 | 11 | 9 | 7 | 5 | 4 | 4 | | | | | |
| 65° | 155 | 109 | 80 | 62 | 49 | 40 | 33 | 28 | 24 | 21 | 18 | 13 | 10 | 7 | 6 | 5 | 4 | 3 | 2 | | | | |
| 70° | 152 | 106 | 78 | 60 | 48 | 38 | 32 | 27 | 23 | 19 | 17 | 12 | 9 | 7 | 5 | 4 | 3 | 2 | 2 | 1 | | | |
| 75° | 150 | 104 | 76 | 58 | 46 | 37 | 30 | 25 | 21 | 18 | 16 | 11 | 8 | 6 | 4 | 3 | 2 | 2 | 1 | 1 | 1 | | |
| 80° | 147 | 102 | 74 | 57 | 45 | 36 | 29 | 24 | 20 | 17 | 15 | 10 | 7 | 5 | 4 | 3 | 2 | 1 | 1 | 1 | 0 | 0 | |
| 85° | 145 | 100 | 73 | 55 | 43 | 34 | 28 | 23 | 19 | 16 | 14 | 10 | 7 | 5 | 3 | 2 | 2 | 1 | 1 | 0 | 0 | 0 | 0 |

| | 90° | 95° | 100° | 105° | 110° | 115° | 120° | 125° | 130° | 135° | 140° | 145° | 150° | 152° | 154° | 156° | 158° | 160° | 162° | 164° | 166° | 168° | 170° |
|---|---|---|---|---|---|---|---|---|---|---|---|---|---|---|---|---|---|---|---|---|---|---|---|
| 90° | 0 | | | | | | | | | | | | | | | | | | | | | | |
| 95° | 0 | 0 | | | | | | | | | | | | | | | | | | | | | |
| 100° | 0 | 0 | 0 | | | | | | | | | | | | | | | | | | | | |
| 105° | 0 | 0 | 0 | 0 | | | | | | | | | | | | | | | | | | | |
| 110° | 1 | 0 | 0 | 0 | 1 | | | | | | | | | | | | | | | | | | |
| 115° | 1 | 1 | 1 | 1 | 1 | 1 | | | | | | | | | | | | | | | | | |
| 120° | 1 | 1 | 1 | 1 | 1 | 1 | 1 | | | | | | | | | | | | | | | | |
| 125° | 2 | 2 | 2 | 2 | 2 | 2 | 2 | 2 | | | | | | | | | | | | | | | |
| 130° | 3 | 3 | 3 | 2 | 2 | 2 | 2 | 3 | 3 | | | | | | | | | | | | | | |
| 135° | 4 | 4 | 4 | 4 | 3 | 3 | 3 | 4 | 4 | 4 | | | | | | | | | | | | | |
| 140° | 6 | 6 | 6 | 5 | 5 | 5 | 5 | 5 | 5 | 5 | 6 | | | | | | | | | | | | |
| 145° | 9 | 9 | 8 | 8 | 7 | 7 | 7 | 7 | 7 | 7 | 8 | 9 | | | | | | | | | | | |
| 150° | 13 | 13 | 12 | 12 | 11 | 11 | 10 | 10 | 10 | 10 | 10 | 11 | 13 | | | | | | | | | | |
| 152° | 16 | 15 | 14 | 14 | 13 | 13 | 12 | 12 | 12 | 12 | 12 | 13 | 15 | 16 | | | | | | | | | |
| 154° | 19 | 18 | 17 | 17 | 16 | 15 | 15 | 14 | 14 | 14 | 14 | 15 | 16 | 17 | 19 | | | | | | | | |
| 156° | 22 | 22 | 21 | 20 | 19 | 19 | 18 | 18 | 17 | 17 | 17 | 17 | 18 | 19 | 21 | 22 | | | | | | | |
| 158° | 27 | 26 | 25 | 25 | 24 | 23 | 22 | 22 | 21 | 21 | 20 | 21 | 21 | 22 | 23 | 25 | 27 | | | | | | |
| 160° | 33 | 32 | 31 | 30 | 30 | 29 | 28 | 27 | 26 | 26 | 25 | 25 | 26 | 26 | 27 | 28 | 30 | 33 | | | | | |
| 162° | 42 | 41 | 40 | 39 | 38 | 37 | 36 | 35 | 34 | 33 | 32 | 32 | 32 | 32 | 33 | 34 | 35 | 38 | 42 | | | | |
| 164° | 54 | 53 | 51 | 50 | 49 | 48 | 46 | 45 | 44 | 43 | 42 | 41 | 40 | 40 | 41 | 42 | 43 | 45 | 48 | 54 | | | |
| 166° | 71 | 70 | 68 | 67 | 65 | 64 | 62 | 61 | 59 | 58 | 56 | 55 | 54 | 53 | 53 | 54 | 54 | 56 | 59 | 63 | 71 | | |
| 168° | 98 | 96 | 95 | 93 | 91 | 89 | 88 | 86 | 84 | 82 | 80 | 77 | 75 | 75 | 74 | 74 | 74 | 74 | 76 | 79 | 86 | 98 | |
| 170° | 143 | 140 | 138 | 136 | 134 | 132 | 129 | 127 | 125 | 122 | 119 | 116 | 112 | 111 | 110 | 108 | 107 | 107 | 107 | 109 | 113 | 122 | 143 |

**Table 10.4**

| Common side | Chain of triangles | Distance angles, deg | | $\delta_A^2 + \delta_A \delta_B + \delta_B^2$ | | R |
|---|---|---|---|---|---|---|
| | | $A_i$ | $B_i$ | Each | $\Sigma$ | |
| AC | ACB | 43 | 60 | 9.8 | 32.0 | 19 |
| | ACD | 40 | 36 | 22.2 | | |
| AD | ADB | 90 | 53 | 2.4 | 7.6 | 5 |
| | ACD | 40 | 104 | 5.2 | | |
| BC | BAC | 77 | 60 | 2.0 | 5.7 | 3 |
| | BCD | 47 | 89 | 3.7 | | |
| BD | BAD | 37 | 53 | 15.2 | 28.0 | 17 |
| | BCD | 47 | 44 | 12.8 | | |

the top. Provision is made for swinging the bottom of the mast to one side when it is desired to place an instrument over the station.

Where a permanent signal that does not need to be moved for setting up the instrument is required, a large tripod or tower can be constructed from round poles or sawed lumber. A vertical mast centered over the station should project upward from the junction of the legs. Figure 10.13 shows such a signal.

Where the instrument must be elevated to secure visibility, a combined observing tower and signal like that shown in Fig. 10.14 is built of wood or steel. A central tripod supports the instrument. Around this, but entirely separate from it, is a three- or four-sided structure supporting the platform upon which the observer stands. Thus the instrument tower is free from the vibrations caused by movements of the observing party. The Bilby steel tower shown in Fig. 10.14 is sectional and can be quickly and easily erected to any height up to 126 ft (38.4 m).

The Lambert tower, invented by A. F. Lambert (see Ref. 11), is a more recently developed tower used for geodetic surveys. This tower is of lightweight tubular construction with an independent interior tube to support the instrument. The tower, which has a maximum height of 60 ft (18.29 m), comes in 15-ft (4.57-m) lengths that can be assembled on the ground, can be hoisted into position by two winches, and is then held in the vertical position

**Fig. 10.13** Tripod signal. (*U.S. National Ocean Survey.*)

**Fig. 10.14** Bilby steel triangulation tower. (*U.S. National Ocean Survey.*)

by guy wires. This tower satisfies the need for a light, easily transported assembly which can be erected by a crew of three in about 2 hours.

When towers are required due to flat terrain, heavy timber, or other factors, it is necessary to use the equation for curvature and refraction [Eqs. (5.5) to (5.7), Art. 5.2] to determine the heights of towers. Thus, if the line of sight is 10 km in length, the necessary height of one tower to permit seeing the surface of the earth at a distance of 10 km is $h = (0.0675)(10)^2 = 6.75$ m. If towers of equal height are placed at both ends of the line, the line of sight will be tangent to the earth at a distance of 5 km, so that $h = (0.0675)(5)^2 = 1.69$ m is the required height of each tower. To minimize the effects of refraction, the line of sight should clear the ground by at least 3.5 m, so that towers 5.2 m in height are needed.

In a small triangulation system where sights are less than 1000 ft or 300 m, temporary signals which are easily moved are all that is necessary. A plumbed line supported from a tripod centered over the station is an adequate signal for a sight a few hundred feet in length. For sights of up to 300 m, a range pole supported by a tripod and plumbed over the station provides a suitable signal. Forced centering (Art. 6.30), where theodolite and sighting target (Fig. 6.67) are compatible with the same tripod, is very useful in triangulation of limited extent. Similarly, tripods equipped with a plumbing rod (Fig. 6.40) have a built-in signal (the plumbing rod) which can be used advantageously for this purpose.

**10.13. Station marks**   For the extensive triangulation systems of the U.S. National Ocean Survey and the U.S. Geological Survey, every triangulation station is permanently marked with a metal tablet (similar to that in Fig. 5.34) which is fastened securely in rock or in a concrete monument. These stations are of great value as reference points for local surveys.

Triangulation stations for networks not related to the national systems should also be marked by a punch mark in a metal plate or rod set in a concrete monument.

Triangulation stations should also be thoroughly referenced and described (Art. 8.14) for future recovery and use.

**10.14. Base lines**   Electronic distance measuring (EDM) equipment (Arts. 4.29 to 4.43) comprises the measuring system most commonly used to determine base line lengths. In a triangulation network, long sides (within limits) are more economical than short sides. Hence, if a short-range EDM device (Fig. 4.12, Art. 4.29) is used or if taping the base line

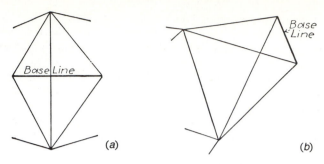

**Fig. 10.15** Base nets.

becomes necessary, the base line is usually considerably shorter than the average length of the triangle sides in the network. To find the required precision in the computed lengths of the main triangles, it is necessary to expand the effect of the base line through a group of smaller triangles called the *base net*.

Figure 10.15*a* is an excellent example of a base net that allows quick and accurate expansion of the base line to the longer sides of the system. The form of the base net suggested by quadrilaterals *GHFE* and *GHJI* (Fig. 10.12) is satisfactory since it can be laid out so as to avoid the small angles present in these figures. This form is also shown in Fig. 10.15*b*.

The number of base lines required will depend on the strength of the figures through which the triangulation computations are being performed. When a maximum allowable cumulated value for *R* has been reached, a new base line is required. These maximum values for *R* for various orders and classifications of triangulation are given in Table 10.2.

When a long-range EDM device is used to measure the base line, there is no need for a base net since the side of a main triangle can be measured directly. EDM instruments having a range comparable to those shown in Figs. 4.13 and 4.14 (Chap. 4) are appropriate for this type of work.

**10.15. Use of EDM equipment to measure base lines**  Utilization of EDM equipment vastly simplifies the measurement of base lines in triangulation operations. The primary requirement for measurement of a base line by an EDM instrument is that an unobstructed measuring path exist between the two ends of the line. Selection of the location of the base line can be made entirely on the basis of strength of figure, since the measurement of the distance is entirely independent of the type of terrain between the two stations.

Care should be exercised to correct for meteorological conditions, systematic instrumental errors, and slope of the line of sight as described in Arts. 4.32 to 4.42. With respect to instrumental errors, the EDM system should be calibrated periodically on a base line of known length (Arts. 4.34 and 4.35). Vertical angles or elevations at the two ends of the line must be obtained in order to allow reduction of slope distances to horizontal distances (Arts. 4.39 and 4.40). At least one elevation is required to permit reduction to sea level.

**10.16. Measurement of the base line with a tape**  For base line measurements of ordinary precision either the steel tape or the invar tape may be employed, but for measurements of higher precision the invar tape is always used. Invar is a nickel-steel alloy for which the coefficient of thermal expansion may be as low as 0.0000002 per 1°F or 0.00000036 per °C (about one-thirtieth that of steel). The invar tapes commonly used have a coefficient of expansion one-eighth to one-tenth that of steel. Often a "long tape" (length 50 m or 200 to 500 ft) is used.

The length of the tape should be precisely determined by comparison with a standard of known length. The National Bureau of Standards at Washington, D.C., for a nominal fee, will compare the tape with a length that has been precisely determined and will issue a certificate showing the actual length of the tape under stated conditions as regards tension, temperature, and supports (see Art. 4.16). It is desirable to have the tape compared with the standard under the conditions of tension and support that will be employed in the field work so that no corrections for sag or tension will be necessary.

A tape that has been compared with the standard at Washington may itself serve as a standard for other tapes in the field, but for work of the greatest precision all the tapes used should be compared with the standard at Washington.

If a tape is kinked in handling, its length will be appreciably changed. Invar is relatively soft and bends easily. Hence, tapes of this metal should be handled with great care and when not in use should be kept on a reel not less than about 15 in (38 cm) in diameter. In the best practice, two or three tapes are provided and, when in use, are compared daily, so as to detect sudden changes in length due to whatever cause. In any case, a tape should have its length again compared with some standard upon the completion of the work.

For second- and third-order accuracy where the base line is along a paved highway or a railroad, measurements can be made with the tape supported over its entire length and at a time when the temperature of the supporting surface (highway or rail) is not appreciably different from that of the surrounding air.

Where the base line is over uneven ground, and for all first-order work end supports for the tape usually consisting of substantial posts, perhaps 2 by 4 in (about 5 by 10 cm) or 4 by 4 in (10 cm by 10 cm), are driven firmly into the ground. These are placed on a transit line at intervals of one tape length, as nearly as can be determined by careful preliminary measurements, and aligned to within 0.50 ft or 0.15 m. A strip of copper or zinc is tacked to the top of the post to receive the markings. Portable tripods or taping bucks (Fig. 10.16) are also used to some extent as tape supports. Profile levels are run over the tops of the end supports to determine the gradient from support to support.

The tape is usually supported at one, two, or three points between the end supports. These intermediate points are placed accurately on the grade line between the tops of the two adjacent end posts, by driving nails at grade in 1 by 2 in (2.5 by 5.0 cm) stakes placed on line at the proper intervals. Preferably, these supports should be provided at the same intervals as those used in the standard comparison, and the nails should be so driven that the tape will not become pinched between nail and stake.

The equipment for base-line measurement includes at least one standardized tape (two for important work); two stretcher devices for applying tension, one of which is equipped with a standardized spring scale (Fig. 10.17) or a weight and pulley; two or three thermometers; a finely divided pocket scale; dividers; and a needle or a marking awl.

**Fig. 10.16** Base-line measurements: making forward contact on a taping tripod. (*U.S. National Ocean Survey.*)

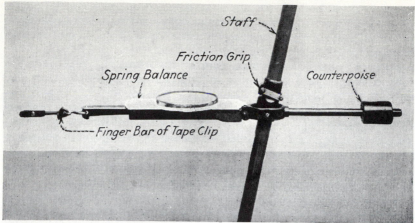

**Fig. 10.17** Tape stretcher and spring scale. (*U.S. National Ocean Survey.*)

The party consists of four to six persons whose duties are as follows. The *head stretcher*[†] applies tension to the head end of the tape and is responsible for seeing that the proper tension is pulled; the *rear stretcher* applies tension on the rear end of the tape; the *rear contact* observes the rear end of the tape to be sure that the rear mark on the tape coincides with the previously established mark; the *front contact* scribes a fine line where the front graduation mark on the tape strikes the metal plate on the head support; if necessary, one person tends the middle support; and the *notekeeper* (usually the chief of party) records the data.

When the rear end of the tape is observed to coincide with the previously established mark and when the proper tension is applied, the position of the forward end of the tape is marked by a fine line engraved by means of a needle or marking awl on the metal strip on top of the post (Fig. 10.16). Thermometers fastened to the tape, one near each end and sometimes one near the middle, are read at the time that the tape length is marked on the forward post.

The tape is then carried forward without allowing it to drag on the ground, and the process is repeated. After a few measurements, the end of the tape will probably fall either beyond or short of the limits of the metal strip of the next forward post because of variations in temperature or because of inaccurate placement of the posts. Accordingly, it will be necessary occasionally to use either *set backs* or *set forwards* as may be necessary to keep the tape ends on top of the posts. These are measurements of small distances made by means of a finely divided pocket scale and a pair of dividers. A record is kept of all conditions and observations.

Long base lines should be divided into sections 1 km in length. Each section is then measured forward and backward, preferably with a different tape on the backward measurement. For second-order triangulation, the two measurements should agree to within $20\sqrt{K}$ mm, where $K$ is the length of the section in km.

Corrections should be made for all systematic errors, including the slope correction (Art. 4.13), incorrect length of tape (Art. 4.16), temperature correction (Art. 4.18), and, if necessary, corrections for sag and change in tension (Arts. 4.19 and 4.20). If the standardized tension and conditions of support for the tape are used, these last two corrections are unnecessary.

The distance corrected for all systematic taping errors is then reduced to sea level as described in Art. 10.17. Thus, it is necessary to have a differential level connection with a bench mark having a known sea-level elevation.

[†]No relationship to "head shrinker."

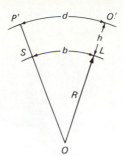

**Fig. 10.18** Reduction to sea level.

The standard deviation of the mean of the measurements of the base line should not exceed the limits given in Table 10.2. The standard deviation is calculated according to Eq. (2.37), Art. 2.17, and is determined using distances corrected for all known systematic errors.

**10.17. Reduction to sea level**    Triangulation networks extending over large areas must be reduced to a common sea-level datum. In Fig. 10.18 distance $d$ is measured from $P'$ to $O'$ at elevation $h$ above sea level. If $R$ is the radius of curvature for the earth's surface at that section, then by proportion

$$\frac{b}{d} = \frac{R}{R+h} \qquad \text{or} \qquad b = d\frac{R}{R+h} \tag{10.24}$$

in which $b$ is the sea-level distance. For most sea-level reductions, an average radius of the earth for the United States of 6,372,160 m or 20,906,000 ft (based on the U.S. Survey Foot; see Art. 1.8) is adequate.

**Example 10.2**    A horizontal distance of 17,690.819 m is measured at an average elevation of 738.22 m above mean sea level. Determine the sea-level distance for the line.

SOLUTION    Use the average $R$ for the United States, as given above. Then, by Eq. (10.24),

$$b = \frac{(17,690.819)(6,372,160)}{6,372,160+738.22} = 17,688.770 \text{ m}$$

Note that the correction in this case is about 2 m, or 1 part in 8000, a significant amount.

**10.18. Measurement of directions and angles**    Directions from which angles are determined for triangulation are usually observed with a direction theodolite (Art. 6.30). First-order triangulation requires a first-order optical-reading direction theodolite with a least count of less than 1 second (Fig. 10.19). A second-order direction theodolite (Figs. 6.34 and 6.35) which has a least count of 1 second is adequate for second- and third-order triangulation. Repeating theodolites and transits of good quality can be used to observe angles directly in third-order triangulation but are not recommended due to the excessive effort required to obtain results of sufficient accuracy.

The use of optical-reading theodolites is developed in Arts. 6.30, 6.40, and 6.41. The details covered in this section concerning the use of these instruments are related to the modifications and refinements in the observational procedures necessary to satisfy first-, second-, and third-order specifications for triangulation as given in Table 10.2.

To achieve the desired accuracy, extra care must be exercised to (1) center the instrument and targets very carefully, (2) eliminate phase in the targets (Art. 10.12), (3) protect the instrument from the heating effects of the sun and vibrations caused by wind, and (4) provide an instrument support of adequate stability. This last requirement is extremely important and frequently leads to use of a concrete pedestal for support of the station and instrument. Scheduling of observations when adverse horizontal refraction is at a minimum

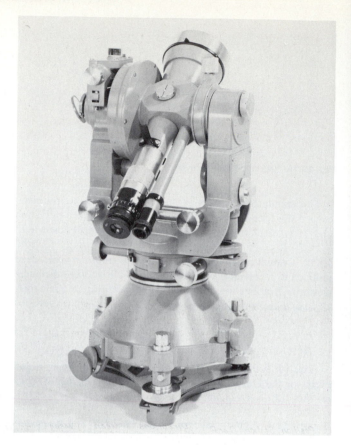

**Fig. 10.19** Wild T-3, first-order theodolite. (*Wild Heerbrugg Instruments, Inc.*)

is also advisable. Operations at night using illuminated signals will help reduce bad atmospheric conditions and assist in achieving optimum results.

Directions should be observed using procedures that minimize collimation, circle, and micrometer errors in the instrument. The procedure for observing one position of directions from a single setup with the telescope direct and reversed is given in Art. 6.41. Depending on the accuracy desired, from 2 to 16 positions may be required (Table 10.2). In order to minimize errors in the horizontal circle and in the micrometer, each position is advanced to a new initial circle reading and micrometer readings should be spread over the range of the micrometer. In general, the horizontal circle is advanced by an interval $I$, where

$$I = \frac{360°}{mn} + \frac{\text{micrometer drum range (in minutes)}}{n} \qquad (10.25)$$

in which $n$ is the number of positions to be observed and $m$ the number of verniers on the instrument. The instrument illustrated in Fig. 6.35 has a micrometer drum range of $10'$. Thus, the interval for four positions is, by Eq. (10.25):

$$I = \frac{360°}{(2)(4)} + \frac{10'}{4} = 45°02'30''$$

and the nominal initial circle readings for each of the four positions are

| Position | Circle reading |
|:---:|:---:|
| 1 | 0°00'10" |
| 2 | 45°02'40" |
| 3 | 90°05'10" |
| 4 | 135°07'40" |

Notes for four positions of directions to three stations are shown in Fig. 10.20. Note that for each direction the micrometer reading was observed twice and the mean is calculated. The mean of the direct and reversed readings are computed and all directions are reduced to a zero azimuth along the initial pointing for each position (from Doris to Clark). These values are recorded in column 7 of the notes. Finally, the means of the four directions for each pointing are calculated yielding an abstract of average directions recorded in column 8 of the notes. This abstract represents the original uncorrelated observations from triangulation station Doris which may be used in any subsequent adjustment of the triangulation. If angles are desired, they can be calculated by subtracting successive directions. The angles are listed in column 9 of the notes. If these angles are to be used as observations, note that they are derived from the directions and are correlated (Art. 6.58). This correlation must be taken into account in any subsequent adjustment involving these angles.

A 20" or 30" engineer's transit or repeating theodolite can be used for third-order triangulation. In this case, each angle is measured by repetition (Art. 6.39). Figure 10.21 shows a portion of the notes (one set) recorded to observe angles $\alpha$, $\beta$, and $\gamma$ between points $B$, $Y$, and $R$ from triangulation point $A$. For this problem, 10 repetitions were observed for each angle. A set comprises all angles about a point, each repeated the requisite number of times. The second set is then initiated with the horizontal circle advanced by an interval determined by Eq. (10.25). For this problem, where four sets are to be observed, the increment is 45°07'30". Using an instrument with a least count of 30", six sets of six repetitions (Art. 6.58) or four sets of 10 repetitions are required for third-order triangulation. Since the horizon has been closed, the condition

$$\alpha_m + \beta_m + \gamma_m = 360° \qquad (10.26)$$

| Position | Station | D or R | Directions ° ' " | Mean " | Mean D & R " | Direction | Average of 4 Positions | Average Angle |
|---|---|---|---|---|---|---|---|---|
| | | | STATION DORIS, INST. WILD T-2 No. 57514 | | | Young Skinner | October 25, 1977 Cool, Clear   30 | |
| 1 | Clark | D | 0 10 16 17 | 16.5 | 21.2 | 00."0 | 0° 00' 00."0 | |
| | | R | 180 10 27 25 | 26.0 | | | | 48° 42' 04."3 |
| | Boyd | D | 48 52 05 08 | 06.5 | 23.5 | 02. 3 | 48 42 04.3 | |
| | | R | 228 52 39 42 | 40.5 | | | | 34° 14' 15."6 |
| | Alice | D | 83 06 33 34 | 33.5 | 40.5 | 19. 3 | 82 56 19.9 | |
| | | R | 263 06 47 48 | 47.5 | | | | N |
| 2 | Clark | R | 45 12 54 53 | 53.5 | 52.5 | 00. 0 | | |
| | | D | 225 12 50 53 | 51.5 | | | | |
| | Boyd | R | 93 55 01 03 | 02.0 | 58.0 | 05. 5 | | |
| | | D | 273 54 53 55 | 54.0 | | | | |
| | Alice | R | 128 09 18 15 | 16.5 | 13.2 | 20. 7 | | |
| | | D | 308 09 10 10 | 10.0 | | | | |
| 3 | Clark | D | 90 07 23 20 | 21.5 | 37.8 | 00. 0 | | |
| | | R | 270 07 55 53 | 54.0 | | | | |
| | Boyd | D | 138 49 34 33 | 33.5 | 42.5 | 04. 7 | | |
| | | R | 318 49 52 51 | 51.5 | | | | |
| | Alice | D | 173 03 56 54 | 55.0 | 56.2 | 18. 4 | | |
| | | R | 353 03 57 58 | 57.5 | | | | |
| 4 | Clark | R | 135 09 40 37 | 38.5 | 31.2 | 00. 0 | | |
| | | D | 315 09 25 23 | 24.0 | | | | |
| | Boyd | R | 183 51 42 41 | 41.5 | 35.8 | 04. 6 | | |
| | | D | 3 51 31 29 | 30.0 | | | | |
| | Alice | R | 218 05 55 57 | 56.0 | 52.5 | 21. 3 | | |
| | | D | 38 05 50 48 | 49.0 | | | | |

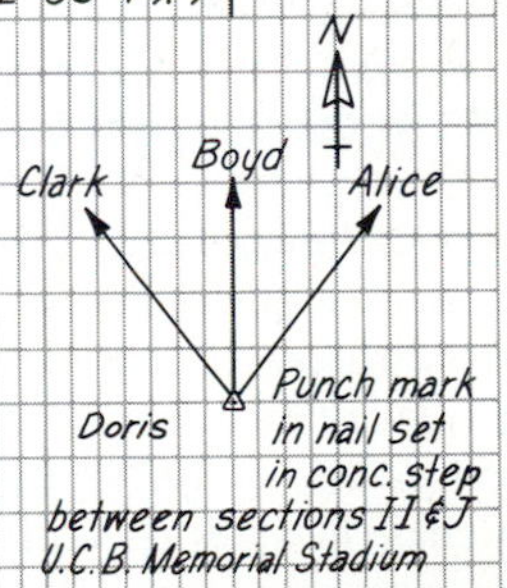

Clark   Boyd   Alice

Punch mark in nail set in conc. step between sections II & J U.C.B. Memorial Stadium

Doris

**Fig. 10.20** Directions with a theodolite from a triangulation station.

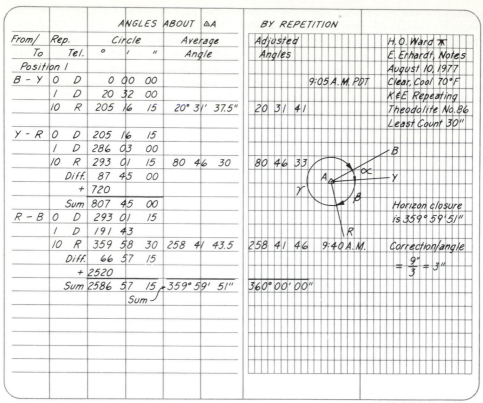

**Fig. 10.21** Angles for triangulation using a repeating instrument.

in which $\alpha_m$, $\beta_m$, and $\gamma_m$ are the average angles, must be satisfied. Applying a least squares adjustment by observations only (Example B.6, Appendix B.8), using condition Eq. (10.26), it can be shown that the adjusted angles are the observed average angles plus $\frac{1}{3}[360° - (\alpha_m + \beta_m + \gamma_m)]$, where $\alpha_m$, $\beta_m$, and $\gamma_m$ are assumed to be uncorrelated observations of equal precision. Thus, in Fig. 10.21 if position 1 is adjusted individually, $[360° - (\alpha_m + \beta_m + \gamma_m)] = 09''$, so that the correction per angle is $09''/3 = 03''$ per angle.

**10.19. Triangulation computations**   The primary objective of triangulation computations is to determine the best possible planimetric positions, usually defined by $X$ and $Y$ coordinates, for the triangulated stations. To achieve this objective it is necessary to (1) correct all base-line measurements for systematic errors and reduce the distances to sea level (Arts. 10.15 to 10.17); (2) reduce observed directions or angles to average values and check triangle closures (Art. 10.18); (3) perform the necessary *reductions to center* when eccentric stations have been occupied (Art. 10.20); (4) compute *spherical excess* where lines and triangles are of sufficient length and size to warrant these corrections (Art. 10.21); (5) compute preliminary values for all lines in the triangulation network (preliminary position computations) when these values are required for subsequent adjustment procedures; (6) adjust the triangulation network by the method of least squares (Art. B.15, Appendix B); and (7) perform final position computations when necessary to determine final coordinates for all triangulated stations. Details concerning computations under (1) and (2) above are discussed in previous sections. Next, consider reduction to center.

**10.20. Reduction to center**   At certain triangulation stations it is difficult, if not impossible, to place the instrument vertically beneath the object which has been observed from adjacent stations. At such a place, the instrument is set over any convenient point near the principal station, and angles to the adjacent stations are measured with the same precision as other angles in the system. These angles will not be the same as those which would be observed if the instrument were occupying the exact location of the station. To obtain the corresponding values for the main station, corrections are computed and applied to the measured angles. This procedure of correcting the observed angles is termed *reduction to center*.

In addition to the measurement of the angles to adjacent stations, measurements are made of (1) the distance from the main station to the occupied station, and (2) the clockwise angle between the occupied station and the adjacent station in the system. Consider the example shown in Fig. 10.22, where $Y$, $G$, and $B$ are triangulation stations; $G$ is the inaccessible station; and $E$ is the eccentric station. The eccentric distance $EG = d$ (as small as is practical) is measured. Eccentric station $E$ is occupied and directions are observed to $G$, $B$, and $Y$. Thus, clockwise angles $\alpha_1$, $\alpha_2$ referenced to line $EG$ can be calculated. To reduce the observations at $E$ to the main triangulation net, it is necessary to calculate corrections $c_1$ and $c_2$. From Fig. 10.22,

$$\sin c_1 = \frac{d \sin \alpha_1}{s_1} \qquad \text{and} \qquad \sin c_2 = \frac{d \sin \alpha_2}{s_2}$$

In general, assuming very small angles, the correction expressed in seconds is

$$c \text{ (seconds)} = \frac{d \sin \alpha}{s \sin 01''} \tag{10.27}$$

where $d =$ distance measured from station to eccentric station
   $\alpha =$ clockwise angle at the eccentric station from the unoccupied to the observed station
   $s =$ length of the side in the main system for which the correction is being calculated
$\sin 01'' = 0.000004848$

Distances between stations in the main system can be calculated from angles and known sides in the balance of the network.

**Example 10.3**   Triangulation station George was inaccessible, so an eccentric station $E$ was set 3.980 m from George as shown in Fig. 10.23. The observed directions reduced to an abstract with zero along line $E$ to George are

| George | $0°00'00.0''$ |
| Rex | $85°10'00.0''$ |
| Mary | $135°25'10.3''$ |
| Exeter | $181°15'26.0''$ |
| Lucy | $261°20'17.4''$ |

Approximate distances from George to Rex, Mary, Exeter, and Lucy were calculated from angles observed at these stations and a known distance in the network. These distances are shown on Fig. 10.23. Calculate corrected directions.

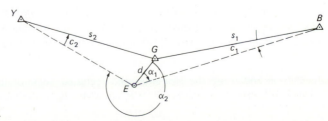

**Fig. 10.22** Reduction to center.

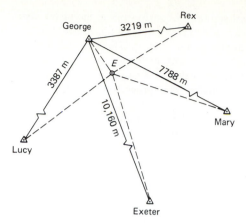

**Fig. 10.23** Example problem, reduction to center.

SOLUTION   Using Eq. (10.27), the correction to each direction is calculated. Data, corrections, and corrected directions are given in the following table.

| | Station | | | |
|---|---|---|---|---|
| | Rex | Mary | Exeter | Lucy |
| $\alpha_i$ | 85°10′00.0″ | 135°25′10.3″ | 181°15′26.0″ | 261°20′17.4″ |
| $s$, m | 3219 | 7788 | 10,160 | 3387 |
| $\sin\alpha$ | 0.99644401 | 0.70191034 | −0.02194091 | −0.98859443 |
| Corr. $c''$ | 254.1 | 74.0 | −1.8 | −239.6 |
| Corrected directions | 85°14′14.1″ | 135°26′24.3″ | 181°15′24.2″ | 261°16′17.8″ |

Note that the sign of the correction follows automatically from Eq. (10.27) when the correct sign is used for $\sin\alpha$.

**10.21. Spherical excess**   A triangulation net having long sides theoretically should be solved as a series of spherical triangles. This problem becomes unnecessarily complicated and may be avoided by applying Legendre's theorem, which is as follows: In a spherical triangle in which the sides are short compared to the radius of the earth and a plane triangle whose sides are equal in length to the corresponding sides of the spherical triangle, the corresponding angles of the two triangles differ by approximately the same amount or one-third of the spherical excess in the triangle.

Figure 10.24 shows a spherical triangle which covers one-eighth of a sphere. From the figure it is obvious that triangle $XYZ$ has a spherical excess of $270° - 180° = 90° = \pi/2$. Since it can be shown by spherical geometry that the spherical excess of a triangle is proportional to the area of the triangle, the following equation can be written from Fig. 10.24:

$$\frac{e}{a} = \frac{\pi/2}{\pi R^2/2} = \frac{1}{R^2}$$

in which $e$ is the spherical excess in radians, $a$ the area of the triangle, and $R$ the radius of curvature for the earth at the average latitude of the triangle. Since in general, the area of a spherical triangle is $(bc \sin\alpha)/2$, the spherical excess in seconds is

$$e \text{ (seconds)} = (bc \sin\alpha)/2R^2 \sin 01'' \qquad (10.28)$$

where $b$, $c$, and $\alpha$ are the two sides and included angle of the triangle and the $\sin 01'' = (4.848)(10^{-6})$. Strictly speaking, $R^2 = R_m N$, where $R_m$ is the radius of curvature of the earth's

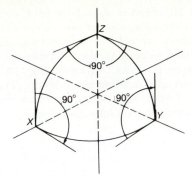

**Fig. 10.24**

meridian and $N$ is the earth's radius of curvature at right angles to the meridian, both at the latitude of the center of the triangle. These values can be calculated as functions of the major and minor axes of the earth and the latitude using Eq. (14.2) for $N$ and Eq. (14.64) for $R_m$. For small triangles common in third-order work as well as for many triangles found in second- and first-order triangulation, use of the average radius of the earth in Eq. (10.28) is adequate. The average radius for the United States is 6,372,160 m or 20,906,000 ft. Using this radius, the spherical excess amounts to about $1''$ per 196.8 km$^2$ or $1''$ per 76 mi$^2$.

**Example 10.4**   The observed angles in triangle $ABC$ are (Fig. 10.25):

$\alpha = 57°53'20.1''$
$\beta = 62°23'32.1''$
$\gamma = 59°43'20.3''$

The distance from $B$ to $C$, designated $d$, as computed from preliminary calculations in the net is 40,320.00 m. Compute the spherical excess in the triangle.

SOLUTION   Solve for the length of side $c$ by the law of sines:

$$c = \frac{d \sin 59°43'20.3''}{\sin 57°53'20.1''} = \frac{(40,320.00)(0.86359190)}{0.84701911} = 41,108.90 \text{ m}$$

Compute the spherical excess, $e$, by Eq. (10.28):

$$e = \frac{(40,230.00)(41,108.90)\sin\beta}{(2)(6,372,160)^2(0.000004848)} = 3.7''$$

The angles corrected for spherical excess are

$\alpha = 57°53'18.9''$
$\beta = 62°23'30.9''$
$\gamma = 59°43'19.1''$
$\overline{\quad 180°00'08.9''\quad}$

The remaining $08.9''$ can be considered random, to be compensated in subsequent adjustment of the triangulation.

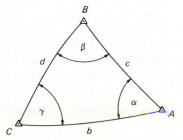

**Fig. 10.25** Spherical excess.

**10.22. Adjustment of triangulation** When all observed data have been compiled and reduced, the next step in the computations is the adjustment of the triangulation network.

Assuming that all systematic errors have been corrected in the measured directions or angles and distances, the observed data can be considered randomly distributed so that the method of least squares is the most appropriate procedure for adjustment. There are two techniques of least squares which are commonly used to adjust triangulation: (1) adjustment of indirect observations, and (2) adjustment of observations only. The concept of the least squares adjustment and details of these two techniques are fully developed in Appendix B.1 to B.8. Only a few brief comments will be made in this chapter.

Adjustment of triangulation by indirect observations (Example B.13, Appendix B.15) requires that the linearized form of Eq. (8.8) be written for each measured distance [Eq. (B.83), Appendix B.14] and the linearized form of the condition equation for each observed angle [Eq. (B.77), Appendix B.12] be written for all measured angles. These equations contain observations (distances and angles) and unknown parameters (corrections to $X$, $Y$ coordinates for which the positions are unknown). Thus, it is necessary to compute approximate coordinates for all stations in the network (Art. 10.23). Output from the least squares adjustment consists of adjusted measurements and adjusted $X$, $Y$ coordinates for all triangulation stations of unknown position.

Adjustment by observations requires that only the appropriate number of angle-condition equations (Art. 10.8) and side-condition equations (Art. 10.9) be formed for the triangulation network. The angle-condition equations are linear, but the side-condition equations [Eq. (10.6) or Eq. (10.10)] must be linearized [Example B.12, Art. B.15, Appendix B]. As implied by the designation of the technique, these equations contain observations only. Thus, the output from the least-squares adjustment consists of adjusted directions or angles. It is then necessary to perform position computations using given coordinates and/or measured base-line distances (which may be assumed correct) to obtain adjusted coordinates for all triangulated stations.

Both techniques also yield the reference standard deviation and the propagated covariance matrices for adjusted coordinates. These statistics are useful for evaluating the quality of the measurements and adjusted positions of the stations. The results obtained by both methods will be identical.

Detailed example solutions for each technique of adjustment can be found in Art. B.15, Appendix B.

Approximate methods for adjustment of triangulation have been developed and are fully documented. Even for small problems, these methods require a substantial effort to formulate and to solve. Since these methods are not rigorous, the answers are not the best possible values, and no rigorous statistics for evaluation of the quality of observations and adjusted positions are available. Consequently, the use of these approximate methods is not recommended for the adjustment of triangulation networks.

**10.23. Position computations** Determination of $X$ and $Y$ coordinates for all triangulated points is necessary to provide approximate coordinates for the subsequent least squares adjustment of indirect observations (Art. 10.22). If the adjustment has been by the least squares technique of observations only, it is necessary to calculate coordinates for all stations using adjusted angles. In either case, the computational procedure is the same. Directions obtained from observed or adjusted angles and distances calculated by the law of sines are utilized in Eqs. (8.7) and (8.8) to compute $X$ and $Y$ coordinates for all stations of unknown position. Thus, the procedure is exactly the same as outlined for traverse computation in Art. 8.15. To illustrate this procedure applied to triangulation computations, consider an example problem.

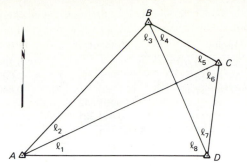

**Fig. 10.26**

**Example 10.5**   The quadrilateral *ABCD* shown in Fig. 10.26 has the following observed interior angles:

$l_1 = 22°01'42.51''$     $l_5 = 86°33'13.45''$

$l_2 = 16°44'31.20''$     $l_6 = 58°46'35.93''$

$l_3 = 57°08'57.10''$     $l_7 = 15°06'52.28''$

$l_4 = 19°33'14.13''$     $l_8 = 84°04'50.66''$

The given coordinates for stations *A* and *D* are

| Station | Coordinates, m | |
|---|---|---|
| | X | Y |
| A | 15,400.812 | 10,425.406 |
| D | 17,901.905 | 10,425.406 |

Thus, the direction of *AD* is due east:

$A_{AD} = 90°00'00.00''$     and     $d_{AD} = 2501.093$ m

To avoid using data containing blunders, the sums of angles in each triangle are checked.

| $l_1 =$ 22°01'42.51'' | $l_1 =$ 22°01'42.51'' | $l_2 =$ 16°44'31.20'' |
|---|---|---|
| $l_6 =$ 58°46'35.93'' | $l_2 =$ 16°44'31.20'' | $l_3 =$ 57°08'57.10'' |
| $l_7 =$ 15°06'52.28'' | $l_3 =$ 57°08'57.10'' | $l_4 =$ 19°33'14.13'' |
| $l_8 =$ 84°04'50.66'' | $l_8 =$ 84°04'50.66'' | $l_5 =$ 86°33'13.45'' |
| = 180°00'01.38'' | = 180°00'01.47'' | = 179°59'55.88'' |

In triangle *ADC*:

| Side | Station | Angles | Distance, m | Sine of angle |
|---|---|---|---|---|
| AD | | | 2501.093 | |
| | A | $l_1$ 22°01'42.51'' | | 0.37506734 |
| | D | $l_7 + l_8$ 99°11'42.94'' | | 0.98714949 |
| | C | $l_6$ 58°46'35.93'' | | 0.85515305 |
| $DC = (AD \sin l_1)/\sin l_6$ | | | 1096.971 | |
| $AC = (AD \sin [l_7 + l_8])/\sin l_6$ | | | 2887.147 | |

In triangle *ADB*:

| Side | Station | Angles | Distance, m | Sine of angle |
|------|---------|--------|-------------|---------------|
| | D | $l_8$ 84°04'50.66" | | 0.99466821 |
| | A | $l_1 + l_2$ 38°46'13.71" | | 0.62620213 |
| | B | $l_3$ 57°08'57.10" | | 0.84008593 |
| $AB = AD \sin l_8 / \sin l_3$ | | | 2961.313 | |
| $BD = AD \sin(l_1 + l_2) / \sin l_3$ | | | 1864.321 | |

In triangle *ABC*:

| Side | Station | Angles | Distance, m | Sine of angle |
|------|---------|--------|-------------|---------------|
| | A | $l_2$ 16°44'31.20" | | 0.28806256 |
| | B | $l_3 + l_4$ 76°42'11.23" | | 0.97319140 |
| | C | $l_5$ 86°33'13.45" | | 0.99819161 |
| $AB = AC \sin l_5 / \sin(l_3 + l_4)$ | | | 2961.315 | |
| $BC = AC \sin l_2 / \sin(l_3 + l_4)$ | | | 854.589 | |

Next, azimuths for all lines are computed.

Triangle *ADC*:

$$A_{AD} = 90°00'00.00" \qquad A_{CA} = 247°58'17.49"$$

$$\text{angle } l_1 = -22°01'42.51" \qquad \text{angle } l_6 = -58°46'35.93"$$

$$A_{AC} = 67°58'17.49" \qquad A_{CD} = 189°11'41.56"$$

$$A_{DC} = 9°11'41.56"$$

$$A_{AD} = 90°00'00.00"$$

$$l_7 + l_8 = 99°11'42.94" \qquad \overline{99°11'41.56"}$$

Triangle *ABD*:

$$A_{AD} = 90°00'00.00" \qquad A_{BA} = 231°13'46.29"$$

$$\text{angle } l_1 + l_2 = -38°46'13.71" \qquad \text{angle } l_3 = -57°08'57.10"$$

$$A_{AB} = \overline{51°13'46.29"} \qquad A_{BD} = \overline{174°04'49.19"}$$

$$A_{DB} = 354°04'49.19"$$

$$\text{angle } l_8 = 84°04'50.66"$$

$$A_{DA} = \overline{269°59'58.53"}$$

Triangle *ABC*:

$$A_{BA} = 231°13'46.29" \qquad A_{CB} = 334°31'35.06"$$

$$\text{angle } l_3 + l_4 = -76°42'11.23" \qquad \text{angle } l_5 = -86°33'13.45"$$

$$A_{BC} = \overline{154°31'35.06"} \qquad A_{CA} = \overline{247°58'21.61"}$$

$$A_{AC} = 67°58'21.61"$$

$$\text{angle } l_2 = -16°44'31.20"$$

$$A_{AB} = \overline{51°13'50.41"}$$

Finally, coordinates for stations $C$ and $B$ are calculated using Eqs. (8.7) and (8.8).

| Station | Distance, m | Azimuth | Coordinates, m | |
|---------|-------------|---------|----------------|---|
| | | | $X$ | $Y$ |
| A | | | 15,400.812 | 10,425.406 |
| | 2887.147 | 67°58′17.49″ | | |
| C | | | 18,077.190 | 11,508.281 |
| | 1096.971 | 189°11′41.56″ | | |
| D | | | 17,901.902 | 10.425.405 |
| | | | [17,901.905] | [10,425.406] |
| A | | | 15,400.812 | 10,425.406 |
| | 2961.313 | 51°13′46.29″ | | |
| B | | | 17,709.632 | 12,279.787 |
| | 1864.321 | 174°04′49.19″ | | |
| D | | | 17,901.906 | 10.425.408 |
| | | | [17,901.905] | [10,425.406] |

As a check, compute $d$ and $A_{BC}$ by Eqs. (8.9) and (8.10).

$$d_{CB} = [(18,077.190 - 17,709.632)^2 + (12,279.787 - 11,508.281)^2]^{1/2}$$

$d_{CB} = 854.588$ m    previously calculated as 854.589

$$A_{CB} = \arctan \frac{18,077.190 - 17,709.632}{12,279.787 - 11,508.281} = \arctan 0.47641626$$

$A_{CB} = 334°31′34.00″$    previously calculated as 334°31′35.06″

The small discrepancies are due to the use of unadjusted observations. The calculated coordinates are more than accurate enough for use as approximations in the least squares adjustment by indirect observations.

Had the adjustment been made by the technique of observations only, the output consists of adjusted angles. In this case, the procedure just outlined is used to compute adjusted coordinates with the adjusted angles. The reader is invited to make these computations with the adjusted angles listed in [Example B.12, Art. B.15, Appendix B].

**10.24. Graphical triangulation**    Graphical triangulation achieves the same results as instrumental triangulation, as described previously. The procedure differs in that the plotted locations of distant signals are determined graphically instead of calculating the positions using a measured base line and observed angles or directions. It is used for control of surveys over a small area and for extension of control for larger surveys. It is especially employed in intermediate-scale or small-scale mapping. As compared with plane-table traversing, graphical triangulation is most advantageous where the terrain offers unobstructed sights, considerable relief, and many well-defined objects.

The particular advantage of the use of the principles of intersection (Art. 9.8) and of resection (Art. 9.9) in plane-table mapping makes it desirable to locate many definite landmarks which are widely visible and suitably situated such as flagstaffs, church spires, and lone trees. Accordingly, distant stations and also land marks which will not be used as instrument stations but that will be useful in subsequent work are located.

Two (preferably three) plane-table stations such as $A$, $B$, and $C$ (Fig. 10.27), the locations of which are known, must be capable of being occupied by the instrument and must be marked by signals. Prior to the beginning of the work, the locations of these stations are plotted on the plane-table sheet at $a$, $b$, and $c$. The field procedure is then as follows. The plane table is set up as at $A$ and is oriented by sighting at $B$ and $C$; and rays are then drawn

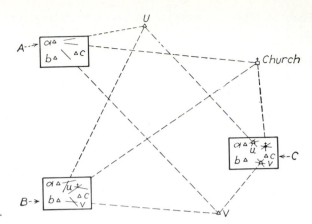

**Fig. 10.27** Graphical triangulation.

toward other stations such as $U$ and $V$ and, say, the church spire. The table is then set up and oriented at stations $B$ and $C$ in succession, and the same objects are sighted again. The correct plotted location of each station is determined by the intersection of two rays, such as by those drawn from stations $A$ and $B$, but the location is improved and checked if the three rays drawn toward a given object are found to pass through a point, as shown by the three rays drawn from station $C$. The series of control stations may be extended as needed.

**10.25. Trilateration**    A trilateration network consists of a system of joined or overlapping triangles in which all lengths are measured and only enough angles or directions are observed to establish azimuth. The continuing development and refinement of EDM equipment has made trilateration feasible and, given suitable geometry, this procedure is now competitive with triangulation for establishing horizontal control. First let us discuss the number of conditions in trilaterated figures.

If one considers a pure trilateration network in which only distances are measured, the number of conditions in the figure must be formulated in terms of these distances. For example, if a single triangle is trilaterated as in Fig. 10.10, where points $A$ and $C$ are fixed, two distances, $a$ and $c$, are measured to fix the position of station $B$. Since the position of $B$ is defined by two parameters, its $X$ and $Y$ coordinates, there are zero extra measurements or no conditions or redundant information to provide a check on the work. Obviously, a chain of single triangles is not a suitable configuration for control extension by trilateration. Next, consider the braced quadrilateral shown in Fig. 10.26. Assuming stations $A$ and $B$ fixed and all lengths measured, there are five distances and four unknown coordinates for $B$ and $C$, resulting in one condition.

At this point, recall that the number of conditions in a triangle and quadrilateral with all angles observed, was one and four, respectively [Eq. (10.23)]. Since strength of figure is a function of the number of conditions in a network, one disadvantage of pure trilateration is that there are not as many conditions possible as can be found in triangulation. The number of conditions for various geometric figures for trilateration and triangulation are listed in Table 10.5. Note that the number of conditions as calculated here for trilateration and

**Table 10.5**

|  | Triangle | Braced quad. | Center-point pentagon | Pentagon (all points | Hexagon intervisible) |
|---|---|---|---|---|---|
| Trilateration | 0 | 1 | 1 | 3 | 6 |
| Triangulation | 1 | 4 | 7 | 9 | 16 |

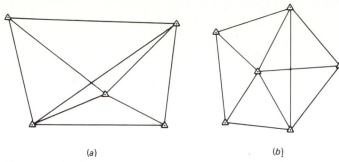

**Fig. 10.28** (*a*) Center-point quadrilateral with brace. (*b*) Center-point pentagon with additional distance.

triangulation is synonomous with the redundancy, $r$, for a least squares adjustment as developed in Art. B.2, Appendix B.

Theoretically, a pentagon or hexagon would be the ideal figure for trilateration. However, from a practical standpoint, establishing even a single pentagon or hexagon in the field would be an almost impossible task.

The U.S. National Ocean Survey recommends use of the traditional braced quadrilateral, a center-point quadrilateral with an additional measurement (Fig. 10.28*a*), or a center-point pentagon with an additional measurement (Fig. 10.28*b*). They further recommend that angles be measured wherever possible to supplement the trilateration, thus increasing the number of conditions.

The procedure for establishing a trilateration network is similar to that required for triangulation. All available data, such as existing maps, references to existing control, and aerial photographs, are assembled for the area in question and preliminary locations for trilateration stations are chosen in the office. A field reconnaissance follows during which the intervisibility of points, the accessibility of stations, and strength of the figure chosen are verified.

Since trilateration has fewer conditions than triangulation, the specifications in Table 10.2 under "Trilateration" should be followed closely. EDM equipment should be chosen that will yield required standard deviations in length and no triangle within the figure should contain an angle less than the specified minimum angles (15 to 25°), particularly for higher-order surveys. Because few, if any, angles will be observed, few signals must be maintained. The construction of towers or stands for instruments is generally unnecessary. Reflectors or antenna units for EDM equipment can be extended up to 13 m above the station without serious effects on the precision of the measurements. In this way distances of up to 50 km can be measured without the need for towers. This represents a real advantage for trilateration, since towers are costly.

Distances measured by EDM equipment must be corrected for all systematic instrumental errors and for the effects of atmospheric conditions. Slope distances must be reduced to horizontal sea-level distances (Arts. 4.37 to 4.41 and 10.17). Vertical angles from both ends of the lines or elevations at the ends of the lines must be obtained. Where the lines are long for an extensive network, elevations can be obtained with sufficient accuracy for this purpose by interpolating from the contours on a topographic map or by use of barometric leveling. Where a relatively small network containing lines of only a few kilometres in length is being surveyed, vertical angles and trigonometric levels (Art. 5.5) are usually employed for slope corrections and reduction to sea level.

**10.26. Adjustment of trilateration**   The observed values in a pure trilateration network are the measured distances. As noted in Art. 10.25, there must be enough measured distances to

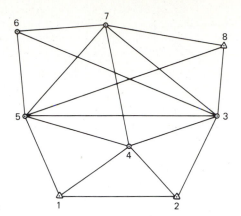

**Fig. 10.29** Trilateration network.

provide redundant observations. With redundant observed distances, the best procedure for adjustment is the method of least squares using the technique of indirect observations (Art. B.7, Appendix B). Application of this technique to trilateration adjustment requires that the linearized form of Eq. (8.7) (Art. 8.15) be written for each measured distance. This linearized or so-called *distance condition equation* [Eq. (B.75)] is developed in Art. B.14, Appendix B.

These distance condition equations contain measured distances and corrections to the $X$ and $Y$ coordinates for stations of unknown position in the network. These corrections have coefficients that are functions of approximate azimuths, distances, and coordinates for all of the measured lines and unknown stations in the network. Consequently, it is necessary to make preliminary calculations so as to obtain approximate values for these quantities.

For example, Fig. 10.29 shows a trilateration network consisting of a center-point quadrilateral 1235 and overlapping braced quadrilaterals 3578 and 3567. An additional distance 4 to 7 was measured to strengthen the figure. Stations 1, 2, and 8 are of known, fixed $X$ and $Y$ coordinates. All distances except 12 have been measured yielding $n = 16$ observations. It is desired to adjust this network by the least squares technique of indirect observations.

First, it is necessary to solve triangles to obtain approximate azimuths and coordinates for stations 3, 4, 5, 6, and 7. As shown in Fig. 10.30, ten measured lines are required to calculate these coordinates using the triangles labeled (1), (2), (3), (4), and (5). Thus, the minimum number of measurements for a determinate figure are $n_0 = 10$ and the redundancy (or conditions as used in Art. 10.24), $r = n - n_0 = 16 - 10 = 6$ [Eq. (B.3), Art. B.4, Appendix B]. Another way to calculate the redundancy is to use Eq. (B.5), (Art. B.5) $r = c - u$, where

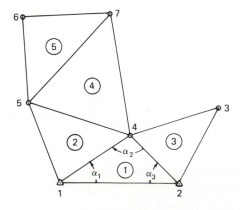

**Fig. 10.30** Trilateration calculations.

$c=$ the number of condition equations written and $u=$ the number of unknown parameters. There are 16 measured lines requiring 16 distance condition equations and five stations with unknown $X$ and $Y$ coordinates so that $u=10$. Thus, $r=16-10=6$.

Calculation of preliminary approximations proceeds as follows:

1. Calculate the azimuth of line 12 using Eq. (8.8), Art. 8.15.

$$A_{12}=\arctan\frac{X_2-X_1}{Y_2-Y_1}$$

2. Calculate the angles in triangles (1), (2), (3), (4), and (5) using the law of cosines.

$$\cos A = \frac{b^2+c^2-a^2}{2bc}$$

Thus, in Fig. 10.30 for triangle (1),

$$\cos\alpha_1 = \frac{(L_{14})^2+(L_{12})^2-(L_{42})^2}{(2)(L_{14})(L_{12})}$$

$$\cos\alpha_2 = \frac{(L_{14})^2+(L_{42})^2-(L_{12})^2}{(2)(L_{14})(L_{42})}$$

$$\cos\alpha_3 = \frac{(L_{12})^2+(L_{42})^2-(L_{14})^2}{(2)(L_{12})(L_{24})}$$

To check:

$$\alpha_1+\alpha_2+\alpha_3=180°+e$$

where $e$ is the spherical excess [Eq. (10.28), Art. 10.21].

This procedure is repeated in triangles (2), (3), (4), and (5) yielding angles in all triangles.

3. Using the azimuth of line 12 calculated in step 1, determine azimuths for all lines in triangles (1), (2),...,(5).
4. Starting with triangle (1), compute the $X$ and $Y$ coordinates for station 4 using measured distances and computed azimuths. Repeat this step for triangles (2), (3), (4), and (5) obtaining approximate coordinates $[X°, Y°]_i$, $i=1, 2, 3, 4, 5$, for each unknown station in the network.

At this point all approximations necessary for the least squares adjustment have been computed. The next step is to form the distance condition equations for each measured line in the network. This procedure is developed in Art. B.16, Appendix B, where Example B.16 contains details for the adjustment of a trilaterated braced quadrilateral. Output from the adjustment consists of adjusted $X$ and $Y$ coordinates for each station of unknown position in the network, residuals in measured distances, the reference standard deviation for the adjustment, and covariance matrices from which error ellipses for each adjusted station can be computed.

**10.27. Combined triangulation and trilateration** Under circumstances that prohibit achieving the necessary strength of figure by pure trilateration or using triangulation alone, it is quite feasible to perform a combination of triangulation and trilateration. This procedure can extend from the addition of a few angles or distances to a trilateration or triangulation network, respectively, to the observation of all angles and all distances in the net. Such an operation can be called a combined control survey and the subsequent adjustment is designated as a combined adjustment.

From an operational standpoint in the field, combined control surveys are certainly practical. This is particularly true for third-order control surveys of smaller extent where electronic tacheometers or surveying systems (Chap. 12) permit observation of distance and direction from a station using a single instrument.

Adjustments by the method of least squares using the technique of indirect observations presents no problems at all. For example, if a combined survey of the quadrilateral in Fig. 10.26 is to be performed, all angles $\ell_1, \ell_2, ..., \ell_8$ and five distances $d_{AB}, d_{AC}, d_{BD}, d_{BC}$, and $d_{CD}$

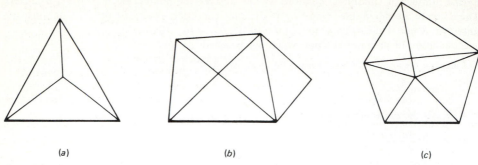

**Fig. 10.31**

are measured (assume that the coordinates are known for $A$ and $D$). For the adjustment, eight angle-condition equations [Eq. (B.70), Art. B.12], and five distance-condition equations [Eq. (B.75), Art. B.14] are formed. Thus, there are 13 condition equations in the system which can be solved by the method of least squares for corrections to four unknown $X$, $Y$ coordinates for stations $B$ and $C$. So the redundancy is $r = c - u = 13 - 4 = 9$, which represents a substantial increase over the one redundancy in pure trilateration and the redundancy four in triangulation adjustments.

Details of the least squares adjustment by indirect observations of combined triangulation and trilateration are similar to the procedures outlined in Appendix B for adjustment of triangulation (Example B.13, Art. B.15) and for adjustment of trilateration (Example B.14, Art. B.16). The systems of equations are simply combined into one set and are solved in single least squares adjustment as outlined in Example B.15, Art. B.17.

## 10.28. Problems

**10.1**  Figure 10.31 shows three configurations for triangulation. Assuming that the heavy line is the measured base line and that all directions are observed, determine the (a) total number of angle conditions; (b) total number of side conditions for each figure.

**10.2**  In Problem 10.1, assuming $X$, $Y$ coordinates are known for each end of the measured base line, calculate the redundancies or statistical degrees of freedom for each figure.

**10.3**  Calculate the total number of conditions for each figure shown in Fig. 10.31, assuming that all directions and distances are measured.

**10.4**  Under the assumptions of Prob. 10.1, determine the values for $C$ and $D$ in Eq. (10.22) (Art. 10.11) for each of the configurations shown in Fig. 10.31.

**10.5**  Determine the strength of figure values $R_1$ and $R_2$ for the triangulation configurations shown in Fig. 10.32. The heavy lines are the known base lines.

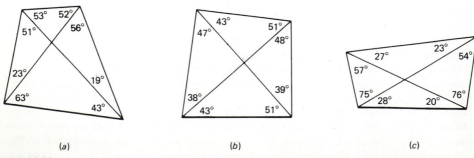

**Fig. 10.32**

**10.6**  In the triangle $ABC$ shown in Fig. 10.10, the line $AC$ is the measured base line, which has an estimated standard deviation of 0.005 m. If the angles at $A$, $B$, and $C$ are observed with respective estimated standard deviations of $10''$, calculate the propagated standard deviation in the calculated length of side $BC$. Assume $\angle A = \angle B = \angle C$.

**10.7**  In Fig. 10.2, line $AC$ or $b$ is the measured base line having a standard deviation of 0.01 ft. All of the angles in this chain of triangles are measured, with each angle having a standard deviation of $05''$. Determine the standard deviation of the line $GH$ or $g$ as calculated through the triangles by the law of sines assuming equilateral triangles and $b = 1,000.000$ m. State the assumptions that must be made in performing this error propagation.

**10.8**  A line of sight for a triangulation network is 20 km in length and must clear vegetation that has a maximum height of 5 m and occurs throughout the line. What heights of towers are necessary at the two ends of the line in order to permit a clear line of sight?

**10.9**  The average elevation at the site of a measured base line is 540 m above mean sea level. The following additional data are given: the tape has a standard length of 50.002 m at 20°C when supported at the 0- and 50-m points with a tension of 9 kg; the coefficient of thermal expansion for the tape is 0.00000036 per 1°C and the tape weighs 1.5 kg; the recorded length of the base line is 1050.241 m; the average temperature was 25.5°C; the cross-sectional area of the tape is 0.06 cm$^2$. the stakes were set on a 3 percent grade at 50-m intervals; the sum of the set forwards was 0.143 m; the sum of the set backs was 0.105 m; and standard interval and tension were used throughout. Compute the length of the base line (Arts. 4.18, 4.19, and 4.20).

**10.10**  For the conditions of Prob. 10.9, assume that the interval between supports in the field is 25 m instead of 50 m. Compute the length of the base line (Art. 4.20).

**10.11**  For the conditions given in Prob. 10.9 compute the normal tension. The modulus of elasticity, $E = 2.1 \times 10^6$ kg/cm$^2$ (Art. 4.21).

**10.12**  A base line measured by an EDM instrument from station 10 at elevation 240.6 m above mean sea level to station 11 at elevation 300.5 m was found to have a slope distance of 21,540.638 m corrected for atmospheric conditions and for EDM system constants. Determine the sea-level distance of the base line.

**10.13**  Eight positions of directions in a quadrilateral are to be observed with a theodolite having a micrometer drum range of $10'$. Compute the nominal initial circle readings for the eight positions.

**10.14**  Triangulation station $A$ was inaccessible, so eccentric station $E$ was set 5.825 m from $A$, as shown in Fig. 10.33. Measured directions reduced to an abstract of directions with zero along the line from $E$ to $A$ are as follows:

| Station | Angle | Line | Distance, m |
|---------|-------|------|-------------|
| A | 0°00′00.0″ | AE | 5.825 |
| B | 130°24′15.1″ | AB | 2530 |
| C | 175°30′50.9″ | AC | 3552 |
| D | 210°15′40.2″ | AD | 2642 |

Correct the directions measured at station $E$ to those which would have been observed had the theodolite been set at station $A$. Present the results as an abstract of directions with zero along the line from $A$ to $B$.

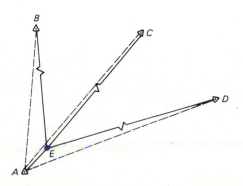

**Fig. 10.33**  $A$

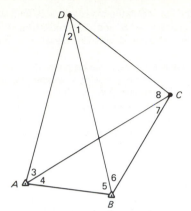

**Fig. 10.34**

**10.15** In measuring the angles at a triangulation station $O$, it was necessary to set the transit over another point $T$, at a distance of 13.25 ft from $O$. The angle measured at $T$ from $O$ to the first distant station $W$, was $95°10'30''$. The angles between the distant stations $W$, $X$, $Y$, and $Z$ were as follows: $WTX = 39°37'48''$; $XTY = 69°04'20''$; $YTZ = 83°16'08''$. The distances to the stations are found to be: $OW = 8,949$ ft; $OX = 14,334$ ft; $OY = 5,647$ ft; and $OZ = 7,326$ ft.

Correct the angles measured at station $T$, to those which would have been measured if the transit had been set at station $O$.

**10.16** For a given triangle $ABC$, the observed angles are $A = 78°30'28''$, $B = 54°17'30''$, and $C = 47°12'16''$. The distance from $A$ to $B$ has been found from preliminary calculations to be 45,300.50 m. Compute the spherical excess in the triangle.

**10.17** The interior angles in triangle $EFG$ are $E = 82°12'45''$, $F = 47°39'58''$, and $G = 50°07'27''$. The length of the line from $E$ to $F$ is 106,250.0 ft long. Compute the spherical excess in the triangle.

**10.18** The unadjusted angles in quadrilateral $ABCD$, with stations arranged and angles numbered as illustrated in Fig. 10.34, are as follows: $1 = 23°44'38''$, $2 = 38°44'06''$, $3 = 75°12'14''$, $4 = 26°25'51''$, $5 = 39°37'48''$, $6 = 69°04'21''$, $7 = 44°52'01''$, $8 = 42°19'09''$. Side $AB$ has a length of 13,000.30 ft and an azimuth from the north of $A_{AB} = 102°35'18''$. The coordinates of $A$ are $X_A = 50,000.00$ ft and $Y_A = 40,000.00$ ft. (a) Calculate preliminary lengths and azimuths for the sides of the quadrilateral. (b) Determine preliminary values for the coordinates of stations $B$, $C$ and $D$.

**10.19** Stations and angles in quadrilateral $ABCD$ are arranged and numbered as shown in Fig. 10.34. Unadjusted angles are as follows: $1 = 33°45'03.6''$, $2 = 63°46'27.5''$, $3 = 44°49'03.8''$, $4 = 34°09'54.9''$, $5 = 37°14'37.6''$, $6 = 44°02'59.4''$, $7 = 64°32'27.5''$, $8 = 37°39'28.2''$. Line $DC$ is 835.186 m long and has an azimuth from the north of $A_{DC} = 82°21'40.0''$. The coordinates of $D$ are $X_D = 10,001.129$ m, $Y_D = 9,999.848$ m. Perform the position computations necessary to provide preliminary estimates for $X$ and $Y$ coordinates for stations $A$ and $B$.

**10.20** The lengths of the sides of triangle $ABC$, as measured with an EDM instrument, are $AB = 1073.007$ m, $BC = 667.373$ m, and $CA = 1174.722$ m. The azimuth from the north of side $CB$ is $160°16'33''$ and the coordinates of station $B$ are $X_B = 11,054.091$ m, $Y_B = 9484.371$ m. Assuming that the given lengths are free from systematic errors, compute preliminary azimuths for sides $AB$ and $AC$ and preliminary coordinates for point $A$ to be used in subsequent adjustment of trilateration by the method of least squares. Assume $A$ is east of $B$.

**10.21** The lengths of sides for quadrilateral $ABCD$ of Prob. 10.19 as measured by an EDM instrument and corrected for all systematic errors are as follows: $AB = 1073.007$, $BC = 667.373$, $BD = 1174.104$, $AC = 1174.695$, and $AD = 723.906$, all distances being in metres. Compute preliminary values for the azimuths of the sides and estimates for the coordinates of points $A$ and $B$ all to be used in a subsequent trilateration adjustment by the method of least squares.

**10.22** If all distances are measured and the coordinates for two stations are known, determine the number of conditions required in a trilateration adjustment by the method of least squares for (a) the center-point quadrilateral with one additional measurement shown in Fig. 10.28a; (b) the center-point pentagon illustrated in Fig. 10.28b.

**10.23**   In Fig. 10.29, assume that stations 1, 2, 6, and 8 have known plane coordinates and that all distances have been measured. Determine the number of conditions (or the statistical degrees of freedom) that exist in a least squares adjustment of trilateration for this network.

## 10.29.  Field problems

### PROBLEM 1    CALIBRATION OF EDM EQUIPMENT

**Object**   To observe the distance using EDM equipment of a base line previously measured over level terrain by an invar or steel tape in order to determine the systematic errors in the EDM system.

**Procedure**

1. Measure the base line at least 10 times using the EDM instrument and the reflector (or combination of reflectors) which comprise the system to be calibrated.
2. Record the slope of the line of sight or elevations of the EDM instrument and reflector and all the necessary meteorological data.
3. Correct each slope distance for meteorological conditions (Art. 4.33) and for slope (Art. 4.37) and compute the average of the several reduced measurements.
4. The difference between the average taped measurement and the average distance measurement by EDM is the system constant $\Delta_s$ (Art. 4.34). If more than one reflector or assembly of reflectors is to be used, calibrate each system separately.

   If the terrain is suitable, the nonlinear, periodic, bounded error (Art. 4.35) can be determined. The procedure is as follows.

1. Using very precise methods, set off from one end of the base line, integral units of the basic measuring unit of the EDM system. For example, if the basic measuring unit is 10 m, then set hubs with tacks at one metre intervals from one end of the base line. Set these points using a standardized tape, with the tape fully supported on a level surface and correcting for all systematic errors.
2. Measure the distance using the EDM instrument from the other end of the base line to the reflector precisely centered over each of the ten points yielding ten distances measured by the EDM system. Correct each distance for meteorological conditions and the EDM system constant $\Delta_s$.
3. Using the known taped distances to each of these points and the distance as measured by EDM, determine the nonlinear periodic error following the procedure outlined in Art. 4.35.
4. Construct a graph of the difference between taped and EDM distances plotted versus the integral units of the basic measuring unit as shown in Fig. 4.22.

### Hints

1. Center the instrument and reflectors very carefully so as to reduce the effects of centering errors on the results.
2. Record meteorological data with great care.
3. Best results will be obtained if the base line is over level terrain. If the terrain is not level, the error for tilt of the reflector must be compensated, either instrumentally using a reflector mount which permits this, or analytically by the methods outlined in Ref. 7 at the end of Chap. 4.

### PROBLEM 2    MEASUREMENT OF DIRECTIONS IN A QUADRILATERAL

**Object**   To observe the directions in a quadrilateral using a theodolite. These directions can be reduced to angles that may then be used in conjunction with a measured base line to adjust the quadrilateral.

**Procedure**

1. Set four points from 500 to 1000 ft (150 to 300 m) apart and located such that all stations are intervisible. At the very least, set hubs with tacks, although it is preferable to set permanent monuments. In either case, the stations should be thoroughly referenced (Art. 8.14). Due regard should be paid to setting points in locations so as to yield strong geometry and optimum strength of figure (Art. 10.11).

2. Occupy each triangulation station with a direction theodolite having a nominal least count of 01″. If sufficient equipment is available, divide the class into four field parties and occupy each station simultaneously. Center the theodolite using the optical plummet and then attach a plumb bob which can be used as a sight by the parties at the other three stations. If the tripods are equipped with plumbing rods, the rod can be used as a target. If four theodolites are not available, targets must be set at the unoccupied stations. For short sights a plumb bob supported by a tripod is adequate. A range pole supported by a tripod is better for sights up to 1000 ft or 300 m. The target illustrated in Fig. 6.67 is excellent and permits forced centering (see the last paragraph of Art. 6.30).
3. Make a careful sketch of the relative locations of all sighted and occupied stations. Keep notes in the form shown in Fig. 10.20.
4. Each member of the party should observe at least one position of directions to the other three stations with a minimum of four positions to be measured at each station. Use the procedure outlined in Art. 6.41. Between each position advance the horizontal circle by the appropriate increment, depending on the total number of positions to be observed. For example, if there are two party members, each will measure two positions, to yield a total of four positions, and the increment will be about 45° (see Art. 10.18).
5. Reduce all directions to an abstract of directions with a zero azimuth along the initial pointing as described in Art. 10.18. The final abstract of directions at each station will consist of the average of the directions for each observed position.
6. When all directions have been observed, calculate angles from the directions. Check the sum of the angles in the quadrilateral and in each triangle. If sufficient care is exercised, the discrepancies should not exceed the specifications for third-order Class II triangulation (Table 10.2).
7. Using the functional relationship between directions and angles as given by Eqs. (10.8) and (10.9), (Art. 10.9) and error propagation equation (2.42), compute the cofactor matrix (Art. 2.14) for the angles [Eqs. (6.15)–(6.19), Art. 6.58]. Assume that the measured directions are uncorrelated and of equal precision.
8. Notes should include a sketch of the quadrilateral, references for the point occupied, a summary of the final abstract of directions, calculated angles for the station, and the propagated cofactor matrix for the angles.

**Hints and precautions**

1. Center the instrument and targets very carefully.
2. After initially leveling the theodolite, check the level bubble periodically. If the bubble goes off center while a position is being measured, finish the position and relevel the instrument. Do not relevel in the middle of a position.

## References

1. Allman, J. S., and Hoar, G. J., "Optimization of Geodetic Networks," *Survey Review*, January 1973.
2. Atia, K. A., "The Adjustment of Trilateration Networks Observed in Paris," *Survey Review*, July 1976.
3. Burke, K. F., "Why Compare Triangulation and Trilateration?" *Proceedings, Annual Meeting, ACSM*, Washington, D.C., March 1971, p. 244.
4. Chrzanowski, A., and Steeves, P., "Control Networks with Wall Monumentation, A Basis for Integrated Survey Systems in Urban Areas," *The Canadian Surveyor*, Vol. 31, No. 3, September 1977, p. 211.
5. Dracup, J. F., "Tests for Evaluating Trilateration Surveys," *Proceedings, Fall Convention, ACSM*, Seattle, Wash., 1976, p. 96.
6. Pinch, M. C., "An Evaluation of the Accuracy of an Autotape Survey," *The Canadian Surveyor*, March 1971.
7. *Proceedings of the International Symposium on Problems Related to the Redefinition of the North American Geodetic Networks*, University of New Brunswick, Fredericton, New Brunswick, Canada, *The Canadian Surveyor*, Vol. 28, No. 5, December 1974.
8. Merry, C. L., and Vanicek, P., "Horizontal Control and the Geoid in Canada," *The Canadian Surveyor*, Vol. 27, No. 1, March 1973, p. 23.

9. Robertson, K. D., "The Use of Atmospheric Models with Trilateration," *Survey Review*, October 1977.
10. Schwartz, W. M., "Control Survey Calculation on a Hand-Held Calculator," *The Canadian Surveyor*, Vol. 30, No. 3, September 1976, p. 181.
11. Smith, W. M., "The Assessment of Angular Measurements with the Lambert Instrument Tower," *The Canadian Surveyor*, Vol. 28, No. 3, September 1974, p. 217.
12. Vincenty, T., "Length Ratios and Scale Unknowns in Trilateration," *Surveying and Mapping*, September 1975.
13. Wolf, P. R., and Johnson, S. D., "Trilateration with Short Range EDM Equipment and Comparison with Triangulation," *Surveying and Mapping*, Vol. 34, No. 4, December 1974, pp. 337–346.

# CHAPTER 11
# Introduction to astronomy

**11.1 General** The surveyor should be familiar with the astronomical and trigonometric principles upon which the observations and computations of field astronomy are based. In this chapter certain fundamentals are given which are applicable to all astronomical observations. However, the discussions are intended to be applied only to surveys of moderate precision.

The science of astronomy offers the surveyor a means of determining the absolute location of any point or the absolute location and direction of any line on the surface of the earth. The absolute location of a point is given by its latitude and longitude, and the absolute direction of a line is defined by the angle which the line makes with the true meridian.

The azimuth of a line is established by angular observations on some celestial body, most commonly on the sun or on Polaris, the North Star or polestar. For the purpose of computing the azimuth from an astronomical observation, it is necessary that the latitude of the place be known. Also, for certain observations it is essential that the longitude be roughly determined. If the survey is through a territory for which there is a reliable map, the latitude and longitude may ordinarily be determined with sufficient precision by scaling from the map.

In *geodetic* surveying it is necessary to determine the latitude and longitude of certain points with great precision, the work involving observations on numerous stars and requiring instruments of high precision. The requirements of *plane* surveying, however, are met if the true or astronomic azimuth of the survey lines is established with a precision at least equal to that with which the angles between survey lines are measured. For plane surveying of ordinary precision, the use of the engineer's transit or theodolite and the methods described herein will yield sufficiently precise results.

**11.2. The celestial sphere** In making observations on the sun and stars, the surveyor is not interested in the distance of these celestial bodies from the earth but merely in their angular position. It is convenient to imagine their being attached to the inner surface of a hollow sphere of infinite radius of which the earth is the center. This imaginary globe is called the *celestial sphere*. It is also helpful to imagine the earth as being fixed, and to consider the celestial sphere as rotating from east to west, its axis being the prolongation of the axis of the earth. Thus, to the naked eye the polestar appears to remain stationary, but the sun (and similarly the stars near the equator) appears above the horizon in the general direction of east, follows a curved path (convex southward) across the heavens, and disappears below the horizon in the general direction of west.

The portion of the celestial sphere seen by the observer is the hemisphere above the plane of his own horizon. The reference plane passes through the center of the earth parallel with

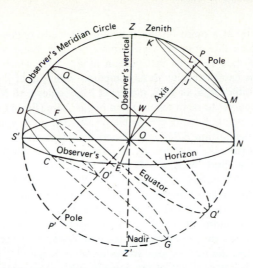

**Fig. 11.1** Celestial sphere.

the observer's horizon plane, but the radius of the earth is so small in relation to the distances to the stars that the error in vertical angle to a star is negligible. In the case of the sun the error produced by this assumption is much larger than for any of the stars, amounting under certain conditions to about 9 seconds of arc and requiring an appropriate correction to the observed vertical angle (see Art. 11.27). In any case, a refraction correction to observed vertical angles is necessary (Art. 11.28).

A vertical line at the location of the observer coincides with the plumb line and is normal to the observer's horizon plane. The point where this vertical line pierces the celestial sphere above the head of the observer is called the *zenith*, and the corresponding point in the opposite hemisphere, directly below the observer, is called the *nadir*.

The *celestial poles* are the points where the earth's axis prolonged pierces the celestial sphere.

The *celestial equator* is the great circle formed by the intersection of the earth's equatorial plane with the surface of the celestial sphere.

Figure 11.1 represents the celestial sphere, the point $O$ being the earth and $NES'W$ being the horizon of an observer, with letters standing for the points of the compass. Figure 11.2 may be taken as an enlarged view of the earth in the same position as that assumed in Fig. 11.1. $A$ is an observer in the northern hemisphere, the line $N_aS_a$ being in the horizon plane at that point. Evidently, this observer views everything above the horizon plane or that portion of the celestial sphere (Fig. 11.1) which is shown by full lines. $B$ is an observer in the

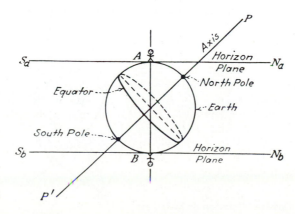

**Fig. 11.2** Observer's horizon.

southern hemisphere, at a point on the earth diametrically opposite $A$; the portion of the celestial sphere which this observer views above horizon plane $N_b S_b$ will be the opposite hemisphere to that seen by $A$, or that portion of Fig. 11.1 which is shown by dash lines. Since the size of the earth is negligible as compared with that of the celestial sphere, it may be considered that either $N_a S_a$ or $N_b S_b$ in Fig. 11.2 coincides with $NS'$ in Fig. 11.1

Assuming the observer to be in the northern hemisphere (Fig. 11.1), $Z$ is the zenith; $P$ and $P'$ are the celestial poles, $P$ being the visible or *elevated* pole; and $EQWQ'$ is the celestial equator, of which the portion $EQW$ is visible to the observer.

Since we are, for the sake of simplicity, assuming that the celestial sphere is rotating and the earth remains stationary, $N$, $E$, $S'$, $W$, and $Z$ are regarded as fixed points with respect to any given station on the surface of the earth. If $S'N$ is a line in the plane of the horizon and also lies in the plane passing through the observer's station, then a vertical plane of which this line is an element cuts the celestial sphere in the great circle $S'ZPNZ'P'$, which is called the meridian circle or, more often, simply the meridian. At a given instant the meridian for one station does not occupy the same position in the celestial sphere as does the meridian for another station, unless the two stations are at the same longitude.

Any star that is below or south of the equator will follow some path, such as $CDFG$. It will become visible at $C$, will pass over the meridian at $D$, and will disappear from view at $F$. It will be above the horizon for a shorter period of time than it will be below, or the angle whose arc is $CDF$ (angle $CO'F$) is less than $180°$. From the figure it is evident that if any star is sufficiently far below the equator, it will never appear above the observer's horizon.

Similarly, any star that is above or north of the equator will be above the horizon for a greater length of time than it is below. If it is far enough above the equator, it will be continuously visible to an observer in a northern latitude and will, during the course of a single revolution of the celestial sphere, follow some path as $JKLM$. When it is at the highest point of its apparent path, at $K$, it is said to be at *upper culmination;* when it is at the lowest point, at $M$, it is said to be at *lower culmination.*

**11.3. Observer's location on earth**   The location, or *position*, of any point on the surface of a sphere may be fixed by angular measurement from two planes of reference at right angles to each other passing through the center of the sphere; these measurements are called the *spherical coordinates* of the point. The spherical coordinates of any station on the surface of the earth are designated as the *latitude* and *longitude* of the station. Figure 11.3 represents the earth, $PP'$ being the axis and $QUVQ'$ being the equator. Let $S$ be the station of an observer. Then $PSUP'$ is a *meridian circle* through the station. Also, $RSR'$ is a *parallel* passing through the station, the plane $RSR'$ being parallel to that of the equator.

The latitude of a place may, for all practical purposes, be defined as the angular distance of the place above or below the equator. When the station is above the equator, the latitude is north and its sign is positive; when below the equator, the latitude is south and its sign is

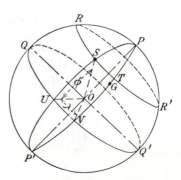

**Fig. 11.3** Observer's location on the earth.

negative. Hence, in the figure the latitude of $S$ is given by the angle $\phi$ or by the corresponding angular distance, measured along any meridian circle, between the equator and the parallel passing through $S$, such as $US$, $VT$, $QR$, etc.; and the latitude is north or positive. The latitude of a place is stated in degrees. Thus, the latitude of the equator is $0°$ and that of the North Pole is $+90°$, or $90°$N.

The longitude of a place is defined as the angular distance measured along the arc of the equator between a reference meridian and the meridian circle passing through the station. The reference meridian is called the *primary meridian*. The primary meridian most generally used is that of Greenwich, England. (Prior to 1912, a Washington, D.C., meridian was extensively used in the United States.) In the figure, if the point $G$ represents Greenwich, $PGP'$ is the primary meridian, and the longitude of $S$ is given by the angle $\lambda$ or by the angular distance $VU$. Longitudes are expressed either in degrees of arc or in hours of time ($15° = 1$ hr.) and are measured either east or west of the Greenwich meridian.

In general, the discussions herein are intended to apply in the northern hemisphere and for longitudes west of Greenwich.

**11.4. Right-ascension equator system**   In Fig. 11.4 is shown the celestial sphere in a position similar to that of the earth in Fig. 11.3, $S$ being a celestial body whose position is to be fixed by spherical coordinates. Comparable with the meridian circles or meridians of longitude of the earth are the *hour circles* of the celestial sphere, all of which converge at the celestial poles. The arc $PSU$ is a portion of the hour circle passing through $S$. Comparable with the parallels of latitude of the earth are the *parallels of declination* of the celestial sphere. $RSR'$ is the parallel of declination passing through $S$. And comparable with the prime meridian through Greenwich is the *equinoctial colure* of the celestial sphere, which passes through the *vernal equinox*, an imaginary point among the stars where the sun apparently crosses the equator on about March 21 of each year. In the figure, $V$ represents the vernal equinox and $PTV$ is the equinoctial colure.

The *right ascension* of the sun or any star is the angular distance measured along the celestial equator between the vernal equinox and the hour circle through the body. It is comparable with the longitude of a station on the earth. Right ascensions are measured *eastward* from the vernal equinox and may be expressed either in degrees of arc ($0°$ to $360°$) or in hours of time ($0^h$ to $24^h$). Thus, in the figure, the right ascension of $S$ is given by the angle $\alpha$ in the plane of the equator or by the arc $VU$.

The *declination* of any celestial body is the angular distance of the body above or below the celestial equator. It is comparable with the latitude of a station on the earth. If the body is above the equator, its declination is said to be north and is considered as positive; if it is below the equator, its declination is said to be south and is considered as negative. Declinations are expressed in degrees and cannot exceed $90°$ in magnitude. In the figure, the

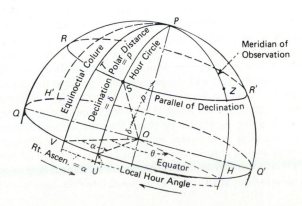

**Fig. 11.4** Equator systems of spherical coordinates.

declination of $S$ is given by the angle $\delta$ or by the arc of any hour circle between the equator and the parallel of declination $RSR'$, such as $US$, $VT$, $QR$, etc.

The *polar distance* or *codeclination* of any celestial body is $p = 90° - \delta$ with due regard to the sign of the declination. In the figure, it is given by the angle $p$ or by the arc $PS$. Polar distances are always positive. For computations referred to the North Pole, when the declination is north, the polar distance is the complement of the declination; but when the declination is south, as in the case of the sun during the winter months, the polar distance is greater than 90°. In defining the position of a star near either pole, often the polar distance is given instead of the declination.

For present purposes it may be considered that the vernal equinox is a fixed point on the celestial equator, just as Greenwich is a fixed point on the earth. But while stations on the earth maintain practically an unvarying location with respect to the equator and the meridian of Greenwich, the coordinates of celestial bodies with respect to the celestial equator and the equinoctial colure change more or less with the passage of time. The fixed stars, or those outside the solar system, alter their positions in the celestial sphere only slightly from month to month and from year to year, the annual change being less than a minute of arc in either right ascension or declination. These variations are due to (1) *precession* or the slow change in the direction of the earth's axis due to attraction of the sun, moon, and planets; and (2) *nutation* or small inequalities in the motion of precession, similar to the oscillation of a spinning top.

As the earth actually travels around the sun but not around the stars, the sun appears to move more slowly than do the stars, making in one year 365 apparent revolutions (approximately) while the stars make 366 apparent revolutions (approximately); thus, the sun apparently makes a complete circuit of the heavens once each year, its right ascension changing from $0^h$ (or 0°) on March 21 to $12^h$ (or 180°) on September 22 and continuing to $24^h$ (or 360°) on the following March 21, when a new cycle begins. Further, as the axis of rotation of the earth is not normal to the plane of the earth's orbit, the path traced by the sun among the stars on the celestial sphere, called the *ecliptic*, is a continuous curved line; each year the sun crosses the equator northward about March 21, reaches a maximum positive declination ($\sim$N$23\frac{1}{2}°$) about June 21, crosses the equator southward about September 22, and reaches a maximum negative declination ($\sim$S$23\frac{1}{2}°$) about December 21.

**11.5. Hour-angle equator system**   In many of the problems of field astronomy it is necessary not only that a star's position in the celestial sphere be known but also that its position with respect to the meridian through a given station on the surface of the earth be determined. In Fig. 11.4, let the arc $PZH$ represent the meridian of the observer stationed on the earth and let $S$ be some celestial body whose position is desired with respect to the meridian $PZH$ and the equator $Q'HUVQ$. The spherical coordinates of the star are given by (1) the angular distance of the star above or below the equator, given in Fig. 11.4 by the arc $US$ defined in Art. 11.4 as the declination $\delta$; and (2) the angular distance $HU$ measured along the equator between the meridian of the observer and the hour angle through the star. This angular measurement is called an *hour angle*. In general, hour angles are measured in a positive sense from east to west (clockwise as viewed from the North Pole) from the reference meridian to the meridian of the celestial body. In Fig. 11.4 the reference meridian is that of the observer and the positive hour angle is the angular distance $HU$. Hour angles are expressed either in hours of time or degrees of arc. When no qualification is stated, it is understood that the hour angle is measured from the upper branch of the meridian, that is, the branch above the station or above the observer's head (meridian $PZH$ in Fig. 11.4).

There are three hour angles which are important in astronomic calculations. To visualize these hour angles, observe the spherical system of Fig. 11.4 from the North Pole (i.e., along line $PO$), as shown by the view from above the North Pole in Fig. 11.5. The local hour angle (L.H.A.) is measured clockwise (westward) from the meridian of observation to the meridian

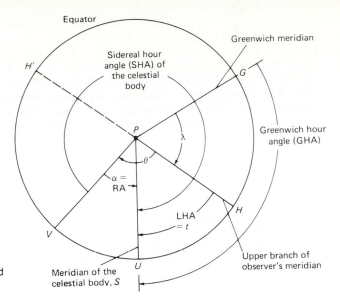

**Fig. 11.5** Local, Greenwich, and sidereal hour angles.

of the celestial body and is angle $HPU = t$ in Fig. 11.5. When the celestial body is west of the observer's meridian (as in Fig. 11.5), the L.H.A. is less than 180° and equals angle $t$. When the celestial body is east of the observer's meridian, the L.H.A. exceeds 180° and

$$t = \text{L.H.A.} - 360° \qquad (11.1a)$$

Thus, $t$ represents a time interval before or after upper culmination of the sun or moon. The *sidereal hour angle* (S.H.A.) is measured clockwise (westward) from the meridian of the vernal equinox, $V$ to the meridian of the celestial body and is the angular distance $VGU$ in Fig. 11.5. It is obvious from the figure that the right ascension,

$$\text{RA} = \alpha = 360° - \text{S.H.A.} \qquad (11.1b)$$

The *Greenwich hour angle* (G.H.A.) is measured clockwise (westward) from the Greenwich meridian to the meridian of the celestial body. In Fig. 11.5 the G.H.A. of the celestial body is the angular distance $GHU$. Also from Fig. 11.5 it can be seen that

$$\text{L.H.A.} = \text{G.H.A.} - \text{observer's west longitude} \qquad (11.1c)$$
$$\text{L.H.A.} = \text{G.H.A.} + \text{observer's east longitude} \qquad (11.1d)$$

In connection with the definition of civil time, hour angles are reckoned from the lower branch of the meridian. In Fig. 11.5, if the hour angle were reckoned from the lower branch, it would be defined by the angular distance $H'GHU$, and would be $12^h$ more than that given by the arc $HU$, which is the hour angle reckoned from the upper branch.

Sometimes the hour angles of celestial bodies east of the meridian are reckoned eastward from the upper branch of the meridian, rather than westward. When an hour angle is expressed in this way, it is preceded by a minus sign. Thus, if the hour angle of $S$ (Fig. 11.5) were reckoned eastward, it would be given by the angular distance $HGH'U$.

**11.6. Equator systems compared**  The system of coordinates described in Art. 11.5 is similar to that described in Art. 11.4 with this difference: that in the hour-angle system the angular distance along the equator is measured (westward) from a *fixed meridian*, while in the right-ascension system the angular distance along the equator is measured (eastward) from the *vernal equinox*, which is a point on the celestial equator that rotates with the celestial sphere. Thus, while right ascensions of fixed stars have annual variations of but a few seconds, hour angles of the stars change rapidly since the celestial sphere apparently rotates (24$^h$ or 360° for each 23$^h$ 56$^m$ of our civil time), and hour angles of the sun change approximately 24$^h$ or 360° for each 24$^h$ of our civil time.

The two systems are called *equator systems of coordinates,* since in each case the primary plane of reference is the celestial equator. The position of a celestial body above or below the equator is given by the declination $\delta$, which is the same in one system as in the other.

Let $\theta$ be the hour angle of the vernal equinox represented in Fig. 11.5 by the angular distance $HUV$ measured along the equator. At any instant of time, if the hour angle of the vernal equinox with respect to a given meridian is known, if the right ascension $\alpha$ of a heavenly body $S$ is known, the hour angle $t$ of the body can be computed, since from the figure

$$t = \theta - \alpha \qquad \text{or} \qquad \theta = t + \alpha$$

This equation is, therefore, an expression by means of which the coordinates of one system may be transformed to those of the other.

**11.7. Astronomical tables used by the surveyor**    By means of astronomical observations and calculations, the positions of many of the celestial bodies are predicted, and values of their right ascensions and declinations for various dates are available in various publications. The position of a celestial body at any time can be obtained by interpolation.

The publication most widely used by *astronomers* in the United States is *The American Ephemeris and Nautical Almanac* (about 500 pages); herein it is called the *American Ephemeris.* It is published one or two years in advance for each year by the Nautical Almanac Office, U.S. Naval Observatory. Formerly, its tables were based on "universal" (civil) time (Art. 11.16), but the tables are now based on "ephemeris" time; it is not convenient for use in surveying.

Most of the astronomical data used by surveyors are presented in *The Nautical Almanac* (about 300 pages), which is also published annually in advance by the Nautical Almanac Office, U.S. Naval Observatory, primarily for use in navigation.

In condensed form is the *Ephemeris of the Sun, Polaris, and Other Selected Stars* (about 30 pages), herein called the *Ephemeris of the Sun and Polaris.* It is published annually in advance by the U.S. Bureau of Land Management. This ephemeris lists for each day of the current year the position of the sun, of Polaris, and selected stars by means of which the surveyor can compute from field observations the latitude and longitude of the point of observation, the time of observation, or the azimuth of a reference line. It is in a form useful for land surveying.

All the foregoing publications are sold by the Superintendent of Documents, Government Printing Office, Washington, D.C. 20402.

Useful condensed tables of data regarding the sun, Polaris and selected stars are furnished for a nominal fee by various manufacturers of surveying instruments.

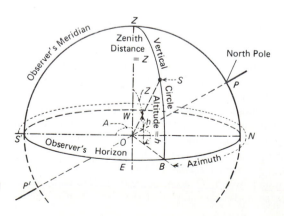

**Fig. 11.6** Horizon system of spherical coordinates.

**11.8. Horizon system of spherical coordinates**   In the ordinary operations of surveying, the angles are measured in horizontal and vertical planes; to use other planes would be inconvenient. Likewise, in astronomical field work the angular location of a celestial body at a given instant is determined by measuring its vertical angle (referred to the horizon plane) and its horizontal angle (referred to a given line on the ground).

Figure 11.6 represents a portion of the celestial sphere in which $O$ represents both the earth and the location of the observer, $NES'W$ the observer's horizon, and $S'ZN$ the meridian plane passing through the observer's location. The point $Z$ on the celestial sphere directly above the observer is called the *zenith*. The point $S$ represents a celestial body, and $BSZ$ is part of a great circle, called a *vertical circle*, through the body and the zenith. In this *horizon system* of spherical coordinates, the angular location of a celestial body is defined by its azimuth and altitude.

The *azimuth* of a celestial body is the angular distance measured along the horizon in a clockwise direction from the meridian to the vertical circle through the body. Azimuths may be reckoned from either the south point or the north point of the meridian; in astronomical work azimuths are usually reckoned from south through 360°, except that for circumpolar stars they are often reckoned from north. In trigonometric computations the azimuths of stars west of north or east of south are often expressed as counter-clockwise angles from the meridian and are considered as negative values. In Fig. 11.6 the azimuth of $S$ reckoned in the customary manner is given by the angle $A$ or by the angular distance $S'NB$, an arc of the horizon. If the azimuth of $S$ were reckoned from north, it would be given by the angle $(A-180°)$ or by the angular distance $NB$. The negative azimuth reckoned from south is given by the arc $S'EB$.

The *altitude* of a celestial body is the angular distance measured along a vertical circle, from the horizon to the body; it corresponds to the vertical angle of ordinary surveying. It is expressed in degrees of arc. The altitude of $S$ (Fig. 11.6) is given by the vertical angle $h$ or by the angular distance $BS$, the arc of a vertical circle passing through the zenith. Except in rare instances, celestial objects are observed when above the true horizon, when the sign of the altitude is positive. It is seen that positive altitudes may vary between 0° and 90°.

The complement of the altitude is called the *zenith distance* or *coaltitude*. It is the angular distance from the zenith to the celestial body measured along a vertical circle. In the figure the zenith distance is given by the angle $z$ or by the angular distance $ZS$. Thus $z = 90° - h$. Zenith distances are always positive.

Since the celestial sphere is apparently rotating about its axis, while the meridian, horizon, and zenith are imagined as remaining fixed in position, it is clear that in general both the azimuth and the altitude of a star are changing continuously.

**11.9. Relations among latitude, altitude, and declination**   Figure 11.7 represents a section of the earth through the poles and the station of an observer. Since the latitude of a place is its angular distance from the equator measured along a meridian of longitude, the latitude of the observer is given by the angle $\phi$ between the equator and a vertical line through the observer's station, the angle being measured in the plane of the meridian. Also, from similar triangles, the latitude is given by the angle $\phi$ between the axis and the horizon, likewise measured in the plane of the meridian; this angle is the altitude of the elevated pole.

Similarly Fig. 11.8 represents a section of the celestial sphere through the celestial poles and the observer's zenith. For reasons just explained,

$$\angle QOZ = \angle NOP = \phi = \text{latitude of observer's place}$$

Hence, the latitude of a place is given by the angular distance $NP$, which is the altitude of the pole, or by the angular distance $QZ$, which is the zenith distance or coaltitude of the equator. The angular distance from the pole to the zenith is $90° - \phi = c$, which is the colatitude of the place.

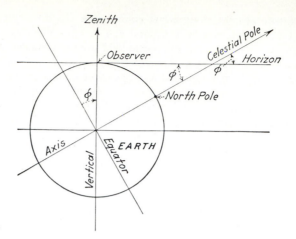

**Fig. 11.7** Latitude of observer.

In a northern latitude if any heavenly body $S$ having declination $\delta$, is on the meridian and south of the zenith, then from Fig. 11.8,

$$\phi = (90° - h) + \delta$$

Similarly, for any star north of the zenith,

$$\phi = h \pm (90° - \delta) = h \pm p$$

in which the sign preceding $p$, the polar distance, is positive or negative according to whether the star is below or above the pole.

**11.10. Horizon and hour-angle equator systems combined**  The relation between the coordinates of the horizon system and those of the hour-angle equator system described in Art. 11.5 is shown, for a star $S$ not on the meridian, by Fig. 11.9. The meridians of the two systems coincide. The place of observation is assumed to be north of the equator at a latitude $\phi$, as given either by the angle between the equator and the zenith or by the angular distance $NP$ between the horizon plane and the celestial axis. The star is at a position east of the meridian and above the celestial equator.

In the horizon system the coordinates of $S$ are $A_N$, the azimuth measured from the north point of the horizon, and $h$, the altitude. In the equator system the coordinates are $t$, the hour angle (also called the local hour angle, L.H.A.) measured westward from the upper branch of the meridian ($360° - t$ is shown in the figure, see also Fig. 11.5), and $\delta$, the declination. The colatitude ($90° - \phi = c$), the zenith distance ($90° - h = z$), and the polar distance ($90° - \delta = p$) define a spherical triangle the vertices of which are the pole $P$, the

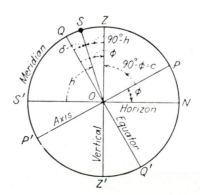

**Fig. 11.8** Relation among latitude, altitude, and declination.

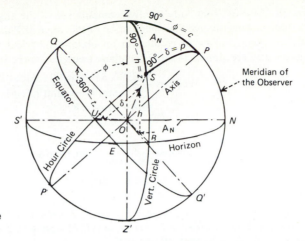

**Fig. 11.9** Horizon and hour-angle equator systems combined.

zenith $Z$, and the celestial body $S$. This triangle is called the *PZS triangle* or the *astronomical triangle*. Most of the problems of field astronomy involve transforming from one system of spherical coordinates to the other and solving the *PZS* triangle for unknown coordinates, having certain coordinates in one or both systems known or observed.

In the figure, the celestial body is shown above the horizon and above the equator. If the body is below the horizon or below the equator, the sides of the *PZS* triangle are defined in a manner similar to that just described but account is taken of the algebraic sign of the altitude and the declination.

In the figure, the celestial body is shown as east of the observer's meridian; the angle $Z$ of the spherical *PZS* triangle is, therefore, its azimuth from the north. Also, the angle $P$ of the *PZS* triangle is equal to $360° - t$. When the body is west of the meridian, $Z = 360° -$ azimuth from north and $P = t$.

**11.11. Solution of the *PZS* triangle**  In surveying, the astronomical triangle is solved in connection with determinations of azimuth. Observations are made on the sun or on some star that can be readily identified. The altitude of the celestial body is measured, its declination at the instant of observation is determined from published tables, and the latitude of the place of observation is known or is determined by separate observation. In Fig. 11.10, *PZS* is the astronomic triangle having the same orientation as the spherical triangle described in Fig. 11.9, where *OP* is the polar axis and *OZ* is the observer's zenith. The three known sides of this triangle are $90 - \phi$, the colatitude; $90 - h$, the coaltitude or zenith distance; and $90 - \delta$, the codeclination or polar distance. Determination of the

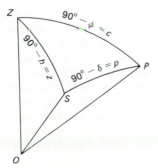

**Fig. 11.10** The astronomic triangle.

azimuth of the celestial body involves computation of the angle at $Z$; and determinations of longitude or time involve the computation of the angle at $P$ as a measure of the hour angle.

By applying the law of cosines for an oblique spherical triangle (the rules of spherical trigonometry are similar to those for plane triangles), there results

$$\cos Z = \frac{\cos(90-\delta) - \cos(90-\phi)\cos(90-h)}{\sin(90-\phi)\sin(90-h)}$$

$$= \frac{\sin\delta - \sin\phi\sin h}{\cos\phi\cos h} \qquad (11.2a)$$

or

$$\cos Z = \frac{\sin\delta}{\cos h\cos\phi} - \tan h\tan\phi \qquad (11.2b)$$

The azimuth from the *north* of the celestial body is obtained from $Z$, as calculated using either Eq. (11.2a) or (11.2b), as follows:

1. When the body is *east* of the observer's meridian, $Z$ is *clockwise from north* and the azimuth from the north is equal to $Z$. If $\cos Z$ in Eq. (11.2a) or (11.2b) is positive, $Z$ will fall between 0° and 90°; when negative, $Z$ will be between 90° and 180°.
2. When the body is *west* of the observer's meridian, the calculated value of $Z$ is *counterclockwise from north* and the azimuth is equal to 360° minus $Z$. Of course, the azimuth is assumed to be always clockwise, and in this case from north. If $\cos Z$ in Eq. (11.2a) or (11.2b) is positive, then $Z$ falls between 0° and 90°, and if negative, then $Z$ falls between 90° and 180°.

When azimuths are reckoned from the *south*, Eq. (11.2b) takes the following form:

$$\cos Z_s = \tan h\tan\phi - \frac{\sin\delta}{\cos h\cos\phi} \qquad (11.3)$$

and the relationships between the azimuth and $Z_s$ are opposite to those above for the azimuth reckoned from north. For example, when the body is west of the observer's meridian, $Z_s$ is *clockwise from south* and is equal to the azimuth (also clockwise from south) with $0° \leqslant Z_s \leqslant 90°$ for positive $\cos Z_s$, and $90° \leqslant Z_s \leqslant 180°$ for negative $\cos Z_s$. When the body is *east* of the observer's meridian, $Z_s$ is counterclockwise from south and the azimuth (which is clockwise from south) is equal to $(360° - Z_s)$. For positive values of $\cos Z_s$, $Z_s$ falls between 0° and 90°, and for negative $\cos Z_s$, $Z_s$ is between 90° and 180°.

Applying the law of cosines for an oblique spherical triangle to the astronomic triangle shown in Fig. 11.10 and solving for the angle $P = t$ yields

$$\cos t = \frac{\sin h - \sin\phi\sin\delta}{\cos\phi\cos\delta} = \frac{\sin h}{\cos\phi\cos\delta} - \tan\phi\tan\delta \qquad (11.4)$$

which is a general expression for determining the hour angle $t$ for any celestial body when the three sides of the astronomical triangle are known. The angle $t$ is defined in Art. 11.5 and is illustrated in Figs. 11.4 and 11.5. When measured westward from the meridian of the observer (clockwise when viewed from the North Pole), $t$ is positive; when measured eastward, $t$ is negative. Equation (11.4) is used to determine time by altitude of the sun (Art. 11.35) and by the altitude of a star (Art. 11.43). When used to determine time by altitude of the sun, angle $t$ by Eq. (11.4) is the time before local noon for a morning observation and the time after local noon for an afternoon observation (see Example 11.15, Art. 11.35).

When the unknown angle in Eqs. (11.1) to (11.4) is either small or near 180°, a relatively small error in the computed value of the cosine will produce a relatively large error in the angle itself, since the magnitude of the cosine is changing slowly. For this reason, *as far as errors in computations are involved*, the foregoing equations are not suitable for precisely computing azimuth and hour angle when the observed celestial body is near the meridian. On the other hand, when the unknown azimuth or hour angle is near 90° or 270°, its cosine is changing rapidly and thus Eqs. (11.1) to (11.4) are quite adequate for precise computation of azimuth or hour angle.

By applying a series of substitutions that can be found in any textbook on spherical trigonometry, Eq. (11.1) can be changed to the form

$$\tan^2\tfrac{1}{2}Z = \frac{\sin(s-h)\sin(s-\phi)}{\cos s \cos(s-p)} \tag{11.5}$$

and Eq. (11.4) may be changed to the form

$$\tan^2\tfrac{1}{2}t = \frac{\cos s \sin(s-h)}{\cos(s-p)\sin(s-\phi)} \tag{11.6}$$

In these two equations $p = 90° - \delta = $ polar distance, $s = \tfrac{1}{2}(h+\phi+p)$, and the remaining letters have the same significance as in Eqs. (11.1) to (11.4). In some cases $(s-p)$ will be negative, but the result will not be affected, since the cosine of a negative angle has the same value and the same sign as the cosine of a positive angle of equal size.

For a given angular value the tangent changes more rapidly than the cosine. Thus, for a given error of computation of the trigonometric function, Eqs. (11.5) and (11.6) will generally render a closer determination of azimuth and hour angle than will Eqs. (11.2) and (11.4). For angles near 90° and 270°, the difference between the rate of change of the tangent and of the cosine is not large and results from the two sets of equations are comparable. However, when the object is near the meridian (azimuth near 0° and 180°), for given computational errors Eqs. (11.5) and (11.6) will provide better determinations of angles than it is possible to obtain by using Eqs. (11.2) and (11.4).

**Azimuths from south**   When azimuths are reckoned from south, Eq. (11.5) takes the following form, $Z_s$ being the angle measured either clockwise or counterclockwise from south:

$$\cot^2\tfrac{1}{2}Z_s = \frac{\sin(s-h)\sin(s-\phi)}{\cos s \cos(s-p)} \tag{11.7}$$

When $\cot\tfrac{1}{2}Z_s$ has been determined, the computations for hour angle are somewhat reduced if Eq. (11.6) is modified as follows:

$$\tan\tfrac{1}{2}t = \frac{\sin(s-h)}{\cot\tfrac{1}{2}Z_s \cos(s-p)} \tag{11.8}$$

**11.12. Azimuth and hour angle at elongation**   The most favorable position for determining azimuth by observation on any star that crosses the upper branch of the meridian north of the zenith occurs when it is farthest east or farthest west of the pole, when the star appears to be traveling vertically for some time. In this position it is said to be at eastern or western *elongation* according to whether it is east or west of the meridian. At the instant of elongation, since the star appears to be traveling vertically, its apparent path in the celestial sphere is tangent to the vertical circle through the observer's zenith, as illustrated at $S$ in Fig. 11.11. Therefore, the angle $S$ between the plane of the hour circle and the plane of the vertical circle is a right angle. For azimuth determinations of this sort, the latitude of the place of observation is known, and either the declination or the polar distance of the star for the given date is obtained from published tables. At the instant of elongation, there are then

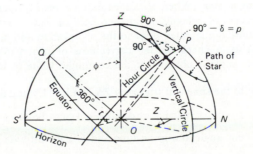

**Fig. 11.11** Star at elongation.

known in the astronomical triangle the side $ZP = 90° - \phi$, the side $PS = 90° - \delta$, and the angle $S = 90°$.

In right spherical triangle $PZS$ in Fig. 11.11, by the law of sines for a spherical triangle,

$$\sin Z = \frac{\sin(90° - \delta)}{\sin(90° - \phi)}$$

This becomes

$$\sin Z = \frac{\sin p}{\cos \phi} \tag{11.9}$$

which is the general expression employed for determining the azimuth of a circumpolar star when at elongation, $Z$ being the azimuth from north when the star is at eastern elongation and $Z = 360° -$ azimuth from north when the star is at western elongation.

By applying the law of cosines to the right spherical triangle $PZS$ (Fig. 11.11), where $S = 90°$ and $t =$ the angle at $P$, the following equation can be derived:

$$\cos t = \tan \phi \tan p \tag{11.10}$$

which is an expression for finding the hour angle of a star at the instant of elongation, the hour angle $t$ being reckoned east or west of the upper branch of the meridian, depending upon the position of the star. The equation is useful in determining the time at which elongation will occur on any given data.

**11.13. Azimuth of a circumpolar star at any position**   In spherical triangle $PZS$ as shown in Fig. 11.9, by the law of sines

$$\frac{\sin(360° - t)}{\sin(90° - h)} = \frac{\sin Z}{\sin(90° - \delta)}$$

or

$$\sin Z = - \frac{\sin t \cos \delta}{\cos h} \tag{11.11}$$

Also, in the same spherical triangle $PZS$, by the law of cosines,

$$\cos h \cos Z = \sin \delta \cos \phi - \cos \delta \sin \phi \cos t \tag{11.12}$$

Division of Eq. (11.11) by Eq. (11.12) yields

$$\tan Z = \frac{\sin t}{\tan \delta \cos \phi - \sin \phi \cos t} \tag{11.13}$$

Equation (11.12) can be used to find the azimuth of the star, $Z$, when the altitude $h$ (corrected for refraction, Art. 11.28) is measured, the declination $\delta$ is obtained from an ephemeris, and the hour angle $t$ (as defined in Art. 11.5) is determined from the observed time of measurement. When the latitude of the place of observation $\phi$ is known or determined from observations (Arts. 11.31 and 11.42), Eq. (11.13) is used to calculate the angle $Z$ from which the azimuth of the star can be determined. In this latter case, the hour angle $t$ can be calculated with Eq. (11.4) or (11.8).

**11.14. Altitude of a star**   When a star cannot be readily identified through the transit telescope, the process of bringing it into the field of view is considerably expedited if its approximate altitude is computed prior to the observation and laid off on the vertical circle of the transit. Also, a check on the correctness of observations and computations for azimuth and hour angle is obtained if the computed value of the altitude agrees with the observed value.

Again referring to the spherical triangle $P\overset{\star}{Z}S$ in Fig. 11.9 and applying the law of cosines,

$$\sin h = \sin \phi \sin \delta + \cos \phi \cos \delta \cos t \tag{11.14}$$

in which $t$ is the hour angle at a given time and $h$ is the altitude at the same instant.

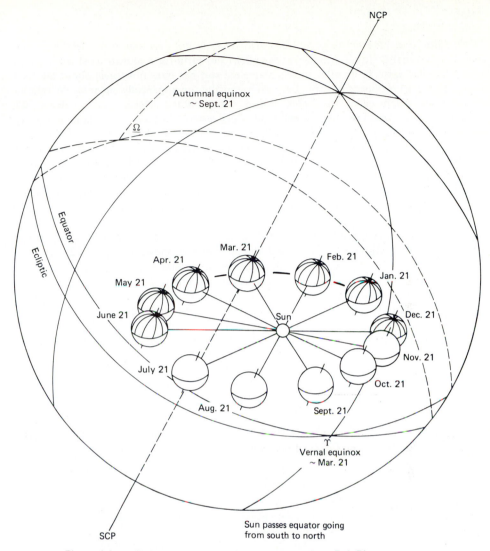

**Fig. 11.12** Plane of the ecliptic and celestial sphere. (*Adapted from Ref. 7.*)

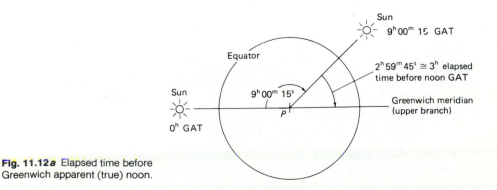

**Fig. 11.12a** Elapsed time before
Greenwich apparent (true) noon.

# TIME

**11.15. Solar and sidereal day**   As the earth rotates about its axis in its travel through space, all celestial bodies apparently rotate about the earth (or about its axis) from east to west. Since the earth in its orbit travels about the sun but does not travel about the fixed stars, which are far outside its orbit, once each year the sun apparently encircles the celestial sphere along a path called the *ecliptic*, which twice cuts the celestial equator during this interval (Fig. 11.12). The point among the stars where the sun in its apparent travel northward cuts the celestial equator on about March 21 of each year is called the *vernal equinox*, a point of reference whose position on the celestial sphere is unchanging. There is no star at that point, but it is helpful to imagine that the vernal equinox is an invisible celestial body rigidly fastened in its position on the celestial sphere, while each of the so-called "fixed" stars slowly moves along a path of extremely small periphery on the surface of the sphere, and the sun travels rapidly along the ecliptic in a direction opposite to that of the rotation of the celestial sphere.

The vernal equinox is referred to in ephemerides as "The First Point of Aries" or simply "Aries"; it is often represented by the zodiacal symbol ♈ as shown in Fig. 11.12.

Because the sun is apparently traveling from west to east among the stars, while the rotation of the celestial sphere about the earth is apparently from east to west, the angular velocity of the sun about the axis of the celestial sphere is less than that of the fixed stars or of the vernal equinox, just as the angular velocity of a passenger walking toward the rear of a train on a circular track is less than that of the train. At a given meridian the hour angle of the sun and that of the vernal equinox will agree at some instant about March 21, but thereafter it will be less for the sun than for the vernal equinox. Six months later, about September 21, when the sun has covered one-half of its annual journey, the hour angle of the sun will be 180° or $12^h$ less than that of the vernal equinox; 1 year later the hour angle of the sun will be 360° or $24^h$ less than that of the vernal equinox; hence the hour angles will again agree.

In the course of a tropical year as measured by the time taken by the sun apparently to make a complete circuit of the ecliptic, there actually occur 366.2422 revolutions of the earth, or apparently a like number of revolutions of the vernal equinox about the earth. For reasons just explained, the sun during this interval will have traveled through a total hour angle of 360° or $24^h$ less than that traversed by the vernal equinox; hence during a tropical year the sun apparently revolves about the earth 365.2422 times.

The interval of time occupied by one apparent revolution of the sun about the earth is called a *solar day*, the unit with which we are all familiar. The interval of time occupied by one apparent revolution of the vernal equinox is called a *sidereal day*, a unit much used by astronomers. Since 366.2422 sidereal days occupy the same period of time as 365.2422 solar days, the sidereal day is a shorter interval of time than the solar day.

When any celestial body, real or imaginary, apparently crosses the upper branch of a meridian, it is said to be at *upper transit* or *upper culmination*; when any celestial body crosses the lower branch of the meridian, it is said to be at *lower transit* or *lower culmination*.

The beginning of a sidereal day at a given place occurs at the instant the vernal equinox is at upper transit.

The solar day is considered as beginning at the instant of lower transit of the sun (midnight), as does the civil day. (Prior to 1925, for astronomical purposes the solar day was considered as beginning at noon.)

Both sidereal and solar days are divided into 24 h each of 60-min duration. For surveying purposes, the hours are reckoned consecutively from 0 to 24.

**11.16. Civil (mean solar) time**   Because of the elliptical shape of the earth's orbit, the apparent angular velocity of the sun that we see, called the *true sun*, is not constant; during four periods of each year it is greater, and during four intervening periods less, than the average velocity. Hence, the days, as indicated by the apparent travel of the true sun about the earth, are not of uniform length. To make our solar days of uniform length, astronomers have invented the *mean sun*, a fictitious body that is imagined to move at a uniform rate along the celestial equator, making a complete circuit from west to east in one year. The time interval as measured by one daily revolution of the mean sun is called a *mean solar day*, which is the same as the civil day. The mean solar day begins at midnight, as does the civil day, and the *mean solar time* at any place is given by the hour angle of the mean sun plus $12^h$. Thus, if the hour angle of the mean sun is $-15° = -1^h$, the mean solar time is $-1^h + 12^h = 11^h$. With regard to time, the terms "mean" and "civil" are interchangeable.

*Civil time* has the same meaning as *mean solar time* or *mean time* or *universal time* and, in the form of *standard time* (Art. 11.21), is the time in general use by the public. *Local civil time* is that for the meridian of the observer. Civil time for any other meridian is designated by name; for example, *Greenwich civil time*. Civil time for any meridian can be converted into terms of civil time for any other meridian by computations involving the longitude of the two meridians, as described in Art. 11.20; $1^h$ civil time corresponds to $1^h$ or $15°$ of longitude.

**11.17. Apparent (true) solar time**  The time interval as measured by one apparent revolution of the true sun about the earth is called an apparent or true solar day. The true solar day begins at midnight and the true solar time at any place is given by the hour angle of the true sun plus $12^h$. Thus, if the hour angle of the true sun is $45° = 3^h$, the true solar time is $3^h + 12^h = 15^h$. The terms "true" and "apparent" are used interchangeably. "True" is a more accurate designation, but since apparent is used in certain ephemerides it must be defined and will be used where necessary. Local true time is the time for the meridian of the observer. True time for any other meridian is designated by name, for example, Greenwich true or apparent time. True time for any meridian can be converted into terms of true time for any other meridian by computations identical with those for civil time, as described in Art. 11.20; $1^h$ true time corresponds to $1^h$ or $15°$ of longitude.

**11.18. Equation of time**  When the true sun is ahead of the mean sun, true time is faster than mean (civil) time; when behind, slower. The difference between true time and civil time at any instant is called the *equation of time*. It is used to convert civil time at any instant into true time, and vice versa.

To formulate the equation of time, it is convenient to utilize the concept of Greenwich hour angle, G.H.A. (Art. 11.5) and Greenwich civil time, G.C.T. (Art. 11.16). Using these terms, the equation of time is

$$\text{equation of time} = \text{G.H.A. of the true sun}$$
$$- \text{G.H.A. of noon at local civil time L.C.T.} \qquad (11.15a)$$

Thus, the equation of time is the elapsed time between an upper transit of the true sun over the observer's meridian and noon L.C.T., or

$$\text{equation of time} = \text{true solar time} - \text{local civil time} \qquad (11.15b)$$

When the equation of time is positive, the true sun precedes noon L.C.T., and is negative when the true sun follows noon L.C.T.

The maximum value of the equation of time is only about $16^m$; hence for work in which its only use is for the determination of change in declination, it is sometimes neglected.

The equation of time may be obtained either from an ephemeris that gives values at given instants of civil time or from one that gives values for apparent time, as follows:

1. In the *Nautical Almanac*, the equation of time, to $01^s$, is given for each day at $0^h$ and $12^h$ Greenwich civil time. The civil time of meridian passage of the apparent (true) sun is also given.

2. In the *Ephemeris of the Sun and Polaris* of the U.S. Bureau of Land Management, the equation of time, to $0.01^s$, is given for each day at the instant of Greenwich apparent noon. The column headings state directly whether the equation of time is to be added to or subtracted from the apparent time when the civil time is desired.

To find the equation of time at any instant other than that for which a value is tabulated, it is necessary to interpolate, adding to or subtracting from the tabulated value of the equation of time the change in the equation of time since the instant to which the tabulated value applies.

**Example 11.1**   It is desired to determine by use of the Nautical Almanac the equation of time at the instant of $3^h30^m45^s$ P.M. Greenwich civil time on December 15, 1978. Greenwich civil time $= 12^h + 3^h30^m45^s = 15.51^h$. Also, calculate the true (apparent) solar time.

SOLUTION   From the *Nautical Almanac* the equation of time at $12^h$ G.C.T. is $04^m57^s$. The change in the equation of time in 12 h ($0^h$ December 16) is determined as follows. From the 1978 *Nautical Almanac*,

Equation of time at $0^h$ Dec. $16 = 04^m43^s$

Equation of time at $12^h$ Dec. $15 = \underline{04^m57^s}$

$\qquad$ Change in equation of time $= \quad -14^s$

The change in the equation of time up to the given instant is

$$\frac{15.51 - 12}{12}(-14) = -4.1^s$$

so that the equation of time at the given instant is

$04^m57^s - 4.1^s = 04^m53^s$

Using Eq. (11.15$b$), we obtain

$04^m53^s =$ true solar time $- 15^h30^m45^s$

or

true solar time $= 15^h30^m45^s + 04^m53^s = 15^h35^m38^s$

This same problem can be worked using an ephemeris distributed by a manufacturer of surveying equipment. In this ephemeris the equation of time is given for $0^h$ G.C.T. for each day of the year. For December 15, 1978, the equation of time at $0^h$ G.C.T. is $05^m11.9^s$ taken directly from the tables. The change in seconds per hour is also tabulated as $1.20^s$. This change is negative since the equation of time for December 16 can be observed to be $04^m43.1^s$. Thus, the equation of time at 15.51 G.C.T. is

equation of time $= (05^m11.9^s) - (15.51^h)(1.20^s) = 04^m53^s$

and the true solar time is calculated as before.

**Example 11.2**   It is desired to determine by use of *Ephemeris of the Sun and Polaris* the Greenwich civil time (G.C.T.) at the instant of $9^h00^m15^s$ Greenwich apparent (true) time (G.A.T.) on October 10, 1978. Reference to Fig. 11.12a illustrates the time relationships and shows that the time which will elapse before Greenwich apparent noon is essentially $3.00^h$.

SOLUTION   From the ephemeris the equation of time at Greenwich apparent noon is:

Equation of time at G.A. noon October $10 = 12^m53.62^s$

$\quad$ Equation of time at G.A. noon October $9 = 12^m37.41^s$

$\qquad$ Change in $24^h$ from October 9 to $10 = + \quad 16.21^s$

The change that will occur in the $3.00^h$ of elapsed time before G.A. noon is

$$\frac{3^h}{24^h}16.21^s = 2.0^s$$

and the equation of time for $9^h00^m15^s$ G.A.T. October 10, 1978, is

$12^m53.62^s - 2.0^s = 12^m51.6^s$

Then, by Eqs. (11.15),

G.C.T. at the instant of $9^h00^m15^s$ G.A.T. $= 9^h00^m15^s - 0^h12^m51.6^s = 8^h47^m23.4^s$

By inspecting the tabulated values of the equation of time as given in the ephemerides, it will be seen that in February the true sun is as much as $14^m$ behind the mean sun and that in November the true sun is more than $16^m$ ahead of the mean sun, while on about the dates April 15, June 15, September 1, and December 25, the equation of time is zero and hence the hour angle of the true sun is for an instant the same as that of the mean sun.

**11.19. Sidereal time**   The *sidereal time* at any place is the hour angle of the vernal equinox at that place; and the beginning of the sidereal day, occurring when the vernal equinox crosses the upper branch of the meridian, is called *sidereal noon*. The sidereal hour angle is defined in Art. 11.5. Twenty-four-hour clocks regulated to keep sidereal time are called *sidereal clocks*. The vernal equinox, also called Aries, is an imaginary point and cannot be observed like the sun; but the right ascensions of stars are referred to the vernal equinox, and therefore the sidereal time can be obtained by determining the hour angle of any star the right ascension of which is known. Then, if $\theta$ is the sidereal time, $\theta = t + \alpha$, as explained in Art. 11.6.

The sidereal day is shorter than the mean solar day by $3^m55.909^s$ mean solar time, or $3^m56.555^s$ sidereal time. The sidereal hour is shorter than the mean solar hour by $3^m55.909^s/24 = 9.830^s$ mean solar time, or $(3^m56.555^s)/24 = 9.856^s$ sidereal time.

The *Nautical Almanac* is useful for converting sidereal time into civil (mean solar) time, and vice versa. The following example illustrates one method of determining the Greenwich sidereal time (G.S.T.) corresponding to a given instant for which the G.C.T. is known.

**Example 11.3**   It is desired to know the Greenwich sidereal time corresponding to $15^h30^m15^s$ G.C.T. on August 1, 1978. The mean solar time interval since $0^h$ G.C.T. $= 15.5042^h$.

SOLUTION   By the *Nautical Almanac* the Greenwich hour angle of the vernal equinox (Aries) is

G.H.A. at $15^h$ G.C.T.                                   $174°51.8'$

G.H.A. at $16^h$ G.C.T.                                   $189°54.3'$

Difference in $1^h$                                       $\overline{15°2.5' = 902.5'}$

Difference in $0.5042^h = (0.5042^h)(902.5') = \quad 455.0' = 7°35.0'$

G.H.A. at $15^h30^m15^s$ G.C.T. $= 174°51.8' + 7°35.0' = 182°26.8'$
The sidereal time expressed in hour units is

$$\frac{182°26.8'}{15} = \frac{182.4467°}{15} = 12.1631^h = 12^h09^m47^s$$

The *Nautical Almanac* is not always available to the practicing surveyor or engineer. The conversion from sidereal to civil time, or vice versa, can be accomplished using the known differences between the times per day and per hour as given above. Consider Example 11.3 using an ordinary ephemeris in which the Greenwich hour angle for the vernal equinox, G.H.A.$_T$, is given only for $0^h$ Greenwich civil time.

G.H.A.$_T$ at $0^h$ G.C.T.                                                     $309°14.9'$

Mean solar interval of $15^h30^m15^s$ converted to arc measure =

$\quad = (15.5042^h)(15°/h) = 232.5630° =$                                      $+232°33.78'$

G.H.A. at $15^h30^m15^s$                                                        $\overline{182°48.68'}$

Correction to sidereal interval $= [(9.856^s/h)(15.5042^h)(15''/s)] (60''/') = + \quad 38.2'$

Sidereal time in angular units                                                 $\overline{182°26.88'}$

Sidereal time converted to hours $= (182.4483°)/(15°/h) = 12.1632^h = \quad 12^h09^m48^s$

Thus, to convert a mean solar interval into a sidereal interval, $9.856^s$ per hour of mean solar interval are added to the solar interval. To convert a sidereal interval into a mean solar interval, $9.830^s$ per hour of sidereal time are subtracted from the sidereal interval.

**11.20. Relation between longitude and time**   As the sun apparently makes a complete revolution (360°) about the earth in one solar day (24 h), and as the longitudes of the earth range from 0° to 360°, it follows that in 1 h the sun apparently traverses $360/24 = 15°$ of longitude. The same statement applies equally well to the sidereal day and the vernal equinox. It follows that at any instant, the *difference in local time* between two places, whether the time under consideration be sidereal, mean solar, or apparent solar, is equal to the *difference in longitude* between the two places, expressed in hours. This relation is used to determine the difference in time when the difference in longitude between two places is known, or vice versa.

Some solar ephemerides are for the meridian of Greenwich, and a problem of frequent occurrence is to find the local time corresponding to a given instant Greenwich time, or vice versa. The local time (L.T.) of a place at a given instant is obtained by adding to or subtracting from the Greenwich time (G.T.) the difference in longitude ($\Delta\lambda$), expressed in hours, between the two places. If the place is east of Greenwich, the difference in longitude is added; if the place is west, the difference in longitude is subtracted.

In these problems concerning time intervals and longitude, conversion from angular units to units of time or the reverse is frequently necessary. These conversions can be accomplished by the following relationships:

| Time   Arc | Arc   Time |
|---|---|
| $24^h = 360°$ | $360° = 24^h$ |
| $1^h = 15°$ | $1° = 4^m$ |
| $1^m = 15'$ | $1' = 4^s$ |
| $1^s = 15''$ | $1'' = 0.067^s$ |

Most ephemerides contain tables for the purpose of making these conversions. With a pocket calculator, the simplest procedure is to convert to decimal parts of the given units and then divide or multiply by 15. Examples 11.3 and 11.6 furnish examples of this type of conversion.

**Example 11.4**   An observation of the sun is taken at $9^h52^m56^s$ local apparent time (L.A.T.). The longitude of the place is $7^h12^m36^s$ west of Greenwich. What is the Greenwich apparent time (G.A.T.)?

SOLUTION

$$\text{G.A.T.} = \text{L.A.T.} + \Delta\lambda = 9^h52^m56^s + 7^h12^m36^s$$

$$= 17^h05^m32^s$$

**Example 11.5**   On a given date the mean sun crosses the lower branch of the Greenwich meridian at $3^h52^m48.6^s$, Greenwich sidereal time. At that instant it is desired to find the local sidereal time (L.S.T.) at a place whose longitude is $5^h12^m24.2^s$ west of Greenwich.

SOLUTION

$$\text{L.S.T.} = \text{G.S.T.} - \Delta\lambda = 3^h52^m48.6^s - 5^h12^m24.2^s + 24^h$$

$$= 22^h40^m24.4^s \text{ of the previous day}$$

**Example 11.6**   At the instant of $18^h48^m15^s$ Greenwich civil time, the local civil time of a place is $10^h37^m42^s$. It is desired to determine the longitude of the place with respect to Greenwich.

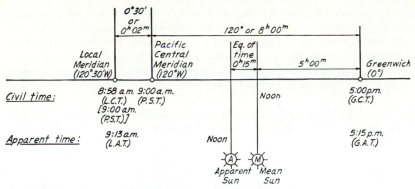

**Fig. 11.13** Relation between longitude and time at a given instant.

SOLUTION

$\Delta\lambda = 18^h48^m15^s - 10^h37^m42^s = 8^h10^m33^s = 8.17583^h$
$= (8.17583)(15) = 122.63750°$
$= 122°38'15''$ west of Greenwich

**Example 11.7** It is desired to find the local civil time at longitude $122°38'15''$W, at the instant of $18^h48^m15^s$ Greenwich civil time.

SOLUTION  The difference in longitude, in hours, is equal to the difference in longitude, in degrees, divided by 15:

$$\text{local civil time} = 18^h48^m15^s - \frac{122°38'15''}{15} = 10^h37^m42^s$$

Sketches are a valuable aid in the solution of problems involving longitude and time, as they enable the surveyor to visualize the relations. A simple "straight-line" type of sketch is shown in Fig. 11.13, for the instant of 9:00 A.M. Pacific standard time. For clarity, values are given only to $01^m$; the actual computations of a surveying problem would be more precise.

**11.21. Standard time**  In order to eliminate the confusion resulting from the use of local time by the public, the United States has been divided into belts, each of which occupies a width of approximately 15° or $1^h$ of longitude. In each belt the watches and clocks that control civil affairs all keep the same time, called *standard time*, which is the local civil time for a meridian near the center of the belt. The time in any belt is a whole number of hours slower than Greenwich civil time, as follows:

| Standard time | Abbreviation | Hours slower than Greenwich civil time | Central meridian | Where used |
|---|---|---|---|---|
| Atlantic | A.S.T. | 4 | 60°W | Maritime provinces of Canada |
| Eastern | E.S.T. | 5 | 75°W | Maine to Indiana |
| Central | C.S.T. | 6 | 90°W | Illinois to central Nebraska |
| Mountain | M.S.T. | 7 | 105°W | Central Nebraska to western Utah |
| Pacific | P.S.T. | 8 | 120°W | West of Utah |
| Yukon | Y.S.T. | 9 | 135°W | Eastern Alaska |
| Alaska | A.H.S.T. | 10 | 150°W | Central Alaska, Hawaii |
| Bering Sea | B.S.T. | 11 | 165°W | Western Alaska |

The exact boundaries of the time belts are irregular and can be determined only from a map.

Correct standard time can be obtained either from a clock known to be closely regulated or from radio signals broadcast by the U.S. Naval Observatory. Time signals are also broadcast by the National Bureau of Standards and radio stations WWV in Fort Collins, Colorado, and WWVH in Kauai, Hawaii, on frequencies of 2.5, 5, 10, and 15 megahertz.

In certain localities, "daylight saving time" is employed during the summer months. Daylight saving time is $1^h$ faster than standard time.

**Computations**   The Greenwich civil (mean) solar time is found by adding to the standard time the longitude (expressed in hours) of the meridian for which standard time is also local mean time.

**Example 11.8**   At a given instant the Central standard time is $9^h00^m$ A.M. It is desired to find the Greenwich civil time. The longitude of the meridian to which Central standard time is referred is $90°$ or $6^h$ west of Greenwich.

SOLUTION   The Greenwich civil time is $9^h00^m + 6^h00^m = 15^h00^m$, or $3^h00^m$ P.M.

If the longitude of a place is known, the standard time of the belt in which the place is situated can be determined by adding algebraically to the local mean time the difference in longitude (expressed in hours) between the given place and the meridian for which standard time is also local mean time.

**Example 11.9**   By observation on a star, the local mean time (L.M.T.) at a given instant is found to be $18^h37^m46^s$. The longitude of the place is $\lambda_l = 89°49'30''W = 5^h59^m18^s$. The standard time at the given instant is to be found. The place is evidently in the Central time belt for which the standard time (C.S.T.) is local time for the 90th meridian. The longitude of this meridian expressed in hours is $\lambda_s = 90°/15 = 6^h$.

SOLUTION

$\lambda_l = 5^h59^m18^s$

$\lambda_s = 6^h00^m00^s$

$\overline{\Delta\lambda = -0^m42^s}$

C.S.T. $=$ L.M.T. $+\Delta\lambda = 18^h37^m46^s + (-0^m42^s) = 18^h37^m04^s = 6^h37^m04^s$ P.M.

## AZIMUTH, LATITUDE, LONGITUDE, AND TIME

**11.22. General**   In the following sections are described the rough methods commonly used in the United States on surveys of ordinary precision where the engineer's transit or the repeating theodolite is employed for angular measurements. For more precise methods, such as those necessary on precise geodetic surveys, the reader is referred to texts on geodesy and engineering astronomy (see the references at the end of the chapter). The methods discussed herein are based upon the relations given in Arts. 11.1 to 11.21.

Since most observations are taken on the sun and Polaris, the discussion is concerned chiefly with these two bodies; but the principles involved are the same for any star.

Measurements to the sun cannot be made with the same degree of precision as to a star, hence the error in computed values is larger than when a fixed star is chosen. However, the sun may be viewed at convenient times, and solar observations are suitable for determinations of azimuth, latitude, and longitude with sufficient precision for most ordinary surveys.

Polaris, being near the pole, changes its position slowly. It is the most favorably located of all bright stars for precise determinations of latitude and azimuth, but owing to its slow change in azimuth, it is not suitable for longitude or time observations.

**11.23. Measurement of angles**   Whenever observations are made to determine azimuth, a part of the field work consists of measuring the horizontal angle between the celestial body and a reference mark on the earth's surface. As the sights to the celestial body are in general steeply inclined, it is highly important that the horizontal axis of the transit or theodolite be in adjustment with respect to the vertical axis and that the instrument be very carefully leveled (Art. 6.46). Even though the horizontal axis is in perfect adjustment, it will be inclined unless the vertical axis is truly vertical; and the error due to such inclination will in general not be eliminated by a reversal of the telescope between sights. For precise observations the transit should be equipped with a sensitive striding level by means of which the horizontal axis may be leveled prior to each sight. With the ordinary transit not so equipped, the plate should be leveled by means of the telescope level when other than rough observations are being made. Also sights should be taken with the telescope in both the direct and the reversed positions in order that the mean of horizontal angles may be free from other instrumental errors.

Whenever altitudes are observed, the index error of the vertical circle should be determined at the time of the observation (see Art. 6.45). The errors due to the line of sight not being parallel to the axis of the level tube and due to the vertical vernier being displaced can be eliminated by double centering (Art. 6.44); however, any error due to inclination of the vertical axis will not be eliminated by double centering and, therefore the transit should be leveled with great care, preferably by means of the telescope level.

The transit should be supported firmly; if the setup is on soft ground, pegs should be driven to support the tripod legs.

## SOLAR OBSERVATIONS

**11.24. Observations on the sun**   To observe the sun directly through the telescopic eyepiece may result in serious injury to the eye. A piece of colored or smoked glass may be held between the eye and the eyepiece. Some instruments are equipped with a colored *sun glass* that may be attached to the eyepiece.

Good observations can be made by bringing the sun's image to a focus on a white card held 3 to 4 in, or 7 to 10 cm, in the rear of the telescopic eyepiece. A rough pointing on the sun is made by sighting over the telescope. The eyepiece is then drawn back, and the objective is focused until the sun's image and the cross hairs are clearly seen on the card. If the eyepiece of the telescope is erecting, the image on the card will be inverted; if the eyepiece is inverting, the image will be erect. The cross hairs are visible only on the image of the sun. As the angle between lines of sight defined by the stadia hairs (on instruments with stadia constant = 100) is 34′, while the diameter of the sun is only 32′, not all of the three horizontal cross hairs are visible on the sun's image at the same time. A common blunder is to mistake one of the stadia hairs for the middle cross hair. This mistake can be avoided by rotating the telescope slightly about the horizontal axis until all three hairs have been seen.

Some transits or theodolites are equipped with a *solar circle*. For this arrangement, the cross-hair reticule has, in addition to the usual vertical, horizontal, and stadia cross hairs, a solar circle inscribed symmetrically about the intersection of the vertical and horizontal cross hairs, as shown in Fig. 11.14. This circle has an angular radius of 15′45″, which is slightly smaller than the sun's semidiameter (Art. 11.25), so that very accurate pointings on the sun are possible.

**11.25. Semidiameter correction**   Since the sun is large (apparent angular diameter about 32′), its center cannot be sighted precisely with the ordinary transit, and it is customary to bring the cross hairs tangent to the sun's image. When a sight is taken directly through an erecting telescope and the horizontal cross hair is brought tangent to the lower edge of the sun, the sight is said to be taken to the sun's *lower limb*. This fact is indicated in the notes by

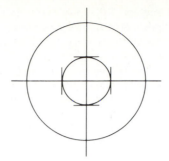

**Fig. 11.14** Reticule with a solar circle.

the symbol $\odot$. Similarly, the symbol $\overline{\odot}$ indicates a sight to the sun's *upper limb*, $\odot|$ a sight with the vertical cross hair to the sun's *right* or *western limb*, and $|\odot$ a sight to the sun's *left*, *eastern*, or *trailing limb*.

When a vertical angle is measured to the sun's upper limb, it is necessary to subtract the sun's semidiameter to obtain the observed altitude of the sun's center; when to the lower limb, add the semidiameter. The *Ephemeris of the Sun and Polaris* and other solar ephemerides give values of the semidiameter of the sun for each day of the year. The semidiameter varies from about 15′46″ in July to about 16′18″ in January; for rough calculations it may be taken as 16′.

When a horizontal angle is measured to the sun's right or left limb, a correction equal to the sun's semidiameter times the secant of the altitude is applied. Thus if the altitude $h$ is 60° and the semidiameter is 16′, the correction to a horizontal angle is 16′ sec $h = 32′$. As the sun approaches the zenith, the correction becomes very large and readings should not be taken to one limb only.

The semidiameter correction to a horizontal angle for the sun at any altitude is illustrated in Fig. 11.15 in which $A$ is the station of an observer on the earth, $S$ is the sun at the horizon, $S'$ is the sun at some altitude $h$ above the horizon, $r$ is the radius of the sun, and $\alpha$ and $\beta$ are the small horizontal angles (semidiameter corrections) subtended by the radius of the sun at $S$ and $S'$, respectively. In the plan view, $\alpha \cdot AC = r$ and $\beta \cdot AB = r$; hence, $\alpha \cdot AC = \beta \cdot AB$, or $\beta = \alpha \cdot AC/AB$. But in the elevation view $AC = AD$ and $AD/AB = \sec h$. Therefore,

$$\beta = \frac{\alpha}{\cos h} \tag{11.16}$$

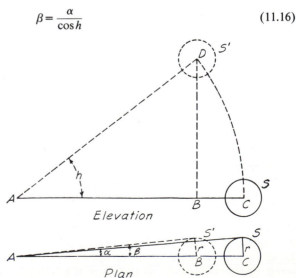

**Fig. 11.15** Semidiameter correction to horizontal angle.

For certain solar observational procedures, an equal number of sights are taken to opposite limbs of the sun; the mean of the horizontal angles and the mean of the vertical angles at the mean of the times are taken, and no corrections for semidiameter are necessary.

**11.26. Procedures for sighting the sun**   When the instrument to be used for sighting the sun is not equipped with a solar circle and colored glass over the eyepiece or with a Roelof's prism (described later in this section), it is necessary to take observations on the sun by making the horizontal and vertical cross hairs tangent to the circumference of the sun. The two methods for accomplishing this task are the *center tangent* method and the *quadrant tangent* method. The procedure for each method will be described assuming an image of the sun reflected on a white card held behind the eyepiece, as described in Art. 11.24.

The center tangent method is somewhat easier to perform and is described first. When observations are made in the morning, the telescope is pointed on the sun so that the image of the sun as reflected on the white card appears as shown in Fig. 11.16a where the horizontal cross hair cuts off a segment of the lower limb of the sun (dashed circle) and the vertical cross hair approximately bisects this segment. In the morning the sun is rising so that with an erecting telescope, the image of the sun on a card will appear to be moving downward and to the right. Thus, the segment will get smaller and smaller as the sun moves in its orbit. The vertical clamp is locked and the horizontal upper tangent-screw is used to adjust the vertical cross hair so as to bisect the disappearing segment of the sun's lower limb. When this segment becomes tangent to the horizontal cross hair (solid circle in Fig. 11.16a), the time is recorded and the horizontal and vertical circle readings are observed. Since the vertical cross hair bisects the sun, no correction is needed for the horizontal circle measurement. However, the horizontal cross hair is tangent to the lower limb of the sun, so that the semidiameter of the sun (Art. 11.25) must be added to the vertical circle reading, which is first corrected for parallax and refraction (Arts. 11.27 to 11.29). Next, the telescope is adjusted so that the horizontal cross hair approximately bisects the disappearing segment of the (trailing or left) limb of the image of the sun reflected on a card as shown by the dashed circle in Fig. 11.16b. This time the horizontal upper clamp is locked and the disappearing segment is bisected by adjusting the vertical tangent motion. When the eastern limb of the sun is tangent to the vertical cross hair, as shown by the solid circle in Fig. 11.16b, the time and horizontal and vertical circle measurements are observed and recorded. The horizontal cross hair bisects the sun so the vertical circle reading, corrected for parallax and refraction, requires no correction for semidiameter of the sun. The vertical cross hair is tangent to the eastern limb of the sun, so that the semidiameter of the sun must be added to the horizontal circle reading. These observations on the sun's lower and eastern limbs are then repeated with the telescope in the reversed position. This constitutes one set of readings.

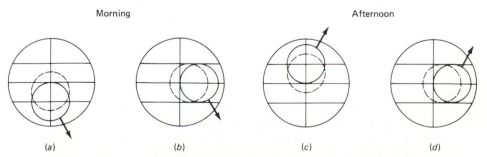

**Fig. 11.16** Image of the sun as reflected on a white card held behind the eyepiece of an erecting telescope for the center tangent method. (*a*) Horizontal cross hair fixed. (*b*) Vertical cross hair fixed. (*c*) Horizontal cross hair fixed. (*d*) Vertical cross hair fixed.

Morning                                                    Afternoon

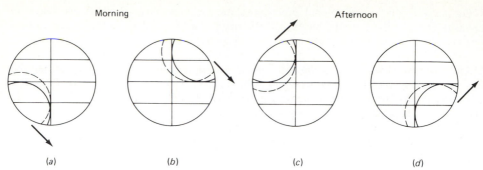

(a)                    (b)                    (c)                    (d)

**Fig. 11.17** Image of the sun reflected on a white card held behind the eyepiece of an erecting telescope for the quadrant tangent method. (a) Horizontal cross hair fixed. (b) Vertical cross hair fixed. (c) Horizontal cross hair fixed. (d) Vertical cross hair fixed.

Observations in the afternoon are performed in a similar manner, but the procedure is modified to accommodate the downward movement of the sun in its orbit that produces an image of the sun which moves upward and to the right when reflected on the card held behind the telescope eyepiece. When the horizontal cross hair becomes tangent to the upper limb of the sun, as shown by the solid circle in Fig. 11.16c, time and vertical and horizontal circle readings are recorded. The semidiameter of the sun must be subtracted from the vertical circle reading corrected for parallax and refraction. The horizontal circle reading requires no correction. Next, the vertical cross hair is made tangent to the eastern limb of the sun, as shown in Fig. 11.16d. This time the vertical circle reading is corrected only for parallax and refraction and the horizontal circle reading is corrected by adding the semidiameter of the sun. As before, these observations should be repeated with the telescope in the reversed position to provide a complete set of readings.

In the quadrant tangent method, both vertical and horizontal cross hairs are brought tangent to the eastern limb and the lower (morning) or upper (afternoon) limb of the sun. The sequence of images observed for morning and afternoon observations by this method is illustrated in Fig. 11.17.

For observations in the morning, the horizontal cross hair is sighted a short distance above the lower limb of the sun, as shown by the dotted circle in Fig. 11.17a. Since the altitude of the sun is increasing, the horizontal cross hair approaches tangency due to the sun's movement in its orbit. At the same time, the vertical cross hair is kept continuously on the sun's western limb by means of the horizontal upper tangent-screw. At the instant when the horizontal and vertical cross hairs are simultaneously tangent to the sun's disk, the motion of the telescope is stopped, the time is observed, and the horizontal and vertical circles are read. A second observation is then taken with the sun in the lower right-hand quadrant that is reflected as the upper right-hand quadrant on the card as shown in Fig. 11.17b. The procedure is as follows. The vertical cross hair is set a short distance to the right of the sun's eastern limb, as shown in Fig. 11.17b by the dashed circle. Since the sun is traveling westward, the vertical cross hair approaches tangency due to the sun's movement. At the same time, the horizontal cross hair is kept continuously on the sun's upper limb by means of the vertical tangent motion. As before, observations are taken when the horizontal and vertical cross hairs are simultaneously tangent to the sun's disk. The procedure is such that the final setting for either observation requires manipulation of only one tangent-screw and that the cross hair which is approaching tangency is visible on the sun's disk. The observations described above should then be repeated with the telescope in the reversed position to yield one set of measurements. Since the horizontal and vertical circle readings are observed in opposite quadrants, the average values represent measurements to the center of the sun. Of course, the vertical circle readings need to be corrected for parallax and

**Fig. 11.18** Image of sun using a Roelof's prism.

refraction, but corrections for the semidiameter of the sun are not required. The disadvantage of the quadrant tangent method is that the two cross hairs must be made simultaneously tangent to the disk of the sun, a difficult task under the best of conditions.

For afternoon observations in northern latitudes the procedure is the same as that just described, except that the sun is sighted first in the upper right-hand quadrant (that appears as the upper left-hand quadrant when the image is reflected on the card; Fig. 11.17c) and then in the lower left-hand quadrant (that appears in the lower right quadrant when the image is reflected on a card; Fig. 11.17d).

In both the center tangent and quadrant methods, it is suggested that several sets of observations be made in rapid succession. When the appropriate corrections have been applied to the horizontal and vertical circle readings, these corrected values can be plotted as a function of the time for each measurement. If the data are consistent, the corrected vertical circle readings and corrected horizontal circle readings will plot as a straight line versus time, assuming that the observations were taken within a 15- or 20-min period of time. In this way blunders can be eliminated before calculations proceed. The procedure for this operation is described in Art. 11.32.

The *Roelof's solar prism* is a device attached to the objective end of the telescope and permits simpler more accurate sighting of the sun. This device produces four over lapping images of the sun simultaneously formed into the pattern illustrated in Fig. 11.18. When the horizontal and vertical cross hairs bisect the small diamond-shaped area at the center of the pattern, the line of sight is on the center of the sun. The image is viewed directly so that observations can be made rapidly and accurately. Vertical circle readings must be corrected for parallax and refraction, but no correction is required for horizontal circle measurements.

**11.27. Parallax correction**  In previous discussions, it was assumed that the celestial sphere is of infinite radius and that a vertical angle measured from a station on the surface of the earth is the same as it would be if measured from a station at the center of the earth. For the fixed stars this assumption yields results that are sufficiently accurate for the work described herein; but the distance between the sun and the earth is relatively small, and for solar observations a *parallax correction* is added to the observed altitude to obtain the altitude of the sun from the center of the earth. The parallax correction is subtracted from the zenith angle to obtain the true zenith distance.

In Fig. 11.19a, $h'$ is the altitude of the sun above the horizon of an observer at $A$, and $h$ is the altitude of the sun above the celestial horizon. The parallax correction is equal to the difference between these two angles. As $h$ is always larger than $h'$, the correction must be added to the observed altitude.

The parallax correction can be computed, since the distance to the sun is known. The magnitude of the correction depends upon the altitude, being zero when the sun is directly overhead, and being a maximum when the sun is on the observer's horizon. When the sun is on the observer's horizon, the correction $C_h$ is called the *horizontal parallax*. It can be readily

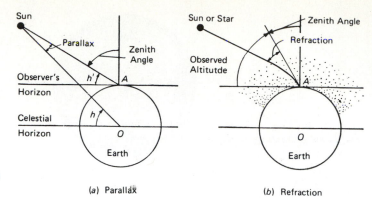

**Fig. 11.19** Parallax and refraction.

(a) Parallax                    (b) Refraction

demonstrated that the parallax correction $C_p$ for any observed altitude $h'$ is

$$C_p = C_h \cos h' \tag{11.17}$$

The horizontal parallax is always slightly less than 09''; hence, the parallax correction at any altitude cannot exceed 09''. Values of the sun's parallax correction at various altitudes are given in the *Ephemeris of the Sun and Polaris*.

Corrections for parallax and refraction are usually made together (see Art. 11.29).

**11.28. Refraction correction**   When a ray of light emanating from a celestial body passes through the atmosphere of the earth, the ray is bent downward, as illustrated in Fig. 11.19b. Hence, the sun and stars appear to be higher above the observer's horizon than they actually are. The angle of deviation of the ray from its direction on entering the earth's atmosphere to its direction at the surface of the earth is called the *refraction* of the ray. A *refraction correction* $C_r$ of an amount equal to the refraction is subtracted from the observed altitude to determine the actual altitude $h'$ above the observer's horizon. When the zenith angle $z'$ is observed, the refraction correction $C_r$ is added to the observation to determine the true zenith angle $z$.

The magnitude of the refraction correction depends upon the temperature and barometric pressure of the atmosphere and upon the altitude of the ray, varying as the cotangent of the altitude. It does not depend upon the distance to the body from which the ray emanates. Under normal conditions the refraction correction is about 34' when the sun or star is on the observer's horizon, about 05' when the altitude is 10°, about 01' when the altitude is 45°, and zero when the altitude is 90°. Table II (Appendix C) gives values of refraction corrections for a barometric pressure of 29.5 in (which may be assumed with sufficient precision for practical purposes), for various temperatures, and for altitudes between 10° and 90°.

Owing to the uncertainties of the refraction correction for low altitudes, observations for precise determinations are never taken on a celestial body which is near the horizon.

**11.29. Combined correction**   For solar observations, refraction and parallax corrections are usually made together. The refraction correction, which is subtractive, is many times larger than the parallax correction, which is additive; hence, the combined correction is of the same sign as the refraction correction. Table I (Appendix C) gives corrections for the combined effect of refraction and parallax, to be subtracted from observed altitudes of the sun to determine the true altitudes above the celestial horizon. Note that this correction should be added to an observed zenith angle to obtain the true zenith distance.

**11.30. Declination of the sun**   For the determination of azimuth, latitude, or longitude by solar observations, it is necessary that the declination of the sun at the instant of sighting be known. The declination at a given instant is obtained by interpolating between values given in a solar ephemeris for the current year. Either of the two common types of ephemeris may be used.

One type gives the apparent declination for each day of the year and is especially adapted for use when the standard time or the Greenwich civil time is known. The *Nautical Almanac* is an example of an ephemeris of this kind. It gives declinations for each hour of Greenwich civil time.

The other type gives the apparent declination for each day of the year at the instant of Greenwich apparent noon and is especially adapted for use when the longitude of the place and the local apparent time of the observation are known. The *Ephemeris of the Sun and Polaris* is of this sort. It is widely used in land surveying.

In the abbreviated solar ephemerides published annually in the form of pamphlets by various manufacturers of surveying instruments, the declination is usually given for $0^h$ G.C.T.

The following examples illustrate the use of each of these types of ephemerides to determine declination.

**Example 11.10**  An observation is taken on the sun at $10^h00^m$ Eastern standard time, on December 15, 1978. It is desired to determine the declination at the given instant.

SOLUTION  The Greenwich civil time at the instant of observation is $10^h00^m + 5^h = 15^h00^m = 15.00^h$. By the ephemeris, for $0^h$ Greenwich civil time, the declination on December 15 is $-23°14.1'$ and the change per hour is $-0.13'$. The change in declination since $0^h$ G.C.T. is $(-0.13)(15.00) = -2.0'$. The declination at the instant of observation is $-23°14.1' - 2.0' = -23°16.1'$. The *Nautical Almanac* gives $-23°16.1'$ directly for $15.00^h$ G.C.T. on December 15, 1978.

**Example 11.11**  An observation is taken on the sun as it crosses the meridian on November 16, 1978, at a place where the longitude is $87°49'30''$ west of Greenwich. It is desired to determine the apparent declination at a given instant.

SOLUTION

$$G.A.T. = \frac{87°49'30''}{15} = 5^h51^m18^s = 5.855^h \text{ P.M.}$$

or

$$G.A.T. = 17.855^h = 17^h51^m18^s$$

It is necessary to compute the change from $0^h$ G.C.T. to $17.855^h$ G.A.T. The equation of time from the ephemeris for November 16 is

$$15'19.8'' + \text{civil time} = \text{apparent time}$$

where the equation of time is changing $-0.45''$ per hour of elapsed time. Thus, at $17.8550^h$ G.A.T. we have

$$15'19.8'' + (17.855^h)(-0.45''/h) + G.C.T. = G.A.T.$$
$$G.C.T. = 17^h51^m18^s - 15^m11.8^s = 17^h36^m06^s = 17.6017^h$$

or the G.C.T of local apparent noon $= 17.6017^h$. From the ephemeris, the declination $= S18°35'30''$ at $0^h$ G.C.T. on Nov. 16, 1978. Also from the ephemeris, the declination changes $-0.63'$ per hour of elapsed time. Consequently, the declination at local apparent noon on November 16, 1978, is

$$\text{declination} = -(18°35'30'') + (-0.63)(17.6017)$$
$$= S18°46'35''$$

If the *Ephemeris of the Sun and Polaris* is used, the declination at Greenwich apparent noon is given in the ephemeris as $S18°42'54''$. The average difference for $1^h$ is $-37.8''$, the minus sign indicating that the declination is increasing. Thus, the change in declination $= (-37.8'')(5.855) = -03'41''$. The declination at local apparent noon at the place of observation $= -(18°42'54'') - 03'41'' = S18°46'35''$.

**Example 11.12**   It is desired to determine the apparent declination of the sun at the instant of $1^h00^m$ P.M. Eastern standard time on November 18, 1976, from a solar ephemeris giving the values for Greenwich apparent noon.

SOLUTION   The difference between Eastern standard time and Greenwich mean time is $5^h$; hence, the instant of observation is $6^h00^m$ after Greenwich mean noon, or $18^h00^m$ G.M.T. At Greenwich apparent noon, the equation of time (from the ephemeris) is $14^m45.59^s$, to be subtracted from Greenwich apparent time to give Greenwich mean time. It follows that apparent time is faster than mean time, and Greenwich apparent time is approximately $18^h15^m$ or $6.25^h$ after Greenwich mean noon. The daily rate of change in the equation of time is given by the difference between the equation of time for November 18 and that for November 19, or $14^m45.59^s - 14^m32.16^s = 13.43^s$. Thus, the change in the equation of time since Greenwich apparent noon is $(6.25)(13.43)/24 = 3.5^s$. The equation of time for the given instant is thus $14^m45.6^s - 3.5^s = 14^m42.1^s$. Consequently, the interval since Greenwich apparent noon is $6^h00^m + 14^m42.1^s = 6.245^h$. At Greenwich apparent noon the declination is given in the ephemeris as $S19°19'08.7''$ and the average difference per hour is $-35.43''$. The change in declination since Greenwich apparent noon is $(-35.53'')(6.245) = 3'42''$ and the declination at the given instant is

declination $= -(19°19'08.7'') - 3'42'' = S19°22'51''$

In Example 11.12 the equation of time has been determined for the given instant. For all practical purposes the equation of time for Greenwich apparent noon might have been employed, since the small error of $3.4^s$ in time would have no effect upon the computed change in the declination unless declinations were carried out to tenths of seconds.

If the equation of time were neglected entirely, the error introduced in the computed value of the apparent declination would be but $09''$, not sufficiently large to be of consequence in rough calculations. Similarly, an observation of time for the sole purpose of determining declination need not be exact.

**11.31. Latitude by observation on the sun at noon**   The latitude of a given station can be determined with a fair degree of precision by observing the altitude of the sun with the engineer's transit or theodolite at local apparent noon, when the sun crosses the meridian. If the longitude of the place is roughly known it is unnecessary to observe the time, but if the longitude is unknown the standard time of the observation must be taken. It is not necessary that the direction of the meridian be known. The problem consists of determining the true altitude $h$ of the center of the sun above the celestial horizon and computing the apparent declination $\delta$ of the sun at the instant of sighting. Then as explained in Art. 11.9 the latitude $\phi$ is (see also Fig. 11.8)

$$\phi = 90° - h + \delta \tag{11.18}$$

The accuracy obtainable ordinarily depends upon the precision of the instrument. As the maximum rate of change of declination is only about $01'$ per hour, a considerable error in time will affect the declination but slightly. With the ordinary transit having a vertical circle reading to $01'$, the latitude may be determined in this manner with an error not greater than $01'$. The mean of a series of observations on different days will, of course, provide a closer result.

The usual procedure is as follows. The instrument is set up and very carefully leveled. The horizontal cross hair is sighted continuously on either the lower limb or the upper limb of the sun until the sun reaches its maximum altitude and begins its apparent descent. At that instant the watch time is observed. The maximum altitude and the watch time are recorded. With the telescope still approximately in the plane of the meridian, the index error is determined, preferably by the method of double centering on a mark as described in Art. 6.45. The watch is compared with a timepiece keeping correct standard time, and the error is noted. The Greenwich civil time of the observation is calculated, and the declination is found in a solar ephemeris as illustrated by Example 11.10, Art. 11.30. The true altitude of the sun's center is determined by applying to the observed altitude the corrections for index error, semidiameter, and refraction and parallax. The latitude is then determined by Eq. (11.18). It should be noted that the sign of the refraction and parallax correction as given in Table I is always negative. The sign of the declination is negative from September to March and is positive for the remainder of the year.

When desired, a second sight may be taken on the opposite limb of the sun with the telescope inverted. The mean of the two vertical angles is taken as the altitude of the sun's center at the mean of the two times of observation, and no correction for index error is necessary. If the time between sights does not exceed 3 or 4 min, the mean altitude may be considered as the altitude at apparent noon. The latitude is then calculated as described in the preceding paragraph.

If an ephemeris giving values at Greenwich apparent noon is to be used, the longitude of the place being unknown and the standard time being known, the Greenwich apparent time of the observation is determined and the declination at the given instant is found as illustrated by Example 11.12, Art. 11.30.

The field notes and computations are made in a form similar to that shown in Fig. 11.20. For these observations the longitude of the place was unknown, and the standard time was recorded. Only the sun's lower limb was sighted. The index error was determined by double centering at a mark; the letters "L" and "R" in the column headed "Circle" indicate whether the vertical circle was left or right and, therefore, whether the telescope was normal or inverted. As the available ephemeris gave values of declination at Greenwich apparent noon, the watch time was converted into Greenwich apparent time. In the line beginning "$\Delta\delta$," the change in declination during the 4.66 h that elapsed since Greenwich apparent noon was computed by multiplying the elapsed time by the variation per hour (57.8″) taken from the ephemeris.

When optical theodolites are used to observe the sun, the zenith angle is measured. In this case, the latitude is computed using

$$\phi = z + \delta \tag{11.19}$$

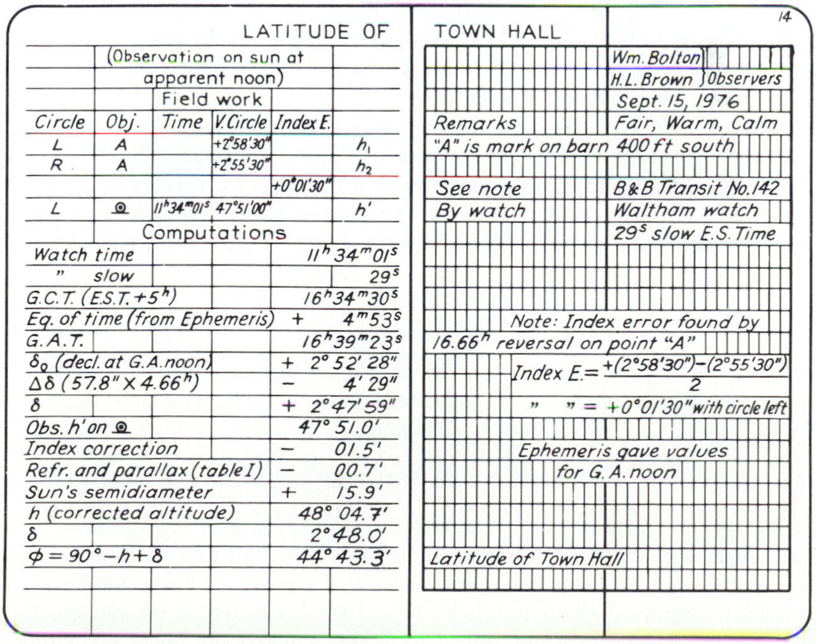

**LATITUDE OF   TOWN HALL**

(Observation on sun at apparent noon)

Field work

| Circle | Obj. | Time | V. Circle | Index E. | |
|--------|------|------|-----------|----------|---|
| L | A | | +2°58′30″ | | $h_1$ |
| R | A | | +2°55′30″ | | $h_2$ |
| | | | | +0°01′30″ | |
| L | ☉ | 11ʰ34ᵐ01ˢ | 47°51′00″ | | $h'$ |

Computations

| | | |
|---|---|---|
| Watch time | | 11ʰ 34ᵐ 01ˢ |
| " slow | | 29ˢ |
| G.C.T. (E.S.T. + 5ʰ) | | 16ʰ34ᵐ30ˢ |
| Eq. of time (from Ephemeris) | + | 4ᵐ53ˢ |
| G.A.T. | | 16ʰ39ᵐ23ˢ |
| δ₀ (decl. at G.A.noon) | + | 2° 52′ 28″ |
| Δδ (57.8″ × 4.66ʰ) | − | 4′ 29″ |
| δ | | + 2°47′59″ |
| Obs. h' on ☉ | | 47° 51.0′ |
| Index correction | − | 01.5′ |
| Refr. and parallax (table I) | − | 00.7′ |
| Sun's semidiameter | + | 15.9′ |
| h (corrected altitude) | | 48° 04.7′ |
| δ | | 2° 48.0′ |
| $\phi = 90° - h + \delta$ | | 44° 43.3′ |

Wm. Bolton ) Observers
H.L. Brown )
Sept. 15, 1976

Remarks   Fair, Warm, Calm
"A" is mark on barn 400 ft south

See note   B & B Transit No.142
By watch   Waltham watch
29ˢ slow E.S. Time

Note: Index error found by
16.66ʰ reversal on point "A"
Index E. = $\dfrac{+(2°58′30″)-(2°55′30″)}{2}$
" " = +0°01′30″ with circle left

Ephemeris gave values
for G.A. noon

Latitude of Town Hall

**Fig. 11.20** Latitude by observation of sun at noon.

**Table 11.1   Latitude by Observation of the Sun Using Zenith Angles**[a]

| | | |
|---|---|---|
| Watch time | | $11^h35^m38^s$ P.S.T. |
| Watch correction | + | $29^s$ |
| Corrected time of observation | = | $11^h36^m07^s = 11.6019^h$ |
| G.C.T. = P.S.T. + $8^h$ = $19.6019^h$ | = | $19^h36^m07^s$ |
| Declination, $\delta$ at $0^h$ G.C.T. (ephemeris) | | N00°24′06″ |
| Change $\Delta\delta$ in $\delta = (0.99'/h)(19.6019^h)$ | = | + 19′24″ |
| where 0.99′/h is the change in declination from the ephemeris | | |
| Declination at instant of observation | | N00°43′30″ |
| Average zenith angle $z'$ to the center of the sun | | 41°51′30″ |
| Correction for refraction and parallax (Table I, Appendix C) or from corresponding tables in ephemeris = 0.94′ | = | + 56″ |
| True zenith angle $z$ | = | 41°52′26″ |
| $\delta$ | = | 00°43′30″ |
| $\phi = z + \delta$ | = | 42°35′56″ |

[a]Average time of observation = $11^h35^m38^s$ = $11.5939^h$ P.S.T. on March 22, 1978; watch correction = $29^s$; temperature = 55°F; average of direct and reversed zenith angles to the upper and lower limbs of the sun = 41°51′30″.

in which $z$ is the observed zenith angle corrected for parallax and refraction (Art. 11.29) and $\delta$ is the declination of the sun at the instant of observation. Table 11.1 shows calculations for latitude when direct and reversed zenith angles are measured to the upper and lower limbs of the sun. Reductions for this example are made using an ephemeris giving the declination at $0^h$ G.C.T.

**11.32. Azimuth by direct solar observation**   The azimuth of a line can be determined by a single observation of the sun at any time when it is visible, provided that the latitude is known. However, as indicated in Art. 11.26, it is advisable to make a series of measurements to the sun in rapid succession with the telescope direct for half the readings and reversed for the remainder (e.g., 4 direct and 4 reversed).

At a known instant of time the sun is observed, and the altitude of the sun and the horizontal angle from the sun to a given reference point are measured. The declination of the sun at the given instant is found from a solar ephemeris. With the declination $\delta$, latitude $\phi$, and altitude $h$ known, the $PZS$ triangle can be solved as described in Art. 11.11, the azimuth of the sun being given by any of several expressions, one of which is Eq. (11.2b), repeated here for convenience.

$$\cos Z = \frac{\sin\delta}{\cos h \cos\phi} - \tan h \tan\phi$$

in which $Z$ is the clockwise or counterclockwise angle from north that is used to calculate the azimuth from the north of the sun, as detailed in Art. 11.11. The angle $Z_s$ from south is given by Eq. (11.3), also repeated for convenience.

$$\cos Z_s = \tan h \tan\phi - \frac{\sin\delta}{\cos h \cos\phi}$$

The azimuth of the line is readily computed from the azimuth of the sun and the observed horizontal angle. The usual procedure is as follows. The transit or theodolite is set up and very carefully leveled over the station at one end of the line. When a transit with a fixed vertical circle is employed, the telescope is pointed approximately at the sun and the vertical circle index error is observed and recorded. When a theodolite equipped with a vertical circle level bubble or with automatic vertical circle indexing is used, this step is not

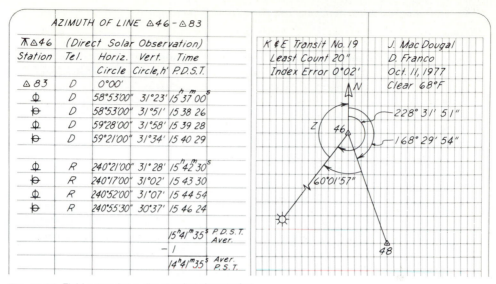

**Fig. 11.21** Field notes for a direct solar observation.

necessary. If a repeating instrument is being used, the horizontal circle index is set to zero and a sight is taken along the given line with the telescope in the direct position. If a direction instrument is being used, a sight is taken along the given line with the telescope in the direct position and the horizontal circle reading is observed and recorded. The upper motion is loosened and a series of sights to the sun is taken first with the telescope direct and then reversed according to the directions given in Art. 11.24. The time, horizontal circle readings, and vertical circle readings should be recorded for each observation.

When the center tangent method is utilized, four pointings on the sun, as illustrated in Fig. 11.16, should be made with the telescope in the direct position. In the morning these measurements would be made pointing at the sun as shown in Fig. 11.16*a, b, a, b*. In the afternoon the pointings would correspond to those shown in Fig. 11.16*c, d, c, d*. Next, the telescope is reversed and the entire group of four pointings is repeated. When the observations are completed, the telescope, now in the reversed position, is again sighted along the line and the circle reading is observed for checking purposes.

When the quadrant tangent method is employed, four pointings, as shown in Fig. 11.17*a, b, a, b*, are made in the morning with the telescope direct. In the afternoon the pointings would correspond to Fig. 11.17*c, d, c, d*. As in the center tangent method, these four pointings are then repeated with the telescope reversed and finally a sight is taken along the original line.

When the instrument is equipped with a solar circle or solar prism, the required number of repetitions with the telescope direct and reversed are taken pointing directly at the center of the sun. As in the other methods, time, vertical circle, and horizontal circle readings are recorded for each pointing.

Figure 11.21 shows notes for a solar observation made from an instrument set up on triangulation station 46 with a sight at triangulation station 48. The point sighted and position of the telescope are recorded in columns 1 and 2. Note that a sketch is made to indicate the position of the sun with respect to the cross hairs for each observation. The procedure followed for this example is the center tangent method using the image of the sun reflected on a card (Art. 11.26), so that the sketches represent the reflected image. The horizontal and vertical circle readings for each pointing are recorded in columns 3 and 4, and the time (P.D.S.T.) is in column 5.

**Table 11.2   Corrections to Horizontal and Vertical Circle Readings for Solar Observation**

| Measured vertical angle $h'$ | Refraction and parallax | Semidiameter of sun | Total correction | Corrected vertical angle $h$ |
|---|---|---|---|---|
| 32°23′ | − 1.33′ | − 16.05′ | − 17′23″ | 32°05′37″ |
| 31°51′ | − 1.36′ | | − 01′22″ | 31°49′38″ |
| 31°58′ | − 1.35′ | − 16.05′ | − 17′24″ | 31°40′36″ |
| 31°34′ | − 1.38′ | | − 01′23″ | 31°32′37″ |
| 31°28′ | − 1.38′ | − 16.05′ | − 17′26″ | 31°10′34″ |
| 31°02′ | − 1.41′ | | − 01′25″ | 31°00′35″ |
| 31°07′ | − 1.40′ | − 16.05′ | − 17′27″ | 30°49′33″ |
| 30°37′ | − 1.43′ | | − 01′26″ | 30°35′34″ |

Average corrected vertical angle = 31°20′35″

| Measured horizontal angle | Correction $\beta = \alpha'/\cos h$ | Corrected horizontal angle |
|---|---|---|
| | | 58°53′00″ |
| 58°53′00″ | $16.05'/\cos 31°49'38'' = +18'53''$ | 59°11′53″ |
| | | 59°28′00″ |
| 59°21′00″ | $16.05'/\cos 31°32'37'' = +18'50''$ | 59°39′50″ |
| | | 60°21′00″ |
| 60°17′00″ | $16.05'/\cos 31°00'35'' = +18'44''$ | 60°35′44″ |
| | | 60°52′00″ |
| 60°55′30″ | $16.05'/\cos 30°35'34'' = +18'39''$ | 61°14′09″ |

Average corrected horizontal angle = 60°01′57″

On completion of the observations, it is necessary to make the corrections to horizontal and vertical circle readings for refraction and parallax and for the semidiameter of the sun, where necessary. The calculations needed for these corrections are summarized in Table 11.2. Each vertical angle is corrected for parallax and refraction (Arts. 11.27 to 11.29) using values from Table I, Appendix C. The arguments in this table are the apparent (observed) latitude and temperature. Most ephemerides have a similar table. In addition, each vertical circle reading to the upper limb of the sun (observations were in the afternoon) is corrected by subtracting the semidiameter of the sun, which is also found from the ephemeris. Similarly, each horizontal circle reading taken on the left or trailing limb of the sun must be corrected by adding a correction for the semidiameter of the sun to the observation (Art. 11.25). Equation (11.16) is used to calculate these corrections. Corrected horizontal and vertical circle readings are listed in Table 11.2.

To detect inconsistent data, the corrected horizontal and vertical circle readings are plotted in a graph where the ordinate is the time of observation and corrected horizontal and vertical angles are plotted on the abscissa. If the observations have been taken over a short period of time of say 10 min, the corrected values for horizontal and vertical angles should plot as straight lines versus time. Any substantial deviation from a straight line indicates a mistake in the observations.

Figure 11.22 shows such a graph plotted using the corrected vertical and horizontal angles in Table 11.2 and corresponding times recorded in Fig. 11.21. Direct and reversed vertical angles plot as two straight parallel lines separated by an increment, $e_h$, which represents twice the uncorrected index error and other instrumental errors in the instrument. Had the index error of 0°02′ been applied to the corrected vertical angles of Table 11.2 and assuming an instrument in perfect adjustment, the eight angles would have defined a single straight line. This correction is unnecessary since the average of direct and reversed angles is free

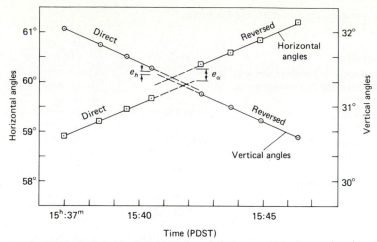

**Fig. 11.22** Vertical and horizontal angles plotted versus time for a solar observation.

from error. Similarly, the direct and reversed horizontal angles plot as two straight parallel lines separated by an amount $e_\alpha$ that represents the collimation error in the instrument (Art. 6.54). The average of the direct and reversed horizontal angles is the angle that will be used in subsequent calculations and is free from collimation error. The results of Fig. 11.22 indicate that there are no blunders in the measurements.

Figure 11.23 is an angle-time graph for a set of afternoon solar observations in which a blunder has occurred. The second direct vertical angle obviously is affected by a mistake. To reduce the effects of systematic instrumental errors, the second direct horizontal angle, the second reversed vertical and horizontal angles, and corresponding times are also rejected. The average values to be used in the azimuth computations include the remaining six vertical and horizontal angles and corresponding times. Computations of azimuth using the data from Fig. 11.21 and Table 11.2 are listed in Table 11.3. Figure 11.24 shows the

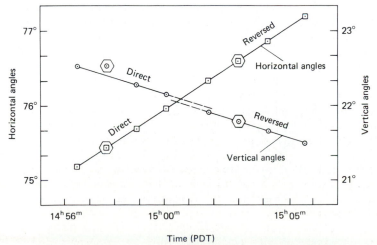

**Fig. 11.23** Angle-time plot revealing a blunder for a solar observation.

**Table 11.3   Calculation of Azimuth from Solar Observations**

Date: October 11, 1978
Latitude = 37°52′20″N
Instrument at △ 46, sight on △ 48
Average corrected vertical angle to center of sun = 31°20′35″
Average corrected horizontal angle (clockwise) to the
   center of the sun = 60°01′57″

| | |
|---|---|
| Average instant of observation | $= 14^h41^m35^s$ P.S.T. |
| Longitude of the 120th meridian | $= 8^h$ |
| Greenwich civil time of observation | $= 22^h41^m35^s$ |
| Declination of the sun at $0^h$ G.C.T. | |
|   from the ephemeris | $= -06°48′12″$ |
| Correction for 22.6931$^h$ of elapsed time | |
|   $= (22.7^h)(-0.94′/h)$ | $= -\quad 21′20″$ |
| Declination, at instant of observation | $= -07°09′32″$ |

Calculate Z using Eq. (11.2a).

$$\cos Z = \frac{\sin\delta - \sin\phi\sin h}{\cos\phi\cos h}$$

$$= \frac{\sin(-07°09′32″) - \sin(37°52′20″)\sin(31°20′35″)}{\cos(37°52′20″)\cos(31°20′35″)}$$

$$= -0.65849723$$

Thus, $Z = 131°11′07″$ counterclockwise from the north.

| | |
|---|---|
| Azimuth of the sun from north | $= 360° - 131°11′07″$ |
| | $= 228°48′53″$   clockwise from north |
| Clockwise angle backsight on | |
|   48 to sun | $=\quad 60°01′57″$   (see Fig. 11.21) |
| Azimuth of 46 to 48 | $= 168°46′56″$ |

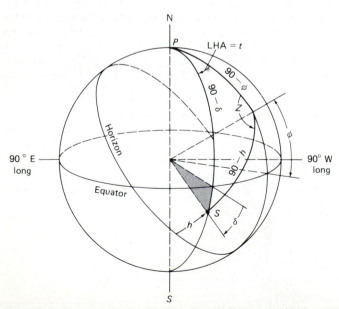

**Fig. 11.24** Spherical triangle for solar observation in Table 11-3.

**Table 11.4   Approximate Change in Calculated Azimuth of Sun for 01′ Change in Latitude, Declination, or Altitude, for Latitude 40°**

| | | | | | Change in azimuth for 01′ change in: | | |
| Months | Decli-nation | Hour angle | Alti-tude | Azi-muth | Lati-tude | Decli-nation | Alti-tude |
| --- | --- | --- | --- | --- | --- | --- | --- |
| November, December, January | −20° | $3^h15^m$ | 15° | $46\frac{1}{2}°$ | 1′10″ | 1′45″ | 1′25″ |
| March, September | 0° | $4^h40^m$ | 15° | 77° | 35″ | 1′25″ | 55″ |
| | 0° | $3^h20^m$ | 30° | 61° | 1′10″ | 1′45″ | 1′20″ |
| | 0° | $1^h30^m$ | 45° | 33° | 3′00″ | 3′20″ | 3′00″ |
| May, June, July | +20° | $5^h45^m$ | 15° | 104° | 04″ | 1′20″ | 45″ |
| | +20° | $4^h30^m$ | 30° | 92° | 35″ | 1′25″ | 50″ |
| | +20° | $3^h10^m$ | 45° | $77\frac{1}{2}°$ | 1′10″ | 1′45″ | 1′00″ |
| | +20° | $1^h45^m$ | 60° | 56° | 2′40″ | 2′55″ | 2′10″ |

astronomic triangle used in this computation. The average corrected horizontal and vertical angles and average time are used for this computation. The average P.D.S.T. is reduced to P.S.T. (Fig. 11.21), which in turn is converted to Greenwich civil time. The declination is obtained from the ephemeris for $0^h$ G.C.T. and is corrected for the elapsed time to the instant of observation. Then the declination $\delta$, the latitude of the place $\phi$ (given), and the true altitude $h$ are used in Eq. (11.2a) to calculate $\cos Z$. Since the observations were made in the afternoon ($14^h41^m35^s$ P.S.T.), the sun is west of the meridian and $Z$ is the counter-clockwise angle from north.

Hence, the azimuth of the sun from the north is $360° - Z$, (Fig. 11.24) [see the explanation following Eq. (11.2b), Art. 11.11]. The desired azimuth of the line is calculated by subtracting the clockwise angle from line 46-48 to the sun from the azimuth of the sun as given in Table 11.3 and illustrated in Fig. 11.21.

Use of average vertical angle, horizontal angle, and time implies that the sun is apparently traveling in a straight line, which of course is not true. Within a period of 10 min, however, the error introduced is so small as to be of little or no consequence.

The watch time does not have to be of high absolute accuracy since it is used only to check the validity of the measurements and to determine declination.


**Precision**   Under ordinary conditions the azimuth of a line can be determined within about 01′. The precision depends not only upon the precision of field observations and the exactitude of the corrections in altitude, but also upon the shape of the astronomical triangle. The $PZS$ triangle becomes weak as the sun approaches the meridian, and the solution becomes indeterminate at the instant of apparent noon. On the other hand, the refraction correction becomes large and very uncertain for low altitudes, particularly for those less than 10°. For these reasons, when possible, observations within the latitudes of the United States are usually taken between the hours of 8 and 10 A.M. or 2 and 4 P.M.

The effect of errors in the sides of the astronomical triangle upon the precision of the computed azimuth of the sun at the instant of observation is given by the accompanying Table 11.4, in which are shown the changes in azimuth of the sun due to changes of 01′ in the latitude, declination, and altitude at the latitude of 40°, which is about the mean for the United States. The changes in azimuth have been computed by means of Eq. (11.3). The values are approximate and are given for comparative purposes only. The months named are those during which the declination is not greatly different from the value given in the second column. Thus, during the period November, December, and January, the declination varies from $-15°$ to $-23°$. The hour angles give the approximate time interval before or after local apparent noon. It will be seen that, when the hour angle is $1^h 30^m$, a 01′ error in latitude, declination, or altitude produces an error of about 03′ in the azimuth. When the hour angle has increased to $3^h$, an error of 01′ in latitude or altitude produces an error of about 01′ in the azimuth. Under the given conditions, the effect of an error in declination is greater than the effect of an error of the same magnitude in either latitude or altitude.

It is important to note that although Table 11.4 is good for comparative purposes, rigorous error propagation techniques may be applied to Eq. (11.3) when a specific data set is given. If Eq. (11.3) is rewritten as

$$\cos Z_s = f(\phi, \delta, h)$$

its linearized form symbolically becomes

$$-\sin Z_s \, dZ_s = \frac{\partial f}{\partial \phi} \, d\phi + \frac{\partial f}{\partial \delta} \, d\delta + \frac{\partial f}{\partial h} \, dh$$

where the partial derivatives may be evaluated from Eq. (11.3). If the three quantities $\phi$, $\delta$, and $h$ are *uncorrelated* measurements, the differentials $dZ_s$, $d\phi$, $d\delta$, and $dh$ may be replaced by the standard deviations $\sigma_{z_s}$, $\sigma_\phi$, $\sigma_\delta$, and $\sigma_h$. In this way the precision of the calculated azimuth can be determined [see Eqs. (2.42) and (2.46), Art. 2.20].

**11.33. Time by observation on the sun at noon**  If the longitude of a station is known and the direction of the meridian has been established, the standard time may be determined accurately by observing the sun as it crosses the meridian at local apparent noon. In determining time in this manner, the transit is set up and carefully leveled over the north end of the meridian line, and a sight is taken along the meridian about one-half hour before noon. The line of sight is elevated to intercept the path of the sun, and at the instant of tangency between the west limb and the vertical cross hair, the time is noted. The telescope is quickly plunged, and a second sight is taken along the meridian. The line of sight is again elevated to intercept the path of the sun, and the time of tangency between the vertical cross hair and the east limb of the sun is observed. The mean of the two times thus observed is the watch time of upper transit of the sun's center, which is local apparent noon.

The longitude of the place expressed in hours, for reasons explained in Art. 11.20, is the Greenwich apparent time reckoned from noon. From the ephemeris giving values for the equation of time at Greenwich apparent noon or $0^h$ Greenwich civil time, the equation of time at the instant of observation is determined as illustrated by Examples 11.1 and 11.2, Art. 11.18. Since the equation of time is the difference between mean and apparent time at the instant of observation, it is clear that $12^h$ plus or minus the equation of time is the local mean time of local apparent noon, the sign being plus or minus depending on whether mean time is faster or slower than apparent time, as indicated by the ephemeris. The local mean time of local apparent noon is changed to standard time by subtracting the difference in longitude (in hours) between the place of observation and the standard-time meridian if the place is east of that meridian and by adding if west. The computed standard time of local apparent noon is the correct watch time of the observation, and a comparison with the observed watch time indicates the correction to be applied to the watch time. Example 11.13 shows the correction for Pacific standard time.

**Example 11.13**  The sun's center is observed to pass the meridian at a given place at $11^h 53^m 50^s$ P.S.T. on November 13, 1978. The longitude of the place is W122°15′40″. Determine the watch correction.

SOLUTION

Longitude W122°15′40″ $= 8^h 09^m 02.7^s = 8.1508^h$

| | |
|---|---|
| L.A.T. of observation | $12^h00^m00.0^s$ |
| Longitude west of Greenwich | $8^h09^m02.7^s$ |
| Greenwich apparent time (G.A.T.) | $20^h09^m02.7^s$ |
| Equation of time at $0^h$ G.C.T. (ephemeris) | $+\quad15^m47.0^s$ |
| Change in equation of time $=(20.1508^h)(0.34^s/h)=$ | $-\qquad06.9^s$ |
| G.C.T. = G.A.T. $-$ equation of time | $=\quad19^h53^m22.6^s$ |
| Longitude to P.S.T. | $8^h$ |
| Pacific standard time of local apparent noon | $11^h53^m23^s$ |
| Watch time | $11^h53^m50^s$ |
| Watch correction | $18^s$ |
| Correction to watch time to get standard time | $-18^s$ |

The length of time taken by the sun in crossing the meridian depends somewhat upon the sun's declination and semidiameter, but it is approximately $2\frac{1}{2}$ min. Therefore, it is clear that no time can be lost in reversing the telescope for a second sight along the meridian preparatory to observing the east limb of the sun. If for any reason it is impractical to observe more than one limb of the sun, the time in seconds earlier or later than the sun's center passes the meridian may be taken as approximately 4(semidiameter of the sun)$/\cos\delta$, where the semidiameter is about $16'$ and $\delta$ is the declination of the sun.

**11.34. Longitude by observation on the sun at noon**   If the standard time is known precisely and the meridian has been established, the longitude of a place may be determined by an observation on the sun at local apparent noon, the field procedure being in all respects identical with that just described for finding time.

With the standard time of passage of the center of the true sun (local apparent noon) known, the Greenwich civil time of local apparent noon can be computed, and the equation of time at the instant of local apparent noon can be found readily from an ephemeris giving values for $0^h$ Greenwich civil time. The standard time of local mean noon (meridian passage of the mean sun) differs from the standard time of local apparent noon by an amount equal to the equation of time, being greater if the mean sun is behind the true sun and less if ahead of it. The difference between the standard time of local mean noon and $12^h$ standard time is the difference in longitude $\Delta\lambda$ (in time units) between the meridian of the place and the standard-time meridian; if local mean noon occurs before $12^h$ standard time, the place is east of the standard meridian; if after, west.

**Example 11.14**   The sun's center is observed to pass the meridian at a given place at $11^h30^m12.2^s$ A.M. P.S.T. on December 2, 1978. Determine the longitude of the place.

SOLUTION

| | |
|---|---|
| P.S.T. December 2, 1978 of local apparent noon | $11^h30^m12.2^s$ |
| Longitude to P.S.T. (120th meridian) | $8^h$ |
| G.C.T. of observation | $19^h30^m12.2^s$ |
| Equation of time at $0^h$ G.C.T. (ephemeris) | $+\quad10^m50.6^s$ |
| Change in equation of time $=(19.5034)(0.96)$ | $-\qquad18.7^s$ |
| G.A.T. of local apparent noon | $19^h40^m44.1^s$ |
| | $-12^h$ |
| G.H.A. of meridian of observation | $7^h40^m44.1^s$ |

Longitude of meridian of observation $=(7.678917^h)(15)=115°11'01''$

**11.35.  Time and longitude by altitude of the sun at any time**   When the azimuth of the sun
is determined by direct solar observation as described in Art. 11.32, the true altitude of the
center of the sun can be obtained by correcting the measured vertical angles for parallax,
refraction, and semidiameter of the sun. Using the true altitude of the sun's center $h$, the
declination of the sun $\delta$, and the latitude of the place of observation determined from
observations (Art. 11.31) or scaled from a map, the local hour angle (L.H.A.), $t$, can be
calculated by any one of Eqs. (11.4), (11.6), or (11.8). This angle $t$ is the hour angle before or
after local apparent noon, which occurs when the sun is over the meridian of the place of
observation or point $Z$ in the astronomic triangle. In Fig. 11.24, the L.H.A. $= t$ converted to
units of time is the time past local apparent noon. Local apparent time plus $12^h$ gives
Greenwich apparent time, which can be converted to Greenwich mean time by applying the
equation of time obtained from the ephemeris for the date and hour of observation. Finally,
Greenwich mean time is converted to standard time by correcting for the longitude of the
place. For best results, several observations should be made to the upper and lower limbs of
the sun. Half the observations should be with the telescope direct and half with the telescope
reversed. With the local hour angle known, the longitude can then be determined.

The solution of the spherical triangle is weak when the altitude of the sun is low or the sun
is too close to the meridian. Consequently, it is advisable to make the measurements between
$8^h$ and $10^h$ and $14^h$ and $16^h$ local civil time to obtain the best results.

**Example 11.15**   The average corrected vertical angle to the center of the sun is $31°11'06''$. The
place of observation has latitude $37°52'20''$ and longitude $122°15'24''$. The approximate observed
time of observation is $14^h42^m40^s$ P.S.T. on October 11, 1977. Determine a watch correction based on
time computed using the altitude of the sun.

SOLUTION

| | |
|---|---|
| P.S.T. of observation (approximate watch time) | $14^h42^m40^s$ |
| Longitude to the 120th meridian | $8^h$ |
| G.C.T. of observation (approximate) | $22^h42^m40^s$ |
| Declination $\delta$ at $0^h$ G.C.T. (ephemeris) | $-6°53'42''$ |
| Correction to time of observation $= (22.7111)(0.94) =$ | $-21'21''$ |
| Declination $\delta$ at instant of observation | $-7°15'03''$ |

Now, calculate the local hour angle (L.H.A.) $= t$ using Eq. (11.4).

$$\cos t = \frac{\sin h - \sin\phi\sin\delta}{\cos\phi\cos\delta}$$

$$= \frac{\sin(31°11'06'') - \sin(37°52'20'')\sin(-7°15'03'')}{\cos(37°52'20'')\cos(-7°15'03'')}$$

$$= 0.76019562$$

$$t = 40°31'06'' = 2^h42^m04^s$$

where the angle $t$ is the hour angle after local apparent noon. Thus,

| | |
|---|---|
| Local apparent time (L.A.T.) | $14^h42^m04^s$ |
| G.A.T. $=$ L.A.T. $+ 8^h$ | $22^h42^m04^s$ |
| Equation of time at $0^h$ G.C.T., October 11, 1977 (ephemeris) | $+\quad 13^m06.0^s$ |
| Change in equation of time $= (22.7161)(0.64) =$ | $+\quad 14.5^s$ |
| Equation of time at $22^h42^m58^s$ G.A.T. | $+\quad 13^m20.5^s$ |
| G.M.T. at meridian of observation $=$ G.A.T. $-$ equation of time $=$ | $22^h28^m43.5^s$ |

This G.M.T. must be corrected for the change in longitude from the meridian of observation to the meridian of P.S.T. Thus,

$\Delta\lambda = 122°15'24'' - 120° = 2°15'24'' =$                        $0^h09^m01.6^s$

G.C.T. at 120th meridian = G.M.T. at meridian of observation $+ \Delta\lambda =$   $22^h37^m45.1^s$

P.S.T.                                                       $14^h37^m45.1^s$

Average watch time                                     $14^h42^m40^s$

Watch correction                                       $-04^m55^s$

The longitude can be determined using the same general procedure when the time of observing the altitude of the sun is sufficiently accurate. The time should be observed using a watch corrected by checking against a time signal on radio station WWV (Art. 11.21). For best results, the altitude of the sun should be measured using a theodolite with a least count of 01".

**Example 11.16** The average corrected zenith angle to the center of the sun was 58°47'04" at an average time of observation of $14^h38^m04^s$ P.S.T., October 11, 1978, in a place where the latitude is known to be 37°52'20"N. Using the methods outlined in Example 11.15, the local hour angle $t$ is found to be $2^h42^m22.1^s$ or 40°35'31.5". Determine the longitude of the place of observation.

SOLUTION

| | |
|---|---:|
| $t$ = time elapsed after local apparent noon | $2^h42^m22.1^s$ |
| | $+12^h$ |
| L.A.T. at meridian of observation | $14^h42^m22.1^s$ |
| Equation of time at $0^h$G.C.T., October 11, 1978 (ephemeris) | $13^m01.7^s$ |
| Change in equation of time to 22 $38^m04^s$ (G.C.T.) = (22.6344)(0.65) = | $14.7^s$ |
| Equation of time at instant of observation | $13^m16.4^s$ |
| Local mean time (L.M.T.) at meridian of observation = L.A.T. − equation of time = | $14^h29^m05.7^s$ |
| $\Delta\lambda$ = P.S.T. of observation − L.M.T. = $14^h38^m04^s - 14^h29^m06^s = 0^h08^m58^s =$ | $2°14'30''$ |
| Longitude to P.S.T. | $+120°$ |
| Longitude of the place of observation | $122°14'30''$W |

## OBSERVATIONS ON STARS

**11.36. General** In general, the methods of determining azimuth, latitude, longitude, and time by direct solar observations are, with slight modifications, applicable to observations on the stars. If a high degree of precision is not required, the same procedure may be followed for stellar observations as for solar observations. Usually, however, it is expected that a higher degree of precision will be obtained; consequently a corresponding degree of refinement is necessary, and special care is taken to eliminate systematic errors.

The measurement of horizontal and vertical angles to celestial bodies is discussed in Art. 11.23. The refraction correction is described in Art. 11.28. For observations on stars, no correction is required for parallax or semidiameter.

As there are fixed stars in all parts of the heavens, it is an easy matter to select a star or stars in a celestial region favorable to a precise determination of the quantity sought. Thus, conditions favorable to a precise determination of latitude by measuring the altitude of a star at culmination are (1) a fairly high altitude, in order that the uncertainty of the refraction correction be small, and (2) a rate of apparent movement that is small, in order that a series of observations may be taken without an appreciable change in the altitude. Within the latitudes of the United States, stars near the pole satisfy these conditions.

Similarly for precise determination, an observation for azimuth by measured altitude and known declination and latitude should be taken on a star in the east or west far enough above the horizon to eliminate the uncertain refraction but not so near the meridian as to produce a weak astronomical triangle. An observation for azimuth with the hour angle, declination, and latitude known should be taken on a circumpolar star—the nearer the pole the better—since the azimuth of such a star changes more slowly in a given length of time than does the azimuth of a star near the equator, and hence any error in time will have less effect.

For determinations of longitude or time, stars should be chosen near the equator because they are apparently traveling more rapidly than those near the pole.

The right ascension (in terms of sidereal hour angle) and declination for many stars are given in the *Nautical Almanac*. In the *Ephemeris of the Sun and Polaris* are given the G.C.T. and the declination for many stars at the instant of their upper transit at the meridian of Greenwich, at intervals of one-half month. Since for the fixed stars these coordinates change very slowly, it is not necessary to determine values for the hour of observation, as with the sun. The sidereal time corresponding to any given solar time can be found, and the hour angle can be computed by the expression $t = \theta - \alpha$, as explained in Art. 11.6.

Stars can be identified by means of charts which show the various constellations. For many stellar observations, however, the published direction and altitude of the star can be set off on the transit with sufficient precision that the star will be brought into the field of view at a given time, and it is not necessary to distinguish the star from among its neighbors. In fact, observations are often taken on stars during daylight hours near the hour of darkness, even when the stars are invisible to the naked eye.

In sighting on a star, the objective should be focused until the star appears as a fine, brilliant point of light. Before looking for a star just before sunset or after sunrise, the objective may be focused approximately by sighting at a distant object in the landscape. The proper position of the objective slide for focus on a star may be permanently marked on the barrel of the transit telescope.

Artificial illumination is required to make the cross hairs of the instrument visible in the darkness. Some instruments are equipped with a reflector sleeve which slips over the objective as a sun shade does. When a flashlight is held to one side of the reflector, the field of view is faintly illuminated and both cross hairs and the star can be seen. On instruments not so equipped, the cross hairs can be illuminated by holding the flashlight several inches in front of the objective and a little to one side of the telescope barrel, thus causing the rays to enter the telescope diagonally. After some experimenting, a position of the light will be found where both cross hairs and star are visible. Modern optical repeating and direction theodolites are equipped with built-in illumination for the reading scales and cross hairs, thus simplifying nighttime operations.

The location of any terrestrial mark used in observing is indicated by a light. The mark may be a strongly illuminated target, the source of illumination being shielded from the observer. Targets adaptable to instrument tripods, such as the one illustrated in Fig. 6.67, can be equipped with lights for nighttime observation.

**11.37. Polaris**   The polestar, Polaris ($\alpha$ Ursa Minor), is the star predominantly used for observations for latitude and azimuth in the latitudes of the United States. Its distance from the pole is approximately 1°. Its annual change in polar distance (or in declination) is less than 01′, and its maximum daily change in polar distance is less than $\frac{1}{2}''$. It is a second-magnitude (or quite bright) star the position of which is readily identified by the neighboring constellations of Ursa Major and Cassiopeia. Figure 11.25 shows the position of Polaris with

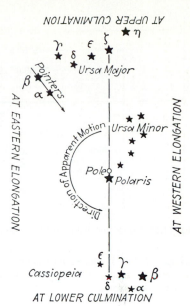

**Fig. 11.25** Positions of constellations near the North Pole when Polaris is at culmination and elongation.

respect to the pole and to these constellations. The seven most brilliant stars in the constellation of Ursa Major are known as the Great Dipper; and the two stars forming the part of the bowl farthest from the handle are called the *pointers* because a line through these stars points very nearly to the celestial north pole. It will be noted that the constellation of Cassiopeia is on the same side of the pole as Polaris, so that when Cassiopeia is above the pole, Polaris is near upper culmination; when Cassiopeia is west of the pole, Polaris is near western elongation; and so on. The position of Polaris relative to the pole may be quite closely estimated by noting the positions of δ Cassiopeia and ζ Ursa Major. A line joining these two stars passes nearly through the pole and Polaris. The line is nearly vertical when the star is at either culmination, and nearly horizontal when the star is at either elongation.

Observations to determine latitude are usually made when Polaris is at upper or lower *culmination* (Art 11.15), when the star appears to be moving almost horizontally for some time. Observations to determine azimuth are usually made when Polaris is at eastern or western *elongation* (Art. 11.12), when it appears to be traveling vertically for some time.

Data concerning Polaris can be found from a variety of sources. The *Ephemeris of the Sun and Polaris* gives the declination of Polaris and the Greenwich mean time of its culmination and elongation at the meridian of Greenwich for each day of the current year. The data in the *American Ephemeris* and the *Nautical Almanac* are not in convenient form for use by the surveyor. Convenient tables are found in the ephemerides published by the manufacturers of surveying instruments.

Since Polaris, in common with other fixed stars, travels at an angular rate more rapid than that of the sun, it follows that at a given meridian it arrives at culmination a little earlier by mean solar (civil) time each day than it did the day before, the amount earlier being approximately equal to the gain of sidereal time on mean solar time for a $24^h$ interval, or $3^m56.6^s$ per day or $9.86^s$ per hour of solar time (Art. 11.19). Furthermore, the time of culmination depends upon the longitude of the place. To determine the local mean time of upper culmination of Polaris at any given meridian on any given date, the value for the meridian of Greenwich is taken from the table and is reduced to the longitude of the place by means of the variation per hour.

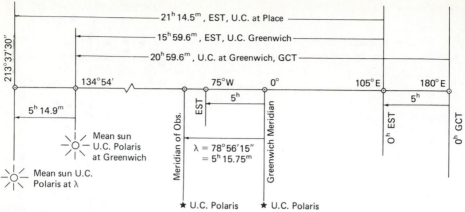

**Fig. 11.26** Upper culmination of Polaris (Example 11.17).

**Example 11.17**   Find the Eastern standard time of upper culmination (U.C.) of Polaris on December 9, 1978 at a place where the longitude is $78°56'15''W = 5^h15^m45^s = 5.2625^h$. Figure 11.26 illustrates the time relationships.

SOLUTION

| | |
|---|---|
| G.C.T. of U.C., December 6, 1978 (ephemeris) | $21^h11.4^m$ |
| Correction to December 9 = (3)(3.95) (ephemeris)   $= -$ | $11.8^m$ |
| G.C.T. of U.C. at Greenwich meridian on December 9 | $20^h59.6^m$ |
| Longitude west of Greenwich to E.S.T.   $= -$ | $5^h$ |
| E.S.T. of U.C. at Greenwich meridian   $=$ | $15^h59.6^m$ |

The longitude to the place of observation is 5.2625 sidereal hours. Polaris must move through this interval to cross the meridian at the place of observation. To convert a sidereal interval into a mean solar interval, $9.83^s$/sidereal hour are subtracted from the sidereal interval (Art. 11.19). The correction is

$(9.83^s/h)(5.2625^h) = -0^m51.7^s = -0.86^m$

The longitude corrected to yield the solar interval between upper culmination (U.C.) at Greenwich and U.C. at place is

corrected longitude $= (5^h15.75^m) - 0.86^m = 5^h14.9^m$

Thus,

| | |
|---|---|
| E.S.T. of U.C. at Greenwich (from above)   $=$ | $15^h59.6^m$ |
| Solar interval U.C. at Greenwich to U.C. at place $= +$ | $5^h14.9^m$ |
| E.S.T. of U.C. at place of observation   $=$ | $21^h14.5^m$ |

To find the time of lower culmination the quantity $12^h$ minus one-half of the variation per day (one-half of $3^m56^s$, or $1^m58^s$) is added to or subtracted from the time of upper culmination. Thus, for the data in this example

| | |
|---|---|
| E.S.T. of U.C. at place of observation | $21^h14.5^m$ |
| Interval between U.C. and L.C. | $-11^h58.0^m$ |
| E.S.T. of lower culmination on December 9, 1978 | $9^h16.5^m$ |

**11.38. Latitude by observation on Polaris at culmination**   As shown in Art. 11.9, the latitude of a place is equal to the altitude of the elevated pole or, if $h$ is the true altitude of any circumpolar star as it crosses the meridian, then the latitude $\phi$ is

$$\phi = h \pm p \qquad\qquad (11.20)$$

in which the sign preceding the polar distance $p$ is positive or negative depending on whether the star is at lower or upper culmination. By this method the latitude of a station is determined by measuring the altitude of Polaris when it is at either upper or lower culmination, and by applying to this altitude, corrected for refraction, the star's polar distance as given in Polaris tables in an ephemeris. Inasmuch as the star is apparently traveling in a horizontal line when at either of these two positions, it is not essential to the precision of the latitude determination that the time of culmination be found, but in any case it facilitates the work of observing if the approximate time is known.

Further, it is not essential that the altitude of the star be observed at the instant it crosses the meridian. For some minutes before and after culmination, the star travels in so nearly a horizontal line that, with the ordinary transit, vertical movement cannot be detected. Within the period $6^m$ before to $6^m$ after culmination, the maximum change in altitude is only $01''$, and within $12^m$ before to $12^m$ after culmination the maximum change in altitude is only $0.1'$.

The procedure to be employed in making an observation depends somewhat upon the precision with which the latitude is to be determined and upon the precision with which the time is known. For an observation with the engineer's transit having a vertical circle reading to minutes, when the watch time or the longitude of the place may be in doubt by a few minutes, the standard time of culmination at the given station is determined from an ephemeris as illustrated in the preceding article. A few minutes before the estimated time of culmination, the transit is set up and is leveled very carefully; as a final test the telescope bubble should remain centered as the transit is revolved about the the vertical axis. The star is found with the naked eye by noting its position with respect to the neighboring constellations shown in Fig. 11.25. The telescope is focused for a star. If the latitude is known approximately, its estimated value, plus or minus the star's polar distance, is set off on the vertical circle to facilitate finding Polaris. The telescope is sighted at Polaris. When the star has been brought within the field of view, the cross hairs are illuminated if necessary, and the star is continuously bisected with the horizontal cross hair. When during a period of 3 or 4 min Polaris no longer appears to move away from the hair but moves horizontally along it, the star is practically at culmination. The vertical angle is read with dispatch, the transit is carefully releveled, the telescope is plunged, and a second observation on the star is taken with the telescope inverted. Usually, the instrument is releveled and a second pair of observations is made. The mean of the observed altitudes, corrected for refraction (Table II, Appendix C) and index error, is taken as the true altitude of the star. The polar distance can be found from any ephemeris giving either declinations or polar distances of Polaris for the days of the current year. Finally, the latitude is computed by applying to the true altitude the polar distance with proper sign. Under ordinary conditions, by this method the latitude can be determined within about $01'$, or less if the mean of several observations is taken.

**Precise determination**    When it is desired to determine the latitude within a few seconds and the standard time and longitude of the place are known within a minute or so, the watch time of culmination may be precisely computed as illustrated in Example 11.17, and a series of observations may be taken on the star when it is near culmination. The observing program is usually arranged so that an equal number of observations will be taken before and after culmination. The observations are begun at a given time interval (usually not more than $10^m$) before the calculated time of culmination, and at each sighting of the star, the watch time and the altitude are observed. Half of the observations are taken with the telescope direct and half with it reversed, and between pairs of observations the instrument is carefully releveled. The observed altitudes of the star for positions other than culmination

**Table 11.5  Corrections to Be Applied to Altitudes of Polaris Near Culmination to Give Altitude at Culmination**

| Interval from culmination, minutes of time | Change in altitude from culmination, seconds of arc |
|:---:|:---:|
| 3 | 00 |
| 6 | 01 |
| 9 | 03 |
| 12 | 06 |
| 15 | 09 |
| 18 | 12 |
| 21 | 17 |
| 24 | 22 |
| 30 | 34 |

are reduced to the altitude at culmination by applying a correction which, for altitudes within the United States, is given approximately in Table 11.5.

When the star is near lower culmination, the correction is subtracted; when near upper culmination, the correction is added. The mean of the altitudes reduced to culmination is corrected for refraction (Table II, Appendix C), the polar distance is found, and the latitude is computed as for the case described in the preceding section.

When culmination of Polaris occurs during the daylight hours, latitude by observation of Polaris at any time can be determined by the following equation:

$$\phi = h - p\cos t + \tfrac{1}{2}p^2\sin^2 t \tan h \sin 1'' \tag{11.21}$$

in which $h$ is the true altitude of Polaris, $p$ the polar distance of Polaris expressed in seconds, and $t$ the local hour angle of Polaris. Thus, it is necessary to measure the vertical or zenith angle to Polaris and record the watch time of each observation. The angle should be observed at least once with the telescope direct and once with the telescope reversed. The average angle is then corrected for refraction to obtain the true altitude. The average watch time is then used in conjunction with the hour angle of Polaris, obtained from the ephemeris, to determine the local hour, $t$. These data are substituted into Eq. (11.21) to solve for the latitude. Consider an example to illustrate the procedure.

**Example 11.18**  The true altitude of Polaris $= 39°40'00''$ observed at $20^h30^m15^s$ E.S.T., July 20, 1978. The longitude of the place is $76°20'00''$W. Determine the latitude for the place of observation.

SOLUTION  The ephemeris used gives the G.H.A. of Polaris at $0^h$ G.C.T. for every day of the year. To compute the L.H.A. $= t$, determine the G.C.T. of observation and correct the G.H.A. of Polaris for the elapsed *sidereal* interval of time. Then $t =$ G.H.A. at the instant of observation minus the west longitude of the place of observation. Figure 11.27 shows hour angles at $0^h$ G.C.T. and at time of observation, respectively.

E.S.T. of observation, July 20, 1978    $22^h30^m15^s$

Longitude west from Greenwich         $5^h$
_____

G.C.T. of observation, July 21, 1978   $3^h30^m15^s = 3.5042^h$

The local hour angle (L.H.A.) is equal to the G.H.A. of Polaris minus the longitude of the place. However, since $0^h$ G.C.T. (Fig. 11.27a), $3.5042^h$ of solar time have elapsed (Fig. 11.27b). During this period of time, Polaris travels faster than the sun by $3.94^m$ per day or $(3.94^m)(60^s)/24^h = 9.86^s/h$, which is the difference between sidereal and solar time (Art. 11.19). Consequently, the elapsed time interval since $0^h$ G.C.T. must be increased by $(3.5042^h)(9.86^s)$, or the sidereal interval, since $0^h$ G.C.T. equals $(3^h30^m15^s) + (3.5042^h)(9.86^s) = 3^h30^m49.6^s = 3.5138^h$. Converting this interval to

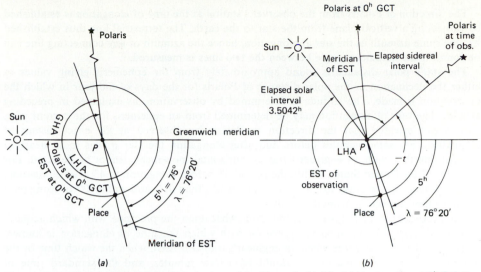

**Fig. 11.27** Latitude from a sight on Polaris at any time (Example 11.18). (*a*) Hour angles at $0^h$ G.C.T. (*b*) Hour angles at time of observation.

degrees, we obtain:

| | | |
|---|---|---|
| Interval since $0^h$ G.C.T. $= (3.5138^h)(15°/h)$ | $=$ | $52°42.4'$ |
| G.H.A. Polaris $0^h$ G.C.T., July 21 (ephemeris) | $=$ | $265°39.9'$ |
| G.H.A. Polaris at instant of observation | $=$ | $318°22.3'$ |
| West longitude | $= -$ | $76°20'$ |
| L.H.A. | $=$ | $242°02.3'$ |
| $t =$ L.H.A. $- 360°$ (Art. 11.6) | $=$ | $-117°57'42''$ |

The latitude is then calculated by Eq. (11.21) using the polar distance, $p$, for Polaris (from the ephemeris).

$$\phi = h - p\cos t + \tfrac{1}{2}p^2\sin^2 t \tan h \sin 1''$$
$$= 39°40'00'' - (3024.6)\cos(242°02.3')$$
$$\quad + \tfrac{1}{2}(3024.6)^2\sin^2(242°02.3')\tan(39°40')(0.000004848)$$
$$= 39°40'00'' + 1418'' + 14'' = 39°40'00'' + 23.9'$$
$$= 40°03.9'$$

**11.39. Azimuth by observation on Polaris at elongation**   The azimuth of a line can be determined conveniently by an observation on Polaris at eastern or western elongation, provided the latitude of the place is known. As shown in Art. 11.12, the angle $Z$ in the spherical triangle containing any star at elongation is given by Equation (11.9), repeated here for convenience.

$$\sin Z = \frac{\sin p}{\cos \phi} \tag{11.9}$$

in which $Z$ is the angle east or west or north depending on whether the star is at eastern or western elongation (Art. 11.12), $p$ is the star's polar distance, and $\phi$ is the latitude of the place. Figure 11.11 shows the astronomic triangle for a Polaris observation at elongation. The azimuth of Polaris at elongation is also given in ephemerides.

The direction of Polaris from the observer's station at the time of elongation is established by projecting a vertical plane from the star to the earth. The terrestrial line thus established has the same azimuth as the star at elongation, hence the azimuth of any connecting line can be found if the horizontal angle between the two lines is measured.

The star's polar distance is found approximately from the ephemeris giving values of either the declination or the polar distance of Polaris for the days of the year in which the observation is made. The latitude is determined by observation, as explained in preceding articles. The time of elongation may be determined from an ephemeris for the current year.

It is not essential that the direction to Polaris be observed at the exact instant of elongation. For some minutes before and after elongation the star travels in so nearly a vertical line that, with an engineer's transit, horizontal movement cannot be detected. For latitudes of the United States, within the period $4^m$ before elongation to $4^m$ after elongation, the maximum change in the azimuth of Polaris is less than $01''$, and within $10^m$ of elongation the maximum change in azimuth is only $0.1'$.

The procedure to be followed depends somewhat upon the precision with which azimuth is to be determined and upon the precision with which the time of elongation is known. First, consider the procedure when an engineer's transit is used, when the watch time or the longitude of the place may be in doubt by a few minutes, and the standard time of elongation is determined by use of an ephemeris. A few minutes before the estimated time of elongation, the transit is set up over a given station and is leveled very carefully. The telescope is focused for a star, the latitude of the place is laid off on the vertical circle to facilitate finding the star, and the transit is revolved about the vertical axis until Polaris comes within the field of view. The horizontal and vertical motions are then clamped, the cross hairs are illuminated if necessary, and the star is continuously bisected with the vertical cross hair. When during a period of 2 or 3 min Polaris no longer appears to move away from the hair but moves vertically along it, the star is practically at elongation. The telescope is depressed, and a point on the line of sight is marked on a stake or other reference monument 300 ft or 100 m or more away. The telescope is then plunged, and another sight is taken on Polaris. The line of sight is again depressed, and a second point is set on the stake beside the first. Usually, the transit is releveled and a second pair of observations is made.

Later the mean of the points is found and marked on the stake. The line joining the occupied station with the established mean point defines the direction of Polaris at elongation. Its azimuth either is computed by Eq. (11.9) or is found directly from tables. The azimuth of any other line through the station can be determined by measuring the horizontal angle between the two lines, by the method of repetition (Art. 6.39).

The precision of azimuth determination by this method necessarily depends upon the quality of the instrument, the care and skill of the observer, and the number of observations. For the procedure described, under ordinary conditions the error should not exceed $10''$.

**Precise determination**   When the azimuth of a line is to be established within $02''$ or $03''$, an optical theodolite with a sensitive striding level for the horizontal axis should be employed, and the standard time and longitude should be known within a minute or so in order that the time of elongation may be computed with precision.

The hour angle of the star when at elongation can be precisely determined as explained in Art. 11.12 by Eq. (11.10) in which $\cos t = \tan\phi \tan p$. The hour angle $t$ expressed in time is practically the sidereal time interval between upper culmination and eastern or western elongation. The corresponding mean solar time interval is found by deducting from the computed hour angle a correction of $9.83^s$ per hour which, as explained in Art. 11.19, is the difference in solar time between the sidereal hour and the mean solar hour. The standard time of upper culmination of the star is found as shown in Example 11.17 of Art. 11.37. The standard time of eastern or western elongation is then determined by adding to or subtracting from the time of culmination, the mean time interval between upper culmination and elongation. Following is a numerical example.

**Example 11.19** It is desired to find the Eastern standard time of western elongation of Polaris occurring in the early morning hours of December 10, 1978, at a place where the longitude is $5^h15^m45^s$ west of Greenwich and the latitude is $50°00'00''$ north.

SOLUTION From an ephemeris, the polar distance, $p = 0°49'45''$. Thus, by Eq. (11.10),

$\cos t = \tan\phi\tan p = \tan(50°00'00'')\tan(0°49'45'')$
$\qquad = 0.01724789$
$\qquad t = 89°00'42''$

Local sidereal hour angle at time of observation

$\qquad = (89°00'42'')/15 = 5.9342^h =$                 $5^h56^m03^s$

Reduction, sidereal to mean solar interval (Art. 11.19)

$\qquad = (5.9341)(9.83) =$                       $-\qquad 58^s$

Mean time interval from upper culmination (U.C.)        $5^h55^m05^s$

From Example 11.17, E.S.T. of U.C. at place of

observation, on December 9, 1978           $+21^h14^m30^s$

E.S.T. of western elongation, December 10, 1978 =     $3^h09^m35^s$

In the *Ephemeris of the Sun and Polaris,* the mean time of elongation for the Greenwich meridian and latitude 40° is given for each day. In other ephemerides, time of elongation is given for selected days for each month for the Greenwich meridian and latitude 40°. Thus, time of elongation can be found by interpolation from the tables and applying a correction for the difference in latitude from 40°.

**Example 11.20** Determine the E.S.T. of western elongation of Polaris for the data of Example 11.19, by using tabular data in the ephemeris.

SOLUTION

G.C.T. of western elongation on December 7, 1978 (ephemeris)   $3^h07.6^m$

Correction to December 10, 1978 = (3.93)(3)         $-\quad 11.8^m$

Western elongation, December 10, 1978, at the Green-

wich meridian at 40°N latitude                  $2^h55.8^m$

Correction for longitude = $(5.263^h)(9.83)^{s/h}$         $-\quad 0.9^m$

Western elongation at meridian of observation      $2^h54.9^m$

Reduction to E.S.T. $= 5^h15^m45^s - 5^h =$         $+\quad 15.8^m$

E.S.T. of western elongation, December 10, 1978

at 40°N latitude                           $3^h10.7^m$

Correction to 50°N latitude (from ephemeris)       $-\quad 1.2^m$

E.S.T. of western elongation at place, December 10, 1978    $3^h09.5^m$

With the time of elongation determined, a series of observations is taken on Polaris, the program being so timed that approximately one-half of the observations will occur in an interval of a few minutes before the instant of elongation, and the remainder will occur in a like period after elongation. If the theodolite is equipped with a striding level, the horizontal axis is leveled each time the star is sighted. If not so equipped, the instrument is carefully leveled with the telescope bubble prior to each set, which consists of two observations, one with the telescope direct and the other with the telescope reversed.

For each set of points marked on the distant reference monument, a mean is taken, and the azimuth of the star at the mean of the times of the two observations comprising each set is found by applying a slight correction to the azimuth at elongation, this correction being found in Table III, Appendix C. It will be noted that for the average latitude of the United States this correction amounts to less than 01″ when the star is $4^m$ from elongation, and about 05″ when the star is $10^m$ from elongation.

The distance from reference monument to transit station is measured. The mean mark for each set of two observations is corrected to give the equivalent mark at the instant of elongation, by calculating the linear offset (for the distance from transit to mark) corresponding to the angular correction found in Table III. If it were not for the random errors connected with the observations, the points thus determined would coincide. The mean of the group is taken as the point which gives the best estimate for the direction of the star when at elongation.

For observations of this character the stability of the instrument during the course of the measurements is of the greatest importance. Preferably the instrument should be removed from the tripod and placed on a concrete pier. Changes in temperature may also seriously affect the relations between the fundamental lines of the instrument, and hence it should be allowed to come to the temperature of the air before observations are begun. Also, the instrument should be protected from wind.

It should be noted that a given error in latitude produces a relatively small error in azimuth. For latitudes of the northern part of the United States, an error of 01′ in latitude produces an error of about 02″ in azimuth, and for lower latitudes the effect is less. Since the latitude can easily be determined within 20″ with an engineer's transit, it is evident that the principal error in azimuth is likely to be due, not to errors in the computed value of the azimuth of Polaris, but to the field operations of projecting the direction of the star to the earth. If the instrument is equipped with a full vertical circle, the procedure is such that practically all instrumental errors of projecting the direction of the star to the ground, except that due to the vertical axis not being truly vertical, are eliminated. Thus, the need arises for extreme care in leveling the instrument and the necessity for a striding level when precise results are to be achieved.

**Example 11.21**   Using the data of Examples 11.19 and 11.20, compute the azimuth of Polaris for the time and place of observations. From Example 11.19, $p = 0°49'45''$ and the latitude, $\phi = 50°00'00''$.

SOLUTION   By Eq. (11.9),

$$\sin Z = \frac{\sin p}{\cos \phi} = \frac{\sin(0°49'45'')}{\cos(50°00'00'')}$$

$$= 0.02251316$$
$$Z = 1°17'24''$$

where $Z$ is the direction of Polaris to the west of north. This same value can also be obtained directly from the ephemeris.

**Example 11.22**   With the conditions as given in Example 11.21, an observation is taken $10^m$ after western elongation. What is the azimuth of Polaris at the given instant?

SOLUTION

From Example 11.2, the direction of Polaris

at western elongation is                          $= -1°17'24''$

From Table III, Appendix C, the correction $= +$          04″

Direction west from north of the star at the

given instant                                    $= -1°17'20''$

**Example 11.23**   For the observation of Example 11.22, the telescope is depressed and a point in the same vertical plane as the star is marked on a monument which is 400 ft (~122 m) from the instrument. What are the amount and direction of linear offset to be measured from this point to establish an equivalent point for Polaris when at western elongation?

SOLUTION   The angular correction is 04″ as found in Example 11.22.

offset = (400)(tan 04″) = (400)(0.0000194) = 0.008 ft

Since the star has reached its most westerly position and is now traveling east, the offset is made to the west of the given mark to indicate the direction of Polaris at western elongation.

**11.40. Azimuth by observation on Polaris at any time**   Although elongation is the most favorable time for precise determination of the azimuth of Polaris, it is often inconvenient or impossible to view the star when it is in this position. Under these circumstances, if the standard time and the longitude of the place are precisely known, the hour angle of the star at any instant can be found and the azimuth of the star at any instant can be determined, as described in Art. 11.13, by Eq. (11.13):

$$\tan Z = \frac{\sin t}{\cos \phi \tan \delta - \sin \phi \cos t} \tag{11.13}$$

or Eq. (11.11):

$$\sin Z = - \frac{\sin t \cos \delta}{\cos h} \tag{11.11}$$

in which $Z$ is the clockwise or counterclockwise angle from the north (Art. 11.11), $t$ is the hour angle of Polaris determined from the observed times as described in Example 11.18, $\phi$ the latitude for the place of observation scaled from a map or determined from observations (Art. 11.38), $h$ the true altitude of Polaris, and $\delta$ the declination of the star obtained from an ephemeris. Values of the azimuth of Polaris for various hour angles, declinations, and latitudes are given in ephemerides. By interpolation, the azimuth for any hour angle and declination can be found.

Field observation consists of measuring the horizontal angle between a terrestrial mark and the star. For azimuth determination to within ±30″, a single set of two observations, one with the telescope direct and the other with it reversed, should be taken with the time of the passing of the star across the vertical cross hair being recorded at each setting. The local hour angle $t$ is computed using the average time of observation and the G.H.A. of Polaris obtained from the ephemeris (Example 11.18, Art. 11.38). Using these data, the azimuth of Polaris is computed by Eq. (11.13). This azimuth combined with the average horizontal angle yields the azimuth of the reference line.

When the latitude of the place of observation is not known, the vertical or zenith angle is measured to Polaris. This angle, corrected for refraction (Table II, Appendix C), is the true altitude, $h$, which can be used in Eq. (11.11) to compute the azimuth of the star.

When a higher degree of precision is desired, a series of measurements is taken. At least six observations of the horizontal angle from a reference line to the star (and also the vertical or zenith angle, if necessary) should be made. Half of the measurements are taken with the telescope direct and half with it reversed. The time is also recorded to the nearest second for each observation. The azimuth of Polaris is computed by Eq. (11.13) using the local hour angle $t$ computed with the average time of observation and Greenwich hour angle of Polaris, and declination taken from the ephemeris. If the latitude is not known, the azimuth of Polaris is computed by Eq. (11.11) using the local hour angle $t$ and the average true altitude of the star. The azimuth of the star determined by one of these two methods is then combined with the average horizontal angle to the reference line to give the azimuth of that line.

**Precision**   The precision to be obtained depends upon the position of the star, the precision of observations of time, the number of observations, the quality of the instrument, and the care and skill of the observer. When the star is near upper or lower culmination, the azimuth changes at a relatively rapid rate, this change amounting to about 01′ of arc in $3^m$ of time for latitude 40°N. For this reason the method should not be expected to give precise results when Polaris is near culmination unless the time is observed precisely. It is also important that systematic errors due to inclination of the vertical axis be eliminated by carefully releveling the instrument. If the instrument is equipped with a striding level, the horizontal axis is leveled just before each observation when the telescope is pointed in the direction of the star. For best results, an optical theodolite equipped with a striding level should be used. Since the stability of the instrument is also extremely important for high precision, the theodolite should be set on a pier or other substantial support.

### 11.41. Observations on other stars

Latitude and azimuth by observation on any circumpolar star can be determined by methods identical with those for Polaris. The other stars near the pole are of less magnitude than Polaris and are therefore not so readily identified; but if the approximate direction of the meridian is known and the hour angle and declination of the star are known, the approximate altitude can be laid off and the telescope can be pointed so that the star will come within the field of view.

Latitude, azimuth, time, and longitude can be determined by observation on stars distant from the pole by methods similar to those described for the sun.

In the ephemerides for each year are tables giving the right ascensions or sidereal hour angles (S.H.A.) and declinations of selected stars for the upper transit at Greenwich. As the right ascension of the fixed stars changes but a small fraction of a second per day, through the relation $\theta = t + \alpha$ (Art. 11.6) it is seen that the hour angle $t$ of a star at any meridian other than Greenwich is readily found if the sidereal time is known, and that the sidereal time of upper culmination or upper transit of the star is equal to its right ascension $\alpha$. Further, knowing the longitude of the place of observation and having given the sidereal time of $0^h$ Greenwich civil time, the relation at a given instant between sidereal time and local civil or standard time can be determined as explained in Example 11.18, Art. 11.38. It is therefore possible to find (by aid of an ephemeris) not only the declination of a star but also its hour angle at any instant of mean solar or standard time. Solutions are expedited by using tables for the conversion of mean solar into sidereal time interval, or vice versa. Such tables are given in the *Nautical Almanac*.

**Example 11.24**  What is the Pacific standard time of the upper transit of Betelgeuse ($\alpha$ Orionis) on February 27, 1976, at a place the longitude of which is 122°15′30″W? To change to Pacific standard time, subtract $8^h$.

SOLUTION   Using the *Ephemeris of the Sun and Polaris*, by interpolation:

G.C.T. of transit of Betelgeuse

at Greenwich meridian                           $19^h26.9^m$

Sidereal interval between transit at place

and transit at Greenwich $= (122°15′30″)/15 = 8^h09.0^m$

Correction to mean solar interval (Art. 11.19)

$= (8.1506^h)(9.83^{s/h})$               $= - \underline{\quad 01.3^m}$

Mean time interval between transit at

Greenwich and transit at place           =      $8^h 07.7^m$

Greenwich mean time of upper transit

at place $= 19^h 26.9^m + 8^h 07.7^m$      =      $27^h 34.6^m$

Correction to Pacific standard time (P.S.T.)      $-$   $8^h$

P.S.T. of upper transit of Betelgeuse

on February 27, 1976               =      $19^h 34.6^m$

**Example 11.25**   What is the hour angle of Betelgeuse at $23^h 00.0^m$ P.S.T. on February 27, 1976, at the same place as in Example 11.24?

SOLUTION

From Example 11.24, the P.S.T. of upper

transit at place is             $19^h 34.6^m$

The mean time interval since upper

transit $= 23^h 00.0^m - 19^h 34.6^m$     $= 3^h 25.4^m$

Sidereal time gains on mean time at the rate of $9.86^s$ per hour (Art. 11.19). Thus, the hour angle of the star is $3^h 25.4^m + (3.42)(9.86) = 3^h 26.0^m$

In other ephemerides, positions of selected stars are given by declination and sidereal hour angle (S.H.A.). The sidereal hour angle is the angle measured westward (or clockwise when viewed from above the north pole) along the celestial equator from the vernal equinox to the meridian of the star (Fig. 11.5, Art. 11.6). Since the position of the vernal equinox is also needed, the Greenwich hour angle of the vernal equinox (G.H.A.$_\Upsilon$) is tabulated for each day of the year at $0^h$ Greenwich civil time (G.C.T.). The G.H.A.$_\Upsilon$ plus the S.H.A. of the star yields the G.H.A. of the star (Fig. 11.28). Knowing the longitude of the place of observation and the G.H.A. of the star, the local hour angle of the star can be computed. With the L.H.A. known, the sidereal interval between $0^h$ G.C.T. and upper transit of the star over the meridian of the place can be calculated. Conversion of this sidereal interval to mean solar time yields the G.C.T. of upper transit. Figure 11.28 shows the relationships of these various

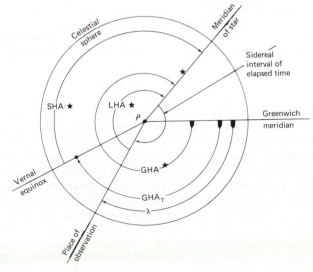

**Fig. 11.28** Hour angles used to determine transit of a star.

hour angles. Consider an example using an ephemeris that gives positions of stars by sidereal hour angle.

**Example 11.26**  Determine the Pacific standard time of upper transit of the star Spica ($\alpha$ Virginis) on February 27, 1978, at a place having a longitude of $122°15'30''$.

SOLUTION   From the ephemeris:

| | |
|---|---|
| G.H.A.$_T$ at $0^h$ G.C.T., February 27, 1978 | $= 156°28.4'$ |
| S.H.A. Spica, February 1978 | $= 158°59.1'$ |
| G.H.A. of Spica at $0^h$ G.C.T. | $= 315°27.5'$ |
| Longitude of place | $= 122°15.5'$ |
| L.H.A. of Spica west of place | $= 193°12.0'$ |
| $360° -$ L.H.A. $=$ sidereal interval between | |
| $0^h$ G.C.T. and upper transit at place | $= 166°48.0'$ |
| | $= 11^h07^m12^s$ |
| Correction to mean solar time $=$ | |
| $(11.120^h)(9.83^{s/h})$ | $= - \ \ 1^m49^s$ |
| G.C.T. of upper transit at place | $= 11^h05^m23^s$ |
| Longitude to P.S.T. | $= -8^h$ |
| P.S.T. of upper transit at place | $= \ \ 3^h05^m23^s$ |

**11.42. Latitude by observation on other stars**   To determine latitude by an observation on a star at upper transit, the star's declination and right ascension or sidereal hour angle are found in an ephemeris, and the approximate standard time of upper transit at the given place is determined as shown in Examples 11.24 and 11.26. Before this time, the transit is set up and the estimated altitude of the star is laid off on the vertical circle. (The latitude will be roughly known, the declination is known, and hence the altitude can be estimated with sufficient precision to bring the star within the field of view.) The telescope is pointed approximately along the meridian, and the instrument is revolved about the vertical axis back and forth through a small angle until the star is sighted. The star is followed with the horizontal cross hair until the maximum altitude is reached; then the vertical angle is read. The latitude is determined as described in Art. 11.31.

In this way, several stars whose times of upper transit differ by short intervals can be observed, and the latitude can be computed by taking the mean of the values thus found.

**11.43. Determination of time by observations on stars**   To determine time by observing the upper transit of any star, the direction of the meridian and the longitude of the place being known, the standard time of upper transit is calculated as described in Examples 11.24 and 11.26, Art. 11.41. The star's declination is found from the ephemeris, and its altitude is roughly calculated, the latitude of the place being at least approximately known. Before the estimated time of upper transit, the instrument is set up, a sight is taken along the meridian, and the horizontal motion is clamped. The estimated altitude of the star is laid off on the vertical circle, and the course of the star is followed until it crosses the vertical hair. At this instant, time is observed. The difference between this time and the calculated time is the error of the timepiece. Time determinations should be made on stars near the celestial equator.

For more precise determinations a succession of observations such as that just described may be made on stars whose calculated times of upper transit differ from each other by only a few minutes. Most instrumental errors will be eliminated if the instrument is plunged between two successive observations. The average clock error thus determined is considered to be the error of the timepiece.

It is evident that time and latitude observations may be made simultaneously if, in addition to the observations just described, the vertical angle to each star as it crosses the meridian is measured.

Time can also be determined by observing the altitude of a star at any time, assuming that the latitude and longitude of the place are known. The procedure is similar to that described in Art. 11.35 for determining time by the altitude of the sun except that the horizontal cross hair bisects the star. To determine the approximate position of a specific star at a given time so that the star may be brought within the field of view, the declination and right ascension (R.A.) or sidereal hour angle (S.H.A.) are obtained from the ephemeris and the local hour angle $t$ is calculated (see Example 11.26). Then the approximate altitude of the star is computed by Eq. (11.14) (Art. 11.14) and the approximate azimuth is determined by Eq. (11.11) or (11.13) (Art. 11.13). With these two values, the star can be brought within the field of view if the direction of the meridian is roughly known. At least two observations, one with the telescope direct and one with it reversed, are made to the star with the time being recorded for each setting. The average vertical angle (or $90° -$ zenith angle) corrected for refraction (Table II, Appendix C), the star's declination (ephemeris), and latitude for the place are then used to calculate the local hour angle $t$ by Eq. (11.4) or Eq. (11.6). The local sidereal time equals the local hour angle ($t$ is + west and − east of meridian) plus the right ascension of the star. The longitude (+ west, − east) added to the local sidereal time yields Greenwich sidereal time. The difference between the Greenwich sidereal time at the place of observation and the Greenwich hour angle of the vernal equinox (G.H.A.$_\Upsilon$, obtained from the ephemeris) at $0^h$ G.C.T. is the sidereal interval of time since $0^h$ G.C.T. This interval, converted to a mean solar time by subtracting $9.83^s$ per hour of sidereal interval, is the Greenwich mean time of observation. The standard time is the Greenwich mean time minus the longitude of the place.

For best results, two stars should be observed, one approximately east and one approximately west of the observer. Each star should be at about the same altitude, preferably at least 30° above the horizon.

**Example 11.27**   The star Arcturus ($\alpha$ Bootis) is to be observed from a place having latitude and longitude of 42°17′30″N and 72°30′00″W, respectively (scaled from a U.S.G.S. quadrangle sheet of the area), on April 1, 1978, at $10^h$ P.M. Eastern standard time. Determine approximate values for azimuth and altitude to assist in locating the star for subsequent observations.

SOLUTION   $10^h00^m00^s$ P.M. $= 22^h00^m00^s$ E.S.T., $\delta = $ N19°17′36″ (ephemeris) G.C.T. of potential observation $= 5^h + 22^h = 27^h = 3^h00^m00^s$ on April 2, 1978 (refer to Fig. 11.29a).

| | |
|---|---|
| S.H.A. Arcturus, April 1978 (ephemeris) | 146°19′42″ |
| G.H.A.$_\Upsilon$ at $0^h$ G. C.T., April 2 (ephemeris) | 189°59′06″ |
| G.H.A. Arcturus, April 2, 1978, at $0^h$ G.C.T. = | 336°18′48″ |
| Longitude to place of observation | − 72°30′00″ |
| L.H.A. at $0^h$ G.C.T. (west of meridian)      = | 263°48′48″ |
| $t$ (east of meridian) = L.H.A. − 360° | = − 96°11′12″ |

Mean solar interval between $0^h$ G.C.T. and time of observation at $3^h$ G.C.T. is $3^h$. Thus, the

| | |
|---|---|
| Sidereal interval $= 3^h + (3^h)(9.86^s/h) = 3^h00^m30^s =$ | 45°07′30″ |
| L.H.A. at $3^h$ G.C.T., April 2, $= t +$ interval | $= -51°03′42″$ |

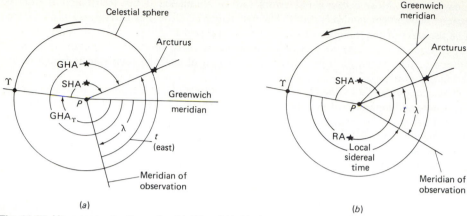

**Fig. 11.29** Hour angles for Examples 11.27 and 11.28. (*a*) Hour angles at $0^h$ G.C.T. (*b*) Hour angles at time of observation.

By Eq. (11.14):

$$\sin h = \sin\phi\sin\delta + \cos\phi\cos\delta\cos t$$

$$= \sin(42°17'30'')\sin(19°17'36'') +$$
$$\cos(42°17'30'')\cos(19°17'36'')\cos(-51°03'42'') = 0.66112891$$

$$h = \quad 41°23'09''$$

$$+ \quad \underline{01'08''} \quad \text{refraction correction}$$

Vertical angle $= \quad 41°24'17''$

By Eq. (11.11):

$$\sin Z = -\frac{\sin t\cos\delta}{\cos h}$$

$$= -\frac{[\sin(-51°03'42'')\cos(19°17'36'')]}{\cos(41°23'09'')}$$

$$= 0.97849492$$

$$Z = 78°05'46'' = \text{azimuth from the north}$$

**Example 11.28**   Using the angles determined in Example 11.27, the true altitude and watch time of observation of Arcturus are $41°23'15''$ and $22^h00^m32^s$ E.S.T., respectively. The latitude and longitude of the place are given in Example 11.27. Determine the time by altitude of the star (refer to Fig. 11.29*b*) and the watch correction.

**SOLUTION**   First calculate the local hour angle by Eq. (11.4) (the declination $\delta$ is obtained from the ephemeris).

$$\cos t = \frac{\sin h - \sin\phi\sin\delta}{\cos\phi\cos\delta}$$

$$= \frac{\sin(41°23'15'') - \sin(42°17'30'')\sin(19°17'36'')}{\cos(42°17'30'')\cos(19°17'36'')}$$

$$= 0.6285112$$

$$t = 51°03'35'' = \frac{51.0596°}{15} = 3.4040^h = 3^h24^m14^s$$

Arcturus is east of the meridian of observation; hence $t$ is negative. R.A. Arcturus, April $2 = 360° -$ S.H.A. Arcturus, where the S.H.A. is obtained from Example 11.27.

R.A. Arcturus $= 360° - 146°19'42'' = 213.6717°$    $=$    $14^h14^m41^s$

Local sidereal time $=$ R.A. $+ t =$

$(14^h14^m41^s) + (-3^h24^m14^s)$    $=$    $10^h50^m27^s$

Longitude to place    $= 72.5°/15$    $= +\ 4^h50^m00^s$

Greenwich sidereal time    $=$    $15^h40^m27^s$

G.H.A._T at $0^h$ G.C.T. $= 189°59'06'' = 12.6656^h$    $= -12^h39^m56^s$

Sidereal interval since $0^h$ G.C.T.    $=$    $3^h00^m31^s$

Correction to mean solar int. $= (3.0086)(9.83)$    $= -$    $30^s$

Greenwich civil time, April 2, 1978    $=$    $3^h00^m01^s$

Greenwich civil time, April 1, 1978    $=$    $27^h00^m01^s$

Longitude to eastern standard time zone    $= -\ 5^h$

Eastern standard time, April 1, 1978    $=$    $22^h00^m01^s$

Watch time    $s =$    $22^h00^m32^s$

Watch correction    $=$    $-31^s$

**11.44. Determination of longitude**    To determine longitude by observing the upper transit of any star, the direction of the meridian and the standard time of the place being known, the standard time of upper transit is calculated for a longitude estimated to be that of the place. The star is found as described in Art. 11.42, and the standard time of its upper transit is observed. The interval between the calculated time and the observed time of upper transit, changed to sidereal time, is the difference between the estimated longitude and the true longitude of the place.

**11.45. Determination of azimuth by observation on other stars**    In some cases it is impossible or impractical to observe circumpolar stars for azimuth, either because of clouds or because the star is too near the horizon or too near the zenith. To determine azimuth by observation on any star other than a circumpolar star, the same general procedure is followed as for solar observations of this character described in Art. 11.32. The approximate position of a given star at a specified time is determined as described in Art. 11.43. When the star is brought within the field of view, the intersection of the cross hairs is sighted on the star, the time is observed, and the horizontal and vertical angles are read.

Where the azimuth is to be determined with a higher degree of precision, a star is chosen that is in a favorable position and several repetitions of horizontal and vertical angles are measured to the star.

Preferably several stars are observed in pairs, one star being nearly east and the other being nearly west of the observer, at about the same altitude, in order to equalize any error in the correction for refraction. The altitude should be not less than about 20° nor more than about 40°.

Still another method of determining azimuth, which requires little computation but which requires pairs of observations several hours apart, is as follows: A star, preferably southward from the observer, is sighted in the eastern sky at the instant it rises to a fixed altitude which should be not less than about 20° nor more than about 50°. Either a mark is set under the star or its azimuth is observed with respect to some reference line on the ground. Later the same star is sighted similarly in the western sky at the instant it descends to the same altitude. The bisector of the horizontal angle between the two directions of pointing lies on

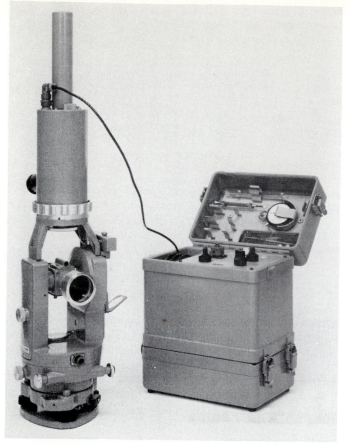

**Fig. 11.30** Wild GAK1 gyro attachment on the T-16 repeating theodolite. Gyro attachment is connected to the battery unit. (*Wild Heerbrugg Instruments, Inc.*)

the meridian of the observer. Several stars may be observed in this manner, and the mean of the observations used.

**11.46. Azimuth by gyro attachment**  A gyro attachment consists of a gyro motor suspended by a thin metal tape from one end of a sealed tube which can be mounted vertically on the standards of a theodolite. Figure 11.30 shows the gyro attachment mounted on a repeating theodolite. The gyro motor is powered by the accompanying battery unit. The tube, enclosure, gyro motor, and suspension are so designed that when the unit is attached to the horizontal axis of a leveled theodolite, the metal tape holding the rotating gyro coincides with the vertical axis of the instrument and also with the direction of gravity. Figure 11.31 illustrates a cross-sectional view of the attachment. Note the suspension strip which supports the gyro motor that has an axis of rotation perpendicular to the vertical or plumb line. The gyro motor spinning at 22,000 revolutions per minute about this horizontal axis tries to maintain in space its initial random spinning plane created by its moment of inertia. The gyro, fixed to the instrument which is earthbound, is pulled out of its original spinning plane by the earth's rotation. This interference causes the gyro to oscillate around the plumb line until the spin axis is oriented in the north-south plane and the rotation of the gyro corresponds to the rotation of the earth. The gyro does not stabilize immediately in the

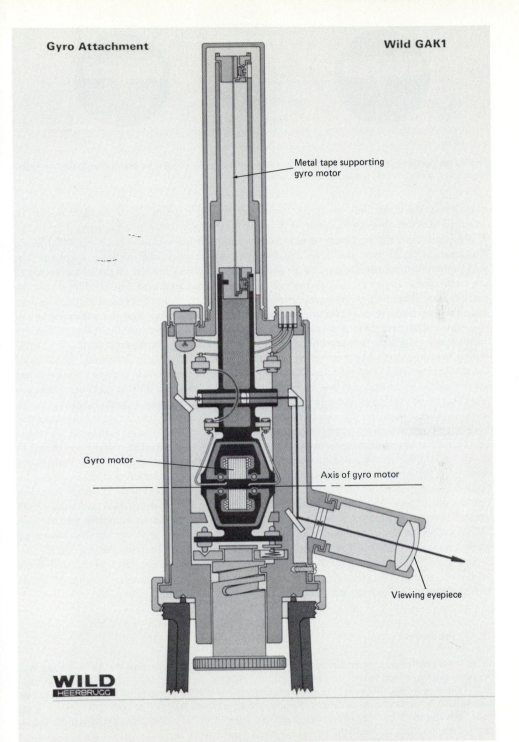

**Gyro Attachment**

**Wild GAK1**

Metal tape supporting
gyro motor

Gyro motor

Axis of gyro motor

Viewing eyepiece

**WILD**
HEERBRUGG

**Fig. 11.31** Cross-sectional view of Wild GAK1 gyro theodolite. (*Wild Heerbrugg Instruments, Inc.*)

(a)                                                                    (b)

**Fig. 11.32** (a) Swing of the gyro mark in the middle of the scale. (b) Gyro mark at elongation—transit method. (*Courtesy Wild Heerbrugg Instruments, Inc.*)

north-south direction, but oscillates about the meridian plane. This oscillation can be observed optically through an eyepiece attached to the tube containing the gyro (Fig. 11.31). These oscillations can be observed as a moving light mark projected on a scale which has a V-shaped central index and that can be observed through the viewing eyepiece. The midposition of these oscillations can be located by measuring the size of the swing period of the oscillations. Figure 11.32a shows the gyro mark centered and Fig. 11.32b shows the mark at one elongation, which is one-half the swing period from the central index. The gyro attachment is oriented on the instrument so that the gyro spin axis and the telescopic line of sight fall in the same vertical plane when the light mark is centered on the index. When this condition is achieved, the telescope is oriented toward true or astronomic north.

The telescope can be approximately oriented toward true north by following the swing of the gyro mark with the upper tangent motion of the theodolite. When the gyro mark reaches an elongation point and reverses direction (a reversal point), the horizontal circle reading is recorded. Next, the gyro mark is again followed using the upper tangent motion, until a second reversal point on the opposite side of the index is reached. At this time a second horizontal circle reading is taken. When the average of these two circle readings is set on the horizontal circle, the telescopic line of sight is approximately directed toward true north to within ±3 to 4′. To obtain a more precise direction, the time intervals of at least three transits of the gyro mark across the central index and amplitudes of elongation points on both sides are measured. An angular correction to the approximate north direction can then be determined as a function of these observed times and half-swings east and west of the central index. The entire operation requires from 20 to 30 min to complete and permits determination of the azimuth with a standard deviation of 20″. Further details concerning determination of azimuth by gyro attachment observations can be found in Art. 20.16.

The gyro method of establishing an azimuth is applicable to any survey where the attainable precision is adequate. It is particularly useful for the transfer of azimuth in the surveys needed for underground tunneling and mining operations (Chap. 20).

## 11.47. Problems

**11.1** When the local apparent time is $8^h17^m12^s$ at a place whose longitude is $96°15′10″$W, what is the Greenwich apparent time?

**11.2** On a given day $0^h$ Greenwich civil time occurs at $4^h17^m32^s$ Greenwich sidereal time. At that instant what is the local sidereal time at a place whose longitude is $7^h17^m43^s$W?

**11.3** When it is $15^h31^m12^s$ Greenwich civil time, it is $10^h16^m37^s$ local civil time at a given place. What is the longitude of the place?

**11.4** What is the Greenwich civil time when it is $3^h15^m$ P.M. Central standard time?

**11.5** If the local civil time at a place is $16^h23^m22^s$ and the longitude of the place is $78°36′20″$W, what is the Eastern standard time?

**11.6**   From an ephemeris find the equation of time for the instant of $4^h15^m00^s$ P.M. Pacific standard time on April 21 of the current year. If the longitude of the place is $7^h46^m03^s$W, calculate the local civil and local apparent times.

**11.7**   From an ephemeris, find the equation of time for the instant of $3^h19^m30^s$ P.M. apparent time on July 4 of the current year, at a place whose longitude is $6^h15^m30^s$W. Compute the corresponding local civil time.

**11.8**   At a given place the hour angle of the true sun at $11^h30^m$ P.M. Greenwich civil time on January 12 of the current year is $42°36'30''$. What is the local sidereal time?

**11.9**   The mean radius of the earth is 3,956 miles (6367 kilometres), and the mean distance to the sun is 92,900,000 miles (149,508,100 kilometres). What is the sun's mean horizontal parallax? What is the parallax correction when the altitude is $30°$?

**11.10**   The observed altitude of a star is $23°15'20''$. The temperature is $90°$F. By Table II (Appendix C) find the refraction correction, and compute the true altitude of the star.

**11.11**   The observed altitude of the sun's center is $15°07'30''$. The temperature is $-10°$C. By Table I (Appendix C) find the correction for parallax and refraction and compute the true altitude of the sun.

**11.12**   Find the apparent declination of the sun for the instant of $9^h00^m$ A.M. Central standard time on February 15 of the current year, using an ephemeris giving values for $0^h$ Greenwich civil time.

**11.13**   Find the apparent declination of the sun for the instant of local apparent noon at a place whose longitude is $5^h52^m54^s$W for the date of July 21 of the current year, using an ephemeris giving values for Greenwich apparent noon.

**11.14**   The observed altitude of the lower limb of the sun as it crosses the meridian at a given place is $55°31'30''$. The observation is made at $11^h34^m20^s$ A.M. Eastern standard time on May 16 of the current year. The temperature is $55°$F. Calculate the latitude of the place.

**11.15**   On August 1 of the current year the observed altitude of the sun at a given place is $30°51'45''$ at $7^h42^m20^s$ A.M. local apparent time. The latitude of the place is $37°18'20''$N, and the longitude is $102°17'30''$W. The temperature is $75°$F ($24°$C). The horizontal angle (measured clockwise) from reference line to sun is $89°39'15''$. What is the azimuth of the sun measured from north, and what is the azimuth of the reference line? Compute the azimuth of the sun by Eq. (11.2b).

**11.16**   Compute the changes in azimuth of sun due to a $01'$ change in latitude, in declination, and in altitude for latitude $50°$, declination $0°$, and altitudes of $15°$ and $30°$. Use Eq. (11.2b) as a basis for computations. Compare the results with the corresponding quantities for latitude $40°$ given in Table 11.4 of Art. 11.32.

**11.17**   Same as problem 11.16 but for latitude $30°$.

**11.18**   At Orono, Maine, on December 5 of the current year the sun's center is observed to cross the meridian at a watch time of $11^h24^m21^s$ A.M. The longitude of the place is $4^h34^m40.3^s$W. What is the watch correction to give local mean time? What is the watch correction to give Eastern standard time?

**11.19**   At a given place the center of the true sun crosses the meridian at $11^h41^m37^s$ A.M. Pacific standard time on September 15 of the current year. What is the longitude of the place?

**11.20**   By a series of observations the true altitude of the sun's center at a given station is $24°28'44''$ at the instant of $4^h13^m12^s$ P.M. Mountain standard time on March 14 of the current year. The clockwise horizontal angle from reference line to sun is $312°16'37''$. The latitude of the place is $39°01'42''$N. By Eqs. (11.5) and (11.6), Art. 11.11, compute the azimuth and hour angle of the sun at the given instant. Determine the longitude of the place and the azimuth of the reference line reckoned from south.

**11.21**   At a given place on January 12 of the current year, the observed altitude of Polaris at upper culmination is $44°36'25''$. The temperature is $15°$F ($-9°$C). What is the latitude of the place?

**11.22**   On September 7 of a given year, upper culmination of Polaris at the meridian of Greenwich occurs at $2^h35^m29^s$ Greenwich civil time. What is the Eastern standard time of upper culmination on September 10 of the same year at a place whose longitude is $78°30'15''$W?

**11.23**   Find the Central standard time of upper culmination of Polaris on December 7 of the current year at Des Moines, Iowa (longitude $6^h14^m30.6^s$W).

**11.24**   The altitude of Polaris is observed $20^m$ after the time of upper culmination and found to be $48°32'20''$. The polar distance is $0°49'25''$. What is the latitude of the place?

**11.25**   Compute the azimuth and hour angle of Polaris when at elongation, the polar distance being $0°49'10''$ and the latitude of the place being $43°00'49''$N.

**11.26**   What is the time of western elongation of Polaris at a given place when upper culmination occurs at $2^h15^m20^s$ P.M. Eastern standard time and the latitude is $42°22'47''$N?

**11.27**   Find the Pacific standard time of eastern elongation of Polaris on August 18 of the current year for latitude 37°52′24″N and longitude 8$^h$9$^m$3$^s$W.

**11.28**   Find the azimuth of Polaris when at elongation on August 18 of the current year for a place whose latitude is 37°52′24″N.

**11.29**   From Table III (Appendix C) determine the azimuth correction to be applied to an observation on Polaris 15$^m$ after elongation to reduce to elongation, the azimuth at elongation being 1°34′12″. Compute the corresponding perpendicular offset to the meridian, at the reference monument beneath the star, when the monument is 180 m from the station occupied.

**11.30**   At a given place upper culmination of Polaris occurs at 3$^h$15$^m$20$^s$ P.M. Central standard time on a given date. On the same date an azimuth observation is made at 7$^h$0$^m$20$^s$ P.M. The latitude of the place is 41°15′30″N, and the polar distance of the star is 0°49′35″. Compute the hour angle and azimuth of the star.

## 11.48. Field problems

### PROBLEM 1   LATITUDE BY OBSERVATION ON SUN AT NOON

**Object**   To determine the latitude of the place by an observation on the sun at local apparent noon, using the transit or theodolite.

**Procedure**   Follow the procedure outlined in Art. 11.31, assuming that the longitude of the place is unknown. Use the method of double-sighting to determine the mean vertical angle to the sun's upper and lower limbs.

#### Hints and precautions

1. See Art. 11.26.
2. Pay particular attention to the algebraic sign of each quantity and of each correction.
3. If the longitude of the place is known approximately, the approximate standard time of upper transit of the sun may be calculated in advance as a guide in observing.

### PROBLEM 2   AZIMUTH BY DIRECT SOLAR OBSERVATION

**Object**   To determine the true azimuth of a line by an observation on the sun with the transit or theodolite.

#### Procedure

1. Follow the procedure outlined in Art. 11.32. Sights must be taken to both right and left limbs of the sun, or the correction for semidiameter (taken as 16′ sec $h$) must be applied to the observed horizontal angle.
2. As a check, observe the magnetic bearing of the line.

#### Hints and precautions

1. See Art. 11.26.
2. Pay particular attention to algebraic signs.

### PROBLEM 3   LATITUDE BY OBSERVATION ON POLARIS AT CULMINATION

**Object**   To determine the latitude of the place by observing Polaris at upper or lower culmination.

**Procedure**   Follow the procedure outlined in Art. 11.38.

#### Hints and precautions

1. See Art. 11.36.
2. Pay particular attention to algebraic signs.
3. As a check, the mean of the times of observation should agree (within a few minutes) with the computed time of culmination.

## PROBLEM 4    AZIMUTH BY OBSERVATION ON POLARIS AT ELONGATION

**Object**   To determine the azimuth of a line by observation on Polaris at eastern or western elongation.

**Procedure**

1. Follow the procedure outlined in Art. 11.39.
2. As a check, observe the magnetic bearing of the established line.

**Hints and precautions**

1. See Art. 11.36.
2. Pay particular attention to algebraic signs.
3. As a check, the mean of the times of observation should agree (within a few minutes) with the computed time of elongation.

## References

1. Adams, L. P., "Astronomical Position and Azimuth by Horizontal Directions: A Rigorous Solution," *Survey Review*, January 1971.
2. Bowker, O. W., "Lightweight Gyro-Azimuth Surveying Instrument," *Proceedings, Annual Meeting, ACSM*, 1972, p. 379.
3. Hoskinson, A. J., and Duerkson, J. A., *Manual of Geodetic Astronomy*, Spec. Publ. 237, U.S. Department of Commerce, Coast and Geodetic Survey (now National Ocean Survey, NOAA), Washington, D.C., 1952.
4. Hosmer, G. L., and Robbins, J. M., *Practical Astronomy*, 4th ed., John Wiley & Sons, Inc., New York, 1948.
5. Husti, G. J., "A Method of Determining Latitude and Azimuth Simultaneously by Star Altitudes," *Survey Review*, April 1977.
6. Keuffel & Esser Company, *K & E Solar Ephemeris for the Year*, _____ , Morristown, N.J., annual (in advance).
7. Kissam, Philip, *Surveying for Civil Engineers*, McGraw-Hill Book Company, New York, 1956.
8. Moffitt, Francis H., and Bouchard, H., *Surveying*, 6th ed., Harper & Row Publishers, New York, 1975.
9. Nassau, J. J., *Textbook of Practical Astronomy*, 4th ed., John Wiley & Sons, Inc., New York, 1948.
10. Robbins, A. R., "Geodetic Astronomy in the Next Decade," *Survey Review*, July 1977.
11. Schwartz, W. M., "Astronomic Azimuth Calculations on a Desk-Top Computer," *The Canadian Surveyor*, Vol. 27, No. 1, March 1973, p. 32.
12. U.S. Bureau of Land Management, *Manual of Instructions for the Survey of the Public Lands of the United States*, Technical Bulletin No. 6, Government Printing Office, Washington, D.C., 1973.
13. U.S. Bureau of Land Management, *Ephemeris of the Sun, Polaris, and Other Selected Stars for the Year*, _____ , Government Printing Office, Washington, D.C., annual (in advance).
14. U.S. Naval Observatory, *The American Ephemeris and Nautical Almanac for the Year*, _____ , Nautical Almanac Office, Government Printing Office, Washington, D.C., annual (in advance).
15. U.S. Naval Observatory, *The Nautical Almanac for the Year*, _____ , Nautical Almanac Office, Government Printing Office, Washington, D.C., annual (in advance).

# CHAPTER 12
# Modern positioning systems

**12.1. Introduction** Position determinations on the surface of the earth have long been accomplished by a combination of directional and distance measurements as well as by astronomic observations (see Chap. 11). These procedures are presently being challenged by more advanced techniques that are based on modern systems and equipment. In this chapter, only an introductory coverage of the following three classes of systems will be given:

1. Electronic positioning systems.
2. Inertial positioning systems.
3. Doppler positioning systems.

Each of the classes will be discussed in a separate section. The electronic positioning systems are an outgrowth of developments in regular surveying equipment where both directional and distance measuring capabilities are combined in one compact instrument. Inertial positioning systems draw on the extensive experience with navigational components used on board aircraft. Such systems are used either on helicopters or on ground vehicles. Finally, doppler positioning systems rely on signals received from artificial satellites.

**12.2. Electronic positioning systems** Electronic positioning systems consist of specially designed short-to medium-range electronic distance measuring (EDM) devices mounted on a theodolite framework which permits determination of horizontal and vertical angles. The net result is a single instrument that can be used to determine distances and directions from a single instrument setup. There are many such systems now available, some of which have minor differences and some that are different in fundamental ways. Generally speaking, electronic positioning systems may be classified into three groups:

1. Optical-reading repeating or direction theodolites with an attached but removable EDM device. These systems are called *combined theodolite and EDM*.
2. Optical-reading theodolites with attached EDM device and built-in electronic computer. These systems are designated as *computerized theodolite and EDM*.
3. *Integrated digitized electronic systems* consisting of a digitized theodolite, microprocessor, and EDM device incorporated into a single system. These systems are designated *electronic tacheometers*. Such a system can also be equipped with a solid-state memory, magnetic tape recorder, or punched-paper tape unit for storage of data.

In addition to the systems listed above, there are also various devices equipped with lasers. Instruments of this type do not permit direct determination of distance or angles but are used primarily for alignment problems, as discussed in Art. 18.5. Each of the three systems listed above is now discussed in more detail.

**1. Combined theodolite and EDM** These systems have specially designed lightweight EDM devices attached to repeating or direction theodolites of the types described in Art. 6.30. Two systems of this class are shown in Fig. 12.1.

On most systems of this group, the EDM transmitter/receiver is a light-wave instrument that utilizes infrared light rays modulated by two or three frequencies (EDM instruments are described in Arts. 4.29 to 4.36). Slope distance between instrument and reflector is determined electronically within the EDM device by phase comparison of the transmitted with the reflected light beam. In most of these systems, this measurement is made by rotating one control knob and centering a needle in one meter. The slope distance is then displayed digitally by a register with light-emitting diodes (L.E.D's). The EDM device is removable so that the theodolite can be used independently if desired. Normally, the entire system is used with distance and direction being measured from a single setup.

The procedure for determining distance and direction is as follows. The instrument is placed over the station and is centered and leveled using the same procedures as described for theodolites in Art. 6.30. A combined reflector/target is placed over the station to be sighted. Figure 12.2 illustrates two designs for combined target/reflectors. Ranges varying from 300 to 700 m are possible with one reflector. To increase the range, reflectors are generally designed so that several can be coupled to a single support, as shown in Fig. 12.2a. With three reflectors and good atmospheric conditions, ranges of from 500 to 1000 m (1600 to 3000 ft) are possible. The operator views through the telescope and centers the cross hairs on the target. This operation also centers the EDM transmitter optics on the reflector for maximum signal return. The necessary controls on the EDM are then manipulated in order to measure the distance electronically and the "measure" button or the proper equivalent control is pressed. The slope distance will be displayed in from 10 to 15 seconds. In the

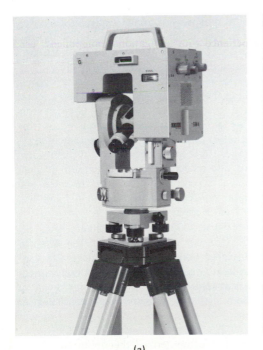

(a)                      (b)

**Fig. 12.1** Combined theodolites and EDM. (a) Zeiss SM4. (*Carl Zeiss, Inc.*) (b) Kern DM SO1. (*Kern Instruments, Inc.*)

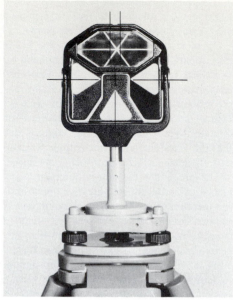

(a)
(b)

**Fig. 12.2** Reflectors/targets for combined theodolite/EDM systems. (a) Three reflectors coupled. (*Kern Instruments, Inc.*) (b) Single reflector. (*Wild Heerbrugg Instruments, Inc.*)

meantime, the operator can observe the horizontal and vertical angles optically, following the procedures given in Art. 6.30 and illustrated in Figs. 6.34b and c and 6.35b. With these data and knowing the heights of instrument and target above the station, the horizontal distance and difference in elevation can be computed directly by hand-held calculator.

In certain systems, the offset between the optics of the EDM device and the sighting telescope must be taken into account when making the slope reduction. In the more recent models, the combined target/reflector is designed so that the sighting target is offset from the reflector center by the same amount as the theodolite optical line of sight is offset from the EDM optics (see Fig. 12.2b). In this way the offset cancels and only a single pointing is needed for distance and direction. Naturally, any difference between height of instrument above the stations at the theodolite/EDM station and reflector must be taken into account for the slope reduction (Art. 4.39). The combined target/reflector can be mounted and force-centered (Art. 6.30) on a compatible tripod or can be connected to a hand-held centering rod when many sights are being taken from a single setup.

The accuracies of these systems for distances of up to 2000 m (~6600 ft) vary from ±5 mm to 10 mm (0.015 ft to 0.030 ft) +2 to 5 parts per million (ppm) of the distance measured. Estimated standard deviations in measured horizontal and zenith angles vary from 3 to 20 seconds of arc and depend on the least count of the horizontal and vertical circles in the theodolite to which the EDM unit is attached.

In order to run the EDM unit, each system contains a power unit consisting of rechargeable nickel-cadmium batteries.

**2. Computerized theodolite and EDM systems**  The computerized theodolite and EDM systems are essentially similar to the combined theodolite and EDM device described in the preceding paragraphs with one important difference. A compact, solid-state computer unit (microprocessor) is integrated into the system. This microprocessor is used for data reduc-

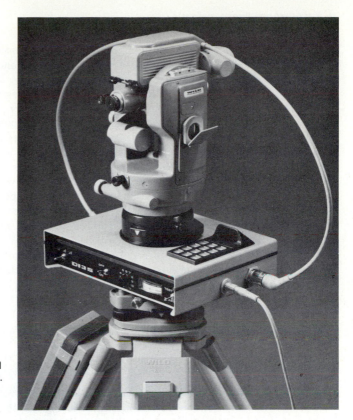

**Fig. 12.3** Computerized theodolite / EDM system. (*Wild Heerbrugg Instruments, Inc.*)

tion. Figure 12.3 shows one such system in which the computer is the flat unit mounted between the leveling head assembly and the base of the theodolite. The vertical axis of the theodolite extends through the computer unit so that optical centering is possible. The adapter on top of the computer unit is compatible with repeating and direction theodolites. There is a keyboard and digital display register (L.E.D.) on top of the computer unit for keying in measured data for subsequent reduction.

These systems are operated in the same way as the combined theodolites and EDM units already discussed. The target is sighted and the keyboard control to start measuring is pressed. While the distance measurement is in progress (10 seconds), the operator reads the angles through the optical train. The horizontal and vertical angles are entered into the computer using the keyboard and the computer reduces the data, providing horizontal distance, vertical difference in elevation, and latitudes and departures for the line measured. It is also possible to introduce constants for sea-level and map projection scale corrections so that these corrections are applied automatically.

The system illustrated in Fig. 12.3 also contains a tracking mode which yields a slope distance every 2 seconds. This feature makes the instrument very useful for field layout operations. The stated measuring accuracy of the tracking mode is 2 to 3 cm.

**3. Integrated electronic surveying systems (electronic tacheometers)** Electronic tacheometers consist of the following components:

a. An electronic digitized theodolite.
b. An EDM unit or distance meter.
c. A microprocessor.

d. Keyboard and display register.

e. Data storage unit or field computer.

Figures 12.4 to 12.6 illustrate three examples of these electronic tacheometers.

Digitized theodolites are described in Art 6.31. Horizontal and vertical (or zenith) angles are not observed directly in these instruments but are sensed by digital encoders fitted to the horizontal and vertical axes of the theodolite. Thus, angular measurements are in the form of digital data that may be easily stored in the microprocessor memory. In this form, the automatically measured angles can be used in subsequent survey calculations or can be displayed on the register in the desired format (degrees, minutes, and seconds, or grads). The theodolite is equipped with the usual tangent-screws found on a direction theodolite, that is, the horizontal clamp and tangent-screw and vertical clamp and tangent-screw. Angular modes possible on the system shown in Fig. 12.4 are average vertical angle, azimuth, or the average of repeated angles turned clockwise or counterclockwise from the backsight. Angular modes available on the system shown in Fig. 12.5 are zenith angle and azimuth, or clockwise angle from the previous sight. In each system both sides of the circle are sensed and averaged automatically.

The EDM unit for the system in Fig. 12.4 is detachable and is equipped with horizontal and vertical clamps and tangent-screws and a seven-power telescope for aiming the device on the reflector. Thus, the theodolite can be used independently. This EDM unit generates a modulated infrared light beam and has a range of up to 2 km (1.2 mi) with a stated accuracy of ±5 mm (0.02 ft)+6 ppm in a temperature range 0 to 41°C (32 to 105°F).

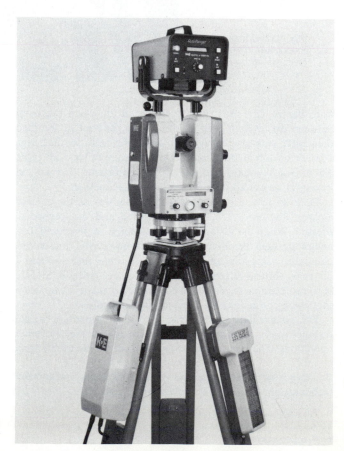

**Fig. 12.4** Electronic surveying system. (*Keuffel & Esser, Inc.*)

**Fig. 12.5** Electronic tacheometer. (*Hewlett Packard, Inc.*)

The EDM unit for the system illustrated in Fig. 12.5 is an integral part of the theodolite and utilizes the sighting telescope of the theodolite as a portion of the optical train for transmitting and receiving a modulated laser beam. This laser beam is generated by a galium arsenide lasing diode and is not comparable to the gas laser generators discussed in Art. 12.3. The range of this unit is up to 5 km (3 mi) with a stated accuracy of $\pm 5$ mm (0.02 ft) + 5 ppm in a temperature range 0 to 41°C (32 to 105°F).

In all three systems (Figs. 12.4 to 12.6), the digitized theodolite and EDM unit are under the control of microprocessor(s), which permit automatic correction of measured slope distances and angles for systematic errors and storage of these reduced values in the microprocessor memory. The system shown in Fig. 12.4 has a microprocessor in each module, so that the theodolite, EDM unit, and field calculator can be used independently. Each system illustrated in Figs. 12.5 and 12.6 is under control of a single microprocessor.

The keyboard is the link between the operator and the microprocessor. By manipulating the proper controls on the keyboard, the operator can display slope distance, horizontal distance, difference in elevation, azimuth or horizontal angle, and zenith or vertical angle. The operator can also store these data in the solid-state memory of the data storage unit and later retrieve a given set of data from storage.

The data storage unit for the system in Fig. 12.4 is also a field computer which can be used independently when provided with its own battery. This computer has sufficient memory to store the measurements from an entire day's work. It may then be disconnected and transported to the office, where the contents are transferred to an electronic calculator for further processing. If desired, the contents of the memory can be read onto magnetic tape in the field or transmitted to the office by telephone. Another option with this field calculator is to perform traverse computations in the field. When the traverse begins and

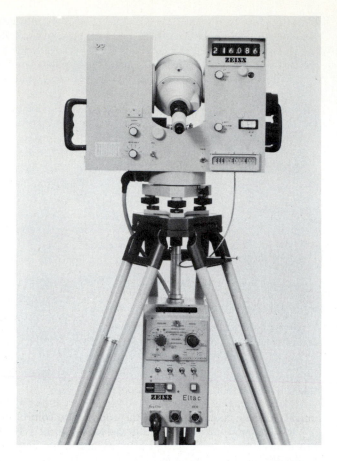

**Fig. 12.6** Electronic tacheometer.
(*Carl Zeiss, Inc.*)

ends on stations of known position or a loop traverse is run, the closure error can be determined before the survey party leaves the field. In the latter case the data stored in the field computer memory consist of $X$, $Y$, $Z$ coordinates from all occupied stations.

The data storage unit for the system in Fig. 12.5 permits storage of from 250 to 4000 lines of data, where a line consists of a four-digit identifier plus 16 characters. The unit can be disconnected and transported to the office, where the contents can be stored on magnetic tape for subsequent processing. Another option is to interface the storage unit with a compatible electronic calculator which can be used to process the data.

With any of the three systems (shown in Figs. 12.4 to 12.6), the data storage unit constitutes a field notebook with numbered lines. Each line contains an identification number (assigned by the computer) and then any one of the following: slope distance, azimuth or horizontal angle, vertical or zenith angle, horizontal distance, or vertical distance, and alphanumeric information such as station number, target number, height of instrument, etc., and other descriptive notes. The final record of these data is a hard-copy listing made either in the field with a portable printing calculator designed for the purpose or in the office after the data have been read onto magnetic tape or into an electronic computer, which, in turn, is interfaced with a printer. The *data storage unit should not be considered a substitute* for the standard field book, only as a supplement. Conventional field notes still serve to provide a record of descriptive information such as sketches, references for traverse stations, and other identification pertinent to the job being performed.

To operate the system shown in Fig. 12.4, it is set over the station, centered with an optical plumb and leveled using a standard plate level bubble (sensitivity of 30″ per 2-mm division). There is automatic compensation to correct vertical angles for mislevel. The telescope is sighted on the signal and the horizontal angle is set to zero or a preselected azimuth is entered via the keyboard. Next, the EDM unit is aimed at the reflector using its own clamp, tangent motion, and telescope. The slope distance is measured by manipulating controls on the face of the EDM unit. There is an electrical connection between the EDM unit and the microprocessor controlling the theodolite so that the slope distance may be automatically transferred to be displayed on the theodolite display register or stored for further processing. At this point, slope distance, horizontal angle, and vertical angle have been measured and are in storage. By proper manipulation of the keyboard and theodolite controls, the operator can now display on the register, or store in the memory of the field calculator, the horizontal distance, difference in elevation, horizontal angle or azimuth, and vertical angle. If desired, angles or directions can be repeated with the telescope direct and reversed. Each measurement is automatically stored and average angles or directions can be displayed on the register of the theodolite or stored in the memory of the field calculator. These data can then be converted to departures, latitudes, and $X$ and $Y$ coordinates which are stored in field calculator memory. When the desired data have been displayed or stored, the signal/reflector is moved to another point or the instrument is moved to another setup.

Operation of the system illustrated in Fig. 12.5 is essentially the same as for the one in Fig. 12.4, with some differences. The instrument is set over the station, centered with an optical plumb, and leveled approximately with a circular level bubble. At this point, a bidirectional electronic level sensor measures the amount by which the instrumental vertical axis fails to coincide with local vertical. These measurements are used in the microprocessor to apply corrections to the sensed horizontal and zenith angles, thus providing analytical compensation for mislevel of the instrument. Fully automatic compensation is possible within a range of 150 seconds of arc ($\sim450^{cc}$) of being level. Next, the telescope is sighted on the target/reflector and by pressing the appropriate control on the keyboard, slope distance, horizontal angle or azimuth, and zenith angle are measured. Again by pressing the proper key, these values as well as horizontal distance and difference in elevation can be displayed and/or stored in the data storage unit. This completes the operation for a single pointing and the next station is sighted or the system is moved to a new station. Angles or directions can be repeated with the telescope direct and reversed, with each entry being stored in the data storage unit. The average angles or directions are calculated in subsequent data-processing operations as described in previous paragraphs in this section. If the azimuth from a stated reference (e.g., true north) for the initial line is known, the horizontal circle can be set to this value. Then by choosing the azimuth mode on the keyboard, all subsequent directions in the survey will be referred to the given reference direction. The system also has a tracking mode for continuous updating of any measurement. The least count in the horizontal and vertical circles is 1 sec ($3^{cc}$) with a specified accuracy of $\pm2″$ in the horizontal angles and $\pm4″$ in zenith angles.

The electronic tacheometer illustrated in Fig. 12.6 is similar in most respects to the system in Fig. 12.5, with a few exceptions. Precise leveling is by a plate level bubble (sensitivity 20″ per 2-mm division) and there is automatic compensation of zenith angles for mislevel. The EDM unit generates a modulated infrared light beam with a range of up to 2 km (1.2 mi). The least count in distance measurement is 5 to 10 mm and in horizontal and vertical angles is 3″ ($10^{cc}$). Measured data can be automatically recorded on punched-paper tape or on a solid-state electronic data storage unit. There is an interface that permits connecting this system to any type of computer for data storage, storage and computing, or just computing. Thus, the potential exists for field evaluation of measured data. The system shown in Fig. 12.6 is equipped with the solid-state storage unit.

***Applications of electronic positioning systems***   The systems described in this section can be applied to practically any surveying task discussed in this text. Naturally, due regard must be given to the precision of the system versus the specified accuracies required for the application. Also, the procedures as detailed for standard surveying equipment may need modifications so as to exploit the features built into these systems. In general, there are applications for all of these systems in control surveys by traverse (Chap. 8) and by combined triangulation/trilateration (Chap. 10); topographic surveying, as mentioned in Art. 15.16, by the controlling point method or by radial survey; and construction layout (Chap. 18). It should be noted that all of these systems provide data that are very adaptable to forming digital terrain models (DTM). In particular, the electronic tacheometers automatically provide data that are directly convertible to a DTM by interfacing the data storage unit with a compatible electronic calculator.

The less sophisticated (and less expensive) systems, such as those shown in Figs. 12.1 to 12.3, will unquestionably become quite common in all surveying operations in the near future. The more sophisticated electronic digitized systems, such as those illustrated in Figs. 12.4 to 12.6, require a substantial initial investment which can be justified only by a high volume of work. Another factor to be considered is the problem of maintenance, particularly of the electronic components.

There is no question that surveying procedures will be modified to allow full exploitation of the characteristics of these systems. Some of these modifications are indicated in the sections where procedural changes are apparent or have already been experienced. Other changes are bound to evolve as experience is gained with these systems on all types of surveying tasks.

**12.3. Inertial positioning systems**   The inertial positioning system is described by its promoters as fulfilling the surveyor's dream of having a "black box" that will instantaneously provide position when set down on any point of the earth's surface. It requires no line-of-sight clearance, no towers, and is unaffected by refraction and other atmospheric conditions. In fact, it can be operated day or night during hot or cold weather, rain or shine, and with the speed of a motor vehicle or a helicopter. All of these apparent advantages should not mislead the student, as it must be realized that an inertial system is a product of the very highly sophisticated advanced aerospace technology. It requires well-trained personnel to operate it, and more important, to service it. Furthermore, to obtain results with the potential high precision of the system, very careful planning and execution of the survey project according to well-defined guidelines is necessary.

In simple terms this type of surveying system is basically an inertial platform with precise *gyroscopes* that keep its three mutually orthogonal axes properly oriented in space. Each axis is controlled by a gyroscope. Briefly, a gyroscope is a balanced mass that spins with a very

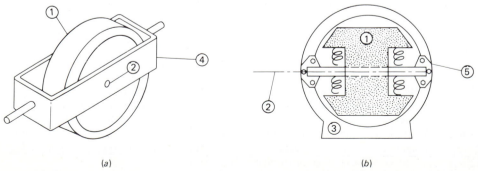

(a)                                                                                          (b)

**Fig. 12.7** Simple gyroscope. (*a*) View. (*b*) Cross section. (*Courtesy L. F. Gregerson.*)

high speed (more than 400 revolutions per second) about an axis of symmetry. This axis must be free to rotate in space to orient the gyroscope to a desired direction. Although there are more sophisticated gyroscopes, such as those employing air-bearing, or electrostatic suspension, the concept is easily understood with the simple gyroscope depicted in Fig. 12.7 where the numbered components are (Ref. 9):

1. The spinning rotor or wheel.
2. The spin axis.
3. The housing with the driving mechanism.
4. The gimbal, which allows rotational freedom of the spin axis.
5. Additional control components (such as thermal control, lateral movement control, etc.).

Once the spinning wheel is rotating at high speed, it attains a tremendous angular momentum and aligns itself in inertial space. In this state, the spinning wheel would resist any torque that attempts to force it out of its plane of rotation. The higher the mass and the spinning speed, the stronger is the inertia or resistance to deviate from the gyroscope plane. However high this inertia is, it is still finite, and unequal forces acting on the spinning mass will cause it to slowly move away from its spin plane. This is called gyroscopic *drift* and is quite important as a source of error to be accounted for. While the gyroscopes keep the three platform axes oriented in space, the acceleration along each of the axes is measured using a very sensitive *accelerometer*. The simplest form of accelerometer is the hanging pendulum or the pendulum balanced by springs. However, these instruments yield a nonlinear output, i.e., at different locations unequal displacements occur as a result of equal acceleration increments and are therefore not suitable. Instead, in an inertial survey system the *sensor-torquer* type of accelerometer shown in Fig. 12.8 is used. The acceleration (as, for example, from the motor vehicle carrying the system) would tend to cause the pendulum to swing out. The *sensor*, however, senses the starting of the movement and immediately sends an electrical signal to the amplifier, which in turn activates the *torquer*. The torquer generates just enough force to counteract the anticipated pendulum movement, thus keeping it in equilibrium. Hence, no actual physical movement of the pendulum occurs, and instead the force generated by the torquer is *quantized* and gives a good linear measure of the acceleration.

It is important to note that an accelerometer measures the component of acceleration in its direction (or the direction normal to the axis on which the pendulum mass is suspended). If the acceleration occurs precisely in that direction, its full value will be measured. On the other hand, if the acceleration takes place in a direction that is normal to the accelerometer's effective direction, it will measure no acceleration.

The essence of an inertial survey system is depicted in Fig. 12.9, where three gyroscopes are used to orient the platform in the east-west, north-south, and up-down directions. Each axis also contains a sensitive accelerometer which measures the acceleration component along that axis. An on-line computer monitors these three components of acceleration by sampling the acceleration along each axis every 17 ms (0.017 s). Since acceleration is the second derivative of distance traveled with respect to time, then by integrating the measured acceleration values twice with respect to time, the three components of the distance traveled

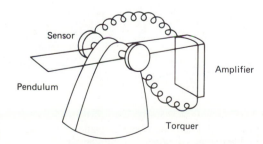

**Fig. 12.8** Accelerometer.
(*Courtesy L. F. Gregerson.*)

Vertical

Vertical
accelerometer

True north

North accelerometer

Vertical
gyroscope

North gyro

Up

East
gyroscope

East-west
accelerometer

N

East

E

**North Alignment**

  Horizontal accelerometers level the
    platform
  Gyroscopes sense earth rate
  Electronics rotates axis to north sensed
    by gyroscopes

**Survey**

  Accelerometers measure acceleration in each
    coordinate axis
  Electronics and computer integrates
    acceleration into distance traveled
  Distance traveled in each axis is resolved
    by computer into latitude, longitude,
    and elevation

**Fig. 12.9** Inertial survey system. (*Litton Guidance and Control Systems.*)

are obtained. It is important to note that the measured accelerations are referred to inertial space. An inertial reference system is fixed in space, while the survey (geodetic) reference coordinate system (with respect to which the results of the survey are expressed, e.g., latitude, longitude, and height) rotates with the earth. Thus, the platform system (which defines the inertial system) must be torqued to follow the curvature of the selected reference spheroid. This is shown in Fig. 12.10, where E-V (east-vertical) represents the inertial platform coordinate system. In Fig. 12.10a, at time $T_0$ the inertial system coincides with the geodetic system, where for example the V axis is along the local vertical, whereas at time $T_1$ it does not. In Fig. 12.10b the inertial system has been torqued to make it coincide with the survey system, also at time $T_1$. In this discussion the change occurs due to the uniform rotation of the earth during the time interval $(T_1 - T_0)$ even when the platform is stationary at one location on the earth's surface. When the platform is moved from one location to another, an additional torquing is also necessary. This torquing is required to compensate for the curvature of the earth as shown in Fig. 12.11. At longitude $\lambda_0$ the inertial system is aligned with the geodetic frame. If we assume that the platform were *instantly* shifted to another longitude $\lambda_1$ (i.e., no time elapsed), Fig. 12.11a, the inertial system would no longer be parallel to the survey system of coordinates. After it is torqued by an amount equivalent to the curvature between $\lambda_0$ and $\lambda_1$, it will remain aligned with the geodetic system (Fig. 12.11b).

As was mentioned previously, the inertial system may be mounted in either a motor vehicle or a helicopter. At the time of this writing (1978), two systems (Honeywell and Ferranti) were still under development, and the earliest system (Litton) was already in use. The following descriptions pertain to the Litton system, whose vehicle-mounted version is schematically depicted in Fig. 12.12; a photograph from the back of the vehicle is shown in Fig. 12.13. It can be seen from the figure that the system is composed of five components:

1. *Inertial measuring unit* (IMU), which is the heart of the system, including the gyroscopes and

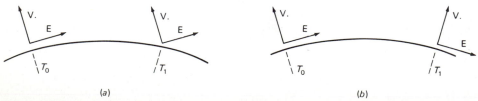

V.

E

V.

E

$T_0$

$T_1$

(a)

V.

E

V.

E

$T_0$

$T_1$

(b)

**Fig. 12.10** Compensation of a stationary inertial platform for rotation of the earth. (a) Not compensated. (b) Compensated. (*Courtesy L. F. Gregerson.*)

**Fig. 12.11** Compensation for earth curvature. (*a*) Without compensation. (*b*) With compensation. (*Courtesy L. F. Gregerson.*)

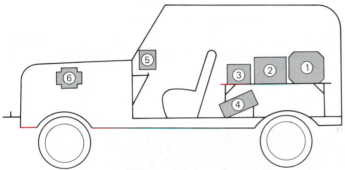

**Fig. 12.12** Schematic of inertial surveying system: (1) inertial measuring unit; (2) power supply; (3) computer; (4) cassette recorder; (5) display and control; (6) power source. (*Courtesy L. F. Gregerson, Ref. 8.*)

**Fig. 12.13** Inertial system mounted in the back of a truck. (*SPAN International, Inc.*)

accelerometers as explained above. The two horizontal accelerometers have a sensitivity of 6 to 8 $\mu$g (1 $\mu$g = 0.98 mgal) while the vertical accelerometer is sensitive to 1 $\mu$g.

2. *Power supply unit* (PSU), where the required current (using 24 V) is derived from an independent circuit using the vehicle's motor. When the engine is stopped, an automatic switchover to a battery occurs (for a maximum of about 1 h).

3. *On-board computer*, which in addition to performing the required arithmetical operations also controls the various functions of the system (i.e., calibration, alignment, adjustment, etc.).

4. *Cassette recorder*, which in addition to recording the data output by the computer, also records the actual biases and changes occurring in the system that are valuable for operational maintenance.

5. *Display and command unit* (Fig. 12.14), which has several functions, including continuous display of the state of the system and visual display of the required measurements; and, through the various switches, operator command and interrogation of the system can be effected. While the various units can be placed anywhere in the vehicle (Fig. 12.13), the display and control unit must logically be close to the operator, as shown in Fig. 12.15.

The operational procedure for inertial surveying proceeds as follows:

1. The platform (IMU) is first aligned into the local horizon and local north (meridian) over the initial point in the survey. Note that when the east-west and north-south accelerometers pick up no acceleration components, the platform must be normal to the directions of gravity. This operation is automatic and requires 1 hour.

2. Once the calibration and alignment are completed, the vehicle may proceed to travel along to the

**Fig. 12.14** Display and command unit. (*SPAN International, Inc.*).

**Fig. 12.15** Location of the display and control unit. (*SPAN International, Inc.*)

next point. The speed is dependent on the condition of the road with a range of 30 to 50 km/h; the technical limitation being maximum acceleration change of no more than 72 m/s². (For a helicopter, the speed range is 120 to 150 km/h.) While moving, E-W and N-S components of the distance traveled are computed every 17 ms, and the platform is made to remain tangent to the *geoid* and the north gyroscope axis made to remain pointing north.

3. After 3 to 5 min the vehicle must be stopped for about 4 s to carry out what is called *zero-velocity update* (ZUPT) (which can be seen in the upper left corner of Fig. 12.14). When the vehicle is stopped, the horizontal accelerometers usually do not register zero. The system then generates torques to make them read zero, and the amount of such torques are noted. These values are used as an indication of change in the *deflection of the vertical*, taking accelerometer drift into consideration.

   The final accuracy depends on ZUPT stops made regularly, preferably at equal time intervals. It is essential that the vehicle supporting the system be stable, particularly during the first 10 s of the ZUPT. In helicopter operations, landed ZUPTs are preferred to those obtained when hovering. When hovered ZUPTs are necessary, they should be followed as soon as possible by a landed ZUPT.

4. When an unknown survey point is reached, first a ZUPT is performed and then the computed data ($\phi, \lambda, h$, etc.) are stored in the computer memory.

5. When the final known control point in the survey is reached, its coordinates are entered into the computer and a *smoothing* subroutine may be called. This is an approximate, *not* a rigorous adjustment, made only to eliminate the large errors. In one commercial inertial survey system (called SPANMARK) this operation is relegated to an off-line minicomputer so as to free the on-line computer to do a better job on other, more necessary functions.

Needless to say, inertial survey systems have been introduced only quite recently and hence will continue to undergo modifications and improvements for several years to come. Therefore, the following accuracy data (adopted from Ref. 9) are applicable only at that time (1977).

Mean errors for any intermediate point in a traverse which runs in both directions between two control points are as follows:

In a vehicle, 60 km (37 mi) maximum length, flat gravity field:

| | |
|---|---|
| Latitude and longitude | 20 cm |
| Elevation | 15 cm |

In mountainous regions, or on longer distances, some deterioration is to be expected.

In a helicopter, 160 km (100 mi) maximum, where the terrain is flat:

| | |
|---|---|
| Latitude and longitude | 60 cm |
| Elevation | 50 cm |

It should be emphasized that these numbers are the result of experience gained up to that time (1977). If the student is interested in a more thorough analysis of errors in inertial survey systems, Ref. 2 should be consulted.

Other factors that are important if maximum accuracy of inertial systems is to be achieved include the following:

1. *Accuracy of basic control*   The positional accuracy can be no better than that of the basic control to which the inertial survey is tied. The highest-order horizontal and vertical control should be used.
2. *Rate of change of latitude and longitude*   Best results are obtained when the rate of change of latitude and longitude are linear with respect to time. Thus, there should be as little meandering as possible between control points, and a uniform rate of travel is necessary. Careful advance planning of the survey is needed to satisfy this requirement, which also results in minimum travel time between points.
3. *Double-run traverses*   Control traverses should be run in both directions between known terminal control points. The results from the two determinations can be averaged.
4. *Gravity anomalies*   Avoid large gravity anomalies when possible. If large gravity gradients are anticipated, zero-velocity-update periods should be taken at shorter intervals—$1\frac{1}{2}$ to 2 min.
5. *Smooth travel*   Travel should be performed in as smooth a manner as possible. Avoid holes in pavements, rough pavement, sudden stops, needless changes in direction, and abrupt accelerations.

**12.4. Applications for inertial positioning systems**   Thus far, experience with inertial systems has permitted identification of those projects where it is most economical. These may be categorized as (1) large projects either in total area or number of points; (2) surveys in remote areas where support is an important factor; (3) surveys where weather, terrain, traffic, environmental restrictions, or the client's desire to remain unnoticed may place restraints on survey personnel and operations; and (4) projects with limited time for completion due to weather, economics, or contract restrictions.

These projects can be further classified according to type of survey: control, property, construction, and geophysical surveys. A brief description of each application follows.

*1. Control surveys*   The capability of utilizing the inertial system to navigate to a point of known coordinates permits rapid, economical recovery and checking of existing control points. Densification of control is another application. Note that it is always necessary to have established geodetic control points from which to initiate the inertial survey. In this connection basic control can be set by doppler satellite methods (Art. 12.5). A major application of the system has been to establish control for aerial photography. Where the terrain lacks identifiable features, targets can be set and coordinated by the inertial system at some time before the acquisition of aerial photography. When the area contains many identifiable features, photography can be taken at the most suitable time and coordinates can be established on photo-identifiable points at a later date.

**2. Property surveys**  Public lands subdivision (Chap. 19) surveys have been successfully surveyed by inertial systems. Use of inertial systems is feasible when the traverse lines between corners do not have to be cleared. There are also applications in establishing control for the staking of mining claims and establishment of right-of-way surveys for route location.

**3. Construction surveys**  Any route location of large extent for pipelines, highways, and transmission lines (Chap. 17) represents a potential application of the inertial positioning system. New city complexes and hydroelectric developments represent other applications. Although the primary function of the inertial system in these projects is to establish control, the use of the system for obtaining data for topographic surveys should not be neglected. This last application is especially pertinent in remote regions covered by heavy rain forests where all other methods of surveying and mapping fail.

**4. Geophysical surveys**  Applications have been made in seismic and gravity studies. In both cases the system is used to establish horizontal position and elevation for the stations where studies occur. In gravity studies, the system can be used to eliminate blunders in reading conventional gravity meters.

**12.5. Doppler positioning systems**  The underlying concept for this class of positioning systems is the *doppler* phenomenon. The classical example usually cited for this phenomenon is the continuously changing pitch of a train whistle as it approaches, travels past, then travels away from a stationary observer. The same principle applies to a continuous tone or frequency signal transmitted from a satellite as it passes over an observer on the earth's surface. As the satellite gets closer to the observer, the received frequency continuously changes. When these frequency changes are properly monitored by the observer, they provide an indirect measure of the changes in distance between the satellite and the observer. If the position of the satellite is known with sufficient accuracy, and if a sufficiently large number of satellite passes is observed, the location of the observer can be determined with a high level of accuracy.

Figure 12.16 depicts schematically the underlying geometry relating the satellite to an observer. It shows that the distance between the satellite and the observer continuously changes during the satellite pass. At time $T_1$, the range between the observer and satellite is $R_1$ and the frequency received by the observer is $F_1$. Usually, instead of measuring the received frequency, a *beat* frequency is generated by subtracting from a locally generated reference frequency $F_0$ the received frequency $F$. The doppler measurements are then the cycle counts $N$ of beat frequency over precisely timed intervals. The received frequency $F$ is the sum of the transmitted frequency $F_T$ and the *doppler shift* resulting from the relative motion of the satellite with respect to the observer. If the satellite were to be assumed stationary, the doppler shift would be zero and the received frequency would be a constant. As the satellite moves closer to the observer, however, more cycles per second than those transmitted must be received by the observer, as shown in Fig. 12.16. For each distance equal to a wavelength that the satellite moves closer to the observation station, an additional cycle must be received. Thus, the doppler count $N$ is a direct measure of the *change* in the range $R$.

The doppler count $N$ consists of two parts: (1) the constant count corresponding to the difference between the transmitted frequency $F_T$ and the reference frequency $F_0$, or $(F_0 - F_T)$; and (2) a count reflecting the doppler shift, which is a function of the change in range or $(R_{i+1} - R_i)$. It can be shown (Ref. 12) that the doppler count $N_{i,i+1}$ during the time interval $T_i$, $T_{i+1}$ is given by

$$N_{i,i+1} = (F_0 - F_T)(T_{i+1} - T_i) + \frac{F_0}{c}(R_{i+1} - R_i) \tag{12.1}$$

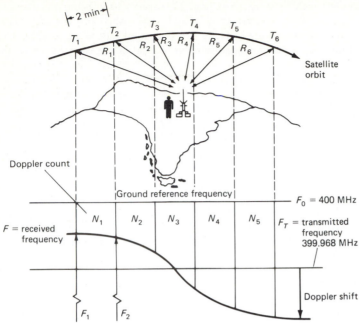

**Fig. 12.16** Each doppler counter measures slant range change. (*Magnavox Company.*)

where  $F_0$ = locally generated reference frequency
$F_T$ = frequency of the signal transmitted by the satellite
$T_i$ = $i$th instant of time
$T_{i+1}$ = $(i+1)$th instant of time
$c$ = velocity of light
$R_i$ = range between the satellite and the observer at time $T_i$
$R_{i+1}$ = range to the satellite at time $T_{i+1}$

It is easily seen from Eq. (12.1) that the first term on the right-hand side is an additive constant which is not pertinent and is usually eliminated during calculation. On the other hand, the second term represents the doppler shift and is the range difference $(R_{i+1} - R_i)$.

The foregoing discussion gives the basic concept of doppler count and its relation to change in range. To determine position accurately, several passes of different satellites are needed. Prior to 1967, the U.S. Navy Navigational Satellite System was used only for military applications by providing accurate position determination for marine vessels. In 1967, the system was made available to nonmilitary users. At the present time there are six satellites in the system, with circular orbits having an altitude of 600 nautical miles (about 1075 km) above the earth (Fig. 12.17). The period of revolution for each satellite is about 107 min. With this configuration, a location near the equator can observe only from 8 to 12 passes, while 40 to 50 passes can be observed from the arctic region. Each satellite continuously transmits the following:

1. Two stable frequencies, one 150 mHz and the other 400 mHz.
2. Precise timing signals.
3. Orbital parameters, which are called the Broadcast Ephemeris, every 2 min predicting the location of the satellite.

An observer needs the following to make use of the foregoing transmitted data:

1. An antenna capable of receiving the signals transmitted by the satellite.
2. A receiver to detect, amplify, and decode the signal.

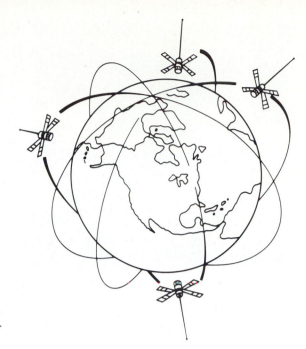

**Fig. 12.17** Navigational satellite orbits.
(*JMR Instruments, Inc.*)

3. A medium suitable for recording the observational data for subsequent processing and analysis.
4. Recording of temperature, humidity, and pressure at the observer's location during each pass in order to apply corrections for systematic errors due to these factors.

Although navigational receivers can be used for doppler surveying, several commercial firms have designed special receivers for the purpose. Such a system is shown in Fig. 12.18. This system is designed to accept both transmitted frequencies, thus allowing for the correction for ionospheric refraction, which is a significant systematic error. The system is essentially composed of an antenna, usually collapsible, receiver, recording medium such as paper or magnetic tape, and a rugged carrying case. The receiver derives its time signal from a very precise internal clock instead of using the satellite data, as in the case of navigational receivers. This allows a considerable reduction in the number of observable passes for a given accuracy requirement.

The accuracy of satellite doppler positioning depends on a number of factors, which according to Brown (Ref. 3) include:

1. Number of passes observed.
2. Number of observational intervals per pass.
3. Accuracy of orbital ephemerides.
4. Design of observational network.
5. Application of precise corrections for atmospheric refraction.
6. Method of data reduction.

Up to a point, accuracy increases in proportion to the square root of the number of passes reduced. This point occurs when 25 to 30 passes are used. Another important factor is whether the *Precise Ephemeris* or the *Broadcast Ephemeris* is used. The Precise Ephemeris is derived from tracking data and is unfortunately not available for all six satellites. The Broadcast Ephemeris is of poorer accuracy than the Precise Ephemeris, but is readily available to the user because of being encoded on the satellite signal itself. The root-mean-square (RMS) errors of the Broadcast Ephemeris are 24 m in track (or along the satellite's

**Fig. 12.18** Doppler receiving system. (*JMR Instruments, Inc.*)

orbit, see Fig. 12.19), 16 m cross-track (or normal to orbit), and 8 m radial (or in the range of the satellite).

Because of these relatively large errors in the Broadcast Ephemeris, there are two basic approaches to doppler positioning: (1) the orbital data from the ephemeris are assumed to be perfect or error-free, and (2) the orbital data from the ephemeris are treated as observational data with errors and are thus allowed to adjust in the data reduction. With these two approaches there are two fundamental procedures for determining point locations: (1) point positioning and (2) translocation. Each of these procedures will be discussed separately.

**Point positioning**   This procedure is characterized by the fact that an independent reduction is performed on all observations taken from a single point or station. Usually in this case, the ephemeris data are assumed to be without error. The end result of this reduction is the set of $X$, $Y$, $Z$ coordinates of the station. Since it is known that the orbital data contain errors, often a large number of passes, such as 100, are used in conjunction with the Broadcast Ephemeris. In such cases one would expect RMS accuracies of 2 to 3 m. Of course to observe this many passes requires about 1 week. So, if this level of accuracy is sufficient, such a procedure is much preferred because the computational effort is quite modest and can even be accomplished in the field on a minicomputer. Furthermore, this technique can be carried out with only one doppler receiver.

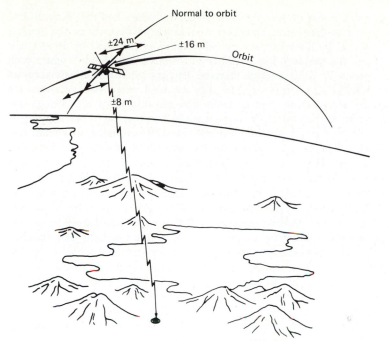

**Fig. 12.19** Root-mean-square errors in Broadcast Ephemeris.

***Translocation***  This procedure is characterized by having *two* receivers, each at different stations, such as *A* and *B* in Fig. 12.20. With this arrangement a *common* set of satellite passes may be observed simultaneously from the stations. When the separation between the two stations is of the same order of magnitude as the altitudes of the satellites, the effect of orbital data errors will be essentially the same. Thus, the relative position between the two stations is determined with higher accuracy than the absolute position. This fact, with proper station combinations, can lead to the determination of a net with high *internal* accuracy.

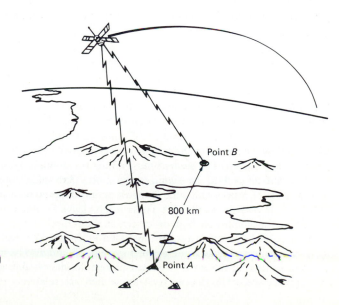

**Fig. 12.20** Doppler positioning by translocation. (*JMR Instruments, Inc.*)

As mentioned above, the design of both the network control points and the observational scheme affect the accuracy of the results. The observational scheme depends on the number of receivers used. Figure 12.21 illustrates several possibilities as proposed by Brown (see Refs. 3, 4, 5 at the end of the chapter). Figure 12.21a represents a net of points where each station is occupied sequentially using a single receiver. This is obviously the least accurate configuration (point positioning). Figure 12.21b may be used when two receivers are available. One receiver occupies the central point and remains there all through the observational period while the second receiver sequentially occupies the other stations. A much more preferred scheme is when four receivers are used, assuming of course that logistic support is possible. Figures 12.21c and d are two possibilities of what is called *singly connected quadrilaterals*, while Fig. 12.21e is of interlocking quadrilaterals. The last figure is by far the strongest, but would require more extensive effort and may therefore be employed when the highest accuracy is required (as for the establishment of a primary geodetic net).

In contrast to the two procedures discussed above (where orbital data are considered errorless), the so-called *short-arc geodetic adjustment* (Ref. 4) is a comprehensive method where all data, including the ephemeris, are given proper a priori weights and thus are allowed to adjust in the reduction program within the constraints imposed by these weights. Such a method of adjustment will necessarily entail a very large number of equations to be solved simultaneously, requiring an extensive computational effort. Comparative results (from a test) between short-arc geodetic adjustment and independent point positioning are given in Tables 12.1 and 12.2. The results relate to a nine-station net sketched in Fig. 12.22. The original design had been planned such that four stations at a time would be occupied during a space of 5 days according to the following schedule:

| Day | Stations to be occupied |
|-----|-------------------------|
| 1 | 1, 2, 4, 5 |
| 2 | 2, 3, 5, 6 |
| 3 | 5, 6, 8, 9 |
| 4 | 4, 5, 7, 8 |
| 5 | 1, 3, 7, 9 |

thus allowing the central station (number 5) to be occupied for 4 days. Because of several unusual factors (although perhaps to be expected), the following passes were used:

| Station | 1 | 2 | 3 | 4 | 5 | 6 | 7 | 8 | 9 |
|---------|---|---|---|---|---|---|---|---|---|
| Number of passes | 22 | 17 | 14 | 5 | 35 | 10 | 20 | 13 | 20 |

Thus, the number of passes observed was significantly less than planned. A minimum of 50 passes from station 5 and 25 passes from each of the other stations had originally been planned. Nevertheless, the results in Tables 12.1 and 12.2 are quite useful. They show the dramatic improvement in the results between independent point positioning and short-arc geodetic adjustment. The student must remember, however, that the field effort in the latter case would in general be significantly more than point positioning with one receiver. Another set of results, this time from an actual doppler surveying project overseas (instead of a test), is given in Table 12.3. These results leave no doubt as to the potential of doppler surveying for establishing high-order nets in the future.

In view of the excellent results illustrated in Tables 12.1 and 12.2, some of the complications that can arise when using doppler systems should also be mentioned. Since many of the most obvious applications are located in remote inaccessible areas, helicopters and camps may be required. The doppler systems, which are relatively expensive, should *not* be left

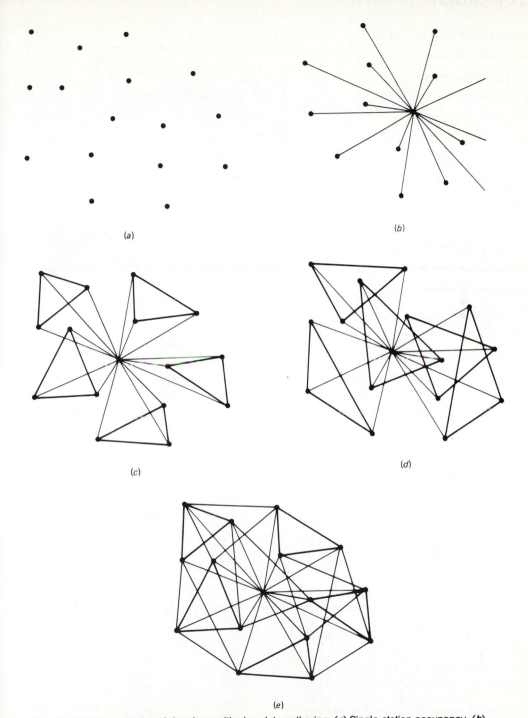

**Fig. 12.21** Various modes of doppler positioning data gathering. (*a*) Single-station occupancy. (*b*) Dual-station, fixed-base occupancy. (*c*) Singly connected quadrilaterals (one possible arrangement). (*d*) Singly connected quadrilaterals (an alternative arrangement). (*e*) Interlocking quadrilaterals [combination of (*c*) and (*d*)]. (*DBA, Inc.*)

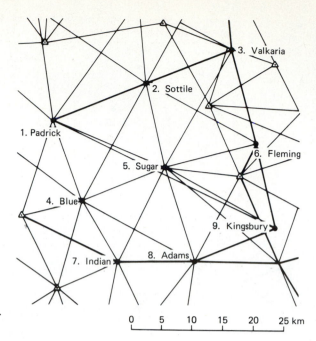

**Fig. 12.22** Nine-station test network. (*DBA, Inc.*)

**Table 12.1  Results from Independent Point Positioning**

(a) Errors in solution in north, east, and up components

| Station | $\Delta N$, m | $\Delta E$, m | $\Delta U$, m |
|---------|---------|---------|---------|
| 1 | 2.48 | 1.07 | 11.65 |
| 2 | − 7.09 | − 14.66 | − 7.68 |
| 3 | 17.30 | − 5.56 | 7.48 |
| 5 | − 9.38 | − 0.77 | − 4.82 |
| 6 | 17.33 | − 4.27 | − 12.51 |
| 7 | − 3.56 | − 10.25 | − 13.93 |
| 8 | 13.74 | − 8.88 | − 11.09 |
| 9 | − 0.09 | − 9.53 | − 13.37 |
| Means: | 3.84 m $\overline{\Delta N}$ | − 6.60 m $\overline{\Delta E}$ | − 8.45 m $\overline{\Delta U}$ |

(b) Errors in solution relative to means (internal errors)

| Station | $\Delta N - \overline{\Delta N}$, m | $\Delta E - \overline{\Delta E}$, m | $\Delta U - \overline{\Delta U}$, m |
|---------|---------|---------|---------|
| 1 | − 1.36 | 7.67 | − 3.20 |
| 2 | − 10.93 | − 8.06 | 0.77 |
| 3 | 13.46 | 1.04 | 15.93 |
| 5 | − 13.22 | 5.83 | 3.63 |
| 6 | 13.49 | 2.33 | − 4.06 |
| 7 | − 7.40 | − 3.65 | − 5.48 |
| 8 | 9.90 | − 2.28 | − 2.64 |
| 9 | − 3.93 | − 2.93 | − 4.92 |
| RMS errors: | 10.17 m | 4.89 m | 6.66 m |

**Table 12.2   Results from Short-Arc Geodetic Adjustment**

(a) Errors in solution in north, east, and up components

| Station | $\Delta N$, m | $\Delta E$, m | $\Delta U$, m |
|---|---|---|---|
| 1 | 7.02 | − 0.05 | − 7.90 |
| 2 | 5.16 | 0.43 | − 7.30 |
| 3 | 2.53 | 1.86 | − 8.14 |
| 5 | 4.39 | 0.58 | − 7.29 |
| 6 | 3.09 | 0.68 | − 6.01 |
| 7 | 5.94 | 0.30 | − 7.31 |
| 8 | 2.28 | − 0.52 | − 7.93 |
| 9 | 4.98 | 0.38 | − 7.47 |
| Means: | 4.42 m $\overline{\Delta N}$ | 0.46 m $\overline{\Delta E}$ | − 7.42 m $\overline{\Delta U}$ |

(b) Errors in solution relative to means (internal errors)

| Station | $\Delta N - \overline{\Delta N}$, m | $\Delta E - \overline{\Delta E}$, m | $\Delta U - \overline{\Delta U}$, m |
|---|---|---|---|
| 1 | 2.60 | − 0.51 | − 0.48 |
| 2 | 0.74 | − 0.03 | 0.12 |
| 3 | − 1.89 | 1.40 | − 0.72 |
| 5 | − 0.03 | 0.12 | 0.13 |
| 6 | − 1.33 | 0.22 | 1.41 |
| 7 | 1.52 | − 0.16 | 0.11 |
| 8 | − 2.14 | − 0.98 | − 0.51 |
| 9 | 0.56 | − 0.08 | − 0.05 |
| RMS errors: | 1.60 m | 0.29 m | 0.38 m |

**Table 12.3   Horizontal Proportional Accuracies Relative to Base Station**
*(Adapted from Ref. 5)*

| Station | Distance from $D$, km | Planimetric $(\sigma_N^2 + \sigma_E^2)^{1/2}$, m | Proportional $(\sigma_N^2 + \sigma_E^2)^{1/2}/D \times 10^3$, 1 part in: |
|---|---|---|---|
| 001 | — | — | — |
| 002 | 531.7 | 0.98 | 543,000 |
| 003 | 189.6 | 0.53 | 358,000 |
| 004 | 351.4 | 0.96 | 366,000 |
| 005 | 639.3 | 1.18 | 542,000 |
| 006 | 523.6 | 0.87 | 602,000 |
| 007 | 230.5 | 0.69 | 334,000 |
| 008 | 623.7 | 1.48 | 421,000 |
| 009 | 717.0 | 1.11 | 646,000 |
| 010 | 598.4 | 0.82 | 730,000 |
| 011 | 818.4 | 1.19 | 688,000 |
| 012 | 820.4 | 1.28 | 641,000 |
| 013 | 609.8 | 0.95 | 642,000 |
| 014 | 432.6 | 0.97 | 446,000 |
| 015 | 524.7 | 0.98 | 535,000 |
| 016 | 456.3 | 0.91 | 501,000 |

unattended. Thus, personnel, transport, and good communications are necessary. If only one set is used, the observation time necessary to obtain a good position may extend to a week or 10 days. When a single doppler set is employed with the Precise Ephemeris (to ensure better results), data must be sent to U.S. Defense Mapping Agency for computation. This process takes 6 months. The other option is to use translocation or short-arc techniques for which a minimum of two doppler sets is needed and four sets are recommended for optimum accuracy and efficiency. Thus, the initial investment is more than doubled or quadrupled, since computer programs to reduce the data must be purchased or developed in house.

Assuming that the finances and work force can be justified for a project, doppler positioning can be used to establish control for (1) surveys to establish right of way, (2) property surveys, and (3) primary and secondary control networks.

Frequently, doppler positioning may best be utilized as a supplement rather than as a substitute for traditional and photogrammetric surveying procedures. For example, in a topographic survey of a large area containing difficult terrain, doppler positioning can be used to establish control at widely separated points. Photogrammetric methods supplemented by ground surveys could then be utilized to complete the control network and perform topographic mapping. Similar examples can be cited for virtually every type of survey.

## 12.6. Problems

**12.1**  Describe the procedure for determining distance and direction using a combined theodolite and EDM.

**12.2**  "An electronic tacheometer may be considered a complete surveying system." Discuss this statement, describing the various components of such a system.

**12.3**  Explain the various applications of electronic positioning systems in which special features of such systems are exploited.

**12.4**  Describe the basic idea of an inertial positioning system and its principal components.

**12.5**  With the aid of a sketch, explain the main parts of a simple gyroscope.

**12.6**  "With the invention of inertial survey systems, classical survey methods will soon be obsolete." Discuss whether this statement is likely to be true or false, expressing your feelings for the role of such systems and their effects on the field of surveying.

**12.7**  Describe in detail the operational steps involved in a survey with a vehicle-mounted inertial survey system.

**12.8**  Discuss concisely four areas of application for inertial survey systems.

**12.9**  What is the doppler phenomenon and how is it exploited in satellite doppler position systems?

**12.10**  Enumerate the factors that affect the accuracy of satellite doppler surveying.

**12.11**  Describe the two methods for determining location by doppler surveying: point positioning and translocation.

**12.12**  Discuss how satellite doppler positioning, inertial survey systems, and electronic survey systems could be utilized to perform the control and topographic surveys required for the design of a dam in a remote region where the terrain is obscured by rain forest.

## References

1. Ball, W. E., "Testing an Airborne Inertial Survey System for B.L.M. Cadastral Survey Application in Alaska," *Proceedings, Annual Meeting, ACSM*, 1975.
2. Ball, W. E., "Adjustment of Inertial Survey System Errors," *Proceedings, Annual Meeting, ACSM*, 1978.
3. Brown, Duane C., *A Primer on Satellite Doppler Surveying*, DBA Systems Technical Note 75-002, April 1975.
4. Brown, D. C., *A Test of Short Arc versus Independent Point Positioning Using Broadcast Ephemeris of Navy Navigational Satellites*, DBA Systems Technical Note 75-001, 1976.
5. Brown, D. C., "Doppler Positioning by the Short Arc Method," Presented to the International

Geodetic Symposium, Physical Science Laboratory, New Mexico State University, Las Cruces, N.M., October 12–14, 1976.

6. Erickson, Kent E., "Electronic Surveying Systems," *Proceedings ACSM, ASP/ACSM Fall Convention*, Little Rock, Ark., October 1977.

7. Gort, Alfred, F., "The Hewlett Packard 3829A Electronic Total Station," *Proceedings ACSM, ASP/ACSM Fall Convention*, Little Rock, Ark., October 1977.

8. Gregerson, L. F., and Carriere, R. J., "Inertial Surveying System Experiments in Canada," *Proceedings, IUGG*, Grenoble, France, August 1975.

9. Gregerson, L. F., "A Description of Inertial Technology Applied for Geodesy," *70th Annual Convention of CIS*, May 1977.

10. Hothem, L. D., Strange, W. E., and White, M., "Doppler Satellite Surveying System," *Journal of the Surveying and Mapping Division, ASCE*, No. SU1, Proceedings Paper 14132, November 1978.

11. Krakiwsky, E. J., Wells, D. E., and Thompson, D. B., "Geodetic Control from Doppler Satellite Observations for Lines Under 200 Km," *The Canadian Surveyor*, Vol. 32, No. 2, June 1972.

12. Laurila, S., *Electronic Surveying and Navigation*, John Wiley & Sons, Inc., New York, 1976.

13. Mancini, A., and Huddle, J., "Gravimetric and Position Determination Using Land Based Inertial System," *Proceedings, Annual Convention, ACSM*, 1975.

14. O'Brien, L. S., "Investigations of Inertial Survey Systems by the Geodetic Survey of Canada," *The Canadian Surveyor*, Vol. 30, No. 5, December 1976, p. 385.

15. Rawlinson, C., "Automatic Angle Measurement in the AGA 700 Geodimeter," *Survey Review*, April 1976.

16. Schellens, Dieter F., "Surveying with the Zeiss REG ELTA 14 in Densely Populated Areas," *Proceedings, Annual Meeting, ACSM*, 1973, p. 451.

# PART IV
# MAPPING

# CHAPTER 13
# Mapping and map drafting

**13.1. General** The various classes, types, and general categories of maps are defined and described in Art. 1.11. It was noted there that the two main types of maps are the line map prepared by a draftsman using traditional drafting procedures from field data and/or a photogrammetrically prepared manuscript, and orthophoto maps produced from a stereo-pair of aerial photographs by photogrammetric methods. An increasing proportion of the line drawn maps are being compiled on a flat-bed plotter under the control of an electronic computer. Thus, compilation and production methods for all kinds of maps are undergoing substantial modifications due to the rapidly changing technologies now adaptable to mapping procedures.

Also defined in Art 1.11 are the two general categories of maps: planimetric maps, which show the plan positions of natural and man-made features; and topographic maps, which show positions of all features as well as the spatial configuration of the terrain called *relief*.

In this chapter emphasis is placed on characteristics of topographic maps; methods of illustrating relief on topographic maps; contours and how contours are plotted; map scales and contour intervals; topographic map construction; types of information that should appear on maps; methods of map reproduction; a brief review of automatic computer-aided map drafting; and tests and standards for the accuracy of maps.

**13.2. Datums for mapping** In surveying and mapping, datums are necessary to correlate measurements used to determine elevations and horizontal positions for points at different locations. A horizontal datum and a vertical datum are used in surveying.

A *horizontal datum* is defined by the latitude and longitude of an initial point, the azimuth of a line from that point, and the two radii needed to define the geometric reference surface (the spheroid) which best approximates the surface of the earth in the region of the surveys. In the conterminous United States and Alaska, the horizontal datum now being used is the North American Datum of 1927 and the reference surface is the Clarke Spheroid of 1866. This datum is scheduled to be revised in 1983 when a new adjustment based on all data gathered to date and satellite observations will be completed.

A *vertical datum* is the surface to which elevations or depths are referred. All surveys and mapping in the conterminous United States are based on the National Geodetic Vertical Datum (NGVD) of 1929, formerly called the Sea Level Datum of 1929. This datum is based on a best fit to 26 mean sea level tide stations in the United States and Canada and is the result of a general adjustment of level networks in the United States and Canada in 1929. This vertical datum will also undergo small changes in the 1983 readjustment. The elevations of the National Vertical Control Network are tied to the National Geodetic Vertical Datum. Thus, the elevations that are normally used in topographic mapping, geodetic surveys, engineering studies, and engineering construction surveys are referred to the NGVD.

The NGVD should not be confused with local mean sea level datums. The national datum was determined by adjusting to a large number of sea level stations located over a broad area. Consequently, small differences exist between the NGVD and local mean sea level for a specific location.

Elevations referred to the NGVD are not used in boundary surveys, which are a function of *tidal datum* in tidal waters or the lake level in the Great Lakes regions. Tidal datums are defined by the phase of the tide and are described as mean high water, mean low water, and mean lower low water.

Other horizontal and vertical datums are employed for Alaska (vertical datum), Puerto Rico, Hawaii, the Virgin Islands, Guam, and other oceanic islands. Specifications for these datums can be obtained by requesting the information from the National Ocean Survey.

**13.3. Topographic maps**   A topographic map shows by the use of suitable symbols (1) the spatial configuration of the earth's surface, which includes such features as hills and valleys; (2) other natural features such as trees and streams; and (3) the physical changes wrought upon the earth's surface by the works of man, such as houses, roads, canals, and cultivation. The distinguishing characteristic of a topographic map, as compared with other maps, is the representation of the terrestrial relief.

Topographic maps are used in many ways. They are a necessary aid in the design of any engineering project that requires a consideration of land forms, elevations, or gradients, and they are used to supply the general information necessary to the studies of geologists, economists, and others interested in the broader aspects of the development of natural resources.

The preparation of general topographic maps is largely in the hands of governmental organizations. The principal example is the topographic map of the United States compiled and published by the U.S. Geological Survey. This map is published in quadrangle sheets, which usually include territory ranging from $7\frac{1}{2}$ by $7\frac{1}{2}$ minutes at a scale of 1:24,000 (1:25,000 in the metric system) to 4° by 12 ° at a scale of 1:1,000,000. A portion of a typical map, the scale of which is 1:24,000, is shown in Fig. 13.1. Note that the figure has been reduced so the scale is no longer 1:24,000. Altogether there are many federal agencies engaged in surveying and mapping. The central source of information regarding all federal maps and aerial photographs is the National Cartographic Information Center, Reston, Virginia 22092. Likewise, many maps are available from state, county, and city agencies.

**13.4. Representation of relief**   Relief may be presented by *relief models, shading, color gradients, hachures, form lines,* or *contour lines.* Of the symbols used on maps, only contour lines indicate elevations directly and quantitatively; they have by far the widest use. They are the principal topic of the next several sections. Form lines are similar to contour lines but are not true to scale and are therefore only qualitative.

Any map that portrays relief by one of the methods above described is called a *hypsometric map.* When the elevations on a map are referred to a sea-level datum, the map may be called a *hypsographic map.* A topographic map may be included in both categories.

**13.5. Relief model**   A relief model is a representation of the terrain done in three dimensions to suitable horizontal and vertical scales; it is a miniature of the terrain it represents. Plastic materials such as wax or clay are used; also laminated models are made by cutting cardboard sheets to the shape of successive contours and then assembling the sheets. The Defense Mapping Agency (DMA) has developed a process for producing three-dimensional contour maps in the form of sheets of plastic which are flat-printed and then molded over a relief model. Cardboard relief maps of certain areas are commercially available.

The relief model is the most legible of all methods of representing relief, and it is of great value for purposes of instruction and public exhibit. It is also an aid in many of the special

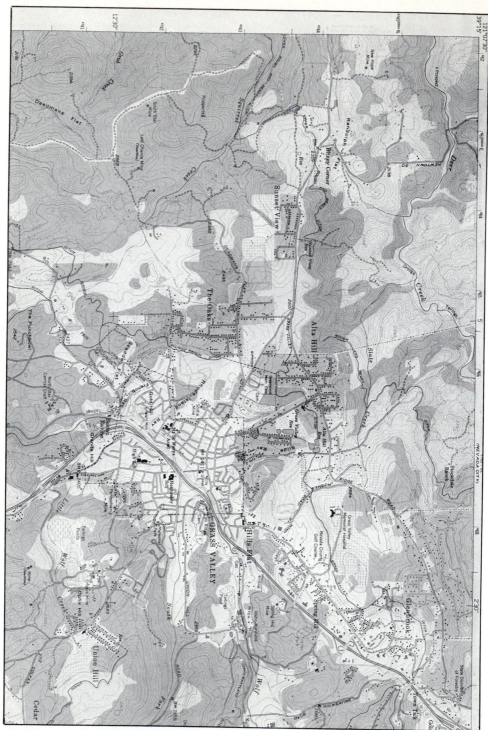

**Fig. 13.1** Typical topographic map of the U.S. Geological Survey. Scale is 1:24,000 (2000 ft/in). Contour interval 20 ft. (*U.S. Geological Survey.*)

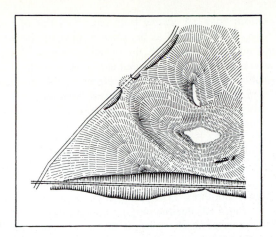

**Fig. 13.2** Hachures.

studies of the geologist, the geographer, and the engineer. However, its use is limited because of its cost and bulk.

**13.6. Shading**   Shading is a method of showing the terrestrial relief roughly in plan as it would appear from a point vertically above and with parallel rays of light flooding the landscape from a given angle causing shadows to lie upon the less-illuminated areas. The method is pictorial and is useful in showing the general features where the relief is high and the slopes are steep. Shading is sometimes used in combination with hachures or contour lines, to render the map more legible.

**13.7. Hachures**   Hachures show relief more definitely but less legibly than does shading. The symbol consists of rows of short, nearly parallel lines whose spacing, weight, and direction produce an effect similar to shading but capable of more definite handling. The lines are drawn parallel to the steepest slopes, and in the best practice a standard scale of lengths and weights of lines is used to represent the various degrees of inclination of slopes. The method is illustrated in Fig. 13.2, which is a representation of a portion of the relief shown by contour lines at right center in Fig. 13.3.

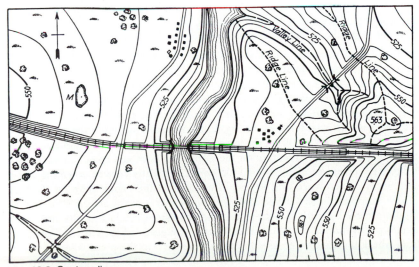

**Fig. 13.3** Contour lines.

**13.8. Contours and contour lines**   A *contour* is an imaginary line of constant elevation on the ground surface. It may be thought of as the trace formed by the intersection of a level surface with the ground surface, for example, the shoreline of a still body of water.

If the locations of several ground points of equal elevation are plotted on a drawing, a line joining these points is called a *contour line*. Thus, contours on the ground are represented by contour lines on the map. Loosely, however, the terms "contour" and "contour line" are often used interchangeably. On a given map, successive contour lines represent elevations differing by a fixed vertical distance called the *contour interval*.

The use of contour lines has the great advantage that it permits the representation of relief with much greater facility, and with far greater definiteness and accuracy, than do other symbols. It has the disadvantage that the map is not as legible to the layman.

**13.9. Characteristics of contour lines**   The principal characteristics of contour lines can be illustrated by reference to Fig. 13.3. For the purpose of this discussion the slope of the river surface is disregarded. The stage of the river at the time of the field survey was at an elevation of 510 ft; hence, the shoreline on the map marks the position of the 510-ft contour line. For this map, the contour interval is 5 ft. If the river were to rise through a 5-ft stage, the shoreline would be represented by the 515-ft contour line; similarly, the successive contour lines at 520 ft, 525 ft, etc., represent shorelines that the river would have if it should rise farther by 5-ft stages.

The principal characteristics of contour lines are as follows:

1. The horizontal distance between contour lines is inversely proportional to the slope. Hence, on steep slopes (as at the railroad and at the river banks in Fig. 13.3) the contour lines are spaced closely.
2. On uniform slopes the contour lines are spaced uniformly.
3. Along plane surfaces (such as those of the railroad cuts and fills in Fig. 13.3) the contour lines are straight and parallel to one another.
4. As contour lines represent level lines, they are perpendicular to the lines of steepest slope. They are perpendicular to ridge and valley lines where they cross such lines.
5. As all land areas may be regarded as summits or islands above sea level, evidently all contour lines must close upon themselves either within or without the borders of the map. It follows that a closed contour line on a map always indicates either a summit or a depression. If water lines or the elevations of adjacent contour lines do not indicate which condition is represented, a depression is shown by a hachured contour line, called a *depression contour*, as shown at *M* in Fig. 13.3.
6. As contour lines represent contours of different elevation on the ground, they cannot merge or cross one another on the map, except in the rare cases of vertical surfaces (see bridge abutments of Fig. 13.3) or overhanging ground surfaces as at a cliff or a cave.
7. A single contour line cannot lie between two contour lines of higher or lower elevation.

Other special features of contours are shown in Figs. 13.4 and 13.5. Figure 13.4 shows contours with a 10-m interval for mountainous terrain. On the right is a well-defined ridge. Note that the V or U shape of the contours points down the slope on a ridge. On the left is a ravine with several small streams leading into it. In contrast to the ridge, the V- or U-shaped contours point up the slope to indicate a ravine, flow line of a stream, or a ditch. Figure 13.5 shows how contours cross artificial features. In Fig. 13.5*a*, the contours cross the paved road built in cut on the upper side and on fill on the lower side. On the upper side there is a ditch, indicated by the V-shaped contours pointing up the slope. The contours cross the pavement as a smooth line (the pavement is relatively smooth) which is slightly convex down the slope since most paved roads are crowned and are slightly higher in the center than on the two edges. In Fig. 13.5*b*, contours cross a paved road with curbs. Most curbs are 0.5 ft or 0.15 m high, so the contour crosses the curb at right angles and runs along the face of the curb until it reaches the contour elevation in the gutter line. Thus, if the grade of the road is 5 percent, the distance from where the contour crosses the top of the curb to the point where it meets the contour elevation in the gutter is 0.5/0.05 = 10 ft, or about 3 m. Then the contour crosses

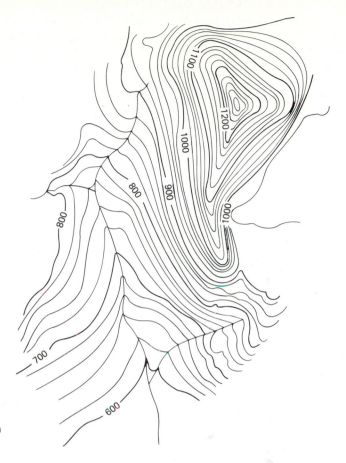

**Fig. 13.4** Mountainous terrain
showing a ravine and a ridge.

the pavement slightly convex down the slope to the gutter line on the other side, follows the
face of the curb until the contour elevation at the top of the curb is reached, and crosses the
curb and sidewalk at right angles (assuming that the sidewalk is horizontal). The contour
interval in Fig. 13.5a is 5 ft and that in Fig. 13.5b is 2 m.

**13.10. Choice of map scale and contour interval**   It is possible to choose a map scale
consistent with the purpose of the survey if the approximate size of plotting error is known.
For example, if it is known that (with reasonable care in plotting) the standard deviation in
measuring a distance between two definite points on the map is $\frac{1}{50}$ in (0.5 mm), and if it is
known that the purpose of the survey will be met if the standard deviation in scaled
distances is 10 ft (3 m), then a map scale of 1:6000 (500 ft/in) satisfies the conditions.

Other factors that influence choice of map scale alone are (1) the clarity with which
features can be shown, (2) the cost (the larger the scale, the higher the cost), (3) the
correlation of map data with related maps, (4) the desired size of the map sheet, and (5)
physical factors such as the number and character of features to be shown, the nature of the
terrain, and the necessary contour interval.

The contour interval may be thought of as the scale by which the vertical distances or
elevations are measured on a map. The choice of a proper contour interval for a topographic
survey and map is based upon four principal considerations: (1) the desired accuracy of
elevations read from the map, (2) the characteristic features of the terrain, (3) the legibility of
the map, and (4) the cost.

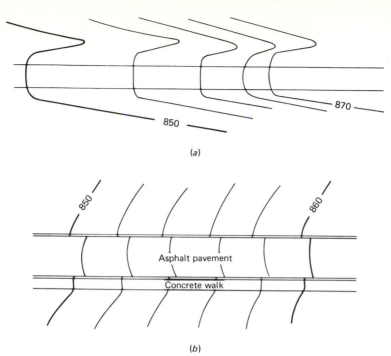

**Fig. 13.5** (*a*) Contours crossing a paved road. (*b*) Contours crossing a paved road with a curb.

1. *Accuracy*   Let it be assumed that two maps are equally accurate, so that the standard deviation in elevations, read from the map, of points chosen at random is one-half of a contour interval. Assume one map to have a contour interval of 5 ft (1 m) and the other 2 ft (0.5 m). It is evident that the standard deviation in elevations of points chosen at random on one map is $2\frac{1}{2}$ ft (0.5 m) and on the other 1 ft (0.25 m). Therefore, the more refined the scale (i.e., the smaller the interval), the more refined should be the measurements of the elevations of chosen points.

2. *Features*   Often field conditions exist where characteristic features require the use of a contour interval that would otherwise be inappropriate. Thus, if the shape of the terrain is such as to show much variation within a small area, or in other words, if the topography is of *fine texture*, then a smaller contour interval is required to show the greater complexity of configuration. On the other hand, if the landscape is composed of large regular forms or is of *coarse texture*, then a larger interval may be used. Where the texture varies considerably within an area to be mapped, or where it is desired to define certain parts of the area more clearly than the general area (as for construction purposes), a smaller contour interval may be used for parts of the area.

3. *Legibility*   A map otherwise excellent may be rendered useless and its appearance disfigured by a mass of contour lines that obscure other essential features. In general, contour lines should not be spaced on the map more closely than 20 or 30 to the inch (8 to 12 per cm), although the legibility of the map depends largely upon the fineness and precision with which the lines are drawn. The lithographed maps of the U.S. Geological Survey and of the U.S. National Ocean Survey yield good results with much closer spacing.

4. *Cost*   The smaller the contour interval, the higher the cost, especially if the usual accuracy of one-half contour interval is maintained.

Obviously, the contour interval and map scale are interrelated. In general, the smaller the scale, the larger the contour interval. For the most part, this interrelationship depends upon the purpose and scale of the map, the character of the terrain represented, and the degree of land development in the area to be mapped.

Traditionally, maps have been classified according to scale as large, intermediate, and small scale with accompanying contour intervals as follows (comparable metric scales shown in parentheses):

*Large scale*          10  to 200 ft/in (1:100 to 1:2000)
  Contour interval   0.2 to 10 ft (0.1 to 2 m)
*Intermediate scale*   200 to 1000 ft/in (1:2000 to 1:10,000)
  Contour interval   1 to 20 ft (0.2 to 5 m)
*Small scale*          1,000 to 5,000,000 ft/in (1:10,000 to 1:100,000,000)
  Contour interval   50 to 5000 ft (5 to 2000 m)

The American Society of Civil Engineers (ASCE) Surveying and Mapping Division has a more detailed classification for map scales and contour intervals.

1. *Design maps*   These maps are used in the design and construction of specific engineering work of all kinds. Scales vary from 10 to 200 ft/in (1:100 to 1:2000) with contour intervals between 0.2 and 10 ft (0.1 and 1 m), depending on the type of project, land use, and terrain characteristics. Two subcategories are given with this group:
   a. *Critical design maps*   These are maps used on projects having critical space, orientation, position, or elevation restrictions. An example is a highway interchange in an urban area.
   b. *General design maps*   These are maps prepared for projects that do not have such rigid restrictions with respect to location. An example is a map prepared for a rural water distribution system.
2. *Planning maps*   These include a large group of maps used in planning engineering work or in overall planning at the urban, regional, national, and international levels. These maps may be used for geological studies, land use, agricultural production, and population studies; for public service planning; and for atlases. Map scales range from 100 to 5,000,000 ft (1000 mi)/in (1:1000 to 1:100,000,000) and contour intervals from 1 to 5000 ft (0.2 to 2000 m).

A summary of the categories of maps, accompanying scale ranges in feet per inch and representative fractions, and corresponding contour intervals in feet and metres is given in Table 13.1, which is taken from the ASCE Task Committee Report (Ref. 3).

The ranges in scale and contour interval for a given category of map in Table 13.1 are broad and do not allow selection of scale or contour interval according to type of project,

**Table 13.1   Categories of Maps with Corresponding Scale Ranges and Contour Intervals**
*(Adapted from Ref. 3, Courtesy ASCE)*

| Category of maps | Scale ranges | | Contour interval, ft (m) |
|---|---|---|---|
| | Feet per inch | Representative fraction, 1: | |
| Design | | | |
| Critical | 10 to 50 | 100 to 500 | 0.2 to 5 (0.1 to 1) |
| General | 40 to 200 | 500 to 2000 | 0.5 to 10 (0.1 to 2) |
| Planning | | | |
| Micro | 100 to 1000 | 1000 to 10,000 | 1 to 20 (0.2 to 5) |
| Local | 400 to 2000 | 5000 to 25,000 | 2 to 50 (0.5 to 10) |
| Regional | 1000 to 10,000 | 10,000 to 100,000 | 5 to 100 (1 to 20) |
| National | 10,000 to 100,000 (2 mi)     (20 mi) | 100,000 to 1,000,000 | 10 to 1000 (2 to 200) |
| International | 100,000 to 5,000,000 (20 mi)     (1,000 mi) | 1,000,000 to 100,000,000 | 50 to 5000 (10 to 2000) |

degree of land development, and terrain roughness. In order to permit this type of selection, the Task Committee classified engineering functions, degrees of land use, and roughness of the terrain.

The engineering functions considered in the classifications are transportation, land development, hydraulic and hydrologic development, urban planning, and cadastral operations. Each function was then separated into specific areas. For example, transportation includes projects on highways, railroads, transmission lines, airports, and waterfront facilities. Other functions are separated into a comparable number of project areas.

Land use is classified as industrial, urban, suburban, farmland, and rangeland. Roughness of the terrain is described as flat, gently rolling, rolling, hilly, and mountainous.

The ASCE Task Committee Report contains three sets of tables in addition to Table 13.1, which permit determination of map scale and contour interval as a function of category of map, engineering function and project area, land use, and roughness of terrain.

As an example, consider the engineering function of transportation, specifically the design and planning of paved highways in a suburban area (present land use) where the terrain varies from flat to rolling to hilly. Table 13.2 shows the ranges of scales and contour intervals for the various categories of maps that might be required for planning and design of these highways.

Tables similar to Table 13.2 could be compiled for any engineering design or planning project of a specific type given the land-use characteristics and roughness of the terrain. Space does not permit inclusion of the additional tables necessary for this purpose. The reader interested in determining map scale and contour interval for a specific job should consult Ref. 3.

**13.11. Topographic map construction**  Normally, the construction of a topographic map consists of three operations: (1) the plotting of the horizontal control, or skeleton upon which the details of the map are placed; (2) the plotting of details, including the map location of points of known ground elevation, called *ground points*, by means of which the relief is to be indicated; and (3) the construction of contour lines at a given contour interval, the ground points being employed as guides in the proper location of the contour lines. A ground point on a contour is called a *contour point*.

Current trends in the production of topographic maps, for use in engineering design and planning and for maps used by the public, are toward photogrammetric compilation of maps (Art. 16.17) and automatic or semiautomatic construction of the map. General details relevant to automated mapping systems can be found in Art. 13.36. Emphasis in the sections that follow is on the conventionally drawn line map. Where appropriate, references are given to the comparable automatic procedure.

**13.12. Methods for plotting horizontal control**  Horizontal control may be plotted by (1) the coordinate method, (2) the tangent method, (3) the chord method, or (4) by use of a protractor and scale. The coordinate method, the most general and accurate procedure for plotting positions of points, is adaptable to plotting by hand or by automatic methods. The latter three methods are of substantially lower accuracy than the coordinate method and are suitable primarily for rough preliminary plotting of control for maps covering small areas. In any case, distances are measured with the engineer's scale; for precise work the points are pricked with a needle and a reading glass is employed. Traverse and triangulation stations are indicated by appropriate symbols, and control lines are drawn carefully with a hard pencil having a fine point. Points are numbered or lettered to conform to the system employed in the field work. Usually, the control is not shown on the finished map.

The precision of plotting the control depends upon the instrument used and upon the care with which the work is done. With a well-sharpened hard pencil and with the use of

Table 13.2   Map Scales and Contour Intervals for Various Categories of Topographic Maps to be Used for Design and Planning of Paved Highways in a Suburban Area (Adapted from Ref. 3, Courtesy ASCE)

| Topographic map subcategory | Scales | | Terrain characteristics | Contour intervals | |
|---|---|---|---|---|---|
| | Feet per inch | Representative fraction, 1: | | Feet | Metres |
| **Design** | | | | | |
| Critical | 40 to 50 | 500 | Flat | 0.5 to 1 | 0.1 |
| | | | Rolling | 2 | 0.5 |
| | | | Hilly | 5 | 1 |
| General | 50 to 200 | 500 to 2000 | Flat | 1 | 0.1 |
| | | | Rolling | 2 to 5 | 0.5 to 1 |
| | | | Hilly | 5 to 10 | 1 to 2 |
| **Planning** | | | | | |
| Local | 100 to 200 | 1000 to 2000 | Flat | 2 to 5 | 0.5 to 1 |
| | | | Rolling | 10 to 20 | 2 to 5 |
| | | | Hilly | 20 to 50 | 5 to 10 |
| Regional | 500 to 1000 | 6000 to 10,000 | Flat | 2 to 5 | 0.5 to 1 |
| | | | Rolling | 20 to 50 | 5 to 10 |
| | | | Hilly | 10 to 50 | 1 to 10 |
| National | 10,000 to 100,000 (2 mi)   (20 mi) | 100,000 to 1,000,000 | | 10 to 1000 | 2 to 200 |

reasonable care, points can be plotted within perhaps $\frac{1}{100}$ in (0.3 mm). With a fine needle and a reading glass and with the use of great care, with the eye directly above the needle, points can be plotted within perhaps $\frac{1}{200}$ in (0.13 mm).

**13.13. Plotting by the coordinate method** The control point positions in horizontal control surveys for topographic maps (Chap. 15) are usually calculated and adjusted in the state plane coordinate system (Chap 14). Thus, data from field operations that have been calculated and adjusted in the office, known to be correct to within specified limits, are available for direct plotting by the coordinate method. This plotting can be performed with high precision, since all measurements are linear distances measured from orthogonal axes.

In order to plot by the coordinate method, a series of grid lines are drawn on the base sheet for the map. These grid lines are sets of $X$ and $Y$ axes of orthogonal coordinate systems. In the state plane coordinate system (and in most other coordinate systems), the $Y$ grid axis is aligned toward grid north. These grid lines are spaced at some regular, uniform interval (50, 100, 200, 500, 1000 ft or m) suitable for the scale of the map being compiled. When drawing the grid lines, extreme care must be exercised to ensure that the lines are straight and of uniform weight and that the $X$ grid lines are perpendicular to the $Y$ grid lines. Each $X$ and $Y$ grid line is labeled at the edges of the map sheet with its coordinate value. Grid lines should be drawn as quickly as possible on stable-base material under uniform temperature conditions so that all measurements are consistent. The positions of the control points are then plotted by laying off the differences between the coordinates for the point and the coordinates of a pair of intersecting grid lines close to the control point.

Figure 13.6 shows a set of coordinate grid lines for plotting the stations of the traverse calculated in Art. 8.16 and adjusted in Art. 8.19. For a small traverse such as this one, it is

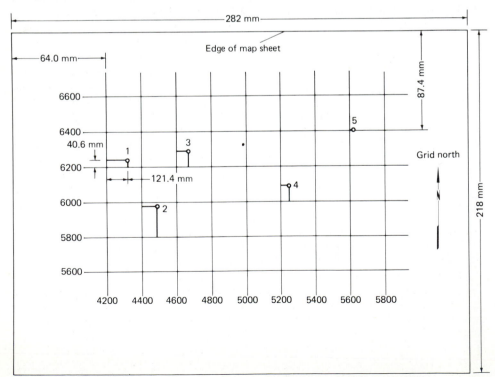

**Fig. 13.6** Traverse stations of Art. 8.16, Table 8.11, plotted by rectangular coordinates.

desirable to locate the grid lines so as to center the traverse approximately on the map sheet. This is done by examining the adjusted coordinates (Table 8.11), noting the maximum and minimum values for $X$ and $Y$ coordinates, and then calculating the maximum changes in $X$ and $Y$ coordinates. These distances reduced to map scale, subtracted from the dimensions of the map sheet and divided by 2, yield the distances that should be measured from the edges of the map sheet to the control points having maximum and minimum $X$ and $Y$ coordinates.

For the example in Fig. 13.6 and using coordinates from Table 8.11, the following data are available:

Map scale $= 1:10,000$
Map sheet dimensions are 282 by 218 mm (11 by $8\frac{1}{2}$ in)
Grid interval $= 200$ m on the ground or 20 mm at map scale
Maximum change in $X = 1297.84$ m on the ground or 129.8 mm at map scale
Maximum change in $Y = 422.34$ m on the ground or 42.2 mm at map scale
Easternmost station is 1, for which $X = 4321.4$ m
Northernmost station is 5, for which $Y = 6405.27$ m

With a grid interval of 200 m and in view of the values of coordinates for the easternmost and northernmost stations, it is necessary to determine the locations for the easternmost $Y$ grid line at $X = 4200$ m and of the northernmost $X$ grid line for $Y = 6400$ m. It is possible to work either at map scale or in metres on the ground. Assume that the best engineer's scale available for the work is graduated only in millimetres, so that working at map scale is more convenient. Since the change in $X$ exceeds that in $Y$, the $X$ axis is aligned with the long dimension of the map sheet. The location of the $Y$ grid line for $X = 4200$ m is determined from the left or western edge of the sheet. Since the maximum change in $X$ coordinates is 129.8 mm at map scale, the distance from the left edge of the sheet to the grid line is $(282 - 129.8)/2 + (4200 - 4321.4)/10 = 64.0$ mm at map scale. This distance is measured from the left edge of the map sheet and the $Y$ grid line for $X = 4200$ m is drawn parallel to the edge of the sheet. The location of the $X$ grid line for $Y = 6400$ m is determined from the top of the map sheet next. The maximum change in $Y$ is 42.2 mm at map scale, so that the distance from the top of the sheet to the $X$ grid line for $Y = 6400$ m is $(218 - 42.2)/2 + (6400 - 6405.3)/10 = 87.4$ mm at map scale. This distance is then scaled from the top of the map sheet (Fig. 13.6) and the $X$ grid line for $Y = 6400$ m is drawn through this point and perpendicular to the $Y$ grid line. The balance of the grid lines are then drawn parallel to these two controlling lines at intervals of 20 mm at map scale or 200 m on the ground. After the grid lines are labeled, the position of station 1 is plotted by subtracting $X = 4200$ and $Y = 6200$ from the $X$ and $Y$ coordinates for 1 to yield increments in the $X$ and $Y$ directions of 121.4 m and 40.6 m, respectively. These distances are scaled along the grid lines and perpendiculars are erected, the intersection of which defines the control point (see Fig. 13.6). The remaining stations are plotted in similar fashion. When all points have been plotted, the distances between points should be scaled as a check against the adjusted distances between the stations of the traverse. In this way the accuracy of the plotting is verified.

Figure 13.7 shows a similar plot of the traverse calculated in Art. 8.17, adjusted in Art. 8.19, and for which adjusted coordinates are given in Table 8.13. These coordinates are in feet, the scale of the plot is 1:4800 (1 in/200 ft), and the coordinate grid interval is 200 ft.

The construction lines drawn to plot the point as indicated in Fig. 13.6 are removed from the finished drawing as shown in the traverse plotted in Fig. 13.7. The grid lines are retained on the finished drawing or map or at the very least, the grid tick intersections are left. These grid lines or grid intersection points are invaluable for scaling the coordinates of points on the map. They can also be useful to evaluate dimensional changes in the map-base material and allow more realistic scaling of distances and locations from a given map sheet.

Rectangular coordinates can be plotted more rapidly and accurately by mechanical methods using a coordinatograph or flat-bed plotter. These devices consist of a flat, smooth

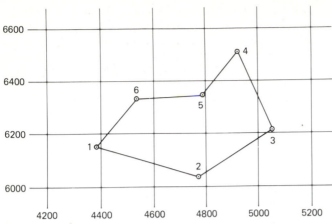

**Fig. 13.7** Traverse of Arts. 8.17 and 8.19, Table 8-13, plotted by rectangular coordinates.

drawing surface mounted on a stand or legs. A beam is then mounted along one edge of the drawing surface in a fixed position. Another beam, at right angles to the first beam, is mounted on wheels or a slide such that it can move back and forth across the entire drawing surface of the table. A chuck for holding a pencil, inking pen, scribing device, needle, or magnifier is mounted on the second beam in such a way that the chuck can be moved the entire length of the beam. In the most elementary form of coordinatograph, the movable beam and pencil chuck are driven by threaded screws turned by hand wheels at the end of each beam. Scales are mounted parallel to the beams, and the hand wheels used to turn the screw are also graduated, so that it is possible to plot grid lines at a specified interval or the positions of points referred to a given origin. The least count on most flat-bed plotters is from 0.001 to 0.004 in (0.025 to 0.10 mm). Figure 13.8 shows a modern flat-bed plotter.

Flat-bed plotters of the most recent design contain servomotors to drive the position of the movable beam and chuck. These plotters are under the control of an electronic computer. The coordinates of the points to be plotted are read into the computer from magnetic tape or punched cards or may be introduced manually using a keyboard. These systems are very adaptable to the automatic plotting of coordinate grid lines and map details

**Fig. 13.8** Flat-bed plotter. (*Wang Laboratories, Inc.*)

when the data are on a suitable input form, such as magnetic tape or punched cards, and a program for processing the data is available. The data would consist of the $X$ and $Y$ coordinates of points to be plotted plus the necessary identifying information for each point. The program is necessary to read the data into the computer and to convert the $X$ and $Y$ coordinates for each point into impulses that drive the plotter to the specified locations. Figure 13.9 shows a drum plotter with the associated computer unit. A plotter, which can be tilted for convenience and to save space, connected to a minicomputer with a cathode ray tube for visual display of program instructions and input is shown in Fig. 13.10.

The advantages of plotting horizontal control by rectangular coordinates are as follows: (1) adjusted coordinates are used to plot the locations so it is known that the data are correct to within the closure of the control survey network; (2) each point is plotted independently so that there is no cumulation of error; (3) high precision and uniform accuracy can be maintained on large map sheets and when multiple map sheets are required; and (4) the method is very adaptable to automatic plotting routines and is compatible with modern data-storage procedures.

**13.14. Other methods for plotting control**   When distances and directions are given between consecutive control points (e.g., in a traverse), it is possible to plot the relative positions of these points by setting off the direction of the line or the angle between two lines and then scaling the distance along the line from one point to the next. Thus, the basic procedure in these methods is that of plotting angles.

***The tangent method***   Here the angle is set off by a linear measurement from the previous line extended 10 or 20 units. The offset is a constant times the natural tangent of the angle.

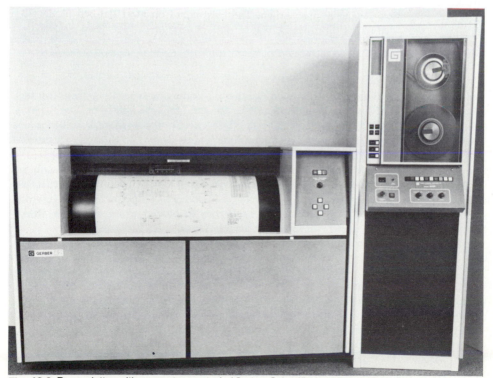

**Fig. 13.9** Drum plotter with computer control. (*Gerber Scientific Instrument Co.*)

**Fig. 13.10** Coordinate plotter, computer, and interactive terminal. (*California Computer Products, Inc.*)

***The chord method*** This method is much like that of plotting by tangents as just described, except that instead of erecting a perpendicular at the end of a base line 10 or 20 units long, an *arc* of 10 or 20 units radius is struck. The chord distance for the given angle is then scaled from the point of intersection between arc and base line to a point on the arc.

***The protractor and scale method*** The protractor is a device for laying off and measuring angles on drawings. The usual form for mapping consists of a full circle or semicircular arc of metal, plastic, or paper graduated in degrees or fractions of a degree. Some have radial scales by means of which a distance and an angle may be plotted at one operation. Others have one or more radial arms with scales.

Figure 13.11 illustrates the most common form of semicircular metal or celluloid protractor.

**13.15. Plotting the details** The work of plotting the details is less refined than that of plotting horizontal control, but the aim is still to plot objects of a definite size within the allowable error.

The method of plotting details is a function of the field procedure used for acquiring topographic details. If details were located by horizontal distance and direction (angle, bearing, or azimuth) from a given control point with a backsight on another control station, the directions can be set off with a protractor and the distances with a scale to position each

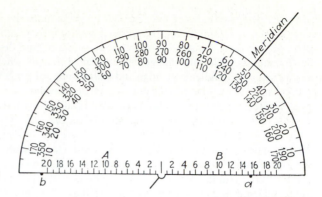

**Fig. 13.11** Protractor for plotting details.

individual point. Data of this type would be available when details are located in the field by stadia (Art. 15.10) or with one of the electronic positioning systems illustrated in Figs. 12.1 and 12.3.

When data are located in the field by the grid or checkerboard method (Art. 15.14) or cross profiles (Art. 15.12), the grid lines or profile center line and cross profile lines must be plotted on the map sheet in the correct relationship to the horizontal control points. Data are then plotted as recorded at the appropriate grid intersections or at the proper offset along the cross-profile lines.

If data are located by radial surveys from a control point using one of the electronic surveying systems (Figs. 12.1 and 12.3 to 12.6, Art. 12.2), the solid-state memory of the data storage unit represents the field record of all measurements to the points so located. When the job is completed, the storage unit is taken to the office and the data can be converted to coordinates by connecting the unit to a compatible electronic calculator. The output, consisting of $X$, $Y$, $Z$ coordinates for each point located, is stored on magnetic tape, disk storage, or punched cards. The output in this form can then be used as input to another electronic computer utilized to drive a drum or flat-bed plotter for automatic plotting by the coordinate method of all the details on the map sheet (Figs. 13.9 and 13.10). Naturally, such an approach requires compatible field and office equipment with the proper programs to process the field data and drive the plotter.

**13.16. Plotting the contours** When the horizontal control points and details have been plotted, the map sheet contains the planimetric details and an array of points for which elevations are known and are plotted on the map sheet. These elevations may be randomly located control points on the terrain or may be positioned on a regular grid or cross-section pattern. The next task in compiling the topographic map is to utilize this array of points in order to plot the contours.

Regardless of the number of ground points whose plotted locations are known, it is evident that any contour line must be drawn, to some degree, by estimation. This condition requires that the draftsman use considerable skill and judgment so that the contour lines may best represent the actual configuration of the ground surface.

Contour lines are shown for elevations that are multiples of the contour interval. They are drawn as fine smooth freehand lines of uniform width. Usually, each fifth contour line is made heavier than the rest, and sometimes these lines are drawn first to facilitate the location of intermediate contour lines.

Elevations of contours are indicated by numbers placed at appropriate intervals; usually, only the fifth or heavier contour lines are numbered. However, in areas of low relief where contours are spaced widely apart, each line may be numbered with its elevation. The line is broken to leave a space for the number. As far as possible, the numbers are faced so as to be read from one or two sides of the map; but on some maps the numbers are faced so that the top of the number is uphill. *Spot elevations* are shown by numbers at significant points, such as road intersections, bridges, water surfaces, summits, and depressions.

Since contours ordinarily change direction most sharply where they cross ridge and valley lines and since the gradients of ridge and valley lines are generally fairly uniform, these lines are important aids to the correct drawing of the contour lines. Special care is taken in the field to locate the ridge and valley lines, and usually these lines are drawn first on the map. Examples of such locations are shown plotted as the ridge line from *a* to *b* and the valley line from *c* to *d* in Fig. 13.12. The stream lines are drawn through those points which represent valleys, and the contour crossings are spaced along them before any attempt is made to interpolate or to draw the contour lines. This procedure aids in the interpretation of the data. For example, in the square bounded by the points *D*-5, *D*-6, *E*-6, and *E*-5, the contour lines are made to show the head of the valley, the existence of which is indicated only by the valley line previously drawn in the square below. In the figure shown, the ridge line is rather indefinite, but sketching the location at the ridge and spacing the contour lines across the ridge does help in drawing the rest of the contours.

**13.17. Drawing contours by interpolation**   The process of spacing the contour lines proportionally between plotted points is called *interpolation*. For example, in Fig. 13.12 the elevations of points *A*-2 and *B*-2 are 848.0 and 852.0 m, respectively. The contour interval for this map has been taken as 2 m, and the 848- and 852-m contour lines pass through the corresponding points. Under the assumption that the slope is uniform, the 850-m contour line passes through a point midway between *A*-2 and *B*-2. If a 1-m interval were used, the additional 849 and 851-m contour lines would be drawn through the quarter points of the line from *A*-2 to *B*-2.

The procedure is usually not so simple as in the case just cited. The elevations at the corners of the other squares in the figure are mostly of such values that the contour lines do not pass through them; further, on many maps the points of known elevation are spaced

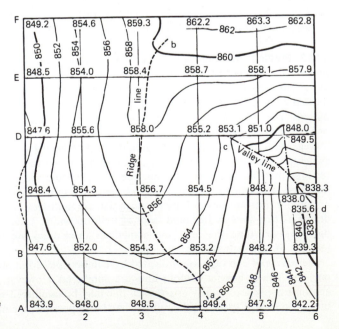

**Fig. 13.12** Contour lines by the grid method.

irregularly. Under such conditions, the interpolations may be made by estimation on the map, by arithmetical computations, or by graphical means, as follows:

**1. Estimation** Since each contour map is the result of more or less interpretation by the draftsman, in many cases it is not inconsistent with the other methods of map construction if the interpolation is made by careful estimation supplemented by approximate mental computations. This method is most commonly used on intermediate-scale and small-scale maps.

**2. Computation** Where considerable precision is desired in the map, the errors of estimation may be eliminated by simple arithmetical computations made with the aid of a hand-held calculator. For example, the elevations of points $E$-6 and $F$-6 (Fig. 13.12) are 857.9 and 862.8 m, respectively. The contour interval is 2 m; hence, the difference in elevation between $E$-6 and the 858-m contour is 0.1 m. Then, since the total difference in elevation is 4.9 m, the proportional part of the distance from $E$-6 to $F$-6 to locate the 858-m contour line is 0.1/4.9 of the map distance between these points. Similarly, the proportional parts for the 860- and 862-m contour lines are, respectively, 2.1/4.9 and 4.1/4.9 of the distance from $E$-6 to $F$-6. The computed map distances are plotted to scale.

**3. Graphical means** The computations indicated in the previous paragraph become laborious if many interpolations are to be made, and accordingly various means of graphical interpolation are in use. One method is illustrated in Fig. 13.13. Parallel lines are drawn at equal intervals on drafting film, each fifth or tenth line being made heavier than, or of a different color from, the rest and being numbered as shown. If it is desired to interpolate the position of, say, the 52- and 54-ft or m contours between $a$ with elevation of 50.7 and $b$ with elevation of 55.1, the line on the drafting film corresponding to 0.7 ft or m (scale at left end) is placed over $a$, and the tracing is turned about $a$ as a center until the line corresponding to 5.1 ft or m (scale at same end) covers $b$. The interpolated points are at the intersections of lines 2.0 and 4.0 (representing elevations 52 and 54) and the line $ab$, and may be pricked through the drafting film. Had the known points been much closer together, the scale at the right end of the tracing would have been used, and thus the value of each space would have been doubled; or if the scale were small, the contour interval large, and the topography rugged, each space might represent 1 ft or 0.2 m. Thus by assigning different values to the spaces, a single piece of drafting film so prepared can be made to suit a variety of conditions.

Another convenient graphical means of interpolation is the use of a rubber band graduated at equal intervals with lines that form a scale similar to that just described for the drafting film. The band is stretched between two plotted points so that these points fall at

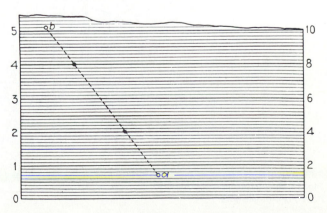

**Fig. 13.13** Graphical interpolation of contour lines.

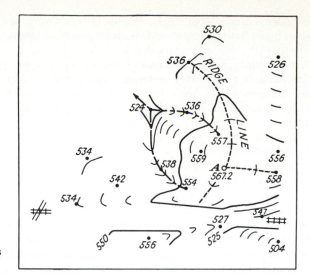

**Fig. 13.14** Interpolation of contours using controlling points.

scale divisions corresponding to their elevations. The intermediate contour points are then marked on the map.

Contours partially plotted by interpolation using data obtained by the controlling point method (Art. 15.9) are illustrated in Fig. 13.14. Figure 13.15 shows contours plotted by interpolation with data from cross profiles (Art. 15.12).

**13.18. Automatic plotting of contours from a digital terrain model**   Modern surveying and photogrammetric equipment enables rapid acquisition of planimetric position and elevation, all referred to a specified datum, for large numbers of points. This dense network of points which defines the configuration of the terrain is called a *digital terrain model* (DTM) or *digital elevation model* (DEM). If the points located in the field are sufficiently close and of large enough numbers, then any one of the methods for acquiring topographic details (Chap.

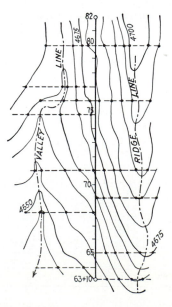

**Fig. 13.15** Interpolation of contours from cross-profile data.

15) yields a DTM. However, in order to be useful in a practical sense, the DTM data must be recorded in "real time" on a medium that is compatible with an electronic computer. Consequently, the DTM data that are usable for mapping are provided by the integrated electronic surveying systems (Art. 12.2) and digitized photogrammetric stereoplotters (Art. 16.16), all of which allow recording the data in real time on magnetic tape, core memory, or disk storage.

The most efficient procedure to utilize this enormous quantity of data (7000 to 10,000 points per photogrammetric stereomodel) is to use an electronic computer to process the data into a form that permits automatic plotting of contours on a flat-bed plotter (Art. 13.12). Some rather complex computer programs are required for this purpose. There are many such programs now available for reducing the DTM to contours. Reference 9 at the end of this chapter contains a number of papers on the subject. Consider a very brief explanation of one approach as an illustration of the concepts involved.

When contours are plotted manually, each contour is located by interpolation between points of known position and elevation (Art. 13.17). Interpolation is also used in the automatic plotting of contours. First, the data must be put into a format that allows interpolation by the computer. Most topographic detail, such as that located by an electronic surveying system consists of randomly located points on the terrain. To get these points into a usable format, a rectangular grid of a specified interval is mathematically interpolated from the area of the DTM. Next, a polynomial surface is fitted to the groups of points within each cell of the grid and elevations for each grid intersection are determined by interpolation with the computer. Output from this stage consists of elevations at all grid intersections of the superimposed rectangular grid. These data are stored in computer memory or written on magnetic tape. The next program reads the elevations at the grid intersections and locates the even unit contour by linear interpolation between the grid intersections. The $X$, $Y$ coordinates for each point on a given contour are then stored on magnetic tape and are identified by the contour elevation. This process is repeated for all cells in the grid and for all integral elevations at the desired contour interval over the range of elevations within the DTM. Output consists of a list of contour points, identified by the contour elevation and with the $X$, $Y$ coordinates for each point all recorded in core memory, on magnetic tape, or in disk storage. This list of contour points is then processed by another computer program which drives an automated plotter. Thus, the $X$, $Y$ coordinates for each point on a given contour are automatically plotted and are connected by either short segments of a straight line or by a third-order curve to yield a contour line. This is repeated for all contours occurring within the DTM, resulting in an automatically plotted set of contours.

Figure 13.16 illustrates two contour maps of the same area plotted at a scale of 1:2500 and with a contour interval of 2.5 m (adapted from Ref. 1). Figure 13.16a was plotted by manual interpolation using about 6100 points located by an electronic tacheometric survey with an instrument similar to the one illustrated in Fig. 12.6. The contour map shown in Fig. 13.16b is of the same area and was plotted automatically from a DTM consisting of the same 6100 points located by the electronic tacheometric survey. The maximum distance between points in the field survey was about 15 m. The interval for the superimposed grid for the DTM was also 15 m on the ground. The data were processed on a large electronic computer. Output from these computations was then used to drive an automatic plotter on which the contours were drawn. Obviously, the two maps agree quite well.

**13.19. Information shown on maps**    Maps are classified variously according to their specific use or type, but in general either they become a part of public records of land division or they form the basis of a study for the works of man. In general, the following information should appear on any map:

1. The direction of the meridian and basis for directions (astronomic, grid, magnetic, etc.).

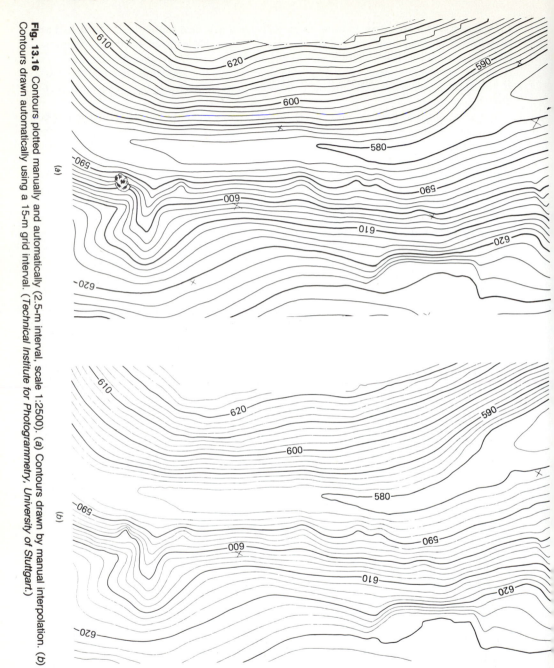

(a)

(b)

**Fig. 13.16** Contours plotted manually and automatically (2.5-m interval, scale 1:2500). (a) Contours drawn by manual interpolation. (b) Contours drawn automatically using a 15-m grid interval. (*Technical Institute for Photogrammetry, University of Stuttgart.*)

2. A graphical scale of the map with a corresponding note stating the scale at which the map was drawn.
3. A legend or key to symbols other than the common conventional signs.
4. An appropriate title.
5. On topographic maps, a statement of the contour interval (Art. 13.8).
6. A statement giving the datum to which horizontal and vertical control are referenced.
7. A statement giving the mapping projection.
8. A statement giving the coordinate systems for which grid lines or grid ticks are published on the map.

In addition, a map that is to become a part of a public record of land division should contain the following information:

1. The length of each line.
2. The bearing or azimuth of each line or the angle between intersecting lines.
3. The location of the tract with reference to established coordinate axes.
4. The number of each formal subdivision, such as section, block, or lot.
5. The location and kind of each monument set, with distances to reference marks.
6. The location and name of each road, stream, landmark, etc.
7. The names of all property owners, including owners of property adjacent to the tract mapped.
8. A full and continuous description of the boundaries of the tract by bearing and length of sides; and the area of the tract.
9. The witnessed signatures of those possessing title to the tract mapped; and if the tract is to be an addition to a town or city, a dedication of all streets and alleys to the use of the public.
10. A certification by the surveyor that the map is correct to the best of his or her knowledge.

Explanatory notes may be employed to give such information as the sources of data for the map and the precision of the survey.

**13.20. Scales**  The map scale should be indicated on the map sheet numerically and graphically. Numerical designations include the representative fraction, where one unit on the drawing represents a stated number of units on the ground (e.g., 1/25,000 or 1:25,000); or a statement that 1 in on the map represents some whole number of feet on the ground, such as 1 in/200 ft. The representative fraction is independent of units and is the method of designating scale that is utilized internationally.

The graphical scale is a line subdivided into map distances corresponding to convenient units of length on the ground. Various forms of graphical scales are shown in Fig. 13.17.

If distances are to be determined accurately from a map, a *graphical scale should always be shown*. The material on which the map is drawn changes dimensions over time, and often the scale of a map is distorted in the reproduction process. Since the graphical scale will change in the same ratio as the map, it provides a means of obtaining accurate distances by scaling from the map.

Choice of the proper scale for a given map is discussed in Art. 13.10.

**13.21. Meridian arrows**  The direction of the meridian is indicated by a needle or feathered arrow pointing north, of sufficient length to be transferred with reasonable accuracy to any

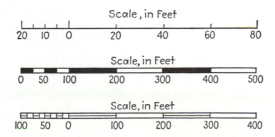

**Fig. 13.17** Graphical scales.

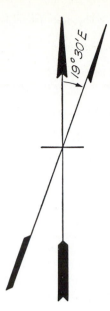

**Fig. 13.18** Meridian arrows.

part of the map. The true meridian is usually represented by an arrow with *full* head; the magnetic meridian by an arrow with *half* head. When both are shown, the angle between them should be indicated. The general tendency is to make needles and arrows too large, blunt, and heavy. A simple design is shown in Fig. 13.18.

Preferably, the top of a map should represent north, although the shape of the area covered or the direction of some principal feature of a project may make another orientation preferable.

**13.22. Lettering**    For *office drawings*, or drawings that are not to be used by the general public, the *Reinhardt* style of single-stroke freehand lettering is employed almost entirely in this country. The letters are constructed rapidly and are easy to read. Reinhardt letters are made either vertical (Fig. 13.19) or inclined (Fig. 13.20). Irregularities of lettering are not so apparent in the slope form as in the vertical form.

In general, letters should be drawn freehand but should be aligned by means of guidelines and slope lines. Commercial devices are available for quickly and uniformly constructing

ABCDEFGHI
JKLMNOPQR
STUVWXYZ
abcdefghijklmn
opqrstuvwxyz
1234567890

NORMAL
Excavation 23 cu. yd.

COMPRESSED
RICHARDSON ESTATE 300 Ac.

EXTENDED
RED RIVER

**Fig. 13.19** Reinhardt letters, vertical form.

ABCDEFGHIJ
KLMNOPQ
RSTUVWXYZ
abcdefghijklmn
opqrstuvwxyz
1234567890
NORMAL
Hickory Tree 10 ft.
COMPRESSED
WASHINGTON Sta. 71+43.8
EXTENDED
NEVADA

**Fig. 13.20** Reinhardt letters, slope form.

these lines. Complete lettering guides are increasing in use because letters of regular form can be made quickly; if the letters are spaced properly, satisfactory lettering can be secured with these guides. Freehand and mechanical lettering should not be used on the same drawing.

**13.23. Titles**    Titles should be so constructed that they will readily catch the eye. The best position for the title is the lower right-hand corner of the sheet, except where the shape of the map makes it advantageous to locate the title elsewhere. The space occupied by the title should be in proportion to the size of the map; the general tendency is to make the title too large. In general, each line should be centered, and the distance between lines should be such that the title as a whole will appear well balanced. The different parts should be weighted in order of their importance, beginning with the principal object of the drawing or the name of the area. Only the common styles of letters should be used. A change in the style of lettering between different parts is permissible to accentuate the important parts of the title, but slope letters and vertical letters should not be included in the same title. The title should include the type or purpose of the map, the name of the tract, location of the tract, the scale and contour interval of the map (unless shown elsewhere), name of the engineer and/or draftsman, and the date. A simple form of title is shown in Fig. 13.21.

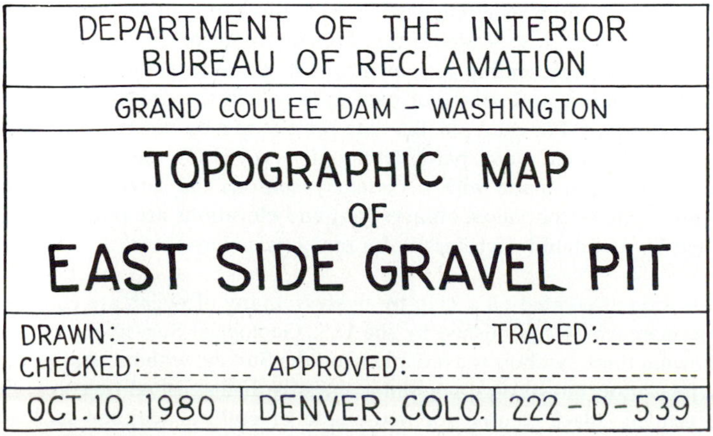

| DEPARTMENT OF THE INTERIOR BUREAU OF RECLAMATION | | |
|---|---|---|
| GRAND COULEE DAM – WASHINGTON | | |
| TOPOGRAPHIC MAP OF EAST SIDE GRAVEL PIT | | |
| DRAWN:_____ TRACED:_____ CHECKED:_____ APPROVED:_____ | | |
| OCT. 10, 1980 | DENVER, COLO. | 222 – D – 539 |

**Fig. 13.21** Title for a map.

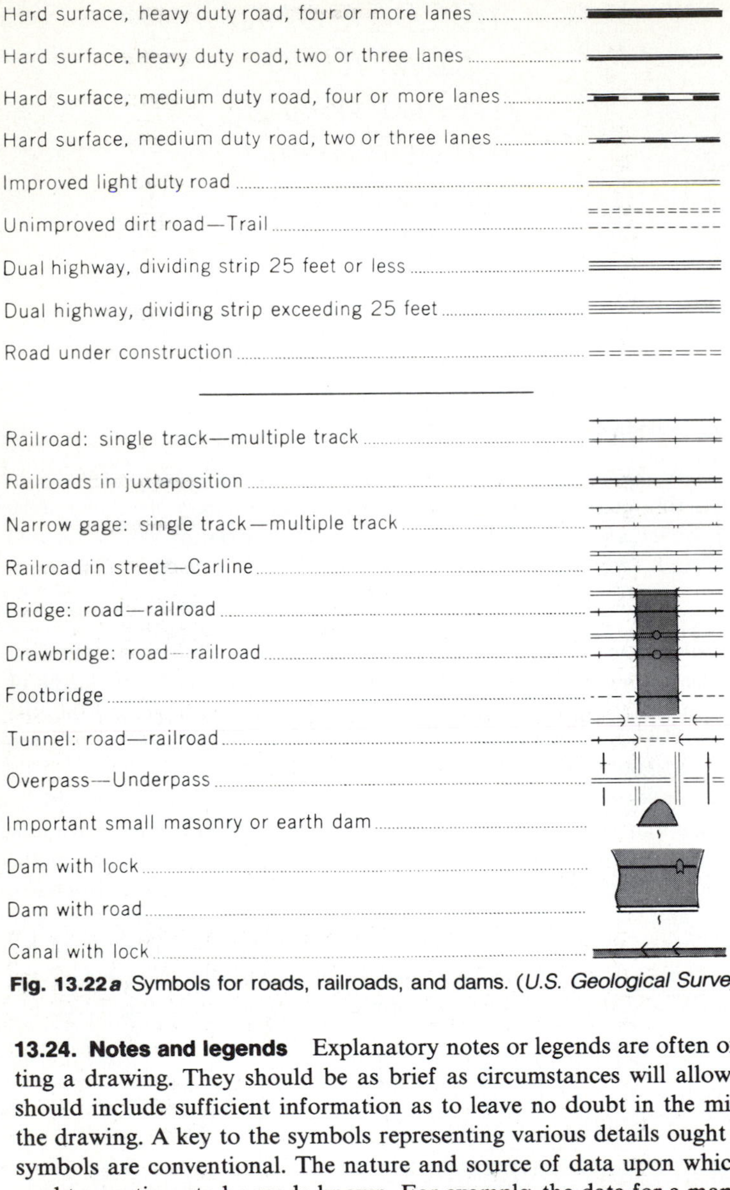

Hard surface, heavy duty road, four or more lanes ..........................

Hard surface, heavy duty road, two or three lanes ........................

Hard surface, medium duty road, four or more lanes.................

Hard surface, medium duty road, two or three lanes .............

Improved light duty road ..........................................................

Unimproved dirt road—Trail..................................................

Dual highway, dividing strip 25 feet or less ........................

Dual highway, dividing strip exceeding 25 feet ..................

Road under construction ..........................................................

Railroad: single track—multiple track ................................

Railroads in juxtaposition ........................................................

Narrow gage: single track—multiple track ........................

Railroad in street—Carline ....................................................

Bridge: road—railroad ............................................................

Drawbridge: road—railroad ....................................................

Footbridge ....................................................................................

Tunnel: road—railroad ............................................................

Overpass—Underpass ..............................................................

Important small masonry or earth dam ................................

Dam with lock..............................................................................

Dam with road ............................................................................

Canal with lock ..........................................................................

**Fig. 13.22a** Symbols for roads, railroads, and dams. (*U.S. Geological Survey.*)

**13.24. Notes and legends**  Explanatory notes or legends are often of assistance in interpreting a drawing. They should be as brief as circumstances will allow, but at the same time should include sufficient information as to leave no doubt in the mind of the person using the drawing. A key to the symbols representing various details ought to be shown unless the symbols are conventional. The nature and source of data upon which the drawing is based ought sometimes to be made known. For example, the data for a map may be obtained from several sources, perhaps partly from old maps, partly from old survey notes, and partly from new surveys; the surveys have been made with a certain precision; the direction of the meridian has been determined by astronomical observation; and elevations are referred to a certain datum as indicated by a certain bench mark of a previous survey.

**13.25. Symbols**  Objects are represented on a map by *symbols*, many of which are conventional. Some topographic map symbols published by the U.S. Geological Survey are shown in Fig. 13.22. A chart showing these symbols is available from the Survey, without charge.

Where feasible, the information shown on topographic maps is distinguished by colors in which the symbols are printed. Black is used for man-made or cultural features such as roads, buildings, names, and boundaries. Blue is used for water or hydrographic features,

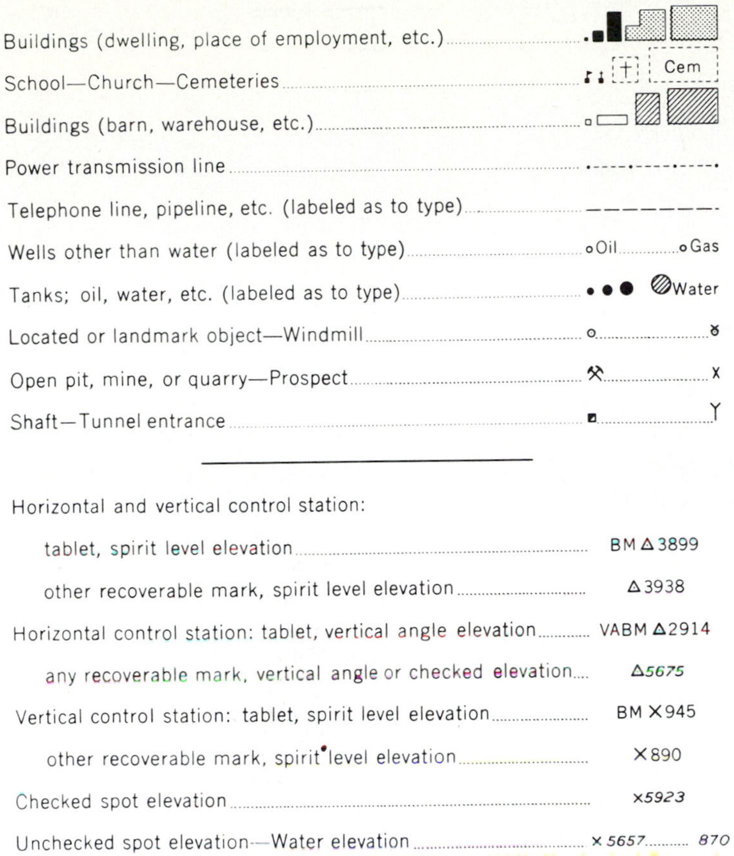

Buildings (dwelling, place of employment, etc.) ...........................

School—Church—Cemeteries ...................................................... Cem

Buildings (barn, warehouse, etc.) ................................................

Power transmission line ............................................................. ·----·----·----·

Telephone line, pipeline, etc. (labeled as to type) ................... ——————-

Wells other than water (labeled as to type) .................... ○Oil ............ ○Gas

Tanks; oil, water, etc. (labeled as to type) .................... • • ● ⊘Water

Located or landmark object—Windmill ............................ ○ ............ ⌇

Open pit, mine, or quarry—Prospect ............................... ⚒ ............ ⤬

Shaft—Tunnel entrance ................................................ ▪ ............ Y

Horizontal and vertical control station:

    tablet, spirit level elevation ........................................ BM △ 3899

    other recoverable mark, spirit level elevation ................. △ 3938

Horizontal control station: tablet, vertical angle elevation .......... VABM △2914

    any recoverable mark, vertical angle or checked elevation .... △5675

Vertical control station: tablet, spirit level elevation .................. BM ✕945

    other recoverable mark, spirit level elevation .................... ✕ 890

Checked spot elevation ............................................................. ✕5923

Unchecked spot elevation—Water elevation ......................... ✕ 5657 ......... 870

**Fig. 13.22b** Symbols for structures and stations. (*U.S. Geological Survey.*)

such as lakes, rivers, canals, and glaciers. Brown is used to show the relief or configuration of the ground surface as portrayed by contours or hachures. Green is used for wooded or other vegetative cover, with typical patterns to show such features as scrub, vineyards, or orchards. Red emphasizes important roads and public-land subdivision lines and shows built-up urban areas. Where colors are not available or where the map is to be reproduced by contact or photographic printing and will therefore be in one color, black is used throughout; the conventional shapes of the symbols are the same as where color is used.

**13.26. Drawing symbols**  The methods described in this section are illustrated in Fig. 13.23.

**Grass**  The symbol for grass consists of a series of lines drawn radially toward a point.

**Marsh**  The symbol for marsh consists of the grass symbol beneath which the water surface is shown by either a single or a double line drawn slightly longer than the base of the grass tuft.

**Trees**  Tree symbols may be drawn either in plan or in elevation. The latter practice is better adapted to reconnaissance sketches or elevation drawings of a terrain, whereas on most topographic drawings the symbol shown in plan is more suitable.

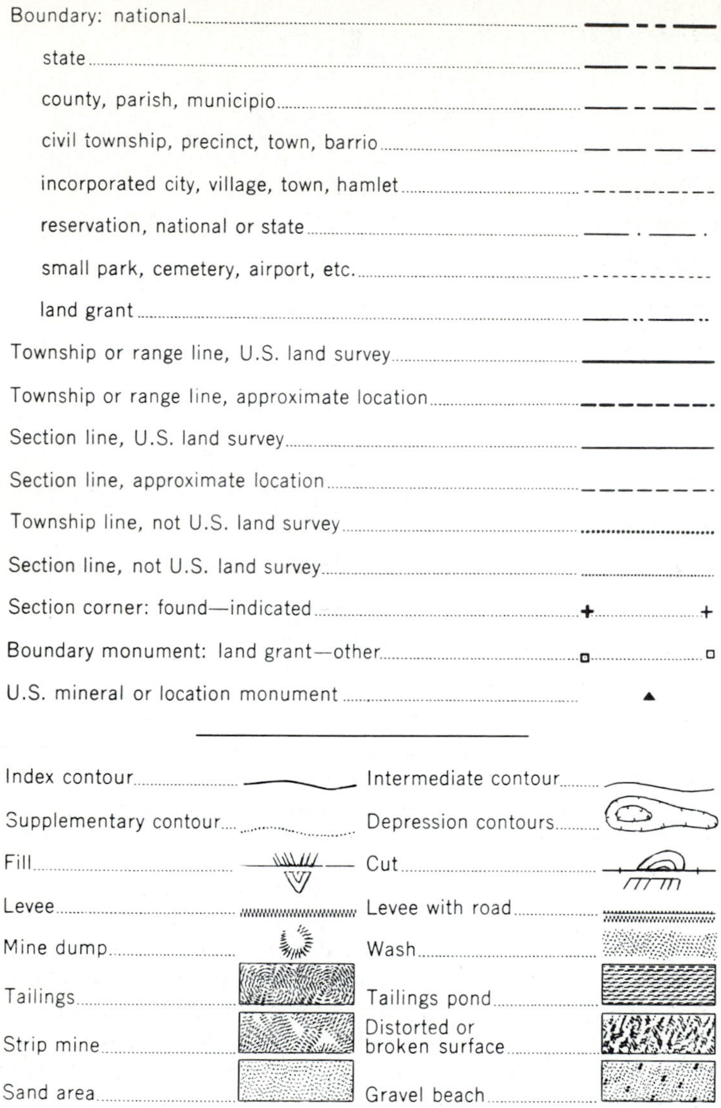

Boundary: national ............................................................... ▬ ▬ ·· ▬

    state ............................................................... ▬ ▬ ·· ▬

    county, parish, municipio ..................................... ▬ ▬ · ▬ ·

    civil township, precinct, town, barrio ..................... ▬ ▬ ▬ ▬

    incorporated city, village, town, hamlet ............... ·▬·▬·▬·▬·

    reservation, national or state ............................... ▬ · ▬ ·

    small park, cemetery, airport, etc. ....................... ------------

    land grant ............................................................... ▬ ·· ▬ ··

Township or range line, U.S. land survey ..................... ▬▬▬▬▬

Township or range line, approximate location .............. ▬ ▬ ▬ ▬ ▬

Section line, U.S. land survey ....................................... ▬▬▬▬▬

Section line, approximate location ................................ ▬ ▬ ▬ ▬ ▬

Township line, not U.S. land survey ............................. ·····················

Section line, not U.S. land survey ................................ ···················

Section corner: found—indicated ................................. +                    +

Boundary monument: land grant—other ....................... ▫                    ▫

U.S. mineral or location monument .............................. ▲

---

Index contour ............... ▬▬▬  Intermediate contour ......... ⌒⌒

Supplementary contour .... ⌇⌇  Depression contours ..........

Fill .............................. ⩗⩗  Cut ...............................

Levee .......................... ⣿⣿⣿  Levee with road ............... ⣿⣿⣿

Mine dump .................. ✸  Wash ............................

Tailings ......................  Tailings pond .................

Strip mine ...................  Distorted or broken surface ...

Sand area ....................  Gravel beach .................

**Fig. 13.22 c** Symbols for boundaries, contours, and excavations. (*U.S. Geological Survey.*)

**Water lines**  To draw the water-line symbol for lakes and ponds, the draftsman begins by sketching the shoreline as a heavy line, then drawing a fine line as close as possible to the shoreline. Fine lines are then drawn successively, each one in close conformity with the preceding line, and the spacing between the lines is increased uniformly outward from the shore.

**Contour lines**  Details concerning the plotting of contours can be found in Art. 13.16.

**13.27. Drafting materials**  Drafting film, tracing papers, and drafting papers constitute the most commonly used drafting materials for drawings and tracings made in engineering offices. As used here, the term "drawing" refers to the original rendition of a map or plan in pencil or ink on opaque or transparent material, such as a base map or map manuscript. A

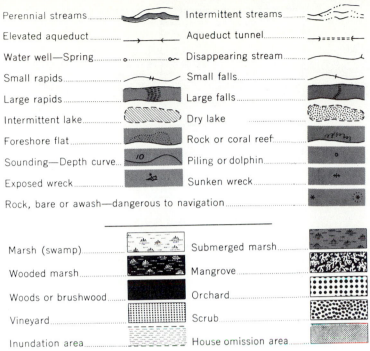

**Fig. 13.22d** Symbols for hydrography and land classification. (*U.S. Geological Survey.*)

tracing is a drawing in ink or pencil on a transparent medium made for the purpose of reproduction. A tracing is generally made by placing the transparent medium over the original drawing (frequently positioned on a light table) and tracing the line work and detail on the transparent material.

Polyester drafting film is the most commonly used material for drawings and tracings because of its high transparency, dimensional stability, tearing strength, and resistance to heat and age. It is insoluble and waterproof. It takes either pencil or ink work, both of which can be erased easily and clearly. It can be coated with semiopaque material of various colors for scribing, a process of forming lines with a sharp-pointed instrument which scratches through the coating. Use of coated polyester film and scribing means that the drawing or tracing is a negative rather than a positive, a feature that can be exploited in the reproduction process (Art. 13.35). The points used in scribing are similar to phonograph needles, with steel or sapphire tips, and are available in various widths; they are fastened to handles similar to penholders or to specially designed scribing tools.

*Tracing paper* comes in several grades, all suitable for pencil drawings. The better grades, usually processed, are also suitable for ink drawings. Ordinary tracing papers will not stand repeated erasures well and will become torn and cracked unless they are handled carefully, but improved tracing papers and papers with a high degree of transparency are available.

Pencil drawings and temporary drawings are often made on a smooth manila *detail paper*, of which there are several grades and weights. For general map work a fairly smooth, tough *drawing paper* of uniform texture is desirable. The paper should take ink well and should stand erasures without its surface becoming fibrous. For permanent drawings a paper should be chosen that will not discolor or become brittle with age.

In consideration of the importance of the map and of the slight expense of the paper in relation to the survey as a whole, it may be considered false economy to use any but high-grade papers for mapping purposes.

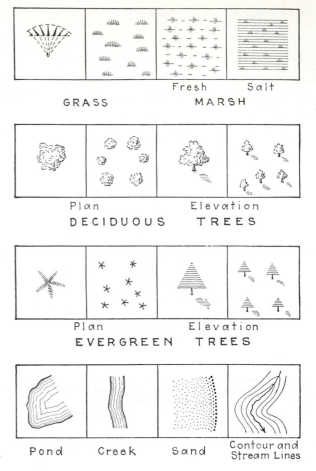

**Fig. 13.23** Details of map symbols.

**13.28. Manuscript preparation** Following the procedures suggested in the preceding sections, the original map manuscript is compiled and usually consists of a drawing in pencil on a worksheet of high-quality drafting film. This worksheet should contain an accurate representation of all the details to be shown, preferably at the final map scale.

When the map consists of a single sheet such as a topographic map for a small subdivision plan or the plan for a property survey, the final map is usually prepared by making an ink tracing on drafting film or a scribed tracing on coated film. If an inked tracing is to be prepared, the translucent stable base film is placed over the worksheet on a light table and all details are traced in ink.

If the final map is to be scribed, the sheet of drafting film is first covered on one side by a uniform thickness of an actinically opaque† coating. This sheet is then placed over the original map manuscript on a light table and a tracing is prepared by scribing, a process in which the coating is removed by scribing tools. This hand-scribed tracing is the reverse of an inked tracing; that is, the lines are the consequence of removing opaque material, whereas on an inked tracing the opaque ink is added to the film to create the lines. Thus, the scribed tracing provides a negative rather than a positive image. This process of hand scribing from the worksheet is usually not done in practice since tracing an image through coated film is difficult due to the diffusion of light. Instead, the line work from the original manuscript (a

_____
†Coating which is opaque to those wavelengths of light by which light-sensitive emulsions are most affected.

positive image on transparent material) is transferred to the coated film, which is treated with a photosensitive material. In this way, the lines to be scribed are reproduced directly on the coated surface so that the process of scribing is greatly simplified. The actual process of scribing is still performed on a light table, since this seems to reduce the amount of eyestrain involved.

The finished tracing produced either by inking or scribing shows all details required, is in one color, and represents the master copy of the map from which reproductions can be made (Art. 13.29).

When maps are produced photogrammetrically, there are original manuscripts or work sheets from the photogrammetric process and from field completion surveys. Thus, details from two work sheets must be combined to create the final map. Also when maps are prepared for publication in color (usually five colors: black, brown, blue, green, and red), color separation sheets are required. Another factor to be considered is that multiple map sheets are frequently needed to cover large areas. Map details at edges common to adjacent map sheets must correspond. Accurate *manuscript registration* is necessary to ensure that details on these various working sheets match properly.

As an example of manuscript registration, consider a topographic map for which nine map sheets are required. The approximate arrangement of the map sheets is illustrated in Fig. 13.24. In order to register and join all map sheets, finely drawn registration crosses are drawn first on the base map sheet so as to appear in each corner of the individual map sheets. These registration marks are then traced on the overlying individual map sheets. The central map sheet, labeled 5, is shown in Fig. 13.24 with registration crosses placed in the margins outside the area of mapping on the sheet. On completing the tracing of sheet 5 and prior to removing it from the table, sheets 1, 2, 3, 4, 6, 7, 8, and 9 are placed on the table so as to overlap sheet 5 by the required amount and registration marks are transferred to the margins of all overlapping sheets. At the same time, details from the completed sheet 5 are drawn on the margins of the yet to be traced sheets so as to ensure a good match between all adjacent manuscript sheets. This last step is called *joining* the edges of the manuscripts.

A more precise method of registration consists of using machined studs of a uniform diameter placed through holes punched in the margins of the manuscript with an ordinary paper punch. First, the stud is placed through a hole punched in a small rectangular piece of heavy drafting film. Holes are then punched in the margins of the manuscript to be traced and in all overlapping sheets. Next, all sheets are removed except the one to be traced, and the studs, previously placed through small pieces of film, are inserted through the holes (stud end up) punched in the margins of the manuscript so that the piece of plastic protrudes beyond the edge of the sheet (Fig. 13.25). The map sheet is placed in the desired location over the original manuscript to be traced, and the projecting edges of the pieces of plastic

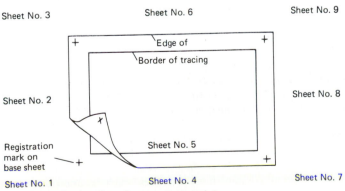

Sheet No. 3    Sheet No. 6    Sheet No. 9

Edge of
Border of tracing

Sheet No. 2    Sheet No. 8

Registration
mark on
base sheet

Sheet No. 5

Sheet No. 1    Sheet No. 4    Sheet No. 7

**Fig. 13.24** Registration of a map manuscript.

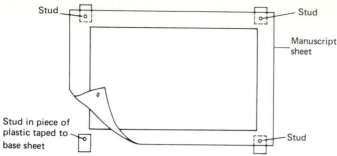

**Fig. 13.25** Registration by using studs taped to the base sheet.

that hold the studs are taped securely to the base sheet. These studs remain throughout the tracing operation. The map manuscript may be removed and replaced without loss of the original registration. When tracings are being prepared for subsequent photographic reproduction, registration crosses should be retained in the corners of the manuscript even though studs are also used. These crosses, reproduced with the drawing, are invaluable for registration of various overlays after reproduction has taken place.

**Stripping**   Areas of solid color such as are used to show water, woodland, and urban regions on multicolor maps are generally reproduced photomechanically by *stripping*. The outline of the areas to be stripped is drawn in black on drafting film or a positive print is made from a scribed negative of these outlines. Then the lines that delineate the desired area are photochemically etched by contact-printing the positive to a sheet of transparent film coated with a specially prepared coating. Using the etched lines as a guide, the coating that covers the desired area is peeled from the film with a sharp blade, leaving an open transparent area surrounded by the actinically opaque coating. This area is referred to as an *open window* through which light can pass allowing preparation of plates for subsequent map reproduction, as described in Art. 13.35.

**Stickup**   Lettering and certain standard symbols can be printed on thin plastic sheets that have a thin coating of adhesive (usually wax) on the reverse side. These letters and symbols, so prepared, are called *stickup*, and when placed in the correct position on the map sheet with the adhesive side down, can be rubbed or burnished so as to make them adhere to the manuscript.

**Color separation**   Multicolor maps are usually printed in five colors, as indicated in Art. 13.25 and illustrated by Fig. 13.22 (see endpapers). Each color requires a separate *color separation guide* to be used in preparation of plates for the reproduction of the map (see Art. 13.35). Scribed sheets, film negatives, or open window sheets may be used for color separation guides. Precise registration is extremely important so that all guides should be made using images reproduced from the compilation manuscript. The major difference between the compilation manuscript and color separation guides is that the manuscript is *right-reading*: that is, all lettering and details appear in the usual left-to-right orientation. On the other hand, the guide is used to prepare plates employed in the map printing process (Art. 13.35). Thus, the guide must be *wrong-reading*, with left and right reversed.

**13.29. Reproduction of drawings and maps**   There are so many different processes and combinations of various methods for reproduction that one can quickly become confused when confronted by the entire array of procedures. In this chapter these procedures are classified into two groups. The first group contains those processes which are appropriate for producing a limited number of copies. The equipment required can range from simple inexpensive devices to rather sophisticated systems, but in general the initial investment is relatively low. The unit cost is low and does not change as the number of copies increases. The equipment is such that it could be operated and maintained in an engineering office. These processes include (1) direct-contact negative prints, (2) direct-contact positive prints, (3) photocopying, (4) film photographs, and (5) xerography.

When a large number of copies is required, some sort of printing process that utilizes plates and ink is used. By these methods the unit costs for copies decrease as the number of copies increases. The equipment required is heavy and requires a substantial initial investment. Printing processes include letterpress and lithography, which are used by most commercial and government agencies for the production of maps reproduced for wide distribution or sale to the public.

First consider the first group of methods for reproducing a limited number of copies.

**13.30. Direct-contact negative prints**  The most common form of direct-contact negative copying at the same scale as the original is the *blueprint*. In this procedure a positive image consisting of black inked or penciled lines on a sheet of transparent film or paper is brought in contact with a sheet of iron-sensitized paper. The side containing the sensitized paper is exposed to a special light (sunlight does the job) for a specified period of time and the exposed sheet is then developed in a bath of water. This procedure produces a white line on a blue background or a *right-reading* negative. "Right-reading negative" means that all images are in the correct left-to-right orientation but that light and dark are interchanged from the original. To produce a blueline print on a white background, the same procedure is employed except that a negative original is used. Thus, a drawing scribed on coated polyester film (Art. 13.27) is a negative from which a blueline print could be made by this process. Blueprints can also be made from negatives produced from the original, as described later.

Alterations on blueprints can be made with a weak solution of caustic soda. This is used as an ink to produce white lines. If a colored line is desired, the solution may be mixed with ink. Several solutions of this nature are on the market in bottled form and are known as erasing fluids. Special inks for alteration of blueprints are available in white, red, and yellow.

If the original consists of a blackline tracing and a blueline or positive reproduction is desired, it is necessary to make a special negative from the original. This is done by placing the positive inked or penciled surface in contact with a special sensitized paper and exposing the sensitized paper to light. The exposed paper is then wet-processed to yield white lines on a dark brown background or a negative copy. Blueline positive prints are then made by placing the brown surface of the negative next to the sensitized sheet of blueprint paper and by exposing and washing in the usual manner.

A blackline contact print on transparent paper, called a *blackline tracing*, is made from a negative produced by a process similar to that for blueline prints. Blackline tracings shrink during the process of manufacture, and the scale is thereby altered appreciably. However, they are economical and are useful for preliminary working drawings from which prints are to be obtained.

A *direct blackline print* is made from the tracing by contact printing in sunlight or artificial light, using a special sensitized paper. The print is developed by applying a chemical furnished by the manufacturer, after which it is thoroughly washed in water and is dried. This type of print has the advantages of printing without a negative, a white background, freedom from excessive shrinkage, and difficulty of alteration without detection.

Prints with a white background are clearer than those with a dark background, and additional notations stand out well. As in the case of blueprints, alterations are evident because erasure of the lines damages the paper.

**13.31. Direct-contact positive prints**  In this process the original drawing on a transparent medium is placed in contact with a special paper sensitized with *diazo* compounds and the sensitized paper is exposed to light. The exposed sensitized sheet is then developed by a dry process where it is placed in a tight container filled with ammonia vapor. The exposed sheet can also be developed in a wet process, in which the sheet is passed through a liquid ammonia solution. Red, blue, brown, or blackline reproductions of the same scale as the original can be made by this process. The image is not absolutely permanent, particularly when exposed to sunlight and the paper has rather poor drafting qualities. *Ozalid* is the trade name for this process.

**13.32. Photocopying**  A drawing on any kind of paper may be reproduced to any scale by the photocopy process, provided that the lines are of a color which photographs well. The process is widely used, especially in the reproduction of pages from books.

The photostat machine is a modified form of camera. The drawing is strongly illuminated by artificial light, and a negative to the desired scale is made, in which black lines of the original appear white. By rephotographing, a positive is produced in which black lines of the original appear black on a white or gray background.

Large reproductions are considerably distorted near the edges, but in the usual sizes the distortion is not great. Reproductions up to 40 by 60 in (100 by 150 cm) may be made by this process.

Photocopy reproduction of a blueprint may be made in one operation, using the blueprint as the negative.

*Photostat* is a trade name that refers to photocopies made by a particular system and paper.

**13.33. Film photography**  This procedure consists of using a camera to make an exposure of the original drawing or map on photographic film. This exposure is then developed into a negative from which prints can be made.

The system consists of a copy camera mounted at one end of a slide and a movable copy board to support the original drawing mounted at the other end of the slide. The camera has a large format and a high-quality lens that can be moved with respect to the plane of the film in the camera (focal plane). The film is held securely against the camera focal plane by applying a vacuum. The original is held tightly against the copy board by applying a vacuum or by using a glass plate. The copy board can be moved with respect to the camera and the lens of the camera with respect to the focal plane so as to produce the desired enlargement or reduction.

Paper prints, film prints, or film negatives can be provided by this process. Prints made on film have the advantages that the copy is on a stable base material that has an excellent drafting surface.

**13.34. Xerography**  Xerography is a dry reproduction process that utilizes photoconductivity and surface electrification to yield copies on nontreated paper. An aluminum drum coated with selenium is given a positive charge which is dissipated when exposed to light. The image of the original to be copied is transmitted to the surface of the drum by an optical train. The light portions of the image reflected to the drum remove the charge and the dark lettering and detail leave a latent image on the surface of the drum which is preserved by a negatively charged powder. Positively charged paper is then brought into contact with the drum, thus receiving the image that is fixed in place by application of heat.

No chemicals or solutions are required, so that excellent copies can be produced in a few seconds on many kinds of paper and on film. When the proper type of lens is introduced into the optical train, enlarged or reduced copies can be made. Reproduction in color on paper and on film is now possible using some of the more recently designed machines.

**13.35. Map printing processes**  Letterpress and lithographic printing are the procedures by which most published maps are reproduced. Both methods require the same basic steps: (1) photographing the original, (2) developing the exposed film to make a negative, (3) preparing a plate from the negative, and (4) using the plates in the printing press to make copies of the original. A brief review of these steps follows. A complete explanation of the details involved is beyond the scope of this book and the interested reader should consult Refs. 5 and 11.

Photography of the original is performed using a copy camera system similar to the arrangement described in Art. 13.33. Great care must be exercised at this point so as to obtain a properly registered, exposed, and focused image on the correct type of film. Reductions in the scale of the images are made during this step.

The exposed film is then processed into a negative. The negative is inspected on a light table and all blemishes in the emulsion are repaired. At this stage it is also possible to add or delete details and to make certain corrections.

When a map has been compiled by scribing on drafting film with a soft opaque coating (Arts. 13.27 and 13.28), the original is a negative and the photographic step is not required.

Up to this point, the reproduction procedure by letterpress and by lithography is the same. The procedures differ in the preparation of the plates and in the way the plates are used in the printing press.

Letterpress plates are made by exposing the sensitized plate to the negative. Light passing through the light portions of the negative hardens the emulsion on the plate. The plate is then developed in an acid bath in which the nonimage areas on the plate are eaten away. Thus, the image areas of the plate stand out in relief. This letterpress plate is put in the press and images are transmitted directly to the paper by the inked, raised images on the plate. Consequently, images on the letterpress plate must be in a reversed or *wrong-reading* position so as to produce right-reading images on the printed map.

Lithographic procedures are the most common methods currently being used for duplicating maps; hence, the lithographic plate is of more interest to the map maker. On lithographic plates the images hold the greasy ink while nonimage areas repel the ink. Image and nonimage areas on the plate are produced by placing the emulsion side of the negative tightly against the sensitized plate and exposing this surface to light. Light transmitted through the light portions of the negative hardens the emulsion in the image areas. The nonimage areas are then washed away in the developing process. The image areas retain the greasy ink while nonimage areas attract water and repel the ink. Different plates are prepared in this way for the various colors on the map, one for each color. Differences in shading may be produced in making the plates by using screens. Screens are meshlike line grids constructed so as to block specific percentages of light, thus producing dotted patterns of color and giving the effect of varying shades.

The lithographic press incorporates an offset arrangement. The inked lithographic plate wrapped around a cylinder or roller can be placed in contact with a rubber blanket around an adjacent roller. In this way, ink on the right-reading images on the plate is transferred to the blanket, where a wrong reading or reversed image is created. Paper passes between the roller containing the blanket and another roller so that the images on the blanket are transferred to the paper as right-reading images. This overall procedure can be described as reproduction using photolithography by *offset printing*.

When multicolored maps are being reproduced, several runs through the same press are required to produce the completed map. Some presses have several printing systems in tandem so that all colors can be applied in sequence during a single run. In either case, it is evident that all plates must be very carefully registered as discussed in Art. 13.28, so that all details and colors appear in the proper locations. This registration is an extremely important aspect of the entire map compilation and reproduction process, all the way from the map manuscript to the production and use of plates in the printing process. Great care must be exercised at all times to maintain proper registration of all materials at all stages of the map compilation and reproduction procedure.

**13.36. Automated mapping** The possibility of automated mapping is created when data from modern ground surveys, existing survey records and maps, and photogrammetric mapping procedures are pooled in a data storage bank under the control of an electronic computer that drives an automated flat-bed plotter.

In Art. 12.2, an electronic surveying system is described which stores the $X$, $Y$ coordinates and elevations of located points in a solid-state device that can later be connected to a computer and read onto magnetic tape, disk storage, or into core memory. Existing maps, photogrammetric stereomodels (Art. 16.19), and the scanned elements of orthophotos (Art. 16.20) can be digitized and stored on tape, disk, or in core memory. These data constitute a digital terrain model (DTM) (Arts. 13.18 and 16.19).

Given a system consisting of an electronic computer interfaced with a flat-bed plotter, the DTM stored in peripheral storage or core memory can be automatically converted into a topographic map. The automatic plotting of contours is discussed in Art. 13.18. Planimetric details can also be plotted from stored $X$, $Y$ coordinates of specified objects. Naturally, an extensive library of computer programs or *software* compatible with the electronic computer system is required for processing and plotting the data. A sufficient number of known ground control points must also be available. Locations for these control points can be established by ground control surveys, analytical photogrammetric triangulation (Art. 16.18), and/or by one of the electronic positioning systems (Chap. 12).

If only a few copies of the map are needed for design and planning purposes, the final product can be a traditional line map with contours, or an orthophoto map with an overlay of contours. All legends, grid lines, and title blocks can be automatically drawn. When the map is for wide distribution, color separation sheets can be individually compiled automatically. These sheets are then used to prepare plates for map printing as described in Art. 13.35.

Map data can be stored in the data bank such that they can be retrieved selectively. Thus, in the foregoing example or preparation of color separation sheets, one could select to plot planimetry, hypsography (contours), forest cover, and hydrography, each separately. On most mapping systems, the capability of selective compilation is controlled by an interactive terminal consisting of a cathode ray tube and keyboard, which gives the operator the option of intervening and controlling the mapping procedure (see Fig. 13.10). This feature not only permits selective plotting but also means that data can be input manually or stored data may be reviewed, edited, and re-stored for compilation. Given the proper programs, calculations with the stored data are also possible. Thus, map revision can be a major function of the system.

An automatically compiled general planning map consisting of an orthophoto map with an overlay of contours is shown in Fig. 13.26. Figure 13.27 illustrates a portion of a subdivision development map that was calculated and plotted using an interactive terminal interfaced with a minicomputer and a flat-bed plotter.

A useful by-product of the DTM which can be produced by an automatic mapping system is the capability of compiling three-dimensional perspective terrain models from a specified vantage point in space. Figure 13.28 illustrates a perspective terrain model automatically plotted from a DTM. Design plans can be incorporated in these models thus providing views from as many points as are necessary of the finished project. Planning group members generally find such a display more comprehensible and useful than topographic maps with design plans included.

**13.37. Tests for accuracy of maps** A topographic map can be tested for accuracy, both in plan and in elevation. In this discussion it is assumed that the errors in field measurement

**Fig. 13.26** Automatically compiled map (orthophoto map with contours). (*Courtesy Riverside County, California, Flood Control District.*)

may be disregarded and that a graphical scale is provided on the map to render negligible any effect of shrinkage of the paper.

*1. Horizontal dimensions*   The test for horizontal dimensions consists of comparing distances scaled from the map to distances measured on the ground between the corresponding points. The precision with which distances may be scaled from a map depends upon the scale of the map and the size of the plotting error. Thus, if for a map scale of 1:1200 or 1 in/100 ft it is known that the error in location of any one point with respect to any other on the map is 1/50 in (0.5 mm), the error represents 2.0 ft or 0.6 m on the ground.

Some surveys are made for the purposes of estimating areas, such as, for example, of a reservoir site. The errors in areas scaled from such maps can readily be determined from a consideration of the errors in the scaled distances and by using error propagation. For example, assume that a rectangular area on a map is 20 cm (7.87 in) by 60 cm (23.62 in), that the map scale is 1:4800 (1 in/400 ft), and the estimated standard deviation in scaling the map distances is 0.08 cm (0.03 in). The area is the product of the length $\ell$ times the width $w$ or

$$A = \ell w$$

and by error propagation Eq. (2.46), the propagated standard deviation in area is (assuming that $\ell$ and $w$ are uncorrelated)

$$\sigma_A = \left[ \left( \frac{\partial A}{\partial \ell} \right)^2 \sigma_\ell^2 + \left( \frac{\partial A}{\partial w} \right)^2 \sigma_w^2 \right]^{1/2}$$

or for the data in the example

$$\sigma_A = \left[ (20 \text{ cm})^2 (0.08 \text{ cm})^2 + (60 \text{ cm})^2 (0.08 \text{ cm})^2 \right]^{1/2}$$

$$= 5.1 \text{ cm}^2 \quad \text{at map scale}$$

or about 0.4 percent of the area.

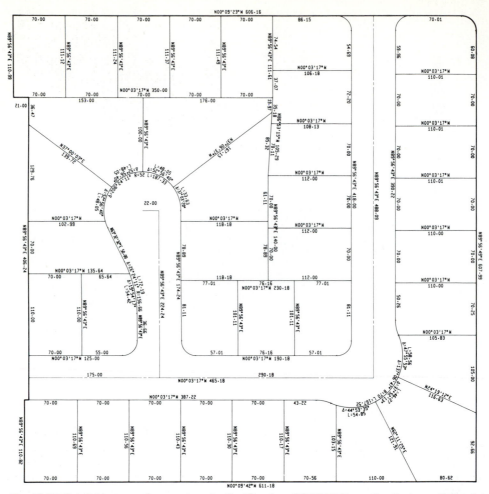

**Fig. 13.27** Subdivision map drawn automatically. (*Courtesy CONCAP Computing Systems, Oakland, California.*)

**2. Elevations** One test for elevations consists of comparing, for selected points, the elevations determined by field levels and the corresponding elevations taken from the map. Usually, the points are taken at 100-ft or 30-m stations along traverse lines crossing typical features of the terrain.

A more searching test is to plot selected profiles of the ground surface as determined by the field levels and the corresponding profiles taken from the map. These profiles provide a graphical record of the agreement between the map profile and the corresponding ground profile. As in the case mentioned above, the comparison between separate 100-ft or 30-m station points can be made; also, the presence of systematic errors will be evidenced if the map profile is above or below the ground profile for an undue proportion of its length. Careless work in spots will be made evident by wide divergences between the profile lines at such places.

**13.38. Standards for the accuracy of maps** The federal agencies engaged in mapping have agreed on minimum requirements which entitle the following statement to be printed on a map, "This map complies with the National Map Accuracy Standards require-

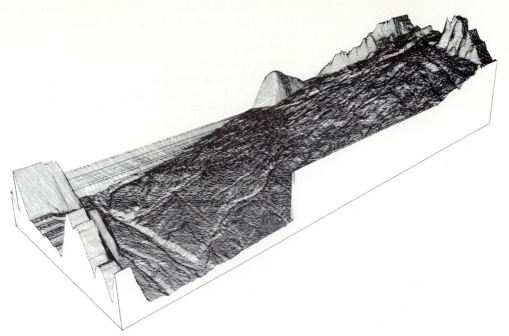

3-D PERSPECTIVE PALM CANYON, PALM SPRINGS, CALIFORNIA - MODELS 1, 2, 3 - X-Y ROTATION=135°

HORIZON ANGLE=20.7° - VIEW ELEVATION=3000' - DISTANCE FROM VIEW POINT TO FOCAL POINT=6000'

**Fig. 13.28** Three-dimensional perspective terrain model, Palm Canyon, Palm Springs, California. (*Courtesy Riverside County, California, Flood Control District.*)

ments," With regard to horizontal accuracy, it is required that for maps on publication scales larger than 1:20,000, not more than 10 percent of the well-defined points tested shall be in error by more than $\frac{1}{30}$ in (0.8 mm), measured on the publication scale; and for maps on publication scales of 1:20,000 or smaller, $\frac{1}{50}$ in (0.5 mm). Well-defined points are those that are easily visible or recoverable on the ground—in general, those which are plottable on the scale of the map within $\frac{1}{100}$ in (0.25 mm). With regard to vertical accuracy, it is required that not more than 10 percent of the elevations tested shall be in error more than one-half the contour interval; the apparent vertical error may be decreased by assuming a horizontal displacement within the permissible horizontal error. The accuracy of the map may be tested by comparing points whose locations or elevations are shown on it with corresponding points as determined by surveys of a higher accuracy.

The National Map Accuracy Standards have provided adequate accuracy criteria for small-scale maps, but are not sufficient for the diversified scales and contour intervals found in medium- and large-scale maps used for engineering design and planning. The American Society of Civil Engineers (ASCE) Task Committee on Selection of Map Types, Scales, and Accuracies for Engineering and Planning has prepared a report on engineering map accuracy standards, which is used as the basis for the following discussion (Ref. 2).[†] These standards are applicable to any map compiled and published at scales of 1:20,000 or larger. The standards do not contain specific numerical values for the parameters recommended for evaluating the accuracy of maps. Rather, the standards constitute a set of recommended parameters that should be used in judging map accuracies. The numerical values to be assigned should be set by the organization involved in the mapping and will be functions of user requirements, feasibility, and the cost of compilation.

[†]See also Merchant, D. C., "Engineering Map Accuracy Standards," unpublished paper for the A.S.C.E. Task Committee Report (Ref. 2), 1978.

The standards are based on estimates for the standard deviations in the $X$, $Y$, and $Z$ components of positions determined from the map. As a rule, this standard deviation would be calculated using discrepancies between map positions and positions determined independently by a survey of adequate accuracy. It is well known that systematic errors are quite frequently present in the measurements taken from the map and on the ground for purposes of the comparison. In an attempt to compensate for this systematic bias, the *mean absolute error* $\bar{\delta}$, defined as the absolute value of the algebraic mean of the discrepancies in each of the coordinate directions, is first estimated and removed. Thus,

$$|\bar{\delta}_x| = \left| \frac{\sum\limits_{i=1}^{n} (\delta_{x_i})}{n} \right| \tag{13.1a}$$

$$|\bar{\delta}_y| = \left| \frac{\sum\limits_{i=1}^{n} (\delta_{y_i})}{n} \right| \tag{13.1b}$$

$$|\bar{\delta}_z| = \left| \frac{\sum\limits_{i=1}^{n} (\delta_{z_i})}{n} \right| \tag{13.1c}$$

where $\delta_{x_i}$, $\delta_{y_i}$, $\delta_{z_i}$ = discrepancies between the position of a well-defined feature determined from map measurements and the position determined by means of an independent survey

$\bar{\delta}_x$, $\bar{\delta}_y$, $\bar{\delta}_z$, = mean algebraic deviations in the $X$, $Y$, $Z$ components, respectively (each + or −)

$|\bar{\delta}_x|$, $|\bar{\delta}_y|$, $|\bar{\delta}_z|$ = absolute values of $\bar{\delta}_x$, $\bar{\delta}_y$, $\bar{\delta}_z$, (always +)

$n$ = number of observed discrepancies ($n \geqslant 30$)

When the mean errors $\bar{\delta}_x$, $\bar{\delta}_y$, $\bar{\delta}_z$ have been determined, the *standard error* (standard deviation $\sigma$) is computed for each component as the standard deviation from the mean error. The standard errors for each component are

$$\sigma_x = \left[ \frac{\sum\limits_{i=1}^{n} (\delta_x - \bar{\delta}_x)_i^2}{n-1} \right]^{1/2} \tag{13.2a}$$

$$\sigma_y = \left[ \frac{\sum\limits_{i=1}^{n} (\delta_y - \bar{\delta}_y)_i^2}{n-1} \right]^{1/2} \tag{13.2b}$$

$$\sigma_z = \left[ \frac{\sum\limits_{i=1}^{n} (\delta_z - \bar{\delta}_z)_i^2}{n-1} \right]^{1/2} \tag{13.2c}$$

where the terms are defined in Eqs. (13.1a) to (13.1c).

Values for mean absolute errors and standard errors are estimated from discrepancies between at least 30 well-defined, widely distributed points which have been withheld as check points during the mapping procedure. Points selected for testing vertical accuracy do not have to be the same as those chosen for testing horizontal accuracy. Points should be located so as to be separated by a minimum of $\frac{1}{12}$ and a maximum of $\frac{1}{4}$ of the diagonal dimension of the area mapped.

Thus, the specifications for a map satisfying these standards could be stated as follows:

1. The *limiting* (or *allowable*) mean absolute errors calculated by Eqs. (13.1a) to (13.1c) shall not exceed _____ feet (metres) in $X$ (plan)

_____ feet (metres) in $Y$ (plan)

_____ feet (metres) in $Z$ (elevation)

2. The limiting standard errors (limiting standard deviation $\sigma$) calculated by Eqs. (13.2a) to (13.2c) shall not exceed

_____ feet (metres) $X$ (plan)

_____ feet (metres) $Y$ (plan)

_____ feet (metres) $Z$ (elevation)

3. Errors exceeding three times the specified limiting standard errors shall be designated blunders and shall be corrected regardless of other accuracy compliance tests.

4. Engineering maps that meet the Engineering Map Standards are to be identified by a note in the map legend as follows:

This map complies with the Engineering Map Accuracy Standards at a scale of _____ with error limits not exceeding:

| Error type | (feet or metres) | | |
| --- | --- | --- | --- |
| | $X$ | $Y$ | $Z$ |
| Standard error $\sigma$ | | | |
| Mean error $|\bar{\delta}|$ | | | |

The specifications as developed have been condensed and somewhat reordered from the format given in the ASCE report. Readers interested in more detail should consult the full text of the Task Committee Report, which is part of Ref. 2.

The limiting standard errors $\sigma_x$ and $\sigma_y$ can be related to the National Map Accuracy Standards by using the concept of circular standard error $\sigma_c$ where (Ref. 8)

$$\sigma_c \cong \tfrac{1}{2}(\sigma_x + \sigma_y) \tag{13.3}$$

This approximation is valid especially for $0.5 < \sigma_y/\sigma_x < 1.0$ (with $\sigma_x > \sigma_y$). Now, the specification for horizontal position stated in the National Map Accuracy Standards can be loosely interpreted as the estimated deviation with a 90 percent probability and can be defined as the circular map accuracy standard (CMAS). This value is the radius of a circle within which 90 percent of the errors are expected to fall. Equation (13.3) can be modified to express a value for the CMAS by using Table 2.4 (Art. 2.19) to give

$$\begin{aligned} \text{CMAS} &\cong \tfrac{1}{2}(\sigma_x + \sigma_y)2.146 \\ &\cong 1.073(\sigma_x + \sigma_y) \end{aligned} \tag{13.4}$$

and assuming that $\sigma_x = \sigma_y$, we have

$$\text{CMAS} \cong 2.146\sigma_x = 2.146\sigma_y$$

or

$$\sigma_x = \sigma_y = 0.466\text{CMAS} \tag{13.5}$$

In a similar fashion, the National Map Accuracy Standard for elevation can be interpreted as the estimated deviation in elevation with a probability of 90 percent. From Eqs. (2.15), Art. 2.11, the constant that relates the standard deviation $\sigma$ with a 90% probability deviation is 1.645. Thus, the vertical map accuracy standard, or VMAS, can be expressed as a function of $\sigma_z$ using

$$\text{VMAS} = 1.645\sigma_z$$

or

$$\sigma_z = 0.608\text{VMAS} \tag{13.6}$$

Note that the foregoing relationships are derived assuming noncorrelation among $X$, $Y$, and $Z$.

**Example 13.1**  A topographic map has a scale of 1:3600 and a contour interval of 2 ft. Accordingly, the National Map Accuracy Standards are:

In position

$$CMAS = \left(\frac{1}{30} \text{ in}\right)\left(\frac{1 \text{ ft}}{12 \text{ ft}} \text{ in}\right)(3600) = 10 \text{ ft}$$

and in elevation

$$VMAS = (\tfrac{1}{2})(2) = 1 \text{ ft}$$

The Engineering Map Accuracy Standards are:

By Eq. (13.5),

$$\sigma_x = \sigma_y = (0.466)(10) = 4.7 \text{ ft}$$

and by Eq. (13.6),

$$\sigma_z = (0.608)(1) = 0.6 \text{ ft}$$

**Example 13.2**  The limiting standard errors for a map are $\sigma_x = 2$ m, $\sigma_y = 2$ m, and $\sigma_z = 0.3$ m. Calculate the corresponding National Map Accuracy Standards.

SOLUTION  By Eq. (13.5), solving for CMAS,

$$CMAS = (2.146)(2) = 4.29 \text{ m}$$

Using Eq. (13.6) for the standard in elevation, we obtain

$$VMAS = (1.645)(0.3) = 0.49 \text{ m}$$

When applying the Engineering Map Accuracy Standards, care should be exercised to ensure that map scale and contour interval are compatible, as discussed in Art. 13.10. Also, due consideration should be given to the degree of slope of the terrain being mapped so that errors in plotted positions of contours do not cause excessive errors in elevations.

Specific values for limiting errors are not suggested for the Engineering Map Accuracy Standards because of the variation in type and applications for maps within this group. Selection of limiting errors for a particular project should be the responsibility of the engineers in charge of the mapping, in consultation with the client who will use the map.

## 13.39. Problems

**13.1**  Describe the various classes, types, and categories of maps.

**13.2**  Discuss the difference between planimetric and topographic maps.

**13.3**  Define horizontal datum and vertical datum as used in the United States.

**13.4**  Discuss the various vertical datums used for different types of surveys, specifying the type of survey that requires a given datum.

**13.5**  What are the different ways in which relief can be represented on a flat sheet of paper?

**13.6**  Discuss the principal characteristics of contour lines.

**13.7**  On a map at a scale of 1:5000 and with a contour interval of 1 m, the distance scaled on the map between two adjacent contours is 15.0 mm. What is the slope of the ground in percent?

**13.8**  The following tabulation gives elevations of points over the area of a 60- by 100-ft city lot. The elevations were obtained by the checkerboard method, using 20-ft squares. Point A-1 is in the northwest corner of the lot, and point F−1 is at the southwest corner. Plot the contours, using a horizontal scale of 1:120 (1 in/10 ft) and a contour interval of 2 ft.

| Point | Elevation, ft | | | |
|-------|-------|-------|-------|-------|
|       | 1 | 2 | 3 | 4 |
| A | 322.9 | 327.0 | 327.5 | 328.4 |
| B | 326.6 | 331.0 | 333.3 | 332.2 |
| C | 327.4 | 333.3 | 335.7 | 333.5 |
| D | 326.6 | 334.6 | 337.0 | 334.2 |
| E | 327.5 | 333.0 | 337.4 | 337.7 |
| F | 328.2 | 333.6 | 338.3 | 341.2 |

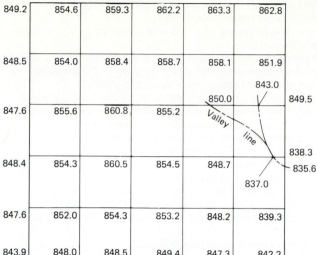

**Fig. 13.29**

**13.9** The grid interval is 50 m for the layout illustrated in Fig. 13.29, and elevations shown at grid intersections are in metres. Draw the layout of grid intersections to a map scale of 1:1500 and plot contours with an interval of 1 m.

**13.10** Plot the planimetric detail and mark elevations on the sheet for the cross-profile data of Fig. 15.8. Draw contours with a 0.5-ft interval, paying particular attention to the way in which the contours cross the pavement.

**13.11** It is possible to plot and scale map details to within 0.5 mm. What scale of map is required to permit scaling distances between well-defined points to within 1 m?

**13.12** In Prob. 13.11, plotting accuracy was used to select a map scale. What other factors affect the choice of map scale?

**13.13** Discuss the criteria that govern the selection of the contour interval for a topographic map.

**13.14** Discuss the relationship between contour interval and map scale. What are the principal factors that affect this relationship?

**13.15** Unadjusted azimuths, distances, and coordinates for traverse *ABCDEF* are as follows:

| Point | Azimuth | Distance, ft | X, ft | Y, ft |
|-------|---------|--------------|-------|-------|
| A     |         |              | 62,480.5 | 49,487.7 |
|       | 0°00′   | 338          |          |          |
| B     |         |              | 62,480.5 | 49,825.7 |
|       | 351°00′ | 307          |          |          |
| C     |         |              | 62,432.5 | 50,128.9 |
|       | 67°45′  | 792          |          |          |
| D     |         |              | 63,165.5 | 50,428.8 |
|       | 142°30′ | 822          |          |          |
| E     |         |              | 63,665.9 | 49,776.7 |
|       | 244°30′ | 624          |          |          |
| F     |         |              | 63,102.7 | 49,508.0 |
|       | 268°00′ | 620          |          |          |
| A     |         |              |          |          |

Plot the positions of the traverse stations by the method of coordinates to a scale of 1:2400. Use a coordinate-grid-line interval of 500 ft. Scale the distances between the plotted points as a check on the plotting accuracy. Since the coordinates are unadjusted, there will be an error of closure which is reflected in the distance and direction of the line *FA*. Scale and record these errors in distance and direction.

**13.16** What is the most desirable form for data that are to be plotted automatically by a computer controlled flat-bed plotter?

**13.17** Discuss the concepts involved in the automatic plotting of contours.

**13.18** What information should appear on any map?

**13.19** What information should appear on a map of a subdivision that is recorded and becomes a part of the public record?

**13.20** Explain the function of color separation sheets in the preparation of a topographic map.

**13.21** Discuss manuscript registration and how it is used in the preparation of maps that are compiled using multiple map sheets and/or color separation sheets.

**13.22** Describe how *stripping* and *stickup* are utilized in the preparation of a map.

**13.23** What is the difference between a *color separation guide* and the *compilation manuscript* if both have been prepared by scribing on coated polyester drafting film?

**13.24** Describe how a letterpress plate is prepared for map reproduction.

**13.25** Discuss how a lithographic plate is prepared for map reproduction.

**13.26** Discuss the differences between *right-reading* and a *wrong-reading* images and how these two types of images are used in map reproduction by photolithographic offset printing.

**13.27** What are the major sources of data for automatic mapping procedures?

**13.28** What are the major components of an automatic mapping system?

**13.29** Discuss the overall concepts involved in the automatic plotting of planimetric details and contours.

**13.30** What are the National Standards of Map Accuracy?

**13.31** Are the National Standards of Map Accuracy adequate for *all* types of maps? Explain.

**13.32** The ASCE has developed accuracy standards for engineering maps compiled at scales of 1:20,000 or larger. What is the basis for these standards and what data must be gathered to evaluate the parameters needed to establish these standards for a given map?

**13.33** Write a sample set of specifications for a map to be compiled according to the ASCE engineering map accuracy standards.

**13.34** A map at a scale of 500 ft/in has a contour interval of 5 ft. Compute the (*a*) National Map Circular and Vertical Map Accuracy Standards (CMAS and VMAS) for position and elevation; (*b*) Engineering Map Accuracy Standards for the same scale map and contour interval.

**13.35** Perform the computations of (*a*) and (*b*) of Prob. 13.34 for a map at a scale of 1:2000 with a contour interval of 5 m.

**13.36** The limiting standard errors for a specific map are $\sigma_x = 1$ m, $\sigma_y = 1$ m, and $\sigma_z = 0.1$ m. (*a*) Calculate the corresponding National Standards of Map Accuracy, and (*b*) Write the statement which includes the specifications necessary to ensure that the Engineering Map Standards will be satisfied.

**13.37** Describe the procedure that is required in order to test a map so as to show that it satisfies the ASCE Engineering Map Accuracy Standards. Write the statement that should be included on the map which has been shown to satisfy the foregoing standards. Use the data from Prob. 13.36.

## 13.40. Office problem

### TOPOGRAPHIC MAP CONSTRUCTION

**Object** To construct a complete topographic map using data collected from field problems in Chap. 15. Relief will be represented by contours.

**Procedure**

1. Calculate adjusted coordinates and elevations for all horizontal and vertical control points.
2. Select a map scale and contour interval compatible with the size and terrain configuration of the area to be mapped and intended purpose of the map (Art. 13.10). For a student project such as this, it is wise to select an area and scale that will allow plotting the map on a single sheet.
3. Plot the control points on the base map sheet by the coordinate method (Art. 13.13). Label each point, mark the elevation next to each vertical point, and mark the distance and azimuth on each line for which these data are available. The base map should be drawn on a good-quality, heavy detail paper or polyester drafting film.
4. The method for plotting details is a function of the procedure employed to collect data in the field. Use one of the methods for plotting details suggested in Art. 13.15.
5. Plot all details, draw the planimetric features, and draw the contours by interpolation (Art. 13.17). Use the standard symbols and include a title on the base map. When the base map has been smooth-drafted (all line work is in pencil at this point), make a tracing of the base map on a

good-quality tracing paper or polyester drafting film. The tracing may be in pencil, ink, or scribed on coated drafting film. Include all the information that is required on a map (Art. 13.19). This final tracing constitutes the map manuscript from which reproductions would be made.

## References

1. Ackermann, F., "Experimental Investigation into the Accuracy of Contouring From DTM," *Photogrammetric Engineering and Remote Sensing*, Vol. 44, No. 12, December 1978.
2. American Society of Civil Engineers, Manual on "Map Uses, Scales, and Accuracies for Engineering and Associated Purposes," Unpublished committee report, 1978.
3. American Society of Civil Engineers, "Selection of Maps for Engineering and Planning," Task Committee for Preparation of a Manual on Selection of Map Types, Scales and Accuracies for Engineering and Planning, *Journal of the Surveying and Mapping Division, ASCE*, No. SU1, July 1972.
4. Collins, S. H., "Terrain Parameters Directly from Digital Terrain Model," *The Canadian Surveyor*, Vol. 29, No. 5, December 1975, p. 507.
5. Keates, J. S., *Cartographic Design and Production*, John Wiley & Sons, Inc., New York, 1973.
6. Kloosterman, B., Norgren, R. R., and Sharp, W. R., "Computer Assisted Color Separation for the Production of Thematic Maps," *The Canadian Surveyor*, Vol. 28, No. 1, March 1974.
7. Lee, K. S., et al., "CRT Light Head for Automatic Drafting Systems," *The Canadian Surveyor*, September 1977.
8. Mikhail, Edward, M., *Observations and Least Squares*, Harper & Row, Publishers, New York, 1976, pp. 32, 33.
9. National Oceanic and Atmospheric Administration and U.S. Geological Survey, Stang, Paul R., and Baxter, Franklin S., eds., *Coastal Zone Mapping Handbook*, March 1976.
10. *Proceedings of the Digital Terrain Models Symposium, American Society of Photogrammetry in cooperation with ASCM*, St. Louis, Mo. May 9–11, 1978, ASP, Falls Church, Va.
11. Robinson, A. H., Sale, R., and Morrison, J., *Elements of Cartography*, 4th ed., John Wiley & Sons, Inc., New York, 1978.
12. Schut, G. H., "Review of Interpolation Methods for Digital Terrain Models," *The Canadian Surveyor*, Vol. 30, No. 5, December 1976, p. 389.
13. Stine, G. E., et al., "Automation in Cartography," *Surveying and Mapping*, December 1971.
14. Thompson, M. M., et al., "On Map Accuracy Specifications," *Surveying and Mapping*, March 1971.
15. Topographic Instructions of the U.S. Geological Survey, *Color-Separation Scribing*, Book 4, Chaps. 4B1–4B3, Government Printing Office, Washington, D.C., 1961.

# CHAPTER 14
# Map projections

**14.1. Introduction**   In plotting a map of a small and limited area, the curvature of the earth need not be considered. A level surface on the earth is assumed to be a plane, and points are plotted on the map in terms of rectangular coordinates from two orthogonal axes representing the east-west and north-south directions.

For maps of larger areas this simple method is not satisfactory because the curvature of the earth can no longer be ignored. The geometric shape of the earth is a *spheroid* with a polar diameter about one-third of 1 percent shorter than the equatorial diameter. Therefore, a plane passing through the equator would cut it in a circle, while a plane through the poles (*meridional* plane) intersects it in an ellipse. However, with that slight difference between polar and equatorial dimensions, the ellipse is very nearly a circle. Consequently, for our purposes here, the earth is assumed to be a *sphere*. This will make it easier for the beginning student to visualize the various projections to be discussed.

Regardless of whether the earth is considered a sphere or spheroid, it is not possible to develop its *surface exactly* onto a plane, just as it is impossible to flatten a section of orange peel without tearing it. It follows, then, that whatever procedure is used to represent a large area on a map, there will *always* be some distortion. To minimize the distortion as well as to develop several possibilities, points on the map are represented in terms of parallels of latitude and meridians of longitude. Position on the earth in terms of latitude and longitude (e.g., New York City is at 40°45′N latitude and 74°00′W longitude) is transformed into scaled linear dimensions on the map. This is accomplished by using the dimensions of the earth and a selected set of criteria for representing the earth on the map. Such a transformation from latitude $\phi$ and longitude $\lambda$ to a map's $X, Y$ coordinates is the function of *map projections*.

Many view the map projection as a mathematical operation in which the map coordinates $X, Y$ are written as a pair of parametric functions of latitude and longitude $\phi, \lambda$, or

$$X = f_x(\phi, \lambda)$$
$$Y = f_y(\phi, \lambda) \tag{14.1}$$

On an *ideal* map *without distortion*, Eq. (14.1) must satisfy the following conditions: (1) all distances and areas on the map would have correct relative magnitude, (2) all azimuths and angles would be correctly shown on the map, (3) all great circles on the earth would appear as straight lines on the map, and (4) geodetic latitudes and longitudes of all points would be correctly shown on the map.

As noted earlier, because of the shape of the earth, it is impossible to satisfy *all* of these conditions in the same map. It is possible, however, to satisfy one or more of the four conditions by imposing them on the transformation Eq. (14.1). Thus, there result several classes of map projection:

1. *Conformal* or *orthomorphic* projection results in a map showing the correct angle between any pair of short intersecting lines, thus making small areas appear in correct *shape*. As the scale varies from point to point, the shapes of larger areas are incorrect.
2. An *equal-area* projection results in a map showing all areas in proper relative *size*, although these areas may be much out of shape and the map may have other defects.
3. In an *equidistant* projection, distances are correctly represented from one central point to other points on the map.
4. In an *azimuthal* projection, the map shows the correct *direction* or azimuth of any point relative to one central point.

**14.2. Types of map projections**   Although all map projections are carried out by computing $X, Y$ for each pair of $\phi, \lambda$ using different forms of Eq. (14.1), there are two methods of projection: geometric and mathematical. In *geometric* projection a surface that can be developed into a plane (such as a plane, a cone, or a cylinder) is selected such that it either cuts or is tangent to the earth. A point is then selected as the projection center from which straight lines are connected to points on the earth and extended until they intersect the selected mapping surface. In *mathematical* projection, there is no one particular projection point; instead, a form of Eq. (14.1) is used to compute the location $X, Y$ of the point on the map from its position $\phi, \lambda$ on the earth. In the following sections the various projections are introduced in groups, beginning with those on a plane tangent to the earth sphere.

**14.3. Map projection to a plane**   The easiest geometric map projection to visualize is that in which the projection surface is a plane tangent to the sphere at any point. The point of tangency becomes the central point in the map. Then there are three possibilities for the point used as projection center: (1) when the center of the sphere is used (Fig. 14.1), the projection is called *gnomonic*; (2) when the end of the diameter opposite to the point of tangency is used (Fig. 14.2), the projection is called *stereographic*; and (3) when the projection center is at infinity, in which case the projection lines become parallel, the projection is called *orthographic*.

**14.4. Gnomonic projection**   This, considered to be the oldest true map projection, is assumed to have been devised by the great abstract geometrist Thales in the sixth century B.C. It is a geometric projection to a plane tangent to the sphere at any point such as $A$ in Fig. 14.1, with the projection center at the center of the sphere. Lines from the earth's center 0 through all points to be mapped, $A, B$, etc., are extended until they intersect the tangent

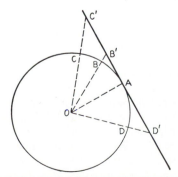

**Fig. 14.1** Gnomonic projection.

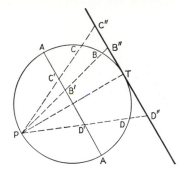

**Fig. 14.2** Stereographic projection.

plane in the map points $A, B', C'$, etc. The new points $A, B', C'$, etc., in the plane represent the gnomonic projection of the original points. This projection has the following properties:

1. Great circles show as straight lines and therefore meridians are straight lines.
2. Shapes and sizes undergo extreme distortion as we go farther from the central tangent point.
3. Azimuths of lines drawn from the tangent point to other map points are correct, because such lines are great circles. Therefore, this is an *azimuthal* projection.
4. Since the point of projection is at the center of the sphere, it is not possible to map an entire hemisphere.

This projection is used mostly for navigational purposes. For other purposes, only a limited area around the central point may be used.

**14.5. Stereographic projection**   This is another projection credited to the Greek astronomer Hipparchus dating back to the second century B.C. It is a geometric projection to a tangent plane with the point of projection $P$ diametrically opposite to the point of tangency $T$ (Fig. 14.2). Rays from $P$ to surface points $B, C$, etc., are extended until they intersect the tangent plane. The points of intersection $B'', C''$, etc., are the map points on the stereographic projection on the tangent plane. On any other intersecting plane, such as $A$-$A$ in Fig. 14.2, an equivalent projection is possible, with only a reduction scale factor from the projection on the tangent plane.

This projection is both azimuthal and conformal. It has the following properties:

1. It is the only projection on which circles on the earth still appear as circles on the map.
2. Like the gnomonic projection, azimuths of lines from the central point are correct because such lines are great circles.
3. The scale increases as we move away from the tangent point and therefore its main defect is that areas are not correctly shown.

This is an excellent projection for general maps showing a hemisphere. It is also used for plotting ranges from central objects and for navigational purposes in higher latitudes, above 80°, when the plane is tangent at the pole.

**14.6. Orthographic projection**   This is another ancient projection, which was used during the Renaissance for artistic representation of the globe. Recent space applications have revived its use. It is again a geometric projection on a tangent plane, with the projection lines parallel to each other and normal to the tangent plane (Fig. 14.3). If the tangent point is one of the poles, each parallel of latitude will be shown correctly to scale as a circle, but the distance between successive parallels becomes rapidly smaller as we move farther from the center of the map. This means that unlike the gnomonic and stereographic projections, the scale in the orthographic projection *decreases* away from the central tangent point. On the other hand, this is an azimuthal projection just like the gnomonic and stereographic, so

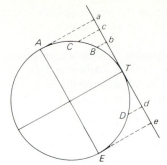

**Fig. 14.3** Orthographic projection.

that the azimuths from the tangent point are correct since the lines through it are great circles. The representation of the globe on this projection appears as it would in a photograph taken from deep space. Therefore, the orthographic projection is also used for map representations of the moon and planets.

**14.7. Conical projections**   Unlike a sphere, both a cone and a cylinder can be developed into a plane without distortion, and therefore both are used for map projection. A cone fits over a sphere, touching it in a small circle of latitude which is called the *standard parallel*, when the cone apex lies on the polar axis (Fig. 14.4*a*). As the height of the cone increases, the standard parallel gets closer to the equator (Fig. 14.4*b*). Finally, when the standard parallel reaches the equator, the elements of the cone become parallel, and the cone becomes a cylinder (Fig. 14.4*c*). When the cone height decreases, the standard parallel moves to higher latitudes (Fig. 14.4*d*), and finally the cone becomes a plane touching the sphere at one point, the pole (Fig. 14.4*e*). Thus, the projection of a sphere to a plane and a cylinder are actually limiting cases (at least geometrically) to conical projection.

Since a tangent cone has one parallel of latitude common to the sphere, its representation on the map will be at true scale, while scale distortions increase as the distance is extended to the north and the south of the parallel. For this reason, many conical projections use a cone that intersects the sphere in *two standard parallels* to minimize scale distortions.

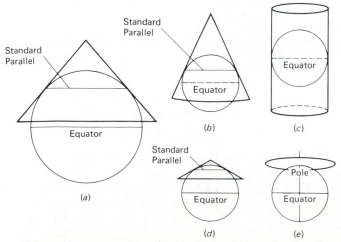

**Fig. 14.4** (*a*) Cone tangent to a sphere. (*b*) Cone tangent to a sphere with standard tangent closer to the equator. (*c*) Cylinder tangent to a sphere at the equator. (*d*) Cone with standard parallel at higher latitudes. (*e*) Plane tangent to a sphere at the pole. (*Courtesy National Geographic Society.*)

Although *geometric* projection on the cone is possible, the more important and useful conical projections are based on *mathematical* projection.

**14.8. Albers equal-area projection**  This is a mathematical projection on a cone with two standard parallels, developed by H. C. Albers in 1805. Its main characteristic is that small areas on the map are equal at the scale of the map, to corresponding small areas on the earth. Other properties of this projection include:

1. Scale is true along the standard parallels, is small between them, and large beyond them along *other parallels*. *Meridional* scale is large between and small beyond standard parallels.
2. Distances and directions measured on the map are reasonably accurate.
3. Meridians are straight lines meeting at a point that is the center of the concentric circular arcs representing the parallels. Therefore, meridians and parallels meet at right angles.
4. Parallel spacing increases away from the standard parallels.

This projection is useful for maps requiring equal-area representation and is used for areas at midlatitudes with extensive east-west dimensions.

**14.9. Polyconic projection**  Instead of a single cone, a series of conical surfaces may be used, points on the surface of the earth being considered as projected to a series of frustums of cones that are fitted together. These conical surfaces are then developed each way from a central meridian. Owing to differences in radii, the resulting strips would not exactly fit together when laid flat, but spaces would appear between them, such spaces increasing in width as the distance from the central meridian increases (Fig. 14.5b). To avoid such spaces, the north-south scale must be modified along the various meridians. Upon such a system of lines points are plotted by latitude and longitude.

In Fig. 14.5 it is seen that each parallel of latitude appears on the map as the arc of a circle having as radius the corresponding tangent distance; the parallel through $A$ has a radius $Aa$, that through $B$ has a radius $Bb$, and so on. The centers of these circles all lie on the central meridian of the map. The length of each tangent distance $Aa$, etc., is $N\cot\phi$, in which $N$ is the length of the normal or vertical at latitude $\phi$ extended to its intersection with the earth's axis. For the assumption that the earth is a sphere, $N$ is equal to the radius of the sphere. More precisely,

$$N = \frac{a}{\sqrt{1 - \varepsilon^2 \sin^2 \phi}} \tag{14.2}$$

in which $a$ is the earth's equatorial radius and $\varepsilon$ is the eccentricity of the ellipse in the meridian section ($\varepsilon^2 = 0.0067686580$). The length of the tangent distance $N\cot\phi$ varies with the latitude.

The distances $Q'A', A'B'$, etc., along the central meridian on the map (Fig. 14.5b) are true scale representations of the corresponding arc distances $QA, AB$, etc., in the meridian section (Fig. 14.5a). The parallels drawn on the map may be selected with as small a difference in

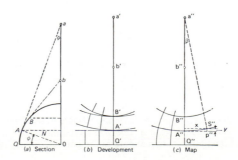

**Fig. 14.5** Polyconic projection.     (a) Section     (b) Development     (c) Map

latitude as may be desired, and each one is drawn with its own particular radius as shown, with the center of the arc on the central meridian at the proper tangent distance (to scale) above the point where the parallel cuts the central meridian.

It should be observed that the method of drawing these arcs of the parallels of latitude on the map is such that each parallel is separately developed as the circumference of the base of its own distinct cone, and that the spacing between them increases with increasing differences of longitude from the central meridian, thereby changing the north-south scale of the map from place to place as the longitude difference increases.

The arc distance $A''S''$ in the map (Fig. 14.5c) represents to true scale the difference in longitude between the points $A''$ and $S''$. The angle $A''a''S''$ is designated $\beta$ and is given by

$$\beta = \lambda \sin \phi$$

in which $\lambda$ is the arc $A''S''$. The rectangular coordinates of the point $S''$ referred to $A''$ as origin are

$$X = A''P'' = a''S'' \sin \beta = N \cot \phi \sin \beta \tag{14.3}$$
$$Y = P''S'' = a''S''(1 - \cos \beta) = N \cot \phi \operatorname{vers} \beta \tag{14.4}$$

and if the chord $A''S''$ is drawn, in the triangle $A''S''P''$, $S''P'' = A''P'' \tan P''A''S''$, or $Y = X \tan(\beta/2)$.

As many points as desired along the parallels on the map, such as point $S''$, are calculated and plotted. Then the meridians are drawn through such points. These meridians are curved, concave toward the central meridian, but if the parallels are drawn close enough together, each meridian may be drawn as a series of straight lines from parallel to parallel. On the network of parallels and meridians so prepared, points are plotted by latitude and longitude. Near the central meridian there is little error in such a map, but the error increases in proportion to the square of the difference in longitude along any one parallel. The variation with difference in latitude is not in direct proportion.

It is to be noted that along the central meridian and along every parallel the map is true to scale; that along the other meridians the scale is somewhat changed; that near the central meridian the parallels and meridians intersect nearly at right angles; and that areas of great extent north and south may be mapped with a very small distortion.

Although better adapted to mapping an area of great extent in latitude than for an area of great extent east and west, the polyconic projection is sufficiently accurate for maps of considerable areas, and it is widely used by the U.S. Geological Survey and the U.S. National Ocean Survey.

**14.10. Conformal mapping** Conformal projection was defined in the introduction as that which preserves the angle between any pair of intersecting short line segments. In this section the mathematical condition for conformal mapping is given. However, instead of expressing the condition as a relation between the map coordinates $X, Y$ and geodetic coordinates $\phi, \lambda$ an intermediate surface is introduced. This surface is called the *isometric plane*, where a new *isometric latitude q* is given by Refs. 9 and 15.

$$q = \ln \left[ \tan \left( \frac{\pi}{4} + \frac{\phi}{2} \right) \left( \frac{1 - \varepsilon \sin \phi}{1 + \varepsilon \sin \phi} \right)^{\varepsilon/2} \right] \tag{14.5}$$

where $\varepsilon$ is the eccentricity; $\varepsilon^2 = (a^2 - b^2)/a^2$, with $a$ and $b$ being the semimajor and semiminor axes of the earth ellipsoid. The map coordinates $X, Y$ are then expressed as functions of $\lambda$ and $q$, or

$$X = f_1(\lambda, q)$$
$$Y = f_2(\lambda, q) \tag{14.6}$$

In order for the relations in Eq. (14.6) to represent a conformal transformation, the following so-called Cauchy-Riemann equations must be satisfied:

$$\frac{\partial X}{\partial \lambda} = \frac{\partial Y}{\partial q}$$
$$\frac{\partial X}{\partial q} = -\frac{\partial Y}{\partial \lambda} \tag{14.7}$$

Equation (14.7) can be used either to derive a conformal mapping or to verify that a mapping is conformal.

**14.11. Lambert conformal conic projection** This projection was developed by Johann Heinrich Lambert in 1772, the same year in which he invented the transverse Mercator projection. The Lambert conformal conic projection is one of the most widely used projections in the United States and even worldwide. It is used for the state plane coordinate systems of states (or zones thereof) of greater east-west than north-south extent. It is a conic projection with two standard parallels, as shown in Fig. 14.6. Therefore, meridians are straight lines meeting in a point outside the map limits; parallels are arcs of concentric circles, and both sets of lines meet at right angles. Scale is true only along the two standard parallels; it is compressed between and expanded beyond them.

The parametric equations for this projection, which must satisfy the conformality conditions of Eq. (14.7), are

$$X = Ke^{-\ell q}\cos \ell\lambda$$
$$Y = Ke^{-\ell q}\sin \ell\lambda \tag{14.8}$$

in which $\ell$ and $K$ are constants computed in terms of the two selected standard parallels. The normal radius of curvature $N$ of the ellipsoid at a particular latitude $\phi$ is given by Eq. (14.2). Denoting by $N_1, N_2$ the radii at the two standard parallels $\phi_1, \phi_2$, the constants $\ell$ and $K$ are given by

$$\ell = \frac{\ln N_1 - \ln N_2 + \ln \cos\phi_1 - \ln \cos\phi_2}{q_2 - q_1} \tag{14.9}$$

$$K = \frac{N_1 \cos\phi_1}{\ell e^{-\ell q_1}} = \frac{N_2 \cos\phi_2}{\ell e^{-\ell q_2}} \tag{14.10}$$

The radius of any circular arc representing a parallel is computed from

$$R^2 = X^2 + Y^2$$

which from Eq. (14.8) becomes

$$R^2 = K^2(e^{-2\ell q})(\cos^2 \ell\lambda + \sin^2 \ell\lambda)$$
$$= K^2 e^{-2\ell q}$$

Then

$$R = Ke^{-\ell q} \tag{14.11}$$

Equations (14.8) may be rewritten in terms of $R$ as

$$X = R\cos \ell\lambda$$
$$Y = R\sin \ell\lambda \tag{14.12}$$

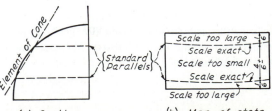

**Fig. 14.6** Lambert conformal conic projection.

(a) *Section*

(b) *Map of state coordinate zone*

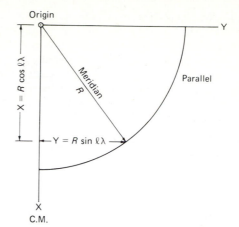

Origin

Y

X = R cos ℓλ

Meridian
R

Parallel

Y = R sin ℓλ

X
C.M.

**Fig. 14.7** Graphical representation of coordinates
on the Lambert conformal projection.

These equations can be represented graphically as shown in Fig. 14.7. The equivalent of the
apex of the cone is taken as the origin of coordinates with the $X$ axis in the direction of the
central meridian. It becomes the center of the concentric circles representing the parallels
and the point in which all the meridians intersect. Unlike the Albers projection, the spacing
of parallels increases away from the standard parallels. For a map of the United States on
this projection, scale errors at any point need not exceed 2 percent. Computation of state
plane coordinates in the Lambert conformal conic projection is explained in detail in Art.
14.17.

**14.12. Mercator projection**   This is a projection on a cylinder tangent to the earth at the
equator (Fig. 14.4c). It is a *mathematical* projection, however, and not geometric in order to
enforce the conformality conditions. It was created by Mercator in 1569 as a result of his
efforts to have the *rhumb* line, or *loxodrome*, or the line of constant bearing on the globe,
appear as a straight line on the map. Thus, with true scale at the equator and taking the
equator as the zero $Y$ value in the map, the mapping equations are

$$X = a\lambda$$
$$Y = aq$$

(14.13)

where $\lambda$ is the longitude and $q$ the isometric latitude given by Eq. (14.5). With these, the
conformality conditions [Eq. (14.7)] are satisfied, the meridians are equally spaced vertical
straight lines, and parallels are unequally spaced horizontal lines. The parallel spacing
increases toward the poles, which in this projection are at infinity on the map. Thus, while
the scale at any one point of intersection of a meridian and a parallel is *equal in all directions
around the intersection* (which is the basis of conformality), the scale expands rapidly at high
latitude. Thus, if $s$ is the scale number at the equator and $s_\phi$ is the scale number at any
latitude $\phi$, then

$$s_\phi = s \cos\phi$$

(14.14)

As an example, if the scale at the equator is $1 : 20,000$, it would be $1 : 10,000$ at a latitude
$\phi = 60°$ because $s_\phi = s \cos 60° = (20,000)(0.5) = 10,000$.

Aside from conformality, the particular feature of the Mercator projection is that the
rhumb line on the earth is plotted as a straight line on the map, a property that renders it
invaluable for purposes of navigation. The shortest course between two points is determined
by drawing on a gnomonic chart a great circle, which there appears as a straight line.
Selected points, at convenient distances apart, of this great circle are then plotted on the
Mercator chart, after making any necessary corrections on account of shoals, wind, currents,

etc. The rhumb line connecting any two adjacent points indicates the true bearing of the course, which is read by means of a protractor. This true bearing, corrected for magnetic declination, gives the compass bearing to be used in steering.

Owing to the rapid variation of scale, maps constructed on the Mercator projection give very inaccurate information as to relative sizes of areas in widely different latitudes. For example, on the map Greenland appears larger than South America, whereas in fact South America is nine times as large as Greenland. Consequently, such a map is not suited to general use, although because of its many other advantages it is widely published.

**14.13. Transverse Mercator projection**    This projection was developed by Lambert in 1772, analytically derived by Gauss 50 years later, and then formulas more suitable for calculations were devised by Kruger in 1912. This is perhaps the reason that it is one of the most widely used conformal map projections.

A transverse Mercator projection is the ordinary Mercator projection turned through a 90° angle so that it is related to a central meridian in the same way that the ordinary Mercator is related to the equator (Fig. 14.8a). Because the cylinder is tangent to the globe at a meridian, the scale is true along that meridian, which is called the central meridian, and

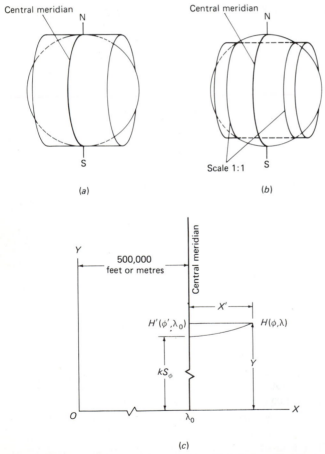

**Fig. 14.8** Transverse Mercator projection. (*a*) Cylinder with one standard line. (*b*) Cylinder with two standard lines. (*c*) Coordinates on the transverse Mercator projection.

is used as the origin of the map $X$ coordinate. The origin of the map $Y$ coordinate is the equator. Although, like the regular Mercator, this projection is conformal, it does not retain the straight-rhumb-line property of the Mercator. Other properties of this projection include: (1) both the central meridian and the normal to it are represented by straight lines; (2) other meridians are complex curves that are concave toward the central meridian; (3) parallels are concave curves toward the pole; and (4) the scale is true only along the central meridian.

This projection is used for the state plane coordinate systems of states (or zones thereof) of greater north-south than east-west extent. For the state systems the Mercator projection cylinder is made to cut the surface of the sphere along two standard lines parallel to the central meridian instead of being tangent to the sphere as in the ordinary Mercator projection (Fig. 14.8$b$). The projection formulas are (Thomas, Ref. 15)

$$\frac{X}{N} = \lambda' \cos\phi + \frac{\lambda'^3 \cos^3\phi}{6}(1 - t^2 + \eta^2)$$

$$+ \frac{\lambda'^5 \cos^5\phi}{120}(5 - 18t^2 + t^4 + 14\eta^2 - 58t^2\eta^2 + 13\eta^4) + \cdots$$

$$\frac{Y}{N} = \frac{S_\phi}{N} + \frac{\lambda'^2}{2}\sin\phi\cos\phi + \frac{\lambda'^4}{24}\sin\phi\cos^3\phi(5 - t^2 + 9\eta^2 + 4\eta^4)$$

$$+ \frac{\lambda'^6}{720}\sin\phi\cos^5\phi(61 - 58t^2 + t^4 + 270\eta^2 - 330t^2\eta^2 + 445\eta^4) + \cdots$$

(14.15)

where $N$ = is given by Eq. (14.2)

$t = \tan\phi$

$\eta^2 = \varepsilon'^2\cos^2\phi = \dfrac{\varepsilon^2\cos^2\phi}{1 - \varepsilon^2} = \dfrac{(a^2 - b^2)\cos^2\phi}{b^2}$

$\lambda' = \lambda - \lambda_0$ = longitude difference from central meridian, $\lambda_0$ in *radians*

$S_\phi$ = length of the meridian arc from the equator to latitude $\phi$ and is given by (Bomford, Ref. 2)

$$S_\phi = \int_0^\phi \frac{a(1 - \varepsilon^2)}{(1 - \varepsilon^2\sin^2\phi)^{3/2}}\, d\phi$$

in which $\varepsilon$ is the eccentricity, where $\varepsilon^2 = (a^2 - b^2)/a^2$ as defined in Art. 14.10, or

$$S_\phi = a(A_0\phi - A_1\sin 2\phi + A_2\sin 4\phi - A_3\sin 6\phi + \cdots)$$

(14.16)

in which

$a$ = semimajor axis of the ellipsoid

$$A_0 = 1 - \frac{1}{4}\varepsilon^2 - \frac{3}{64}\varepsilon^4 - \frac{5}{256}\varepsilon^6 - \cdots$$

$$A_1 = \frac{3}{8}\varepsilon^2 + \frac{3}{32}\varepsilon^4 + \frac{45}{1024}\varepsilon^6 + \cdots$$

$$A_2 = \frac{15}{256}\varepsilon^4 + \frac{45}{1024}\varepsilon^6 + \cdots$$

$$A_3 = \frac{35}{3072}\varepsilon^6 + \cdots$$

The terms used in Eq. (14.15) are sufficient to yield $X$ and $Y$ values that are accurate to 0.01 m for zones within $\pm 3°$ of longitude about the central meridian. Computation of state plane coordinates on a transverse Mercator projection is explained in Art. 14.19.

The convergence of meridians (difference between grid north and geodetic north) is

$$\Delta\alpha = \lambda'\sin\phi\left[1 + \frac{(\lambda')^2\cos^2\phi(1 + 3\eta^2)}{3} + \frac{(\lambda')^4\cos^4\phi(2 - t^2)}{15}\right]$$

(14.16$a$)

in which $\lambda'$ and $\Delta\alpha$ are in radians.

The scale determined from geographic coordinates (latitude and longitude) is

$$k = 1 + \frac{(\lambda')^2 \cos^2 \phi}{2}(1+\eta^2)$$

$$+ \frac{(\lambda')^4 \cos^4 \phi}{24}(5 - 4t^2 + 14\eta^2 + 13\eta^4 - 28t^2\eta^2 + 4\eta^6 - 48t^2\eta^4 - 24t^2\eta^6)$$

$$+ \frac{(\lambda')^6 \cos^6 \phi}{720}(61 - 148t^2 + 16t^4) \tag{14.17}$$

Latitude and longitude are obtained from the $X, Y$ rectangular map coordinates (Fig. 14.8c) by the following equations (Thomas, Ref. 15, for projections 2° on either side of the central meridian):

$$\Delta\phi = \phi - \phi' = t_1 \left[ -\frac{X^2}{2R_1 N_1} + \frac{X^4}{24 R_1 N_1^3}(5 + 3t_1^2) \right]$$

$$\Delta\lambda = \lambda - \lambda_0 = \sec\phi' \left[ \frac{X}{N_1} - \frac{1}{6}\left(\frac{X}{N_1}\right)^3(1 + 2t_1^2 + \eta^2) \right.$$

$$\left. + \frac{1}{120}\left(\frac{X}{N_1}\right)^5(5 + 28t_1^2 + 24t_1^4) \right] \tag{14.17a}$$

where $\phi' =$ footpoint latitude (Fig. 14.8c)

$$R_1 = \frac{a(1 - \epsilon^2)}{(1 - \epsilon^2 \sin^2 \phi')^{3/2}}$$

$N_1$ is given by Eq. (14.2) for latitude $\phi'$

$t_1 = \tan\phi'$

$$\eta_1^2 = \frac{\epsilon^2}{1 - \epsilon^2} \cos^2 \phi'$$

The meridian convergence from rectangular coordinates is

$$\Delta\alpha = t_1 \left[ \frac{X'}{N_1} - \frac{1}{3}\left(\frac{X'}{N_1}\right)^3(1 + t_1^2 - \eta_1^2) + \frac{1}{15}\left(\frac{X'}{N_1}\right)^5(2 + 5t_1^2 + 3t_1^4) \right] \tag{14.17b}$$

The scale is given by

$$k = 1 + \frac{1}{2}\left(\frac{X'}{N_1}\right)^2(1 + \eta_1^2) + \frac{1}{24}\left(\frac{X'}{N_1}\right)^4(1 + 6\eta_1^2) \tag{14.17c}$$

Computation of state plane coordinates on the transverse Mercator projection is described in Art 14.19.

**14.14. Universal transverse Mercator (UTM)** This projection is based entirely on the transverse Mercator projection discussed in the preceding section. Its specifications are (UTM 1958):

1. Transverse Mercator projection is in zones that are 6° wide.
2. The reference ellipsoid is Clarke 1866 in North America.
3. The origin of longitude is at the central meridian.
4. The origin of latitude is at the equator.
5. The unit of measure is the metre.
6. For the southern hemisphere, a false northing of 10,000,000 m is used.
7. A false easting of 500,000 m is used for the central meridian of each zone.
8. The scale factor at the central meridian is 0.9996.
9. The zones are numbered beginning with 1 for zone between 180°W and 174°W meridians and increasing to 60 for the zone between meridians 174°E and 180°E (Fig. 14.9).
10. The latitude for the system varies from 80°N to 80°S.

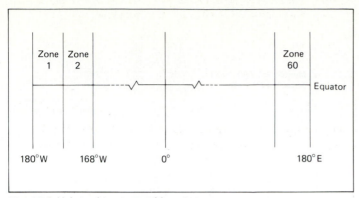

**Fig. 14.9** Universal transverse Mercator zones.

The UTM zones in the United States are shown in Fig. 14.10. An individual zone is illustrated in Fig. 14.11.

**14.15. Transverse Mercator in three degree zones**   The UTM projection system is the transverse Mercator projection in 6° zones. It is also possible to have a transverse Mercator projection in 3° zones. Since the zone width is only 3° and the scale factor of 0.9999 is assigned to the central meridian, a smaller scale error can be expected throughout the zone. This scale error is 1/10,000 on the central meridian as compared to 1/2500 for the UTM.

**14.16. State plane coordinate systems**   Each state in the United States has a plane coordinate system which is based on one or more zones of either the Lambert conic conformal projection (Art. 14.11) or the transverse Mercator projection (Art. 14.13). The Lambert projection is used in states having a large east-west dimension; the transverse Mercator projection is employed for states with a longer north-south dimension. If the north-south dimension of the Lambert projection and the east-west dimension of

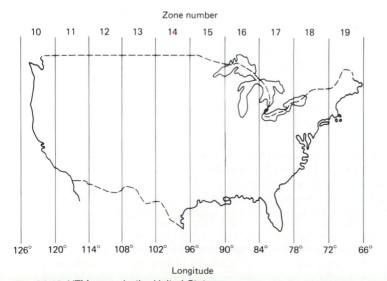

**Fig. 14.10** UTM zones in the United States.

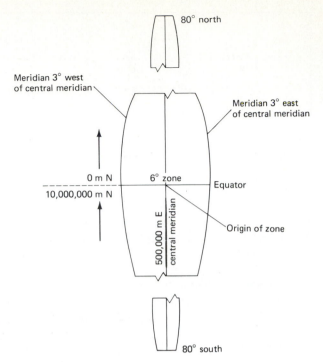

**Fig. 14.11** $X$ and $Y$ coordinates of the origin of a UTM grid zone. (From Ref. 17e.)

the transverse Mercator projection are held to 158 mi or 254 km, the maximum distortion in the projection is 1 part in 10,000. In states that exceed these dimensions, more than one zone of the same projection or perhaps two different projections are used. For example, New York has three transverse Mercator zones and one Lambert zone which covers Long Island; two transverse Mercator zones cover the peninsula of Florida, while a Lambert zone covers the western portion of the state; California has seven Lambert zones; and Vermont is covered by one transverse Mercator zone. There are a total of 126 zones (including Alaska and Hawaii) in the state plane coordinate systems in the United States. These systems are ratified by legislation in 35 states.

The present state plane coordinate systems, introduced in the early 1930s and tied to the 1927 North American Datum, utilize the foot as the basic unit.

When the North American Datum is redefined (scheduled to occur in 1983) two plane coordinate systems are to be published: (1) the universal transverse Mercator (UTM 1958, as described in Art. 14.14), and (2) the state plane coordinate (SPC) system. The constants used to define both of these systems will be in metric units and both systems will be referred to the North American Datum. Thus, the present state plane coordinate systems are to be retained with only a change in units and a slight shift in the respective origins.

Since new parameters and projection tables are not yet available, all state plane coordinate computations demonstrated in the following sections are in feet, using constants defined according to the 1927 North American Datum. The basic theory related to computing $X$ and $Y$ coordinates from projection parameters will remain the same.

**14.17. State plane coordinates using the Lambert conformal conic projection**  The coordinate system illustrated in Fig. 14.7 is not suitable for a state plane coordinate system. Instead, Fig. 14.12 shows that the origin of coordinates is shifted by a $Y$ value of $R_b$ and the central meridian is given an $X$ value designated by $C$. It follows from the figure that the

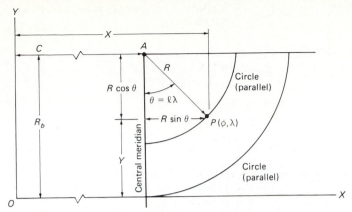

**Fig. 14.12** State plane coordinates on the Lambert projection.

state plane coordinates are given by

$$X = C + R \sin \ell\lambda = C + R \sin \theta$$
$$Y = R_b - R \cos \ell\lambda = R_b - R \cos \theta \qquad (14.18)$$

Values of $C$, $\ell$, and $R_b$ are given for each state plane coordinate system or zone thereof. In addition, values for $R$ are tabulated in terms of the latitude $\phi$, and $\theta$ is tabulated as a function of the longitude $\lambda$. It is, of course, possible to compute these quantities ($R$ and $\theta$) directly, using the relations given in Art. 14.11 and the latitudes for the two standard parallels, $\phi_1$ and $\phi_2$. For example, the procedure is as follows:

1. Compute $N_1$ and $N_2$ with Eq. (14.2), using $\phi_1$ and $\phi_2$ for the projection.
2. Compute the isometric latitudes $q_1$ and $q_2$ for the two standard parallels, using Eq. (14.5).
3.. Compute $\ell$, a constant for the projection, using Eq. (14.9).
4. Using Eq. (14.10) compute $K$, another constant for the projection.
5. The isometric latitude $q$ for the point of interest is computed by Eq. (14.5).
6. Equation (14.11) is used to compute $R$.

Then

$$\theta = \ell(\lambda_0 - \lambda) \qquad (14.19)$$

where $\lambda_0$ is the longitude of the central meridian which is given for the projection. Finally, Eqs. (14.18) are used to calculate the projection coordinates.

If the latitude $\phi$ and the longitude $\lambda$ are required from given coordinates $X, Y$ (called the *inverse transformation*), they may be obtained by rearranging Eqs. (14.18) or

$$R \sin \theta = X - C$$
$$R \cos \theta = R_b - Y$$

Division of the first by the second equation yields

$$\tan \theta = \frac{X - C}{R_b - Y}$$

or

$$\theta = \arctan\left(\frac{X - C}{R_b - Y}\right) \qquad (14.20)$$

Also,

$$R = \frac{R_b - Y}{\cos \theta} \qquad (14.21)$$

and from Eq. (14.19) compute

$$\lambda = \lambda_0 - \frac{\theta}{\ell} \tag{14.22}$$

To compute the latitude $\phi$, recall Eq. (14.11):

$$R = Ke^{-\ell q}$$

or

$$\ln R = \ln K - \ell q$$

from which

$$q = \frac{\ln K - \ln R}{\ell} \tag{14.23}$$

With the value of the isometric latitude $q$ known, the latitude $\phi$ can be evaluated from Eq. (14.5). Since $\phi$ cannot be found directly from this equation, an iterative procedure (such as Newton-Raphson) must be applied to Eq. (14.5) to solve for $\phi$.

Of course, if state plane coordinate tables are available, the value of $\phi$ is obtained directly as a function of $R$ except for linearly interpolating for the seconds portion of the angle.

The United States National Ocean Survey publishes tables for each state plane projection (Ref. 4). These tables can be obtained from the U.S. Government Printing Office. A portion of the Plane Coordinate Projection Tables for California, Special Publication No. 253 (1952), is illustrated in Table 14.1. Table I in this publication gives the value for $R$ in feet for every minute of latitude within the zone. The tabular difference for 01″ of latitude is provided for the purpose of obtaining the values sought by interpolation. Thus, given $\phi$ to some decimal part of a second, $R$ can be determined to the nearest hundredth of a foot. Conversely, given $R$, the latitude can be found. The scale of the projection is given in units of the seventh place of logarithms and as a ratio for each minute of latitude. Table II contains the mapping angle, $\theta$ [Eq. (14.19)], for each minute of longitude. The constants listed at the bottom include $\ell, C, R_b$ discussed previously and $1/2\rho_0^2 \sin 1″$ used in converting from grid to geodetic or geodetic to grid azimuth for long lines.

**Table 14.1   Lambert Projection for California I**
Table I

| Latitude | R, ft | Y′, y value on central meridian feet | Tabular difference for 1″ of latitude | Scale in units of seventh place of logs | Scale expressed as a ratio |
|---|---|---|---|---|---|
| 41°16′ | 24,088,135.79 | 704,300.44 | 101.20383 | − 334.6 | 0.9999230 |
| 17 | 24,082,063.56 | 710,372.67 | 101.20433 | − 324.9 | 0.9999252 |
| 18 | 24,075,991.30 | 716,444.93 | 101.20483 | − 314.8 | 0.9999275 |
| 19 | 24,069,919.01 | 722,517.22 | 101.20533 | − 304.4 | 0.9999299 |
| 20 | 24,063,846.69 | 728,589.54 | 101.20583 | − 293.6 | 0.9999324 |
| 41°21′ | 24,057,774.34 | 734,661.89 | 101.20650 | − 282.4 | 0.9999350 |
| 22 | 24,051,701.95 | 740,734.28 | 101.20717 | − 270.9 | 0.9999376 |
| 23 | 24,045,629.52 | 746,806.71 | 101.20767 | − 259.0 | 0.9999404 |
| 24 | 24,039,557.06 | 752,879.17 | 101.20817 | − 246.7 | 0.9999432 |
| 25 | 24,033,484.57 | 758,951.66 | 101.20883 | − 234.1 | 0.9999461 |
| 41°26′ | 24,027,412.04 | 765,024.19 | 101.20950 | − 221.1 | 0.9999491 |
| 27 | 24,021,339.47 | 771,096.76 | 101.21000 | − 207.7 | 0.9999522 |
| 28 | 24,015,266.87 | 777,169.36 | 101.21067 | − 193.9 | 0.9999554 |
| 29 | 24,009,194.23 | 783,242.00 | 101.21133 | − 179.8 | 0.9999586 |
| 30 | 24,003,121.55 | 789,314.68 | 101.21183 | − 165.3 | 0.9999619 |

Table II

| Longitude | | $\theta$ | |
|---|---|---|---|
| 120°56′ | +0 | 41 | 50.9158 |
| 57 | +0 | 41 | 11.6827 |
| 58 | +0 | 40 | 32.4497 |
| 59 | +0 | 39 | 53.2166 |
| 121°00′ | +0 | 39 | 13.9836 |
| 121°01′ | +0 | 38 | 34.7505 |
| 02 | +0 | 37 | 55.5174 |
| 03 | +0 | 37 | 16.2844 |
| 04 | +0 | 36 | 37.0513 |
| 05 | +0 | 35 | 57.8183 |
| 121°06′ | +0 | 35 | 18.5852 |
| 07 | +0 | 34 | 39.3521 |
| 08 | +0 | 34 | 00.1191 |
| 09 | +0 | 33 | 20.8860 |
| 10 | +0 | 32 | 41.6530 |

$C = 2,000,000$  $Y_0 = 547,078.17$
Central meridian $= 122°00′$  $\ell = 0.6538843192$

$R_b = 24,792,436.23$ ft  $\dfrac{1}{2\rho_0^2 \sin 1''} = (2.38)(10^{-10})$

More recently the U.S. National Ocean Survey has published a different set of tables with constants for use in computations with automatic data-processing equipment. Table 14.2 shows these constants for California zone I. The formulas to be used with these constants are as follows (Ref. 3 at the end of this chapter):

Given:

$\phi =$ latitude of the station
$\lambda =$ longitude of the station

$$s = 101.2794065\{(60)(L_7 - \phi') + L_8 - \phi'' + [1052.893882$$
$$- (4.483344 - 0.023520\cos^2\phi)\cos^2\phi]\sin\phi\cos\phi\}  \tag{14.24}$$

where $\phi' =$ latitude, expressed as whole minutes
$\phi'' =$ remainder of $\phi$, expressed in seconds

**Table 14.2    Constants for Lambert State Plane Coordinates by Automatic Data Processing, California Zone I**

| Constant | California zone I |
|---|---|
| $L_1$ | 2,000,000.00 |
| $L_2$ | 439,200.00 |
| $L_3$ | 24,245,358.05 |
| $L_4$ | 24,792,436.23 |
| $L_5$ | 0.9998946358 |
| $L_6$ | 0.6538843192 |
| $L_7$ | 2441. |
| $L_8$ | 26.75847 |
| $L_9$ | 3.80992 |
| $L_{10}$ | 3.93575 |
| $L_{11}$ | 0. |

then

$$R = L_3 + sL_5\left\{1 + \left(\frac{s}{10^8}\right)^2\left[L_9 - \left(\frac{s}{10^8}\right)L_{10} + \left(\frac{s}{10^8}\right)^2 L_{11}\right]\right\}$$  (14.25)

$$\theta = L_6(L_2 - \lambda)$$  (14.26)

in which $\theta$ and $\lambda$ are in seconds. Then the plane coordinates are

$$X = L_1 + R\sin\theta$$  (14.27a)

$$Y = L_4 - R + 2R\sin^2\frac{\theta}{2}$$  (14.27b)

and the scale factor is

$$k = \frac{L_6 R(1 - 0.00676\ 86580\sin^2\phi)^{1/2}}{20{,}925{,}832.16\cos\phi}$$  (14.28)

Equations and steps for the inverse solution are as follows (Ref. 3):
   Given: $X$ and $Y$ of the state plane coordinates of the Lambert projection:

$$\theta = \arctan\frac{X - L_1}{L_4 - Y}$$  (14.29)

$$\lambda = L_2 - \frac{\theta}{L_6}$$  (14.30)

where $\theta$ and $\lambda$ are in seconds. Then

$$R = \frac{L_4 - Y}{\cos\theta}$$  (14.31)

and

$$s_1 = \frac{L_4 - L_3 - Y + 2R\sin^2(\theta/2)}{L_5}$$  (14.32a)

$$s_2 = \frac{s_1}{1 + (s_1/10^8)^2 L_9 - (s_1/10^8)^3 L_{10} + (s_1/10^8)^4 L_{11}}$$  (14.32b)

$$s_3 = \frac{s_1}{1 + (s_2/10^8)^2 L_9 - (s_2/10^8)^3 L_{10} + (s_2/10^8)^4 L_{11}}$$  (14.32c)

$$s = \frac{s_1}{1 + (s_3/10^8)^2 L_9 - (s_3/10^8)^3 L_{10} + (s_3/10^8)^4 L_{11}}$$  (14.33)

$\omega' = L_7 - 600$   (degrees and minutes of $\omega$ in whole minutes)  (14.34)
$\omega'' = 36{,}000 + L_8 - 0.009873675553\ s$   (remainder of $\omega$ in seconds)  (14.35)
$\omega = \omega' + \omega''$  (14.36)
$\phi' = L_7 - 600$   (degrees and minutes of $\phi$ in whole minutes)  (14.37)
$\phi'' = \omega'' + [1047.546710 + (6.192760 + 0.050912\cos^2\omega)\cos^2\omega]\sin\omega\cos\omega$
(remainder of $\phi$ in seconds)  (14.38)

Finally, the latitude is found by

$$\phi = \phi' + \phi''$$  (14.39)

Consider an example problem.

**Example 14.1**   The latitude and longitude for a control point 9 in northern California zone I are
$\phi = 41°20'10.''403$
$\lambda = 121°05'20.''541$
Compute $X$ and $Y$ plane coordinates for the point on the California Lambert projection zone I, using the data in Tables 14.1 and 14.2 as well as the rigorous equations developed previously in Art. 14.13.

SOLUTION    Obtain $R$ by interpolation from Table I, Special Publication 253 (Table 14.1):

$$
\begin{aligned}
R \text{ for } 41°20' &= \quad 24{,}063{,}846.69 \\
(101.20583)(10.403'') &= - \quad\quad 1{,}052.84 \\
\hline
R \text{ for } 41°20'10.403'' &= \quad 24{,}062{,}793.85
\end{aligned}
$$

Find $\theta$ from Table II by interpolation (Table 14.1):

$$
\begin{aligned}
\theta \text{ for } 121°05' &= +0°35'57.8183'' \\
(\ell = 0.65388432)(20.541'') &= - \quad\quad 13.4314'' \\
\hline
(\theta \text{ for } \lambda \text{ of } 121°05'20.541'' &= 0°35'44.3869''
\end{aligned}
$$

$\sin\theta = 0.01039609370 \qquad \cos\theta = 0.9999459591$

Then, by Eq. (14.18),

$R\sin\theta = +25{,}015.906 \qquad\qquad R\cos\theta = 24{,}061{,}493.47$

$\quad C = 2{,}000{,}000 \qquad\qquad\qquad R_b = 24{,}792{,}436.23$

$\quad X_9 = \overline{2{,}250{,}159.06} \text{ ft} \qquad\quad Y_9 = \overline{\quad 730{,}942.76} \text{ ft}$

Also, from Table 14.1, using $\phi$ as the argument, the scale factor $= 0.9999329$.

Next, find the $X$ and $Y$ coordinates of point 9 using the constants listed in Table 14.2 and Eqs. (14.24) to (14.28):

$\phi' = (41)(60) + 20 = 2480'$
$\phi'' = 10.403''$

Solve for $s$ by Eq. (14.24).

$$
\begin{aligned}
s = 101.2794065 \{ &(60)(2441 - 2480) + 26.75847 - 10.403 \\
&+ [1{,}052.893882 - (4.483344 - (0.023520)(0.750846722)^2) \\
&\times (0.750846722]^2)(0.660476494)(0.750846722) \} \\
&= -182{,}581.1185
\end{aligned}
$$

Calculate $R$ using Eq. (14.25).

$$
R = (24{,}245{,}358.05) + (-182{,}581.1185)(0.9998946358)
$$

$$
\times \left\{ 1 + \left( \frac{-182{,}581.1185}{10^8} \right)^2 \left[ (3.80992) - \left( \frac{-182{,}581.1185}{10^8} \right)(3.93975) \right] \right\}
$$

$R = 24{,}062{,}793.85$, which corresponds to the value obtained from Table 14.1 using $\phi$ as the argument. Continuing, the mapping angle $\theta$ is computed by Eq. (14.26):

$$
\begin{aligned}
\theta &= (0.6538843192)(439{,}200.00 - 435{,}920.541) \\
&= 2144.386816'' \\
&= 0°35'44.386816''
\end{aligned}
$$

The coordinates of 9 are calculated by Eqs. (14.27a) and (14.27b):

$X = 2{,}000{,}000.00 + (24{,}062{,}793.85)(0.0103960937) = 2{,}250{,}159.05$ ft
$Y = 24{,}792{,}436.23 - 24{,}062{,}793.85 + 2(24{,}062{,}793.85)(0.000027020) = 730{,}942.75$ ft

These coordinates differ by 0.01 ft from those calculated using Table 14.1.

The scale factor is determined by Eq. (14.28):

$$
k = \frac{(0.6538843192)(24{,}062{,}793.85)[1 - (0.006768658)(0.436229199)]^{1/2}}{(20{,}925{,}832.16)(0.750846723)}
$$

$= 0.999932835$

Finally, compute the $X$ and $Y$ coordinates using the basic equations. For this solution the parameters of the Clarke ellipsoid of 1866 and the latitudes for the south and north standard parallels are:

$a = 6{,}378{,}206.4$ m $=$ semimajor axis $\Big\}$ of the
$b = 6{,}356{,}583.8$ m $=$ semimajor axis $\Big\}$ ellipsoid

$\varepsilon^2 = 0.006768658$
$\phi_1 = 40°00'$ latitude of south standard parallel
$\phi_2 = 41°40'$ latitude of north standard parallel

SOLUTION Using Eq. (14.2), compute $N_1$ and $N_2$.

$$N_1 = \frac{6{,}378{,}206.4}{[1-(0.006768658)(\sin^2 40°)]^{1/2}} = \frac{6{,}378{,}206.4}{0.9986006978} = 6{,}387{,}143.945 \text{ m}$$

$$N_2 = \frac{6{,}378{,}206.4}{[1-(0.006768658)(\sin^2 41°40')]^{1/2}} = \frac{6{,}378{,}206.4}{0.9985031635} = 6{,}387{,}767.844 \text{ m}$$

Compute $q_1$ and $q_2$, the isometric latitudes for the south and north standard parallels. Use Eq. (14.5):

$$q_1 = \ln\left[\tan\left(\frac{\pi}{2}+\frac{40°}{2}\right)\left(\frac{1-\varepsilon\sin 40°}{1+\varepsilon\sin 40°}\right)^{\varepsilon/2}\right] = \ln[(\tan 65°)(0.8995457012)^{\varepsilon/2}]$$

$$= \ln[(2.144506921)(0.9956545965)] = \ln(2.135188173) = 0.7585547800 \text{ rad}$$

$$q_2 = \ln\left[\tan\left(\frac{\pi}{2}+\frac{41°40'}{2}\right)\left(\frac{1-\varepsilon\sin 41°40'}{1+\varepsilon\sin 41°40'}\right)^{\varepsilon/2}\right] = \ln[(2.2285672597)(0.8962846298)^{\varepsilon/2}]$$

$$= \ln[(2.228567597)(0.995505858)] = \ln(2.218552098) = 0.7968547750 \text{ rad}$$

Compute $\ell$ using $q_1$ and $q_2$ above in Eq. (14.9).

$$\ell = \frac{\ln N_1 - \ln N_2 + \ln\cos\phi_1 - \ln\cos\phi_2}{q_2 - q_1} = \frac{15.66979777 - 15.66989545 - 0.2665150909 + 0.2916565319}{0.7968547750 - 0.7585547800}$$

$$= \frac{0.02504376100}{0.03829999500} = 0.6538841846$$

Note that calculated $\ell$ varies in the last four places from the value given in the projection tables where $\ell = 0.6538843192$. Compute the isometric latitude $q$ for the point 9. Use Eq. (14.5).

$$q_9 = \ln\left[\tan\left(\frac{\pi}{2}+\frac{41°20'10.403''}{2}\right)\left(\frac{1-\varepsilon\sin\phi}{1+\varepsilon\sin\phi}\right)^{\varepsilon/2}\right]$$

$$= \ln[(2.211471987)(0.8969237692)^{\varepsilon/2}] = \ln[(2.211471987)(0.9955350504)]$$

$$= \ln[2.20159786] = 0.7891834040 \text{ rad}$$

Next, calculate $K$ with Eq. (14.10) using the value of $\ell$ computed above.

$$K = \frac{N_1\cos\phi_1}{\ell}e^{\ell q_1} = \frac{N_2\cos\phi_2}{\ell}e^{\ell q_2}$$

$$= \frac{(6{,}387{,}143.945)(0.7660444433)}{0.6538841846}(e^{\ell q_1}) = (7{,}482{,}725.907)(1.642151010) = 12{,}287{,}765.91 \text{ m}$$

Compute $R$ for the point using calculated $\ell$ in Eq. (14.11).

$$R = Ke^{-\ell q} = \frac{12{,}287{,}765.91}{1.675370855} = 7{,}334{,}355.778 \text{ m} = 24{,}062{,}798.88 \text{ ft} \quad (1 \text{ ft} = 0.30480061 \text{ m})$$

Calculate $\theta$ with Eq. (14.19).

$$\theta = \ell(\lambda_0 - \lambda) = (0.6538841846)[122° - (121°.0890391)]$$

$$= (0.6538841846)(0.9109609)$$

$$= 0°.5956629253$$

$$= 0°35'44.3865''$$

Finally, $X$ and $Y$ coordinates are computed with the values of $R$ and $\theta$ just calculated and using Eqs. (14.18).

$$X = 2{,}000{,}000 + (24{,}062{,}798.88)(0.01039609200) = 2{,}250{,}159.07 \text{ ft}$$

$$Y = 24{,}792{,}436.23 - (24{,}062{,}798.88)(0.9999459591)$$

$$= 24{,}792{,}436.23 - 24{,}061{,}498.50 = 730{,}937.73 \text{ ft}$$

where the $Y$ coordinate is seen to vary by 5.02 ft from the values calculated previously. This difference can be traced to the calculated constant $\ell$, which differed from tabulated $\ell$ for the projection in the seventh place. This apparently small discrepancy was sufficient to cause a difference between the calculated and tabulated values of $R$ of 5.03 ft.

The published values for $\ell$ were computed using logarithms, the best available method for performing this task at the time the tables were originally compiled. These tabulated values for $\ell$ are consistent throughout the tables for a given projection and provide *relative* positions for locations of points that are entirely adequate for users of the state plane coordinate systems.

The use of the rigorous equations is of primary value as an educational device to furnish an example of a theoretical basis for the projection equations and to illustrate the origin of the tabulated constants. However, it must be recognized that use of the rigorous equations throughout will yield answers different from those obtained with the tables. Consequently, if the rigorous equations are used, the tabulated value of $\ell$ for the projection must be utilized in Eqs. (14.10), (14.11), and (14.19) in the procedure [just as $R_b$ is used in Eq. (14.18)] in order to obtain reasonable agreement with coordinates calculated by the tables or the equations of Ref. 3.

**Example 14.2**   The $X$ and $Y$ California state plane coordinates, zone I, for control point 10 are $X = 2,269,500.94$ ft and $Y = 778,680.42$ ft. Compute the latitude and longitude for the point. First determine these values using the tables in Special Publication 253 (Table 14.1).

SOLUTION   First, calculate the longitude. From Table 14.1, $\ell = 0.6538843192$, $R_b = 24,792,436.23$ ft, $C = 2,000,000$ ft. From Fig. 14.12 and using Eq. (14.20),

$$\theta = \arctan \frac{X-c}{R_b - Y}$$

Thus,

$$\theta = \arctan \left[ \frac{(2,269,500.94) - (2,000,000)}{(24,792,436.23) - (778,680.42)} \right] = \arctan 0.01122277340 = 0.°6429905562$$

or

$\theta = 0°38'34.7660''$

With this value of $\theta$ and $\ell$ from the tables, compute $\lambda_0 - \lambda$ with Eq. (14.19), in which $\theta$ is expressed in seconds.

$$\lambda_0 - \lambda = \frac{\theta}{\ell} = \frac{0.°6429905562}{0.6538843192}$$

$$\lambda = 122° - 0.°9833399232 = 121.°0166601$$

$$= 121°00'59.9764''$$

Next compute the latitude. From the computation above,

$R_b - Y = 24,013,755.81$ ft

$\cos\theta = 0.9999370306$

Solve for $R$ using Eq. (14.21).

$$R = \frac{R_b - Y}{\cos\theta} = 24,015,268.04 \text{ ft}$$

Using the extract from Special Publication 253 in Table 14.1, determine the latitude $\phi$, by interpolation from Table I as follows:

$R$ for $41°27' = 24,021,339.47$

$R$ for $\phi$ $\qquad = 24,015,268.04$

$\qquad$ Difference $\qquad$ 6,071.43

Also from the tables, column 4, the tabular difference for $1''$ is 101.21000, so that

$$\text{difference in seconds} = \frac{6071.43}{101.21000''}$$

$$= 59.9884''$$

and the latitude is

$\phi = 41°27'59.9884''$

Next, consider the solution to this inverse problem using Eqs. (14.29) to (14.39) and constants in Table 14.2. Compute $\theta$ by Eq. (14.29):

$$\theta = \arctan \frac{(2,269,500.94 - 2,000,000.00)}{24,792,436.23 - 778,680.42} = \arctan(0.011222773) = 0°38'34.7660022''$$

Next, calculate $\lambda$ by Eq. (14.30).

$$\lambda = 439,200.00'' - \frac{21,314.766002''}{0.6538843192} = 435,659.9763'' = 121°00'59.9764''$$

which checks the value calculated previously. From Eq. (14.31),

$$R = \frac{24,792,436.23 - 778,680.42}{0.999937031} = 24,015,268.04 \text{ ft}$$

To obtain $s$, first $s_1$, $s_2$, and $s_3$ must be calculated sequentially as follows, using Eqs. (14.32a) to (14.32c):

$$s_1 = \frac{(24,792,436.23) - (24,245,358.05) - (778,680.42) + (2)(24,015,268.04)(0.000031484686)}{0.9998946358}$$

$$= -230,114.2595$$

$$s_2 = \frac{-230,114.2595}{1 + (-0.002301143)^2(3.80992) - (-0.002301143)^3(3.93575)}$$

$$= \frac{-230,114.2595}{1.000020223} = -230,109.6060$$

$$s_3 = \frac{-230,114.2595}{1 + (-0.002301096)^2(3.80992) - (-0.002301096)^3(3.93575)}$$

$$= \frac{-230,114.2595}{1.000020222} = -230,109.6062$$

Then, from Eq. (14.33),

$$s = \frac{230,114.2595}{1 + (-0.002301096)^2(3.80992) - (-0.002301095)^3(3.93575)}$$

$$= \frac{230,114.2595}{1.000020222} = -230,109.6062$$

Compute $\omega'$, the degrees and minutes of $\omega$ in whole minutes by Eq. (14.34).

$$\omega' = 2441 - 600 = 1841'$$

and

$$\omega'' = 36,000 + 26.75847 - 0.009873675553s$$

$$= 38,298.78606 \quad \text{(the remainder of } \omega \text{ in seconds)}$$

$$\omega = 1841' + 38,298.78606''$$

$$= 41°19'18.786036'' = 41°32188501$$

$$\phi' = 2441 - 600 = 1841'$$

$$\phi'' = 38,298.78606 + [1047.546710 + (6.19760 + 0.050912\cos^2\omega)\cos^2\omega]\sin\omega\cos\omega$$

$$= 38,819.98845$$

$$\phi = 1841' + \frac{38,819.98845''}{60} = 2487.999808'$$

$$= 41°27'59.98849''$$

This checks the value obtained from Table 14.1 by 0.0001''.

In the alternative procedure, without the tables, $\ell$ is computed as in Example 14.1 using Eqs. (14.2), (14.5), and (14.9). The mapping angle, $\theta$, is computed by Eq. (14.20) as in this example and $\lambda$ is calculated with Eq. (14.19), also as in this example. Thus, the procedure is the same to obtain $\lambda$ except a calculated value for the constant $\ell$ is employed. Solving for $\lambda$ using calculated $\ell$ in Eq. (14.20) yields 121°00'59.9756'' different from that previously computed using tabular values by 0.0008''.

To determine the latitude, $\phi$, without tables, the constant $K$ is computed with Eq. (14.10) as shown in Example 14.1. This value of $K$ converted to feet, and $R$ for the point calculated using Eq. (14.21) (computed previously in this example), are used in Eq. (14.23) to obtain the isometric latitude $q$ for the point. Thus,

$$q = \frac{\ln K - \ln R}{\ell}$$

$$= \frac{\ln\left(\dfrac{12,287,765.91}{0.30480061}\right) - \ln(24,015,268.04)}{\ell} = \frac{0.5180117900}{0.6538843192} = 0.7922070843 \text{ rad}$$

There is no direct computational procedure to get $\phi$ from the isometric latitude, $q$, since Eq. (14.5) cannot be inverted directly. Therefore, an iterative procedure such as the Newton-Raphson method must be applied to Eq. (14.5) to obtain $\phi$.

As in the direct solution for $X$ and $Y$ coordinates, the tabulated value of $\ell$ for the projection should be used in Eqs. (14.10), (14.19), and (14.23) if reasonable correspondence with latitudes and longitudes calculated with the tables or the equations from Ref. 3 is to be expected.

Obviously, the Lambert plane coordinate projection tables provide the most efficient direction solution, by electronic hand calculator, for $X$ and $Y$ coordinates given $\phi$ and $\lambda$ and for the inverse transformation.

Equations (14.24) to (14.28) and (14.29) to (14.39), when employed with constants in Table 14.2, are designed for use on an electronic computer or programmable desk or pocket calculator. The procedure using these equations is not necessarily the most efficient method for state plane coordinate computations using a hand-operated calculator.

The basic theory was developed and applied to demonstrate the origin of the constants in Table 14.1 and also to provide a guide for solving the problem on a programmable calculator or electronic computer. The tabular value of $\ell$ from the projection tables should be used in these computations.

Those interested in the origin of constants in Table 14.2 and method of deriving the equations should refer to Ref. 3.

### 14.18. Grid and geodetic azimuths and distances on the Lambert conformal projection

In Fig. 14.13, the line $AP$ from the apex of the cone to point $P$ is a meridian and all lines parallel to the central meridian are grid-north lines. The azimuth of a line between two points can be referred to a grid-north line by *grid azimuth* or to a meridian by *geodetic azimuth*. For lines of up to 5 mi or 8 km in length, grid and geodetic azimuths differ by the mapping angle $\theta$, as shown in Fig. 14.13. As demonstrated in Art. 14.17, $\theta$ can be obtained as a function of $\lambda$ from Table II in the state plane coordinate tables (Example 14.1); can be calculated as a function of $\lambda$ using Eqs. (14.26) and (14.29) with constants in Ref. 3; can be calculated with Eq. (14.19) (Example 14.1); or can be computed by Eq. (14.20) when the $X$ and $Y$ coordinates for the point are given (Example 14.2). The difference between geodetic and grid azimuth expressed as an equation is

$$\text{geodetic azimuth} - \text{grid azimuth} = \theta \tag{14.40}$$

in which the sign of $\theta$ is determined from Eq. (14.19).

When lines exceed 5 mi or 8 km in length a second term must be included in Eq. (14.40) so as to achieve the same accuracy as would be possible in geodetic computations. The modified equation is

geodetic azimuth − grid azimuth

$$= \theta - \frac{X_2 - X_1}{2\rho_0^2 \sin 1''}\left(Y_1 - Y_0 + \frac{Y_2 - Y_1}{3}\right) \tag{14.41}$$

for azimuth from point 1 to point 2, with map coordinates $(X_1, Y_1)$ and $(X_2, Y_2)$, respectively. The remaining quantities used in the second term of Eq. (14.41) are illustrated in Fig. 14.14.

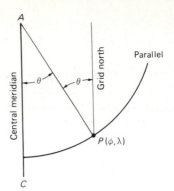

**Fig. 14.13** Geodetic and grid
azimuths on the Lambert projection.

The line labeled $\phi_0$ is a parallel of latitude slightly north of the midpoint between the two
standard parallels. The intersection of this parallel with the central meridian has a $Y$
coordinate of $Y_0$. East-west lines $GH$ and $LK$ lie north and south, respectively, of the
parallel of latitude $\phi_0$. Note that the geodetic line $GH$ curves outward (northward) from the
parallel $\phi_0$ when projected on the grid. The geodetic line $LK$ also curves away from
the parallel $\phi_0$. The second term in Eq. (14.41) allows for this curvature. The quantity $\rho_0$ is
the mean radius of the ellipsoid at latitude $\phi_0$. The constants $1/(\rho_0^2 \sin 1'')$ and $Y_0$ are given
in the projection tables for each zone.

Note that north-south lines $GL$ and $HK$ project onto the projection as straight lines. When
a line is close to or in a grid north or south direction, $X_2 - X_1$ in Eq. (14.41) is either very
small or zero, so that the correction term is negligible or vanishes.

Grid distances or distances computed on the map projection are calculated using the given
$X$ and $Y$ coordinates in Eq. (8.8) in Art. 8.15. In order to obtain the geodetic distance at sea
level, the scale factors $k$ are taken from the projection tables (Table 14.1, Table I) or are
calculated by Eq. (14.28) for the end points of the line. The relationship between grid and
geodetic distance at sea level is then

$$\text{grid distance} = (\text{geodetic distance})(\text{scale factor})$$

As a rule, the average of the two scale factors at the two ends of the line, $k_{av}$, is used in this
computation.

If it is necessary to convert grid distances to geodetic distances on the terrain or reduce
measured distances to grid distances, the elevation of the line above sea level must be known
and a sea-level correction is made [Eq. (10.24), Art. 10.17].

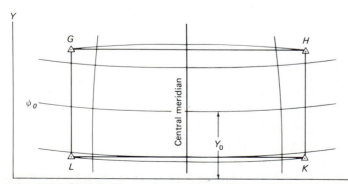

**Fig. 14.14** Geodetic lines on the Lambert projection. (*Adapted from Ref. 11.*)

A combined scale and sea-level correction factor may also be calculated where

$$\text{combined factor} = (\text{scale factor})(\text{sea-level correction factor})$$

or

$$\text{combined factor} = k_{\text{av}} \frac{R}{R+h}$$

in which in the second term, the sea-level correction factor, $R$, is the average radius of the earth (20,906,000 ft, 6,372,000 m) and $h$ the average elevation of the line or area.

**Example 14.3**   The coordinates for points 9 and 10 as calculated in Example 14.1 (using Table 14.1) and given in Example 14.2 are as follows:

| Point | X, ft | Y, ft |
|-------|-------|-------|
| 9 | 2,250,159.06 | 730,942.76 |
| 10 | 2,269,500.94 | 778,680.42 |

Determine the grid and the geodetic azimuth from both ends of the line. Also, compute the grid distance and geodetic sea-level distance between points 9 and 10.

SOLUTION   Since these are plane coordinates, the grid azimuth is given by Eq. (8.6). Thus, the grid azimuth from the south of the line from 9 to 10 is

$$\tan A = \frac{X_9 - X_{10}}{Y_9 - Y_{10}} = 0.4051702576$$

so that the azimuth from south is

$$A_{9\text{-}10} = 202°03'22.8''$$

The grid azimuth from south of the line from 10 to 9 is exactly 180° different from the value above. Thus,

$$A_{10\text{-}9} = 22°03'22.8''$$

Next, compute the geodetic azimuth of the line from 9 to 10 from the south by Eq. (14.41). From Example 14.1, $\theta = 0°35'44.3869''$. From Table 14.1,

$$\frac{1}{2\rho_0^2 \sin 1''} = (2.38)(10^{-10}) \quad \text{and} \quad Y_0 = 547,078.17$$

Thus, the second correction term in Eq. (14.41) is

$$= (19,341.88)(2.38)(10^{-10})\left(730,942.76 - 547,078.17 + \frac{47,737.66}{3}\right) = 0.92''$$

and by Eq. (14.41),

$$\text{geodetic azimuth} = (202°03'22.8'') + (0°35'44.39'') - (0.92'')$$

$$= 202°39'06.3''$$

Similarly, the geodetic azimuth of line 10 to 9 from the south is

$$\text{geodetic azimuth} = (22°03'22.8'') + (0°38'34.7660'')$$

$$- (-19,341.88)(2.38)(10^{-10})\left(778,680.42 - 547,078.17 - \frac{47,737.66}{3}\right)$$

$$= 22°03'22.8''$$

$$+ 0°38'34.77''$$

$$\underline{+ 0°00'00.99''}$$

$$\text{geodetic azimuth } 10\text{-}9 = 22°41'58.6''$$

Note that the geodetic azimuth from 9 to 10 does not differ from the geodetic azimuth from 10 to 9 by 180° as do corresponding grid azimuths. Figure 14.15 illustrates the grid and geodetic azimuths from

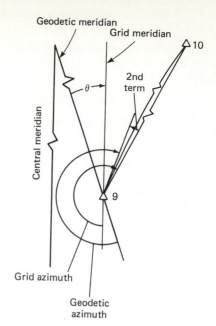

**Fig. 14.15** Geodetic azimuth with the correction term.

9 to 10 for Example 14.3. The student can make a similar sketch for the geodetic and grid azimuths from 10 to 9 as an exercise to verify, graphically, the signs of $\theta$ and the correction term in Eq. (14.41) for this part of Example 14.3.

The grid distance from 9 to 10 calculated by Eq. (8.8) is

$$d_{9\text{-}10} = [(X_{10} - X_9)^2 + (Y_{10} - Y_9)^2]^{1/2} = 51{,}507.208 \text{ ft}$$

The scale factor for station 9 calculated by Eq. (14.28) in Example 14.1 was found to be 0.9999328. The scale factor for 10 also calculated by Eq. (14.28) using $R$ and $\phi$ for 10 from Example 14.2 is

$$k_{10} = \frac{(0.6538843192)(24{,}015{,}268.04)[1 - 0.006768658 \sin^2(41°27'59.9885'')]^{1/2}}{(20{,}925{,}832.16)\cos(41°27'59.9885'')}$$

$$= 0.9999553$$

Thus, the average scale factor, $k_{av} = (0.9999328 + 0.9999553)/2 = 0.9999441$ and the geodetic distance at sea level is

$$d_{9\text{-}10}(\text{geod}) = \frac{51{,}507.208}{0.9999441} = 51{,}510.09 \text{ ft}$$

Assuming an average elevation of 2500 ft above mean sea level for the line, a sea-level correction factor can be calculated using Eq. (10.24):

$$\text{sea-level factor} = \frac{R}{R+h} = \frac{(20.906)10^6}{(20.906)10^6 + 2500} = 0.9998804$$

and the geodetic distance at average terrain elevation is

$$d_{9\text{-}10}(\text{terrain}) = \frac{51{,}510.09}{0.9998804} = 51{,}516.25 \text{ ft}$$

A combined factor could have been calculated as follows:

combined factor $= (0.9999441)(0.9998804) = 0.9998245$

and applied directly to the grid distance to yield

$$d_{9\text{-}10}(\text{geod}) = \frac{51{,}507.208}{0.9998245} = 51{,}516.25 \text{ ft}$$

Note that for a line of length cited in Ex. 14.3 (in excess of 5 mi or 8 km), a single average scale factor is not adequate. The line should be subdivided into shorter sections with a separate scale factor being calculated for each section (see Art. 14.23 and Ref. 17a).

**14.19. Transverse Mercator state plane coordinate projection**    The theory of the transverse Mercator projection is developed in Art. 14.13. Since the basic projection formulas [Eqs. (14.15)] are not usually used directly in the computation of state plane coordinates, some additional developments are required here to clarify use of the projection tables provided for each state plane system by the National Geodetic Survey. For the state systems, the Mercator projection cylinder cuts the surface of the sphere at two standard lines, parallel to and equally spaced from the central meridian as shown in Fig. 14.8b. The projection distance between points along these standard lines equals the distance between corresponding points on the sphere. Between the two standard lines, the projection distance is less than the corresponding distance on the sphere and the scale of the projection is too small. Beyond the two standard lines, the projection distance exceeds the corresponding distance on the sphere and the scale of the projection is too large. Thus, the scale on an east-west line varies from point to point while the scale on a north-south line is constant throughout even though it may not be 1:1. If the two standard circles are set so the distance between them is about 96 km (60 mi), and the total width of the projection is about 254 km (158 mi), then errors due to scale change throughout the projection will not exceed 1:10,000. Since the scale is constant along the meridian, this type of projection is used in states that have a long north-south dimension.

For the state plane transverse Mercator projections, the central meridian is taken as the $Y$ axis, which has an arbitrarily assigned $X$ coordinate of $X_0=500,000$ ft (current projection tables are in feet so all calculations must be in feet). In this manner there will be no negative $X$ coordinates within the system. The $X$ coordinate axis is along a line perpendicular to the central meridian at a latitude selected such that there will be no negative $Y$ coordinates within the system. Still assuming a sphere, the $Y$ coordinate for a point $H$ (refer to Fig. 14.16) would be the length of meridian from the origin to a perpendicular from the central meridian passing through $H$. The $X$ coordinate would be the length of this perpendicular from the central meridian to the point.

Now, it is necessary to recognize that the earth is not a sphere but is represented by an ellipsoid of revolution so the parameters for this figure of the earth must be taken into account. Theoretically, given the latitude and longitude $(\phi,\lambda)$ for $H$, Eqs. (14.15) could be used for direct calculation of $X$ and $Y$. However, when the state plane coordinate systems were established, these equations were not considered feasible due to the difficulty of the computations, particularly those required in the inverse solution for $\phi$ and $\lambda$ given $X$ and $Y$. Thus, an empirical approach based on these rigorous equations was adopted.

Several sets of tables have evolved over the years for use in the computations of state plane coordinates on a transverse Mercator projection as in the case of the Lambert

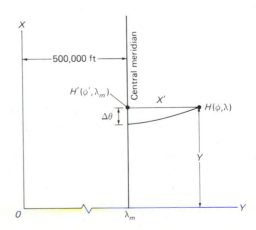

**Fig. 14.16** Coordinates on the transverse Mercator projection.

conformal projection discussed in Arts. 14.17 and 14.18 (Ref. 4). It should be emphasized that each set of tables was quite effective at the time of its use particularly because of the computational aids available. For our purposes here, we shall present the most recent tables, recognizing that perhaps in the not-too-distant future, the basic projection equations (14.15) to (14.17c) will be programmed on a digital computer and used directly.

National Ocean Survey Publication 62-4 for state plane coordinate processing (Ref. 3) contains formulas suitable for computing on the transverse Mercator projection using a programmable electronic calculator. For the forward computation of coordinates from latitude and longitude, the following equations are employed:

$$S_1 = \frac{30.92241720\cos\phi}{(1-0.0067686580\sin^2\phi)^{1/2}}\left[T_2-\lambda-3.9174\left(\frac{T_2-\lambda}{10^4}\right)^3\right] \qquad (14.42)$$

in which $\lambda$ is in seconds.

$$S_m = S_1 + 4.0831\left(\frac{S_1}{10^5}\right)^3 \qquad (14.43)$$

$$X = T_1 + 3.28083333\,S_m\,T_5 + \left(\frac{3.28083333\,S_m\,T_5}{10^5}\right)^3 T_6 \qquad (14.44)$$

$$\phi_1 = \phi + \frac{25.52381}{10^{10}}S_m^2(1-0.006768658\sin^2\phi)^2\tan\phi \qquad (14.45)$$

in which $\phi$ is in seconds.

$$\phi_2 = \phi + \frac{25.52381}{10^{10}}S_m^2(1-0.006768658\sin^2\phi_1)^2\tan\phi_1 \qquad (14.46)$$

$$Y = 101.2794065\,T_5\{60(\phi'-T_3)+\phi''-T_4$$
$$-\left[1052.893882-(4.483344-0.023520\cos^2\phi_2)\cos^2\phi_2\right]\sin\phi_2\cos\phi_2\} \qquad (14.47)$$

in which $\phi'$ is degrees and minutes of $\phi_2$ expressed in whole minutes and $\phi''$ is the remainder of $\phi_2$ in seconds. The convergence of the meridians, $\Delta\alpha$, is (Art. 14.20)

$$\Delta\alpha = (T_2-\lambda)\left[\sin\frac{\phi+\phi_2}{2} + \frac{1.9587}{10^{12}}(T_2-\lambda)^2\sin\left(\frac{\phi+\phi_2}{2}\right)\cos\left(\frac{\phi+\phi_2}{2}\right)\right] \qquad (14.48)$$

in which $\lambda$ is in seconds. The scale factor is

$$k = T_5\left[1 + \frac{(1+0.0068147849\cos^2\phi)^2}{881.749162\,T_5^2}\left(\frac{X-T_1}{10^6}\right)^2\right] \qquad (14.49)$$

The constant values for the $T$'s are given for the Arizona transverse Mercator projection eastern zone in Table 14.3. Readers interested in the details of these constants should refer to the Appendix of Publication 62-4 of the National Ocean Survey.

**Example 14.4** Compute $X$ and $Y$ coordinates on the transverse Mercator projection, Arizona eastern zone, for station Orbit, and calculate the convergence of the meridian at Orbit where

$\phi = 32°10'31.505''$
$\lambda = 109°40'10.230''$
Use Eqs. (14.42) to (14.47).

**Table 14.3   Constants for Transverse Mercator Projection Arizona Eastern Zone**

| | |
|---|---|
| $T_1$ | 500,000.00 |
| $T_2$ | 396,600.00 |
| $T_3$ | 1852. |
| $T_4$ | 16.62358 |
| $T_5$ | 0.9999 |
| $T_6$ | 0.3816485 |

SOLUTION   Compute $S_1$ by Eq. (14.42).

$$S_1 = \left[ \frac{30.92241720\cos(32°10'31.505)}{\{1 - 0.0067686580\sin^2(32°10'31.505'')\}^{1/2}} \right]$$
$$\times \left[ 396,600.00 - 394,810.230 - 3.9174\left( \frac{396,600.00 - 394,810.230}{10^4} \right)^3 \right]$$

Compute $S_m$ by Eq. (14.43).

$$S_1 = 46,888.80832$$

$$S_m = 46,888.80832 + 4.0831\left( \frac{46,888.80832}{10^5} \right)^3 = 46,889.22924$$

Save $S_m$, as it will be used in subsequent calculations.
   Compute $X$ by Eq. (14.44).

$$X = 500,000.00 + 3.28083333(46,889.22924)(0.9999)$$
$$+ \left[ \frac{(3.28083333)(46,889.22924)(0.9999)}{10^5} \right]^3 (0.3816485)$$
$$= 500,000.00 + 153,820.36 + 1.39 = 653,821.75 \text{ ft}$$

Next, determine $\phi_1$ and $\phi_2$ with Eqs. (14.45) and (14.46).

$$\phi_1 = 115,831.505 + \frac{25.52381}{10^{10}}(46,889.22924)^2(1 - 0.006768658\sin^2\phi)^2\tan\phi$$
$$= 115835.0220'' = 32°.17639500$$

Save this value.

$$\phi_2 = 115,831.505'' + \frac{25.52381}{10^{10}}(46,889.22924)^2(1 - 0.006768658\sin^2\phi_1)^2\tan\phi_1$$
$$= 115,835.0221 = 32.17639503$$

Save this value. Compute $Y$ by Eq. (14.47).

$$Y = (101.2794065)(0.9999)\{ 60(1930 - 1852) + 35.022108 - 16.62358$$
$$- [1052.893882 - (4.483344 - 0.023520\cos^2\phi_2)\cos^2\phi_2]\sin\phi_2\cos\phi_2\}$$
$$= (101.2794065)(0.999900000)(4225.261442)$$
$$= 427,889.18 \text{ ft}$$

The convergence of the meridian is computed by Eq. (14.48) using $\phi$ as given and $\phi_2$ as calculated previously.

$$\Delta\alpha = (396,600.00 - 394,810.23)\left[ \sin\left( \frac{32.17541806 + 32.17639503}{2} \right) \right.$$
$$+ \frac{1.9587}{10^{12}}(396,600.00 - 394,810.23)^2$$
$$\left. \times \sin\left( \frac{32.17541806 + 32.17639503}{2} \right)\cos^2\left( \frac{32.17541806 + 32°.17639503}{2} \right) \right]$$

$$\Delta\alpha = (1789.77)[(0.532520396) + (0.000006274)(0.381509342)]$$
$$= (1789.77)(0.532522790) = 953.093 = 0°15'53.093''$$

The use of $\Delta\alpha$ is explained in Art. 14.20.

The equations for calculating geographic coordinates from plane coordinates are:

$$Sg_1 = X - T_1 - T_6 \left( \frac{X - T_1}{10^5} \right)^3 \tag{14.50}$$

$$S_m = \frac{0.3048006099}{T_5} \left[ X - T_1 - T_6 \left[ \frac{Sg_1}{10^5} \right]^3 \right] \tag{14.51}$$

$$\omega' = T_3 \quad \text{(degrees and minutes of } \omega \text{ in whole minutes)} \tag{14.52}$$

$$\omega'' = T_4 + \left( \frac{0.009873675553}{T_5} \right) Y \tag{14.53}$$

$$\omega = \omega' + \omega''$$
$$(\phi')' = T_3 \quad \text{(degrees and minutes of } \phi \text{ and } \phi' \text{ in whole minutes)}$$
$$(\phi')'' = \omega'' + [1047.546710 + (6.192760 + 0.050912 \cos^2 \omega) \cos^2 \omega] \sin \omega \cos \omega \tag{14.54}$$

where $(\phi')''$ is the remainder of $\phi'$ in seconds.

$$\phi' = (\phi')' + (\phi')'' \tag{14.55}$$

$$(\phi)'' = (\phi')'' - 25.52381(1 - 0.006768658 \sin^2 \phi')^2 \left( \frac{Sm}{10^5} \right)^2 \tan \phi' \tag{14.56}$$

$$\phi = (\phi')' + (\phi)'' \tag{14.57}$$

$$Sa = Sm - 4.0831 \left( \frac{Sm}{10^5} \right)^3 \tag{14.58}$$

$$S_1 = Sm - 4.0831 \left( \frac{Sa}{10^5} \right)^3 \tag{14.59}$$

$$\Delta\lambda_1 = \frac{S_1(1 - 0.006768658 \sin^2 \phi)^{1/2}}{30.92241724 \cos \phi} \tag{14.60}$$

$$\Delta\lambda_a = \Delta\lambda_1 + 3.9174 \left( \frac{\Delta\lambda_1}{10^4} \right)^3 \tag{14.61}$$

$$\lambda'' = T_2 - \Delta\lambda_1 - 3.9174 \left( \frac{\Delta\lambda_a}{10^4} \right)^3 \tag{14.62}$$

**Example 14.5** Compute latitude and longitude for station Roger, where
$X = 692,435.81$
$Y = 353,400.22$
on the transverse Mercator projection for Arizona eastern zone. Use Eqs. (14.50) to (14.62).

SOLUTION

$$Sg_1 = 692,435.81 - 500,000.00 - 0.3816485 \left( \frac{692,435.81 - 500,000.00}{10^5} \right)^3 = 192,433.0903$$

$$S_m = \frac{0.3048006099}{0.9999} \left[ 692,435.81 - 5,000,000.00 - 0.3816485 \left( \frac{192,433.090}{10^5} \right)^3 \right] = 58,659.5893$$

$$\omega' = 1852'$$

$$\omega'' = 16.62358 + \left( \frac{0.009873675553}{0.9999} \right)(353,400.22) = 3506.331668''$$

$$\omega = 1852' + 3506.331660'' = 1910.438861' = 31°84064768$$
$$(\phi')' = T_3 = 1852'$$

$$(\phi')'' = 3506.3317 + [1047.546710 + (6.192760 + 0.050912 \cos^2 \omega) \cos^2 \omega] \sin \omega \cos \omega$$
$$= 3506.331664 + (1052.042429) \sin(31°84064768) \cos(31.84064768)$$
$$= 3506.331664 + 471.49477 = 3977.82643''$$
$$\phi' = 1852' + 3977.82643'' = 31.97161845°$$

$$(\phi)'' = 3977.82643'' - 25.52381[1 - 0.006768658 \sin^2(31.97161845)]^2$$
$$\times \left( \frac{58,659.5893}{10^5} \right)^2 \tan(31.97161845°) = 3977.82643 - 5.46114911 = 3972.36528''$$

$$\phi = 1852' + 3972.36528'' = 31.97010147° = 31°58'12.365''$$

Longitude is computed using Eqs. (14.58) to (14.62) as follows:

$$S_a = 58{,}659.5893 - 4.0831\left(\frac{58{,}659.5893}{10^5}\right)^3 = 58{,}659.5893 - 0.8241515 = 58{,}658.76515$$

$$S_1 = 58{,}659.5893 - 4.0831\left(\frac{58{,}658.76515}{10^5}\right)^3 = 58{,}659.5893 - 0.8241168 = 58{,}658.76518$$

$$\Delta\lambda_1 = \frac{(58{,}658.76518)[1 - 0.006768658\sin^2(31°58'12.365)]^{1/2}}{(30.92241724)\cos(31°58'12.365'')} = 2234.009547''$$

$$\Delta\lambda_a = 2234.009547 + 3.9174\left(\frac{2234.009547}{10^4}\right)^3 = 2234.0523224$$

$$\lambda'' = 396{,}600.00 - 2234.009547 - 3.9174\left(\frac{2234.0523224}{10^4}\right)^3$$
$$= 394{,}365.9468'' = 109.5460963° = 109°32'45.947''$$

Computation of state plane coordinates on the transverse Mercator projection by the presented method is adaptable to programmable calculators and electronic computers, since only a limited number of constants (six) are required. These constants are listed in Coast and Geodetic Survey Publication 62-4 for all states that have the transverse Mercator projection. The method is also adaptable for nonprogrammable pocket calculators, but the equations are long and susceptible to mistakes.

From a theoretical standpoint, the basic projection equations (14.15) (Art. 14.13), which are adaptable to automatic data processing, should provide the best results. However, if Eqs. (14.15) were to be programmed, the results obtained would not correspond exactly to those obtained by the outlined method. The differences would be due to the empirical determination of constants in the original projection tables. This is no cause for concern, because these tables and formulations provide results that are well within the relative accuracies of 1 part in 10,000 specified for the state plane coordinate systems.

**14.20. Grid and geodetic azimuth on the transverse Mercator projection**   Grid and geodetic azimuths are illustrated for the transverse Mercator projection in Fig. 14.17. Station $H$ has a latitude and longitude of $\phi$ and $\lambda$. The projection of $H$ on the central meridian is at $H'$, which has latitude of $\phi'$ and longitude $\lambda_m$. The parallel of latitude passing through $H$ strikes the central meridian at $J$ with geographic coordinates of $\phi$ and $\lambda_m$. The straight line passing through $H$ and parallel to the central meridian is a grid-north line from which grid azimuths are measured. The curved line passing through $H$ concave toward the central meridian is a true meridian from which geodetic azimuths are reckoned. A line tangent to the true meridian at $H$ makes an angle with the grid meridian of $\Delta\alpha$ which is calculated by Eq. (14.48) using the constants given in Publication 62-4 for the projection. This factor $\Delta\alpha$, frequently designated as the *convergence* of the meridians for this projection, is used in the conversion from grid to geodetic and from geodetic to grid azimuths.

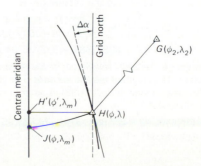

**Fig. 14.17** Geodetic and grid azimuths on the transverse Mercator projection.

The equation for azimuth on the transverse Mercator projection is

$$\text{geodetic azimuth} = \text{grid azimuth} + \Delta\alpha$$
$$+ \frac{(Y_2 - Y_1)(2X_1' + X_2')}{6\rho_0^2 \sin 1''} \tag{14.63}$$

in which $\Delta\alpha$ is computed by Eq. (14.48); $X_1', X_2'$ (referred to the projection central meridian, Fig. 14.16) and $Y_1, Y_2$ are coordinates of the origin and terminus of the line; and $\rho_0^2 = R_m N_m$. The value for $N_m$ is calculated by Eq. (14.2) and

$$R_m = \frac{a(1 - \varepsilon^2)}{(1 - \varepsilon^2 \sin^2 \phi_0)^{3/2}} \tag{14.64}$$

where $\phi_0$ is the average latitude for the area of the projection, and $a$ and $\varepsilon$ are the semimajor axis and eccentricity, as defined previously. The value of $\phi_0$ is not given in Publication 62-4, but the constant $T_6$ is found by

$$T_6 = \left(\tfrac{1}{6} R_m N_m T_5^2\right)(10^{15}) \tag{14.65}$$

so that

$$\frac{1}{6 R_m N_m \sin 1''} = \frac{T_6 (T_5)^2}{10^{15} \sin 1''}$$

or

$$\frac{1}{6\rho_0^2 \sin 1''} = \frac{(T_6)(T_5)^2}{10^{15} \sin 1''} \tag{14.66}$$

The third term of Eq. (14.63) is insignificant for lines that do not exceed 5 mi or 8 km in length. It reaches a maximum for long lines which are predominantly of a north-south direction. The third term is zero for an east-west line ($Y_2 - Y_1 = 0$) and is very small even for long lines where the line is predominantly east-west.

**Example 14.6** Determine grid and geodetic azimuths (from the south) of the line from station Orbit (Example 14.4) to station Roger (Example 14.5) which have state plane coordinates on the transverse Mercator projection of Arizona east zone as follows:

| Station | X, ft | Y, ft |
|---------|-----------|------------|
| Orbit | 653,821.75 | 427,889.18 |
| Roger | 692,435.81 | 353,400.22 |

Also, calculate the grid and geodetic distances between the two stations.

SOLUTION   Orbit is the origin (station 1) and Roger the terminus of the line (station 2). Thus, the grid azimuth from the south of the line Orbit-Roger is

$$A_{12}(\text{south}) = \arctan \frac{X_1 - X_2}{Y_1 - Y_2} = \frac{653,821.75 - 692,435.81}{427,889.18 - 353,400.22}$$
$$= \arctan(-0.5183863488) = 332°35'54.215''$$

The convergence $\Delta\alpha$ was calculated in Example 14.4 using Eq. (14.48) and was found to be $0°15'53.093''$. The length of the line exceeds 5 mi (it is 15.9 mi or 25.6 km), so that it is necessary to evaluate the second term of Eq. (14.63) in order to calculate the correct geodetic azimuth. Evaluation of this term follows:

$$Y_2 - Y_1 = (353,400.22 - 427,889.18) = -74,488.96$$
$$2X_1' + X_2' = (2)(653,821.75 - 500,000.00) + (692,435.81 - 500,000.00) = 500,079.31$$

and by Eq. (14.66) using $T_6$ and $T_5$ from Table 14.3,

$$\frac{1}{6\rho_0^2 \sin 1''} = \frac{(0.3816485)(0.9999)^2}{(10^{15})(4.848)(10^{-6})} = (0.78707)(10^{-10})$$

so that the second term in Eq. (14.63) is

$$\frac{(Y_2 - Y_1)(2X_1' + X_2')}{6\rho_0^2 \sin 1''} = (-74,488.96)(500,079.31)(0.78707)(10^{-10}) = -2.932''$$

By Eq. (14.63) the geodetic azimuth of the line Orbit-Roger from the south is then

geodetic azimuth $= 332°35'54.215'' + 0°15'53.093'' - 0°00'02.932'' = 332°51'44.376''$

Figure 14.18 illustrates the relationship between grid and geodetic azimuths for the line from Orbit to Roger. The grid line on the projection is represented by the straight line from Orbit to Roger. Thus, the grid azimuth is measured from grid south to the grid line. The geodetic line is a curve concave toward the central meridian. The geodetic azimuth is measured from the geodetic meridian (south) to a tangent to the curving geodetic line at station Orbit. The third term in Eq. (14.63) is the angle from the grid line between the points to a tangent to the geodetic line at the station which is the origin for the azimuth.

The grid distance between Orbit and Roger is

$$d_{12} = [(X_1 - X_2)^2 + (Y_1 - Y_2)^2]^{1/2} = [(653,821.75 - 692,435.81)^2$$
$$+ (427,889.18 - 353,400.22)^2]^{1/2} = 83,902,63 \text{ ft}$$

In order to obtain the geodetic distance between the two points, the grid distance must be modified by the average of the scale factors $k_1$ and $k_2$ computed for Orbit and Roger by Eq. (14.49). Using $T_1$ and $T_5$ from Table 14.3 and $\phi$ for Orbit from Example 14.4:

$T_1 = 500,000.00$      $\phi = 32°10'31.505''$

$T_5 = 0.9999$      $\cos^2\phi = 0.716429713$

$k$ for Orbit by Eq. (14.49) is

$$k_1 = 0.9999\left[ 1 + \frac{(1 + 0.0068147849\cos^2\phi)^2}{(881.749162)(0.9999)^2}\left( \frac{653,821.75 - 500,000.00}{10^6} \right)^2 \right]$$
$$= 0.9999270973$$

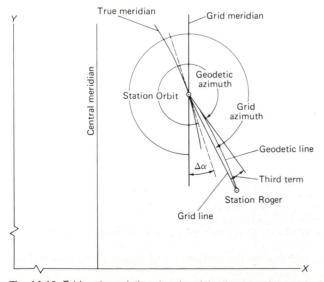

**Fig. 14.18** Grid and geodetic azimuths of the line from Orbit to Roger on the transverse Mercator projection.

The latitude for station Roger from Example 14.5 is

$$\phi = 31°58'12.365'' \qquad \text{and} \qquad \cos^2\phi = 0.7196544710$$

Thus, $k$ for Roger is

$$k_2 = 0.9999\left[1 + \frac{(1 + 0.0068147849\cos^2\phi)^2}{(881.749162)(0.9999)^2}\left(\frac{692{,}435.81 - 500{,}000.00}{10^6}\right)^2\right]$$

$$= 0.9999424148$$

$$\frac{k_1 + k_2}{2} = 0.999934756$$

The grid distance is divided by this average $k$ to yield the geodetic distance at sea level. Thus, geodetic distance Orbit to Roger $= 83{,}902.63/0.9999347560 = 83{,}908.10$ ft at sea level.

For a line of this length and orientation, the difference between grid and geodetic lengths is substantial. When a geodetic distance is reduced to the grid, the scale factor $k$ is multiplied times the sea level geodetic distance.

Since the state plane coordinate systems are in general computed at sea level, the geodetic distance from Orbit to Roger is a scaled sea-level distance. (All the state plane coordinate systems are referred to sea level except the Michigan Lambert zones, which are computed for the 800-ft elevation above mean sea level.)

**14.21. Computation of coordinates on the universal transverse Mercator projection**   The basic equations for the transverse Mercator projection are given in Art. 14.13, and the universal transverse Mercator (UTM) projection is described in Art. 14.14. Coordinates using the UTM projection can be calculated on any one of five spheroids (International, Clarke 1866, Clarke 1880, Everest, and Bessel), so this system is applicable anywhere in the world. In the United States, the Clarke 1866 spheroid is almost always specified.

The conversion from latitude and longitude to rectangular coordinates and the inverse solution can of course be determined by Eqs. (14.15) and (14.17a). Tabular values based on these equations have been compiled to simplify the computational procedure. The use of these tables is illustrated in this section by example transformation from geographic positions to rectangular coordinates and the inverse solution. The results of these computations are then checked by solving Eqs. (14.15) and (14.17a) directly.

First the appropriate spheroid must be chosen. A different set of tables is required for each spheroid. Example problems in this section are computed on the Clarke 1866 spheroid.

Next, the zone number and longitude of the central meridian are chosen from a list of zone numbers, central, and bounding meridians, as shown in Table 14.4. For example, a point with longitude $77°25'25.452''$ is in zone 18, where the longitude of the central meridian $\lambda_0$ is 75°W. In case this table is not available, the zone number and $\lambda_0$ (longitude of the central meridian) can be computed using

$$\text{zone number} = \frac{180° - \lambda}{6} \tag{14.67}$$

Thus, for the example given,

$$\text{zone number} = \frac{180° - 77°25'}{6} = \frac{102°35'}{6} = 17 +$$

So, zone 18 must be used. Zone 18 is $(6°)(18)$, or $108°$ east of $180°$ longitude and ranges from $72°$ to $78°$ (UTM zones are $6°$ wide). Consequently, the central meridian is 75°W.

**Transformation of geographic to rectangular coordinates**   The equations for determining $X, Y$ coordinates, given $\phi$ and $\lambda$, are [Ref. 17(c)]

$$X = 500{,}000.000 + (\text{IV})p + (\text{V})p^3 + B_5$$

$$Y = (\text{I}) + (\text{II})p^2 + (\text{III})p^4 + A_6 \qquad \text{(north of equator)}$$

**Table 14.4 UTM Zone Numbers with Bounding and Central Meridians (*Adapted from Ref. 17a*)**

| Zone | Central meridian | Bounding meridians | Zone | Central meridian | Bounding meridians | Zone | Central meridian | Bounding meridians |
|---|---|---|---|---|---|---|---|---|
| | | 180° | | | 60°W | | | 60°E |
| 1 | 177°W | 174°W | 21 | 57°W | 54°W | 41 | 63°E | 66°E |
| | | 174°W | | | 54°W | | | 66°E |
| 2 | 171°W | 168°W | 22 | 51°W | 48°W | 42 | 69°E | 72°E |
| | | 168°W | | | 48°W | | | 72°E |
| 3 | 165°W | 162°W | 23 | 45°W | 42°W | 43 | 75°E | 78°E |
| | | 162°W | | | 42°W | | | 78°E |
| 4 | 159°W | 156°W | 24 | 39°W | 36°W | 44 | 81°E | 84°E |
| | | 156°W | | | 36°W | | | 84°E |
| 5 | 153°W | 150°W | 25 | 33°W | 30°W | 45 | 87°E | 90°E |
| | | 150°W | | | 30°W | | | 90°E |
| 6 | 147°W | 144°W | 26 | 27°W | 24°W | 46 | 93°E | 96°E |
| | | 144°W | | | 24°W | | | 96°E |
| 7 | 141°W | 138°W | 27 | 21°W | 18°W | 47 | 99°E | 102°E |
| | | 138°W | | | 18°W | | | 102°E |
| 8 | 135°W | 132°W | 28 | 15°W | 12°W | 48 | 105°E | 108°E |
| | | 132°W | | | 12°W | | | 108°E |
| 9 | 129°W | 126°W | 29 | 09°W | 06°W | 49 | 111°E | 114°E |
| | | 126°W | | | 06°W | | | 114°E |
| 10 | 123°W | 120°W | 30 | 03°W | 00° | 50 | 117°E | 120°E |
| | | 120°W | | | 00° | | | 120°E |
| 11 | 117°W | 114°W | 31 | 03°E | 06°E | 51 | 123°E | 126°E |
| | | 114°W | | | 06°E | | | 126°E |
| 12 | 111°W | 108°W | 32 | 09°E | 12°E | 52 | 129°E | 132°E |
| | | 108°W | | | 12°E | | | 132°E |
| 13 | 105°W | 102°W | 33 | 15°E | 18°E | 53 | 135°E | 138°E |
| | | 102°W | | | 18°E | | | 138°E |
| 14 | 99°W | 96°W | 34 | 21°E | 24°E | 54 | 141°E | 144°E |
| | | 96°W | | | 24°E | | | 144°E |
| 15 | 93°W | 90°W | 35 | 27°E | 30°E | 55 | 147°E | 150°E |
| | | 90°W | | | 30°E | | | 150°E |
| 16 | 87°W | 84°W | 36 | 33°E | 36°E | 56 | 153°E | 156°E |
| | | 84°W | | | 36°E | | | 156°E |
| 17 | 81°W | 78°W | 37 | 39°E | 42°E | 57 | 159°E | 162°E |
| | | 78°W | | | 42°E | | | 162°E |
| 18 | 75°W | 72°W | 38 | 45°E | 48°E | 59 | 165°E | 168°E |
| | | 72°W | | | 48°E | | | 168°E |
| 19 | 69°W | 66°W | 39 | 51°E | 54°E | 59 | 171°E | 174°E |
| | | 66°W | | | 54°E | | | 174°E |
| 20 | 63°W | 60°W | 40 | 57°E | 60°E | 60 | 177°E | 180° |

and

$$Y = 10,000,000 - \left[ I + (II)p^2 + (III)p^4 + A_6 \right] \quad \text{(south of equator)} \qquad (14.68)$$

in which factors (I), (II), (III), (IV), and (V) are values tabulated as functions of latitude in tables prepared for a specific spheroid; $p = 0.0001\Delta\lambda''$; $B_5$ is a correction factor taken from a nomogram with $\Delta\lambda$ and $\phi$ as arguments; and $A_6$ is a correction factor taken from a bar graph using $\Delta\lambda$ as the argument. Note that $\Delta\lambda = \lambda'$, as defined in Art. 14.13. Table 14.5 illustrates an extract from tables of UTM factors for the Clarke 1866 spheroid [Ref. 17(c)].

Factor (I) is the *grid* distance along the central meridian from the equator to the intersection of the parallel of latitude through the point with the central meridian (point $J$, Fig. 14.17). As latitude increases, this distance becomes greater.

## Table 14.5 UTM Projection Tables: Clarke 1866 Spheroid, Metres
### (Adapted from Ref. 17c)

$$p = 0.0001\Delta\lambda'' \qquad X' = (IV)p + (V)p^3 + B_5$$

| Latitude | (IV) | Diff. 1" | (V) | Diff. 1" |
|---|---|---|---|---|
| 43°00' | 226,418.474 | − 1.02007 | 6.360 | − 0.00089 |
| 01 | 226,357.270 | 1.02039 | 6.307 | 0.00089 |
| 02 | 226,296.046 | 1.02071 | 6.253 | 0.00089 |
| 03 | 226,234.803 | 1.02103 | 6.200 | 0.00089 |
| 04 | 226,173.541 | 1.02136 | 6.147 | 0.00089 |
| 43°05' | 226,112.260 | − 1.02168 | 6.094 | − 0.00089 |
| 06 | 226,050.959 | 1.02200 | 6.040 | 0.00089 |
| 07 | 225,989.640 | 1.02232 | 5.987 | 0.00089 |
| 08 | 225,928.301 | 1.02264 | 5.934 | 0.00089 |
| 09 | 225,866.942 | 1.02296 | 5.881 | 0.00089 |
| 43°10' | 225,805.565 | − 1.02328 | 5.828 | − 0.00089 |
| 11 | 225,744.168 | 1.02360 | 5.775 | 0.00088 |
| 12 | 225,682.752 | 1.02392 | 5.721 | 0.00088 |
| 13 | 225,621.317 | 1.02424 | 5.668 | 0.00088 |
| 14 | 225,559.862 | 1.02456 | 5.615 | 0.00088 |
| 43°15' | 225,498.389 | − 1.02488 | 5.562 | − 0.00088 |
| 16 | 225,436.896 | 1.02520 | 5.509 | 0.00088 |
| 17 | 225,375.384 | 1.02552 | 5.456 | 0.00088 |
| 18 | 225,313.853 | 1.02584 | 5.403 | 0.00088 |
| 19 | 225,252.303 | 1.02616 | 5.351 | 0.00088 |
| 43°20' | 225,190.733 | − 1.02648 | 5.298 | − 0.00088 |
| 21 | 225,129.145 | 1.02680 | 5.245 | 0.00088 |
| 22 | 225,067.537 | 1.02712 | 5.192 | 0.00088 |
| 23 | 225,005.910 | 1.02744 | 5.139 | 0.00088 |
| 24 | 224,944.264 | 1.02775 | 5.086 | 0.00088 |
| 43°25' | 224,882.599 | − 1.02807 | 5.034 | − 0.00088 |
| 26 | 224,820.914 | 1.02839 | 4.981 | 0.00088 |
| 27 | 224,759.211 | 1.02871 | 4.928 | 0.00088 |
| 28 | 224,697.488 | 1.02903 | 4.876 | 0.00088 |
| 29 | 224,635.746 | 1.02935 | 4.823 | 0.00088 |

$\Delta^2$(IV)
Units of (IV)

```
        ┌ .003
  30" ─┤   +
20" 40" ┼ .002
 10" 50" ┤
        ├ .001
        │   +
   0" ─┴ .000
```

$A_6$, metres

```
        ┌ .004
        │   +
        ┤ .003
        │
        ┤ .002
        │
 3°30' ┤ .001
 3°00' ┴ .000
```

$B_5$

(Chart with axes labeled 43°, 30', 44° across the top and bottom; vertical axis 1°00', 2°00', 2°20', 2°30', 2°40', 2°50', 2°55', 3°00', 3°05', 3°10', 3°15', 3°20', 3°22', 3°24', 3°26', 3°28', 3°30'; right-side scale .000, —, .010, .020, .030, .040, .050, .060, .070, .080, .090, —, .100. Horizontal axis labeled Metres.)

North of Equator: $Y = (I) + (II)p^2 + (III)p^4 + A_6$

South of Equator: $Y = 10{,}000{,}000 - [(I) + (II)p^2 + (III)p^4 + A_6]$

$$p = 0.0001\Delta\lambda''$$

| Latitude | (I) | Diff. 1" | (II) | Diff. 1" | (III) |
|---|---|---|---|---|---|
| 43°00' | 4,760,599.818 | 30.84642 | 3743.174 | 0.00259 | 1.633 |
| 01 | 4,762,450.603 | 30.84652 | 3743.330 | 0.00257 | 1.631 |
| 02 | 4,764,301.395 | 30.84661 | 3743.484 | 0.00255 | 1.630 |
| 03 | 4,766,152.191 | 30.84671 | 3743.637 | 0.00253 | 1.629 |
| 04 | 4,768,002.994 | 30.84679 | 3743.788 | 0.00250 | 1.628 |
| 43°05' | 4,769,853.801 | 30.84689 | 3743.938 | 0.00248 | 1.627 |
| 06 | 4,771,704.614 | 30.84697 | 3744.087 | 0.00246 | 1.625 |
| 07 | 4,773,555.433 | 30.84706 | 3744.235 | 0.00244 | 1.624 |
| 08 | 4,775,406.256 | 30.84716 | 3744.382 | 0.00242 | 1.623 |
| 09 | 4,777,257.085 | 30.84726 | 3744.527 | 0.00240 | 1.622 |
| 43°10' | 4,779,107.921 | 30.84734 | 3744.671 | 0.00238 | 1.620 |

Factors $(II)p^2$, $(III)p^3$, and $A_6$ represent the *grid* distance along the central meridian from the intersection of the parallel with the meridian to the foot of a perpendicular to the meridian also passing through the point (distance $JH'$, Fig. 14.17). The term $A_6$ is not necessary for all problems.

Term $(IV)p$ is an approximation to the grid distance from the central meridian to the point. Terms $(V)p^3$ and $B_5$ are corrections to this approximation to give the correct $X$ coordinate. Term $(IV)$ becomes smaller as the latitude increases. Term $(V)$ also decreases with increasing latitude until $\phi = 63°55'$, when it increases with increasing latitude.

Terms $(I) \cdots (V)$ can also be correlated with basic equations (14.15) as follows:

$$(I) = S_\phi k_0$$

$$(II)p^2 = \left( N\frac{\lambda'^2}{2} \sin\phi\cos\phi \right) k_0$$

$$(III)p^4 = \left[ N\frac{\lambda'^4}{24} \sin\phi\cos^3\phi(5 - t^2 + 9\eta^2 + 4\eta^4) \right] k_0$$

$$A_6 = \left[ N\frac{\lambda'^6}{720} \sin\phi\cos^5\phi(61 - 58t^2 + t^4 + 270\eta - 330t^2\eta^2 + 445\eta^4) \right] k_0$$

in which $N$, $\lambda'$, $t$, and $\eta^2$ are defined for Eqs. (14.15), Art. 14.13, and $k_0 = 0.9996$ is the scale factor of the UTM projection along the central meridian. Similarly,

$$(IV)p = [N\lambda'\cos\phi]k_0$$

$$(V)p^3 = \left[ N\frac{\lambda'^3\cos^3\phi}{6}(1 - t^2 + \eta^2) \right] k_0$$

$$B_5 = \left[ N\frac{\lambda'^5}{120}(5 - 18t^2 + t^4 + 14\eta^2 - 58t^2\eta^2 + 13\eta^4) \right] k_0$$

In order to illustrate use of the tables and calculation of UTM coordinates directly by the equations of Art. 14.13, consider an example problem.

**Example 14.7**   Determine the $X, Y$ coordinates on the UTM projection (Clarke 1866 spheroid) for a station (Astro) having latitude and longitude of

$\phi = 43°09'01.0054''$
$\lambda = 77°25'25.4521''$

SOLUTION   From Table 14.4 or using Eq. (14.67), this station falls in zone 18, which has $\lambda_0 = 75°$W. First compute $\lambda'$.

$\lambda' = \lambda_0 - \lambda = (75°00'00'') - 77°25'25.4521''$
$\quad = 2.4237367° = -2°25'25.4521'' = -8725.4521''$

where $\lambda' = \Delta\lambda$, as used in the UTM projection tables. Note that the sign is the reverse of that used in computing $\Delta\lambda$ for the transverse Mercator projection in the state plane coordinate systems. In the tables, $\Delta\lambda$ is always taken as positive. First compute the $X$ coordinate by the first of equations (14.68) and using Table 14.5 with $\phi$ as the argument (units are metres).

| | |
|---|---|
| Tabular (IV)   (even minute of $\phi$) = | 225,866.942 |
| Interpolation for seconds $= (-1.02296)(1.0054'') = -$ | 1.028 |
| $\Delta^2$(IV)   (from graph, Table 14.5) = | 0.000 |
| (IV) = | 225,865.914 |

| | |
|---|---|
| Tabular (V)   (even minutes of $\phi$) | $= 5.881$ |
| Interpolation for seconds $= (0.00089)(1.0054'') = 0.000$ | |
| (V) | $= 5.881$ |

$B_5$ from graph, Table 14.5; use $\lambda'$ and $\phi$ as arguments $= -0.015$

$$p = \quad 0.0001\Delta\lambda'' = (0.0001)(-8725.4521) = -0.87254521$$
$$p^2 = \quad 0.761335144$$
$$p^3 = -0.664299333$$
$$p^4 = \quad 0.579631201$$
$$\text{(IV)}p = -197{,}078.221$$
$$\text{(V)}p^3 = - \qquad 3.907$$
$$B_5 = - \qquad 0.015$$
$$X' = -197{,}082.143$$
$$X_0 = \quad 500{,}000.000$$
$$\overline{X = X_0 + X' = \quad 302{,}917.857 \text{ m}}$$

Next, calculate $Y$ with the second of equations (14.68) using Table 14.5 to obtain (I), (II), and (III) with $\phi$ as the argument.

$$\text{(I) to nearest minute of } \phi = 4{,}777{,}257.085$$
$$\text{Interpolation to seconds} = (30.84726)(1.0054) = \qquad 31.014$$
$$\overline{\text{(I)} = 4{,}777{,}288.099}$$
$$\text{(II) to nearest minute} = \qquad 3744.527$$
$$\text{Interpolation to seconds} = (0.00240)(1.0054) = \qquad 0.002$$
$$\overline{\text{(II)} = \qquad 3744.529}$$
$$\text{(III)} = \qquad 1.622$$
$$\text{(II)}p^2 = \qquad 2{,}850.842$$
$$\text{(III)}p^4 = \qquad 0.940$$
$$\overline{A_6 \text{ from graph} = \qquad 0.000}$$
$$Y = \text{(I)} + \text{(II)}p^2 + \text{(III)}p^4 + A_6 = 4{,}780{,}139.881 \text{ m}$$

In order to demonstrate the correlation between the tabular values and the rigorous equations, the problem will also be solved by Eqs. (14.15), Art. 14.13. The equations are repeated here for convenience.

$$\frac{X}{N} = \lambda' \cos\phi + \frac{\lambda'^3 \cos^3\phi}{6}(1 - t^2 + \eta^2)$$

$$+ \frac{\lambda'^5 \cos^5\phi}{120}(5 - 18t^2 + t^4 + 14\eta^2 - 58t^2\eta^2 + 13\eta^4) + \cdots$$

$$\frac{Y}{N} = \frac{S_\phi}{N} + \frac{\lambda'^2}{2}\sin\phi\cos\phi + \frac{\lambda'^4}{24}\sin\phi\cos^3\phi(5 - t^2 + 9\eta^2 + 4\eta^4)$$

$$+ \frac{\lambda'^6}{720}\sin\phi\cos^5\phi(61 - 58t^2 + t^4 + 270\eta^2 - 330t^2\eta^2 + 445\eta^4) + \cdots$$

By Eq. (14.2),

$$N = \frac{a}{(1 - \varepsilon^2\sin^2\phi)^{1/2}} \qquad \phi = 43°09'01.0054''$$

$a = 6{,}378{,}206.4$ m

$\varepsilon^2 = 0.006768658$  for the Clarke 1866 spheroid

$$N = \frac{6{,}378{,}206.4}{[1 - 0.006768658(\sin^2\phi)]^{1/2}} = 6{,}388{,}327.021 \text{ m}$$

$\lambda' = -8725.4521'' = 0.04230218553$ rad

$t = \tan\phi = 0.937430794 \qquad t^2 = 0.8787764937$

$$\eta^2 = \frac{\varepsilon^2\cos^2\phi}{1 - \varepsilon^2} = \frac{(0.006768658)\cos^2\phi}{1 - 0.006768658} = 0.003627246226$$

By definition, the scale of the UTM projection is fixed at $k_0 = 0.9996$ on the central meridian. Thus, each term in Eqs. (14.15) must be multiplied by $k_0$. First solve for $X'$.

$$N\lambda' \cos\phi k_0 \qquad\qquad\qquad\qquad = (IV)p \quad = -197,078.221$$

$$\frac{N(\lambda')^3 \cos^3\phi}{6}(1 - t^2 + \eta^2)k_0 \qquad = (V)p^3 \quad = -\qquad\quad 3.906$$

$$5 - 18t^2 + t^4 + 14\eta^2 - 58t^2\eta^2 + 13\eta^4 = -10.1797$$

$$\frac{N(\lambda')^5 \cos^5\phi}{120}(-10.1797)k_0 \qquad = B_5 \qquad = -\underline{\qquad 0.015}$$

$$X' \qquad\qquad\qquad\qquad = -197,082.142$$

$$X = X_0 + X' \qquad\qquad\qquad = \quad 302,917.858 \text{ m}$$

In the conversion to the $Y$ coordinate, it is necessary to evaluate $S_\phi$, where

$$S_\phi = a(A_0\phi - A_1\sin 2\phi + A_2\sin 4\phi - A_3\sin 6\phi + \cdots)$$

in which

$$A_0 = 1 - \frac{\varepsilon^2}{4} - \frac{3}{64}\varepsilon^4 - \frac{5}{256}\varepsilon^6 = 0.998305682$$

$$A_1 = \frac{3}{8}\varepsilon^2 + \frac{3}{32}\varepsilon^4 + \frac{45}{1024}\varepsilon^6 \;= 0.002542556$$

$$A_2 = \frac{15}{256}\varepsilon^4 + \frac{45}{1024}\varepsilon^6 \qquad = 0.000002698$$

$$A_3 = \frac{35}{3072}\varepsilon^6 \qquad\qquad\qquad = 0.000000004$$

$$\phi = 0.753114447 \text{ rad}$$

$$S_\phi = 6,378,206.4(0.749301525)(0.9996) = (I) = 4,777,288.100$$

$$N\frac{(\lambda')^2}{2}\sin\phi\cos\phi(0.9996) = (II)p^2 = \qquad 2850.842$$

$$5 - t^2 + 9\eta^2 + 4\eta^4 = 4.153921350$$

$$N\frac{(\lambda')^4}{24}\sin\phi\cos^3\phi(4.153921350)(0.9996) = (III)p^4 = \qquad\quad 0.940$$

$$Y = \overline{4,780,139.882}$$

    The answers obtained by both methods are the same to within 1 mm. The tables are more convenient for use with hand-operated, nonprogrammable calculators. Equations (14.15) should be used for programmable calculators and electronic computers.

**Transformation from rectangular to geographic coordinates**    When $X, Y$ UTM grid coordinates are to be transformed into latitude and longitude, the procedure is essentially the reverse of the solution just outlined for conversion of geographic to rectangular coordinates. Equations (14.17a) provide the basic relationship for this transformation. The equations for use with the tables (Table 14.6) are

$$\phi = \phi' - (VII)q^2 + (VIII)q^4 - D_6 \qquad\qquad (14.69a)$$

$$\Delta\lambda = (IX)q - (X)q^3 + E_5 \qquad\qquad\qquad (14.69b)$$

$$\lambda = \lambda_0 \pm \Delta\lambda \qquad\qquad\qquad\qquad\quad (14.69c)$$

in which $\phi'$ is the latitude of the foot of the perpendicular, from the station to the meridian. Thus, $\phi'$ is the latitude for which tabular factor (I) equals $Y$ for the station and, consequently, $\phi'$ can be determined by interpolation in Table 14.6 using $Y$ as the argument. The value of $q = 0.000001X'$, where $X' = X - 500,000.000$. The value of $X'$ is always taken as being positive. Factors (VII) and (VIII) are obtained from the tables using $\phi'$ as the argument. These values multiplied by $q^2$ and $q^4$, together with $D_6$ obtained from the graph

**Table 14.6   Universal Transverse Mercator Grid 43° (*Adapted from Ref. 17d*)**

$$q = 0.000001\,X' \qquad \phi = \phi' - (VII)q^2 + (VIII)q^4 - D_6$$

| Latitude | (I) | Diff. 1″ | (VII) | Diff. 1″ | (VIII) |
|---|---|---|---|---|---|
| 43°00′ | 4,760,599.818 | 30.84642 | 2367.075 | 0.02286 | 36.69 |
| 01 | 4,762,450.603 | 30.84652 | 2368.447 | 0.02287 | 36.72 |
| 02 | 4,764,301.395 | 30.84661 | 2369.819 | 0.02288 | 36.76 |
| 03 | 4,766,152.191 | 30.84671 | 2371.192 | 0.02290 | 36.79 |
| 04 | 4,768,002.994 | 30.94679 | 2372.566 | 0.02291 | 36.83 |
| 43°05′ | 4,769,853.801 | 30.84689 | 2373.940 | 0.02292 | 36.87 |
| 06 | 4,771,704.614 | 30.84697 | 2375.316 | 0.02293 | 36.90 |
| 07 | 4,773,555.433 | 30.84706 | 2376.692 | 0.02294 | 36.94 |
| 08 | 4,775,406.256 | 30.84716 | 2378.068 | 0.02296 | 36.97 |
| 09 | 4,777,257.085 | 30.84726 | 2379.446 | 0.02297 | 37.01 |
| 43°10′ | 4,779,107.921 | 30.84734 | 2380.824 | 0.02298 | 37.05 |
| 11 | 4,780,958.761 | 30.84742 | 2382.203 | 0.02299 | 37.08 |
| 12 | 4,782,809.607 | 30.84752 | 2383.583 | 0.02301 | 37.12 |

$$q = 0.000001\,X' \qquad \Delta\lambda = (IX)q - (X)q^3 + E_5$$

| Latitude | (IX) | Diff. 1″ | (X) | Diff. 1″ |
|---|---|---|---|---|
| 43°00′ | 44,166.007 | 0.19903 | 495.124 | 0.00832 |
| 01 | 44,177.949 | 0.19920 | 495.623 | 0.00833 |
| 02 | 44,189.901 | 0.19937 | 496.123 | 0.00834 |
| 03 | 44,201.864 | 0.19954 | 496.623 | 0.00835 |
| 04 | 44,213.837 | 0.19972 | 497.125 | 0.00837 |
| 43°05′ | 44,225.819 | 0.19989 | 497.627 | 0.00838 |
| 06 | 44,237.813 | 0.20006 | 498.129 | 0.00839 |
| 07 | 44,249.816 | 0.20023 | 498.633 | 0.00840 |
| 08 | 44,261.830 | 0.20040 | 499.137 | 0.00841 |
| 09 | 44,273.854 | 0.20057 | 499.642 | 0.00843 |
| 43°10′ | 44,285.888 | 0.20074 | 500.147 | 0.00844 |
| 11 | 44,297.933 | 0.20092 | 500.654 | 0.00845 |
| 12 | 44,309.988 | 0.20109 | 501.161 | 0.00846 |

$E_5$

$\Delta^2(IX)$
Units of (IX)

$D_6$, seconds

$q$

included with tables using $q$ as the argument, provide corrections to $\phi'$ to give the correct latitude $\phi$.

The change of longitude between the station and the central meridian $\Delta\lambda$ is determined by interpolating factors (IX) and $X$ from the tables using $\phi'$ as the argument. Table 14.6 contains a sample of this section of the tables (Ref. 17$d$). These terms multiplied by $q$ and $q^3$, respectively, are then cumulated with $E_5$ from the chart (using $\phi'$ and $q$ as arguments) to calculate $\Delta\lambda$ in seconds.

When $X$ exceeds 500,000, the station is east of the central meridian. In this case $\Delta\lambda$ is subtracted from west longitude and added to east longitude. When $X$ is less than 500,000, the station is west of the central meridian and $\Delta\lambda$ is added to west longitude and subtracted from east longitude.

**Example 14.8**    Determine latitude and longitude for station Astro of Example 14.7, for which

$X = 302,917.857$ m
$Y = 4,780,139.881$ m

SOLUTION    Compute $X'$.

$X' = 500,000.000 - 302,917.857$
   $= 197,082.143$

Compute $q = (0.000001)X' = 0.197082143$.

$q^2 = 0.038841371$

$q^3 = 0.007654941$

$q^4 = 0.001508652$

Using (I) = $Y = 4,780,139.881$, obtain $\phi'$ from Table 14.6.

$\qquad$ (I) = 4,779,107.921 $\qquad$ for 43°10′

$\qquad\qquad$ Y = 4,780,139.881

| | |
|---|---|
| Difference = | 1 031.960 |
| Diff. / second = | 30.84734 |

Seconds of $\phi'$ = $\dfrac{1031.960}{30.84734}$ $\qquad = \qquad$ 33.4538″

$\qquad\qquad \phi'' = 43°10'33.4538''$

With $\phi'$ as the argument, obtain (VII) and (VIII) from Table 14.6.

$\qquad\qquad$ (VII) for 43°10′ = 2380.824

Correction for seconds of $\phi' = (33.4538'')(0.02298/\text{second}) =$ $\qquad$ 0.769

$\qquad\qquad$ (VII) for 43°10′33.4538″ = 2381.593

$\qquad\qquad$ (VIII) for 43°10′ = $\qquad$ 37.05

$\qquad$ Correction for seconds of $\phi' = (33.4538/60)(0.03) =$ $\qquad$ 0.02

$\qquad\qquad$ (VIII) = $\qquad$ 37.07

Obtain $D_6$ from the bar scale in Table 14.6 using $q = 0.197$ as the argument. $D = 0.0000$. Calculate $\phi$ using Eq. (14.69a).

$\phi = \phi' - (\text{VII})q^2 + (\text{VIII})q^4 - D_6$

$\phi' =$ $\qquad\qquad\qquad\qquad\qquad$ 43°10′33.4538″

$-(\text{VII})q^2 = (2381.593)(0.038841371) = 92.5043'' =$ $\qquad$ −1′32.5043″

$\qquad (\text{VIII})q^4 = (37.07)(0.001508652) =$ $\qquad$ + $\qquad$ 0.0559″

$\qquad\qquad\qquad\qquad -D_6$ $\qquad$ − $\qquad$ 0.0000″

$\qquad\qquad\qquad\qquad \phi = 43°09'01.0054''$

Next, compute the longitude. Determine factors (IX) and (X) in Eq. (14.69b) using $\phi'$ as the argument.

$\qquad\qquad$ (IX) for 43°10′ = 44,285.888

Correction for seconds of $\phi' = (33.4538'')(0.20074) =$ $\qquad$ 6.715

$\Delta^2(\text{IX})$ from bar scale using seconds of $\phi'$ as argument = $\qquad$ 0.001

$\qquad\qquad$ (IX) for $\phi' = 44,292.604$

$\qquad\qquad$ (X) for 43°10′ = $\qquad$ 500.147

Correction for seconds of $\phi' = (33.4538'')(0.00844) =$ $\qquad$ 0.282

$\qquad\qquad$ (X) for $\phi' =$ $\qquad$ 500.429

$E_5$ from the chart in Table 14.6 using $q$ and $\phi$ as arguments $\qquad$ 0.0030″

Cumulate the combined factors in Eq. (14.69b).

$$\Delta\lambda = (IX)q - (X)q^3 + E_5$$

$(IX)q = (44,292.598)(0.197082143) = \quad +8729.2813''$

$(X)q^3 = (500.429)(0.0076554941) = \quad - \quad 3.8308$

$$E_5 = \quad + \quad 0.0030$$

$$\Delta\lambda = \quad 8725.4535''$$

$$\Delta\lambda = 2°25'25.4535''$$

$$\lambda = \lambda_0 + \Delta\lambda = 77°25'25.4535''$$

Transformation from rectangular coordinates to latitude and longitude on the UTM projection is also possible by direct solution of Eqs. (14.17a). A value for $\phi'$ can be obtained by using the first term in the second of equations (14.15), where

$$Y = S_\phi k_0$$

in which $Y$ is the given $Y$ coordinate for the station, $k_0 = 0.9996$, and $S_\phi$ is defined by Eq. (14.16). Solve for $S_\phi = Y/k_0$ and substitute this value for $S_\phi$ into Eq. (14.16) to determine $\phi'$. Since direct solution of Eq. (14.16) for $\phi'$ is not possible, an iterative procedure such as the Newton-Raphson method can be used to obtain $\phi'$. When $\phi'$ has been determined, Eqs. (14.17a) are used to compute $\Delta\phi$ and $\Delta\lambda$, from which $\phi$ and $\lambda$ for the station can be computed. In order to demonstrate this procedure, the coordinates given in Example 14.7 are transformed to latitude and longitude by Eqs. (14.17a).

**Example 14.9** Determine $\phi$ and $\lambda$ given $X$ and $Y$ for station Astro from Example 14.7. Assume that $\phi'$ has been calculated by an iterative solution of Eq. (14.16) and is found to be $43°10'33.4538''$. From Example 14.7, $X = 302,917.857$ m, $Y = 4,780,139.881$ m. First compute $\Delta\phi$ using the first of Eqs. (14.17a).

SOLUTION

$$\Delta\phi = t_1 \left[ \frac{-(X')^2}{2R_1 N_1} + \frac{(X')^4}{24 R_1 N_1^3} (5 + 3t_1^2) \right]$$

in which the variables are defined in Art. 14.13. Compute $X'$, $R_1$, $N_1$, and $t_1$.

$$R_1 = \frac{a(1 - \varepsilon^2)}{(1 - \varepsilon^2 \sin^2\phi')^{3/2}} = \frac{(6,378,206.4)(1 - 0.006768658)}{(1 - 0.006768658 \sin^2\phi')^{3/2}} = 6,365,267.731 \text{ m}$$

$$N_1 = \frac{a}{(1 - \varepsilon^2 \sin^2\phi')^{1/2}} = \frac{6,378,206.4}{(1 - 0.006768658 \sin^2\phi')^{1/2}} = 6,388,336.723 \text{ m}$$

$$t_1 = \tan\phi' = 0.938273220$$

$$\eta_1^2 = \frac{\varepsilon^2}{(1 - \varepsilon^2)} \cos^2\phi' = \frac{0.00676858}{1 - 0.006768658} = 0.003624198$$

$$X' = 500,000 - 302,917.857 = 197,082.143 \quad \text{grid distance}$$

$$\frac{X'}{0.9996} = 197,161.007 \text{ m}$$

$$\frac{-t_1 (X'/k_0)^2}{2R_1 N_1} = -0.000448474 \text{ rad} = -92.5043''$$

$$\frac{t_1 (X'/k_0)^4}{24 R_1 N_1^3} (5 + 3t_1^2) = 0.000000272 \text{ rad} = 0.0561''$$

$$\Delta\phi = -92.4482''$$

$$\Delta\phi = -0°01'32.4482''$$

$$\phi = \phi' - \Delta\phi = (43°10'33.4538'') - (0°01'32.4482'')$$
$$= 43°09'01.0056'' \quad \text{(discrepancy of } 0.0002'')$$

Longitude is calculated by the second of Eqs. (14.17a) using $R_1$, $N_1$, $t_1$, and $\eta_1$ computed in determining $\phi$.

$$\Delta\lambda = \sec\phi'\left[\frac{X'}{N_1} - \frac{1}{6}\left(\frac{X'}{N_1}\right)^3(1 + 2t_1^2 + \eta_1^2) + \frac{1}{120}\left(\frac{X'}{N_1}\right)^5(5 + 28t_1^2 + 24t_1^4)\right]$$

$$\frac{X'/k_0}{\cos\phi'N_1} \qquad\qquad = \quad 0.042320749 \text{ rad} = \quad +8729.2810''$$

$$\frac{-1}{(6\cos\phi')}\left(\frac{X'/k_0}{N_1}\right)^3(1 + 2t_1^2 + \eta_1^2) \quad = -0.000018572 \text{ rad} = \quad -3.8308''$$

$$\frac{1}{120\cos\phi}\left(\frac{X'/k_0}{N_1}\right)^5(5 + 28t_1^2 + 24t_1^4) = \quad 0.000000016 \text{ rad} = \quad +0.0034''$$

$$8725.4536''$$
$$\Delta\lambda = \quad 2°25'25.4536''$$

$$\lambda = \lambda_0 + \Delta\lambda = 75°00'00'' \qquad\qquad +2°25'25.4536'' = 77°25'25.4536''$$

$$(25.4521'') \quad \text{original } \lambda$$

Equations (14.17a) are for projections with a 4° band width. The UTM has a 6° band width. Consequently, the results obtained by direct computation do not correspond to the third decimal place of seconds with those obtained using the tables. Additional terms are required in the equations in order to obtain identical results to four places. Equations with the additional terms adequate for projections band widths of up to 10 to 12 degrees of arc can be found in Ref. 15.

**14.22. Azimuth on the UTM projection**   When *geographic coordinates* are given for a station, the angle between grid north and geodetic north (meridian convergence) is given by Eq. (14.16a), Art. 14.13. The terms in this equation are also tabulated in the UTM projection tables. With these tabulated factors, convergence is given by

$$\Delta\alpha = (\text{XII})p + (\text{XIII})p^3 + C_5 \qquad\qquad (14.70)$$

Factors (XII) and (XIII) are given in the tables with $\phi$ as the argument. Factor $C_5$ is taken from a graph using $\Delta\lambda$ as the argument. Factors (XII) and (XIII) from a portion of the UTM projection tables are shown in Table 14.7. When $\Delta\alpha$ has been calculated,

$$\text{grid azimuth} = \text{geodetic azimuth} - \Delta\alpha \qquad\qquad (14.71)$$

In the northern hemisphere, $\Delta\alpha$ is positive when the station is east of the central meridian and negative when the station is west of the central meridian. The reverse is true in the southern hemisphere.

Convergence of meridians when *rectangular coordinates* are given is calculated by Eq. (14.17b), Art. 14.13. This convergence can also be determined from projection tables, an extract from which is shown in Table 14.8. Using factors tabulated according to the foot-point latitude $\phi'$, the equation for convergence is

$$\Delta\alpha = (\text{XV})q - (\text{XVI})q^3 + F_5 \qquad\qquad (14.72)$$

Factors (XV) are (XVI) are listed in Table 14.8 with $\phi'$ as the argument. Factor $F_5$ is taken from a graph using $q$ as the argument, where $q = 0.000001X'$. When the convergence has been calculated, the grid or geodetic azimuth is calculated using Eq. (14.71).

**Example 14.10**   From Example 14.8 the plane coordinates for Astro are

$X = 302,917.857$ m
$Y = 4,780,139.881$ m

Calculate the convergence at this station.

**Table 14.7 UTM Projection Tables—Scale Factor Using Geographic Coordinates: Clarke 1866 Spheroid Metres** (*Adapted from Ref. 17a*)

$$p = 0.0001\Delta\lambda'' \qquad \Delta\alpha = (XII)p + (XIII)p^3 + C_5$$

| Latitude $\phi$ | (XII) | Diff. 1″ | (XIII) | $C_5$ |
|---|---|---|---|---|
| 43°00′ | 6819.984 | 0.03545 | 2.889 | |
| 01 | 6822.111 | 0.03543 | 2.889 | |
| 02 | 6824.237 | 0.03543 | 2.888 | |
| 03 | 6826.363 | 0.03543 | 2.887 | |
| 04 | 6828.489 | 0.03540 | 2.887 | |
| 43°05′ | 6830.613 | 0.03542 | 2.886 | |
| 06 | 6832.738 | 0.03538 | 2.885 | |
| 07 | 6834.861 | 0.03538 | 2.885 | |
| 08 | 6836.984 | 0.03538 | 2.884 | |
| 09 | 6839.107 | 0.03537 | 2.883 | |
| 43°10′ | 6841.229 | 0.03535 | 2.882 | |
| 11 | 6843.350 | 0.03535 | 2.882 | |
| 12 | 6845.471 | 0.03533 | 2.881 | |
| 13 | 6847.591 | 0.03533 | 2.880 | |
| 14 | 6849.711 | 0.03532 | 2.880 | |
| 43°15′ | 6851.830 | 0.03530 | 2.879 | |
| 16 | 6853.948 | 0.03530 | 2.878 | |
| 17 | 6856.066 | 0.03530 | 2.878 | |
| 18 | 6858.184 | 0.03527 | 2.877 | |
| 19 | 6860.300 | 0.03527 | 2.876 | |

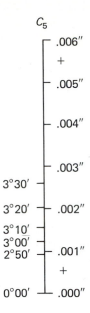

**Table 14.8 Scale Factor Using Rectangular Coordinates: Clarke 1866 Spheroid Metres** (*Adapted from Ref. 17a*)

$$q = 0.000001 X' \qquad \Delta\alpha = (XV)q - (XVI)q^3 + F_5$$

| Latitude $\phi'$ | (XV) | Diff. 1″ | (XVI) | $F_5$ |
|---|---|---|---|---|
| 43°00′ | 30,121.15 | 0.2924 | 459.4 | |
| 01 | 30,138.69 | 0.2925 | 459.9 | |
| 02 | 30,156.24 | 0.2927 | 460.5 | |
| 03 | 30,173.80 | 0.2928 | 461.0 | |
| 04 | 30,191.37 | 0.2930 | 461.5 | |
| 43°05′ | 30,208.95 | 0.2932 | 462.0 | |
| 06 | 30,226.54 | 0.2933 | 462.5 | |
| 07 | 30,244.14 | 0.2935 | 463.1 | |
| 08 | 30,261.74 | 0.2936 | 463.6 | |
| 09 | 30,279.36 | 0.2938 | 464.1 | |
| 43°10′ | 30,296.99 | 0.2940 | 464.6 | |
| 11 | 30,314.63 | 0.2941 | 465.2 | |
| 12 | 30,332.27 | 0.2943 | 465.7 | |
| 13 | 30,349.93 | 0.2944 | 466.2 | |
| 14 | 30,367.60 | 0.2946 | 466.7 | |

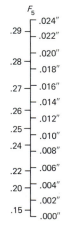

SOLUTION   From Example 14.8, $\phi' = 43°10'33.4538''$ and $X' = 197,082.143$ m. Thus, $q = 0.197082143$ and $q^3 = 0.007654941$. From Table 14.8, using $\phi'$ as the argument:

$$\text{(XV) for } 43°10' = 30,296.99$$

Corr. for 33.4538'' = (0.2940)(33.4538) = 9.835

$$\text{(XV) for } 43°10'33.4538'' = 30,306.825$$

$$\text{(XVI) for } 43°10'0.8 = 464.6$$

$$\text{Corr.} = \frac{33.4538}{60}(0.6) = 0.3$$

$$\text{(XVI)} = 464.9$$

By Eq. (14.72),

$$\Delta\alpha = \text{(XV)}q - \text{(XVI)}q^3 + F_5$$

$$\text{(XV)}q = (30,306.825)(0.197082143) = 5972.9340''$$

$$- \text{(XVI)}q^3 = (464.9)(0.007654941) = - \quad 3.5590''$$

$F_5$ from the chart using $q$ as argument = 0.0036''

$$\Delta\alpha'' = 5969.3786''$$

$$\Delta\alpha = -1°39'29.379''$$

The sign of $\Delta\alpha$ is negative since the station is west of the central meridian.

The convergence at this station can also be computed, using Eq. (14.17b) where

$$\Delta\alpha = t_1\left[\frac{X'/k_0}{N_1} - \frac{1}{3}\left(\frac{X'/k_0}{N_1}\right)^3(1 + t_1^2 - \eta_1^2) + \frac{1}{15}\left(\frac{X'/k_0}{N_1}\right)^5(2 + 5t_1^2 + 3t_1^4)\right]$$

in which the terms are defined in Art. 14.13 and $k_0 = 0.9996$ is the scale factor needed to convert grid to arc distance. From Example 14.9,

$$\frac{X'}{0.9996} = 197,161.007 \text{ m} \qquad N_1 = 6,388,336.723 \text{ m}$$

$$t_1 = \tan\phi' = 0.938273220 \qquad \eta_1^2 = 0.003624198$$

$$t_1\frac{X'/k_0}{N_1} = 0.028957599 \text{ rad} = 5972.934''$$

$$-\frac{1}{3}t_1\left(\frac{X'/k_0}{N_1}\right)^3(1 + t_1^2 - \eta_1^2) = 0.000017255 \text{ rad} = - \quad 3.559''$$

$$\frac{1}{15}t_1\left(\frac{X'/k_0}{N_1}\right)^5(2 + 5t_1^2 + 3t_1^4) = 0.000000015 \text{ rad} = 0.003''$$

$$\Delta\alpha'' = 5969.378''$$

$$\Delta\alpha = -1°39'29.378''$$

which agrees with the results from the tables to within 0.001''. Note that factors (XV)$q$, (XVI)$q^3$, and $F_5$ of Eq. (14.72) correspond to the first, second, and third terms, respectively, of Eq. (14.17b). Since $X$ is less than 500,000, the station is west of the central meridian and $\Delta\alpha$ is negative.

**Example 14.11**   Determine the convergence of meridians at station Astro using geographic coordinates as given in Example 14.7, where

$$\phi = 43°09'01.0054''$$
$$\lambda = 77°25'25.4521''$$

SOLUTION   From Example 14.7,

$\lambda' = \Delta\lambda$ (in tables) = 8725.4521''    (note: $\Delta\lambda$ is taken as positive in tables)
$p = 0.87254521 \qquad p^3 = 0.664299333$

Equation (14.70) is used to evaluate $\Delta\alpha$. Factors (XII) and (XIII) are obtained from Table 14.7 using $\phi$ as the argument.

| | |
|---|---:|
| (XII) for 43°09′ = | 6839.107 |
| Correction for seconds of $\phi = (1.0054)(0.03537) =$ | 0.036 |
| (XII) = | 6839.143 |
| (XIII) for 43°09′01.0054″ = | 2.883 |
| (XII)$p$ = | 5967.461 |
| (XIII)$p^3$ = | 1.915 |
| $C_5$ from chart using $\Delta\lambda = 2°25′$ as the argument = | 0.001 |
| $\Delta\alpha''$ = | 5969.377 |
| $\Delta\alpha = -1°39.''29.377''$ | |

which is the same as obtained using rectangular coordinates in Example 14.9. Again $\Delta\alpha$ is negative because the station is west of the central meridian.

Using Eq. (14.16a),

$$\Delta\alpha = \lambda' \sin\phi \left[ 1 + \frac{(\lambda')^2 \cos^2\phi}{3}(1 + 3\eta^2) + \frac{(\lambda')^4 \cos^4\phi}{15}(2 - t^2) \right]$$

where $\Delta\alpha$ and $\lambda'$ are in radians. From Example 14.7,

$\lambda' = -0.0423021855$ radians     $\eta^2 = 0.0036272462$
$t^2 = 0.8787764937$

| | | |
|---|---|---:|
| $\lambda' \sin\phi = -0.028931068$ radians = | | $-5967.461''$ |
| $\dfrac{(\lambda)^3 \sin\phi \cos^2\phi(1 + 3\eta^2)}{3} = -0.000009285$ rad = | | $-1.9152''$ |
| $\dfrac{(\lambda')^5 \sin\phi \cos^4\phi(2 - t^2)}{15} = 0.000000002$ rad = | | $-0.0004''$ |
| $\Delta\alpha =$ | | $-5967.3766''$ |
| $\Delta\alpha = -1°39'29.377''$ | | |

If the grid azimuth from station Astro to station Orient from the south has been determined from UTM plane rectangular coordinates as 135°45′20.59″, then by Eq. (14.71),

geodetic azimuth = grid azimuth + $\Delta\alpha$
$\qquad = (135°45'20.59'') - (1°39'29.38'')$
$\qquad = 134°05'51.21''$

For lines greater than 8 km in length, additional terms are required to determine the correct azimuth. This correction term for azimuth between stations 1 and 2 is

$$\text{correction} = \frac{(Y_2 - Y_1)(2X_1' - X_2')}{6R_m^2 \sin 1''} \tag{14.73}$$

in which $R_m = (R_1 N_1)^{1/2}$, where $N_1$ is calculated by Eq. (14.2) and $R_1$ is defined by Eq. (14.64). The procedure for calculating and applying this correction can be found in Example 14.6 for the transverse Mercator state plane projection system.

**14.23. Computation of the scale factor on the UTM projection** Measured distances corrected for systematic errors and reduced to sea level must be multiplied by a scale factor before being used in plane coordinate computations on the UTM projection.

For surveys having relative accuracies of from 1 : 1000 to 1 : 10,000, the scale factor can be determined from a graph of scale factors plotted versus the $X$ coordinate (Fig. 14.19). Usually, an average $X$ coordinate is computed for the area surveyed and an average scale factor is determined for the region.

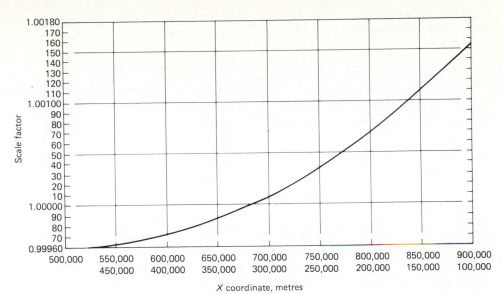

**Fig. 14.19** Universal transverse Mercator grid, scale factor. (*Adapted from Ref. 17c.*)

When a survey is to have a relative accuracy in excess of $1:10,000$, the scale factor can be computed by Eq. (14.17) (for geographic coordinates), by Eq. (14.17c) (rectangular coordinates) or by the following equation using factors tabulated in the projection tables:

$$k = k_0[1 + (XVIII)q^2 + (0.00003)q^4]$$   (14.74)

in which $k_0 = 0.9996$ is the scale factor at the central meridian and (XVIII) is taken from the projection tables using $Y$ as the argument. Table 14.9 illustrates scale factors for the UTM projection.

For surveys that do not exceed 8 km between extremities, it is usually adequate to calculate a scale factor based on an average of the $X$ and $Y$ coordinates for the area.

**Example 14.12**   Maximum and minimum $X$ and $Y$ coordinates on the UTM projection (zone 18) for a traverse are

|         | X, m       | Y, m         |
|---------|------------|--------------|
| Maximum | 303,450.58 | 4,800,451.25 |
| Minimum | 309,125.92 | 4,795,000.34 |

Calculate the scale factor for the area covered by this traverse.

SOLUTION

$\bar{X} = 306,288$     $\bar{Y} = 4,797,725$

Calculate $X'$ and $q$.

$X' = 500,000 - \bar{X} = 193,712$     $q = 0.193712$

$q^2 = 0.037524339$     $q^4 = 0.001408076$

From Table 14.9, using $\bar{Y}$ as the argument,

$(XVIII) = 0.012305$

Then by Eq. (14.74), the scale factor is

$k = 0.9996[1 + (0.012305)(0.037524339) + (0.00003)(0.001408076)]$
  $= 1.000061594$

**Table 14.9   UTM Projection Tables—Factors for Determining Scale: Clarke 1866 Spheroid, Units in Metres (*Adapted from Ref. 17a*)**

$$q = 0.000001\,X'$$
$$k_0 = 0.9996$$

$$k = k_0[1 + (XVIII)q^2 + (0.00003)q^4]$$

| Y Coordinate | | | Y Coordinate | | |
|---|---|---|---|---|---|
| Southern hemisphere | Northern hemisphere | (XVIII) | Southern hemisphere | Northern hemisphere | (XVIII) |
| 10,000,000 | 000,000 | 0.012384 | 5,500,000 | 4,500,000 | 0.012313 |
| 9,900,000 | 100,000 | 0.012384 | 5,400,000 | 4,600,000 | 0.012311 |
| 9,800,000 | 200,000 | 0.012384 | 5,300,000 | 4,700,000 | 0.012308 |
| 9,700,000 | 300,000 | 0.012384 | 5,200,000 | 4,800,000 | 0.012305 |
| 9,600,000 | 400,000 | 0.012384 | 5,100,000 | 4,900,000 | 0.012303 |
| 9,500,000 | 500,000 | 0.012383 | 5,000,000 | 5,000,000 | 0.012300 |
| 9,400,000 | 600,000 | 0.012383 | 4,900,000 | 5,100,000 | 0.012297 |
| 9,300,000 | 700,000 | 0.012382 | 4,800,000 | 5,200,000 | 0.012295 |
| 9,200,000 | 800,000 | 0.012382 | 4,700,000 | 5,300,000 | 0.012292 |
| 9,100,000 | 900,000 | 0.012381 | 4,600,000 | 5,400,000 | 0.012290 |
| 9,000,000 | 1,000,000 | 0.012380 | 4,500,000 | 5,500,000 | 0.012287 |
| 8,900,000 | 1,100,000 | 0.012379 | 4,400,000 | 5,600,000 | 0.012284 |
| 8,800,000 | 1,200,000 | 0.012378 | 4,300,000 | 5,700,000 | 0.012282 |
| 8,700,000 | 1,300,000 | 0.012377 | 4,200,000 | 5,800,000 | 0.012279 |
| 8,600,000 | 1,400,000 | 0.012376 | 4,100,000 | 5,900,000 | 0.012277 |
| 8,500,000 | 1,500,000 | 0.012375 | 4,000,000 | 6,000,000 | 0.012274 |
| 8,400,000 | 1,600,000 | 0.012374 | 3,900,000 | 6,100,000 | 0.012272 |
| 8,300,000 | 1,700,000 | 0.012372 | 3,800,000 | 6,200,000 | 0.012269 |
| 8,200,000 | 1,800,000 | 0.012371 | 3,700,000 | 6,300,000 | 0.012267 |
| 8,100,000 | 1,900,000 | 0.012370 | 3,600,000 | 6,400,000 | 0.012265 |
| 8,000,000 | 2,000,000 | 0.012368 | 3,500,000 | 6,500,000 | 0.012262 |
| 7,900,000 | 2,100,000 | 0.012366 | 3,400,000 | 6,600,000 | 0.012260 |
| 7,800,000 | 2,200,000 | 0.012365 | 3,300,000 | 6,700,000 | 0.012258 |
| 7,700,000 | 2,300,000 | 0.012363 | 3,200,000 | 6,800,000 | 0.012256 |
| 7,600,000 | 2,400,000 | 0.012361 | 3,100,000 | 6,900,000 | 0.012253 |
| 7,500,000 | 2,500,000 | 0.012359 | 3,000,000 | 7,000,000 | 0.012251 |
| 7,400,000 | 2,600,000 | 0.012358 | 2,900,000 | 7,100,000 | 0.012249 |
| 7,300,000 | 2,700,000 | 0.012356 | 2,800,000 | 7,200,000 | 0.012247 |
| 7,200,000 | 2,800,000 | 0.012354 | 2,700,000 | 7,300,000 | 0.012245 |
| 7,100,000 | 2,900,000 | 0.012352 | 2,600,000 | 7,400,000 | 0.012243 |
| 7,000,000 | 3,000,000 | 0.012349 | 2,500,000 | 7,500,000 | 0.012241 |
| 6,900,000 | 3,100,000 | 0.012347 | 2,400,000 | 7,600,000 | 0.012240 |
| 6,800,000 | 3,200,000 | 0.012345 | 2,300,000 | 7,700,000 | 0.012238 |
| 6,700,000 | 3,300,000 | 0.012343 | 2,200,000 | 7,800,000 | 0.012236 |
| 6,600,000 | 3,400,000 | 0.012340 | 2,100,000 | 7,900,000 | 0.012234 |
| 6,500,000 | 3,500,000 | 0.012338 | 2,000,000 | 8,000,000 | 0.012233 |
| 6,400,000 | 3,600,000 | 0.012336 | 1,900,000 | 8,100,000 | 0.012231 |
| 6,300,000 | 3,700,000 | 0.012333 | 1,800,000 | 8,200,000 | 0.012230 |
| 6,200,000 | 3,800,000 | 0.012331 | 1,700,000 | 8,300,000 | 0.012229 |
| 6,100,000 | 3,900,000 | 0.012329 | 1,600,000 | 8,400,000 | 0.012227 |
| 6,000,000 | 4,000,000 | 0.012326 | 1,500,000 | 8,500,000 | 0.012226 |
| 5,900,000 | 4,100,000 | 0.012323 | 1,400,000 | 8,600,000 | 0.012225 |
| 5,800,000 | 4,200,000 | 0.012321 | 1,300,000 | 8,700,000 | 0.012224 |
| 5,700,000 | 4,300,000 | 0.012318 | 1,200,000 | 8,800,000 | 0.012223 |
| 5,600,000 | 4,400,000 | 0.012316 | 1,100,000 | 8,900,000 | 0.012222 |

For surveys in which the extremities of the region cover distances in excess of 8 km, considerable judgment must be exercised in computing scale factors in order to achieve specified accuracies. Thus, lines running north and south at $X = 160,000$ m or $X = 840,000$ m have constant scale factors near unity. On the other hand, lines running east and west in these portions of the projection have rapidly changing scale factors. In the first case, a single scale factor would be adequate for lines predominantly of north-south directions. In the latter case, a line may have to be divided into sections with a separate scale factor calculated for each section.

## 14.24. Problems

**14.1** "Any map projection has some distortions." Is this statement true or false, and why?

**14.2** What are the four conditions that an ideal distortion-free map must satisfy?

**14.3** Describe briefly four classes of map projections.

**14.4** Explain the basic differences between geometric map projection and mathematical map projection.

**14.5** The Lambert conformal conic projection and the transverse Mercator projection are two of the most commonly used map projections; discuss the principal differences between these two projections.

**14.6** Enumerate the specifications of the universal transverse Mercator projection.

**14.7** "The shape of a state, or its north-south extent relative to its east-west extent, influences the type of map projection used for its state plane coordinate system." Discuss this statement fully.

**14.8** The latitude and longitude for station 1 are 41°20'45.618"N and 121°05'50.623"W. Compute on the Lambert projection, California zone I): (a) the mapping angle $\theta$ for station 1; (b) $X$ and $Y$ state plane coordinates for the station; (c) the scale factor at station 1. Use the projection tables (Table 14.1).

**14.9** Solve Prob. 14.8 using the constants given in Ref. 3 for the California Lambert projection, zone I (Table 14.2).

**14.10** The state plane coordinates on the Lambert projection for California zone I are $X_2 = 2,300,000.10$ ft and $Y_2 = 770,421.89$ ft. The average elevation of the terrain between point 2 and point 10 of Example 14.2 is 5000 ft above mean sea level. Calculate (a) the grid azimuth from station 2 to station 10; (b) the geodetic azimuth from station 2 to station 10 and from station 10 to station 2; (c) the grid distance, geodetic distance at mean sea level, and geodetic distance at average terrain elevation between stations 2 and 10. Assume an average scale factor between 2 and 10 of 0.99995315.

**14.11** In Prob. 14.10, compute the geographic coordinates for station 2, using the projection tables in Table I.

**14.12** In Prob. 14.11, calculate geographic coordinates for station 2 using the constants and equations for the Lambert projection for California zone I as found in Ref. 3 and shown in Table 14.2.

**14.13** Geographic coordinates for station Glenn are $\phi = 33°40'20.500"$N and $\lambda = 110°10'05.250"$W. Using Eqs. (14.42) to (14.49) and the constants of Table 14.3, compute (a) the state plane coordinates for station Glenn on the transverse Mercator projection, Arizona eastern zone; (b) the convergence of the meridian and the scale factor for station Glenn.

**14.14** The state plane coordinates for station Gail on the transverse Mercator projection, Arizona eastern zone, are $X = 610,005.05$ ft and $Y = 465,650.50$ ft. (a) Compute the grid distance and grid azimuth from station Orbit (Example 14.4) to Gail; (b) determine the geodetic azimuth from the south of the line from Orbit to Gail.

**14.15** Compute geographic coordinates for station Gail in Prob. 14.14.

**14.16** Calculate the convergence of the meridians and the scale factor for station Gail in Prob. 14.14. Also compute the sea-level geodetic distance and the geodetic azimuth from the south of the line from Gail to Orbit.

**14.17** The distance between two stations calculated from their state plane coordinates is 10,500.825 m. The scale factor and average terrain elevation are 1.0004 and 1000 m above mean sea level, respectively. What is the horizontal distance between these two stations as it would be measured on the ground?

**14.18** In the area described for Prob. 14.17, the state plane coordinates for two control points $A$ and $B$ are $X_A = 2,140,982.13$ ft, $Y_A = 640,283.85$ ft; and $X_B = 2,141,116.82$ ft, $Y_B = 640,496.27$ ft. The coordinates for a construction control point $C$, which is to be established from $A$, are $X_C = 2,139,885.06$ ft and $Y_C = 639,480.09$ ft. Compute (a) the clockwise angle to be turned from an instrument set up at $A$ with a backsight on $B$ in order to direct the line of sight toward point $C$, which is to be established; (b) the distance that must be measured over level terrain from $A$ to set point $C$.

**14.19** Control point Arch has geographic coordinates of $\phi = 43°10'30.205''$N and $\lambda = 121°30'45.638''$W. Determine the UTM zone in which this point falls. Compute the plane coordinates on the UTM projection for this point.

**14.20** Compute the convergence of the meridian and the scale factor at point Arch for which data are given in Prob. 14.19.

**14.21** Determine the plane coordinates on the UTM projection for station 1 in Prob. 14.8. Use Eqs. (14.15), Art. 14.13.

**14.22** Calculate the grid and geodetic azimuths on the UTM projection for a line from station 1 in Prob. 14.8 to station Arch in Prob. 14.19.

**14.23** The plane coordinates on the UTM projection (Zone 14) for station Able are $X = 379,272.805$ m and $Y = 4,765,501.354$ m. Compute the latitude and longitude for station Able.

**14.24** The plane coordinates for station Plum on the UTM projection (Zone 14) are $X = 401,560.354$ m and $Y = 4,782,465.550$ m. Compute the geographic coordinates for Plum.

**14.25** Calculate the scale factor and convergence of the meridian on the UTM for station 1 in Prob. 14.8. Use Eqs. (14.16a) and (14.17).

**14.26** Calculate the convergence of the meridian and the scale factor for station Able in Prob. 14.23.

**14.27** Compute the convergence of the meridian and the scale factor for station Plum in Prob. 14.24. Use the rigorous equations for this solution.

**14.28** Compute the grid and geodetic azimuths from the south on the UTM projection for the line from station Able (Prob. 14.23) to station Plum (Prob. 14.24). Assume $R_m = 4.063311271 \times 10^{13}$.

**14.29** Compute the geodetic azimuth from the south on the UTM projection for the line from Plum (Prob. 14.24) to Able (Prob. 14.23).

**14.30** A microwave transmission antenna is to be erected at station Able in Prob. 14.23, so that the axis of the antenna is directed toward station Plum in problem 14.24. From an instrument set up over station Able, a backsight is possible on control point $Z$ which has plane coordinates of $X_Z = 379,071.621$ m and $Y_Z = 4,765,100.052$ m. Compute the clockwise horizontal angle to be turned from an instrument set up over Able with a backsight on point $Z$ so as to place the instrument line of sight in the proper direction for aligning the antenna.

## References

1. Adams, O. S., and Claire, C. N., *Manual of Plane Coordinate Computation*, Spec. Publ. No. 193, U.S. Department of Commerce, Coast and Geodetic Survey (now National Ocean Survey, NOAA), Rockville, Md., 1935; reprinted 1971.

2. Bomford, G., *Geodesy*, 2nd ed., Oxford University Press, London, 1962.

3. Claire, C. N., *State Plane Coordinates by Automatic Data Processing*, Publ. 62-4, U.S. Department of Commerce, Environmental Sciences Services Administration (now National Ocean Survey, NOAA), Rockville, Md., 1968; reprinted with corrections 1973/1976.

4. Coast and Geodetic Survey, *Plane Coordinate Projection Tables*, Spec. Publ. series, U.S. Department of Commerce, National Ocean Survey, NOAA, Rockville, Md., also Government Printing Office, Washington, D.C., for some states.

5. Dracup, J. F., *Fundamentals of the State Plane Coordinate Systems*, National Geodetic Survey Information Center, C18, NOAA, Rockville, Md., October 1974.

6. Dracup, J. F., "The New Adjustment of the North American Datum." *ACSM Bulletin*, November 1977.

7. Ewing, C. E., and Mitchell, M. M., *Introduction to Geodesy*, American Elsevier Publishing Company, Inc., New York, 1970.

8. Hradilek, L., and Hamilton, A. C., "A Systematic Analysis of Distortions in Map Projections," University of New Brunswick Lecture Notes No. 34, August 1973.

9. Krakiwsky, E. J., "Conformal Map Projections in Geodesy," University of New Brunswick Lecture Notes No. 37, September 1973.
10. Meade, B. K., "Coordinate Systems for Surveying and Mapping," *Surveying and Mapping*, Vol. 33, No. 3, September 1973.
11. Moffitt, F. H., and Bouchard, H., *Surveying*, 6th ed., Harper & Row, Publishers, New York, 1975.
12. National Geographic Society, *The Round Earth on Flat Paper*, National Geographic Society, Monograph, 1947.
13. Raisz, Erwin, *Principles of Cartography*, McGraw-Hill Book Company, New York, 1962.
14. Simmons, Lansing G., *Philippine Islands Plane Coordinate Projection*, G-59, U.S. Department of Commerce, Coast and Geodetic Survey (now National Ocean Survey, NOAA), Rockville, Md., 1947.
15. Thomas, Paul D., *Conformal Projections in Geodesy and Cartography*, Spec. Publ. 251, U.S. Department of Commerce, Coast and Geodetic Survey (now National Ocean Survey NOAA), Rockville, Md., 1952.
16. Whitten, C. A., "A Suggested Plan for Improving State Coordinate Systems Referenced to a New National Geodetic Datum," *Surveying and Mapping*, Vol. 30, No. 3, September 1970.
17. U.S. Department of the Army Technical Manuals: (a) *The Universal Grid Systems*, TMS-241, TO116-1-233, 1951; (b) *Universal Transverse Mercator Grid*, TMS-241-8, 1958; (c) *Transverse Mercator Grid Tables for Latitudes 0°–80°*, Vol. 1, *Transformation of Coordinates from Geographic to Grid*, TM 5-241-4/1, July 1958, rev. May 1971; (d) *Transverse Mercator Grid Tables for Latitudes 0°–80°, Transformation of Coordinates from Grid to Geographic*, Vol. 2, TM-5-241-4/2, July 1958, rev. May 1971, Washington, D.C.; (e) *Field Manual, Map Reading*, FM 21–26, 1973.

# PART V
# TYPES OF SURVEYS

# CHAPTER 15
# Control and topographic surveying

**15.1. General** The distinguishing feature of a topographic survey is the determination of the location, both in plan and elevation, of selected ground points which are necessary for plotting contour lines and the planimetric location of features on the topographic map. A topographic survey consists of (1) establishing over the area to be mapped a system of horizontal and vertical *control*, which consists of key stations connected by measurements of high precision; and (2) locating the details, including selected ground points, by measurements of lower precision from the control stations.

Topographic surveys fall roughly into three classes, according to the map scale employed, as follows:

*Large scale*  1:1200 (1 inch to 100 ft) or larger
*Intermediate scale*  1:1200 to 1:12,000 (1 in to 100 ft to 1 in to 1000 ft)
*Small scale*  1:12,000 (1 in to 1000 ft) or smaller

Because of the range in uses of topographic maps and variations in the nature of the areas mapped, topographic surveys vary widely in character.

Topographic surveys can be performed by aerial photogrammetric methods, by ground survey methods, or by some combination of these two procedures. The largest portion of almost all of the small and intermediate scale as well as some large-scale topographic mapping is now performed by photogrammetric methods (Art. 16.17). This photogrammetric operation includes establishing portions of the horizontal control in addition to compilation of the topographic map. However, ground survey methods are still applicable for large-scale topographic mapping of small areas and for field completion surveys which are usually needed for photogrammetrically compiled topographic maps. The discussions in this chapter are directed primarily toward the various procedures for topographic surveys by ground survey methods.

**15.2. Planning the survey** The choice of field methods for topographic surveying is governed by (1) the intended use of the map, (2) the area of the tract, (3) the map scale, and (4) the contour interval.

**1. Intended use of map** Surveys for detailed maps should be made by more refined methods than surveys for maps of a general character. For example, the earthwork estimates to be made from a topographic map by a landscape architect must be determined from a map which represents the ground surface much more accurately in both horizontal and vertical dimensions than one to be used in estimating the storage capacity of a reservoir. Also, a survey for a bridge site should be more detailed and more accurate in the immediate vicinity of the river crossing than in areas remote therefrom.

**2. Area of tract**   It is more difficult to maintain a desired precision in the relative location of points over a large area than over a small area. Control measurements for a large area should be more precise than those for a small area.

**3. Scale of map**   It is sometimes considered that, if the errors in the field measurements are not greater than the errors in plotting, the former are unimportant. But since these errors may not compensate each other, the errors in the field measurements should be considerably less than the errors in plotting at the given scale. The ratio between field errors and plotting errors should be perhaps one to three.

The ease with which precision may be increased in plotting, as compared with a corresponding increase in the precision of the field measurements, points to the desirability of reducing the total cost of a survey by giving proper attention to the excellence of the work of plotting points, of interpolation, and of interpretation in drawing the map.

The choice of a suitable map scale is discussed in Art. 13.6.

**4. Contour Interval**   The smaller the contour interval, the more refined should be the field methods. The choice of a suitable contour interval is discussed in Art. 13.6.

**15.3. Establishment of control**   Control consists of two parts: (1) *horizontal control*, in which the planimetric positions of specified control points are located by trilateration, triangulation, intersection, resection, or traverse; and (2) *vertical control*, in which elevations are established on specified bench marks located throughout the area to be mapped. This control provides the skeleton, which is later clothed with the *details*, or locations of such objects as roads, houses, trees, streams, ground points of known elevations, and contours.

On surveys of wide extent, a relatively few stations distributed over the tract are connected by more precise measurements, forming the *primary control*; within this system, other control stations are located by less precise measurements, forming the *secondary control*. For small areas only one control system is necessary, corresponding in precision to the secondary control used for larger areas.

**15.4. Horizontal control**   Horizontal control can be established by triangulation, trilateration, traverse, aerial photogrammetric methods, and inertial and doppler positioning systems.

Triangulation and trilateration (Chap. 10) can be used to establish primary and secondary control for relatively large topographic surveys. These methods are also utilized in areas of lesser extent when field conditions are appropriate (hilly, urban, or rugged mountainous regions). Specifications for triangulation and trilateration are given in Table 10.1, Art. 10.2.

Traverse (Chap. 8) with a theodolite and electronic distance measurement (EDM) instrument can also be used for primary and secondary control surveys. Specifications for various orders and classes of traverse are given in Table 8.15, Art. 8.23.

Horizontal control determination by aerial photogrammetric methods is feasible and particularly applicable to small-scale mapping of large areas. Note that this procedure requires a basic framework of horizontal control points established by traditional methods (trilateration and/or triangulation) or by inertial or doppler positioning systems.

Inertial and doppler electronic positioning systems are most appropriate for establishing primary horizontal control when the topographic survey is of very large extent, covers an inaccessible region or where the conduct of the survey is governed by special conditions. A description of these modern positioning systems is given in Arts. 12.3 to 12.5.

**15.5. Vertical control**   The purpose of vertical control is to establish bench marks at convenient intervals over the area to serve (1) as points of departure and closure for leveling operations of the topographic parties when locating details, and (2) as reference marks during subsequent construction work.

Vertical control is usually accomplished by direct differential leveling (Art. 5.35), but for small areas or in rough country the vertical control is frequently established by trigonometric leveling (Art. 5.5).

All elevations for topographic mapping should be tied to bench marks which are referred to the sea-level datum (Art. 13.2).

Specifications for first-, second-, and third-order differential levels are given in Table 5.3, Art. 5.43. These specifications may be relaxed somewhat depending on map scale, character of the terrain to be mapped, the contour interval desired, and ultimate use of the survey. Table 15.1 gives ranges of approximate closures applicable to intermediate- and large-scale topographic mapping surveys. The smaller error of closure for a given map and type control is used for very flat regions where a contour interval of 1 ft (0.5 m) or less is required and on surveys which are to be used for determination of gradients of streams or to establish the grades of proposed drainage or irrigation systems. The higher errors of closure apply to surveys in which no more exact use is made of the results other than to determine the elevations of ground points for contours having 2-, 5-, and 10-ft or 0.5-, 2-, and 3-m intervals. Where the largest error of closure is specified [0.5 ft (miles)$^{1/2}$ and 240 mm (kilometres)$^{1/2}$], stadia leveling is adequate.

**15.6. Horizontal and vertical control by three-dimensional traverse**   A three-dimensional stadia survey may be adequate for establishing control of a small area where the contour interval exceeds 1 m or 5 ft and a relative accuracy (Art. 8.23) of less than 1 part in 500 or 600 is satisfactory. The method for performing a stadia traverse with elevations required is described in Art. 8.12.

When higher accuracy is required, a three-dimensional traverse using a theodolite and EDM device or an electronic tacheometer (Chap. 12) can be performed. The procedures involved for a traverse in three dimensions are outlined in Art. 8.39.

**15.7. Location of details: general methods**   It is assumed in the following sections that horizontal and vertical control has been established and that the field party is concerned

**Table 15.1   Topographic Survey Vertical Control Specifications**

| Scale of map | Type of control | Length of circuit | | Maximum error of closure | |
|---|---|---|---|---|---|
| | | Miles | Kilometres | ft | mm |
| Intermediate | Primary | 1–20 | 2–30 | 0.05–0.3 × √mi | 12–72 × √km |
| | Secondary | 1–5 | 2–8 | 0.1–0.5 × √mi | 24–120 × √km |
| Large | Primary | 1–5 | 2–8 | 0.05–0.1 × √mi | 12–24 × √km |
| | Secondary | $\frac{1}{2}$–3 | 1–5 | 0.05–0.1 × √mi | 12–24 × √km |

only with the location of details. If the plane-table method is to be used, horizontal control is plotted on the plane table sheet (Art. 7.20).

The adequacy with which the resultant map sheet meets the purpose of the survey depends largely upon the task of locating details. Thus, the topographer should be completely informed as to the uses of the map so that the proper emphasis is placed on each part of the work.

The principal instruments used are the engineer's transit, self-reducing tacheometer, the engineer's level (preferably self-leveling), plane table and alidade, hand level, and clinometer. The transit and self-reducing tacheometer have advantages where there are many definite points to be located or where ground cover limits visibility and requires many setups. Conditions favorable to the plane table are open country and many irregular lines to be mapped; the plane table is also advantageous for small-scale mapping. Another class of instruments, electronic surveying systems (Art. 12.2), can also be employed to advantage on topographic detail surveys. These systems provide direction, horizontal distance, and difference in elevation between stations. When available, these devices can be used for almost any phase of a topographic survey.

The four principal methods for acquiring topographic detail in the field are the *controlling-point*, *cross-profile*, *checkerboard*, and *trace-contour* methods. These methods are analogous to the systems of points used to plot contours as described in Chap. 13. Each of the methods is described in subsequent sections in this chapter. For a given survey, a combination of the methods may be used; for example the checkerboard method can be used to obtain detail in open, slightly rolling terrain, while the controlling point method is employed to get data in irregular, wooded portions of the region being mapped. The aim is to locate details with minimum time and effort.

**15.8. Precision**    The precision required in locating such definite objects as buildings, bridges, and boundary lines should be consistent with the precision of plotting, which may be assumed to be a map distance of about 0.5 mm or $\frac{1}{50}$ in. Such less definite objects as shorelines, streams, and edges of woods are located with a precision corresponding to a map distance of perhaps 0.9 to 1.3 mm or $\frac{1}{30}$ to $\frac{1}{20}$ in. For use in maps of the same relative precision, more located points are required for a given area on large-scale surveys than on intermediate-scale surveys; hence, the location of details is relatively more important on large-scale surveys.

**Contours**    The accuracy with which contour lines represent the terrain depends upon (1) the accuracy and precision of the observations, (2) the number of observations, and (3) the distribution of the points located. Although ground points are definite, contour lines must necessarily be generalized to some extent. The error of field measurement in plan should be consistent with the error in elevation, which in general should not exceed one-fifth of the horizontal distance between contours. The error in elevation should not exceed one-fifth of the vertical distance between contours. The purpose of a topographic survey will be better served by locating a greater number of points with less precision, within reasonable limits, than by locating fewer points with greater precision. Thus, if for a given survey the contour interval is 5 ft, a better map will be secured by locating with respect to each instrument station perhaps 50 points whose standard deviation in elevation is 1 ft than by locating 25 points whose standard deviation is only 0.5 ft. Similarly, if the contour interval is 2 m, it is better to have 50 points with a standard deviation of 0.4 m than 25 points with a standard deviation of 0.2 m.

A general principle which should serve as a guide in the selection of ground points may be noted. As an example, let it be supposed that a given survey is to provide a map which shall be accurate to the extent that if a number of well-distributed points is chosen at random on

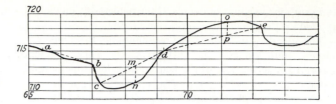

**Fig. 15.1** Effect of omission of significant ground points.

the map, the average difference between the map elevations and ground elevations of identical points shall not exceed one-half of a contour interval (Art. 13.37). Under this requirement, the attempt is made in the field to choose ground points such that a straight line between any two adjacent points will in no case pass above or below the ground by more than one contour interval. Thus, in Fig. 15.1, if the ground points were taken only at *a*, *b*, *c*, *d*, and *e* as shown, the resulting map would indicate the straight slopes *cd* and *de*; the consequent errors in elevation of *mn* and *op* on the profile amount to two contour intervals and show that additional readings should have been taken at the points *n* and *o*. The corresponding displacement of the contours on the map is shown by dotted and full lines in Fig. 15.2.

**Angles**  The precision needed in the field measurements of angles to details may be readily determined by relating it to the required precision of corresponding vertical and horizontal distances. Thus, for a sight at a distance of 1,000 ft (300 m), a permissible error of 0.3 ft (0.09 m) in elevation corresponds to a permissible error of 01′ in the vertical angle; likewise, a permissible error of 0.3 ft (0.09 m) in azimuth (measured along the arc from the point sighted) corresponds to a permissible error of 01′ in the horizontal angle. Values for other lengths of sight or degrees of precision are obtained in a similar manner; thus, if it is desired to locate a point to the nearest 2 ft in azimuth (or elevation) and if the length of the sight is 500 ft, the corresponding permissible error in the angle is 2/500 = 0.004 rad = 14′.

**15.9. Details by the controlling-point method**  Details may be located by the controlling-point method employing the transit or self-reducing tacheometer and stadia, the plane table and alidade, or electronic systems (Art. 12.2). This method is applicable to practically every type of terrain and condition encountered in topographic mapping. Since only the controlling points that govern the configuration of the land and location of details are located, a measure of economy is achieved that is not possible in the other techniques. However, the topographer must be experienced, since the success of the method depends on selection of the key controlling points. Procedures for use of this method are outlined in subsequent articles.

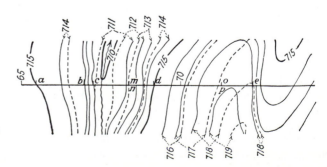

**Fig. 15.2** Errors in contour lines due to insufficiency of ground points.

**15.10. Transit stadia**  The personnel of the topography party using the transit or self-reducing tacheometer usually consists of an instrument operator, a notekeeper, and one or two people operating the rods.

In locating ground points, the vertical angles are measured more precisely than the horizontal angles. Accordingly, the vertical circle of the instrument assumes greater importance than the horizontal circle; but because the vertical angles are measured with respect to a horizontal plane, it is important that the horizontal plate be truly horizontal and remain so without need for constant releveling. In this respect, the modern optical repeating theodolites with automatic self-indexing of the vertical circle are a definite asset in this type of work.

**Procedure**  The transit is set up on either a primary or secondary control station, the elevation and position of which are known. The height of instrument (h.i.) of the horizontal axis of the instrument above the station over which it is set is measured with a rod or tape or observed on the plumbing rod if the tripod is so equipped. The vertical circle is checked for index error and the transit is oriented by backsighting along a line the azimuth of which is known, this azimuth having been set off on the clockwise horizontal circle. The upper motion is unclamped and sights to the desired points are taken.

The notekeeper records all stadia intervals, vertical or zenith angles, and azimuths, and describes all points by remarks or sketches so that the draftsman can interpret the data correctly and draw all features properly on the map. Figure 15.3 shows notes of observations taken from station $C$ of a traverse, the elevation of the station having been previously determined as 419.5 ft above the datum. The h.i. of the instrument is 4.4 ft. The transit is oriented by sighting to $B$, the azimuth of the line $CB$ being set off on the horizontal circle prior to taking the sight. In the first column are given the numbers of the side shots, their locations being shown on the sketch. In the second column are the azimuths of the several points sighted; in the third column are the rod intervals. The measured vertical angles are recorded in the fourth column, and in the following columns are shown, respectively, the

| | | | TOPOGRAPHIC | | DETAILS, BLACK ESTATE | | 32 |
|---|---|---|---|---|---|---|---|
| | Inst. at C; El. 419.5; h.i. =4.4 | | | | Lietz Transit No.1532 | G. Burke, ⚐ | |
| Obj. | Az. | Rod Int | Vert. Ang | Hor. Dist | (f+c)=0 ; K=100 | M.D. Rand, Notes | |
| B | 176°14′ | | | | Elev. | F.J. & K.D., Rods | |
| | | | | Diff. El. | | Apr. 4, 1978 | |
| 1 | 10°21′ | 7.23 | -3°11′ | 721 | -40.1 | 383.8 Water's Edge-Corner | Cloudy, Cold |
| 2 | 3°14′ | 7.02 | -3°17′ | 700 | -40.1 | 383.8 " " -On Line | |
| 3 | 352°45′ | 5.64 | -4°11′ | 561 | -41.0 | 382.9 " " | |
| 4 | 7°18′ | 5.76 | -4°04′ | 573 | -40.7 | 383.2 " | |
| 5 | 349°10′ | 7.17 | -2°33′ | 716 | -31.9 | 392.0 Line (Intervals) | |
| 6 | 16°55′ | 5.50 | -2°50′ | 549 | -27.2 | 396.7 " | |
| 7 | 315°20′ | 7.86 | -2°41′ | 784 | -36.8 | 387.1 " | Grass |
| 8 | 349°15′ | 4.13 | -5°46′ | 409 | -41.3 | 382.6 Water's Edge | |
| 9 | 339°30′ | 5.40 | -4°22′ | 537 | -41.0 | 382.9 Bank Brook 6'wide | |
| 10 | 0°05′ | 3.71 | -4°12′ | 369 | -27.1 | 396.8 | |
| 11 | 344°40′ | 4.85 | -4°54′ | 482 | -41.3 | 382.6 " " 15' | |
| 12 | 25°00′ | 2.86 | 0°on 3.2 | 286 | +1.2 | 420.7 Direct Levels | |
| 13 | 307°45′ | 4.88 | -4°56′ | 484 | -41.8 | 382.1 Water's Edge | |
| 14 | 319°10′ | 4.02 | -5°56′ | 398 | -41.3 | 382.6 " " | |
| 15 | 309°45′ | 5.80 | -3°00′ | 578 | -30.3 | 393.6 | |
| 16 | 318°25′ | 3.27 | -4°36′ | 325 | -26.1 | 397.8 | |
| B | 176°15′ | c k. | | | | | |
| 17 | 340°00′ | 6.34 | -3°08′ | 632 | -34.6 | 389.3 | |
| 18 | 278°35′ | 2.51 | -5°43′ | 249 | -24.9 | 399.0 | |
| 19 | 276°20′ | 3.07 | -7°56′ | 301 | -42.0 | 381.9 Water's Edge | |
| 20 | 277°40′ | 4.24 | -5°40′ | 420 | -41.7 | 382.2 " " | |

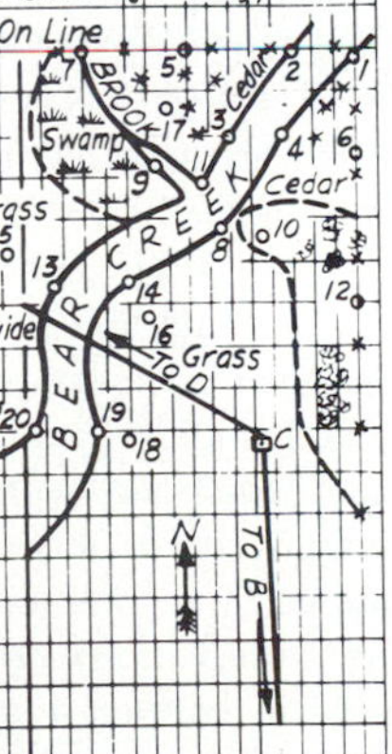

**Fig. 15.3** Stadia notes for location of details, with elevations (units are feet).

computed horizontal distances, differences in elevations, and elevations. Methods for reducing stadia data to horizontal distances and differences in elevation are outlined in Art. 7.9.

In measuring vertical angles it is customary, when practical, to sight at a rod reading equal to the height of instrument above the station over which the transit is set. In this way, the difference in elevation between the center of instrument and the h.i. on the rod is the same as the difference in elevation between the station over which the transit is set and the point on which the rod is held. When the line of vision is obstructed so that the h.i. cannot be sighted, the sight is taken on some other graduation of the rod, usually a whole number of feet or decimetres above or below the h.i., and this difference in rod readings is recorded in the notes.

When the detail to be observed is at nearly the same elevation as the point over which the transit is set, there is a marked advantage in determining difference in elevation by direct leveling. The notes of Fig. 15.3 show that object 12 was observed in this manner.

Where measurements are made solely for plotting a map, usually horizontal angles or directions are estimated to 05′ without reading the vernier. Vertical angles are usually measured to minutes, and differences in elevation are computed to tenths of feet or to centimetres. Where elevations to the nearest foot or nearest 3 dm are sufficiently precise, the vertical angles (except for long shots) may properly be read by estimation without use of the vernier.

The usual procedure of observing is to sight on the rod, the lower stadia hair being set on an even foot or decimetre mark such that the horizontal cross hair falls somewhere near the h.i., and to read the stadia interval. The horizontal cross hair is then set on the h.i., the horizontal angle is read and the vertical angle is observed.

When numerous observations are to be taken from a single station, sights to some object the azimuth of which is known are taken at intervals in order to make sure that there is no undetected movement of the lower motion of the transit. The notes of Fig. 15.3 show a check measurement of this kind taken to station $B$ after observations to object 16 had been completed.

The rodmen choose ground points along valley and ridge lines and at summits, depressions, important changes in slope, and definite details. The selection of points is important, and the rodmen should be instructed and trained for their work. They should follow a systematic arrangement of routes such that the entire area is covered and that no important objects are overlooked. They should observe the terrain carefully (often with the aid of a hand level) and report any important features which cannot be seen from the transit station.

When an optical repeating theodolite is used to locate controlling points by stadia, the zenith angle (Fig. 6.50, Art. 6.43) is measured instead of a vertical angle. For example, if such an instrument were used for the measurements in Fig. 15.3, the observations on point 1 would have a zenith angle of 93°11′, on point 2 it would be 93°17′,...etc. Thus, the vertical angle $= 90° - 93°11' = -3°11'$ and the stadia reductions are performed as indicated by Example 7.1, Art. 7.9. Note that the zenith angle can be used directly to calculate horizontal distance and difference in elevation by using the relationship that cos(90° − zenith angle) = sin(zenith angle) and sin(90° − zenith angle) = cos(zenith angle) in Eqs. (7.3) and (7.4a).

If an instrument equipped with a Beaman stadia arc (Art. 7.11) or a self-reducing tacheometer is employed for locating details by stadia, the note form and observational procedure is modified to accommodate the differences in the instrument. Figure 15.4 shows notes recorded for observations using an instrument with a Beaman stadia arc. The point number is recorded in column 1 and the interval is observed and recorded in column 2. When reading the stadia interval there is no need to have the middle cross hair near the value of the h.i. Next, the index on the $V$ scale is set to the nearest integral $V$ graduation using the vertical tangent motion. The values of $H$ and $V$ are observed and recorded in columns 3 and 4 and the rod reading or rod correction is observed and recorded in column 8. The azimuth is observed and recorded in column 6. Then the horizontal distance is

| Pt. | Int. (S) | H 木 | V @C | Horiz. Dist. | Az. | (V-50)(I) h.i. = | Rod Corr.(-) 4.4 ft. | ΔElev. | H.I. 428.3 | Elev. ft. 423.9 | K=100; f+c=0 |
|---|---|---|---|---|---|---|---|---|---|---|---|
| B | | | | | 176°14′ | | | | | | |
| 21 | 2.90 | 3 | 32 | 281 | 276° 40′ | -52.2 | 3.2 | -55.4 | | 372.9 | Flow Line Stream |
| 22 | 1.50 | 5 | 72 | 142 | 175° 20′ | +33.0 | 11.0 | +22.0 | | 450.3 | Top of hill |
| | | | | | | | | | | | |

**Fig. 15.4** Note form for details about a point using an instrument equipped with a Beaman stadia arc (units are feet).

calculated and recorded in column 5. As an example, the horizontal distance between $B$ and $21 = (K)(s) - (H)(I) = (100)(2.90) - (3)(2.9) = 281$ ft. Next, calculate the difference in elevation between the instrument and point sighted on the rod, and record in column 7. Again, for point $B$ to point 21, this value is $(V - 50)(s) = (32 - 50)(2.90) = -52.2$. Continuing for the line $B$-21, the change in elevation between the horizontal axis of the instrument and the ground point $= \Delta$ el. $= (V - 50)(s) -$ rod reading $= -52.2 - 3.2 = -55.4$ and is recorded in column 9. Note that the height of instrument above point $C$ (4.4 ft) has been added to the elevation of $C$ and recorded as the height of instrument above the datum or H.I. in column 10. The value for change in elevation or $\Delta$ el. can be applied directly to this H.I. to yield the elevation of ground point 21 recorded in column 11. Thus, elevation of $21 = 428.3 - 55.4 = 372.9$ ft.

When using one of the self-reducing tacheometers such as the one illustrated in Fig. 7.9, a note form similar to that shown in Fig. 15.4 but modified to accommodate the special features of the instrument can be employed. These instruments yield horizontal distance and difference in elevation simply by multiplication of the stadia interval times a constant of 100. Thus, when an experienced operator is available, there would be no need to record the cross-hair readings. The intervals could be observed and mentally multiplied by 100 so that horizontal distance and the change in elevation could be recorded directly. Instruments such as these frequently have an automatic vertical circle index compensator, and there is no need to center a level bubble prior to observing the intervals. Consequently, use of such instruments can yield substantial savings in time when locating details by stadia.

**15.11. Details by plane table**   The personnel of the plane-table party for mapping details by the controlling-point method usually consists of the plane-table operator, computer, and one or more rodmen. The equipment usually includes a plane table, telescopic alidade (preferably of the optical-reading, self-reducing type), scale, small triangles, 6H or 8H pencil, and electronic pocket calculator.

Before the party goes into the field, the horizontal control, the coordinate system, and the outline of the map sheet are adjusted and plotted on the plane-table sheet (Art. 7.20) according to the method developed in Art. 13.13. The elevations of all bench marks either are recorded on the sheet or are in the hands of the computer. It is also advisable to indicate the distance and direction for each course of the control survey on the plane table sheet.

The instrument operator sets up the table at a convenient station, orients it usually by backsighting on an adjacent station and measures the h.i. of the alidade above point $B$. The rodmen are directed to the controlling points of the terrain, as just described for the transit. When a rodman holds the rod on a ground point, the instrument operator sights on the rod and observes the data necessary in order to determine horizontal distance and difference in elevation from the instrument to the ground point sighted. These data are called out to the computer, who enters the information into the pocket calculator and makes the necessary reductions to provide horizontal distance to the point and its elevation. The plane-table operator motions to the rodman to move ahead, draws a short portion of the ray on the plane-table sheet from the station to the point sighted at the approximate location of the point, and then plots the point, to the scale of the plane-table sheet, using the distance provided by the computer. Finally, the point is marked with a small circle and the elevation is recorded on the map next to the point. As rapidly as sufficient data are secured, the plane-table operator sketches the contour lines. Other objects of the terrain are located and are drawn either in finished form or with sufficient detail so that they may be completed in the office.

The specific procedure used in observing the rod for each sighting varies somewhat depending on the type of alidade employed. There are three principal types of alidades currently in use: (1) those with a Beaman stadia arc equipped with a vertical circle control level bubble (Art. 7.11); (2) instruments with optically observed scale readings analogous to the Beaman arc, some of which have vertical circle control level bubbles and others that are equipped with automatic vertical circle indexing (Art. 7.19); and (3) alidades with optically observed self-reducing scales with variable stadia intervals which lead directly to the horizontal distance and difference in elevation when multiplied by a constant (Art. 7.17). These last instruments have either a vertical circle control bubble or automatic vertical circle indexing. The methods for using these instruments and the computations to reduce the measurements to distance and elevation are detailed in the sections indicated above in Chap. 7 and need not be repeated here.

Generally in practice, no record is made of the data observed and used to plot points on the plane table. Since all plotting is done in the field, mistakes or omissions are apparent at once and can be corrected immediately. However, for instructional purposes a record of the data observed can be useful for future reference. When notes of plane-table operations are recorded, a form similar to the ones indicated in Figs. 15.3 and 15.4 for transit stadia can be employed. Naturally, the column for azimuth is deleted and column headings are modified to suit the type of alidade being used.

Figure 15.5 illustrates such a note form designed for observations taken with the alidade shown in Fig. 7.18, an optical-reading instrument with a vertical circle control level bubble. The readings on point 1 are for the setting illustrated in Fig. 7.19. The procedure for obtaining these data for point 1 is as follows. The alidade is sighted on the rod and the stadia interval $s$ is read by setting the bottom stadia cross hair on an integral decimetre or foot graduation. This interval is recorded in column 2. Next, the vertical circle index bubble is centered, the $V$ scale is set to $-8.0$ using the vertical tangent motion, and the rod is read at the middle cross hair (1.25 m). The values of $H$, $V$, and the rod readings are recorded in columns 3, 4, and 7, respectively. The computer calculates the horizontal distance, $(V)(s)$, difference in elevation, and elevation of the point which are recorded in columns 5, 6, 8, and 10, respectively. It must be emphasized that these notes are *not* to be used for plotting the data at a later time. The notes provide a training device which helps systematize party operation during the initial stages of learning to use the plane table. The plotting and drawing of all contours and planimetric detail is done in the field on the plane-table sheet.

The utmost skill of the topographer is used in judging the features of the terrain and

| PLANE | TABLE | SURVEY | | | | ARMSTRONG PROPERTY | | May 6, 1978 | | |
|---|---|---|---|---|---|---|---|---|---|---|
| Inst. @ A | h.i. = 1.22 m | | | | | | | | Sunny 70°F | |
| Point | S | H | V | Horiz. Dist. | V×S | Rod Corr. (-) | Δ Elev. | H.I. | Elev.,m | ⊼ -Jones |
| B | | | | | | | | 31.22 | 30.000 | Notes -Arnold |
| | | | | | | | | | | Rod -Adams |
| 1 | 0.348 | 0.6 | -8 | 99.8 | -2.78 | 1.25 | -4.03 | | 27.19 | Top of inlet |
| 2 | 0.375 | 0 | +2 | 37.5 | 0.75 | 1.12 | -0.37 | | 30.85 | Edge of sidewalk |

**Fig. 15.5** Notes for details with a plane table.

representing them on the map with the required precision and with the least expenditure of time.

Many objects are located by the method of intersection, the elevations being determined by trigonometric leveling.

Because the plane table permits a ready solution of the three-point problem (Art. 9.10), use is made of this technique to enable the topographer to utilize advantageous instrument stations which have not been included in the control surveys, especially where control has been initially established by triangulation. The elevation of such a station is determined either by stadia leveling or by trigonometric leveling.

**Transit and plane table**   For large-scale maps and where many details are to be sighted, sometimes it is advantageous to use both the transit or self-reducing tacheometer and the plane table. This method saves time in the field, but it may not reduce the total cost, as a larger party is required than for the plane table alone.

The transit or tacheometer is set up and oriented at the control station, the location of which is plotted on the plane-table sheet. The plane table is set up and oriented nearby, and its location is plotted on the map in its correct relation to the transit station. (In some cases the map distance between the transit station and the plane-table station is negligible, and the two points are regarded as identical.) When a rodman has selected a ground point, the transit operator observes the stadia distance and vertical angle to it; the plane-table operator sights in the direction of the point, draws a ray toward it from the plotted location of the plane-table station, plots the point at the correct distance scaled from the plotted location of the transit station, and records on the map the elevation (computed by the transit operator or the computer) of the plotted point.

**15.12. Details by the cross-profile method**   In the cross-profile method of locating details, the ground points are on relatively short lines transverse to the main traverse or base line. The control points are stations on the transit traverse. These stations are set along the traverse lines at an interval which is a function of the irregularity of the terrain and the desired contour interval. When the unit is the foot, a full station is 100 ft and stations are usually set at 100- or 50-ft intervals. In the metric system, a full station is 100 m, so that stations are generally set at 10- or 20-m intervals. This horizontal control can be established by transit and EDM equipment, transit and tape, or transit stadia (between control stations established by taping or EDM equipment). The elevations of these stations are determined by differential profile leveling and are available to the topography party. Cross profiles are particularly appropriate when the area to be mapped is a long, narrow band such as a route survey (Chap. 17).

The cross profiles (or cross sections as they are sometimes called, particularly in route surveying) are taken using a level rod plus a transit and tape, level and tape, transit stadia, hand level and tape or plane table and stadia. The party consists of a topographer and two people to work with the rod and tape. Sometimes a Jacob's staff or other rod about 5 ft long is used as support for the hand level while sights are being taken.

When the transit, rod, and tape are used for cross profiles, the transit is set over the specified station and the height of instrument (h.i.) is determined by holding the rod alongside the transit. A sight is taken on another station on the same line and an angle of 90° is turned, establishing the line of the cross profile. The rodman moves out along this line to the first break in the terrain. The distance is measured from the center line with the tape and a rod reading is taken with the telescope leveled. Distance and elevation are recorded and the procedure is repeated at the next and all subsequent breaks in terrain along the cross profile to the limit of the area to be mapped. If the terrain slope is too steep to allow a rod reading with the telescope level, a vertical angle is turned to a point on the rod equal to the height of the instrument above the center-line station and a slope distance is measured to the point. The difference in elevation equals the slope distance times the sine of the vertical angle. The horizontal distance equals the slope distance times the cosine of the vertical angle. Use of the transit and tape is feasible for cross profiles if the strip to be mapped is not too wide and the terrain is not excessively steep and irregular. If it becomes necessary to move the instrument ahead to complete the profile, considere an alternative method.

Figure 15.6 shows one form of field notes for cross-profile data taken from station 23+00 to 24+00 with 20-m stations. The stations are recorded from the bottom of the page to the top and cross-profile data are on the left and right of the page, corresponding to the left and right sides of the center line when facing toward the next higher station. As an example, consider station 24+00, where the h.i. at the center line measured from the ground point to the instrumental horizontal axis is 1.65 m and the center-line elevation is 426.60 m, taken from the profile notes. Thus, the instrument H.I. is 426.60 + 1.65 = 428.25 m, recorded above the center-line reading and enclosed in brackets. The offset distance to the first break point on the right is 10 m where the rod reading, taken with the telescope level, is 0.40 m and is recorded below the 10-m offset distance. The elevation of this point = 428.25 − 0.40 = 427.8 m, which is calculated immediately and recorded beneath the rod reading. At the next break point to the right of center line, the ground is higher than the instrumental H.I., so that a vertical angle of +2°52′ was observed to 1.65 m on the rod at a slope distance of 15.0 m, resulting in a change in elevation of +0.75 m and a horizontal distance of 15 m from the center line. This procedure is then repeated to obtain readings 20 and 30 m to the right. Similarly, rod readings are taken with the telescope level at 10, 17, 20, and 30 m to the left of the center line.

|  |  | Left |  |  |  | Center Line |  |  | Right |  |  |
|---|---|---|---|---|---|---|---|---|---|---|---|
|  | H.I. |  |  |  |  | [428.25] |  |  |  |  |  |
|  | Offset | 30 | 20 | 17 | 10 | 0 | 10 | 15 | 20 | 30 |  |
| 24+00 |  | 2.65 | 3.45 | 4.05 | 3.15 | 1.65 | 0.40 | +0.75 | +2.75 | +3.25 |  |
|  | Elev. | 425.60 | 424.80 | 424.20 | 425.10 | 426.60 | 427.80 | 429.00 | 431.00 | 431.20 |  |
|  |  |  | F.L.Stream |  |  | [428.05] |  |  | Ridge |  |  |
|  |  | 30 | 20 |  | 10 | 0 | 10 | 20 | 25 | 30 |  |
| +80 |  | 3.55 | 3.85 |  | 2.95 | 1.55 | 0.25 | +1.55 | +3.05 | +2.15 |  |
|  |  | 424.50 | 424.20 |  | 425.10 | 426.50 | 427.80 | 429.60 | 431.10 | 430.20 |  |
|  |  |  | F.L.Stream |  |  | [427.50] |  |  |  |  |  |
|  |  | 30 | 24 | 20 | 10 | 0 | 10 | 20 | 24 | 30 |  |
| +60 |  | 4.50 | 4.30 | 3.60 | 2.40 | 1.60 | 0.30 | +1.40 | +1.80 | +1.10 |  |
|  |  | 423.00 | 423.20 | 423.90 | 425.10 | 425.90 | 427.20 | 428.90 | 429.30 | 428.60 |  |
|  | F.L.Stream |  |  |  |  | [427.10] |  | Ridge |  |  |  |
|  | 35 | 32 | 30 | 20 | 10 | 0 | 10 | 18 |  | 30 |  |
| +40 | 4.90 | 5.30 | 4.20 | 3.60 | 2.50 | 1.50 | 0.10 | +1.00 |  | +0.10 |  |
|  | 422.20 | 421.80 | 422.90 | 423.50 | 424.60 | 425.60 | 427.00 | 428.10 |  | 427.00 |  |
|  | F.L.Stream |  |  |  |  | [426.45] |  | Ridge |  |  |  |
|  | 40 | 36 | 30 | 20 | 10 | 0 | 10 | 20 |  | 30 |  |
| +20 | 5.35 | 4.00 | 3.70 | 2.00 | 1.20 | 1.65 | 0.25 | +0.65 |  | 0.55 |  |
|  | 421.20 | 420.80 | 421.10 | 422.80 | 423.60 | 424.80 | 426.20 | 427.10 |  | 425.90 |  |
|  |  |  |  |  |  | [425.20] |  | Ridge |  |  |  |
|  |  |  | 30 | 20 | 10 | 0 | 10 | 17 | 20 | 30 |  |
| 23+00 |  |  | 4.40 | 2.30 | 2.30 | 1.30 | +0.30 | +0.90 | +0.80 | 0.30 |  |
|  |  |  | 420.80 | 422.90 | 422.90 | 423.90 | 425.50 | 426.10 | 426.00 | 424.90 |  |

**Fig. 15.6** Notes for the cross-profile method (units are metres).

Where specific planimetric details fall within the area being mapped, they are located by plus and offset. For example, the flow line of the stream is located 20 m to the left of station 23 + 80, 24 m left of station of 23 + 60, etc. Similarly, any type of planimetric features, such as buildings, utilities, trees, fences, etc., can be located in the cross-profile notes by means of plus and offset, as described at the end of this section.

One of the most common methods for taking cross profiles is by using a hand level, rod, and tape. For this operation, the right angle is established previously by the transit, by estimation, or by using a *pentaprism*, a hand-held device for setting off a right angle. The center-line profile is run first using a rod and level. Consequently, elevations for each center-line station are available. First, the direction of the cross profile is established and the rod is held on the center line. The hand-level operator is stationed along this cross profile in a position which allows comfortable viewing of the rod. Since the hand level has no magnification, lengths of sights should be no more than 15 m or about 50 ft. If the ground is rising, the center-line reading should be as high as possible. If the terrain slopes away from the center line, the reading on the rod at center line should be as low as possible. For example in Fig. 15.7, the original rod reading in the profile notes for the center line was 1.50 m, assuming that the profile was run independently. In order to run the cross profile up the slope the hand-level operator is positioned along the line of the cross profile so that a backsight of 3.50 m can be taken on the rod held at the center-line station. Thus, the hand level is "up 2 m" and it is the responsibility of the notekeeper to subtract 2 m from all subsequent rod readings taken from that position. In the example, a reading of 2.40 m is taken 10 m right and is recorded as $(2.40-2)=0.40$ at 10 m right. The next break point is 20 m right, where the rod reading is 1.00 m, which is recorded as +1.00 m, 20 m right. The rod reading is positive since the bottom of the rod is now 1 m above the original H.I. of the instrument used to obtain the center-line reading of 1.50 m, to which all rod readings on the cross profile are referred. At this point, the hand-level operator moves ahead to a position which permits a backsight of 3.00 m on the rod still held 20 m right. Now the hand level is "up 4 m" and 4 m must be subtracted from all subsequent rod readings so as to make them consistent with the rest of the cross profile. The next reading at 30 m right is 1.60 m, recorded as +2.40 m, 30 m right; and the final reading to the right of center line is 1.40 m, recorded as +2.60 m, 35 m right. The note form shown below the sketch is similar to that employed in Fig. 15.6. When running a cross profile with a hand level, a stake is placed at the position of the last reading on each profile. Then each profile is closed on the previous line to that side of the center line and mistakes are avoided. The closures should be within 3 dm or 1 ft. Since all rod readings have been referred to the profile H.I. (recorded above the center-line data and enclosed in brackets [315.58]), the readings are subtracted from or added to this H.I. to yield ground-point elevations as shown in the sample cross-profile notes of Fig. 15.7. The same procedure can be applied to terrain that slopes downward from the center line. In this case the height of the hand level will be "down" a given number of metres or feet, which will be added to all readings taken along that profile.

In regions of slight relief or in urban areas where it is necessary to obtain elevations and locations of pavement, curb lines, sidewalks, utilities, etc., the engineer's level, tape, and rod are frequently used for cross profiles. When this procedure is followed, the center-line profile can be run concurrently with the cross profiles or as a separate operation. In either case, center-line elevations are available. The procedure is as follows. The level is set up at a location convenient for viewing to the left and right of the center or base line and the rod is held on a center-line station where a reading is taken. The right angle to set the direction of the cross profile may have been previously established by a transit or is now estimated by the rodman. The rodman then moves out along the cross profile. Rod readings are taken and distances are measured to significant features and break points along the cross profile to the left and right of the center line.

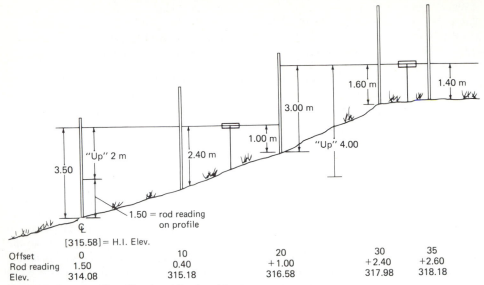

| Offset | 0 | | 10 | 20 | 30 | 35 |
|---|---|---|---|---|---|---|
| Rod reading | 1.50 | | 0.40 | +1.00 | +2.40 | +2.60 |
| Elev. | 314.08 | | 315.18 | 316.58 | 317.98 | 318.18 |

**Fig. 15.7** Cross profile with a hand level and tape.

Figure 15.8 illustrates a note form for cross profiles using a level, tape, and rod in an urban area. English units are employed in these notes. A plan view of the area involved is shown in Fig. 15.9a. The plan positions for the two instrument stations and two turning points utilized in the notes of Fig. 15.8 are also shown on Fig. 15.9a. Locations of instrument setups are usually chosen so as to allow a maximum number of readings from a single setup and also to balance backsight and foresight distances between turning points. For this example, stations are set along the center line of the existing pavement and the center-line profile and cross profiles are being run concurrently. Thus, the notes are recorded from the top of the page down, as is generally done for regular profile levels, in contrast to the example in Fig. 15.6, where the cross profiles are run independently of the center-line profile. Note that all objects are described and rod readings on semipermanent features such as pavement, curbs, sidewalks, etc., are read to the nearest hundredth of a foot while readings on the ground are to the nearest tenth of a foot. A sectional view of the cross profile at station 11 + 00 is shown in Fig. 15.9b oriented as one would see it when standing at station 11 + 00 and looking toward station 11 + 50. For convenience, the level is shown in this sectional view even though it does not fall on the line of the cross profile, as can be seen in Fig. 15.9a. Since the center-line profile is also being run, all rod readings (foresights) are subtracted from the H.I. established at T.P.$_3$ (turning point 3) in the profile levels. A new turning point, T.P.$_4$, is taken in order that the level can be positioned so as to run the cross profile at the next station, station 11 + 50. Frequently, when lack of visibility or irregular terrain prevents completing a cross profile, a hand level can be used to finish the cross profile, assuming that rod readings to the nearest tenth of a foot or to the nearest 0.03 m are sufficiently accurate.

Cross profiles with transit stadia are essentially the same as those run using the transit, tape, and rod, except that distances and differences in elevations to break points and other features are determined by the stadia method. This method is particularly advantageous when the terrain is exceptionally steep and irregular.

When cross profiles are run using a plane table and alidade, the center line is plotted on the plane-table sheet. The plane table is then set over a station and the plotted center line is aligned with the center line in the field. The alidade is aligned with a perpendicular drawn

|  | B.S. | H.I. | F.S. |  |  |
|---|---|---|---|---|---|
| TP3 | 1.45 | 877.90 | 5.24 | 876.45 |  |
|  |  | 25 | E.S.W. 16.2 | Top Curb 12 | Gutter 12 |
| 10+50 |  | 5.0 | 4.20 | 4.30 | 4.80 |
|  |  | 872.9 | 873.70 | 873.60 | 873.10 |

| ¢ | Gutter | Top Curb | Edge Side Walk |  |  |
|---|---|---|---|---|---|
| 0 | 12 | 12 | 16 | 25 | 50 |
| 4.50 | 4.75 | 4.25 | 4.30 | 5.3 | 7.4 |
| 873.40 | 873.15 | 873.65 | 873.60 | 872.6 | 870.5 |

|  |  | E.S.W. | T.C. | G |
|---|---|---|---|---|
|  | 25 | 16 | 11.9 | 11.9 |
| 11+00 | 4.2 | 3.20 | 3.20 | 3.70 |
|  | 873.70 | 874.70 | 874.70 | 874.20 |

| ¢ | G | T.C. | E.S.W. | 2 | Ground at Side of House |
|---|---|---|---|---|---|
| 0 | 12.1 | 12.1 | 16.1 | 25 | 45 |
| 3.45 | 3.74 | 3.25 | 3.30 | 4.8 | 5.5 |
| 874.45 | 874.16 | 874.65 | 874.60 | 873.1 | 872.4 |

| ¢ |  | Floor Utility Room | Ground S.W.Cor. I.-S.-F. |
|---|---|---|---|
| 0 |  | 32.5 | 32.5 |
| 11+04 | 3.37 | 3.55 | 4.0 |
|  | 874.53 | 874.35 | 873.9 |

| ¢ | Grd. Cor. I-S-Fr. 32.2 |  |
|---|---|---|
| 0 |  |  |
| 11+07.8 | 3.29 | 4.4 |
|  | 874.61 | 873.5 |

| ¢ | Ground N.W. Cor. 39.5 | First Floor I.-S.-Fr. 39.5 |
|---|---|---|
| 0 |  |  |
| 11+38.4 | 2.85 | 4.1 | 3.50 |
|  | 875.05 | 873.8 | 874.40 |

| TP4 | 3.45 | 878.85 | 2.50 | 875.40 |
|---|---|---|---|---|
| 11+50 |  |  |  |  |

¢
1.52
877.33

**Fig. 15.8** Cross-profile notes in an urban area using an engineer's level and rod (units are feet).

through the station and the cross profile is run along this line. Distances and differences in elevation to break points and features are determined by stadia with the point and elevation being plotted directly on the sheet. Thus, there is no need to keep a separate set of notes. One advantage of utilizing a plane table for cross profiles is that all additional planimetric detail can be easily located by the controlling-point method and plotted directly on the plane-table sheet.

The completed topographic map should also show the planimetric position of all significant natural and constructed features in the area. Some of these features may not be obtained by the cross profiles where the station interval is 20 m or 50 ft or larger, so that a supplementary operation is required to locate the additional detail. When cross profiles are taken by transit or plane-table stadia, these methods are also employed to locate details between the cross profiles by distance and direction.

If cross profiles are run by engineer's level or a hand level and tape, additional details are usually located by *plus* and offset from center or base line. A plus is the numerical value for the center-line station of a line perpendicular to the center line and also passing through the object to be located. The distance from the station to the object is the offset. Figure 15.9a illustrates the location of a building by plus and offset. Thus, the northwest corner of the building is located by an offset of 39.5 ft to the right of station 11+43.1. A station on the intersection of a *range* line from the north end of the building with the center line is also indicated as being 11+53.0. This provides a check on the plus and offset location. For intermediate-scale and many large-scale maps, the plus is determined by estimation or using a pentaprism. When higher precision is required, such as in a property survey where it is

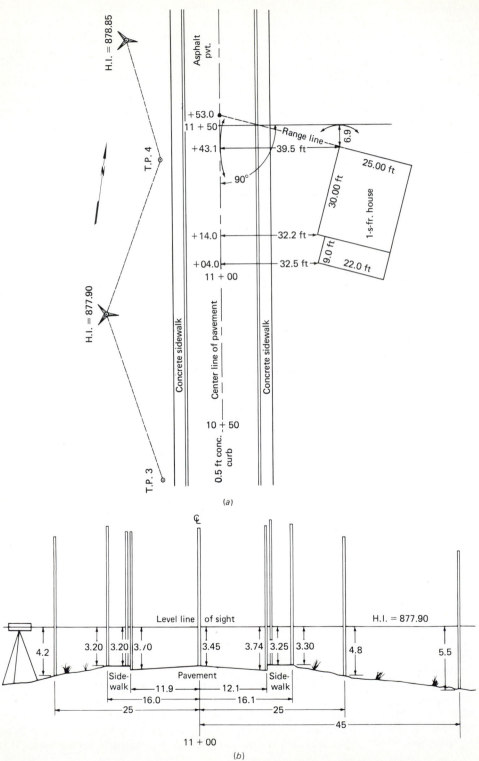

**Fig. 15.9** (a) Plan view for cross profiles in an urban area and location of details by plus and offset. (b) Section through cross profile of 11 + 00 from the notes of Fig. 15.8 (units are feet).

desired to show the location of a structure with respect to property lines, a transit and tape are used. For example, in Fig. 15.9a the transit is set over station $11+50$ and sighted on $10+50$. The zero end of the tape is held on the northwest corner of the building and the shortest distance between the corner and center line is determined by *swinging* the tape. Swinging the tape is analogous to swinging (or waving) the rod as described in Art. 5.34, except that the tape is swung in a horizontal plane to and from the transit. In this way the transit operator can observe the shortest distance between the line of sight, directed on the tape, and the object being located at the zero end of the tape. A continuously graduated tape with legible numbered graduations is necessary for this purpose. On long sights the transit operator aligns a pencil placed against the face of the tape and used as a target. In Fig. 15.9a the offset distance determined this way is 39.5 ft. Next, an angle of 90° is turned to the left from the center line and the offset between the northwest corner of the building and a line normal to the center line at the station occupied is determined by swinging the tape. Again referring to Fig. 15.9a, this offset is recorded as 6.9 ft. The precision of this operation can be to within the least count of the tape but definitely can be no better than the precision of the control survey. Plus and offset notes can be recorded with data for cross profiles as indicated in Fig. 15.8. However, when much detail must be located, a separate sketch similar to the one in Fig. 15.9a should be made.

The data from the cross profiles are plotted on a master sheet to the scale desired, and the contours are plotted by interpolation as described in Chap. 13. The cross-profile data from Fig. 15.6 are plotted in Fig. 15.10. The stations are indicated along the plotted center line, perpendiculars are drawn through these stations, and the points located are plotted on these perpendiculars. The elevations of these points would be recorded on the sheet lightly and then the contours are drawn by interpolation. This portion of the job could be done directly in the field when plane-table stadia are used to collect cross-profile data.

## 15.13. Contour points with hand level

The party proceeds from station to station along the traverse. At each station the topographer notifies the hand-level operator and rodman of the elevation of the station. The head rodman carrying the rod moves out on a line estimated to be at right angles with the traverse line until the rod is on the next contour (either higher or lower) from the station, as determined by the hand-level operator; the distance out to the contour is then measured with the tape. The hand-level operator then goes out to the point occupied by the rod, and the rodman again moves out until the next contour is reached; and so the process is repeated until all contour points are located out to the edge of the strip being surveyed. A similar procedure is followed on the other side of the traverse line. Usually the trends or directions of the contours are sketched at each crossline and along ridge and valley lines, but on the field sheets the contour lines are not sketched for their full length. Definite details are located with relation to the transit line by tape measurements. If the topography is regular, sometimes the sketches are omitted, and the distances from traverse to contour points are recorded numerically.

**Example 15.1**   The method is illustrated by reference to Figs. 15.11 and 15.12, which represent the field notes and the finished map, respectively, of a preliminary route survey. The contour interval is 5 ft. Suppose that the topography party has reached station $9+00$, where the topographer notifies the party members that the elevation of that station is 821.1 ft. To the left of the traverse line the ground slopes downward. To locate the 820-ft contour, the rodman carrying the rod (which is graduated from the bottom) and the zero end of the tape moves out to the left until the hand-level operator by the use of the hand level supported, say, on a 5.0-ft staff, reads 6.1 on the rod ($821.1 + 5.0 - 6.1 = 820.0$). The horizontal distance out from the station is read on the tape and called to the topographer who plots its location (9 ft from the traverse line). To locate the 815-ft contour, the hand-level operator moves out to the 820-ft contour, and the rodman moves out until the hand-level operator reads 10.0 on the rod ($820.0 + 5.0 - 10.0 = 815.0$); the horizontal distance between the two contours (26 ft) is read from the tape, the second point is plotted; etc. A similar procedure is followed in going uphill.

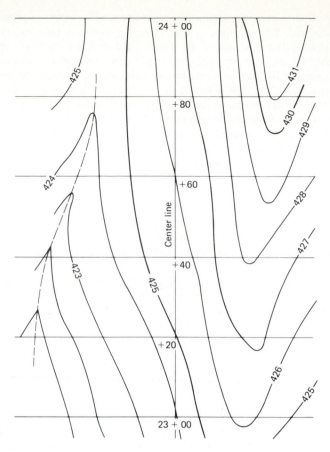

**Fig. 15.10** Contour lines
(1-m interval) interpolated from
cross-profile data in Fig. 15.6.

**15.14. Details by the checkerboard method**   The checkerboard method of locating details
is well adapted for large-scale surveys, as the points are located in plan by tape measurements. It is also useful where the tract is wooded, where the topography is smooth, and on
urban surveys where blocks and lots are rectangular. The tract is staked off into squares or
rectangles—usually 50- or 100-ft squares (15-, 20-, or 30-m squares). The ground points and
other details are then located with reference to the stakes and connecting lines.

The usual procedure is first to run a traverse near the perimeter of the tract such as
traverse $A$-6-6'-$K$ shown in Fig. 15.13. Next, one side of the traverse is chosen as a base line
and is marked off in stations of the desired interval. In Fig. 15.13, if line $AK$ is selected as
the base line and the interval chosen is 20 m, then $A=0+00$, $B=0+20$, $C=0+40$, etc.
Stakes are set at each 20-m station along the base line. Next each station along the base line
is occupied with the transit, an angle of 90° is turned from the base line $AK$, and stakes are
set at 20-m intervals along lines $BB', CC',\ldots, GG'$. Points $B'$, $C'$, $D'$, $E'$, and $F'$ are set by
intersecting the respective grid line with line 66', usually by estimation. Points 1 through 6
and 1' through 6' can be set by ranging in along the appropriate interior grid line and
intersecting the grid line with the traverse line. All points are labeled by number and letter
which designate row and column, respectively. Thus, along the grid line from 1 to 1' there
are stations $1A, 1B,\ldots, 1H$. Several different methods for setting the grid intersections are
possible, all of which would be satisfactory if the accuracy required for the survey is
satisfied. The method outlined would provide fairly precise results for horizontal position of
the points.

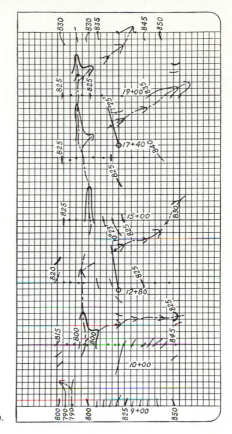

**Fig. 15.11** Notes for contour points by cross profile.

When all stations have been set and marked, levels are run and an elevation is determined at each grid-line intersection and the intersections of the grid lines with the perimeter traverse lines. An engineer's level or an engineer's transit can be used for the leveling. In addition, all irregularities and significant planimetric features are located from the nearest pair of intersecting grid lines. When all elevations have been determined and details located, the data are plotted on a map sheet on which the traverse and grid lines have been plotted to the desired scale of the map. Contours are drawn by interpolation and all planimetric features are plotted to complete the topographic map of the area.

**15.15. Details by the trace-contour method**    The trace-contour method of locating contour points on the ground is commonly used on large-scale surveys, and sometimes on intermediate-scale surveys where the ground is irregular. Under these conditions, if visibility is good, the trace-contour method is more accurate than the checkerboard method.

Although the transit may be used in this work, either alone or with the engineer's level, the plane table is commonly used because it requires fewer points to be observed and because of the saving of time in plotting.

A control survey is run and plotted on the plane-table sheet. The plane table is set up over a control point of known position and elevation, such as point *A* (plotted as *a* on the plane-table sheet) in Fig. 15.14, and the plane table is oriented by sighting control point *B*. The height of the instrument above point *A* is measured using the rod and found to be 1.25 m. If the elevation of *A* is known to be 352.40 m, the H.I. = 352.40 + 1.25 = 353.65 m. With the H.I. known, rod readings for specific contours can be calculated. For example, to locate

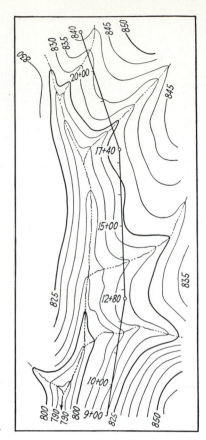

**Fig. 15.12** Preliminary map for route survey.

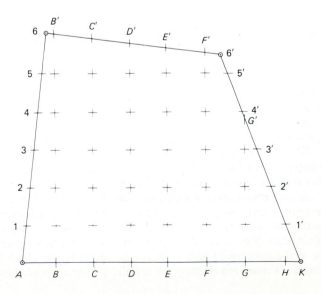

**Fig. 15.13** Checkerboard system
for topographic survey.

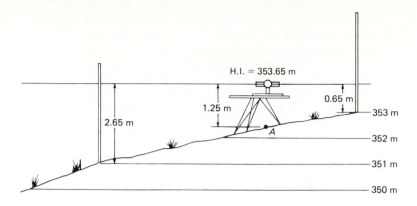

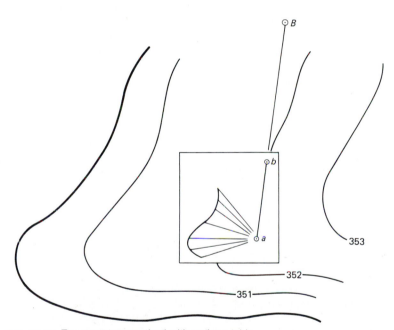

**Fig. 15.14** Trace-contour method with a plane table.

the 351-m contour, the rod reading with the telescope of the alidade leveled should be
$353.65 - 351 = 2.65$ m. The plane-table operator informs the rodman what reading is required
to locate the desired contour, and the rodman estimates where the rod should be held to
obtain this reading. The plane-table operator focuses the telescope on the rod, centers the
vertical circle level bubble, sets the vertical circle index to zero, and then directs the rodman
up or down the slope until the line of sight cuts the rod at 3.65 m. The plane-table operator
checks the vertical circle bubble and index reading, observes the stadia interval, draws a ray
along the alidade blade, and plots the calculated horizontal distance to the location of the
contour on the plane-table sheet. The rodman then moves along the slope to another
estimated location of the contour and the procedure is repeated until the contour has been
traced throughout the range of the control on the plane-table sheet. As soon as at least three
points have been plotted, the plane-table operator should draw and label the contour line

interpreting all the irregularities that can be observed as the contour is traced. Readings should not be taken too close together in fairly regular terrain. Where the terrain is highly irregular, rod readings should be taken on all break points in the terrain. When the 351-m contour has been located, the rod reading for the 352-m contour, calculated to be 353.65 − 352 = 1.65 m, is traced. From the setup shown in Fig. 15.14, the 353-m contour could also be traced, and then a new plane-table setup would have to be made. The trace-contour method is quite accurate but tends to be slow and costly. Thus, this procedure should not be used unless it is absolutely necessary in order to meet rigid accuracy requirements.

When contours are traced using transit stadia or a self-reducing tacheometer, the procedure is the same except that the H.I., stadia interval, rod-reading horizontal distance, and azimuth to the contour locations are recorded in a field notebook. The instrument is set over a known control point and a backsight is taken along a line of known azimuth using the lower motion. The azimuth of the known line is set off on the clockwise horizontal circle before taking this sight. All sights to locate the contour being traced are taken with the upper horizontal motion and with the telescope in a horizontal position.

The engineer's level can also be used to trace contours provided that the telescope has stadia hairs and the level has a horizontal circle. The self-leveling instrument shown in Fig. 5.17 is so equipped.

When contours are traced using transit stadia, self-reducing tacheometer, or engineer's level, the data are plotted in the office. First, a base map must be prepared on which all control points are plotted. Then the contour points are plotted by distance and azimuth from the given control points using a protractor and scale. Finally, the contour points are connected to form contour lines.

**15.16. Electronic positioning systems for topographic surveys**   Reference to Art. 12.2 shows that the combined theodolite and EDM systems and electronic tacheometers automatically provide horizontal distance and difference in elevation between the station occupied and the ground point sighted. Since these are the basic quantities involved in all of the outlined methods for topographic surveying, electronic positioning systems are ideally suited for gathering data for topographic surveys, particularly by the controlling-point method. The procedures for use of such a system for a topographic survey by the controlling-point method are similar to those suggested for acquiring topographic details by transit or self-reducing tacheometer in Art. 15.10. The major differences occur in the operation of the instruments and methods for recording data.

When the electronic tacheometer is used for a controlling-point topographic survey, the instrument is set up over a known control point and oriented by a backsight on another known station. Then the target or signal is placed at a desired ground point and is sighted by the theodolite. On the simplest of these systems, proper manipulation of the controls on theodolite and EDM (Art. 12.2) yields visual output consisting of horizontal distance, direction, and difference in elevation to the point. These data are recorded in the field notebook for subsequent plotting in the office. The more sophisticated electronic tacheometers not only have a visual display of the positional data but also allow for storage of the data in solid-state core memory, on punched-paper tape, or on magnetic tape. Voice data entry, described in Art. 3.14, provides a method potentially useful for recording verbal descriptions automatically. Thus, the need for complete field notes is reduced and the bulk of the data are automatically stored for processing in the office. Consequently, the time needed for the operation is limited to the time required to move the signal from one point to another and that necessary to operate the instrument.

Under certain conditions, use of the electronic positioning systems with a plane table is appropriate. The procedure would be similar to the method described in Art. 15.11 for the transit and plane table. Acquisition of data would be substantially faster in this operation so that the limiting time factor would be the ability of the plane-table operator to plot and draw topographic detail. The advantage of this method is that a more reliable map is produced because of the proximity of the map compiler to the terrain.

Electronic positioning systems could also be utilized for cross-profile topographic surveys, especially where the cross profiles or cross sections are being run for route alignment in rugged, steep terrain. The procedure would be similar to the method outlined for cross profiles using the transit, tape, and rod in Art. 15.12. The transit and tape are replaced by the electronic positioning instrument and the rod is replaced by the reflector or signal. Data are then recorded or stored as previously described in this section for controlling-point topography.

**15.17. Topographic surveys and digital terrain models**   The digital terrain model (DTM) or digital elevation model (DEM) is described in Arts. 13.18 and 16.19 along with some of its uses and applications in topographic mapping. Since the DTM is a numerical ($X$ and $Y$ coordinates and elevations) representation of the configuration of the terrain, it can be stored in the memory of a computer and may be processed by a computer program which permits plotting planimetric detail and contours. Thus, the DTM is an extremely flexible method for storing data which describes terrain configuration.

The reason for inserting these comments at this time is to emphasize that the data from each of the methods for obtaining topographic detail constitutes a DTM in the most basic form.

The methods for acquiring topographic detail described in this chapter are designed primarily for the small or special purpose topographic survey or map completion survey. The traditional methods for plotting and compiling maps of this type will probably continue to be used well into the future. However, the surveyor or engineer involved with these special mapping problems needs to be aware of the potential of the data being collected. In this way field methods and data recording and storage procedures can be designed so as to be compatible with possible automatic data processing in the future.

**15.18. Problems**

**15.1**   Describe the major steps involved in performing a topographic survey.
**15.2**   Differentiate among large-, intermediate-, and small-scale topographic surveys.
**15.3**   What factors govern the choice of surveying method used for topographic surveys?
**15.4**   Discuss the various surveying methods appropriate for establishing horizontal and vertical control for topographic mapping and the conditions that favor each method.
**15.5**   What sort of equipment would be required for running a three-dimensional traverse to establish horizontal and vertical control for a topographic survey?
**15.6**   List the four principal methods for acquiring topographic details.
**15.7**   The controlling-point method is appropriate for locating details in what sort of terrain? What are the advantages of this method?
**15.8**   The cross-profile method of locating details is best adapted to what type of topographic survey?
**15.9**   Describe how a cross profile may be run using a transit and a tape. Assume that the terrain is steep but of constant slope on both sides of the center line. Use a sketch to illustrate your explanation.
**15.10**   Use a sketch to show how details can be located from a base or center line by plus and offset.
**15.11**   Describe the principal features of the checkerboard method for locating topographic details.
**15.12**   What are the advantages and disadvantages of locating topographic details by the trace-contour method?
**15.13**   Describe how electronic positioning systems can be used for locating topographic details.
**15.14**   Discuss the relationship between surveys to obtain topographic details and digital elevation models.

## 15.19. Field problem

ACQUIRING TOPOGRAPHIC DETAILS BY THE CONTROLLING-POINT METHOD

**Object**  To obtain a sufficient amount of data for making a topographic map of a small area having a moderate amount of relief by using the controlling-point method (Art. 15.9). Data from this problem can be used in Prob. 1 (Art. 13.40).

**Procedure**

1. Make a rapid reconnaissance of the tract, selecting the most advantageous points for instrument stations from which areas comprising the entire area can be observed.
2. Run a closed traverse through the selected points, observing the stadia intervals and vertical angles. Follow the procedures suggested in Field Problem 3, Chap. 8, using a transit or tacheometer and stadia. If an electronic surveying system (Art. 12.2) is available, this may be used to run the traverse.
3. If details are to be located by transit stadia (Art. 15.10), occupy each of the traverse stations and with the instrument correctly oriented observe the azimuth, stadia distance, and vertical angle to all changes in ground slope and to other natural and artificial features which are within range of the instrument.
4. Include in the notes a sketch drawn approximately to scale. Use a note form similar to Fig. 15.3.
5. Reduce the stadia readings to horizontal distances and differences in elevation. Plot control points and details as directed in the office problem of Chap. 13.
6. If topographic details are to be located by a plane table, adjusted coordinates (Arts. 8.18 and 8.19) and elevations (Art. 5.50) are calculated.
7. Plot the positions of horizontal control points (see the instructions in Art. 13.13) on a plane-table sheet (Art. 7.20) of heavy detail paper or drafting film.
8. Occupy each station with the plane table, locating detail according to the instructions given in Art. 15.11.
9. Draw planimetric features and contours *in the field*. On student projects it is also advisable to keep notes using a form similar to Fig. 15.5.

### Hints and precautions

1. If the elevation of a point is not required for mapping, often the point can be located advantageously by the method of intersection, the azimuth being observed from two or more traverse stations.
2. When using transit stadia, it is sometimes advantageous, particularly if there are a large number of details, to plot the map in the field as the work progresses. This is most easily accomplished by using a plane table and a transit as suggested at the end of Art. 15.11.
3. When details are close to the horizontal control station, it is often faster to get the horizontal distance by measuring with a tape rather than reading the stadia interval.

### References

1. American Society of Civil Engineers, *Definitions of Surveying and Associated Terms*, Manual of Engineering Practice No. 34, 1978 (rev.).
2. American Society of Civil Engineers, *Technical Procedures for City Surveys*, Manual of Engineering Practice, No. 10, 1963.
3. Bere, C. G. T., "Direct Field Plotting Using Short Range EDM," *Survey Review*, January 1976.
4. Chrzanowski, A., et al., "Control Networks with Wall Monumentation: A Basis for Integrated Survey Systems in Urban Areas," *The Canadian Surveyor*, September 1977.
5. Dearden, J. D., et al., "The BOMB Project, A Control Survey in Southern Ontario," *The Canadian Surveyor*, March 1973.
6. Emerson, J. R., and Schultz, R. J., "A County and University Cooperative Control Survey," *Surveying and Mapping*, June 1972.
7. Wolf, P. R., Wilder, B. P., and Mahun, G., "An Evaluation of Accuracies and Applications of Tacheometry," *Surveying and Mapping*, Vol. 38, No. 3, 1978.

# CHAPTER 16
# Photogrammetric surveying and mapping

**16.1. Introduction** The discipline of photogrammetry involves obtaining information about an object *indirectly* by measuring photographs taken of the object. Therefore, unlike surveying procedures, where measurements are usually made *directly* on the object in the field, in photogrammetry the object is first recorded on an intermediate medium, mostly photographs, and the measurements are carried out later in the office. Thus, photogrammetry requires the following operations: (1) planning and taking the photography, (2) processing the photographs, and (3) measuring the photographs and reducing the measurements to produce the end results, such as point coordinates or maps.

Two broad categories are involved in photogrammetry: *metrical* or *quantitative* activities and *interpretive* or *qualitative* work. Metrical photogrammetry involves all quantitative work, such as the determination of ground positions, distances, differences in elevations, areas, volumes, and various types of maps. In the second category, classically called *photo interpretation*, photographs are analyzed qualitatively for purposes of identifying objects and assessing their significance. Photo interpretation relies on human ability to assimilate and correlate such photographic elements as sizes, shapes, patterns, tones, textures, colors, contrasts, and relative location. Accurate and reliable photo interpretation requires extensive training and experience. The photo interpreter is called upon in many fields of application, such as ecology, environmental analysis, forestry, geology, engineering site selection, resource inventory, planning, and of course military intelligence. In recent years, records from other imaging systems have been used for interpretive purposes, and therefore the more general name *remote sensing* is used. Civilian applications of remote sensing, however, still employ aerial photography extensively.

In photo interpretation, some measurements are necessary, and in metric photogrammetry some interpretation is required. Therefore, photogrammetry combines both activities and can be broadly defined to include acquisition, measurement, interpretation, and evaluation of photographs, imageries, and other remotely sensed data. Because of this broad definition, photogrammetry has found applications in a large number of fields, ranging from biomedicine and dentistry to aerospace engineering and astronomy, from cloud-chamber measurements in physics to all types of engineering, from tailoring to transportation and urban planning, from accident analysis to forestry, and so on. But by far the most common application is in surveying and mapping. Almost all of the topographic maps produced by federal, state, and private organizations are compiled from aerial photographs. Photogrammetric procedures are also used in deriving supplemental ground control, both horizontal and vertical. Therefore, the surveying and engineering students must be aware of the various facets of this field and the ways photogrammetry can assist them in their future professional activities.

Metric photogrammetry has been classically divided into terrestrial and aerial types. In *terrestrial photogrammetry*, the photographs are taken from fixed, often known, points on or near the ground. In *aerial photogrammetry*, a high-precision camera is mounted in an aircraft and photographs are taken in an organized manner as the aircraft flies over the terrain. A more recent branch is *space photogrammetry*, which deals with extraterrestrial photography and imagery where the camera may be fixed on earth, mounted on board an artificial satellite, or located on a moon or a planet. *Close-range photogrammetry* involves applications where the camera is relatively close to the object photographed. Such applications exist in many fields where direct measurement of the object is either impractical, uneconomical, or simply impossible. Examples are found in archaeology, architecture, medicine, and many aspects of experimental engineering laboratory investigations.

This chapter will be devoted to the principles of aerial photogrammetry and its applications in surveying and mapping. An important fact should be established at the outset that an aerial photograph is *not* equivalent to a map except in very unusual circumstances. If the terrain is flat and level, and if the aerial photograph is exposed with the camera pointing perfectly downward (i.e., its optical axis is truly vertical), and assuming no image aberrations, the resulting photograph will be a map with a constant scale. When any of the foregoing restrictions are not met, the photograph will then only *approximate* a map, the degree of approximation depending upon the amount of distortion in the photograph. For example, when the terrain contains relief, the photographic scale at high points will be larger than that at lower points because the higher points are closer to the camera. In order to derive a proper map of the terrain, it is necessary that the ground is imaged in at least two successive aerial photographs. Photogrammetric techniques may then be used to reconstruct, to a scale, a faithful replication of the ground in all three dimensions from the overlapping photographs. Once this is accomplished, horizontal and vertical ground point positions, planimetric maps, topographic maps, cross sections, and a variety of other products can be extracted. The objective of this chapter is to introduce the student to the various operations and products of photogrammetry.

**16.2. Cameras and accessories**   The camera is obviously the first component in the total photogrammetric system since it is used to obtain the photography from which the data can be extracted. The camera is similar in function to the surveying instruments since it is used to gather information about the object in the field. It is, however, quite different with regard to the amount of data it gathers. One sighting by a transit or theodolite yields *one* direction, while one photograph makes possible the determination of an essentially limitless number of directions or as many directions as the number of points identified in the photograph. Consequently, the design of the photogrammetric camera is rather important to guarantee that the photographs obtained are of good metric quality.

Cameras used for terrestrial photogrammetry differ substantially from those used for aerial photography. Since in terrestrial photogrammetry the camera is usually stationary, its design is much simpler than the fast-moving aerial camera. In fact, a camera is combined with a theodolite, thus called a *phototheodolite* (Fig. 16.1), and is used to obtain photographs from known control points. The camera is usually pointed horizontally during exposure. Figure 16.2 is an example of a terrestrial photograph. Note that there are four marks, called *fiducial marks*, one in the middle of each side of the picture. The lines joining each pair of opposite fiducials intersect in the point representing the geometric center of the photograph, called the *principal point*. This point is important in the proper reconstruction of the geometry of the photograph, as will be explained later.

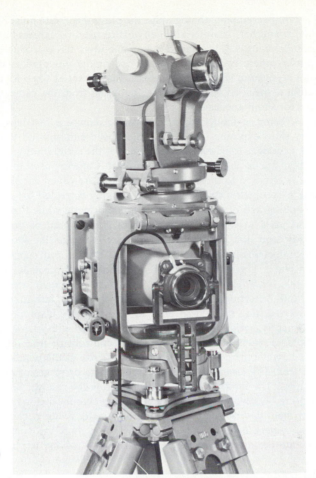

**Fig. 16.1** Wild P-30 phototheodolite.

201,89 +        05

**Fig. 16.2** Terrestrial photograph.
(*Courtesy Carl Zeiss, Inc.*)

In order to obtain overlapping photographs, the phototheodolite must be moved from one exposure station to another. Another alternative is to use two cameras and expose the two photographs simultaneously. Consequently, manufacturers have produced what is called a *stereometric camera*, where two metric cameras are rigidly mounted at the ends of a base tube with known separation (Fig. 16.3). The cameras are set at fixed focus, thus producing sharp images from a specified near distance (e.g., 2.5 m from the cameras) to infinity, and are synchronized for simultaneous exposure.

Aerial cameras, being in motion during exposure, require a fast lens, a reliable shutter system for a very short exposure time, and a high-speed emulsion on the film. All of these requirements are necessary to guarantee the quality of the aerial photograph, particularly by minimizing the blur due to the motion of the aircraft. Cameras used to gather photographs for metric work, as opposed to those for interpretation, are called *cartographic* cameras and are characterized by geometric stability and lenses highly corrected for geometric distortion. Figure 16.4 is a photograph of a modern cartographic camera, while Fig. 16.5 is a schematic of the component parts. It is seen that the camera is composed of three main parts: lens cone, camera body, and film magazine. The *lens cone* contains the *lens assembly*, which is composed of a multielement lens, shutter, and diaphragm, and supports the frame that defines the focal plane. The lens is made of several elements, provides for a large angular coverage (about 90° for wide-angle and 120° for super-wide-angle), and is highly corrected for aberrations and distortions. Because the light from the terrain travels through the atmosphere before reaching the lens, it ends up containing a disproportionate amount of blue light. Consequently, aerial cameras are usually provided with filters which prevent some of that blue light from reaching the film. The *shutter* controls the time interval during which the film is exposed , while the *diaphragm* determines the size of the bundle of light allowed to pass through the lens. The *camera body* houses the camera drive motor and mechanism and a recording chamber. The motor and mechanism operate the shutter, film flattening device, and advance the film between exposures. The recording chamber allows for printing on each frame several pertinent data such as level bubble, altimeter reading, clock, date, photo and mission numbers, camera number, etc. As shown in Fig. 16.5, the cone and camera body upper surface define the *focal plane*, where a sharp image of the terrain is formed on the

**Fig. 16.3** Stereometric camera. (*Courtesy Wild Heerbrugg Instruments, Inc.*)

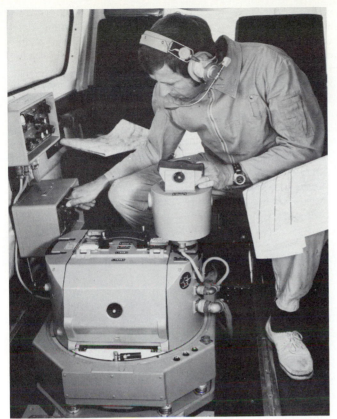

**Fig. 16.4** Aerial cartographic camera. (*Courtesy Wild Heerbrugg Instruments, Inc.*)

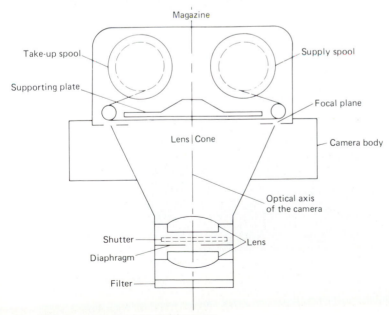

**Fig. 16.5** Components of an aerial camera.

emulsion of the film. In this plane, a number of so-called *fiducial marks*, usually four, are registered on each frame, as shown in Fig. 16.6a, which illustrates *side* fiducial marks. These marks are most important since they define the coordinate system of the photograph. Furthermore, in well-adjusted cameras, the lines connecting opposite pairs of fiducials intersect in a point that falls on the optical axis of the lens. In this way, the photo system can be tied to the terrain system once the camera position and orientation (i.e., the direction of its optical axis) are determined. Although there are other sizes and shapes (e.g., rectangular), the most common type of aerial photographic *format* is that which is square and measures 9 in (23 cm) on the side.

The light entering an aerial camera lens is considered, for all practical purposes, as coming from infinity, owing to the distance between the camera and terrain. Consequently, the image of the terrain will be formed at the focal plane of the camera lens. This is why the emulsion plane of the film is located at a distance from the lens equal to the focal length. Focal lengths of aerial camera lenses are nominally 3.5 in (88 mm), 6 in (153 mm), 8.25 in (210 mm), and 12 in (305 mm), although the 6-in lens is by far the most common.

(a)　　　　　　　　　　　　　　　　　　(b)

(c)

**Fig. 16.6** (*a*) Vertical aerial photograph with four side fiducial marks. (*Courtesy Pacific Aerial Surveys, Oakland, California.*) (*b*) Low oblique aerial photograph. (*Courtesy Wild Heerbrugg Instruments, Inc.*) (*c*) High oblique aerial photograph. (*Courtesy Carl Zeiss, Inc.*)

The *magazine* is light-tight and holds the unexposed and exposed film spools. Magazines vary in capacity from 180 ft (55 m) to 500 ft (150 m) of $9\frac{1}{2}$ in (24 cm) wide thin-base film.

Vertical aerial photography is exposed with the intention that the camera optical axis be as truly vertical as possible during exposure. Therefore, it is necessary that the camera be supported in a *stabilized mount* such that it is isolated from aircraft vibrations and perturbations. These mounts employ torquer motors controlled by gyroscopes. Other auxiliary systems used with aerial cameras include viewfinder, intervalometer, and $V/H$ computer. The *viewfinder* provides the photographer with a clear unobstructed view of the terrain below and ahead of the airplane. The *intervalometer* is a timing device that can be set to trigger the camera shutter at a specified time interval between exposures. The $V/H$ (velocity/height) computer develops a voltage that is directly proportional to the aircraft velocity, $V$, and inversely proportional to its altitude, $H$. It may be used to determine the exposure interval as well as the film shift in the direction of flight during exposure to minimize image blur.

**16.3. Acquisition and processing of aerial photographs**    Once a project area has been specified, the first step in the photogrammetric project is to plan the acquisition of the photography in a manner suitable for the desired purpose. For topographic mapping and the production of mosaics or orthophotos, vertical aerial photography is most common. A *vertical* photograph is that taken with the camera axis intended to be vertical, although unavoidable aircraft motion may cause it to tilt a few degrees from the vertical (usually a maximum of 5°, although the average is often 1° or less). An example of a vertical aerial photograph is shown in Fig. 16.6a. In small-scale mapping, the camera axis may be intentionally tilted in order to increase the area covered by one photograph. In this case, if the tilt angle, is say, 20°, the photograph is called a *low oblique*, an example of which is shown in Fig. 16.6b. When the tilt angle is large enough (50° or 60°) such that the horizon appears in the photograph, it is called a *high oblique*, as shown in Fig. 16.6c. The existence of the horizon line in the photograph can be used to advantage when deriving metric information from the high-oblique photograph.

In the case of vertical aerial photography with square format [usually 9×9 in (228.6× 228.6 mm)] the ground-area coverage of a single photograph is accordingly a square. As the airplane flies over the ground, successive photographs are exposed in such a way that each two adjacent photographs cover a common area which is more than half the single photo coverage. This common area is called *forward overlap* or simply *overlap* and is usually 60 percent, as shown in Fig. 16.7. Thus, if photographs along one pass from one end of the project area to the other are laid down on a table, each two successive photographs would overlap by about 60 percent and each three successive photographs by about 20 percent.

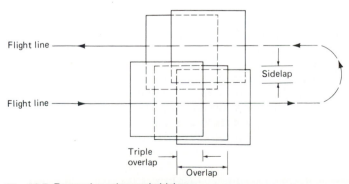

**Fig. 16.7** Forward overlap and sidelap.

This type of coverage is necessary to ensure that each area on the ground is covered at least twice so that its three-dimensional geometry may be recovered. The 20 percent *triple overlap* (see Fig. 16.7) is provided to assure the proper connection from one pair of photographs to the next. In some special situations, such as mosaic of rugged terrain, overlap may be increased to say 80 percent. The nominal line passing through the middle of successive photographs is called a *flight line*, and the set of photographs in one line is often referred to as a *strip*.

Once the airplane has traversed the length of the project area, it turns around and returns in the opposite direction as shown in Fig. 16.7. The second strip of photography is exposed in such a way that there is at least 20 percent *sidelap* with the first. This again assures no gaps and allows for the tie-in between strips so that the required maps or mosaics can be continuously produced.

After an aerial photography mission, the film is brought to the photo lab for processing. The images exposed on the film are called *latent* images because they are not visible until they have been processed. The degree of complexity in the processing of film depends on its type. There are in general four types of emulsions used in aerial photography: (1) *panchromatic black and white* (B & W), which is the most widely used type in metric photogrammetry and interpretation; (2) *color*, which is composed of three emulsion layers to render the different color hues originally in the object, and is used more for interpretation than in mapping; (3) *infrared black and white* (B & W IR) is sensitive to near-infrared light (longer wavelength) and thus penetrates haze and is used mainly for interpretation and intelligence work as it permits detection of camouflage; and (4) *infrared color* (*false color IR*), in which different colors are arbitrarily used to code images from different portions of the light spectrum (thus the name false color) and is employed in various facets of interpretation, such as crop disease detection, pollution monitoring, etc., as well as intelligence.

After processing in the photo lab, the result is either a film negative or a film positive, depending upon the processing procedure used. On a *film negative*, light portions of the original scene appear dark, while dark scene portions appear light, so the scene illumination is reversed on a film negative. On a *film positive*, the original scene light and dark distribution is preserved. It is much more common for B & W films (both panchromatic and IR) to be processed to negatives, while both color and false color films are processed either to positives or negatives. When the result of film processing is a negative, positive-tone prints are usually made both for initial inspection, as well as for planning and execution of the various photogrammetric operations. It is a common and good practice to lay all the photographs in a project together in order to determine whether there are any gaps. Also, the tilts of the different photographs are checked in case some are unacceptable. In either case, a re-flight may be required to obtain more suitable photography. Once the photographs are judged as meeting the original specified requirements, the photogrammetric project may then proceed.

**16.4. Stereoscopy**   *Stereoscopy* refers to the ability of the individual to perceive the object space in three dimensions through using *both* eyes. Each human eye represents a single camera, and therefore *monocular viewing* or viewing with one eye results in *flat perspective* and the person's ability to perceive depth is hampered. *Binocular viewing*, on the other hand, allows the person to view an object from two different locations due to the separation between the eyes. The student can see the similarity between the function of the human eyes and the pair of camera stations when taking overlapping photographs. This is the reason for studying the principles and techniques of stereoscopy in support of photogrammetry.

Viewing with one eye fixes only *one* direction from the eye to the object, which is insufficient for fixing the object's distance from the viewer. When the other eye is utilized, a second direction is fixed and its intersection with the first locates the point. The closer the

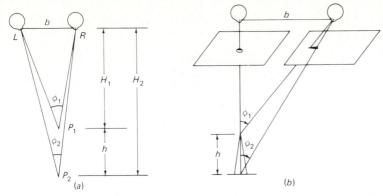

**Fig. 16.8** (*a*) Convergence angle in stereoscopic vision. (*b*) Stereoscopic viewing with a pair of overlapping photographs.

point to the eyes, the larger is the convergence angle between the two directions, as depicted in Fig. 16.8*a*. Since the angle $\phi_1$ is larger than $\phi_2$, the observer will perceive point $P_1$ as closer than point $P_2$. In fact, the difference in distance from the observer, $H_2 - H_1 = h$, is a function of the difference in convergence angles, $\phi_1 - \phi_2$. The closest distance for distinct comfortable vision (without aids) is 25 cm (10 in), for which the convergence angle is about 15°. This can be computed on the basis of an average eye separation of $b = 65$ mm. The lower limit of the convergence angle for unaided vision is 10 to 20 seconds of arc.

Figure 16.8*b* as compared to Fig. 16.8*a* shows the correspondence between natural binocular vision and stereoscopic viewing of a pair of overlapping photographs. An idealized tower is assumed to have been photographed with the camera first directly overhead and second after having traveled a distance past the tower. If it were possible to view the first photograph with the left eye only and the second photograph with the right eye only, the observer would perceive the tower in three dimensions, as shown schematically in Fig. 16.8*b*. The lines joining the eyes and the two images of the top of the tower will intersect at an angle $\phi_1$, while those passing through the images of the tower base make an angle $\phi_2$. Since $\phi_1$ is larger than $\phi_2$ the top of the tower will appear closer to the viewer than its base, and the tower will then be perceived in three dimensions. Stereoscopic viewing of objects and terrain from overlapping photographs is very important to most all operations of both interpretation and metric photogrammetry, as will be shown in subsequent sections of this chapter.

**16.5. Analysis of the single photograph**   As noted earlier, an aerial photograph will be the same as a map only in the very special situation when the terrain is flat and level and the photograph is truly vertical. In general, the photograph is neither taken precisely vertical nor is it common that the terrain be flat and level. Figure 16.9 shows schematically the analysis of both factors. In Fig. 16.9*a* the terrain is assumed to be flat and level between points $A$ and $B$ and between points $D$ and $E$, and of variable relief in between. Thus, $h_A = h_B$ and $h_D = h_E$. The photograph is *assumed* to be truly vertical. The scale *at* any point in the photograph such as $a$ is the ratio of the distance from the camera location, $C$, to the image point, divided by the distance from $C$ to the object point, $A$, or $S_a = Ca/CA$. From the two similar triangles $Cao$ and $CAO$, $S_a = Ca/CA = Co/CO$. Since $Co = f =$ focal length and $CO = H - h_A =$ the flying height above the object point of interest, the scale at any point on a truly vertical photograph is

$$S = \frac{f}{H - h} \qquad (16.1)$$

in which $H$, the flying height, and $h$, the elevation of the point, are both referred to the same

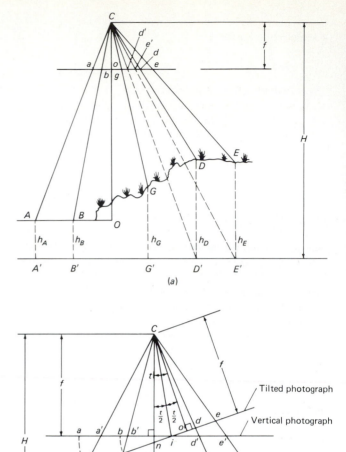

**Fig. 16.9** (*a*) Geometry of a single vertical aerial photograph. (*b*) Geometry of a tilted photograph over level terrain.

datum. It can be seen from Fig. 16.9*a* that the scale is the same for the area between points *a* and *b* and is equal to $f/(H-h_A)$. Similarly, the scale is constant between points *d* and *e* and is equal to $f/(H-h_D)$. In a region with a constant scale on a vertical photograph, the scale can be derived by dividing an image distance by the corresponding ground distance. Hence, $S_a = ab/AB$ and $S_d = de/DE$. Since $h_D$ is larger than $h_A$, $DE$ is closer to the camera than $AB$ and the scale along *de* is larger than along *ab*. In fact, the two ground distances $AB$ and $DE$ are equal, and the student can see that the image distance *de* is longer than *ab*. It is important, then, to note that unlike a map whose scale *must* by definition be constant, the scale on an aerial photograph is in general not. Therefore, it is better to speak of a photographic scale at a *point* unless a nominal, an average, or an overall scale is specified. These latter scales are approximate and do not apply necessarily to a particular point or area on the photograph. They are usually obtained from a relation similar to Eq. (16.1) using an average value $h_{av}$ for $h$, or

$$S_{av} = \frac{f}{H - h_{av}} \qquad (16.2)$$

The variability of the scale on an aerial photograph, even when truly vertical, is due to the existence of relief. It is clear from Fig. 16.9a that if the object points $D$ and $E$ were on the datum at $D'$ and $E'$, they would have been imaged at point $e'$ and $d'$. The ratio between $d'e'$ and $D'E'$ is equal to $f/H$ and is often called the datum scale of the photograph. It is usually the smallest scale unless there are points below the datum. It can be seen that the distance $d'e'$ is, in fact, smaller than the image distance $de$. The effect of relief on the location of image points is called *relief displacement* and is discussed in Art. 16.6.

**Example 16.1**   Referring to Fig. 16.9a, the elevation of point $A$ is $h_A = 123.2$ m and the scale at that point on the aerial photograph is $1:32,000$. The elevation of point $E$ is 275.6 m. Compute the scale at the image for point $E$ and the average scale of the photograph if the focal length is $f = 152.4$ mm.

SOLUTION   The scale at point $A$, according to Eq. (16.1), is

$$S_A = \frac{f}{H - h_A} \quad \text{or} \quad \frac{1}{32,000} = \frac{152.4}{(H - 123.2)1000}$$

or

$H - 123.2 = 152.4 \times 32 = 4876.8$ m

Then

$H = 4876.8 + 123.2 = 5000$ m

The scale at point $E$ is

$$S_E = \frac{f}{H - h_E}$$

or

$$S_E = \frac{152.4}{(5,000 - 275.6)1000} = \frac{1}{31,000} \quad \text{or} \quad 1:31,000$$

Because of the shape of the terrain in Fig. 16.9a, the mean terrain will be taken as the average of the elevations of points $A$ and $E$. Thus, $h_{av} = \frac{1}{2}(123.2 + 275.6) = 199.4$ m and the average scale, according to Eq. (16.2), is

$$S_{av} = \frac{152.4}{(5,000 - 199.4)1000} = \frac{1}{31,500} \quad \text{or} \quad 1:31,500$$

Figure 16.9b shows the effect of the second factor, the tilt in the photograph. To isolate and dramatize this effect, the terrain is assumed to be flat and level, with the distances $AB$ and $DE$ being equal. If a truly vertical photograph is exposed from $C$, it will have a constant scale equal to $f/H$. This is borne out by observing that the image distance $a'b'$ and $d'e'$ are equal. An aerial photograph exposed from the same point $C$, but with its optical axis tilted an angle $t$ from the vertical, contains images of the four object points at $a$, $b$, $d$, and $e$. It is clear that the scale is variable along the tilted photograph. The image distance $ab$ is larger than $de$, even though both represent the same distance on the ground. The effect of tilt on image location on a tilted photograph is called *tilt displacement* and is discussed in Art. 16.7.

**16.6. Relief displacement on a vertical photograph**   Figure 16.10a shows a section of a vertical plane passing through the camera station $C$ of a truly vertical photograph and an object point $A$. The datum position of point $A$ is $A'$, and their image points are $a$ and $a'$, respectively. Of course, point $a$ is what actually appears on the photograph, while $a'$ is an imaginary point representing the image of $A'$ if it were possible for $A'$ to be seen by the camera. The distance $d$ between $a'$ and $a$ is the *relief displacement on the photograph*. The extension of line $CaA$ intersects the datum in point $A''$, which represents the datum point that would have been imaged at $a$. The distance $D$ between $A'$ and $A''$ is the *relief displacement in the datum plane*. The ratio between $d$ and $D$ is the datum scale of the photograph, or $f/H$; thus,

$$d = \frac{Df}{H} \tag{16.3}$$

The *principal point o* of a truly vertical photograph is the only point that undergoes no relief displacement. It is seen from Fig. 16.10a that no matter how high the ground point $O$ is above the datum, both $O$ and its datum position $O'$ are imaged at $o$, since both lie on the vertical line through $C$. Consequently, all other points are displaced due to relief radially from $o$ as depicted in Fig. 16.10b. The amount of relief displacement depends on the radial distance $r$ between the image point and the principal point $o$. The farther the image point from $o$, the larger is the relief displacement, as shown from the relation to be derived.

The two triangles $AA'A''$ and $Coa$ (Fig. 16.10a) are similar and therefore

$$\frac{D}{r} = \frac{h}{f} \qquad \text{or} \qquad D = \frac{rh}{f}$$

which when substituting from Eq. (16.3) leads to

$$d = \frac{rhf}{fH}$$

or

$$d = \frac{rh}{H} \tag{16.4}$$

From Eq. (16.4), the relief displacement is seen to be directly proportional to $r$ and the elevation of the point $h$, and inversely proportional to camera altitude $H$; both $h$ and $H$ must be measured above the same datum. One must watch for units in applying Eq. (16.4). In general, $d$ and $r$ may be in the same units, say $mm$, while $h$ and $H$ are in metres or feet.

Equation (16.4) gives the total relief displacement for a given image point computed when a scale is used to measure directly the distance $r$ from $o$ to the image. Sometimes, instead of $r$, the coordinates $x,y$ of the image are measured as shown in Fig. 16.10b. Here, opposite fiducial marks are connected by straight lines to intersect in the principal point $o$. One line, usually that in the general direction of the flight line, is taken as the $x$ axis, and the other as

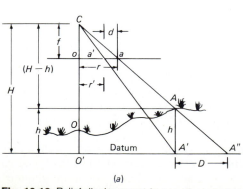

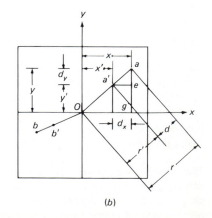

(a)                               (b)

**Fig. 16.10** Relief displacement in a vertical photograph.

the $y$ axis, with $o$ as the origin of the coordinate system. One way is to compute $r$ from $r = \sqrt{x^2 + y^2}$ and apply Eq. (16.4) to compute $d$ and scale it from or toward $o$ to get $a'$. Another way is to compute instead the two components $d_x, d_y$ of the relief displacement. In Fig. 16.10*b* the two triangles $aa'e$ and $aog$ are similar; hence,

$$\frac{aa'}{ao} = \frac{a'e}{og} = \frac{ae}{ag} \qquad \text{or} \qquad \frac{d}{r} = \frac{d_x}{x} = \frac{d_y}{y}$$

But from Eq. (16.4), $d/r = h/H$, so that

$$d_x = \frac{xh}{H}$$
$$d_y = \frac{yh}{H} \qquad (16.5)$$

Thus, if $x, y$ are given, $d_x$ and $d_y$ can be evaluated directly from Eq. (16.5). Then the coordinates of $a'$ are simply $x' = x - d_x$ and $y' = y - d_y$. If one is interested in the total relief displacement, then $d = \sqrt{d_x^2 + d_y^2}$ . These relations are demonstrated in the following numerical example.

**Example 16.2**    A vertical photograph is taken with a camera having $f = 151.52$ mm. Two images $a$ and $b$ appear on the photograph, such that $a$ falls on the $x$ axis with $x_a = +78.70$ mm, and $b$ falls on the $y$ axis, with $y_b = -91.30$ mm. The two points $A, B$ on the ground have the same elevation 98 m above datum. Compute the image coordinates of the datum point images $a', b'$ if the datum scale is 1 : 16,500.

SOLUTION    The datum scale is $f/H$; therefore,

$$H = \frac{(16,500)(151.52)}{1000} = 2500 \text{ m}$$

Since point $a$ lies on the photo $x$ axis, point $a'$ will also be on the $x$ axis, and the radial distance is equal to $x_a$. Thus, the relief displacement is [Eq. (16.5)]

$$d_a = d_{xa} = \frac{x_a h}{H} = \frac{(78.70)(98)}{2500} = 3.09 \text{ mm}$$

and

$$x_a' = x_a - d_{xa} = 78.70 - 3.09 = 75.61 \text{ mm}$$

Similarly, since $b$ lies on the $y$ axis, then [Eq. (16.5)]

$$d_b = d_{yb} = \frac{y_b h}{H} = \frac{(-91.30)(98)}{2500} = -3.58 \text{ mm}$$

and the coordinate of $b'$ is

$$y_b' = y_b - d_{yb} = -91.30 - (-3.58) = -87.72 \text{ mm}$$

**16.7. Tilt displacement**    Figure 16.9*b* shows the relation between a truly vertical photograph and a tilted photograph both taken from the same camera station. The plane of the figure, in which the tilt angle is measured, is called the *principal plane* and is the vertical plane through the optical axis when the tilted photograph is exposed. The trace of that plane on the tilted photograph is a line that passes through the principal point and is called the *principal line*, as shown in Fig. 16.11. Only one line is common to both the vertical and tilted photographs. This line is represented by point $i$ in Fig. 16.9*b* and by $i'i'$ in Fig. 16.11. It is called the *isoline* or sometimes the *axis of tilt*, and its intersection with the principal line at $i$ is called the *isocenter*. As can be seen from Fig. 16.9*b*, point $i$ lies on the bisector of the tilt angle $t$. Therefore, the distance along the principal line from principal point $o$ to the isocenter $i$ is

$$oi = f \tan \frac{t}{2} \qquad (16.6)$$

The vertical or plumb line through $C$ pierces the tilted photograph at point $n$ which is called the nadir point and also lies on the principal line. The distance $on$ from the principal point to the nadir point is given by (from Fig. 16.9$b$)

$$on = f \tan t \tag{16.7}$$

The principal line is located on the photograph by an angle $s$, called the *swing angle* and defined as the clockwise angle from the photographic $+y$ axis to the line $on$, *in that direction*, as shown in Fig. 16.11.

Imagine rotating the tilted photograph about its common line with the truly vertical photograph until they coincide. When this is done, images of the same object point do not match, the difference being the *tilt displacement*. Points $a$ and $b$, being on the downside of the isoline, are displaced radially *away* from the isocenter, $i$, while points $d$ and $e$ are shifted radially *toward* $i$ because they are above the isoline. The amount of tilt displacement is therefore a function of the distance $\ell$ from the image point to the isocenter. It can be shown that the tilt displacement $d_u$ is given by

$$d_u = \frac{\ell^2 \cos I \sin t}{f - \ell \cos I \sin t} = \frac{\ell \ell_p \sin t}{f - \ell_p \sin t} \tag{16.8}$$

in which $d_u$ is tilt displacement, $\ell$ the distance between the image and isocenter $i$, $\ell_p$ the component of $\ell$ along the principal line, $f$ the focal length of the camera lens, $t$ the tilt angle, and $I$ the angle that the line between $i$ and the image makes with the principal line. The quantities $d_u$, $\ell$, and $f$ must be in the same units. Equation (16.8) applies to points in that portion of the photograph above the isoline. For those points below the isoline, the sign in

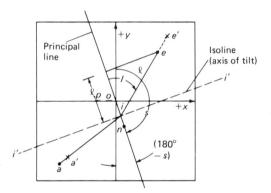

**Fig. 16.11** Principal line and tilt axis on a tilted photograph.

the denominator is changed to positive, or

$$d_L = \frac{\ell^2 \cos I \sin t}{f + \ell \cos I \sin t} = \frac{\ell \ell_p \sin t}{f + \ell_p \sin t} \tag{16.9}$$

Clearly, points on the isoline undergo no tilt displacement, since they already fall on the associated truly vertical photograph. This is verified by both Eqs. (16.8) and (16.9), since in this case the angle $I$ is $\pi/2$ and $\cos I$ is zero, thus reducing both $d_u$ and $d_L$ to zero.

To compute tilt displacement, first the fiducial lines are drawn to locate the principal point. Next, the swing angle, which is assumed to be known, is used to locate the downside of the principal line (the side containing points $i$ and $n$). Equation (16.6) is then used to locate point $i$ and the distances $\ell$ and $\ell_p$ are scaled off the photograph for the points for which tilt displacement is to be evaluated. This assumes that both the focal length $f$ and tilt angle $t$ are known. The following example demonstrates these calculations.

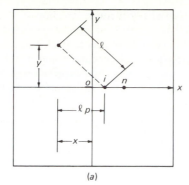

**Fig. 16.11** (a)                    (a)

**Example 16.3**    For a tilted photograph the swing angle is $s = 90°$, and the tilt is $1°40'$. Compute the tilt displacement for a point whose image coordinates are $x = -20$ mm and $y = 40$ mm, given $f = 152.2$ mm (Fig. 16.11a).

SOLUTION    Since $s = 90°$, the $x$ axis is the principal line with $i$ and $n$ falling on the $+x$ side. According to Eqs. (16.6) and (16.7),

$oi = 152.20 \tan(50') = 2.21$ mm
$on = 152.20 \tan(1°40') = 4.43$ mm

Since the image point has a negative $x$ coordinate, it falls above the axis of tilt and Eq. (16.8) applies (see Fig. 16.11a). The distance $\ell$ from the isocenter $i$ to the image point is

$$\ell = [(oi - x)^2 + y^2]^{1/2} = [(2.21 + 20)^2 + (40)^2]^{1/2} = 45.75 \text{ mm}$$

The distance $\ell_p$ (see Fig. 16.11a) is 22.21 mm, and hence the tilt displacement according to Eq. (16.8) is

$$d_u = \frac{(45.75)(22.21)\sin(1°40')}{152.20 - (22.21)\sin(1°40')} = 0.20 \text{ mm}$$

**16.8. Uses and products of aerial photogrammetry**    Aerial photographs are used in one of two primary ways, metric and interpretive. However, many of the applications of photogrammetry employ both interpretive and metric techniques. But by far the most extensive use of aerial photographs is in the compilation of topographic maps and orthophotographs, as will be discussed in later sections of this chapter. There are many other uses of aerial photographs. Practically all phases of modern highway design, location, construction, and maintenance make use of aerial photographs. Earthwork quantities, for example, are often computed from a photogrammetric model composed of two overlapping aerial photographs. Maintenance operations, such as those relating to pavement condition and possible erosion of highway banks are determined from the interpretation of photographs. Photographs are also used in a multitude of engineering applications, such as dam settlement and other structural deformation, wave action, channel sedimentation, and traffic studies. In hydrology, analysis of slopes, ground coverage, watershed areas, and snow depth determined from photographs assist in the determination of quantities of runoff water.

   In geology, interpretation of photographs for structural form are a valuable supplement to field methods. In agriculture, crop inventory and crop disease analysis are effectively performed using aerial photographs. Aerial photographs are also used for archaeological purposes to explore old sites for new and buried finds.

   Aerial photographs are used in land surveying for identification and location of boundary lines and corners (Art. 19.49), and the determination of soil and vegetation cover types. Analytical photogrammetry can be used to determine the location of land boundary corners if they have been well targeted before the aerial photographs are flown. Alternatively, the positions of arbitrarily targeted points can be determined by photogrammetry and other points located by ground surveys using the targeted points as control. Land subdivision can be readily staked out using the photogrammetrically determined points as references.

The uses enumerated above are just a few examples of the extensive range of applications in photogrammetry. In the course of these applications, a variety of products are derived. Some of the more common products include:

1. *Paper-print copies* of the aerial photographs themselves, which may be used for general planning and for field operations.
2. *Mosaics*, which are composed of segments of photographs assembled to give the appearance of a continuous picture of the terrain. Since each is a segment of a perspective view, the mosaic contains distortions and should not be considered as an accurate map.
3. *Orthophotos*, specially prepared photographic representations of the terrain *without* distortions (relief and tilt) which can be used as a planimetric map. In some cases contour lines are superimposed on an orthophoto, resulting in an *orthophotomap* which is then the equivalent of a topographic map.
4. *Planimetric maps* (Art. 13.1), photogrammetrically prepared, contain only horizontal position information of terrain features. It is usually compiled from a stereomodel reconstructed in a plotter, as explained in a later section.
5. *Topographic maps* (Chap. 13), which show both planimetric detail and contour lines, compiled photogrammetrically on a stereoplotter.
6. *Thematic maps*, which are portrayals of different "themes" of interest, such as soil type, drainage patterns, transportation network, vegetation, etc. Modern scanning imagery systems are quite suitable for this product, particularly using computer classification techniques.
7. *Digital data*. The photogrammetric model can be used to produce either a listing of ground coordinates of a limited number of required control points or a very dense network of points to represent the total terrain surface. The latter is called the *digital terrain model* (DTM) (see also Art. 16.19) and is usually in the form of a regular $XY$ grid with the elevation of each grid intersection given.

**16.9. Elementary operations in photogrammetry: general**    As mentioned previously, a photograph is a perspective projection of the terrain. Therefore, to obtain accurate metric information about the terrain from photographs, rigorous principles must be applied. However, many elementary photogrammetric operations can be based on some simplifying assumptions. Such assumptions include (1) accepting the fiducial center as the principal point of the photograph; (2) lens distortion, earth curvature, and atmospheric refraction are relatively of no significance and can be ignored; (3) only linear corrections for film and paper deformations are considered, since elementary operations occur almost totally with paper prints; and (4) the calibrated focal length is used as the focal length, $f$. In the course of discussion in various sections to follow, other assumptions may be made and will be specifically pointed out.

**16.10. Planimetric information from aerial photographs**    When overlapping photographs are used, three-dimensional information which may be contained in the photographed object can be recovered. The same applies to terrain photographed by an aerial camera. However, when elementary operations are considered, it is much easier to extract horizontal or planimetric information in a separate operation from that concerned with the height information. Planimetric data can be obtained from overlapping aerial photographs either in the form of horizontal point coordinates, or planimetric line maps, or mosaics. The photogrammetric determination of horizontal control points is accomplished by an operation called *radial line triangulation*, which is fundamentally similar to survey triangulation (see Chap. 10). Since the determination of such horizontal control is necessary before either plotting planimetric detail or producing a mosaic, radial triangulation is discussed first.

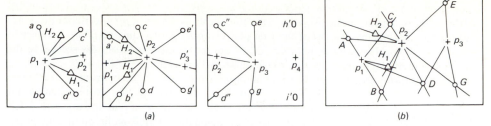

**Fig. 16.12** (*a*) Locations of points on three overlapping photographs for a radial line plot. (*b*) Map sheet showing control points for a radial line plot.

**16.11. Radial line triangulation**   In addition to the assumptions made in Art. 16.9, radial line triangulation presumes that the photographs are *truly vertical*. If the terrain were flat and level, angles measured about any point in a truly vertical photograph would be true horizontal angles. On the other hand, for terrain with relief, only angles about the principal point of a truly photograph are horizontal angles. This fact is basic to radial line triangulation, and even when the aerial photographs contain some tilt (with about 3° maximum), the effect on this elementary procedure is not significant.

Figure 16.12*a* shows a schematic of three successive photographs with two known horizontal control points $H_1, H_2$ appearing in the first two photographs; Fig. 16.12*b* is a map sheet containing the two control points. The principal points (fiducial centers )$P_1, P_2, P_3$ are marked at the center of each photograph. Using a stereoscope such as that shown in Fig. 16.13, the first pair of photographs is properly aligned such that the overlap area is perceived in stereo under the stereoscope. While viewing in stereo the conjugate principal points $p_1$ and $p_2$ are marked on the second (right) and first (left) photograph, respectively. Also selected and marked in stereo are the two control points $H_1$ and $H_2$ and the four *pass points* $a, b, c, d$ and their conjugates $a', b', c', d'$. Pass points are those image points the coordinates of which are to be determined by the triangulation. They are also involved in the geometric construction of the method. Therefore, they must be sharp, well-defined points located in the outer areas of the photographs to guarantee strong intersection of the rays from different

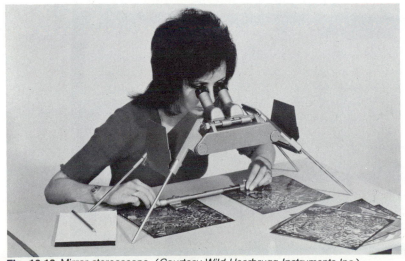

**Fig. 16.13** Mirror stereoscope. (*Courtesy Wild Heerbrugg Instruments Inc.*)

photographs. To ensure that the triangulation may continue from one photographic overlap to the next, a pair of pass points, such as $c, d$, are selected on opposite sides of the principal point $p_2$ so as to appear on the two adjacent photographs to the left ($p_1$) and right ($p_3$). Once points are marked in the overlap of the first pair of photographs, the procedure is repeated for the second pair. In this case, points $p_2', c'', d''$ are first transferred from the second to the third photo, and then points $e, e', g, g'$ are selected and marked.

A tracing paper is placed over the first photograph and rays are drawn from $p_1$ to all other points of interest $a, H_2, c', p_2', H_1, d'$, and $b$, as depicted in Fig. 16.12$a$. Under the assumption of truly vertical photography, all angles about $p_2$ in the tracing paper will represent horizontal angles as if a theodolite occupied the ground nadir point and angles were measured to the corresponding ground points. Similarly, a tracing paper is prepared for the second and third photographs as shown in Fig. 16.12$a$. All the rays in the overlays, $p_1 p_2'$ and $p_1' p_2$ represent the same *base line* between the first and second exposure stations, and $p_2 p_3'$ and $p_2' p_3$ the base line between the second and third photographs. Thus, when the three overlays are placed over the map sheet (Fig. 16.12$b$), rays representing the same base line must coincide. The overlays are then shifted relative to each other along the base lines (to effect a change in scale) and also rotated until the rays to the two horizontal control points $H_1, H_2$ pass through their plotted map positions as shown in Fig. 16.12$b$. At this point rays from different principal points drawn toward the same point (e.g., $C$) will usually intersect in a point that represents its map position. When the assembly of overlays has been properly positioned on the map sheet, all principal points $p_1, p_2, p_3$ and pass points $A, B, C, D, E, G$ are marked by pressing a needle through the points of intersection. These points then represent the map locations of the ground points, and their horizontal coordinates can be extracted by reference to the map grid.

Tracing paper was considered in the foregoing discussion only to facilitate the explanation. In practice, paper is used for only a limited number of photographs; otherwise, it would become difficult to handle. Instead, each photograph is represented by a segment of cardboard called a *slotted templet*. The principal point of the photograph is represented by a circular hole in the center of the templet, while the radial lines are represented by slots that are cut radially from the central hole. When several slots overlap, a circular hole will result in which a stud can be inserted. A needle or a pin inserted along a thin hole along the axis of the stud would lead to marking a point on the base map which represents the point of

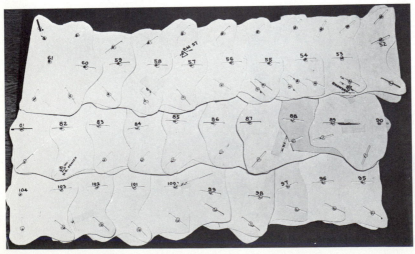

**Fig. 16.14** Slotted-templet assembly.

intersection of the corresponding ray. The assembly of the slotted templets from three short adjacent strips of photographs is shown in Fig. 16.14.

An alternative to the slotted templets is the slotted-arm templet, an assembly of which is shown in Fig. 16.15. In this case each templet is formed using a set of slotted strips of steel, one for each ray, radiating from a center bolt which corresponds to the principal point of the photograph.

The results of radial triangulation are used to control the transfer of planimetric data from photographs to a map, as discussed next.

**16.12. Plotting of planimetry** While radial line triangulation yields the horizontal position (within accuracies limited by the assumptions made) of only a finite number of points, it is possible to plot planimetric detail and produce a line map. The simplest procedure is to trace directly on transparent paper (representing the map base) placed over the photographic print. To minimize the effect of tilt and relief displacements, the map sheet should contain as many control points as can be determined by radial triangulation (at least nine well-distributed points per photograph; see Fig. 16.12a). One control point on the map sheet is made to coincide with its image on the photograph and the map sheet properly oriented. Planimetric detail is then traced around that one point. The sheet is shifted such that another control point coincides with its image and details around it traced, and the process is repeated. Mismatch will occur but can be eliminated by the draftsman. Obviously, the lower the amount of relief (and tilt, in general) the better is the accuracy of this method. Of course, the scale of the planimetric map must be the same as the average scale of the aerial photograph.

Direct tracing is the simplest form of data transfer from photograph to map. To accommodate some change in scale and compensate to some extent for photo tilts, the *camera lucida* principle shown in Fig. 16.16a is used. Light reflected off the photo is deflected by the mirror $M$ to the beam splitter (or half-silvered mirror) $BS$, which directs it to a single eye. Light from the map passes through $BS$ to the eye, which then perceives both the photo and map as if coincident on each other. Thus, the operator can trace on the map the superimposed photo detail. The ratio between the lengths of the optical paths between photo and eye, and map and eye, determines the scale. The sketch master (Fig. 16.16b) is a direct implementation of the camera lucida principle. Figure 16.17a is a diagram of a more

**Fig. 16.15** Spider-templet laydown. (*Courtesy Abrams Aerial Survey Corporation.*)

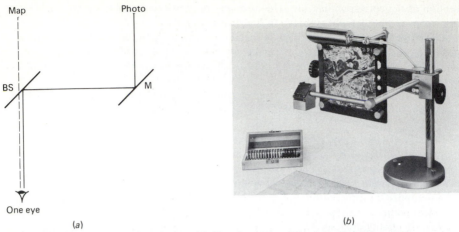

**Fig. 16.16** (*a*) Camera lucida principle. (*b*) Sketch master. (*Courtesy Carl Zeiss, Inc.*)

sophisticated arrangement of the principle used in the *zoom transfer scope* (ZTS), illustrated in Fig. 16.17*b*. The ZTS allows for a continuous change in magnification from 1 to 14 to accommodate a large range of photo and map scale ratios. As shown in the figures, the viewing is binocular, which minimizes eye fatigue. It also allows for *anamorphic* correction, which affects a scale change in only one direction, thus compensating to some extent for geometric distortions in the photograph due to tilt and relief.

**16.13. Mosaics**   A mosaic is an assembly of adjacent aerial photographs such that a continuous pictorial display of the terrain is obtained (Fig. 16.18). There are usually three types of mosaics: *uncontrolled*, *controlled*, and *semicontrolled* mosaics, depending upon whether and how much horizontal control has been used in constructing the mosaic. When little or no control is used, the mosaic is uncontrolled. The simplest form of an uncontrolled mosaic is the *index* mosaic (sometimes called index "map"), where overlapping photographs are placed down consecutively and stapled to a board. It is used as an index for the mapping project, usually to monitor its progress.

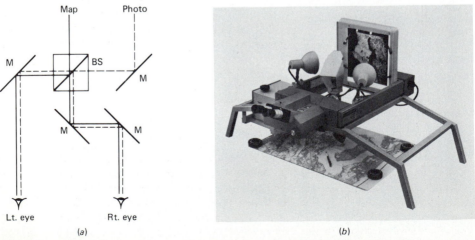

**Fig. 16.17** (*a*) Optical train of zoom transfer scope (ZTS). (*b*) Zoom transfer scope (ZTS). (*Courtesy Bausch & Lomb Inc.*)

**Fig. 16.18** Aerial mosaic. (*Courtesy Pacific Aerial Surveys, Oakland, California.*)

Controlled mosaics are made both for pictorial display and the extraction of some metric data, particularly for planning. The accuracy of the mosaic depends on minimizing the effects of relief and tilt displacements. Considering relief, the smaller the area around the center of each photograph to be used in the mosaic, the less is the relief displacement, which depends on the radial distance from the center, $r$ [see Eq. (16.4)]. Therefore, long focal length, higher-altitude flying, and increased overlap and sidelap are preferable for photography to be used in a mosaic. With regard to tilt displacement, the regular aerial photographs are usually replaced by *rectified* prints produced by instruments called *rectifiers*. Rectification produces an equivalent truly vertical photograph from the same exposure station. Since the attitude of the camera also varies somewhat from one photograph to another, the rectification process also allows for bringing all rectified prints to the same scale.

Although small central segments of the rectified prints are used, each is still a perspective projection of the terrain. Therefore, horizontal control points are utilized to *control* the mosaic. Such control is obtained either by ground survey methods or (more frequently) by radial line triangulation (Art. 16.11). The control is plotted at the required scale on the board on which the mosaic is to be assembled. Each control point must also be identified on the photographic segment to be used. Only the central portion of a photograph is used with its edges being brought to a feather edge using sandpaper. Some small overlap is allowed between adjacent segments. Each segment is then pasted on the board, matching control, if it exists, and other details on adjacent segments as much as possible. Once all segments are pasted and the assembly is finished, the mosaic is then photographed by a large copy camera and reproductions are made at the desired scale. Although the mosaic is an excellent product that offers wealth of detail and information, it is only a map substitute and not the same as a planimetric map since each small area would still contain perspective distortions which may be minimized but not totally eliminated. A semicontrolled mosaic is then a mosaic classified between controlled and uncontrolled.

Radial line triangulation, plotting of planimetry, and preparing mosaics are elementary procedures for obtaining *horizontal* information from aerial photographs. Next, consider procedures that permit obtaining not only horizontal but also vertical information.

**16.14. Topographic information from aerial photographs**   As we mentioned in Art. 16.4, a three-dimensional view may be perceived from a pair of overlapping photographs of the object (see Fig. 16.8*b*). For this stereo perception, it is necessary that each photograph be viewed by one eye, a function that is fulfilled through the use of a stereoscope (Fig. 16.13). Although high and low points in the terrain can be ascertained when viewing stereoscopically through a stereoscope, a quantitative measure is not possible until we define *parallax* and show how it can be measured and then converted into differences in elevation. In simple terms parallax is a linear measure on the photographs that corresponds to the parallactic angle $\phi$ (see Fig. 16.8). The larger the parallactic angle, the larger is the parallax, and the closer is the object to the exposure stations (i.e., the higher the elevation of the terrain point).

The quantitative definition of parallax depends on the simplifying assumptions that the photographs have *zero tilts* and that both photographs are exposed from the same altitude. Any rotation of the camera about its (vertical) axis from one photograph to the other is accommodated by replacing the fiducial system of axes by another. The new system (Fig. 16.19) has its abscissa along the projection of the air base on the planes of the photographs. Then the $x$ parallax (or simply parallax) is defined by

$$p = x - x' \tag{16.10}$$

where $x$ and $x'$ are the coordinates on the left and right photographs, respectively, with respect to the new axis-of-flight, or base-line, system, and $p$ is the parallax. The algebraic values (i.e., with the proper sign) of both $x$ and $x'$ must be used in Eq. (16.10). The value of $p$ can be used to determine the elevation of terrain points appearing in the overlap area of two aerial photographs. Before this is done, first consider how parallax is measured directly from the photographs.

The fiducial center of each photograph is first determined and assumed to be its principal point. Under the stereoscope, the two photographs are manipulated until stereoviewing is achieved. Conjugate principal points are then transferred. Lines connecting the pair of principal points on each photograph represent the base and therefore the lines parallel to which parallax is measured. Thus, as shown in Fig. 16.20, the two photographs are laid down such that the base or flight lines are coincident. A simple instrument, called a *parallax bar* and schematically depicted in Fig. 16.20, is used to yield the values of $p$ at various points. The bar has two identical dots, called *half-marks* and identified in the figure, which fuse under the stereoscope and appear as a *floating* dot in the space occupied by the terrain model. When the micrometer $m$ is rotated, the distance between the two half-marks is changed and the dot appears to go up and down. When it appears to be just in contact with the terrain at the point of interest, the reading of the parallax bar would lead to the required value of parallax $p$, for that point.

Figure 16.21 shows a schematic of two photographs and a terrain point, drawn in a plane through the flight line. The altitude of each photograph is $H$ above datum, while the elevation of the point, $A$, is $h$ above the same datum. The two coordinates $x$ and $x'$ are shown for the images $a$ and $a'$ on the left and right photographs, respectively. A line $Eb$ is

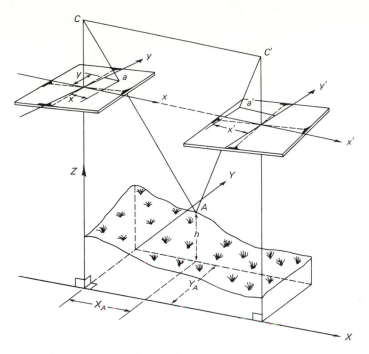

**Fig. 16.19** Parallax in stereophotography.

drawn parallel to $E'A$ by construction; thus, $c_1 b$ is equal to $x'$. Since from Eq. (16.10) the parallax is $(x - x')$, the line segment $ab$ is the parallax $p$. From the fact that the two triangles $abE$ and $EE'A$ are similar,

$$\frac{H-h}{f} = \frac{B}{p} \quad \text{or} \quad H - h = \frac{Bf}{p}$$

so that

$$h = H - \frac{Bf}{p} \qquad (16.11)$$

Thus, in order to determine the elevation of a point $h$ from a pair of aerial photographs, it is

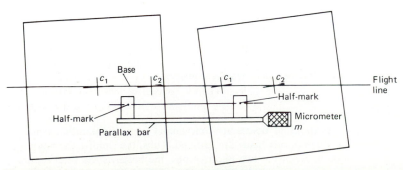

**Fig. 16.20** Measurement of parallax with a parallax bar.

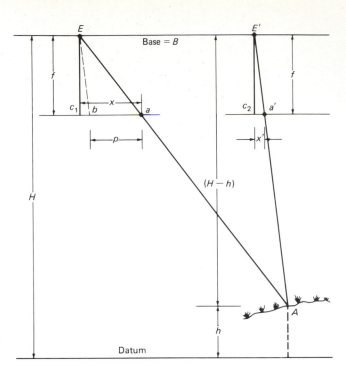

**Fig. 16.21** Relation between parallax and point elevation.

necessary to have the flying height above datum $H$, the value of the air base $B$, and the focal length $f$. Usually, $f$ is known from camera calibration data, $B$ is determined from a radial line plot, and $H$ is either given from prior computation or determined on the basis of a known *vertical* control point in the overlap area. In the latter case, $h$ would be known and Eq. (16.11) is rearranged to give $H$, or

$$H = h + \frac{Bf}{p} \tag{16.12}$$

**Example 16.4**  A pair of photographs are taken with a camera the focal length of which is $f = 152.70$ mm. A radial line plot prepared at a scale of 1:10,000 was used to determine the distance between the two principal points as 5.65 cm. The measured parallax for a vertical control point having an elevation of 153.2 m is 91.12 mm. Compute the elevation of another point $A$ whose measured parallax is 92.38 mm.

SOLUTION  The air base is

$B = (5.65)(10,000) = 56,500$ cm $= 565$ m

Using Eq. (16.12), the flying height above datum, $H$, is

$H = 153.2 + \dfrac{(565)(152.7)}{91.12} = 1100.0$ m

Using Eq. (16.11), the elevation of the new point may be determined.

$h = 1100.0 - \dfrac{(565)(152.7)}{92.38} = 166.1$ m

The accuracy with which elevations of points are determined from the parallax equations depends on how closely the assumptions made are fulfilled. Thus, the smaller the amount of tilt and difference in altitude between the two photographs, the better is the final accuracy. In general, the obtainable accuracy is consistent with the methodology and material used, particularly if paper prints subject to significant shrinkage are used.

Coordinates in the ground system can also be obtained in terms of parallax. Recall from Eq. (16.1) that the scale, $S$, at the point of elevation $h$ is

$$S = \frac{f}{H-h}$$

Substitution of $H$ from Eq. (16.12) yields

$$S = \frac{p}{B} \tag{16.13}$$

in which $p$ is the parallax of the point in question and $B$ the air base. If, as shown in Fig. 16.19, a coordinate system is established in the terrain with the $X$ axis parallel to the photographic axis-of-flight $x$ axis, then

$$X = \frac{x}{S} = \frac{xB}{p}$$

$$Y = \frac{y}{S} = \frac{yB}{p} \tag{16.14}$$

in which $X$ and $Y$ are the ground coordinates in the specific system shown in Fig. 16.19 and $x$ and $y$ are the coordinates of the image point in the *left* photograph with respect to the axis-of-flight coordinate system. On the basis of Eq. (16.14), the distance between two points could be determined if their elevations are known, or at least the elevation of one point is known. In this case, the measured parallax values are used together with image coordinates from Eq. (16.14). Then the distance is

$$D = \left[ (X_1 - X_2)^2 + (Y_1 - Y_2)^2 \right]^{1/2} \tag{16.15}$$

where the subscripts 1 and 2 refer to the point numbers. Directions between points could also be determined in a like manner. The use of Eq. (16.14) is not so much for determining absolute location as for extracting relative information between points.

In addition to determining vertical information on a point to point basis, it is possible to plot contour lines, *in perspective*, using a combination mirror stereoscope and parallax bar. The basis for this is Eq. (16.12), which can be rearranged to the form

$$p = \frac{Bf}{H-h} \tag{16.16}$$

Thus, given $B, f$, and $H$, the parallax value $p$ for each contour $h$ (i.e., the plane at elevation $h$ in the terrain) can be calculated. The corresponding reading on the parallax bar is then set, fixing the floating dot in the space where the overlap is perceived in stereo. If the stereoscope–parallax bar assembly is moved about such that the floating dot remains in contact with the perceived terrain surface, this motion would describe a contour line. A drawing pencil attached to the assembly would draw such a contour. It is important to note these contours are not orthographic because they are drawn in relation to one of the photographs, which is of course a perspective. There are several simple plotters designed on this principle, such as the *stereocomparagraph*, the *contour finder*, and the K.E.K. plotter, more details on which are given in the references at the end of this chapter.

**16.15. Rigorous operations in photogrammetry: general**   As indicated in Art. 16.9, several assumptions regarding the geometry of aerial photographs can be made which will then allow the deriving of products using relatively simple and elementary procedures. On the other hand, if high-accuracy products are required, the previous simplifications are unsatisfactory and the geometry of the photographs must be rigorously recovered. At the instant a photograph is exposed from an airplane, the aerial camera location may be expressed by three coordinates in the ground space. Also, the attitude of the camera at the time of exposure may be expressed by three angles, usually called *orientation* angles. Thus, attached to each aerial photograph are six elements, usually unknown, designated by the term *exterior orientation*. These elements, when determined, specify the location and attitude of the bundle of light rays that enter the aerial camera lens at the time of exposure. The shape of the bundle, on the other hand, is determined by what is called *interior orientation* elements, which are *three* in number. They express the geometric relationship between the exposure station and the photograph plane. In aerial photography, the interior orientation elements are usually known from camera calibration and consist of the principal point coordinates $x_0$, $y_0$ and camera calibrated focal length, $f$.

Given a pair of photographs, their 12 exterior orientation elements (assuming interior orientation is known) must be recovered before the geometry of the photographed terrain can be accurately determined. There are a number of different techniques of accomplishing this, the most common being through the use of analog instruments called *stereoplotters*. They employ optical-mechanical components to simulate and help recover rigorously the geometry of the photographs. There are also digital procedures for this operation, but discussion of these methods will be deferred until the analog techniques are introduced.

**16.16. Photogrammetric stereoplotters**   A photogrammetric stereoplotter is an instrument of high precision which rigorously reconstructs the geometry of overlapping photographs such that accurate three-dimensional information about the photographed object may be obtained. In the case of aerial photography, planimetric and topographic maps may be compiled, or coordinate listings, both horizontal and vertical, may be derived. Although there are many classifications of stereoplotters, it is sufficient for the purpose of this chapter to consider two broad classes: *direct-projection* plotters and *indirect-viewing* plotters. The first class is easier to understand and will therefore be presented first.

Figure 16.22 is a schematic of a direct-projection stereoplotter. The aerial camera is replaced by a *projector* similar in function to a slide projector. Two such projectors are needed so that the pair of overlapping photographs may be projected simultaneously. Each bundle of rays emerging from the projector lens should duplicate, as nearly as possible, the bundle that entered the aerial camera during the photographic exposure. This duplication is accomplished by properly recovering the three elements of interior orientation. If the projector accepts photographs, usually called *diapositives*, which are the same size as the original format used in the aerial camera, the distance between the photograph and the projector lens (i.e., its principal distance) is set equal to the aerial-camera calibrated focal length. If, on the other hand, the plotter's projector accepts only a reduced-size photograph, the principal distance in the plotter should be the focal length *reduced by the same factor*. For example, if the plates used in the plotter are *one-third* the size of the original format, the corresponding principal distance would be $f/3$, where $f$ is the calibrated focal length of the camera. The remaining elements of interior orientation are recovered by placing the plate such that its center (principal point) is directly over the optical axis of the projector. This is usually accomplished by matching the plate's fiducial marks to reference marks in the diapositive holder of each projector.

Once interior orientation is accomplished and the plates are illuminated from above, the ray bundles emerging from the two projectors will each be a replica of the bundle that originally entered the camera. However, the relative position and orientation will not duplicate the original situation during photography because the 12 exterior orientation elements have not yet been recovered. Each projector has six degrees of freedom, including three translations along $X$, $Y$, and $Z$, and three rotations about the $X$, $Y$, and $Z$ axes. These are used to recover the exterior orientation elements, as explained in Art. 16.17.

When exterior orientation is properly recovered, a stereoscopic model will result, as shown in Fig. 16.22. In order for the human operator to perceive the model in three dimensions, the image from each projector should be viewed with one eye. One common procedure is to use

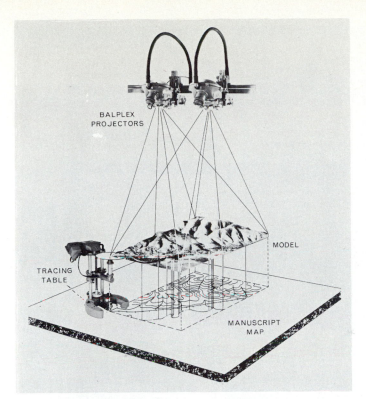

BALPLEX
PROJECTORS

MODEL

TRACING
TABLE

MANUSCRIPT
MAP

**Fig. 16.22** Model representation.

two glass filters, one *blue-green* and the other *red*, placed in the path of the projecting light rays. The operator then views the model through a pair of spectacles with corresponding blue-green and red lenses. The resulting stereoscopic model is measured by using a *tracing table* (Fig. 16.22), which moves through the model space in all three directions $X, Y, Z$. It slides freely over the map table (the $X$ and $Y$ motions) and its top surface, called the *platen*, is moved up and down by a knob (the $Z$ motion). All three motions can be monitored and recorded either manually or through the use of automatic devices. The reference measuring point is a small hole in the center of the platen illuminated by a small bulb directly underneath it. Thus, when viewing the model stereoscopically the operator sees within the model a sharp circular white point which can be moved about and brought in apparent contact with the model surface at any desired spot. Directly underneath the measuring mark is a pencil which can be lowered so as to contact the map sheet when plotting.

A good example of a direct-projection stereoscopic instrument is the Kelsh plotter shown in Fig. 16.23. It is used basically for plotting maps at medium to large scales. It is built on essentially the same principle explained above. The basic features of the Kelsh plotter are: (1) diapositives the same size as the original negative are used; (2) only a small area of each plate is illuminated, enough to cover the platen, by small lamps shown in the figure (two rods connect the tracing table to the lamps so that as the table moves the lamps rotate to illuminate the proper image areas); and (3) the model scale is usually five times that of the original negative.

In the *indirect-viewing* instruments, the operator looks at the photographic plates and consequent stereomodel through optical trains. Thus, these instruments, as far as viewing is concerned, are essentially very sophisticated stereoscopes. However, since there is no direct

**Fig. 16.23** Direct-projection stereoscopic plotter (Kelsh plotter). (*Courtesy Danko Arlington, Inc., Kelsh Instrument Division.*)

projection with light rays, the relative geometry of the photographs is accomplished through optical-mechanical simulation. For example, each pair of conjugate light rays from a terrain point to the camera lens is represented by two *space rods*, as can be seen in Fig. 16.24 of one such plotter. Instead of the tracing table, a *base carriage* is used, at which the two rods intersect. The virtual point at which the axes of the two rods intersect represents the model point. As the carriage is moved in model space, each rod rotates about a pivot point that represents the original camera location, or exposure station. Viewing is accomplished through a pair of binoculars which pick up the corresponding images via optical trains. The objectives of the optical trains are close to the photographic plates and are connected to the upper ends of the space rods, thus moving relative to the plates, when the rods move, to pick up the proper areas of the photographs. Within each objective is inserted a half-mark, similar to those used in the parallax bar (Fig. 16.20), so that a floating dot appears in the model.

Plotting in this type of instrument is usually not done in the model space as in the direct projection instruments. Instead, the movements of the base carriage are transferred through either pantographs or gear boxes to a plotting table, called a *coordinatograph*. This added feature allows for some additional scale change between the model and the map.

The Zeiss Planicart E3 (Fig. 16.24) is an example of the indirect-viewing mechanical type of projection plotter. It uses full-size diapositives (the same size as the original negative) from both wide-angle and super-wide-angle cameras. Each diapositive can be rotated about

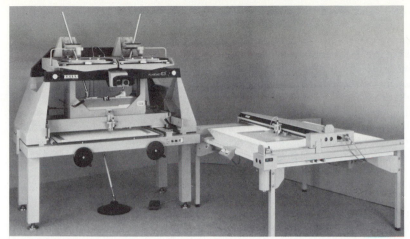

**Fig. 16.24** Indirect-viewing mechanical projection plotter, Zeiss Planicart E-3. (*Courtesy Carl Zeiss, Inc.*)

each of the three axes, but the projectors can be displaced relative to each other only in one direction, representing the total air base. The model is usually at a scale which is about twice that of the original photography. A linear pantograph can effect an additional scale change of 2.5 times.

The two plotters mentioned here are but one example from a multitude of photogrammetric plotters on the market. The student should consult textbooks on photogrammetry (see the references at the end of the chapter) for more extensive treatment of other instruments.

**16.17. Orientation and map compilation**    In the preceding section it was shown that in a photogrammetric stereoplotter both interior and exterior orientation must be properly recovered for the pair of photographs before plotting of a map from the stereomodel can begin. While the procedure for interior orientation was explained, the method for recovering the 12 elements of exterior orientation was not discussed. Referring to Fig. 16.22, note that the plane of the table top where the map manuscript rests represents the $XY$ plane of the plotter; the line normal to it (i.e., the vertical direction of travel of the platen) represents the $Z$ axis. It is possible to bring the two bundles of rays in proper relative registration such that each pair of conjugate rays intersects in a point and one can perceive a model in stereo. The resulting stereomodel, however, may be at an arbitrarily unknown scale and may also be tipped and tilted relative to the plotter's coordinate system. Since, at the time of map plotting the plotter's system represents the object coordinate system, the model must be rotated, translated, and its scale changed until it sits properly into the instrument system just as the terrain was relative to the object coordinate system and the model is at the scale of the map. This operation of manipulating an established (or *restituted*) stereomodel is called *absolute orientation* and relies on having an adequate number of survey control points. Since a model at an unknown scale can be translated along each of the three axes, and rotated about each of the three axes, and its scale changed, absolute orientation involves the recovery of *seven* parameters. This leaves *five* parameters to make the total of 12 exterior orientation elements that must be recovered before one can compile a map from a stereomodel.

The five-parameter operation by which a model is established at an arbitrary scale and arbitrary absolute orientation is called *relative orientation*. It in effect causes each pair of corresponding rays to intersect at a point by recovering the relative tilts of two projectors with respect to each other. There are a number of specific procedures of relative orientation depending upon the type of plotter available and to some extent on the nature of the model (as, for example, the existence of water bodies covering parts of the model). Interested students should consult the references at the end of the chapter for more details.

Absolute orientation is usually accomplished in two separate operations: *scaling*, and *leveling* the model. In scaling, two horizontal control points are plotted on the map manuscript at the scale at which the map is to be compiled. These control points are then used to make the horizontal distance between the corresponding model points equal to the map distance by changing the model base in the instrument. For leveling, an absolute minimum of three vertical control points is needed, since three points determine a plane. In practice, however, the elevation of four points at the four corners of the stereomodel used, and frequently a fifth point in the center is considered desirable. Usually, two rotations, one front to back, and the other side to side, are effected to make the stereomodel level. Because leveling affects scaling, and vice versa, these two operations are usually repeated alternately to refine the results sufficiently.

Once the model is absolutely oriented relative to a map manuscript that is firmly fastened to the plotting table, map compilation can begin. The plotting of planimetry is performed separately from, and usually before, the drawing of contour lines. The planimetry is drawn totally before contours are added to make sure that the locations of the contour lines fall in the proper relationship to the horizontal features (such as streams and roads). A planimetric feature, for example a road, is plotted by keeping the floating dot always in contact with the terrain. Thus, if the road is ascending or descending, the elevation of the floating dot should be continuously changed to keep it in contact with the apparent terrain surface in the model. It is advisable that the operator compile like features at the same time: all roads first, then all streams, then structures, etc. Whenever the operator encounters difficulty, the manuscript should be annotated with notes to be used later in the completion stage.

Contour plotting is somewhat more difficult than planimetry, particularly for the inexperienced. The floating dot is *fixed* at the elevation representing the contour value and the tracing table, or base carriage, is moved in the *XY* plane *only*, keeping the dot in contact with the terrain surface in the model. It is here that the difficulty is encountered and often a contour line drawn by an operator with little experience will be in the form of a zigzagging line. It is therefore advisable that the operator scan the entire model first to become familiar with its various areas, and then begin with those details that are simpler to compile. Once these areas are finished and as more familiarity with the model terrain is gained, the operator may proceed with the remaining more difficult areas, such as those which are relatively flat and those with ground cover. In fact, it is in such areas that photogrammetric mapping is at a disadvantage compared to field mapping. Thus, when the plotting is finished on the instrument, the compiled manuscript must be checked. Wherever difficulties are encountered, the respective areas are field-completed. Then the manuscript is checked for mistakes, names are added, and its accuracy is determined by comparing photogrammetric measurements to those correspondingly determined in the field. At this point the manuscript is ready for cartographic and reproduction operations, as explained in Art. 13.29.

**16.18. Photogrammetric triangulation**   It was explained in the preceding two sections that a stereomodel can be established such that it becomes a faithful reproduction of a segment of the earth's surface at a desired scale. Consequently, the horizontal and vertical coordinates of any terrain point can be extracted from the stereoplotter. This means that given two horizontal and three (but usually four) vertical control points to absolutely orient a stereomodel, the ground coordinates of any desired number of points in the model can be determined. This analysis can be extended by visualizing a stereoplotter with *three* projectors. Three overlapping photographs can be used and a stereomodel composed of the total overlap (which is equal to the area of the middle photograph) can be relatively oriented. To absolutely orient this large model (composed of two adjacent stereopairs) still involves seven parameters and requires two horizontal and three (preferably four) vertical control points. Once this is accomplished, supplemental control, both horizontal and vertical, can be obtained. When map compilation is required, each stereopair is used separately on a plotter such as the Kelsh (Fig. 16.23) or the Zeiss Planicart (Fig. 16.24). For each stereopair, enough control will be required to perform absolute orientation. Part of this control would have been field-determined, but the other part is determined photogrammetrically through restituting the three photographs together. This concept can obviously be extended to more photographs than just three and is called *photogrammetric triangulation*. It may be defined as the procedure of establishing the geometric relationships among overlapping and sidelapping photographs for the purpose of determining the positions for supplemental horizontal and vertical control points. In general, the supplemental control thus obtained is used for map compilation. Photogrammetric triangulation is a very powerful tool, since it reduces substantially the control required by field methods, thus markedly improving the economic aspects of photogrammetric mapping.

Photogrammetric triangulation can be performed by one of three techniques: (1) analog triangulation, (2) semianalytical or independent model triangulation, and (3) analytical triangulation. Each technique will be discussed briefly.

*1. Analog triangulation*  In this procedure stereoplotters are used to form a continuous strip model at one scale from an overlapping strip of photographs. Operations involved include relative orientation of successive models and scale transfer between consecutive models, and adjustment of the resulting strip to fit the available ground control. The procedure is easiest to visualize on an instrument with a long bar from which are suspended a large number of projectors. The first pair of photographs in the strip, say photos 1/2, are relatively oriented at some arbitrary scale. Then the light illuminating photo 1 is turned off and photo 3 is inserted in the third projector so that model 2/3 may be relatively oriented. It is *imperative* that the relative orientation of model 2/3 be carried out by manipulating the *third projector only*, leaving the second projector completely undisturbed, an operation usually called *dependent relative orientation*. This procedure is required so as not to destroy the first model and guarantee the continuity of the strip model. Once the relative orientation of model 2/3 is completed, it is usually at a scale that may in general be different from that of model 1/2. The operation of *scale transfer* between the two models is performed so that the scale of the new model, 2/3, is the same as that of the preceding model, 1/2. The processes of dependent relative orientation and scale transfer are repeated for all other models down the strip. At the end, a continuous-strip model at an arbitrary uniform scale results. The data from the plotter are then transformed and adjusted to fit available ground control as discussed at the end of this section.

Long-bar multiprojector instruments are not frequently used in analog triangulation; use of *universal* plotters is more common. The latter are instruments that have only two projectors, with a feature called *base-in base-out*. This feature enables the viewing of overlap areas of succeeding photographs as either the inner or outer areas of the photographs. The procedure is similar to the long-bar case with one notable exception. Unlike the long-bar, a universal plotter has only one viewable model in the instrument at one time. Consequently, it is mandatory that the model coordinates of all points of interest (pass points and control points) be recorded before proceeding to the following model. At the end, instead of having a physical strip model as in a multiprojector plotter, the strip is represented by a listing of all needed model coordinates. These coordinates are then used in strip transformation and adjustments, as will be explained shortly.

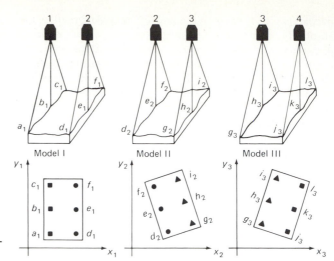

**Fig. 16.25** Three consecutive independent models and their coordinate systems. (*Courtesy California Department of Transportation.*)

**2. Semianalytical or Independent model triangulation**   In this procedure model relative orientation is performed instrumentally on a plotter, but model connection, including scale transfer between successive models, is accomplished analytically; hence the *semianalytical* designation. The method is also called *independent model* because each model 1/2, 2/3, 3/4, etc., is relatively oriented on the plotter completely independently of other models. After relative orientation, each model will have its own coordinate system, as schematically depicted for three consecutive models in Fig. 16.25. Also, each model will be represented by a list of the coordinates of all model points, such as $a_1, b_1, c_1, d_1, e_1, f_1$ for model I in Fig. 16.25, for example. Each two successive models will in general have at least three points in common, as for example $d_1, e_1, f_1$ and $d_2, e_2, f_2$ between models I and II in the figure. Since the extent of terrain relief is usually considerably less than the dimension of the model in the $y$ direction, the use of such model points may lead to weak geometric ties between models. In fact, the determination of the angle about the $y$ axis may be so weak that the model tie becomes unstable. For this reason, the coordinates of the perspective center of both projectors must be determined in the same coordinate system as the model points in each model. These perspective center coordinates can then be used in the model connection to provide depth and guarantee a strong geometric tie. This tie is accomplished by a computer program which implements a transformation involving seven parameters: three translations, three rotations, and one scale change. When all independent models are assembled into a continuous strip, it would be much the same as that obtained from analog triangulation. Semianalytical triangulation has become more common in practice than analog triangulation because of its flexibility. Plotters used for independent model triangulation are usually less expensive than universal plotters. Also, they may be used for map compilation when the triangulation task is finished. Furthermore, analytical strip assembly is in principle more accurate than analog and with the rapid increase in availability and capacity of computers, the computational task is no longer difficult.

**3. Analytical triangulation**   In this method no plotter is needed; and instead, an instrument called a *comparator* is used to measure the plate coordinates $x,y$ in the plane of the photograph of each pass point and control point in each photograph. The ideal locations of the pass points are the same as those used in radial line triangulation and are schematically depicted in Fig. 16.26. These photographic coordinates are then used together with other camera parameters in a relatively extensive program to analytically enforce the intersection

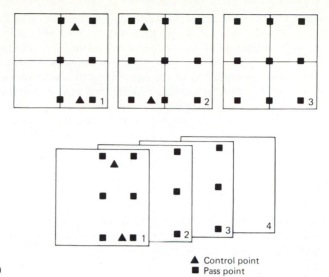

**Fig. 16.26** Ideal pass point locations. (*Courtesy California Department of Transportation.*)

▲ Control point
■ Pass point

of conjugate rays and compute the strip coordinates of all pass points. Thus, unlike analog and semianalytical methods, in analytical triangulation no physical model is constructed. What makes this procedure potentially the most accurate is the ability to correct computationally for all possible systematic errors, such as film shrinkage, lens distortion, and atmospheric refraction. Another advantage of the analytical technique is that triangulation can be, and often is, performed directly in the ground coordinate system using the available ground control, thus eliminating strip adjustment as a separate step.

Strips formed by the analog and semianalytical methods must be transformed and corrected for various deformations in order to fit the ground control that exists in it. As the process progresses from model to model, either in an instrument or computationally, errors build up causing scale variation, increased azimuth error, and twist, bend, and bow of the strip. It is therefore necessary to have both horizontal and vertical control points at regular intervals (usually four to six models) to help correct for these errors. In the past, graphical means were used, but at the present time strip adjustment is almost always performed analytically. Polynomials of various degrees are used, the type depending upon the length of the strip and amount and distribution of the control points available. The end result is a list of the ground coordinates for all points in the strip.

**16.19. Digital terrain models**  As indicated in the preceding two sections, the ground coordinates of any point in a model or strip can be determined once the exterior parameters of the constituent photographs are recovered. The density of points of supplemental control has been relatively sparse, on the order of 10 or at most 20 points per model. These points are usually determined in order to be used for other operations, most notably producing a graphic representation of the terrain in the form of a topographic map. In recent years, however, a totally different product is being photogrammetrically derived, in which the terrain surface is numerically represented by a very dense network of points of known $X, Y, Z$ coordinates. This product is now given the standard name digital terrain model (DTM) or digital elevation model (DEM). The DTM may be produced in a photogrammetric plotter by supplementing it with special processing components that make model digitization possible. The $X, Y, Z$ coordinates are usually stored on magnetic tapes. The data may be in profile form with equal spatial interval or time interval during digitization, or selectively at terrain break points as designated. The data may also be obtained as $Z$ values for an $X, Y$ grid with equal $X$ and $Y$ spacing. Another possibility is to digitize along contour lines, where a large set of $X, Y$ values are developed for each given $Z$ value. These various forms of DTM are produced for a variety of purposes. One of the areas where DTM has been successfully applied is in the automation of road location. Highway engineers have created a DTM, stored it into a computer, and used it over and over to try out different alignments and grades in order to arrive at optimum design factors (Art. 17.47). The increased significance of terrain data banks makes the use of DTM all the more important. During the next decade or two a considerable increase in DTM production, processing, and applications will occur, particularly in relation to automated cartography. In this regard photogrammetric instrumentation and methodology will play an important role.

**16.20. Orthophotography**  One of the more recent products of photogrammetry that is being used increasingly is the orthophoto. It can be defined as a pictorial depiction of the terrain derived from aerial photography in such a way that there are no relief or tilt displacements. Therefore, an orthophoto is equivalent to a planimetric map except that instead of lines and symbols, image tonal variations convey the information. If the photographed terrain is flat and level, a *rectified* aerial photograph would be the same as an orthophoto.

The principle of orthophoto generation is depicted schematically in Fig. 16.27. First a stereomodel is relatively and absolutely oriented as if for plotting a topographic map. Visualizing the plotting a point of detail $A$, the tracing table is elevated to point $A$ in the model and its map location $A_1$ is plotted *orthographically* directly underneath on the map plane. When another model point $B$ is to be plotted, the platen is lowered until the floating mark is at $B$ and the point dropped at $B_1$, on the map. Similarly, point $C_1$ is plotted when the plane of the tracing-table surface is at $C$. It is clear then that the relationship between the *horizontal* positions of model points $A, B, C$ is identical to that on the map sheet. If it were convenient, the entire map-sheet plane could have been raised to the level of the point under consideration and the map position recorded. The result in this case would be identical. This is exactly what is done when generating an orthophoto optically. The tracing table is replaced by a large enough sheet of film to cover the total model area. The film is covered by a light-tight screen and rests on a table that can be moved up and down by the operator just as the tracing table. Similar to the floating mark, a small rectangular slit is open in the screen and can be moved in the $X$ and $Y$ directions. When the slit is at point $A$ in the model (Fig. 16.27), the light from one projector, usually the blue-green, exposes the film to the point detail within the slit. The film is normally not sensitive to red light and therefore is not affected by the image projected from the second projector. When at another point, $B$, the total film plane is moved down to eliminate the effect of relief, and the slit area is exposed. It can be seen that the exposed film would be equivalent to a planimetric map.

In actual orthophoto systems, the slit travels automatically either in the $X$ or $Y$ direction of the model at a selected constant speed, thus continuously exposing a film strip as wide as the longer dimension of the slit. This feature frees the operator to concentrate on raising or lowering the film plane to keep the slit always at the model surface. When one strip is finished, the slit is stepped over and an exactly adjacent film strip is exposed until the entire model area is covered. The film is then developed into a negative from which orthophoto prints can be made.

The on-line optical rectification explained above is but one of many methods for the production of an orthophoto. Another method that is widely used employs electronic techniques. In such an approach, the two conjugate images are electronically correlated and the output signal from the correlation circuits are sent to an orthoprinter. In the printer, a cathode ray tube (CRT) is used to display an image which is then orthogonally printed as an

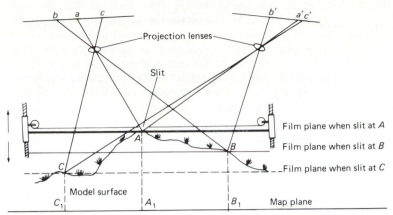

**Fig. 16.27** Orthophoto generation.

orthophoto. Interested students should consult recent textbooks on photogrammetry [such as the Moffitt and Mikhail book (Ref. 11)] for more detail.

Compared to a planimetric map, an orthophoto offers several advantages. Since the details are reproduced in pictorial form, there is a wealth of information on the orthophoto, whereas in a line map the operator has already performed the classification task and selected only those features deemed important. Being an accurate map and rendering abundance of detail makes an orthophoto useful for more applications than either a planimetric map or a regular aerial photograph. Then, if contour lines are added to the orthophoto, it becomes even more useful than a topographic map. Figure 16.28a shows an example of an orthophoto with added contours. For comparison purposes, Fig. 16.28b illustrates a conventional topographic map of a portion of the same area at a larger scale.

**16.21. Planning of photogrammetric project**   There are three interrelated phases involved in the planning for a photogrammetric survey: (1) the design of a flight plan which must be followed by the pilot of the airplane while taking the aerial photography; (2) deciding on the amount and location of necessary horizontal ground control and performing the field surveys to obtain such control to the designated accuracy measures; and (3) estimation of

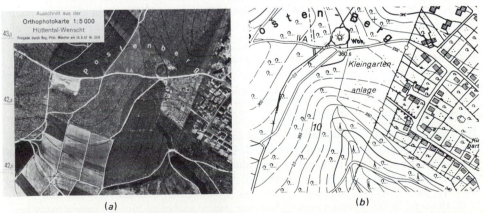

**Fig. 16.28** (a) Orthophoto with contours added (scale 1:5000). (b) Topographic of a portion of the same area as shown in part (a) at a larger scale. (*Courtesy Carl Zeiss, Inc.*)

costs where the relative costs of various operations are determined so that the most economical combination may be selected without compromising the accuracy requirements. Analysis of all three phases must be made in relation to the particular end product of the project, which may include topographic maps of a wide range of scales, orthophotomaps, mosaics, and lists of supplemental control. Usually, the factors affecting the acquisition of photography for one type of end product may be substantially different from those affecting photography for another product. In general, the most important factors to be considered in planning aerial photography include the following:

1. The intended use of the photography (quantitative or interpretive).
2. The desired product (map, orthophoto, mosaic, numerical data, etc.).
3. Accuracy specifications.
4. Size and shape of the area to be covered by photography.
5. The amount and disposition of relief in the area to be photographed.
6. The scale of photography, which in turn depends on the scale of the final product, the camera to be used, the compilation instruments available, etc.
7. Tilt, crab, and drift as they influence the effective overlap and sidelap between adjacent photographs.

When these factors are carefully analyzed for a given project, the basic elements of the flight plan can be resolved. These elements include the flying height above datum (usually mean seal level), the ground distance between successive photo exposures, and the ground spacing between flight lines. Except for applications requiring only one flight line (such as for highway design), the resulting flight lines are usually laid out on the best available map of the area to be photographed.  The general scheme for photographic coverage has been discussed and is shown in Fig. 16.7. Although overlap is most commonly taken at 60 percent, it may be increased for some applications (e.g., mosaic). Also, sidelap can be as high as 60 percent, particularly for block triangulation to produce supplemental control. The actual details of the steps involved in designing a flight plan and the computations associated with it are outside the scope of this chapter.

**16.22. Automated photogrammetric systems**   This chapter has covered the various aspects of photogrammetry involved in the regular production of maps, orthophotos, mosaics, etc. Recently, there has been a significant emphasis in photogrammetry toward automating various functions which are normally performed by a human being. Obviously, details of such aspects are well beyond the scope of this chapter. However, the student should be aware of these developments and their significance in order to be able to call on specialists whenever the use of such systems becomes feasible in future professional work.

Although there are many more possible classes, here we consider automated systems as falling in three categories:

1. Systems where a *correlator* is used to automatically match corresponding imagery on overlapping photographs (thus replacing stereoperception by the human operator).
2. Photogrammetric plotters which operate with the assistance of a digital computer, called *analytical plotters*.
3. Totally digital systems in which photographs are scanned and digitized and all photogrammetric operations performed on the digital data.

The earliest attempts at automation were in the first category, where the function of the eye-brain combination was being synthesized. Electronic, optical, and digital cross-correlation techniques are used with some success, although they still remain less capable than the far superior ability of the human observer to accomplish stereoviewing. In the third category, very extensive computer capabilities are required, and activities are still in the research stage. In the second category, however, many analytical plotters have already been commercially produced and sold. In an analytical plotter, an on-line digital computer is used to implement on a real-time basis the geometric conditions that exist between the images on the plates and model points. The stereomodel is never fully developed as a projection model as in regular plotters; instead, it is incrementally perceived by the operator. The computer continuously computes the locations of conjugate images and transmits $X$ and $Y$ photo shifts to the plates, which are then translated by servomotors to bring the image over the viewing optics. This is done so fast that the operator is never aware of the incremental nature of model formation.

The most distinct advantage of automated systems is flexibility. Analytical plotters usually accommodate a variety of photographic types and can develop several products rapidly. For example, orthophotos could be produced either on-line or off-line, usually automatically without any human intervention. Also, the system can be programmed to automatically produce high-density digital terrain model (DTM) data. These data can themselves be used for producing other useful information (areas, volumes, line of sight, etc.).

## 16.23.  Problems

**16.1**   Describe concisely the following: (a) photo interpretation; (b) remote sensing; (c) aerial photogrammetry; (d) terrestrial photogrammetry; (e) close-range photogrammetry.

**16.2**   Explain the basic differences between a phototheodolite, a stereometric camera, and an aerial camera.

**16.3**   Draw a simple sketch showing the main components of an aerial camera and briefly discuss each component.

**16.4**   "Overlap and sidelap are necessary for proper aerial photographic coverage." Discuss this statement using a sketch to illustrate the two cases.

**16.5**   Explain the correspondence between stereo perception by the pair of human eyes and the geometry of overlapping vertical photographs.

**16.6**   Describe fully, with the aid of sketches, the two types of displacements that make an aerial photograph of the terrain different from a map of such terrain.

**16.7**   Enumerate five different photogrammetric products. Explain each briefly.

**16.8**   What is the main objective of radial line triangulation? Discuss the assumptions underlying the method.

**16.9**   Explain the following operations for a stereopair of dispositives to be used in a photogrammetric plotter: (a) interior orientation; (b) exterior orientation; (c) relative orientation; (d) absolute orientation.

**16.10**   How is a topographic map compiled on a stereoplotter?

**16.11**   Describe briefly three procedures for photogrammetric triangulation.

**16.12**   Explain the principal differences between two photogrammetric products: the digital terrain model and the orthophotograph.

**16.13**   List the most important factors to be considered in planning vertical aerial photography.

**16.14**   What is the height of a tower above terrain which appears on a truly vertical photograph with the following data:

flying height above base of tower $= 3200$ m
distance between principal point and the image of the tower base $= 75.11$ mm
distance between principal point and the image of the top of the tower $= 82.54$ mm

**16.15**   The distance between two well-defined points on a vertical aerial photograph is 72.05 mm. The corresponding distance measured on a U.S. quadrangle map is 55.17 mm. If the map scale is $1:24,000$, calculate the approximate scale of the photograph. Calculate the flying height above terrain if the focal length of the camera used is 150.00 mm.

**16.16**   A vertical photograph is taken with a lens having a focal length of 152.40 mm from a flying height of 2670 m above an airport where the elevation is 380 m above sea level. Determine the representative fraction, expressing the scale of the photograph at a point at which the elevation is 221 m above sea level.

**16.17**   How high above sea level must an aircraft fly in order that photographs at a scale of $1:9600$ may be obtained, if the focal length is 8.25 in and the average elevation of the terrain is 800 ft?

**16.18**   A distance measured on a vertical photograph between two points, both lying at ground elevation of 366 m, scales 82.677 mm. The focal length of the camera is 152.908 mm. The distance between the same pair of points measures 25.451 mm on a 1:24,000 quadrangle sheet. Compute the flying height above sea level at which the photograph was taken. Calculate the datum scale of the photograph.

**16.19**   Photographs are to be taken for preparing a highway design map. The lowest elevation in the area to be photographed is 144 m and the highest elevation is 282 m. The minimum photographic scale is to be 1:6000. What must be the flying height above sea level if the camera to be used contains a lens with a focal length of 134.63 mm? What will be the maximum scale?

**16.20**   The distance between two section corners is assumed to be exactly 1 mi. Both corners lie at 450 ft above sea level. If the distance between the images of these two corners scales 2.342 in on a vertical photograph taken with a lens having an 8.262-in focal length, what is the flying height at which the photograph was taken?

**16.21**   The datum scale of a vertical photograph taken with a lens having a focal length of 152.400 mm is 1:8000. A hilltop lies at an elevation of 800 m above sea level and the image of the hilltop is 71.679 mm from the principal point of the photograph. Compute the relief displacement of the hilltop. If this photo should happen to be tilted 2° such that the hilltop is now on the "up side" of the tilted photo, explain the effects of tilt on the displacement caused by relief.

**16.22**   Two overlapping vertical photographs are taken with a camera having a focal length of 152.00 mm. It is assumed that both photographs are exposed at a flying height of 3500 m above mean terrain, and that the overlap is 60 percent. A control point with elevation of 100.22 m is used as a basis for calculating the elevations of two other points. If the differences in parallax between the control point and the two points are 4.11 mm and −1.73 mm, respectively, calculate the elevations of the two points.

## References

1. American Society of Photogrammetry, *Manual of Photogrammetry*, 3rd. ed., Vols. 1 and 2, George Banta, 1966.
2. Anderson, J. M., et al., "Analytic Block Adjustment," *Photogrammetric Engineering*, Vol. 39, No. 10, 1973.
3. Danko, J. O., "Color, the Kelsh, and the PPV," *Photogrammetric Engineering*, Vol. 38, No. 1, 1972.
4. Hallert, B., *Photogrammetry*, McGraw-Hill Book Company, New York, 1960.
5. Hughes, T., et al., "U.S.G.S. Automatic Orthophoto System," *Photogrammetric Engineering*, Vol. 37, No. 10, 1971.
6. Jensen, N., *Optical and Photographic Reconnaissance Systems*, John Wiley & Sons, Inc., New York, 1968.
7. LaPrade, G. L., "Stereoscopy—A More General Theory," *Photogrammetric Engineering*, Vol. 38, No. 12, 1972.
8. Meyer, D., "Mosaics You Can Make," *Photogrammetric Engineering*, Vol. 28, No. 1, 1962.
9. Mikhail, E. M., "A Study in Numerical Radial Triangulation," *Photogrammetric Engineering*, Vol. 34, No. 4, 1968.
10. Moffitt, F. H., and Bouchard, H., *Surveying*, 6th ed., Harper & Row, Publishers, New York, 1975, Chapter 15.
11. Moffitt, F. H., and Mikhail, E. M., *Photogrammetry*, 3rd ed., Harper & Row, Publishers, New York, 1980.
12. Nash, A. J., "Use a Mirror Stereoscope Correctly," *Photogrammetric Engineering*, Vol. 38, No. 12, 1972.
13. Radlinski, W., "Orthophotomaps versus Conventional Maps," *The Canadian Surveyor*, Vol. 22, No. 1, 1967.
14. Stanton, B. T., "Education in Photogrammetry," *Photogrammetric Engineering*, Vol. 32, No. 3, 1971.
15. Thompson, E. H., "Aerial Triangulation by Independent Models," *Photogrammetria*, Vol. 19, No. 7, 1964.
16. U.S. Department of Transportation, Federal Highway Administration, *Reference Guide Outline*

*Specifications for Aerial Surveys and Mapping by Photogrammetric Methods for Highway,* Washington, D.C., 1968.

17. Whiteside, A. E., et al., "Recent Analytical Stereoplotter Developments," *Photogrammetric Engineering,* Vol. 38, No. 4, 1972.
18. Wolf, P. R., *Elements of Photogrammetry,* McGraw-Hill Book Company, New York, 1974.
19. Woodward, L. A., "Survey Project Planning," *Photogrammetric Engineering,* Vol. 36, No. 6, 1970.
20. Yacoumelos, N., "The Geometry of the Stereomodel," *Photogrammetric Engineering,* Vol. 38, No. 8, 1972.

## Periodicals

21. *Photogrammetria,* published by the International Sociey of Photogrammetry, Delft, Holland.
22. *Photogrammetric Engineering and Remote Sensing,* published by the American Society of Photogrammetry, Washington, D.C.
23. *Photogrammetric Record,* published by the British Society of Photogrammetry, London.
24. *Surveying and Mapping,* published by the American Congress on Surveying and Mapping, Washington, D.C.
25. *The Canadian Surveyor,* published by the Canadian Institute of Surveying and Photogrammetry, Ottawa, Canada.

# CHAPTER 17
# Route surveying

**17.1. Planning the route alignment: general**  The expression "route surveying" used in a very general sense can be applied to the surveys required to establish the horizontal and vertical alignment for transportation facilities. In the most general case, the transportation facilities are assumed to comprise a network that includes the transport of people or goods on or by way of highways, railways, rapid transit guideways, canals, pipelines, and transmission lines. For the past three decades in the United States, highways have been the most highly developed form of transportation facility in the overall network. As a result, route surveys for highways are well defined and widely practiced. Most of the methods that have been developed for highway surveys are equally applicable to the other specified means of transport. Consequently, the emphasis in this chapter is on route surveys for highway alignment.

Surveys of some type are required for practically all phases of route alignment planning, design, and construction work. For small projects involving widening or minor improvement of an existing facility, the survey may be relatively simple and may include only the obtaining of sufficient information for the design engineer to prepare plans and specifications defining the work to be done. For more complex projects involving multilane highways on new locations, the survey may require a myriad of details, including data from specialists in related fields to determine the best location; to prepare plans, specifications, and estimates for construction; and to prepare deed descriptions and maps for appraisal and acquisition of the necessary rights of way.

A description of all aspects and the various stages involved for the planning of route alignments is beyond the scope of this book. Details concerning these processes can be found in the references at the end of the chapter.

However, it is the function of the survey or project engineer to plan the surveys and gather all survey data that may be needed to execute the design of a route alignment for a particular project. This process includes obtaining the necessary information regarding terrain and land use, making surveys to determine detailed topography, and establishing horizontal and vertical control required for construction layout.

In order to plan and perform the surveys needed to acquire these types of data, the survey engineer must be familiar with (1) the geometry of horizontal and vertical curves and how they are used in the route alignment procedure, (2) the methods of acquiring terrain data utilized in the route design procedure, (3) the procedures followed in processing terrain data to obtain earthwork volumes, and (4) the earthwork distribution processes. These topics are covered in the sections that follow.

**17.2 Route curves for horizontal and vertical alignment: general**  In highway, railway, canal, and pipeline location, the horizontal curves employed at points of change in direction are arcs of circles. The straight lines connecting these *circular curves* are tangent to them and are therefore called *tangents*. For the completed line, the transition from tangent to circular curve and from circular curve to tangent may be accomplished gradually by means of a segment in the form of a *spiral* (Arts. 17.13 to 17.18).

Vertical curves (Art. 17.19) are usually arcs of parabolas. Horizontal parabolic curves are occasionally employed in route surveying and in landscaping; they are similar to vertical curves and will not be discussed in this book.

The subject of route curves is extensive, as indicated by the list of references at the end of the chapter. Herein are discussed only some of the simpler relationships.

**17.3. Circular curves: general**   The stationing of a route progresses around a curve in the same manner as along a tangent, as indicated in Fig. 17.1. The point where a circular curve begins is commonly called the *point of curve*, written P.C.; that where the curve ends is called the *point of tangent*, written P.T.; and that where two tangents produced intersect is called the *point of intersection* or the *vertex*, written P.I. or V. A point on the curve is written P.O.C. Other notations are also used; for example, the point of curve may be written T.C., signifying that the route changes from tangent to circular curve, whereas the point of tangent is written C.T. Or the beginning of the curve may be written B.C. and the end of curve E.C. The point of change from tangent to spiral is written T.S., and the point of change from spiral to circular curve S.C.

In the field the distances from station to station on a curve are necessarily measured in straight lines, so that essentially the curve consists of a succession of chords. In this text, a full station is a function of the system of measurement units employed. Thus, for the metric system a full station is 100 m. For the foot system a full station is 100 ft. Where the curve is of long radius, as in railroad practice, the distances along the arc of the curve are considered to be the same as along the chords. In highway practice and along curved property boundaries, the distances are usually considered to be along the *arcs*, and the corresponding chord lengths are computed for measurement in the field.

The sharpness of curvature may be expressed in any of three ways:

**1. Radius**   The curvature is defined by stating the length of radius. This method is often employed in subdivision surveys and sometimes in highway work. The radius is usually taken as a multiple of 100 ft or 100 m.

**2. Degree of curve, arc definition**   Here the curvature is expressed by stating the "degree of curve" $D_a$, which has traditionally been defined as the angle subtended at the center of the curve by an arc 100 ft long. In the metric system, $D_a$ is defined as the angle subtended by a 100-m arc and, in general, $D_a$ is the angle subtended by an arc equal in length to one full station. The arc definition for degree of curve is the method most frequently followed in

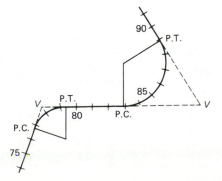

**Fig. 17.1** Route stationing.

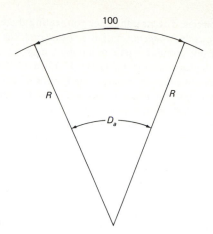

**Fig. 17.2** Degree of curve $D_a$, arc definition.

highway practice. In Fig. 17.2, $D_a$ is the degree of curve by the arc definition and $R$ is the radius, so that

$$\frac{D_a}{100} = \frac{360°}{2\pi R}$$

or

$$D_a = \frac{36{,}000}{2\pi R} = \frac{5729.578}{R} \tag{17.1}$$

Equation 17.1 is applicable in both systems of measurement. For example, the radius of a 1° curve is 5729.578 m in the metric system and 5729.578 ft in the foot system, and a 5° curve would have a radius of 1145.916 units in the respective systems.

Since 1 m equals 3.28084 ft, degrees of curve in the two systems differ by the same proportion. Thus, a 1° curve in the metric system is the same curve as a $0°3048$ curve in the foot system. Conversely, a 1° curve in the foot system is equivalent to a $3°28084$ curve in the metric system. Comparable values in both systems rounded to the nearest tenth of a degree are shown in Table 17.1 (see Ref. 8).

In design practice, $D_a$ is usually selected on the basis of design speed, allowable superelevation, and friction factor and is rounded *down* to the nearest integral number. The radius is then calculated using Eq. (17.1) for subsequent curve calculations.

**3. Degree of curve, chord definition**  Degree of curve according to the chord definition, $D_c$, is defined as the angle subtended by a *chord* having a length of one full station or 100 ft in the foot system. In the metric system, $D_c$ would be the angle subtended by a chord of 100 m. This definition in the foot system has been followed almost invariably in railroad practice. From Fig. 17.3,

$$\sin\left(\frac{D_c}{2}\right) = \frac{50}{R} \tag{17.2}$$

**Table 17.1  Degree-of-Curve Conversions**

| Metric | Foot | Foot | Metric |
|--------|------|------|--------|
| 1 | 0.3 | 0.5 | 1.6 |
| 5 | 1.5 | 1.0 | 3.3 |
| 10 | 3.0 | 1.5 | 4.9 |
| 15 | 4.6 | 2.0 | 6.6 |
| 20 | 6.1 | 2.5 | 8.2 |
| 25 | 7.6 | 3.0 | 9.8 |

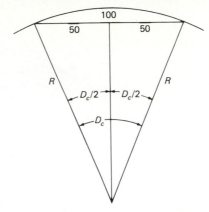

**Fig. 17.3** Degree of curve $D_c$, chord definition.

Note that the radius of curvature varies inversely as the degree of curve; for example, according to the arc definition, the radius of a 1° curve is 5729.578 ft or m and the radius of a 10° curve is 572.958 ft or m. According to the chord definition, the radius of a 1° curve is 5729.651 ft or m and the radius of a 10° curve is 573.686 ft or m.

Field measurements of the curve with the tape must, of course, be made along the chords and not along the arc. When the arc basis is used, either a correction is applied for the difference between arc length and chord length or the chords are made so short as to reduce the error to a negligible amount. In the latter case, usually 100-ft chords are used for curves up to about 3°, 50-ft chords from 3° to 8°, 25-ft chords from 8° to 25°, and 10-ft chords for curves sharper than 25°. In the metric system, 100-m chords can be used for curves up to 1°, 50-m chords from 1° to 3°, 25-m chords from 3° to 5°, and 10-m chords for curves sharper than 5°. Note that curve layout using 100-m or 50-m chords could prove impractical under field conditions where terrain configuration and vegetation present obstacles.

**17.4. Geometry of the circular curve**    In discussing circular curves, the following geometrical facts are employed:

1. An inscribed angle is measured by one-half its intercepted arc, and inscribed angles having the same or equal intercepted arcs are equal. Thus, in Fig. 17.4, the angle $ACB$ (at any point $C$ on the circumference) subtending an arc $AB$ is one-half the central angle $AOB$ subtending the same arc $AB$; and the angles at the points $C$ and $C'$ are equal.
2. An angle formed by a tangent and a chord is measured by one-half its intercepted arc. Thus, in Fig. 17.4, the angle at the point $A$ between $AD$, the tangent to the curve at that point, and the chord $AB$, is one-half the central angle $AOB$ subtending the same arc $AB$. This is a special case of the proposition above, when the point $C$ moves to $A$.

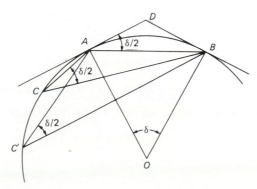

**Fig. 17.4** Geometry of circular curve.

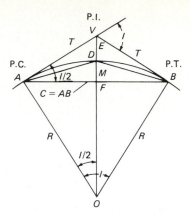

**Fig. 17.5** Basis for curve formulas.

3. The two tangent distances to a circular curve, from the point of intersection of the tangents to the points of tangency, are equal. In Fig. 17.4, lines $AD$ and $DB$ are tangent to the curve at points $A$ and $B$, respectively. Thus, distance $AD$ = distance $DB$.

**17.5. Curve formulas**   Figure 17.5 represents a circular curve joining two tangents. In the field the intersection angle $I$ between the two tangents is measured. The radius of the curve is selected to fit the topography and the proposed operating conditions on the line. The line $OV$ bisects the angles at $V$ and at $O$, bisects the chord $AB$ and the arc $ADB$, and is perpendicular to the chord $AB$ at $F$. From the figure, $\angle AOB = I$ and

$$\angle AOV = \angle VOB = \angle VAB = \angle VBA = \frac{I}{2}$$

The chord $AB = C$ from beginning to end of curve is called the *long chord*. The distance $AV = BV = T$ from vertex to P.C. or P.T. is called the *tangent distance*. The distance $DF = M$ from midpoint of arc to midpoint of chord is called the *middle ordinate*. The distance $DV = E$ from midpoint of arc to vertex is called the *external distance*.

Given the radius of the curve $OA = OB = R$ and the intersection angle $I$, then in the triangle $OAV$

$$\frac{T}{R} = \tan\frac{I}{2}$$

$$T = R\tan\frac{I}{2} = \text{tangent distance} \tag{17.3}$$

$$E = R\sec\frac{I}{2} - R = R\left[\frac{1}{\cos(I/2)} - 1\right] \tag{17.4}$$

$$E = R\operatorname{exsec}\frac{I}{2} = \text{external distance}$$

From the triangle $AOF$, in which $AF = C/2$,

$$C = 2R\sin\frac{I}{2} = \text{long chord} \tag{17.5}$$

$$M = R - R\cos\frac{I}{2} = R\left(1 - \cos\frac{I}{2}\right)$$

$$M = R\operatorname{vers}\frac{I}{2} = \text{middle ordinate} \tag{17.6}$$

From the triangle $AVF$, in which $\angle VAF = I/2$ and $AF = C/2$,

$$\frac{C}{2} = T\cos\frac{I}{2}$$

$$C = 2T\cos\frac{I}{2} \tag{17.7}$$

From the triangle $ADF$, in which $\angle DAF = I/4$,

$$M = \frac{C}{2}\tan\frac{I}{4}$$
(17.8)

**17.6. Length of curve**  The length of the circumference of a circle is $2\pi R$; this is the arc length for a full circle or 360°. As the arc length corresponding to a given radius varies in direct proportion to the central angle subtended by the arc, the length of arc for any central angle $I$ is

$$arc = \left(\frac{I°}{360°}\right)2\pi R$$
(17.9)

in which angle $I°$ is expressed in degrees. This solution is simplified by the use of a table of arc lengths for various angles and for unit radius.

If the degree of curvature is expressed on the *arc* basis, from Eqs. (17.1) and (17.9) the length of curve $L_a$ is

$$L_a = 100\frac{I}{D_a}$$
(17.10)

If the degree of curvature is expressed on the *chord* basis, the length of curve is considered to be the sum of the lengths of the chords, normally each 100 ft long. In the metric system, the chord lengths would be 100 m. For these cases, the length of curve (on the chords) is

$$L_c = 100\frac{I}{D_c}$$
(17.11)

which is somewhat less than the actual arc length. Thus, if the central angle $I$ of the curve $AD$ (Fig. 17.6) is equal to three times the degree of curve $D$, as shown, then there are three 100-ft chords between $A$ and $D$, and the length of "curve" on this basis is 300 ft. Similarly, in the metric system, there would be three 100-m chords between $A$ and $D$, and $L_c$ is 300 m.

**17.7. Laying out a curve by deflection angles**  Curves are staked out usually by the use of deflection angles turned at the P.C. from the tangent to stations along the curve together with the use of chords measured from station to station along the curve. The method is illustrated in Fig. 17.7, in which $ABC$ represents the curve, $AX$ the tangent to the curve at $A$, and angles $XAB$ and $XAC$ the deflection angles from the tangent to the chords $AB$ and $AC$.

Assume the transit to be set up at $A$. Given $R$, $\delta$, and $\theta$ it is required to locate points $B$ and $C$. Considering point $B$,

$$\angle XAB = \frac{\delta}{2}$$
(17.12)

and by Eq. (17.5)

$$AB = 2R\sin\frac{\delta}{2}$$
(17.13)

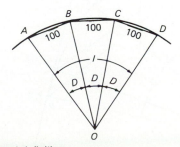

**Fig. 17.6** Length of curve, chord definition.

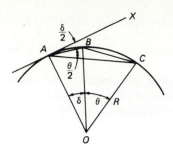

**Fig. 17.7** Curve layout by deflection angles.

In the field, point $B$ is located as follows. The deflection angle $XAB = \delta/2$ is set off from the tangent, the distance $AB$ is measured from $A$, and the forward end of the tape at $B$ is lined in with the transit.

Considering point $C$,

$$\angle BAC = \frac{\theta}{2} \tag{17.14}$$

$$BC = 2R \sin \frac{\theta}{2} \tag{17.15}$$

$$\angle XAC = \frac{\delta + \theta}{2} \tag{17.16}$$

In Fig. 17.8, let points $a$, $b$, $c$, $d$ represent station points on a simple curve. Point $a$ is an odd distance from the P.C. and the distance $dB$ is also an odd increment. The deflection angles are

$$VAa = \frac{d_1}{2}$$

$$VAb = \frac{d_1}{2} + \frac{D}{2}$$

$$VAc = \frac{d_1}{2} + \frac{D}{2} + \frac{D}{2} = \frac{d_1}{2} + D$$

$$VAd = \frac{d_1}{2} + D + \frac{D}{2} = \frac{d_1}{2} + 3\frac{D}{2}$$

$$VAB = \frac{d_1}{2} + \frac{3D}{2} + \frac{d_2}{2} = \frac{I}{2}$$

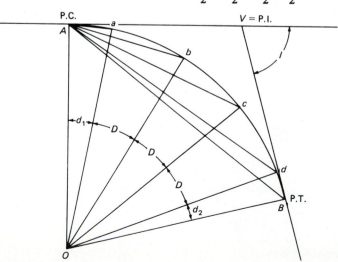

**Fig. 17.8** Deflection angles on a simple curve.

Note that the sum of all deflection angles must equal $I/2$, a check on the calculations.

For a curve having degree of curve $D$ (arc or chord) deflection angles per station, half-station, quarter-station, and one-tenth station are $D/2$, $D/4$, $D/8$, and $D/20$, respectively.

For odd lengths of arc or odd stations, it is convenient to calculate the deflection of the arc in minutes per unit of arc. If $D=9°30'$, then $D/2=4°45'=285'$ per station. Thus, the deflection equals $285'/100$ units $=2.85'/$unit of arc or station. If the radius is given, then

$$\frac{d}{L}=\frac{360}{2\pi R}$$

so that

$$\frac{d}{2}\ (\text{deflection in minutes/unit})=\frac{1718.873}{R}$$

For $D=9°30'$, $R=5729.578/9.5=603.113$ units and

$$\frac{d}{2}=\frac{1718.873}{R}=2.85'/\text{unit of arc or station}$$

A curve is normally located in the field as follows. The P.C. and P.T. are marked on the ground. The deflection angle from the P.C. is computed for each full station on the curve and for any intermediate stations that are to be located. The transit is set up at the P.C., a sight is taken along the tangent, and each point on the curve is located by the deflection angle and by distance measured from the preceding station. The following examples illustrate the procedure and give the usual form of field notes.

**Example 17.1**   In Fig. 17.9, assume that stations have been set as far as $88+00$ at $C$. The directions of tangents $CV$ and $MV$ have been fixed by stakes, but the tangent distances have not been measured.

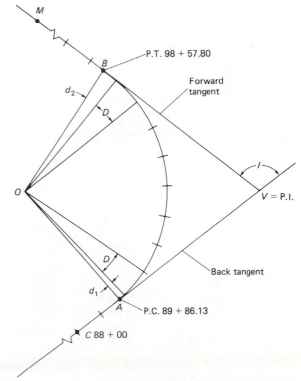

**Fig. 17.9** Laying out a simple curve.

The degree of curve $D_a$ is to be $12°$ according to the arc definition. It is desired to stake full stations on the curve using feet as the unit of measurement.

SOLUTION    The tangents $CA$ and $MB$ are produced to an intersection at $V$ and the distance from $C$ to $V$ is found to be 803.90 ft. Then the transit is set at $V$ and the angle $I$ is observed by double deflection to be $104°36'$. Since $D_a$ is given, the radius of the curve, tangent distance, and length of curve using Eqs. (17.1), (17.3), and (17.10) can be computed as follows:

$$R = \frac{5729.578}{12} = 477.46 \text{ ft}$$

$$T = R \tan\frac{I}{2} = 617.77 \text{ ft}$$

$$L = \frac{100I}{D} = 871.67 \text{ ft}$$

Thus, the stations at the P.C. and P.T. are:

| | |
|---|---|
| Station at P.I. = 88 + 00 + 803.90 = | 96 + 03.90 |
| Tangent distance = | − 6   17.77 |
| Station at the P.C. | 89 + 86.13 |
| Length of curve | + 8   71.67 |
| Station at the P.T. | 98 + 57.80 |

The deflection angle for one full 100-ft station is $D/2 = 6°00$. The distance from the P.C. to the first full station is $(90 + 00) - (89 + 86.13) = 13.87$ ft, and the distance from the last full station on the curve to the P.T. is $(98 + 57.80) - (98 + 00) = 57.80$ ft. The deflection in minutes per unit of arc equals $(12°)(60'/°)/200 = 3.6'/\text{ft}$, so that deflection angles for the odd increments at the beginning and end of the curve are:

$$\frac{d_1}{2} = (13.87)(3.6) = 0°49'56'' = 0°49.93'$$

$$\frac{d_2}{2} = (57.80)(3.6) = 208.08 = 3°28'05'' = 3°28.08'$$

Chord distances for the initial and final odd increments of arc and the full stations are

$$c_1 = 2R\sin\frac{d_1}{2} = (2)(477.46)\sin 0°49'56'' = 13.87 \text{ ft}$$

$$c_2 = (2)(477.46)\sin 3°28'05'' = 57.76 \text{ ft}$$

$$c_{100} = (2)(477.46)\sin 6°00'00'' = 99.82 \text{ ft}$$

Deflection angles are found by cumulating the individual deflections to full stations from the P.C. to P.T. as follows:

| Station | Deflection angles | Chord | Curve data |
|---|---|---|---|
| P.T. 98 + 57.80 | 52°18.0′ | 57.76 | |
| 98 | 48°49.9′ | 99.82 | |
| 97 | 42°49.9′ | 99.82 | |
| 96 | 36°49.9′ | 99.82 | $\Delta = 104°36'$ |
| 95 | 30°49.9′ | 99.82 | $D_a = 12°00'$ |
| 94 | 24°49.9′ | 99.82 | $R = 477.46$ ft |
| 93 | 18°49.9′ | 99.82 | $T = 617.77$ ft |
| 92 | 12°49.9′ | 99.82 | $L = 871.67$ ft |
| 91 | 6°49.9′ | 99.82 | Sta. at P.I. = 96 + 03.90 |
| 90 | 0°49.9′ | 99.82 | |
| P.C. 89 + 86.13 | 0°00.0′ | 13.87 ft | |

Note that the total deflection angle should equal $I/2$ which provides a positive check on the cumulated deflection angles.

**Example 17.2**   The directions of two tangents on an existing road are established from the curb lines. The deflection angle to the right is $44°00'00''$. Because the existing curbs are to be retained, a curve must be determined for which the tangent distance does not exceed 50 m but is equal to or greater than 45 m. Determine a radius and integral value for $D_a$ to satisfy this specification and calculate the parameters and deflection angles needed to set one-tenth stations on this curve. The station of the P.I. is $8 + 43.892$.

SOLUTION   A trial value for $R$ can be found using Eq. (17.3).

$$R = \frac{T}{\tan(I/2)} = \frac{45}{\tan 22°} = 111.379 \text{ m}$$

which yields

$$D_a = \frac{5729.578}{111.379} = 51°4$$

Round this value of $D_a$ to 50°, recompute $R$, and calculate $T$.

$$R = \frac{5729.578}{50°} = 114.592$$

$$T = (114.592)\tan 22° = 46.298 \text{ m}$$

This satisfies the stated specification. The balance of the curve computations are included in the sample note form in Fig. 17.10.

**17.8. Transit setups on the curve**   Because of obstacles, great lengths of curve, and so on, often it is impractical or impossible to run all of a given curve with the transit at the P.C. In such cases, one or more setups are required along the curve between P.C. and P.T.

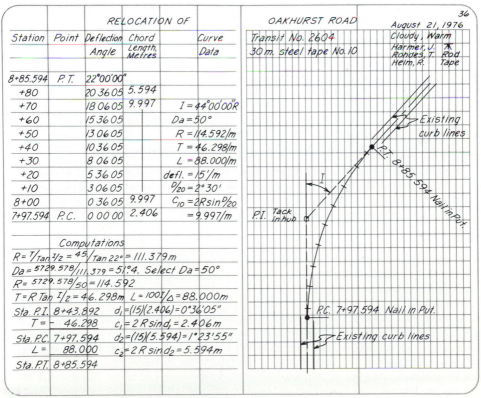

**Fig. 17.10** Notes for layout of a simple curve.

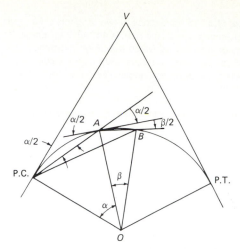

**Fig. 17.11** Transit setup on a simple curve.

Figure 17.11 illustrates the case where the transit is set up at an intermediate point $A$. The curve is begun at the P.C. and is located as far as $A$, where a hub is set. The transit is then set up at $A$. A backsight (with the $A$ vernier set on 180° and the telescope direct) is taken on the last preceding station at which the transit was set up, in this case the P.C. The upper motion is loosened and the telescope is turned toward $B$ with the angle $\alpha/2 + \beta/2$ being set off on the horizontal circle. The line of sight is thus directed along the chord $AB$. The angle $(\alpha + \beta)/2$, turned from the line P.C. to $A$ extended, is equal to the deflection angle at the P.C. for point $B$; therefore, the vernier setting to locate point $B$ from transit station $A$ is the same as that which would have been used had the transit remained at the P.C. According to this method, the following procedure may be used to orient the transit at any point on the curve:

1. Compute deflection angles as for use at the P.C.
2. When set up at any point on the curve, backsight at any preceding transit station with the telescope direct and the $A$ vernier set on 180° + the deflection angle to the station for curves deflecting to the right and 180° − the deflection angle to the station sighted for curves deflecting to the left.
3. To locate stations in a forward direction along the curve, loosen the upper motion, turn the instrument through 180°, and turn the previously computed deflection angles.

If the instrument is in good adjustment, an alternative procedure is possible. Backsight the preceding station with the telescope reversed and with the vernier set on the deflection angle for the station sighted. To locate subsequent stations, plunge the telescope and use the computed deflection angles. If the backsight is to the P.C., the backsight vernier reading is zero.

**17.9. Layout of a curve by tape alone**  Often it is convenient or necessary to lay out a circular curve by means of tape alone. Of the various methods employed, three of the more useful are briefly described here.

**1. Offsets from tangent**  Occasionally, it is convenient to establish the various points on the curve by perpendicular offsets from the tangents. For example, it is desired to establish point $E'$ at station $4+00$ on the curve shown in Fig. 17.12, the intersection angle and degree of curve being known. The central angle $d_1$ is equal to the distance from P.C. to $E'$ (station $3+45.00$ to station $4+00$), in stations, multiplied by the degree of curve. The distance along the tangent from P.C. to $E$, the foot of the perpendicular offset, is equal to $R\sin d_1$; and the

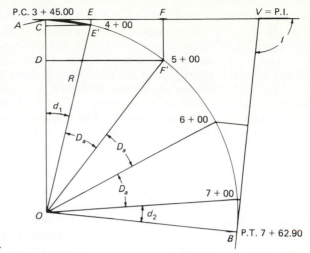

**Fig. 17.12** Offsets from the tangent.

length of the offset $EE'$ is equal to $R$ vers $d_1$ or $R(1-\cos d_1)$. In the field, point $E$ is established by measuring along the tangent from the P.C. If the offset distance $EE'$ is very short, the perpendicular may be established with sufficient precision by estimation; otherwise, point $E$ is established by measuring from $E$ and the P.C. with two tapes. In this case, the chord $AE' = 2R\sin(d_1/2)$ must be calculated. Other points on the curve are established similarly, those on the second half of the curve being located by offsets from the forward tangent. A transit may be used to establish the offsets. Figure 17.12 illustrates graphically how stations on a curve can be established using tangent offsets.

**2. Chord offsets**  Offsets from the chord produced can be employed in conjunction with tangent offsets using two tapes, two range poles, and plumb bobs. Figure 17.13 illustrates the

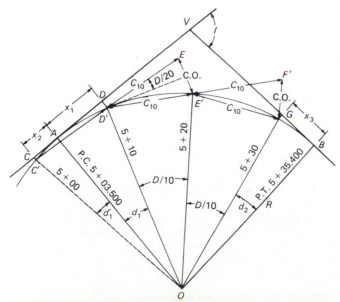

**Fig. 17.13** Curve layout by chord offsets.

procedure. First, points $D'$ at station $5+10$ and $C'$ at station $5+00$ produced backward from the P.C. are set by tangent offsets as described in the previous paragraph. Thus, $x_1 = R\sin d_1$, $x_2 = R\sin d_1'$, $DD' = R(1 - \cos d_1)$, and $CC' = R(1 - \cos d_1')$. Range poles are set at $C'$ and $D'$. In order to set station $5+20$, the chord $C_{10} = 2R\sin(D/20)$ is measured from $D'$ and point $E$ is visually set on line with the range poles at $C'$ and $D'$. Point $E$ is marked on the ground. Next, the *chord offset*, C.O. $= EE' = 2C_{10}\sin(D/20)$ is calculated. Finally, station $5+20$ at $E'$ is set by intersection, measuring the chord offset $EE'$ from $E$ and the chord for the one-tenth station from $D'$. This procedure is repeated for station $5+30$, where the closure can be checked, as illustrated in Fig. 17.13, by tangent offsets from the P.T. at $5+35.400$.

**3. Middle ordinates**  Another method of laying out a circular curve with a tape involves the location of successive stations by use of a middle ordinate of a two-station chord. Such a layout is illustrated on Fig. 17.14, where one-tenth stations on a metric curve are located by a combination of tangent offsets and middle ordinates. The first even tenth station $(5+10)$ ahead of the P.C. is located by distance $x_1$ and tangent offset $DD'$. Next, the first even tenth station $(5+00)$ back from the P.C. is also set by tangent offset $CC'$ and distance $x_2$.

Now, the middle ordinate for a two-station chord or distance $D'D''$ is calculated using Eq. (17.6) so that $D'D'' = R[1 - \cos(D/10)]$. In the field, the distance $D'D''$ is set off from $D'$ along a line the direction of which is estimated to be that of the radius $D'O$ of the curve. Points $C'$ and $D''$ are marked by range poles. One end of the tape is held at $D'$ and the chord $D'E'$ for a one-tenth station is swung until it is on line with points $D''$ and $C'$. Station $5+20$ is marked on the ground. In a similar manner, the next station, $5+30$, is established by middle ordinate $E'E''$ and chord $E'F'$. The work is checked by offset from the tangent $FF'$ at a distance $x_3$ from the P.T. along the forward tangent. The chord distance from $5+30$ to the P.T. should also be checked.

These methods for location of a curve by tapes are most appropriate when extra personnel are available, the transit is being used for other work, and the terrain is relatively level and open. Under other conditions, layout of curves using the transit and deflection angles is preferable.

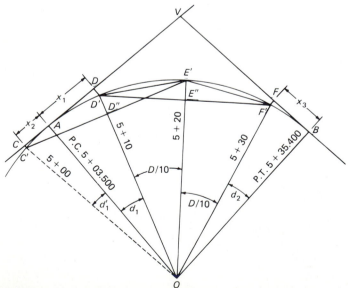

**Fig. 17.14** Curve layout by middle ordinates.

**17.10. String-lining of curves**   Railroad track, particularly on curves, is eventually thrown out of alignment by the action of trains. *String-lining* is a simple method of determining and applying the amounts by which the track must be moved laterally at various points to restore proper curvature. It involves the use of middle ordinates from chord to curve (see Art. 17.9); it is described in detail in various texts on route surveying. Briefly, the method is as follows. At regular intervals along the outer rail, a chord of length equal to two intervals is stretched, and the middle ordinate is measured with a scale and is recorded. For the circular portion of the curve, all middle ordinates should be equal; for the portion along which a gradual transition is made from curve to tangent, the middle ordinates should be progressively smaller by uniform increments. Irregularities in the tabulated values of middle ordinate are noted, and for each point of measurement the amount necessary to move the track is computed. Stakes are set in the ballast to serve as reference points and the track is moved to conform with computed values.

**17.11. Compound curves**   A compound curve consists of two or more simple curves which deflect in the same direction, are tangent to one another, and have two or more centers on the same side of the curve.

Use of compound curves permits better fitting of highway and railroad center lines to difficult topographic conditions. Compound curves should not be used where a simple curve satisfies alignment conditions.

Figure 17.15 illustrates a two-center compound curve. There are several variables: $I_1$, $I_2$, $T_1$, $T_2$, $R_1$, $R_2$, and $I$, of which six are independent since $I = I_1 + I_2$. It is beyond the scope of this treatment to deal with all solutions to compound curves. Two methods are discussed: the vertex triangle procedure, and the traverse method.

*Vertex triangle method*   In Fig. 17.15, assume that $R_1$, $R_2$, $I_1$, $I_2$, $I$, and the station of the P.I. at $V$ are known. The line $A'B'$ is tangent to the curve at the point of compound curve or P.C.C. and forms the vertex triangle $A'B'V$. The distance $A'D = T_1 = R_1 \tan(I_1/2)$ and $DB' = T_2 = R_2 \tan(I_2/2)$, yielding $A'B' = T_1 + T_2$. The triangle $A'B'V$ is solved by the law of sines for $p$ and $q$. Then $T_L = T_1 + p$ and $T_R = T_2 + q$.

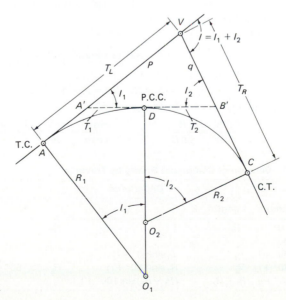

**Fig. 17.15** Compound curve.

***Traverse method*** The procedure for applying this method to compound curves is as follows:

1. Draw a sketch of the curve to be computed. Form a closed traverse, including all independent variables, and label variables. In Fig. 17.15, polygon $AVCO_2O_1A$ comprises the closed traverse.
2. Assume a direction of zero azimuth to be parallel or perpendicular to one unknown line. Thus, in Fig. 17.15, $O_1A$ can be assumed due north.
3. Calculate azimuths for each line, including unknown directions expressed as functions of unknown angles. For example, in Fig. 17.15, the azimuth of $O_1O_2$ is $I_1^\circ$, assuming that $O_1A$ is due north.
4. Tabulate data as for traverse calculations and compute departures and latitudes, expressing those for unknown lines as functions of unknowns.
5. Take $\Sigma$ latitudes $=0$ and the $\Sigma$ departures $=0$ to form two equations which can be solved for the unknowns.

The tabular form used to set up the traverse solution of the compound curve in Fig. 17.15 is illustrated in Table 17.2. Taking $\Sigma$ latitudes $=0$ yields

$$R_1 - (R_1 - R_2)\cos(180^\circ + I_1) - T_R\cos(90^\circ + I) + R_2\cos(180^\circ + I) = 0$$

which can be solved for $T_R$:

$$T_R = \frac{R_1 - (R_1 - R_2)\cos(180^\circ + I_1) + R_2\cos(180^\circ + I)}{\cos(90^\circ + I)} \tag{17.17}$$

Setting $\Sigma$ departures $=0$ and substituting $T_R$ from Eq. (17.17) into this equation yields a value for $T_L$:

$$T_L = R_2\sin(180^\circ + I) + (R_1 - R_2)\sin(180^\circ + I_1) + T_R\sin(90^\circ + I) \tag{17.18}$$

Many different solutions to various compound curve problems are possible depending on the combinations of known and sought parameters. Reference to a route surveying text is desirable to become familiar with all the possibilities (Ref. 6).

### 17.12. Intersection of a curve and a straight line

A problem which frequently occurs in route alignment is that of locating the intersection of a straight line with a curve. In Fig. 17.16, the line $XP'$ of known direction intersects the curve at $P$. It is desired to locate point $P$ so as to fix the corner of property bounded by the arc $BP$ and the line $PX$. Thus, the angle $\theta$ and distance $PP'$ must be calculated. The angle $\alpha$, distance $AP'$, $I$, and $R$ are known. One solution to this problem is as follows.

In triangle $NAP'$:

$$NA = AP'\tan\alpha \quad \text{and} \quad NP' = \frac{AP'}{\cos\alpha}$$

so that

$$ON = R - NA = R - AP'\tan\alpha$$

Since $\beta = 90 - \alpha$, $\sin\beta = \cos\alpha$. In triangle $ONP$,

$$\frac{ON}{\sin\phi} = \frac{R}{\sin\beta} \quad \text{or} \quad \frac{\sin\phi}{\cos\alpha} = \frac{ON}{R} = \frac{R - AP'\tan\alpha}{R}$$

**Table 17.2  Solution of Compound Survey by Traverse**

| Side | Azimuth | Length | E | W | N | S |
|------|---------|--------|---|---|---|---|
| | | | | Departures | | Latitudes |
| $O_1A$ | $0^\circ$ | $R_1$ | | | $R_1$ | |
| $AV$ | $90^\circ$ | $T_L$ | $T_L$ | | | |
| $VC$ | $90^\circ + I$ | $T_R$ | | $T_R\sin(90^\circ + I)$ | | $T_R\cos(90^\circ + I)$ |
| $CO_2$ | $180^\circ + I$ | $R_2$ | | $R_2\sin(180^\circ + I)$ | $R_2\cos(180^\circ + I)$ | |
| $O_2O_1$ | $180^\circ + I_1$ | $R_1 - R_2$ | | $(R_1 - R_2)\sin(180^\circ + I_1)$ | | $(R_1 - R_2)\cos(180^\circ + I_1)$ |

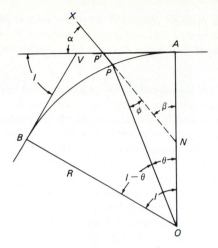

**Fig. 17.16** Intersection of a curve by a straight line.

so that

$$\sin\phi = \cos\alpha - \frac{AP'}{R}\sin\alpha$$

from which $\phi$ may be determined and used to calculate $\theta$, where

$$\theta = \beta - \phi = 90 - \alpha - \phi$$

In triangle $ONP$, by the law of sines

$$NP = \frac{R\sin\theta}{\cos\alpha}$$

and

$$PP' = NP' - NP = \frac{AP'}{\cos\alpha} - \frac{R\sin\theta}{\cos\alpha}$$

This problem can also be solved by the traverse method as outlined in Art. 17.11. The closed traverse would consist of $PP'AOP$. A bearing of due north could be assumed for line $PP'$. Traverse data are shown in Table 17.3, from which

$$\Sigma\ \text{departures} = P'A\sin\alpha + R\cos(\alpha+\theta) - R\cos\alpha = 0$$

which can be solved for

$$\cos(\alpha+\theta) = \frac{R\cos\alpha - P'A\sin\alpha}{R} \tag{17.19}$$

to yield $\theta$ since $\alpha$ is known.

$$\Sigma\ \text{latitudes} = PP' + R\sin(\alpha+\theta) - P'A\cos\alpha - R\sin\alpha = 0$$

from which

$$PP' = P'A\cos\alpha + R\sin\alpha - R\sin(\alpha+\theta) \tag{17.20}$$

This completes the solution.

**Table 17.3   Intersection of a Curve by a Straight Line**

| | | | Departures | | Latitudes | |
|---|---|---|---|---|---|---|
| Line | Direction | Distance | E | W | N | S |
| PP' | N0°E | PP' | | | PP' | |
| P'A | S α E | P'A | P'A sin α | | | P'A cos α |
| AO | S(90° − α)W | R | | R cos α | | R sin α |
| OP | N[90° − (α + θ)]E | R | R cos(α + θ) | | R sin(α + θ) | |

A third solution to this problem consists of using the method of coordinates according to the procedures developed in Art. 8.21 and illustrated in Example 8.3. To apply this method, coordinates for the center of the circular curve and for a point on the property line must be available.

**17.13. Superelevation**   On high-speed highways and railroad curves, the velocity of movement of the vehicle or train develops a horizontal centrifugal force. So that the plane of the pavement or rails may be normal to the resultant of the horizontal and vertical forces acting on the vehicle, the outer edge of pavement or outer rail of the track is *superelevated*, or elevated above the inner edge of the pavement or inner rail of track. In railway work, the amount of superelevation is made equal to approximately $0.00067 V^2 D$ expressed in inches, where $V$ is the train speed in miles per hour and $D$ the degree of curve according to the chord definition. The amount of this superelevation should not exceed 7 or 8 in (0.18 to .20 m) because of the use of the track by slow trains. For a speed of 40 mi (64 km) per hour, the superelevation equals (in inches) a fraction more than the degree of curve in degrees. The elevation of the inner rail is maintained at grade.

The superelevation of highway curves is usually combined with a friction factor $f$ between the vehicle wheels and pavements. The amount of pavement superelevation plus the side friction factor is

$$e + f = \frac{V^2}{15R} \qquad (17.21)$$

where $e$ = superelevation, feet per foot
$f$ = side friction factor
$V$ = vehicle speed, miles per hour
$R$ = radius of curve, feet

Substituting for $R$ using Eq. (17.1) (rounded to one decimal place) and solving the resulting equation for $D$ yields

$$D_a = \frac{85,900(e + f)}{V^2} \qquad (17.22)$$

in which $D_a$ is the degree of curve according to the arc definition. Maximum values for $e$ and assumption for $f$ depend on climatic conditions, terrain configuration, type of area (urban vs. rural), and frequency of slow-moving vehicles. Depending on these factors, the American Association of State Highway Officials (AASHO) recommends values of $e + f$ which vary from 0.19 to 0.24. Thus, with a knowledge of local conditions, Eq. (17.22) can be used to calculate a maximum safe degree of curve for specified vehicle speeds and values for $(e + f)$. For example, assuming that $e = 0.08$, $f = 0.14$, and $V = 50$ mi/h, the minimum safe radius is 758 ft and the maximum $D_a = 7°6$. Tables for selecting $D_a$ and $R$ given the design speed and $e + f$ can be found in many route surveying textbooks and in the AASHO *A Policy on Geometric Design of Rural Highways* (Ref. 1).

Superelevation and side friction factor for highway curves in the metric system are given approximately (modified on the safe side) by (Ref. 8)

$$e + f = \frac{DV^2}{730,000} \qquad (17.23)$$

or

$$D = \frac{730,000(e + f)}{V^2} \qquad (17.24)$$

where $D$ = angle subtended by an arc of 100 m
$e$ = superelevation, metres/metre
$f$ = side friction factor
$V$ = design speed, km/h

Equation (17.24) can be used to determine maximum degree of curve given $(e+f)$ and vehicle speeds. For example, assuming that $V=80$ km/h, $e=0.08$ m/m, and $f=0.16$, computation with Eqs. (17.24) and (17.1) yields $R=209.3$ m and $D=27.38°$. These values could be rounded to 210 m and 27°, respectively, without sacrificing safety.

Superelevation on highway pavements is generally introduced by gradually rotating the pavement cross section about the center line. In freeways and in interchange design, special problems often require that the rotation occurs about the inner edge or some intermediate point on the pavement cross section. Obviously, full superelevation cannot be introduced immediately at the beginning of a curve but must be developed gradually. In order to provide smooth transition from normal pavement crown on a tangent to full superelevation on the circular curve a spiral curve could be used and is introduced in the next Article.

**17.14. Spirals**   Transition curves were first utilized by railways as early as 1880 to provide easement between tangents and circular curves. The curve chosen for this purpose was a *clothoid*, in which the curvature varies inversely as the radius and increases linearly from zero, at the tangent to spiral or T.S., to the degree of curvature of the simple curve at the point where the spiral is tangent to the curve. Spiral curves are also called easement or transition curves. They are used as easement curves in both railway and highway alignments.

**17.15. Geometry of the spiral**   Figure 17.17 illustrates a spiral with a length of $L_s$. Since the degree of curvature for a spiral increases from zero at the T.S. to $D_a$ at the S.C., the rate of change of curvature of a spiral in degrees per station is

$$K = \frac{100D}{L_s} \tag{17.25}$$

in which $L_s$ is in the units of measurement and $D$ is in degrees. For example, if $L_s=800$ ft and $D_a$ at the S.C. is $8°$, then $K=(100)(8)/800=1$ degree per station. Thus, if $L_s$ and $D_a$ or $R$ are known for the circular curve at the S.C., then $D_p$ or the radius $r$ at any point $p$ a distance $\ell_s$ from the T.S. (Fig. 17.17) can be determined as follows:

$$D_p = \frac{\ell_s K}{100}$$

$$r = \frac{5729.578}{D_p} = \frac{(5729.578)(100)}{\ell_s K} \tag{17.26}$$

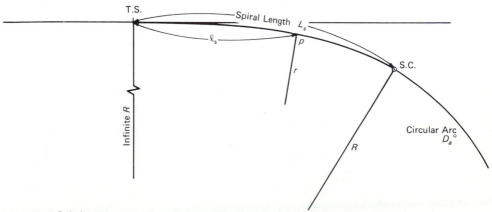

**Fig. 17.17** Spiral curve.

The radius at the S.C. is

$$R = \frac{5729.578}{D_a} = \frac{(5729.578)(100)}{L_s K} \qquad (17.27)$$

Division of Eq. (17.26) by Eq. (17.27) yields

$$\frac{r}{R} = \frac{L_s}{\ell_s} \qquad (17.28)$$

which indicates that the radius of a spiral varies inversely as the length of the spiral times a constant.

To provide room for the spiral, the original circular curve is shifted inward from the main tangent as to position $KCC'K''$ in Fig. 17.18. The portion $CC'$ is retained and spirals are introduced from $A$ to $C$ and from $C'$ to $B$.

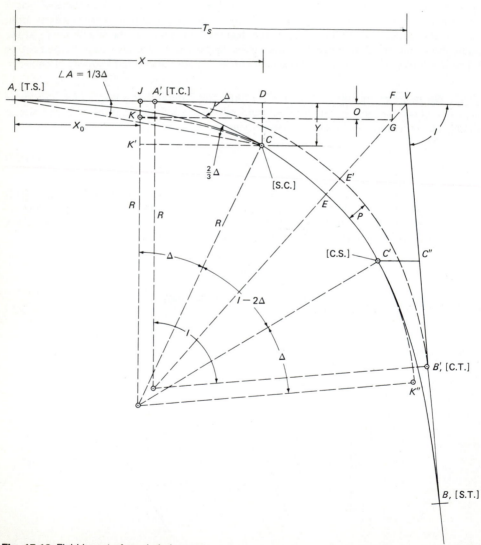

**Fig. 17.18** Field layout of a spiraled curve.

The central angle of a spiral is a function of the average degree of curvature of the spiral or $D_a/2$ and is given by

$$\Delta = \frac{L_s D_a}{200} \tag{17.29}$$

Note that from here on, $\Delta$ refers to the central angle of a spiral while $I$ is used to designate the central angle of a simple curve. So if a simple curve having a central angle $I$ has equal spirals with central angles of $\Delta$ introduced at both ends of the curve, as shown in Fig. 17.18, the central angle of the remaining simple curve from $C$ to $C'$ is $I - 2\Delta$.

Referring to Fig. 17.19, consider the equations required to evaluate deflection angles to points on spirals. For a segment of the spiral $d\ell$ at point $p$ located a distance $\ell_s$ from the T.S.,

$$d\delta = \frac{d\ell}{r} \tag{17.30}$$

From Eq. (17.28),

$$r = \frac{RL_s}{\ell_s} \tag{17.31}$$

Substitution of $r$ from Eq. (17.31) into Eq. (17.30) gives

$$d\delta = \frac{d\ell\, \ell_s}{RL_s} \tag{17.32}$$

Integration of Eq. (17.32) from 0 to $\ell_s$ yields $\delta$, or

$$\delta_p = \frac{1}{RL_s} \int_0^{\ell_s} \ell\, d\ell = \frac{\ell_s^2}{2RL_s} \tag{17.33}$$

At the S.C., $\Delta$ in radians equals $\delta_{S.C.}$, so that

$$\Delta = \frac{L_s}{2R} \tag{17.34}$$

Division of Eq. (17.33) by Eq. (17.34) leads to

$$\delta_p = \frac{\ell_s^2}{L_s^2} \Delta \tag{17.35}$$

Thus, the angle from the T.S. to any point on the spiral is proportional to the square of the distance of the point from the T.S.

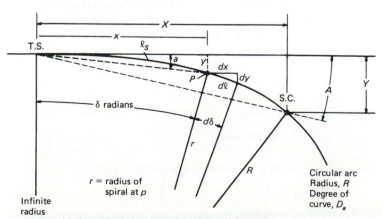

**Fig. 17.19** Mathematical development of a spiral curve.

Using the equations derived, several approximations useful for field layout of spirals can be developed. Referring to Fig. 17.19, we obtain

$$\sin \delta = \frac{dy}{d\ell} = \delta \qquad \text{(approx.)} \quad (17.36)$$

and

$$dy = \delta d\ell \qquad \text{(approx.)} \quad (17.37)$$

Substituting Eq. (17.33) into Eq. (17.37) and integrating from 0 to $\ell$ yields

$$y = \frac{\ell^3}{6RL_s} \qquad \text{(approx.)} \quad (17.38)$$

or $y$, which is equal to the tangent offset to any point on the spiral, varies closely as the cube of the distance from the T.S. to the point.

Next, let $a$ be the deflection angle to any point $p$ on the spiral where

$$\sin \alpha = \frac{y}{\ell_s} = a \qquad \text{(approx.)} \quad (17.39)$$

Substitution of $y$ from Eq. (17.38) into Eq. (17.39) yields

$$a = \frac{\ell_s^2}{6RL_s} \qquad \text{(approx.)} \quad (17.40)$$

Applying similar logic to the total deflection angle, $A$, from the T.S. to the S.C., results in

$$A = \frac{L_s}{6R} \qquad \text{(approx.)} \quad (17.41)$$

From Eq. (17.34), $R = L_s/2\Delta$, so that

$$A = \frac{\Delta}{3} \qquad \text{(approx.)} \quad (17.42)$$

Division of Eq. (17.40) by Eq. (17.41) yields

$$a = \frac{l_s^2}{L_s^2} A \qquad \text{(approx.)} \quad (17.43)$$

Also, it can be shown by using Eqs. (17.42), (17.35), and (17.43) that

$$a = \frac{\delta}{3} \qquad \text{(approx.)} \quad (17.44)$$

The approximations given by Eqs. (17.36) to (17.44) are sufficiently accurate for field work in most practical situations.

Several parameters remain to be determined before field layout of a spiral is possible. In order to locate the S.C. from the T.S. or the C.S. from the S.T., $X$ and $Y$ must be calculated for the spiral (Fig. 17.18). Also, to locate the T.S. and S.T. with respect to the P.I. or point $V$, the distance $o = FG = JK$ must be determined.

Using Eqs. (17.30) and (17.25), it can be shown that

$$\delta = \frac{K\ell_s^2}{20,000} \qquad (17.45)$$

in which $\delta$ is in radians, $K$ in radians/station, and $\ell_s$ in the units of the curve. From Fig. 17.19,

$$dx = d\ell \cos \delta \qquad (17.46a)$$
$$dy = d\ell \sin \delta \qquad (17.46b)$$

Given $\delta$, the sin and cos can be found in a power-series expansion to yield

$$dx = d\ell \left( 1 - \frac{\delta^2}{2!} + \frac{\delta^4}{4!} - \frac{\delta^6}{6!} + \cdots \right) \qquad (17.47a)$$

$$dy = d\ell \left( \delta - \frac{\delta^3}{3!} + \frac{\delta^5}{5!} - \frac{\delta^7}{7!} + \cdots \right) \qquad (17.47b)$$

Substitution of $\delta$ from Eq. (17.45) into Eq. (17.47) and integration of the resulting modified Eqs. (17.47$a$) and Eq. (17.47$b$) from 0 to $\ell_s$ yields

$$x = \ell_s \left[ 1 - \frac{\delta^2}{(5)(2!)} + \frac{\delta^4}{(9)(4!)} - \frac{\delta^6}{(13)(6!)} + \cdots \right] \qquad (17.48a)$$

$$y = \ell_s \left[ \delta/3 - \frac{\delta^3}{(7)(3!)} + \frac{\delta^5}{(11)(5!)} - \frac{\delta^7}{(15)(7!)} + \cdots \right] \qquad (17.48b)$$

For the S.C., $\Delta = \delta$ and $L_s = \ell_s$. Thus, given $\Delta$ in radians and $L_s$, exact values for $X$ and $Y$ can be calculated with Eqs. (17.48$a$) and (17.48$b$).

With a value for $Y$ calculated, the "throw" or distance $o = JK$ (see Fig. 17.18) is

$$o = Y - KK' = Y - R(1 - \cos \Delta) \qquad (17.49)$$

and the external distance $EV$ is

$$EV = EG + GV = R \left[ \frac{1}{\cos(I/2)} - 1 \right] + \frac{o}{\cos(I/2)} \qquad (17.50)$$

**17.16. Calculation and field layout of a spiral**    In the usual case, the station of the P.I. (point $V$), direction of the tangent(s), $I$ and $D_a$ are known. A brief review of the computational details involved follows.

1. Select $L_s$ so as to fit the existing conditions. (Refer to the AASHO *A Policy on Geometric Design of Rural Roads*.)
2. Calculate $\Delta$ using Eq. (17.29).
3. Calculate $(I - 2\Delta)$, $R$, and $L_a$ (length of circular arc), using Eqs. (17.1) and (17.10) for the last two.
4. Compute $X$ and $Y$ using Eqs. (17.48$a$) and (17.48$b$) or take values from the tables found in most route surveying textbooks.
5. Calculate $o$ using Eq. (17.49).
6. Calculate $T_s$ where $T_s = AJ + KG + FV$ (Fig. 17.18), in which $AJ = X - R \sin \Delta = X_0$, $KG = R \tan(I/2)$, and $FV = o \tan(I/2)$ to yield

$$T_s = X - R \sin \Delta + (R + o) \tan \frac{I}{2} \qquad (17.50a)$$

7. Calculate the stations of control points as follows:

    station T.S. = station P.I. $- T_s$
    station S.C. = station T.S. $+ L_s$
    station C.S. = station S.C. $+ L_a$
    station S.T. = station of C.S. $+ L_s$

8. Calculate spiral deflection angles using Eqs. (17.42) and (17.43).
9. Calculate deflection angles to desired stations from the S.C. to the C.S. on the simple curve.

    Field layout of the equally spiraled curve is as follows:

1. Occupy the P.I. and sight along the back tangent setting the T.S. and point $D$ (Fig. 17.18).
2. Establish the direction of the forward tangent by turning $I$ by double deflection (Art. 6.42). Measure to set $C''$ and the S.T.
3. Occupy points $D$ and $C''$, setting the S.C. and C.S. by right-angle offset where $DC = C'C'' = Y$.
4. Occupy the T.S. and set stations on the approach spiral to the S.C. by deflection angles and chords. Usually, the spiral is divided into an equal number of chords. Chord lengths can be taken equal to the nominal difference in stationing for relatively flat spirals.
5. Occupy the S.C., backsight the T.S. with $180 \pm 2A$ set on the horizontal circle. Loosen the upper motion, turn through $180°$, and set stations on the simple curve by deflection angles and chords. Check the closure on the C.S.
6. Occupy the S.T. and set stations on the leaving spiral from the S.T. to the the the C.S. On an equally spiraled curve, the same station interval and deflection angles calculated for the approach spiral can be used. When even stations are being set, new deflection angles must be calculated for the leaving spiral.

7. Stations on the leaving spiral can be set from the C.S. with a backsight on the S.C. where $180° \pm (I - 2\Delta)/2$ is set on the horizontal circle. In this case, deflection angles from the C.S. to the T.S. may be calculated using the *osculating-circle principle* (Art. 17.17).

At this point, an example problem is useful.

**Example 17.3** It is desired to place spirals on an existing metric curve for which the station of the P.I. is $54 + 61.460$ and $I$ is $40°00'$ right. The design speed is 80 km/h and $e + f = 0.08 + 0.15 = 0.23$. Using Eq. (17.24), $D_a = (730,000)(0.23)/(80)^2 = 26°2$. This value is rounded down to 26°, yielding a slightly longer radius, to be on the safe side. The length of spiral $L_s$ is generally selected on the basis of design speed and superelevation. Using tables from the AASHO *A Policy on Geometric Design of Rural Highways*, the length of spiral required for superelevation runoff for $e = 0.08$ and a design speed of 50 mi/h (~80 km/h) for a two-lane highway is 190 ft, so that $L_s$ is chosen as 60 m. Thus, by Eq. (17.29),

$$\Delta = \frac{L_s D_a}{200} = \frac{(60)(26)}{200} = 7°8 = 7°48'$$

By Eq. (17.1),

$$R = \frac{5729.578}{26°} = 220.368 \text{ m}$$

The central angle of the spiraled circular curve is

$$I - 2\Delta = 40° - (2)(7°8) = 24°4 = 24°24'$$

From Eq. (17.10), the length of circular curve is

$$L_a = \frac{100(I - 2\Delta)}{D_a} = \frac{(100)(24.4)}{26} = 93.846 \text{ m}$$

Calculate $X$ and $Y$ using Eqs. (17.48$a$) and (17.48$b$) (using only the first two terms), where $\delta = 0.136136$ rad.

$$X = L_s\left(1 - \frac{\delta^2}{10} + \frac{\delta^4}{216} - \frac{\delta^6}{9360} + \cdots\right)$$
$$= 60(0.998148) = 59.889 \text{ m}$$

$$Y = L_s\left(\frac{\delta}{3} - \frac{\delta^3}{42} + \frac{\delta^5}{132} - \frac{\delta^7}{75,600} + \cdots\right)$$
$$= 60(0.045319) = 2.719 \text{ m}$$

The throw, $o$, is calculated by Eq. (17.49).

$$o = Y - R(1 - \cos\Delta) = 2.719 - (220.368)(1 - 0.990748)$$
$$= 0.680 \text{ m}$$

Using Eq. (17.50$a$), $T_s$ is computed.

$$T_s = X - R\sin\Delta + (R + o)\tan\frac{I}{2}$$
$$= 59.889 - (220.368)\sin 7°48' + (220.368 + 0.680)\tan 20°$$
$$= 110.437 \text{ m}$$

The stations at T.S., S.C., C.S., and T.S. are:

| | | |
|---|---|---|
| Station at P.I. | 54 + 61.460 | |
| $- T_s$ | 1 | 10.437 |
| Station at T.S. | 53 + 51.023 | |
| $L_s$ | | 60.000 |
| Station at S.C. | 54 + 11.023 | |
| $L_a$ | | 93.846 |
| Station at C.S. | 55 + 04.869 | |
| $L_s$ | | 60.000 |
| Station at S.T. | 55 + 64.869 | |

**Table 17.4   Computation of Spiral Deflection Angles**

| Station | $\ell/10$ | $\ell^2/100$ | $L_s^2/100$ | $\ell^2/L_s^2$ | $a_i = (\ell^2/L_s^2)A$ |
|---|---|---|---|---|---|
| | | | | | $i = (1, 2, \ldots, 6)$ |
| S.C. 54 + 11.023 | 6 | 36 | 36 | 1 | 2°36.0′ |
| + 01.023 | 5 | 25 | 36 | 25/36 | 1°48.3′ |
| + 91.023 | 4 | 16 | 36 | 4/9 | 1°09.3′ |
| + 81.023 | 3 | 9 | 36 | 1/4 | 0°39.0′ |
| + 71.023 | 2 | 4 | 36 | 1/9 | 0°17.3′ |
| + 61.023 | 1 | 1 | 36 | 1/36 | 0°04.3′ |
| T.S. 53 + 51.023 | | | | | 0°00.0′ |

The total deflection angle from the T.S. to the S.C. is calculated by Eq. (17.42).

$$A = \frac{\Delta}{3} = 2°6 = 2°36' = 156'$$

Deflection angles are determined using Eq. (17.43). Computations for spiral deflection angles on the approach are accomplished most easily by arranging them in tabular form, as shown in Table 17.4.

Deflection angles to 20-m stations on the simple curve are computed in the usual way and are set from an instrument setup at the S.C. A backsight is taken on the T.S. with the clockwise horizontal circle set on $180° - \frac{2}{3}\Delta = 180° - 2A = 174°48'$. The leaving spiral is set from the S.T. using the same uniformly spaced chord intervals and the same deflection angles as were calculated from the T.S. to the S.C.

Figure 17.20 illustrates the field notes for the spiraled curve. Note that the chord lengths on the spiral can be taken equal to the length of spiral between stations without introducing significant errors since the station interval is small and the spiral is relatively short and flat.

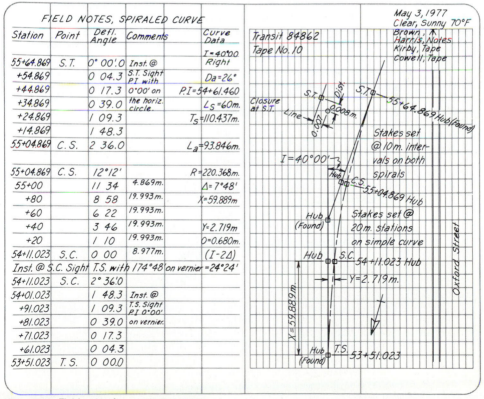

**Fig. 17.20** Field notes for a spiraled curve.

**17.17. Principle of the osculating circle**   The osculating circle is utilized for intermediate instrument setups on a spiral, calculating deflection angles to points on a spiral from the C.S. to the S.T., and the calculations necessary to place a spiral between two simple curves.

At any point $p$ on the spiral illustrated in Fig. 17.21, the radius $r$ is defined by Eq. (17.1), in which $D_a$ is the degree of curve at that point. The circular arc having radius $r$ and tangent to the spiral at $p$ is called the *osculating circle*. The following significant properties govern the relationship between the osculating circle and spiral:

1. The osculating circle departs inside the spiral toward the T.S. and outside the spiral toward the S.C.
2. The rate of departure between the spiral and the osculating circle in both directions is the same as the rate of departure of the same spiral from the tangent to the spiral at the T.S.

The latter statement pertains only to rate of departure of degree of curve and is not exact with respect to rate of departure of deflection angles and offsets. However, the approximations which result from applying the theory to deflection angles and offsets are sufficiently close for the relatively flat spirals employed in highway and railway alignment. Consider how the osculating circle can be used to calculate deflection angles from the C.S. to the S.T. of a spiral.

**Example 17.4**   Calculate deflection angles from the C.S. to the S.T., using the data of Example 17.3.

SOLUTION   As illustrated in Fig. 17.22, the osculating circle at the C.S. ($R = 220.368$ m and $D_a = 26°$) departs inside the spiral as stations increase from the C.S. toward the S.T. With the transit set at the C.S., the telescope can be oriented to a zero setting along the local tangent $V'D'$ by setting $180° - [(I - 2\Delta)/2] = 167°48'$ on the clockwise horizontal circle and taking a sight with the telescope direct on the S.C. Deflection angles can be calculated from the local tangent to stations at 10-m intervals on the osculating circle (a simple curve $D_a = 26°$). These angles are tabulated in column 2 of Table 17.5. Now, the rate of departure between the osculating circle and spiral is the same as that between the tangent and the spiral from the T.S. Consequently, the deflection angles to 10-m station intervals from the T.S. to the spiral calculated in Example 17.3 and listed in Table 17.5 are equal to the deflections between the spiral and osculating circle at corresponding intervals from the C.S. toward the S.T. Thus, the deflection from the local tangent at the C.S. at $55 + 14.869$ equals the deflection from the local tangent to the osculating circle ($b = 1°18'$) minus the deflection between spiral and osculating circle ($a = 0°04.3'$) equals $1°13.7'$. Calculations to all the equal 10-m stations on the leaving spiral are shown in Table 17.5. For convenience, stations are labeled 1, 2,...,6 so that the final angle can be expressed as $c_i = b_i - a_i$.

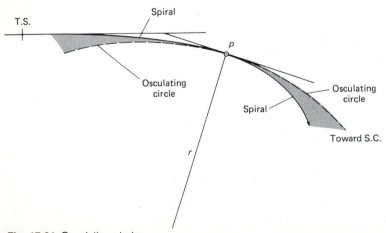

**Fig. 17.21** Osculating circle.

**Table 17.5    Deflection Angles C.S. to the T.S.**

| Station | Deflection to osculating circle, $b_i(i=1,2,\ldots,6)$ | Deflection between osculating circle and spiral, $a_i(i=1,2,\ldots,6)$ | Deflection angles local tangent to spiral, $c_i = b_i - a_i$ |
|---|---|---|---|
| S.T. 55 + 64.869 (6) | 7°48′ | 2°36.0′ | 5°12.0′ |
| + 54.869 (5) | 6°30′ | 1°48.3′ | 4°41.7′ |
| + 44.869 (4) | 5°12′ | 1°09.3′ | 4°02.7′ |
| + 34.869 (3) | 3°54′ | 0°39.0′ | 3°15.0′ |
| + 24.869 (2) | 2°36′ | 0°17.3′ | 2°18.7′ |
| + 14.869 (1) | 1°18′ | 0°04.3′ | 1°13.7′ |
| C.S. 55 + 04.869 | | | 0°00.0′ |

The osculating circle is also used when an intermediate setup is necessary on a spiral.

**Example 17.5**    Assume that in setting stations for the spiral of Example 17.3, it is necessary to move ahead to station 53 + 71.023 in order to complete laying out the spiral. In this case, the spiral departs inside the osculating curve as stationing progresses toward the S.C. (Fig. 17.23 illustrates the relationships involved). Thus, the deflection angles from the local tangent at 53 + 71.023 to subsequent stations will be the *sum* of the deflection angle to the osculating curve, $b_i$, and the deflection angle from the osculating curve to the spiral, $a_i$. First, it is necessary to calculate the degree of curve of the spiral at 54 + 71.023. Again, for convenience in notation, label the stations on the spiral 1, 2, 3, 4, 5, 6, as shown in Table 17.6. Then

$$D_2 = \left(\frac{20}{60}\right)26° = 8.6667°$$

or the deflection per metre = 2.6′/metre and the deflection per one 10-m station of osculating curve is 0°26′. Table 17.6 contains the deflection angles from 53 + 71.023 to subsequent points on the spiral. With the instrument set at 53 + 71.023, a backsight is taken on the T.S. with the telescope direct and the horizontal circle set on 180° − 2$a_2$, or 180° − (2)(0°17.3′) = 179°25.4.

Should another intermediate setup be required to complete the spiral layout, a backsight must be made to a station on the spiral. In this case, the osculating circle must be used to calculate the backsight setting as well as the deflection angles to subsequent stations. Suppose, in the previous

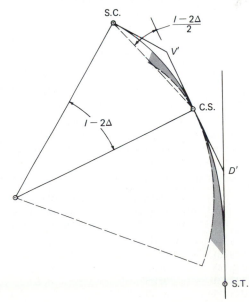

**Fig. 17.22** Deflection angles from C.S. to S.T.

T.S. | 53 + 51.023

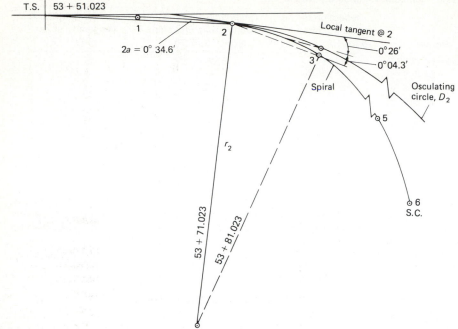

**Fig. 17.23** Intermediate setup on a spiral.

example, that an instrument setup is necessary at 53 + 91.923 or chord point (4) and a backsight is made on chord point (2) (Fig. 17.24). The degree of curve of the osculating circle at (4), $D_4$, is

$$D_4 = \left(\frac{4}{6}\right)26° = 17.333°$$

yielding a deflection of 0°52′ per 10-m station. As illustrated in Fig. 17.24, the osculating circle departs inside the spiral toward the T.S. For two 10-m stations, the deflection between the local tangent and osculating circle is 1°44′. The deflection between osculating circle and spiral for two stations is 0°17.3 ($a_2$, Table 17.5). Thus, the angle between local tangent and spiral is (1°44) − (0°17.3) = 1°26.7′ and the backsight from 4 to 2 is made with the telescope direct and the horizontal circle set on 180° − (1°26.7′) = 178°33.3′. Deflection angles to stations 5 and 6 (the S.C.) are calculated as in Example 17.5, using $D_4$ = 17.333° for the osculating circle.

In the previous examples, the assumption has been made that the total length of spiral, $L_s$, equals the sum of the chords used to set the stations. This assumption is valid for relatively flat spirals of moderate length. Chord corrections are necessary only for long chords near the end of a long sharp spiral. When necessary, the chord correction can be calculated using the methods employed for simple curves as outlined in Art. 17.5.

**Table 17.6  Intermediate Setup on a Spiral**

| Station | Deflection to osculating circle, $b_i$ ($i=3,4,5,6$) | Deflection between osculating circle and spiral, $a_i$ ($i=3,4,5,6$) | Deflection angle local tangent to spiral, $c_i = a_i + b_i$ |
|---|---|---|---|
| S.C 54 + 11.023 (6) | 1°44′ | 1°09.3′ | 2°53.3′ |
| 54 + 01.023 (5) | 1°18′ | 0°39.0′ | 1°57.0′ |
| + 91.023 (4) | 0°52′ | 0°17.3′ | 1°09.3′ |
| + 81.023 (3) | 0°26′ | 0°04.3′ | 0°30.3′ |
| 53 + 71.023 (2) | 0°00′ | | 0°00.0′ |

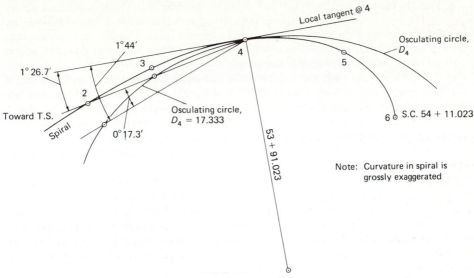

**Fig. 17.24** Backsight from an intermediate station on a spiral.

**17.18. Corrections to spiral deflection angles**   The approximate equations derived in Art. 17.15 for deflection angles are valid for relatively flat spirals of moderate length. For long, sharp spirals, it may be necessary to calculate a correction to deflection angles for stations near the end of the spiral. In Fig. 17.19, let $a$ be the deflection angle in radians to point $p$ on the spiral. Then

$$\tan a = \frac{y}{x} \tag{17.51}$$

Substitution of Eqs. (17.48$a$) and (17.48$b$) into Eq. (17.51) yields

$$\tan a = \frac{\delta}{3} + \frac{\delta^3}{105} + \frac{26\delta^5}{155,925} - \frac{17\delta^7}{3,378,375} + \cdots \tag{17.52}$$

When the tangent of an angle is known and the angle is desired, it can be found by a series expansion. Thus,

$$a = \tan a - \frac{(\tan^3 a)}{3} + \frac{(\tan^5 a)}{5} - \frac{(\tan^7 a)}{7} + \cdots \tag{17.53}$$

Substitution of Eq. (17.52) into Eq. (17.53) and collecting terms gives

$$a = \frac{\delta}{3} - \frac{8\delta^3}{2835} - \frac{32\delta^5}{467,775} - \cdots \tag{17.54}$$

Consequently,

$$a = \frac{\delta}{3} - \text{correction} \tag{17.55}$$

**Example 17.6**   Let $\delta = \Delta = 20° = 0.34907$ rad.

$$\text{correction} = \frac{(8)(0.34907)^3}{2835} = 0.000120 \text{ rad}$$
$$= 0°00'25''$$

In a 100-m spiral, this correction amounts to an offset at the S.C. of only 0.012 m. In general for values of $\Delta$ less than 15°, the correction is negligible. For long sharp spirals, this correction term should be evaluated and applied to deflection angles for stations near the end of the spiral.

**17.19. Vertical curves** On highways and railways, in order that there may be no abrupt change in the vertical direction of moving vehicles, adjacent segments of differing grade are connected by a curve in the vertical plane called a *vertical curve*. Usually, the vertical curve is the arc of a parabola, as this shape is well adapted to gradual change in direction and as elevations along the curve are readily computed.

Figure 17.25 illustrates crest and sag parabolic and equal tangent vertical curves. Point $A$ is the beginning of the vertical curve or $BVC$, $V$ is the P.I. or intersection of the tangents, and $B$ is the end of the vertical curve or $EVC$. The length of the curve $AB$ is designated $L$ and is measured in horizontal units. The two grades in the direction of stationing along tangents $AV$ and $VB$ expressed in percent are $G_1$ and $G_2$, respectively. When the tangent rises in the direction of stationing, the grade is positive, and when the tangent slopes downward, the grade is negative.

Design of crest and sag vertical curves is a function of the grades of the intersecting tangents, stopping or passing sight distance, which in turn are functions of vehicle design speed and height of the driver's eye above the roadway, and drainage. In addition to these factors, design of a sag vertical curve also depends on headlight beam distance, rider comfort, and appearance. Details governing design of vertical curves are beyond the scope of this text and may be found in route surveying textbooks as well as in the AASHO Policies on Rural and Urban Highways (Refs. 1 and 2).

Elevations on vertical parabolic curves can be computed by using the (1) equation of the parabola directly, (2) geometric properties of the parabola to calculate vertical offsets from the tangent, and (3) geometric properties of the parabola as exemplified by the chord gradient method.

**17.20. Vertical curves by equation of the parabola** (Ref. 4)  Consider the plane parabolic curve, $AB$, of Fig. 17.25a with the $y$ axis passing through the $BVC$ and the $x$ axis corresponding to the datum for elevations. Let $L$ be the length of the vertical curve in stations. The slope of the curve at the $BVC$ is $G_1$ and the slope at the $EVC$ is $G_2$. Since the rate of change in slope of a parabola is constant, the second derivative of $y$ with respect to $x$ is a constant, or

$$\frac{d^2y}{dx^2} = \text{constant} = r \qquad (17.56)$$

Integration of Eq. (17.56) yields the first derivative or the slope of the parabola,

$$\frac{dy}{dx} = rx + H \qquad (17.57)$$

When $x = 0$, the slope $= G_1$ and when $x = L$, the slope $= G_2$. Thus,

$$G_1 = 0 + H \qquad \text{and} \qquad G_2 = rL + H$$

from which

$$r = \frac{G_2 - G_1}{L} \qquad (17.58)$$

The value of $r$ is the rate of change of the slope in percent per station. Substitution of Eq. (17.58) into Eq. (17.57) gives

$$\frac{dy}{dx} = \left(\frac{G_2 - G_1}{L}\right)x + G_1 \qquad (17.59)$$

To obtain $y$, integrate Eq. (17.59) to yield

$$y = \left(\frac{G_2 - G_1}{L}\right)\frac{x^2}{2} + G_1 x + c \qquad (17.60)$$

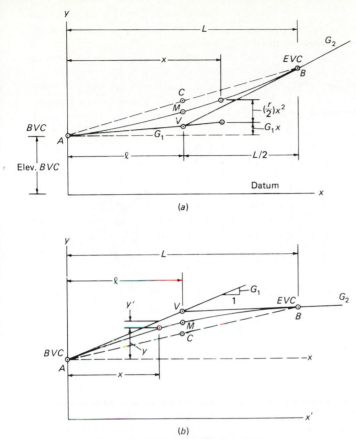

**Fig. 17.25** (a) Sag vertical curve. (b) Crest vertical curve.

which becomes

$$y = \left(\frac{r}{2}\right)x^2 + xG_1 + (\text{elevation of } BVC) \qquad (17.61)$$

that is the equation of the parabolic curve $AB$ and can be used to calculate elevations on the curve given $G_1$, $G_2$, $L$ and the elevation of the $BVC$. Note that the first term, $(r/2)x^2$, is the vertical offset between the curve and a point on the tangent to the curve at a distance $x$ from the $BVC$. The second term $G_1 x$ represents the elevation on the tangent at a distance $x$ from the $BVC$, and the third term is the elevation of the $BVC$ above the datum (see Fig. 17.25a). Any system of units can be employed in Eq. (17.61). The grades $G_1$ and $G_2$ are dimensionless ratios and expressed as percentages, but $r$ [Eq. (17.58)] must be determined in units compatible with the units of $x$ and $y$. If distances are measured in feet, $L$ in Eq. (17.58) must be in stations where 100 ft = one station. If distances are in metres, $L$ is in metric stations where 100 m = one station.

The high or low point of a vertical curve is frequently of interest for drainage purposes. At the high or low point, the tangent to the vertical curve is zero. Equating the first derivative of $y$ with respect to $x$ [Eq. (17.59)] to zero gives

$$xr + G_1 = 0 \qquad (17.62)$$

or

$$x = -\frac{G_1}{r} \qquad (17.63)$$

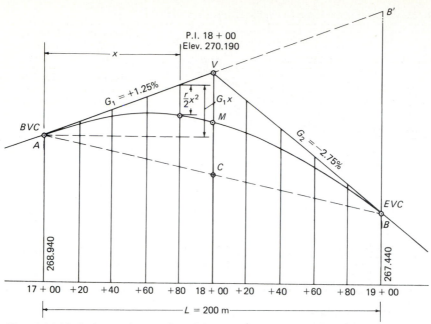

**Fig. 17.26** Vertical curve by equation of the parabola and by vertical offsets.

**Example 17.7** A 200-m equal tangent parabolic vertical curve is to be placed between grades of $G_1 = 1.25$ percent, and $G_2 = -2.75$ percent intersecting at station $18 + 00$, which has an elevation of 270.190 m above mean sea level (m.s.l.). Calculate elevations at even 20-m stations on the vertical curve and determine the station and elevation of the high point on the vertical curve. Figure 17.26 shows the geometry involved.

SOLUTION    The elevation of the $BVC = 270.190 - (0.0125)(100) = 268.940$ m and the elevation of the $EVC = 270.190 - (0.0275)(100) = 267.440$ m above m.s.l. Next compute the rate of change in grade by Eq. (17.58), where $L = 2$ stations. Thus,

$$r = \frac{-2.75 - 1.25}{2} = -2.000$$

so that

$$\frac{r}{2} = -1.000$$

From Eq. (17.61), the equation of the parabola for this vertical curve is

$$y = -1.000x^2 + 1.25x + 268.940$$

in which $y$ equals the elevation of a point on the vertical curve $x$ stations from the $BVC$. Calculation of elevations on the vertical curve by this equation is best illustrated by Table 17.7. Note that elevations on the curve for a given station are the sum of columns 3, 4, and 6 of the table.
By Eq. (17.63), the station of the high point is

$$x = -\frac{G_1}{r} = \frac{1.25}{2} = 0.625 \text{ station or } 62.500 \text{ m}$$

so that the

station of high point $= 17 + 00 + 62.500 = 17 + 62.500$

Using the equation of the parabola, the elevation of this station is

$$\text{elevation at } 17 + 62.500 = (-1.0)(0.625)^2 + (1.25)(0.625) + 268.940$$
$$= 269.331 \text{ m above m.s.l.}$$

**Table 17.7   Elevations on a Vertical Curve by Equation of Parabola**

| Station | $x$ | $(r/2)x^2$ | $G_1x$ | $(r/2)x^2 + G_1x$ | Elevation BVC (m) | Elevation on curve (m) |
|---|---|---|---|---|---|---|
| BVC 17 + 00 | 0. | 0. | 0. | 0. | 268.940 | 268.940 |
| + 20 | 0.2 | − 0.04 | 0.250 | 0.210 | 268.940 | 269.150 |
| + 40 | 0.4 | − 0.16 | 0.500 | 0.340 | 268.940 | 269.280 |
| + 60 | 0.6 | − 0.36 | 0.750 | 0.390 | 268.940 | 269.330 |
| + 80 | 0.8 | − 0.64 | 1.000 | 0.360 | 268.940 | 269.300 |
| 18 + 00 | 1.0 | − 1.00 | 1.250 | 0.250 | 268.940 | 269.190 |
| + 20 | 1.2 | − 1.44 | 1.500 | 0.060 | 268.940 | 269.000 |
| + 40 | 1.4 | − 1.96 | 1.750 | − 0.210 | 268.940 | 268.730 |
| + 60 | 1.6 | − 2.56 | 2.000 | − 0.560 | 268.940 | 268.380 |
| + 80 | 1.8 | − 3.24 | 2.250 | − 0.990 | 268.940 | 267.950 |
| EVC 19 + 00 | 2.0 | − 4.00 | 2.500 | − 1.500 | 268.940 | 267.440 |

**17.21. Vertical curves by vertical offsets from the tangent**   In Fig. 17.25b, translate the $x$ coordinate axis so that it passes through $BVC$ at $A$. Then by similar triangles,

$$\frac{y + y'}{x} = G_1$$

or

$$y = xG_1 - y' \tag{17.64}$$

Substituting Eq. (17.64) into Eq. (17.60) yields $(c = 0)$

$$xG_1 - y' = \left(\frac{G_2 - G_1}{2L}\right)x^2 + G_1x$$

which can be solved for $y'$ to give

$$y' = -\left(\frac{G_2 - G_1}{2L}\right)x^2 = -\left(\frac{r}{2}\right)x^2 \tag{17.65}$$

or vertical offsets from the tangent vary as the square of the distance from the point of tangency. Now, at point $V$, the P.I. of the vertical curve, $y' = VM$ and $x = L/2$, so that

$$y' = VM = -\left(\frac{G_2 - G_1}{2L}\right)\frac{L^2}{4} = \frac{-(G_2 - G_1)L}{8} \tag{17.66}$$

Divide Eq. (17.65) by Eq. (17.66) and solve for $y'$ to give

$$y' = \left(\frac{x^2}{\ell^2}\right)VM \tag{17.67}$$

in which $\ell = L/2$.

In Fig. 17.25b, point $C$ bisects the chord $AB$ and since the midpoint $M$ on the vertical curve is halfway between $V$ and $C$, $VM = \frac{1}{2}VC$. Thus, given the elevations of the $BVC$ and $EVC$, $VM$ can be calculated. On sag vertical curves, $VM$ is positive and on crest curves $VM$ is negative, so that $y'$ is positive on sag (Fig. 17.25a) and negative on crest (Fig. 17.25b) vertical curves.

To determine elevations on the vertical curve by vertical offsets, calculate the elevations for desired stations along the tangent to the vertical curve from the $BVC$ to $V$ using $G_1$ and from $V$ to the $EVC$ using $G_2$. Next, calculate the vertical offsets [Eq. (17.67)] and apply to the elevations on the tangent. Since an equal tangent parabolic vertical curve is symmetrical, vertical offsets need to be computed for only one-half the curve when stations are equally spaced and symmetrical about the P.I., as in Example 17.7. Then these same vertical offsets can be applied to elevations along the tangent at the $EVC$, allowing determination of

**Table 17.8   Elevations on a Vertical Curve by Vertical Offsets**

| Station | Grade along tangent | $x$ in stations | $(x^2/\ell^2)VM^a$ | Elevation on curve | Difference First | Second |
|---|---|---|---|---|---|---|
| $17+00$ *BVC* | 268.940 | 0. | 0. | 268.940 | | |
| | | | | | > 0.210 | |
| $+20$ | 269.190 | 0.2 | −0.040 | 269.150 | | >0.08 |
| | | | | | > 0.130 | |
| $+40$ | 269.440 | 0.4 | −0.160 | 269.280 | | >0.08 |
| | | | | | > 0.050 | |
| $+60$ | 269.690 | 0.6 | −0.360 | 269.330 | | >0.08 |
| | | | | | > −0.030 | |
| $+80$ | 269.940 | 0.8 | −0.640 | 269.300 | | >0.08 |
| | | | | | > −0.110 | |
| $18+00$ | 270.190 | 1.0 | −1.000 | 269.190 | | >0.08 |
| | | | | | > −0.190 | |
| $+20$ | 269.640 | 0.8 | −0.640 | 269.000 | | >0.08 |
| | | | | | > −0.270 | |
| $+40$ | 269.090 | 0.6 | −0.360 | 268.730 | | >0.08 |
| | | | | | > −0.350 | |
| $+60$ | 268.540 | 0.4 | −0.160 | 268.380 | | >0.08 |
| | | | | | > −0.430 | |
| $+80$ | 267.990 | 0.2 | −0.040 | 267.950 | | >0.08 |
| | | | | | > −0.510 | |
| $19+00$ *EVC* | 267.440 | 0. | 0. | 267.440 | | |

$^aVM = -1.0$; $\ell = 1$ station.

elevations on the second half of the curve. Table 17.8 shows the calculations necessary to compute the vertical curve of Example 17.7 by vertical offsets from the tangent.

*VM* is calculated as follows:

| Point | Elevation |
|---|---|
| *BVC* | 268.940 |
| *EVC* | 267.440 |
| | 536.380 |
| *C* | 268.190 |

$$VM = -\frac{\text{el. } V - \text{el. } C}{2}$$
$$= -\frac{270.190 - 268.190}{2}$$
$$= -1.000$$

Elevations along the tangent are computed from *BVC* to *V* using $G_1$. Elevations along the tangent *V* to the *EVC* are calculated with $G_2$. In this case where points are equally spaced, second differences of the elevations must be equal. This is an important check which can be applied to all vertical curves calculated by the equation of the parabola (as in Example 17.7) or by vertical offsets.

If stations are not equally spaced and symmetrical about the P.I., grades are calculated for the desired stations along the tangent *AV* extended to *B′* (Fig. 17.26) and vertical offsets are computed with Eq. (17.67), where $x$ is always measured from the *BVC*.

**17.22. Elevations on vertical curves by chord gradients**   The rate of change in slope of a tangent to the vertical curve, $r$, has been defined by Eq. (17.58). Using this equation, the grade of chords can be calculated. Knowing these grades, elevations on the vertical curve can be computed by the *chord gradient* method.

Figure 17.27 illustrates chord gradients for full stations on a 500-ft vertical curve. Since $r$ is defined as the change in grade between tangents to the vertical curve, the change between the tangent at the *BVC* and the first station is $r/2 = a$, the change between chords of successive full stations is $r = 2a$, and the change between the last chord and the tangent at the *EVC* is $a$. Thus, for the example in Fig. 17.27, the chord gradients to full stations are as follows: $G_1 + a$, $G_1 + 3a$, $G_1 + 5a$, $G_1 + 7a$, $G_1 + 9a$, and as a check $G_1 + 10a = G_2$. Elevations at full stations are: elevation of the $BVC + G_1 + a$; elevation of $16 + 00 + G_1 + 3a; \cdots$; elevation of $19 + 00 + G_1 + 9a$. Since elevations are calculated cumulatively, the method is self-checking.

Chord gradients for subchords are determined as follows:
1. The change in gradient between a tangent and a subchord is "$a$" times the length of the chord in stations.
2. The change in gradient between two adjacent subchords is $a$ times the sum of the lengths of the subchords in stations.

In the vertical curve of Example 17.7, a 200-m curve with 20-m stations, $r = 2a = -2.000$, so that $a = -1.000$. Consequently, for the first 20-m subchord, $a_{20} = (-1.0)(0.2) = -0.2$ percent and for subsequent consecutive 20-m subchords $2a_{20} = (-1.0)(0.4) = -0.4$ percent.

**Example 17.8**   In the vertical curve illustrated by Fig. 17.27, $G_1 = 1.50$ percent, $G_2 = -3.00$ percent, and the elevation of the P.I. is 850.85 ft above m.s.l. Calculate elevations at half station points on the vertical curve.

SOLUTION   Elevation of the $BVC = 850.85 - (250)(0.015) = 847.10$. The elevation at the $EVC = 850.85 - (250)(0.03) = 843.35$ ft above m.s.l. The rate of change in gradient is

$$r = \frac{G_2 - G_1}{L} = \frac{-3.00 - 1.50}{5} = \frac{-4.50}{5} = -0.9\% / \text{station}$$

so that $a = -0.45$ percent. According to rule (1) for subchords, the change in gradient between the tangent at the $BVC$ and a chord to the first half-station is equal to $(-0.45)(0.5) = -0.225$ percent, while the change between two successive 50-ft subchords is $(-0.45)(1.0) = -0.45$ percent. Changes in gradients, and elevations for half-stations, are listed in Table 17.9.

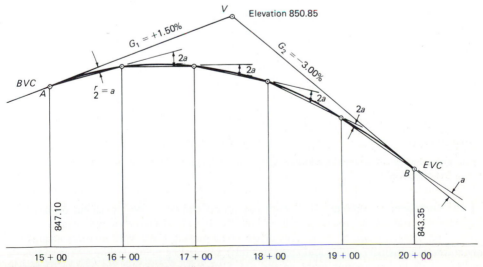

**Fig. 17.27** Vertical curve by chord gradients.

**Table 17.9  Vertical Curve by Chord Gradients[a]**

| From/to station | Chord gradient, % $= G_1 + na$ $n = (0.5, 1.5, \ldots, 9.5)$ | $\Delta$ el. = $(G_1 + na)(0.5)$ | Elevation, ft above m.s.l. |
|---|---|---|---|
| *BVC* 15 + 00 | | | 847.10 |
| 15 + 00 to 15 + 50 | $G_1 + 0.5a =$  1.275 | 0.638 | 847.74 |
| 15 + 50 to 16 + 00 | $G_1 + 1.5a =$  0.825 | 0.413 | 848.15 |
| 16 + 00 to 16 + 50 | $G_1 + 2.5a =$  0.375 | 0.188 | 848.34 |
| 16 + 50 to 17 + 00 | $G_1 + 3.5a = -0.075$ | $-0.038$ | 848.30 |
| 17 + 00 to 17 + 50 | $G_1 + 4.5a = -0.525$ | $-0.263$ | 848.04 |
| 17 + 50 to 18 + 00 | $G_1 + 5.5a = -0.975$ | $-0.488$ | 847.55 |
| 18 + 00 to 18 + 50 | $G_1 + 6.5a = -1.425$ | $-0.713$ | 846.84 |
| 18 + 50 to 19 + 00 | $G_1 + 7.5a = -1.875$ | $-0.938$ | 845.90 |
| 19 + 00 to 19 + 50 | $G_1 + 8.5a = -2.325$ | $-1.163$ | 844.74 |
| *EVC* 19 + 50 to 20 + 00 | $G_1 + 9.5a = -2.775$ | $-1.388$ | 843.35 |
| | $G_1 + 10.a = -3.000$ (check) | | |

[a] $r = -0.90\%$; $a = -0.45\%$; $G_1 = 1.50\%$; $G_2 = -3.00\%$.

The high point on the curve by Eq. (17.63) is

$$x = -\frac{G_1}{r} = \frac{1.50}{0.90} = 1.6667 \text{ stations} = 166.67 \text{ ft}$$

so that the station of the high point is 16 + 66.67. The elevation of the high point by the chord gradient method is as follows:

| | | |
|---|---|---|
| Chord gradient 16 + 00 to 16 + 50 | = | 0.375 |
| Change in gradient at 16 + 50 | | |
| $= (0.5 + 0.16667)(-0.45)$ | = | $-0.300$ |
| Subchord gradient 16 + 50 to 16 + 66.67 | = | 0.075 |
| Elevation at 16 + 50 | = | 848.34 |
| $\Delta$ El. 16 + 50 to 16 + 66.67 = (0.075)(0.16667) = | | 0.01 |
| Elevation at 16 + 66.67 | = | 848.35 |

To check this elevation into station 17 + 00, proceed as follows:

| | | |
|---|---|---|
| Chord gradient 16 + 50 to 16 + 66.67 | = | 0.075 |
| Change in gradient at 16 + 66.67 | | |
| $= (0.16667 + 0.3333)(-0.45)$ | = | $-0.225$ |
| Subchord gradient 16 + 66.67 to 17 + 00 | = | $-0.150$ |
| Elevation 16 + 66.67 | = | 848.35 |
| $\Delta$ El. 16 + 66.67 to 17 + 00 = (−0.150)(0.333) = | | $-0.050$ |
| Elevation 17 + 00 | = | 848.30  checks |

Thus, the chord gradient method, although somewhat more difficult to visualize, has the advantage of being self-checking.

**17.23. Earthwork operations**  Earthwork operations consist of the movement of materials in order to establish a predetermined surface for the construction of public and private works and determination of the volume of materials moved. Field work involves acquisition of terrain data (usually profile and cross sections) and setting points (grade stakes and slope stakes) to guide construction work on the site. Office work involves acquisition of terrain data from maps or by photogrammetric methods, processing terrain data, calculating volumes of excavated or embanked materials, and determination of the most economic procedure for performing the excavation and embankment. Earthwork operations and computations are key elements in the overall route location procedure.

For large projects, most of the terrain data can be extracted more accurately and at a lower cost from large-scale photogrammetrically compiled topographic maps or by applying photogrammetric methods to remotely sensed digital data such as can be obtained from aerial photographs. The traditional field methods described in the following sections are still appropriate for projects of small extent, for terrain that is relatively level, and for field conditions where the ground surface is concealed by heavy vegetation.

Although the main emphasis in this chapter is on earthwork operations for route surveying, the field and office methods developed are also applicable for determining volumes of stockpiles of materials, reservoir volumes, and volumes of concrete structures.

**17.24. Cross-section levels**  Frequently, in connection with problems of drainage, irrigation, grading of earthwork, location and construction of buildings, and similar enterprises, the shape of the surface of a parcel of land is desired. This may be obtained by staking out the area into a system of squares and then determining the elevations of the corners and of other points where changes in slope occur. The sides of the squares are usually 100, 50, or 25 ft (30, 15, or 10 m). Directions of the lines may be obtained with either the tape or the transit, distances may be laid off either with the tape or by stadia, and elevations may be determined with either the engineer's level or the hand level, all depending upon the required precision. The elevations are carried forward and the rod readings on ground points are determined as in profile leveling. The data may be employed in the construction of a contour map (see Art. 15.14).

Figure 17.28 illustrates a suitable form of notes. The sketch on the right-hand page shows the area divided into 10-m squares, the lines running in one direction being numbered and those running the other direction being lettered. The coordinates of a given corner may then be stated as the letter and number of the two lines intersecting at the corner. Thus, $B2$ is a point at the intersection of lines $B$ and 2.

Surveyors who own EDM equipment with real-time display of horizontal distance, in particular those who have access to an electronic surveying system (Art. 12.2), should consider optional methods of approaching this problem as outlined in Art. 15.16.

**17.25. Route cross sections**  Surveys for highways, railroads, or canals are often made by establishing a center line along the proposed route. This center line may be a location on paper using a topographic map, may consist of points set on the ground in the field, or may be a line of numerically known positions with respect to a digital terrain model (DTM). In any case, station points (a full station is 100 ft in the English and 100 m in the metric systems) are designated or set at full, half, or one-tenth stations along the route center line. Odd station points are set at breaks in the terrain configuration. Then, in order to furnish data for location studies and for estimating volumes of earthwork, a profile is run along the center line (Art. 5.53), and cross profiles (Art. 15.12) are taken along lines passing through each station and at right angles to the center line. The elevations along these cross lines and respective distances to the left and right (as one looks toward the next numerically higher station) constitute terrain cross-section data. The solid line in Fig. 17.29 is the center line, the dashed lines represent cross-section lines, and the tick marks are points where elevations are obtained to the left and right of center line, which is usually designated as ℄.

There are four major methods by which terrain cross-section data can be obtained: (1) field surveys; (2) from topographic maps compiled by field surveys or photogrammetric methods, (3) from photogrammetric stereomodels, and (4) from digital terrain models. The

| CROSS-SECTION LEVELS | | | BORROW | | PIT PROPERTY OF A. N. HAWK | 27 |
|---|---|---|---|---|---|---|
| Sta. | B.S. | H.I. | IFS | Elev. | K&E Transit No. 1245  Rod No. 5 | B.C. Gunther ⚲  R.N. Wald  Rod |
| BM 10 | 3.320 | 127.778 | | 124.458 | R.R. Spike in 18 in. Oak | Clear, Hot, 85°F |
| A-1 | | | 3.12 | 124.66 | 1 m. N.E. of N.E. Corner | |
| A-2 | | | 3.00 | 124.78 | R.N. Hawk property (Page 10) | |
| A-3 | | | 2.48 | 125.30 | | |
| A-4 | | | 2.12 | 125.66 | | |
| B-1 | | | 2.00 | 125.78 | | |
| B-2 | | | 2.85 | 124.93 | | |
| B-3 | | | 2.60 | 125.18 | | |
| B-4 | | | 2.50 | 125.28 | | |
| C-1 | | | 2.00 | 125.78 | | |
| C-2 | | | 2.45 | 125.33 | | |
| C-3 | | | 3.00 | 124.78 | | |
| C-4 | | | 3.05 | 124.73 | | |
| D-1 | | | 0.50 | 127.78 | | |
| D-2 | | | 1.00 | 126.78 | | |
| D-3 | | | 1.20 | 126.58 | | |
| D-4 | | | 1.50 | 126.28 | Base line A-F set | |
| E-1 | | | 1.00 | 126.78 | parallel to West | |
| E-2 | | | 2.00 | 125.78 | boundary of A.B. | |
| E-3 | | | 2.50 | 125.28 | Henderson (see | |
| F-1 | | | 0.80 | 126.98 | page 15 for property | |
| F-2 | | | 0.40 | 127.38 | survey notes) | |
| BM 10 | | | 3.320 | 124.458 | | |

Note: Stakes set on 10-m squares

C.R. Tobey

Base line A-F set parallel to West boundary of A.B. Henderson (see page 15 for property survey notes)

**Fig. 17.28** Cross-section (checkerboard) notes.

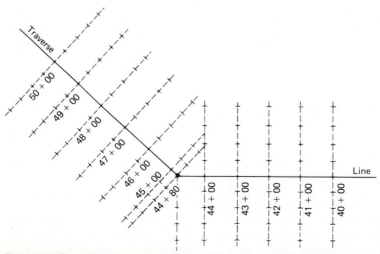

**Fig. 17.29** Cross lines for cross sections by the plus and offset method.

data will ultimately be in the form described above in defining the cross section no matter which method is used. The method selected depends on the size and nature of the project under study, and the equipment, personnel, and instrumentation available.

**17.26. Terrain cross sections by field methods**   This method constitutes the traditional approach and resembles the procedure for obtaining data for topographic mapping by the cross-profile method (Art. 15.12). The center line is set in the field and a profile is run over the center-line stations by differential leveling (Art. 5.53). The directions of short cross lines are set off by eye; that of long cross lines by transit or right-angle prism. Usually, elevations are determined with an engineer's level and rod in level terrain and by a hand level and rod in rough, irregular country. For each cross section, the height of instrument is established by a backsight on the center line station. The rod is then held on the cross line at specified distances or at breaks in the surface slope, where rod readings are observed and distances measured with a metallic tape. In very steep, irregular terrain, the use of self-reducing tacheometers and the stadia method (Art. 7.17) or EDM equipment and theodolites to obtain distance and elevation would be advantageous.

A note form for recording cross-section data by the traditional method is illustrated in Fig. 17.30. The elevations for center-line stations are obtained from the profile-level notes. This form of notes would be utilized when profile levels and cross sections are run separately and recorded in separate notebooks. The right-hand page of the notebook represents the traverse line, and to the right or left of this line are recorded the observed distances, rod readings, and the computed elevations. When a second H.I. is required to secure the necessary observations for a given cross line, a second line for that station is shown in the notes; thus station 405 occupies two lines in the notes, and station 406 occupies three. The location of fences, streams, etc., may be indicated by appropriate symbols or abbreviations.

Figure 17.31 shows a note form where profile levels and cross-section data are recorded on the same form, with all readings referenced to the same height of instrument (H.I.). The actual operations of profile and cross sections could be performed concurrently or separately. For example, profile leveling could be run separately, yielding the notes on the left-hand page, and at another time, cross sections could be run with a hand level, tape, and

| CROSS-SECTIONS C. & R. EXTENSION | | | | | | | O.H. Ellis, 杰  C.O. Lord, Rod   Jan.20,1979  J.A. Crum, Tape   Cold, Snow | | |
|---|---|---|---|---|---|---|---|---|---|
| Sta. | B.S. | H.I. | F.S. | Elev. | | Left | ¢ | Right | |
| 405 | 12.4 | 633.0 | | 620.6 | (Dist.) (Elev.) (Rod) | 300 210 123 80  632.1 630.8 627.0 626.7 620.6  0.9 2.2 6.0 6.3 | | | |
| 405 | 0.6 | 621.2 | | 620.6 | | | | 50 160 250 350  617.0 612.3 610.0 609.7  4.2 8.9 11.2 11.5 | |
| 406 | 12.1 | 628.9 | | 616.8 | | | 90 50  628.2 624.3 616.8  0.7 4.6 | | |
| 406 | 11.5 | 639.7 | 0.7 | 628.2 | | 280 200 120  638.3 635.6 632.2  1.4 4.1 7.5 | | | |
| 406 | 1.9 | 618.7 | | 616.8 | | | | 60 100 200 300  615.5 614.2 610.9 609.3  3.2 4.5 7.8 9.4 | |
| 407 | 4.7 | 615.9 | | 611.2 | | 300 180 100  615.6 614.7 612.6 611.2  0.3 1.2 3.3 | | 75 155 270  606.2 604.5 602.9  9.7 11.4 13.0 | |
| 408 | 10.6 | 615.9 | | 605.3 | | 280 200 100  611.6 609.2 607.7 605.3  4.3 6.7 8.2 | | 100 200 300  604.1 603.2 603.0  11.8 12.7 12.9 | |

**Fig. 17.30** Route cross-section notes.

**PRELIMINARY CROSS SECTIONS**
**HALF MOON BAY**

| Sta. | B.S. | H.I. | F.S. | Elev. |
|---|---|---|---|---|
| BM 24 | 2.750 | 213.200 | | 210.450 metres |
| 43+00 | | | | (Dist.) (Rod) (Elev.) |
| TP1 | 0.815 | 210.263 | 3.752 | 209.448 |
| 44+00 | | | | |

O.H. Ellis,        June 5, 1978
C.O. Lord, Rod     Fair, Warm
J.A. Crum, Tape    75°F                      28

Level No. 4211  Rod No. 10
Brass plate set in concrete. See page 10.

Station 43+00:

| (Dist.) | 25 | 16 | 8 | ₵ | 8 | 16 | 24 |
|---|---|---|---|---|---|---|---|
| (Rod) | 3.05 | 3.60 | 4.40 | 2.51 | 2.05 | 1.45 | 1.00 |
| (Elev.) | 210.15 | 209.60 | 208.80 | 210.69 | 211.15 | 211.75 | 212.20 |

| 36 | | | | | | 35 |
|---|---|---|---|---|---|---|
| 3.95 | | | | | | 0.30 |
| 209.25 | | | | | | 212.90 |

Station 44+00:

| 25 | 15 | 8 | ₵ | 7 | 15 | 25 |
|---|---|---|---|---|---|---|
| 3.42 | 3.15 | 2.95 | 2.25 | 2.04 | 1.25 | 0.95 |
| 206.84 | 207.11 | 207.31 | 208.01 | 208.22 | 209.01 | 209.31 |

| 35 | | | | | | 35 |
|---|---|---|---|---|---|---|
| 4.01 | | | | | | 0.25 |
| 206.25 | | | | | | 210.01 |

**Fig. 17.31** Notes for cross sections taken concurrently with profile leveling (metric units).

rod as described for the cross profile in Art. 15.12. The notes are recorded on the right-hand pages in space left for that purpose when the profile was run.

Since most cross-section data are processed by computer programs, it is frequently advantageous to record field notes on a note form that is directly usable by a key-punch operator for punching terrain data in a specified format. Figure 17.32 illustrates such a form using the data from cross sections given in Fig. 17.30 for stations 405 and 406.

Cross sections by field methods are applicable on small projects, in areas where the terrain is level, in regions where heavy vegetation inhibits use of photogrammetric methods, and where final cross sections and slope staking are combined.

Code 1, ₵ Elevation in col's. 17–24; Code 2, H.I. in col's. 17–24; Offset left is minus, Offset right is plus.

| Code | Whole station | Plus | ₵ Elevation or H.I. | Offset | Rod | Offset | Rod | Offset | Rod | Offset | Rod | Offset | Rod |
|---|---|---|---|---|---|---|---|---|---|---|---|---|---|
| 1 | 405. | | 620.6 | 0. | -12.4 | -80. | -6.3 | -123. | -6. | -210. | | | -2.2 |
| 1 | 405. | | 620.6 | 0. | -12.4 | -300. | -0.9 | | | | | | |
| 1 | 405. | | 620.6 | 0. | -0.6 | 50. | -4.2 | 160. | -8.9 | 250. | | | -11.2 |
| 1 | 405. | | 620.6 | 0. | -0.6 | 350. | -11.5 | | | | | | |
| 2 | 406. | | 628.9 | 0. | -12.1 | -50. | -4.6 | -90. | -0.7 | | | | |
| 2 | 406. | | 639.7 | 0. | -11.5 | -120. | -7.5 | -200. | -4.2 | 280. | | | -2.4 |
| 2 | 406. | | 618.7 | 0. | -1.9 | 60. | -3.2 | 100. | -4.5 | 200. | | | -7.8 |
| 2 | 406. | | 618.7 | 0. | -1.9 | 300. | -9.4 | | | | | | |

**Fig. 17.32** Format for field cross sections to be processed by electronic computer.

**17.27. Terrain cross sections from topographic maps**   During the successive stages of highway mapping for route alignment (reconnaissance, project planning, and design), trial locations for proposed highway center lines are plotted on topographic maps for evaluation of grades and earthwork volumes. One method of obtaining cross-section data for evaluating earthwork volumes is to plot the cross-section lines on the topographic map at each full or half-station and all breaks in the terrain. Offset distances from the center line are then scaled from the map and associated elevations are interpolated from the contours. The simplest application of this procedure involves scaling offset distances to contours with an engineer's scale and recording offset distances and elevations as terrain notes.

A portion of a topographic map at a scale of $1:480$ and with a contour interval of 2 ft is shown in Fig. 17.33 (scale in the figure is reduced). A trial location for a route center line and cross-section lines at full stations are also plotted on the map. Cross-section data are obtained by scaling from the center line to the contours that intersect the cross line. If the cross line is parallel to the contour, elevations are interpolated at specified intervals left and right of the center line. Cross-section notes determined from station $51+00$ in Fig. 17.33 are shown in Fig. 17.34$a$. The plotted cross section for station $51+00$ with a horizontal scale of $1:480$ and vertical scale of $1:12$ is illustrated in Fig. 17.34$b$ (scale in the figure is reduced).

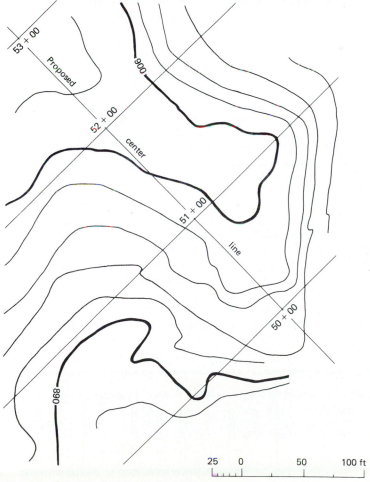

**Fig. 17.33** Cross-section data from topographic maps.

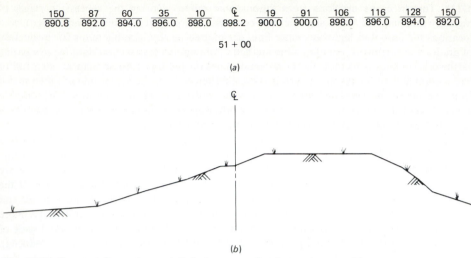

$$\frac{150}{890.8} \quad \frac{87}{892.0} \quad \frac{60}{894.0} \quad \frac{35}{896.0} \quad \frac{10}{898.0} \quad \frac{\text{¢}}{898.2} \quad \frac{19}{900.0} \quad \frac{91}{900.0} \quad \frac{106}{898.0} \quad \frac{116}{896.0} \quad \frac{128}{894.0} \quad \frac{150}{892.0}$$

51 + 00

(a)

¢

(b)

**Fig. 17.34** Cross-section notes and plotted cross section from a topographic map.

A note form compatible with punching the data on cards for electronic computer processing is shown in Fig. 17.35.

Cross-section data from a topographic map can be observed and recorded automatically by means of an analog-to-digital converter. The basic elements of the system are an electronic unit for storing digital data, a scaling bar with a movable pointer that can be aligned with a cross-section line, a keyboard to permit entering station numbers and elevations into the storage unit, and a card- or tape-punching device or magnetic tape recorder on which digital data in the storage unit can be recorded when a button is pressed.

To record data, the operator aligns the scaling bar with the cross line, indexes the bar to zero distance, and enters the station number and center-line elevation into the storage unit using the keyboard. These data are then put on the recording medium by pressing a button. Next, the operator moves the pointer along the scaling bar to a contour or point of desired elevation and introduces this elevation into the storage unit by way of the keyboard. The distance from the center line to the point is automatically cumulated in the storage unit as the pointer is moved along the scaling bar. The operator only needs to indicate the proper sign of + to the right and − to the left of center line and then records the information in the storage unit on the recording medium by pressing the button. This operation is repeated for each point desired on the cross section and for all cross sections along the center line. The procedure is efficient and provides data on any desired input medium and in the necessary format so as to be compatible with the computing equipment available.

*Terrain Data*
*Plus distances are right of centerline, minus distances are left of centerline, in the order of increasing stations.*

| ID Code | Whole station | Pluses | Elev | x-dist | Elev | x-dist | Elev | x-dist | Elev | x-dist |
|---|---|---|---|---|---|---|---|---|---|---|
| 30 | 51+ | . | 898.2 | 0 | 898.0 | −10 | 896.0 | −35 | 894.0 | −60 |
| 30 | 51+ | . | 892.0 | −87 | 890.8 | −150 | 900.0 | +19 | 900.0 | +91 |
| 30 | 51+ | . | 898.0 | +106 | 896.0 | +116 | 894.0 | +128 | 892.0 | +150 |
| 30 | + | . | | | | | | | | |

**Fig. 17.35** Format for cross-section data taken from a topographic map or photogrammetric stereomodel.

**17.28. Terrain cross-section data from photogrammetric stereomodels**   Topographic maps prepared for the planning and design stages of route locations are usually compiled by photogrammetric methods. Consequently, aerial photography of the study area is available almost from the very beginning of an alignment study. In order to compile a topographic map by photogrammetric methods, overlapping aerial photographs are placed in a stereoplotter and are relatively and absolutely oriented according to the methods outlined in Art. 16.17. The stereomodel that results from these operations provides a scaled, three-dimensional model of the photographed terrain. Given the proper instrumentation, it is possible to obtain cross sections directly from the stereomodel. Since fewer operations are involved, cross-section data from the stereomodel are more accurate than those taken from a topographic map compiled from the same stereomodel.

The first step in obtaining cross sections from the stereomodel is to plot the positions of known horizontal control points and the location of the proposed center line on a base sheet using a coordinatograph (Fig. 13.8, Art. 13.12). All center-line stations are also plotted and lines perpendicular to the center line are drawn through each station. The stereomodel is then absolutely oriented with respect to horizontal control points plotted on the base sheet and known vertical control points. At this point, the profile and cross-section data may be obtained by centering the measuring mark in the platen on the tracing table of the stereoplotter over a given center-line station and placing this mark on the apparent surface of the terrain in the stereomodel. The elevation of the terrain is read from the counter on the tracing table and the station and elevation are recorded on a specially prepared form. This procedure is then repeated at each break in the terrain to the left and right of the station along the perpendicular plotted through the center-line station. The offset distance is scaled and offset and elevation are recorded for each point. Figure 17.35 shows cross-section data obtained in this manner. This process is tedious, time-consuming, and requires manual

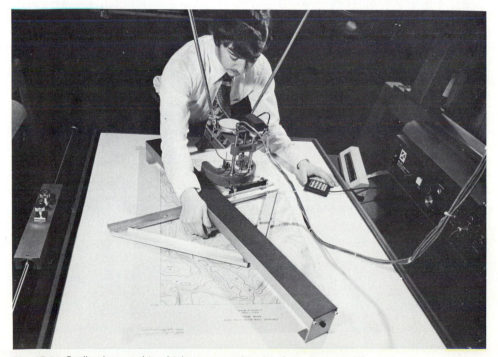

**Fig. 17.36** Scaling bar used to obtain cross-section data from stereomodel.

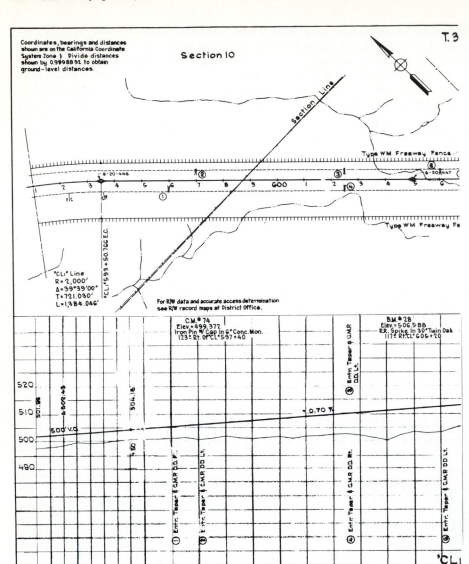

**Fig. 17.37** Typical location plan and profile. (*Courtesy California Department of Transportation.*)

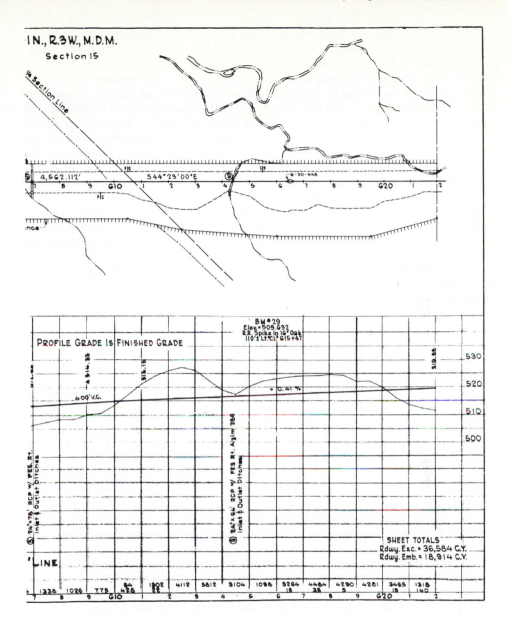

scaling of offset distances. Feasible practical application of this technique requires that the stereoplotter be equipped with a cross-section guide, keying device, and electronic storage unit similar to that described for obtaining data from a topographic map in Art. 17.27.

Figure 17.36 illustrates a stereoplotter with keying device and a digitized tracing table attached to a cross-section guide. This guide, consisting of a channel beam and attachment to the tracing table, can be oriented so that the digitized measuring mark in the tracing table moves in line with a given plotted cross section. First, the operator aligns the guide and measuring mark with a plotted cross-section line. Then the measuring mark is centered over a center-line station and is placed on the apparent terrain surface of the stereomodel. The station number is recorded using the keying device and the digitizing unit is indexed to zero. When the proper button on the keying device is depressed, station number, offset, and terrain elevation are automatically recorded on magnetic tape, punched cards, or punched-paper tape. The operator then moves the floating mark along the cross line to the first break in the terrain, places the floating dot on the apparent terrain, and presses the button again recording elevation and offset distance. This process is repeated for all breaks in terrain and at specified distances from the center line for the entire cross line. The data recorded constitute a cross section of the terrain at the given station. The entire procedure is repeated for all cross lines on the base sheet.

Photogrammetric cross sections have been shown to be more accurate than the usual field methods in rugged, steep terrain where the ground is not concealed by heavy vegetation. Field methods are more accurate and less expensive when the terrain is level.

**17.29. Preliminary cross sections**  Preliminary cross sections are those terrain cross sections obtained by any one of the three methods described in Arts. 17.26 to 17.28, which are then used for studying various alignments and for making preliminary estimates of earthwork.

When a preliminary alignment and grade are being studied, a trial grade is selected and the preliminary grade line is plotted on the profile sheet (Fig. 17.37). Details concerning plotting the profile can be found in Art. 5.55. This trial grade is based on maximum permissible rates of grade and elevations at controlling points such as termini, stream, and railway and roadway crossings. A crude effort is also made at this time to balance the amount of cut and fill on the center line. When a grade line is temporarily fixed, grade elevations are computed for all center-line stations and the finished roadbed templets superimposed on the terrain cross sections are plotted to a scale of from $1:60$ to $1:120$. Figure 17.38 shows typical cross sections in cut, in fill, and a sidehill cross section for a highway.

The roadbed templet consists of a subgrade which is usually a plane surface, transversely level but on highway curves perhaps superelevated, and side slopes at a given rate of slope. On a given road, the subgrade is of one width in cut (Fig. 17.38$a$); of a second and smaller width in fill (Fig. 17.38$b$); and of still a third width where the section is partly in cut and partly in fill (Fig. 17.38$c$). The finished cross section may be sloped in various ways to provide shoulders, drainage, and rounded corners, as illustrated in Fig. 17.38$c$. The rate of the side slope is stated in terms of the number of units measured horizontally to one unit measured vertically; thus a 2 to 1 slope indicates a slope that in a horizontal distance of 2 units rises (or falls) 1 unit. The slope most commonly employed for cuts or fills through ordinary earth is $1\frac{1}{2}$ to 1. For coarse gravel, the slope is often made 1 to 1; for loose rock, $\frac{1}{2}$ to 1; for solid rock, $\frac{1}{4}$ to 1; for soft clay or sand, 2 or 3 to 1.

After study of the tentative profile and the cross-section data, areas of the cross sections are determined for each station according to the methods given in Art. 17.31, and cumulated volumes of earth work are computed by the procedures outlined in Art. 17.32.

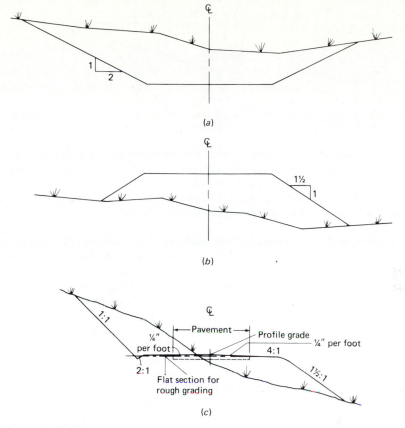

**Fig. 17.38** Preliminary cross sections. (*a*) Cross section in cut. (*b*) Cross section in fill. (*c*) Side-hill cross section.

For major highway alignment projects where earthwork volumes are determined by computation on an electronic computer, cross-sectional areas are not plotted manually. Terrain cross-section data and roadbed templet information, plus alignment and grade parameters, are used as input to a computer program which computes cumulated earthwork volumes, positions for slope stakes (Art. 17.30), ordinates for a mass diagram (Art. 17.39), and will plot the profile, mass diagram, and the outline of the cross-section areas at specified stations.

In either case, earthwork volumes and mass diagram for the tentative alignment are studied and whatever modifications necessary are made to provide the most economic solution and at the same time satisfy all the standards specified for horizontal and vertical alignment for the proposed highway.

**17.30. Final cross sections and slope stakes**   When the highway project reaches the final design stage, the horizontal and vertical alignments are fairly well fixed. During this stage final design plans are prepared and construction of the roadway begins.

Prior to the actual construction, final cross sections are taken and slope stakes marking the intersection of the side slopes with the natural ground surface (also called *catch points*) are set opposite each center-line station.

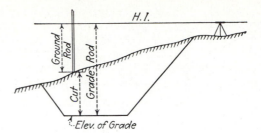

**Fig. 17.39** Road cross section in cut.

Final cross-section data, which are used to calculate earthwork quantities for payment, can be determined by field surveys or by photogrammetric methods. Distances to slope stakes can be determined by trial and error in the field concurrently with running the final cross sections, or can be calculated from the photogrammetric cross section. In the latter case, slope stakes are then set in the field during construction layout.

**1. Final cross sections and slope stakes by field methods**  Level readings for final cross sections are generally taken with an engineer's level, and distances to the right or to the left of the center line are measured with a metallic tape, all to tenths of feet or to centimetres. In rough, steep terrain, final cross sections may be taken by measuring slope distances and vertical angles. EDM equipment can be used to advantage in such situations.

Prior to going to the field, the surveyor secures a record of elevations of ground points as obtained from the profile levels, and also the elevation of the established grade at each station. In the field, the instrument is set up in any convenient location, and the H.I. is obtained by a backsight on a bench mark. At each station the rod is held on the ground (sometimes on a short wooden peg driven flush with the ground), a foresight is taken, and the ground elevation is checked against that obtained by profile leveling. Next, the computed cut or fill is marked on the back of the center stake, and a cross line through the station is established as described in Art. 17.26.

Figure 17.39 shows the engineer's level in position for taking rod readings at a section *in cut*. The height of instrument (H.I.) has been determined; the elevation of grade at the particular station is known. The leveler computes the difference between the H.I. and the grade elevation, a difference known as the *grade rod*; that is, H.I. − elevation of grade = grade rod. The rod is held at any point for which the cut is desired, and a reading called the *ground rod* is taken. The difference between the grade rod and the ground rod is equal to the cut designated $c$.

Figure 17.40 is a similar illustration for a cross section *in fill*. If the H.I. is *above* grade (as at $A$), the fill is the *difference* between the grade rod and the ground rod; if the H.I. is *below* grade (as at $B$), the fill is the *sum* of the grade rod and the ground rod. Fill is designated by $f$.

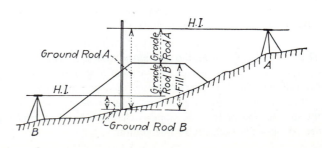

**Fig. 17.40** Road cross section in fill.

If the ground is level in a direction transverse to the center line, the only rod reading necessary is that at the center stake, and the distance to the slope stake can be calculated once the center cut or fill has been determined; such a cross section is called a *level section*. When rod readings are taken at each slope stake in addition to the reading taken at the center, as will normally be done where the ground is sloping, the cross section is called a *three-level section*. When rod readings are taken at the center stake, the slope stakes, and at points on each side of the center at a distance of half the width of the roadbed, the cross section is called a *five-level section*. A cross section for which observations are taken to points between center and slope stakes at irregular intervals is called an *irregular section* (Fig. 17.38b). Where the cross section passes from cut to fill, it is called a *side-hill section* (Fig. 17.38c), and an additional observation is made to determine the distance from center to the grade point, that is, the point where the subgrade will intersect the natural ground surface. A peg is usually driven to grade at this point, and its position is indicated by a guard stake marked "grade." In this case, cross sections are also taken at additional plus stations where the center line, the left edge, and the right edge of the roadway pass from cut to fill or from fill to cut, as at station $C$, $D$, and $F$ in Fig. 17.41.

The process of setting the slope stakes requires some additional explanation. If $w$ is the width of roadbed or canal bed, $d$ the measured distance from center to slope stake, $s$ the side-slope ratio (ratio of horizontal distance to drop or rise), $c$ the cut (or $f$ the fill) at the

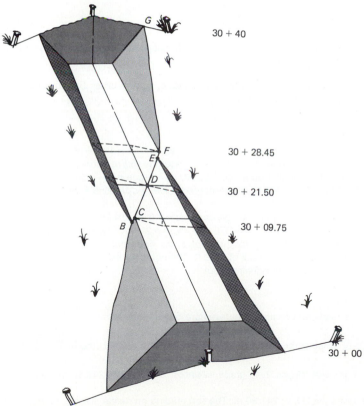

30 + 40

30 + 28.45

30 + 21.50

30 + 09.75

30 + 00

**Fig. 17.41** Cross sections in a transition area.

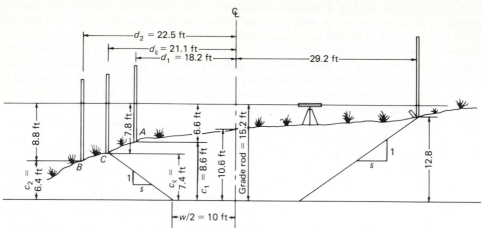

**Fig. 17.42** Slope stakes for a cross section in cut.

slope stake, then from Fig. 17.42, when the slope stake is in the correct position (at $C$),

$$d = \frac{w}{2} + cs \qquad (17.68)$$

and

$$d = \frac{w}{2} + fs \qquad (17.69)$$

The following numerical example for a cut illustrates the steps involved in establishing the correct location for a slope stake in the field.

**Example 17.9** Let $w = 20$ ft, side slope $s = 1\frac{1}{2}$ to 1, and grade rod $= 15.2$ ft. Suppose that a slope stake is to be set on the left of the center stake (Fig. 17.42). As a first trial, the rod is held at $A$; ground rod $= 6.6$ ft.

$c_1 =$ grade rod $-$ ground rod $= 15.2 - 6.6 = 8.6$ ft

The computed distance for this value of $c_1$ by Eq. (17.68) is $w/2 + c_1 s = 10.0 + (8.6)(\frac{3}{2}) = 22.9$ ft. Measurement from the center stake shows $d_1$ to be 18.2 ft; hence, the rodman should go farther out.

For a second trial, the rod is held at $B$; ground rod $= 8.8$ ft; $c_2 =$ grade rod $-$ ground rod $= 15.2 - 8.8 = 6.4$ ft; $w/2 + (c_2)(s) = 10.0 + 9.6 = 19.6$ ft. The measured value of $d_2$ is 22.5 ft; hence, the rod is too far out.

Eventually, by trial, the rod will be held at $C$; ground rod $= 7.8$ ft; $c = 15.2 - 7.8 = 7.4$ ft. The computed distance for this value of $c$ is $w/2 + cs = 10.0 + (7.4)(\frac{3}{2}) = 21.1$ ft. The measured value of $d$ is also 21.1 ft; hence, this is the correct location for the slope stake. The slope stake on the right, also set by trial and error, is 29.2 ft right and has a cut of 12.8 ft. The notes for this final cross section are recorded as follows, where the symbol $c$ designates a cut:

| $c7.4$ | $c8.6$ | $c10.6$ | $c12.8$ |
|---|---|---|---|
| 21.1 | 18.2 | 0.0 | 29.2 |

The reading at 18.2 ft left is recorded because it represents a break in the terrain.

Consider an example of slope stakes for a cross section in fill, as illustrated in Fig. 17.43.

**Example 17.10** Let $w = 7$ m; side slopes 2:1; grade elevation at $32 + 00 = 240.36$ m; center-line elevation $= 239.25$ m.

A backsight on a B.M. gives the H.I. of 241.90 m. The rod reading on center line is 2.65 m, which verifies the given center-line elevation. The grade rod $=$ H.I. $-$ grade elevation $= 241.90 - 240.36 = 1.54$ m. In order to set the right slope stake the rod is held, as a first trial, at point 1, where the ground

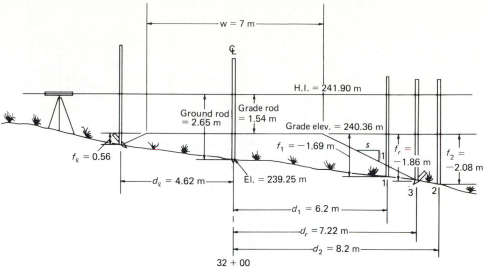

**Fig. 17.43** Slope stakes for a cross section in fill.

rod = 3.23. Thus, $f_1$ = grade rod − ground rod = 1.54 − 3.23 = − 1.69 m, where the negative sign indicates a fill. The computed distance for the absolute value of $f_1$ by Eq. (17.69) is 3.5 + (1.69)(2) = 6.88 m. The measured value for $d_1$ is 6.2 m, so that the rod must be moved farther from the center line.

A second trial is made at point 2, where the ground rod is 3.62, $f_2$ = 1.54 − 3.62 = − 2.08 m, and the computed distance to $f_2$ is 3.5 + (2.08)(2) = 7.66 m. Since the measured distance $d_2$ is 8.2 m, the rod is out too far.

Eventually, by trial, the rod is held at 3, where the ground rod = 3.40, $f_r$ = 1.54 − 3.40 = − 1.86 m, and the computed distance is 3.5 + (1.86)(2) = 7.22 m. This distance agrees with the measured distance $d_r$; hence, point 3 is the correct location for the slope stake. The notes for this final cross section are:

$$
\begin{array}{llll}
\text{H.I.} \\
241.90 \\
32+00 & \dfrac{f0.56}{4.62} & \dfrac{f1.11}{0.00} & \dfrac{f1.86}{7.22}
\end{array}
$$

where the left slope stake is also set by trial and error and the symbol $f$ designates a fill.

Slope stakes can also be set in the field by slope distance and vertical or zenith angle. This procedure is particularly appropriate where the cuts are deep and fills or embankment are high, so that setting slope stakes from a single setup of a level is not possible.

In Fig. 17.44 the instrument is set over the center-line station with a height above the ground equal to the h.i. A sight is taken on the rod at $A$ (the correct position for the slope stake) such that the rod reading $AB$ equals the h.i. and the vertical angle $\alpha$ or zenith angle $z$ is observed. The difference in elevation between the center-line station at $D$ and the ground at the slope stake at $A$ is $BC = V$, where

$$V = \ell \sin \alpha = \ell \cos z \qquad (17.70)$$

and the horizontal distance from $D$ to $A$ is

$$d = \ell \cos \alpha = \ell \sin z \qquad (17.71)$$

If $c$ is the cut to grade elevation at the center line, taken from the profile and grade plans, the difference in elevation between the center line and the ground at the slope stake is $(c + V)$ for an uphill sight and $(c - V)$ for a downhill sight. Thus, the calculated distance

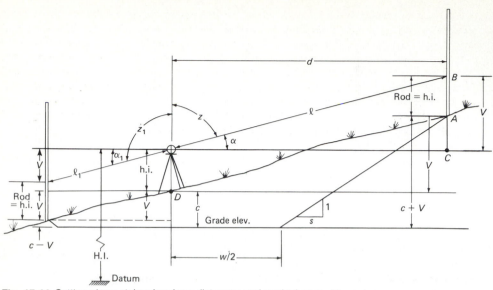

**Fig. 17.44** Setting slope stakes by slope distances and vertical or zenith angles.

from center line to slope stake is

$$d(\text{calc.}) = \frac{w}{2} + s(c \pm V) \tag{17.71a}$$

When the value $d(\text{calc.})$ from Eq. (17.71$a$) equals $d$ obtained by Eq. (17.71) using field measurements, the rod is in the correct position to set the slope stake. If $d$ is greater or less than $d(\text{calc.})$, the rod must be moved toward or away from the center line and another set of measurements taken. The correct position for the slope stake is found by trial.

For cross sections in fill, the calculated distance is

$$d(\text{calc.}) = \frac{w}{2} + s(f \pm V) \tag{17.72}$$

where $f$ is the fill at center line (taken as a positive value) and the height of instrument (H.I.) above the datum is assumed to be below the center-line grade elevation. The value of $V$ is negative when the sight is to the uphill side and $V$ is positive when the sight is to the downhill side of the center line. Equation (17.72) can also be used if the H.I. is above the grade elevation at the center line, but it is usually more convenient to set slope stakes with a horizontal line of sight and distance measurement as described in Example 17.10.

Transit stadia, theodolite and steel tape, an EDM instrument mounted on a theodolite, or electronic surveying system (Art. 12.2) can be used to obtain the necessary measurements for setting slope stakes by this method.

Slope stakes are set side to the line, sloping outward in fill and inward in cut. On the back of the stake the station number is marked with crayon. On the front (side nearest the center line) are marked the cut or fill at the stake and the distance from center line to slope stake. The numbers read down the stake.

Some organizations do not set stakes at the top or bottom of the slope (frequently referred to as the *catch point*) but set a reference point at a fixed distance back from the catch point. This distance is from 10 to 20 ft (3 to 7 m) from the catch point. Figure 17.41 shows this method for stations 30+00 and 30+40. In this way the stakes are not so likely to be destroyed during roadway construction. For further details concerning this procedure, Chap. 18. Some organizations set stakes at both the catch points and reference points.

**2. Final cross sections and slope stakes by photogrammetric methods**   Final cross sections by photogrammetric methods (Art. 17.28) are feasible when the terrain is steep and fairly free of heavy vegetation or if the right of way has been cleared of vegetation prior to beginning construction. In the latter case, photography is taken specifically for determining final cross sections photogrammetrically. The photogrammetric cross sections are combined with design templet data (roadbed width, side slopes, etc.) for calculating areas of cross sections (Art. 17.31) and volumes of earthwork (Art. 17.32). The computations are performed using an electronic computer program which also provides as output, the data needed for field layout of the slope stakes. These data include roadbed templet notes, side-slope ratios, and distances right and left of center line to the slope stakes. Further details concerning this procedure and the setting of slope stakes from *random control* points can be found in Art. 18.7.

**3. Location of cross sections**   When measurements are in feet, cross sections are taken at all full or half-stations and at all intermediate stations where significant breaks in the terrain occur. When the metric system is used, stations are generally set at 10- or 20-m intervals (one-tenth and two-tenth station points). If grading is very heavy or rock is present, cross sections should be taken at closer intervals.

Additional cross sections are required when there is a transition from cut to fill on a side-hill location. Figure 17.41 shows a transition from fill to cut from station $30+00$ to station $30+40$. Additional cross sections are required at $B$, $C$, $D$, $E$, and $F$. Generally, $B$ is so close to $C$ and $E$ is so close to $F$ that these points are omitted. Consequently, three additional cross sections are required: (1) at the fill-base grade point at $C$, (2) at the center-line grade point at $D$, and (3) at the cut-base grade point at $F$.

**17.31. Areas of cross sections**   Regular cross sections are those for which levels are taken at one point on each side of the center line. Level and three-level sections as defined in Art. 17.30 are regular cross sections.

The area of a level cross section (Fig. 17.45*a*) is

$$A = f(w + sf) \tag{17.73}$$

in which $f$ is the center-line fill, $w$ the roadway width, and $s$ the side-slope ratio. If the section is in cut, the center-line cut $c$ is substituted for $f$ in Eq. (17.73).

A three-level section is illustrated in Fig. 17.45*b*. As shown in the figure, this section can be divided into four parts, for which the respective areas are

$$A_1 = \tfrac{1}{2}(f_\ell)\left(\frac{w}{2}\right); \quad A_2 = \tfrac{1}{2}(f_r)\left(\frac{w}{2}\right); \quad A_3 = f\left(\frac{d_\ell}{2}\right); \quad A_4 = f\left(\frac{d_r}{2}\right)$$

in which $f_r$ and $f_l$ are the fills at the right and left catch points, and $d_r$ and $d_l$ are the distances from center line to right and left catch points. Taking the sum of the four areas and collecting terms yields

$$A_{3\ell} = \tfrac{1}{2}f(d_\ell + d_r) + \frac{w}{4}(f_\ell + f_r) \tag{17.74}$$

If the section is in cut, substitute $c = f$, $c_\ell = f$, and $c_r = f_r$ in Eq. (17.74).

A five-level section is shown in Fig. 17.45*c*. This cross section can be divided into five triangles (where $A_5 = A_6$, so $2A_5 = A_5 + A_6$) as indicated in the figure. The area of each triangle can then be determined as a function of $c$, $w/2$, $c'_\ell$, $c'_r$, $d_\ell$, and $d_r$. For example, $A_1 = \tfrac{1}{2}c'_\ell(d_\ell - w/2)$, $A_2 = \tfrac{1}{2}c'_r(d_r - w/2)$, etc. Taking the sum of these areas and collecting terms gives

$$A_{5\ell} = \tfrac{1}{2}(c'_\ell d_\ell + c'_r d_r + cw) \tag{17.75}$$

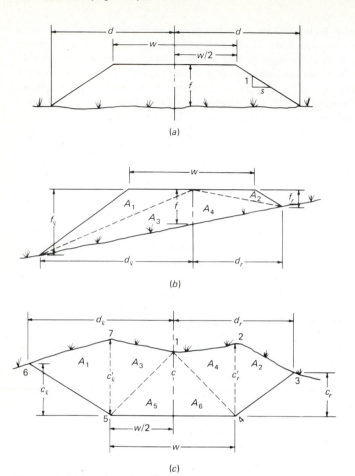

**Fig. 17.45** (*a*) Level section. (*b*) Three-level section. (*c*) Five-level section.

An irregular cross section is illustrated in Fig. 17.46. The area of this type of section can be determined by dividing the section into trapezoids and triangles. However, the most efficient procedure is to use the method of coordinates as developed to determine the area within a closed traverse in Eq. (8.41), Art. 8.28. To compute the area of a section by Eq. (8.41), let the center line in the plane of the cross section define the plus $Y$ axis and the finished grade line define the $X$ axis. The coordinates of slope stake positions or catch points, break points, and *hinge points* (hinge point is the angle point between the grade line and side slope) in the cross section are the offset distances from center line ($X$ coordinates) and differences in elevation between the finished grade and the terrain ($Y$ coordinates). Thus, the coordinates of these points can be determined from the cross-section notes (Figs. 17.31 and 17.34*a*) with due regard for sign of offsets and differences in elevation between road bed templet and terrain. The area of a cross section having $n$ sides by Eq. (8.41) (Art. 8.28) is

$$A = \tfrac{1}{2}[Y_1(X_2 - X_n) + Y_2(X_3 - X_1) + Y_3(X_4 - X_2) + \cdots + Y_{n-1}(X_n - X_{n-2}) + Y_n(X_1 - X_{n-1})]$$

For the cross section shown in Fig. 17.46*c* with seven sides,

$$A = \tfrac{1}{2}[Y_1(X_2 - X_8) + Y_2(X_3 - X_1) + Y_3(X_4 - X_2) + \cdots + Y_6(X_7 - X_5) + Y_7(X_1 - X_6)]$$

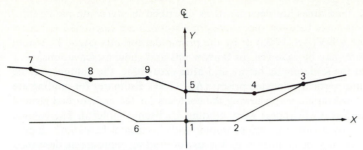

**Fig. 17.46** Irregular cross section.

A convenient way to systematize the calculation of cross-section areas by the coordinate method is to divide the section at the center line, taking the intersection of the grade line and center line (point 1 in Fig. 17.46) as the origin of the coordinate system. Then applying expression (8.44) (Art. 8.28), the area of the right half of the section in Fig. 17.46 is

$$A_R = \frac{1}{2}\left[ \frac{Y_1}{X_1} \diagdown \frac{Y_2}{X_2} \diagdown \frac{Y_3}{X_3} \diagdown \cdots \frac{Y_5}{X_5} \diagdown \frac{Y_1}{X_1} \right]$$

where the coordinates are listed for each point with $Y$ in the numerator and $X$ in the denominator starting and ending with the first point and moving in a counterclockwise direction around the loop. Similarly, the area to the left of the center line beginning with point 1 and moving in a clockwise direction around the loop is

$$A_L = \frac{1}{2}\left[ \frac{Y_1}{X_1} \diagdown \frac{Y_6}{X_6} \diagdown \frac{Y_7}{X_7} \diagdown \cdots \frac{Y_5}{X_5} \diagdown \frac{Y_1}{X_1} \right]$$

The total area enclosed by the cross section is the sum of the absolute values of $A_R$ and $A_L$.

**Example 17.11**   Compute the area enclosed by the cross section and roadbed templet of Example 17.10. Use the three-level equation (17.74) and check by the coordinate method. The final cross-section notes are as follows (units are in metres):

$$\frac{f0.56}{4.62} \quad \frac{f1.11}{0.00} \quad \frac{f1.86}{7.22}$$

SOLUTION   By Eq. (17.74), the area is

$$A_{3\ell} = \tfrac{1}{2}(1.11)(4.62 + 7.22) + \tfrac{7}{4}(0.56 + 1.86)$$

$$= 10.81 \text{ m}^2$$

Checking by the coordinate method yields

$$A_L = \frac{1}{2}\left[ \frac{0.00}{0.00} \quad \frac{1.11}{0.00} \quad \frac{0.56}{4.62} \quad \frac{0.00}{3.50} \quad \frac{0.00}{0.00} \right]$$

$$= \tfrac{1}{2}[(0.00)(0.00) + (1.11)(4.62) + (0.56)(3.50) + (0.00)(0.00)$$
$$- (1.11)(0) - (0.56)(0) - (0)(4.62) - (0)(3.50)]$$

$$= 3.54 \text{ m}^2$$

$$A_R = \frac{1}{2}\left[ \frac{0.00}{0.00} \quad \frac{1.11}{0.00} \quad \frac{1.86}{7.22} \quad \frac{0.00}{3.50} \quad \frac{0.00}{0.00} \right]$$

$$= \tfrac{1}{2}[(0)(0) + (1.11)(7.22) + (1.86)(3.50) + (0)(0) - (1.11)(0) - (1.86)(0) - (0)(7.22) - (0)(3.50)]$$

$$= 7.26 \text{ m}^2$$

$$A = A_L + A_R = 3.54 + 7.26 = 10.80 \text{ m}^2$$

**17.32. Volumes of earthwork**   Volumes of earthwork are calculated by a variety of methods, depending upon the nature of the excavation and of the data. If cross sections have been taken along a route, their areas are determined as described in preceding paragraphs, and the volumes of the prismoids between successive cross sections are calculated either by the method of average end areas (Art. 17.33) or by the prismoidal formula (Arts. 17.34 and 17.35). The same procedure may be followed for borrow pits and similar excavations, or if elevations are observed at the same points before and after excavating, the volume may be computed by dividing it into vertical truncated prisms (Art. 17.41). Estimates for grading are frequently based upon a topographic map showing the contours for the undisturbed ground and contours for the ground as it will appear when grading has been completed. The volume is conveniently determined by dividing it into prismoids with horizontal bases and sloping sides. Volumes of earthwork may be computed by the use of grading contours as described in Art. 17.42. Total volumes are expressed in cubic yards or cubic metres.

Because of the repetitive nature of computations for earthwork, calculation by electronic computer is highly desirable.

**17.33. Volumes by average end area**   The common method of determining volumes of excavation along the line of highways, railroads, canals, and similar works is that of *average end areas*. It is assumed that the volume between successive cross sections is the average of their areas multiplied by the distance between them, or

$$V = \frac{\ell}{2}(A_1 + A_2) \tag{17.76}$$

in which $V$ is the volume (cubic feet or cubic metres) of the prismoid of length $\ell$ (feet or metres) between cross sections having areas (square feet or square metres) $A_1$ and $A_2$.

If $\ell$ is in feet and cross sections at successive stations are in square feet, the volume between the stations in cubic yards, and $V_y$ is

$$V_y = \frac{\ell}{54}(A_1 + A_2) \tag{17.77}$$

Formulas (17.76) and (17.77) are valid only when $A_1 = A_2$ but are approximate for $A_1 \neq A_2$. As one of the areas approaches zero, as on running from cut to fill on side-hill work, a maximum error of 50 percent would occur if the formulas were followed literally. In this case, however, the volume is usually calculated as a pyramid; that is,

$$\text{volume} = \tfrac{1}{3}(\text{area of base})(\text{length})$$

Considering the fact that cross sections are usually a considerable distance apart and that minor inequalities in the surface of the earth between sections are not considered, the method of average end areas is sufficiently precise for ordinary earthwork.

Where heavy cuts or fills occur on sharp curves, the computed volume of earthwork may be corrected for curvature (Art. 17.36), but ordinarily the correction is not large enough to be considered.

**17.34. Prismoidal formula**   It can be shown that the volume of a prismoid is

$$V = \frac{\ell}{6}(A_1 + 4A_m + A_2) \tag{17.78}$$

in which $\ell$ is the distance between end sections, $A_1$ and $A_2$ the areas of the end sections, and $A_m$ the middle area or area halfway between the end sections. $A_m$ is determined by averaging the corresponding linear dimensions of the end sections and *not* by averaging the end areas $A_1$ and $A_2$. Equation (17.78) is one application of Simpson's rule (Art. 8.32).

**Example 17.12**   In Table 17.10 are shown the three-level cross-section notes for two stations 100 ft apart. The width of the roadbed is 20 ft, and the side-slope ratio is $1\frac{1}{2}$ to 1.

**Table 17.10   Earthwork Data**

| Station | Cross section | | | Area, ft$^2$ | Volume, yd$^3$ |
|---|---|---|---|---|---|
| | $\mathcal{C}$ | $\mathcal{C}$ | R | | |
| 115 | $\dfrac{c\,4.0}{16.0}$ | $\dfrac{c\,6.0}{0}$ | $\dfrac{c12.0}{28.0}$ | 212 | |
| | | | | | 575 |
| 116 | $\dfrac{c2.0}{13.0}$ | $\dfrac{c\,3.0}{0}$ | $\dfrac{c8.0}{22.0}$ | 103 | |
| Midsection | $\dfrac{c3.0}{14.5}$ | $\dfrac{c\,4.5}{0}$ | $\dfrac{c10.0}{25.0}$ | 154 | |

The volume of earthwork between the two stations is to be computed by the prismoidal formula. Below the regular cross-section notes are shown those for the midsection obtained by averaging the values given for sections at stations 115 and 116. In the column headed "Area, ft$^2$" are areas of cross sections computed by Eq. (17.74) (Art. 17.31). Then by the prismoidal formula given above,

$$V = \frac{100}{6}[212.0 + (4)(154.0) + 103.0] = 15{,}520 \text{ ft}^3 \text{ or } 575 \text{ yd}^3$$

For the foregoing example, the volume computed by average end areas is 583 yd$^3$, and the difference between the results obtained by the two methods is 8 yd$^3$ or about 1.4 percent.

As far as volumes of earthwork are concerned, the use of the prismoidal formula is justified only if cross sections are taken at short intervals, if small surface deviations are observed, and if the areas of successive cross sections differ widely. Usually, it yields smaller values than those computed from average end areas. For excavation under contract, the basis of computation should be understood in advance; otherwise, the contractor will usually claim (and obtain) the benefit of the common method of average end areas.

**17.35. Prismoidal correction**   It can be shown that the difference between the volume computed by the prismoidal formula and that computed by average end areas for the prismoids defined by three-level sections is

$$C_V = \frac{\ell}{12}(c_0 - c_1)(d_0 - d_1) \tag{17.79}$$

where $C_V$ = difference in volume, or correction, for a prismoid $\ell$ feet or metres in length, in cubic feet or cubic metres

$c_0$ = center height at one end section, feet or metres

$c_1$ = center height at the other end section, feet or metres

$d_0$ = distance between slope stakes at the end section where the center height is $c_0$, feet or metres

$d_1$ = distance between slope stakes at the other end section, feet or metres

$C_V$ is known as the *prismoidal correction*; it is *subtracted algebraically* from the volume as determined by the average-end-area method to give the more nearly correct volume as determined by the prismoidal formula.

All units in Eq. (17.79) must be the same. When all distances are in feet, the prismoidal correction in cubic yards is

$$C_{Vy} = \frac{\ell}{324}(c_0 - c_1)(d_0 - d_1) \tag{17.79a}$$

**Example 17.13**   For the prismoid of Example 17.12, the prismoidal correction by Eq. (17.79a) is

$$C_{Vy} = \frac{100}{324}(6.0 - 3.0)(44.0 - 35.0) = 8.33 \text{ yd}^3$$

which is consistent with the difference $(583 - 575) = 8$ yd$^3$ between the volumes obtained by the average-end-area method and the prismoidal equation, respectively.

**17.36. Earthwork curvature correction**   The average-end-area method for calculating volumes is based on parallel end areas. Volumes computed on curves will have discrepancies if this correction is neglected.

According to a theorem of Pappus, a plane area revolved about an axis generates a volume equal to the product of the revolving area and the path generated by the center of gravity of the revolving area. Let

$C_e$ = curvature correction = corrected volume − volume by the average-end-area method
$R$ = radius of the highway center line
$\bar{e}$ = eccentricity of the cross-section area

The corrected volume, $V_{corr}$, is

$$V_{corr} = \frac{\ell}{2}(A_1 + A_2) \pm C_e \tag{17.80}$$

where $A_1$ and $A_2$ have respective eccentricities of $\bar{e}_1$ and $\bar{e}_2$, and $\ell$ is the horizontal center-line distance between stations. Then $V_1 = A_1 \ell_1$, where $\ell_1$ is the length of arc at $R + e_1$, so that by proportion

$$\ell_1 = \ell\left(\frac{R + \bar{e}_1}{R}\right)$$

and

$$V_1 = A_1\left(\frac{R + \bar{e}_1}{R}\right)\ell \tag{17.80a}$$

Similarly,

$$V_2 = A_2\left(\frac{R + \bar{e}_2}{R}\right)\ell \tag{17.80b}$$

Next, assume that

$$V_{corr} = \frac{V_1 + V_2}{2} \tag{17.80c}$$

Substitution of Eqs. (17.80a) and (17.80b) into Eq. (17.80c) and then the result of this manipulation into Eq. (17.80) yields

$$C_e = \frac{\ell}{2R}(\bar{e}_1 A_1 + \bar{e}_2 A_2) \tag{17.81}$$

The curvature correction is positive when the excess area is on the outside of the curve and negative when the excess area is on the inside.

The centroid of the cross section is determined by taking moments about the center line. For example, the three-level section shown in Fig. 17.47a is divided into four triangles having respective areas of $a_1$, $a_2$, $a_3$, and $a_4$, each with a centroid at distances of $r_1$, $r_2$, $r_3$, and $r_4$ from the center line. Taking moments about the center line of the section yields

$$\bar{e} = \frac{r_1 a_1 + r_2 a_2 + r_3 a_3 + r_4 a_4}{a_1 + a_2 + a_3 + a_4} \tag{17.82}$$

For a three-level section such as that illustrated in Fig. 17.47a, the areas of the triangles can be calculated using the cuts or fills and offsets taken from the cross-section notes and the dimensions of the roadbed templet. The respective centers of gravity can be scaled from the cross section plotted at a scale of 1 : 120 (10 ft per 1 in) and recalling that the centroid of any triangle falls at the intersection of its medians.

**Example 17.14**   Assume that the center line containing stations 115 and 116 of Example 17.13 is a 10° circular curve (arc definition; see Art. 17.3) which turns to the left. Calculate the curvature correction to the earthwork volume between these two stations.

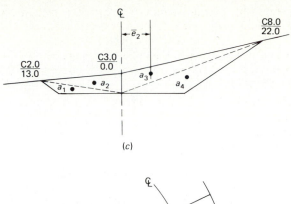

(c)

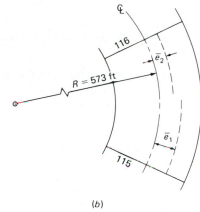

(b)

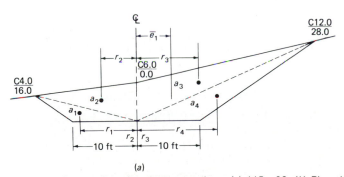

(a)

**Fig. 17.47** Center of gravity of cross sections. (a) 115 + 00. (b) Plan view. (c) 116 + 00.

SOLUTION   The cross sections for stations 115 and 116 are plotted in Fig. 17.47. From Fig. 17.47a and the cross-section notes of Example 17.13, the centroid for 115 + 00 is found as follows:

$$a_1 = (\tfrac{1}{2})(10.0)(4.0) \ = 20 \text{ ft}^2 \qquad r_1 = \ 8.8 \text{ ft}$$

$$a_2 = (\tfrac{1}{2})(6.0)(16.0) \ = 48 \text{ ft}^2 \qquad r_2 = \ 5.5 \text{ ft} \quad \text{scaled from}$$

$$a_3 = (\tfrac{1}{2})(6.0)(28.0) \ = 84 \text{ ft}^2 \qquad r_3 = \ 9.9 \text{ ft} \quad \text{cross section}$$

$$a_4 = (\tfrac{1}{2})(10.0)(12.0) = 60 \text{ ft}^2 \qquad r_4 = 12.5 \text{ ft}$$

$$\Sigma a_i \qquad\qquad = 212 \text{ ft}^2$$

$$\bar{e}_1 = \frac{(20)(-8.8) + (48)(-5.5) + (84)(9.9) + (60)(12.5)}{212} = 5.4 \text{ ft}$$

Similarly, for station $116+00$,

$$\bar{e}_2 = \frac{-\left(\frac{1}{2}\right)(10)(2.0)(7.8) - \left(\frac{1}{2}\right)(3)(13)(4.4) + \left(\frac{1}{2}\right)(3)(22)(7.6) + \left(\frac{1}{2}\right)(10)(8)(10.2)}{103}$$

$$= 4.8 \text{ ft}$$

From Eq. (17.1), the radius of the center line is

$$R = \frac{5729.58}{10} = 573 \text{ ft}$$

By Eq. (17.81), the curvature correction is

$$C_e = \frac{100}{(2)(573)}[(5.4)(212) + (4.8)(103)] = 143 \text{ ft}^3 = 5 \text{ yd}^3$$

Since the center of gravity falls outside the center line, the correction is positive, so that

$$V = 583 \text{ yd}^3 + 5 \text{ yd}^3 = 588 \text{ yd}^3$$

Note that this correction is only 1 percent of the total. Unless side-hill cuts or fills are very heavy, this correction is frequently neglected.

**17.37. Volumes in transitional areas**   The cross sections required in the transition from cut to fill or fill to cut are discussed in Art. 17.30, section 3, and are illustrated in Fig. 17.41. In that illustration, additional sections are necessary at $30+09.75$ (point $C$), where the cut runs out; $30+21.50$ (point $D$), the grade point at center line; and $30+28.45$ (point $F$), where the fill runs out. These cross sections are plotted and shown in Fig. 17.48 together with the sections for $30+00$ and $30+40$. Note that these sections are plotted assuming a straight line for the finished grade in cut.

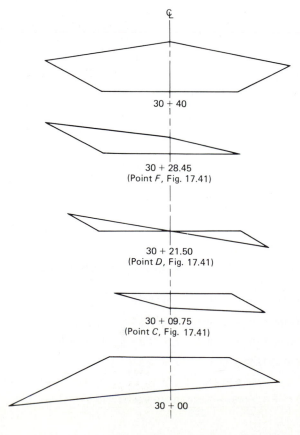

**Fig. 17.48** Cross sections in transition from cut to fill.

From $30+00$ to $30+09.75$, the volume is all fill. From $30+09.75$ to $30+21.50$, the volume consists of fill on the right and cut on the left. Note that the volume on the left between $30+09.75$ (where cut runs out) and $30+21.50$ must be calculated using the equation for volume of a pyramid (Art. 17.33). From $30+21.50$ to $30+28.45$, the volume on the left is cut and on the right is fill, which also must be calculated using the equation for a pyramid. In calculating the volumes as pyramids, the bases are the cross-sectional areas at $30+21.50$ and the altitudes of the pyramids are the distances from $30+21.50$ to the stations where fill or cut runs out. From $30+28.45$ to $30+40$, the volume is all cut.

**Example 17.15**  Table 17.11 contains cross-section data from which the cross sections of Fig. 17.48 were plotted. Compute the volumes of cut and fill in this transitional area from $30+00$ to $30+40$. The units are metres. The roadbed templet is $w=21$ m in cut and 18 m in fill with 2:1 side slopes.

**Table 17.11  Cross-Section Data for Transition Area**

| Station | Surface elevation | Grade elevation | Left | ℄ | Right |
|---|---|---|---|---|---|
| $30+00$ | 149.56 | 154.565 | f9.10 / 27.20 | f5.00 / 0.00 | f3.60 / 16.20 |
| $30+09.75$ | 152.87 | 155.073 | 0.00 / 9.00 | f2.20 / 0.00 | f2.28 / 13.56 |
| $30+21.50$ | 155.69 | 155.686 | c2.64 / 15.78 | 0.00 / 0.00 | f2.35 / 13.70 |
| $30+28.45$ | 158.20 | 156.048 | c4.10 / 18.70 | c2.15 / 0.00 | 0.00 / 10.50 |
| $30+40$ | 164.10 | 156.650 | c4.15 / 18.80 | c7.45 / 0.00 | c3.84 / 18.18 |

$30+00$:   $A = \frac{1}{2}(5.00)(27.20+16.20) + \frac{18}{4}(9.10+3.60)$

$\qquad = 165.65$ m$^2$    fill

$30+09.75$:  $A = \frac{1}{2}(2.20)(9+13.56) + \frac{18}{4}(0+2.28)$

$\qquad = 35.08$ m$^2$    fill

$30+21.50$: calculate as triangles, first on the left:

$A_{\text{left}} = (\frac{1}{2})(10.5)(2.64) = 13.86$ m$^3$    cut

$A_{\text{right}} = \frac{1}{2}(9)(2.35) = 10.58$ m$^2$    fill

$30+28.45$:  $A = \frac{1}{2}(2.15)(18.7+10.5) + \frac{21}{4}(4.1+0.0)$

$\qquad = 52.92$ m$^2$    cut

$30+40$:   $A = \frac{1}{2}(7.45)(18.80+18.18) + \frac{21}{4}(4.15+3.84)$

$\qquad = 179.70$ m$^2$    cut

These results and the accompanying volumes are listed in Table 17.12.

Volumes are computed as follows: from $30+00$ to $30+09.75$ by Eq. (17.76):

$V = \frac{9.75}{2}(165.65+35.08) = 978.6$ m$^3$    fill

From $30+09.75$ to $30+21.50$ the volume in fill by Eq. (17.76) is

$V = \frac{11.75}{2}(35.08+10.58) = 268.3$ m$^3$    fill

**Table 17.12   Summary of Areas and Volumes**

| Station | Area, m$^2$ | | Volume, m$^3$ | |
|---|---|---|---|---|
| | Cut | Fill | Cut | Fill |
| 30 + 00 | | 165.65 | | |
| | | | | 978.6 |
| 30 + 09.75 | | 35.08 | | |
| | | | 54.3 | 268.3 |
| 30 + 21.50 | 13.86 | 10.58 | | |
| | | | 232.1 | 24.5 |
| 30 + 28.45 | 52.92 | | | |
| | | | 1343.4 | |
| 30 + 40 | 179.70 | | | |
| | | | 1629.8 | 1271.4 |

and the pyramid of cut is

$$V = \frac{1}{3}(13.86)(11.75) = 54.3 \text{ m}^3 \qquad \text{cut}$$

From 30 + 21.50 to 30 + 28.45 there is a pyramid of fill on the right,

$$V = \frac{1}{3}(10.58)(6.95) = 24.5 \text{ m}^3 \qquad \text{fill}$$

and the volume of cut is computed by Eq. (17.76) for this pair of stations and between 30 + 28.45 and 30 + 40:

$$V = \frac{6.95}{2}(13.86 + 52.92) = 232.1 \text{ m}^3 \qquad \text{cut}$$

and

$$V = \frac{11.55}{2}(52.92 + 179.70) = 1343.4 \text{ m}^3 \qquad \text{cut}$$

The total volumes are 1629.8 m$^3$ of cut and 1271.4 m$^3$ of fill.

**17.38. Distribution analysis of earthwork**   Earthwork contracts usually allow payment for earthwork by lump sum or on the basis of the volume of materials moved. Thus, for a given job the contractor needs to know the volumes of cut and fill, where these quantities are located, and the distances materials must be moved. Preparatory to detailed discussions concerning distribution analysis, first consider some definitions related to the movement of cut and placing of fill or embankment.

*Excavation* is a pay quantity consisting of materials in cut which are transported to another location and placed in fill or embankment. The distance that a cubic unit (cubic yard or cubic metre, where 1 yd$^3$ = 0.765 m$^3$ and 1 m$^3$ = 1.308 yd$^3$) of material is transported from cut to fill is called a *haul distance* or simply *haul*. The *free-haul distance* (usually specified in the contract) is the distance a contractor can haul a cubic unit of excavated material and place it in fill without extra cost above the cost for excavation. Any haul distance beyond free haul is called *overhaul*, for which there is an extra charge. Material excavated but not used for fill is called *waste*. Material needed for fill but not obtained from the roadway grading is called *borrow*. Some types of soils shrink when placed in fill and others swell. Consequently, *shrinkage* or *swell* of the materials must be taken into account when calculating volumes used in the earthwork analysis.

*Distribution analysis* of earthwork involves determining balance points along the roadway center line between cut and fill. On a simple job one could make separate subtotals of cuts and fills. The balance points would occur where these subtotals are equal. On larger projects, this method is inadequate and the analysis is performed by the station-to-station method or by study of the mass diagram. Details concerning the station-to-station method can be found in Ref. 6. The mass diagram method is more general and is discussed in the sections that follow.

**17.39. The mass diagram**[†]   The earthwork mass diagram is a continuous graphical display of cumulated cut and fill volumes plotted as ordinates versus stations along the center line as abscissas. Cut is positive and fill is negative. Usually, the initial ordinate of the diagram is translated so that all ordinates are positive. The mass diagram is plotted above or below the center-line profile so that both can be used to analyze the earthwork operation. The abscissas are generally plotted at the same scale as the profile (say, 1 in to 400 ft or 1:4000 in the metric system), and ordinates are plotted at as large a scale as possible considering the range in the cumulated cuts and fills. Figure 17.49 shows a portion of a profile plotted over the mass diagram for the corresponding stations.

Reference to Fig. 17.49 reveals the following features of the mass diagram: (1) a rising curve indicates a cut, as from $A$ to $C$; (2) a falling curve indicates fill, as from $C$ to $B$; (3) a maximum point on the curve as at $C$ indicates a change from cut to fill in the direction of stationing (note that if there is extensive side-hill cut and fill, this maximum point and the grade point $c$ on the profile may not coincide exactly); (4) a minimum point as at $D$ indicates a change from fill to cut in the direction of stationing; and (5) a horizontal line intersects the curve of the mass diagram at balance points between which there is a balance of cut and fill volumes. In Fig. 17.49$b$, the line $AB$ is called a *balance line*.

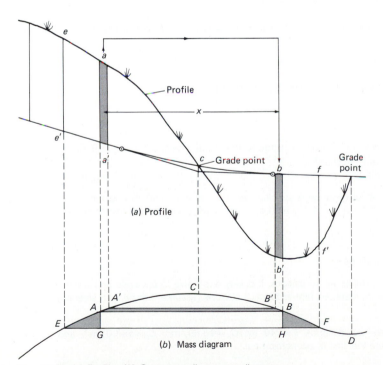

**Fig. 17.49** (*a*) Profile. (*b*) Corresponding mass diagram.

[†]Foote, F. S., "Notes on the Mass Diagram as Applied to Earthwork Computations and Operations," unpublished notes, University of California Institute for Transportation and Traffic Engineering.

In Fig. 17.49a, if *aa′* represents 1 cubic unit of cut which is transported a distance of *x* stations to be placed as a fill of 1 cubic unit at *bb′*, this volume times the distance transported in stations constitutes the number of *station units* of haul that is represented by the area *AA′B′B* under the mass diagram in Fig. 17.49b. In the English system, haul is expressed in station yards, where one station equals 100 ft. In the metric system, haul is expressed in station metres, where one station equals 100 m. The total haul resulting from placing the entire cut *aa′c* in the fill *cbb′* (Fig. 17.49a), is the area between the curve of the mass diagram *ACB* and the balance line *AB* (Fig. 17.49b). If the balance line *AB* equals the free-haul distance at the scale of the abscissa of the mass diagram, material in *aa′c* can be excavated and placed in fill at *cbb′* at the unit price for excavation.

**Overhaul**   Again referring to Fig. 17.49 if it is desired to place excavation *eaa′e′* in fill at *bff′b′*, the volumes so handled and transported constitute *overhaul* in station units. An additional charge is assigned for overhaul. The total amount of overhaul in Fig. 17.49b is represented by the areas *EAG* and *BFH*.

**Limit of economic overhaul**   Under certain conditions, it may pay to borrow or waste. In other words, there is a limit to the overhaul distance which is profitable. Let

$$C_E = \text{cost to excavate 1 unit volume (includes free-haul)}$$
$$C_{EW} = \text{cost to excavate and waste 1 unit volume}$$
$$C_B = \text{cost to borrow 1 unit volume}$$
$$C_{OH} = \text{cost for overhaul per station unit}$$
$$F = \text{free-haul distance, stations}$$
$$\text{L.E.H.} = \text{limit of economic haul, stations}$$

The cost to excavate and waste plus the cost to borrow equals the cost of excavation plus the cost for overhaul, or

$$C_{EW} + C_B = C_{OH}(\text{L.E.H.} - F) + C_E \tag{17.83}$$

Next, assume that $C_{EW} = C_E$, so that

$$\text{L.E.H.} = \frac{C_B}{C_{OH}} + F \tag{17.84}$$

Equation (17.84) governs the limiting economic overhaul given the free-haul distance and unit costs for borrow and overhaul.

**Example 17.16**   The free-haul distance is specified as 10 stations (1000 ft) and the unit costs are $C_E = \$1.80/\text{yd}^3$, $C_B = \$2.00/\text{yd}^3$, and $C_{OH} = \$0.40/\text{station yd}$. Compute the limit of economic haul.

SOLUTION   By Eq. (17.84),

$$\text{L.E.H.} = \frac{2.00}{0.40} + 10 = 15 \text{ stations}$$

Thus, if *AB* (Fig 17.49b) is taken as 10 stations, it is less expensive to excavate and waste at *E* and borrow at *F* than to haul the materials farther than 15 stations (assume that *EF* ⩾ 15 stations).

**Example 17.17**   The free-haul distance $F = 240$ m, $C_E = \$3.60/\text{m}^3$, $C_B = \$4.00/\text{m}^3$, and $C_{OH} = \$0.50/\text{station m}$. Compute the limit of economic haul.

SOLUTION   By Eq. (17.84),

$$\text{L.E.H.} = \frac{4.00}{0.50} + 2.4 = 10.4 \text{ stations}$$

beyond which it is less expensive to excavate and waste and borrow.

***Shrinkage and swell*** Shrinkage is small in granular soils (sands and gravels) but can be substantial (up to 30 percent) in fine-grained silts and clays. Shrinkage occurs when excavated soils are compacted into fills. Thus, 1 cubic unit of excavated material will not provide 1 cubic unit of compacted fill. In contrast to fine-grained soils, rock when excavated will occupy a larger volume in fill than when in place. In this case swell occurs and 1 cubic unit of excavated rock will occupy more than 1 cubic unit of fill. Swell of up to 30 percent is not uncommon in excavated rocky materials.

When excavated materials have uniform shrinkage or swell factors, compensation is accomplished by increasing or decreasing the calculated fill quantities before computing the mass-diagram ordinates. For example, if the excavated material shrinks 20 percent when placed in fill, 125 m³ of cut is required to provide 100 m³ of fill. Consequently, all fills must be increased by 20 percent to have a proper balance.

When excavated materials are nonhomogeneous and several different shrinkage factors are involved, the proper factor is applied to each cut volume and actual fills are utilized. Thus, if the shrinkage factor is 20 percent, a measured cut of 100 m³ will occupy a fill of 80 m³ when placed in fill. Compensation is achieved by decreasing all cut volumes by 20 percent prior to computing the cumulated cuts and fills and the ordinate of the mass diagram represents actual volumes of fill and volumes of cut available for fill.

***Determination of overhaul*** The amount of overhaul can be determined from the mass diagram. Portions of one loop of a mass diagram and the corresponding profile are shown in Fig. 17.50. Assume that balance line $AB$ is the free-haul distance, at the scale of the mass diagram, and balance line $A'B'$ is the limit of economic haul (L.E.H.). Thus, material from cut $aa'd'd$ excavated, hauled, and placed in fill at $bb'e'e$ is overhaul which is represented under the mass diagram by areas $AA'D$ and $BB'D'$. When $AA'$ and $BB'$ are relatively straight, the sum of these two areas equals $AD(EE'-AB)$, where $EE'$ lies midway between $AB$ and $A'B'$. Consequently, the overhaul in station units (station yd or station m) is found by multiplying the difference in ordinates between the free-haul balance line and the overhaul balance line by the difference between the overhaul distance and free-haul distance

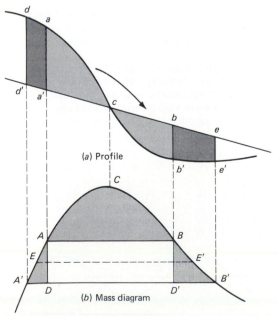

(a) Profile

(b) Mass diagram

**Fig. 17.50** Free-haul and overhaul.

in stations. When the mass diagram is very curved or irregular, the areas that represent overhaul can be measured by a planimeter, an analog device for measuring area.

**17.40. Balancing procedures**  So far, only the case involving one loop of the mass diagram has been studied. In practice, several or many loops of varying magnitude may occur consecutively and the mass diagram is very useful in determining the most economical solution by applying certain balancing procedures.

When two loops occur as in Fig. 17.51$a$, the most economical position for the balance line $ABC$ is that in which $AB = BC$ providing the distances $AB$ and $BC$ are each less than the limit of economic haul.

If three consecutive loops occur, as in Fig. 17.51$b$, the most economical position for balance line $ABCD$ is that in which $(AB + CD) - BC = $ L.E.H., provided that $AB$, $CD$, and $BC$ are each less than the limit of economic haul (L.E.H.).

In general, the most economical position for a balance line cutting an even number of loops is the one where the sum of the segments cutting concave loops equals the sum of the segments cutting convex loops. Each segment should be less than the limit of economical haul. The most economical position for a balance line cutting an odd number of loops is that in which the sum of the segments cutting concave (or convex) loops minus the sum of the segments cutting loops in the opposite direction equals the limit of economic haul. As before, each segment should be less than the limit of economic haul.

**Example 17.18**  Figure 17.52 shows the mass diagram and corresponding center-line profile for a small project. The unit costs for this project are cost of excavation $C_E = \$4.00/\text{m}^3$, cost of borrow $C_B = \$4.50/\text{m}^3$, cost of overhaul $C_{OH} = \$0.45/\text{station m}$, and the free-haul distance $F = 800 \text{ m} = 8$ stations. The total excavation from the volume summary is 24,848 m³. Since variable shrinkage factors were involved (1.00 and 0.80, see Fig. 17.52), excavated volumes adjusted for shrinkage and actual fills were used to compute ordinates for the mass diagram.

SOLUTION  By Eq. (17.84), the limit of economic haul is

$$\text{L.E.H.} = \frac{4.50}{0.45} + 8 = 18 \text{ stations}$$

Referring to Fig. 17.52, balance lines $AB$ (free-haul distance) and $CD$ (limit of economic haul) are placed on the first large loop. The cut from $A'B$ can be placed in fill at $A'A$ at the cost of excavation. Since balance line $CD$ is the limit of economic haul, the cut $DB$ placed in fill at $AC$ is overhaul and from the origin to $C$ requires borrow.

The balance line $EFG$ for the two consecutive loops is placed such that $EF = FG$. Note that both $EF$ and $FG$ are less than the free-haul distance. Thus, excavation from $D$ to $E$ and from $G$ to $H$ is wasted. The cut from $E$ to $E'$ can be placed in fill from $E'$ to $F$ and the cut from $G$ to $G'$ can be placed in the fill from $G'$ to $F$ at the cost for excavation since $EF = FG < $ the free-haul distance of 8 stations.

From the mass diagram the distance $JK$ is scaled and found to be 15 stations, so that

overhaul $= (15 - 8)(6170 - 3730) = 17,080$ station m

and the total costs are

Excavation $= (24,848)(4.00)$  $= \$99,392$

Borrow $= (10,000 - 6170)(4.50) =$  $17,235$

Overhaul $= (17,080/0.80)(0.45) =$  $9,608$

Total cost  $= \$126,235$

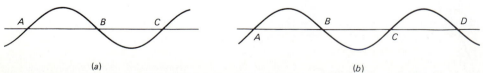

(a)  (b)

**Fig. 17.51** Balancing procedures. (a) Two loops. (b) Three loops.

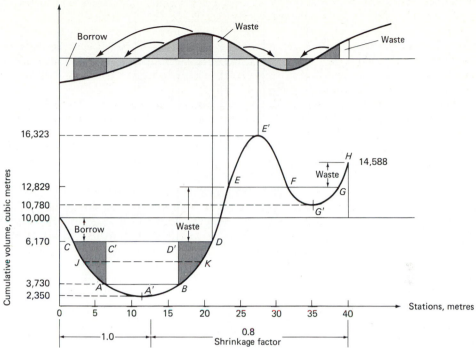

**Fig. 17.52** Mass diagram and profile for a small project (Example 17.17).

Note that the total cost of excavation must be calculated from the earthwork volume summary and the overhaul is adjusted by the shrinkage factor. These steps are necessary for this example, since ordinates on the mass diagram reflect cut volumes reduced by the shrinkage factor.

Certain checks can be applied to the mass-diagram ordinates: (1) the difference between the final and initial ordinates is equal to the difference between the total volume of waste and total volume of borrow; and (2) the final ordinate is equal to the difference between the total volume of cut and the total volume of fill taken from the earthwork volume summary.

Applying the first check to the mass diagram of Fig. 17.52, the total volume of borrow from the origin to $C$ is 3830 m³ and the total volume of waste from $D$ to $E$ and from $G$ to $H$ is $(6659 + 1759) = 8418$ m³. Thus, the total waste minus the total borrow $= 8418 - 3830 = 4588$ m³, which equals the difference between the final and initial ordinates, or $14{,}588 - 10{,}000 = 4588$ m³.

Application of the second check requires data from the volume summary. The total volume of cut is 24,848 m³ and the total volume of fill is 10,260 m³, yielding a difference of 14,588 m³. This value corresponds to the final ordinate of the mass diagram as taken from Fig. 17.52.

**17.41. Borrow pit**   The common method of determining the volume of a borrow pit is to cross-section the area before and after excavating (Art. 17.24). The pit is then plotted in plan, and its area is divided into rectangles, triangles, and trapezoids, thus dividing the volumes into truncated prisms. Actually, the upper and lower surfaces of the prisms are warped, but for earthwork computations they are assumed to be plane.

Figure 17.53 illustrates the plan view of a borrow pit, observations having been taken at the intersections of full lines. The numbers written diagonally are the cuts in feet.

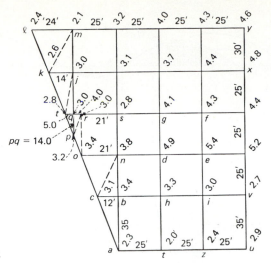

**Fig. 17.53** Borrow pit.

The volume of a triangular truncated prism (such as *abc* in Fig. 17.53) is

$$V = \frac{A}{3}(h_1 + h_2 + h_3) \tag{17.85}$$

in which $A$ is the horizontal sectional area and $h_1$, $h_2$, and $h_3$ are the corner heights at $a$, $b$, and $c$ in this figure.

The volume of any rectangular prism (as *defg* in Fig. 17.53) can be found using the equation

$$V = A\left(\frac{h_1 + h_2 + h_3 + h_4}{4}\right) \tag{17.86}$$

in which $A$ is the area of the base of the rectangular prism and $h_1, h_2, h_3, h_4$ are the corner depths of the prism from original terrain surface to the finished grade of the borrow pit.

When a uniform grid is used, the base area for all rectangular prisms is the same and it is possible to derive a single equation to determine the volume within the area included by the uniform grid. It is possible to accomplish this by summing the common corner heights. For example, in rectangular prism *ivuz* the height at $u$ is used once, at $v$ and $z$ twice, and at $i$ four times. Let

$$h_1 = \text{sum of corner heights used once}$$
$$h_2 = \text{sum of corner heights used twice}$$
$$h_3 = \text{sum of corner heights used three times}$$
$$h_4 = \text{sum of corner heights used four times}$$

Then the sum of the volumes of the rectangular prisms all having a base area of $A$ is

$$V = \frac{A}{4}\left(\sum h_1 + 2\sum h_2 + 3\sum h_3 + 4\sum h_4\right) \tag{17.87}$$

The volumes of marginal prisms must be calculated separately and added to the volume determined by Eq. (17.87) to get the total volume of the borrow pit.

**Example 17.19** Compute the volume of the borrow pit shown in Fig. 17.53. First compute the volume of the rectangular prisms having uniform base areas of $(25)(25) = 625 \text{ ft}^2$ enclosed by *xjqsbv*.

SOLUTION   Use Eq. (17.87).

| $h_1$ | $h_2$ | $h_3$ | $h_4$ |
|---|---|---|---|
| 4.8 | 4.4 | 2.8 | 4.1 |
| 3.0 | 3.7 | $\Sigma h_3 = 2.8$ | 4.3 |
| 3.0 | 3.1 | | 4.9 |
| 3.4 | 3.8 | | 5.4 |
| 2.7 | 3.3 | | $\Sigma h_4 = 18.7$ |
| $\Sigma h_1 = 16.9$ | 3.0 | | |
| | 5.2 | | |
| | 4.4 | | |
| | $\Sigma h_2 = 30.9$ | | |

By Eq. (17.87), the volume of *xjqsbv* is

$$V = \frac{625}{4}[16.9 + (2)(30.9) + (3)(2.8) + (4)(18.7)] = 25,297 \text{ ft}^3$$

Next, compute the volumes of *bvua* and *xymj*, both composed of rectangular prisms.

| | *bvua* | | | *xymj* | |
|---|---|---|---|---|---|
| | $h_1$ | $h_2$ | | $h_1$ | $h_2$ |
| | 2.7 | 3.0 | | 4.8 | 4.3 |
| | 3.4 | 3.3 | | 4.6 | 4.0 |
| | 2.3 | 2.0 | | 2.1 | 3.2 |
| | 2.9 | 2.4 | | 3.0 | 3.1 |
| | $\Sigma h_1 = 11.3$ | $\Sigma h_2 = 10.7$ | | $\Sigma h_1 = 14.5$ | 3.7 |
| | | | | | 4.4 |
| | | | | | $\Sigma h_2 = 22.7$ |

$$V_{bvua} = \frac{(25)(35)}{4}[11.3 + (2)(10.7)] = 7153 \text{ ft}^3 \qquad V_{xymj} = \frac{(30)(25)}{4}[14.5 + (2)(22.7)] = 11,231 \text{ ft}^3$$

The volumes of the remaining triangular truncated prisms and the one remaining rectangular prism are calculated by Eqs. (17.85) and (17.86), respectively. The results are tabulated below together with the volumes already computed.

| Prism | $\Sigma$ Corner heights, ft | Area, ft² | Times | Volume, ft³ |
|---|---|---|---|---|
| *abc* | 8.8 | (6)(35) | $\frac{1}{3}$ | 616 |
| *bcn* | 10.3 | (6)(25) | $\frac{1}{3}$ | 515 |
| *con* | 10.3 | (10.5)(25) | $\frac{1}{3}$ | 901 |
| *por* | 9.6 | (2)(25) | $\frac{1}{3}$ | 160 |
| *pqr* | 9.2 | (2)(14) | $\frac{1}{3}$ | 86 |
| *pqt* | 9.0 | (2.5)(14) | $\frac{1}{3}$ | 105 |
| *tqj* | 8.8 | (2.5)(25) | $\frac{1}{3}$ | 183 |
| *kjt* | 8.4 | (7)(25) | $\frac{1}{3}$ | 490 |
| *kjm* | 7.7 | (7)(30) | $\frac{1}{3}$ | 539 |
| *klm* | 7.1 | (12)(30) | $\frac{1}{3}$ | 852 |
| *nors* | 13.0 | (21)(25) | $\frac{1}{4}$ | 1,706 |
| *xjqsbv* | | | | 25,297 |
| *bvua* | | | | 7,153 |
| *xymj* | | | | 11,231 |
| Total | | | | 49,834 ft³ |
| or | | | | 1,846 yd³ |

**17.42. Earthwork volumes by grading contours** Contours plotted on a topographic map to portray the finished surface of a proposed grading operation are called *grading contours*. Since the finished surface is usually smooth and of constant slope, grading contours are generally smooth, equally spaced, parallel lines which may be straight or curved, depending on the character of the final surface. Volumes from grading contours are determined by the method of horizontal planes and the method of equal-depth contours.

**1. Horizontal planes** For preliminary estimates for grading areas, especially where the graded surface is itself more or less irregular, the common practice is to use the topographic map directly as a basis for calculations of volume. On the map are shown contours for the natural ground and contours for the proposed graded surface. This method consists of determining the volumes of earth to be moved between the horizontal planes at the elevations of successive contours.

The light full lines of Fig. 17.54 represent contours of the original ground, and the dashed lines represent contours of the proposed graded surface. The heavy full lines are drawn through points of no cut or fill. Thus, the line *abcdefa* bounds an area that is entirely in fill, and the line *dghjked* bounds an area that is entirely in cut. The "no cut or fill" lines are seen to pass through the points of intersection between full contours and the corresponding dash contours, as at *a*, *b*, *d*, *e*, *h*, and *j*. The conditions surrounding the problem make it possible to estimate the position of the lines where the cut or fill runs out between contours, as the lines *bcd*, *efa*, and *jke*. The crosshatched portions are the horizontal sections of earth cut or fill at the contour elevations; thus, $F_1$ represents the horizontal section of earth filled at elevation 98 m. The volume of earthwork between the two horizontal planes at the elevations of successive contours is a solid the altitude of which is the contour interval and the top and bottom bases of which are the horizontal projections of the cut or fill at the contour elevations (as for the fill between the 97- and 98-m contours where the height is 1 m and the bases are $F_2$ and $F_1$ the areas of which may be determined by use of a planimeter). Where the cut or fill runs out between contours (as along line *bcd*), the height of the end volume will be less than the contour interval. This height may be estimated by assuming the slope of the ground to be uniform between contours; thus, point *c* is estimated to be at elevation 98.6 m, and the volume above the 98-m contour is a solid the base of which is $F_1$ and the altitude of which is $98.6 - 98 = 0.6$ m. The end volumes may be considered as pyramids.

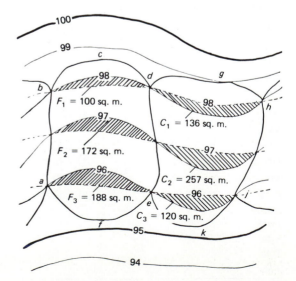

**Fig. 17.54** Volume by horizontal planes.

**Example 17.20**    It is desired to determine the volume of earthwork in fill bounded by the line *abcdefa* (Fig. 17.54). The intermediate volumes are to be calculated by the method of average end areas; the end volumes are to be considered as pyramids. The areas of fill at the contours are as shown in the figure. Point *c* is estimated to lie 0.6 m above the 98-m contour, and point *f* is estimated to be 0.8 m below the 96-m contour.

SOLUTION

| Elevation | Base area, m$^2$ | Altitude, m | | Volume m$^3$ |
|---|---|---|---|---|
| *c* = 98.6 | 0 | | | |
| | | 0.6 | $(\frac{1}{3})(0.6)(100)$ | 20 |
| 98 | 100 | | | |
| | | 1.0 | $(\frac{1}{2})(1.0)(272)$ | 136 |
| 97 | 172 | | | |
| | | 1.0 | $(\frac{1}{2})(1.0)(360)$ | 180 |
| 96 | 188 | | | |
| | | 0.8 | $(\frac{1}{3})(0.8)(188)$ | 50 |
| *f* = 95.2 | 0 | | | |
| | | | Total | 386 m$^3$ |

**2. Equal-depth contours**    This method consists of determining volumes between irregularly inclined upper and lower surfaces bounding certain increments of cut or fill. Horizontal projections of the inclined areas are taken from the map, usually with the planimeter, and the volume between any two successive areas is determined by multiplying the average of the two areas by the depth between them.

Figure 17.55 represents the topographic map of a portion of a tract that is to be graded by filling. The light full lines represent contours of the original ground surface, and the dashed lines represent contours of the proposed graded surface. Above the dashed 102-ft contour the fill drops abruptly to the natural ground. Along the bank thus formed just above and paralleling the 102-ft contour, actually there would be 100-, 98-, and 96-ft contour lines, but

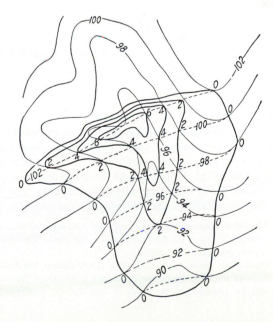

**Fig. 17.55**  Volume by equal-depth con-tours.

to avoid confusion of lines these are not shown. At the intersection of each light full line with each of the dashed lines the depth of fill (or cut) is recorded.

The heavy full lines drawn through points of equal fill are called lines of equal fill (or cut). The heavy outer line passes through points of zero fill and marks the limit of the fill. The next heavy line encloses the area over which the fill is a minimum of one contour interval and passes through points of intersection between a full contour and a dashed contour the elevation of which is 2 ft greater; and so on. Along the side of the bank above the dashed 102-ft contour, the heavy lines are seen to be close together and nearly parallel.

The fill between the graded surface and the surface 2 ft below is represented by the solid the altitude of which is 2 ft and the upper and lower surfaces of which are shown in horizontal projection by the line of zero fill and the line of 2-ft fill, respectively. Similarly, the lines of 2- and 4-ft fill define the volume of fill between the depths of 2 ft and 4 ft from the graded surface. The volume below the innermost line of equal fill may be considered as a pyramid the base area of which is that bounded by the line and the altitude of which is estimated, being always less than the full contour interval. Volumes are usually determined by multiplying the contour interval by the average of the areas of successive surfaces of equal cut or fill. When there is a large difference between successive areas the prismoidal, Eq. (17.78) (Art. 17.34) or the equation (17.79) for prismoidal correction is sometimes used.

**Example 17.21** An estimate of volume of earthwork in fill is to be made from a contour map similar to that of Fig. 17.55. Lines of equal fill are drawn, and the areas of the horizontal projections of surfaces of equal fill are determined by measurement with a planimeter. The altitude of the pyramid below the innermost surface of equal fill is estimated to be 1 ft.

SOLUTION

| Fill, ft | Area, ft$^2$ | Altitude, ft | | Volume, ft$^3$ |
|---|---|---|---|---|
| 0 | 101,000 | | | |
| | | 2.0 | $(\frac{1}{2})(2)(134,000)$ | 134,000 |
| 2 | 33,000 | | | |
| | | 2.0 | $(\frac{1}{2})(2)(50,000)$ | 50,000 |
| 4 | 17,000 | | | |
| | | 2.0 | $(\frac{1}{2})(2)(22,000)$ | 22,000 |
| 6 | 5,000 | | | |
| | | 1.0 | $(\frac{1}{3})(1)(5000)$ | 2,000 |
| 7 | 0 | | | |
| | | | Total | 208,000 ft$^3$ |
| | | | or | 7700 yd$^3$ |

**17.43. Earthwork for roadway by grading contours** Figure 17.56 shows the contour lines for a proposed roadway (the grade line of which has been drawn on the profile) drawn dotted over the existing contour map. Above the contour map are shown a profile of the ground along the center line and the grade of the proposed roadway. The side slopes of the earthwork are $1\frac{1}{2}$ to 1. The width of the roadway is 36 ft in cut and 30 ft in fill. From a study of these two drawings, the following observations may be made:

1. The 840-ft contour line of the proposed roadway crosses the roadway at a point on the map vertically beneath the point on the profile where the grade line crosses the 840-ft elevation line; and similarly for the other gradient contours.
2. On the side slopes of the earthwork at any station, the distance out from the edge of the roadway to a contour line is given by the difference in elevation (between that which the contour line represents and the elevation of the grade at that station), multiplied by the side-slope ratio. Thus, at station

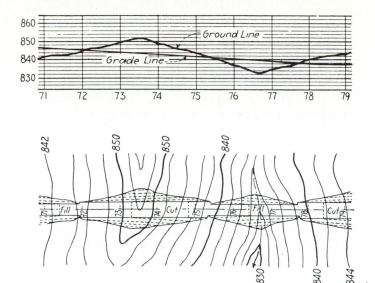

**Fig. 17.56** Earthwork for roadway by grading contours.

76 + 40 the elevation of grade is 840.0 ft and the elevation of the first contour line out from the edge of the fill is 838.0 ft; hence, the distance out is $2 \times 1\frac{1}{2} = 3$ ft (or 18 ft from ℄). (For clearness, in the illustration the lateral scale is exaggerated.)

3. As the grade is not level, the contour lines on the earthwork slopes are not parallel to the roadway. Thus, the 844-ft dotted contour line which crosses the roadway at station 73 + 30 is so inclined in direction that at station 74 + 80, where the elevation of grade is 842 ft the 844-ft contour line is out from the edge of the roadway a distance of $2 \times 1\frac{1}{2} = 3$ ft.
4. The top or toe of a slope is drawn on the contour map by connecting the points where the dotted lines intersect the corresponding full lines.

The volume of earthwork is estimated by means of horizontal planes as described in Art. 17.42, section 1. Figure 17.57, an enlargement of the portion of Fig. 17.56 from stations 72 + 00 to 75 + 00 illustrates the method. As an example, consider the volume contained between the 848 and 850 grading contours. The area *abdc* enclosed by the 850 grade contours and the 850-ft contour line is determined using a planimeter. Next, the area *efgh* enclosed by the 848-ft grade contours and the 848-ft contour is also measured by the planimeter. These two areas constitute end areas separated by the contour interval, or 2 ft in this case. Thus, the volume of this solid can be calculated by the average-end-area equation (17.77) (Art. 17.33). Volumes of the remaining segments are calculated in a similar manner. The volume of the segment bounded by the 852 grading contour, the 852 contour, and the top of the slope at elevation 853 (interpolated from the map) is determined as the volume of a pyramid having a base area of *jkℓ* and an altitude of 1 ft. The total volume is the sum of the volumes of all segments with the area of interest.

**17.44 Reservoir areas and volumes**  A contour map can be employed to determine the capacity of a reservoir, the location of the flow line, the area of the reservoir, and the area of the drainage basin. The procedure may be illustrated by reference to the fill across the valley in Fig. 17.56, the fill being considered as a dam which extends from the 841-ft elevation on the left to about the 838-ft elevation on the right. If water is imagined to stand at the elevation of 834 ft., the water surface is represented by that within the full and dotted 834-ft contour line. If the water were to rise through a 2-ft stage to the elevation of 836 ft, the water

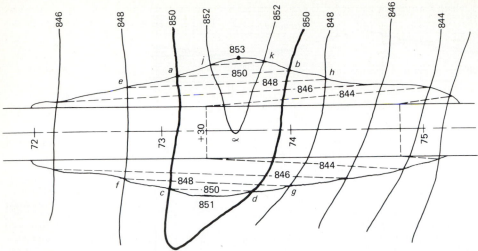

**Fig. 17.57** Roadway earthwork volumes by grading contours.

surface would be represented by that within the full and dotted 836-ft contour line (the full line being continued until the two parts meet). The volume of water that caused the 2-ft rise is given by the average of the two surface areas multiplied by the vertical distance of 2 ft. Similarly, the volume of water required to cause a rise of the water surface from 836 to 838 ft may be found, and so on. By a similar procedure, the volume of any reservoir can be calculated.

The outline of the submerged area of a proposed reservoir is given by the contour line representing the maximum stage of the impounded water. The drainage area may be estimated by sketching the watershed line on the map and measuring the extent of the watershed with a planimeter.

**17.45. Error propagation in volume computation**  It is useful to consider the propagated errors in earthwork volume computation. As an example, consider three-level cross sections, computation of the area by Eq. (17.74) (Art. 17.31), and calculation of volumes by the average-end-area method using Eq. (17.76) or (17.77). Equation (17.74) for area of a three-level section is repeated here for convenience:

$$A = \tfrac{1}{2}c(d_l + d_r) + \frac{w}{4}(c_l + c_r)$$

Measured values are the center-line cut $c$, the cut at the left slope stake $c_l$, the cut at the right slope stake $c_r$, and $d_l$ and $d_r$ the horizontal distances to the left and right slope stakes, respectively. Applying the error-propagation equation (2.46) (Art. 2.20) to Eq. (17.74) yields an estimated variance in the area $A$ of

$$\sigma_A^2 = \left(\frac{\partial A}{\partial c}\right)^2 \sigma_c^2 + \left(\frac{\partial A}{\partial d_l}\right)^2 \sigma_{d_l}^2 + \left(\frac{\partial A}{\partial d_r}\right)^2 \sigma_{d_r}^2 + \left(\frac{\partial A}{\partial c_l}\right)^2 \sigma_{c_l}^2 + \left(\frac{\partial A}{\partial c_r}\right)^2 \sigma_{c_r}^2$$

in which $c, d_l, \ldots, c_r$ are assumed to be uncorrelated. Assuming that $\sigma_c^2 = \sigma_{c_r}^2 = \sigma_{c_l}^2$ and $\sigma_{d_l}^2 = \sigma_{d_r}^2 = \sigma_d^2$, then

$$\sigma_A^2 = \left[\left(\frac{d_l + d_r}{2}\right)^2 + \frac{w^2}{8}\right]\sigma_c^2 + \left(\frac{c^2}{2}\right)\sigma_d^2 \tag{17.88}$$

Equation (17.76) for volume by the average-end-area method is

$$V = \frac{\ell}{2}(A_1 + A_2)$$

in which $\ell$ is the measured horizontal distance between stations and $A_1$ and $A_2$ are derived from measured elevations and distances using Eq. (17.74). The propagated variance in the volume by applying Eq. (2.46) to Eq. (17.76) is

$$\sigma_V^2 = \left(\frac{\partial V}{\partial \ell}\right)^2 \sigma_\ell^2 + \left(\frac{\partial V}{\partial A_1}\right)^2 \sigma_{A_1}^2 + \left(\frac{\partial V}{\partial A_2}\right)^2 \sigma_{A_2}^2$$

$$\sigma_V^2 = \left(\frac{A_1 + A_2}{2}\right)^2 \sigma_\ell^2 + \left(\frac{\ell}{2}\right)^2 \left(\sigma_{A_1}^2 + \sigma_{A_2}^2\right) \tag{17.89}$$

in which $\sigma_{A_1}^2$ and $\sigma_{A_2}^2$ are estimated by Eq. (17.88) and $\sigma_\ell^2$ is estimated on the basis of the equipment and methods used. Application of error propagation to Eq. (17.77), where volumes are expressed in cubic yards, yields

$$\sigma_V^2 = \left(\frac{A_1 + A_2}{54}\right)^2 \sigma_\ell^2 + \left(\frac{\ell}{54}\right)^2 \left(\sigma_{A_1}^2 + \Sigma_{A_2}^2\right) \tag{17.90}$$

When slope stakes are set in the field, a metallic tape, level, and rod or EDM equipment is used. Estimated standard deviations of 0.05 ft or 0.015 m in horizontal distance and 0.05 ft or 0.015 m in elevation are reasonable with these types of equipment.

Propagated errors for areas and volumes computed by Eqs. (17.74) and (17.77) for several three-level cross sections of varying size are given in Table 17.13. The assumptions used are: roadway width $w = 24$ ft, $\ell = 100$ ft, slopes are 2 to 1, $\sigma_c = 0.05$ ft, and $\sigma_t = \sigma_d = 0.05$ ft.

An inspection of the tabulation shows (1) that the percentage of error in the area and in the volume varies inversely with the depth of the cut or fill, (2) that the magnitude of the estimated standard deviation is not important as compared with the errors due to variations over the ground surface, and (3) that the estimated standard deviations indicate an uncertainty of one or more in the last unit of the computed quantities. Hence, it will be consistent to carry one decimal place in intermediate computations of areas and volumes; but it is absurd to record values beyond the last whole unit, either of areas or of volumes.

The procedures in the preceding paragraphs apply equally to volumes using irregular cross sections and borrow-pit excavation. Thus, if volumes are calculated with irregular cross-sections, the variance in areas is propagated by applying Eq. (2.46) to Eq. (8.41) for calculating area by coordinates. Estimated standard deviations in volumes computed with irregular cross sections obtained by field methods which satisfy $\sigma_c = 0.05$ ft and $\sigma_d = 0.05$ ft would be approximately the same as those given in Table 17.13. The critical item in the error

**Table 17.13   Propagated Errors In Volume Computations**[a]

| Left slope stake | $\underline{\ell}$ | Right slope stake | Area, $A$ ft$^2$ | $\sigma_A^2$, ft$^4$ | $\sigma_A$, ft$^2$ | $\sigma_A/A$ Percent | Volume, yd$^3$ | yd$^3$ | $\sigma_v/V$, Percent |
|---|---|---|---|---|---|---|---|---|---|
| $\dfrac{c1.0}{14}$ | $\dfrac{c2.3}{0}$ | $\dfrac{c3.0}{18}$ | 61 | 0.83 | 0.91 | 1.5 | | | |
| | | | | | | | 350 | 2.5 | 0.7 |
| $\dfrac{c2.0}{16}$ | $\dfrac{c5.1}{0}$ | $\dfrac{c4.0}{20.0}$ | 128 | 1.02 | 1.01 | 0.8 | | | |
| | | | | | | | 952 | 3.3 | 0.3 |
| $\dfrac{c5.0}{22}$ | $\dfrac{c11.6}{0}$ | $\dfrac{c9.0}{30}$ | 386 | 2.04 | 1.43 | 0.4 | | | |
| | | | | | | | 1576 | 3.9 | 0.2 |
| $\dfrac{c6.0}{24}$ | $\dfrac{c13.9}{0}$ | $\dfrac{c9.0}{30}$ | 465 | 2.24 | 1.50 | 0.3 | | | |

[a] $\ell = 100$ ft, $w = 24$ ft, $s$ is 2 to 1, $\sigma_t = \sigma_d = 0.05$ ft, $\sigma_c = 0.05$ ft.

propagation is in selecting valid estimates for standard deviations in elevation and horizontal distance.

Suppose that cross-section data are obtained from a topographic map at a scale of $1:600$ (1 in to 50 ft) and with a 2-ft contour interval. According to National Map Accuracy Standards (NMAS) as discussed in Art. 13.38, errors in planimetric position should not exceed $\frac{1}{50}$ in or 0.5 mm of map scale for 90 percent of the well-defined images checked. Similarly, errors in elevation should not exceed one-half the contour interval for 90 percent of the points checked. These standards correspond to a circular map accuracy standard (CMAS) $= (1/50 \text{ in})(50 \text{ ft/in}) = 1 \text{ ft } (0.3 \text{ m})$ and a vertical map accuracy standard (VMAS) $= (\frac{1}{2})(2 \text{ ft}) = 1 \text{ ft } (0.3 \text{ m})$. Using Eqs. (13.5) and (13.6) from Art. 13.38, these values for CMAS and VMAS correspond to $\sigma_d = \sigma_l = 0.5 \text{ ft } (0.15 \text{ m})$ and $\sigma_c = 0.6 \text{ ft } (0.18 \text{ m})$. Propagation of these errors using Eqs. (17.88) and (17.90) for the three cross sections of Table 17.13 yields errors in volumes which are 9, 4, and 3 percent, respectively. Obviously, cross-section data obtained from topographic maps should not be used to compute volumes used as a basis for payment since the errors are likely to be excessive. If such an approach is necessary, a contour interval of 0.5 ft or 0.2 m and a map scale of $1:480$ (1 in to 40 ft) or $1:400$ is required.

When cross-section data are determined photogrammetrically using photography taken over cleared terrain, $\sigma_c = 0.4 \text{ ft}$ or 0.12 m and $\sigma_l = \sigma_d = 0.2 \text{ ft}$ or 0.06 m are reasonable estimates. Using these values propagated standard deviations in volumes for the three cross sections in Table 17.13 are 7, 3, and 2 percent of their respective volumes. It is apparent that if photogrammetric cross sections are to be utilized for evaluating payment quantities of earthwork, the strictest control must be exercised over the photogrammetric operation in order to achieve the desired accuracy.

**17.46. Use of digital terrain models (DTM) in route location**  The digital terrain model [DTM or digital elevation model (DEM)] is described in detail in Art. 16.19. For the purposes of this section, it is sufficient to define the DTM as being a numerical representation of terrain relief and planimetric detail. This numerical representation usually consists of $X, Y, Z$ coordinates in a known system (state plane coordinates and elevations, for example) for a dense array of points within the area to be studied. The data that comprise the DTM are generally recorded on magnetic tape but can be punched on cards or stored in the solid-state memory of an electronic computer.

It is possible to use data from all sources to form the DTM. For example, field survey data (Art. 15.17), data taken from topographic maps (Art. 13.36), and photogrammetrically determined data (Art. 16.19) can be digitized and combined to form a DTM. In this way all existing techniques and information are employed to the fullest extent.

In order to use the DTM for route alignment, it is necessary to acquire or develop electronic computer programs which permit extraction of profiles and cross-section data from the DTM stored on magnetic tape. These extracted data can then be used in the route alignment procedure in the same way as terrain data obtained by traditional methods. Thus, a proposed alignment and many optional locations can be thoroughly studied without access to additional data other than that obtained from the DTM.

The primary requirements for utilization of the DTM in this way are (1) a system for compiling and recording the data used to form the DTM, (2) sufficient ground control data to permit absolute orientation of the DTM, and (3) a set of electronic computer programs, normally called *software*, which allows the design engineer(s) to extract the information needed for the route location.

At the time of this writing, use of the DTM for route location is still in the experimental stage. However, the potential of the DTM is enormous and extensive use of this source of information for route location and design is inevitable in the near future.

## 17.47. Problems

**17.1**  Given: $I = 34°30'$, $D$ (chord basis) $= 3°00'$, and P.C. $=$ station $74 + 30.0$. Required: $R$, $L$, $T$, and $E$; also deflection angles arranged in notebook form for staking out this curve, using 100-ft stations.

**17.2**  Given: $I = 60°40'$, $E = 125.5$ ft. Required: $R$, $D$ (arc basis), $C$, $T$, $M$, and $L$.

**17.3**  Given: $I = 37°30'$, $D_a = 22°00'$, the station of the P.I. $= 14 + 80.000$ m. Compute (a) $T$, $E$, $L$, the middle ordinate for the entire curve, and the middle ordinate for one full station or 100 m; (b) the station of the P.C., the station of the P.T., and deflections angles for staking the curve at even one-tenth station points; (c) the true lengths of one-tenth station chords and the true lengths for the subchords at either end of the curve. (d) Write the necessary field notes for staking this curve in the form in which they should appear in the field book.

**17.4**  Given the data of Prob. 17.3. This curve is to be staked by the method of chord offsets and then checked by tangent offsets. Points are to be set at even two-tenths station points (20 m). (a) Compute those chord offsets and tangent offsets required for this operation. (b) Describe in detail, with the aid of suitable sketches, the necessary field work procedures for staking and field checking of this curve by chord offsets and tangent offsets.

**17.5**  Two tangents $AV$ and $BV$ have an intersection angle of $45°00'$. A point $C$ is located by the coordinates $VH = 270.2$ ft and $HC = 157.4$ ft, $VH$ being measured along the tangent $VA$, and $HC$ being measured perpendicular thereto. It is desired to connect the two tangents with a curve passing through the point $C$. Required: $R$, $D$ (chord basis), $T$, $L$, and $E$. Station of P.I. is $17 + 40.00$.

**17.6**  In Prob. 17.5, change the value of $D$ to $2°00'$ (chord basis) and all other elements to agree with the new value of $D$. Compute deflection angles to full stations.

**17.7**  Given the data of 17.1. Make the necessary computations for the insertion of a spiral of length 250 ft at each end of the curve. Calculate deflection angles to 50 ft stations for the entire curve.

**17.8**  Given $I = 53°30'$ to the right, $D_a = 28°00'$, and the station of the P.I. $= 30 + 14.143$. The sharpness of this curve requires that a spiral be used at each end of the circular curve. The rate of change of curvature of the spiral is to be $20°00'$ per station (100 m). Using the rigorous equations for $X$ and $Y$, compute (a) the stations for the $TS$, $SC$, $CS$, and $ST$ points on the spiraled curve; (b) deflection angles for staking the curve with 20-m chords on the spirals and one-tenth station points (10 m) on the circular curve. The spiral from the $TS$ to the $SC$ is to be laid out from an instrument setup at the $TS$; and the second spiral is to be staked by deflection angles from an instrument set at the $CS$ point. Tabulate the results in the form of field notes, showing therein the necessary information for orienting the transit when set up at the points specified.

**17.9**  The layout of the spirals in Prob. 17.8 is to be checked by means of offsets from the appropriate tangent to the spiral. Compute the offsets from the tangent to the spiral between the $TS$ and $SC$ points and between the $CS$ and $ST$ points (a) using the approximate equation (17.38); and (b) using the rigorous equations (17.48a) and (17.48b). Show the results of parts (a) and (b) by a neat sketch and compare the offsets computed by the approximate and rigorous methods. Tabulate the results.

**17.10**  Given the following finished grade elevations for the center line of a roadway: $A$ at $0 + 00 = 901.200$ m, $E$ at $20 + 96.143 = 990.000$ m, and $D$ at $41 + 75.470 = 950.000$ m, each above mean sea level. An equal tangent parabolic vertical curve with a total length of 150 m and centered about point $E$ is to be calculated for subsequent construction layout. Compute the: (a) grade elevations for the BVC and EVC and at points spaced at 20-m intervals throughout the vertical curve; (b) station and elevation of the high (or low) point on the vertical curve. Use the equation of the parabola to solve this problem.

**17.11**  In Prob. 17.10, do the computations for parts (a) and (b) using the method of vertical offsets from the tangent.

**17.12**  Do the computations for Prob. 17.10, parts (a) and (b), using the chord gradient method.

**17.13**  Following are the notes for cross sections at stations 109 and 110. The width of the roadbed is 24 ft, and the side slopes are 2 to 1. Plot these two cross sections at a scale of $1:120$ (1 in per 10 ft). Compute the areas of the two sections.

| Station | Cross section | | |
|---|---|---|---|
| 109 | $\dfrac{c2.4}{16.8}$ | $\dfrac{c1.2}{0.0}$ | $\dfrac{c0.4}{12.8}$ |
| 110 | $\dfrac{c12.2}{36.4}$ | $\dfrac{c9.2}{0.0}$ | $\dfrac{c4.8}{21.6}$ |

**17.14** Plot the preliminary cross section for station 44 + 00 in Fig. 17.31 using a horizontal scale of 1 : 200 and vertical scale of 1 : 100. Assuming a roadbed width of 10 m, side slopes of 2 to 1, and a center-line grade elevation of 207.300 m, plot the roadbed templet and compute the cross-section area by the coordinate method.

**17.15** Using the roadbed templet specifications given in Prob. 17.14, compute the cross-section area in square metres for the data of station 43 + 00 in Fig. 17.31.

**17.16** Compute the volume in cubic metres between stations 43 + 00 and 44 + 00 of Probs. 17.14 and 17.15 by the average-end-area method.

**17.17** The width of roadbed for a highway is 8 m with side slopes of 2 to 1; the grade elevation at station 48 + 00 is 345.40 m, and the center-line elevation is 344.29 m. The H.I. for obtaining the final cross section at this station is 346.940 m. The rod readings at the right and left slope stakes are 4.60 and 7.10 m, respectively. Compute the grade rod, the center-line cut or fill, and the cut or fill and distance to each slope stake.

**17.18** Compute the volume in cubic yards between stations 109 and 110 of Prob. 17.13. Use both the average-end-area method and the prismoidal formula. Note the discrepancy expressed as a percentage between volumes as determined by the two methods.

**17.19** What error in volume between station 109 and station 110 of Prob. 17.24 would be introduced if the recorded cuts at centers and slope stakes were 0.1 ft too great? What is the error in terms of percentage of the volume by average end areas?

**17.20** Given: a roadway width of 8 m, side slopes of 1.5 to 1, grade elevation at station 102 + 00 = 642.156 m, and ground elevation at the center line of 644.38 m. A theodolite is set over the center line station 102 + 00 with an h.i. = 1.45 m. The vertical angle and slope distance to a point 1.45 m above the ground on a rod held at the approximate location of the slope stake are + 14°40′ and 11.50 m, respectively. In order to set the slope stake in the proper position, should the rod be moved toward or away from center line, and by about how much?

**17.21** Assume that the center line for the cross-sectional data in Prob. 17.13 turns to the right with a simple circular curve having $D_a = 10°$ (arc definition). Compute the curvature correction in cubic yards to the earthwork volume between these two stations and the percentage of the total volume of earthwork which this correction constitutes.

**17.22** Compute the volume in cut and in fill for the given cross sections between stations 62 and 64. The roadbed is 24 ft wide in cut and 20 ft wide in fill, and the side slopes are $1\frac{1}{2}$ to 1. Tabulate the data in the following form: "Station," "Cross section," "Area," and "Volume." Use the prismoidal formula.

| Station | Cross section | | | |
|---|---|---|---|---|
| 62 | $\dfrac{c2.6}{15.9}$ | $\dfrac{c4.8}{0.0}$ | $\dfrac{c6.4}{21.6}$ | |
| 63 | $\dfrac{0.0}{12.0}$ | $\dfrac{c3.1}{0.0}$ | $\dfrac{c4.4}{18.6}$ | |
| 63 + 25 | $\dfrac{f4.6}{16.9}$ | $\dfrac{0.0}{0.0}$ | $\dfrac{c2.6}{15.9}$ | |
| 64 | $\dfrac{f7.2}{20.8}$ | $\dfrac{f4.8}{0.0}$ | $\dfrac{0.0}{6.0}$ | $\dfrac{c1.8}{14.7}$ |

**17.23** Solve Prob. 17.22 by the method of average end areas, computing the volume of pyramids by the relation, volume = $\frac{1}{3}$ (area of base times length).

**17.24** In plan, a borrow pit is 75 by 135 ft. Before and after excavation, levels are run and offsets are measured from stations along one of the 135-ft sides. The computed cuts are shown in the following table:

| | Cut, ft | | | | | | |
|---|---|---|---|---|---|---|---|
| Offset | Sta. 0 | Sta. 0 + 30 | Sta. 0 + 50 | Sta. 0 + 75 | Sta. 1 + 00 | Sta. 1 + 15 | Sta. 1 + 35 |
| 0 | 0.0 | 1.5 | 0.0 | 4.5 | 6.2 | 4.7 | 0.0 |
| 25 | 1.2 | 2.9 | 10.6 | 9.7 | 7.9 | 8.4 | 2.5 |
| 50 | 2.5 | 3.7 | 8.7 | 8.7 | 9.4 | 8.4 | 3.6 |
| 75 | 0.0 | 0.0 | 1.9 | 7.6 | 6.8 | 6.3 | 0.0 |

Compute the volume of excavation in cubic yards using the method outlined in Art. 17.41.

**17.25** Compute the volume in cut and fill in cubic metres for the given cross sections (all distances and elevations in metres) between stations 10 + 00 and 10 + 65. The roadbed is 8 m wide in cut and 6 m wide in fill and the slopes are 1 to 1.5. Tabulate as directed in Prob. 17.22. Use the method of average end areas.

| Station | Cross section | | |
|---------|---------------|---|---|
| 10 + 00 | $\dfrac{c2.00}{7.00}$ | $\dfrac{c1.50}{0.00}$ | $\dfrac{c0.80}{5.20}$ |
| 10 + 30 | $\dfrac{c1.60}{6.40}$ | $\dfrac{c1.00}{0.00}$ | $\dfrac{0.00}{4.00}$ |
| 10 + 40 | $\dfrac{c0.85}{5.28}$ | $\dfrac{0.00}{0.00}$ | $\dfrac{f1.40}{5.10}$ |
| 10 + 65 | $\dfrac{c1.0}{5.50}$ | $\dfrac{0.00}{2.00}$ | $\dfrac{f1.62}{0.00}$ | $\dfrac{f2.54}{6.81}$ |

**17.26** In Fig. 17.28, the area enclosed by $F1$, $A1$, $A4$, $D4$, and $F2$ is excavated to a grade of 120 m above mean sea level. Assuming vertical slopes for all sides of the excavation, compute the volume removed in cubic metres.

## 17.48 Field Problem

### LOCATION OF A SIMPLE CURVE BY DEFLECTION ANGLES

**Object** To calculate and stake a simple circular curve in the field by the method of deflection angles (Art. 17.7). Suggested parameters for the curve are as follows: in the metric system $\Delta = 30°00'$, $D_a = 33°$, station at the P.I. $= 16 + 34.500$; and using the foot as the measuring unit, $\Delta = 30°00'$, $D_a = 10°$, station of the P.I. as given above.

**Procedure** Before the beginning of the field session, compute (a) length of curve; (b) radius; (c) tangent distance; (d) deflection angles from the P.C. to each half-station (one-tenth station for metric units); (e) chord lengths between the P.C. and the first curve station, between each half station (tenth stations when metric units are used) and between the last station on the curve and the P.T. All calculations are to be to the nearest hundredth of a foot (nearest millimetre).

In the field, the instructor will designate the P.I. and a point on the back tangent. Each party will occupy the assigned P.I. and locate the P.C. and P.T. according to the instructions given below. Each party should make one setup on an intermediate station near the midpoint of the curve to complete the curve layout. Consult the text (Art. 17.8) for procedures. The procedure for curve layout is:

1. Set the transit over the P.I. and take a backsight on the P.O.T. Measure the tangent distance from the P.I. and set a hub at the P.C. Measure to the nearest hundredth of ft (the nearest mm).
2. Lay off the direction of the forward tangent by turning a double deflection angle according to the method outlined in Art. 6.42.
3. Measure the tangent distance from the P.I. along the forward tangent to set a hub at the P.T. Measure to the nearest hundredth of ft (mm).
4. Occupy the P.C. with the transit. With the $A$ vernier or horizontal index mark on zero, sight the P.I.
5. Set stations by deflection angles and chord distances.
6. About halfway along the curve, set a hub to mark the curve station. (This point will be designated by the instructor.) Occupy this point with the transit; set the $A$ vernier or horizontal index mark to zero, invert the telescope, and backsight on the P.C. using the lower motion. Replunge the telescope to its direct position and continue setting curve stations.
7. Measure the distance error and offset error at the P.T. Record discrepancies in the field book.
8. Record data for curve and sketch the curve in your field book. Use the note form illustrated in Fig. 17.10.

## References

1. American Association of State Highway Officials, *A Policy on Geometric Design of Rural Highways* (Blue book), Washington, D.C., 1966.

2. American Association of State Highway Officials, *A Policy on Design of Urban Highways and Arterial Streets* (Red book), Washington, D.C., 1973.

3. California Division of Highways, *Highway Design Manual of Instructions*, Sacramento, Calif., 1972.

4. Hurdle, V. F., "Vertical Curves," unpublished notes, University of California, Institute of Transportation and Traffic Engineering, Berkeley, Calif., 1974.

5. Leeming, J. J., "Road Curvature and Superelevation," *Survey Review*, January 1973.

6. Meyer, C. F. and Gibson D. W., *Route Surveying*, 5th ed., Harper & Row, Publishers, New York, 1980,

7. Moffitt, F. H., and Bouchard, H., *Surveying*, 6th ed., Harper & Row, Publishers, New York, 1975.

8. Pryor, W. T., "Metrication," *Surveying and Mapping*, Vol, 35, No. 3, September 1979, pp. 229–237.

9. Sanyaolu, A., "New Analytic Formulas and Automatic Checks for Earthwork," *The Canadian Surveyor*, Vol. 32, No. 2, June 1978.

10. State of California, Business and Transportation Agency, Department of Transportation, *Action Plan for Transportation Development*, Sacramento, Calif., 1973.

11. State of California, Business and Transportation Agency, Department of Public Works, Division of Highways, *Surveys Manual*, Sacramento, Calif., 1976.

12. Weisner, P. C., "Highway Earthwork Computer Program, User's Manual," unpublished report for special studies under direction of Professor P. R. Wolf, University of California at Berkeley, 1970.

# CHAPTER 18
# Construction surveying

**18.1. General** Surveys for construction involve the following: (1) establishing on the ground a system of stakes or other markers, both in plan and in elevation, from which measurement of earthwork and structures can be taken conveniently by the construction force, (2) giving line and grade as needed either to replace stakes disturbed by construction or to reach additional points on the structure itself, and (3) making measurements necessary to verify the location of completed parts of the structure (the as-built survey) and to determine the volume of work actually performed up to a given date (usually each month), as a basis of payment to the contractor.

Prior to construction, a topographic survey (Chap. 15) of the site is performed and maps are prepared to be used in the development of plans for the project. As soon as approval of the project is assured, property line surveys are initiated to be used for acquisition of lands or rights of way. The control network established for these topographic and property surveys contains many of the horizontal and vertical control points which will eventually form the basis for subsequent construction surveys. Consequently, the survey engineer in charge of planning the surveys has the responsibility to organize the initial basic control surveys so as to provide the maximum number of horizontal and vertical control points which are useful not only for topographic and property surveys but also for the construction surveys that follow.

New developments in equipment, such as electronic surveying systems and laser equipped alignment instruments (Art. 18.5), provide powerful tools which have many applications in construction surveys. Although the application of these new devices does result in somewhat altered operations, the basic ideas concerning construction surveying as outlined above remain the same.

The detailed methods employed on construction surveys vary greatly with the type, location, and size of structure and with the preference of the engineering and construction organizations. Much depends on the ingenuity of the surveyor so that the correct information is given without confusion or needless effort. Consequently, the discussions that follow are of a general nature with a minimum number of specific examples and procedures set forth.

**18.2. Alignment** Temporary stakes or other markers are usually set at the corners of the proposed structure, as a rough guide for beginning the excavation. Outside the limits of excavation or probable disturbance but close enough to be convenient, are set permanent stations which are established with the precision required for the measurement of the structure itself. These permanent stations should be well referenced (Art. 8.14), with the reference stakes in such number and in such position that the loss of one or two will not invalidate any portion of the location survey. Permanent targets or marks called *foresights* may be erected as convenient means of orienting the transit on the principal lines of the structure and for sighting along these lines by eye.

Stakes or other markers are set on all important lines in order to mark clearly the limits of the work. The number of such markers should be sufficient to avoid the necessity for many measurements by the workmen but should not be so great as to cause confusion. A simple and uniform system of designating the various points, satisfactory to the construction superintendent, should be adopted. Also, the exact points, lines, and planes from which and to which measurements are to be made should be well understood.

In many cases, line and grade are given more conveniently by means of *batter boards* than by means of stakes. A batter board is a board [usually 1 by 6 in (2.5 by 15 cm)] nailed to two substantial posts [usually 2 by 4 in (5 by 10 cm)] with the board horizontal and its top edge preferably either at grade or at some whole number of feet above or below grade. The alignment is fixed by a nail driven in the top edge of the board. Between two such batter boards a stout cord or wire is stretched to define the line and grade. Batter boards for sewer construction are shown in Fig. 18.1. Note that the string line provides the reference line for alignment and grade between the surveyed stations.

Laser-equipped instruments, such as those described in Art. 18.5, provide a means of establishing a laser beam which can be used in the same way as the string line shown in Fig. 18.1.

Often it is impractical to establish permanent markers on the line of the structure. Thus, the face of a bridge abutment may be beyond the shore line and therefore inaccessible. Also,

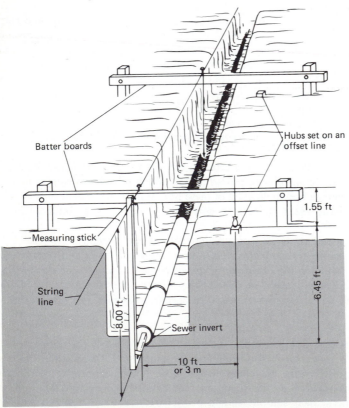

**Fig. 18.1** Batter boards for sewer construction.

stakes placed at the edge of a concrete pavement would interfere with grading and with setting the forms. In such cases the survey line is established parallel to the structure line, as close as practical and with the offset distance some whole number of feet, decimetres, or metres.

**18.3. Grade**   A system of bench marks is established near the structure in convenient locations that will probably not be subject to disturbance, either of the bench mark itself or of the supporting ground. From time to time these bench marks should be checked against one another to detect any disturbance. Every care should be taken to preserve existing bench marks of state and federal surveys; if construction necessitates the removal of such marks, the proper organization should be notified and the marks transferred in accordance with its instructions.

The various grades and elevations are defined on the ground by means of stakes and/or batter boards, as a guide to the workers. The grade stakes may or may not be the same as the stakes used in giving line. When stakes are used, the vertical measurements may be taken from the top of the stake, from a crayon mark or a nail on the side of the stake, or (for excavation) from the ground surface at the stake; in order to avoid mistakes, only one of these bases for measurement should be used for a given kind of work, and the basis should be made clear at the beginning of construction. When batter boards are used, the vertical measurements are taken from the top edge of the board, which is horizontal. The stake or the batter board may be set either at grade or at a fixed whole number of feet (metres) above or below grade.

On some jobs, grade stakes consisting of hubs driven flush with the ground are set, elevations are obtained on the tops of the hubs, and cuts or fills to finished grade are calculated in whole and fractions of units from the top of the hub. These hubs are also referred to as grade stakes and are usually guarded by a flat stake. It is then the responsibility of the construction engineer to set batter boards or other alignment devices as desired using the given stakes, cuts, and fills.

Use of instruments equipped with lasers to establish grades is discussed in Art. 18.5.

**18.4. Alignment and grade by electronic surveying systems**   When construction layout is performed using transit or theodolite and tape, alignment is usually established in the field by direct measurement from the highway center line, the exterior dimensions of a structure, or from the property line. This means that center line, side line, or property line must be established prior to setting offset stakes or batter boards. This procedure has definite advantages in that it provides a check on office calculations and furnishes proof that the structure as designed is consistent with basic control points in the field. Also, the process for setting offset stakes or batter boards from center line or building corners is straightforward, involves right-angle offsets or parallel lines, and provides ample visual checks for the validity of the layout. The disadvantages are that the procedure is time-consuming, expensive, and during initial stages of construction (e.g., rough excavation) most of the points set will be lost and will have to be replaced for the final layout.

Use of electronic surveying systems with direct readout of distance and a tracking mode (Art. 12.2) in conjunction with coordinate geometry make construction layout from randomly located control points feasible. In order to perform this operation, horizontal and vertical control points are positioned at locations appropriate for layout. These locations can be outside the area of construction, but must provide good visibility for construction layout. All random control points must be coordinated and be part of a closed survey network. It is also advantageous to have elevations on these control points. The accuracy maintained is a function of the type of work being done.

The points to be staked must also be coordinated. Thus, if stakes on right-angle offsets from the corner of a building or offsets from the catch point in slope staking (Art. 17.30) are to be set, the offset point is coordinated. Next, the inverse solution between the random control point and coordinated offset point is computed to yield distance and direction (azimuth or bearing). Most surveying organizations have standard electronic computer programs for this computation, given the coordinates for both ends of the line. The minimum amount of data provided for field operations consists of azimuth or bearing and horizontal distance from the random control point to the offset point. When offset stakes are to be set for alignment and grade (as for slope stakes in highway excavation), azimuth or vertical angle, slope distance, horizontal distance, and elevation of the offset point are given. Data are usually precalculated and furnished to field personnel in the form of a hard-copy listing.

In the field, the operation consists of occupying the designated random control point, taking a backsight on an intervisible random control station, turning off the direction, and measuring the distance to set the specified offset stake. Ideally, a series of offset stakes are set from a single well-located random control point. To verify calculations and field work, the locations of the stakes as set are checked by measuring distance and direction from another random control point or by measurements between offset stakes on the site.

Obviously, the locations for random control stations must be chosen carefully so as to achieve maximum benefits from this method. In preparation for layout from random stations, enough inverse data must be generated from several control stations as to provide for checking and to guard against lack of visibility along certain lines. In case insufficient inverses are generated to complete the job, ordinary or programmable pocket calculators provide enough computing power to allow calculation of the necessary inverses on the site.

The choice as to whether construction layout should be performed in the traditional manner or from random controlling points depends on many factors related to characteristics of the job and site. These factors are generally unique for every project. In all probability a combination of both methods will be used on most projects. For example, in highway construction, slope stakes may be set from random control points while the final grade stakes are established from the center line, which of course is set from random control points. Examples of setting slope stakes from random control points are given in Art. 18.7.

**18.5. Laser leveling and alignment equipment for construction layout**   As noted in Art. 4.29, laser beams are used in certain EDM instruments as the carrier beam for modulated light in distance measurement, thus increasing the range of these instruments substantially. Laser light is of a single color or *monochromatic*, is *coherent* (i.e., the light waves are in step with each other), and is highly collimated, so that the beam spreads only very slightly as the distance from generator to target increases. These characteristics make the laser a useful device for surveying instruments to be utilized in various types of construction layout.

As an example of the use of lasers in surveying instruments, consider application of a helium neon laser or gas laser to an ordinary theodolite or transit (Ref. 13). The laser is a separate unit equipped with a portable power supply. This laser can be strapped to the leg of the tripod. The parallel rays of the beam emitted by the laser are focused by an objective lens down to a very small diameter (0.08 mm) for transmission by a flexible fiber optic cable to the telescope of the theodolite equipped with a special eyepiece attachment. Within the telescope the light is focused on a reticule which masks out a cross. The laser beam then passes through a beam-splitting cube which directs the laser along the optical axis of the telescope. The surface of the beam splitting cube, visible through the eyepiece, contains a second reticule, designed to permit the telescope to be used as usual under good lighting conditions. A filter is installed to absorb any laser light directed toward the eyepiece. The

**Fig. 18.2** Kern DKM 2 theodolite equipped with a gas laser. (*Courtesy Kern Instruments, Inc.*)

beam directed toward the telescope objective is a narrow beam containing a projected cross which may be brought to a focus on a target by focusing the telescope on the target.

Figure 18.2 shows a theodolite equipped with a gas laser, power supply, and eyepiece attachment. Figure 18.3 is a schematic drawing which illustrates the way in which the laser is projected into the telescope using the attachment shown in Fig. 18.2.

Laser eyepieces, now manufactured as accessories, can also be attached to levels and the laser plummet. A laser plummet which projects a vertical laser beam up and down is illustrated in Fig. 18.4.

The system just described and shown in Fig. 18.2 is a *single-beam* laser that projects a visible "string line" which can be seen on targets under all lighting conditions. Thus, if a sight is taken on a distant point in the conventional manner, alignment can be achieved at any intermediate point by moving a special target about until the projected cross is centered on the target. This alignment can be established at any intermediate point without instructions from an operator behind the instrument.

The laser beam can be projected at any inclination or may be used as a level beam of light. For leveling, the graduated rod would have a movable target equipped with a vernier that is moved up or down until the beam or projected cross is centered on the target. The

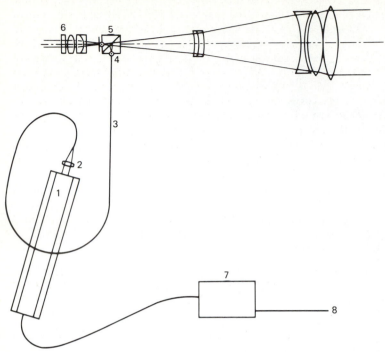

**Fig. 18.3** Schematic of the transmission of laser light to the theodolite: (1) laser light source; (2) objective; (3) light tube (glass fiber optics); (4) reticule; (5) light-splitting cube; (6) filter; (7) laser power converter; and (8) power source. (*Courtesy Kern Instruments, Inc.*)

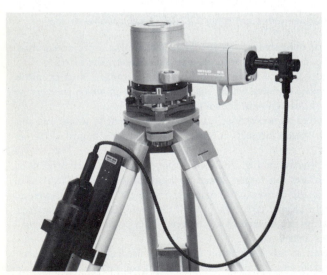

**Fig. 18.4** Laser plummet, Wild ZL with eyepiece attachment GLO2. (*Courtesy Wild Heerbrugg Instruments, Inc.*)

operator of the rod would then read the rod to the least count of the scale. The range of the laser attachment shown in Fig. 18.2 is from 10 to 400 m (30 to 1300 ft) with a maximum range of 300 m (1000 ft) under moderate dust conditions as might be common in mining and tunneling operations. Naturally, the range of the laser beam is a function of the power of the laser. Using an output beam with a diameter of 1.7 mm (0.005 ft), the estimated standard deviation in alignment is about 6 to 10 seconds of arc or a displacement of from 6 to 10 mm (0.019 to 0.033 ft) in 200 m (~650 ft). This accuracy is most easily achievable by utilizing a laser beam with a dark spot or cross in the center of the beam for more precise centering of the target. A stable instrument setup is extremely important and therefore whenever possible permanent stands with built-in centering plates or trivets should be used.

The system described is quite elementary but serves to illustrate the principles involved and application of laser technology to existing equipment.

Experiments have been performed using prototypes of laser-equipped theodolites and levels for high-precision alignment and leveling. The reader should consult Refs. 3 and 4, listed at the end of the chapter, for details.

As laser technology developed, an impressive array of special purpose laser-equipped instruments has become available. Most of these instruments have been designed for construction layout and can be classified into two general groups:

1. Instruments which project single beams that are visible on targets under all lighting conditions. These devices are usually employed for horizontal and vertical linear alignment. Included in this group are transits or theodolites equipped with lasers (Fig. 18.2), lasers for alignment of tunneling machinery, and lasers for alignment of pipes and drains.
2. Instruments in which the laser beam is rotated by rapidly spinning optics so as to provide a reference plane in space over open areas and trace reference lines which are visible indoors. The speed of rotation can be varied and at zero revolutions per minute a single beam is available.

A single-beam laser alignment instrument is shown in Fig. 18.5. This laser is mounted on a transit-like framework which has horizontal and vertical motions, a spirit level parallel to the

**Fig. 18.5** Laser alignment instrument. (*Courtesy Spectra-Physics, Inc.*)

axis of the laser, a vertical circle (not visible in figure), and a horizontal circle. Horizontal and vertical circles are graduated in full degrees with double, direct-reading verniers having a least count of 5 minutes. Mounted on top of the laser enclosure is a telescope which allows the operator to sight the location of the transmitted laser spot. The spot diameter of the beam is 0.016 m (0.053 ft), at the laser, and due to spread becomes 0.062 m at 1 km (0.33 ft at 1 mi), and 0.50 m at the maximum range of 8 km (1.67 ft at 5 mi). This range is effective in mild haze and increases somewhat at night.

A fanning lens can be attached to the emission end of the laser housing. This attachment allows the beam spot to be converted into a horizontal or vertical line beam shape (Fig. 18.6). The dimensions of this beam are 0.062 m by 161 m at a distance of 1 km (0.33 ft by 850 ft in 1 mi), increasing to 0.50 m by 1288 m at 8 km (1.67 ft by 4250 ft in 5 mi). This instrument can be used for horizontal and vertical control of dredging operations, control of tunneling machines, and azimuth and grade control for pipe-laying operations.

A self-leveling, rotating laser is shown in Fig. 18.7. In this instrument, the laser unit is mounted vertically on a platform containing two orthogonally mounted sensors which act

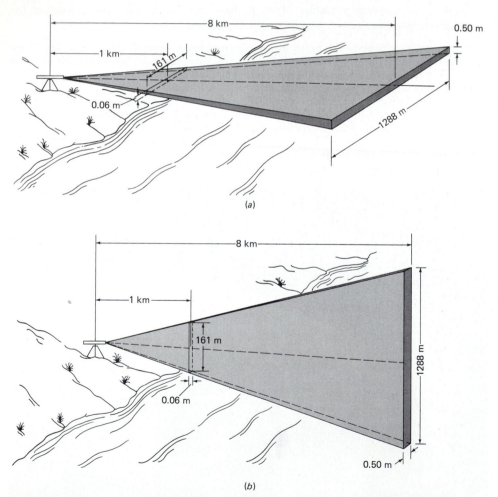

(a)

(b)

**Fig. 18.6** Horizontal and vertical line beams produced by fanning a laser beam. (*Courtesy Spectra-Physics, Inc.*) (a) Horizontal line beam. (b) Vertical line beam.

like spirit levels and deviate from center when the platform is not level. The amount of deviation is detected electronically and the consequent electrical impulses drive servomotors which automatically level the base and make the axis of the laser vertical. The laser beam is emitted at an angle of 90° to the axis of the laser by an optical train and the optics rotate to form a horizontal reference plane. This device can also be side-mounted so that the axis of the laser is in a horizontal position and a vertical plane can be formed by the rotating beam. There is also an electronic sensing device parallel to the axis of the laser to allow self-plumbing of the rotating beam. The instrument is self-leveling and self-plumbing within a range of 8°. The tolerance specified for the position of the reference plane with respect to

**Fig. 18.7** Rotating laser level. (*Courtesy Spectra-Physics, Inc.*)

true level or true vertical is 20 seconds of arc. Thus, in a distance of 100 m (330 ft) from the instrument, a deviation of 10 mm (0.03 ft) is possible.

A laser rod equipped with a laser detector (Fig. 18.8) is used in conjunction with the rotating laser. This rod contains a sliding battery-powered sensor on the front face of the rod. When within 0.14 m (0.45 ft) above or below the rotating laser beam, this sensor "locks on" to the beam and emits a "beep" which indicates that a reading should be taken. The operator then reads the rod directly to the nearest 5 mm (0.01 ft) at an index on the sensor which allows reading the scale on the front of the rod. There are two modes for the sensor. The *lock* mode means the sensor will seek the beam, average the position of the beam, and then lock onto it, giving a beep so that the operator may read the scale. The *float* mode enables the sensor to fix on the laser beam and continue reading the beam as the rod is moved up and down. The lock mode is used for normal leveling and determination of elevations or position. The float mode is useful when forms or stakes must be adjusted. The sensor is controlled by a mode switch at the top of the rod.

**Grade and alignment by laser equipped instruments**   The basic idea in the application of laser devices to construction layout is that the laser plane replaces the horizontal line of sight of the engineer's level and the laser beam replaces the string line. This plane or beam can be located in space at any intermediate point by using a small plastic target or an electronic laser detector which automatically seeks the laser plane or beam when it is within a specified range. Consequently, no operator must be stationed at the instrument when it is desired to get on line or obtain a rod reading.

When using a horizontal laser beam or rotating laser beam to form a horizontal reference line or plane, the operation of setting grade stakes to a given elevation is the same as described in Art. 18.3 except that there is no need for instructions from the operator of the instrument. The horizontal laser beam is visible as a red dot on a target which slides up and

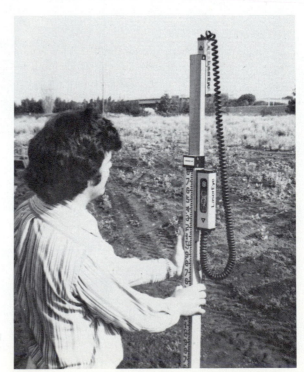

**Fig. 18.8** Laser level rod equipped with a laser detector. (*Courtesy Spectra-Physics, Inc.*)

down along the face of the rod. The horizontal laser plane is detected by a sliding sensor on the face of the rod which emits a shrill sound or beep when the sensor index mark coincides with the laser plane. Thus, the operator of the rod must know what rod reading is required to set the grade stake and then can set the stake or form accordingly without assistance from the other end of the line at the instrument. When the laser beam is used as a substitute for the string line, the laser transmitting device is set over a control point at one end of the line and sighted at a control point at the other end of the line. The laser beam then serves the same function as the string line described in Art. 18.2.

Another possibility involves attaching the laser target to the machine which is grading, paving, tunneling, etc. The operator of the machine can observe the target and by keeping the visible red dot centered on the target can maintain the machine on the proper alignment and grade. A more sophisticated approach to this operation consists of attaching an electronic laser detector to the machine. In this way, the machine can be guided automatically in the proper direction.

A simple example consists of the grading operations for an area which is to be graded horizontally at a specified elevation. Assume that the laser level of Fig. 18.7 is set up near the area to be graded, permitted to level, and a reading is taken on a nearby bench mark. This reading establishes the height of the laser level above the datum (H.I.) so that the cut to the given elevation desired for the area can be calculated. Assume that this cut is 6.05 ft (1.84 m), so that the blade of the grader must be adjusted to 6.05 ft below the laser plane. Special graduated devices which support laser detectors and can be attached to various types of machinery are available for this purpose. Figure 18.9 shows a laser level mounted on a tripod and a power grader with laser detectors mounted at each end of the blade. When the laser detector is within about 0.5 ft (0.15 m) of the beam, it will automatically energize display lights in the cab which indicate low, on beam, or high positioning of the blade. There is also a pointer display to show how much adjustment is needed to get the detector on the beam. When the on-beam light is illuminated, the cutting edge of the blade is on grade. The same principles can be used to guide tunneling, paving, and ditch-digging machinery.

**Fig. 18.9** Grading machine controlled by laser level and laser detector. (*Courtesy Spectra-Physics, Inc.*)

There is a precaution (in addition to the usual steps to ensure safety) which must be observed when using any of the laser alignment devices. Theoretically, one of the advantages of these types of systems is that once the laser transmitter is set up and aligned, no operator needs to stand by the instrument. However, laser transmitters are subject to the same disturbances that cause conventional optical alignment telescopes to become misaligned. Settlement, vibration, and accidental bumping are common factors around any construction job which can cause a perfectly aligned laser beam to be deflected. Consequently, periodic checks of the laser-beam alignment are absolutely necessary. Such checks can consist of having a party member check the alignment device at regular intervals or the checking device may be built into the system. The laser level illustrated in Fig. 18.7 shuts off when the laser beam deflects from horizontal. This constitutes a built-in check on the system.

Laser alignment devices and systems have many applications, particularly in construction layout. However, it should be emphasized that many of these devices are not general-purpose surveying instruments having the flexibility in application of an engineer's transit or theodolite (with the exception of the laser attached to a theodolite shown in Fig. 18.2). Consequently, laser systems are more likely to be useful to contractors with a high volume of construction layout operations.

Various applications of laser alignment devices are described in the sections that follow.

**18.6. Precision**   For purposes of excavation only, usually elevations are given to the nearest 0.1 ft (3 cm). For points on the structure, usually elevations to 0.01 ft (3 mm) are sufficiently precise. Alignment to the nearest 0.01 ft (3 mm) will serve the purpose of most construction, but greater precision may be required for prefabricated steel structures or members.

It is still desirable to give dimensions to the workers in feet, inches, and fractions of an inch. Ordinarily, measurements to the nearest $\frac{1}{4}$ or $\frac{1}{8}$ in (6 or 3 mm) are sufficiently precise, but certain of the measurements for the construction of buildings and bridges should be given to the nearest $\frac{1}{16}$ in (1.5 mm). Often, it is convenient to use the relation that $\frac{1}{8}$ in equals approximately 0.01 ft ($\sim$3 mm).

**18.7. Highways**   As a general rule, numerous horizontal and vertical control points are available along a highway right of way prior to construction. These control points are set during the planning and design stages for the highway and usually provide a good basic network of control points for construction surveys.

There are two basic approaches to highway construction surveys. These two approaches consist of layout from (1) center-line control, and (2) random control stations. The method chosen depends on terrain configuration, amount of vegetation present, equipment available, construction requirements, and other factors unique to the particular project.

*1. Center-line control*   Usually, just prior to the beginning of construction of a section of highway, the center line is rerun as a base line for construction, missing stakes are replaced, and hubs are referenced. Borrow pits (if necessary) are staked out and cross-sectioned. Lines and grades are staked out for bridges, culverts, and other structures. Slope stakes are set except where clearing is necessary; in that case they are set when the right of way has been cleared. For purposes of clearing, only rough measurements from the center-line stakes are necessary.

The method of setting slope stakes is described in detail in Art. 17.30 for the case where an engineer's level, tape, and rod are used to set the stakes at the catch points. When modern equipment is available, it has become common practice to set slope stakes by measuring slope distances and determining differences in elevation by trigonometric leveling. This procedure is particularly efficient when combined theodolite and EDM instruments (Art.

12.2) are available. The general details of this procedure are also given in Art. 17.30 and are illustrated by Fig. 17.44. When this method is employed, the field party is frequently provided with an electronic computer listing which contains roadbed templet notes, slope ratios, and theoretical catch point or slope stake locations.

A typical set of slope stake notes derived from a computer listing for field layout, is shown in Fig. 18.10a. The data in the computer listing could have been determined from field notes, topographic map, or from photogrammetric cross sections. In this particular example, the data were derived photogrammetrically. Figure 18.10b shows the plotted cross section for station 46+00. At this station the grade at center line is 220.48 and the cut at center line is 15.52 ft. The center line of the ditch 35 ft right has a grade elevation of 218.88 ft and the catch point 65.23 ft right has an elevation of 228.95 ft. The slope stake data are given in the column farthest to the right where the $C = (30.23)(0.333) = 10.07$ ft is for a point 30.23 ft to the right of the center line of the ditch (see Fig. 18.10b). This catch point is calculated by the program as the intersection of a 3 to 1 slope (0.333) and the terrain surface from the photogrammetric cross section. Similar data are given for the left side of the cross section. The hand-lettered entries in Fig. 18.10a are inserted during field operations.

In the field, the center-line station is occupied, the direction of the cross-section line is established, and an offset stake or reference point (R.P.) is set 80.23 ft right of center line or 15 ft from the catch point. In this way the reference point is outside the area of construction and will not be destroyed. The elevation of the reference point is determined trigonometrically as being 227.4 ft. Thus, the catch point is 1.5 ft higher and 15 ft toward the center line from the R.P. (see Fig. 18.10b). These data concerning offset distance and elevation of the

| Slope stake left | | | Catch point left | Center line ditch | Center line | Center line ditch | Catch point right | | | Slope stake right |
|---|---|---|---|---|---|---|---|---|---|---|
| | *R.P.* | *R.P.* | | | C15.52 | | | *R.P.* | *R.P.* | |
| C30.19 @ 90.57 | *El. 251.1* | *-2.0* | 249.07 | 218.88 | 220.48 | 218.88 | 228.95 | *+1.5* | *El. 227.4* | C10.07 @30.23 |
| s = 0.333 | *₵ 140.6* | *15.0* | 125.57 | 35.00 | 0.00 | 35.00 | 65.23 | *15.0 ₵* | *80.2* | s = 0.333 |
| | | | | | 46 + 00 | | | | | |
| | | | | | C23.34 | | | | | |
| C36.16 @ 108.47 | | | 254.22 | 218.06 | 219.66 | 218.06 | 235.52 | | | C17.46 @ 52.37 |
| s = 0.333 | | | 143.47 | 35.00 | 0.00 | 35.0 | 87.37 | | | s = 0.333 |
| | | | | | 45 + 50 | | | | | |

(a)

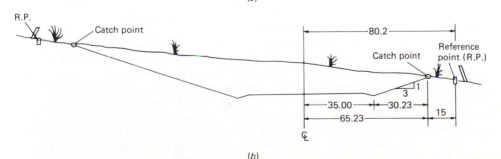

(b)

**Fig. 18.10** (a) Listing of slope stake notes derived from computer output for field layout. (b) Plotted cross section for station 46 + 00.

R.P. and distance and difference in elevation from R.P. are entered on the listing by hand in the field, as shown in Fig. 18.10a. The difference in elevation and offset distance between catch point and R.P. (+1.50/15) are also marked on the center-line side of the face of a stake set to guard the R.P. hub. Next, the position of the catch point is set and the elevation is checked against that given on the computer listing. If the elevation of the catch point as determined in the field does not agree with the theoretical location given in the listing, the true position of the catch point is determined by trial according to the methods given in Art. 17.30 and the reference point and slope stake data are adjusted accordingly. Any changes should be shown directly on the listing.

Where the depth of cuts and fills does not average more than about 3 ft or 1 m, the slope stakes may be omitted; in this case the line and grade for earthwork may be indicated by a line of hubs (with guard stakes) along one side of the road and offset a uniform distance such that they will not be disturbed by the grading operations. Hubs are usually placed on both sides of the road at curves, and may be so placed on tangents; when this is done, measurements for grading purposes may be taken conveniently by sighting across the two hubs or by stretching a line or tape between them.

When rough grading has been completed, a second set of grade stakes is set on both sides of the roadway. It should be understood whether the cuts and fills from the tops of these stakes are to subgrade or to finished grade (final grade). If the slopes of cuts are terraced to provide drainage, finishing stakes are set along the terraces.

In order to give line and grade for the pavement of concrete highways, a line of stakes is set along each side, offset a uniform distance (usually 2 ft or 1 m) from the edge of the pavement. The grade of the top of the pavement, at the edge, is indicated either by the top of the stake or by a nail or line on the side of the stake. The alignment is indicated on one side of the roadway only, by means of a tack in the top of each stake. The distance between stakes in a given line is usually 100 or 50 ft (30 or 10 m) on tangents at uniform grade and half the normal distance on horizontal or vertical curves. The dimensions of the finished subgrade and of the finished pavement are checked by the construction inspector, usually by means of a templet.

**2. Random control stations**    Construction layout from random control stations is described in Art. 18.4. Three procedures are possible: (a) staking directly from random control stations, (b) establishing a random line along or near the work area tied into existing random control points, and (c) establishing the center line or a line offset from the center line tied to existing random control points.

For each of these procedures, the order of establishing offset stakes for grading is the same as the center-line method described in the first paragraph of this section. It is the method of establishing the positions of the stakes that differs.

As an example, consider staking directly from random control points. The coordinates for all control points, reference points, and catch points are calculated. Inverses between conveniently located random control points and the reference points to the slope stakes are then calculated using an electronic computer program. A listing from the computer giving inverse data from random control point CM 131 to reference points to slope stakes (RPSS) for stations 130+00 through 133+00 is shown in Fig. 18.11. Given in this listing are the offset distance of the catch point (or slope stake) left or right of the center line, azimuth and zenith angle from the control point to the reference point, slope and horizontal distance from the control point to the reference point in metres and feet, and elevation and coordinates of the reference point. Figure 18.12 shows a plan view of the reference-point layout from control station CM 131 to reference points from slope stakes for stations 131 through 133.

Consider as an example the setting of the reference point for station 130+00 (140 ft left) using a combined theodolite and EDM. The theodolite is set on station CM 131 and a

```
13 OG 01 01                                                      DATE 04/17/80
TRANSIT AT    PT                                           ELEVATION     NORTH COORD      EAST COORD      PAGE 2
CM 131        02                                           1,228.944     550,190.755      1,999,254.589
```

| DESCRIPTION (POINT SIGHTED) | PT | AZIMUTH (NORTH) | ZENITH ANGLE | SLOPE DISTANCE (METERS) | (FEET) | HORIZONTAL DISTANCE (METERS) | (FEET) | ELEVATION | NORTH COORD | EAST COORD |
|---|---|---|---|---|---|---|---|---|---|---|
| CM 124 | 01 | 198 54 19 | 90 40 51 | 288.265 | 945.751 | 288.245 | 945.684 | 1,217.708 | 549,296.085 | 1,998,948.184 |
| CM 134 | 03 | 333 15 37 | 92 04 50 | 221.642 | 727.169 | 221.496 | 726.690 | 1,202.545 | 550,839.733 | 1,998,927.624 |
| 130+00 RPSS 140 LT | 04 | 237 22 34 | 92 01 30 | 223.793 | 734.226 | 223.653 | 733.767 | 1,203.000 | 549,795.165 | 1,998,636.590 |
| 130+00 RPSS 373 RT | 05 | 213 56 58 | 126 25 28 | 95.196 | 312.321 | 76.598 | 251.306 | 1,043.500 | 549,982.289 | 1,999,114.244 |
| 130+50 RPSS 145 LT | 06 | 240 39 01 | 91 59 17 | 223.551 | 733.433 | 223.416 | 732.992 | 1,203.500 | 549,831.487 | 1,998,615.681 |
| 130+50 RPSS 368 RT | 07 | 230 05 53 | 127 31 15 | 86.250 | 282.971 | 68.407 | 224.433 | 1,056.600 | 550,046.787 | 1,999,082.417 |
| 131+00 RPSS 152 LT | 08 | 243 52 27 | 92 09 55 | 225.427 | 739.589 | 225.266 | 739.061 | 1,051.000 | 549,865.313 | 1,998,591.039 |
| 131+00 RPSS 370 RT | 09 | 248 41 01 | 127 45 20 | 84.100 | 275.919 | 66.492 | 218.149 | 1,060.000 | 550,111.454 | 1,999,051.364 |
| 131+50 RPSS 154 LT | 10 | 247 00 46 | 92 35 38 | 227.253 | 745.577 | 227.020 | 744.813 | 1,195.200 | 549,899.885 | 1,998,568.920 |
| 131+50 RPSS 372 RT | 11 | 266 08 25 | 125 28 43 | 88.985 | 291.944 | 72.463 | 237.739 | 1,059.200 | 550,174.752 | 1,999,017.389 |
| 132+00 RPSS 133 LT | 12 | 250 30 10 | 91 23 13 | 223.446 | 733.088 | 223.380 | 732.873 | 1,211.200 | 549,946.149 | 1,998,563.741 |
| 132+00 RPSS 359 RT | 13 | 277 17 02 | 118 11 00 | 100.124 | 328.490 | 88.253 | 289.544 | 1,073.800 | 550,227.465 | 1,998,967.381 |
| 132+50 RPSS 107 LT | 14 | 254 21 59 | 89 45 27 | 220.150 | 722.274 | 220.148 | 722.267 | 1,232.000 | 549,996.116 | 1,998,559.042 |
| 132+50 RPSS 365 RT | 15 | 286 42 46 | 115 37 57 | 114.667 | 376.201 | 103.382 | 339.178 | 1,066.200 | 550,288.294 | 1,998,929.738 |
| 133+00 RPSS 82 LT | 16 | 258 24 48 | 88 08 56 | 219.440 | 719.944 | 219.325 | 719.568 | 1,252.200 | 550,046.229 | 1,998,549.684 |
| 133+00 RPSS 353 RT | 17 | 290 49 05 | 111 19 15 | 132.909 | 436.052 | 123.813 | 406.208 | 1,070.400 | 550,335.122 | 1,998,874.901 |

```
ZENITH ANGLES ARE CALCULATED TO A ROD READING = (HI + 0.000 FEET)

          TRANSIT AT         POINT         PAGE 2
          CM 131             02
```

Fig. 18.11 Computer listing for direction and distances from a random control station to reference points to slope stakes. (Courtesy California Department of Transportation.)

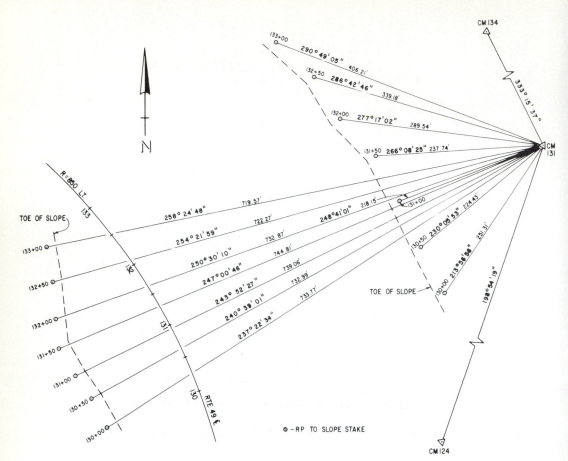

**Fig. 18.12** Slope staking from random control points. (*Courtesy California Department of Transportation.*)

backsight is taken on station CM 124. An angle of 237°22′34″ − 198°54′19″ = 38°28′15″ (Fig. 18.11) is turned to the right and a zenith angle of 92°01′30″ is set on the vertical circle. The reflector is placed on line at the approximate location of the reference point and with the EDM in the tracking mode, the reflector is moved until the EDM registers the slope distance of 734.226 ft (223.793 m) and a hub is set. The measurements are then repeated, including a backsight on CM 124, and the distance and elevation are determined to the hub just set. The field party should also have the slope stake listing (Fig. 18.13) so that the elevation of the catch point can be checked. The catch point is set and checked from the reference point using a tape, hand level, and rod. Note that data for the reference-point location is entered on the computer listing (Fig. 18.13) for station 130+00 on the left as RP El. = 1203.0@140 ₵.. The catch point is set by measuring 15 ft toward the center line and is found to be − 8.7 ft lower than the reference point. The field elevation of the catch point is 1203.0 − 8.7 = 1194.3, which checks the precomputed elevation for the catch point also given in the listing. As a rule for earthwork operations, this discrepancy should not exceed 5 ft or about 2 m. Reference points to slope stakes can be checked from a second instrument setup, at random control point CM 134 (Fig. 18.12), for example. However, if the elevations of reference and

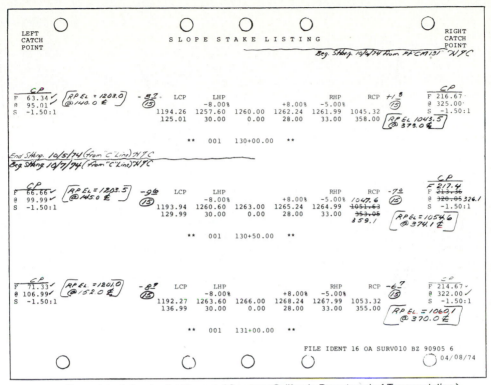

**Fig. 18.13** Slope stake listing for field layout. (*Courtesy California Department of Transportation.*)

catch points as determined in the field check the theoretical values in the computer listings (Figs. 18.11 and 18.13, respectively), these checks are assumed sufficient and a second setup is considered unnecessary.

Referring to Fig. 18.13, at station 130+50 the field measurements do not check the elevations and offsets in the listing (see the voided entries on the right). Consequently, the correct position of the catch point is determined in the field and adjusted values are entered in the listing as shown.

When a random line is established near the working area, it may be a line connecting two random control points or may be a series of supplementary points established between existing random control points. These supplementary points are used as angle points on the random line. The random line should be located so as to be as close as possible to the points to be set. The relationship between the center line and the random line must be determined and then all reference points to slope stakes are set from the stations on the random line. Figure 18.14 illustrates an example of a random line established to set reference points to slope stakes for a portion of curved center line for stations 129 through 134. As in the case of staking directly from random points, this procedure is simplified by electronic computer calculations to determine the distances and directions from the random line to the reference line. Slope stake listings and offset listings are used in the field to assist in the setting of the points. The method is adaptable to the use of combined theodolite EDM instruments and in general involves measuring shorter distances than when staking directly from the control stations.

When excavation is heavy (cuts greater than 30 ft or 10 m), intermediate slope stakes may

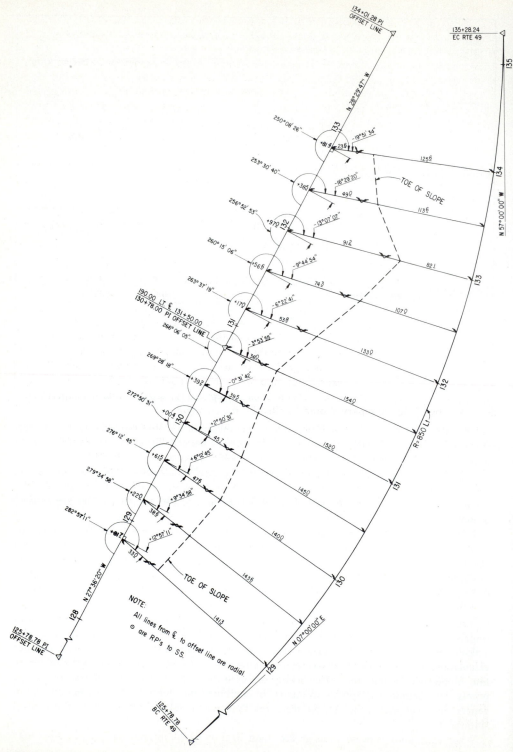

**Fig. 18.14** Slope staking from a random offset line. (*Courtesy California Department of Transportation.*)

be required after grading has been partially completed. These stakes can be set from random control points but are more conveniently established from a random offset line or line of constant offset (offset line parallel to the center line).

The center line of the alignment is established from ties to random control points when rough excavation is completed. The balance of the layout work then becomes a combination of staking from random control points and staking from the center line. Prior to setting final grade stakes, bench-mark levels are run throughout the graded alignment and bench marks are set to provide vertical control for all subsequent operations. These bench marks should be established so as to satisfy third-order specifications (Table 5.2, Art. 5.43) and should have a minimum density of three per mile (two per kilometre).

Final grade stakes for subgrade excavation and paving are usually set from the center line. However, grade stakes in the vicinity of an interchange may be more efficiently set from random control points.

As construction proceeds, monthly estimates are made of the work completed to date. A quantity survey is made near the close of each month, and the volumes of earthwork, etc. are classified and summarized as a basis for payment.

**18.8. Streets**    For street construction the procedure of surveying is similar to that just described for highways. Ordinarily, the curb is built first. The line and grade for the top of each curb are indicated by hubs driven just outside the curb line, usually at 50-ft (15-m) intervals.

On alignments containing horizontal and vertical curves, grade stakes should be set at closer intervals commensurate with the curvature of the center line and grade line. Differential levels are run obtaining an elevation on the top of each hub. Cuts or fills are calculated from the top of each hub to finished grade at the top of the curb. These cuts and fills are marked on guard stakes set next to each hub and are also tabulated in the field notebook. Frequently, a separate summary of cuts and fills is prepared for the construction superintendent. A copy of this summary should be filed by the surveyor.

When the curb is completed, levels should be run over the finished curb to check the accuracy of the construction. The grade for the edge of the pavement is then marked on the face of the completed curb; or for a combined curb and gutter it is indicated by the completed gutter. Hubs are set on the center line of the pavement, either at the grade of the finished subgrade (in which case holes are dug when necessary to place hubs below the ground surface) or with the cut or fill indicated on the hub or on an adjacent stake. Where the street is wide, an intermediate row of hubs may be set between center line and each curb. It is usually necessary to reset the hubs after the street is graded. Where driving stakes is impractical because of hard or paved ground, nails or spikes may be driven or marks may be cut or painted on the surface.

The surveys for street location and construction should determine the location of all surface and underground utilities that may affect the project; and notification of necessary changes should be given well in advance.

On paving projects where the rate of grade is constant, the possibility of using a laser level and rod such as are illustrated in Figs. 18.7 and 18.8 should be investigated. The use of laser levels for setting grade stakes is discussed in Art. 18.5. There must be a bench mark within the work area and care must be exercised to avoid sights of excessive lengths. The laser level plane is accurate to within $\pm 0.03$ ft or about 10 mm in a sight of 330 ft (100 m).

The automatic laser guidance equipment described in Art. 18.5 can be used to guide graders in establishing subgrade elevations and paving machines utilized for placing concrete to finished grade elevations.

**18.9. Railroads**    Construction surveys for railroads are similar to those described in Art. 18.7 for highways. Prior to construction, the located center line is rerun, missing stakes are replaced, control station hubs are referenced, borrow pits are staked out, slope stakes are set, and lines and grades for structures are established on the ground. When rough grading is completed, final grade stakes are set to grade at the outer edges of the roadbed, as a guide in trimming the slopes.

The foregoing operations can also be performed from random control points or using a random offset line as described in Arts. 18.4 and 18.7.

When the roadbed has been graded, alignment is established precisely by setting tacked stakes along the center line at full stations on tangents and usually at fractional stations on horizontal and vertical curves. Spiral curves are staked out at this time. An additional line of hubs is set on one side of the track and perhaps 3 ft (1 m) from the proposed line of the rail, with the top of the hub usually at the elevation of the top of the rail. Track is usually laid on the subgrade and is lifted into position after the ballast has been placed and compacted around the ties.

**18.10. Sewers and pipelines**   The center line for a proposed sewer is located on the ground with stakes or other marks set usually at 50-ft or 15-m intervals where the grade is uniform and as close as 10 ft or 3 m on vertical curves. At one side of this line, just far enough from it to prevent being disturbed by the excavation, a parallel line of hubs is set, with the hubs at the same intervals as those on the center line. A guard stake is driven beside each hub, with the side to the line; on the side of the guard stake farthest from the line is marked the station number and offset, and on the side nearest the line is marked the cut [to the nearest $\frac{1}{8}$ in (0.01 ft or 3 mm)]. In paved streets or hard roads where it is impossible to drive stakes and pegs, the line and grade are marked with spikes (driven flush), chisel marks, or paint marks.

When the trench has been excavated, batter boards are set across the trench at the intervals used for stationing. The top of the board is set at a fixed whole number of feet or decimetres above the sewer invert (inside surface of bottom of sewer pipe); and a measuring stick of the same length is prepared. A nail is driven in the top edge of each batter board to define the line. As the sewer is being laid, a cord is stretched tightly between these nails, and the free end of each section of pipe is set at the proper distance below the cord as determined by measuring with the stick (see Fig. 18.1).

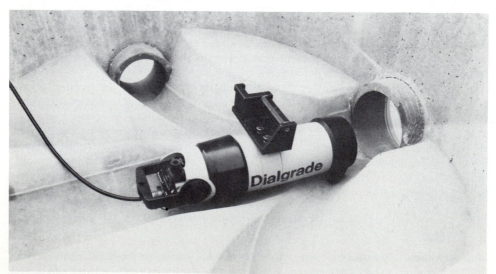

**Fig. 18.15** Laser device for sewer alignment. (*Courtesy Spectra-Physics, Inc.*)

For pipelines, the procedure is similar to that for sewers, but the interval between grade hubs or batter boards may be greater, and less care need be taken to lay the pipe at the exact grade.

For both sewers and pipelines, the extent of excavation in earth and in rock is measured in the trench, and the volumes of each class of excavation are computed as a basis of payment to the contractor.

The records of the survey should include the location of underground utilities crossed by, or adjacent to, the trench.

Laser alignment instruments can be used instead of the batter boards and stringline described above. The simplest approach consists of utilizing a laser-equipped theodolite to establish a laser beam parallel with the flow line of the sewer. This alignment beam would be used first for rough grading to get the ditch excavated and then for placing the pipe on line and at grade in the ditch.

There are also several special purpose laser alignment devices designed specifically for sewer construction. Figure 18.15 shows such an instrument placed in the invert of a manhole. This type of device can also be mounted on a tripod. These instruments are self-leveling and transmit a laser beam that can be adjusted from a $-10$ to a $+30$ percent grade. Instruments such as the laser level (Fig. 18.7) and this one are not useful for general surveying purposes but would be appropriate for a general contractor engaged in substantial amounts of excavation and sewer construction.

Note that the laser alignment instruments serve only to establish a line between previously surveyed control points and bench marks. Thus, the sewer alignment still needs to be established initially as in the traditional approach.

**18.11. Tunnels**   Tunnel surveys are run to determine by field measurements and computations the length, direction, and slope of a line connecting given points, and to lay off this line by appropriate field measurement. The methods employed naturally vary somewhat with the purpose of the tunnel and the magnitude of the work. A coordinate system is particularly appropriate for tunnel work.

For a short tunnel such as a highway tunnel through a ridge, a traverse and a line of levels are run between the terminal points; and the length, direction, and grade of the connecting line are computed. Where practical, the surface traverse between the terminals takes the form of a straight line. Outside the tunnel, on the center line at both ends, permanent monuments are established. Additional points are established in convenient surface locations on the center line, to fix the direction of the tunnel on each side of the ridge. As construction proceeds, the line at either end is given by setting up at the permanent monument outside the portal, taking a sight at the fixed point on line, and then setting points along the tunnel, usually in the roof. Grade is given by direct levels taken to points in either the roof or the floor, and distances are measured from the permanent monuments to stations along the tunnel. If the survey line is on the floor of the tunnel, it is usually offset from the center line to a location relatively free from traffic and disturbance; from this line a rough temporary line is given as needed by the construction force.

The dimensions of the tunnel are usually checked by some form of templet transverse to the line of the tunnel, but may be checked by direct measurement with the tape. Photogrammetric methods may also be used to check the cross-sectional dimensions of the tunnel (see Ref. 8).

Railroad and aqueduct tunnels in mountainous country are often several miles in length and are not uniform in either slope or direction. Tunnels of this character are usually driven not only from the ends but also from several intermediate points where shafts are sunk to intersect the center line of the tunnel (see Mine Surveying, Chap. 20). The surface surveys for the control of the tunnel work usually consist of a precise triangulation and/or trilateration system tied to monuments at the portals of the main tunnel and at the entrances

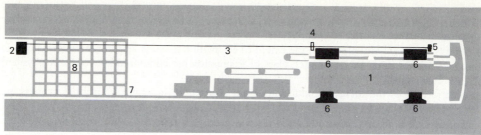

**Fig. 18.16** Laser alignment to control a tunneling machine: (1) tunneling machine; (2) laser theodolite station; (3) laser beam; (4) glass target; (5) second target; (6) hydraulic bracing; (7) tunnel invert (lining); (8) concrete forms. (*Courtesy Kern Instruments, Inc.*)

of shafts and adits, and a precise system of differential levels connecting the same points. With these data as a basis, the length, direction, and slope of each of the several sections of the tunnel are calculated; and construction is controlled by establishing these lines and grades as the work progresses.

Alignment of tunneling machinery was one of the earlier applications of laser equipped instruments. The procedure for establishing control and the center line of the tunnel is the same as described previously in this section except that a laser beam is projected ahead to control the direction of the tunneling machine (see Refs. 10 and 14). As the center line of the tunnel is produced, a laser-equipped theodolite (Fig. 18.2) is suspended from a support at a center-line station on the roof of the tunnel. The line of sight is produced ahead by double centering from a backsight. The laser transmitter is turned on to create a laser beam which is parallel to the center line of the tunnel. The tunneling machine has two targets placed on the top of its framework. The first target is glass with a center cross and the second is an opaque target with a center mark. The machine is adjusted until the laser beam passes through the center of the first target and strikes the center of the second target. When these two conditions are met, the tunneling machine is on line and work can progress. The operator of the machine can detect and correct for any deviation from this alignment. The laser beam is also used to control setting of forms for the tunnel lining. Figure 18.16 illustrates in schematic form the arrangement for control of the tunneling operation.

The range of the laser beam is only about 1000 ft (300 m), owing to the dust created by tunneling operations. Thus, periodically the line is produced ahead by double centering and a target is centered on a new station attached to the tunnel roof and immediately behind the tunnel machine. The target and theodolite are interchanged, the theodolite is centered by forced centering, a backsight is taken on the target, and the line is again produced by double centering. As the tunneling progresses, the backsight should be checked periodically to ensure that the laser beam does not deviate from the correct position.

When the alignment of the tunnel is curved, the machine must be driven ahead along chords on the curve. In order to get directed along a chord from the point of curvature, the targets on top of the machine are shifted laterally by a precomputed amount so as to produce the desired amount of deflection of the cutting heads in the direction of the chord. When the machine has moved ahead enough to provide clearance, the chord point is set on the roof of the tunnel, the theodolite is moved ahead, a backsight is taken on the previous station, and the deflection angle to the next chord point is turned. This procedure establishes the direction of the tunnel until the cutting heads of the machine advance to the next chord point, where the entire procedure is repeated. The machine is driven forward from chord point to chord point until the end of the curve is reached. Obviously, to maintain correct direction of the tunnel, very careful attention must be given to centering the theodolite and producing the laser beam by double centering. Further details concerning use of lasers for tunnel alignment can be found in Refs. 10 and 11.

Electronic distance measuring equipment (EDM) is also very useful in establishing the distances to theodolite stations, points of curvature, and chord points as the tunnel is driven.

To establish absolute directions (azimuth), a gyroscopic theodolite can be used. This instrument is designed to determine direction and is described in Arts. 11.46 and 20.15 to 20.18. Periodic determinations of the azimuth of the tunnel center line provide a check on the direction established by producing the line by double centering.

**18.12. Bridges**   Normally, the location survey will provide sufficient information for use in the design of culverts and small bridges; but for long bridges and for grade-separation structures usually a special topographic survey of the site is necessary. This survey should be made as early as possible in order to allow time for design and—in the case of grade crossings or navigable streams—to permit approval of the appropriate governmental agencies to be secured. The site map should show all the data of the location survey, including the line and grade of the roadway and the marking and referencing of all survey stations.

A scale of 1:1200 (100 ft/in) with a 5-ft (2-m) contour interval is usually suitable for large bridges in rough terrain. Where the terrain is level and in urban regions, a scale of 1:600 with a contour interval of 2 ft (1 m) is desirable.

Where the bridge crosses rivers or other bodies of water, a hydrographic survey (Chap. 21) which will provide a continuous profile of the bottom of the river or bay at 10- to 25-ft (3- to 8-m) intervals is required.

When planning the horizontal and vertical control for design mapping, due consideration should be given to establishing control points suitably located for the construction survey. Long crossings over water require elaborate triangulation and/or trilateration networks of first-order accuracy (Tables 10.1 and 10.2) which must be planned well in advance of the actual construction. For a short bridge with no offshore piers, first the center line of the roadway is established, the stationing of some governing line such as the abutment face is established on the located line, and the angle of intersection of the face of the abutment with the located line is turned off. This governing cross line may be established by two well-referenced control points at each end of the cross line offset beyond the limits of excavation or, if the face of the abutment is in the stream, by a line parallel to the face of the abutment offset on the shore. Similarly, governing lines for each of the wing walls are established on shore beyond the limits of excavation, with two stations on the line prolonged at one, or preferably both, ends of the wing-wall line. If the faces are battered, usually one line is established for the bottom of the batter and another for the top. Stakes are set as a guide to the excavation and are replaced as necessary. When the foundation concrete has been cast, line is given on the footings for the setting of forms and then by sighting with the theodolite for the top of the forms. As the structure is built, grades are carried up by leveling, with marks on the forms or on the hardened portions of the concrete. Also, the alignment is established on completed portions of the structure. The data are recorded in field books kept especially for the structure, principally by means of sketches.

For long sights or for work of high precision, as in the case of offshore piers, various control points are established on shore and/or on specially built towers by a system of triangulation and/or trilateration, such that favorable intersection angles and checks will be obtained for all parts of the work. Electronic distance measuring (EDM) equipment and first-order theodolites should be employed in establishing this control network. For ordinary river crossings, a single quadrilateral or two adjacent quadrilaterals with the common side coincident with the center line of the bridge are appropriate figures and provide well-located stations for construction layout. For river and bay crossings, vertical control is established by simultaneous, reciprocal leveling (Art. 5.45). For long crossings, special illuminated targets are used on the rod and all measurements are made at night.

To establish the location of offshore structures, simultaneous sights are taken from two coordinated control points which have favorable geometric locations with respect to the point being set. Thus, the points are located by intersection. When combined theodolite and EDM systems are available, points can be located by distance and direction so that triangulation and trilateration are involved. The survey is usually accomplished in stages with the pier locations being established with lower precision than those of the fittings subsequently established on top of the pier.

Laser-equipped instruments can be used for establishing the alignment and vertical positions for cofferdams and pile-driving equipment used in the construction of foundations for piers. The laser alignment device shown in Fig. 18.5 used in conjunction with EDM equipment provides an excellent system for this purpose.

As soon as the pier foundation is placed, survey control is transferred to the foundation so as to verify the location and provide control for construction of the rest of the pier. Elevations are also transferred to the pier at this time. When the pier is completed, the precise locations of anchor bolts and other fastenings to which the bridge structure is connected are set out on the top of the pier. All center-line points, anchor-bolt locations, and control monuments are coordinated in a system referred to a common datum. Thus, it is possible to locate anchor bolt positions, etc., by triangulation and trilateration from at least two control monuments that have good geometric positions with respect to the points being located. Note that this procedure is an application of location by random control points (Arts. 18.4 and 18.7).

When anchor bolts have been placed, their positions should be verified by triangulation and trilateration from control monuments. In addition, distances between anchor bolts and other pier connections on adjacent piers should be measured using EDM equipment.

On completion of the structure, permanent survey points are established and referenced for use in future surveys to determine the direction and extent of any movement of the piers and bridge structure.

**18.13. Culverts**    At the intersection of the center line of the culvert with the located survey line of the highway or railway, the angle of intersection is turned off, and a survey line defining the direction of the culvert is projected for a short distance beyond its ends and is well referenced. At (or offset from) each end of the culvert, a line defining the face is turned off and referenced. If excavation is necessary for the channel to and from the culvert, it is staked out in a manner similar to that for a roadway cut. Bench marks are established nearby, and hubs are set for convenient leveling to the culvert. Line and grade are given as required for the particular type of structure.

**18.14. Buildings**    At the beginning of excavation, the corners of the building are marked by stakes, which will of course be lost as excavation proceeds. Sighting lines are established and referenced on each outside building line and line of columns, preferably on the center line of wall or column. A batter board is set at each end of each outside building line, about 3 ft or 1 m outside the excavation. If the ground permits, the tops of all boards are set at the same elevation; in any event the boards at opposite ends of a given line (or portion thereof) are set at the same elevation so that the cord stretched between them will be level. The elevations are chosen at some whole number of feet or metres above the bottom of the excavation, usually that for the floor rather than that for the footings, and are established by holding the level rod on the side posts for the batter board and marking the grade on the post. When the board has been nailed on the posts, a nail is driven in the top edge of the board on the building line, which is given by the transit. Carpenter's lines stretched between opposite batter boards define both the line and the grade, and measurements can be made conveniently by the workers for excavation, setting forms, and aligning masonry and framing. If the distance between batter boards is great enough that the sag of the carpenter's line is appreciable, the grades must be taken as approximate only. Laser alignment and leveling devices provide an excellent substitute for the string line (Art. 18.5) and are adaptable for establishing line and grade. Use of these devices is particularly convenient when obstructions prohibit the erection and use of batter boards.

When excavation is completed, grades for column and wall footings are given by hubs driven in the ground to the elevation of either the top of the footing or the top of the floor. Lines for footings are given by batter boards set in the bottom of the excavation. Column bases and wall plates are set to grade directly by a level and level rod. The position of each column or wall is marked in advance on the footing; and when a concrete form, a steel member, or a first course of masonry has been placed on the footing, its alignment and grade are checked directly.

In setting the form for a concrete wall, the bottom is aligned and fixed in place before the top is aligned.

As before, laser alignment and leveling devices can also be utilized at this stage to establish line and grade.

Similarly, at each floor level the governing lines and grade are set and checked, except that for prefabricated steel framing, the structure as a whole is plumbed by means of the transit, theodolite, or plumbed laser beam at every second or third story level. Notes are kept in a field book used especially for the purpose, principally by means of sketches.

Whenever the elevation of a floor is given, it should be clearly understood whether the value refers to the bottom of the base course, the top of the base course, or the top of the finished floor.

Throughout the construction of large buildings, selected key points are checked by means of stretched wires, plumb lines, optical plummet, plumbed laser beam, plumbed laser plane, transit, theodolite, or level in order to detect settlement, excessive deflection of forms or members, or mistakes. Bench marks are checked to detect any disturbance.

**18.15. Dams** Prior to the design of a dam, a topographic survey is made to determine the feasibility of the project, the approximate size of the reservoir, and the optimum location and height of the dam. To provide information for the design, a topographic survey of the site similar in many respects to that for a bridge (Art. 18.12), is made usually by photogrammetric methods supplemented by field surveys where necessary. Extensive soundings and borings are made, and topography is taken in detail sufficient to define not only the dam itself but also the appurtenant structures, necessary construction plant, roads, and perhaps a branch railroad. A property-line survey is made of the area to be covered by, or directly affected by, the proposed reservoir.

When feasibility and location studies are being made for dams in remote areas, due consideration should be given to using doppler positioning and inertial positioning systems (Chap. 12) for establishing the basic control network. Inertial systems can also be used to obtain data for preliminary topographic mapping where heavy ground cover (such as rain forest) combined with bad weather conditions make ground surveys and photogrammetric topographic surveys impractical.

Prior to construction, horizontal control stations, sighting points, and bench marks are permanently established and referenced upstream and downstream from the dam at advantageous locations and elevations for sighting on the various parts of the structure as work proceeds. These reference points are usually established by triangulation and/or trilateration and all points are referred to a system of rectangular coordinates. In the United States, the state plane coordinate systems should be used.

Precise levels (Art. 5.48) should be run to establish vertical control. For major dam projects, first-order specifications should be satisfied in establishing both horizontal and vertical control (Tables 10.1, 10.2, and 5.2). A dam is a vast project in which construction may last for a period of several years. Comprehensive planning of the survey control is essential from the beginning of the project. Careful attention must be given to setting horizontal and vertical control points in locations suitable for subsequent construction layout.

To establish the horizontal location of a point on the dam, as for the purpose of setting concrete forms or of checking the alignment of the dam, simultaneous sights are taken from two theodolites set up at horizontal control points, each instrument being sighted in a direction previously computed from the coordinates of the control station and of the point to be established. The method of staking from random control points as described in Art. 18.4 can be used to advantage in dam construction. Thus, coordinated points on the dam structure are set from at least two coordinated horizontal control points by triangulation using combined theodolite and EDM or an electronic surveying system (Art. 12.2).

Vertical control is usually established by direct leveling. When warranted, trigonometric leveling (Art. 5.5) is used to establish elevation by using measurements obtained with one of the electronic surveying systems from two coordinated vertical control stations.

Wherever possible, laser alignment instruments should be utilized. For example, lasers can be used in controlling tunneling operations in the underground excavation associated with the construction of the dam.

A traverse is run around the reservoir above the proposed shoreline and monuments are set for use in connection with property-line surveys and for future reference. Similarly, bench marks are established at points above the shoreline. The shoreline may be marked out by trace contour leveling (Art. 15.15), with stakes set at intervals along the specified contour line. The area to be cleared is defined with reference to these stakes. The area and volume of the reservoir may be computed as described in Art. 17.44.

When the dam is completed, permanent horizontal control monuments and bench marks are placed on the dam structure appropriate for subsequent studies related to deformation of the dam. A network of control points suitable for combined triangulation/trilateration (Art. 10.27) provides optimum control for monitoring possible movement of the dam. Observations on this control network should be made with first-order theodolites (Art. 10.18) and EDM equipment (Arts. 4.29 through 4.43). Elevations should be determined by geodetic leveling (Art. 5.48). First-order specifications should govern all horizontal and vertical control surveys (Tables 10.1 and 10.2, Art. 10.2, and Table 5.2, Art. 5.43).

**18.16. As-built surveys**  The *as-built survey* is performed on completion of a construction project for the purpose of reestablishing the principal horizontal and vertical controlling points and to locate the positions of all structures and improvements. As-built surveys should be performed after the construction of highways, railroads, bridges, buildings (including private homes), dams, and underground facilities.

Essentially, the as-built survey consists of rerunning center lines and/or property lines, setting permanent monuments at controlling points, and locating all improvements relative to these lines. These data are then plotted on a map to provide a permanent record or inventory of the work done and its precise location. A record of these data filed in a data bank (magnetic tape, disk storage, or core memory) according to a coordinated position in the state plane coordinate system is perhaps the most efficient method for storing this information. Such a record is easily accessible using a computer program and permits efficient retrieval and automatic plotting of the desired data.

## References

1. American Society of Civil Engineers Task Committee Report on "Engineering Surveying, Chapters 2, 5, 6, 8, 16, and 20," *Journal of the Surveying and Mapping Division, ASCE*, Vol. 102, No. SU1, December 1976, p. 59.
2. California Department of Transportation, *Surveying Manual*, Sacramento, Calif., 1977, Chaps. 7, 11.
3. Chrzanowski, A., Jarzymowski, A., and Kaspar, M., "A Comparison of Precision Alignment Methods," *The Canadian Surveyor*, Vol. 30, No. 2, June 1976.
4. Chrzanowski, A., and Janssen, H.-D., "Use of Laser in Precision Leveling," *The Canadian Surveyor*, Vol. 26, No. 4, September 1972.
5. Cooney, Austin, "Laser Alignment Techniques in Tunneling," *Journal of the Surveying and Mapping Division, ASCE*, Vol. 94, No. SU2, September 1968, p. 203.
6. Kahn, Michael, "Lasers, Today's Great Levelers in Construction," *World Construction*, February 1977.
7. Kahn, Michael, "Lasers as Construction Tools," *Constructor*, April 1976.
8. Katibah, G. P., "Photogrammetry in Highway Practice," Unpublished report, California Department of Transportation, 1968.
9. More, Norman L., and Begell, R. G., "Control Surveys for Major Bridges," *Journal of the Surveying and Mapping Division, ASCE*, Vol. 98, No. SU1, May 1971.
10. Peterson, Edward W., and Frobenius, Peter, "Tunnel Surveying and Tunneling Machine Control," *Journal of the Surveying and Mapping Division, ASCE*, Vol. 99, No. SU1, September 1973, p. 21.
11. Thompson, Bruce J., "Planning Economical Tunnel Surveys," *Journal of the Surveying and Mapping Division, ASCE*, Vol. 100, No. SU2, November 1974, p. 95.
12. Willis, Maynard J., "Planning Concrete Dam Construction Control," *Journal of the Surveying and Mapping Division, ASCE*, Vol. 98, No. SU1, July 1972, p. 27.
13. "Kern DKM2-A Laser Theodolite," *Kern Bulletin 15*, Kern Instruments Inc., 5001 Aarau, Switzerland.
14. "Laser Control in Tunnel Driving," *Kern Bulletin 15*, Kern Instruments Inc., 5001 Aarau, Switzerland.

# CHAPTER 19
# Land surveys

**19.1. General**  Land surveying deals with the laying off or the measurement of the lengths and directions of lines forming the boundaries of real or landed property. Land surveys are made for one or more of the following purposes:

1. To secure the necessary data for writing the legal description and for finding the area of a designated tract of land, the boundaries of the property being defined by visible objects.
2. To reestablish the boundaries of a tract for which a survey has previously been made and for which the description as defined by the previous survey is known.
3. To subdivide a tract into two or more smaller units in accordance with a definite plan which predetermines the size, shape, and location of the units.

Whenever real estate is conveyed from one owner to another, it is important to know and state the location of the boundaries, particularly if there is a possibility of encroachment by structures or roadways.

The functions of the land surveyor are to carry out field surveys, to calculate dimensions and areas, to prepare maps showing the lengths and directions of boundary lines and areas of lands, and to write descriptions by means of which lands may be legally conveyed, by deed, from one party to another.

The land surveyor must be familiar not only with technical procedures but also with the legal aspects of real property and boundaries. Usually, the land surveyor is required to be licensed by the state, either directly or as a civil engineer.

In this chapter, practices as applied to both rural and urban properties are described, and some of the legal aspects of land surveying are discussed. The United States system of subdividing the public lands is also outlined. Methods of calculating and subdividing areas are discussed in Chap. 8.

**19.2. Kinds of land surveys**  In accordance with the purposes listed in Art. 19.1, land surveys may be classified as follows:

1. *Original surveys*, to measure the unknown lengths and directions of boundaries already established and in evidence. Surveys of this character are usually of rural lands. For example, Adams may purchase from Brown a certain parcel of land bounded or defined by features or objects such as fences, roads, or trees. In order that the deed may contain a definite description of the tract, a survey is necessary.
2. *Resurveys*, to reestablish the boundaries of a tract for which a survey has previously been made. The surveyor is guided by a description of the property based upon the original survey, and by evidence on the ground. The description may be in the form of the original survey notes, an old deed, or a map or plat on which are recorded the measured lengths and bearings of sides and other pertinent data. When land is transferred by deed from one party to another, often a resurvey is made.
3. *Subdivision surveys*, to subdivide land into more or less regular tracts in accordance with a

prearranged plan. The division of the public lands of the United States into townships, sections, and quarter sections is an example of the subdivision of rural lands into large units. The laying out of blocks and lots in a city addition or subdivision is an example of the subdivision of urban lands.

**19.3. Equipment and methods**   Land surveys can be run with a transit and tape, transit or theodolite and electronic distance measurement (EDM) device, or electronic surveying system (Art. 12.2). In some states, land surveys have been performed using photogrammetric methods to establish the control points from which property corners have been located and set. Ample evidence exists to indicate that photogrammetric methods provide adequate accuracy for property surveys (Ref. 4). Acceptance of photogrammetry as one of the tools useful for performing property surveys is bound to come as more surveyors become familiar with the procedures (Chap. 16) and as the legality of the method is recognized in the courts.

Directions of lines for property surveys are usually referred to the true or astronomic meridian, and angular measurements are transformed into bearings. Distances are still given in feet and decimal parts of the foot, but with metrication on the horizon, it is recommended that distances be given in feet and in metres. Angles are measured to minutes or fractions thereof. On the U.S. public-land surveys, all distances are in Gunter's chains (66 ft) with 100 links per chain as prescribed by law.

Formerly, the surveyor's compass and 66-ft link chain were used extensively, particularly in rural surveying; and the directions and lengths of lines contained in many old deeds are given in terms of magnetic bearings and Gunter's chains. In retracing old surveys of this character, allowance must be made for change in magnetic bearing since the time of the original survey. Also, it must be kept in mind that the compass and link chain used on old surveys were relatively inaccurate instruments and that great precision was not regarded as necessary since usually the land values were low. Further, for many years the United States public lands were surveyed under contract, at the low price of a few dollars per mile. Many of the lines and corners established by old surveys are not where they theoretically should be; nevertheless, these boundaries legally remain fixed as they were originally established.

Wherever possible, the field procedure is such that the lengths of boundary lines and the angles between boundaries are determined by direct measurement. Therefore, the land survey is in general a traverse, the stations being at corners of the property, and the traverse lines coinciding with property lines. Where obstacles render direct measurement of boundaries impossible, a traverse is run as near the property lines as practical, and measurements are made from the traverse to property corners; the lengths and directions of the property lines are then calculated. Where the boundary is irregular or curved, the traverse is established in a convenient location, and offsets are taken from the traverse line to points on the boundary; the length of the boundary is then calculated.

In general, the required precision of land surveys depends upon the value of the land, being higher in urban than in rural areas. The possibility of increase in land values should also be considered.

Wherever possible, the survey utilized to establish property corners should be tied to monuments for which state plane coordinates are available (Arts. 14.16 to 14.23). Use of state plane coordinates integrates all surveys into a common system, facilitates replacing lost corners, and provides a means of parcel identification that is compatible with filing survey data in electronic computer data banks.

**19.4. Legal aspects of land surveys and definitions**   Property lines are established and indicated by acts on the ground or by legal documents. Thus, old boundaries are located on the basis of acts and documents exercised in the past. New boundaries are created by current acts or new documents. Consequently, the present positions of property lines or boundaries depend on work performed on the ground and documents executed in the past as well as in the present.

The positions of boundaries are controlled by two principles of law: (1) the *intent* of the parties involved in originally establishing the boundaries, and (2) acceptance of conditions as they have existed over a period of years. Intent is judged on the basis of the acts and documents of the involved parties. If old evidence of intent is obscure and misleading, then conditions as defined by existing evidence are accepted. The longer the period of acceptance, the stronger the evidence becomes.

Two types of law govern the foregoing principles: *common law*, which consists of the body of laws that have been inherited from our predecessors; and *statutory law*, which is established by governing bodies. Much of the common law in the United States is based on laws originating in England. Statutory laws *usually take precedence over common law*.

The evidence used to establish property lines consists of title transfers, transfer of rights such as in easements for right of way, by agreement, by marks on the ground, and by acts leading to adverse possession. Consider each of these types of evidence briefly.

**Title transfer**   Title is transferred by legal instruments such as deeds or wills, inheritance without a will, or by adverse possession (Art. 19.12). Deeds are usually recorded in some public office (Art. 19.6) and contain a description of the property (Art. 19.7). This description generally contains references to marks on the ground. The surveyor makes much use of this type of evidence in relocating old boundary lines.

In connection with deeds, it is necessary to define *senior deed*. When land is partitioned, in most cases the first parcel sold is the senior deed. Subsequent deeds related to partitioning of the remainder of the land have seniority in the order of their execution. In a dispute, the owner of the senior deed has rights over those with less senior rights.

**Easements for rights of way**   The right to use the land of another for some specific purpose constitutes an *easement*. An easement that grants the right to pass over the land of another is a *right of way*. An easement can be established by an owner through deed or by *dedication*. The state has the right of *eminent domain*, which is the right to use a person's property for a specific use (such as a highway) provided that the owner of the property involved receives fair remuneration for the land taken. *Condemnation* proceedings are usually instituted to acquire land by eminent domain. These proceedings are recorded and become part of the public record.

Although most transfers of land are by written documents, such transfers can also be accomplished by *oral agreement* between the parties involved. Dedication of land for public use is one example of such a transfer. Another example arises when owners of adjacent properties agree as to the location of a property line acceptable to both parties.

Visual evidence of property-line location consists of marks on the ground. These marks include monuments (Art. 19.5), iron pipes and rods, stakes, fences, and structures. Evidence of this type can be extremely valuable when attempting to reestablish old property lines.

**Adverse possession**   Land acquired by the open and hostile use of another's land with the full knowledge of the owner is said to have been obtained by adverse possession. Details concerning adverse possession can be found in Art. 19.12.

Additional legal terms quoted from an early edition of Bouvier's *Law Dictionary* (Ref. 2 at the end of the chapter) of a few of the more common legal terms having to do with the transfer of land are as follows:

**Color of title**   Color of title, for the purposes of adverse possession under the statute of limitations, is that which has the semblance or appearance of title, legal or equitable, but which in fact is no title.

A writing which upon its face professes to pass title but which does not in fact do so, either from a want of title in the person making it or from the defective conveyance used, is also known as color of title. The term is also applied to a title that is imperfect, but not so obviously that it would be apparent to one not skilled in the law.

**Fee**   The word "fee" signifies that the land or other subject of property belongs to its owner and is transmissible, in the case of an individual, to those whom the law appoints to succeed him under the appellation of heirs.

**Fee simple**   An estate of inheritance is a fee simple. The word "simple" adds no meaning to the word fee standing by itself, but it excludes all qualifications or restrictions as to the persons who may inherit it as heirs.

**Parol**   Parol is a term used to distinguish contracts which are made verbally, or in writing not under seal, which are called parol contracts, as distinguished from those which are under seal, which bear the name of deeds or specialties.

**Patent**   A patent is the title deed by which a government, either state or federal, conveys its lands.

**19.5. Monuments**   Monuments are classified as *natural, artificial, record,* or *legal.*

Examples of natural monuments are trees, large stones, or other substantial, naturally occurring objects in place before the survey was made.

Artificial monuments can consist of an iron pipe or bar driven in the ground; a concrete or stone monument with drill hole, cross, or metal plug marking the exact corner; a stone with identifying mark, placed below the ground surface; charcoal placed below the surface; a mound of stones; a mound of earth above a buried stone; and a metal marker set in concrete below the surface, reached through a covered shaft. Monuments for city lots are usually set nearly flush with the ground. Subsurface stones were commonly used for corner monuments in localities where roads followed section lines. On many old governmental surveys, through wooded country where stones were not available, corners were established by building up a mound of earth over a quart of charcoal or a charred stake, or by building a mound around a tree at which the corner fell. The U.S. Bureau of Land Management has adopted as the standard for the monumenting of the public-land surveys a post made of iron pipe filled with concrete, the lower end of the pipe being split and spread to form a base, and the upper end being fitted with a brass cap with identifying marks (Art. 19.43).

A record monument exists because of a reference in a deed or description. For example, the phrase in a deed "to the side line of a street" is a call for a record monument (the street) that could be marked by improvements, stakes, a fence, or may not be marked in any way.

A legal monument is one that is controlling in the description. Thus, the statement in a deed "to a concrete post" is a call to a legal artificial monument. Similarly, the statement "to Johnson's property line" is a call to a legal record monument.

Original, natural monuments control over artificial monuments and record monuments. Original, artificial monuments control over record monuments. Further details concerning the principles that govern conflicts between monuments can be found in Ref. 4.

If there is a possibility that a corner monument will become displaced, the corner should be *referenced*, or connected to nearby objects of more or less permanent character in such manner that it may be readily replaced in case of loss (Art. 8.14). Usually, the recorded measurement is called a *connection*, and the object is called a *reference mark* or a *corner accessory*; essentially it is a part of the monument. Examples of corner accessories are trees, mounds, pits, large stones, and buildings. In many large cities, systems of permanent monuments are established and all surveys are referred to them. On public-land surveys, the bearing and distance from a corner to a tree are taken where possible, the tree being blazed and so marked as to identify the section on which the tree stands, the mark terminating with the letters "B.T." signifying bearing tree. The Bureau of Land Management specifies that every corner established in the public-land surveys shall be referenced by one or more objects of any of the following classes: (a) "bearing trees, or other natural objects...; (b) permanent improvements and memorials; (c) mound of stone; and (d) pits."

If the location of a corner within reasonable accuracy can be determined beyond reasonable doubt, the corner is said to *exist*; otherwise, it is said to be *lost*. If the monument marking an existing corner cannot be found, the corner is said to be *obliterated*, but it is not necessarily lost.

Where a corner falls in such location as to make it impossible or impractical to establish a monument in its true location, it is customary to set a point on one or more of the boundary lines leading to the corner, as near to the true corner as practical. A point thus established is called a *witness corner*. Everything that has been said concerning monuments at the true corners also applies to witness corners. Witness corners are necessary where the true corner falls in a roadway or body of water, within a building, or upon a precipitous slope. Under certain circumstances, as when boundaries are in roads, it is impossible to place the witness corner on any of the property lines approaching the true corner, in which case the witness corner is established in any convenient location.

The field notes should give detailed information concerning the character, size, and location of all monuments and reference marks; and the data should be recorded in such manner that there will be no possibility of misinterpretation. As far as possible, all points established in the field should be clearly marked to indicate the object which they represent.

**19.6. Boundary records** Descriptions of the boundaries of real property may be found from deeds, official plats or maps, or notes of original surveys. Typical descriptions are given in Arts. 19.8 and 19.9.

Records of the transfer of land from one owner to another are usually kept in the office of the county registry of deeds, exact copies of all deeds of transfer being filed in deed books. These files are open to the public and are a frequent source of information for the land surveyor in search of boundary descriptions when it is inconvenient or impossible to secure permission of the owner to examine the original deed.

Originally, all official records were copied by hand into the record books. The first step toward modernization of this system was to make photocopies of each record, which were still filed in the books. As records proliferated, many recording offices switched to making microfilm copies of new and old records so as to save space. The microfilm copies can be stored on cards (microfiche) and then are filed in a system that permits manual or automatic retrieval.

No matter what system of filing the records is employed, some method for indexing is required. The system that has and still is in most general use in the United States consists of an alphabetic index, usually kept by years, giving in one part the names of the *grantors* or persons selling property and in the other the names of the *grantees* or persons buying property. Thus, if either or both parties to the transfer and the approximate date of the transfer are known, a given deed can be located. Usually, the preceding transfer of the same property is noted on the margin of the deed.

In most cases the deeds of transfer of city lots give only the lot or block number and the name of the addition or the subdivision. The official plat or map showing the dimensions of all lots and the character and location of permanent monuments is on file either in the office of the city clerk or in the county registry of deeds; copies are also on file in the offices of city and county assessors. Filing land transfer records according to these data is called indexing by *lot*, *block*, *and parcel*, or more simply indexing by *land parcel*.

Some organizations, usually called *title companies*, for a fee will search the records for boundary descriptions and will guarantee the title against possible defects in description, legal transfer, and certain types of claims such as those for right of way. Title insurance does not necessarily mean that the property corners are correctly marked on the ground; and if assurance is desired, a survey should be made. The files of title companies are usually based on land parcel indexing.

As U.S. public lands are subdivided, official plats are prepared showing the dimensions of subdivisions and the character of monuments marking the corners. When the surveys within a state have been completed, records are given to the state. States in possession of records have them on file at the state capital. Usually, information concerning these records can be secured from the state secretary of state. Photographic copies of the official plats are obtainable at nominal cost. Land transfer records maintained in this fashion are said to be filed according to the federal rectangular survey system.

Several states have passed land registration acts which specify the procedures to be followed in the transfer of land. These acts usually designate the state plane coordinate systems as the reference framework to be used. The state of Massachusetts has a special "land court" where title to land can be confirmed by a simple procedure and at nominal cost (see Ref. 25).

Current trends are toward storing legal records related to land transfers in computer data banks (see Arts. 1.11 and 3.15). Such a procedure has the advantage that title searches can be performed more rapidly and at lower cost. Most title companies already have automated or semiautomated search procedures for land records. Electronic computer storage and retrieval methods and interactive computer techniques are utilized in these procedures (Ref. 22). The problems encountered in such systems are mostly related to the fact that the grantor-grantee alphabetic listing is not exactly compatible with efficient computer usage and a unique identifier for each parcel involved in a land transfer is necessary.

One proposal is to pass legislation requiring that all property surveys be tied into the state plane coordinate system. In this way, records for each property transaction would have a parcel identifier consisting of the state plane coordinates for the approximate center of the parcel, visually determined from a scaled map (Ref. 7). These coordinates would be preceded by identifier digits which specify the state, county, and other necessary municipal subdivision. Naturally, the success of such a system would hinge on stringent but realistic accuracy requirements for all property surveys so as to avoid ambiguities in coordinates of the visual centers for adjacent parcels.

The North American Institute for Modernization of Land Data Systems (MOLDS) was organized in 1966 to study problems related to land data systems and to investigate methods of improving these systems. The first meeting of MOLDS was a conference on a Comprehensive Unified Land Data System (CULDATA) held at the University of Cincinnati in December 1966. Subsequent meetings of MOLDS have been held in Atlanta, Georgia, in 1972 on Compatible Land Identifiers—The Problems, Prospects and Payoffs (CLIPP), and at Washington, D.C., in 1975 on Modernization of Land Data Systems (MOLDS). Proceedings for these three meetings were published and are available (Ref. 18).

**19.7. Property descriptions**   As noted in Art. 19.4, a description of the property is an essential item in the legal document referring to the transfer of the land. The major purpose of this description is to identify the property for title purposes and to describe its size, shape, and location. The description should be precise, clear, and concise. It should be worded with sufficient legal terms so as to perpetuate the intent of the parties in a legal sense. The dimensions given to describe the property lines should be mathematically correct, and there should be no conflicts in the description of the property or with respect to adjoining areas.

There are two basic methods for describing property: (1) metes and bounds descriptions, and (2) subdivision descriptions. Each of these methods for describing land is discussed in more detail in the sections that follow.

**19.8. Metes and bounds descriptions**   In the older portions of the United States, nearly all of the original land grants were of irregular shape, many of the boundaries following stream and ridge lines. Also, in the process of subdivision the units were taken without much regard for regularity, and it was thought sufficient if lands were specified as bounded by natural or artificial features of the terrain and if the names of adjacent property owners were given. An example of a description of a tract as recorded in a deed reads:

Bounded on the north by Bog Brook, bounded on the northeast by the irregular line formed by the southwesterly border of Cedar Swamp of land now or formerly belonging to Benjamin Clark, bounded on the east by a stone wall and land now or formerly belonging to Ezra Pennell, bounded on the south and southeast by the turnpike road from Brunswick to Bath, and bounded on the west by the irregular line formed by the easterly fringe of trees of the wood lot now or formerly belonging to Moses Purington.

As the country developed and land became more valuable, and as many boundaries such as those listed in the preceding description ceased to exist, land litigations became numerous. It then became the general practice to determine the lengths and directions of the boundaries of land by measurements with the link chain and surveyor's compass, and to fix the locations of corners permanently by monuments. The lengths were ordinarily given in rods, perches, or chains, and the directions were expressed as bearings usually referred to the magnetic meridian. Surveys of this character are now usually made with the transit and tape, with theodolite and EDM device, or electronic surveying system, distances being recorded in feet, chains, or metres and directions being given in astronomic, geodetic, or grid bearings computed from angular measurements. In describing a tract surveyed in this manner the lengths and bearings of the several courses are given in order, and the objects marking the corners are described; if any boundary follows some prominent feature of the terrain, the fact is stated; and the calculated area of the tract is given. When the bearings and lengths of the sides are thus given, the tract is said to be described by *metes and bounds*. Within the limits of the precision of the original survey, it is possible to relocate the boundaries of a tract if its description by metes and bounds is available, provided that at least one of the original corners can be identified and the true direction of one of the boundaries can be determined.

It should be noted that conveyances of land having senior and junior rights, as discussed in Art. 19.4, will come under the classification of metes and bounds descriptions. This is in contrast to the subdivision description in which all lots within the subdivision have equal rights (Art. 19.9).

There are several types of metes and bounds descriptions, the most common of which is the method of *successive bounds*, as discussed above. This type of metes and bounds description is sometimes regarded as the true metes and bounds description.

**Example 19.1**   Alan Tart has sold a portion of his property to Richard Merlin. Tart's land was obtained in a previous conveyance from Biggert to Tart. There is no recorded plan of lots (subdivision), so that a metes and bounds description is necessary in the deed that is the legal document in which the transaction from Tart to Merlin will be recorded. Prior to the sale, the land to be sold is surveyed and the corners are marked as illustrated in Fig. 19.1. The following is one form of a deed description with a metes and bound description by successive bounds:

A parcel of land in the City of _____, County of _____, State of _____, being a part of the land conveyed by John Biggert to Alan Tart by deed dated June 14, 1956, and recorded in Book 201, page 131, at the _____ County Recorder's Office, and described as follows:

Beginning at a concrete monument on the northerly line of Woodlawn Avenue, said monument bearing S47°40′E a distance of 100.00 ft (30.480 m) along the northerly line of Woodlawn Avenue from the intersection of the northerly line of Woodlawn Avenue with the easterly line of Hawthorne Avenue and running: (1) thence along the northerly line of Woodlawn Avenue S47°40′E a distance of 102.88 ft (31.358 m) to an iron pipe (2 in or 51 mm inside diameter) at the line of the land of John

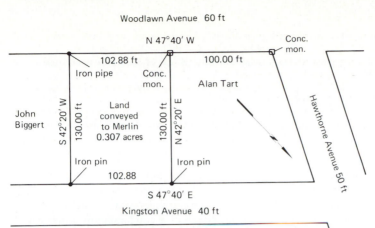

**Fig. 19.1** Plan of property described in Example 19.1.

Biggert; (2) thence along said land N42°20′E a distance of 130.00 ft (39.624 m) to an iron pin ($\frac{1}{2}$ in or 13 mm outside diameter) on the southerly line of Kingston Avenue; (3) thence along the southerly line of Kingston Avenue N47°40′W a distance of 102.88 ft (31.358 m) to an iron pin ($\frac{1}{2}$ in or 13 mm outside diameter) at the line of the land of the grantor; (4) thence along the land of the grantor S42°20′W a distance of 130.00 ft (39.624 m) to the point of beginning.

All directions are based on the stated bearing of the northerly line of Woodlawn Avenue.

This description was written on July 1, 1978, by _____ , Professional Land Surveyor, Registration No. _____ , State of _____ , and is based on data gathered in the field by _____ , on June 15, 1978.

The deed contains a general statement which indicates the approximate location of the land. The body of the deed includes a specific description of the land conveyed by the method of successive bounds. Note that the description has a point of beginning which is referenced to the block corner. Such a reference is necessary to fix the position of the land being conveyed. Then each course is described in turn and the description ends with a return to the point of beginning.

Following the description are given the basis for bearings, the name and registration number of the surveyor, and the dates when the deed was written and field work performed.

The basis for bearings should always be stated clearly. Thus, bearings should be described as being astronomic, grid, or as being referenced to a stated line. The basis for bearings can be stated indirectly. In Example 19.1, the initial statement "said monument bearing S47°40′E along the northerly line of Woodlawn Avenue..." fixes the basis for bearings along that line. Hence, the statement on basis for bearings, following the body of the deed, is redundant only in situations such as this example.

Note that the deed conveying land from Biggert to Tart has senior rights over the deed in Example 19.1 conveying land from Tart to Merlin. Thus, if the monuments along the northerly line of Woodlawn Avenue are destroyed at some later date and a subsequent survey reveals that the record distance of 202.88 ft is 201.88 ft to the iron pin, then Tart must get his 100.00 ft of frontage along Woodlawn Avenue even at the expense of Merlin, who has the junior deed. When senior and junior rights are involved, there is no proportioning of errors in surveys (Arts. 19.18 and 19.47).

Several other types of metes and bounds descriptions exist, as follows:

1. *Strip conveyances* are employed to describe a road easement or right of way. The form used is "a right of way for highway purposes over and across a strip of land lying 25 ft (7.620 m) on either side

of the following described center line...." Stationing is used to denote distance along the center line.

2. *Conveyance by division line*: "All of lot 62 lying south of Rock Creek" is a description by division line.

3. *Conveyance by distance*: "The westerly 20 m of lot 32" constitutes the very brief description by distance.

4. *Proportional conveyances* are those which describe a fractional part of the whole tract, such as "the eastern 1/8 of lot 11."

5. Land may be conveyed by *exception*, such as "all except the westerly 100 ft of lot 61," or by *acreage*, such as "the west 2 acres of lot 15."

Further details concerning the variations on the true metes and bounds descriptions listed above can be found in Ref. 4.

**Descriptions for condominiums**  In general, the term *condominium* refers to a method of ownership. This term may be applied to a multiunit building or to a unit or units within the building. Thus, the owner of an individual unit or of several units with a multiunit building is called a condominium owner. The limits of ownership of a condominium owner are the top surface of the floor, the surfaces of the walls, and the bottom surface of the ceilings of the specified unit. Condominium ownership also includes a fractional interest in the common elements associated with each unit such as swimming pools, tennis courts, and other common areas in the building. A grant of interest may also be made for items such as garage, patio, and storage space.

Since the potential condominium owner is buying a volume of space as defined above, it is necessary to provide a description of the unit in three dimensions. To describe each condominium unit in three dimensions by a metes and bounds description would be a challenging task, to say the least. A simpler approach is to prepare a map of the land on which the building is located and a set of plans of the building referenced to the map. These plans of the building show each individual unit with an identifying label. In this way any specified unit can be described by the identifying label and a reference to the map.

The land on which the condominium building is constructed can be described by a metes and bounds description, a lot or parcel on a map already recorded, or by a new subdivision plan of one or more lots. The most convenient instrument of record is a map which shows the boundaries of the land and the position of the exterior walls of the condominium building with respect to these boundaries. The condominium property acts of some states require that this type of map be recorded in order to establish a condominium. Along with the map there should be a set of plans of the building. These plans show all horizontal dimensions of the interior walls and the relationships of these walls with the exterior walls of the structure. Elevations are also required for the top of floor and bottom of ceiling of each unit. These elevations can be shown on the plans, indicated on a cross section of the building, or tabulated for each unit. All elevations should be based on bench marks referred to a specified datum.

The dimensions shown on the plan should be certified by the surveyor as conforming to the physical building. Thus, the surveyor needs to perform an as-built survey (Chap. 18) of the structure after it is completed. If this is not possible and measurements in the field are made to the unfinished elements of the structure, the interior dimensions can be based on the architect's plans, a fact that must be indicated on the plans. When the building is finished, the surveyor should resurvey the building, including the interior horizontal and vertical dimensions. Any differences between dimensions determined from this final as-built survey and those projected from the architect's plans should be indicated on the plans. If major differences occur, a new plan should be filed.

The following example description for a condominium unit is one of several forms recommended by the California Title Association.

**Example 19.2**  Assume that the condominium building is on land described by a recorded subdivision plan and that all units are within one lot.

"A condominium comprised of:
PARCEL 1: An undivided _____ interest in and to Lot _____ of Tract No. _____ , in the City of _____ County of _____ , State of _____ , as per map recorded in Book _____ , of Maps, at Page __ , in the Office of the County Recorder of said County.
PARCEL 2: Unit _____ , as shown upon the map referred to in PARCEL 1."

If the condominium building rests on more than one lot, this fact must be noted in the description of Parcel 1. Where the condominium buildings rest on land described by metes and bounds, the description for Parcel 1 in the example is modified as follows:
An undivided _____ interest in and to the land described as follows:... insert the metes and bounds description here and add the description of Parcel 2.

For further details concerning condominiums and their descriptions, consult Ref. 16 and 23.

**Descriptions by coordinates**  As noted in Art. 19.6, there is a trend toward requiring all property surveys to be performed in the state plane coordinate systems. In some states, the locations of land corners are legally described by state plane coordinates. Although practice is not as yet uniform in this regard, the following description by the Tennessee Valley Authority illustrates the description of land both by metes and bounds and by coordinates, with further reference to corners and lines of the U.S. public-land survey. The public-land survey is referred to the Huntsville principal meridian. A map of the tract is shown in Fig. 19.2.

**Example 19.3**  A tract of land lying in Jackson County, State of Alabama, in the South Half (S $\frac{1}{2}$) of the Northwest Quarter (NW $\frac{1}{4}$) of Section Three (3), Township Six (6) South, Range Five (5) East of the Huntsville principal meridian, and more particularly described as follows:
Beginning at a fence corner at the southwest corner of the Northwest Quarter (NW $\frac{1}{4}$) of Section Three (3) (coordinates X416,239; Y1,470,588), said corner being North six degrees twenty-four minutes West (N6°24'W) twenty-six hundred (2600) feet from the southwest corner of Section Three (3) (X416,529; Y1,468,004), and a corner to the land of T. E. Morgan; thence with Morgan's line, the west line of Section Three (3), and a fence line, North five degrees thirty-three minutes West (N5°33'W) thirteen hundred four (1304) feet to a fence corner (X416,113; Y1,471,886), a corner of the lands of T. E. Morgan, and the G. T. Cabiness Estate... thence with Weeks' line, the south line of the Northwest Quarter (NW $\frac{1}{4}$) of Section Three (3), and a fence line, North eighty-nine degrees eleven minutes West (N89°11'W) two thousand five hundred fifty (2550) feet to a point on the ground shown by S. L. Cobler, a corner of the lands of H. O. Weeks and T. E. Morgan; thence with Morgan's line, the south line of the Northwest Quarter (NW $\frac{1}{4}$) of Section Three (3), and the fence line North eighty-nine degrees eleven minutes West (N89°11'W), one hundred twenty-five (125) feet to the point of beginning.
The above described land contains seventy-nine and six-tenths (79.6) acres more or less, subject to the rights of a county road which affects approximately five-tenths (0.5) acres, and is known as Tract No. GR 275, as shown on Map No. 8-4159-45, prepared by the Engineers of the Tennessee Valley Authority.
The coordinates referred to in the above description are for the Alabama Mercator (East) Coordinate System as established by the U. S. Coast and Geodetic Survey, 1934. The Central Meridian for this coordinate system is Longitude eight-five degrees (85°) fifty minutes (50') no seconds (00").

An outstanding advantage of recording the coordinates of the corners in the deed description is that lost corners can be replaced at any subsequent time without having to resurvey the entire property. As discussed in Art. 19.6, coordinates for the approximate

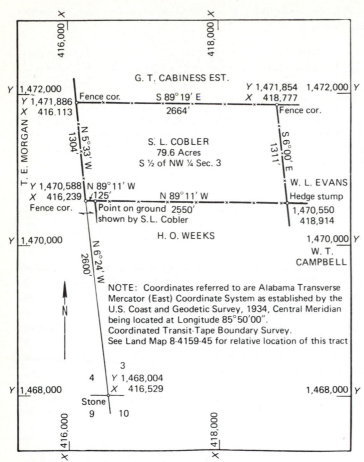

**Fig. 19.2** Land map.

center of the property can be utilized as an identifier for the parcel when filing the deed. For this example, the approximate center is taken as the average values of the coordinates for the corners, which are $X = 417,510$ ft and $Y = 1,471,219$ ft.

Note that both the deed and map for this example fail to indicate the date of the survey or name of the surveyor. Normally, these items of information are included in both the description and on the map.

In Example 19.3, coordinates are given for each corner. Theoretically, a single coordinated point in a figure which closes mathematically is adequate to position (but not to orient) the property. As a rule, coordinates for at least two points should be given. Consider the deed of Example 19.1 and Fig. 19.1 using coordinates for two points:

**Example 19.4** To illustrate the use of two coordinated points in this description only the necessary portions of the deed are repeated as follows:

A parcel of land in the City of _____ , County of... and described as follows:

Beginning at a concrete monument on the northerly line of Woodlawn Avenue, having established grid coordinates of $X =$ _____ , $Y =$ _____ , North Zone of the _____ State Coordinate System, said monument bearing S47°40′E a distance of 100.00 ft (30.480 m) along the northerly line of... (3) thence along the southerly line of Kingston Avenue N47°40′W a distance of 102.88 ft (31.358 m) to an iron pin ($\frac{1}{2}$ in or 13 mm outside diameter), having grid coordinates of $X =$ _____ , $Y =$ _____ , of said North Zone, thence along the land of the grantor S42°20′W a distance of 130.00 ft (39.624 m) to the point of beginning.

All directions are based on the stated ... data gathered in the field by _____ on June 15, 1978.

Whenever possible, the coordinates cited in the description should be for control points of a higher-order survey. Note that in each description the state plane coordinate system and zone are designated.

Coordinates do provide an excellent method for identifying and reestablishing the location of a point and an unambiguous means for parcel identification compatible with computer technology. As such, coordinates should be employed in all descriptions. However, coordinates alone do not provide a legal description for a property corner except where laws have been established to recognize coordinated points as the sole criteria for property lines. In most regions, court decisions in the past have firmly established monuments as the best evidence of property ownership.

Properly written legal descriptions provide checks and ties to which property lines must correspond. Coordinates in a given system referred to a known datum certainly provide an invaluable supplement to the description but do not furnish sole evidence of property ownership.

It is the responsibility of the surveyor to become familiar with the state, county, and local laws of the area in which the surveying practice occurs so as to know the legal status of coordinates in a description.

**19.9. Descriptions by subdivisions**   When land is partitioned into lots and a map showing the manner in which the land has been divided and also satisfying state, county, and local laws is placed on record, a subdivision has been created. The significant aspects of land descriptions by subdivision are (1) a plan that shows all dimensions, directions, areas, and monuments is recorded; and (2) all lots within a given subdivision are created simultaneously so that there are no senior rights involved among lots within a subdivision. Sectionalized land surveyed according to the Public Land Surveys in the United States (Arts. 19.21 to 19.48) and private subdivisions fall within this definition.

Most states have laws that control subdivision development. In the more densely populated regions, counties and local municipal governments also have laws regulating subdivisions. Details concerning the development of subdivision plans and their subsequent approval and recording are given in Art. 19.18.

When the boundaries of the property to be described coincide exactly with a lot within a subdivision for which there is an official recorded map, the lot may be described legally by a statement giving the lot and block numbers and the name, date, and place of filing the map. Most city property and many lots in urban developments are described in this way. The following are descriptions of this type:

**Example 19.5**   Lot 29, Block 0, Map of Berkeley Highlands Addition, filed November 20, 1912, in Map Book 8, Page 194, Contra Costa County Records, State of California.

Lot 15 in Block 5 as said lots and blocks are delineated and so designated upon that certain map entitled *Map of Thousand Oaks, Alameda County, California*, filed August 23, 1909, in Liber 25 of Maps, page 2, in the office of the County Recorder of the said County of Alameda.

If the boundaries of a given tract within a subdivision for which there is a recorded map do not conform exactly to boundaries shown on the official map, the tract is described by metes and bounds (Art. 19.8), with the point of beginning referred to a corner shown on the official map. Also, the numbers of lots of which the tract is composed are given. Following is an example of a description of this kind:

**Example 19.6**   Beginning at the intersection of the Northern line of Escondido Avenue, with the Eastern boundary line of Lot 16, hereinafter referred to; running thence Northerly along said Eastern boundary line of Lot 16, and the Eastern boundary line of Lot 17, eighty-nine (89) feet; thence at right angles Westerly, fifty-one (51) feet; thence South 12°06′ East, seventy-five (75) feet to the Northern line of Escondido Avenue; thence Easterly along said line of Escondido Avenue, fifty-three and $\frac{13}{100}$ (53.13) feet, more or less, to the point of beginning.

Being a portion of Lots 16 and 17, in Block 5, as said lots and blocks are delineated and so designated upon that certain map entitled *Map of Thousand Oaks, Alameda County, California*, filed August 23, 1909, in Liber 25 of Maps, page 2, in the office of the County Recorder of the said County of Alameda.

Note that if land is divided within a subdivision and metes and bounds descriptions are written into the deed, senior rights must be observed in the order in which the division of the land occurs.

**19.10. Legal interpretation of deed descriptions**    The descriptions of the boundaries of a tract include the objects that fix the corners, the lengths and directions of lines between the corners, and the area of the tract. A deed description may contain errors or mistakes of measurement, or mistakes of calculation or record, thus introducing inconsistencies which cannot be reconciled completely when retracement becomes necessary. In such cases, where uncertainty has arisen as to the location of property lines, it is a universal principle of law that the endeavor is to make the deed effectual rather than void, and to execute the intentions of the contracting parties. The following general rules are pursuant to this principle:

**1. Monuments**    It is presumed that the visible objects which marked the corners when a conveyance of ownership was made indicated best the intentions of the parties concerned; hence it is agreed that a corner is established by an existing material object or by conclusive evidence as to the previous location of the object. A corner thus established will prevail against all other conflicting evidence, provided there is reason to believe that the monument was set in accordance with the original intention and that its location has not been disturbed. The kinds of evidence which are valid in relocating obliterated corners are stated in Art. 19.48. Specifically, monuments will control over distance, bearing, and area.

**2. Distance, direction, and area**    In case of conflicts among "calls" in the deed (or dimensions on a recorded subdivision map) for distance, bearings, and area, the following order of importance is observed: distance controls over bearings; and bearings control over area.

**3. Mistakes**    It is a well-established principle that a deed description which taken as a whole plainly indicates the intentions of the parties concerned will not be invalidated by evident mistakes or omissions. For example, such obvious mistakes as the omission of a full tape length in a dimension or the transposition of the words "northeast" for "northwest" will have no effect on the validity of a description, provided it is otherwise complete and consistent or provided its intention is manifest.

**4. Purchaser favored**    In the case of a description that is capable of two or more interpretations, that one will prevail which favors the purchaser (the "grantee").

**5. Ownership of highways**    Land described as being bounded by a highway or street often conveys ownership to the center of the highway or street. Any variation from this interpretation must be explicitly stated in the description.

**6. Original government surveys presumed correct**    Errors found in original government surveys do not affect the location of the boundaries established under those surveys, and the boundaries remain fixed as originally established.

**19.11. Riparian rights**  An owner of property that borders on a body of water is a riparian proprietor and has riparian rights (pertaining to the use of the shore or of the water) which may be valuable. Because of the difficulties arising from the irregularity of such boundaries, it is important that the surveyor be familiar with the general principles relating to riparian rights and with the statutes and precedents established in the particular state where work is being performed. For example, as regards the ownership of the bed of a navigable river, the two states of Iowa and Illinois bordering on the same river have very different laws. Clark (Ref. 6) states: "It is a rule of property in Illinois, that the fee of the riparian owner of land in that state bordering on the Mississippi River extends to the middle line of the main channel of the river," whereas the Iowa courts hold "that the bed of the Mississippi River and the banks to the high-water mark belong to the state, and that the title of the riparian proprietor extends only to that line."

Before developing the principles that govern riparian rights, consider some frequently used definitions.

**Alluvium**  That increase of earth on a bank of a river, or on the shore of the sea, by the force of the water, as by a current or by waves, or from the recession of water in a navigable lake, which is so gradual that no one can judge how much is added in a given interval, is known as alluvium. The proprietor of the bank which is increased by alluvium is entitled to the addition, this being regarded as the equivalent for the loss he or she might sustain from the encroachment of the waters upon the land. The process by which alluvium is formed is called *acretion* or *alluvion*.

**Avulsion**  The removal of a considerable quantity of soil from the land of one owner and its deposit upon or annexation to the land of another, suddenly and by the perceptible action of water, is avulsion. In such case the property belongs to the first owner. Avulsion by the Missouri River, the middle of whose channel forms the boundary line between the states of Missouri and Nebraska, works no change in such boundary, but leaves it in the center line of the old channel. Avulsion is the opposite of acretion and is sometimes called *revulsion*.

**Bed**  The bed of a lake or river is generally that portion covered by water for a long enough time to prevent vegetation from growing.

**High water**  The high-water mark is wherever the presence of the water is so common as to mark on the soil a character, in respect to vegetation, distinct from that of the banks; it does not include low lands which, though subject to periodic overflow, are valuable for agricultural purposes.

That part of the shore of the sea to which the waves ordinarily reach when the tide is at its highest is also known as the high-water mark.

**Low-water mark**  Low-water mark is that part of the shore of the sea to which the waters recede when the tide is lowest, that is, the line to which the ebb tide usually recedes; or it is the ordinary low-water mark unaffected by drought. It has been said to be the point to which a river recedes at its lowest stage.

**Meander line**  A meander line is run by the surveyor for the purpose of determining the shape and size of the body of water. In the U. S. public-land surveys where regular corners fall in water, traverses called *meander lines* are run roughly following the bank of stream or shore of lake (see also Art. 19.38). The process of establishing such a line is called *meandering*. Meander lines are for surveying and mapping purposes only and are not property lines except in the rare cases where they are specifically stated as property lines in a deed.

***Thread of a river***   The line formed an equal distance from the shores is called the *thread of the river*. The thread of a lake is the line connecting the thread of the inlet with the thread of the outlet. If the lake has no inlet or outlet, the thread passes through the longest axis of the lake. The thread of a river or lake is determined at normal water level.

***Reliction***   The increase of the land by retreat or recession of the sea or a river is known as reliction.

In establishing the property lines of riparian owners many dissimilar and complex situations are encountered, but the principles that usually apply are stated below under six general cases.

***1. Meander lines***   It is a well-established principle that government patents of land bordering on meandered streams or lakes convey ownership, not to the meander line, but to the thread of a nonnavigable stream, or to the bank of a navigable stream, or to the shore of a lake.

***2. Origin of dividing lines***   There are two opposing lines of decisions. Under one it is held that a dividing line has its origin at the high-water line of a river, or at the shore line of a lake, and not at the meander line. Thus, in Fig. 19.3, the dividing line between lots 2 and 3 would be made perpendicular to the thread of the stream, beginning at $e$ (on the high-water line) and not at $E$ (on the meander line). Under the other line of decisions, the reverse is held.

***3. Alluvion and reliction***   The direction of the property lines dividing areas created by alluvium or by reliction is determined by the proportional lengths of the old and of the new shore lines. The extremities of these lines are fixed either by definite bends, as $A$, $F$, $A'$ and $F'$ (Fig. 19.4) or by the intersections of the old and new lines as $A$ and $F$ (Fig. 19.5). The general rule is to measure along the old shore line between the old extremities, as $A$ and $F$; measure along the new shore line between the new extremities, as $A'$ and $F'$; and divide the new line ($A'F'$) into parts proportional in length to those of the old line ($AF$). Thus, for lots 4 and 12 by proportion $B'C'/BC = A'F'/AF$. The area $BB'C'C$ represents the area of alluvium added by acretion or alluvion to lots 4 and 12.

***4. Bays or coves***   Property lines fixing riparian rights in bays or coves sometimes are established by lines beginning at the extremities of the property lines on shore and having a direction perpendicular to a line connecting the adjacent headlands of the bay or cove. Thus, the lines $BB'$, $CC'$, etc., for lots 1, 2, and 3 (Fig. 19.6) are established perpendicular to the line $AF$, which connects the two headlands $A$ and $F$.

Other court decisions have fixed the lines according to the following rule: Divide the straight line joining the headlands ($AF$ in Fig. 19.6) into parts proportional to the lengths of

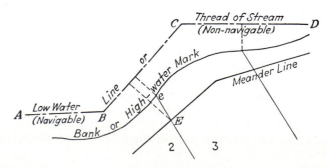

**Fig. 19.3** Origin of riparian boundaries.

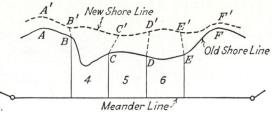

**Fig. 19.4** Riparian boundaries for areas added by alluvion or acretion between definite bends in a shoreline.

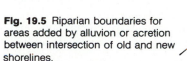

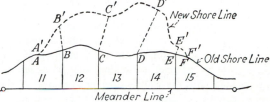

**Fig. 19.5** Riparian boundaries for areas added by alluvion or acretion between intersection of old and new shorelines.

the shoreline held by each owner; the property line of inundated land is determined by joining the extremities of the property lines on shore and the corresponding points of subdivision on the line between headlands. These are shown in the figure by lines $BB''$, $CC''$, etc.

**5. Streams and rivers**   The lines fixing the riparian rights of owners of property bordering on streams and rivers are established by lines perpendicular to the thread of the stream if nonnavigable, or to the low-water line (sometimes to the middle of the channel) if navigable. Thus, the lines for lots 2 and 3 of Fig. 19.3 are established perpendicular to the line $ABCD$.

**6. Lakes**   The riparian property lines are established perpendicular to the center line of the lake; or, in the case of a circular shoreline, by lines to the thread of the lake. Thus, in Fig. 19.7, the lines for lot 7 are established by the boundary $ABCDE$, and for lot 4 by the boundary $FGCHI$. Where the shoreline is circular at the end of the lake, the land lines terminating at $J$ and $K$ are drawn to $O$, the center of the circular shoreline.

**19.12. Adverse possession**   The many legal aspects of adverse possession cannot be treated here, but it is desirable to direct attention to the important fact that property lines may be fixed by continued possession and use of the land (usually for 20 years) as against original survey boundaries. The conditions and the period of time necessary to gain title are fixed by statute in the various states.

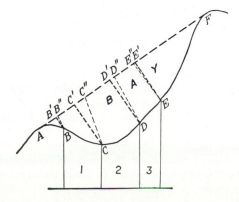

**Fig. 19.6** Riparian boundaries in a bay.

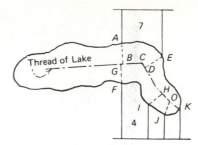

**Fig. 19.7** Riparian boundaries in a lake.

Adverse possession, to become effective, must be plainly evident to the owner, without his or her permission, to his or her exclusion, and hostile to his or her interests. Such possession may be evidenced by fencing, cultivation, erection of buildings, etc.

Right to title by adverse possession may be acquired by individuals, corporations, and even by the state. But the statute does not run *against* the state; that is, property in a street or highway cannot be acquired by adverse possession.

Under this principle, if a person should use the land up to a fence and should recognize it as a boundary line, to the exclusion of the owner, for the statutory period, the fence then becomes the legal property line even though it may be shown later that it is not on the true and original line. However, if the possession of the land has not been held adversely, that is, to the exclusion of the owner, and if the fence has merely served the convenience of the persons concerned, both parties recognizing that it was probably not on the true line, title cannot be claimed.

It is therefore clear that the application of the principle of adverse possession is entirely a matter of intention and belief. If land is held openly and notoriously with the intent to acquire title, or with the belief that the occupation is proper and right, title will be granted if and when the statutory requirements are fulfilled. But if by parol agreement or by actions it is manifest that the parties concerned had no intention to occupy beyond the true line, at the same time knowing that the location of the true line was uncertain, title cannot be gained adversely.

Adverse possession under "color of title" will "ripen into title" under the statute of limitations in some jurisdictions in half the time required without color of title; for example, if title may be gained without color of title in 20 years, it may be gained in 10 years with it.

**19.13. Legal authority of the surveyor**   A resurvey may be run to settle a controversy between owners of adjoining property. The surveyor should understand that, although it is possible to act as an arbiter in such cases, it is not within the power of the surveyor legally to fix boundaries without the mutual consent and authority of all interested parties. In the event of a dispute involving court action, the surveyor may present evidence and argument as to the proper location of a boundary, but has no authority to establish such a boundary against the wishes of either party concerned. A competent surveyor by wise counsel can usually prevent litigation; but if no agreement occurs, the boundaries in dispute become valid and defined only by a decision of the court. In boundary disputes the surveyor is an expert witness, not a judge.

The right to enter upon property for the purpose of making public surveys is usually provided by law, but there is no similar provision regarding private surveys. The surveyor (or employer, whether public or private) is liable for damage caused by cutting trees, destroying crops or fences, etc.

Permission should be obtained before entering private property, in order to avoid a possible charge of trespassing. If permission is not granted, it may be necessary to traverse around the property or to resort to aerial photographs. In certain cases, as for example, where a boundary is in doubt or dispute, a court order may be obtained.

**19.14. Liability of the surveyor**   It has been held in court decisions that county surveyors and surveyors in private practice are members of a learned profession and may be held liable for incompetent services rendered. Thus, Clark (Ref. 6), quoting from court decisions, states: "If a surveyor is notified of the nature of a building to be erected on a lot, he may be held liable for all damages resulting from an erroneous survey; and he may not plead in his defense that the survey was not guaranteed." Similarly, it has been held that in any case where the surveyor knows the purpose for which the survey is made, he or she is liable for damages resulting from incompetent work.

The general principle invoked in such cases is that the surveyor is bound to exhibit that degree of prudence, judgment, and skill which may reasonably be expected of a member of the profession. Thus, in the following quotations from Clark, a Connecticut court says "the gist of the plaintiff's cause of action was the negligence of the defendant in his employment as a civil engineer. Having accepted that service from the plaintiff, the defendant... was bound to exercise that degree of care which a skilled civil engineer would have exercised under similar circumstances." Also, a Kansas court declared, "reasonable care and skill is the measure of the obligation created by the implied contract of a surgeon, lawyer, or any other professional practitioner." But Ruling Case Law says, "... yet a person undertaking to make a survey does not insure the correctness of his work, nor is absolute correctness the test of the amount of skill the law requires. Reasonable care, honesty, and a reasonable amount of skill are all he is bound to bring to the discharge of his duties."

**19.15. Original survey**   The need for an original survey usually arises when one person desires to transfer to another a tract of land which has not been previously surveyed but which is defined by certain natural or artificial features of the terrain.

With the desired boundaries of the land given, the surveyor establishes monuments at the corners and runs a closed traverse around the property, measuring the lengths of lines and the angles between intersecting lines. Whenever possible the traverse should be tied to monuments in the state plane coordinate systems established by the National Ocean Survey. Where boundaries are not straight, offsets from transit line to curved boundary are measured at known intervals; and where obstructions make direct measurement along boundaries impossible, the traverse is run as close to the boundary as convenient and measurements are taken from transit stations to corners of the tract. Angular measurements may be taken by any of the methods described in Chap. 6, but usually the interior angles are observed. Preferably the corners should be referenced to permanent objects. Also the direction of the astronomic meridian should be determined, usually by a solar observation (see Art. 11.32).

The information thus obtained is recorded in the surveyor's notebook, the angles and distances of the main traverse being tabulated, and the remaining data being recorded in the form of a sketch. The bearings of the sides are then computed with respect to the true or astronomic rather than the magnetic meridian.

A description of the tract, by metes and bounds and/or by coordinates, is prepared (Arts. 19.7 and 19.8). Usually, a plat is drawn, the boundaries being plotted by one or another of the methods described in Chap. 13, and details being shown as suggested by Fig. 19.2 and Art. 13.19. The area is calculated as described in Chap. 8. In the process of calculation, the error of closure of the traverse is determined and thus a check on the reliability of the survey is obtained. A copy of the description and a tracing and/or prints of the plat are submitted to the person for whom the survey is made.

When state, county, or other local laws require it, the map or plat must be approved and filed in the county recorder's office. These regulations vary from state to state and from county to county. Thus, it is the responsibility of the surveyor to become familiar with the rules governing division of land in the region where the work occurs.

**19.16. Resurveys** The resurvey of lands is attended with greater difficulty than is usually appreciated by those inexperienced in work of this character. This is particularly true in the older sections of the United States where the early surveys were not of the rectangular system and were not under the control of the U.S. Bureau of Land Management. The proper relocation of old lines calls for greater ingenuity and broader experience on the part of the surveyor than does any other kind of surveying.

The purpose of the resurvey is to reestablish boundaries in their original locations. As a guide the surveyor has available the description contained in the deed or obtained from old records, and descriptions of adjoining property.

The first step in a resurvey is to obtain a copy of the deed description from the owner. It should be established that the document is the deed, not a tax statement or other incomplete document. Next, the surveyor should obtain copies of deed descriptions and maps of all adjoining properties, city or county surveyor's records, township plats, utility maps, field notes (if any), and record of surveys. Depending on the region being surveyed and the local regulations in effect, many of these records may not be available or even exist. However, a thorough search should be performed to ensure that no records potentially useful for the survey have been neglected. Copies of deeds are generally found in the county recorder's office, wills at the recorder's office, and maps or plans at the county surveyor's office or in city hall. These arrangements vary from community to community and it is necessary to become familiar with the situation in the region where the surveying takes place.

When all records are assembled, the surveyor examines the descriptions, maps, and plats for gross errors, computes departures and latitudes for the courses given in the description, determines the closure error, and plots the boundaries of the tract to scale. When a metes and bounds description is involved, the seniority of the deeds must be determined. A copy of the title insurance policy gives the order of seniority of the deeds related to the property being surveyed and should be used if available.

Next, data concerning the locations of all marks that show the positions of streets, roads, or other public properties near the tract to be surveyed should be ascertained. The state, county, township, or city engineer may have this information. If the property is near a newly constructed state highway, the right of way for the highway is generally marked by monuments which have state plane coordinates and ties to the nearest property corners. In many states, this information would be filed as a record of survey and is available to the public. Many large cities have established systems of control monuments which have been coordinated in either an arbitrary or the state plane coordinate system. Quite frequently, these monuments are on a line parallel to but offset from the property line and are located in the sidewalks at the corners of the city blocks. The city engineer would have data for these monuments.

Only when all available data have been gathered and analyzed should field surveys commence. The importance of a thorough search for all records, followed by an equally thorough analysis of the data, cannot be overemphasized.

In rural regions, initial survey operations should be concentrated on locating monuments. When monuments are not available or cannot be found, old fence lines, corner posts, stakes, pipes, iron pins, and hedge lines are some of the features that provide indications of the existing boundary locations. Old aerial photographs are very beneficial in some cases.

In urban regions or in the city, monuments should be sought initially. In the absence of monuments, property corners marked by iron pins, metal survey markers, iron pipes, stakes, fence lines, walls, and hedge rows provide a method for establishing the boundaries. When there are sidewalks and curbs, cross marks or drill holes are frequently chiseled into the walks or on the tops of curbs. These marks are generally on the lot lines produced to a line offset an even number of feet or metres from the side line of the street. In the absence of all visible marks on the ground, the center line of the street can be established approximately by splitting the distance between the curb lines or the edges of the pavement. In the city, buildings are frequently constructed on the property lines or may be offset a few tenths of a foot or a few centimetres inside of the line. By measuring the distances between buildings and between curb lines and taking an average of one-half of these measurements, it is possible to establish an approximate center line for the street. This procedure provides a means of getting started in the search for evidence of old monuments or property corners which are not visible.

When the approximate positions for boundaries of the property to be surveyed have been located and in the absence of positive visible evidence of property corners, a control traverse is run around the tract. The traverse lines should be established approximately parallel and as close as is practical to the estimated positions for the property lines. If possible, traverse lines can coincide with the estimated positions for the property lines when existing survey markers are found and there are no obstacles on the line. Permanent markers should be set at all traverse points and each station should be referenced for subsequent relocation. For city and urban surveys, the control traverse may extend around the entire block, particularly if this is the first survey in the area. In rural surveys, the control traverse is generally restricted to the outline of the property being surveyed plus the necessary ties to control monuments not in the immediate vicinity of the survey. Whenever possible in city, urban, or rural surveys the property survey control traverse should be tied to control points in the national geodetic network so that the survey can be coordinated on the state plane coordinate system.

As the control traverse is being run, ties should be measured to all details relevant to the boundaries. Locate property corners, fence lines, hedge rows, walls, and all buildings on the lot being surveyed and on adjoining properties.

The data collected in the field are then reduced, calculated, and plotted in the office. Coordinates are determined for the traverse stations and potential property corners and by inverse computations, the distance and bearings between located, possible property corners can be calculated. The control traverse and all evidence of property lines and corners are then plotted to scale.

The deed information and the physical locations of all evidence in the field are then compared and evaluated and a solution is sought which will best fit all the data, written and physical. This procedure requires substantial skill, experience, and judgment on the part of the surveyor. Frequently, a preliminary examination and computations of all the evidence allows calculation of a bearing and distance from a traverse control station to the possible location of a property corner or monument for which there is a call in the deed. If the called for corner was not found during the initial survey, a return trip should be made to the field with every effort being expended to locate the corner. The bearing and distance can be set off from the designated traverse station and the area examined very carefully to locate any evidence pertaining to the old corner. If no evidence exists on the surface, the soil should be carefully sliced away in increments in an effort to detect the rotted remains of an old corner stake or post or the remains of an old monument.

When a final solution has been reached, the property corners chosen as those which best fit all the data are coordinated and ties by direction and distance to the nearest traverse control stations are computed. Using these computed ties, the property corners, as determined from the final evaluation, are set off in the field from the traverse control stations.

With the property corners established, structures on the tract are located by perpendicular offsets from the nearest property line and a map is prepared showing the property corners, property lines, fence lines, and buildings. Figure 19.8 illustrates such a map, sometimes referred to as a *plat* or plot plan.

The map is submitted to the client along with a report stating exactly what evidence was found and what procedure was followed in reestablishing the property corners.

Depending on the state, county, or local regulations, it may also be necessary for the surveyor to file a record of survey in the appropriate governmental office. Many of the more

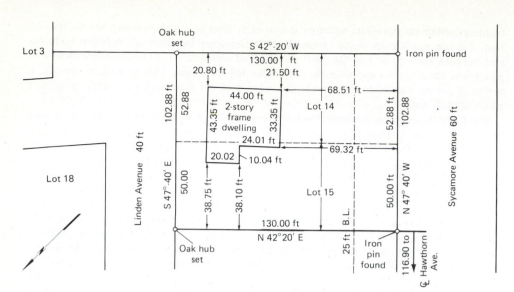

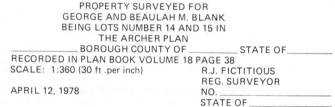

PROPERTY SURVEYED FOR
GEORGE AND BEAULAH M. BLANK
BEING LOTS NUMBER 14 AND 15 IN
THE ARCHER PLAN
_____ BOROUGH COUNTY OF _____ STATE OF_____
RECORDED IN PLAN BOOK VOLUME 18 PAGE 38
SCALE: 1:360 (30 ft .per inch)     R.J. FICTITIOUS
                                   REG. SURVEYOR
      APRIL 12, 1978               NO. _____
                                   STATE OF_____

**Fig. 19.8** Plat or plot plan for a property survey.

densely populated counties in several states now have statutes which require that any division of land be approved by the county surveyor and that the map of the survey be filed with the county recorder.

Obviously, the procedure as outlined is an extensive and expensive project. The cost of such a survey would be more than one could reasonably expect to charge a client for a lot survey. However, the method as described is necessary when doing a first job in an area with which the surveyor is unfamiliar. Subsequent surveys in the same vicinity can be accomplished with much less expenditure of time and money. Thus, the initial investment in a complete search and survey of the area will be returned with dividends. Surveyors frequently must indulge in a certain amount of speculation in order to become familiar with the conditions that govern the locations of boundaries in a given region. In order to exploit the efforts invested in an initial survey, field notes and office records for the survey should be rigorously maintained and all relevant monuments and control traverse stations must be very carefully referenced for future use.

**19.17. Subdivision survey of rural land**   A subdivision survey of rural land implies a survey which is conducted for the purpose of subdividing into two or more tracts, in accordance with some prearranged plan, an area whose boundaries are already established. In such cases, a resurvey of the tract is run, new monuments are established on the new boundary lines, and a new plat and description are prepared as in the case of an original survey.

***Public lands***   The public lands are divided into townships, sections, and quarter sections by United States land surveyors, in a manner prescribed by law. In general the United States surveys establish the boundaries of sections and establish quarter-section corners on section lines; and any further subdivision is made after the lands have passed into the hands of private individuals, the work being carried out by surveyors in private practice. Subdivisions of this kind are described in detail in Arts. 19.35 through 19.37.

**Irregular subdivisions**   Surveys of this class are conducted for a variety of purposes. The following examples serve to illustrate the procedure for certain cases:

**Example 19.7**   A railroad is to traverse the land belonging to Black, and the railroad company desires to secure title to a right of way of a definite width on either side of the center line which has already been surveyed and marked with stakes. A description of Black's tract has been secured.

The right-of-way surveyor reruns the boundaries of Black's tract that are intersected by the railroad line, establishes the directions of the right-of-way boundaries parallel with the center line, and sets monuments at the intersections of these boundaries with those of Black's tract. The surveyor then makes a survey of the tract thus defined, securing sufficient data so that the lengths and directions of the boundaries of the right-of-way tract now within Black's tract are obtained. He or she also ties right-of-way corners which he (or she) has established to the nearest old corners of Black's property.

With these data the area of the right-of-way tract is calculated, and a description of the tract is prepared as for an original survey (see Arts. 19.7, 19.8, and 19.15). The point of beginning is referred to one of the old monuments marking the original tract, and not to the center line of the railroad.

**Example 19.8**   It is stipulated in the will of Green that his New England farm is to be divided equally among his three sons, each to have an equal frontage on the highway which forms one of the boundaries of the tract. The farm is of irregular shape and has not been surveyed for many years.

In a case of this kind, the surveyor first makes a resurvey of the entire tract, angles being measured from $\frac{1}{10}$ to the nearest $\frac{1}{2}$ minute, and distances being measured probably to hundredths of feet. From the data thus obtained the area is calculated. In connection with the resurvey, subdivision corners are established on the highway.

The simplest division is one for which the subdividing lines are straight, each cutting off the required area from the given tract. With the area of the entire tract known, the area each son is to receive is calculated, and the length and bearing of each of the lines rendering the subdivision are computed as described in Art. 8.36.

Finally, from each of the two subdivision corners already established on the highway frontage, the surveyor lays off the computed direction of the subdividing line through that point and establishes the remaining unknown corner at the point where this subdividing line intersects the opposite boundary. The distances from this latter corner to adjacent corners are measured, and the survey is considered as checked if these measured distances agree closely with the computed lengths of the same courses. A plat is drawn as for an original survey, the lengths and bearings of all lines and the area of each subdivision being shown; and a description of each of the three subdivisions is prepared.

**19.18. Subdivision survey of urban land**   As a city or town develops, unimproved lands are subdivided into lots which are placed on sale as residential or business property. In most instances such extensions are the result of the activities of real-estate operators who acquire a tract of rural land of considerable area, develop a plan of subdivision which is approved by the authorities of the municipality to which the tract is to be attached, and cause surveys to be made for the purpose of establishing the boundaries of individual lots. A tract thus divided according to an acceptable plan is known as an *addition* or *subdivision*.

For large and important developments the work of originating the general plan is often carried out by persons specializing in city planning and landscape architecture, under whose direction the surveyor works. Such developments require a high degree of skill, and usually extensive surveys (particularly in hilly sections) are carried out before the actual plan of subdivision can be decided upon. Problems of this character can be adequately discussed only in treatises on city planning. However, it is appropriate to state here that the preliminary studies should consider the probable future character of the district; the probable location of business sections; the probable magnitude, direction, and character of future traffic; the topography of the land; the location, width, grade, and character of paving of streets; the size and shape of lots and blocks; the location and size of storm and sanitary sewers; and the disposition of electric and telephone wires and cables.

For the ordinary real-estate development the owner usually calls for the services of an engineer or surveyor who has had experience in such work. The surveyor confers with the owner, and they discuss a general plan. The surveyor makes a resurvey of the entire property; and if the character of the topography is irregular, certain preliminary surveys for the purpose of finding the location and elevation of the governing features of the terrain are made. In some cases a complete topographic survey may be made. With the general plan fixed, and having studied the results of the field investigation and having considered the items listed in the previous paragraph, the surveyor works out a detailed plan on paper, showing on the drawing the names of all streets and the numbers of all blocks and lots, the dimensions of all lots, the width of streets, the length and bearing of all street tangents, and the radius and length of all street curves. The surveyor also prepares a report which, in addition to a discussion of the plan of subdivision, may consider the cost of subdividing, including not only the establishing of boundaries but also the work of grading, paving, constructing sewers, and landscaping.

This detailed plan, when approved by the owner, is submitted to the governing body in the municipality. If it meets with the requirements of this body, it is approved.

Upon the authority of the owner, the surveyor then proceeds to execute the necessary subdivision surveys, including the laying out of roads, walks, blocks, and lots. Often the lot and block corners are marked with permanent monuments; but in many cases, contrary to what may be considered good practice, the lot corners are marked by wooden stakes. When the surveys are completed, the map of the subdivision is revised to show minor changes made during the survey, together with the location and character of permanent monuments. A tracing is submitted to the municipality, and this, when duly signed by those in authority, becomes the official map of the subdivision. It then becomes a part of the public records and is usually filed in the registry of deeds of the county in which the municipality lies. Upon this approval, if the subdivision is outside the corporate limits of the municipality, they are extended to include it.

Sample portions of typical subdivision plans are shown in Fig. 19.9. The line work, distances, and directions on the subdivision plan illustrated in Fig. 19.9*b* were plotted automatically (see Art. 13.36). The latter plan is in a form suitable for recording as the final official map of the subdivision.

***Resurveys within a subdivision***   When land is bought and sold within a subdivision, often a resurvey is required to relocate lost property corners. These surveys are in general performed according to the procedures set forth in Art. 19.16. Since there is an official recorded map containing all distances and directions which are referenced to monuments and lot corners, many of which exist on the ground, the problems related to subdivision resurveys are not so complex as those found in metes and bounds resurveys. However, problems do occur. Many older subdivision plans are incomplete and contain incorrect information. Furthermore, incorrect information can occur on any such plan, even one that was approved and presumably satisfied all requirements.

Prior to performing the survey, always calculate a mathematical closure using recorded map data not only for the lot to be surveyed but also for the entire block within which the lot falls. When the field survey is performed, always close around the block in which the lot falls. In this way the surveyor can be certain that the lot corners established by the resurvey do not conflict with any other property owner in the block.

Since all lots set off according to a recorded map for a given subdivision are created simultaneously, all property owners have equal rights and there are no senior deeds. Thus if the surveyor finds an excess or shortage of distance within the block in which a lot is being surveyed, it is distributed proportionally to all the lots within the block except where there is an obvious mistake in the exterior survey, which is corrected where it occurs.

For example, in Fig. 19.9a, assume that the corners for lot 6 in block 5 are to be resurveyed. The center lines of Highland Road and Amanda and Hopewell Streets are established from monuments and original property corners referenced in the original survey notes for the subdivision. Next, the distance of 285.00 ft (86.868 m) is measured from the center-line intersection of Hopewell and Amanda Streets to the east boundary of lot 6. Temporary points are set on the center line of Amanda Street at the east and west lot lines of lot 6 extended. Then the distance is measured from the west boundary of lot 6 along the center line of Amanda Street to the center line of Highland Road. If this distance measured in the field compares with the map distance so as to yield an acceptable relative accuracy for the block, the property corners for lot 6 are set from the temporary points on the center line of Amanda Street.

However, suppose that an excess of 0.60 ft in the dimension of block 5 was detected on closing at the center line of Highland Road. After double checking to be sure there are no errors in measurement such as uncorrected systematic errors, mistakes, etc.; that the plan dimensions close for block 6; and that the center-line intersections are correctly established, the excess of 0.60 ft is distributed proportionately among all the lots in block 5. Since the distance corner to corner is 570 ft, lots 1 and 11 receive (60/570)(0.60)=0.063 ft (0.019 m) each and the remaining lots receive (50/570)(0.60)=0.053 ft (0.016 m) each. Consequently, the distance from the center line of Hopewell Street to the east boundary of lot 6 is 285.274 ft (79.332 m), the frontage for lot 6 is 50.053 ft (15.256 m), and the distance from the west boundary of lot 6 to the center line of Highland Road should check with a measurement of 285.274 ft (86.952 m). Distances are carried to the third place to avoid round-off error in cumulating the proportional change in lot dimensions. In practice, all distances would be rounded to the second place using feet or to the third place using metres.

The foregoing procedure is called *single proportionate measurement. Double proportionate measurement* (Art. 19.47) occurs when a corner is reestablished from four known points and is applicable when land is subdivided in a checkerboard pattern as in the U.S. public-land surveys system.

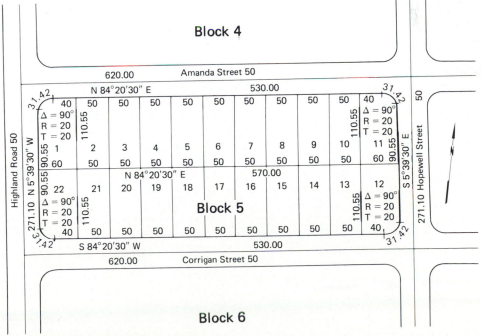

**Fig. 19.9a** Portion of a subdivision lot plan.

**19.19. City surveying**　The term *city survey* refers to an extensive coordinated survey of the area in and near a city for the purposes of fixing reference monuments, locating property lines and improvements, and determining the configuration and physical features of the land for a city lot. In Ref. 1, Manual No. 10, it is characterized as "an inventory of the physical facts relating to the land and its occupation." Such a survey is of value for a wide variety of purposes, particularly for planning city improvements. The technical procedure for city surveys is described in detail in Ref. 1. Briefly, the work consists of:

1. Establishing *horizontal and vertical control*, as described for topographic surveying. The primary horizontal control is usually by triangulation and/or trilateration, supplemented as desired by precise traversing. Secondary horizontal control is established by traversing of appropriate precision. Photogrammetric methods for establishing horizontal control are feasible for setting control points in the secondary network. All horizontal control should be tied to the state plane coordinate system. Primary vertical control is by precise leveling and should be referred to the National Geodetic Vertical Datum (NGVD) of 1929 (Art. 13.2).

2. Making a *topographic survey* and a *topographic map*. Usually, the scale of the map may range from 1 : 1200 to 1 : 2400 (1 : 1000 to 1 : 2500 in the metric system) with contour intervals of from 1 ft to 5 ft (0.5 m to 2 m). Photogrammetric methods for map compilation should be utilized.

3. *Monumenting* a system of selected points at suitable locations such as street corners, for reference in subsequent surveys. These monuments are referred to the state plane coordinate system and to the national datum.

4. Making a *property map*. The survey for the map consists of (*a*) collecting recorded information regarding property; (*b*) determining the location on the ground of street intersections, angle points, and curve points; (*c*) monumenting the points so located; and (*d*) determining the coordinates of the monuments. Usually, the scale of the property map is 50 ft/in or 1 : 600 (1 : 500 in metric units). The property map shows the length and bearing of all street lines and boundaries of public property, coordinates of governing points, control, monuments, important structures, natural features of the terrain, etc., all with appropriate legends and notes.

5. Making a *wall map*, which shows essentially the same information as the topographic map but which is drawn to a smaller scale; the scale should be not less than 2000 ft per in or 1 : 24,000 (1 : 25,000 in metric units). The wall map is reproduced in the usual colors—culture in black, drainage in blue, wooded areas in green, and contours in brown.

6. Making a *map of underground utilities*. Usually the scale and the size of the map sheets are the same as those for the property map. The underground map shows street and easement lines, monuments, surface structures and natural features affecting underground construction, and underground structures and utilities (with dimensions), all with appropriate legends and notes.

**19.20. Cadastral surveying**　Cadastral surveying is a general term referring to extensive surveys relating to land boundaries and subdivisions made to create units suitable for transfer or to define limitations of title. The expression is derived from the word "cadastre," meaning register of the real property of a political subdivision with details of area, ownership, and value (see Ref. 1, Manual No. 32). The term is applied to the U.S. public-land surveys (Arts. 19.21 to 19.48) by the U.S. Bureau of Land Management and may also be used to describe corresponding surveys outside the public lands. However, the term property, land, or boundary surveys is usually used by preference.

A cadastral map shows individual tracts of land with corners, length and bearing of boundaries, acreage, ownership, and sometimes the cultural and drainage features. The surveying methods are the same as those described for topographic surveying for maps of intermediate and large scale (Chap. 15).

**19.21. U.S. public-lands surveys: general** The following sections deal with the methods of subdividing the public lands of the United States in accordance with regulations imposed by law. The public lands are subdivided into townships, sections, and quarter sections—in early years by private surveyors under contract, later by the Field Surveying Service of the General Land Office, and currently by civil service employees of the Bureau of Land Management which succeeded the General Land Office in 1946. Further subdivision of such lands is made after the lands have passed into the hands of private owners, the work being carried out by surveyors in private practice.

The methods described herein are those now in force, but with minor differences they have been followed in principle since 1785, when the rectangular system of subdivision was inaugurated. Under this system, the public lands of 30 states have been or are in progress of being surveyed. In general, these methods of subdividing land do not apply in the 13 original states and in Hawaii, Kentucky, Tennessee, Texas, and West Virginia. As the progress of the public-land surveys has been from east to west, the details in states east of the Mississippi River differ somewhat from those of present practice.

The laws regulating the subdivision of public lands and the surveying methods employed are fully described in the *Manual of Instructions for the Survey of the Public Lands of the United States*, published by the Bureau of Land Management (Ref. 20). From the manual is drawn much of the material for this chapter.

Field notes and plats of the public-land surveys may be examined in the regional offices of the Bureau, and copies may be procured for a nominal fee.

**19.22. General scheme of subdivision** The regulations for the subdivision of public lands have been altered from time to time; hence, the methods employed in surveying various regions of the United States show marked differences, depending upon the dates when the surveys were made. In general principle, however, the system has remained unchanged, the primary unit being the *township*, bounded by meridional and latitudinal lines and as nearly as may be 6 mi (9.66 km) square. The township is divided into 36 secondary units called *sections*, each as nearly as may be 1 mi (1.61 km) square. Because the meridians converge (Art. 19.29), it is impossible to lay out a square township by such lines; and because the township is not square, not all the 36 sections can be 1 mi square even though all measurements are without error.

**19.23. Standard lines** Since the time of the earliest surveys, the townships and sections have been located with respect to principal axes passing through an origin called an *initial point*; the north-south axis is a true or astronomic meridian called the *principal meridian*, and the east-west axis is an astronomic parallel of latitude called the *base line*.

The principal meridian is given a name to which all subdivisions are referred. Thus, the principal meridian which governs the rectangular surveys (wholly or in part) of the states of Ohio and Indiana is called the First Principal Meridian; its longitude is 84°48′11″W, and the latitude of the base line is 40°59′22″N. The extent of the surveys which are referred to a given initial point may be found by consulting a map, published by the Bureau of Land Management, entitled "United States, Showing Principal Meridians, Base Lines, and Areas Governed Thereby," or from Ref. 20.

Secondary axes are established at intervals of 24 mi (38.62 km) east or west of the principal meridian and at intervals of 24 mi north or south of the base line, thus dividing the tract being surveyed into quadrangles bounded by astronomic meridians 24 mi long and by true or astronomic parallels, the south boundary of each quadrangle being 24 mi long, and the north boundary being 24 mi long less the convergence of the meridians in that distance. [In some early surveys, these distances were 30 (48.28 km) or 36 (57.94 km) mi.] The secondary parallels are called *standard parallels* or *correction lines*, and each is continuous throughout its length. The secondary meridians are called *guide meridians*, and each is broken at the base line and at each standard parallel.

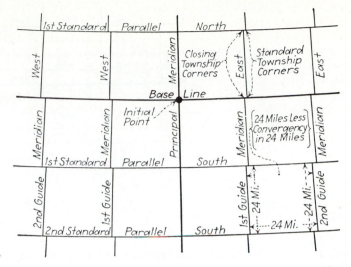

**Fig. 19.10** Standard lines.

The principal meridian, base line, standard parallels, and guide meridians are called *standard lines*.

A typical system of principal and secondary axes is shown in Fig. 19.10. The base line and standard parallels, being everywhere perpendicular to the direction of the meridian, are laid out on the ground as curved lines, the rate of curvature depending upon the latitude. The principal meridian and guide meridians, being astronomic north and south lines, are laid out as straight lines but converge toward the north, the rate of convergence depending upon the latitude.

Standard parallels are counted north or south of the base line; thus, the *second standard parallel south* indicates a parallel 48 mi (77.25 km) south of the base line. Guide meridians are counted east or west of the principal meridian; thus the *third guide meridian west* is 72 mi (154.50 km) west of the principal meridian.

**19.24. Townships** The division of the 24-mi (38.62 km) quadrangles into townships is accomplished by laying off astronomic meridional lines called *range lines* at intervals of 6 mi (9.66 km) along each standard parallel, the range line extending north 24 miles to the next standard parallel; and by joining the township corners established at intervals of 6 miles (9.66 km) on the range lines, guide meridians, and principal meridian with latitudinal lines called *township lines*.

The plan of subdivision is illustrated by Fig. 19.11. A row of townships extending north and south is called a *range*; and a row extending east and west is called a *tier*. Ranges are counted east or west of the principal meridian and tiers are counted north or south of the base line. Usually, for purposes of description the word "township" is substituted for "tier." A township is designated by the number of its tier and range and the name of the principal meridian. For example, T7S, R7W (read *township seven south, range seven west*) designates a township in the seventh tier south of the base line and the seventh range west of the principal meridian.

**19.25. Sections** The division of townships into sections is performed by establishing, at intervals of 1 mi (1.61 km), "meridional" lines parallel to the east boundary of the township and by joining the section corners established at intervals of 1 mi with straight latitudinal lines. (These lines are not exactly meridional, but are parallel to the east boundary of the township, which is a meridional line.) These lines, called *section lines*, divide each township into 36 sections, as shown in Fig. 19.12. The sections are numbered consecutively from east to west and from west to east, beginning with No. 1 in the northeast corner of the township

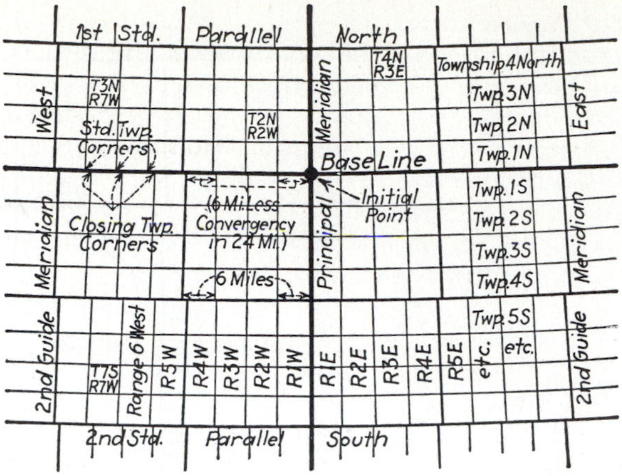

**Fig. 19.11** Township and range lines.

and ending with No. 36 in the southeast corner. Thus section 16 is a section whose center is $3\frac{1}{2}$ mi (5.63 km) north and $3\frac{1}{2}$ mi (5.63 km) west of the southeast corner of a township.

A section is legally described by giving its number, the tier and range of the township, and the name of the principal meridian; for example, Section 16, T7S, R7W, of the Third Principal Meridian.

On account of the convergence of the range lines (true or astronomic meridians) forming the east and west boundaries of townships, the latitudinal lines forming the north and south boundaries of townships are less than 6 mi (9.66 km) in length, except for the south boundary of townships that lie just north of a standard parallel. As the north-south section lines are run parallel to the *east* boundary of the township, it follows that all sections except those adjacent to the west boundary will be 1 mi (1.61 km) square, but that those adjacent to the west boundary will have a latitudinal dimension less than 1 mi by an amount equal to the convergence of the range lines within the distance from the section to the nearest standard parallel to the south.

The subdivision of sections is described in Arts. 19.35 through 19.37.

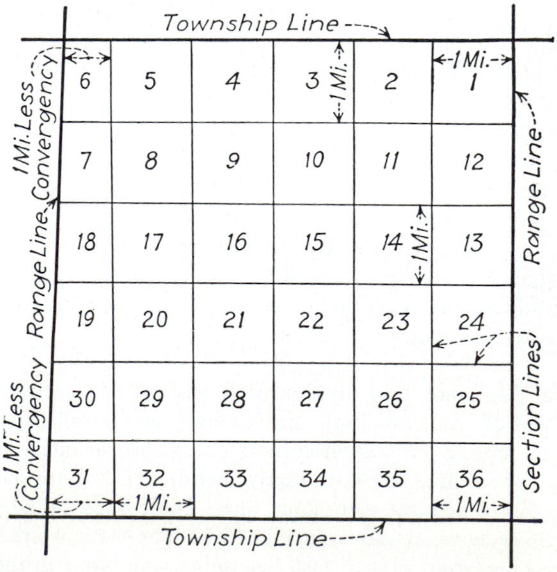

**Fig. 19.12** Numbering of sections.

**19.26. Standard corners**  Corners called *standard corners* are established on the base line and standard parallels at intervals of 40 chains or 2640 ft (804.67 m); these standard corners govern the meridional subdivision of the land lying between each standard parallel and the next standard parallel to the north. Other corners called *correction corners* or *closing corners* are later established on the base line and standard parallels during the process of subdivision; these corners fall at the intersection of the base line or standard parallel either with the meridional lines projected from the standard township corners of the next standard parallel to the south (see Fig. 19.11) or with the intermediate section and quarter-section lines. Standard parallels are also called *correction lines*.

**19.27. Irregularities in subdivision**  It should be understood that the plan of subdivision just described is the one which is carried out when conditions allow. Of course, errors of measurement are always present so that the actual lengths and directions established in the field do not entirely agree with the theoretical values. But in addition, conditions met in the field often make it inexpedient or impossible to establish the lines of the survey in exact accordance with the specified plan. Thus, there are numerous instances of standard parallels and guide meridians having been originally established at intervals of 30 and 36 mi (48.27 and 57.92 km), under old regulations; and of regions having been only partly surveyed. Later, under present regulations, meridians have been established between the old guide meridians; and recent subdivisions are, therefore, referred to standard lines many of which are less than 24 mi (38.62 km) apart. Also, the presence of large bodies of water, mountain ranges, Indian reservations, etc., may greatly modify the method of division, many townships and sections being made fractional.

**19.28. Establishing standard lines**

*Principal meridian*  The principal meridian is established as an astronomic meridian through the initial point, either north or south, or in both directions, as conditions require. Permanent quarter-section and section corners are established alternately at intervals of 40 chains ($\frac{1}{2}$ mi or 0.80 km), and regular township corners are placed at intervals of 480 chains (6 mi or 9.66 km).

Independent linear measurements are taken either by two sets of tapemen or, when this is not possible, by the duplication of each measurement by one set of tapemen. When the discrepancy between two sets of measurements taken in the prescribed manner exceeds 20 links (13.02 ft or 4.02 m) per mile, it is required that the line be remeasured to reduce the difference. The use of electronic distance measurement (EDM) equipment is feasible and simplifies setting corners substantially. The corners are set at the mean distances. When successive independent tests of the alignment, as determined by astronomical observations, indicate that the line has departed from the astronomic meridian by more than 03′, it is required that the necessary correction be made to reduce the deviation in azimuth.

*Base line*  From the initial point the base line is extended east and west on a true or astronomic parallel of latitude, standard quarter-section and section corners being established alternately at intervals of 40 chains ($\frac{1}{2}$ mi or 0.80 km) and standard township corners being placed at intervals of 480 chains (6 mi or 9.66 km). The manner of taking the linear measurements of the base line and the required precision of both linear measurements and alignment are the same as for the survey of the principal meridian. Any of the three methods described in Art. 19.30, for laying out the astronomic latitude curve, may be used.

*Standard parallels*  At intervals of 24 mi (38.62 km) north and south of the base line, true parallels of latitude called *standard parallels* or *correction lines* are run east and west from the principal meridian, these lines being established in a manner identical with that prescribed for the survey of the base line.

**Guide meridians**  The guide meridians are extended north from the base line and standard parallels at intervals of 24 mi (38.62 km) east and west of the principal meridian. Each guide meridian is established as a true meridian in a manner identical with that employed in laying off the principal meridian. The guide meridians terminate at the points of their intersection with the standard parallels, and hence are broken lines, each segment being theoretically 24 mi (38.62 km) long. Errors of measurement are placed in the most northerly half mi (0.80 km) of each 24-mi segment. At the point of intersection of the guide meridian and standard parallel, a township corner, called a *correction corner* or *closing corner*, is established by retracing the standard parallel between the first standard corners to the east and to the west of the point for the closing corner; and the distance from the closing corner to the nearest standard corner on the standard parallel is measured.

**19.29. Convergence of meridians**  The angular convergence of two meridians is a function of the distance between the meridians, the latitude, and the dimensions of the reference ellipsoid. In Fig. 19.13, $d$ is the distance between the meridians along parallel of latitude $AB$ that lies at latitude $\phi$. The angular convergence in seconds is

$$\hat{\theta} = \frac{d \tan \phi (1 - \epsilon^2 \sin^2 \phi)^{1/2}}{a \sin 1''} \tag{19.1}$$

where    $a$ = semimajor axis of the ellipsoid = 3963.3 mi = 6378.3 km
$\quad\quad\quad \epsilon$ = eccentricity of the ellipsoid (Clarke 1866)
$\quad\quad\quad\quad$ = 0.0822718948
$\quad\quad \sin 1'' = (4.8481)(10^{-6})$

The linear convergence, $c$, of two meridians having length $\ell$ and separated by a distance $d$ (Fig. 19.13) is given by

$$c = \frac{\ell d \tan \phi (1 - \epsilon^2 \sin^2 \phi)^{1/2}}{a} \tag{19.2}$$

in which $c$, $\ell$, $d$, and $a$ are in the same units.

**Example 19.9**  Find the angular convergence of two guide meridians 24 mi (38.62 km) apart at latitude 43°20′.

SOLUTION  By Eq. (19.1),

$$\hat{\theta} = \frac{(24)(\tan 43°20')[1 - (0.0822718948)^2 \sin^2 43°20']^{1/2}}{(3963.3)(0.000004848)}$$

$\quad$ = 1176.5″
$\quad$ = 0°19′36.5″

**Example 19.10**  Find the linear convergence for the data in Example 19.9.

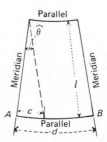

**Fig. 19.13** Convergence between two meridians.

SOLUTION   By Eq. (19.2),

$$c = \frac{(24)^2(\tan 43°20')[1 - (0.0822718948)^2 \sin^2 43°20']^{1/2}}{3963.3} = 0.136896 \text{ mi}$$
$$= 10.952 \text{ chains} = 722.81 \text{ ft} = 220.313 \text{ m}$$

Note in Example 19.9 that the convergence of the two guide meridians is nearly one-third of a degree. In Example 19.10, the linear convergence is nearly 11 chains (726 ft or 220 m) in 24 mi; this amount represents the jog at the correction line in the first guide meridian east or west, or one-half of the jog in the second guide meridian.

In Table V (Appendix C) are given, for each degree of latitude, the linear and angular convergence of meridians 6 mi long and 6 mi apart. The linear convergence represents the correction to be applied to the north boundary of a regular township in computing the error of closure about the township. This value likewise represents double the amount of the offset from the tangent to the parallel at a distance of 6 miles from the point of tangency (see Art. 19.30).

Table V also gives for the various latitudes the difference in longitude for 6 mi in both angle and time, and the difference in latitude for both 1 and 6 mi in angular measure.

**Meridional section lines**   In the subdivision of townships into sections, the establishment of section lines parallel to the east boundary of the township necessitates a correction in azimuth of these section lines on account of the angular convergence of the meridians. While meridional section lines are being run north, they are made to deflect to the left or west of the true meridian by an angle equal to the convergence in the distance to the section line from the east boundary. Hence, $\frac{1}{6}$, $\frac{1}{3}$, $\frac{1}{2}$, $\frac{2}{3}$, and $\frac{5}{6}$ of the angles of convergence given in Table V represent, respectively the deflections from the true meridian for section lines respectively, 1, 2, 3, 4, and 5 mi (1.61, 3.22, 4.83, 6.44, and 8.05 km) west of the east boundary of the township.

**19.30. Laying off a parallel of latitude**   As the base line, standard parallels, and latitudinal township lines are astronomic parallels of latitude, they are curved lines when established on the surface of the earth. This is evident from the fact that meridians converge and that a parallel of latitude is a line whose direction at any point is perpendicular to the direction of the meridian at that point. It is defined by a plane at right angles to the earth's polar axis cutting the earth's surface on a circle whose radius is less for higher latitudes. The rate of curvature within the latitudes of the United States is so small that two points $\frac{1}{4}$ mi (0.4 km) apart on the same parallel of latitude will, for all practical purposes, define the direction of the curve at either point; but the continuation of a line so defined in either direction would describe a great circle of the earth, gradually departing southerly from the astronomic parallel. The great circle tangent to the parallel at any point along the parallel is called the *tangent to the parallel*, and it coincides with the astronomic latitude curve only at the point of tangency.

Though the tangent to the parallel is a straight line in plan, its bearing is not constant but varies with the distance from the point of tangency, the deflection from true east or true west being equal to the angle of convergence of the meridians within the distance from the point of tangency to the given point. Hence, the angles of convergence given in Table V also represent the deviation in azimuth of the tangent from the parallel in a distance of 6 mi (9.66 km), and $\frac{1}{6}$, $\frac{1}{3}$, $\frac{1}{2}$, $\frac{2}{3}$, and $\frac{5}{6}$ of the tabulated angles represent the changes in azimuth of the tangent in distances respectively 1, 2, 3, 4, and 5 mi (1.61, 3.22, 4.83, 6.44, and 8.05 km) from the point of tangency. Note that these values can be calculated by Eq. (19.1).

Within the limits of precision necessary in land surveying, the offset from tangent to parallel at any distance from the point of tangency is one-half of the linear convergence of the meridians within the same distance. Referring to Fig. 19.14, let $a_i$ be the tangent offset,

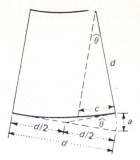

**Fig. 19.14** Relation between tangent offset and linear convergence.

where $i = (\frac{1}{2}, 1, 1\frac{1}{2}, \dots, 5\frac{1}{2}, 6)$; $c$ the linear convergence calculated by Eq. (19.2); $\hat{\theta}$ the angular convergence calculated by Eq. (19.1); and $d$ the distance along the parallel between the meridians. Then from Fig. 19.14,

$$c = \hat{\theta}d$$
$$a = \hat{\theta}\frac{d}{2} \qquad (19.3)$$
$$a = \frac{c}{2} \quad \text{(approx.)}$$

Hence, values one-half as great as the values of the linear convergence in 6 mi (9.66 km) given in Table V or calculated by Eq. (19.2) represent the offset from tangent to parallel, measured along the meridian, at a distance of 6 mi from the point of tangency. With small error a parallel of latitude may, within the limits of distance here considered, be assumed to behave as a parabola. Hence, the offset from tangent to curve at any point, for all practical purposes, may be said to vary as the square of the distance from the point of tangency; and the offsets at $\frac{1}{2}$, 1, $1\frac{1}{2}$, 2, etc., miles from the point of tangency would bear to the offset at 6 mi the ratios $\frac{1}{144}$, $\frac{1}{36}$, $\frac{1}{16}$, $\frac{1}{9}$, etc., respectively.

There are three general methods of establishing a true or astronomic parallel of latitude, which may be employed independently to arrive at the same result: (1) the solar method, (2) the tangent method, and (3) the secant method. The secant method is most commonly employed in heavily wooded areas since offsets from the secant line to the parallel are small, thus eliminating a lot of brush cutting.

**Solar method** By this method a solar attachment to the engineer's transit is employed (see Ref. 20, pp. 45–51). If the instrument is in good adjustment, the true meridian may be established with sufficient precision at each transit station, and the true parallel may be established by turning an angle of 90° in either direction from the meridian. If sights taken with the telescope pointing in the latter direction are not longer than 20 to 40 chains (1320 to 2640 ft or 402.3 to 804.7 m), the line thus defined will not depart appreciably from the true parallel.

**Tangent method** This method consists in determining the true meridian at the point of tangency, from which the tangent to the parallel is established by laying off an angle of 90°. The tangent is extended in a straight line for a distance of 6 miles, and as each distance of 40 chains ($\frac{1}{2}$ mi) is laid off along the tangent, the corresponding section or quarter-section corner is established on the parallel by laying off along the meridian the appropriate offset from tangent to parallel.

At the end of 6 mi a new tangent is laid off, and the process just described is repeated. The values of the offsets may be found from Table V, as previously suggested.

**Example 19.11** Compute tangent offsets for a standard parallel at latitude 44°30′. Figure 19.15 shows the configuration of a tangent to the parallel.

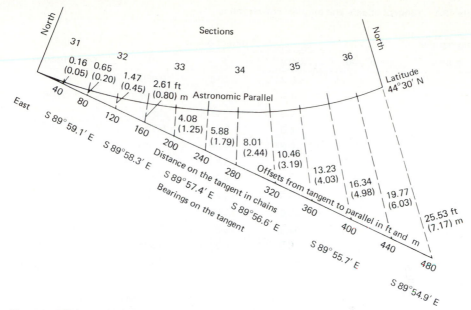

**Fig. 19.15**  Tangent method for establishing a parallel of latitude (44°30′N).

SOLUTION   Calculate the linear and angular convergence at 6 mi by Eqs. (19.2) and (19.1), respectively.

$$c = \ell d \frac{\tan(44°30′)[1-(0.0822718948)^2\sin^2 44°30′]^{1/2}}{3963.3}$$

$$= (6)(6)(0.000247537) = 0.008911 \text{ mi} = 47.05 \text{ ft} = 14.341 \text{ m} = 71.3 \text{ links}$$

$$\hat{\theta}'' = \frac{(6)\tan(44°30′)[1-(0.0822718948)^2\sin^2 44°30′]^{1/2}}{(4.848)(10^{-6})3963.3}$$

$$= \frac{(6)(0.000247537)}{(4.848)(10^{-6})} = 306.4''$$

$$= 0°05′06.4''$$

The tangent offset at $d = 6$ mi is approximately $c/2 = 23.53$ ft $= 7.17$ m $= 35.6$ links. Tangent offsets at the $\frac{1}{2}$, 1, 1$\frac{1}{2}$, 2, 2$\frac{1}{2}$, 3,…, etc., mile points are proportional to the squares of distances from the point of tangency. Thus, the tangent offset from the $\frac{1}{2}$-mi point $= (\frac{1}{144})(23.53$ ft$) = 0.163$ ft, etc. Results of these computations for each $\frac{1}{2}$ mi are given in Table 19.1 in links, feet, and metres.

The angles of convergence for each 1-mi point are $\frac{1}{6}$, $\frac{1}{3}$, $\frac{1}{2}$, $\frac{2}{3}$, and $\frac{5}{6}$ of the convergence at the 6-mi point. These values are also listed in Table 19.1, in column 6, to the nearest tenth of a second and are used to compute bearings along the tangent to the nearest tenth of a minute, as shown in Fig. 19.15.

The tangent offsets and values for angular convergence could have been determined using the U.S. Bureau of Land Management *Standard Field Tables and Trigonometric Tables* (Ref. 21).

**Secant method**   This is a modification of the tangent method, in which the secant is a straight line 6 mi in length forming the arc of a great circle, which intersects the true parallel at the end of the first and fifth miles from the point of beginning, as illustrated by Fig. 19.16. For the latitude of the given parallel, the offsets (in feet and metres) from secant to parallel are given in the figure, at intervals of $\frac{1}{2}$ mi. From the figure it is clear that the secant is parallel with a tangent to the parallel at the end of the third mile (240 chains); hence, the offset south from the third-mile point on the secant line to the corner on the true parallel is

**Table 19.1 Tangent offsets and angular convergence**

| Distance along parallel $d$, mi | $d^2/36$ | Tangent offset | | | Angular convergence |
|---|---|---|---|---|---|
| | | links | ft | m | |
| $\frac{1}{2}$ | $\frac{1}{144}$ | 0.2 | 0.16 | 0.050 | |
| 1 | $\frac{1}{36}$ | 1.0 | 0.65 | 0.199 | 51.1″ |
| $1\frac{1}{2}$ | $\frac{1}{16}$ | 2.2 | 1.47 | 0.448 | |
| 2 | $\frac{1}{9}$ | 4.0 | 2.61 | 0.797 | 1′42.1″ |
| $2\frac{1}{2}$ | $\frac{25}{144}$ | 6.2 | 4.08 | 1.245 | |
| 3 | $\frac{1}{4}$ | 8.9 | 5.88 | 1.793 | 2′33.2″ |
| $3\frac{1}{2}$ | $\frac{49}{144}$ | 12.1 | 8.01 | 2.440 | |
| 4 | $\frac{4}{9}$ | 15.8 | 10.46 | 3.187 | 3′24.2″ |
| $4\frac{1}{2}$ | $\frac{9}{16}$ | 20.0 | 13.23 | 4.033 | |
| 5 | $\frac{25}{36}$ | 24.8 | 16.34 | 4.980 | 4′15.3″ |
| $5\frac{1}{2}$ | $\frac{121}{144}$ | 30.0 | 19.77 | 6.025 | |
| 6 | 1 | 35.6 | 23.53 | 7.171 | 5′06.3″ |

the same as the offset from the tangent to the parallel in a distance of 2 mi. Also, it is evident that the offset south of the point of beginning to the initial point on the secant, and the offset north of the secant to the true parallel at the end of the sixth mile, is equal to the difference between the tangent offset in a distance of 3 mi and the tangent offset in a distance of 2 mi.

In general, offsets from the secant line to corners on the parallel at $\frac{1}{2}$-mi intervals are calculated by combining the appropriate tangent offsets computed using Eqs. (19.2) and (19.3) or taken from the standard field tables (Ref. 21). Let the secant offset be designated $(so)_i$, where $i=(0,\frac{1}{2},1,\ldots,5\frac{1}{2},6)$. The determination of the secant offsets for the example illustrated in Fig. 19.16, where the latitude of 44°30′N is the same as for Example 19.11, is as follows (note: use the tangent offsets $a_{\frac{1}{2}},a_1,\ldots,a_6$ from Table 19.1):

$$(so)_0 = a_3 - a_2 = 5.88 - 2.61 = 3.27 \text{ ft } (1.00 \text{ m})$$
$$(so)_0 = (so)_6$$
$$(so)_{\frac{1}{2}} = a_{2\frac{1}{2}} - a_2 = 4.08 - 2.61 = 1.47 \text{ ft } (0.45 \text{ m})$$
$$(so)_{\frac{1}{2}} = (so)_{5\frac{1}{2}}$$
$$(so)_1 = 0 = (so)_5$$
$$(so)_{1\frac{1}{2}} = a_2 - a_{1\frac{1}{2}} = 2.61 - 1.47 = 1.14 \text{ ft } (0.35 \text{ m})$$
$$(so)_{1\frac{1}{2}} = (so)_{4\frac{1}{2}}$$
$$(so)_2 = a_2 - a_1 = 2.61 - 0.65 = 1.96 \text{ ft } (0.60 \text{ m})$$
$$(so)_2 = (so)_4$$
$$(so)_{2\frac{1}{2}} = a_2 - a_{\frac{1}{2}} = 2.61 - 0.16 = 2.45 \text{ ft } (0.75 \text{ m})$$
$$(so)_{2\frac{1}{2}} = (so)_{3\frac{1}{2}}$$
$$(so)_3 = a_2 = 2.61 \text{ ft } (0.80 \text{ m})$$

Owing to the convergence of meridians, the azimuth of the secant—a straight line in plan—varies along its length. If the secant is laid off toward the east, the direction of the secant from the point of beginning to the end of the third mile is north of astronomic east, and beyond the end of the third mile is south of astronomic east, the variation from astronomic east increasing directly with the distance in either direction from the third-mile point. At the third-mile point the secant bears astronomic east; at the point of beginning the

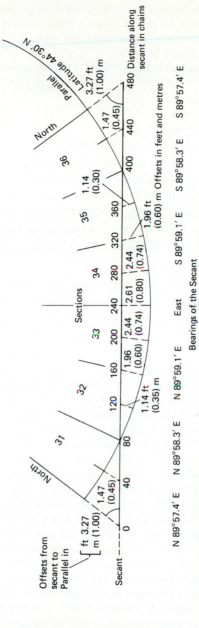

**Fig. 19.16** Secant-line method for establishing parallel of latitude (latitude 44° 30' N).

secant bears north of east by an amount equal to the angular convergence of meridians 3 mi apart; and at the end of the sixth mile the secant bears south of east by the same amount.

For the data of Fig. 19.16, where $\phi = 44°30'$N, the values for angular convergence by Eq. (19.1) for distances $d$ of 1, 2, and 3 mi are:

$$\hat{\theta}_1'' = (1)\frac{\tan(44°30')\left[1-(0.0822718948)^2\sin^2 44°30'\right]^{1/2}}{(4.848)(10^{-6})(3963.3)}$$

$$= 51.06''$$

$$\hat{\theta}_2'' = (2)(51.06)$$
$$= 102.2'' = 01'42.12''$$
$$\hat{\theta}_3'' = (3)(51.06) = 153.18'' = 02'33.18''$$

The bearings $B_i$, $i = (0, 1, \ldots, 6)$ for segments of the secant line at 1-mi intervals are as follows: First, from the point of beginning to the 3-mi point,

$$B_0 = (90°) - \hat{\theta}_3 = (90°) - (0°02'33.18'') = \text{N}89°57'26.82''\text{E} = \text{N}89°57.4'\text{E}$$
$$B_1 = B_0 + \hat{\theta}_1 = (89°57'26.82'') + (51.06'') = \text{N}89°58'17.88''\text{E} = \text{N}89°58.3'\text{E}$$
$$B_2 = B_0 + \hat{\theta}_2 = B_1 + \hat{\theta}_1 = (89°58'17.88'') + (51.06'') = \text{N}89°59'08.94''\text{E} = \text{N}89°59.1'\text{E}$$
$$B_3 = B_0 + \hat{\theta}_3 = B_2 + \hat{\theta}_1 = (89°59'08.94'') + (51.06'') = \text{due east}$$

Next from the 3-mi point to the 6-mi point:

$$B_4 = (90) - \hat{\theta}_1 = (90°) - (0°00'51.06'') = \text{S}89°59'08.94''\text{E} = \text{S}89°59.1'\text{E}$$
$$B_5 = B_4 - \hat{\theta}_1 = (89°59'08.94'') - (0°00'51.06'') = \text{S}89°58'17.88''\text{E} = \text{S}89°58.3'\text{E}$$
$$B_6 = B_5 - \hat{\theta}_1 = (89°58'17.88'') - (0°00'51.06'') = \text{S}89°57'26.82''\text{E} = \text{S}89°57.4'\text{E}$$
$$B_6 = 90° - \hat{\theta}_3 = (90°) - (0°02'33.18'') = \text{S}89°57'26.82''\text{E} = \text{S}89°57.4'\text{E}$$

as a check.

Bearings are calculated to two places so as to avoid round-off error. Final values should be rounded to tenths of minutes.

The azimuths of the secant for intervals of 0, 1, 2, and 3 mi (0, 1.61, 3.22, and 4.83 km) are given for various latitudes in Table VI (Appendix C). Offsets in links from the secant line to the parallel are given for various latitudes in Table VII (Appendix C). To convert these offsets to feet, multiply by 0.66; and to metres, multiply by 0.201. Azimuths and latitudes are also given for 1° intervals from 25° to 70°N latitude in the *Standard Field Tables* (Ref. 21).

The procedure for establishing a true parallel by the secant method is as follows. The initial point on the secant is located by measuring south of the beginning corner a distance equal to the secant offset for 0 mi determined by calculations as described previously in this section or given in Table VII, Appendix C (3.27 ft = 1.00 m = 4.97 links, Fig. 19.16). The transit is set up at this point, and the direction of the secant line is established by laying off from astronomic north the azimuth or bearing calculated as discussed previously in this section or given in Table VI in the column headed 0 mi; for the conditions illustrated by Fig. 19.16, the bearing of the secant which extends east from the point of beginning is N89°57.4'E. The secant is then projected in a straight line for 6 mi (9.66 km); and as each 40 chains ($\frac{1}{2}$ mi or 0.805 km) is laid off along the secant, the proper offset is taken to establish the corresponding section or quarter-section corner on the true parallel.

At the end of 6 mi, if it is not convenient to determine the true meridian, the succeeding secant line may be established by laying off, at the sixth-mile point, a deflection angle from the prolongation of the preceding secant to the succeeding secant line, the angle being equal to the convergence of meridians 6 mi apart. The angular convergence can be calculated using Eq. (19.1), or the deflection angle for 6 mi can be obtained from the last column of Table VI. When the direction of the new secant line has been thus defined, the process of measurement to establish corners on the true parallel is continued as before.

**19.31. Establishing township exteriors** When practical, the township exteriors are surveyed successively through a 24-mi quadrangle in ranges, beginning each range with the township on the south (Fig. 19.17). The range lines or meridional boundaries of the townships take precedence in the order of survey and are run from south to north on astronomic meridians.

For example, in Fig. 19.17, the meridional line from *a* is run due north to *b*, setting quarter-section and section corners alternately at intervals of 40 chains (2640 ft or 804.67 m). At the end of 6 mi a temporary township corner is set at *b*, pending latitudinal measurements necessary to close the township exterior and to calculate the error of closure.

Next, a random line is run from the corner at *b* west to intersect the principal meridian (or guide meridian). Temporary monuments are set at quarter-section and section corners alternately at intervals of 40 chains (2640 ft or 804.67 m) as this line is being run. The closure on the corner at *c* is noted, and this amount (assuming it is within the permissible value) is used to determine the direction of an improved estimate for the true line between *b* and *c*. This correction line is run from *c* to *b*, correcting the temporary monuments at quarter-section and section corners and setting permanent monuments. Owing to the convergence of the meridians, this northern boundary of the township will be less than 6 mi. All errors due to convergence and errors in measurement are thrown into the westernmost $\frac{1}{2}$ mi.

The meridional or range line is then extended from the corner at *b* north to the corner at *d* setting quarter-section and section corners alternately at 40-chain (2640-ft or 804.67-m) intervals. From the corner at *d*, a random line is run west to the principal meridian or range line, where the closure is noted at corner *e*. A correction line is run from *e* to *d*, setting permanent quarter corners and section corners alternately at 40-chain intervals. Similarly, the range line is extended to *f*, a random line is run west to *g*, and a correction line is run from *g* to *f*. Finally, the range line is run from *f* to *h*, where it is terminated at its intersection with the standard parallel, the excess or deficiency between standard parallels being placed in the most northerly half-mile. At the point of intersection between the range line and the standard parallel, a closing township corner (correction corner) is established. To determine

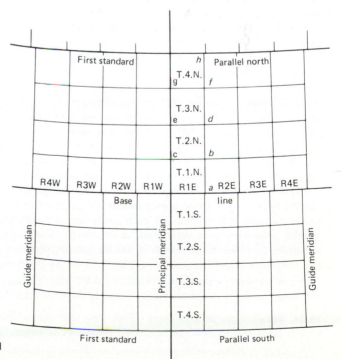

**Fig. 19.17** Four quadrangles of 16 townships, each bounded by standard lines.

the alignment of the line closed upon, the standard parallel is retraced between the two standard corners adjacent to the closing corner. The distance from the closing corner to the nearest standard corner is measured in order that the error of closure may be calculated.

In a similar manner, the boundaries of townships forming the next or second range line to the east are established. The range lines are extended north, setting quarter-section and section corners alternately at 40-chain ($\frac{1}{2}$-mi or 0.805-km) intervals, and random lines are run west to close on the previously established township corners on the first range line.

When the third range line is extended north from the southwest corner of the southeast township, quarter-section corners are set as before, but random latitudinal lines at the 6-, 12-, and 18-mi points are run (1) to the west to connect with corresponding township corners, and (2) to the east to connect with the first, second, and third regular township corners north of the standard parallel on the guide meridian.

As the line run to the east is corrected back to the west, with the initial measurement being made from the guide meridian, quarter-section and section corners are set alternately at 40-chain ($\frac{1}{2}$-mi or 0.805-km) intervals and any error is thrown into the westerly half-mile of the township as was done for the others.

**19.32. Allowable limits of error of closure**   According to the *Manual of Instructions for the Survey of Public Lands of the United States*, 1973 (Ref. 20), the maximum *limit of closure* (relative accuracy; Art. 8.23) is $\frac{1}{905}$ provided that the limit of closure in either the sum of the departures or sum of the latitudes does not exceed 1/1280. Where a survey qualifies under the latter limit, the first must be satisfied. A cumulative error of $6\frac{1}{4}$ links (4.13 ft or 1.26 m) should not be exceeded in either departure or latitude.

The departures and latitudes of a normal section should each close within 25 links (16.5 ft or 5.03 m); of a normal range or tier of sections to within 88 links (58.1 ft or 17.70 m); and of a normal township within 150 links (99.0 ft or 30.18 m). The boundaries of all fractional sections, irregular claim lines or meanders, and all broken or irregular boundaries should have a limit of closure (relative accuracy) of 1/1280 to be determined separately for the sum of the departures and the sum of the latitudes.

Whenever a closure is effected, the departures, latitudes, and error of closure of the lines composing the figure (quadrangle, township, section, meander, etc.) must be calculated, and corrective steps must be taken whenever the test discloses an error in excess of the allowable value.

**19.33. Rectangular limits**   Before considering further the methods employed in the subdivision of townships, the legal requirement relative to the rectangular surveys of the public lands should be stated. Of the 36 sections in each normal township (Fig. 19.18), 25 are returned as containing 640 acres (259 ha) each; 10 sections adjacent to the north and west boundaries (comprising sections 1–5, 7, 18, 19, 30, and 31) each containing regular subdivisions totaling 480 acres (194 ha) and in addition 4 *fractional lots* each containing 40 acres (16 ha) plus or minus definite differences to be determined in the survey; and 1 section (section 6) in the northwest corner contains regular subdivisions totaling 360 acres (146 ha) and in addition 7 fractional lots each containing 40 acres (16.2 ha) plus or minus certain definite differences to be determined in the survey. The aliquot[†] parts of 640 acres (259 ha), or regular subdivisions of a section, are the quarter section ($\frac{1}{2}$ mi or 0.805 km square), the half-quarter or eighth section ($\frac{1}{4}$ by $\frac{1}{2}$ mi or 0.402 by 0.805 km), and the quarter-quarter or sixteenth section ($\frac{1}{4}$ mi square or 0.402 by 0.402 km), the last containing 40 acres (16.2 ha) and being the legal minimum for purposes of disposal under the general land laws.

**19.34. Subdivision of townships**   Following the normal plan for subdividing townships with regular boundaries, the subdivisional survey is begun on the south boundary of the

---

[†] A part of the distance which divides the distance without a remainder (Ref. 1, Manual No. 32).

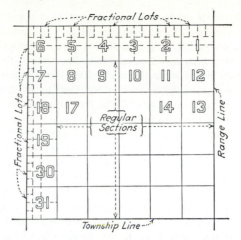

**Fig. 19.18** Subdivision of a township.

township at the section corner between sections 35 and 36 (see Fig. 19.19). The line between sections 35 and 36 is run in a northerly direction parallel to the east boundary of the township, the quarter-section corner between 35 and 36 being set at 40 chains (2640 ft or 804.67 m) and the section corner common to sections 25, 26, 35, and 36 being set at 80 chains (5280 ft or 1609.34 m). From the latter corner a random line is run eastward on a course calculated to be parallel with the south boundary of section 36, a temporary quarter-section corner being set at 40 chains. If this random line intersects the east boundary of the township at the corner of sections 25 and 36, it is blazed (a blaze is a mark made with an axe on a living tree, see Ref. 20 and Art. 19.40) and established as the true line, and if the linear error of closure is within the allowable limits, the temporary quarter-section corner is made permanent by shifting it to a position midway between adjacent section corners, as determined by field measurements.

If the point of intersection between the random line and east boundary falls to the north or to the south of the section corner on the township boundary, as will generally be the case, the falling is measured and from the data thus obtained the bearing of the true return course is calculated and the true line joining the section corners is blazed and established, the quarter-section corner common to sections 25 and 36 being placed midway between section corners, as described above.

This process is repeated for the successive meridional and latitudinal lines in the eastern range of sections until the north boundary of section 12 is established, the order in which the lines are surveyed being as indicated by the numbers on the section lines in Fig. 19.19.

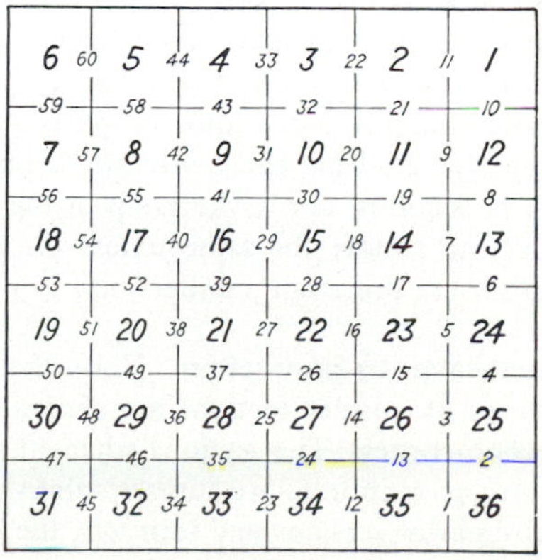

**Fig. 19.19** Order of establishing section lines.

When the northern boundary of the township is not a base line or standard parallel, the line between sections 1 and 2 is run north as a random line parallel to the east boundary, the distance to its points of intersection with the northern boundary of the township being measured. If the random line intersects the northern boundary at the corner of sections 1 and 2 and the linear error of closure of the tier of sections is within the allowable limit, the random line is blazed back and established as the true line, the fractional measurement being thrown into that portion of the line between the quarter-section corner and the north boundary of the township.

If, as is usually the case, the random line intersects the north boundary to the east or to the west of the corner of sections 1 and 2, the falling is measured, the bearing of the true return course is calculated, and the true line joining the section corners is established, the permanent quarter-section corner common to sections 1 and 2 being placed a full 40 chains (2640 ft or 804.67 m) from the south boundary of these sections. In this way the excess or deficiency in linear measurement is, as before, placed in that portion of the line between the permanent quarter-section corner and the north boundary of the township.

When the north boundary of the township is a base line or a standard parallel, the line between sections 1 and 2 is run as a true line parallel to the east boundary of the township, a permanent quarter-section corner being set at 40 chains (2640 ft or 804.7 m), a closing section corner being established at the point of intersection of the section line and base line or standard parallel, and the distance from this closing corner to the nearest standard corner being measured.

The successive ranges of sections from east to west are surveyed in a manner identical with the procedure described in the preceding paragraphs for the most easterly range until the two most westerly ranges are reached.

The west and north boundaries of section 32 are established as for corresponding sections to the east. A random line parallel to the south boundary of the township is then run west from the corner of sections 29, 30, 31, and 32, and the point of intersection between the random line and the west boundary of the township is determined. The falling of the intersection from the true corner is then measured, the course of the true line is calculated, and the true line is blazed and established, the permanent quarter-section corner being placed on the true line at a full 40 chains (2640 ft or 804.7 m) from the corner of sections 29, 30, 31 and 32. Thus, the deficiency due to convergence of the meridians and the excess or deficiency due to errors in linear measurements are thrown in the most westerly half-mile.

The survey of the other sections comprising the two most westerly ranges is continued in similar manner, the order in which the lines are surveyed being indicated by the numbers shown in Fig. 19.19.

**19.35. Subdivision of sections**   The function of the United States surveyor is to establish the official monuments so that the officially surveyed lines may be identified and the subdivision of the section may be controlled as contemplated by law. There the duties of the United States surveyor cease, and those of the surveyor in private practice begin. In the work of subdividing sections into the parts shown on the official plat the local surveyor cannot properly serve his client without being familiar with the land laws regarding the subdivision of sections, nor in the event of the loss of original monuments can the surveyor expect legally to restore the same unless the principles employed in the execution of the original survey are thoroughly understood.

**19.36. Subdivision by protraction**   Upon the official government township plats the interior boundaries of quarter sections are shown as dashed straight lines connecting opposite quarter-section corners. The sections adjacent to the north and west boundaries of a normal township, except section 6, are further subdivided by protraction into parts containing two regular half-quarter sections and four *lots*, the latter containing the fractional areas resulting

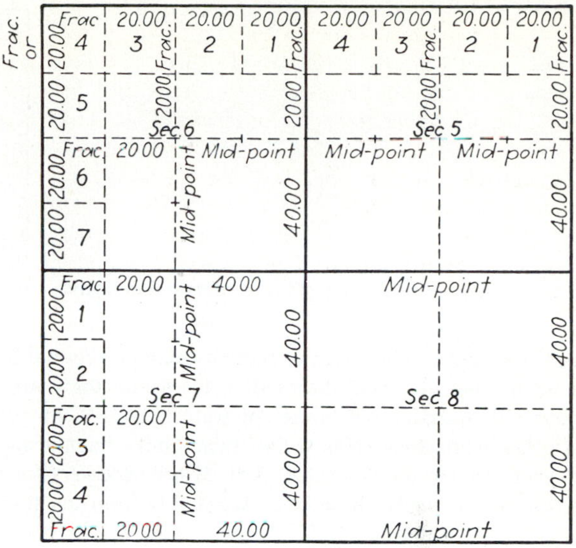

**Fig. 19.20** Subdivision of sections.

from the plan of subdivision of the normal township. This process of plotting the interior, unsurveyed boundaries on the official plat is called *subdivision by protraction*. Figure 19.20 illustrates the plan of the normal subdivision of sections. The regular half-quarter sections are protracted by laying off a full 20 chains (1320 ft or 402.34 m) from the line joining opposite quarter-section corners. The lines subdividing the fractional half-quarter sections into the fractional lots are protracted from midpoints of the opposite boundaries of the fractional quarter sections.

In section 6 the two quarter-quarter-section corners on the south and east boundaries of the fractional northwest quarter are similarly fixed, one at a point 20 chains north and the other at a point 20 chains west of the center of the section, from which points lines are protracted to corresponding points on the west and north boundaries of the section. Hence, the subdivision of the northwest quarter of section 6 results in one regular quarter-quarter section and three lots.

In all sections bordering on the north boundary the fractional lots are numbered in succession beginning with No. 1 at the east. In all sections bordering on the west boundary the fractional lots are numbered in succession beginning with No. 1 at the north, except section 6 which, being common to both north and west boundaries, has its fractional lots numbered in progression beginning with No. 1 in the northeast corner and ending with No. 7 in the southwest corner, all as illustrated by Fig. 19.20.

Figure 19.21 illustrates a typical plat of section 6 on which the protracted areas are shown. Figure 19.22 is a similar section giving the calculated dimensions of the protracted areas.

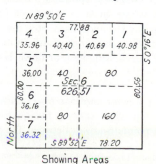

**Fig. 19.21** Subdivisonal areas of section 6 in acres.

Showing Areas

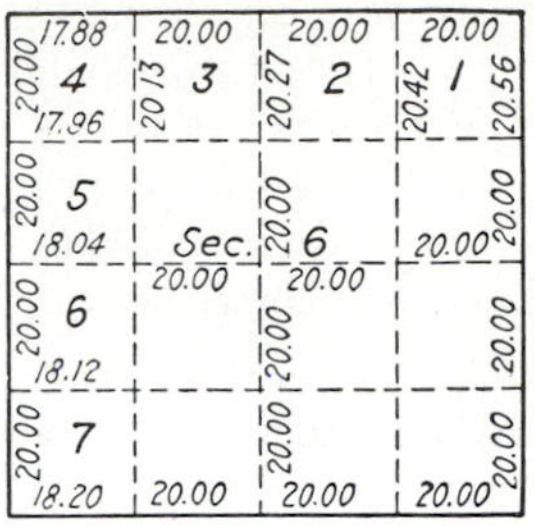

**Fig. 19.22** Subdivisional dimensions of section 6 in chains.  Showing Calculated Distances

**Fractional lots**   In addition to sections made fractional by reason of their being adjacent to the north and west boundaries of a township, there are also sections made fractional on account of meanderable bodies of water (Art. 19.38), mining claims, Indian Reservations and other segregated areas within their limits. Such sections are subdivided by protraction into such regular and fractional parts as are necessary for the entry of the undisposed public lands and to describe these lands separately from the segregated areas.

**19.37. Subdivision by survey**   The rules for the subdivision of sections given in the following paragraphs are based upon the general land laws. When an entryman[†] has acquired title to a certain legal subdivision, he becomes the owner of the identical ground area represented by the same subdivision on the official plat. Preliminary to subdivision it is necessary to identify the actual boundaries of the section, as it cannot be legally subdivided until the section and exterior quarter-section corners have been found or have been restored to their original locations, and the resulting courses and distances have been redetermined in the field. When the opposite quarter-section corners have been located, the legal center of the section, or interior quarter-section corner, may be placed. If the boundaries of quarter-quarter sections or of fractional lots are to be established on the ground, it is necessary to measure the boundaries of the quarter section and to fix thereon the quarter-quarter-section corners at distances in proportion to those given upon the official plat; then the interior quarter-quarter-section corner may be placed.

**Subdivision of sections into quarter sections**   According to law, the procedure to be followed in the subdivision of a section into quarter sections is to run straight lines between the established opposite quarter-section corners. The point of intersection of lines thus run is the quarter-section corner common to each of the four quarter sections into which the section is divided. It is called the *interior quarter-section corner* and is the legal center of the section.

**Subdivision of fractional sections**   Where opposite corresponding quarter-section corners of a section have not been or cannot be fixed, as is frequently the case when sections are made fractional by streams, lakes, etc., the lines of sectional subdivisions are run on courses with the mean of the bearings of adjoining established section lines, or are run on courses parallel to the east, south, west, or north boundary of the section where there is no opposite section line.

**Subdivision of quarter sections into quarter-quarter sections**   Preliminary to the subdivision of regular quarter sections, the quarter-quarter-section (sixteenth-section) corners are established at points midway between the section and exterior quarter-section corners and

---

[†]One who initiates procedures to acquire title to public lands under public land laws (Ref. 1, Manual No. 34).

between the exterior quarter-section corners and the center of the section. The quarter-quarter-section corners having thus been established, the center lines of the quarter section are run as straight lines between opposite corresponding quarter-quarter-section corners on the boundaries of the quarter section. The intersection of these lines is common to the four quarter-quarter sections into which the quarter section is divided. It is called the *interior quarter-quarter-section corner*, and it marks the legal center of the quarter section.

1. *Irregular quarter sections*  This case arises (1) when the quarter section is adjacent to the north or west boundary of a regular township and (2) when the quarter section adjoins any irregular boundary of an irregular township. The procedure is the same as that outlined in the preceding paragraph, except that the quarter-quarter-section corners on the boundaries of the quarter section which are normal to the township exterior are placed at 20 chains (1320 ft or 402.34 m), proportionate measurement (Art. 19.47), counting from the regular quarter-section corner.

2. *Fractional quarter sections*  The subdivisional lines of fractional quarter sections are run from properly established quarter-quarter-section corners, with courses governed by the conditions represented upon the official plat, to the lake, water course, or reservation which renders such quarter sections fractional.

**19.38. Meandering**  In the process of surveying the public lands, all navigable bodies of water and other important rivers and lakes below the line of mean high water are segregated from the lands which are open to private ownership. In the process of subdivision, the regular section lines are run to an intersection with the mean high-water mark of such a body of water, at which intersection corners called *meander corners* are established. The traverse which is run between meander corners, approximately following the margin of a permanent body of water, is called a *meander line*, and the process of establishing such lines is called *meandering*. The mean high-water mark is taken as the line along which vegetation ceases. The fact that an irregular line must be run in tracing the boundary of a reservation does not entitle such a line to be called a meander line except where it follows closely the shore of a lake or the bank of a stream.

Meander lines are not boundaries but are lines run for the purpose of locating the water boundaries approximately, and although the official plats show fractional lots as bounded in part by meander lines, it is an established principle that ownership does not stop at such boundaries. A Supreme Court decision reads as follows:

Meander lines are run in surveying fractional portions of the public lands bordering on navigable rivers, not as boundaries of the tract, but for the purpose of defining the sinuosities of the banks of the stream and as the means of ascertaining the quantity of land in the fraction subject to sale, which is to be paid for by the purchaser. In preparing the official plat from the field notes, the meander line is represented as the border line of the stream, and shows to a demonstration that the water-course, and not the meander line as actually run on the land, is the boundary.

In running a meander line, the surveyor begins at a meander corner and follows the bank or shore line, as closely as convenience permits, to the next meander corner, the traverse being a succession of straight lines. The true length and bearing of each of the courses of the meander line are observed with precision, but for convenience in plotting and computing areas the intermediate courses are laid off to the exact quarter degree and each intermediate transit station is placed a whole number of chains, or at least a multiple of 10 links (6.60 ft or 2.01 m), from the preceding station. Inasmuch as meander lines are not true boundaries, this procedure defines the sinuosities of the mean high-water line with sufficient accuracy. When a meander line is "closed" on a second meander corner, the departures and latitudes of the courses bounding the fractional lot are computed and the error of closure is determined. If this exceeds the allowable value, the line is rerun until an error in bearing or distance is discovered which will bring the closure within the specified limits (maximum error in either departure or latitude 1/1280).

**Rivers**   Proceeding downstream, the river bank on the left hand is termed the left bank and that on the right hand the right bank. Navigable rivers and bayous as well as all rivers not embraced in the class denominated "navigable," the right-angle width of which is 3 chains (198 ft or 60.35 m) and upward, are meandered on both banks, at the ordinary mean high-water mark, by taking the general courses and distances of their sinuosities.

**Lakes**   Regulations provide for the meandering of all lakes having an area of 25 acres (101,171.4 m$^2$) or greater, the procedure being the same as for the meandering of streams. If the lake lies entirely within a section, there will be obviously no regular meander corners, and a *special meander corner* is established at the intersection of the shore of the lake with a line run from one of the quarter-section corners on a theoretical course to connect with the opposite quarter-section corner, the distance from the quarter-section corner to the special meander corner being measured. The lake is then meandered by a line beginning and ending at the special meander corner. If a meanderable lake is found to lie entirely within a quarter section, an *auxiliary meander corner* is placed at any convenient place on its margin, and this is connected by traverse with one of the regular corners established on the boundary of the section.

**Islands**   In the progress of the regular surveys, every island of any meanderable body of water, except those islands which have formed in navigable streams since the admission of a state to the union, is located with respect to regular corners on section boundaries and is meandered and shown upon the official plat. Also in the survey of lands fronting on any nonnavigable body of water, any island opposite such lands is subject to survey.

**19.39. Field notes**   The field notes taken in connection with the survey of public lands are required to be in narrative form and are designed to furnish not only a record of the exact surveying procedure followed in the field but also a report showing the character of the land, soil, and timber traversed by the line of subdivision and a detailed schedule of the topographic features adjacent to the lines, together with reference measurements showing the location of the lines with respect to natural objects, to improvements, and to the lines of other surveys. In this way the notes serve three purposes: (1) the field procedure is made a matter of official record, (2) the general characteristics of the territory served by the subdivision surveys are secured, and (3) the reference measurements to objects along the surveyed lines furnish evidence by which the established points and lines become practically unchangeable.

**19.40. Marking lines between corners**   As a final step in the survey of the public lands, it is the aim to fix the location of the legal lines of subdivision permanently with reference to objects on the surface of the earth. This aim is accomplished (1) by setting monuments, of a character later to be defined, at the regular corners, (2) by finding the location of the officially surveyed lines with respect to natural features of the terrain, and (3) by indicating the position of the regular lines through living timber by *blazing* and by *hack marks*.

The last method of fixing the location of the regular subdivisional lines is required by law just as definitely as is the establishment of monuments at the corners. All legal lines of the public-land surveys through timber are marked in this manner. Those trees which lie on the line, called *line trees*, are marked with two horizontal notches, called *hack marks*, on each side of the tree facing the line; and an appropriate number of trees on either side of the line and within 50 links (33 ft or 10.06 m) thereof are marked by flat axe marks, called *blazes*, a single blaze on each of two sides quartering toward the line.

**19.41. Corners**   In the subdivision of the public lands as described in the preceding sections, it is required that the United States surveyors shall permanently mark the location of the township, section, exterior quarter section, and meander corners, as well as those quarter-quarter-section corners necessary in connection with the subdivision of fractional sections. Monuments of a character specified by regulations of the Bureau of Land Management are employed for this purpose. The location of every such corner monument is, in accordance with definite rule, referred to such nearby objects as are available and suitable for this purpose; and where the corner itself cannot be marked in the ordinary manner an appropriate witness corner is established (Art. 19.42).

At the appropriate place in the field notes of the survey a record of each established monument is introduced, this record including the character and dimensions of the monument itself, the manner in which it is placed, the significance of its location, its markings, and the nature of the objects to which reference measurements are taken, together with the reference measurements.

**19.42. Witness corners**   Where a true corner point falls within an unmeandered stream or lake, within a marsh, or in an inaccessible place, a witness corner is established in a convenient location nearby, preferably on one of the surveyed lines leading to the location of the regular corner. Also, where the true point falls within the traveled limits of a road, a cross-marked stone is deposited below the road surface, and a witness corner is placed in a suitable location outside the roadway.

The witness corner is placed on any one of the surveyed lines leading to a corner, if a suitable place within a distance of 10 chains (660 ft or 201.168 m) is available; but if there is no secure place to be found on a surveyed line within the stated limiting distance, the witness corner may be located in any direction within a distance of 5 chains (330 ft or 100.584 m).

**19.43. Corner monuments**   The Bureau of Land Management has adopted a standard iron post for monumenting the public-land surveys. This post is to be used unless exceptional circumstances warrant the use of other material (Ref. 20).

Where the procedure is duly authorized, durable native stone may be substituted for the model iron post described above, provided that the stone is at least 20 in (51 cm) long and at least 6 in (15.2 cm) in its least lateral dimension. Stone may not be used as a monument for a corner whose location is among large quantities of loose rock. The required corner markings are cut with a chisel, and usually the stone is set with about three-fourths of its length in the ground.

Where the ground is underlaid with rock close to the surface and it is impractical to complete the excavations for monuments to the regular depth, the monument is placed as deep as practical and is supported above the natural ground surface by a mound of stone. Where the solid rock is at the surface, the exact corner point is marked by a cross cut in the rock; and if it is practical to do so, the corner monument is established in its proper location and is supported by a mound of stones.

Where the corner point falls within the trunk of a living tree which is too large to be removed readily, the tree becomes the corner monument and, as such, is scribed with the proper marks of identification.

Legal penalties are prescribed for damage to government survey monuments or marked trees.

**19.44. Marking corners**   A complete treatment of the system of marking corner monuments established in the survey of the public lands is beyond the scope of this text. However a brief description of the general features of the system is given. For further details the reader is referred to the *Manual of Instructions for the Survey of Public Lands of the United States* of the Bureau of Land Management.

All classes of monuments are marked in accordance with a system which has been designed to provide a ready identification of the location and character of the monument on which the markings appear. Iron posts and tree corners are marked with capital letters which are themselves keys to the character of the monument and with arabic figures giving the section and township and range numbers of the adjacent subdivisions and the year in which the survey was made.

In the case of stone monuments, certain marks in the form of *notches* and *grooves* are placed on the vertical edges or faces; for an exterior corner the number of marks is made equal to the distance in miles from the adjoining township corner along the township or range line to the monument, and for an interior corner the number of marks is made equal to the distance in miles from the adjoining township boundary along section lines to the monument. These marks furnish a means of determining the number of the adjoining sections.

A witness corner and its accessories are constructed and marked similarly to a regular corner for which it stands, with the additional letters "WC" to signify *witness corner* and with an arrow pointing to the true corner.

Following is an index of the ordinary markings common to all classes of corners:

| Mark | Meaning | Mark | Meaning |
|------|---------|------|---------|
| AMC | Auxiliary meander corner | S | Section |
| BO | Bearing object | S | South |
| BT | Bearing tree | SC | Standard corner |
| C | Center | SMC | Special meander corner |
| CC | Closing corner | T | Township |
| E | East | W | West |
| MC | Meander corner | WC | Witness corner |
| N | North | WP | Witness point |
| R | Range | $\frac{1}{4}$ | Quarter section |
| RM | Reference monument | $\frac{1}{16}$ | Quarter-quarter section |

**19.45. Corner accessories**   When a corner is referred by direction and distance to some other more-or-less permanent object nearby, and the operation becomes a matter of record, it is possible to relocate the corner with respect to the object. In land surveying a recorded measurement of this kind is often called a *connection*, and the object thus located is called a *corner accessory*. It is specified that the United States surveyors in the survey of the public lands shall employ at least one accessory for every corner established. The character of the accessories is to fall within the following groups: (a) bearing trees, or other natural objects such as notable cliffs and boulders, permanent improvements, and reference monuments; (b) mounds of stone; and (c) pits and memorials. Essentially, such an accessory is a part of the monument.

The marks on a bearing tree are made on the side nearest the corner, in the manner already described for tree-corner monuments. The mark includes the section number in which the tree stands and is terminated by the letters "BT."

Where a bearing object is of rock formation, the point to which measurements are taken is indicated by a cross, and it is marked with the letters "BO" and the section number, all marks being cut with a chisel.

Where it is impossible to make a single connection to a bearing tree or other bearing object and where a mound of stone or a pit is impracticable, a *memorial*, or durable article such as glassware, stoneware, a cross-marked stone, a charred stake, a quart of charcoal, or piece of metal is deposited alongside the base of the monument.

Where native stone is at hand, a mound of stones of sufficient size to be conspicuous is employed as an accessory.

Where accessories such as those mentioned in the preceding paragraphs are not available, pits may be used if conditions are favorable to their permanence. Where the ground is covered with sod, the soil is firm, and the slope is not steep, the pit will gradually fill with a material different in color or in texture from the original soil; and often a new species of vegetation springs up. Thus, it may be possible to identify the location of a pit after the lapse of many years.

**19.46. Restoration of lost corners**   Although it has been the aim of the Bureau of Land Management in the subdivision of the public lands to monument the established corners so that there will always be physical evidence of their location, it is a matter of common experience that many corner marks become obliterated with the progress of time. It is one of the important duties of the local or county surveyor, in the relocation of property lines or in the further subdivision of lands, to examine all available evidence and to identify the official corners if they exist. Should a search of this kind result in failure, it is the duty of the surveyor to employ a process of field measurement which will result in the obliterated corner being restored to its most probable original location.

As here employed, the term *corner* is used to designate a point established by a survey, while the term monument is used to indicate the object placed to mark the corner point upon the surface of the earth.

A corner is said to *exist* when its location, within reasonable accuracy, can be determined beyond reasonable doubt, either by means of the original monument, by means of the accessories to which connections were made at the time of the original survey, by the expert testimony of surveyors who may have identified the original corner and recorded connections to other accessories, or even by land owners who have indisputable knowledge of the exact location of the original monument. If the original location of a corner cannot be determined beyond reasonable doubt, the corner is said to be *lost*. If the monument of an existent corner cannot be found, the corner is said to be *obliterated*, but it is not necessarily lost.

In the absence of an original monument, either a line tree or a definite connection to natural objects or to improvements may fix a point of the original survey for both departure and latitude. The mean location of a line marked by blazed trees, when identified as the original line, may sometimes help to fix a meridional line for departure, or a latitudinal line for latitude. Other calls of the original field notes in relation to various items of topography may assist materially in the recovery of the locus of the original survey. Such evidence may be developed in infinite variety.

A lost corner is restored to its original location, as nearly as possible, by processes of surveying that involve the retracement of lines leading to the corner. Restoration of a corner does not insure that it is placed exactly in its original location, and when a corner is restored the record of the survey should so state.

**19.47. Proportionate measurements**   It is essential that the laying off of a given distance at the time of a resurvey to restore a lost corner should render the same absolute distance between two points on the ground as was measured during the original survey. For reasons which have been discussed in earlier chapters (Arts. 4.24 to 4.28), the measurement of a given known line at the time of a resurvey will not in general agree with the length of the line as recorded in the original survey. Thus, where linear measurements are necessary to the restoration of a lost corner, the principle of *proportionate measurement* must be employed. Proportionate measurement distributes an excess or deficiency in an overall remeasured distance so that each of the remeasured parts will have the same ratio to the remeasured distance as the corresponding original parts had to the originally measured distance.

Single proportionate measurement consists of first comparing the resurvey measurement with the original measurement between two existing corners on opposite sides of the lost corner, and then laying off a proportionate distance from one of the existing corners to the lost corner. Double proportionate measurement consists of single proportionate measurement on each of two such lines perpendicular and intersecting at the lost corner.

**19.48. Field process of restoration** Following are the field procedures to be followed in a few of the simpler cases of the restoration of lost corners. In any event the restorative process must be in harmony with the methods employed in originally establishing the lines involved, and the preponderant lines must be given the greater weight in determining whether a corner should be relocated by single or double proportionate measurement or by some other method. Thus, standard parallels are given precedence over township exteriors, the latter are given precedence over subdivisional lines, and quarter-section corners are relocated after adjoining section corners have been restored. Detailed instructions for the relocation of lost corners are given in Ref. 20.

1. *Township corner common to four townships* Where all the connecting lines have been established in the field, retracement is made between the nearest existing corners on the meridional line, one north and one south of the lost corner, and a temporary stake is set at the proportionate distance for the lost corner; this defines the latitude of the lost corner. Similarly, measurement is made between the nearest existing corners on the latitudinal line through the point, and at the proper proportionate distance a second temporary stake is set; this marks the departure (or longitude) of the lost corner. The location of the lost corner is then found at the intersection of an east-west line through the first stake and a north-south line through the second; the corner is thus relocated by double proportionate measurement.

2. *Section corner common to four sections in interior of township* Where all lines have been run, the section corner common to four sections in the interior of a township is restored by double proportionate measurement, in the manner described in (1).

3. *Regular corner on range line but not at corner of township* The range line is straight between township corners. Two original corners on the 6-mi (9.66 km) segment of the range line, one north and one south of the point sought, are identified and a line is run between them. The lost corner is relocated by a single proportionate measurement along this line. This procedure applies either to section or quarter-section corners.

4. *Regular corner on township line but not at corner of township* The township line was originally run as a parallel of latitude for 6 mi (9.66 km). A parallel is rerun between the nearest existing corners to the east and west of the point sought, and the corner is relocated by single proportionate measurements along this line.

5. *Standard corner* The standard corner includes any township, section, quarter-section, or meander corner, established on a base line or standard parallel at the time the line was originally run. The corner is relocated by the process explained in (4), that is, by single proportionate measurement along the parallel reestablished between the nearest existing standard corners on opposite sides of the point sought.

6. *Quarter-section corner on either meridional or latitudinal section line but not on range or township line* The corner is relocated by single proportionate measurement along the straight line joining the adjacent section corners of the same section. If these section corners cannot be identified, they must be restored, as previously explained, before the quarter-section corner can be reestablished.

7. *Quarter-section corner at center of section* The corner is relocated at the intersection of meridional and latitudinal lines between opposite quarter-section corners on the boundaries of the section.

8. *Closing corner on standard parallel* The parallel is reestablished between the nearest existing corners on opposite sides of the corner sought. The lost corner is relocated by single proportionate measurement along the parallel from the nearest *standard* corners on opposite sides of the point sought.

9. *Quarter-quarter-section corner on section and quarter-section lines* The corner is relocated by single proportionate measurement between quarter-section and section corners on opposite sides of the point sought.

10. *Quarter-quarter-section corner at center of quarter section* The corner is relocated at the intersection of the meridional and latitudinal lines between opposite quarter-quarter-section corners on the exterior of the quarter section.

**19.49. Photogrammetric methods applied to property surveys** Instrumental and analytical photogrammetric triangulation (Art. 16.18) are methods for determining positions of photographed objects. Qualitative interpretation of terrain features is also possible with overlapping aerial photographs. These two aspects of photogrammetry, quantitative and qualitative, make it a powerful tool for performing property surveys.

The procedure involved in preparing and executing a photogrammetric property survey is similar in many respects to the methods described for original surveys, resurveys, and subdivision surveys described in Arts. 19.15 to 19.18. The primary differences are in the methods employed to determine horizontal positions of points. Generally, the surveyor and photogrammetrist work as a team on a photogrammetric surveying project. Few individuals possess sufficient experience and expertise to do the job alone. Consider, as an example, a subdivision survey for a large tract of land.

As in any property survey, all available maps, aerial photographs, descriptions of the property to be surveyed, descriptions of adjacent lands, descriptions and references to all existing horizontal and vertical control points, and all field notes should be assembled and analyzed prior to any field operations or photogrammetric work. Assume for purposes of this discussion that a topographic map adequate for planning is available for the area and that a tentative plan of streets and lots has been prepared for the subdivision. The objective of the photogrammetric survey is to establish horizontal positions of control points for two purposes: (1) in order to fix the positions of boundaries around the perimeter of the tract; and (2) to provide horizontal control points around the perimeter and throughout the interior of the area with sufficient density to permit locating subdivision monuments and center-line control points for subsequent layout of all street center lines and property corners in the subdivision.

Assume that analytical photogrammetric triangulation is to be used to establish positions for horizontal control points (Art. 16.18). First, photography must be planned so as to provide complete stereoscopic coverage of the area at a scale compatible with the specified accuracy requirements (Art. 16.21). During this stage, ground control points are located around the perimeter in order to provide control for the block of aerial photographs. These points should be positioned so as to be usable in a subsequent theodolite and EDM traverse around the perimeter. In addition, points should be located as closely as possible to existing property corners and other areas where evidence of boundary positions may exist. These latter points may or may not be included in the perimeter traverse. At the same time control-point locations are also selected throughout the interior of the tract. These interior or *photogrammetric control points* should be in locations suitable for photogrammetric positioning. Thus, points must be in the overlap of at least two aerial photographs and ought not to be under trees, next to steep cliffs, etc. These points should also be chosen in locations convenient for subsequent layout of subdivision controlling points, such as center-line intersections, points of intersections for curves, angle points in lot lines, etc. Each interior photogrammetric control point must be placed so as to be intervisible with at least one other point in order to provide azimuth in subsequent field surveys. Selection of locations for perimeter ground control points and photogrammetric control points (perimeter and interior) can be expedited by use of existing aerial photography. Stereoscopic examination of this photography allows locating perimeter control points near presumed property corners, fence lines, or other evidence of boundary position as interpreted from studying the three-dimensional model. This type of analysis is also helpful in choosing strategic locations for interior photogrammetric control points. The selected locations can be marked directly on the photographs for later use in the field. Vertical control points used in the photogrammetric triangulation must also be planned. The highest density of vertical control should be around the perimeter of the tract. Trigonometric leveling (Art. 5.5), performed concurrently with the perimeter traverse, provides adequate accuracy in elevation.

Prior to taking the aerial photography, all ground control points, photogrammetric control points, existing monuments and bench marks, and all potential and identified property corners are marked with targets. These targets should be of a size, configuration, and color suitable for the analytical photogrammetric triangulation. In choosing the size of the target, the scale of the photography to be flown must be taken into account.

After all points have been marked, or concurrently with the placing of targets, a traverse is run around the tract. This traverse should be a closed loop tied to at least two monuments in the national control system or the individual sections of the traverse must be closed on monuments of known position. *No point should remain unchecked.* Ties should be made to all U.S. National Ocean Survey, U.S. Geological Survey, and county monuments in the area. Computations should be performed in the state plane coordinate system and all elevations should be referred to the National Geodetic Vertical Datum (NGVD) (Art. 13.2).

When marking of all points and field surveys are completed, the data are reduced and adjusted yielding state plane coordinates and elevations for all ground points. After aerial photography has been acquired, the images of all ground control and photogrammetric control points are measured using a comparator. The measurements and the $X$, $Y$ coordinates and elevations for known ground control points are utilized in an electronic computer program which performs the calculations required for photogrammetric triangulation. This program yields as output the $X$, $Y$ coordinates and elevations for all unknown perimeter and interior photogrammetric control points included in the adjustment. These data are retained for future computations.

The control point locations around the perimeter of the tract determined by both field survey and photogrammetric methods are now used to make final determination of boundary locations around the property. Evidence located by stereoscopic examination of aerial photographs permits defining the area of search within rather narrow limits so as to minimize field examination for evidence of property lines. Azimuths and distances are computed from ground and photogrammetric control points to positions of potential property corners and monuments so as to facilitate the search in the field. When all data from examining photography and field searches are consolidated, analyzed, and correlated with the deed description of the property, the exterior boundaries of the tract are fixed and coordinates are computed for each property corner and angle point. Next, coordinates are computed for all center-line and lot-line controlling points and all lot corners within the subdivision. Finally, the azimuths and distances from photogrammetric control points to the nearest subdivision monument or control point are calculated. A listing of these azimuths and distances is prepared for use in the field. It is also possible to plot automatically (Art. 13.13) the positions of photogrammetric control points and the nearest subdivision monuments on work sheets which can also be used in the field.

Field layout surveys consist of occupying the photogrammetric control points, backsighting on another photogrammetric control point, and setting the nearest subdivision monument by angle and distance from the data sheets prepared in the office. When a sufficient number of subdivision monuments have been established, the center lines of streets and the balance of the lot corners should be set so as to provide a check on the layout. No lot corner or subdivision monument should remain unchecked.

As the monumentation and lot survey proceeds, spot checks on the accuracy of the photogrammetric survey can be made. Independent field traverses are run from the perimeter control traverse for this purpose. These traverses should be well distributed throughout the tract so as to provide an adequate sample of check points.

When the monumentation survey and checking have been completed to the satisfaction of the surveyor, the photogrammetrist, and the county engineer or surveyor, the photogrammetric survey is completed. All control points and monuments should be thoroughly referenced for subsequent use and the subdivision plan should be submitted to the proper authorities for approval and recording.

The procedure as outlined for a new subdivision can be applied in essentially the same way to the resurvey of an old subdivision, a group of lots located by metes and bounds, and land subdivided according to the U.S. public-land surveys. To summarize, some of the principal factors to be remembered in any photogrammetric property survey are:

1. The surveyor and photogrammetrist must work as a team. Careful advance planning to correlate field and photogrammetric work is absolutely necessary.

2. Whenever possible, locate all visible evidence of property lines and corners by examination of existing photography and field searches prior to acquiring the aerial photography to be used in the photogrammetric survey.
3. Set targets on all ground control points, photogrammetric control points, and existing monuments and bench marks. Set targets on or near all identified and potential property corners.

Finally, it is necessary to remember that the photogrammetrist can determine the position of a signalized control point with an accuracy more than adequate for most property surveys. However, to locate a property corner properly from this control point requires the services of a surveyor with substantial experience in property surveying and an adequate knowledge of the laws and customs regulating property subdivision in the region where the work is being performed.

References 5 and 15 contain additional details concerning accuracies possible and application of photogrammetric methods to property surveys.

## 19.50. Problems

**19.1** Name and discuss the different types of property or land surveys.

**19.2** Define the following terms: senior deed, easement, eminent domain, oral agreement, and adverse possession.

**19.3** Define and give examples for natural, artificial, and record monuments.

**19.4** Describe three methods by which land transfer records may be filed.

**19.5** List the major elements that should be included in a land description.

**19.6** What is a metes and bounds description? Describe five different types of metes and bounds descriptions.

**19.7** Write a metes and bounds description for the property shown in Fig. 19.8. Do not use coordinates in the description.

**19.8** Give examples of descriptions by subdivision. What is a significant difference between a description by subdivision and one by metes and bounds?

**19.9** Discuss how a description can be written for the property owned in a condominium.

**19.10** Assume that you have bought unit 10 in a condominium constructed on the property shown in Fig. 19.2 and described in Example 19.3. Write the description for this condominium unit.

**19.11** The coordinates (Lambert projection, south zone, Pennsylvania) for the southernmost corner of the property shown in Fig. 19.8 are $X = 1,005,431.05$ ft and $Y = 405,110.85$ ft. All directions shown are referred to grid north for the projection. Write a metes and bounds description for this property using coordinates.

**19.12** Write a lot and block description for lot 22 in the plan shown in Fig. 19.9. Assume that this subdivision is part of the Artesian Heights Addition recorded in Plan Book 140, page 32, in the Office of the Recorder of _____ County, State of _____ .

**19.13** Give the order of precedence for calls in a deed when conflicts exist.

**19.14** Define the following terms which are related to riparian rights: thread of a river, meander line, alluvium, alluvion, bed, high-water mark, low-water mark, and reliction.

**19.15** List the conditions that must be satisfied in order to acquire possession of property by adverse possession.

**19.16** Outline the steps that should be taken by a registered surveyor in preparing to execute a resurvey of property located in a region unfamiliar to the surveyor.

**19.17** Brown owns a farm containing about 60 acres (about 243 ha). In 1965 he sold a 10-acre (40.5-ha) portion to Boyd, and in 1970 sold a 20-acre portion (80.9-ha) to Wilson. Brown continues to live on and farm the remaining land. Among the three property owners, who has senior rights? Explain.

**19.18** In block 5 of the subdivision illustrated in Fig. 19.9a, lot 7 was sold in March 1972. Thereafter, lots 20, 8, 13, and 11 were sold in April, July, August, and October of 1973. Among these five property owners, which has senior rights? Explain.

**19.19** Joyce, who owns a lot 200 m in width along Route 82, sold the west half, a lot 100 m wide, to Eliot in 1967. Both properties were surveyed at the time and metes and bounds descriptions were recorded. Ten years later both of these properties were resurveyed and a shortage of 1 m was discovered in the frontage along Route 82. How is this discrepancy proportioned in the resurvey? Support your conclusions.

**19.20**  A subdivision contains twenty 60-ft lots numbered 1, 2,...,20 from east to west in one block. During a resurvey to set the corners of lot 5, the total distance between block corners (proven to be correct) is shown to be 1200.80 ft. What distance should be measured from the eastern corner of the block to set the eastern corner of lot 5, and how much frontage should be allocated to this lot? Support your reasons for proportioning the discrepancy.

**19.21**  Describe, briefly, how photogrammetric methods can be used for property surveys.

**19.22**  In performing a property survey using photogrammetric methods, describe the responsibilities of (a) the photogrammetric engineer; (b) the registered land surveyor. Assume that both are involved in a joint effort in the task.

**19.23**  Find the angle of convergence between two meridians 6 mi apart at a mean latitude of 32°20'. Compute the linear convergence, measured along a parallel of latitude, in a distance of 6 mi.

**19.24**  Find the length of 1° longitude at a latitude of 40°06'20".

**19.25**  Find the offsets between the tangent and the parallel at intervals of $\frac{1}{2}$ mi over a distance of 6 mi at a latitude of 40°06'20".

**19.26**  Find the azimuth of the secant and the offsets from the secant to the parallel at intervals of $\frac{1}{2}$ mi over a distance of 6 mi at a latitude of 40°06'20".

**19.27**  Show the dimensions and areas of the protracted subdivisions of section 7, as required by law to be shown on the official plat, when the north, east, south, and west boundaries are respectively 76.84, 80.00, 76.64, and 80.00 chains.

**19.28**  Show the dimensions and areas of the protracted subdivisions of section 6, as required by law to be shown on the official plat, when the north, east, south, and west boundaries are respectively 76.36, 80.44, 76.60, and 80.00 chains.

**19.29**  A lost interior section corner is to be restored by a resurvey. The nearest corners which can be identified are regular section corners 1 mi north, 2 mi east, 3 mi south, and 1 mi west of the point sought. The records show the corresponding original measured distances to be 80.40, 160.56, 240.00, and 78.32 chains. The resurvey measurement between the nearest existing monuments on the meridional line through the lost corner is 320.16 chains, and that along the latitudinal line between the nearest existing corners is 238.48 chains. Calculate the proportionate measurements to be used in the relocation of the lost corner and state the procedure to be employed in its reestablishment.

**19.30**  A lost section corner on a range line is to be restored by a resurvey. One mile to the south the township corner is identified, and $2\frac{1}{2}$ mi to the north the quarter-section corner is found. According to the records the corresponding distances measured at the time of the original survey were 80.00 and 200.00 chains. The resurvey distance between the existing corners is 279.64 chains. State the procedure to be followed in restoring the lost corner, and calculate the proportionate distances to be employed.

## 19.51. Office problem

### DEED DESCRIPTION

**Object**  To obtain the legal description and other data needed to perform a property survey for a specific lot.

### Procedure

1. Go to the county recorder's office (or other public office used for this purpose) and obtain the deed description for the specified property. This property may be the lot on which the house where you live is located or some other appropriate lot.
2. If the deed contains a reference to a recorded plan of lots (usually by plan book volume and page number), locate the plan and get a copy of those portions of the plan you believe necessary to perform the survey.
3. Using the calls from the deed or data from a subdivision map, calculate departures and latitudes for the sides of the property, compute the closure error, and calculate coordinates for the property corners (Art. 8.16). Plot the outline of the property, at an appropriate scale, on an $8\frac{1}{2}$ by 11 in sheet (Art. 13.13).
4. Rewrite the description by the metes and bounds method using coordinates.

5. The final report should consist of copies of the deed description obtained at the recorder's office; plan of lots in which the property lies; map of the lot on the $8\frac{1}{2}$ by 11 in sheet, and the deed written using coordinates.

## References

1. American Society of Civil Engineers, Manuals of Engineering Practice, *Technical Procedures for City Surveys*, No. 10, 1963; *Urban Planning Guide*, No. 49, 1969; *Accommodation of Utility Plant within the Rights of Way of Urban Streets and Highways*, No. 14, 1974; and *Definitions of Surveying Mapping and Related Terms*, No. 32, 1978 (rev.).
2. Bouvier, John, *Bouvier's Law Dictionary*, William E. Baldwin, Ed., Banks-Baldwin Publishing Company, Cleveland, Ohio, 1948.
3. Brown, C. M., "Identifying Monuments," *Surveying and Mapping*, September 1976.
4. Brown, Curtis M., *Boundary Control and Legal Principles*, 2nd ed., John Wiley & Sons, Inc., New York, 1969.
5. Brown, Duane C., "Accuracies of Analytical Triangulation in Applications to Cadastral Surveys," *Surveying and Mapping*, September 1973.
6. Clark, F. F., *Law of Surveying and Boundaries*, 3rd ed., The Bobbs-Merrill Company, Inc., Indianapolis, Ind., 1959.
7. Cook, Robert N., "Multi-purpose Land Data System, the Legal Parcel," *Proceedings, 37th Annual Meeting, ACSM*, 1977, pp. 612–617.
8. Danial, N. F., "Some Guidelines for Laying Out Subdivisions," *Surveying and Mapping*, June 1972.
9. Dean, D. R., and McEntyre, J. G., "Surveyor's Guide to the Use of a Law Library," *Surveying and Mapping*, Vol. 35, No. 3, September 1975.
10. Grimes, J., *Clark on Surveying and Boundaries*, 4th ed., The Bobbs-Merrill Company, Inc., Indianapolis, Ind., 1976.
11. Greulich, G., "Use of Metric System in Metes and Bounds Surveys," *Proceedings, Fall Convention, ACSM*, 1974.
12. Greulich, G., "CLIPP, RESPA, and Cadastre," *Journal of the Surveying and Mapping Division, ASCE*, No. SU1, Proceedings Paper 14145, November 1978.
13. Hill, J. M., "Riparian Rights—From Estuary to Tidal Headwaters," *Surveying and Mapping*, June 1973.
14. Kissam, P., *Surveying for Civil Engineers*, McGraw-Hill Book Company, New York, 1956.
15. Lafferty, M. E., "Photogrammetric Control for Subdivision Monumentation," *Proceedings, Fall Convention, ACSM*, 1971.
16. Laundry, M. E., "The Condominium Survey: Its Legal and Practical Aspects under Illinois law," *Surveying and Mapping*, Vol. 34, No. 3, September 1974.
17. Moyer, D. D., and Fisher, P. K., "Land Parcel Identifiers for Information Systems," *Proceedings CLIPP*, sponsored by the American Bar Foundation, Atlanta, Ga, 1972.
18. North American Conference on Modernization of Land Data Systems (MOLDS), *Proceedings*, Washington, D.C., 1975.
19. Robillard, W. G., "The Surveyor and the Law," *Surveying and Mapping*, June 1978.
20. U.S. Bureau of Land Management, U.S. Department of the Interior, *Manual of Instructions for the Survey of Public Lands of the United States, 1973*, Technical Bulletin No. 6, Government Printing Office, Washington, D.C.
21. U.S. Bureau of Land Management, *Standard Field Tables and Trigonometric Tables*, 8th ed., Government Printing Office, Washington, D.C., 1956.
22. Vorhies, J. H., "A Computerized Land Title Plant in Joint Use," *Proceedings Annual Meeting, ACSM*, 1977.
23. Wattles, G. H., *Writing Legal Descriptions*, Gurdon H. Wattles Publications, Orange, Calif., 1976.
24. Wattles, W. C., *Land Survey Descriptions*, 10th ed., Gurdon H. Wattles Publications, Inc., Orange, Calif., 1974.
25. Woodbury, R. L., "The Surveyor and the Law (Massachusetts Land Court History and Procedure)," *Surveying and Mapping*, Vol. 33, No. 2, June 1973.

# CHAPTER 20
# Mining surveys[†]

**20.1. Introduction**  Mine surveying includes underground surveying as practiced in mining and tunneling as well as the surface operations associated with underground work and open-pit mining.

Conditions underground are very different from those on the surface. Traverses may contain very short legs and run along narrow, dusty corridors. Levels to establish elevations may have to be brought into the workings through deep shafts. Astronomic observations are not possible, so that underground orientation must be controlled by plumbing wires in a shaft or by means of a gyro theodolite. Rock movement can affect the stability of survey marks and may also cause more serious problems associated with cave-ins, property damage, or perhaps loss of life. The mine surveyor must monitor these rock movements and cooperate closely with geologists and other related specialists.

Because of the expanding role of surveyors in the mining industry (Ref. 3), it is not possible to cover all aspects of mine surveys in one chapter. Emphasis is concentrated on a few of the more basic tasks in underground and tunnel surveys, such as mine orientation procedures and the control of tunneling. However, many other equally important aspects, such as open-pit surveying, ground deformation monitoring, and three-dimensional mine modeling and mapping have been omitted.

**20.2. Mining terminology**  The composite sketch in Fig. 20.1 illustrates some mining terms the definition of which follow:

*Adit*  A horizontal or nearly horizontal passage driven from the surface for working or dewatering a mine.

*Back*  The top of a drift, cross cut, or slope. Also called a roof.

*Back fill*  Waste rock or other material used to fill a mined out stope to prevent caving.

*Bedded deposit*  An ore deposit of tabular form that lies horizontally or slightly inclined and is commonly parallel to the stratification of the enclosing rocks.

*Cage*  An elevator for workers and material in a mine shaft.

*Chute*  A channel or trough underground, or inclined trough above ground, through which ore falls or is shot by gravity from a higher to a lower level (also spelled shoot).

*Collar*  The term applied to the timbering or concrete around the mouth or top of a shaft and the mouth of a drill hole.

*Cross cut*  A horizontal opening driven from shaft to a vein across the course of the vein in order to reach the ore zone.

*Dip*  The angle at which a bed, stratum, or vein is inclined from the horizontal.

[†]This chapter was written by Dr. Adam Chrzanowski, University of New Brunswick, Canada, and Dr. A. J. Robinson, University of New South Wales, Australia.

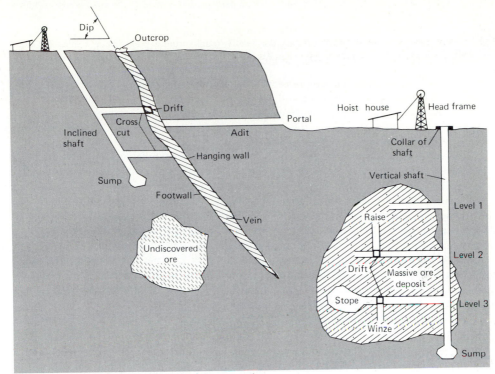

**Fig. 20.1** Cross section of a typical mining operation.

*Drift*   A horizontal opening in or near a mineral deposit and parallel to the course of the vein or long dimension of the deposit.

*Entry*   Manway, haulage way, or ventilation way below ground, of a permanent nature (i.e., not in an ore to be removed).

*Face*   End wall of a drift or cross cut or of bedded deposits.

*Foot wall*   The wall or rock under a vein or under other steeply inclined mineral formations.

*Gangue*   Undesired minerals associated with ore.

*Gangway*   A main haulage road underground.

*Hanging wall*   The wall or rock on the upper side of steeply inclined deposits. It is called a roof in bedded deposits.

*Headframe*   A construction at top of a shaft which houses hoisting equipment.

*Level*   Mines are customarily worked from shafts through horizontal passages or drifts called levels. These are commonly spaced at regular intervals in depth and are either numbered from the surface in regular order or are designated by their actual elevation below the top of a shaft.

*Ore pass*   Vertical or diagonal opening between levels to permit movement of ore by gravity.

*Pillars*   Natural rock, or ore supports, left in stopes to avoid or to decrease the roof subsidence as mining progresses.

*Raise*   A vertical or inclined opening driven upward in ore from a level.

*Rib*   Wall in an entry. Also simply wall.

*Roof*   *See* Back.

*Shaft*   A vertical or inclined excavation in a mine extending downward from the surface or from some interior point as a principal opening through which the mine is exploited.

*Shoot*   *See* Chute.

*Sill*   Synonymous with floor.

*Stope*   Underground "room" or working area from which ore is removed.

*Strike*   The horizontal course, bearing, or azimuth of an inclined bed, stratum, or vein.

*Sump*   An excavation made at the bottom of a shaft to collect water.

*Tunnel* A horizontal or nearly horizontal underground passage that is open to the atmosphere at both ends.

*Waste* Mined rocks that do not contain useful minerals.

*Winze* A vertical or inclined opening driven downwards (sunk) from a point inside a mine for the purpose of connecting with a lower level or of exploring the ground for a limited depth below a level.

**20.3. Design of horizontal control networks in underground mines** Control networks consist of traverses (frequently open-end traverses) that must follow the existing net of mining workings and excavations. Since open-end traverses may often serve as basic control, they must be executed with the utmost care and are usually independently checked by a second resurvey.

Distances between the survey stations are generally very short, ranging from 10 to 20 ft (a few metres) to an average of 160 ft (50 m). Only in the main transportation roads may the distances be increased to about 1000 ft or a few hundred metres.

The control network consists of (1) first-order loops which serve as basic control and are run in the permanent mine workings, (2) second-order traverses run into headings and development areas, and (3) third-order stations (short traverses) used for detailed mapping of excavated areas and daily checks of mining progress in stopes and headings.

The establishment of the underground control network is done in a reversed sequence from that used on the surface. The lowest-order control is established first and is subsequently replaced by the higher-order control once the developed area allows for longer sights and for a loop closure of the traverses.

Typically, the following maximum errors in relative positions of the control points are permitted:

*first-order control*   (*a*) 1:10,000 in small and medium-size mines
                           (*b*) 1:20,000 in large mines extended over areas of several kilometres in diameter
*second-order control*    1:5000
*third-order control*      1:1000

The relative accuracies given above are usually interpreted as the ratio of the semimajor axis of the relative error ellipse (Art. 2.19), at the 95 percent probability level, to the distance between the points of interest.

**20.4. Monumentation and marking of points** The stations of the horizontal control network are usually marked in the roof (back) or walls of the mining workings. A hole is drilled, a wooden plug is inserted, and into this a spad (Fig. 20.2*a*) or a metal plug (Fig. 20.2*b*), with a hole for the string of the plumb bob, is driven. The markers may also be cemented directly in the drilled holes, using, for instance, an epoxy glue.

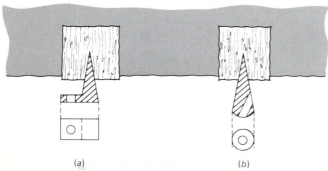

(*a*)                        (*b*)

**Fig. 20.2** Roof markers.

The wall markers require either a special type of a portable bar (Fig. 20.3) which is inserted into the marker during the survey procedure, or the markers are used as eccentric stations to which the position of the survey instruments is referenced by measurements of short distances and/or angles (see Ref. 9). The latter method, although requiring some additional measurements and trigonometric calculations, has the advantage that the survey instruments can be set up in any convenient place without the time-consuming task of centering under the marker.

**20.5. Angle measurements**   The old-type vernier transits, although still in use in some mines in North America, are being replaced by much smaller and lighter modern theodolites with the optical micrometer readout (Art. 6.30). The theodolites are equipped with electric illumination of the horizontal and vertical circles and the cross hairs. Vertical axes of the theodolite are marked on the top of the telescopes for centering under the roof stations.

Owing to the generally short sights in underground traversing, accurate centering of the instruments is very crucial. Centering under the roof markers is more difficult than the conventional centering above the marked points. If one adds to it the cramped conditions, darkness, and difficulties for setting the legs of the tripod on an uneven floor, the centering procedure requires a lot of experience. It is usually done by means of a string plumb bob. The telescope with a centering marker must, of course, be set in a horizontal position during the centering procedure. The tips of the bobs should be very sharp and protected against any damage to ensure a good accuracy of centering. Optical zenith plummets attached on the top of the telescopes or interchangeable with the theodolite in the same tribrach, available as an optional accessory with some models (e.g., Wild Heerbrugg) of theodolites, are also used in centering under the roof markers.

First-order traversing is usually done by means of forced-centering traversing equipment, using interchangeable theodolite and targets fitted with detachable tribrachs (Art. 6.30).

Repeating theodolites with direct micrometer readouts of 20″ to 1′ are usually sufficient for most of the control surveys except first-order traversing in very long headings and tunnels when precision theodolites are required.

Plumb-bob strings lighted from behind by means of a mining head lamp (covered with a piece of tracing paper) serve as the targets in third- and sometimes in second-order traversing. Traversing equipment with lighted targets (see Fig. 6.67) are used in first-order surveys. Special parabolic reflectors with 6-volt bulbs and with a changeable aperture have

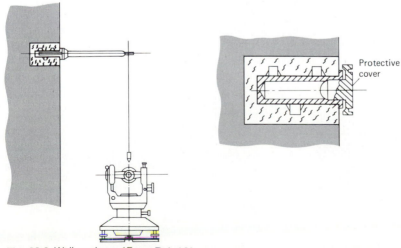

Protective
cover

**Fig. 20.3** Wall markers. (*From Ref. 16.*)

to be used as the targets when distances exceed a few hundred metres (Ref. 21). A small helium–neon laser mounted on the telescope of a precision theodolite and small corner reflectors used as the targets have been successfully tested at the Department of Surveying Engineering of the University of New Brunswick in Canada in precision angle measurements in long tunnels.

Frequently, very steep sights are encountered when traversing through steep raises or other inclined openings. The influence of an error $\varepsilon_\beta$ (expressed in seconds of arc) in leveling the theodolite on the accuracy of angle measurement can be very dangerous and is expressed by the approximate formula (Ref. 10)

$$\varepsilon_\beta = \varepsilon_L \sqrt{(\tan \gamma_1)^2 + (\tan \gamma_2)^2} \tag{20.1}$$

where $\gamma_1$ and $\gamma_2$ are vertical angles of the sights and $\varepsilon_L$ is the error of leveling the horizontal plate of the theodolite. For instance, if $\gamma_1 = \gamma_2 = 40°$ and a theodolite with a spirit level of a sensitivity of 30″ is misleveled by one division only (i.e., $\varepsilon_L = 30″$), then the error of the measured angle $\varepsilon_\beta = 36″$. Additional striding levels of a higher sensitivity must then be used in leveling the theodolite.

Setting and reading the micrometer on very steep sights may be most difficult or even impossible without using additional diagonal eyepieces which are available as optional accessories with most models of modern theodolites. Some companies, for example Breithaupt, Keuffel & Esser, and Berger, produce theodolites with additional eccentric side telescopes for sighting downwards on very steeply inclined traverse legs. Kern Co. of Switzerland produces small additional telescopes which fit on the top of the main telescope of their model DKM-1 theodolite. When using the side telescope, the mean of an angle turned with the telescope direct and reversed is free of the eccentricity influence of the side telescope. Use of the top telescope is inferior in this respect because it does not allow for direct and reversed measurements and, therefore, requires a very careful instrument adjustment prior to the measurements. Vertical angles measured with eccentric telescopes must be corrected for the eccentricity. Details of the adjustment and use of eccentric telescopes are given in Refs. 19 and 23.

In some cases the cramped space in the mining workings or an unstable and steeply inclined floor do not allow the use of tripods in angle measurements. The theodolite must then be set up on special bars (Fig. 20.4) or supporting arms equipped with a bracket screwed or bolted to the timber. Also, telescopic beams stretched between the walls of the headings are used. The Wild and Zeiss companies, among others, produce special supporting bars for their theodolites for mining and construction survey applications. Any mining workshop should be able to make supporting arms at the request of the mine surveyor.

Reversed hanging theodolites (Fig. 20.5a), which fit on supporting bars screwed into wall timbering, are popularly used in the third-order surveys in coal mines in central and eastern European countries. They are very convenient to use in narrow and cramped conditions.

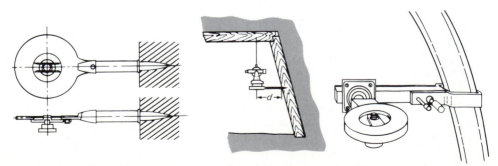

**Fig. 20.4** Supporting bars for theodolites. (*From Ref. 16.*)

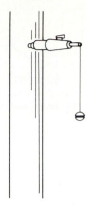

**Fig. 20.5** (*a*) Hanging theodolite.
(*Courtesy Academy of Mining and
Metalurgy, University of Krakow, Krakow,
Poland.*) (*b*) Hanging spherical target.

(*a*)                                         (*b*)

They allow for very fast setting up and leveling. Small targets in a form of small spheres
(Fig. 20.5*b*) or cones suspended on a short chain are interchangeable with the theodolite on
the portable supporting arms, thus allowing for an automatic (forced) centering. Model
Temin made by Breithaupt weighs only 2.5 kg, has 1 centigrade (about 30″) interpolated
accuracy of the readout, and 1.1 m shortest focusing distance.

**20.6. Distance measurements**   Steel tapes are still the most popular tools in distance
measurement. Lightweight, short-range electronic distance measurement (EDM) instruments
(Arts. 4.29 to 4.35) have become popular in first-order traverse, but their use is still
uneconomical for a daily application in traverses with very short distances. Mining regulations require that only fire- and damp-proof EDM instruments, such as the Zeiss Eldi-2
(Mining), may be used in gaseous mines.

Steel tapes are used mainly supported at the two ends (in catenary) unless the floor is flat
and the procedure and corrections to be applied are practically the same as on the surface.

Optical distance measurements (stadia) with short fluorescent tacheometric rods or optical
range finders, such as Zeiss-Jena BRT 006 telemeter, are very useful in stope measurements
and in detail underground mapping. Their use in mining surveying, due to the specific
conditions, will probably survive a much longer time than in surface measurements, where
electronic tacheometry is rapidly replacing optical measurements.

**20.7. Traverse computations**   Mining operations must be based on a framework of coordinated points. Coordinates of underground stations should be calculated in the surface
coordinate system so that positions of details on the surface can be analytically correlated
with details on individual levels and sublevels of the mine. Of course, this requirement also
implies that both surface and underground systems have the same orientation. First-order
and possibly second-order underground traverse loops should be simultaneously adjusted by
the method of least squares (see Art. B.14, Appendix B).

**20.8. Error propagation in open-end traverses**   As mentioned previously, most of the
control surveys in the underground development areas are based on open-end traverses. A
thorough analysis of the expected positional accuracy of the last point of the traverse should
always be done before starting a new development survey.

Figure 20.6 shows a traverse $O - K$ with known coordinates of points 0 and 1 treated as
errorless (belonging to the higher-order traverse). If angles $\beta_i$ and distances $d_i$ are measured,

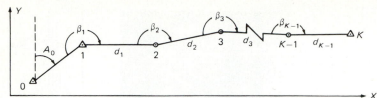

**Fig. 20.6** Traverse with measured angles and distances.

the coordinates of point $K$ are calculated from

$$X_k = X_1 + d_1 \sin(A_0 + \beta_1 - 180°) + d_2 \sin[A_0 + \beta_1 + \beta_2 - (2)(180)] + \cdots$$
$$+ d_{k-1} \sin[A_0 + \beta_1 + \beta_2 + \cdots + \beta_{k-1} - (k-1)180] \tag{20.2}$$
$$Y_k = Y_1 + d_1 \cos(A_0 + \beta_1 - 180) + d_2 \cos[A_0 + \beta_1 + \beta_2 - (2)(180)] + \cdots$$
$$+ d_{k-1} \cos[A_0 + \beta_1 + \beta_2 + \cdots + \beta_{k-1} - (k-1)180] \tag{20.3}$$

where $A_0$ is calculated from the coordinates of points 0 and 1.

The propagated variances and covariances of the coordinates for point $K$, $\sigma_{x_k}^2$, $\sigma_{y_k}^2$, and $\sigma_{xy_k}$ can be found by applying Eqs. (8.33), Art. 8.23, to Eqs. (20.2) and (20.3). Standard deviations of angles, $\sigma_\beta$ (in radians), and distance, $\sigma_d$, are estimated for this purpose.

The variances and covariances determined using Eqs. (8.33) allow calculation of the parameters of the error ellipse for point $K$ using Eqs. (2.39) to (2.41), Art 2.19.

If azimuths of each traverse leg are measured (Fig. 20.7) using a gyrotheodolite, the coordinates of $K$ are calculated from

$$X_k = X_1 + d_1 \sin A_1 + d_2 \sin A_2 + \cdots + d_{k-1} \sin A_{k-1} \tag{20.4}$$

and

$$Y_k = Y_1 + d_1 \cos A_1 + d_2 \cos A_2 + \cdots + d_{k-1} \cos A_{k-1} \tag{20.5}$$

In this case the error propagation will result in

$$\sigma_{x_k}^2 = \sum_1^{k-1} (Y_{i+1} - Y_i)^2 \sigma_{A_i}^2 + \sum_1^{k-1} \left( \frac{X_{i+1} - X_i}{d_i} \right)^2 \sigma_{d_i}^2 \tag{20.6}$$

$$\sigma_{y_k}^2 = \sum_1^{k-1} (X_{i+1} - X_i)^2 \sigma_{A_i}^2 + \sum_1^{k-1} \left( \frac{Y_{i+1} - Y_i}{d_i} \right)^2 \sigma_{d_i}^2 \tag{20.7}$$

and

$$\sigma_{xy_k} = - \sum_1^{k-1} (X_{i+1} - X_i)(Y_{i+1} - Y_i)\sigma_{A_i}^2 + \sum_1^{k-1} (X_{i+1} - X_i)(Y_{i+1} - Y_i)\frac{\sigma_{d_i}^2}{d_i^2} \tag{20.8}$$

Comparing the Eqs. (8.33), Art. 8.23, with Eqs. (20.6) to (20.8) it can be seen that there is a significant difference in the propagation of errors of angle measurements versus errors of azimuth measurements. In long traverses with many stations, the positional accuracy of the last point may be smaller in a traverse with measured azimuths, compared to a traverse with measured angles even when the angles are measured with a much higher accuracy than the azimuths.

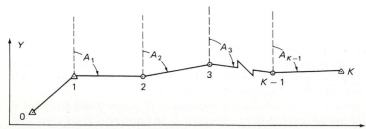

**Fig. 20.7** Traverse with measured azimuths and distances.

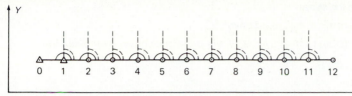

**Fig. 20.8** Error propagation of angles vs. azimuths (Example 20.1).

**Example 20.1**   A traverse 5.5 km long (Fig. 20.8) is run from fixed, higher order stations 0 and 1 to station No. 12. All the distances are equal to 500 m. An option is available either to measure angles $\beta$ with $\sigma_\beta = 4''$ or azimuth with $\sigma_A = 15''$. Which method will give smaller $\sigma_{y_{12}}$? (Note that in this example the X-coordinate of 12 is not affected by the errors of angles or azimuths because the traverse is parallel to the X axis).

SOLUTION   Using angle measurements and applying Eqs. (8.33) to Eq. (20.3),

$$\sigma_{y_{12}}^2 = \sum_1^{11} (X_{12} - X_i)^2 \left( \frac{4''}{\rho''} \right)^2 = 0.0476 \text{ m}^2 \qquad \text{and} \qquad \sigma_{y_{12}} = 0.218 \text{ m}$$

Using azimuth measurements [Eq. (20.7)]:

$$\sigma_{y_{12}}^2 = \sum_1^{11} (X_{i+1} - X_i)^2 \left( \frac{15''}{\rho''} \right)^2 = 0.0145 \text{ m}^2 \qquad \text{and} \qquad \sigma_{y_{12}} = 0.121 \text{ m}$$

where $\rho'' = 206,265''$.

The results show that the azimuth measurements would give almost twice the positional accuracy of the angle measurements even though the azimuths would be measured with almost four times lower accuracy than angles.

**Example 20.2**   Given are approximate coordinates of traverse stations points 1 to 4 (Fig. 20.9):

| Point | X, m | Y, m |
|---|---|---|
| 1 | + 300.00 | + 100.00 |
| 2 | 400.00 | 200.00 |
| 3 | 500.00 | 200.00 |
| 4 | 600.00 | 200.00 |

Points 0 and 1 are known and are treated as errorless. Find the standard error ellipse of point 4 if estimated standard deviations of measured angles and distances are $\sigma_\beta = 10''$, $\sigma_d = 10$ mm. First apply Eqs. (8.33) to Eqs. (20.2) and (20.3) to obtain propagated variances and covariances.

SOLUTION

$$\sum_1^3 (Y_4 - Y_i)^2 \sigma_{\beta_i}^2 = 23 \text{ mm}^2 \qquad\qquad \sum_1^3 (X_4 - X_i)^2 \sigma_{\beta_i}^2 = 329 \text{ mm}^2$$

$$\sum_1^3 \frac{(X_{i+1} - X_i)^2}{d_i^2} \sigma_{d_i}^2 = 250 \text{ mm}^2 \qquad\qquad \sum_1^3 \frac{(Y_{i+1} - Y_i)^2}{d_i^2} \sigma_{d_i}^2 = 50 \text{ mm}^2$$

$$\sum_1^3 (Y_4 - Y_i)(X_4 - X_i)\sigma_{\beta_i}^2 = 71 \text{ mm}^2 \qquad\qquad \sum_1^3 \frac{(X_{i+1} - X_i)(Y_{i+1} - Y_i)}{d_i^2} \sigma_{d_i}^2 = 50 \text{ mm}^2$$

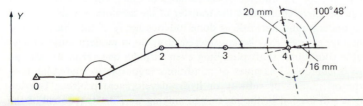

**Fig. 20.9** Positional error in an open traverse (Example 20.2).

where $\sigma_{B_i}$ is in radians. From the above:

$$\sigma_{X_4}^2 = 273 \text{ mm}^2 \qquad \sigma_{Y_4}^2 = 379 \text{ mm}^2 \qquad \sigma_{X_4 Y_4} = -21 \text{ mm}^2$$

Placing these values in Eqs. (2.39) to (2.41), Art. 2.19, the parameters of the error ellipse are

$a = 20$ mm
$b = 16$ mm
$\theta = 100°48'$

**20.9. Mine orientation surveys: basic principles and classification**  If the mine is accessible by means of adits or inclined transportation roads, the orientation process is comparatively simple and limited to running a traverse between the surface geodetic network and points of the underground control net. Very often, however, the only access to the mine is by way of vertical shafts and then direct traversing from the surface is impossible. In these cases, one of the following three methods of mine orientation can be applied:

1. Shaft plumbing with two or more plumb lines in one vertical shaft.
2. Shaft plumbing through two or more vertical shafts with one plumb line in each shaft.
3. Gyro orientation with one plumb line.

The process of orientation is supposed to give coordinates of at least one point and azimuth of one line of the underground network in the surface coordinate system. In the first two methods, the two plumb lines serve for a simultaneous transfer of the coordinates and of the azimuth directly from the surface, assuming that the plumb lines are truly vertical. This assumption is particularly critical in the first method in which the distance between the two plumb lines is comparatively short. This distance is usually not longer than 2 to 4 m even when the diameter of the shaft is larger because there are always many obstacles in the shaft such as cages, pipes, cables, etc. In this case two small random deflections, $e_1$ and $e_2$, one for each plumb line separated by a distance $b$, will produce an error $\varepsilon''$ of the transferred azimuth, $A$:

$$\varepsilon_A'' = \frac{206{,}265''}{b} \sqrt{e_1^2 + e_2^2} \tag{20.9}$$

For example, random deflections, $e_1 = e_2 = 1$ mm, of the plumb lines which are separated by a distance of 3 m will produce an error of $\varepsilon_A'' = 97''$. Therefore, the method of shaft plumbing orientation through one vertical shaft requires the utmost care and experience in establishing the plumb lines in the shaft as it is discussed in detail in Ref. 3 and summarized in the next section.

The error of shaft plumbing caused by the possible deflections of the plumb lines is, of course, much smaller when two plumb lines are used in the two separate shafts. In that case the distance $b$ in Eq. (20.9) is usually several hundred metres long and even large errors in the verticality of the plumb lines may be tolerated. The method of orientation through two or more vertical shafts is also called the *fitted traverse method*. This method usually gives much higher accuracy of mine orientation than shaft plumbing through one vertical shaft, and very often it may also give a better accuracy than the gyro orientation. Unfortunately, not every mine has an access to the surface through two or more vertical shafts from the mining levels which require the orientation.

Use of the gyro attachments has revolutionized mine orientation surveys. In this method, the shaft plumbing is used only for the transfer of coordinates of one point, and an error of even a few centimetres can be tolerated because the transfer of the azimuth, which is critical for the orientation of the underground network, is done independently of the shaft plumbing. Although shaft-plumbing methods may be considered obsolete in modern mine surveying under certain favorable conditions, shaft plumbing even through one vertical shaft can be competitive with gyro orientation. Examples of conditions favorable to shaft plumbing are shallow shafts, or the orientation of subway or hydro-development projects. Besides, precision shaft plumbing is still needed in the process of sinking new shafts (control of the shaft construction) and in periodic deformation measurements of shafts, which is mandatory in many mines. Therefore, the gyro method and the shaft-plumbing methods are both discussed in more detail in the following sections.

There are no general specifications for the accuracy requirements for mine orientation. In each individual case the chief surveyor has to decide, depending on the importance of the orientation, what accuracy is needed. Generally, the relative position of the underground points in respect to the surface and other mining levels should be known with an accuracy better than 1 m. The accuracy requirements may be much higher, of the order of 0.2 m, when two mine workings from two different levels, or when two different mines are supposed to meet each other.

The process of mine orientation is one of the most responsible tasks of the mine surveyor and should not be entrusted to persons who do not have good experience and knowledge in all aspects of mine orientation, including error analysis.

**20.10. Shaft plumbing through one vertical shaft: general methodology**   The basic idea of the method is shown in Fig. 20.10. Plumb lines $P_1$ and $P_2$ serve as intermediate traverse stations between known (coordinated) points $A$ and $B$ on the surface and $C$ and $D$ underground. The only problem in the method is the determination of the angles $\beta_1$ and $\beta_2$. They cannot be measured directly because of difficulties in setting up theodolites in the shaft opening and centering them precisely in the locations of the plumb lines. One possibility is to establish points $B$ and $C$ exactly in line of the vertical plane of the two plumb lines. This method is time consuming and if high accuracy is required, special equipment such as micrometric sliding devices for the theodolites is necessary to bring them precisely in the plane of the plumb lines.

There are several other methods of connecting surveys which allow the determination of angles $\beta_1$ and $\beta_2$. The most popular are Weissbach's method (or Weissbach's triangle method) and the quadrilateral method, which is also known in the literature either as the Hause or the Weiss method.

**20.11. Shaft-plumbing procedure**

*Selection of the plumb bob*   Thin steel wire with a heavy suspended plumb bob is the most popularly used plumb line in the orientation process. Precision optical or laser plummets will be briefly discussed later, but their application is not as popular as the mechanical plumb bobs.

Steel wires with very high tensile strength (200 kg/mm² or larger) (piano wires) should be used for shaft plumbing. As a rule, the wire should be as thin as possible and the load (the bob) should be as heavy as possible. As a compromise, the weight of the bob is usually selected as equal to $H/3$ in kilograms, where $H$ is the depth of plumbing in metres. For safety reasons the load should not exceed half of the maximum (breaking) load of the wire. Therefore, if wires with tensile strength of 200 kg/mm² are used in a depth of $H=600$ m, a

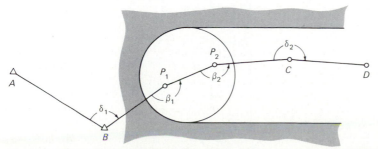

**Fig. 20.10** Mine orientation using shaft plumbing (angles $\beta_1$ and $\beta_2$ must be determined indirectly).

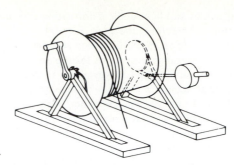

**Fig. 20.11** Wire reel for shaft plumbing.

bob of 200 kg should be suspended on a wire with cross section of 2 mm² (this corresponds to a diameter of 1.6 mm). The wire should be stored on special drums (Fig. 20.11) which allow for a slow lowering of the plumb bob to a desired mining level. The plumb bob consists of a rod and removable lead disks (Fig. 20.12), usually of 20 kg each.

**Lowering and stabilizing the plumb bobs**    The wire drums are located near the shaft (Fig. 20.13) and are securely fastened to the ground. Safety platforms are built across the shaft opening on the surface and at the oriented level. The wires are led over beams of the shaft head frame and over guiding pulleys (Fig. 20.14) in the preselected locations. Only small weights, of about 11 lb (5 kg), are attached to the wires during the lowering procedure. The main loads are suspended when the wires reach the bottom. The distance between the two safety platforms at the bottom should be designed according to the precalculated extension of the wire when the full load is suspended:

$$\Delta H = \frac{PH}{aE} \qquad \text{m or ft} \qquad\qquad (20.10)$$

where $P$ = weight of the plumb bob, kg
  $a$ = cross section of the wire, cm² or in²
  $H$ = depth, m or ft
  $E = (2.1)10^6$ kg/cm² (28 to 30 million lb/in²)

For example, if $H = 600$ m, $P = 200$ kg, and $a = 0.02$ cm², then $\Delta H = 2.9$ m.
    The plumb line should be checked to ensure that it does not touch any obstacles in the

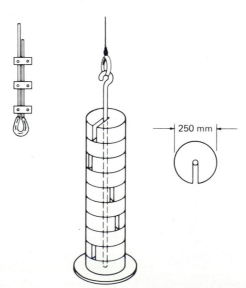

**Fig. 20.12** Typical plumb bob with removable weight disks.

250 mm

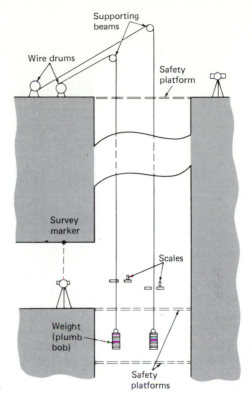

**Fig. 20.13** Shaft-plumbing procedure.

shaft. For that purpose a small ring of wire is wrapped around the plumb line and allowed to drop. Another check is to compare the actual and theoretical period, $T$, of oscillations of the plumb line using the formula

$$T = 2\pi \sqrt{\frac{H}{g}} \tag{20.11}$$

where $g = 980$ cm/sec$^2$ and $H$ is the depth of shaft in centimetres ($g = 32$ ft/sec$^2$ and $H$ is in feet).

In shallow shafts, when the weight is small, the plumb bob should be submerged in a container of oil to dampen the oscillations of the plumb line, which are produced by air currents and dropping water in the shaft. In very deep shafts it is impossible to completely dampen the oscillations, and the plumb line is in a continuous swing along ellipses a few centimetres in diameter. The vertical position of the wire may be found by taking readings of the extreme left and right deflections of the wire on a millimetre scale placed behind the wire at the oriented level. The readings are taken with a telescope of the theodolite, which

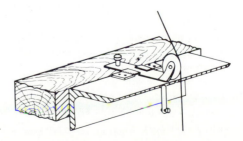

**Fig. 20.14** Guiding pulley for plumb wire.

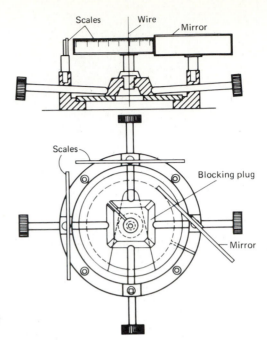

**Fig. 20.15** Typical apparatus for setting plumb wire in its vertical position.

should be set up a few metres from the wire. Fractions of a millimetre are estimated when taking the readings of the left and right edges of the wire at its turning points.

Usually, a set of 10 readings of the turning points of the swinging wire is sufficient to calculate the mean "vertical" position of the plumb line. Care must be taken so that the plumb bob swings in a plane parallel to the scale, which should be perpendicular to the line of sight of the theodolite. This is done by holding the plumb bob near its center of gravity and gently deflecting it in the desired direction. The calculated mean reading on the scale serves as a target for angle measurements in the connecting surveys at the oriented level. Sometimes, for example in the aforementioned quadrilateral method of the connecting surveys, the plumb line must be clamped in its vertical position. Two perpendicular scales must then be used for observing the swings of the wire from two perpendicular directions, or a mirror is attached (Fig. 20.15) which allows observations on two scales from one station. More details on the procedure and instrumentation used in lowering and stabilization of the wires in their vertical positions is given in Ref. 3.

**Influence of air currents**   The movement of air in the shaft exerts a force on the hanging wire (Fig. 20.16). If the plumb bob itself is screened from the air influence, as it should be, the approximate value of the deflection of the plumb line may be calculated (see Ref. 3) from the equation

$$e = \frac{30(h)(H)(d)(v)^2}{P} \text{ mm} \tag{20.12}$$

where $h$ = portion of the wire in metres exposed to the side stream of air, m; it is taken as approximately equal to the height of the shaft opening at the shaft entrance

$H$ = depth of plumbing, m
$d$ = diameter of the wire, m
$v$ = velocity of air, m/s at the cross section $h$
$P$ = weight of the plumb bob, kg

For example, if $h = 5$ m, $H = 600$ m, $d = 0.0016$ m, $v = 1.5$ m/s, and $P = 200$ kg, the expected deflection would be $e = 1.6$ mm.

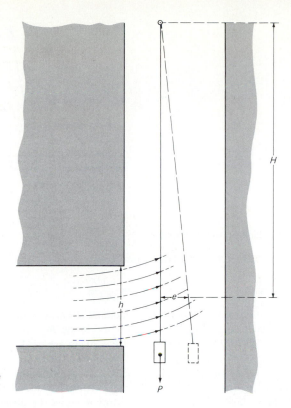

**Fig. 20.16** Deflection of the plumb line by air current.

The direction of the wire deflection depends on the location of the wire in the horizontal cross section of the shaft. Generally, the direction coincides with the direction of the axis of the shaft opening, but it can differ by $\pm 45°$, so that the plumb bobs hanging in different positions may be deflected in different directions causing a large error in the orientation. Therefore, the air current should be dampened during the orientation process at all levels entering the shaft by shutting all the ventilation doors. The complete damping of the air stream in the shaft is not usually possible. The disconnection of the forced draft ventilation is not a sufficient precaution, as the speed of the natural air stream may be greater than 1 m/sec.

When the influence of the air current is too strong, a method of wire plumbing with two or more different loads may be used. Two positions $r_1$ and $r_2$ of the deflected wire are then determined on the scale (Fig. 20.17) using two different loads on the same wire (usually at the ratio $P_1/P_2 = 1:2$). The vertical position ($r_0$ reading on the scale) of the wire can then be extrapolated from:

$$r_0 = r_2 - \frac{P_1(r_1 - r_2)}{P_2 - P_1} \tag{20.13}$$

The double-weight method has, however, some disadvantages:

1. The point of the suspension of the wire on the surface may move when changing the weights of the plumb bob.
2. The smaller weight $P_1$ may be too small to counteract the spiral shape of the wire (see below).

**Influence of the spiral shape of the wires**   During manufacture, and later during storage on small-diameter reels, the wire becomes permanently deformed and assumes a spiral shape. Even a very heavy load may fail to straighten the wire completely, and what appears to the naked eye as a long, straight wire may in fact be a long, small-diameter spiral (Fig.

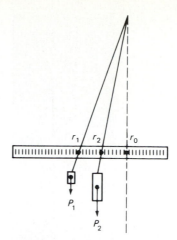

**Fig. 20.17** Shaft plumbing with two weights.

20.18), with a radius $r$ (cm) calculated from Ref. 3:

$$r = \frac{\pi d^4 E}{64 RP} \tag{20.14}$$

where $R$ is the radius of the spiral of the unloaded (free) wire in centimetres, $d$ the diameter of the wire in cm, $P$ the weight of the plumb bob in kg. $R$ is usually between 10 and 25 cm. For example, if $d = 0.16$ cm, $P = 200$ kg, and $R = 20$ cm are used, $r = 0.17$ mm. The same wire if loaded only with 50 kg would give $r = 0.7$ mm.

The spiral shape of the hanging wire gives an error in the vertical projection of the suspension point at the head of the shaft; the maximum value of this error can be $2r$ if the observer on the surface sights the wire in point $A$ while the observer at the oriented level uses point $B$ when observing the scale.

**20.12. Connecting surveys using the Weisbach method**  In this method the orientation angles $\beta_1$ and $\beta_2$ (see Fig. 20.10) are determined from measurements of the angle $\alpha_1$ and

**Fig. 20.18** Spiral shape of the plumb wire.

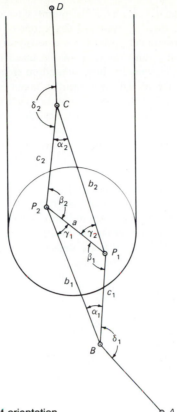

**Fig. 20.19** Weisbach method of orientation.

distances $a$, $b_1$, and $c_1$ in the triangle on the surface (Fig. 20.19) and measurements of $\alpha_2$ and distances $b_2$ and $c_2$ in the underground triangle. The distance $a$ between the plumb wires is also measured underground to approximately check the verticality of the wires. The measurements on the surface are made directly to the plumb wires. The measurement of the underground angle $\alpha_2$ is made to the predetermined vertical positions of the wires on the scales (one scale for each wire) perpendicular to the lines of sight. The distances $c_2$ and $b_2$ are taped directly to the swinging wires, averaging the readings on the tape. To complete the orientation measurements, the angles $\delta_1$ and $\delta_2$ are measured at stations $B$ and $C$.

The angles $\beta_1$ and $\beta_2$ are calculated from the simple trigonometric relationships

$$\sin \beta_1 = \frac{b_1 \sin \alpha_1}{a} \qquad \text{and} \qquad \sin \beta_2 = \frac{b_2 \sin \alpha_2}{a} \tag{20.15}$$

As a check, the angles $\gamma_1$ and $\gamma_2$ are calculated from

$$\sin \gamma_1 = \frac{c_1 \sin \alpha_1}{a} \qquad \text{and} \qquad \sin \gamma_2 = \frac{c_2 \sin \alpha_2}{a} \tag{20.16}$$

and the closure to $180°$ should be obtained in both triangles.

Once the angles $\beta_1$ and $\beta_2$ are determined, the mine orientation procedure is completed by calculating the azimuth of the line $CD$ and coordinates of $C$ from the traverse $A$-$B$-$P_1$-$P_2$-$C$-$D$ as shown in Fig. 20.10, in which stations $A$ and $B$ on the surface have known coordinates.

The variance of the azimuth of the underground line $CD$ with respect to the surface line $AB$ can be calculated from

$$\sigma^2_{A_{CD}} = \sigma^2_{\delta_1} + \sigma^2_{\beta_1} + \sigma^2_{\beta_2} + \sigma^2_{\delta_2} + \varepsilon^2_A \tag{20.17}$$

where $\varepsilon_A$ is the influence of the deflections of the plumb lines calculated from Eq. (20.9) if the deflections $e_1$ and $e_2$ are treated as standard deviations of the verticality of the plumb lines. They are estimated approximately from the previously discussed errors due to the air current, spiral shape of the wires, and the error of the mean positions determined on the scales (adding them, of course, in square values, as random errors). Usually, only half of the calculated [Eq. (20.12)] influence of the air current is taken as the part of the estimated standard deviation, owing to the fact that both plumb lines are most probably deflected in a similar direction. The standard deviation of positioning the wires on the scales can be kept equal to or smaller than 0.2 mm if the plane of the oscillations of the plumb bobs is parallel within $\pm 10°$ to the scale, the amplitude is smaller than 10 cm, and if at least 10 readings (with an estimation to 0.2 mm) of the left and right reversal positions are taken for the calculation of the mean position.

Standard deviations of angles $\delta_1$ and $\delta_2$ can be estimated in the usual way from an examination (Ref. 10) of the survey method and instruments used. They can easily be kept within a few seconds if the angles are measured in four to six positions with a theodolite with $20''$ or smaller nominal values of the micrometer divisions.

The variances of the calculated angles $\beta$ are determined by applying the rule of error propagation [Eq. (2.42)] to Eqs. (20.15), obtaining (for $\beta_1$ or $\beta_2$)

$$\sigma_\beta^2 = \frac{\tan^2 \beta}{b^2}\sigma_b^2 + \frac{\tan^2 \beta}{a^2}\sigma_a^2 + \left(\frac{b^2}{a^2\cos^2\beta} - \tan^2\beta\right)\sigma_\alpha^2 \qquad (20.18)$$

Equation (20.18) allows for an easy optimization of the Weisbach method; the best accuracy will be obtained when $\beta \approx 180°$ because then the errors of the distances can be neglected and then

$$\sigma_\beta \approx \frac{b}{a}\sigma_\alpha \qquad (20.19)$$

or, analogically,

$$\sigma_\gamma = \frac{c}{a}\sigma_\alpha \qquad (20.20)$$

if angles $\gamma$ are used in the traverse computations.

For optimum results, theodolite stations at $B$ and $C$ should be as close as possible to the nearer wire, and almost in line with both wires, the distance between the wires should be as long as possible and the angles $\alpha$ should be measured with a high degree of precision.

Errors in distance as large as 10 mm may be neglected if the angle $\alpha \leqslant 30'$. Generally, the Weisbach method is not applied when the angle $\alpha$ must be larger than $10°$, because the influence of the errors of the measured distances becomes critical. In this case other methods of connecting surveys, for example the quadrilateral method, are recommended.

**20.13. Connecting surveys using the quadrilateral (Hause) method** Typical configurations for connecting surveys by the quadrilateral method are shown in Fig. 20.20.

Generally, the following values are measured in the quadrilateral method: angles $\delta_1$, $\delta_2$, $\gamma_1$, and $\gamma_2$ and distances $P_1 P_2$ and $CD$. The errors of the distances have no influence on the accuracy of the transferred azimuth; therefore, they do not need to be measured precisely.

Since the angles to each wire are measured from two different stations, the "vertical" position of each wire must be determined on two scales perpendicular to each other using the apparatus shown in Fig. 20.15. The blocking plug is removed to allow the wire to swing in directions perpendicular to each scale, changing the direction of the swing after observations on the first scale are completed, and a set of observations is taken on each scale. The mean "vertical" positions are then calculated and the wire is fixed in this position by means of the blocking plug and slow-motion screws. The angles are then measured directly to the fixed wires.

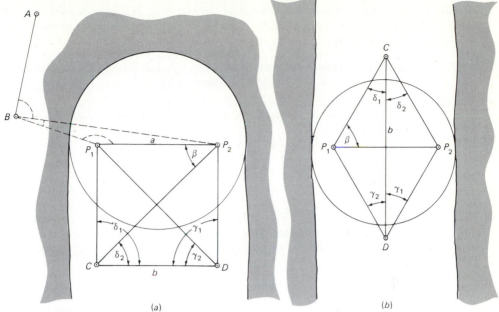

**Fig. 20.20** Quadrilateral (Hause) method of orientation.

Since the distance $CD$ is usually just a few metres, the accuracy of centering the theodolite and the target is critical. Forced centering is recommended or else two theodolites should be used simultaneously at stations $C$ and $D$, each pointing at the cross hairs of the other (telescopes focused to infinity).

Calculation of the orientation angle $\beta$ can be done as follows:

1. A local coordinate system is arbitrarily chosen for the calculation of the angle $\beta$, taking, for instance, point $C$ as the beginning of the system and the line $CD$ as the $+X$ axis.
2. The coordinates of $P_1$ and $P_2$ are calculated in the local system by simple intersections from the base $CD$ using angles $\delta_1$ and $\gamma_2$ for point $P_1$ and angles $\delta_2$ and $\gamma_1$ for point $P_2$. An error in the distance $b$ will produce only a scale change of the figure without changing the shape (the angles) so that the accuracy of $b$ is not critical in the calculation of $\beta$.
3. Angle $\beta$ is calculated from the known coordinates of points $P_1$, $P_2$, and $C$ (in the local system).
4. Angles $\delta_2$ and $\gamma_2$ are calculated from the obtained coordinates and as a check are compared with their original values.

Equations (9.3) and (9.4) (Art. 9.2) derived from the point slope [Eq. (8.13), Art. 8.21] are recommended for use in calculating the intersections. When coordinates $X_{P_1}, Y_{P_1}, X_{P_2}, Y_{P_2}$ have been computed, the angle $\beta$ is found using

$$\tan\beta = \frac{(Y_C - Y_{P_2})(X_{P_1} - X_{P_2}) - (X_C - X_{P_2})(Y_{P_1} - Y_{P_2})}{(Y_C - Y_{P_2})(Y_{P_1} - Y_{P_2}) + (X_C - X_{P_2})(X_{P_1} - X_{P_2})} \qquad (20.21)$$

Once the angle $\beta$ is calculated, the orientation calculations proceed the same way as described in the preceding section.

Error analysis (see Ref. 13) of the quadrilateral method shows that (1) the best geometrical shape for the connecting quadrilateral is a square; (2) given a square figure, better results are obtained if the shaft can be entered from the sides (Fig. 20.20$b$) because in this case $\sigma_\beta = \sigma_\alpha$, where $\sigma_\alpha$ is the average standard deviation of the measured angles $\delta_1$, $\delta_2$, $\gamma_1$, and $\gamma_2$, and for the case where the shaft can only be entered from one side (Fig. 20.20$a$), $\sigma_\beta = 2.4\sigma_\alpha$; and (3) by increasing the ratio of distance $CD$ to distance $P_1P_2$ the standard deviation of the orientation is also increased.

**20.14. Orientation through two vertical shafts**   If two vertical shafts are sunk to the mining level, the orientation process is performed in the following steps:

1. One plumb line is established in each shaft and coordinates $X_1, Y_1$ and $X_2, Y_2$ of the plumb lines $P_1$ and $P_2$ (Fig. 20.21) are determined at the surface by means of a connecting survey to the nearest points of the geodetic control network.
2. The azimuth $A_{1,2}$ and distance $d_{1,2}$ between the plumb lines are calculated from

$$A_{1,2} = \arctan \frac{X_2 - X_1}{Y_2 - Y_1} \tag{20.22}$$

and

$$d_{1,2} = \sqrt{(X_2 - X_1)^2 + (Y_2 - Y_1)^2} \tag{20.23}$$

3. An underground traverse is measured from $P_1$ to $P_2$ using the shortest possible route.
4. Distances in the underground traverse are reduced to the reference level of the surface coordinate system by adding corrections:

$$\Delta d_i = \frac{d_i H}{R} \tag{20.24}$$

where $H$ is the vertical distance to the reference surface and $R$ the mean radius of the earth.
5. The underground traverse is calculated in a local $\overline{XY}$ coordinate system having the coordinates of $P_1$ in the surface system as an origin and with the $+\overline{X}$ axis aligned with the first traverse leg (Fig. 20.21). Thus, coordinates of $D$, $E$, and $P_2$ (Fig. 20.21) are calculated in the $\overline{XY}$ system.
6. The distance $\overline{d}_{1,2}$ and azimuth $\overline{A}_{1,2}$ are calculated from the local coordinates of $P_1$ and $P_2$ and compared with previously calculated $d_{1,2}$ and $A_{1,2}$ on the surface, giving the rotation angle $\omega$ of the local coordinate system:

$$\omega = A_{1,2} - \overline{A}_{1,2} \tag{20.25}$$

and the scale

$$\lambda = \frac{d_{1,2}}{\overline{d}_{1,2}} \tag{20.26}$$

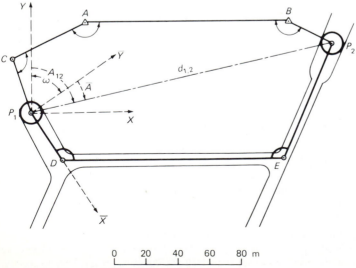

0   20   40   60   80 m

**Fig. 20.21**  Mine orientation through two shafts.

7. Coordinates for points in the underground traverse are now calculated in the $XY$ surface coordinate system using distances multiplied by $\lambda$ and the azimuths of the first and subsequent legs rotated by the angle $\omega$ (similarity transformation).

A least-squares adjustment may be used in the adjustment of the underground traverse at this point. However, with only one redundant observation in the traverse, the value of such an adjustment is questionable.

The accuracy of the orientation angle $\omega$ can be determined by finding the standard deviations of $A$ and $\bar{A}$ and applying the rule of error propagation to Eq. (20.25). One should note that the error of $\bar{A}$ has two components: one component is the result of random errors in angle and distance measurements in the underground traverse; and the second component occurs as the result of the possible deflections of the plumb lines expressed by the error $\varepsilon_A$ [Eq. (20.9)]. If $\sigma_{\bar{A}}$ is defined as a standard deviation of $\bar{A}$ caused only by the errors of the underground traverse, the total value of the standard deviation $\sigma_\omega$ can be calculated from

$$\sigma_\omega^2 = \sigma_A^2 + \sigma_{\bar{A}}^2 + \varepsilon_A^2 \tag{20.27}$$

If variances and covariances of the coordinates of $P_1$ and $P_2$ on the surface are known, the value of $\sigma_A$ is found by applying error propagation Eq. (2.42) to Eq. (20.22), obtaining (in radians²)

$$\sigma_A^2 = \frac{\Delta X^2}{d^4}\left(\sigma_{Y_1}^2 + \sigma_{Y_2}^2 - 2\sigma_{Y_1 Y_2}\right) + \frac{\Delta Y^2}{d^4}\left(\sigma_{X_1}^2 + \sigma_{X_2}^2 - 2\sigma_{X_1 X_2}\right)$$
$$+ 2\frac{\Delta X \Delta Y}{d^4}\left(\sigma_{Y_1 X_2} + \sigma_{Y_2 X_1} - \sigma_{Y_1 X_1} - \sigma_{Y_2 X_2}\right) \tag{20.28}$$

where $\Delta X = X_2 - X_1$ and $\Delta Y = Y_2 - Y_1$.

The value of $\sigma_{\bar{A}}$ underground is obtained in a similar manner, except that in this case the coordinates of $P_1$ have to be treated as errorless because the errors in $\bar{A}$ are caused only by relative positional errors of $P_2$ with respect to $P_1$ in the local system. Therefore,

$$\sigma_{\bar{A}}^2 = \frac{(\Delta \bar{X})^2}{\bar{d}^4}\sigma_{\bar{Y}_2}^2 + \frac{(\Delta \bar{Y})^2}{\bar{d}^4}\sigma_{\bar{X}_2}^2 - 2\frac{\Delta \bar{X} \Delta \bar{Y}}{\bar{d}^4}\sigma_{\bar{Y}_2 \bar{X}_2} \tag{20.29}$$

The variances and covariances of the coordinates $\bar{X}_2$ and $\bar{Y}_2$ can be calculated using Eq. (8.33), Art. 8.23, with estimated accuracies for angle and distance measurements.

The calculated standard deviation of $\omega$ gives the accuracy of the determination of the azimuth of the first leg of the underground traverse. In order to calculate the accuracy of any leg in the traverse, the full variance covariance matrix for coordinates of all the traverse points would have to be known. This can be obtained by a simultaneous error analysis of the combined surface and underground connecting surveys, including the squares of the estimated deflections of the plumb lines as additional variances of the coordinates of $P_1$ and $P_2$. A detailed description of the combined error analysis is beyond the scope of this chapter. If more than two vertical shafts with plumb lines are connected by an underground network, a simultaneous least-squares adjustment using the technique of indirect observations is recommended (Example B.10, Art. B.14). Variances and covariances for adjusted coordinates are a by-product of this adjustment, so that error analysis of the orientation is possible.

Orientation through two or more vertical shafts can give an accuracy better than 20 seconds of arc in transferring the azimuth from the surface to the mine if the accuracy and survey methodology of the connecting surveys are properly designed.

**Example 20.3**   A design of a mine orientation through two shafts is shown in Fig. 20.21. Two plumb lines, $P_1$ and $P_2$, are to be established in the shafts, at a distance $d$ of about 195 m and connected by two independent traverses to the reference control stations $A$ and $B$ on the surface. Underground, a traverse $P_1 - D - E - P_2$ will be measured. Approximate coordinates (scaled graphically from a large-scale plan) of the points are as follows:

| Point | X, m | Y, m |
|-------|------|------|
| $P_1$ | 515  | 235  |
| C     | 505  | 270  |
| A     | 548  | 290  |
| B     | 677  | 293  |
| $P_2$ | 705  | 278  |
| E     | 675  | 208  |
| D     | 535  | 206  |

Determine the accuracy of the orientation angle $\omega$ if:

1. Points A and B are treated as errorless.
2. Standard deviations of plumbing $e_1 = e_2 = 5$ mm.
3. All angles are measured with $\sigma_\alpha = 5''$.
4. All distances are measured with $\sigma_d = 10$ m.

First, the three component errors used in Eq. (20.27) are calculated. Next, the influence of the shaft plumbing is calculated using Eq. (20.9):

$$\varepsilon''_A = \frac{\rho''}{d} \sqrt{e_1^2 + e_2^2} = 7.5''$$

where $\rho'' = 206,265.''$ The influence of the connecting surveys on the surface, $\sigma_A$, is calculated from Eq. (20.28). The variances and covariances of coordinates of $P_1$ and $P_2$ are calculated by propagating errors in the open-end traverses $A - C - P_1$ and $B - P_2$ using Eqs. (8.33), (Art. 8.23), yielding:

$$\sigma_{X_1}^2 = 2.5 + 89.9 = \quad 92.3 \text{ mm}^2 \qquad \sigma_{X_2}^2 = 0.1 + 77.5 = \quad 77.8 \text{ mm}^2$$
$$\sigma_{Y_1}^2 = 0.7 + 110.3 = 111.0 \text{ mm}^2 \qquad \sigma_{Y_2}^2 = 0.5 + 22.3 = \quad 22.8 \text{ mm}^2$$
$$\sigma_{X_1 Y_1} = -0.9 + 11.9 = 11.0 \text{ mm}^2 \qquad \sigma_{X_2 Y_2} = 0.3 - 41.5 = -41.2 \text{ mm}^2$$

Finally, using Eq. (20.28)

$$\sigma_A^2 = 142 + 9 + 15 = 166 \qquad \text{and} \qquad \sigma_A = 12.9''$$

(Note that traverses $A - C - P_1$ and $B - P_2$ are uncorrelated and, therefore, the covariances between coordinates of $P_1$ and $P_2$ are zero.) Influence of the errors in the underground traverse is calculated from Eq. (20.29), holding point $P_1$ and direction $P_1 - D$ as errorless (arbitrarily selected underground coordinate system). Using given variances $\sigma_\alpha^2$ and $\sigma_d^2$ in Eqs. (8.33) yields

$$\sigma_{\bar{X}_2}^2 = \quad 7.0 + 147.8 = 154.7 \text{ mm}^2$$
$$\sigma_{\bar{Y}_2}^2 = 38.7 + 152.3 = 191.0 \text{ mm}^2$$

and

$$\sigma_{\bar{X}_2 \bar{Y}_2} = -13.2 - 9.1 = -22.3 \text{ mm}$$

obtaining

$$\sigma_{\bar{A}}^2 = 203 + 8 + 11 = 222 ''^2 \qquad \text{and} \qquad \sigma_{\bar{A}} = 14.9''$$

The total orientation error (error of the azimuth of the direction $P_1 - D$)

$$\sigma_\omega = \sqrt{\sigma_A^2 + \sigma_{\bar{A}}^2 + \varepsilon_A^2} = \sqrt{166 + 222 + 56} = 21''$$

If the distances and angles are measured with twice the accuracy ($\sigma_d = 5$ mm and $\sigma_\alpha = 2.5''$), the orientation error would become

$$\sigma_\omega = \sqrt{41.5 + 55.5 + 56} = 12''$$

## 20.15. Gyroscopic methods of mine orientation: Introduction
Orientation may be determined by means of the gyrotheodolite. The basic theory of the gyro attachment is applicable to the gyrotheodolite and is discussed briefly in Art. 11.46. The reader is referred

to Refs. 1 and 14 for an account of the interesting historical development of the gyrotheodolite and for a detailed mathematical treatment of the theory. The gyrotheodolite is available
in two forms, one in which the gyro is a separate unit and is attached above the theodolite,
such as the Wild GAK1 manufactured by Wild Heerbrugg Instruments, Inc. This instrument
is described in Art. 11.46 (Figs. 11.30 and 11.31) and will also be discussed in this chapter.
The other form available is one in which the gyro is mounted below and forms an integral
part of the theodolite, such as the GYMO Gi/Bl, marketed by Gyro, a division of Plessey
Canada (Fig. 20.22). The accuracy of the azimuth determination of this latter type is usually
greater than for the attachment, a standard deviation of 3 seconds of arc being quoted for
the Gymo Gi/Bl. Consequently, instruments of the latter type have been used to determine
azimuth for geodetic control (Ref. 11).

There are two basic approaches for azimuth determination, one in which the gyro is
allowed to precess about the meridian while the observer reads the horizontal circle of the
theodolite or the time of oscillation and the amplitude of the swing. With this technique the
damping of the gyro movement is very small and the gyro usually spins about 22,000
rev/min. The other approach is to use the torque acting on the spinning gyro when it is not
aligned in the meridian and, from several observations of this torque, east and west of the
meridian, determine the azimuth. This principle is applied in the P.I.M. (Precision Indicator
of the Meridian, British Aircraft Corporation, Ref. 23).

The latest gyrotheodolite to be developed makes use of a heavier gyro which rotates at a
slower speed and is heavily damped. An example of this type of instrument is the
Meridianweiser MW77, which was developed by Westfälische Berggewerkschaftskasse, In-
stitut für Markschiedewesen, Bochum, West Germany. It is not yet available in North
America.

The gyro attachment is very popular because of its compact size and because it is an
attachment for a theodolite. The theodolite on which it is mounted is usually a direct-reading type (direct to nearest minute, estimation to 0.1 minute of arc), and it is used in other
survey work associated with the mine. The standard deviation for an azimuth determined
with the Wild GAK 1 is given as 20 seconds (this value is considered to be comfortably

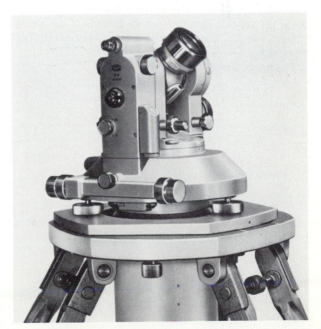

**Fig. 20.22** Gyrotheodolite, GYMO
Gi/B1. (*Courtesy Gymo, a Division
of Plessey Canada Ltd.*)

attainable, as noted in Refs. 1 and 12) and hence is suitable for most mining and tunneling work.

**20.16. Azimuth determination**   Azimuth determinations with the gyro attachment are usually classified as approximate and accurate. The *approximate* or *quick method* is employed when there is no azimuth available from the existing survey records. The results of this determination are then used as the initial setting for the second classification type, the *accurate methods*.

**Quick method**   The basic principle of this method is to observe two reversal points.

The gyro theodolite is set up and carefully leveled, the gyro is run up to its operating speed, lowered, and allowed to precess. The observer must track the gyro (use the upper tangent motion of the theodolite) so that the gyro mark remains in the center of the V notch (see Fig. 11.32). When a reversal point is reached, the gyro mark slows down and momentarily stops before moving in the opposite direction. At the reversal point the observer stops tracking and reads the horizontal circle of the theodolite, reading $U_1$. The gyro mark is then tracked until a second reversal point is reached, where the observer stops tracking and reads the horizontal circle of the theodolite, reading $U_2$. The approximate orientation is given by (see Fig. 20.23)

$$N = \frac{U_1 + U_2}{2} \tag{20.30}$$

**Accurate methods**   Two methods are considered, the reversal-point and transit methods. The reversal-point method requires an experienced observer to track the gyro mark accurately, keeping it in the V notch, and to read the horizontal circle of the theodolite at the reversal points (i.e., where the gyro mark changes its direction of movement). Generally, eight reversal points are recorded, taking the observer about 20 to 30 min, depending on the latitude of the place of observation.

**The transit method**   This method was invented by H. R. Schwendener (Ref. 20) and is based on the fact that the first 20 percent of the swing curve (which is a sine curve) is a straight line and hence there is a linear relationship between time and change in direction (Fig. 20.24).

The north direction $N$ is given by

$$N = N' + \Delta N$$

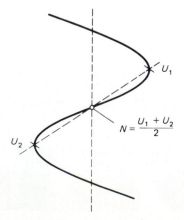

**Fig. 20.23**

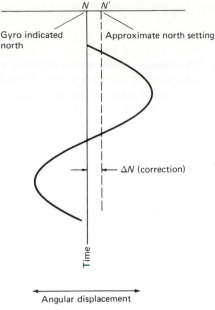

**Fig. 20.24**

$$N = N' + \Delta N$$

where $N'$ is the approximate north setting and $\Delta N$ is a correction to $N'$ to give $N$. It can be shown (see Ref. 20) that

$$\Delta N = ca\Delta t \tag{20.31}$$

where $c$ = constant (and is dependent on latitude, see Art 20.17)

      $a$ = average amplitude of swing right and left in scale divisions

    $\Delta t$ = time difference in seconds between the time for a swing to the right and the time for a swing to the left

The method of observation is as follows. The gyrotheodolite is centered over the station and carefully leveled, direct and reversed observations are taken on the reference object, and the theodolite is oriented to the north direction. This orientation must be within 10 to 20 minutes of arc or closer to the north direction. The horizontal circle reading at this setting is read and recorded. Since any error in this reading enters directly into the resulting azimuth determination it is important to check this observation. The nonspinning gyro readings are taken, then the gyro is run up to its operating speed and carefully released. The amplitude of the gyro mark is adjusted to be about +10 to −10 scale divisions. (If the setting is exactly true north, both amplitudes would be the same.) A stopwatch with a trailing hand is started when the gyro mark passes through the center of the V notch and the amplitude of the swing to the left or right is observed and recorded. When the gyro mark passes through the V notch the time is again taken, and the amplitude in scale divisions is also observed and recorded. Times are observed for eight transits of the gyro mark through the V notch, and at least two left and two right amplitudes are recorded. The gyro is then clamped, allowed to run down or braked, and the final nonspinning readings are observed. Finally, direct and reversed observations are taken to the reference object. The recorder calculates the swing times to the left, which should agree to about 0.2″, and the swing times to the right, which should also agree to 0.2″. The recorder also calculates the individual $\Delta N$ values [$\Delta N = (c)(a_1)\Delta t_1$], which when averaged ought to equal $(c)(a)_{av}(\Delta t)_{av}$.

The value of $\Delta N$ is added to or subtracted from the initial setting $N'$ to obtain the circle reading for the gyro indicated north (G.I.N.). The calibration value $E$ and the mean horizontal circle reading to the reference object must be applied to the value for $N$ to obtain the azimuth to the reference object.

An example of this method is shown in Fig. 20.25 and Table 20.1 (after Ref. 1).

**20.17. Determination of the constant c**  The constant $c$, used in the calculation of $\Delta N$, can be determined by observations. Two determinations of north are necessary for the calculation; one determination is carried out with a setting of about 20 minutes of arc west of north and the other 20 minutes of arc east of north. Thus, the direction of the meridian must be known to determine the constant $c$. True or astronomic north is given by each determination:

$$N = N_1 + \Delta N_1 = N_2 + \Delta N_2$$

where $N_1$ and $\Delta N_1$ refer to the first determination and $N_2$ and $\Delta N_2$ refer to the second determination. Thus,

$$N = N_1 + ca_1\Delta t_1 = N_2 + ca_2\Delta t_2 \tag{20.32}$$

so that

$$c = \frac{N_2 - N_1}{a_1\Delta t_1 - a_2\Delta t_2} \tag{20.33}$$

An example of the determination follows:

| First determination | Second determination |
|---|---|
| $N_1 = 0°10'00$ | $N_2 = 359°55'00$ |
| $a_1 = 10.45$ | $a_2 = 10.30$ |
| $\Delta t_1 = -23.9''$ | $\Delta t_2 = 7.4''$ |
| $c = \dfrac{359°55' - 360°10'}{(10.45)(-23.9) - (10.30)(7.4)}$ | |
| $= 0.0460'/\text{division}/\text{second}$ | |

It was stated previously that the initial orientation of the theodolite should be within $20'$ for the transit method.

Using the calculated value of $c = 0.046$ and letting $a = 10$ divisions and $\Delta t = 45''$ (these are about the maximum values within the $20'$ limit),

$$\Delta N = (0.046)(10)(45)$$
$$= 20.70''$$
$$= 0°20'42''$$

If the value of $c$ is changed to 0.0459, then $\Delta N$ is $0°20'39.3''$; that is, a change in $c$ of 0.0001 gives a change in azimuth of about $3''$.

The constant $c$ is dependent on latitude and it can be shown (Ref. 20) that a change in latitude of about $1°$ will cause the factor $c$ to change by 0.0001. Therefore, the same $c$ factor may be used within a radius of about 100 km, the resulting azimuth error being about $3''$.

**20.18. Calibration value E**  The calibration value $E$ is the horizontal angle between the plane of the meridian and the direction of the line of sight of the theodolite's telescope determined by the exact symmetry of the gyro oscillations (Ref. 24). The manufacturer cannot always set the $E$ value to zero but attempts to construct the instrument so that the $E$ value is stable. It is recommended that the value of $E$ be regularly checked, particularly if the gyro has been in storage, has been transported over long distances, or is suspected of having been bumped.

The $E$ value is determined by a direct comparison on a line, the azimuth of which has been determined from astronomical observations (see Arts. 11.32 and 11.39). If the azimuth of the line determined by gyrotheodolite (using one of the accurate methods) is designated

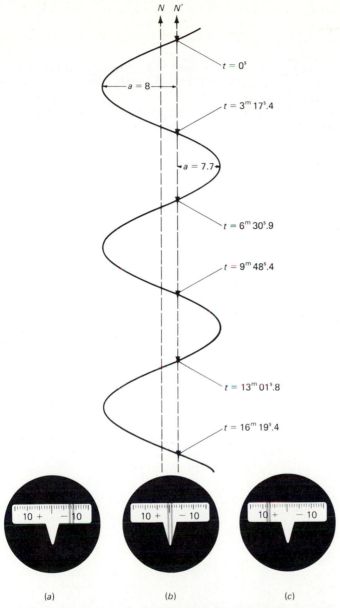

**Fig. 20.25** The transit method. ($a$) Right (west) reversal point, amplitude $a_w = -7.7$. ($b$) Transit, take the time. ($c$) Left (east) reversal point, amplitude $a_e = 8.0$. (Note: The north index mark moves in the opposite sense to the *north-seeking end* of the axis.) (*Courtesy Ref. 3.*)

$A_G$ and the astronomic azimuth is $A$, then

$$E = A - A_G \tag{20.34}$$

When a line of known azimuth in a plane coordinate system is available, it may be used for the determination of $E$ provided that correction for convergence of the meridians is applied to convert the grid direction to a geodetic azimuth. In this case

$$E = \text{grid azimuth} \pm \text{convergence of meridians} - A_G$$

**Table 20.1 (After Ref. 1)**

Gyro-theodolite Survey — Transit Method

Place *Erindale College*
Date *April 7/78*    Time *1440 E.S.T.*    Line *ASTRO*          to *STAR*
Observer *A.J.R.*    Recorder *A.C.*    Battery *Int.*
Theod. *T16.205404*  Gyro *GAK1. 25818*    Run up time *66 secs*    Braking Time *55 secs*

Ref. Obj. Circle Readings

| | Start | | Finish | | | Means |
|---|---|---|---|---|---|---|
| F.L. | F.R. | F.L. | F.R. | | | |
| *348°50'00"* | *168°50'00"* | *348°50'06"* | *168°50'06"* | Start | *348°00'00"* | |
| *50'00"* | *50'00"* | *50'06"* | *50'06"* | Finish | *348°50'06"* | |
| *50'00"* | *50'00"* | *50'06"* | *50'06"* | Mean | *348°50'03"* (4) | |
| Means *348°50'00"* | *168°50'00"* | *348°50'06"* | *168°50'06"* | | | |

Non-Spinning Gyro Readings

| | Start | | | Finish | | |
|---|---|---|---|---|---|---|
| Left + | Right − | Mean | Left + | Right − | Mean | |
| | *−9.8* | | *15.4* | | | |
| *10.0* | *(−9.75)* | *0.125* | *(15.35)* | *−14.5* | *−0.075* | |
| *(9.95)* | *−9.7* | *0.125* | *15.3* | *(14.45)* | *−0.075* | |
| *9.9* | *(−9.7)* | *0.100* | *(15.3)* | *−14.4* | *−0.050* | |
| *(9.9)* | *−9.7* | *0.100* | *+15.3* | *(−14.35)* | *−0.025* | |
| *9.9* | *(9.65)* | *0.125* | *(15.25)* | *−14.3* | *−0.025* | |
| *(9.7)* | *−9.6* | *0.050* | *15.2* | *(−14.3)* | *−0.050* | |
| *9.8* | | | | *−14.3* | | |
| | | Mean  *0.104* | | | Mean *−0.050* | |

Spinning Gyro Readings

| | | | | | Auxiliary Scale | | |
|---|---|---|---|---|---|---|---|
| | Transit Time | Swing Time | Δt | Left | Right | ΔN = c.a. Δt | |
| Swinging to left | | | | | | | |
| Swinging to Right | *0ᴹ 00ˢ0* | *+* | | | *7.7* | | |
| | *3 17.4* | *−3ᴹ 17ˢ4* | *−3.9* | *8.0* | | *−1.41* | |
| | *6 30.9* | *+3 13.5* | *−4.0* | | *7.7* | *−1.44* | |
| | *9 48.4* | *−3 17.5* | *−4.1* | *8.0* | | *−1.48* | |
| | *13 01.8* | *+3 13.4* | *−4.2* | | | *−1.52* | |
| | *16 19.4* | *−3 17.6* | *−4.3* | | | *−1.55* | |
| | *19 32.7* | *+3 13.3* | *−4.2* | | | *−1.52* | |
| | *22 50.2* | *−3 17.5* | | | | | |
| | | Mean *−4.117* | | | Mean *−1.49* (2) | | |

c. = *0.046*    c.a. = *0.361*         a. = *7.85*

Calculation of Azimuth

| | | |
|---|---|---|
| Circle Setting N' | *360° 00' 00"* | 1 |
| Mean ΔN | *−01 29* | 2 |
| N = N' + ΔN | *359 58 31* | 3 = 1 + 2 |
| Mean R.O. Circle Reading | *348 50 03* | 4 |
| Gyro Azimuth | *348 51 32* | 5 = 4 − 3 |
| E | *−12 30* | 6 |
| Azimuth of R.O. | *348 39 02* | 7 = 5 + 6 |

**20.19. Meridian convergence**   The orientation of a mine (see Art. 20.9) may be considered to be the angular difference between the gyroazimuth of the surface line and the gyroazimuth of the underground line plus or minus a correction for convergence of meridians. Referring to Fig. 20.26, this angular difference, $\omega$, can be determined as follows:

$$\omega = (A_{CD} - \hat{\theta}_C) - (A_{AB} - \hat{\theta}_A) = A_{CD} - A_{AB} - (\hat{\theta}_C - \hat{\theta}_A)$$
$$= A_{CD} - A_{AB} - \Delta\hat{\theta}_{CA} \tag{20.35}$$

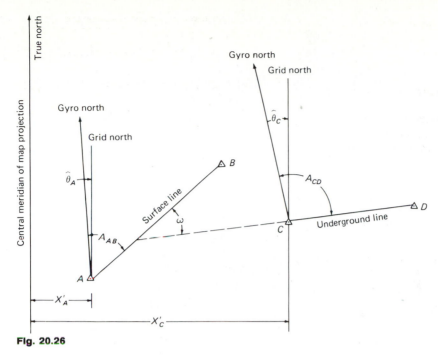

**Fig. 20.26**

where $A_{AB}$ = gyroazimuth at $A$
$\quad A_{CD}$ = gyroazimuth at $C$
$\quad \hat{\theta}_A$ = convergence of the meridian at $A$
$\quad \hat{\theta}_C$ = convergence of the meridian at $C$
$\quad \Delta\hat{\theta}_{CA}$ = convergence of the meridians from $C$ to $A$

Convergence should always be calculated and applied to azimuth determinations even when the distance between two stations is very short because its value may be significant.

For example, suppose that station $A$ is 200 km from the central meridian at latitude $45°.0$ and station $C$ is 205 km from the central meridian at latitude $45.1°$. Then $\Delta\hat{\theta}''_{CA} = \hat{\theta}''_C - \hat{\theta}''_A = 6650.1'' - 6465.3'' = 184.8''$ by Eq. (19.1), where $\hat{\theta}_C$ and $\hat{\theta}_A$ were computed individually using their respective latitudes and distances from the central meridian and assuming that $R = 6730$ km.

**20.20. Optical and laser plummets**   Several models of precision optical plummets (zenith and/or nadir) are available using tubular spirit levels, self-compensating leveling systems, or a mercury surface as a reference for setting the line of sight in the vertical direction. Table 20.2 shows characteristics of four optical plummets.

**Table 20.2   Precision Optical Plummets**

| Manufacturer | Model | Type | Magnification | Achievable accuracy |
|---|---|---|---|---|
| Zeiss-Jena | PZL | Self-compensating zenith | 31.5 | 1:100,000 (2″) |
| Kern | OL | Spirit level, zenith and nadir | 22.5 | 1:100,000 (2″) |
| Breithaupt | TELIM | Spirit level, nadir | 40 | 1:50,000 (4″) |
| Wild | GLQ | Mercury horizon, nadir | 40 | 1:200,000 (1″) |

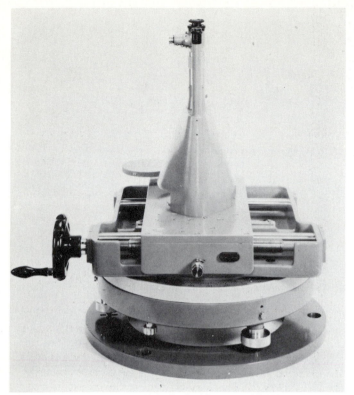

**Fig. 20.27** Optical plummet Wild GLQ. (*Courtesy Wild Heerbrugg Instruments, Inc.*)

The Wild GLQ model (Fig. 20.27) sets the line of sight in a vertical direction with an autocollimating telescope by achieving coincidence of the cross hair with its image reflected from a surface of mercury. The pool of mercury is removed from the field of view of the telescope after setting the line of sight vertical. Use of optical plummets in shaft plumbing is limited to a short range only (100 to 200 m) because of the poor visibility usually found in the shaft atmosphere.

Slightly longer ranges can be achieved by using a collimated laser beam as the plumb line.

A laser optical plummet (Fig. 20.28*a*) and a laser interference plummet (Fig. 20.28*b*), both utilizing mercury reference surfaces, have been developed at the University of New Brunswick in Canada. Practical tests with laser plummets indicate a repeatability in setting the vertical lines with a standard deviation of better than 0.5 seconds of arc. Details of these plummets are given in Refs. 4 and 7.

Other models of precision laser plummets are being developed, for example, by the National Physical Laboratory in England and by the optical industry in Poland. Some simple laser applications in plumbing are also possible by projecting a laser beam through telescopes of existing optical plummets or through precision automatic levels (see Ref. 5) equipped with 90° pentaprisms placed in front of the objective lenses. A commercially available laser plummet is illustrated in Fig. 18.4, Art. 18.5.

Laser plummets may be very useful in controlling shaft-sinking procedures and in transferring coordinates (shaft plumbing) when using the gyro or the two-shaft method of mine orientation. Their accuracy is, however, not sufficient for shaft-plumbing orientation through one vertical shaft.

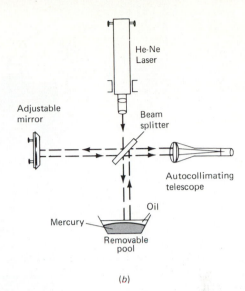

He-Ne
Laser

Adjustable
mirror

Beam
splitter

Autocollimating
telescope

Oil

Mercury

Removable
pool

(a)

(b)

**Fig. 20.28** Optical laser plummet. (*a*) Prototype. (*Courtesy University of New Brunswick.*) (*b*) Schematic cross section.

**20.21. Vertical control surveys and leveling**    Special steel tapes of lengths up to 1000 m (see Refs. 13 and 15) stored on large reels (Fig. 20.29) are available for the transfer of heights from the surface to the underground workings. The principle of the height transfer is shown in Fig. 20.30. The tape is slowly lowered to the required level and a weight is suspended at its end. The weight should preferably be equal to the tension used during the standardization of the tape (usually 20 to 45 lb or 10 to 20 kg). A bench mark, tied to the existing leveling network on the surface, is established near the collar of the shaft. Underground bench marks of the same type as those on the surface are cemented in the walls of the oriented levels near the shaft opening. Two survey crews with spirit levels take simultaneous readings on the tape at the surface and at the level. Usually, a set of 10 readings is taken changing the position of the tape (lowering or raising) by a few centimetres between the readings. The tapes are usually marked every 10 cm. Therefore, additional short scales with 1-mm divisions are clamped to the tape at the reading heights.

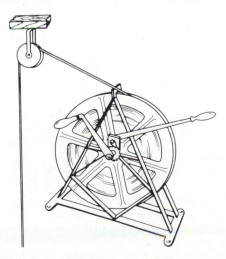

**Fig. 20.29** Shaft tape for height transfer. (*Courtesy Ref. 10.*)

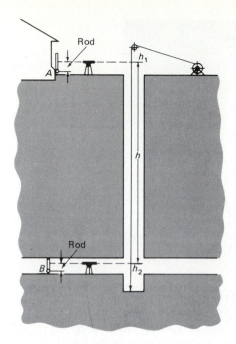

**Fig. 20.30** Procedure for the height transfer with a special tape.

The elevation of the bench mark $B, H_B$, at the oriented level is calculated from

$$H_B = H_A - h + \text{rod}_A - \text{rod}_B \tag{20.36}$$

where $H_A$ is the elevation at $A$, $\text{rod}_A$ and $\text{rod}_B$ are the respective rod readings at $A$ and $B$, and $h$ is the mean difference between readings $h_2 - h_1$ on the tape corrected by (1) standardization correction $\Delta h_d$; (2) temperature correction $\Delta h_t$; (3) stretch by the tape's own weight, correction $\Delta h_w$; and (4) stretch by the applied weight (if different from the standard tension), correction $\Delta h_p$.

The first correction is analogous to the tape correction $C_d$ described in Art. 4.16 and does not require any explanation. The temperature correction is the same as $C_t$ in Art. 4.18 but is more complicated because of the nonlinear change of temperature in the mining shaft. Temperatures should be measured at different levels (100 to 150 ft or every 30 to 50 m) in the shaft just before the height-transfer procedure and a weighted mean temperature is then calculated from (see Ref. 16)

$$T = \frac{\sum\limits_{i=1}^{n} [(h_{i+1} - h_i)(T_i + T_{i+1})]}{2(h_n - h_1)} \tag{20.37}$$

where $T_i$ is the temperature at a depth $h_i$.

The correction $\Delta h_t$ is calculated using Eq. (4.13), repeated here for convenience

$$\Delta h_t = h\alpha(T - T_0) \tag{20.38}$$

where $T_0$ is the temperature of standardization and $\alpha$ the thermal coefficient of expansion, which is equal to $(11.6)10^{-6}$ per $1°C$ $[(6.45)10^{-6}$ per $1°F]$ for steel.

The stretch correction

$$\Delta h_w = \frac{w}{aE}\left(Lh - \frac{h^2}{2}\right) \tag{20.39}$$

where $w$ = weight of one unit of length of the tape
$a$ = cross-sectional area of the tape, $\text{cm}^2$ or $\text{in}^2$
$E$ = modulus of elasticity, which for steel is usually given as $2.1 \times 10^6$ kg/cm² (28 to 30 million lb/in²)
$L$ = total length of the tape freely suspended, m or ft (same units as $h$)

The last correction, $\Delta h_p$, is analogous to $C_p$ given by Eq. (4.14) in Art. 4.19, repeated here for convenience:

$$\Delta h_p = \frac{(P - P_0)L}{aE} \tag{20.40}$$

where    $P$ = applied weight, kg or lb
    $P_0$ = standardization tension, kg or lb
  $a, L, E$  are as defined for Eq. (20.39)

The value of the product $aE$ may be checked experimentally (it is recommended to do this) by measuring $\Delta h$ for two different weights, say 20 lb and 100 lb (10 kg and 50 kg) and calculating $aE$ from Eq. (20.40).

Transfer of the heights to the mine through a vertical shaft may also be made with electro-optical distance measuring (EDM) instruments (Arts. 4.29 to 4.35) if the visibility conditions are favorable. In this case the EDM instruments should be clamped in a vertical position above the shaft opening. It requires some ingenuity on behalf of the surveyor and the cooperation of the mining workshop to make the necessary adaptors and brackets to fasten the instrument in this manner. The heights of the center of the instrument and of the reflector must be carefully determined by means of spirit leveling from bench marks. Another possibility in the use of EDM equipment is to use the instrument in its upright position near the shaft opening using a good quality mirror (first surface coating) to direct the electromagnetic signal down the shaft. Laser instruments with visible radiation should be used to facilitate the search for the reflector at the bottom of the shaft and to find a reference light spot on the mirror so that it can be referenced to the bench mark by spirit leveling.

The leveling network in the mine is divided into three orders of accuracy similar to the horizontal network. Height measurement in the third-order network is carried by trigono-metric leveling (Art. 5.5) simultaneously with the traverse measurements in the horizontal control surveys. The roof markers serve as bench marks. Vertical angles are measured to a mark made on a plumb-bob string which usually serves as the target and vertical distances are measured with pocket tapes from the mark and from the horizontal axis of the theodolite to the bottom of the roof mark. Trigonometric leveling is also frequently used in second- and first-order networks when running the leveling traverses through raises and other inclined openings. Because of the comparatively short distances and usually quite stable atmospheric conditions, the accuracy of trigonometric leveling in the mine is competitive with spirit leveling if proper precautions are made in the measurements of the height of the instruments and the targets.

Spirit leveling is usually done between wall bench marks. When a connecting survey to a roof station is required, special inverted leveling rods have to be used.

Similar to the theodolites, spirit levels sometimes have to be used on supporting arms fixed to the wall lining or timbering. Detailed descriptions of different types of levels and special adaptors used in the mines are given in Refs. 13 and 15.

## 20.22. Problems

**20.1**   A tunnel 11 km long is driven from reference point 1 to point 12 as shown in Fig. 20.8. The tunneling procedure is controlled by the straight traverse $0-1-2\cdots 12$. All distances between successive traverse points equal 1 km. Coordinates of reference point 1 and azimuth of the reference line $0-1$ are fixed and errorless. What misclosure in the $Y$ coordinate of point 12 would you expect if the angles in the traverse are measured with a standard deviation $\sigma_\beta = 2''$?

**20.2**   Answer Prob. 20.1 if the angle measurements are replaced by azimuth measurements of each traverse leg using a gyrotheodolite with standard deviations of the gyroazimuths of $\sigma_A = 20''$. Which of the two traverse measurements (angles vs. gyroazimuths) would give a smaller lateral deviation of the tunnel at point 12?

**20.3** A mine orientation survey is to be done using two mechanical plumb lines in one vertical shaft. The depth of the oriented level $H = 300$ m. The distance between the two plumb lines is 4 m. Steel wires of tensile strength 200 kg/mm² are available for plumbing. The height of the shaft opening to the oriented level $h = 5$ m and the average air velocity in the cross section of the opening $v = 1$ m/s. There are no other openings to intermediate levels between the surface and the oriented level. Answer the following questions:

(a) What diameter ($d$) of the plumb wires and what weight ($P$) of the plumb bobs would you use for the orientation purpose?

(b) What should be the distance between the safety platforms at the oriented level?

(c) What will be the period of swing of the plumb lines (for checking purposes)?

(d) What error of the transferred azimuth would you expect as a result of the air current and spiral shape of the wires [take $R = 15$ cm and use values of $d$ and $P$ as obtained from part (a)].

**20.4** Mine orientation of the level $H = 300$ m has been performed using the Weisbach method as shown in Fig. 20.19. The plumbing procedure has been performed according to the previous design (see Prob. 20.3). The following values of the measured angles and distances have been obtained:

On the surface: $a = 4.001$ m, $b_1 = 8.030$ m, $c_1 = 4.035$ m, $\delta_1 = 160°20'20''$, $\alpha_1 = 0°08'10''$ with standard deviations of distances $\sigma = 2$ mm and angles $\sigma = 4''$.

Underground: $a = 3.997$ m, $b_2 = 7.005$ m, $c_2 = 3.015$, $\alpha_2 = 0°16'00''$, $\delta_2 = 178°25'40''$ with standard deviations of distances $\sigma = 5$ mm and angles $\sigma = 6''$.

Calculate coordinates of point $C$, azimuth of the line $CD$, and standard deviation of the azimuth $CD$. Coordinates of point $B$ on the surface are $Y_B = +360.320$ m, $X_B = +538.435$ m, and the azimuth $AB = 310°15'20''$. The coordinates of $B$ and azimuth $AB$ are treated as errorless. (*Note:* In the calculations of the error of the azimuth of the line $CD$ you should include the error of shaft plumbing as calculated in Prob. 20.3.)

**20.5** Grid azimuth of a line $AB$ (Fig. 20.26) is equal to $26°16'30''$. A gyrotheodolite was calibrated on the line $AB$ giving the gyroazimuth of the line equal to $27°14'00''$. The same gyrotheodolite was used at station $C$ in order to determine the grid azimuth of the line $CD$. The gyroazimuth of $CD$ was $72°20'00''$. What is the grid azimuth of the line $CD$ if $X'_A = 101250$ m, $X'_C = 102416$ m, and the latitudes are $\phi_A = 43°20'30''$ and $\phi_C = 43°21'00''$?

### References

1. Bennett, G. G., "New Methods of Observation with the Wild GAK-1 Gyrotheodolite," *Unisurve Report No. 15*, University of New South Wales, Australia, 1969.
2. Chrzanowski, A., and Derenyi, E., "Role of Surveyors in the Mining Industry," *ASP–ACSM Semi-annual Convention*, St. Louis, Mo., October 1967.
3. Chrzanowski, A., Derenyi, E., and Wilson, P., "Underground Survey Measurements, Research for Progress," *The Canadian Mining and Metallurgical Bulletin*, June 1967.
4. Chrzanowski, A., "New Techniques in Mine Orientation Surveys," *The Canadian Surveyor*, Vol. 24, March 1970.
5. Chrzanowski, A., and Jansses, H.D., "Use of Laser in Precision Leveling," *The Canadian Surveyor*, December 1972.
6. Chrzanowski, A., and Masry, S., "Tunnel Profiling Using a Polaroid Camera," *The Canadian Mining and Metallurgical Bulletin*, March 1969.
7. Chrzanowski, A., Jarzymowski, A., and Kaspar, M., "A Comparison of Precision Alignment Methods," *The Canadian Surveyor*, Vol. 30, June 1976.
8. Chrzanowski, A., Ahmed, F., and Kurz, B., "New Laser Applications in Geodetic and Engineering Surveys," *Applied Optics*, Vol. 2, February 1972.
9. Chrzanowski, A., and Steeves, P., "Control Surveys with Wall Monumentation," *The Canadian Surveyor*, Vol. 31, June 1977.
10. Chrzanowski, A., "Design and Error Analysis of Surveying Projects," Department of Surveying Engineering, University of New Brunswick, Lecture Notes No. 47, November 1977.
11. Gregerson, L. F., "An Investigation of Gyroscopic Theodolites," Paper presented to the 62nd Annual Meeting of the Canadian Institute of Surveying, Ottawa, Canada, 1969.
12. Hodges, D. J., and Brown, J., "Underground and Surface Orientation Measurements with Gyrotheodolite Attachments," *The Mining Engineer*, October/November 1972.
13. Kowalczyk, Z., *Miernictwo Gornicze*, Vol. 2, Wyd. Slask., Katowice, 1965; Vol. 3, 1968 (in Polish).

14. Lauf, G. B., "The Gyrotheodolite and Its Application in the Mining Industry of South Africa," *Journal of the South African Institute of Mining and Metallurgy*, Vol. 63, March 1963.
15. Neset, K., *Dulni Merictvi*, SNTL, Praha, 1966.
16. Richardus, P., *Project Surveying*, North-Holland Publishing Company, Amsterdam, 1966.
17. Sheehan, J. F., "Mine Surveying at Mount Isa," Presented at the 11th Congress of Institution of Surveyors Australia, Brisbane 1968 (*Australian Surveyor*, March 1969).
18. Smith, R. C. H., "A Modified GAK1 Gyro Attachment," *Survey Review*, Vol. 24, No. 183, January 1977.
19. Staley, W. W., *Introduction to Mine Surveying*, 2d. ed., Stanford University Press, Stanford, Calif., 1964.
20. Strasser, G. J., and Schwendener, H. R., "A North Seeking Gyro Attachment for the Theodolite as a New Aid to the Surveyor," Wild Heerbrugg, Switzerland.
21. Wasserman, W., "Underground Survey Procedures," *New Zealand Surveyor*, March 1967.
22. Williams, H. S., "A 'New' Method for Gyrotheodolites Operable in the Non-tracking Mode," *The South African Survey Journal*, April 1978.
23. Winiberg, F., *Metalliferous Mine Surveying*, 5th ed., Mining Publications Ltd., London, 1966.
24. Wild Heerbrugg, *Handbook GAK1*, Gyro Attachment.

# CHAPTER 21
# Hydrographic surveying†

**21.1. General** Hydrographic surveys are those made of a body of water such as a bay, harbor, lake, or river. These surveys are made for the purposes of (1) determination of channel depths for navigation; (2) determination of quantities of bottom excavation; (3) location of rocks, sand bars, navigational aids; (4) measurement of areas subject to scour or silting; and (5) for offshore structure siting. In the case of rivers, surveys are made for flood control, power development, navigation, water supply, bridges, pipeline crossings, underground cable crossings, and water storage.

A certain amount of shoreline is generally included or shown on most hydrographic surveys. These shoreline data can be determined by (1) photogrammetric surveys, (2) plane-table methods, or (3) azimuth and distance measurements from ground control points. A shoreline is necessary to make the proper junction between the water and land interface.

**21.2. Datum** Hydrographic surveys must be referenced to both a horizontal and a vertical datum (Art. 13.2). In most cases these datums will be the accepted national horizontal and vertical datums but need not be in all cases. A "floating" or independent datum is acceptable (but not recommended) in cases where connecting the survey to the national datum is too costly to justify.

The most important feature of connecting surveys to the national datums is the ability to resurvey the area at a later time and make a legitimate direct comparison between the results of the two surveys. The minimum information to position but not orient a given survey is a single permanent survey control marker from which the relative $X$, $Y$, and $Z$ coordinates of other temporary control points can be determined. Connecting the survey to the national datum ensures that resurveys can be properly repeated even if the survey control points are inadvertently destroyed.

The vertical datum for hydrographic surveys is generally determined by measuring the water level in the project area over a period of time. The water levels are then averaged so that the mean low-water, mean water, mean high-water, or other average water level is used as the reference level of the survey. The important factor in using a vertical datum is to ensure that all depths are referenced to the same datum elevation.

**21.3. Controlling vessel position** The term "vessel" is used here in the general sense. It might be a ship, launch, small boat, or even a person in waders. This vessel is used as the platform from which depths of water or *soundings* are measured. The horizontal location ($X, Y$ coordinates) of each sounding must be determined to produce a hydrographic survey. This is accomplished by using an electronic positioning system or by measuring angles and/or distances between the vessel and fixed (shore) control points.

†This chapter was written by Commander James Collins, NOAA, National Ocean Survey.

The survey methods, or the methods employing the measurement of angles and distances, are varied in number. The basic criterion is that two independent quantities be measured and that the measurements produce lines of position at the vessel which intersect with no less than a 30° acute angle. The following are some of the more common techniques:

1. Measuring an angle, either to or from a vessel positioned along a known range line.
2. Measuring distances between the vessel and a control point when the vessel is positioned along a known range line.
3. Measuring the azimuths to the vessel from two (fixed shore) control points.
4. Measuring the azimuth and distance to the vessel from one (fixed shore) control point.

The angles and distances can be measured with theodolites, transits, sextants, alidades, stadia, electronic ranging, surveying tapes, and specially constructed measuring wires or cables. The choice and combination should be dictated by the desired accuracy and economy of the survey.

The direct measurement (survey) method, preferred for open bodies of water where the vessel operates quite far from shore, is to measure two sextant angles to three control points from the vessel. The economy of this method is that all survey operations are accomplished aboard the vessel and cumbersome communications with shore personnel are eliminated. This procedure is an application of three-point resection which is developed in Arts. 9.5 to 9.7. A full treatment of this method is given in Ref. 6 listed at the end of this chapter. When the method of observing two angles from the vessel is used, two angles are simultaneously observed from the boat to three fixed points on shore whose relative positions are known, as illustrated by Fig. 21.1. In Fig. 21.1, $\alpha$ and $\beta$ are the angles read by the observer in the boat to the known points $B$, $A$, and $C$ on shore. Since a boat is too unstable to support a transit, the angles are read with the sextant. The sextant is shown in Figs. 6.69 and 6.70. A complete description of the instrument and operating instructions can be found in Arts. 6.59 and 6.60. Two angles are sufficient to locate a sounding unless the boat happens to be on the circumference of a circle passing through $A$, $B$, and $C$ as shown in Fig. 9.11; in such a case the location of the sounding is indeterminate.

The accuracy of the plotted location will vary with the relative location of the known points $A$, $B$, and $C$, as follows:

1. If $A$, $B$, and $C$ are in a straight line, or if $A$ is nearer the boat than $B$ and $C$, the location is strong unless one of the angles $\alpha$ or $\beta$ is small.
2. Extremely long sights will give small values for $\alpha$ or $\beta$ and the location will be weak.
3. On long or short sights, small angles should be avoided, as they are difficult to plot and may give weak locations.
4. The error in the plotted location of the point due to errors in plotting the angles increases with the length of sight, and better results can be obtained with shorter sights.
5. The accuracy of location is poor when the point occupied approaches the circle through the three fixed points.

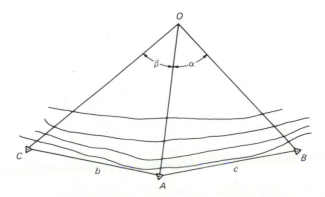

**Fig. 21.1** Three-point fix.

The method may be used in combination with the time-interval method with satisfactory results. The boat is propelled at a uniform rate along a range line, and at intervals the position of the sounding boat is located by two angles read with the sextant. Soundings taken between sextant readings are plotted in proportion to the time intervals. This method reduces the labor of plotting and speeds up the work of observing in the field.

Surveys of rivers, harbors, lakes, or other inland bodies of water generally require that a combination of survey positioning procedures be employed. For example, when cross sections are required for a river, the tag-line method is often used. This method involves establishing a number of control points on opposite banks of the river, such as points $A$ through $H$ shown in Fig. 21.2. A special wire cable is constructed with points marked along its length (e.g., every 10 ft). This is most easily accomplished by crimping special copper sleeves on the wire. These sleeves can either be arranged in a coded manner or they can be used to secure coded leather strips to the wire (tag line). The tag line is stretched between points on opposite banks of the river such as between $A$ and $E$, $B$ and $F$, $C$, and $G$, and $D$ and $H$. Soundings are taken at marked intervals along the tag line, and cross sections can be plotted from the sounding and tag-line distance measurements.

Variations of the tag-line procedure, such as establishing ranges on the same side of the river bank, can be employed. Signals or targets erected over points $A$ and $A'$, and $B$ and $B'$ (Fig. 21.2) will permit the hydrographer to position the vessel along the respective range line. The position of the vessel along the cross section is then determined by measuring either an angle to two shore points or measuring the distance from one shore point. For example, the angle $AVB$ or the distance $AV$ (Fig. 21.2) measured by the hydrographer will uniquely determine the position of the vessel at point $V$. Another variation of this method is to measure the azimuth of one or more lines from shore control points. Azimuths to point $V$ measured from points $A$ and $B$ would also uniquely locate the position of the survey vessel.

One additional positioning method that is of particular value in making hydrographic surveys of narrow bodies of water is the location by azimuth and distance. For example, measuring the length and azimuth of line segment $BV$ (Fig. 21.2) provides an excellent geometric determination of the vessel's position. This is due to the fact that the azimuth and distance measurements always generate lines of position perpendicular to one another. The

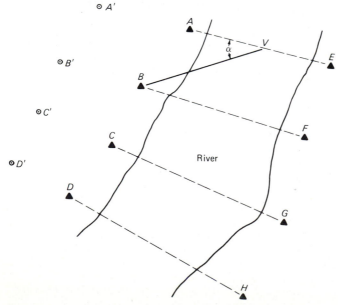

**Fig. 21.2** River range lines for acquisition of cross-section data.

other methods of location (except tag line) generate lines of position whose intersection forms an acute angle.

Angle (azimuth) and distance measurements can be made with a variety of instruments. Transits or theodolites are more than adequate for measuring angles for cartographic purposes. The plane table and sextant, although of lesser accuracy, generally suffice for most cartographic angle measurements. Single distances can be made with a variety of standard surveying electronic distance measurement equipment. The choice of instruments is more often dictated by economics rather than by technical considerations.

**21.4. Electronic positioning**   Beginning in the late 1960s, short-range (line-of-sight) electronic positioning equipment has been increasingly used to provide horizontal positioning for hydrographic surveys. The advantages of electronic positioning equipment over direct measurement methods are:

1. Fewer shore control points are required to cover a given area.
2. Survey data are in digital format and are compatible with electronic computer processing.
3. Position coordinates $(X, Y)$ are instantly obtainable for each sounding.
4. Fewer launch personnel are required to conduct a survey.

A number of electronic positioning systems of both domestic and foreign manufacture are currently available. These systems are generally divided into three categories: (1) long-range, (2) intermediate-range, (3) and short-range. The long-range systems, such as Loran, have accuracies of a fraction of a kilometre; intermediate-range systems, such as Raydist, have accuracies in the tens of metres; and short-range systems, such as Del Norte, have accuracies of several metres. Short-range electronic positioning systems are limited to line-of-sight operations and are the type most often used for near-shore hydrographic surveys. Other types of short-range EDM instruments used for general surveying are described in Art. 4.29.

All electronic positioning systems measure the time of transit or difference in transit times of electromagnetic waves in the atmosphere. The accuracy of these electronic systems is therefore highly dependent upon atmospheric temperature, pressure, and humidity at the time of measurement. One method of correcting for these atmospheric parameters is to calibrate the electronic systems by comparing electronically determined positions with positions determined by survey methods. For example, a three-point fix can be taken simultaneously with an electronic position determination. The difference between the $X, Y$ coordinates from this simultaneous determination is then plotted as a correction to the electronic determination versus the distance from the respective electronic shore stations. Calibration curves can be developed from a number of these comparisons to correct for various atmospheric conditions. As a general rule, electronic positioning systems should be thoroughly calibrated prior to beginning a project and whenever there are major changes in the electronic components. Daily, and sometimes more frequent, checks should be made of the electronic system by verifying that the calibration curve correction factors are still valid. This is accomplished by electronic position comparison with a three-point fix or comparison at a point of known position. Calibration of electronic distance measuring (EDM) equipment is discussed in Arts. 4.32 through 4.35.

There is presently a rather large selection of electronic positioning systems available which are suitable for hydrographic surveying. A listing of some of these systems, their accuracies, and operating characteristics is contained in Ref. 2.

Some examples of short-range electronic positioning equipment are the Autotape, Miniranger, Tellurometer, and Trisponder. These systems are all of the ranging type or circular systems. They generally have a ranging repeatability of $\pm 1$ m, so that accuracies of $\pm 5$ m in position are readily obtainable under actual operating conditions.

Examples of medium-range positioning systems are ARGO, HI-FIX, LORAC, and RAYDIST. These systems can be used either in a ranging mode (circular lane pattern) or in a phase differencing mode (hyperbolic lane pattern). Medium-range systems are capable of operating up to 200 km from shore and have repeatabilities of $\pm 1$ m. System accuracies under field conditions are highly dependent upon shore station geometric configuration with respect to the vessel and the atmospheric conditions. Under favorable conditions vessel position accuracies of less than $\pm 10$ m are readily obtainable.

**21.5. Computation of position**   Electronic positioning systems that are known as ranging systems measure the distance from fixed shore stations to the vessel. The typical geometric arrangement of a ranging system is shown in Fig. 21.3. Points $A$ and $B$ represent the location of the shore stations having known $X$ and $Y$ coordinates and point $O$ is the location of the vessel which is to be determined. The distances $R_A$ and $R_B$ are measured by the ranging system. Thus, the three sides of triangle $AOB$ are known and the plane coordinates for point $O$ (the position of the vessel) can be computed. First, calculate the angle $\alpha$ by the law of cosines:

$$\cos \alpha = \frac{R_A^2 + (AB)^2 - R_B^2}{2(R_A)(AB)}$$

in which the length $AB$ is calculated by inverse computation using Eq. (8.8) (Art. 8.15). The azimuth of line $AB$ can be calculated by inverse computation with Eq. (8.9), so that the azimuth of line $AO$ is

$$A_{AO} = A_{AB} + \alpha$$

allowing computation of the coordinates $X_0$, $Y_0$ of the vessel position by Eq. (8.7) as follows:

$$X_0 = R_A \sin A_{AO} + X_A$$
$$Y_0 = R_A \cos A_{AO} + Y_A$$

The position of a vessel located by measuring angles to at least three stations having known coordinates and located on the shore is found by solving the three-point resection problem as outlined in Art. 9.6 and demonstrated in Example 9.3.

An alternative solution to the three-point resection problem follows. Referring to Fig. 21.1, compute the base-line lengths and azimuths from north for the lines between stations on the shore. Thus, the distances $c$ and $b$ are

$$c = \left[ (X_B - X_A)^2 + (Y_B - Y_A)^2 \right]^{1/2}$$
$$b = \left[ (X_C - X_A)^2 + (Y_C - Y_A)^2 \right]^{1/2}$$

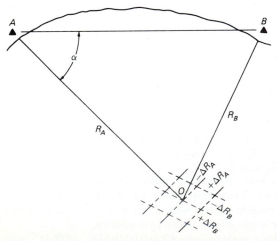

**Fig. 21.3** Position determination by electronic ranging system.

and the azimuths $A_{AB}$ and $A_{AC}$ are

$$A_{AB} = \arctan \frac{X_B - X_A}{Y_B - Y_A}$$

$$A_{AC} = \arctan \frac{X_C - X_A}{Y_C - Y_A}$$

Now, translate the coordinates of the ground stations into an $X', Y'$ system with an origin at $A$ as shown in Fig. 21.4. Next, compute coordinates for the centers $O_1$ and $O_2$ of the circles which circumscribe triangles $OAB$ and $OAC$, respectively. For the circle centered at $O_1$,

$$X_1' = \frac{X_B'}{2} - \frac{c \cos A_{AB}}{2 \tan \alpha}$$

$$Y_1' = \frac{Y_B'}{2} + \frac{c \sin A_{AB}}{2 \tan \alpha} \tag{21.1}$$

For the circle centered at $O_2$,

$$X_2' = \frac{X_C'}{2} + \frac{b \cos A_{AC}}{2 \tan \beta}$$

$$Y_2' = \frac{Y_C'}{2} - \frac{b \sin A_{AC}}{2 \tan \beta} \tag{21.2}$$

With these values, the coordinates of $O$ in the $X', Y'$ system are

$$X_0' = \frac{2(Y_2' - Y_1')(X_1' Y_2' - X_2' Y_1')}{d^2}$$

$$Y_0' = \frac{(-2)(X_2' - X_1')(X_1' Y_2' - X_2' Y_1')}{d^2} \tag{21.3}$$

where $d^2 = (X_2' - X_1')^2 + (Y_2' - Y_1')^2$.

Finally, plane coordinates for point $O$, in the system in which they were originally given, are

$$X_0 = X_A + X_0'$$

$$Y_0 = Y_A + Y_0' \tag{21.4}$$

As an exercise, students should solve the problem of Example 9.3 (Art. 9.6) by this method.

Coordinates of control points and sounding positions for line-of-sight areas are performed on a standard map projection in the case of state plane coordinate computations (Chap. 14).

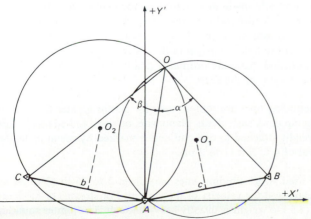

**Fig. 21.4** Position by three-point resection.

However, when surveys are so extensive that they extend beyond line of sight from land, a more rigorous computational method must be used. Ellipsoidal or geodetic formulae in which earth curvature and convergence of the meridians are taken into consideration must be used for computing the positions of vessels when long distances are involved. To avoid making geodetic computations, a graphic procedure is often used to plot the vessel's position on a suitable projection (e.g., polyconic). This graphical procedure requires that the electronic rate lines (circles or hyperbolae) be drawn on the map projection. The position of the vessel is found by interpolating between rate lines with a special protractor.

**21.6. Accuracy of positions** The accuracy of a given determined horizontal position is directly calculable by applying the standard laws of error propagation [Eqs. (2.42) and (2.46), Art. 2.20] to the equations used to compute the position. The error at any determined point is represented by an error ellipse aligned so that its major axis lies in the direction representing the weakest geometry at the given point (Art. 2.19). A graphical approximation of the error ellipse at a point can be made by plotting the lines of position at that point generated by the positioning system being used. For example, an electronic ranging system would generate two intersecting circles passing through the vessel (circle centers at two respective shore stations). The distance from station $A$ (Fig. 21.3) to the vessel will have errors $\pm \Delta R_A$ and the errors in the distance from station $B$ will have errors $\pm \Delta R_B$. The true position of the vessel will lie somewhere within the parallelogram at point $O$ bounded by the errors in ranges to the vessel.

If sufficient observations are made to provide redundant data, a least-squares adjustment can be performed to yield a best estimate for the position of the vessel (e.g., Example B.14, Art. B.16, Appendix B). In this case error ellipses can be derived from error propagation data obtained as a by-product of the adjustment.

**21.7. Soundings** The speed and precision of making soundings depend greatly upon the character of the equipment used. As mentioned previously, if the survey requires relatively few soundings, and if the depths to be sounded are relatively shallow, the soundings are made with a sounding rod or a sounding line. For surveys of greater scope and depth, echo sounders are used.

*Sounding units* The present convention is to express water depths in feet or fathoms. This convention is gradually changing and in the future soundings will be expressed in metres.

*Sounding rods* With depths of about 12 ft or less and with low current velocities, rods can be used to advantage. The rod is usually made in 4-ft sections for convenience in carrying and must be of sufficient thickness to withstand the pressure of the current. The edges are usually rounded to give minimum resistance to the flowing water. The lower end is fitted with a metal shoe of sufficient weight to hold it upright in the water and with area enough to prevent its sinking into the mud or sand. Rods are ordinarily graduated on both sides to feet and tenths, the zero being at the bottom of the shoe.

*Hand sounding lines* Hand sounding lines are seldom used for depths greater than 150 ft. The lines may be of cotton or hemp cord, sash chain, piano wire, or small linked steel chain. The sounding line has a lead weight attached to one end and is marked by metal or leather tags at some appropriate interval of 4 to 5 ft (1 to 2 m).

*Sounding leads* The weights used with sounding lines vary from 5 to 75 lb (2 to 30 kg) depending upon the depth of water and the velocity of the current. For streams of moderate depth and low velocity a 10-lb (4.5-kg) weight is usually heavy enough. The lighter sounding leads are usually made similar in shape to a window sash weight, with a slight taper toward

**Fig. 21.5** Portable echo sounder. (*Courtesy Raytheon Corporation.*)

the top or "eye" end. They are circular in cross section and three to four times as long as their average diameter. The heavier weights are often "torpedo"-shaped, with stabilizing fins to offer less resistance to the current. Sounding lines are infrequently used today. They are, however, still used to calibrate echo sounders.

**Echo sounders**  Continuous-recording echo sounders are currently widely used. They range in size from highly portable, battery-operated models (Fig. 21.5) to high-powered permanent installations that measure the deepest oceans. All fathometers[†] operate on the basic principle that sound produced near the water's surface will travel to the bottom and be reflected to the surface as an echo. Echo sounders employ electromechanical means to produce the sound, receive and amplify the echo, and convert the elapsed time to units of depth. Some echo sounders display the depth in digital form and others produce an analog graph or strip chart. Deep-water fathometers use low-frequency transmissions, owing to the higher absorption of the high-frequency transmissions by the water column. Shallow-water fathometers operate at high or medium frequencies at a sounding rate of 600 pulses/min. A portable digital survey fathometer is illustrated in Fig. 21.5.

[†] A fathometer is a device which measures water depth by echo soundings.

Airborne lasers and multispectral scanners (MSS) are being developed to replace conventional vessel hydrography. The airborne laser system consists of a pulse-type laser that emits a burst of light which is reflected at the water surface and again at the bottom of the water body. The transit time between these two reflected pulses is accurately measured and converted into water depth. Laser systems can be used in the scanning mode so that a long narrow area along the aircraft's track is covered. Multispectral scanner systems are being developed to remotely (satellite or aircraft) measure water depths. These systems work on the densitometric principle, where less energy is returned from deep water than from shallow water. The MSS system can be calibrated by comparing returned energy (density) with known water depths. Both the laser and MSS systems show promise for providing substantial economies in the future.

Another airborne hydrographic system currently being used is photobathymetry. This system involves photographing underwater areas with an aerial cartographic camera and using standard aerial mapping techniques (Art. 16.14). Natural color film is used for this underwater mapping and depths up to 20 m can be mapped in clear water. Photobathymetry is significantly more economical than conventional vessel hydrography; however, it cannot be used in all areas. Where water clarity is low and in areas of low bottom contrast, conventional hydrography must be used.

Profiles of bottom sediment and geologic structuring are often as important as bathymetry. This information is obtained using special low-frequency fathometers, or sound devices that generate a low-frequency echo with towed air gun or electric sparker.

**21.8. Reduction of soundings** Observed soundings must be corrected for departure from true depths. Most corrections vary with the type of sounding system being used. One correction that is commonly applied to all observed soundings is the correction for water level at the instant of sounding. This correction is determined by continuously measuring the water-level elevation in the vicinity of the hydrography. The water elevation will vary predictably in saltwater areas, being affected by tidal and meteorological forces. However, in rivers, lakes, and other freshwater bodies, the water elevation will vary due to runoff, natural spring discharge, and a number of man-made causes. In any case, soundings must be reduced or increased by the amount that the water surface elevation differs from the elevation of the chosen sounding datum.

Lead lines and sounding poles require little or no additional correction. Lead lines should, however, be checked frequently to ensure that they are of the proper length.

Fathometer soundings require a number of corrections. The largest correction is due to the variability of sound in water.

Survey fathometers are generally calibrated for a sound velocity of 4800 ft/s. However, the sound velocity in seawater varies with the temperature, salinity, and depth (pressure) of the water. The indicated depth given by the fathometer needs to be corrected for the difference between the calibrated velocity of 4800 ft/s and the actual velocity determined by the water's temperature and salinity. This can be accomplished by measuring the temperature and salinity of the water at various depths and making corrections based on tables and graphs. A more direct and simpler method is to construct calibration (correction) curves from data gathered by making bar checks or lead-line comparisons.

A *bar check* is a long, narrow sheet of metal that is suspended beneath the fathometer's transducer. The metal sheet is raised and lowered by means of two marked lines similar to lead lines. In this manner, the metal sheet can be lowered to exact distances below the water surface and the corresponding fathometer reading made. The difference between the true (bar check) and indicated (fathometer) depths is plotted versus the depth tested. The curve generated (Fig. 21.6) is a calibration curve which can be used to correct future fathometer readings. Bar checks are taken to a limited depth. Calibration curves are extended beyond the bar-check values by taking lead-line comparisons. For depths beyond the reach of lead

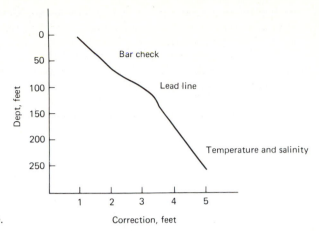

**Fig. 21.6** Velocity correction curve.

lines, the curves are extended by continuing the curve linearly with a slope determined by the temperature and salinity of the deeper water.

Additional corrections to soundings are made for heave and for dynamic draft (settlement and squat). The heave error is due to sea conditions. Often in rough seas the soundings vessel is appreciably displaced by waves. This correction does not, of course, apply in protected waters. Dynamic draft is caused by the displacement of a vessel as it proceeds through the water at a given speed. The dynamic draft of the sounding transducer is found by sighting on a survey level rod placed directly above the transducer as the vessel passes by a pier or other location where a surveyor's level can be set up. This correction is often ignored, because it is quite small for moderate speeds of vessel operation.

Fathometers with a variable initial fathometer adjustment for the transducer's depth, often require an additional correction. If the initial is set for a given value and maintained at that setting for all bar checks, no additional correction is necessary. However, if the initial varies from the bar check value, a correction must be made to the observed sounding.

Sounding spacing on the final hydrographic sheet (Fig. 21.7) is mainly a matter of individual requirements. The depths displayed on the hydrographic sheet should not be so crowded that individual soundings cannot be easily read. The National Ocean Survey plots single-digit soundings 5 mm apart on the office plot of field data or smooth sheets. If more information than can be shown using a 5-mm interval is desired, consideration should be given to increasing the scale of the survey or using overlays. The selection of soundings that appear on the smooth sheet will vary with the purpose of the survey. Hydrography performed primarily for charting (Fig. 21.8) or navigation purposes attempts to develop the shoal soundings. Therefore, shoal soundings are shown on the smooth sheet. Bathymetric surveys are conducted primarily to show bottom contours. On bathymetric smooth sheets (Fig. 21.9) soundings that delineate the depth contours should be displayed.

**21.9. Display of hydrographic data**   The work sheet upon which the original vessel's positions and approximate water depths are shown is called a *boat sheet*. Generally, a durable, stable-base material such as Mylar is used for the boat sheet. This sheet contains the plotted graticule, representing the latitude and longitude lines, and the basic control points. Additional data, such as the shoreline, floating and fixed aids to navigation, and landmarks, are often shown on boat sheets. When photobathymetry is accomplished prior to vessel hydrography, photo soundings and offshore rocks, coral heads, reefs, and sand bars are shown on the boat sheet.

*Smooth sheets* are office plots of all field data gathered during a survey. These sheets represent the data after all corrections have been applied. For example, smooth sheet

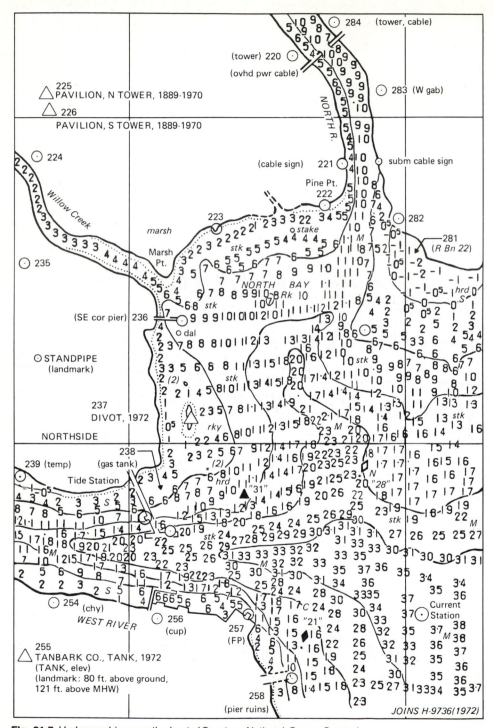

**Fig. 21.7** Hydrographic smooth sheet. (*Courtesy National Ocean Survey.*)

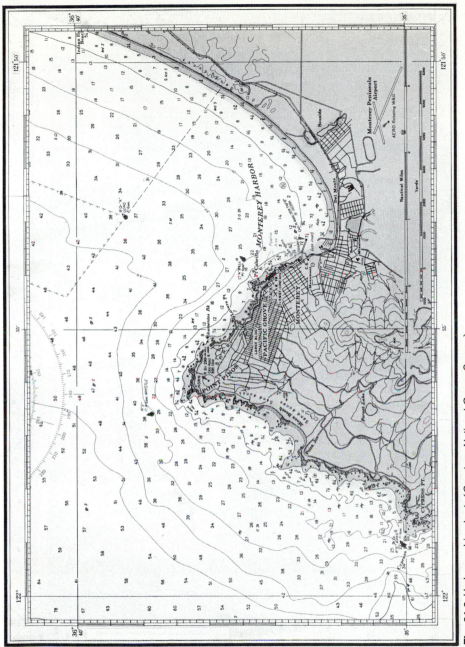

**Fig. 21.8** Hydrographic chart. (*Courtesy National Ocean Survey.*)

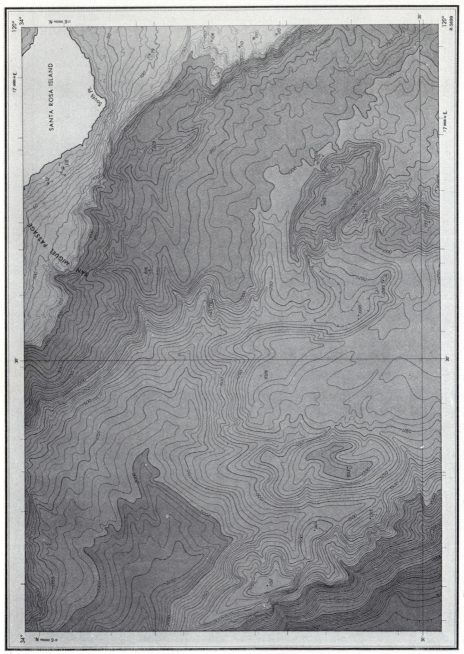

**Fig. 21.9** Bathymetric map. (*Courtesy National Ocean Survey.*)

soundings include final water-level corrections, instrumental corrections, and dynamic draft corrections. Smooth sheets include shoreline data, hydrographic data, bottom sample information, and contours depicting the bottom topography. In addition, they may include floating and fixed aids to navigation and as much land planimetric data as are necessary for the purpose of the sheet.

**21.10. Special hydrographic surveys**   *Wire-drag surveys* are conducted to locate sunken wrecks, or other underwater obstructions too small to be located by conventional hydrography. This type of survey is performed by towing a submerged cable between two vessels. The depth of the submerged wire is kept constant by a series of weights and surface floats connected to the underwater (ground) wire by a series of vertical wires (uprights). Obstructions in the path of the ground wire will "hang" the drag so that a "V" formed in the ground wire (Fig. 21.10) is transmitted to the surface floats by way of the uprights. The position of the hang is determined by intersecting the azimuths of observations from both vessels to the V visible in the surface buoys.

Wire-drag hangs are shown on the smooth sheets and charts by special symbols. Areas cleared by wire drag are shown as shaded regions with the cleared depth indicated.

In recent years, side-scanning sonar has been used to locate submerged objects. This equipment works on the same principle as the fathometer except that a swath is scanned on either side of the vessel.

**21.11. Currents**   It is often quite important to know the current vector and its periodic characteristics particularly for engineering projects. In tidal areas, currents ebb and flow with the tide; and in streams the currents vary in magnitude in response to discharge. Two basic current types are measured: surface and subsurface. Surface currents are measured by floating targets which are tracked over a period of time. These surface targets are periodically located by survey techniques and the velocity vectors computed. Surface currents can also be determined photogrammetrically. A special dye marker is dropped on the water surface and the marked point is tracked on aerial photographs.

Current meters are used to measure subsurface currents. These meters vary greatly in complexity and cost. Generally, the more costly meters measure both the current velocity and direction. Less expensive, simple meters, such as the Price meter (Fig. 21.11), measure only the current velocity. In streams current meters are lowered into the water from a boat, bridge, or special overhead cable car. A heavy weight is attached directly below the meter and the meter is gimballed so that the horizontal component of the current is measured.

In open bodies of water the procedure is more complicated. In these areas meters are suspended from a surface buoy or tethered to an anchored, buoyed vertical cable. Generally, several meters are suspended at one location. Several spaced vertical meters permit the determination of flow characteristics of the body of water. In many tidal areas there is a bottom counter flow associated with the tidal current cycle.

**21.12. Water levels**   Most saltwater areas are affected by the tides even if to a minor degree. The National Ocean Survey uses the mean low-water and mean lower-low-water

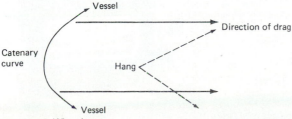

**Fig. 21.10** Wire drag.

**Fig. 21.11** Price current meter.

levels as vertical datums to which water depths are referenced. The choice of a datum, however, will depend upon the purpose of the survey. For example, a convenient vertical datum is the zero elevation of the nearest national geodetic vertical bench mark. If this zero elevation is used as a sounding datum, the water and land contours will be referenced to the same datum. However, care must be exercised, since the mean sea-level surface is not an equipotential surface (level surface), especially, on inland, constricted waterways. The suitability of the water-level surface can be determined by locating water-level gaging stations throughout the project area. If all gages record the same water elevation at the same time, the water surface can be represented by an equipotential surface.

In saltwater areas, water-level datums are determined by referencing tide or water-level gages to longer period standard gages (Fig. 21.12). Approximately 200 primary tide gages are located along the coasts of the United States. Water levels are continuously measured at these primary gages for a minimum period of 19 years. This period of time is called a *tidal epoch* and accounts for one complete 18.6-year luni-solar cycle. Shorter-period tide gages, which measure the tide cycle continuously for 1 year, are referenced to the primary gages. These secondary (one-year) gages permit the computation of a tidal datum with a degree of accuracy nearly equal to the 19-year primary gages. Less accurate tertiary gages are generally installed at specific sites where hydrographic surveys are being conducted. These gages operate for the duration of the project or for a minimum of 30 days. Even shorter-period gages operating for a day or two to a week are installed in tidal marshes and up tidal

**Fig. 21.12** Water-level gage. (*Courtesy Fischer Porter, Inc.*)

streams. These very short period gages are used to establish tidal datums in constricted inland water areas. Tide gages operating for 1 week are referenced to the 30-day gages, which are in turn referenced to the 1-year gages, which are ultimately referenced to the 19-year primary gages. This successive referencing of short-to-long-period gages, when properly done, permits a more accurate datum computation for the short-period gages than would be possible if referencing were not done. A complete treatment of this method and the accuracies of each gage in the chain is given in Ref. 5. A complete discussion of the various water-level gages is contained in the NOS *Hydrographic Manual*, 4th ed. (Ref. 6).

## 21.13. Problems

**21.1**  Outline the purposes of hydrographic surveys.

**21.2**  What information is required to position and orient a hydrographic survey horizontally?

**21.3**  In the absence of a bench mark referenced to the national vertical datum, how can a vertical datum be established for a given hydrographic survey?

**21.4**  Outline the method by which vessels involved in hydrographic surveys are generally positioned horizontally.

**21.5**  When a vessel is in open water but within sight of shore, what is a method for determining the horizontal position without using electronic equipment? What equipment is utilized for this procedure, and what additional data are required?

**21.6**  Describe the most economical method for surveying cross sections of a stream 100 m wide where the area to be covered is 1 km in extent and sections are to be taken at 100-m intervals. Also list the equipment needed.

**21.7**  Outline the different categories of electronic positioning systems used for hydrographic surveys, giving examples of each and the estimated accuracies possible with each system.

**21.8**  Outline a procedure for calibrating an electronic positioning system on board a vessel.

**21.9**  Compute the coordinates of a vessel given the following information. Coordinates of shore signals in metres:

$X_A = 10,000.00$    $Y_A = 10,000.00$
$X_B = 13,000.00$    $Y_B = 9,000.00$
$X_C = 7,000.00$     $Y_C = 9,500.00$

Angles taken to signals from vessel:

$\alpha = 27°47'$
$\beta = 37°15'$

Compute the values of coordinates of points $B$ and $C$ when $A$ is the origin of local coordinate system. Use Eqs. (21.1) to (21.4). Assume that the vessel lies north of points $A$, $B$, and $C$.

**21.10**  Compute the position of a vessel located by ranging equipment with the following values (see Fig. 21.3):

| Station A, m | Station B, m |
|---|---|
| $X_A = 1141.2$ | $X_B = 5463.6$ |
| $Y_A = 7035.7$ | $Y_B = 8154.7$ |
| Range $A_0 = 3230.0$ m | |
| Range $B_0 = 4150.0$ m | |

**21.11**  In Prob. 21.9, the standard deviation in measuring the angles is 01′ for each angle. Assuming that the coordinates of $A$, $B$, and $C$ are errorless, determine the standard error ellipse for the position of the vessel.

**21.12**  In Prob. 21.9, the standard deviations in the measurements and in the coordinates of $A$, $B$, and $C$ are $\sigma_\alpha = \sigma_\beta = 01′$ and $\sigma_{X_A} = \sigma_{Y_A} = 0.05$ m. Assume that the $X$, $Y$ coordinates of $A$, $B$, and $C$ are of equal precision and uncorrelated. Compute the propagated standard error ellipse for the position of the vessel.

**21.13**  For the data in Prob. 21.10, assume that the coordinates of points $A$ and $B$ are of equal precision and uncorrelated with $\sigma_{X_A} = \sigma_{Y_A} = 0.05$ m, while $\sigma_{A_0} = \sigma_{B_0} = 1$ m. Compute the standard error ellipse for the position of the vessel.

**21.14** Describe the various methods by which soundings may be measured. Compare the merits of these methods.

**21.15** What corrections should be applied to soundings measured using electronic fathometers? Describe how these corrections are made.

**21.16** In compiling a hydrographic chart, what are the differences between a boat sheet and a smooth sheet?

**21.17** What are the differences between a hydrographic chart for navigation and a bathymetric map?

**21.18** Describe two methods for locating underwater objects that are not generally found by the usual hydrographic survey.

## References

1. Collins, J., "Formulas for Positioning at Sea by Circular, Hyperbolic, and Astronomic Methods," NOAA Technical Report NOS81, U.S. Dept. of Commerce, Washington, D.C., February 1980.
2. Freeman, N., Haras, W. S., and Wigen, S. O., "Hydrodynamic Surveys and the New Technology," *The Canadian Surveyor,* Vol. 28, No. 3, September 1974.
3. Munson, R. C., "Positioning Systems," *Report on the Work of WG 41b,* presented at the 15th International Congress of Surveyors, Stockholm, Sweden, June 1971.
4. Ritchie, G. S., "Technological Advances and the Sea Surveyor," *The Canadian Surveyor,* Vol. 27, No. 4, December 1973.
5. Swanson, R. L., *Variability of Tidal Datums and Accuracy in Determining Datums from Short Series of Observations,* NOAA Technical Report NOS-64, Rockville, Md., October 1974.
6. Umbach, M. J., *Hydrographic Manual,* 4th ed., NOAA, National Ocean Survey, Government Printing Office, Washington, D.C., 1976.

# APPENDIX A
# An introduction to matrix algebra

**A.1. Definitions** A matrix is simply a collection of numbers or symbols collected in an array form. The following are examples of matrices:

$$\begin{bmatrix} 1 & 2 & 0 \\ 6 & 4 & 3 \end{bmatrix} \qquad \begin{bmatrix} 1 \\ 5 \end{bmatrix} \qquad [7 \quad 9 \quad 3] \qquad \begin{bmatrix} a & b \\ c & d \end{bmatrix}$$

$$(1) \qquad\qquad (2) \qquad\quad (3) \qquad\qquad (4)$$

Every matrix has a specified number of rows and columns. Thus, the matrix in (1) has 2 rows and 3 columns and is said to be a $2 \times 3$ matrix. Similarly, (2) is a $2 \times 1$ matrix, (3) is a $1 \times 3$ matrix, and (4) is a $2 \times 2$ matrix. The two numbers representing the rows and columns are referred to as the *matrix dimensions*.

A matrix is designated by a boldface capital Roman letter. Thus, if **A** is an $m \times n$ matrix, it can be written symbolically as

$$\mathbf{A} = \begin{bmatrix} a_{11} & a_{12} & \cdots & a_{1n} \\ a_{21} & a_{22} & \cdots & a_{2n} \\ \vdots & \vdots & & \vdots \\ a_{m1} & a_{m2} & \cdots & a_{mn} \end{bmatrix}$$

A lowercase letter with a double subscript designates an element in a matrix, so $a_{ij}$ represents a typical element of the matrix **A**. The first subscript, $i$, refers to the number of the row in which $a_{ij}$ lies, starting with 1 at the top and proceeding down to $m$ at the bottom. The second subscript, $j$, refers to the number of the column containing $a_{ij}$, starting with 1 on the left and proceeding to $n$ at the right. Thus, $a_{ij}$ lies at the intersection of the $i$th row and $j$th column. As an example, $a_{23}$ in matrix (1) is 3, while $a_{12}$ in matrix (3) is equal to 9. The smallest matrix size is the $1 \times 1$, which is then called a *scalar*.

## A.2. Types of matrices

**A.2.1. Square matrix** A *square matrix* is a matrix with an equal number of rows and columns. In this case $\underset{m,m}{\mathbf{A}}$ is a square matrix of order $m$. The principal (or main, or leading) diagonal is that composed of the elements $a_{ij}$ for $i = j$. The following are examples of the square matrix:

$$\mathbf{A} = \begin{bmatrix} 1 & 2 \\ 3 & 4 \end{bmatrix} \qquad \mathbf{B} = \begin{bmatrix} a & b & c \\ d & e & f \\ g & h & k \end{bmatrix}$$

The main diagonal of **A** is composed of 1 and 4, while that for **B** contains $a$, $e$, and $k$. Two special cases of the square matrix are the *symmetric* and *skew-symmetric* ones. The case of a symmetric matrix will be introduced later.

**A.2.2. Row matrix**  A *row matrix*, or *row vector*, is a matrix composed of only one row. It will be designated by a lowercase boldface Roman letter: for example,

$$\underset{1,n}{\mathbf{a}} = [a_1 \quad a_2 \quad \cdots \quad a_n] \qquad \text{and} \qquad \underset{1,3}{\mathbf{c}} = [1 \quad 2 \quad 4]$$

**A.2.3. Column matrix**  A *column matrix* or *column vector*, is composed of only one column, such as

$$\underset{m,1}{\mathbf{b}} = \begin{bmatrix} b_1 \\ b_2 \\ \vdots \\ b_m \end{bmatrix} \qquad \text{and} \qquad \mathbf{d} = \begin{bmatrix} -1 \\ 3 \end{bmatrix}$$

**A.2.4. Diagonal matrix**  A *diagonal matrix* is a square matrix such that all elements above and below the main diagonal are zero:

$$\mathbf{D} = \begin{bmatrix} d_{11} & & & 0 \\ & d_{22} & & \\ & & \ddots & \\ 0 & & & d_{mm} \end{bmatrix}$$

where

$$\begin{aligned} d_{ij} &= 0 \qquad \text{for all } i \neq j \\ d_{ij} &\neq 0 \qquad \text{for some or all } i = j \end{aligned}$$

For example,

$$\mathbf{G} = \begin{bmatrix} 1 & 0 & 0 \\ 0 & 0 & 0 \\ 0 & 0 & -3 \end{bmatrix} \qquad \text{and} \qquad \mathbf{H} = \begin{bmatrix} p & 0 & 0 \\ 0 & q & 0 \\ 0 & 0 & r \end{bmatrix}$$

**A.2.5. Scalar matrix**  A *scalar matrix* is a diagonal matrix whose diagonal elements are *all* equal to the same scalar; hence,

$$\mathbf{A} = \begin{bmatrix} a & 0 & & 0 \\ & a & & \\ & & \ddots & \\ 0 & & & a \end{bmatrix} \qquad \begin{aligned} a_{ij} &= 0 \quad \text{for all } i \neq j \\ a_{ij} &= a \quad \text{for all } i = j \end{aligned}$$

and

$$\mathbf{H} = \begin{bmatrix} 2 & 0 & 0 \\ 0 & 2 & 0 \\ 0 & 0 & 2 \end{bmatrix}$$

are scalar matrices.

**A.2.6. Unit or identity matrix**  A *unit* or *identity matrix* is a diagonal matrix whose diagonal elements are all equal to 1. It will always be referred to by

$$I = \begin{bmatrix} 1 & 0 & & 0 \\ 0 & 1 & & \\ & & \ddots & \\ 0 & & & 1 \end{bmatrix} \qquad \begin{array}{l} a_{ij} = 0 \quad \text{for all } i \neq j \\ a_{ij} = 1 \quad \text{for all } i = j \end{array}$$

**A.2.7. Null or zero matrix**   A *null* or *zero matrix* is a matrix whose elements are *all* zero. It is denoted by a boldface zero, **0**.

**A.2.8. Triangular matrix**   A *triangular matrix* is a square matrix whose elements above (or below), but not including, the main diagonal are all zero. An upper-triangular matrix takes the form

$$A = \begin{bmatrix} a_{11} & a_{12} & \cdots & a_{1m} \\ 0 & a_{22} & \cdots & a_{2m} \\ \cdots & \cdots & \cdots & \cdots \\ 0 & 0 & \cdots & a_{mm} \end{bmatrix} \qquad \text{with } a_{ij} = 0, \quad \text{for } i > j$$

The matrix

$$A = \begin{bmatrix} -1 & 3 & 4 \\ 0 & 1 & 0 \\ 0 & 0 & 7 \end{bmatrix}$$

is an example of an upper-triangular matrix of order 3.

On the other hand, a lower-triangular matrix is of the form

$$A = \begin{bmatrix} a_{11} & 0 & \cdots & 0 \\ a_{21} & a_{22} & \cdots & 0 \\ \cdots & \cdots & \cdots & \cdots \\ a_{m1} & a_{m2} & \cdots & a_{mm} \end{bmatrix} \qquad \text{where } a_{ij} = 0, \quad \text{for } i < j$$

For example,

$$B = \begin{bmatrix} 18 & 0 \\ 2 & -11 \end{bmatrix}$$

is a lower-triangular matrix of order 2.

## A.3. Matrix operations

**A.3.1. Equality**   Two matrices **A** and **B** of the *same dimensions* are *equal* if each element $a_{ij} = b_{ij}$ for all $i$ and $j$. Matrices of different dimensions cannot be equated. Some relationships which apply to matrix equality include:

If $A = B$, then $B = A$ for all $A$ and $B$             (A.1)

If $A = B$ and $B = C$, then $A = C$ for all $A$, $B$, and $C$      (A.2)

As an example, let

$$A = \begin{bmatrix} 1 & 2 \\ 3 & 4 \end{bmatrix} \qquad \text{and} \qquad B = \begin{bmatrix} b_{11} & b_{12} \\ b_{21} & b_{22} \end{bmatrix}$$

If $A = B$ (noting that both are $2 \times 2$ matrices); then

$$b_{11} = 1 \qquad b_{12} = 2 \qquad b_{21} = 3 \qquad \text{and} \qquad b_{22} = 4$$

**A.3.2. Sums**   The sum of two matrices **A** and **B**, of the *same* dimensions, is a matrix **C** of the same dimensions, the elements of which are given by $c_{ij} = a_{ij} + b_{ij}$ for all $i$ and $j$. Matrices of different dimensions cannot be added. The following relationships apply to matrix addition:

$$\mathbf{A} + \mathbf{B} = \mathbf{B} + \mathbf{A} \tag{A.3}$$
$$\mathbf{A} + (\mathbf{B} + \mathbf{C}) = (\mathbf{A} + \mathbf{B}) + \mathbf{C} = \mathbf{A} + \mathbf{B} + \mathbf{C} \tag{A.4}$$

With the null or zero matrix **0**, we have

$$\mathbf{A} + \mathbf{0} = \mathbf{0} + \mathbf{A} = \mathbf{A} \tag{A.5}$$
$$\mathbf{A} + (-\mathbf{A}) = \mathbf{0} \tag{A.6}$$

where $(-\mathbf{A})$ is the matrix composed of $(-a_{ij})$ as elements. As an example, if

$$\mathbf{A} = \begin{bmatrix} 1 & 2 & 0 \\ 0 & 3 & 5 \end{bmatrix} \quad \mathbf{B} = \begin{bmatrix} 6 & 4 & 2 \\ 3 & 2 & 7 \end{bmatrix} \quad \mathbf{C} = \begin{bmatrix} x & y & z \\ u & v & w \end{bmatrix}$$

and $\mathbf{C} = \mathbf{B} - \mathbf{A}$, compute the values of the six variables $x, y, z, u, v,$ and $w$. First, compute $\mathbf{B} - \mathbf{A}$ as

$$\begin{bmatrix} 6 & 4 & 2 \\ 3 & 2 & 7 \end{bmatrix} - \begin{bmatrix} 1 & 2 & 0 \\ 0 & 3 & 5 \end{bmatrix} = \begin{bmatrix} 5 & 2 & 2 \\ 3 & -1 & 2 \end{bmatrix}$$

and then from

$$\mathbf{C} = \begin{bmatrix} x & y & z \\ u & v & w \end{bmatrix} = \begin{bmatrix} 5 & 2 & 2 \\ 3 & -1 & 2 \end{bmatrix}$$

we get $x = 5$, $y = 2$, $z = 2$, $u = 3$, $v = -1$, and $w = 2$.

**A.3.3. Scalar multiplication**   Multiplication of a matrix **A** by a scalar $\alpha$ is another matrix **B** whose elements are $b_{ij} = \alpha a_{ij}$, for all $i$ and $j$. (A scalar is generally denoted by a lowercase Greek letter.) Thus, $\mathbf{B} = \alpha \mathbf{A}$. As an example, for

$$\mathbf{A} = \begin{bmatrix} 7 & 2 \\ 3 & -4 \end{bmatrix} \quad \mathbf{B} = 3\mathbf{A} = \begin{bmatrix} 21 & 6 \\ 9 & -12 \end{bmatrix}$$

The following relations hold for scalar multiplication:

$$\alpha(\mathbf{A} + \mathbf{B}) = \alpha \mathbf{A} + \alpha \mathbf{B} \tag{A.7}$$
$$(\alpha + \beta)\mathbf{A} = \alpha \mathbf{A} + \beta \mathbf{A} \tag{A.8}$$
$$\alpha(\mathbf{A}\mathbf{B}) = (\alpha \mathbf{A})\mathbf{B} = \mathbf{A}(\alpha \mathbf{B}) \tag{A.9}$$
$$\alpha(\beta \mathbf{A}) = (\alpha \beta)\mathbf{A} \tag{A.10}$$

Some examples are

$$\mathbf{A} = \begin{bmatrix} 1 & 2 \\ 2 & 1 \\ 3 & 0 \end{bmatrix} \quad \text{and} \quad \mathbf{B} = \begin{bmatrix} 0 & 3 \\ 2 & 0 \\ 1 & 1 \end{bmatrix}$$

To verify Eq. (A.7), compute $2(\mathbf{A} + \mathbf{B})$ both ways:

$$\mathbf{A} + \mathbf{B} = \begin{bmatrix} 1 & 5 \\ 4 & 1 \\ 4 & 1 \end{bmatrix} \quad \text{and} \quad 2(\mathbf{A} + \mathbf{B}) = \begin{bmatrix} 2 & 10 \\ 8 & 2 \\ 8 & 2 \end{bmatrix}$$

On the other hand,

$$2\mathbf{A} = \begin{bmatrix} 2 & 4 \\ 4 & 2 \\ 6 & 0 \end{bmatrix} \quad \text{and} \quad 2\mathbf{B} = \begin{bmatrix} 0 & 6 \\ 4 & 0 \\ 2 & 2 \end{bmatrix}$$

and then

$$2\mathbf{A} + 2\mathbf{B} = \begin{bmatrix} 2 & 10 \\ 8 & 2 \\ 8 & 2 \end{bmatrix}$$

which checks with the answer above.

To demonstrate Eq. (A.8),

$$(2+1)A = 3A = \begin{bmatrix} 3 & 6 \\ 6 & 3 \\ 9 & 0 \end{bmatrix}$$

Also,

$$2A + A = \begin{bmatrix} 2 & 4 \\ 4 & 2 \\ 6 & 0 \end{bmatrix} + \begin{bmatrix} 1 & 2 \\ 2 & 1 \\ 3 & 0 \end{bmatrix} = \begin{bmatrix} 3 & 6 \\ 6 & 3 \\ 9 & 0 \end{bmatrix}$$

The relation in Eq. (A.9) could not be applied until matrix multiplication is defined, which is given in the next section. Using $\alpha = 2$ and $\beta = 3$, the student should be able to verify Eq. (A.10) as an exercise.

**A.3.4. Matrix multiplication**   The product of two matrices is in general another matrix. A relation between the dimensions of both matrices must exist before multiplication can be performed. The relation is that the number of columns of the first matrix must equal the number of rows of the second matrix. Thus, if $A$ is $m \times k$ and $B$ is $k \times n$, the product $AB$ *in that order* is another matrix $C$ with $m$ rows (as in $A$) and $n$ columns (as in $B$). Each element $c_{ij}$ in $C$ is obtained taking the $k$ elements of the $i$th row in $A$ and multiplying each by the corresponding element in the $j$th column in $B$ and adding the results. Algebraically, this is written as

$$c_{ij} = a_{i1}b_{1j} + a_{i2}b_{2j} + \cdots + a_{ik}b_{kj} \tag{A.11}$$

This process may be shown schematically as

$$
\begin{bmatrix}
a_{11} & a_{12} & \cdots & a_{1k} \\
& & & \\
a_{i1} & a_{i2} & \cdots & a_{ik} \\
& & & \\
a_{m1} & a_{m2} & \cdots & a_{mk}
\end{bmatrix}
\begin{bmatrix}
b_{11} & \cdots & b_{1j} & \cdots & b_{1n} \\
b_{21} & \cdots & b_{2j} & \cdots & b_{2n} \\
& & & & \\
b_{k1} & \cdots & b_{kj} & \cdots & b_{kn}
\end{bmatrix}
$$

$$
= \begin{bmatrix}
c_{11} & \cdots & & c_{1j} & \cdots & c_{1n} \\
& & & & & \\
c_{i1} & \cdots & & c_{ij} = \sum_{r=1}^{r=k} a_{ir}b_{rj} & \cdots & c_{in} \\
& & & & & \\
c_{m1} & \cdots & & c_{mj} & \cdots & c_{mn}
\end{bmatrix} \tag{A.12}
$$

From this definition it follows that if $A$ is a row matrix and $B$ is a column matrix, then $AB$ is a scalar. For example, if

$$A = \begin{bmatrix} 1 & 2 & 3 \end{bmatrix} \quad \text{and} \quad B = \begin{bmatrix} 4 \\ 5 \\ 6 \end{bmatrix}$$

then

$$\underset{1,3}{A} \; \underset{3,1}{B} = \underset{1,1}{C} = (1)(4) + (2)(5) + (3)(6) = 32$$

As examples for multiplication of matrices, let

$$A = \begin{bmatrix} 1 & 1 \\ 2 & 0 \end{bmatrix} \quad B = \begin{bmatrix} 1 & 2 \\ 0 & 1 \\ 1 & 0 \end{bmatrix} \quad C = \begin{bmatrix} 1 & 0 & 1 \\ 0 & 1 & 2 \\ 1 & 0 & 1 \end{bmatrix}$$

Then

$$\begin{matrix} \mathbf{B} & \mathbf{A} & = \mathbf{D} = \\ 3,2 & 2,2 & 3,2 \end{matrix} \begin{bmatrix} 1 & 2 \\ 0 & 1 \\ 1 & 0 \end{bmatrix} \begin{bmatrix} 1 & 1 \\ 2 & 0 \end{bmatrix} = \begin{bmatrix} (1)(1)+(2)(2) & (1)(1)+(2)(0) \\ (0)(1)+(1)(2) & (0)(1)+(1)(0) \\ (1)(1)+(0)(2) & (1)(1)+(0)(0) \end{bmatrix} = \begin{bmatrix} 5 & 1 \\ 2 & 0 \\ 1 & 1 \end{bmatrix}$$

$$\begin{matrix} \mathbf{C} & \mathbf{B} & = \mathbf{E} = \\ 3,3 & 3,2 & 3,2 \end{matrix} \begin{bmatrix} 1 & 0 & 1 \\ 0 & 1 & 2 \\ 1 & 0 & 1 \end{bmatrix} \begin{bmatrix} 1 & 2 \\ 0 & 1 \\ 1 & 0 \end{bmatrix} = \begin{bmatrix} (1)(1)+(0)(0)+(1)(1) & (1)(2)+(0)(1)+(1)(0) \\ (0)(1)+(1)(0)+(2)(1) & (0)(2)+(1)(1)+(2)(0) \\ (1)(1)+(0)(0)+(1)(1) & (1)(2)+(0)(1)+(1)(0) \end{bmatrix}$$

$$= \begin{bmatrix} 2 & 2 \\ 2 & 1 \\ 2 & 2 \end{bmatrix}$$

Note that because of the given matrix dimensions the multiplications $\begin{matrix}\mathbf{A}&\mathbf{B}\\2,2&3,2\end{matrix}$, $\begin{matrix}\mathbf{B}&\mathbf{C}\\3,2&3,3\end{matrix}$, and so on are not possible. Therefore, it is a general rule that $\mathbf{TS} \neq \mathbf{ST}$, even if both matrices are square. For example, if

$$\mathbf{T} = \begin{bmatrix} 1 & 2 \\ 5 & 0 \end{bmatrix} \quad \text{and} \quad \mathbf{S} = \begin{bmatrix} 3 & 4 \\ 0 & 2 \end{bmatrix}$$

then

$$\mathbf{TS} = \mathbf{P} = \begin{bmatrix} 1 & 2 \\ 5 & 0 \end{bmatrix} \begin{bmatrix} 3 & 4 \\ 0 & 2 \end{bmatrix} = \begin{bmatrix} 3 & 8 \\ 15 & 20 \end{bmatrix}$$

while

$$\mathbf{ST} = \mathbf{Q} = \begin{bmatrix} 3 & 4 \\ 0 & 2 \end{bmatrix} \begin{bmatrix} 1 & 2 \\ 5 & 0 \end{bmatrix} = \begin{bmatrix} 23 & 6 \\ 10 & 0 \end{bmatrix}$$

with the obvious result that $\mathbf{P} \neq \mathbf{Q}$ and thus $\mathbf{TS} \neq \mathbf{ST}$.

The following relationships regarding matrix multiplication hold:

$$\mathbf{AI} = \mathbf{IA} = \mathbf{A} \quad \text{with } \mathbf{I} = \text{identity matrix} \tag{A.13}$$
$$\mathbf{A(BC)} = \mathbf{(AB)C} = \mathbf{ABC} \tag{A.14}$$
$$\mathbf{A(B+C)} = \mathbf{AB} + \mathbf{AC} \tag{A.15}$$
$$\mathbf{(A+B)C} = \mathbf{AC} + \mathbf{BC} \tag{A.16}$$

In all these relations the sequence of the matrices is *strictly* preserved, since if the order is reversed, the results will be different, as has just been shown.

An important property of matrix multiplication, which distinguishes it from scalar multiplication, is that the product of two matrices can be the null or zero matrix, without either matrix being the zero matrix. The following are three examples:

1. $\begin{bmatrix} 1 & 1 \\ 0 & 0 \end{bmatrix} \begin{bmatrix} 2 & 3 \\ -2 & -3 \end{bmatrix} = \begin{bmatrix} 0 & 0 \\ 0 & 0 \end{bmatrix}$

2. $\begin{bmatrix} 1 & 2 & 3 \\ 1 & -1 & 0 \end{bmatrix} \begin{bmatrix} 3 & -5 \\ 3 & -5 \\ -3 & 5 \end{bmatrix} = \begin{bmatrix} 0 & 0 \\ 0 & 0 \end{bmatrix}$

3. $\begin{bmatrix} 2 & 2 & -2 \end{bmatrix} \begin{bmatrix} 1 & 2 & 1 \\ 0 & -2 & 1 \\ 1 & 0 & 2 \end{bmatrix} = \begin{bmatrix} 0 & 0 & 0 \end{bmatrix}$

**A.3.5. Matrix transpose**   The transpose of a matrix $\mathbf{A}$ of dimensions $m$ by $n$ is an $n \times m$ matrix formed from $\mathbf{A}$ by interchanging rows and columns such that the $i$th row of $\mathbf{A}$ becomes the $i$th column of the transposed matrix. We denote the transpose by adding a superscript $t$ to the matrix, or $\mathbf{A}^t$. If $\mathbf{B} = \mathbf{A}^t$, it follows that $b_{ij} = a_{ji}$ for all $i$ and $j$. For example,

$$\text{if} \quad \mathbf{A} = \begin{bmatrix} 3 & 2 \\ -1 & 1 \end{bmatrix} \quad \text{then} \quad \mathbf{A}^t = \begin{bmatrix} 3 & -1 \\ 2 & 1 \end{bmatrix}$$

$$\mathbf{B} = \begin{bmatrix} -1 & 6 \\ 0 & 4 \\ 5 & 0 \end{bmatrix} \qquad \mathbf{B}^t = \begin{bmatrix} -1 & 0 & 5 \\ 6 & 4 & 0 \end{bmatrix}$$

$$\mathbf{C} = \begin{bmatrix} a & b & c \end{bmatrix} \qquad \mathbf{C}^t = \begin{bmatrix} a \\ b \\ c \end{bmatrix}$$

The following relationships apply to a matrix transpose:

$$(\mathbf{A} + \mathbf{B})^t = \mathbf{A}^t + \mathbf{B}^t \tag{A.17}$$

$$(\mathbf{AB})^t = \mathbf{B}^t\mathbf{A}^t \tag{A.18}$$

$$(\alpha\mathbf{A})^t = \alpha\mathbf{A}^t \tag{A.19}$$

$$(\mathbf{A}^t)^t = \mathbf{A} \tag{A.20}$$

The first relationship (A.17) can be readily verified by recalling that matrix addition is element by element; therefore, whether you sum first and transpose, or transpose first then add, the result will be the same. The second relation (A.18) is rather important, since transposing a matrix product leads to transposing each matrix, then *reversing* the sequence before performing the multiplication. This becomes logical when we recall that if $\mathbf{A}$ is $m \times k$ then $\mathbf{A}^t$ is $k \times m$ and if $\mathbf{B}$ is $k \times n$, then $\mathbf{B}^t$ is $n \times k$ and multiplication is defined only for $\mathbf{B}^t\mathbf{A}^t$. Furthermore, since $\mathbf{AB}$ is $m$ by $n$, then $(\mathbf{AB})^t$ is by definition an $n \times m$ matrix, which is the same as the dimensions of $\underset{n,k}{\mathbf{B}^t}\ \underset{k,m}{\mathbf{A}^t}$. As a demonstration of Eq. (A.18), let

$$\underset{2,3}{\mathbf{A}} = \begin{bmatrix} 1 & 1 & 0 \\ 0 & 2 & 3 \end{bmatrix} \quad \text{and} \quad \underset{3,1}{\mathbf{B}} = \begin{bmatrix} 1 \\ 1 \\ 2 \end{bmatrix}$$

then

$$\underset{2,3}{\mathbf{A}}\ \underset{3,1}{\mathbf{B}} = \underset{2,1}{\mathbf{C}} = \begin{bmatrix} 1 & 1 & 0 \\ 0 & 2 & 3 \end{bmatrix}\begin{bmatrix} 1 \\ 1 \\ 2 \end{bmatrix} = \begin{bmatrix} 2 \\ 8 \end{bmatrix} \quad \text{and} \quad \mathbf{C}^t = \begin{bmatrix} 2 & 8 \end{bmatrix}$$

$$\mathbf{B}^t\mathbf{A}^t = \begin{bmatrix} 1 & 1 & 2 \end{bmatrix}\begin{bmatrix} 1 & 0 \\ 1 & 2 \\ 0 & 3 \end{bmatrix} = \begin{bmatrix} 2 & 8 \end{bmatrix}$$

which is equal to $\mathbf{C}^t$. The last two relationships (A.19) and (A.20) are rather straightforward and the student should verify them by numerical examples.

When the original matrix is square, the operation of transpose does not affect the elements of the main diagonal. For example, for

$$\mathbf{A} = \begin{bmatrix} a & b \\ c & d \end{bmatrix}$$

the transpose is

$$\mathbf{A}^t = \begin{bmatrix} a & c \\ b & d \end{bmatrix}$$

It follows, then, that if the matrix has zero elements above and below the main diagonal, then it is equal to its transpose. We have introduced three matrices with this property: the identity matrix $\mathbf{I}$, the diagonal matrix $\mathbf{D}$, and the scalar matrix $\mathbf{K}$. Hence, $\mathbf{I}^t = \mathbf{I}$, $\mathbf{D}^t = \mathbf{D}$, and $\mathbf{K}^t = \mathbf{K}$. The scalar matrix $\mathbf{K}$ can also be written as $\mathbf{K} = k\mathbf{I}$, where $k$ is the value of each element along the main diagonal. Therefore, $\mathbf{K}^t = (k\mathbf{I})^t = k\mathbf{I}^t = k\mathbf{I} = \mathbf{K}$. If $\mathbf{x}$ is a column matrix, which is usually called *column vector*, then $\mathbf{x}^t\mathbf{x}$ is a positive scalar which is equal to the sum of the squares of the vector components, or the square of its length. For example,

$$\text{if} \quad \mathbf{x}=\begin{bmatrix} x_1 \\ x_2 \\ x_3 \end{bmatrix} \quad \text{then} \quad \mathbf{x}'\mathbf{x}=\begin{bmatrix} x_1 & x_2 & x_3 \end{bmatrix}\begin{bmatrix} x_1 \\ x_2 \\ x_3 \end{bmatrix}=x_1^2+x_2^2+x_3^2$$

On the other hand, $\mathbf{xx}'$ is a *square-symmetric* matrix, as explained in the following section.

**A.3.6. Symmetric matrix**   A square matrix is called *symmetric* if it is equal to its transpose, or $\mathbf{A}'=\mathbf{A}$. Since transposing a matrix does not change the elements of the main diagonal, the elements above the main diagonal of a symmetric matrix are "mirror images" of those below the diagonal. For example,

$$\begin{bmatrix} 3 & 2 & 1 \\ 2 & 5 & 6 \\ 1 & 6 & 4 \end{bmatrix}, \quad \begin{bmatrix} a & b \\ b & c \end{bmatrix} \quad \text{and} \quad \begin{bmatrix} 6 & 3 & 0 & 1 \\ 3 & 7 & 2 & 0 \\ 0 & 2 & 4 & 0 \\ 1 & 0 & 0 & 5 \end{bmatrix}$$

are symmetric matrices.

For any matrix $\mathbf{A}$ (not necessarily square), both $\mathbf{AA}'$ and $\mathbf{A}'\mathbf{A}$ are symmetric. The proof is direct. For example, if $\mathbf{C}=\mathbf{AA}'$, then $\mathbf{C}'=[\mathbf{AA}']'=(\mathbf{A}')'(\mathbf{A})'=\mathbf{C}$, which means that $\mathbf{C}$ is symmetric. If $\mathbf{B}$ is a symmetric matrix of suitable dimensions, then for any matrix $\mathbf{A}$, both $\mathbf{ABA}'$ and $\mathbf{A}'\mathbf{BA}$ are symmetric. Let

$$\mathbf{A}=\begin{bmatrix} 1 & 1 & 0 \\ 0 & 2 & 1 \end{bmatrix} \quad \text{and} \quad \mathbf{B}=\begin{bmatrix} 3 & 1 \\ 1 & 4 \end{bmatrix}$$

Then

$$\mathbf{A}'\mathbf{BA}=\begin{bmatrix} 1 & 0 \\ 1 & 2 \\ 0 & 1 \end{bmatrix}\begin{bmatrix} 3 & 1 \\ 1 & 4 \end{bmatrix}\begin{bmatrix} 1 & 1 & 0 \\ 0 & 2 & 1 \end{bmatrix}=\begin{bmatrix} 3 & 1 \\ 5 & 9 \\ 1 & 4 \end{bmatrix}\begin{bmatrix} 1 & 1 & 0 \\ 0 & 2 & 1 \end{bmatrix}=\begin{bmatrix} 3 & 5 & 1 \\ 5 & 23 & 9 \\ 1 & 9 & 4 \end{bmatrix}$$

which is obviously symmetric. The sum and difference of symmetric matrices are also symmetric. The product, however, is in general different.

**A.3.7. Matrix Inverse**   Unlike scalars, *division* of matrices is *not defined*. In fact, the relationship $\mathbf{AB}=\mathbf{AC}$ may exist without having $\mathbf{B}=\mathbf{C}$. This implies that the operation of "dividing" by $\mathbf{A}$, even if $\mathbf{A}\neq\mathbf{0}$, is not possible. As an example, let

$$\mathbf{A}=\begin{bmatrix} 2 & 0 \\ 4 & 0 \end{bmatrix} \quad \mathbf{B}=\begin{bmatrix} 2 & -2 \\ 5 & 3 \end{bmatrix} \quad \mathbf{C}=\begin{bmatrix} 2 & -2 \\ 1 & 4 \end{bmatrix}$$

where obviously $\mathbf{B}\neq\mathbf{C}$. Computation of both $\mathbf{AB}$ and $\mathbf{AC}$ yields

$$\mathbf{AB}=\begin{bmatrix} 4 & -4 \\ 8 & -8 \end{bmatrix}=\mathbf{AC}$$

In place of division, the concept of matrix *inverse* is used. It is symbolized by $\mathbf{A}^{-1}$, for a matrix $\mathbf{A}$ (similar to the reciprocal $\alpha^{-1}$ of the scalar $\alpha$).

The *inverse* of a square matrix $\mathbf{A}$, *if it exists*, is the unique matrix $\mathbf{A}^{-1}$ with the following property:

$$\mathbf{AA}^{-1}=\mathbf{A}^{-1}\mathbf{A}=\mathbf{I} \tag{A.21}$$

where $\mathbf{I}$ is the identity matrix. Thus, for

$$\mathbf{A}=\begin{bmatrix} 3 & 1 \\ 2 & 1 \end{bmatrix} \quad \text{the matrix } \mathbf{A}^{-1}=\begin{bmatrix} 1 & -1 \\ -2 & 3 \end{bmatrix}$$

is the inverse because

$$\begin{bmatrix} 1 & -1 \\ -2 & 3 \end{bmatrix}\begin{bmatrix} 3 & 1 \\ 2 & 1 \end{bmatrix}=\begin{bmatrix} 1 & 0 \\ 0 & 1 \end{bmatrix} \quad \text{and} \quad \begin{bmatrix} 3 & 1 \\ 2 & 1 \end{bmatrix}\begin{bmatrix} 1 & -1 \\ -2 & 3 \end{bmatrix}=\begin{bmatrix} 1 & 0 \\ 0 & 1 \end{bmatrix}$$

Some of the properties of the inverse are

$$(\mathbf{AB})^{-1} = \mathbf{B}^{-1}\mathbf{A}^{-1} \tag{A.22}$$

$$(\mathbf{A}^{-1})^{-1} = \mathbf{A} \tag{A.23}$$

$$(\mathbf{A}^t)^{-1} = (\mathbf{A}^{-1})^t \tag{A.24}$$

$$(\alpha\mathbf{A})^{-1} = \frac{1}{\alpha}\mathbf{A}^{-1} \tag{A.25}$$

assuming that all inverses exist. The square matrix for which an inverse exists is called *nonsingular*, while that which does not have an inverse is called *singular*.

As shown previously, the product $\mathbf{AB}$ can equal $\mathbf{0}$ without either $\mathbf{A}=\mathbf{0}$ or $\mathbf{B}=\mathbf{0}$. On the other hand, if either $\mathbf{A}$ or $\mathbf{B}$ is nonsingular, the other matrix must be a null matrix. Hence, the product of two nonsingular matrices cannot be a null matrix.

There are several procedures for computing the inverse of a square matrix. It is beyond the scope of this text to develop these methods. Details of matrix inversion procedures can be found in the references at the end of this appendix.

**A.4. Solution of linear equations**  Linear equations are encountered frequently in surveying and mapping adjustment and computation problems. When the number of equations is equal to the number of unknowns, the system of equations is said to be *unique*. The following is a set of $n$ equations:

$$\begin{aligned}
a_{11}x_1 + a_{12}x_2 + \cdots + a_{1n}x_n &= b_1 \\
a_{21}x_1 + a_{22}x_2 + \cdots + a_{2n}x_n &= b_2 \\
&\cdots \\
a_{n1}x_1 + a_{n2}x_2 + \cdots + a_{nn}x_n &= b_n
\end{aligned} \tag{A.26}$$

In Eq. (A.26), the symbols $a_{ij}$ and $b_i$ are numerical coefficients and constant terms, while the $x_j$ are the unknowns. The solution to Eq. (A.26) is a unique set of numerical values for $x_j$ such that when substituted into the equations, they would be simultaneously satisfied. Equation (A.26) may be expressed in matrix form as

$$\underset{n,n}{\mathbf{A}} \ \underset{n,1}{\mathbf{x}} = \underset{n,1}{\mathbf{b}} \tag{A.27}$$

where $\mathbf{A}$ is a square matrix of order $n$. Consideration here is limited to the more practical unique case for which $\mathbf{A}$ must be nonsingular. Thus, premultiplying both sides of Eq. (A.27) by $\mathbf{A}^{-1}$ yields

$$\mathbf{A}^{-1}\mathbf{A}\mathbf{x} = \mathbf{A}^{-1}\mathbf{b}$$

or

$$\mathbf{x} = \mathbf{A}^{-1}\mathbf{b} \tag{A.28}$$

since $\mathbf{A}^{-1}\mathbf{A} = \mathbf{I}$ by definition. Consequently, the solution to linear equations is equivalent to finding the inverse of the coefficient matrix. If the equations contain only two or three unknowns (i.e., $\mathbf{A}$ is of dimensions $2 \times 2$ or $3 \times 3$), it may be practical to compute the inverse by a simple procedure such as the adjoint method described in Ref. 3. However, computing the inverse by the adjoint method becomes quite tedious and inefficient for matrices of order more than three. Instead, other procedures are employed for both finding the inverse as well as the direct solution of linear equations. Many of these procedures make use of what are called primary row or column operations, which are given in the next section.

**A.5. Primary row or column operations**  These operations, which do not change the order of the matrix being operated on, are:

1. The interchange of any two rows (or two columns).
2. The multiplication of all the elements of any row (or column) by the same nonzero constant.
3. The addition to any row (or column) of an arbitrary multiple of any other row (or column).

Two procedures for solving linear equations, which apply to these operations, are given in the next two sections.

**A.6. The Gauss method**   In this method the elementary operations are used to reduce the coefficient matrix to an upper-triangular form (preferably with diagonal elements of 1). This operation is called the *forward solution*. The subsequent computation of **x** is called *backward solution*. As an example, solve the system

$$x_1 + 2x_2 = 3$$
$$-3x_1 + 4x_2 = 1$$

or

$$\begin{bmatrix} 1 & 2 \\ -3 & 4 \end{bmatrix} \begin{bmatrix} x_1 \\ x_2 \end{bmatrix} = \begin{bmatrix} 3 \\ 1 \end{bmatrix}$$

The first equation may be written as

| $x_1$ | $x_2$ | $b$ |
|---|---|---|
| 0 | 10 | 10 |

which needs no modification since the coefficient of $x_1$ is already 1. The next step is to reduce the coefficient of $x_1$ in the second equation to zero. This is done by multiplying the first equation by 3 and adding to the second. Hence, the new second equation is

| $x_1$ | $x_2$ | $b$ |
|---|---|---|
| 0 | 10 | 10 |

which already completes the forward solution. The backward solution begins with the new second equation, since it has only one unknown, $x_2$, and therefore $x_2 = 1$. Using this in the first equation, compute $x_1 = 3 - 2x_2 = 3 - 2 = 1$. As another example, consider the following set of three equations:

$$x_1 + \quad - x_3 = -1$$
$$x_2 + 3x_3 = \quad 7$$
$$-x_1 + \quad +2x_3 = \quad 3$$

or

$$\begin{bmatrix} 1 & 0 & -1 \\ 0 & 1 & 3 \\ -1 & 0 & 2 \end{bmatrix} \begin{bmatrix} x_1 \\ x_2 \\ x_3 \end{bmatrix} = \begin{bmatrix} -1 \\ 7 \\ 3 \end{bmatrix}$$

In this system, the initial coefficients of $x_1$ are 1 in each equation. Thus, in order to achieve the desired upper-triangular form it is necessary only to add the first and third equations as follows:

| | $x_1$ | $x_2$ | $x_3$ | $b$ | Remarks |
|---|---|---|---|---|---|
| Eq. 1 | 1 | 0 | -1 | -1 | The original data |
| Eq. 3 | -1 | 0 | 2 | 3 | |
| | 1 | 0 | -1 | -1 | First and second equations unaltered since |
| | 0 | 1 | 3 | 7 | $a_{11} = a_{22} = 1$ already; add original first equation |
| | 0 | 0 | 1 | 2 | to original third equation to get a revised third equation; this completes the forward solution |

The backward solution proceeds as follows:

$$x_3 = \quad 2$$
$$x_2 = \quad 7 - 3x_3 = 1$$
$$x_1 = -1 + \quad x_3 = 1$$

**A.7. The Gauss-Jordan method**    In this procedure, the primary operations are continued until the coefficient matrix reduces to the identity matrix. At this point, **b** becomes the answer vector. If all the operations are simultaneously performed on an identity it will become $A^{-1}$ at the end. To show that such is the case, solve the two-equation problem of Art. A.6.

| Line | $x_1$ | $x_2$ | b | I | | Remarks |
|---|---|---|---|---|---|---|
| 1 | 1 | 2 | 3 | 1 | 0 ⎫ | Original data augmented by **I** |
| 2 | −3 | 4 | 1 | 0 | 1 ⎭ | |
| 3 | 1 | 2 | 3 | 1 | 0 ⎫ | First equation (line 1) unaltered; |
| 4 | 0 | 10 | 10 | 3 | 1 ⎭ | 3 times line 1 added to line 2 |
| 5 | 1 | 0 | 1 | 0.4 | −0.2 ⎫ | Divide line 4 by 10 to get line 6; subtract |
| 6 | 0 | 1 | 1 | 0.3 | 0.1 ⎭ | 2 times line 6 from line 3 to get line 5 |
| | | **I** | | **x** | **A⁻¹** | |

It can be seen that **A** reduced to **I**, **b** reduced to **x**, and **I** reduced to $A^{-1}$. Next, work the three-equation problem of Art. A.6:

| Line | $x_1$ | $x_2$ | $x_3$ | b | I | | | Remarks |
|---|---|---|---|---|---|---|---|---|
| 1 | 1 | 0 | −1 | −1 | 1 | 0 | 0 ⎫ | |
| 2 | 0 | 1 | 3 | 7 | 0 | 1 | 0 ⎬ | Given data |
| 3 | −1 | 0 | 2 | 3 | 0 | 0 | 1 ⎭ | |
| 4 | 1 | 0 | −1 | −1 | 1 | 0 | 0 | Same as line 1 |
| 5 | 0 | 1 | 3 | 7 | 0 | 1 | 0 | Same as line 2 |
| 6 | 0 | 0 | 1 | 2 | 1 | 0 | 1 | Line 1 + line 3 |
| 7 | 1 | 0 | 0 | 1 | 2 | 0 | 1 | Line 4 + line 6 |
| 8 | 0 | 1 | 0 | 1 | −3 | 1 | −3 | (Line 5) − (3)(line 6) |
| 9 | 0 | 0 | 1 | 2 | 1 | 0 | 1 | Same as line 6 |
| | | **I** | | **x** | | **A⁻¹** | | |

An important note concerns the possibility of having to divide by zero. For example, if the coefficient of $x_1$ in the first equation is zero, the procedure for both the Gauss and Gauss-Jordan methods could not progress unaltered. In such a case the equations are rearranged such that the first equation will have a nonzero coefficient for $x_1$. This should also be done for $x_2$ in the second step, for $x_3$ in the third step, and so on.

**A.8. Eigenvalues and eigenvectors**    For each square matrix **A** there exists a set of scalars $\lambda_i$ and vectors $x_i$, one vector for each scalar, such that the following relationship holds:

$$Ax_i = \lambda_i x_i \tag{A.29}$$

The scalar $\lambda_i$ is called an *eigenvalue* and $x_i$ is called an *eigenvector*. If the matrix **A** is symmetric, all eigenvalues are real and all eigenvectors are mutually orthogonal to each other. This is the case most often encountered in surveying and mapping, where this technique is used to determine error ellipses from known covariance matrices. This will be shown in conjunction with quadratic forms in the following section.

Equation (A.29) may be rearranged, and the subscript *i* dropped to allow for determining all eigenvalues; thus,

$$Ax - \lambda x = 0$$

or

$$(A - \lambda I)x = 0 \tag{A.30}$$

Equation (A.30) represents a set of $n$ homogeneous equations (assuming that $\mathbf{A}$ is of order $n$) which will have a nontrivial solution if the determinant[†] of the coefficient matrix $(\mathbf{A} - \lambda\mathbf{I})$ is zero. Expanding such a determinant leads to an $n$th-degree polynomial in $\lambda$, the $n$ roots of which would be the eigenvalues. From a practical standpoint, consider the two-dimensional case. Therefore (designating a determinant by two bars $|\cdot|$),

$$|\mathbf{A} - \lambda\mathbf{I}| = 0$$

or

$$\begin{vmatrix} a_{11} - \lambda & a_{12} \\ a_{21} & a_{22} - \lambda \end{vmatrix} = 0$$

or

$$(a_{11} - \lambda)(a_{22} - \lambda) - a_{12}a_{21} = 0$$
$$\lambda^2 - (a_{11} + a_{22})\lambda + (a_{11}a_{22} - a_{12}a_{21}) = 0$$
$$\lambda^2 - \mathrm{tr}(\mathbf{A})\lambda + |\mathbf{A}| = 0 \tag{A.31}$$

in which $\mathrm{tr}(\mathbf{A})$ means the *trace* of $\mathbf{A}$ and is equal to the sum of its diagonal elements, and $|\mathbf{A}|$ is the determinant of $\mathbf{A}$. In general, two values of $\lambda$ are obtained from Eq. (A.31). For each $\lambda$ the corresponding eigenvector is derived using Eq. (A.30). Note that because Eq. (A.30) represents a set of homogeneous equations in the components of the eigenvector, it can be determined only by direction. The concept is illustrated by the numerical computation of the eigenvalues and eigenvectors of the matrix

$$\mathbf{B} = \begin{bmatrix} 2 & 1 \\ 1 & 2 \end{bmatrix}$$

First, $\mathrm{tr}(\mathbf{B}) = 4$ and $|\mathbf{B}| = (2)(2) - (1)(1) = 3$. Then, substituting in Eq. (A.31) gives $\lambda^2 - 4\lambda + 3 = 0$ or $(\lambda - 3)(\lambda - 1) = 0$, from which $\lambda_1 = 3$ and $\lambda_2 = 1$. For $\lambda_1 = 3$, the eigenvector is evaluated from [see Eq. (A.30)]

$$(\mathbf{B} - 3\mathbf{I})\mathbf{x} = 0$$

or

$$\left[ \begin{bmatrix} 2 & 1 \\ 1 & 2 \end{bmatrix} - \begin{bmatrix} 3 & 0 \\ 0 & 3 \end{bmatrix} \right] \begin{bmatrix} x_1 \\ x_2 \end{bmatrix} = \begin{bmatrix} 0 \\ 0 \end{bmatrix}$$

or

$$\begin{bmatrix} -1 & 1 \\ 1 & -1 \end{bmatrix} \begin{bmatrix} x_1 \\ x_2 \end{bmatrix} = \begin{bmatrix} 0 \\ 0 \end{bmatrix}$$

for which $x_1 = x_2$. This means that for $\lambda_1 = 3$ the eigenvector makes $45°$ with the $x_1$ axis and can be taken as $(1, 1)$. For $\lambda_2 = 1$,

$$\begin{bmatrix} 1 & 1 \\ 1 & 1 \end{bmatrix} \begin{bmatrix} x_1 \\ x_2 \end{bmatrix} = \begin{bmatrix} 0 \\ 0 \end{bmatrix}$$

from which $x_1 = x_2$. This means that for $\lambda_1 = 3$ the eigenvector makes $45°$ with the $x_1$ axis and can be taken as $(1, 1)$. For $\lambda_2 = 1$,

**A.9. Quadratic forms**    If $\mathbf{A}$ is a symmetric matrix of order $n$ and $\mathbf{x}$ is an $n \times 1$ vector of variables, the scalar

$$u = \mathbf{x}^t\mathbf{A}\mathbf{x} \tag{A.32}$$

is called a *quadratic form* because it is the sum of second-order terms in the elements of $\mathbf{x}$. As a demonstration, the two-dimensional case is given by

$$u = \begin{bmatrix} x_1 & x_2 \end{bmatrix} \begin{bmatrix} a_{11} & a_{12} \\ a_{12} & a_{22} \end{bmatrix} \begin{bmatrix} x_1 \\ x_2 \end{bmatrix} = a_{11}x_1^2 + 2a_{12}x_1x_2 + a_{22}x_2^2 \tag{A.33}$$

[†] For definition and evaluation of a determinant, see Ref. 2.

A good example of quadratic forms is the sum of weighted residuals squared, which is minimized in least squares. It takes the general form

$$\phi = v'Wv$$

in which $v$ is the vector of observational residuals and $W$ is the symmetric weight matrix of the observations.

If the matrix $A$ in Eq. (A.33) represents the inverse of a covariance matrix, that equation will express a family of ellipses the axes of which are inclined with respect to the $x_1, x_2$ system. If the directions of the eigenvectors of the covariance matrix are designated by $y_1, y_2$, it can be shown that these coincide with the semimajor and semiminor axes of the ellipses. Furthermore, if $u = 1$ in Eq. (A.33), the resulting ellipse is called the *standard error ellipse* with semimajor and semiminor axes which are equal to $\sqrt{\lambda_1}$ and $\sqrt{\lambda_2}$, respectively. The quantities $\lambda_1$ and $\lambda_2$ are the eigenvalues of the covariance matrix. Consider an example. The covariance matrix is

$$B = \begin{bmatrix} 2 & 1 \\ 1 & 2 \end{bmatrix} \quad \text{and} \quad A = B^{-1} = \frac{1}{3} \begin{bmatrix} 2 & -1 \\ -1 & 2 \end{bmatrix}$$

The quadratic form for the standard ellipse is

$$[ x_1 \quad x_2 ] \frac{1}{3} \begin{bmatrix} 2 & -1 \\ -1 & 2 \end{bmatrix} \begin{bmatrix} x_1 \\ x_2 \end{bmatrix} = 1$$

$$\tfrac{2}{3} x_1^2 - \tfrac{2}{3} x_1 x_2 + \tfrac{2}{3} x_2^2 = 1 \tag{A.34}$$

Equation (A.34) represents the equation of the ellipse in Fig. A.1 with respect to the $x_1, x_2$ axis system. If the dimensions of that ellipse are computed then the semimajor axis, $a = \sqrt{3}$ and the semiminor axis, $b = 1$, and the axes are oriented at 45° as shown in the figure. This will agree with the fact that $\lambda_1 = 3$ and $\lambda_2 = 1$ for the covariance matrix $B$ as was computed in the preceding example. Furthermore, $y_1$ and $y_2$ do, in fact, represent the directions of the eigenvectors of the covariance matrix $B$. The equation of the standard ellipse with respect to the $y_1, y_2$ system is

$$\frac{y_1^2}{3} + \frac{y_2^2}{1} = 1 \tag{A.35}$$

The student should prove this equality by substituting, for example, the $x_1, x_2$ coordinates for one or two points on the ellipse, such as $c$ and $d$, and see that they satisfy Eq. (A.34), and correspondingly the $y_1, y_2$ coordinates and see that they satisfy Eq. (A.35).

**A.10. Differentiation of matrices and quadratic forms**    The following are useful relations that are encountered in surveying and mapping computation and adjustment.

1. If the elements of a matrix $\underset{m,n}{A}$ are functions of a (scalar) variable $u$, then the

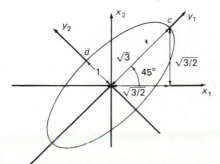

**Fig. A.1**

derivative $dA/du$ of the matrix with respect to $u$ is another $m$ by $n$ matrix given by

$$\frac{dA}{\underset{m,n}{du}} = \begin{bmatrix} \dfrac{da_{11}}{du} & \cdots & \dfrac{da_{1n}}{du} \\ \vdots & & \vdots \\ \dfrac{da_{m1}}{du} & \cdots & \dfrac{da_{mn}}{du} \end{bmatrix} \tag{A.36}$$

2. If a (column) vector $\underset{m,1}{y}$ is composed of $m$ functions of some or all of the elements of another variable (column) vector $\underset{n,1}{x}$, then the partial derivative $\partial y/\partial x$ of $y$ with respect to $x$ is an $m$ by $n$ matrix, which is called the *jacobian matrix*, whose elements are

$$J_{yx} = \frac{\partial y}{\partial x} = \begin{bmatrix} \dfrac{\partial y_1}{\partial x_1} & \dfrac{\partial y_1}{\partial x_2} & \cdots & \dfrac{\partial y_1}{\partial x_n} \\ \vdots & & & \vdots \\ \dfrac{\partial y_m}{\partial x_1} & \dfrac{\partial y_m}{\partial x_2} & \cdots & \dfrac{\partial y_m}{\partial x_n} \end{bmatrix} \tag{A.37}$$

For example, if $y_1 = x_1 + 2x_2 + 3x_3^2$ and $y_2 = 7 - 2x_2^2 + 5x_3$, then the jacobian $J_{yx}$ is

$$J_{yx} = \begin{bmatrix} \partial y_1/\partial x_1 & \partial y_1/\partial x_2 & \partial y_1/\partial x_3 \\ \partial y_2/\partial x_1 & \partial y_2/\partial x_2 & \partial y_2/\partial x_3 \end{bmatrix} = \begin{bmatrix} 1 & 2 & 6x_3 \\ 0 & -4x_2 & 5 \end{bmatrix}$$

3. Assume first that $A$ is not symmetric in the quadratic form $u = x'Ax$ and is also independent of $x$. If both $A$ and $x$ are of order $n$, then $\partial u/\partial x$ must be a 1 by $n$ row vector according to item (2) above. That partial derivative may be evaluated by visualizing partially differentiating first with respect to $x$ and then with respect to $x'$ and adding the results. Keeping in mind that the result is $1 \times n$, then

$$\frac{\partial u}{\partial x} = \left( \underset{1,n}{x'} \underset{n,n}{A} \right) + \left( \underset{1,n}{x'} \underset{n,n}{A'} \right) = x'(A + A') \tag{A.38}$$

Now, if $A$ is symmetric which is the more practical case, Eq. (A.38) becomes

$$\frac{\partial u}{\partial x} = 2x'A \tag{A.39}$$

We demonstrate this for the two-dimensional case by referring to Eq. (A.33):

$$\frac{\partial u}{\partial x_1} = 2a_{11}x_1 + 2a_{12}x_2 = 2[\begin{matrix} x_1 & x_2 \end{matrix}] \begin{bmatrix} a_{11} \\ a_{12} \end{bmatrix}$$

$$\frac{\partial u}{\partial x_2} = 2a_{12}x_1 + 2a_{22}x_2 = 2[\begin{matrix} x_1 & x_2 \end{matrix}] \begin{bmatrix} a_{12} \\ a_{22} \end{bmatrix}$$

Thus,

$$\frac{\partial u}{\partial x} = 2[\begin{matrix} x_1 & x_2 \end{matrix}] \begin{bmatrix} a_{11} & a_{12} \\ a_{12} & a_{22} \end{bmatrix} = 2x'A$$

## References

**1.** Eves, Howard, *Elementary Matrix Theory*, Allyn and Bacon, Inc., Boston, Mass., 1966.
**2.** Fuller, Leonard E., *Basic Matrix Theory*, Prentice-Hall, Inc., Englewood Cliffs, N.J., 1964.
**3.** Mikhail, Edward M., *Observations and Least Squares*, Harper & Row, Publishers, New York, 1976.

# APPENDIX B
# Least squares adjustment

**B.1. Concept of adjustment** It was explained in Chap. 2 that many of the variables in surveying are random variables which would take different values if the experiments are repeated. It was also emphasized (in Arts. 2.21 and 2.22) that a single measurement for each variable of interest would provide estimates that are consistent with the model. If, however, there are more measurements than necessary to uniquely determine the model, *redundant* observations are present and an *adjustment* is required in order to have a set of estimates that are consistent with the model. Consider the very simple example of measuring an angle $\alpha$. The model here is simply composed of one random variable $\alpha$ for which the minimum number of observations is one. If the angle is measured twice, with two different values, $\alpha_1, \alpha_2$, there will be one redundant observation and the apparent inconsistency of having two different values must be resolved. Let $v_1$ and $v_2$ represent two "residuals" which when added to $\alpha_1$ and $\alpha_2$, respectively, yield one unique value $\hat{\alpha}$ for the angle $\alpha$. In other words,

$$\hat{\alpha} = \hat{\alpha}_1 = \alpha_1 + v_1 = \hat{\alpha}_2 = \alpha_2 + v_2 \tag{B.1}$$

represents the new estimate for $\alpha$ *which is consistent with the given model.* To obtain $\hat{\alpha}$, $v_1$ and $v_2$ are required; and in order to determine $v_1$ and $v_2$, the method of least squares adjustment is applied as explained in detail in the subsequent sections of this appendix. It is clear from Eq. (B.1) that there is an infinite number of possible pairs of values for $v_1$ and $v_2$, and the choice of a particular pair is made on the basis of the least squares criterion which leads to certain optimum properties.

**B.2. The least squares criterion** The least squares criterion was introduced in Art. 2.22, and therefore only a summary is given here for the sake of completeness. With $\ell$, $v$, and $\hat{\ell}$ representing the given observations, residuals, and least squares estimates for the observations, respectively,

$$\hat{\ell} = \ell + v \tag{2.52}$$

Next, if $\mathbf{W}$ denotes the weight matrix of the observations, the general least squares criterion is

$$\phi = v^t \mathbf{W} v \rightarrow \text{minimum} \tag{2.53}$$

If $\mathbf{W}$ is a diagonal matrix (uncorrelated observations),

$$\phi = (w_1 v_1^2 + w_2 v_2^2 + \cdots + w_n v_n^2) \rightarrow \text{minimum} \tag{2.54}$$

and if $\mathbf{W}$ is the identity matrix (uncorrelated observations and of equal precision), the criterion becomes

$$\phi = (v_1^2 + v_2^2 + \cdots + v_n^2) \rightarrow \text{minimum} \tag{2.55}$$

The last relationship would apply to $v_1$ and $v_2$ in the example for the angle $\alpha$ (in Art. B.1) if

**Table B.1**

| $v_1$ | $v_2$ | $\hat{\alpha}_1 = \alpha_1 + v_1$ | $\hat{\alpha}_2 = \alpha_2 + v_2$ | $\phi = v_1^2 + v_2^2$ |
|---|---|---|---|---|
| $+1.5'$ | $-0.5$ | $30°27'.5$ | $30°27'.5$ | $2.5$ |
| $+2'$ | $0$ | $30°28'$ | $30°28'$ | $4.0$ |

both $\alpha_1$ and $\alpha_2$ are measured with equal precision (or weight). Thus, if $\alpha_1 = 30°26'$ and $\alpha_2 = 30°28'$, it will be shown later that $v_1 = +1'$ and $v_2 = -1'$ and therefore

$$\hat{\alpha} = \hat{\alpha}_1 = \alpha_1 + v_1 = (30°26') + 1' = 30°27'$$

and

$$\hat{\alpha} = \hat{\alpha}_2 = \alpha_2 + v_2 = (30°28') - 1' = 30°27'$$

Note that $\hat{\alpha}_1$ and $\hat{\alpha}_2$ estimated from least-squares adjustment are equal and therefore are consistent with the given model. Furthermore, computation of $\phi$ according to Eq. (2.55) yields

$$\phi = (+1)^2 + (-1)^2 = 2(\text{min of arc})^2$$

which is the minimum value possible, keeping $\hat{\alpha}_1 = \hat{\alpha}_2$. To show that this is true, consider two other possibilities, as given in Table B.1.

This rather simple example demonstrates the optimum property of least squares: when two observations are of equal weight, the final estimate is their mean. Similarly, if the two observations are of unequal precision, the best estimate is the weighted mean (see Ex. B.5).

**B.3. Geometric demonstration of the least squares principle** In order for the reader to have a better understanding of the basic concept of the criterion of least squares, simple geometric demonstrations may be used. For example, consider the case of measuring the angle $\alpha$ twice as discussed in the preceding two sections. As before, $\hat{\alpha}_1 = \alpha_1 + v_1$ and $\hat{\alpha}_2 = \alpha_2 + v_2$ must be equal. This means that there is a conditional relationship corresponding to the one redundant observation which may be written from Eq. (B.1) as

$$\hat{\alpha}_1 - \hat{\alpha}_2 = 0 \tag{B.2}$$

The condition in Eq. (B.2) represents a straight line in a two-dimensional cartesian coordinate system, with one axis for $\hat{\alpha}_1$ and the other for $\hat{\alpha}_2$, as depicted in Fig. B.1. That

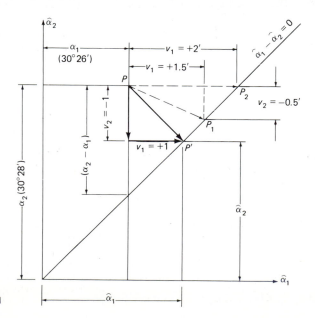

**Fig. B.1**

line bisects the angle between the $\hat{\alpha}_1$ and $\hat{\alpha}_2$ axes. The two actual observations, $\alpha_1 = 30°26'$ and $\alpha_2 = 30°28'$, define a point $P$ which falls above the condition line because $\alpha_1 < \alpha_2$. Adjustment in general sense is to adhere to the model by enforcing the condition, and therefore point $P$ should be replaced by another point $P'$ which lies on the condition line. In other words, the new estimates $\hat{\alpha}_1$, $\hat{\alpha}_2$ (the coordinates of $P'$) are consistent with the model. It is clear that there is an infinite number of possible points $P'$ which lie on the line. From all these possibilities, the least squares principle selects the one point $P'$ such that the distance $PP'$ is a minimum. Consequently, the line $PP'$ is perpendicular to the condition line, satisfying the intuitive property that the new estimated observations ($P'$) be as close as possible to the given observations ($P$).

If the given numerical values for $\alpha_1$ and $\alpha_2$ are considered, the diagram in Fig. B.1 leads to

$$v_1 = +\tfrac{1}{2}(\alpha_2 - \alpha_1) = +1'$$

$$v_2 = -\tfrac{1}{2}(\alpha_2 - \alpha_1) = -1'$$

The two other possibilities given in Table B.1 are represented in the figure by points $P_1$ and $P_2$. It should be clear from Fig. B.1 that the length of $PP_1$ is $\sqrt{2.5}$ and of $PP_2$ is $\sqrt{4} = 2$, and both lengths are longer than the minimum distance $PP'$. This should demonstrate graphically that the least squares solution is, in fact, optimum in that it yields estimates which are closest to the given observations.

If the two observations $\alpha_1$, $\alpha_2$ were of different weights, $w_1$ and $w_2$, respectively, the geometric demonstration would still be possible. In such a case, the $\hat{\alpha}_1$ axis would have divisions which are equal to $\sqrt{w_1}$, while the divisions along the $\hat{\alpha}_2$ axis would be multiples of $\sqrt{w_2}$. If the observations are in addition correlated, geometric representation is in principal possible, except that in such cases the situation gets sufficiently complex that it defeats the purpose of the demonstration.

## B.4. Redundancy and the model

Before planning the acquisition phase of surveying data, a general model is usually specified either explicitly or implicitly. Such a model is determined by a certain number of variables and a possible set of relationships among them. Whether or not an adjustment of the survey data is necessary depends on the amount of observational data acquired. There is always a *minimum number* of independent variables needed to determine the selected model uniquely. Such a minimum number is designated by $n_0$. If $n$ measurements are acquired ($n > n_0$) with respect to the specified model, then the *redundancy*, or (statistical) degrees of freedom, is specified as the amount by which $n$ exceeds $n_0$. Denoting the redundancy by $r$,

$$r = n - n_0 \tag{B.3}$$

As illustrations, consider the following examples.

1. The shape of a plane triangle is uniquely determined by a minimum of two interior angles, or $n_0 = 2$. If three interior angles are measured, then with $n = 3$, redundancy is $r = 1$.
2. The size and shape of a plane triangle require a minimum of three observations, at least one of which is the length of one side; or $n_0 = 3$. If three interior angles and all three lengths are available, then with $n = 6$ the redundancy is $r = 3$.
3. In addition to size and shape of the plane triangle, its location and orientation with respect to a specified cartesian coordinate system $xy$, are also of interest (Fig. B.2).

   In this case, the minimum of variables necessary to determine the model is $n_0 = 6$, which can be explained in one of two ways. From statement 2 above, the size and shape requires that $n_0 = 3$; then the location of one point (e.g., $x_1$ and $y_1$ in the Fig. B.2) and the orientation of one side (e.g., $\alpha$ in the figure) add three more to make a minimum total of six. Another way to determine $n_0$ is to express the model as simply locating three points (1, 2, 3 in the figure) in the two-dimensional coordinate system $xy$ which obviously requires six coordinates. If observations $x_1$, $y_1$, $\alpha$ are known in addition to the three interior angles and three sides, then with $n = 9$, the redundancy is $r = 3$.

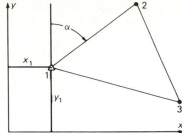

**Fig. B.2**

The success of a survey adjustment depends to a large measure on the proper definition of the model and the correct determination of $n_0$. Next, the acquired measurements must relate to the specified model and must have a set that is sufficient to determine the model. If this is not the case, a deficient situation will exist and the adjustment would not be meaningful. This could be illustrated by having three different measurements of *one* interior angle in a plane triangle. In this case even though $n=3$ and $n_0=2$, it is clear that the shape of the triangle could not be determined from these data.

**B.5. Condition equations—parameters**   After the redundancy $r$ is determined, the adjustment proceeds by writing equations that relate the model variables in order to reflect the existing redundancy. Such equations will be referred to either as *condition equations* or simply as *conditions*. The number of conditions to be formulated for a given problem will depend on whether only observational variables are involved in them, or the conditions contain other unknown variables. To illustrate this point, recall the case of having two measurements $\alpha_1$ and $\alpha_2$ for the angle $\alpha$. If no additional unknown variables are introduced, there will be, as explained in Art. B.3, only *one condition equation* corresponding to the one redundancy, or $\hat{\alpha}_1 - \hat{\alpha}_2 = 0$. Once the adjustment is performed, the least squares estimate of the angle, $\hat{\alpha}$, is obtained from another relationship, namely, $\hat{\alpha} = \hat{\alpha}_1$ (or $\hat{\alpha} = \hat{\alpha}_2$). Note that this relationship is almost self-evident. Nevertheless, such additional relations are required in order to evaluate other variables, as will be shown in the following example. As an alternative, $\hat{\alpha}$ could be carried in the adjustment as an additional unknown variable. In such a case, one more condition must be written in addition to the one corresponding to $r=1$ (i.e., there must be two conditions). These may be written as

$$\hat{\alpha} - \hat{\alpha}_1 = 0$$
$$\hat{\alpha} - \hat{\alpha}_2 = 0$$

The additional unknown variable, which is also a random variable like the observations, will be called a *parameter*. The one thing that distinguishes a parameter from an observation is that the parameter does not have an a priori sample value while the observation does. After the adjustment, both the observations and the parameters will have new least squares estimates, as well as estimates for their cofactor matrices, as will be explained in later sections of this appendix.

To summarize, then, if the redundancy is $r$, there exist $r$ independent condition equations which can be written in terms of the given $n$ observations. If $u$ unknown parameters are included in the adjustment, a total of

$$c = r + u \tag{B.4}$$

independent condition equations in terms of both the $n$ observations and $u$ parameters must be written. In order for the parameters to be functionally independent, their number $u$ should not exceed the minimum number of variables, $n_0$, necessary to specify the model. Hence, the following relation must be satisfied:

$$0 \leqslant u \leqslant n_0 \tag{B.5}$$

Similarly, for the formulated condition equations to be independent, their number $c$ should not be larger than the total number of observations, $n$. Hence,

$$r \leqslant c \leqslant n \tag{B.6}$$

To demonstrate these relations as well as elaborate further on the concept of a parameter, consider another example.

**Example B.1**   Figure B.3 is a sketch of a small level net which contains a bench mark, B.M., and three points, the elevations of which are needed. To determine these three elevations, five differences in elevation, $\ell_1, \ell_2, \ldots, \ell_5$, are measured. The arrow along a line in Fig. B.3 leads (by convention) from a low point to a higher point.

The model involves the elevation of four points, of which one is a bench mark having a known elevation. Therefore, the minimum number of variables needed to fully specify the model is $n_0 = 3$. Given $n = 5$ observations, it follows that the redundancy is $r = 2$. Consequently, for the least squares adjustment, two independent condition equations are written in terms of five observations. One possibility is to write one condition for each of the two loops $a$ and $b$, shown in Fig. B.3 as follows:

$$\ell_1 + \ell_2 - \ell_3 = 0$$
$$\ell_3 + \ell_4 - \ell_5 = 0 \tag{B.7}$$

After the adjustment, new estimates, $\hat{\ell}_1, \hat{\ell}_2, \ldots, \hat{\ell}_5$, for the five observations are obtained. From these new values the elevations of points 1, 2, and 3 can be uniquely computed no matter which combination of estimated observations are used. For instance, the elevation of point 2 may be computed in any one of the following ways and the value would be identical:

elevation of point $2 =$ B.M. $+ \ell_1 + \ell_2$
$\qquad\qquad\qquad = $ B.M. $+ \ell_3$
$\qquad\qquad\qquad = $ B.M. $+ \ell_5 - \ell_4$

An alternative least squares procedure is possible if the elevations of points 1, 2, 3 are carried as parameters in the adjustment. In such a case, $u = 3$ and according to Eq. (B.4), the number of conditions would be $c = 2 + 3 = 5$. Here, $u$ takes on its upper value of $n_0 = 3$, and likewise $c$ is equal to $n$ [see Eqs. (B.5) and (B.6)]. Denoting the parameters by $x_1$, $x_2$, $x_3$, the five condition equations may be written as

B.M. $+ \ell_1 - x_1 = 0$
$\quad x_1 + \ell_2 - x_2 = 0$
B.M. $+ \ell_3 - x_2 = 0$
$\quad x_2 + \ell_4 - x_3 = 0$
B.M. $+ \ell_5 - x_3 = 0 \tag{B.8}$

The estimates $x_1$, $x_2$, $x_3$ computed from the direct least squares adjustment would be identical to those computed from $\ell_1, \ell_2, \ldots, \ell_5$ in the previous technique.

This example also demonstrates that least squares adjustment can be performed by at least one of two techniques: one where both observations and parameters are involved, and the other where only observations and no parameters are included in the condition equations. The two techniques are discussed in the following section.

**B.6.  Techniques of least squares**   Although there are several techniques for least squares adjustment, consider only two techniques. It is important to point out that for any given

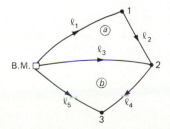

**Fig. B.3**

survey adjustment problem, *all* least squares techniques yield *identical* results. Therefore, the choice of a technique depends mainly on the type of problem, the required information, and to some extent the computing equipment available. Further discussion of this point will be given after the actual techniques are presented.

The first technique is called *adjustment of indirect observations* and is characterized by the following properties:

1. The condition equations include both observations and parameters.
2. There are as many conditions as the number of observations, or $c = n$
3. Each condition equation contains *only one observation* with the specific stipulation that its coefficient is unity.

With these properties the condition equations take the functional form

$$v_1 + b_{11}\delta_1 + b_{12}\delta_2 + \cdots + b_{1u}\delta_u = f_1$$
$$v_2 + b_{21}\delta_1 + b_{22}\delta_2 + \cdots + b_{2u}\delta_u = f_2 \qquad \text{(B.9)}$$
$$\cdots \cdots \cdots \cdots \cdots \cdots$$
$$v_n + b_{n1}\delta_1 + b_{n2}\delta_2 + \cdots + b_{nu}\delta_u = f_n$$

where $v_1, v_2, \ldots, v_n =$ residuals for the $n$ observations
$\delta_1, \delta_2, \ldots, \delta_u = u$ unknown parameters
$b_{11}, b_{12}, \ldots, b_{nu} =$ numerical coefficients of the parameters
$f_1, f_2, \ldots, f_n =$ constant terms for the $n$ conditions, which will usually contain the a priori numerical values of the observations.

The set of equations in (B.9) can be collected into matrix form as

$$\begin{bmatrix} v_1 \\ v_2 \\ \vdots \\ v_n \end{bmatrix} + \begin{bmatrix} b_{11} & b_{12} & \cdots & b_{1u} \\ b_{21} & b_{22} & \cdots & b_{2u} \\ \vdots & & & \\ b_{n1} & b_{n2} & \cdots & n_{nu} \end{bmatrix} \begin{bmatrix} \delta_1 \\ \delta_2 \\ \vdots \\ \delta_u \end{bmatrix} = \begin{bmatrix} f_1 \\ f_2 \\ \vdots \\ f_n \end{bmatrix} \qquad \text{(B.10)}$$

or, more concisely,

$$\underset{n,1}{\mathbf{v}} + \underset{n,u}{\mathbf{B}} \underset{u,1}{\boldsymbol{\Delta}} = \underset{n,1}{\mathbf{f}} \qquad \text{(B.11)}$$

In linear problems, the constant-term vector $\mathbf{f}$ in Eq. (B.11) is usually the vector of given observations $\boldsymbol{\ell}$ subtracted from a vector of numerical constants $\mathbf{d}$:

$$\mathbf{f} = \mathbf{d} - \boldsymbol{\ell} \qquad \text{(B.12)}$$

and Eq. (B.11) is, in fact,

$$\mathbf{v} + \mathbf{B}\boldsymbol{\Delta} = \mathbf{d} - \boldsymbol{\ell} \qquad \text{(B.13)}$$

**Example B.2** To demonstrate Eqs. (B.9), write the conditions in Eq. (B.8) for the level net of Example B.1 as follows:

$$\begin{aligned} v_1 - x_1 \quad &= -\ell_1 - \text{B.M.} = f_1 \\ v_2 + x_1 - x_2 \quad &= -\ell_2 \quad = f_2 \\ v_3 \quad - x_2 \quad &= -\ell_3 - \text{B.M.} = f_3 \\ v_4 \quad + x_2 - x_3 &= -\ell_4 \quad = f_4 \\ v_5 \quad - x_3 &= -\ell_5 - \text{B.M.} = f_5 \end{aligned} \qquad \text{(B.14)}$$

or, in matrix form,

$$\underset{5,1}{\mathbf{v}} + \underset{5,3}{\mathbf{B}} \underset{3,1}{\boldsymbol{\Delta}} = \underset{5,1}{\mathbf{f}} \qquad \text{(B.15)}$$

with

$$
\mathbf{v}=\begin{bmatrix} v_1 \\ v_2 \\ \vdots \\ v_5 \end{bmatrix} \quad
\mathbf{B}=\begin{bmatrix} -1 & 0 & 0 \\ 1 & -1 & 0 \\ 0 & -1 & 0 \\ 0 & 1 & -1 \\ 0 & 0 & -1 \end{bmatrix} \quad
\mathbf{\Delta}=\begin{bmatrix} x_1 \\ x_2 \\ x_3 \end{bmatrix} \quad
\mathbf{f}=\begin{bmatrix} f_1 \\ f_2 \\ \vdots \\ f_5 \end{bmatrix}
\tag{B.16}
$$

It is clear from Eq. (B.14) that the vector f is

$$
\mathbf{f}=\begin{bmatrix} -\text{B.M.} \\ 0 \\ -\text{B.M.} \\ 0 \\ -\text{B.M.} \end{bmatrix} - \begin{bmatrix} \ell_1 \\ \ell_2 \\ \ell_3 \\ \ell_4 \\ \ell_5 \end{bmatrix} = \mathbf{d}-\boldsymbol{\ell}
$$

which demonstrates Eq. (B.12).

The second least squares technique to be considered is called *adjustment of observations only*. As its name implies, no parameters are included in the condition equations, which must therefore be equal in number to the redundancy $r$. They take the general functional form

$$
\begin{aligned}
a_{11}v_1 + a_{12}v_2 + \cdots + a_{1n}v_n &= f_1 \\
a_{21}v_1 + a_{22}v_2 + \cdots + a_{2n}v_n &= f_2 \\
\cdots\cdots\cdots\cdots\cdots\cdots\cdots & \\
a_{r1}v_1 + a_{r2}v_2 + \cdots + a_{rn}v_n &= f_r
\end{aligned}
\tag{B.17}
$$

which in matrix notation becomes

$$
\begin{bmatrix} a_{11} & a_{12} & \cdots & a_{1n} \\ a_{21} & a_{22} & \cdots & a_{2n} \\ \cdots\cdots\cdots\cdots\cdots \\ a_{r1} & a_{r2} & \cdots & a_{rn} \end{bmatrix}
\begin{bmatrix} v_1 \\ v_2 \\ \vdots \\ v_n \end{bmatrix} =
\begin{bmatrix} f_1 \\ f_2 \\ \vdots \\ f_r \end{bmatrix}
\tag{B.18}
$$

or, more concisely,

$$
\underset{r,n}{\mathbf{A}}\ \underset{n,1}{\mathbf{v}} = \underset{r,1}{\mathbf{f}}
\tag{B.19}
$$

For linear adjustment problems, the constant-term vector f in Eq. (B.19) is given by

$$
\mathbf{f}=\mathbf{d}-\mathbf{A}\boldsymbol{\ell}
\tag{B.20}
$$

in which d is the vector of numerical constants as in the case of Eq. (B.12). Then, Eq. (B.19) becomes

$$
\mathbf{A}\mathbf{v}=\mathbf{d}-\mathbf{A}\boldsymbol{\ell}
\tag{B.21}
$$

**Example B.3**   As an illustration, rewrite the two conditions in Eq. (B.7) for the level net in Example B.1.

$$
v_1 + v_2 - v_3 = -\ell_1 - \ell_2 + \ell_3 = f_1
\tag{B.22}
$$

$$
v_3 + v_4 - v_5 = -\ell_3 - \ell_4 + \ell_5 = f_2
$$

or

$$
\underset{2,5}{\mathbf{A}}\ \underset{5,1}{\mathbf{v}} = \underset{2,1}{\mathbf{f}}
\tag{B.23}
$$

in which

$$
\mathbf{A}=\begin{bmatrix} 1 & 1 & -1 & 0 & 0 \\ 0 & 0 & 1 & 1 & -1 \end{bmatrix} \quad
\mathbf{v}=[\,v_1\ \ v_2\ \ \cdots\ \ v_5\,]^t \quad
\mathbf{f}=\begin{bmatrix} f_1 \\ f_2 \end{bmatrix}
\tag{B.24}
$$

The vector $f$ can be written from Eq. (B.22) as

$$f = \begin{bmatrix} 0 \\ 0 \end{bmatrix} - \begin{bmatrix} 1 & 1 & -1 & 0 & 0 \\ 0 & 0 & 1 & 1 & -1 \end{bmatrix} \begin{bmatrix} \ell_1 \\ \ell_2 \\ \ell_3 \\ \ell_4 \\ \ell_5 \end{bmatrix}$$

which according to Eq. (B.24) becomes

$$f = d - A\ell$$

thus demonstrating Eq. (B.21).

Once the condition equations are formulated, the least squares solution is obtained through the application of the minimum criterion given in Art. B.2. The reason for having to apply an additional criterion is that the set of condition equations will always contain more unknowns than equations. For instance, Eq. (B.9) [or (B.10)] is a set of only $n$ equations but in terms of $(n + u)$ unknowns: $n$ residuals and $u$ parameters. Similarly, Eq. (B.17) [or (B.18)] is a set of only $r$ equations but in terms of $n$ unknown residuals, where $n$ is always larger than $r$ according to Eq. (B.3). The application of the minimum criterion will yield an additional set of equations which when combined with the set of conditions results in a unique situation with as many linear equations as unknowns. The mechanics of enforcing the minimum principle and deriving the additional equations vary slightly for the two techniques, and thus each technique will be addressed in a separate section.

**B.7. Adjustment of indirect observations**   In order to appreciate the matrix derivation to follow, first work with the level net of Examples B.1 and B.2. Also, for the purpose of keeping the introductory treatment uncomplicated, assume that the five observed differences in elevation are uncorrelated and with the weights $w_1, w_2, \ldots, w_5$, respectively. Under this assumption, the minimum criterion from Eq. (2.54) applies and thus

$$\phi = w_1 v_1^2 + w_2 v_2^2 + w_3 v_3^2 + w_4 v_4^2 + w_5 v_5^2$$

which, in view of the equations in (B.14), becomes

$$\phi = w_1(f_1 + x_1)^2 + w_2(f_2 - x_1 + x_2)^2 + w_3(f_3 + x_2)^2 + w_4(f_4 - x_2 + x_3)^2 + w_5(f_5 + x_3)^2 \tag{B.25}$$

For $\phi$ in Eq. (B.25) to be a minimum, its partial derivative with respect to each unknown variable (i.e., with respect to $x_1, x_2, x_3$) must equal zero. Hence,

$$\frac{\partial \phi}{\partial x_1} = 2w_1(f_1 + x_1) - 2w_2(f_2 - x_1 + x_2) \qquad\qquad = 0$$

$$\frac{\partial \phi}{\partial x_2} = 2w_2(f_2 - x_1 + x_2) + 2w_3(f_3 + x_2) - 2w_4(f_4 - x_2 + x_3) = 0 \tag{B.26}$$

$$\frac{\partial \phi}{\partial x_3} = 2w_4(f_4 - x_2 + x_3) + 2w_5(f_5 + x_3) \qquad\qquad = 0$$

Clearing and rearranging, the set of equations in (B.26) becomes

$$(w_1 + w_2)x_1 \qquad - w_2 x_2 \qquad\qquad = -w_1 f_1 + w_2 f_2$$
$$-w_2 x_1 + (w_2 + w_3 + w_4)x_2 - w_4 x_3 \qquad = -w_2 f_2 - w_3 f_3 + w_4 f_4 \tag{B.27}$$
$$-w_4 x_2 + (w_4 + w_5)x_3 = -w_4 f_4 - w_5 f_5$$

In order to express Eq. (B.27) in matrix form recall that $\Delta$ and $f$ are vectors given in Eq.

(B.16) and that the weight matrix of the observations is a diagonal matrix

$$
\begin{bmatrix}
w_1 & & & & \mathbf{0} \\
& w_2 & & & \\
& & \ddots & & \\
\mathbf{0} & & & & w_5
\end{bmatrix}
\tag{B.28}
$$

Then the equations in (B.27) can be written in matrix form as

$$
\begin{bmatrix}
-1 & 1 & 0 & 0 & 0 \\
0 & -1 & -1 & 1 & 0 \\
0 & 0 & 0 & -1 & -1
\end{bmatrix}
\begin{bmatrix}
w_1 & & & & \mathbf{0} \\
& w_2 & \cdots & & \\
& & \ddots & & \\
\mathbf{0} & & & & w_5
\end{bmatrix}
\begin{bmatrix}
-1 & 0 & 0 \\
1 & -1 & 0 \\
0 & -1 & 0 \\
0 & 1 & -1 \\
0 & 0 & -1
\end{bmatrix}
\begin{bmatrix}
x_1 \\
x_2 \\
x_3
\end{bmatrix}
$$

$$
=
\begin{bmatrix}
-1 & 1 & 0 & 0 & 0 \\
0 & -1 & -1 & 1 & 0 \\
0 & 0 & 0 & -1 & -1
\end{bmatrix}
\begin{bmatrix}
w_1 & & & & \mathbf{0} \\
& w_2 & \cdots & & \\
& & \ddots & & \\
\mathbf{0} & & & & w_5
\end{bmatrix}
\begin{bmatrix}
f_1 \\
f_2 \\
\vdots \\
f_5
\end{bmatrix}
\tag{B.29}
$$

The reader should multiply the matrices in Eq. (B.29) to ascertain that the equations in (B.27) will result. Recalling the matrix **B** from Eq. (B.16), Eq. (B.29) can be written more concisely as

$$
\left(\underset{3,5}{\mathbf{B}'}\ \underset{5,5}{\mathbf{W}}\ \underset{5,3}{\mathbf{B}}\right)\underset{3,1}{\Delta} = \underset{3,5}{\mathbf{B}'}\ \underset{5,5}{\mathbf{W}}\ \underset{5,1}{\mathbf{f}}
\tag{B.30}
$$

The relation in Eq. (B.30), although derived for an example, is in fact general and applies to any problem for which the condition equations are of the general form given in Eq. (B.11). Furthermore, there is no restriction on the structure of the weight matrix **W**. The following derivation shows that this is true.

The least squares criterion is

$$
\phi = \mathbf{v}'\mathbf{W}\mathbf{v}
$$

which, upon substituting for **v** from Eq. (B.11), becomes

$$
\phi = (\mathbf{f} - \mathbf{B}\Delta)'\mathbf{W}(\mathbf{f} - \mathbf{B}\Delta)
$$
$$
\phi = \mathbf{f}'\mathbf{W}\mathbf{f} - \Delta'\mathbf{B}'\mathbf{W}\mathbf{f} - \mathbf{f}'\mathbf{W}\mathbf{B}\Delta + \Delta'\mathbf{B}'\mathbf{W}\mathbf{B}\Delta
\tag{B.31}
$$

Since all terms on the right-hand side of Eq. (B.31) are scalars, the second and third terms are equal and thus

$$
\phi = \mathbf{f}'\mathbf{W}\mathbf{f} - 2\mathbf{f}'\mathbf{W}\mathbf{B}\Delta + \Delta'\mathbf{B}'\mathbf{W}\mathbf{B}\Delta
\tag{B.32}
$$

For $\phi$ to be a minimum, $\partial\phi/\partial\Delta$ must be zero, or

$$
-2\mathbf{f}'\mathbf{W}\mathbf{B} + 2\Delta'\mathbf{B}'\mathbf{W}\mathbf{B} = 0
$$

or

$$
\left(\underset{u,n}{\mathbf{B}'}\ \underset{n,n}{\mathbf{W}}\ \underset{n,u}{\mathbf{B}}\right)\underset{u,1}{\Delta} = \underset{u,n}{\mathbf{B}'}\ \underset{n,n}{\mathbf{W}}\ \underset{n,1}{\mathbf{f}}
\tag{B.33}
$$

which is identical to Eq. (B.30). Note that **W** could be a full matrix and therefore no restriction is placed on the statistical properties of the observations. Of course, certain assumptions about the structure of **W** may be made in practice. If the auxiliaries

$$
\underset{u,u}{\mathbf{N}} = \mathbf{B}'\mathbf{W}\mathbf{B}
$$

$$
\underset{u,1}{\mathbf{t}} = \mathbf{B}'\mathbf{W}\mathbf{f}
\tag{B.34}
$$

are used, Eq. (B.33) becomes

$$\mathbf{N\Delta = t} \tag{B.35}$$

the solution of which is

$$\mathbf{\Delta = N^{-1}t} \tag{B.36}$$

The set of equations in (B.33) [or (B.35)] are usually called the reduced normal equations, or simply the normal equations in the parameters. The matrices $\mathbf{N}$ and $\mathbf{t}$ [in (B.34)] are called the normal equations coefficient matrix and the normal equations constant term vector, respectively.

The precision of the estimated parameters is the cofactor matrix $\mathbf{Q}_{\Delta\Delta}$.

This matrix is obtained by applying the relationships of error propagation developed in Art. 2.20 to Eq. (B.36). Using Eq. (B.34) for $\mathbf{t}$ and Eq. (B.12) for $\mathbf{f}$, Eq. (B.36) becomes

$$\mathbf{\Delta = N^{-1}B'W(d-\ell)} \tag{B.37}$$

The only vector of random variables on the right-hand side of Eq. (B.37) is $\boldsymbol{\ell}$, since $\mathbf{d}$ is a vector of numerical constants. The matrix $\mathbf{Q}$ is used to designate the cofactor of the observations (i.e., in place of $\mathbf{Q}_{\ell\ell}$), and therefore when Eq. (2.43) is applied to Eq. (B.37), the following results:

$$\mathbf{Q}_{\Delta\Delta} = \mathbf{J}_{\Delta\ell}\mathbf{Q}_{\ell\ell}\mathbf{J}'_{\Delta\ell} = \mathbf{J}_{\Delta\ell}\mathbf{Q}\mathbf{J}'_{\Delta\ell}$$
$$= (-\mathbf{N^{-1}B'W})\mathbf{Q}(-\mathbf{N^{-1}B'W})'$$

or

$$\mathbf{Q}_{\Delta\Delta} = \mathbf{N^{-1}B'WQWBN^{-1}}$$

noting that $\mathbf{N^{-1}}$ and $\mathbf{W}$ are symmetric matrices. Since $\mathbf{W = Q^{-1}}$, using Eq. (B.34) for $\mathbf{N}$, then

$$\mathbf{Q}_{\Delta\Delta} = \mathbf{N^{-1}(B'WB)N^{-1} = N^{-1}NN^{-1}}$$

or

$$\mathbf{Q}_{\Delta\Delta} = \mathbf{N^{-1}} \tag{B.38}$$

Once $\mathbf{\Delta}$ is computed [from Eq. (B.35)], the observational residuals may be computed using Eq. (B.11), or

$$\mathbf{v = f - B\Delta} \tag{B.39}$$

and the least squares estimate of the observations, $\hat{\boldsymbol{\ell}}$, is evaluated from Eq. (2.52), or

$$\hat{\boldsymbol{\ell}} = \boldsymbol{\ell} + \mathbf{v} \tag{B.40a}$$

In a manner similar to that used to obtain $\mathbf{Q}_{\Delta\Delta}$ a relationship for the cofactor matrix $\mathbf{Q}_{\tilde{v}\tilde{v}}$ may be derived as

$$\mathbf{Q}_{\tilde{v}\tilde{v}} = \mathbf{BN^{-1}B'} \tag{B.40b}$$

If, originally, the covariance matrix of the observations $\mathbf{\Sigma}$ was given and used in the least squares solution instead of the cofactor matrix $\mathbf{Q}$, then $\mathbf{\Sigma}_{\Delta\Delta} = \mathbf{N^{-1}}$ and $\mathbf{\Sigma}_{\tilde{v}\tilde{v}} = \mathbf{BN^{-1}B'}$, instead of Eqs. (B.38) and (B.40b). On the other hand, if only relative covariances and variances were given a priori, Eqs. (B.38) and (B.40b) may be used to compute $\mathbf{Q}_{\Delta\Delta}$ and $\mathbf{Q}_{\tilde{v}\tilde{v}}$. Then, in order to get $\mathbf{\Sigma}_{\Delta\Delta}$ and $\mathbf{\Sigma}_{\tilde{v}\tilde{v}}$ an estimate $\hat{\sigma}_0^2$ of the reference variance may be computed from the adjustment using the relationship

$$\hat{\sigma}_0^2 = \frac{\mathbf{v'Wv}}{r} \tag{B.41}$$

in which $r$ is the redundancy [see Eq. (B.3), Art. B.4], $\mathbf{v}$ is the vector of residuals computed from Eq. (B.39), and $\mathbf{W}$ is the a priori weight matrix of the observations. Then, according to Eq. (2.31),

$$\mathbf{\Sigma}_{\Delta\Delta} = \hat{\sigma}_0^2 \mathbf{Q}_{\Delta\Delta} \tag{B.42}$$
$$\mathbf{\Sigma}_{\tilde{v}\tilde{v}} = \hat{\sigma}_0^2 \mathbf{Q}_{\tilde{v}\tilde{v}} \tag{B.43}$$

**Example B.4**  Suppose that $\ell_1$, $\ell_2$, $\ell_3$ are three different measurements of a distance (which is a random variable) and that these measurements are uncorrelated and of equal precision; find the least squares estimate of the distance, $\bar{\ell}$.

SOLUTION    If $\bar{\ell}$ is the final estimate of the distance, it will be equal to the sum of each distance and its corresponding residual, or

$$\ell_1 + v_1 = \bar{\ell}$$
$$\ell_2 + v_2 = \bar{\ell}$$
$$\ell_3 + v_3 = \bar{\ell}$$

These three condition equations may be rearranged and put in matrix form,

$$\begin{bmatrix} v_1 \\ v_2 \\ v_3 \end{bmatrix} + \begin{bmatrix} -1 \\ -1 \\ -1 \end{bmatrix} \bar{\ell} = \begin{bmatrix} -\ell_1 \\ -\ell_2 \\ -\ell_3 \end{bmatrix}$$

which is of the general form $\mathbf{v} + \mathbf{B}\Delta = \mathbf{f}$, with

$$\mathbf{B} = \begin{bmatrix} -1 \\ -1 \\ -1 \end{bmatrix} \qquad \Delta = \bar{\ell} \qquad \text{and} \qquad \mathbf{f} = \begin{bmatrix} -\ell_1 \\ -\ell_2 \\ -\ell_3 \end{bmatrix}$$

Since the observations are uncorrelated and of equal precision, then $\mathbf{W} = \mathbf{I}$ and the normal equations coefficient matrix $\mathbf{N}$ and constant-term vector $\mathbf{t}$ are computed from Eq. (B.34) as

$$\mathbf{N} = \mathbf{B}^t\mathbf{B} = \begin{bmatrix} -1 & -1 & -1 \end{bmatrix} \begin{bmatrix} -1 \\ -1 \\ -1 \end{bmatrix} = 3$$

$$\mathbf{t} = \mathbf{B}^t\mathbf{f} = \begin{bmatrix} -1 & -1 & -1 \end{bmatrix} \begin{bmatrix} -\ell_1 \\ -\ell_2 \\ -\ell_3 \end{bmatrix} = \ell_1 + \ell_2 + \ell_3$$

According to Eq. (B.36), the least squares estimate of the distance is given by

$$\bar{\ell} = \Delta = \mathbf{N}^{-1}\mathbf{t} = (3)^{-1}(\ell_1 + \ell_2 + \ell_3) = \frac{\ell_1 + \ell_2 + \ell_3}{3}$$

This result is rather interesting, since it shows that the least squares estimate of three uncorrelated observations of equal weight is their arithmetic mean, which is what one would have intuitively taken as the "best estimate" of the distance. Although only three observations were considered, the result is general and applies to any number of observations. Thus, for $n$ observations which are uncorrelated and of equal precision (weight),

$$\mathbf{N} = n \tag{B.44a}$$

$$\mathbf{t} = \ell_1 + \ell_2 + \cdots + \ell_n = \sum_{i=1}^{n} \ell_i \tag{B.44b}$$

$$\bar{\ell} = \frac{1}{n} \sum_{i=1}^{n} \ell_i \tag{B.44c}$$

and from Eq. (B.38),

$$q_{\bar{\ell}\bar{\ell}} = \mathbf{Q}_{\Delta\Delta} = \mathbf{N}^{-1} = \frac{1}{n} \tag{B.44d}$$

Thus, the weight of the arithmetic mean of $n$ uncorrelated observations each of unit weight is equal to $n$, because

$$w_{\bar{\ell}\bar{\ell}} = q_{\bar{\ell}\bar{\ell}}^{-1} = n \tag{B.44e}$$

**Example B.5**  Compute the least squares estimate, $\bar{\ell}$, for the distance measured three times $(\ell_1, \ell_2, \ell_3)$ of Example B.4 if the observations are uncorrelated but have different weights, $w_1$, $w_2$, $w_3$, respectively.

SOLUTION   The condition equations are the same as those given in Example B.4. However, the normal equations coefficient matrix is in this case

$$N = B'WB = \begin{bmatrix} -1 & -1 & -1 \end{bmatrix} \begin{bmatrix} w_1 & 0 & 0 \\ 0 & w_2 & 0 \\ 0 & 0 & w_3 \end{bmatrix} \begin{bmatrix} -1 \\ -1 \\ -1 \end{bmatrix} = w_1 + w_2 + w_3$$

recognizing of course that the weight matrix $W$ is a diagonal matrix. The constant-term vector, $t$, is computed as

$$t = B'Wf = \begin{bmatrix} -1 & -1 & -1 \end{bmatrix} \begin{bmatrix} w_1 & 0 & 0 \\ 0 & w_2 & 0 \\ 0 & 0 & w_3 \end{bmatrix} \begin{bmatrix} -\ell_1 \\ -\ell_2 \\ -\ell_3 \end{bmatrix} = w_1 \ell_1 + w_2 \ell_2 + w_3 \ell_3$$

and thus the least-squares estimate is

$$\bar{\ell} = N^{-1} t = \frac{w_1 \ell_1 + w_2 \ell_2 + w_3 \ell_3}{w_1 + w_2 + w_3}$$

which is the weighted mean of the given observations.

As before, the result from Example B.5 is also general and may be extended to $n$ observations:

$$N = w_1 + w_2 + \cdots + w_n = \sum_{i=1}^{n} w_i \tag{B.45a}$$

$$t = w_1 \ell_1 + w_2 \ell_2 + \cdots + w_n \ell_n = \sum_{i=1}^{n} w_i \ell_i \tag{B.45b}$$

$$\bar{\ell} = \frac{\sum_{i=1}^{n} w_1 \ell_i}{\sum_{i=1}^{n} w_i} \tag{B.45c}$$

$$q_{\bar{\ell}\bar{\ell}} = Q_{\Delta\Delta} = N^{-1} = \left( \sum_{i=1}^{n} w_i \right)^{-1} \tag{B.45d}$$

$$w_{\bar{\ell}\bar{\ell}} = q_{\bar{\ell}\bar{\ell}}^{-1} = \sum_{i=1}^{n} w_i \tag{B.45e}$$

Equation (B.45e) implies that the weight of the weighted mean of a set of uncorrelated observations (with different weights) is equal to the sum of their weights. It can be seen then that Eq. (B.44e) is the special case of Eq. (B.45e) when the weights are all equal to unity.

**Example B.5a**   The three measured interior angles of a plane triangle are $\ell_1 = 45°25'01''$, $\ell_2 = 65°20'00''$, and $\ell_3 = 69°15'02''$. Compute the least squares estimates for the three angles assuming that the measurements are uncorrelated and of equal precision.

SOLUTION   Let $x_1$ and $x_2$ be the adjusted measurements for angles $\ell_1$ and $\ell_2$. Three condition equations can be formed as follows:

$$v_1 + \ell_1 = x_1$$
$$v_2 + \ell_2 = x_2$$
$$v_3 + \ell_3 = 180° - x_1 - x_2$$

which become

$$v_1 - x_1 = -\ell_1$$
$$v_2 - x_2 = -\ell_2$$
$$v_3 + x_1 + x_2 = 180° - \ell_3$$

which may be written in matrix form as

$$v + B\Delta = f$$

where

$$v = \begin{bmatrix} v_1 \\ v_2 \\ v_3 \end{bmatrix} \quad B = \begin{bmatrix} -1 & 0 \\ 0 & -1 \\ 1 & 1 \end{bmatrix} \quad \Delta = \begin{bmatrix} x_1 \\ x_2 \end{bmatrix} \quad f = \begin{bmatrix} -45°25'01'' \\ -65°20'00'' \\ 110°44'58'' \end{bmatrix}$$

Since noncorrelated observations of equal precision have been assumed, $W = I$ and normal equations (B.33) become

$$(B^t B)\Delta = B^t f$$

or

$$\begin{bmatrix} -1 & 0 & 1 \\ 0 & -1 & 1 \end{bmatrix} \begin{bmatrix} -1 & 0 \\ 0 & -1 \\ 1 & 1 \end{bmatrix} \begin{bmatrix} x_1 \\ x_2 \end{bmatrix} = \begin{bmatrix} -1 & 0 & 1 \\ 0 & -1 & 1 \end{bmatrix} \begin{bmatrix} -45°25'01'' \\ -65°20'00'' \\ 110°44'58'' \end{bmatrix}$$

$$\begin{bmatrix} 2 & 1 \\ 1 & 2 \end{bmatrix} \begin{bmatrix} x_1 \\ x_2 \end{bmatrix} = \begin{bmatrix} 156.166389° \\ 176.082778° \end{bmatrix}$$

Solution of these equations yields

$x_1 = 45°25'00''$

$x_2 = 65°19'59''$

Substitution of these values into the original condition equations gives

$v_1 = 45°25'00'' - 45°25'01'' = -01''$

$v_2 = 65°19'59'' - 65°20'00'' = -01''$

$v_3 = 180° - (45°25'00'' + 65°19'59'' + 69°15'02'') = -01''$

The propagated cofactor matrix for the adjusted observations by Eq. (B.40b) is

$$Q_{\tilde{v}\tilde{v}} = B(B^t B)^{-1} B^t = B N^{-1} B^t$$

Thus,

$$Q_{\tilde{v}\tilde{v}} = \begin{bmatrix} -1 & 0 \\ 0 & -1 \\ 1 & 1 \end{bmatrix} \begin{bmatrix} \frac{2}{3} & -\frac{1}{3} \\ -\frac{1}{3} & \frac{2}{3} \end{bmatrix} \begin{bmatrix} -1 & 0 & 1 \\ 0 & -1 & 1 \end{bmatrix}$$

and

$$Q_{\tilde{v}\tilde{v}} = \begin{bmatrix} \frac{2}{3} & -\frac{1}{3} & -\frac{1}{3} \\ -\frac{1}{3} & \frac{2}{3} & -\frac{1}{3} \\ -\frac{1}{3} & -\frac{1}{3} & \frac{2}{3} \end{bmatrix}$$

Note that the precision of each adjusted angle is higher than that for the unadjusted, original observation, since $\hat{q} = \frac{2}{3}$ as compared to $q = 1.0$. In addition, although the original observations were uncorrelated, the adjusted observations are correlated as indicated by the off-diagonal elements in $Q_{\tilde{v}\tilde{v}}$.

**B.8. Adjustment of observations only**   Similar to the case of indirect observations, this technique is introduced by working the level-net problem discussed in Examples B.1 and B.3 and assuming as before that the five observations are uncorrelated and have the weights $w_1$, $w_2$, $w_3$, $w_4$, and $w_5$. With this information the minimum criterion from Eq. (2.54) is

$$\phi = w_1 v_1^2 + w_2 v_2^2 + w_3 v_3^2 + w_4 v_4^2 + w_5 v_5^2 \rightarrow \min.$$

Unlike the case of adjustment of indirect observations, it is not possible here to substitute for the residuals from the condition equations since there are five residuals and only two condition equations [in (B.22)]. Therefore, in this case a minimum for the function $\phi$ is sought under the constraint imposed by the condition equations (B.22). This makes the problem that of seeking a *constrained minimum* instead of a *free minimum*, as it is termed in mathematics. Such a constrained minimum is most conveniently obtained by adding (algebraically) to $\phi$ each of the condition equations multiplied by a factor $\lambda$. These factors are called *Lagrange multipliers* after the great French analyst Lagrange (Ref. 12). It is numerically more convenient for our later development to use $-2k$ instead of $\lambda$, and therefore the function to be minimized becomes

$$\phi' = w_1 v_1^2 + w_2 v_2^2 + w_3 v_3^2 + w_4 v_4^2 + w_5 v_5^2$$
$$- 2k_1(v_1 + v_2 - v_3 - f_1)$$
$$- 2k_2(v_3 + v_4 - v_5 - f_2) \tag{B.46}$$

Note that after the adjustment, the quantities within parenthesis in (B.46) vanish because the two condition Eqs. (B.22) are fully satisfied after the adjustment. Consequently, the minimum of $\phi'$ corresponds to the minimum of the original function $\phi$. Taking the partial derivatives of $\phi'$ with respect to each of the five residuals and equating to zero leads to

$$\frac{\partial \phi'}{\partial v_1} = 2w_1 v_1 - 2k_1 \qquad = 0$$

$$\frac{\partial \phi'}{\partial v_2} = 2w_2 v_2 - 2k_1 \qquad = 0$$

$$\frac{\partial \phi'}{\partial v_3} = 2w_3 v_3 + 2k_1 - 2k_2 = 0 \tag{B.47}$$

$$\frac{\partial \phi'}{\partial v_4} = 2w_4 v_4 - 2k_2 \qquad = 0$$

$$\frac{\partial \phi'}{\partial v_5} = 2w_5 v_5 + 2k_2 \qquad = 0$$

Partial differentiation of $\phi'$ with respect to $k_1$ and $k_2$ and equating the result to zero yields the two condition equations in (B.22). Therefore, combining Eqs. (B.47) and (B.22) results in seven linear equations in seven unknowns: $v_1, v_2, \ldots, v_5, k_1, k_2$. Solving Eq. (B.47) for the five residuals yields

$$v_1 = \frac{1}{w_1} k_1$$

$$v_2 = \frac{1}{w_2} k_1$$

$$v_3 = \frac{1}{w_3}(-k_1 + k_2)$$

$$v_4 = \frac{1}{w_4} k_2$$

$$v_5 = \frac{1}{w_5}(-k_2)$$

which in matrix form becomes

$$
\begin{bmatrix} v_1 \\ v_2 \\ v_3 \\ v_4 \\ v_5 \end{bmatrix}
=
\begin{bmatrix}
1/w_1 & & & & \mathbf{0} \\
& 1/w_2 & & & \\
& & 1/w_3 & & \\
& & & 1/w_4 & \\
\mathbf{0} & & & & 1/w_5
\end{bmatrix}
\begin{bmatrix}
1 & 0 \\
1 & 0 \\
-1 & 1 \\
0 & 1 \\
0 & -1
\end{bmatrix}
\begin{bmatrix} k_1 \\ k_2 \end{bmatrix}
\tag{B.48}
$$

The first (diagonal) matrix on the right-hand side of Eq. (B.48) is the inverse of the weight matrix or $\mathbf{W}^{-1}$, which is equal to the cofactor matrix of the observations $\mathbf{Q}$. The second matrix is $\mathbf{A}'$ as can be seen by reference to Eq. (B.24). Thus, Eq. (B.48) may be written more concisely as

$$\mathbf{v} = \mathbf{Q}\mathbf{A}'\mathbf{k} \tag{B.49}$$

in which all the terms have been defined except that $\mathbf{k}$ is the vector of Lagrange multipliers, or $\mathbf{k}=[k_1 \quad k_2]'$. Substituting for $\mathbf{v}$ from Eq. (B.49) into the condition equations (B.23) gives

$$\mathbf{AQA'k}=\mathbf{f} \qquad (\text{B.}50a)$$

which may be solved for $k$ as

$$\mathbf{k}=(\mathbf{AQA'})^{-1}\mathbf{f} \qquad (\text{B.}50b)$$

Finally, substitute the value of $\mathbf{k}$ computed from Eq. (B.50$b$), into Eq. (B.48) to get values for the residuals.

The relations (B.50$a$) and (B.50$b$) are not specific for this particular example, but are rather general for this technique of least squares adjustment of observations only as shown in the following derivation. Let $\underset{r,1}{\mathbf{k}}$ be the vector of $r$ Lagrange multipliers, one for each of the $r$ condition equations in (B.19). Then the function to be minimized is

$$\phi'=\mathbf{v'Wv}-2\mathbf{k'}(\mathbf{Av}-\mathbf{f})$$

For $\phi'$ to be a minimum, $\partial\phi'/\partial\mathbf{v}$ must be zero, or

$$\frac{\partial\phi'}{\partial\mathbf{v}}=2\mathbf{v'W}-2\mathbf{k'A}=0$$

which after, transposing and rearranging, becomes

$$\mathbf{Wv}=\mathbf{A'k}$$

and solving for $\mathbf{v}$ yields

$$\underset{n,1}{\mathbf{v}}=\mathbf{W}^{-1}\mathbf{A'k}=\underset{n,n}{\mathbf{Q}}\ \underset{n,r}{\mathbf{A'}}\ \underset{r,1}{\mathbf{k}} \qquad (\text{B.}51)$$

Substituting Eq. (B.51) into Eq. (B.19) yields

$$(\mathbf{AQA'})\mathbf{k}=\mathbf{f}$$

which when using the auxiliary

$$\underset{r,r}{\mathbf{Q}_e}=\underset{r,n}{\mathbf{A}}\ \underset{n,n}{\mathbf{Q}}\ \underset{n,r}{\mathbf{A'}} \qquad (\text{B.}52)$$

leads to

$$\mathbf{Q}_e\mathbf{k}=\mathbf{f}$$

or

$$\mathbf{k}=\mathbf{Q}_e^{-1}\mathbf{f}=\mathbf{W}_e\mathbf{f} \qquad (\text{B.}53)$$

The matrix $\mathbf{Q}_e$ can be considered as the cofactor matrix for an equivalent set of observations, $\underset{r,1}{\ell_e}$, containing as many observations as there are condition equations. Since $r<n$, the number of equivalent observations is always less than the number of original observations. Each equivalent observation is a linear combination of the original observations, as can be seen from either Eq. (B.21) or (B.22). The linear relations are expressed by the matrix $\mathbf{A}$ and therefore

$$\ell_e=\mathbf{A}\ell \qquad (\text{B.}54)$$

By error propagation, then, the cofactor matrix $\mathbf{Q}_e$ for $\ell_e$ may be evaluated as

$$\mathbf{Q}_e=\mathbf{J}_{\ell_e\ell}\mathbf{QJ'}_{\ell_e\ell}=\mathbf{AQA'}$$

which is identical to Eq. (B.52). The inverse of $\mathbf{Q}_e$ is designated $\mathbf{W}_e$, as shown in Eq. (B.53). The final relation for $v$ is obtained by substituting for $\mathbf{k}$ from Eq. (B.53) into Eq. (B.51), or

$$\mathbf{v}=\mathbf{QA'W}_e\mathbf{f} \qquad (\text{B.}55)$$

Precision estimation after the adjustment may be performed using the rules of propagation developed in Art. 2.20. The estimated observations, $\hat{\ell}$, are given by Eq. (B.40), or

$$\hat{\ell}=\ell+\mathbf{v}=\ell+\mathbf{QA'W}_e\mathbf{f}$$

which, from Eq. (B.20), becomes

$$\hat{\ell}=\ell+\mathbf{QA'W}_e(\mathbf{d}-\mathbf{A}\ell) \qquad (\text{B.}56)$$

Applying Eq. (2.43) to Eq. (B.56) results in

$$Q_{\tilde{\ell}\tilde{\ell}} = J_{\tilde{\ell}\ell} Q J_{\tilde{\ell}\ell}'$$

$$= (I - QA'W_e A)Q(I - QA'W_e A)'$$

$$= (Q - QA'W_e AQ)(I - A'W_e AQ)$$

or

$$Q_{\tilde{\ell}\tilde{\ell}} = Q - QA'W_e AQ - QA'W_e AQ + QA'W_e AQA'W_e AQ$$

From the definition of $Q_e$ in Eq. (B.52) and the fact that $W_e = Q_e^{-1}$, the last term reduces to the negative of the third term, and therefore the two cancel out and the final expression for the cofactor matrix of the estimated observations becomes

$$Q_{\tilde{\ell}\tilde{\ell}} = Q - QA'W_e AQ \tag{B.57}$$

As an exercise, the reader should evaluate the cofactor matrix of the residuals $Q_{vv}$, from Eq. (B.55) and verify that it is the negative of the last term in Eq. (B.57). Thus, an alternative to Eq. (B.57) is

$$Q_{\tilde{\ell}\tilde{\ell}} = Q - Q_{vv} \tag{B.58}$$

**Example B.6**  The interior angles of a plane triangle are measured to be $\ell_1 = 45°25'01''$, $\ell_2 = 65°20'00''$, and $\ell_3 = 69°15'02''$ (the same data as used for Example B.5a). Compute the least squares estimates of the three angles if the measurements are uncorrelated and of equal precision.

SOLUTION  Since it takes a minimum of two angles to fix the shape of a plane triangle, given three measured angles, the redundancy according to Eq. (B.3) is

$r = n - n_0 = 3 - 2 = 1$

Therefore, *one* condition equation exists which relates the observations. This equation is

$\ell_1 + \ell_2 + \ell_3 = 180°$

meaning that the sum of the interior angles in a plane triangle must equal 180°. This equation may be rewritten in the form of $Av = f$, or

$v_1 + v_2 + v_3 = 180° - \ell_1 - \ell_2 - \ell_3 = -3''$

or

$$[1 \quad 1 \quad 1] \begin{bmatrix} v_1 \\ v_2 \\ v_3 \end{bmatrix} = -3''$$

Since the observations are uncorrelated and of equal weight, $Q = W = I$, and from Eqs. (B.52), (B.53), and (B.55), we obtain

$$W_e = (AA')^{-1} = \tfrac{1}{3}$$

$$v = A'W_e f = \begin{bmatrix} 1 \\ 1 \\ 1 \end{bmatrix} \tfrac{1}{3}(-3'') = \begin{bmatrix} -1'' \\ -1'' \\ -1'' \end{bmatrix}$$

The least squares estimates of the observations are

$$\hat{\ell} = \ell + v = \begin{bmatrix} 45°25'01'' \\ 65°20'00'' \\ 69°15'02'' \end{bmatrix} + \begin{bmatrix} -1'' \\ -1'' \\ -1'' \end{bmatrix} = \begin{bmatrix} 45°25'00'' \\ 65°19'59'' \\ 69°15'01'' \end{bmatrix}$$

which when added together yield 180°, thus satisfying the required condition.

As expected, these adjusted angles correspond to the angles obtained by the least squares adjustment by indirect observations in Example B.5a. The propagated cofactor matrix for adjusted observations is obtained as follows by Eq. (B.58):

$$\mathbf{Q}_{\tilde{u}\tilde{u}}=\mathbf{Q}-\mathbf{QA^tW_eAQ}$$

$$=\begin{bmatrix}1&0&0\\0&1&0\\0&0&1\end{bmatrix}-\begin{bmatrix}1&0&0\\0&1&0\\0&0&1\end{bmatrix}\begin{bmatrix}1\\1\\1\end{bmatrix}[1/3][1\ \ 1\ \ 1]\begin{bmatrix}1&0&0\\0&1&0\\0&0&1\end{bmatrix}=\begin{bmatrix}\frac{2}{3}&-\frac{1}{3}&-\frac{1}{3}\\-\frac{1}{3}&\frac{2}{3}&-\frac{1}{3}\\-\frac{1}{3}&-\frac{1}{3}&\frac{2}{3}\end{bmatrix}$$

which is identical to the propagated cofactor matrix for adjusted angles calculated in Example B.5a.

## B.9. Linearization of nonlinear equations

So far the discussion of least squares has been based on the assumption that the condition equations are linear in the observations and the parameters. In practice, surveying problems yield condition equations which frequently are nonlinear in the observations and/or the parameters. The direct use of nonlinear equations in least squares is very complex and is rarely done. Instead, the equations are linearized using Taylor's series expansion and solving the resulting linear equations, then iterating until the effect of the neglected higher-order terms is minimized.

Let $y=f(x)$ be a function that is nonlinear in $x$. The Taylor series expansion is

$$y=f(x^0)+\frac{dy}{dx}\bigg|_{x^0}\Delta x+\frac{1}{2!}\frac{d^2y}{dx^2}\bigg|_{x^0}(\Delta x)^2+\cdots \tag{B.59}$$

in which $x^0$ is the approximate value of the variable at which the function is evaluated. The first term on the right-hand side of Eq. (B.59) is the zero-order term, which is equal to the value of the function evaluated at $x=x^0$; the second term is the first-order term, which contains the first derivative evaluated at $x=x^0$; the third term is the second-order term, which includes the second derivative evaluated at $x=x^0$; and so on. Thus, in order to have a linear form, only the zero- and first-order terms are used from the series expansion. If $y=f(x_1,x_2)$ is a nonlinear function of the two variables $x_1,x_2$, the linearized form is

$$y=f(x_1^0,x_2^0)+\frac{\partial y}{\partial x_1}\bigg|_{x_1^0,x_2^0}\Delta x_1+\frac{\partial y}{\partial x_2}\bigg|_{x_1^0,x_2^0}\Delta x_2$$

$$=f(x_1^0,x_2^0)+j_1\Delta x_1+j_2\Delta x_2$$

$$=y^0+[j_1\ j_2]\begin{bmatrix}\Delta x_1\\\Delta x_2\end{bmatrix}$$

or

$$y=y^0+\mathbf{J}_{yx}\Delta\mathbf{x} \tag{B.60}$$

in which $\mathbf{J}_{yx}$ is the jacobian matrix containing the partial derivatives of $y$ with respect to $x_1$ and $x_2$, respectively, as described in Art. A.10. The relation in Eq. (B.60) can be generalized to the case of $m$ functions $\mathbf{y}$, each in terms of some or all of $n$ variables $\mathbf{x}$ [i.e., $\mathbf{y}=f(\mathbf{x})$]; thus,

$$\mathbf{y}=\mathbf{y}^0+\mathbf{J}_{yx}\Delta\mathbf{x} \tag{B.61}$$

where

$$\mathbf{y}^0=\begin{bmatrix}y_1^0\\y_2^0\\\vdots\\y_m^0\end{bmatrix}=\begin{bmatrix}f_1(x_1^0,x_2^0,\ldots,x_n^0)\\f_2(x_1^0,x_2^0,\ldots,x_n^0)\\\cdots\cdots\cdots\\f_m(x_1^0,x_2^0,\ldots,x_n^0)\end{bmatrix}\qquad \Delta\mathbf{x}=\begin{bmatrix}\Delta x_1\\\Delta x_2\\\vdots\\\Delta x_n\end{bmatrix}$$

$$\mathbf{J}_{yx}=\frac{\partial\mathbf{y}}{\partial\mathbf{x}}=\begin{bmatrix}\dfrac{\partial y_1}{\partial x_1}&\dfrac{\partial y_1}{\partial x_2}&\cdots&\dfrac{\partial y_1}{\partial x_n}\\[2mm]\dfrac{\partial y_2}{\partial x_1}&\dfrac{\partial y_2}{\partial x_2}&\cdots&\dfrac{\partial y_2}{\partial x_n}\\[1mm]\cdots&\cdots&\cdots&\cdots\\[1mm]\dfrac{\partial y_m}{\partial x_1}&\dfrac{\partial y_m}{\partial x_2}&\cdots&\dfrac{\partial y_m}{\partial x_n}\end{bmatrix}$$

evaluated at $\mathbf{x}=\mathbf{x}^0$.

**Fig. B.4**

**B.10. Adjustment of various survey problems**  In the preceding sections of this appendix the tools necessary for the least squares adjustment of any survey problem have been presented. However, some problems may be solved using either of the two techniques derived, while for other problems it is considerably easier to use one of the two techniques. In the following sections the least squares adjustment of level nets, resection, intersection, traverse, triangulation, and trilateration are discussed. In each case, the smallest element or unit is analyzed and the appropriate condition equations derived. This is followed by a suitable numerical example to demonstrate the techniques.

**B.11. Adjustment of level nets**  Adjustment of a level net is very straightforward and is ideal for use in the introductory examples leading to the derivation of the least squares techniques. Both techniques as used in level-net adjustments are summarized.

Let $\ell_{ij}$ represent the measured difference in elevation between two points $i$ and $j$ and in the particular direction $i$ to $j$ (see Fig. B.4). If $x_i$ and $x_j$ represent the elevations of point $i,j$, respectively (either or both known or unknown), *the condition equation for the adjustment of indirect observations* is

$$x_i + \ell_{ij} + v_{ij} = x_j$$

or

$$v_{ij} + x_i - x_j = -\ell_{ij} \tag{B.62}$$

in which $x_i$, $x_j$ are the parameters.

In the case of *adjustment of observations only*, the elevations of the different points do not appear in the condition equations. For every closed *loop* composed of points for which elevations are to be determined (see Fig. B.5), one condition equation is written. For example, going from point $i$ to $j$ to $k$ and back to $i$ (Fig. B.5), the sum of the differences in elevation must add up to zero. Noting that the arrow implies the direction in which the elevation increases from a lower point to a higher point (by convention) the condition is

$$\ell_{ij} + v_{ij} - \ell_{jk} - v_{jk} + \ell_{ki} + v_{ki} = 0$$

or

$$v_{ij} - v_{jk} + v_{ki} = -\ell_{ij} + \ell_{jk} - \ell_{ki} \tag{B.63}$$

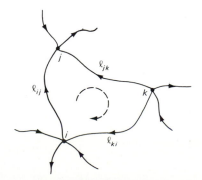

**Fig. B.5**

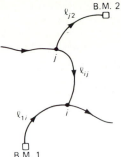

**Fig. B.6**   B.M. 1

In some cases the loop is not closed on the original bench mark, but rather begins and ends at bench marks for which the elevations are known. Let $x_1$ and $x_2$ represent the elevations of the two bench marks in Fig. B.6. The condition equation for the adjustment of observations only is

$$x_1 + \ell_{1i} + v_{1i} - \ell_{ij} - v_{ij} + \ell_{j2} + v_{j2} = x_2$$

or

$$v_{1i} - v_{ij} + v_{j2} = x_2 - \ell_{1i} + \ell_{ij} - \ell_{j2} - x_1 \qquad (B.64)$$

**Example B.7.   Adjustment of a level net**   Figure B.7 shows a simple level net which contains two known bench marks $A$ and $D$, whose elevations are 237.15 ft and 246.05 ft, respectively. The method of least squares is to be used to determine the adjusted elevations of points $B$, $C$, and $E$. Lengths and differences in elevations of each line are indicated on the figure (with the arrows showing the direction in which levels are run) and are summarized in the following tabulation:

| From | To | Length, mi | Observation | Difference in elevation, ft | Weight |
|---|---|---|---|---|---|
| A | B | 15 | $\ell_1$ | − 22.93 | $w_1 = \frac{1}{15}$ |
| B | C | 12 | $\ell_2$ | + 10.94 | $w_2 = \frac{1}{12}$ |
| C | D | 28 | $\ell_3$ | + 21.04 | $w_3 = \frac{1}{28}$ |
| D | A | 26 | $\ell_4$ | − 8.92 | $w_4 = \frac{1}{26}$ |
| E | B | 17 | $\ell_5$ | − 5.23 | $w_5 = \frac{1}{17}$ |
| E | A | 11 | $\ell_6$ | + 17.91 | $w_6 = \frac{1}{11}$ |
| D | E | 13 | $\ell_7$ | − 27.15 | $w_7 = \frac{1}{13}$ |

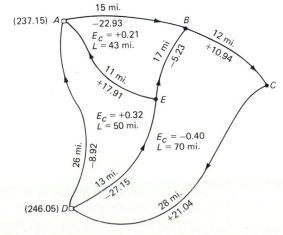

**Fig. B.7**

The seven observations are assumed to be *uncorrelated*. Because it was not possible to determine variances, each observation is assumed to have an empirical weight that is inversely proportional to the length of line in miles. Under the assumption of lack of correlation, the weights are computed directly as given in the last column in the table of data given above. The weight matrix is then the following diagonal matrix, containing the elements in that last column:

$$W = \begin{bmatrix} \frac{1}{15} & & & & & & \mathbf{0} \\ & \frac{1}{12} & & & & & \\ & & \frac{1}{28} & & & & \\ & & & \frac{1}{26} & & & \\ & & & & \frac{1}{17} & & \\ & & & & & \frac{1}{11} & \\ \mathbf{0} & & & & & & \frac{1}{13} \end{bmatrix}$$

*Adjustment by the method of indirect observations*—Denoting the elevations of points $B$, $C$, and $E$ by $x_1$, $x_2$, and $x_3$, respectively, the seven condition equations according to Eq. (B.62) are

$$
\begin{aligned}
v_1 + 237.15 - x_1 &= -\ell_1 & \text{or} \quad v_1 - x_1 &= -214.22 \\
v_2 + x_1 - x_2 &= -\ell_2 & \text{or} \quad v_2 + x_1 - x_2 &= -10.94 \\
v_3 + x_2 - 246.05 &= -\ell_3 & \text{or} \quad v_3 + x_2 &= 225.01 \\
v_4 + 246.05 - 237.15 &= -\ell_4 & \text{or} \quad v_4 &= 0.02 \quad \text{(ft)} \\
v_5 + x_3 - x_1 &= -\ell_5 & \text{or} \quad v_5 - x_1 + x_3 &= 5.23 \\
v_6 + x_3 - 237.15 &= -\ell_6 & \text{or} \quad v_6 + x_3 &= 219.24 \\
v_7 + 246.05 - x_3 &= -\ell_7 & \text{or} \quad v_7 - x_3 &= -218.90
\end{aligned}
$$

which in matrix form become

$$
\begin{bmatrix} v_1 \\ v_2 \\ v_3 \\ v_4 \\ v_5 \\ v_6 \\ v_7 \end{bmatrix} + \begin{bmatrix} -1 & 0 & 0 \\ 1 & -1 & 0 \\ 0 & 1 & 0 \\ 0 & 0 & 0 \\ -1 & 0 & 1 \\ 0 & 0 & 1 \\ 0 & 0 & -1 \end{bmatrix} \begin{bmatrix} x_1 \\ x_2 \\ x_3 \end{bmatrix} = \begin{bmatrix} -214.22 \\ -10.94 \\ 225.01 \\ 0.02 \\ 5.23 \\ 219.24 \\ -218.90 \end{bmatrix} \quad \text{(ft)}
$$

which is of the general form $v + B\Delta = f$.

With the given diagonal weight matrix, the normal equations are

$$(B^t W B)\Delta = (B^t W f)$$

or

$$
\begin{bmatrix} 0.209 & -0.083 & -0.059 \\ -0.083 & 0.119 & 0 \\ -0.059 & 0 & 0.227 \end{bmatrix} \begin{bmatrix} x_1 \\ x_2 \\ x_3 \end{bmatrix} = \begin{bmatrix} 13.062 \\ 8.948 \\ 37.077 \end{bmatrix}
$$

The solution of these equations yields the three adjusted elevations, or

$\hat{x}_1 = $ elevation of point $B = 214.07$ ft
$\hat{x}_2 = $ elevation of point $C = 225.01$ ft
$\hat{x}_3 = $ elevation of point $E = 219.14$ ft

For purposes of comparison with the following procedure of adjustment, the vector of residuals is computed from $v = f - B\Delta$ and the adjusted observations from $\hat{\ell} = \ell + v$ [see Eqs. (B.11) and (B.40a)].

Thus,

$$\mathbf{v} = \begin{bmatrix} -0.146 \\ -0.001 \\ -0.003 \\ 0.020 \\ 0.163 \\ 0.099 \\ 0.241 \end{bmatrix} \quad \text{and} \quad \hat{\ell} = \begin{bmatrix} -23.076 \\ 10.939 \\ 21.037 \\ -8.900 \\ -5.067 \\ 18.009 \\ -26.909 \end{bmatrix} \quad \text{(ft)}$$

According to Eq. (B.38), the cofactor matrix for the adjusted elevations of the three points is

$$\mathbf{Q}_{\Delta\Delta} = \mathbf{Q}_{xx} = (\mathbf{B}^t\mathbf{W}\mathbf{B})^{-1} = \mathbf{N}^{-1} = \begin{bmatrix} 7.395 & 5.176 & 1.919 \\ & 12.024 & 1.343 \\ \text{symmetric} & & 4.910 \end{bmatrix}$$

Since only relative precisions for the observations were given in the form of weights, evaluation of the a posteriori reference variance, $\hat{\sigma}_0^2$, is necessary in order to calculate the corresponding covariance matrix $\Sigma_{xx}$. Thus, from Eq. (B.41),

$$\hat{\sigma}_0^2 = \frac{\mathbf{v}^t\mathbf{W}\mathbf{v}}{r} = \frac{0.00836}{7-3} = 0.0021 \ \text{ft}^2$$

in which $r = n - n_0$ by Eq. (B.3). Finally, according to Eq. (B.42), the covariance matrix for adjusted observations is

$$\Sigma_{xx} = \hat{\sigma}_0^2 \mathbf{Q}_{xx} = \begin{bmatrix} 0.0155 & 0.0108 & 0.0040 \\ & 0.0251 & 0.0028 \\ \text{symmetric} & & 0.0103 \end{bmatrix} \quad \text{(ft}^2\text{)}$$

*Adjustment of observations only*   Since there are seven observations and only a minimum of three is required for a unique solution, the redundancy is [see Eq. (B.3)]

$$r = n - n_0 = 7 - 3 = 4$$

Thus, there exist four condition equations that relate the observations together. In a manner similar to Eq. (B.63), the following four conditions are written:

Loop $A \rightarrow B \rightarrow E \rightarrow A$: $\quad \ell_1 + v_1 - \ell_5 - v_5 + \ell_6 + v_6 = 0$

Loop $B \rightarrow C \rightarrow D \rightarrow E \rightarrow B$: $\quad \ell_2 + v_2 + \ell_3 + v_3 + \ell_7 + v_7 + \ell_5 + v_5 = 0$

Loop $A \rightarrow E \rightarrow D \rightarrow A$: $\quad -\ell_6 - v_6 - \ell_7 - v_7 + \ell_4 + v_4 = 0$

Loop $A \rightarrow B \rightarrow C \rightarrow D$: $\quad 237.15 + \ell_1 + v_1 + \ell_2 + v_2 + \ell_3 + v_3 - 246.05 = 0$

Thus,

$$v_1 - v_5 + v_6 = -\ell_1 + \ell_5 - \ell_6 = -(-22.93) + (-5.23) - 17.91 = -0.21$$
$$v_2 + v_3 + v_5 + v_7 = -\ell_2 - \ell_3 - \ell_5 - \ell_7 = -10.94 - 21.04 + 5.23 + 27.15 = 0.40$$
$$v_4 - v_6 - v_7 = -\ell_4 + \ell_6 + \ell_7 = 8.92 + 17.91 - 27.15 = -0.32$$
$$v_1 + v_2 + v_3 = 8.9 - \ell_1 - \ell_2 - \ell_3 = 8.9 + 22.93 - 10.94 - 21.04 = -0.15$$

or

$$\begin{bmatrix} 1 & 0 & 0 & 0 & -1 & 1 & 0 \\ 0 & 1 & 1 & 0 & 1 & 0 & 1 \\ 0 & 0 & 0 & 1 & 0 & -1 & -1 \\ 1 & 1 & 1 & 0 & 0 & 0 & 0 \end{bmatrix} \begin{bmatrix} v_1 \\ v_2 \\ \vdots \\ v_7 \end{bmatrix} = \begin{bmatrix} -0.21 \\ 0.40 \\ -0.32 \\ -0.15 \end{bmatrix} \quad \text{(ft)}$$

which is of the form $\mathbf{Av} = \mathbf{f}$. Applying Eq. (B.50a), we get $(\mathbf{AQA}^t)\mathbf{k} = \mathbf{f}$, in which $\mathbf{Q} = \mathbf{W}^{-1}$, or

$$\mathbf{Q} = \begin{bmatrix} 15 & & & & & & 0 \\ & 12 & & & & & \\ & & 28 & & & & \\ & & & 26 & & & \\ & & & & 17 & & \\ & & & & & 11 & \\ 0 & & & & & & 13 \end{bmatrix}$$

Then

$$
\begin{bmatrix}
43 & -17 & -11 & 15 \\
 & 70 & -13 & 40 \\
 & & 50 & 0 \\
\text{symmetric} & & & 55
\end{bmatrix}
\begin{bmatrix}
k_1 \\ k_2 \\ k_3 \\ k_4
\end{bmatrix}
=
\begin{bmatrix}
-0.21 \\ 0.40 \\ -0.32 \\ -0.15
\end{bmatrix}
\quad \text{(ft)}
$$

and from Eqs. (B.49) and (B.50b) the residuals are

$$
v = QA^t(AQA^t)^{-1}f =
\begin{bmatrix}
-0.146 \\
-0.001 \\
-0.003 \\
0.020 \\
0.163 \\
0.099 \\
0.241
\end{bmatrix}
\quad \text{and} \quad
\hat{\ell} = \ell + v =
\begin{bmatrix}
-23.076 \\
10.939 \\
21.037 \\
-8.900 \\
-5.067 \\
18.009 \\
-26.909
\end{bmatrix}
\quad \text{(ft)}
$$

Both $v$ and $\hat{\ell}$ are identical to those computed by the preceding technique.

Let the adjusted elevations of points $B$, $C$, and $E$ be designated by $\hat{X}^t = [x_B \; x_C \; x_E]$ and known elevations of points $A$ and $D$ be indicated by $X_0^t = [x_A \; x_A \; x_D]$. Thus, adjusted elevations are (referring to Fig. B.7)

$$
\begin{bmatrix} x_B \\ x_C \\ x_E \end{bmatrix}
=
\begin{bmatrix} x_A \\ x_A \\ x_D \end{bmatrix}
+
\begin{bmatrix}
1 & 0 & 0 & 0 & 0 & 0 & 0 \\
1 & 1 & 0 & 0 & 0 & 0 & 0 \\
0 & 0 & 0 & 0 & 0 & 0 & 1
\end{bmatrix}
\begin{bmatrix}
\hat{\ell}_1 \\ \hat{\ell}_2 \\ \hat{\ell}_3 \\ \hat{\ell}_4 \\ \hat{\ell}_5 \\ \hat{\ell}_6 \\ \hat{\ell}_7
\end{bmatrix}
$$

which can be written more compactly as

$$
\hat{X} = X_0 + J\hat{\ell}
$$

Substitution of known elevations and adjusted differences in elevations into the equations above yields the following adjusted elevations:

$$
\begin{aligned}
x_A &= x_A + \hat{\ell}_1 = 237.15 - 23.076 & &= 214.07 \text{ ft} \\
x_C &= x_A + \hat{\ell}_1 + \hat{\ell}_2 = 237.15 - 23.076 + 10.939 &&= 225.01 \text{ ft} \\
x_E &= x_D + \hat{\ell}_7 = 246.05 - 26.909 & &= 219.14 \text{ ft}
\end{aligned}
$$

Obviously, results from both techniques of least squares adjustment are identical.

As indicated above, adjusted elevations are functions of the adjusted measurements. Consequently, to determine the cofactor matrix for adjusted elevations, it is first necessary to evaluate the cofactor matrix for the adjusted measurements, $Q_{\hat{\ell}\hat{\ell}}$. According to Eqs. (B.57) and (B.58),

$$
Q_{\hat{\ell}\hat{\ell}} = Q - QA^t W_e AQ = Q - Q_{vv}
$$

in which $W_e = (AQA^t)^{-1}$. For this problem,

$$
Q_w =
\begin{bmatrix}
7.6049 & 2.2185 & 5.1766 & 0.0 & -5.4759 & 1.9193 & -1.9193 \\
 & 2.9344 & 6.8470 & 0.0 & 1.6428 & -0.5758 & 0.5758 \\
 & & 15.9764 & 0.0 & 3.8331 & -1.3435 & 1.3435 \\
 & & & 26.0000 & 0.0 & 0.0 & 0.0 \\
 & & & & 8.5333 & -2.9908 & 2.9908 \\
 & & & & & 6.0899 & 4.9101 \\
\text{symmetric} & & & & & & 8.0899
\end{bmatrix}
\quad \text{(ft}^2\text{)}
$$

and

$$Q_{\tilde{t}\tilde{t}}= \begin{bmatrix} 7.3951 & -2.2185 & -5.1766 & 0.0 & 5.4759 & -1.9193 & 1.9193 \\ & 9.0656 & -6.8470 & 0.0 & -1.6428 & 0.5758 & -0.5758 \\ & & 12.0236 & 0.0 & -3.8331 & 1.3435 & -1.3435 \\ & & & (6)10^{-13} & 0. & 0. & 0. \\ & & & & 8.4667 & 2.9908 & -2.9908 \\ & & & & & 4.9101 & -4.9101 \\ \text{symmetric} & & & & & & 4.9101 \end{bmatrix} \text{(ft}^2)$$

Application of error propagation equation (2.43), Art. 2.20, to the equation used to evaluate adjusted elevations yields

$$Q_{xx}=JQJ^t$$

or the cofactor matrix for adjusted elevations is

$$Q_{xx}= \begin{bmatrix} 7.3952 & 5.1766 & 1.9193 \\ & 12.0236 & 1.3435 \\ \text{symmetric} & & 4.9101 \end{bmatrix}$$

Using Eq. (B.41), the a posteriori reference variance is

$$\hat{\sigma}_0^2 = 0.002090 \text{ ft}^2$$

so that the propagated covariance matrix for adjusted elevations [Eq. (B.42)] is

$$\Sigma_{xx}=\sigma_0^2 Q_{xx}= \begin{bmatrix} 0.0155 & 0.0108 & 0.0040 \\ & 0.0251 & 0.0028 \\ \text{symmetric} & & 0.0103 \end{bmatrix} \text{(ft}^2)$$

Note that the elements in this matrix are identical to the corresponding elements in the covariance matrix propagated in the adjustment of this same level net by indirect observations.

**B.12. Resection adjustment**    Resection is the operation of determining the position of an unknown point from angular measurements made at the point and sighting at known points (refer to Arts. 9.5 to 9.7). From the geometric standpoint, a minimum of two angles, such as $\alpha, \beta$ in Fig. B.8, are necessary for a unique determination of the unknown point. Each additional observation contributes one redundancy to the adjustment. Consequently, it is considerably easier to *use the least squares adjustment of indirect observations for resection problems*. In order to derive the condition equation for the angle $\ell_{ij}$ in Fig. B.8, first write the two azimuth condition equations for the azimuths (from the north) $A_i$ and $A_j$;

$$A_i = \arctan \frac{X_i - X_P}{Y_i - Y_P} \tag{B.65}$$

$$A_j = \arctan \frac{X_j - X_P}{Y_j - Y_P} \tag{B.66}$$

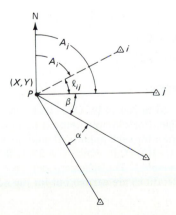

**Fig. B.8**

Then, the *angle condition* equation for the angle $\ell_{ij}$ is

$$v_{ij} + \ell_{ij} = A_j - A_i = \arctan\frac{X_j - X_P}{Y_j - Y_P} - \arctan\frac{X_i - X_P}{Y_i - Y_P}$$

or

$$v_{ij} + \ell_{ij} + \arctan\frac{X_i - X_P}{Y_i - Y_P} - \arctan\frac{X_j - X_P}{Y_j - Y_P} = 0 \tag{B.67}$$

Equation (B.67) is obviously nonlinear in the parameters and must be linearized. Recall that

$$\frac{\partial}{\partial x}\arctan u = \frac{1}{1+u^2}\frac{\partial u}{\partial x} \tag{B.68}$$

and then write Eq. (B.67) in functional form:

$$v_{ij} + \ell_{ij} + F_{ij}(X_P, Y_P, X_i, Y_i, X_j, Y_j) = 0 \tag{B.69}$$

Thus, the linearized form of the angle condition equation is

$$v_{ij} + \ell_{ij} + F_{ij}(X_P, Y_P, X_i, Y_i, X_j, Y_j)^0 + \frac{\partial F_{ij}}{\partial X_P}\delta X_P + \frac{\partial F_{ij}}{\partial Y_P}\delta Y_P$$

$$+ \frac{\partial F_{ij}}{\partial X_i}\delta X_i + \frac{\partial F_{ij}}{\partial Y_i}\delta Y_i + \frac{\partial F_{ij}}{\partial X_j}\delta X_j + \frac{\partial F_{ij}}{\partial Y_j}\delta Y_j = 0$$

or

$$v_{ij} + b_1\delta X_P + b_2\delta Y_P + b_3\delta X_i + b_4\delta Y_i + b_5\delta X_j + b_6\delta Y_j$$
$$= -\ell_{ij} - F_{ij}\big[(X,Y)^0_P, (X,Y)^0_i, (X,Y)^0_j\big] = f_{ij} \tag{B.70}$$

Equation (B.67) is for the case where coordinates for the three points $P$, $i$, and $j$ are unknown (Fig. B.8). This general case was *derived* for use later in the adjustment of triangulation (Example B.13, Art. B.15). For the special case of resection, coordinates for points $i$ and $j$ are known so that $b_3 = b_4 = b_5 = b_6 = 0$ and the angle equation for each measured angle contains only two unknown corrections, $\delta X_P$ and $\delta Y_P$. Thus, the linearized angle condition equation for resection is

$$v_{ij} + b_1\delta X_P + b_2\delta Y_P = -\ell_{ij} - F_{ij}(X^0, Y^0)_P = f_{ij} \tag{B.71a}$$

in which

$$b_1 = \frac{\partial F_{ij}}{\partial X_P} = \frac{-(Y_i - Y_P^0)}{(Y_i - Y_P^0)^2 + (X_i - X_P^0)^2} - \frac{-(Y_j - Y_P^0)}{(Y_j - Y_P^0)^2 + (X_j - X_P^0)^2} \tag{B.71b}$$

$$b_2 = \frac{\partial F_{ij}}{\partial Y_P} = \frac{X_i - X_P^0}{(Y_i - Y_P^0)^2 + (X_i - X_P^0)^2} - \frac{X_j - X_P^0}{(Y_j - Y_P^0)^2 + (X_j - X_P^0)^2} \tag{B.71c}$$

$$f_{ij} = -\ell_{ij} - \arctan\frac{X_i - X_P^0}{Y_i - Y_P^0} + \arctan\frac{X_j - X_P^0}{Y_j - Y_P^0} \tag{B.71d}$$

or

$$f_{ij} = (\text{calculated angle}) - (\text{measured angle}).$$

The remainder of the coefficients $b_3, \ldots, b_6$ are derived in Art. B.15. One condition equation of the type (B.71a) is written for each observed angle. For example, three conditions would be written for the case given in Fig. B.8: one condition for each angle $\alpha$, $\beta$, and $\ell_{ij}$. The total set of linearized conditions would be of the form $\mathbf{v} + \mathbf{B\Delta} = \mathbf{f}$, in which $\mathbf{v}$ is $3\times1$, $\mathbf{B}$ is $3\times2$, and $\mathbf{\Delta}$ is $2\times1$, which contains $\delta X_P$ and $\delta Y_P$ as parameters. The scheme is demonstrated by the following example, which also shows how the iterations are carried out for the nonlinear problem.

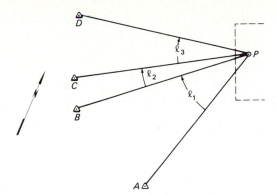

**Fig. B.9**

**Example B.8   Problem on resection adjustment**   Figure B.9 shows a sketch for the resection of the position of point $P$ from three observed angles $\ell_1$, $\ell_2$, $\ell_3$, to four known control points $A, B, C, D$, whose coordinates are as follows:

| Station | California state plane coordinates, Zone III | |
| --- | --- | --- |
| | X, ft | Y, ft |
| A   Flagpole on top of Oakland City Hall | 88,237.92 | 80,132.03 |
| B   Top of Durkee tank | 82,279.10 | 97,418.58 |
| C   Top of Peet Bros. stack | 81,802.35 | 98,696.21 |
| D   Top of Spenger's Fish Grotto tank | 80,330.69 | 102,911.40 |

The observed directions (mean of two positions), their standard deviations, and the corresponding angles are as follows:

| Instrument at point $P$ (mean of two positions): abstract of directions | $\sigma_D$ | Average angle |
| --- | --- | --- |
| A     0°00'00" | 5" | |
| | | $\ell_1 = 44°55'30.5"$ |
| B   44°55'30.5" | 5" | |
| | | $\ell_2 = 5°56'19.4"$ |
| C   50°51'49.9" | 5" | |
| | | $\ell_3 = 19°56'18.3"$ |
| D   70°48'08.2" | 5" | |

Compute the coordinates $X, Y$ of the unknown point $P$.

SOLUTION   Because the angles are computed from directions, which are assumed to be uncorrelated, their covariance matrix must be computed from error propagation. Thus,

$$\ell_1 = D_{PB} - D_{PA}$$
$$\ell_2 = D_{PC} - D_{PB}$$
$$\ell_3 = D_{PD} - D_{PC}$$

which in matrix form become

$$\begin{bmatrix} \ell_1 \\ \ell_2 \\ \ell_3 \end{bmatrix} = \begin{bmatrix} -1 & 1 & 0 & 0 \\ 0 & -1 & 1 & 0 \\ 0 & 0 & -1 & 1 \end{bmatrix} \begin{bmatrix} D_{PA} \\ D_{PB} \\ D_{PC} \\ D_{PD} \end{bmatrix} \qquad \text{or} \qquad \boldsymbol{\ell} = \mathbf{JD}$$

Since the standard deviation of each direction is 5″ or the variance is 25 seconds$^2$ or 5.876107 × $10^{-10}$ rad$^2$, the variance matrix of the directions is $\Sigma_{DD}=(5.876107)10^{-10}I$. Then the covariance matrix of the angles is [by Eq. (2.42)]

$$\Sigma_{tt}=J\Sigma_{DD}J^t=(5.876107)10^{-10}\begin{bmatrix} -1 & 1 & 0 & 0 \\ 0 & -1 & 1 & 0 \\ 0 & 0 & -1 & 1 \end{bmatrix}\begin{bmatrix} -1 & 0 & 0 \\ 1 & -1 & 0 \\ 0 & 1 & -1 \\ 0 & 0 & 1 \end{bmatrix}\ (\text{rad}^2)$$

$$=(5.876107)10^{-10}\begin{bmatrix} 2 & -1 & 0 \\ -1 & 2 & -1 \\ 0 & -1 & 2 \end{bmatrix}\ (\text{rad}^2)$$

The three angle condition equations (B.67) are

$$F_1=\ell_1-A_{PB}+A_{PA}=\ell_1-\arctan\frac{X_B-X}{Y_B-Y}+\arctan\frac{X_A-X}{Y_A-Y}=0$$

$$F_2=\ell_2-A_{PC}+A_{PB}=\ell_2-\arctan\frac{X_C-X}{Y_C-Y}+\arctan\frac{X_B-X}{Y_B-Y}=0$$

$$F_3=\ell_3-A_{PD}+A_{PC}=\ell_3-\arctan\frac{X_D-X}{Y_D-Y}+\arctan\frac{X_C-X}{Y_C-Y}=0$$

The linearized equations according to the form of Eq. (B.71a) for the adjustment are:

$$v_1+b_{11}\,\delta X+b_{12}\,\delta Y=f_1$$
$$v_2+b_{21}\,\delta X+b_{22}\,\delta Y=f_2$$
$$v_3+b_{31}\,\delta X+b_{32}\,\delta Y=f_3$$

which may be expressed in matrix form as

$$v+B\Delta=f$$

Denoting the approximations for unknown parameters by $X^0$, $Y^0$, the coefficients and constant terms are

$$b_{11}=\frac{\partial F_1}{\partial X}=-\frac{Y_A-Y^0}{L_{AP}^{0^2}}+\frac{Y_B-Y^0}{L_{BP}^{0^2}}$$

$$b_{12}=\frac{\partial F_1}{\partial Y}=\frac{X_A-X^0}{L_{AP}^{0^2}}-\frac{X_B-X^0}{L_{BP}^{0^2}}$$

$$b_{21}=\frac{\partial F_2}{\partial X}=-\frac{Y_B-Y^0}{L_{BP}^{0^2}}+\frac{Y_C-Y^0}{L_{CP}^{0^2}}$$

$$b_{22}=\frac{\partial F_2}{\partial Y}=\frac{X_B-X^0}{L_{BP}^{0^2}}-\frac{X_C-X^0}{L_{CP}^{0^2}}$$

$$b_{31}=\frac{\partial F_3}{\partial X}=-\frac{Y_C-Y^0}{L_{CP}^{0^2}}+\frac{Y_D-Y^0}{L_{DP}^{0^2}}$$

$$b_{32}=\frac{\partial F_3}{\partial Y}=\frac{X_C-X^0}{L_{CP}^{0^2}}-\frac{X_D-X^0}{L_{DP}^{0^2}}$$

$$f_1=-\ell_1+\arctan\frac{X_B-X^0}{Y_B-Y^0}-\arctan\frac{X_A-X^0}{Y_A-Y^0}$$

$$f_2=-\ell_2+\arctan\frac{X_C-X^0}{Y_C-Y^0}-\arctan\frac{X_B-X^0}{Y_B-Y^0}$$

$$f_3=-\ell_3+\arctan\frac{X_D-X^0}{Y_D-Y^0}-\arctan\frac{X_C-X^0}{Y_C-Y^0}$$

where $L_{AP}^{0^2}=(X_A-X^0)^2+(Y_A-Y^0)^2$
$\phantom{where}L_{BP}^{0^2}=(X_B-X^0)^2+(Y_B-Y^0)^2$
$\phantom{where}L_{CP}^{0^2}=(X_C-X^0)^2+(Y_C-Y^0)^2$
$\phantom{where}L_{DP}^{0^2}=(X_D-X^0)^2+(Y_D-Y^0)^2$

The value of the azimuth angle computed from the arctan function in evaluating the elements of the f vector will be correct only if that function is properly derived in the computer. If, however, the value given is less than 90° in magnitude, with either a positive or negative sign, the azimuth angle should be determined according to the signs of both the numerator and denominator of the arctan function. Denoting the numerator by $\Delta X$ and denominator by $\Delta Y$, the azimuth A (from the north) is given as follows (refer also to Example B.10, Art. B.14):

(a) $+\Delta X, +\Delta Y$    $A = \arctan \dfrac{\Delta X}{\Delta Y}$

(b) $+\Delta X, -\Delta Y$    $A = \pi - \left| \arctan \dfrac{\Delta X}{\Delta Y} \right|$

(c) $-\Delta X, -\Delta Y$    $A = \pi + \left| \arctan \dfrac{\Delta X}{\Delta Y} \right|$

(d) $-\Delta X, +\Delta Y$    $A = 2\pi - \left| \arctan \dfrac{\Delta X}{\Delta Y} \right|$

We repeat, however, that if a given computer yields the azimuth directly as a positive angle between 0° and 360°, the foregoing relations become unnecessary to program.

Using the approximations $X^0 = 93{,}600$ ft, $Y^0 = 104{,}000$ ft as determined from the geometry of the figure, the numerical values of the B and f matrices are

$$\mathbf{B} = 10^{-6} \begin{bmatrix} 1.50 & 57.06 \\ 6.68 & 4.49 \\ 25.56 & 4.35 \end{bmatrix} \quad \text{and} \quad \mathbf{f} = 10^{-3} \begin{bmatrix} 39.12349 \\ 0.45342 \\ -7.35596 \end{bmatrix}$$

Using the covariance matrix $\Sigma_{\ell\ell}$ evaluated above, the normal equations are formed as $(\mathbf{B}'\Sigma_{\ell\ell}^{-1}\mathbf{B})\Delta_1 = \mathbf{B}'\Sigma_{\ell\ell}^{-1}\mathbf{f}$ and solved to yield

$$\Delta_1 = \begin{bmatrix} -405.436 \text{ ft} \\ 696.625 \text{ ft} \end{bmatrix}$$

The updated approximations for the parameters become $93{,}600 - 405.436 = 93{,}194.564$ and $104{,}000 + 696.625 = 104{,}626.625$. Re-forming and solving the normal equations gives the correction vector:

$$\Delta_2 = \begin{bmatrix} -41.072 \text{ ft} \\ -11.353 \text{ ft} \end{bmatrix}$$

Continuing the process to four iterations gives a $\Delta_4$ which is essentially zero, and the final least squares estimates for the coordinates of point P are

$\hat{X} = 93{,}153.645$ ft    and    $\hat{Y} = 104{,}685.246$ ft

**B.13. Intersection adjustment**    Intersection is the operation of determining the position of an unknown point using observed angles taken at known points (refer to Arts. 9.1 to 9.4). As shown in Fig. B.10, a unique determination of the unknown point P requires a minimum of two angles, such as $\alpha$ and $\beta$. Each additional *measured angle* contributes one redundancy to the adjustment. On the other hand, when one additional control point, such as i in Fig. B.10, is used, two angles $\ell_i$ and $\ell_i'$ are added. For each of the measured angles one condition

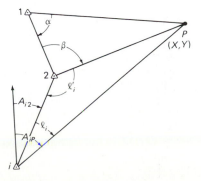

**Fig. B.10**

equation may be written, (B.67). For example, the condition equation for the angle $\ell_i$ in Fig. B.10 is

$$v_i + \ell_i = A_{iP} - A_{i2} = \arctan\frac{X_P - X_i}{Y_P - Y_i} - \arctan\frac{X_2 - X_i}{Y_2 - Y_i}$$

or

$$v_i + \ell_i - \arctan\frac{X_P - X_i}{Y_P - Y_i} + \arctan\frac{X_2 - X_i}{Y_2 - Y_i} = 0 \tag{B.72}$$

The linearization of Eq. (B.72) is very similar to that of Eq. (B.67) and is not included here. However, note that for this special case, the stations occupied and sighted have known coordinates, while the station to which the angle is turned is of unknown position. This is the reverse of the resection. In linearizing Eq. (B.72), partials are taken with respect to the coordinates of the station to which the angle is turned ($X_P, Y_P$ in Fig. B.10) and the resulting coefficients are not the same as $b_1$ and $b_2$ derived in Eqs. (B.71b) and (B.71c). Coefficients for the intersection adjustment are derived in Example B.9.

Since the angle condition is written as the difference between two azimuths, the sign convention is very important and must be carefully observed. Thus, if the azimuth is defined as *clockwise* from north, the measured angles must also be defined as *clockwise* from the first line encountered from the north to the second line. Figure B.11 depicts four different cases. In Fig. B.11a the azimuths $A_a$ and $A_b$ are both in the first quadrant and $\ell$ is easily seen to be equal to $(A_b - A_a)$. In Fig. B.11b the azimuth $A_a$ is in the first quadrant while $A_b$ is in the second quadrant because $A_b = \arctan[(X_b - X_0)/(Y_b - Y_0)]$ and $(X_b - X_0)$ is positive while $(Y_b - Y_0)$ is negative. [Note that the sign of $(X_b - X_0)$ corresponds to the sign of $\sin A_b$, while that for $(Y_b - Y_0)$ corresponds to the sign of $\cos A_b$.] Here, too, $\ell = A_b - A_a$. In Fig. B.11c the azimuth $A_b$ is in the third quadrant and $\ell = A_b - A_a$. In Fig. B.11d the azimuth $A_b$ is in the

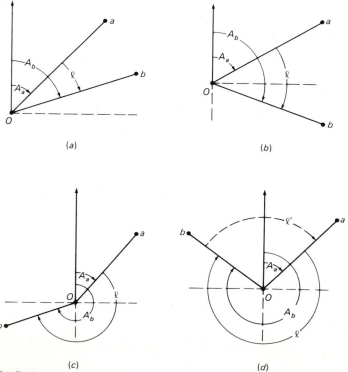

(a)          (b)

(c)          (d)

**Fig. B.11**

fourth quadrant. There are two possibilities for the observed angle, either $\ell$ or $\ell'$ (see Fig. B.11d). If $\ell$ is given as the clockwise angle from $Oa$ to $Ob$, it will be directly equal to $(A_b - A_a)$. On the other hand, if $\ell'$ is given as the clockwise angle from $Ob$ to $Oa$, then $(A_b - A_a)$ will actually be $(\ell' - 360°)$, which is a negative value. Therefore, whenever the difference between two directions is negative, 360° should be added to yield the observed angle.

**Example B.9     Intersection adjustment problem**   Figure B.12 shows a sketch for the intersection of the position of point $A$ from angles observed at three horizontal control points $B$, $C$, and $D$, the coordinates of which are as follows:

| Point | X, ft | Y, ft |
|-------|-------|-------|
| B | 11,054.091 | 9,484.370 |
| C | 10,827.622 | 10,112.150 |
| D | 10,000.000 | 10,000.00 |

The measured angles, also shown in the figure, are

$\ell_1$ = angle $ABC$ = 81°17'37.5"
$\ell_2$ = angle $BCA$ = 64°32'27.5"
$\ell_3$ = angle $ACD$ = 37°39'28.2"
$\ell_4$ = angle $CDA$ = 97°31'31.1"

These angles are derived from directions taken with a standard deviation of 2". Compute the least squares estimates of the coordinates $X_A$, $Y_A$ of the unknown point $A$.

SOLUTION   To evaluate the precision of the four given angles, their relationship to the seven directions involved, $d_1, d_2, \ldots, d_7$, shown in Fig. B.12 are:

$$\ell_1 = d_2 - d_1 \qquad \ell_2 = d_4 - d_3 \qquad \ell_3 = d_5 - d_4 \qquad \ell_4 = d_7 - d_6$$

In matrix form, these relations become

$$\begin{bmatrix} \ell_1 \\ \ell_2 \\ \ell_3 \\ \ell_4 \end{bmatrix} = \begin{bmatrix} -1 & 1 & 0 & 0 & 0 & 0 & 0 \\ 0 & 0 & -1 & 1 & 0 & 0 & 0 \\ 0 & 0 & 0 & -1 & 1 & 0 & 0 \\ 0 & 0 & 0 & 0 & 0 & -1 & 1 \end{bmatrix} \begin{bmatrix} d_1 \\ d_2 \\ \vdots \\ d_7 \end{bmatrix} = \mathbf{Jd}$$

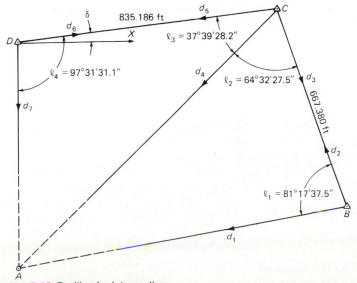

**Fig. B.12** Position by intersection.

The variance matrix for the directions is $\Sigma_{dd} = 41$ seconds$^2 = (9.401772)10^{-11}\mathbf{I}$ rad$^2$. Then from the rules of error propagation (Eq. 2.42), the covariance matrix for the four angles is

$$\Sigma_{\ell\ell} = \mathbf{J}\Sigma_{dd}\mathbf{J}^t = (9.401772)10^{-11}\begin{bmatrix} -1 & 1 & 0 & 0 & 0 & 0 & 0 \\ 0 & 0 & -1 & 1 & 0 & 0 & 0 \\ 0 & 0 & 0 & -1 & 1 & 0 & 0 \\ 0 & 0 & 0 & 0 & 0 & -1 & 1 \end{bmatrix}\begin{bmatrix} -1 & 0 & 0 & 0 \\ 1 & 0 & 0 & 0 \\ 0 & -1 & 0 & 0 \\ 0 & 1 & -1 & 0 \\ 0 & 0 & 1 & 0 \\ 0 & 0 & 0 & -1 \\ 0 & 0 & 0 & 1 \end{bmatrix}$$

or

$$\Sigma_{\ell\ell} = (9.401772)10^{-11}\begin{bmatrix} 2 & 0 & 0 & 0 \\ 0 & 2 & -1 & 0 \\ 0 & -1 & 2 & 0 \\ 0 & 0 & 0 & 2 \end{bmatrix} \quad \text{(rad}^2\text{)}$$

Before writing the condition equations, first compute approximations for the two unknown parameters $X_A$, $Y_A$. As shown in Fig. B.12, the line $DC$ makes an angle $\delta$ with the $X$ axis which may be computed from $\delta = \arctan[(Y_C - Y_D)/(X_C - X_D)] = \arctan 112.150/827.622 = 7.°7$. With the given value of $\ell_4$, the line $DA$ is very nearly parallel with the $Y$ axis. Thus, a good approximation for $X_A$ is to take it equal to $X_D$ or $X_A^0 = 10,000.0000$ ft. To get $Y_A^0$, the length of the side $AD$ is computed by applying the sine law to the triangle $ACD$, realizing that the length of $CD$ is

$$[(X_C - X_D)^2 + (Y_C - Y_D)^2]^{1/2} = [(827.622)^2 + (112.150)^2]^{1/2} = 835.186 \text{ ft.}$$

Thus,

$$\overline{AD} = 835.186\frac{\sin \ell_3}{\sin(180 - \ell_3 - \ell_4)} \approx 700 \text{ ft} \quad \text{and} \quad Y_A^0 = Y_D - \overline{AD} \approx 10,000 - 700 = 9300 \text{ ft}$$

The condition equations are [ from Eq. (B.72)]

$$F_1 = \ell_1 - \arctan\frac{X_C - X_B}{Y_C - Y_B} + \arctan\frac{X_A - X_B}{Y_A - Y_B} = 0$$

$$F_2 = \ell_2 - \arctan\frac{X_A - X_C}{Y_A - Y_C} + \arctan\frac{X_B - X_C}{Y_B - Y_C} = 0$$

$$F_3 = \ell_3 - \arctan\frac{X_D - X_C}{Y_D - Y_C} + \arctan\frac{X_A - X_C}{Y_A - Y_C} = 0$$

$$F_4 = \ell_4 - \arctan\frac{X_A - X_D}{Y_A - Y_D} + \arctan\frac{X_C - X_D}{Y_C - Y_D} = 0$$

Using $X_A^0$, $Y_A^0$ to represent the approximations for the coordinates of point $A$, linearization of these equations to the form $\mathbf{v} + \mathbf{B\Delta} = \mathbf{f}$ yields the following coefficients:

$$b_{11} = \frac{\partial F_1}{\partial X_A} = \frac{Y_A^0 - Y_B}{L_{AB}^{0^2}} \qquad\qquad b_{12} = \frac{\partial F_1}{\partial Y_A} = -\frac{X_A^0 - X_B}{L_{AB}^{0^2}}$$

$$b_{21} = \frac{\partial F_2}{\partial X_A} = -\frac{Y_A^0 - Y_C}{L_{AC}^{0^2}} \qquad\qquad b_{22} = \frac{\partial F_2}{\partial Y_A} = \frac{X_A^0 - X_C}{L_{AC}^{0^2}}$$

$$b_{31} = \frac{\partial F_3}{\partial X_A} = \frac{Y_A^0 - Y_C}{L_{AC}^{0^2}} = -b_{21} \qquad b_{32} = \frac{\partial F_3}{\partial Y_A} = -\frac{X_A^0 - X_C}{L_{AC}^{0^2}} = -b_{22}$$

$$b_{41} = \frac{\partial F_4}{\partial X_A} = -\frac{Y_A^0 - Y_D}{L_{AD}^{0^2}} \qquad\qquad b_{42} = \frac{\partial F_4}{\partial Y_A} = \frac{X_A^0 - X_D}{L_{AD}^{0^2}}$$

where

$$L_{AB}^{0^2} = (X_A^0 - X_B)^2 + (Y_A^0 - Y_B)^2$$

$$L_{AC}^{0^2} = (X_A^0 - X_C)^2 + (Y_A^0 - Y_C)^2$$

$$L_{AD}^{0^2} = (X_A^0 - X_D)^2 + (Y_A^0 - Y_D)^2$$

$$L_{BC}^2 = (X_B - X_C)^2 + (Y_B - Y_C)^2 = 445,395.9364 \text{ ft}^2$$

$$L_{CD}^2 = (X_C - X_D)^2 + (Y_C - Y_D)^2 = 697,535.7974 \text{ ft}^2$$

Unless the computer used, yields the value of the azimuth in the proper quadrant (i.e., between 0° and 360°) from the arctan function, the quadrant should be determined from the signs of the numerator and denominator as explained in Example B.8. The elements of the vector **f** are

$$f_1 = -\ell_1 + \arctan\frac{X_C - X_B}{Y_C - Y_B} - \arctan\frac{X_A^0 - X_B}{Y_A^0 - Y_B}$$

$$f_2 = -\ell_2 + \arctan\frac{X_A^0 - X_C}{Y_A^0 - Y_C} - \arctan\frac{X_B - X_C}{Y_B - Y_C}$$

$$f_3 = -\ell_3 + \arctan\frac{X_D - X_C}{Y_D - Y_C} - \arctan\frac{X_A^0 - X_C}{Y_A^0 - Y_C}$$

$$f_4 = -\ell_4 + \arctan\frac{X_A^0 - X_D}{Y_A^0 - Y_D} - \arctan\frac{X_C - X_D}{Y_C - Y_D}$$

Using the approximations above, the **B** and **f** matrices are

$$\mathbf{B} = 10^{-4}\begin{bmatrix} -1.6101 & 9.2052 \\ 6.0403 & -6.1554 \\ -6.0403 & 6.1554 \\ 14.2857 & 0 \end{bmatrix}\left(\frac{1}{\text{ft}}\right) \quad \text{and} \quad \mathbf{f} = 10^{-2}\begin{bmatrix} -2.110557 \\ 1.459653 \\ -1.597851 \\ 0.334695 \end{bmatrix}\text{(rad)}$$

With the covariance matrix $\Sigma$ computed above, the normal equations are formed as $(\mathbf{B}'\Sigma^{-1}\mathbf{B})\Delta_1 = \mathbf{B}'\Sigma^{-1}\mathbf{f}$ and solved to yield

$$\Delta_1 = \begin{bmatrix} 2.345 \text{ ft} \\ -22.524 \text{ ft} \end{bmatrix}$$

Adding these corrections to the original approximations gives the new set of approximate coordinates $10,000 + 2.345 = 10,002.345$ and $9300 - 22.524 = 9277.476$. With these new approximations the normal equations are re-formed and solved for a new vector of corrections. After a total of three iterations, the correction vector becomes essentially zero and the final least squares estimates of the coordinates of point $A$ are

$$\hat{X}_A = 10,002.445 \text{ ft} \qquad \text{and} \qquad \hat{Y}_A = 9277.390 \text{ ft}$$

**B.14.  Traverse adjustment**   As explained in Art. 8.1, a traverse is composed of consecutive distance and angle measurements. Figure B.13 shows a traverse between two horizontal control points 1 and 5, at each of which the azimuths $A_1$ and $A_5$ are also known. The observations are five angles, $\alpha_1$ to $\alpha_5$, and 4 distances, $\ell_{12}$ to $\ell_{45}$. A traverse can be adjusted using either of the two techniques of least squares presented in this appendix. The *technique of least squares adjustment of indirect observations* is more frequently applied in practice and is therefore presented first.

There are obviously two types of condition equations: the *angle condition*, such as that used in the preceding two sections, [Eq. (B.67)] and the *distance condition*. As an example, the angle condition for $\alpha_2$ is

$$\alpha_2 = A_{23} - A_{21} + 360$$
$$= \arctan\frac{X_3 - X_2}{Y_3 - Y_2} - \arctan\frac{X_1 - X_2}{Y_1 - Y_2} + 360°$$

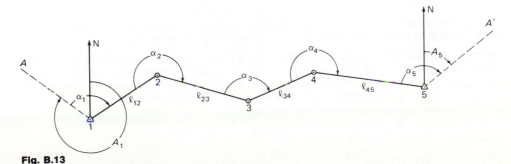

**Fig. B.13**

The distance condition expresses the distance between two points $i$ and $j$ as a function of their coordinates, or

$$\ell_{ij} + v_{ij} - \left[(X_i - X_j)^2 + (Y_i - Y_j)^2\right]^{1/2} = 0 \tag{B.73}$$

Since Eq. (B.73) is nonlinear, it must be linearized using Taylor series by first writing it in the form

$$\ell_{ij} + v_{ij} + F_{ij}(X_i, Y_i, X_j, Y_j) = 0 \tag{B.74}$$

and then applying Eq. (B.60) to (B.74). Thus,

$$v_{ij} + b_1\,\delta X_i + b_2\,\delta Y_i + b_3\,\delta X_j + b_4\,\delta Y_j = -\ell_{ij} + \ell_{ij}^0 = f_{ij} \tag{B.75}$$

in which

$$b_1 = \frac{\partial F_{ij}}{\partial X_i} = -\frac{X_i^0 - X_j^0}{\ell_{ij}^0} \tag{B.76a}$$

$$b_2 = \frac{\partial F_{ij}}{\partial Y_i} = -\frac{Y_i^0 - Y_j^0}{\ell_{ij}^0} \tag{B.76b}$$

$$b_3 = \frac{\partial F_{ij}}{\partial X_j} = \frac{X_i^0 - X_j^0}{\ell_{ij}^0} = -b_1 \tag{B.76c}$$

$$b_4 = \frac{\partial F_{ij}}{\partial Y_j} = \frac{Y_i^0 - Y_j^0}{\ell_{ij}^0} = -b_2 \tag{B.76d}$$

with

$$\ell_{ij}^0 = \left[(X_i^0 - X_j^0)^2 + (Y_i^0 - Y_j^0)^2\right]^{1/2} \tag{B.76e}$$

The superscript "o" designates an approximate value for the parameters. Thus, for the adjustment of the traverse in Fig. B.13, with the technique of adjustment of indirect observations, five angle and four distance conditions need to be written. The nine condition equations would include as unknown parameters the six coordinates of points 2, 3, and 4. Six normal equations are formed and solved for corrections to the approximate values of the parameters. The corrections are added to the approximations to update their values, and the solution is repeated until the last set of corrections is insignificantly small. The following example demonstrates this technique of adjusting traverses by the method of indirect observations.

**Example B.10**  Figure B.14 shows a traverse for which the following data are given:

| Angle | Value | $\sigma$ | Distance | Value, m | $\sigma$, m |
|---|---|---|---|---|---|
| $\alpha_1$ | 172°53′34″ | 2″ | $d_1$ | 281.832 | 0.016 |
| $\alpha_2$ | 185°22′14″ | 2″ | $d_2$ | 271.300 | 0.016 |
| $\alpha_3$ | 208°26′19″ | 2″ | $d_3$ | 274.100 | 0.016 |
| $\alpha_4$ | 205°13′51″ | 2″ | Point | $X$, m | $Y$, m |
| $A_B$ | 68°15′20.7″ | 0 | $B$ | 8478.139 | 2483.826 |
| $A_E$ | 300°11′30.5″ | 0 | $E$ | 7709.336 | 2263.411 |

The observations are assumed to be uncorrelated. It is required to compute the coordinates of points $C$ and $D$ using the least-squares technique of adjustment of indirect observations.

SOLUTION  For each of the seven observations given, a condition equation is written in terms of the observation and the coordinates of points $C$ and $D$ carried as parameters and denoted by $X_C, Y_C, X_D, Y_D$. Using the regular symbols $\ell_1, \ell_2, \ldots, \ell_7$ to represent the observations $\alpha_1, \alpha_2, \alpha_3, \alpha_4, d_1, d_2, d_3$, respectively, the condition equations are developed as follows:

$$\alpha_1 + A_{BA} - A_{BC} = 0 \quad \text{or} \quad F_1 = \alpha_1 + A_B - \arctan\frac{X_C - X_B}{Y_C - Y_B} = 0$$

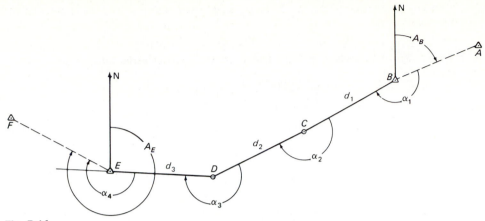

**Fig. B.14**

$$\alpha_2 + A_{CB} - A_{CD} = 0 \quad \text{or} \quad F_2 = \alpha_2 + \arctan\frac{X_B - X_C}{Y_B - Y_C} - \arctan\frac{X_D - X_C}{Y_D - Y_C} = 0$$

$$\alpha_3 + A_{DC} - A_{DE} = 0 \quad \text{or} \quad F_3 = \alpha_3 + \arctan\frac{X_C - X_D}{Y_C - Y_D} - \arctan\frac{X_E - X_D}{Y_E - Y_D} = 0$$

$$\alpha_4 + A_{ED} - A_{EF} = 0 \quad \text{or} \quad F_4 = \alpha_4 + \arctan\frac{X_D - X_E}{Y_D - Y_E} - A_E = 0$$

$$F_5 = d_1 - [(X_B - X_C)^2 + (Y_B - Y_C)^2]^{1/2} = 0$$

$$F_6 = d_2 - [(X_C - X_D)^2 + (Y_C - Y_D)^2]^{1/2} = 0$$

$$F_7 = d_3 - [(X_D - X_E)^2 + (Y_D - Y_E)^2]^{1/2} = 0$$

The linearized equations are

$$v_1 + b_{11}\,\delta X_C + b_{12}\,\delta Y_C + b_{13}\,\delta X_D + b_{14}\,\delta Y_D = f_1$$
$$v_2 + b_{21}\,\delta X_C + b_{22}\,\delta Y_C + b_{23}\,\delta X_D + b_{24}\,\delta Y_D = f_2$$
$$v_3 + b_{31}\,\delta X_C + b_{32}\,\delta Y_C + b_{33}\,\delta X_D + b_{34}\,\delta Y_D = f_3$$
$$v_4 + b_{41}\,\delta X_C + b_{42}\,\delta Y_C + b_{43}\,\delta X_D + b_{44}\,\delta Y_D = f_4$$
$$v_5 + b_{51}\,\delta X_C + b_{52}\,\delta Y_C + b_{53}\,\delta X_D + b_{54}\,\delta Y_D = f_5$$
$$v_5 + b_{61}\,\delta X_C + b_{62}\,\delta Y_C + b_{63}\,\delta X_D + b_{64}\,\delta Y_D = f_6$$
$$v_7 + b_{71}\,\delta X_C + b_{72}\,\delta Y_C + b_{73}\,\delta X_D + b_{74}\,\delta Y_D = f_7$$

which may be expressed in matrix form as

$$\mathbf{v} + \mathbf{B}\Delta = \mathbf{f}$$

in which the coefficients in **B** for angle and distance equations are determined by partial differentiation of the condition equations with respect to the unknown coordinates. The coefficients for this adjustment are

$$b_{11} = \frac{\partial F_1}{\partial X_C} = -\frac{Y_C^0 - Y_B}{(X_C^0 - X_B)^2 + (Y_C^0 - Y_B)^2} = -\frac{Y_C^0 - Y_B}{L_{BC}^{0^2}}$$

$$b_{12} = \frac{\partial F_1}{\partial Y_C} = \frac{X_C^0 - X_B}{(X_C^0 - X_B)^2 + (Y_C^0 - Y_B)^2} = \frac{X_C^0 - X_B}{L_{BC}^{0^2}}$$

$$b_{13} = \frac{\partial F_1}{\partial X_D} = 0$$

$$b_{14} = \frac{\partial F_1}{\partial Y_D} = 0$$

$$b_{21} = \frac{\partial F_2}{\partial X_C} = -b_{11} + \frac{Y_D^0 - Y_C^0}{(X_D^0 - X_C^0)^2 + (Y_D^0 - Y_C^0)^2} = -b_{11} + \frac{Y_D^0 - Y_C^0}{L_{CD}^{0^2}}$$

$$b_{22} = \frac{\partial F_2}{\partial Y_C} = -b_{12} - \frac{X_D^0 - X_C^0}{(X_D^0 - X_C^0)^2 + (Y_D^0 - Y_C^0)^2} = -b_{12} - \frac{X_D^0 - X_C^0}{L_{CD}^{0^2}}$$

$$b_{23} = \frac{\partial F_2}{\partial X_D} = -\frac{Y_D^0 - Y_C^0}{(X_D^0 - X_C^0)^2 + (Y_D^0 - Y_C^0)^2} = -(b_{11} + b_{21})$$

$$b_{24} = \frac{\partial F_2}{\partial Y_D} = \frac{X_D^0 - X_C^0}{(X_D^0 - X_C^0)^2 + (Y_D^0 - Y_C^0)^2} = -(b_{12} + b_{22})$$

$$b_{31} = \frac{\partial F_3}{\partial X_C} = b_{23}$$

$$b_{32} = \frac{\partial F_3}{\partial Y_C} = b_{24}$$

$$b_{33} = \frac{\partial F_3}{\partial X_D} = -b_{23} + \frac{Y_E - Y_D^0}{(X_E - X_D^0)^2 + (Y_E - Y_D^0)^2} = -b_{23} + \frac{Y_E - Y_D^0}{L_{DE}^{0^2}}$$

$$b_{34} = \frac{\partial F_3}{\partial Y_D} = -b_{24} - \frac{X_E - X_D^0}{[(X_E - X_D^0)^2 + (Y_E - Y_D^0)^2]} = -b_{24} - \frac{X_E - X_D^0}{L_{DE}^{0^2}}$$

$$b_{41} = \frac{\partial F_4}{\partial X_C} = 0$$

$$b_{42} = \frac{\partial F_4}{\partial Y_C} = 0$$

$$b_{43} = \frac{\partial F_4}{\partial X_D} = -(b_{23} + b_{33})$$

$$b_{44} = \frac{\partial F_4}{\partial Y_D} = -(b_{24} + b_{34})$$

$$b_{51} = \frac{\partial F_5}{\partial X_C} = \frac{X_B - X_C^0}{L_{BC}^0}$$

$$b_{52} = \frac{\partial F_5}{\partial Y_C} = \frac{Y_B - Y_C^0}{L_{BC}^0}$$

$$b_{53} = \frac{\partial F_5}{\partial X_D} = 0$$

$$b_{54} = \frac{\partial F_5}{\partial Y_D} = 0$$

$$b_{61} = \frac{\partial F_6}{\partial X_C} = -\frac{X_C^0 - X_D^0}{L_{CD}^0}$$

$$b_{62} = \frac{\partial F_c}{\partial Y_C} = -\frac{Y_C^0 - Y_D^0}{L_{CD}^0}$$

$$b_{63} = \frac{\partial F_6}{\partial X_D} = -b_{61}$$

$$b_{64} = \frac{\partial F_6}{\partial Y_D} = -b_{62}$$

$$b_{71} = \frac{\partial F_7}{\partial X_C} = 0$$

$$b_{72} = \frac{\partial F_7}{\partial Y_C} = 0$$

$$b_{73} = \frac{\partial F_7}{\partial X_D} = -\frac{X_D^0 - X_E}{L_{DE}^0}$$

$$b_{74} = \frac{\partial F_7}{\partial Y_D} = -\frac{Y_D^0 - Y_E}{L_{DE}^0}$$

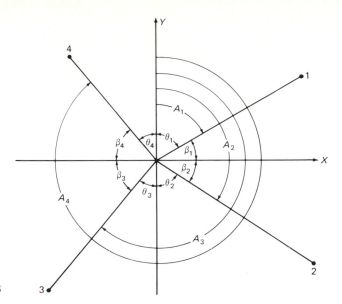

**Fig. B.15**

To evaluate the first four elements of **f**, it is necessary to determine the quadrant for each azimuth as implied by the signs associated with the numerator and denominator of the arctan functions. In general, the angles $\theta$ computed (in the computer) from the arctan functions in the condition equations are either positive or negative and are less than 90°. Therefore, the value of the azimuth angle is obtained on the basis of the signs of $\Delta X$ and $\Delta Y$ depicted in Fig. B.15 as follows [azimuth, $A$, is clockwise from $+Y$ (north), thus consistent with angles turned to the right in the problem]:

(a) $+\Delta X, +\Delta Y$     $A = \theta = \arctan \dfrac{\Delta X}{\Delta Y}$

(b) $+\Delta X, -\Delta Y$     $A = 180° - |\theta| = 180° - \left|\arctan \dfrac{\Delta X}{\Delta Y}\right|$

(c) $-\Delta X, -\Delta Y$     $A = 180° + |\theta| = 180° + \left|\arctan \dfrac{\Delta X}{\Delta Y}\right|$

(d) $-\Delta X, +\Delta Y$     $A = 360° - |\theta| = 360° - \left|\arctan \dfrac{\Delta X}{\Delta Y}\right|$

If the computer yields acute positive angles with the signs of $\Delta X, \Delta Y$ as in cases (a) and (c) and acute negative angles in cases (b) and (d) using the arctan function, the following scheme is simpler and more efficient to program (see Fig. B.15):

1. $+\Delta X$, $A = 90° - \beta = 90° - \arctan(\Delta Y/\Delta X)$; thus if $\Delta Y$ is positive, $A$ is less than 90°, and if it is negative, $A$ is between 90° and 180°.
2. $-\Delta X$, $A = 270° - \beta = 270° - \arctan(\Delta Y/\Delta X)$; again, with $-\Delta Y$, then $180° < A < 270°$; and with $+\Delta Y$, $270° < A < 360°$.

In larger computers there are special arctan functions (such as ATAN 2 in the CDC system) which directly evaluate the total angle with a value from $-180°$ to $+180°$. When such functions are available, neither of the foregoing two schemes is necessary.

   Let us assume the following approximations for the four unknown parameters.

$X_C^0 = 8200$ m     $Y_C^0 = 2340$ m
$X_D^0 = 7980$ m     $Y_D^0 = 2230$ m

From Eq. (B.71$d$), the rules set forth in (a) to (d) above, and (B.76$e$), the constant terms in the **f** vector are:

$$f_1 = -\alpha_1 + [A_{BC} - A_{BA}]$$

$$= -\alpha_1 + \left[180° + \arctan\frac{\Delta X_{BC}}{\Delta Y_{BC}} - A_{BA}\right] = 0.026318430 \text{ rad}$$

$$f_2 = -\alpha_2 + [A_{CD} - A_{CB}]$$

$$= -\alpha_2 + \left[180° + \arctan\frac{\Delta X_{CD}}{\Delta Y_{CD}} - \arctan\frac{\Delta X_{CB}}{\Delta Y_{CB}}\right] = -0.080146734 \text{ rad}$$

$$f_3 = -\alpha_3 + [A_{DE} - A_{DC}]$$

$$= -\alpha_3 + \left[360° - \arctan\frac{\Delta X_{DE}}{\Delta Y_{DE}} - \arctan\frac{\Delta X_{DC}}{\Delta Y_{DC}}\right] = 0.09011977 \text{ rad}$$

$$f_4 = -\alpha_4 + [A_{EF} - A_{ED}]$$

$$= -\alpha_4 + \left[A_{EF} - (180° - \arctan\frac{\Delta X_{ED}}{\Delta Y_{ED}})\right] = -0.036234261 \text{ rad}$$

$$f_5 = -d_1 + [\Delta X_{BC}^2 + \Delta Y_{BC}^2]^{1/2} = -d_1 + L_{BC} = \quad 31.2929 \text{ m}$$

$$f_6 = -d_2 + [\Delta X_{CD}^2 + \Delta Y_{CD}^2]^{1/2} = -d_2 + L_{CD} = -25.3325 \text{ m}$$

$$f_7 = -d_3 + [\Delta X_{DE}^2 + \Delta Y_{DE}^2]^{1/2} = -d_3 + L_{DE} = -1.3817 \text{ m}$$

The variance of each measured angle is $9.401772217 \times 10^{-11}$ rad$^2$. The variance of each measured distance is $2.56 \times 10^{-4}$ m$^2$. Thus, assuming a reference variance of 1.0 and uncorrelated measurements, the covariance matrix [Eq. (2.29), Art. 2.14] of the measurements is

$$\Sigma = 2.56 \times 10^{-4} \text{ diag}\{(3.672567)10^{-7} \quad (3.672567)10^{-7} \quad (3.672567)10^{-7}$$

$$(3.672567)10^{-7} \quad 1.0 \quad 1.0 \quad 1.0\}$$

By Eq. (2.32), Art. 2.14, the weight matrix is $\mathbf{W} = \Sigma^{-1}$, or

$$\mathbf{W} = 3906.25 \text{ diag}\{2,722,890.898 \quad 2,722,890.898 \quad 2,722,890.898 \quad 2,722,890.898 \quad 1.0 \quad 1.0 \quad 1.0\}$$

With the approximations selected above, the **B** matrix is

$$\mathbf{B} = 10^{-3}\begin{bmatrix} 1.466905 & -2.836786 & 0 & 0 \\ -3.285087 & 6.473150 & 1.818182 & -3.636364 \\ 1.818182 & -3.636364 & -1.368960 & 7.275529 \\ 0 & 0 & 0.4492217 & -3.639165 \\ 888.2685 & 459.3247 & 0 & 0 \\ -894.4272 & -447.2136 & 894.4272 & 447.2136 \\ 0 & 0 & -992.4672 & 122.5110 \end{bmatrix}$$

The normal equations coefficient matrix is $\mathbf{N} = \mathbf{B'WB}$, [Eq. (B.30)], or

$$\mathbf{N} = 10^5\begin{bmatrix} 1.7904 & -3.361 & -0.9313 & 2.6620 \\ & 6.7352 & 1.7657 & -5.3254 \\ & & 0.6421 & -1.5778 \\ \text{symmetric} & & & 8.4536 \end{bmatrix}$$

and its inverse is

$$\mathbf{N}^{-1} = 10^{-5}\begin{bmatrix} 11.6976 & 5.1548 & 3.1756 & 0.1566 \\ & 2.8713 & 0.0680 & 0.1983 \\ & & 6.6941 & 0.2923 \\ \text{symmetric} & & & 0.2485 \end{bmatrix}$$

The correction vector is found by solving Eq. (B.36) for $\mathbf{\Delta}$, or

$$\mathbf{\Delta}_1 = \mathbf{N}^{-1}\mathbf{t}$$

The vector of corrections after the first iteration is

$$\mathbf{\Delta}_1 = [32.5209 \quad 7.8617 \quad 4.8214 \quad 9.1344]^t \text{ (m)}$$

which leads to the following improved approximations: $X_{C_1}^0 = 8232.5209$, $Y_{C_1}^0 = 2347.8617$, $X_{D_1}^0 = 7984.8214$, and $Y_{D_1}^0 = 2239.1344$. Using these values, the solution is iterated. After four iterations, the correction vector is zero to four decimal places, and the final estimates of the coordinates are

$X_C = 8231.263$ m $\qquad Y_C = 2347.818$ m
$X_D = 7982.404$ m $\qquad Y_D = 2239.714$ m

According to Eq. (B.38), the covariance matrix of the parameters (when the reference variance is unity as in this example) is equal to the inverse of the normal equations coefficient matrix, or $\Sigma_{\Delta\Delta} = N^{-1}$. At the end of four iterations the inverse was

$$
N^{-1} = \begin{bmatrix}
(1.077237)10^{-4} & (5.016788)10^{-5} & ( \ 2.456713)10^{-5} & (4.457744)10^{-6} \\
& (2.808001)10^{-5} & (-4.638273)10^{-7} & (3.091250)10^{-6} \\
\hline
& & ( \ 7.525994)10^{-5} & (4.779247)10^{-6} \\
\text{symmetric} & & & (2.685166)10^{-6}
\end{bmatrix}
$$

$$
= \begin{bmatrix}
\Sigma_{CC} & \Sigma_{CD} \\
\hline
\Sigma_{CD}^t & \Sigma_{DD}
\end{bmatrix}
$$

where $\Sigma_{CC}$ is the 2 × 2 covariance matrix for the coordinates of point $C$, $\Sigma_{DD}$ is the 2 × 2 covariance matrix for the coordinates of point $D$, and $\Sigma_{CD}$ is a 2 × 2 cross-covariance matrix between the coordinates of point $C$ and point $D$. It is used only when the precision of relative information between $C$ and $D$ is desired. The derivation of such precision is beyond the scope of this book.

The covariance matrices $\Sigma_{CC}$ and $\Sigma_{DD}$ express the precision with which points $C$ and $D$ are located. Each matrix can be used to give the standard deviation for each coordinate. For example, the standard deviation of the $X$ coordinate of point $C$ is $\sigma_{X_c} = [(1.077237)(10^{-4})]^{1/2} = 0.010$ m. Alternatively, each covariance matrix could be utilized to establish a standard error ellipse about the point as explained in Art. 2.19 and demonstrated by Example 2.9.

A traverse may also be adjusted using the technique of *least squares adjustment of observations only* (Art. B.8). Usually, the redundancy is three and the following three conditions must be satisfied: (1) the angles must close; (2) the sum of the departures must equal zero; and (3) the sum of the latitudes must equal zero. The angle closure condition is expressed by starting from one end of the traverse and summing the starting azimuth and measured angles through to the other end of the traverse [see Eq. (8.5), Art. 8.9]. Thus, the total sum minus multiples of 180° must be equated to the azimuth at the end of the traverse. As an example, the angle closure condition for the traverse in Fig. B.13 is

$$A_1 + (\alpha_1 + v_1) + (\alpha_2 + v_2) + (\alpha_3 + v_3) + (\alpha_4 + v_4) + (\alpha_5 + v_5) - (4)(180°) - A_5 = 0$$

To write the departure and latitude conditions, the accumulated azimuth must be evaluated at each station in the traverse. Let $x_{ij}$ and $y_{ij}$ represent the departure and latitude for one traverse leg $\ell_{ij}$ shown in Fig. B.16. If $A_i$ represents the accumulated azimuth at point $i$, then

$$x_{ij} = \ell_{ij} \sin A_i$$
$$y_{ij} = \ell_{ij} \cos A_i \tag{B.77}$$

and the coordinates at point $j$ would be

$$X_j = X_i + x_{ij}$$
$$Y_j = Y_i + y_{ij}$$

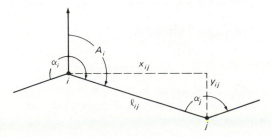

**Fig. B.16**

Thus, for the traverse in Fig. B.13, the departure and latitude conditions are

$$X_1 + x_{12} + x_{23} + x_{34} + x_{45} - X_5 = 0$$
$$Y_1 + y_{12} + y_{23} + y_{34} + y_{45} - Y_5 = 0$$

where $(X_1, Y_1)$ and $(X_5, Y_5)$ are the coordinates of the beginning and ending control points. The following example shows how the technique of least squares adjustment of observations only is applied to a numerical traverse problem.

**Example B.11** Using the data for the traverse in Example B.10 (Fig. B.14), compute the coordinates of points C and D using the least squares technique of adjustment of observations only.

SOLUTION The three conditions that relate the seven observations are:

1. Angle closure:

$$F_1 = A_B + (\alpha_1 - \pi) + (\alpha_2 - \pi) + (\alpha_3 - \pi) + \alpha_4 - A_E = 0$$

2. Sum of departures is equal to zero:

$$F_2 = X_B - d_1 \sin(A_B + \alpha_1 - \pi) - d_2 \sin(A_B + \alpha_1 - \pi + \alpha_2 - \pi) - d_3 \sin(\alpha_4 + \pi - A_E) - X_E = 0$$

3. Sum of latitudes is equal to zero:

$$F_3 = Y_B - d_1 \cos(A_B + \alpha_1 - \pi) - d_2 \cos(A_B + \alpha_1 - \pi + \alpha_2 - \pi) + d_3 \cos(\alpha_4 + \pi - A_E) - Y_E = 0$$

Linearization yields the following three equations in terms of coefficients, unknown residuals, and constant terms:

$$a_{11} v_1 + a_{12} v_2 + a_{13} v_3 + a_{14} v_4 + a_{15} v_5 + a_{16} v_6 + a_{17} v_7 = f_1$$
$$a_{21} v_1 + a_{22} v_2 + a_{23} v_3 + a_{24} v_4 + a_{25} v_5 + a_{26} v_6 + a_{27} v_7 = f_2$$
$$a_{31} v_1 + a_{32} v_2 + a_{33} v_3 + a_{34} v_4 + a_{35} v_5 + a_{36} v_6 + a_{37} v_7 = f_3$$

which may be expressed in matrix form as

$$Av = f$$

where

$$a_{11} = \frac{\partial F_1}{\partial \alpha_1} = 1$$

$$a_{12} = \frac{\partial F_1}{\partial \alpha_2} = 1$$

$$a_{13} = \frac{\partial F_1}{\partial \alpha_3} = 1$$

$$a_{14} = \frac{\partial F_1}{\partial \alpha_4} = 1$$

$$a_{15} = \frac{\partial F_1}{\partial d_1} = 0$$

$$a_{16} = \frac{\partial F_1}{\partial d_2} = 0$$

$$a_{17} = \frac{\partial F_1}{\partial d_3} = 0$$

$$a_{21} = \frac{\partial F_2}{\partial \alpha_1} = -d_1 \cos(A_B + \alpha_1 - \pi) - d_2 \cos(A_B + \alpha_1 + \alpha_2 - 2\pi) = d_1 a_{35} + d_2 a_{36} = -244.093 \text{ m}$$

$$a_{22} = \frac{\partial F_2}{\partial \alpha_2} = -d_2 \cos(A_B + \alpha_1 + \alpha_2 - 2\pi) = d_2 a_{36} = -108.098 \text{ m}$$

$$a_{23} = \frac{\partial F_2}{\partial \alpha_3} = 0$$

$$a_{24} = \frac{\partial F_2}{\partial \alpha_4} = -d_3 \cos(\alpha_4 + \pi - A_E) = -d_3 a_{37} = -23.697 \text{ m}$$

$$a_{25} = \frac{\partial F_2}{\partial d_1} = -\sin(A_B + \alpha_1 - \pi) = -0.87583538$$

$$a_{26} = \frac{\partial F_2}{\partial d_2} = -\sin(A_B + \alpha_1 + \alpha_2 - 2\pi) = -0.91719283$$

$$a_{27} = \frac{\partial F_2}{\partial d_3} = -\sin(\alpha_4 + \pi - A_E) = -0.99625383$$

$$a_{31} = \frac{\partial F_3}{\partial \alpha_1} = -d_1 a_{25} - d_2 a_{26} = 495.684 \text{ m}$$

$$a_{32} = \frac{\partial F_3}{\partial \alpha_2} = -d_2 a_{26} = 248.834 \text{ m}$$

$$a_{33} = \frac{\partial F_3}{\partial \alpha_3} = 0$$

$$a_{34} = \frac{\partial F_3}{\partial \alpha_4} = d_3 a_{27} = -273.073 \text{ m}$$

$$a_{35} = \frac{\partial F_3}{\partial d_1} = -\cos(A_B + \alpha_1 - \pi) = -0.48254072$$

$$a_{36} = \frac{\partial F_3}{\partial d_2} = -\cos(A_B + \alpha_1 + \alpha_2 - 2\pi) = -0.39844360$$

$$a_{37} = \frac{\partial F_3}{\partial d_3} = \cos(\alpha_4 + \pi - A_E) = 0.08647715$$

$f_1 = -A_B - \alpha_1 - \alpha_2 - \alpha_3 - \alpha_4 + 3\pi + A_E = (5.72066569)10^{-5} \text{ rad}$
$f_2 = -X_B + d_1 \sin(A_B + \alpha_1 - \pi) + d_2 \sin(A_B + \alpha_1 + \alpha_2 - 2\pi) + d_3 \sin(\alpha_4 + \pi - A_E) + X_E = -0.046217 \text{ m}$
$f_3 = -Y_B + d_1 \cos(A_B + \alpha_1 - \pi) + d_2 \cos(A_B + \alpha_1 + \alpha_2 - 2\pi) - d_3 \cos(\alpha_4 + \pi - A_E) + Y_E = -0.025221 \text{ m}$

The covariance matrix $\Sigma$ for the measurements is given in Example B.10. Since the reference variance is unity, the cofactor matrix $Q = \Sigma$ and Eqs. (B.50$b$) and (B.49) (Art. B.8) become $k = (A\Sigma A^t)^{-1}f$ and $v = \Sigma A^t k$, which lead to $v = \Sigma A^t(A\Sigma A^t)^{-1}f$. Substitution of the elements of $A$ and $f$ above and $\Sigma$ from Example B.10 into this equation yields the vector of residuals $v$, from which adjusted measurements are found by Eq. (2.52) or $\hat{l} = l + v$.

From a computer program written to perform these operations, the adjusted observations are:

$$\hat{l}_1 = \hat{\alpha}_1 = 3.01755395 \text{ rad} = 172°53'35.2''$$

$$\hat{l}_2 = \hat{\alpha}_2 = 3.23533786 \text{ rad} = 185°22'16.3''$$

$$\hat{l}_3 = \hat{\alpha}_3 = 3.63795702 \text{ rad} = 208°26'22.5''$$

$$\hat{l}_4 = \hat{\alpha}_4 = 3.58197693 \text{ rad} = 205°13'55.8''$$

$$\hat{l}_5 = \hat{d}_1 = 281.862 \text{ m}$$

$$\hat{l}_6 = \hat{d}_2 = 271.325 \text{ m}$$

$$\hat{l}_7 = \hat{d}_3 = 274.095 \text{ m}$$

From these values, the adjusted coordinates of points $C$ and $D$ are computed as follows:

$$\hat{X}_C = X_B - \hat{d}_1 \sin(A_B + \hat{\alpha}_1 - \pi) = 8478.139 - (281.862)\sin(61°08'55.9'') = 8231.263 \text{ m}$$

$$\hat{Y}_C = Y_B - \hat{d}_1 \cos(61°08'55.9'') = 2483.826 - 136.008 = 2347.818 \text{ m}$$

$$\hat{X}_D = X_E + \hat{d}_3 \sin(\hat{\alpha}_4 + \pi - A_E) = 7709.336 + (274.095)\sin(85°02'25.3'') = 7982.405 \text{ m}$$

$$\hat{Y}_D = Y_E - \hat{d}_3 \cos(85°02'25.3'') = 2263.411 - 23.697 = 2239.714 \text{ m}$$

These adjusted coordinates correspond to within 0.001 m of the values computed in Example B.10 by the technique of indirect observations. Should error propagation be desired, the propagated covariance matrix for the adjusted coordinates may be calculated by Eq. (B.57)(Art.B.8), in which $Q = \Sigma$.

**B.15. Adjustment of triangulation**   As presented in Chap. 10, triangulation is a means of establishing horizontal position by measuring mainly angles and the lengths of a few lines. Each measured line is called a base line, which has to be measured with care since its errors will propagate throughout the triangulation net. If the ends of a line are two horizontal control points, that line can be considered as a base line. Consequently, both resection and intersection presented in Arts. B.12 and B.13 may be considered as special cases of triangulation. In practice, however, the single triangle with measured internal angles is usually the main unit in triangulation (hence the name). With this unit, chains of quadri-laterals and central point figures can be constructed. The least squares adjustment of triangulation may be performed by either of the techniques of observations only (Art. B.8) or indirect observations (Art. B.7). To illustrate the procedures, an example will be worked using both techniques. First, consider adjustment of triangulation by observations only.

**Example B.12**   In the adjustment of triangulation by the method of observations only, no $X, Y$-coordi-nate parameters are carried in the adjustment. Instead, only the observations are involved in the condition equations. This usually allows for the reduction in the number of condition equations to be written, the number of normal equations to be solved, and does not require the computation of approximate values (since there are no unknown parameters). In this example, consider the *shape* of the quadrilateral *ABCD* shown in Fig. B.17. The observations are the eight interior angles shown with the following values:

$\ell_1 = 22°01'42.51''$

$\ell_2 = 16°44'31.20''$

$\ell_3 = 57°08'57.10''$

$\ell_4 = 19°33'14.13''$

$\ell_5 = 86°33'13.45''$

$\ell_6 = 58°46'35.93''$

$\ell_7 = 15°06'52.28''$

$\ell_8 = 84°04'50.66''$

The two angles at each point are derived from three *observed* directions, which are assumed to be uncorrelated and of equal precision ($\mathbf{Q}_{\text{directions}} = \mathbf{I}$). Thus, every two angles at a point will be correlated. For example, if $d_{AB}, d_{AC}, d_{AD}$ represent the directions at $A$, then

$$\ell_1 = d_{AD} - d_{AC} \qquad \text{and} \qquad \ell_2 = d_{AC} - d_{AB}$$

or

$$\begin{bmatrix} \ell_1 \\ \ell_2 \end{bmatrix} = \begin{bmatrix} 0 & -1 & 1 \\ -1 & 1 & 0 \end{bmatrix} \begin{bmatrix} d_{AB} \\ d_{AC} \\ d_{AD} \end{bmatrix}$$

The cofactor matrix for $\ell_1$ and $\ell_2$ can be readily obtained from error propagation [see Eq. (2.43)] using the identity matrix as the cofactor matrix of the directions, or

$$\mathbf{Q}_{\text{angles}} = \begin{bmatrix} 0 & -1 & 1 \\ -1 & 1 & 0 \end{bmatrix} \begin{bmatrix} 0 & -1 \\ -1 & 1 \\ 1 & 0 \end{bmatrix} = \begin{bmatrix} 2 & -1 \\ -1 & 2 \end{bmatrix}$$

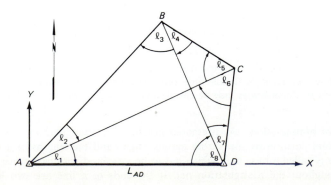

**Fig. B.17**

Thus, the cofactor matrix for all eight angles is the following block diagonal 8 × 8 matrix:

$$
\mathbf{Q} =
\begin{bmatrix}
2 & -1 & 0 & 0 & 0 & 0 & 0 & 0 \\
-1 & 2 & 0 & 0 & 0 & 0 & 0 & 0 \\
0 & 0 & 2 & -1 & 0 & 0 & 0 & 0 \\
0 & 0 & -1 & 2 & 0 & 0 & 0 & 0 \\
0 & 0 & 0 & 0 & 2 & -1 & 0 & 0 \\
0 & 0 & 0 & 0 & -1 & 2 & 0 & 0 \\
0 & 0 & 0 & 0 & 0 & 0 & 2 & -1 \\
0 & 0 & 0 & 0 & 0 & 0 & -1 & 2
\end{bmatrix}
$$

Since the shape of the quadrilateral is of primary interest, start with the length of one side, such as $AD$ (the known side for this example), in order to determine the minimum number of observations, $n_0$. At point $A$, angles $\ell_1$, $\ell_2$ are used to fix the directions $AC$ and $AB$, respectively; at point $D$, angles $\ell_8$, $\ell_7$ are used to fix the directions $DB$ and $DC$, respectively, thus locating points $B$ and $C$. Since it took four angles to uniquely determine the shape of the quadrilateral and there are a total of eight measured angles, the redundancy by Eq. (B.3) is $r = n - n_0 = 8 - 4 = 4$. The four conditions relating the eight observed angles include angle and side conditions (Arts. 10.8 and 10.9). The number of conditions of each type can be determined separately using Eq. (10.1) (Art. 10.8) for angles to give $C_A = A - L + 1 = 8 - 6 + 1 = 3$ angle conditions and Eq. (10.3) (Art. 10.9) for the sides to yield $C_s = L - 2s + 3 = 6 - 8 + 3 = 1$ side condition (where $A =$ total number of angles, $L =$ number of sides, and $s =$ number of stations in the quadrilateral). Note that the total number of conditions determined by $C_A + C_s = 4$, corresponds to the redundancy, $r$, as found by Eq. (B.3).

The three *independent* angle condition equations chosen for this adjustment are that the angles in triangles $ABD$, $BCD$, and $ACD$ each sum to $180°$. The side condition ensures that the calculated side (e.g., $BC$) computed from two different sets of triangles must be of the same length. The side condition Eq. (10.6) as developed in Art. 10.9 is utilized for this purpose. The four condition equations are

$$F_1 = \ell_1 + \ell_2 + \ell_3 + \ell_8 - \pi = 0$$
$$F_2 = \ell_4 + \ell_5 + \ell_6 + \ell_7 - \pi = 0$$
$$F_3 = \ell_1 + \ell_6 + \ell_7 + \ell_8 - \pi = 0$$
$$F_4 = \frac{\sin \ell_1 \, \sin \ell_3 \, \sin \ell_5 \, \sin \ell_7}{\sin \ell_2 \, \sin \ell_4 \, \sin \ell_6 \, \sin \ell_8} - 1 = \frac{T}{U} - 1 = 0$$

LInearization to the form $\mathbf{Av} = \mathbf{f}$ yields the following values:

$$
\mathbf{A} =
\begin{bmatrix}
1 & 1 & 1 & 0 & 0 & 0 & 0 & 1 \\
0 & 0 & 0 & 1 & 1 & 1 & 1 & 0 \\
1 & 0 & 0 & 0 & 0 & 1 & 1 & 1 \\
a_{41} & a_{42} & a_{43} & a_{44} & a_{45} & a_{46} & a_{47} & a_{48}
\end{bmatrix}
$$

in which

$$a_{41} = \frac{\partial F_4}{\partial \ell_1} = \frac{T}{U} \cot \ell_1 = 2.4716$$

$$a_{42} = \frac{\partial F_4}{\partial \ell_2} = -\frac{T}{U} \cot \ell_2 = -3.3244$$

$$a_{43} = \frac{\partial F_4}{\partial \ell_3} = \frac{T}{U} \cot \ell_3 = 0.6457$$

$$a_{44} = \frac{\partial F_4}{\partial \ell_4} = -\frac{T}{U} \cot \ell_4 = -2.8156$$

$$a_{45} = \frac{\partial F_4}{\partial \ell_5} = \frac{T}{U} \cot \ell_5 = 0.0602$$

$$a_{46} = \frac{\partial F_4}{\partial \ell_6} = -\frac{T}{U} \cot \ell_6 = -0.6062$$

$$a_{47} = \frac{\partial F_4}{\partial \ell_7} = \frac{T}{U} \cot \ell_7 = 3.7025$$

$$a_{48} = \frac{\partial F_4}{\partial \ell_8} = -\frac{T}{U} \cot \ell_8 = -0.1037$$

and
$$\mathbf{f}=[-1.47 \quad 4.21 \quad -1.38 \quad -6.267\,]^t \quad (\text{sec})$$

With these values the least squares solution is obtained as follows: From Eq. (B.52),

$$\mathbf{Q_e}=\mathbf{AQA'}=\begin{bmatrix} 6 & -2 & 2 & -\ 0.6557 \\ -2 & 6 & 2 & 0.6859 \\ 2 & 2 & 6 & 10.5939 \\ -0.6557 & 0.6859 & 10.5939 & 100.1021 \end{bmatrix}$$

$$\mathbf{W_e}=\mathbf{Q_e^{-1}}=\begin{bmatrix} 0.2773 & 0.1490 & -0.1765 & 0.0195 \\ 0.1490 & 0.2712 & -0.1703 & 0.0171 \\ -0.1765 & -0.1703 & 0.3471 & -0.0367 \\ 0.0195 & 0.0171 & -0.0367 & 0.0139 \end{bmatrix}$$

and from Eq. (B.53), the vector of Lagrange multipliers is

$$\mathbf{k}=\mathbf{W_e f}=[0.3414 \quad 1.0502 \quad -0.7066 \quad 0.0072\,]^t$$

The vectors of residuals and adjusted observations are computed next from Eq. (B.49) and (2.52):

$$\mathbf{v}=\mathbf{QA'k}=\begin{bmatrix} -1.01'' \\ 0.98'' \\ -0.34'' \\ 1.71'' \\ 1.76'' \\ -0.37'' \\ 1.11'' \\ -1.10'' \end{bmatrix} \quad \text{and} \quad \hat{\boldsymbol{\ell}}=\boldsymbol{\ell}+\mathbf{v}=\begin{bmatrix} 22°01'41.50'' \\ 16°44'32.18'' \\ 57°08'56.76'' \\ 19°33'15.84'' \\ 86°33'15.21'' \\ 58°46'35.56'' \\ 15°06'53.39'' \\ 84°04'49.56'' \end{bmatrix}$$

Once the adjusted angles are computed, each triangle in Fig. B.12 closes to within 0.01'', and therefore any elements of $\hat{\boldsymbol{\ell}}$ can be used to compute point locations uniquely.

In this example problem, stations $A$ and $D$ are of known position and have the following coordinates:

| Station | X, m | Y, m |
|---|---|---|
| A | 15,400.812 | 10,425.406 |
| D | 17,901.905 | 10,425.406 |

By inspection of the coordinates it can be seen that the direction of line $AD$ is due east and the distance from $A$ to $D$ by inverse computation [Eq. (8.8)] is 2501.093 m. In order to calculate adjusted coordinates for stations $B$ and $C$, position computations using adjusted angles are performed following the procedure set forth in Example 10.5 (Art. 10.23). The results of position computations for quadrilateral $ABCD$ (Fig. B.17) using adjusted angles yields coordinates for stations $B$ and $C$ as follows:

| Station | X, m | Y, m |
|---|---|---|
| B | 17,709.633 | 12,279.787 |
| C | 18,077.199 | 11,508.269 |

Note that the adjusted coordinates differ only slightly from those obtained using un-adjusted angles (Example 10.5). This is due to the excellence of the measurements and small corrections that result from the adjustment. Next, the same quadrilateral is adjusted by the least squares technique of indirect observations.

**Example B.13** Adjustment of triangulation by indirect observations requires that estimates be calculated for all stations to be adjusted. Thus in this example, approximate coordinates are computed for stations $B$ and $C$. Double position computations using unadjusted angles can be found in Example 10.5 (Art. 10.23). From that example, fixed coordinates for $A$ and $D$ and approximate values for $B$ and

*C* are:

| Station | *X*, m | *Y*, m | |
|---------|--------|--------|---|
| A | 15,400.812 | 10,425.406 | } Fixed |
| D | 17,901.905 | 10,425.406 | |
| B | 17,709.632 | 12,279.787 | } Approximate |
| C | 18,077.190 | 11,508.281 | |

To perform this adjustment, an angle condition equation must be written for each measured angle. Angle condition equation (B.67) is repeated here for convenience:

$$v_{ij} + \ell_{ij} + \arctan\frac{X_i - X}{Y_i - Y} - \arctan\frac{X_j - X}{Y_j - Y} = 0$$

For the example problem, the following eight angle condition equations of the form of (B.67) are formed:

$$F_1 = \ell_1 - \arctan\frac{X_D - X_A}{Y_D - Y_A} + \arctan\frac{X_C - X_A}{Y_C - Y_A} = 0$$

$$F_2 = \ell_2 - \arctan\frac{X_C - X_A}{Y_C - Y_A} + \arctan\frac{X_B - X_A}{Y_B - Y_A} = 0$$

$$F_3 = \ell_3 - \arctan\frac{X_A - X_B}{Y_A - Y_B} + \arctan\frac{X_D - X_B}{Y_D - Y_B} = 0$$

$$F_4 = \ell_4 - \arctan\frac{X_D - X_B}{Y_D - Y_B} + \arctan\frac{X_C - X_B}{Y_C - Y_B} = 0$$

$$F_5 = \ell_5 - \arctan\frac{X_B - X_C}{Y_B - Y_C} + \arctan\frac{X_A - X_C}{Y_A - Y_C} = 0$$

$$F_6 = \ell_6 - \arctan\frac{X_A - X_C}{Y_A - Y_C} + \arctan\frac{X_D - X_C}{Y_D - Y_C} = 0$$

$$F_7 = \ell_7 - \arctan\frac{X_C - X_D}{Y_C - Y_D} + \arctan\frac{X_B - X_D}{Y_B - Y_D} = 0$$

$$F_8 = \ell_8 - \arctan\frac{X_B - X_D}{Y_B - Y_D} + \arctan\frac{X_A - X_D}{Y_A - Y_D} = 0$$

The linearized angle condition equations [see Eq. (B.70)] are

$$v_1 + b_{11}\,\delta X_B + b_{12}\,\delta Y_B + b_{13}\,\delta X_C + b_{14}\,\delta Y_C = f_1$$
$$v_2 + b_{21}\,\delta X_B + b_{22}\,\delta Y_B + b_{23}\,\delta X_C + b_{24}\,\delta Y_C = f_2$$
$$v_3 + b_{31}\,\delta X_B + b_{32}\,\delta Y_B + b_{33}\,\delta X_C + b_{34}\,\delta Y_C = f_3$$
$$v_4 + b_{41}\,\delta X_B + b_{42}\,\delta Y_B + b_{43}\,\delta X_C + b_{44}\,\delta Y_C = f_4$$
$$v_5 + b_{51}\,\delta X_B + b_{52}\,\delta Y_B + b_{53}\,\delta X_C + b_{54}\,\delta Y_C = f_5$$
$$v_6 + b_{61}\,\delta X_B + b_{62}\,\delta Y_B + b_{63}\,\delta X_C + b_{64}\,\delta Y_C = f_6$$
$$v_7 + b_{71}\,\delta X_B + b_{72}\,\delta Y_B + b_{73}\,\delta X_C + b_{74}\,\delta Y_C = f_7$$
$$v_8 + b_{81}\,\delta X_B + b_{82}\,\delta Y_B + b_{83}\,\delta X_C + b_{84}\,\delta Y_C = f_8$$

which may be expressed in matrix form as

**v + BΔ = f**

in which the elements in **B** are determined by partial differentiation of the condition equations with respect to the unknown coordinates of stations *B* and *C* to yield

$$b_{11} = b_{12} = b_{33} = b_{34} = b_{61} = b_{62} = b_{83} = b_{84} = 0$$

$$b_{13} = \frac{\partial F_1}{\partial X_C} = \frac{Y_C^0 - Y_A}{(L_{AC}^0)^2}$$

$$b_{14} = \frac{\partial F_1}{\partial Y_C} = -\frac{X_C^0 - X_A}{(L_{AC}^0)^2}$$

$$b_{21} = \frac{\partial F_2}{\partial X_B} = \frac{Y_B^0 - Y_A}{(L_{AB}^0)^2}$$

$$b_{22} = \frac{\partial F_2}{\partial Y_B} = -\frac{X_B^0 - X_A}{(L_{AB}^0)^2}$$

$$b_{23} = \frac{\partial F_2}{\partial X_C} = -\frac{Y_C^0 - Y_A}{(L_{AC}^0)^2} = -b_{13}$$

$$b_{24} = \frac{\partial F_2}{\partial Y_C} = \frac{X_C^0 - X_A}{(L_{AC}^0)^2} = b_{14}$$

$$b_{31} = \frac{\partial F_3}{\partial X_B} = -\frac{Y_D - Y_B^0}{(L_{BD}^0)^2} + \frac{Y_A - Y_B^0}{(L_{BA}^0)^2}$$

$$b_{32} = \frac{\partial F_3}{\partial Y_B} = \frac{X_D - X_B^0}{(L_{BD}^0)^2} - \frac{X_A - X_B^0}{(L_{BA}^0)^2}$$

$$b_{41} = \frac{\partial F_4}{\partial X_B} = -\frac{Y_C^0 - Y_B^0}{(L_{BC}^0)^2} + \frac{Y_D - Y_B^0}{(L_{BD}^0)^2}$$

$$b_{42} = \frac{\partial F_4}{\partial Y_B} = \frac{X_C^0 - X_B^0}{(L_{BC}^0)^2} - \frac{X_D - X_B^0}{(L_{BD}^0)^2}$$

$$b_{43} = \frac{\partial F_4}{\partial X_C} = \frac{Y_C^0 - Y_B^0}{(L_{BC}^0)^2}$$

$$b_{44} = \frac{\partial F_4}{\partial Y_C} = -\frac{X_C^0 - X_B^0}{(L_{BC}^0)^2}$$

$$b_{51} = \frac{\partial F_5}{\partial Y_B} = \frac{Y_B^0 - Y_C^0}{(L_{BC}^0)^2} = -b_{43}$$

$$b_{52} = \frac{\partial F_5}{\partial Y_B} = -\frac{X_B^0 - X_C^0}{(L_{BC}^0)^2} = -b_{44}$$

$$b_{53} = \frac{\partial F_5}{\partial X_C} = -\frac{Y_A - Y_C^0}{(L_{AC}^0)^2} + \frac{Y_B^0 - Y_C^0}{(L_{CB}^0)^2}$$

$$b_{54} = \frac{\partial F_5}{\partial Y_C} = \frac{X_A - X_C^0}{(L_{CA}^0)^2} - \frac{X_B^0 - X_C^0}{(L_{CB}^0)^2}$$

$$b_{63} = \frac{\partial F_6}{\partial X_C} = -\frac{Y_D - Y_C^0}{(L_{DC}^0)^2} + \frac{Y_A - Y_C^0}{(L_{AC}^0)^2}$$

$$b_{64} = \frac{\partial F_6}{\partial Y_C} = \frac{X_D - X_C^0}{(L_{CD}^0)^2} - \frac{X_A - X_C^0}{(L_{AC}^0)^2}$$

$$b_{71} = \frac{\partial F_7}{\partial X_B} = \frac{Y_B^0 - Y_D}{(L_{DB}^0)^2}$$

$$b_{72} = \frac{\partial F_7}{\partial Y_B} = -\frac{X_B^0 - X_D}{(L_{BD}^0)^2}$$

$$b_{73} = \frac{\partial F_7}{\partial X_C} = -\frac{Y_C^0 - Y_D}{(L_{DC}^0)^2}$$

$$b_{74} = \frac{\partial F_7}{\partial Y_C} = \frac{X_C^0 - X_D}{(L_{DC}^0)^2}$$

$$b_{81} = \frac{\partial F_8}{\partial X_B} = -\frac{Y_B^0 - Y_D}{(L_{DB}^0)^2} = -b_{71}$$

$$b_{82} = \frac{\partial F_8}{\partial Y_B} = \frac{X_B^0 - X_D}{(L_{DB}^0)^2} = -b_{72}$$

where

$$(L_{AC}^0)^2 = (X_C^0 - X_A)^2 + (Y_C^0 - Y_A)^2$$
$$(L_{AB}^0)^2 = (X_B^0 - X_A)^2 + (Y_B^0 - Y_A)^2$$
$$\vdots$$
$$(L_{DC}^0)^2 = (X_C^0 - Y_D)^2 + (Y_C^0 - Y_D)^2$$

The elements of the matrix **f** formulated according to Eq. (B.71$d$) (Art. B.12) are as follows:

$$f_1 = -\ell_1 + \arctan\frac{X_D - X_A}{Y_D - Y_A} - \arctan\frac{X_C^0 - X_A}{Y_C^0 - Y_A}$$

$$f_2 = -\ell_2 + \arctan\frac{X_C^0 - X_A}{Y_C^0 - Y_A} - \arctan\frac{X_B^0 - X_A}{Y_B^0 - Y_A}$$

$$f_3 = -\ell_3 + \arctan\frac{X_A - X_B^0}{Y_A - Y_B^0} - \arctan\frac{X_D - X_B^0}{Y_D - Y_B^0}$$

$$f_4 = -\ell_4 + \arctan\frac{X_D - X_B^0}{Y_D - Y_B^0} - \arctan\frac{X_C^0 - X_B^0}{Y_C^0 - Y_B^0}$$

$$f_5 = -\ell_5 + \arctan\frac{X_B^0 - X_C^0}{Y_B^0 - Y_C^0} - \arctan\frac{X_A - X_C^0}{Y_A - Y_C^0}$$

$$f_6 = -\ell_6 + \arctan\frac{X_A - X_C^0}{Y_A - Y_C^0} - \arctan\frac{X_D - X_C^0}{Y_D - Y_C^0}$$

$$f_7 = -\ell_7 + \arctan\frac{X_C^0 - X_D}{Y_C^0 - Y_D} - \arctan\frac{X_B^0 - X_D}{Y_B^0 - Y_D}$$

$$f_8 = -\ell_8 + \arctan\frac{X_B^0 - X_D}{Y_B^0 - Y_D} - \arctan\frac{X_A - X_D}{Y_A - Y_D}$$

Using the approximations for coordinates and measured angles, the elements for the **B** and **f** matrices respectively, are

$$\mathbf{B} = 10^{-4}\begin{bmatrix} 0 & 0 & 1.29909 & -3.21077 \\ 2.11461 & -2.63282 & -1.29909 & 3.21077 \\ 2.70246 & 3.18600 & 0 & 0 \\ 5.22866 & 4.47963 & -10.56394 & -5.03283 \\ -10.56394 & -5.03283 & 11.86303 & 1.82206 \\ 0 & 0 & 7.69978 & 1.75412 \\ 5.33528 & 0.55320 & -8.99889 & 1.45665 \\ -5.33528 & -0.55320 & 0 & 0 \end{bmatrix}\left(\frac{1}{m}\right)$$

$$\mathbf{f} = 10^{-7}\begin{bmatrix} 1.77 \\ -1.46 \\ -11.30 \\ 62.43 \\ 150.06 \\ 26.42 \\ -34.81 \\ -60.28 \end{bmatrix}(\text{rad})$$

The normal equations [Eq. (B.33)] are

$$(\mathbf{B}^t\mathbf{WB})\Delta = \mathbf{B}^t\mathbf{Wf}$$

in which $\Delta^t = [\Delta X_B \quad \Delta Y_B \quad \Delta X_C \quad \Delta Y_C]^t$, **B** and **f** have been defined, and $\mathbf{W} = \mathbf{Q}^{-1}$ [according to Eq. (2.32)], where **Q** is defined for this same quadrilateral in Example B.12. The normal equation coefficient matrix $\mathbf{N} = \mathbf{B}^t\mathbf{WB}$ is

$$\mathbf{N} = 10^{-6}\begin{bmatrix} 1.28871 & 0.64644 & -1.73919 & -0.36232 \\ & 0.51371 & -0.96007 & -0.31980 \\ & & 3.23750 & 0.58952 \\ \text{symmetric} & & & 0.31569 \end{bmatrix}\left(\frac{1}{m^2}\right)$$

for which the inverse is

$$\mathbf{N}^{-1}=10^6 \begin{bmatrix} 3.73108 & -2.88114 & 1.36624 & -1.18780 \\ & 10.03030 & 0.27060 & 6.34898 \\ & & 1.19345 & -0.38646 \\ \text{symmetric} & & & 8.95786 \end{bmatrix} (\text{m}^2)$$

The column matrix of constant terms, $\mathbf{t}=\mathbf{B}^t\mathbf{Wf}$, is

$$\mathbf{t}=10^{-8} \begin{bmatrix} -0.87157 \\ -0.33023 \\ 1.80319 \\ 0.05992 \end{bmatrix} (\text{rad}/\text{m})$$

Solution of the normal equations yields corrections

$$\boldsymbol{\Delta}= \begin{bmatrix} 0.001 \\ 0.001 \\ 0.009 \\ -0.012 \end{bmatrix} (\text{m})$$

Generally, more than one iteration is required for convergence of the solution. In this example, the initial approximations for coordinates are very good, so that one iteration is considered adequate. The very small corrections support this assumption. Application of corrections to the approximate coordinates for stations $C$ and $D$ yields the following adjusted values:

| Station | X, m | Y, m |
|---------|------|------|
| B | 17,709.633 | 12,279.788 |
| C | 18,077.199 | 11,508.269 |

Comparison of these coordinates with the final adjusted values calculated in Example B.12 shows that the answers are the same except for $Y_B$, which differs by 1 mm.

The residuals are computed using Eq. (B.39):

$$\mathbf{v}=\mathbf{f}-\mathbf{B}\boldsymbol{\Delta}= \begin{bmatrix} -1.0'' \\ 1.0'' \\ -0.3'' \\ 1.7'' \\ 1.7'' \\ -0.4'' \\ 1.1'' \\ -1.1'' \end{bmatrix}$$

and the adjusted angles are found by Eq. (B.40$a$):

$$\hat{\boldsymbol{\ell}}=\boldsymbol{\ell}+\mathbf{v}= \begin{bmatrix} 22°01'41.5'' \\ 16°44'32.2'' \\ 57°08'56.8'' \\ 19°33'15.8'' \\ 86°33'15.2'' \\ 58°46'35.5'' \\ 15°06'53.4'' \\ 84°04'49.6'' \end{bmatrix}$$

The estimated reference variance determined using Eq. (B.41) is

$$\hat{\sigma}_0^2=\frac{\mathbf{v}^t\mathbf{Wv}}{r}=\frac{4.84}{4}=1.21 \text{ seconds}^2=(0.2854)10^{-10} \text{ rad}^2$$

Recall that angles were derived from the original directions, which were assumed noncorrelated and of equal precision, so that $\mathbf{Q}=\mathbf{I}$. Therefore, $\sigma_0^2$ is assumed unknown and $\sigma_0^2$ must be used to evaluate the propagated covariance matrix for the adjusted coordinates. Thus, using Eqs. (B.38) and (B.42), we obtain

$$\Sigma_{AA} = \sigma_0^2 N^{-1}$$

$$= \begin{bmatrix} (1.06485)10^{-4} & (-0.82228)10^{-4} & (0.38993)10^{-4} & -(0.33900)10^{-4} \\ & (0.28627)10^{-3} & (0.77228)10^{-5} & (1.81200)10^{-4} \\ \hline & & (0.34061)10^{-4} & -(1.10296)10^{-5} \\ \text{symmetric} & & & (0.25566)10^{-3} \end{bmatrix}$$

or

$$\Sigma_{AA} = \begin{bmatrix} \Sigma_{BB} & \Sigma_{BC} \\ \Sigma_{BC}^t & \Sigma_{CC} \end{bmatrix}$$

which has characteristics similar to the covariance matrix propagated and described in Example B.10.

**B.16. Trilateration adjustment**    In Chap. 10, trilateration was explained as the procedure for determining horizontal position through the measurement (most frequently by EDM equipment) of only distances between points. A trilateration net would include several points with a large number of interconnecting lines to provide for redundancy. To fix the orientation of the net, either two horizontal control points or one control point and the azimuth of one line must be known. The least squares adjustment of trilateration nets is best carried out by the method of indirect observations using the distance condition of Eq. (B.73) for each measured line. This adjustment is demonstrated by the following example.

**Example B.14**    Figure B.18 shows a quadrilateral $ABCD$ in which $A$ and $B$ are two horizontal control points with the following California Zone III state plane coordinates:

| Point | $X$, ft | $Y$, ft |
|-------|---------|---------|
| $A$ | 1,495,316.983 | 503,991.197 |
| $B$ | 1,495,056.547 | 504,269.054 |

An HP 3800 distance meter was used to measure all five lines between $A, B$ and $C, D$, and the distances were corrected for reflector and instrument constants and nonlinearity. The distances, reduced to sea level and plane coordinate grid, are:

| Line | Distance, ft | Standard deviation, ft |
|------|--------------|------------------------|
| $AD$ | $\ell_1 = 542.899$ | $\sigma_1 = 0.02543 \approx 0.026$ |
| $AC$ | $\ell_2 = 678.904$ | $\sigma_2 = 0.02678 \approx 0.026$ |
| $BD$ | $\ell_3 = 676.289$ | $\sigma_3 = 0.02676 \approx 0.026$ |
| $BC$ | $\ell_4 = 509.192$ | $\sigma_4 = 0.02509 \approx 0.026$ |
| $DC$ | $\ell_5 = 479.820$ | $\sigma_5 = 0.02479 \approx 0.026$ |

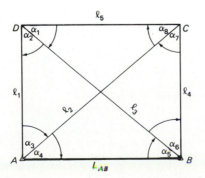

**Fig. B.18**

The uncertainty in EDM measurements can be assumed as 0.02 ft + 10 ppm. Using the least squares method of indirect observations, compute the coordinates of points $C$ and $D$.

SOLUTION Applying the uncertainty criterion for the measured distances, the standard deviation for each distance is computed. For example, $\sigma_1 = 0.02 + 10(542.899)10^{-6} = 0.02543$ ft. The values of all five $\sigma$'s are given in the third column of the preceding table. It can be seen that all five values are very nearly the same, since the five distances do not vary substantially. Consequently, it is quite reasonable to consider that all observations are uncorrelated and of equal precision with a standard deviation of 0.026 ft. Hence, the variance matrix is $(0.026)^2 I$ or $(6.76)10^{-4} I$. Thus, $\sigma_0^2$ is assumed equal to unity.

The five distance condition equations are

$$F_1 = \ell_1 - [(X_A - X_D)^2 + (Y_A - Y_D)^2]^{1/2} = 0$$
$$F_2 = \ell_2 - [(X_A - X_C)^2 + (Y_A - Y_C)^2]^{1/2} = 0$$
$$F_3 = \ell_3 - [(X_B - X_D)^2 + (Y_B - Y_D)^2]^{1/2} = 0$$
$$F_4 = \ell_4 - [(X_B - X_C)^2 + (Y_B - Y_C)^2]^{1/2} = 0$$
$$F_5 = \ell_5 - [(X_D - X_C)^2 + (Y_D - Y_C)^2]^{1/2} = 0$$

The linearized distance equations are

$$v_1 + b_{11} \Delta X_C + b_{12} \Delta Y_C + b_{13} \Delta X_D + b_{14} \Delta Y_D = f_1$$
$$v_2 + b_{21} \Delta X_C + b_{22} \Delta Y_C + b_{23} \Delta X_D + b_{24} \Delta Y_D = f_2$$
$$v_3 + b_{31} \Delta X_C + b_{32} \Delta Y_C + b_{33} \Delta X_D + b_{34} \Delta Y_D = f_3$$
$$v_4 + b_{41} \Delta X_C + b_{42} \Delta Y_C + b_{43} \Delta X_D + b_{44} \Delta Y_D = f_4$$
$$v_5 + b_{51} \Delta X_C + b_{52} \Delta Y_C + b_{53} \Delta X_D + b_{54} \Delta Y_D = f_5$$

in which the coefficients in matrix **B** and elements in the **f** matrix are defined in Eqs. (B.76a) to (B.76e). For this example, the coefficients in **B** and elements of the **f** matrix are

$$b_{11} = \frac{\partial F_1}{\partial X_C} = 0 \qquad\qquad b_{12} = \frac{\partial F_1}{\partial Y_C} = 0$$

$$b_{13} = \frac{\partial F_1}{\partial X_D} = \frac{X_A - X_D^0}{L_{AD}^0} \qquad\qquad b_{14} = \frac{\partial F_1}{\partial Y_D} = \frac{Y_A - Y_D^0}{L_{AD}^0}$$

$$b_{21} = \frac{\partial F_2}{\partial X_C} = \frac{X_A - X_C^0}{L_{AC}^0} \qquad\qquad b_{22} = \frac{\partial F_2}{\partial Y_C} = \frac{Y_A - Y_C^0}{L_{AC}^0}$$

$$b_{23} = \frac{\partial F_2}{\partial X_D} = 0 \qquad\qquad b_{24} = \frac{\partial F_2}{\partial Y_D} = 0$$

$$b_{31} = \frac{\partial F_3}{\partial X_C} = 0 \qquad\qquad b_{32} = \frac{\partial F_3}{\partial Y_C} = 0$$

$$b_{33} = \frac{\partial F_3}{\partial X_D} = \frac{X_B - X_D^0}{L_{BD}^0} \qquad\qquad b_{34} = \frac{\partial F_3}{\partial Y_D} = \frac{Y_B - Y_D^0}{L_{BD}^0}$$

$$b_{41} = \frac{\partial F_4}{\partial X_C} = \frac{X_B - X_C^0}{L_{BC}^0} \qquad\qquad b_{42} = \frac{\partial F_4}{\partial Y_C} = \frac{Y_B - Y_C^0}{L_{BC}^0}$$

$$b_{43} = \frac{\partial F_4}{\partial X_D} = 0 \qquad\qquad b_{44} = \frac{\partial F_4}{\partial Y_D} = 0$$

$$b_{51} = \frac{\partial F_5}{\partial X_C} = \frac{X_D^0 - X_C^0}{L_{DC}^0} \qquad\qquad b_{52} = \frac{\partial F_5}{\partial Y_C} = \frac{Y_D^0 - Y_C^0}{L_{DC}^0}$$

$$b_{53} = \frac{\partial F_5}{\partial X_D} = -b_{51} \qquad\qquad b_{54} = \frac{\partial F_5}{\partial Y_D} = -b_{52}$$

$$f_1 = -\ell_1 + L_{AD}^0 \qquad\qquad f_2 = -\ell_2 + L_{AC}^0$$
$$f_3 = -\ell_3 + L_{BD}^0 \qquad\qquad f_4 = -\ell_4 + L_{BC}^0$$
$$f_5 = -\ell_5 + L_{DC}^0$$

where

$$L_{AD}^0 = [(X_A - X_D^0)^2 + (Y_A - Y_D^0)^2]^{1/2}$$
$$L_{AC}^0 = [(X_A - X_C^0)^2 + (Y_A - Y_C^0)^2]^{1/2}$$
$$L_{BD}^0 = [(X_B^0 - X_D^0)^2 + (Y_B - Y_D^0)^2]^{1/2}$$
$$L_{BC}^0 = [(X_B - X_C^0)^2 + (Y_B - Y_C^0)^2]^{1/2}$$
$$L_{DC}^0 = [(X_D^0 - X_C^0)^2 + (Y_D^0 - Y_C^0)^2]^{1/2}$$

When computations are performed by a computer program on an electronic computer, coarse values for the coordinates of unknown stations scaled from an accurate drawing of the quadrilateral are adequate to use for estimates. If a hand calculator is to be used to form coefficients, fairly good estimates are desirable. In this example, the procedure described in Art. 10.26 is followed and the two plane triangles ABD and ABC are solved by the law of cosines to obtain angles used to determine preliminary azimuths of the lines. These azimuths are then employed with the measured distances to calculate approximate coordinates for stations C and D. In triangle ABD,

$$\alpha_5^0 = \cos^{-1} \frac{(380.830)^2 + (676.289)^2 - (542.899)^2}{(2)(380.830)(676.289)} = 53°19'29''$$

$$\alpha_3^0 + \alpha_4^0 = \cos^{-1} \frac{(380.830)^2 + (542.899)^2 - (676.289)^2}{(2)(380.830)(542.899)} = 92°26'20''$$

$$\alpha_2^0 = \cos^{-1} \frac{(542.899)^2 + (676.289)^2 - (380.830)^2}{(2)(542.899)(676.289)} = 34°14'11''$$

$$\overline{\qquad\qquad 180°00'00''}$$

and in triangle ABC,

$$\alpha_4^0 = \cos^{-1} \frac{(380.830)^2 + (678.904)^2 - (509.192)^2}{(2)(380.830)(678.904)} = 47°54'04''$$

$$\alpha_5^0 + \alpha_6^0 = \cos^{-1} \frac{(380.830)^2 + (509.192)^2 - (678.904)^2}{(2)(380.830)(509.192)} = 98°23'32''$$

$$\alpha_7^0 = \cos^{-1} \frac{(509.192)^2 + (678.904)^2 - (380.830)^2}{(2)(509.192)(678.904)} = 33°42'24''$$

$$\overline{\qquad\qquad 180°00'00''}$$

Preliminary azimuths from the north for all lines in the quadrilateral are calculated using the fixed azimuth of line AB determined from the coordinates of these stations.

$$A_{AB} = \arctan \frac{X_B - X_A}{Y_B - Y_A} = \frac{-260.436}{277.857} = \arctan(-0.937302281)$$
$$= 316°51'13.1''$$

Computations for azimuths are as follows:

| | Triangle ABD | | Triangle ABC |
|---|---|---|---|
| $A_{AB}$ | 316°51'13.1'' | $A_{BA}$ | 136°51'13.1'' |
| $-(\alpha_3^0 + \alpha_4^0)$ | 92°26'20.0'' | $+(\alpha_5^0 + \alpha_6^0)$ | 98°23'31.6'' |
| $A_{AD}^0$ | 224°24'53.1'' | $A_{BC}^0$ | 235°14'44.7'' |
| $-\alpha_2^0$ | 34°14'11.0'' | $A_{CB}^0$ | 55°14'44.7'' |
| $A_{BD}^0$ | 190°10'42.1'' | $+\alpha_7^0$ | 33°42'24.2'' |
| $-\alpha_5^0$ | 53°19'29.0'' | $A_{CA}^0$ | 88°57'08.9'' |
| $A_{BA}$ | 136°51'13.1'' Checks | $A_{AC}^0$ | 268°57'08.7'' |
| | | $+\alpha_4^0$ | 47°54'04.3'' |
| | | Check | 316°51'13.2'' |

The preliminary azimuths just calculated and measured lengths of lines are next used to calculate approximate coordinates for the adjustment. The results of these computations are given in Table B.2.

**Table B.2   Computation of Approximate Coordinates**

| Station | Distance, ft | Azimuth | Coordinates, ft | |
|---|---|---|---|---|
| | | | X | Y |
| | | Triangle ADB | | |
| A | | | 1,495,316.983 | 503,991.197 |
| | 542.899 | 224°24′53.1″ | | |
| D | | | 1,494,937.037 | 503,603.408 |
| | 676.289 | 10°10′42.1″ | | |
| B | | | 1,495,056.546 | 504,269.054 |
| | | Triangle BCA | | |
| B | | | 1,495,056.547 | 504,269.054 |
| | 509.192 | 235°14′44.7″ | | |
| C | | | 1,494,638.192 | 503,978.785 |
| | 678.904 | 88°57′08.9″ | | |
| A | | | 1,495,316.983 | 503,991.197 |

The length and azimuth of line $DC$ are

$$L_{DC}^0 = [(375.377)^2 + (298.845)^2]^{1/2} = 479.808 \text{ ft}$$

$$A_{DC}^0 = \arctan \frac{298.845}{-375.377} = 321°28′34″$$

When computations are programmed for an electronic computer, the approximate coordinates can be used directly to form coefficients for the condition equations. If a hand calculator is being used, it is more convenient to calculate the coefficients with the sine and cosine of the approximate azimuths. Preliminary approximate values for distances and azimuths are listed in Table B.3.

**Table B.3   Preliminary Values for Lines**

| Line | Calculated length $L^0$ | Approximate azimuth $A^0$ |
|---|---|---|
| AD | 542.899 | 224°24′53.1″ |
| AC | 678.903 | 268°57′08.7″ |
| BD | 676.289 | 190°10′42.4″ |
| BC | 509.192 | 235°14′44.7″ |
| DC | 479.808 | 321°28′34.0″ |

Using these estimates, measured angles, and distances, the numerical values for the elements of the **B** and **f** matrices are

$$\mathbf{B} = \begin{bmatrix} 0 & 0 & 0.699847248 & 0.714292538 \\ 0.999832875 & 0.018282758 & 0 & 0 \\ 0 & 0 & 0.176714458 & 0.984262161 \\ 0.821604656 & 0.570057707 & 0 & 0 \\ 0.622840883 & -0.782348537 & -0.622840883 & 0.782348537 \end{bmatrix}$$

$$\mathbf{f}' = [0 \quad -0.001 \quad 0 \quad 0 \quad -0.012] \quad \text{ft}$$

As noted previously, the covariance matrix for measured distances is

$$\Sigma_{tt} = (6.76)10^{-4}\mathbf{I} \quad (\text{ft}^2)$$

and the weight matrix is

$$\mathbf{W} = \Sigma_{tt}^{-1} \quad \left( \frac{1}{\text{ft}^2} \right)$$

The normal equations are

$$(\mathbf{B'WB})\Delta = \mathbf{B'Wf}$$

or

$$\mathbf{N}\Delta = \mathbf{t}$$

from which the inverse of the coefficient matrix and vector of constant terms are

$$\mathbf{N}^{-1} = 10^{-4} \begin{bmatrix} 4.96982 & -4.67851 & 5.24397 & -2.99539 \\ & 20.15700 & -14.50300 & 8.28421 \\ & & 18.91870 & -7.15497 \\ \text{symmetric} & & & 6.99401 \end{bmatrix} (\text{ft}^2)$$

$$\mathbf{t}' = [-12.5354 \quad 13.8608 \quad 11.5603 \quad -13.8878] \left(\frac{1}{\text{ft}}\right)$$

Solution of the normal equations yields corrections

$$\Delta' = [\delta X_C \quad \delta Y_C \quad \delta X_D \quad \delta Y_D]$$

as follows:

$$\Delta' = [-0.0027 \quad 0.0063 \quad 0.0042 \quad -0.0024](\text{ft})$$

Application of these corrections to approximate values gives the following adjusted coordinates for stations $C$ and $D$:

|   | X, ft | Y, ft |
|---|---|---|
| C | 1,494,638.190 | 503,978.791 |
| D | 1,494,937.041 | 503,603.406 |

Since the reference variance $\sigma_0^2$ was initially assumed equal to unity, the propagated covariance matrix $\Sigma_{\Delta\Delta}$ for the adjusted coordinates equals $\mathbf{N}^{-1}$ according to Eqs. (B.38) and (B.42) and as described for the traverse adjustment in Example B.10.

**B.17. Adjustment of combined triangulation and trilateration**    The need for this adjustment (described in Art. 10.27) arises when both angles and distances are measured in a horizontal control survey. Equation (B.67) for angles and Eq. (B.73) for distances are formed to yield one set of normal equations that are solved in a least squares adjustment by indirect observations. The procedures required for this adjustment are illustrated by the following example.

**Example B.15**    Quadrilateral $ABCD$ (Fig. B.18), for which the five measured distances with respective estimated standard deviations are given in Example B.14, also has the eight angles $\alpha_1, \alpha_2, \ldots, \alpha_8$ derived from observed directions. These directions are of equal precision and noncorrelated with an estimated standard deviation of 10″. The angles determined from the directions are listed in Table B.4.

As in the previous adjustments, approximate values are necessary for the coordinates of stations $C$ and $D$ and the distances. Approximate angles, azimuths, distances, and coordinates previously

**Table B.4    Measured and Approximate Angles**

|  | Measured angles | Calculated angles (from Example B.14) | $\alpha_i^0 - \alpha_i$ | |
|---|---|---|---|---|
|  | $\alpha_i$ | $\alpha_i^0$ | seconds | radians |
| 1 | 48°42′06.0″ | 48°42′08.4″ | 2.4 | 0.000011636 |
| 2 | 34°14′03.6″ | 34°14′10.5″ | 6.9 | 0.000033452 |
| 3 | 44°32′06.0″ | 44°32′15.6″ | 9.6 | 0.000046542 |
| 4 | 47°54′00.6″ | 47°54′04.4″ | 3.8 | 0.000018423 |
| 5 | 53°19′16.4″ | 53°19′29.5″ | 13.1 | 0.000063511 |
| 6 | 45°04′00.8″ | 45°04′02.1″ | 1.3 | 0.000006303 |
| 7 | 33°42′29.4″ | 33°42′24.0″ | -5.4 | -0.000026180 |
| 8 | 52°31′13.1″ | 52°31′25.5″ | 12.4 | 0.000060117 |

calculated in Example B.14 are used for this adjustment. Angles calculated in Example B.14 are also tabulated in Table B.4. The differences between calculated and measured angles constitute the elements of the **f** matrix of the angle condition equations and are listed in column 3 of Table B.4.

Coefficients for the eight angle condition equations (B.67) are calculated as described in Example B.13 using approximate azimuths and distances from Table B.3 (Art. B.16) or given and approximate coordinates for the triangulation stations as provided in Example B.14, Table B.2.

Coefficients for the five distance equations (B.73) and elements of the **f** matrix as determined in Example B.14 are used for this adjustment. Thus, eight angle and five distance equations, in that order, or a total of 13 condition equations are formed for this adjustment by indirect observations $(\mathbf{v}+\mathbf{B}\Delta=\mathbf{f})$ to form the following **B** and **f** matrices:

$$\mathbf{B}=10^{-3}\begin{bmatrix} 1.630545 & 1.298104 & -0.175158 & -1.559405 \\ 0 & 0 & -0.139685 & -1.027791 \\ 0.026931 & -1.472718 & -1.315702 & 1.289092 \\ -0.026931 & 1.472718 & 0 & 0 \\ 0 & 0 & 1.455387 & -0.261300 \\ 1.119535 & -1.613545 & -1.455387 & 0.261300 \\ -1.092604 & 0.140827 & 0 & 0 \\ -1.65748 & 0.174614 & 1.603545 & 1.298104 \\ 0 & 0 & 699.8466 & 714.2932 \\ 999.8329 & 18.2827 & 0 & 0 \\ 0 & 0 & 176.7146 & 984.2622 \\ 821.6046 & 570.0577 & 0 & 0 \\ 622.8409 & -782.3485 & -622.8409 & 782.3485 \end{bmatrix}$$

$$\mathbf{f}'=10^{-5}[\ 1.1636 \quad 3.3452 \quad 4.6542 \quad 1.8423 \quad 6.3511 \quad 0.6303 \quad -2.6180$$
$$6.0117 \quad 0 \quad 100 \quad 0 \quad 0 \quad -1200\ ]$$

Assuming a variance in measured directions of 100 seconds$^2$ or $2.3504(10)^{-9}$ rad$^2$, the propagated covariance matrix for angles from a single station [see Example B.12 and propagate with Eq. (2.43)] is

$$\Sigma_{angles}=\begin{bmatrix} 0 & -1 & 1 \\ -1 & 1 & 0 \end{bmatrix}(10)^{-9}\begin{bmatrix} 2.35044 & 0 & 0 \\ 0 & 2.35044 & 0 \\ 0 & 0 & 2.35044 \end{bmatrix}\begin{bmatrix} 0 & -1 \\ -1 & 1 \\ 1 & 0 \end{bmatrix}$$

$$=(2.35044)10^{-9}\begin{bmatrix} 2 & -1 \\ -1 & 2 \end{bmatrix}$$

Measured distances are of equal precision and noncorrelated with respective variances of 0.000676 ft$^2$/line. Assuming a reference variance, $\sigma_0^2$, of unity, the covariance matrix for the eight measured angles and five distances is

$$\Sigma_{\ell\ell}=(2.3504)10^{-9}$$

$$\times\begin{bmatrix} 2 & -1 & 0 & & & & & & & & & & \\ -1 & 2 & 0 & 0 & & & & & & & & & \quad \mathbf{0} \\ 0 & 0 & 2 & -1 & 0 & & & & & & & & \\ 0 & -1 & 2 & 0 & 0 & & & & & & & & \\ & & 0 & 2 & -1 & 0 & & & & & & & \\ & & 0 & -1 & 2 & 0 & 0 & & & & & & \\ & & & & 0 & 2 & -1 & & & & & & \\ & & & & 0 & -1 & 2 & 0 & & & & & \\ & & & & & & 0 & (2.876)10^5 & 0 & & & & \\ & & & & & & & 0 & (2.876)10^5 & 0 & & & \\ & & & & & & & & 0 & (2.876)10^5 & 0 & & \\ & & & & & & & & & 0 & (2.876)10^5 & 0 & \\ \mathbf{0} & & & & & & & & & & 0 & (2.876)10^5 \end{bmatrix}$$

and the weight matrix becomes $\mathbf{W}=\Sigma_{\ell\ell}^{-1}$. The normal equations are

$$\mathbf{N}\Delta=\mathbf{t}$$

in which $\mathbf{N}=\mathbf{B}'\mathbf{WB}$, $\Delta'=[\delta X_C \quad \delta Y_C \quad \delta X_D \quad \delta Y_D]$, and $\mathbf{t}=\mathbf{B}'\mathbf{Wf}$. The inverse of the normal equation coefficient matrix and the vector **t** are

$$\mathbf{N}^{-1} = 10^{-4} \begin{bmatrix} 2.25526 & -0.51482 & 1.56597 & 0.10381 \\ & 5.50016 & -2.93210 & 2.05263 \\ & & 5.49383 & -1.18173 \\ \text{symmetric} & & & 2.64349 \end{bmatrix} (\text{ft}^2)$$

$$\mathbf{t}^t = [-10.593 \quad 3.4984 \quad 20.4850 \quad 2.3002] \left(\frac{1}{\text{ft}}\right)$$

Solution of the normal equations gives corrections

$$\Delta^t = [0.001 \quad -0.004 \quad 0.008 \quad -0.002] \quad (\text{ft})$$

These corrections applied to the approximate coordinates for stations $C$ and $D$ yield adjusted coordinates as follows:

$$\hat{X}_C = X_C^0 + \delta X_C = 1{,}494{,}638.192 + 0.001 = 1{,}494{,}638{,}193$$
$$\hat{Y}_C = Y_C^0 + \delta Y_C = \phantom{1{,}49}503{,}978.785 - 0.004 = \phantom{1{,}49}503{,}978{,}781$$
$$\hat{X}_D = X_D^0 + \delta X_D = 1{,}494{,}937.037 + 0.008 = 1{,}494{,}937{,}045$$
$$\hat{Y}_D = Y_D^0 + \delta Y_D = \phantom{1{,}49}503{,}603.408 - 0.002 = \phantom{1{,}49}503{,}603{,}406$$

# References

1. Aguilar, A. M., "Principles of Survey Error Analysis and Adjustment," *Surveying and Mapping*, September 1973.
2. Baker, J. R., "Adjusting Radial Surveys," *Surveying and Mapping*, November 1972.
3. Bird, R. G., "Least Squares Adjustment of E.D.M. Traverse," *Survey Review*, July 1972.
4. Cross, P. A., "The Effect of Errors in Weights," *Survey Review*, July 1972.
5. Greggor, K. N., "Computation and Adjustment by Geometric Network," *Survey Review*, October 1973.
6. Jones, P. B., "A Comparison of the Precision of Traverses Adjusted by the Bowditch Rule and by Least Squares," *Survey Review*, April 1972.
7. Mikhail, E. M., *Observations and Least Squares*, Harper & Row Publishers, New York, 1976.
8. Mikhail, E. M., and Gracie, G., *Survey Computations and Adjustment*, Van Nostrand Reinhold Company, New York, 1980.
9. Murphy, B. T., "The Adjustment of Single Traverses," *Australian Surveyor*, December 1974.
10. Pinch, M. C., et al., "A Method for Adjusting Survey Networks in Sections," *Canadian Surveyor*, March 1974.
11. Richardus, P., *Project Surveying*, North-Holland Publishing Company, Amsterdam, 1966.
12. Sokolnikoff and Redheffer, *Mathematics of Physics and Modern Engineering*, McGraw-Hill Book Company, New York, 1958, pp. 249–257.
13. Thompson, E. H., "The Theory of the Method of Least Squares," *Photogrammetric Record*, April 1962.
14. Veress, S. A., *Adjustment by Least Squares*, Monograph, ACSM, Washington, D.C., 1974.
15. Wolf, P. R., "Horizontal Position Adjustment," *Surveying and Mapping*, December 1969.

# APPENDIX C
# General tables

**Table I   Correction for Refraction and Parallax, to Be Subtracted from the Observed Altitude of the Sun (Barometric pressure, 29.5 In )**

| App't alt. | Temperature | | | | | | | | | | App't alt. |
|---|---|---|---|---|---|---|---|---|---|---|---|
| | 10°C 14°F 1+ | 5°C 23°F 1+ | 0°C 32°F 0+ ++ | 5°C 41°F ++ | 10°C 50°F ++ | 15°C 59°F ++ | 20°C 68°F ++ | 25°C 77°F ++ | 30°C 86°F ++ | 35°C 95°F ++ | |
| ° | ' | ' | ' | ' | ' | ' | ' | ' | ' | ' | ° |
| 10 | 5.52 | 5.42 | 5.30 | 5.20 | 5.10 | 5.00 | 4.92 | 4.83 | 4.75 | 4.67 | 10 |
| 11 | 5.02 | 4.92 | 4.82 | 4.73 | 4.63 | 4.55 | 4.47 | 4.38 | 4.32 | 4.23 | 11 |
| 12 | 4.60 | 4.50 | 4.42 | 4.33 | 4.25 | 4.17 | 4.10 | 4.03 | 3.97 | 3.88 | 12 |
| 13 | 4.23 | 4.15 | 4.07 | 4.00 | 3.92 | 3.85 | 3.78 | 3.72 | 3.65 | 3.58 | 13 |
| 14 | 3.92 | 3.83 | 3.77 | 3.70 | 3.62 | 3.55 | 3.50 | 3.45 | 3.37 | 3.32 | 14 |
| 15 | 3.65 | 3.58 | 3.50 | 3.43 | 3.37 | 3.32 | 3.25 | 3.20 | 3.13 | 3.08 | 15 |
| 16 | 3.43 | 3.35 | 3.30 | 3.23 | 3.17 | 3.12 | 3.07 | 3.00 | 2.95 | 2.90 | 16 |
| 17 | 3.22 | 3.15 | 3.10 | 3.03 | 2.98 | 2.92 | 2.88 | 2.82 | 2.77 | 2.72 | 17 |
| 18 | 3.02 | 2.95 | 2.90 | 2.85 | 2.80 | 2.75 | 2.70 | 2.65 | 2.60 | 2.55 | 18 |
| 19 | 2.83 | 2.78 | 2.73 | 2.68 | 2.63 | 2.58 | 2.53 | 2.48 | 2.43 | 2.40 | 19 |
| 20 | 2.68 | 2.63 | 2.58 | 2.53 | 2.48 | 2.43 | 2.38 | 2.33 | 2.30 | 2.27 | 20 |
| 21 | 2.53 | 2.48 | 2.43 | 2.38 | 2.35 | 2.30 | 2.27 | 2.22 | 2.17 | 2.13 | 21 |
| 22 | 2.38 | 2.35 | 2.30 | 2.25 | 2.22 | 2.18 | 2.13 | 2.08 | 2.05 | 2.02 | 22 |
| 23 | 2.28 | 2.25 | 2.20 | 2.15 | 2.12 | 2.08 | 2.03 | 1.98 | 1.95 | 1.93 | 23 |
| 24 | 2.17 | 2.13 | 2.08 | 2.05 | 2.02 | 1.98 | 1.93 | 1.88 | 1.87 | 1.83 | 24 |
| 25 | 2.07 | 2.03 | 1.98 | 1.95 | 1.92 | 1.88 | 1.83 | 1.80 | 1.77 | 1.75 | 25 |
| 26 | 1.99 | 1.95 | 1.90 | 1.87 | 1.83 | 1.80 | 1.75 | 1.72 | 1.70 | 1.67 | 26 |
| 27 | 1.88 | 1.85 | 1.82 | 1.78 | 1.75 | 1.72 | 1.68 | 1.63 | 1.62 | 1.60 | 27 |
| 28 | 1.80 | 1.77 | 1.72 | 1.70 | 1.67 | 1.63 | 1.60 | 1.57 | 1.53 | 1.52 | 28 |
| 29 | 1.72 | 1.68 | 1.65 | 1.63 | 1.60 | 1.57 | 1.53 | 1.50 | 1.47 | 1.46 | 29 |
| 30 | 1.65 | 1.62 | 1.58 | 1.57 | 1.53 | 1.50 | 1.47 | 1.45 | 1.42 | 1.40 | 30 |
| 32 | 1.53 | 1.50 | 1.47 | 1.45 | 1.42 | 1.38 | 1.35 | 1.33 | 1.30 | 1.28 | 32 |
| 34 | 1.41 | 1.37 | 1.35 | 1.32 | 1.30 | 1.27 | 1.25 | 1.23 | 1.20 | 1.18 | 34 |
| 36 | 1.30 | 1.27 | 1.25 | 1.22 | 1.20 | 1.18 | 1.15 | 1.13 | 1.10 | 1.08 | 36 |
| 38 | 1.20 | 1.18 | 1.15 | 1.13 | 1.12 | 1.10 | 1.07 | 1.05 | 1.02 | 1.02 | 38 |
| 40 | 1.11 | 1.10 | 1.07 | 1.05 | 1.03 | 1.02 | 0.98 | 0.97 | 0.95 | 0.93 | 40 |
| 42 | 1.03 | 1.00 | 0.98 | 0.97 | 0.95 | 0.93 | 0.90 | 0.88 | 0.87 | 0.87 | 42 |
| 44 | 0.96 | 0.93 | 0.92 | 0.90 | 0.88 | 0.87 | 0.85 | 0.83 | 0.82 | 0.80 | 44 |
| 46 | 0.89 | 0.88 | 0.87 | 0.85 | 0.83 | 0.82 | 0.80 | 0.78 | 0.77 | 0.75 | 46 |
| 48 | 0.83 | 0.82 | 0.80 | 0.78 | 0.77 | 0.75 | 0.73 | 0.72 | 0.70 | 0.68 | 48 |
| 50 | 0.77 | 0.75 | 0.73 | 0.72 | 0.70 | 0.68 | 0.67 | 0.67 | 0.65 | 0.63 | 50 |
| 55 | 0.63 | 0.62 | 0.60 | 0.60 | 0.58 | 0.57 | 0.57 | 0.55 | 0.53 | 0.52 | 55 |
| 60 | 0.52 | 0.52 | 0.50 | 0.50 | 0.48 | 0.47 | 0.47 | 0.45 | 0.45 | 0.43 | 60 |
| 65 | 0.42 | 0.40 | 0.40 | 0.40 | 0.38 | 0.38 | 0.37 | 0.37 | 0.35 | 0.33 | 65 |
| 70 | 0.32 | 0.32 | 0.32 | 0.30 | 0.30 | 0.30 | 0.28 | 0.28 | 0.28 | 0.27 | 70 |
| 75 | 0.23 | 0.23 | 0.23 | 0.22 | 0.22 | 0.22 | 0.20 | 0.20 | 0.20 | 0.18 | 75 |
| 80 | 0.15 | 0.15 | 0.13 | 0.13 | 0.13 | 0.13 | 0.13 | 0.12 | 0.12 | 0.12 | 80 |
| 85 | 0.07 | 0.07 | 0.07 | 0.07 | 0.07 | 0.07 | 0.07 | 0.05 | 0.05 | 0.05 | 85 |
| 90 | 0.00 | 0.00 | 0.00 | 0.00 | 0.00 | 0.00 | 0.00 | 0.00 | 0.00 | 0.00 | 90 |

## Table II   Correction for Refraction, to Be Subtracted from the Observed Altitude of a Star (Barometric pressure, 29.5 In )

| App't alt. | Temperature | | | | | | | | | | App't alt. |
|---|---|---|---|---|---|---|---|---|---|---|---|
| | °C 10° °F 14° 1+ | °C 5° °F 23° 1+ | °C 0° °F 32° 0+ | °C 5° °F 41° ++ | °C 10° °F 50° ++ | °C 15° °F 59° ++ | °C 20° °F 68° ++ | °C 25° °F 77° ++ | °C 30° °F 86° ++ | °C 35° °F 95° ++ | |
| ° | ′ | ′ | ′ | ′ | ′ | ′ | ′ | ′ | ′ | ′ | ° |
| 10 | 5.67 | 5.57 | 5.45 | 5.35 | 5.25 | 5.15 | 5.07 | 4.98 | 4.90 | 4.82 | 10 |
| 11 | 5.17 | 5.07 | 4.97 | 4.88 | 4.78 | 4.70 | 4.62 | 4.53 | 4.47 | 4.38 | 11 |
| 12 | 4.75 | 4.65 | 4.57 | 4.48 | 4.40 | 4.32 | 4.25 | 4.18 | 4.12 | 4.03 | 12 |
| 13 | 4.38 | 4.30 | 4.22 | 4.15 | 4.07 | 4.00 | 3.93 | 3.87 | 3.80 | 3.73 | 13 |
| 14 | 4.06 | 3.97 | 3.91 | 3.84 | 3.76 | 3.69 | 3.64 | 3.59 | 3.51 | 3.46 | 14 |
| 15 | 3.79 | 3.72 | 3.64 | 3.57 | 3.51 | 3.46 | 3.39 | 3.34 | 3.27 | 3.22 | 15 |
| 16 | 3.57 | 3.49 | 3.44 | 3.37 | 3.31 | 3.26 | 3.21 | 3.14 | 3.09 | 3.04 | 16 |
| 17 | 3.36 | 3.29 | 3.24 | 3.17 | 3.12 | 3.06 | 3.02 | 2.96 | 2.91 | 2.86 | 17 |
| 18 | 3.16 | 3.09 | 3.04 | 2.99 | 2.94 | 2.89 | 2.84 | 2.79 | 2.74 | 2.69 | 18 |
| 19 | 2.97 | 2.92 | 2.87 | 2.82 | 2.77 | 2.72 | 2.67 | 2.62 | 2.57 | 2.54 | 19 |
| 20 | 2.82 | 2.77 | 2.72 | 2.67 | 2.62 | 2.57 | 2.52 | 2.47 | 2.44 | 2.41 | 20 |
| 21 | 2.67 | 2.62 | 2.57 | 2.52 | 2.49 | 2.44 | 2.41 | 2.36 | 2.31 | 2.27 | 21 |
| 22 | 2.52 | 2.49 | 2.44 | 2.39 | 2.36 | 2.32 | 2.27 | 2.22 | 2.19 | 2.16 | 22 |
| 23 | 2.42 | 2.39 | 2.34 | 2.29 | 2.26 | 2.22 | 2.17 | 2.12 | 2.09 | 2.07 | 23 |
| 24 | 2.31 | 2.27 | 2.22 | 2.19 | 2.16 | 2.12 | 2.07 | 2.02 | 2.01 | 1.97 | 24 |
| 25 | 2.21 | 2.17 | 2.12 | 2.09 | 2.06 | 2.02 | 1.97 | 1.94 | 1.91 | 1.89 | 25 |
| 26 | 2.12 | 2.08 | 2.03 | 2.00 | 1.96 | 1.93 | 1.88 | 1.85 | 1.83 | 1.80 | 26 |
| 27 | 2.01 | 1.98 | 1.95 | 1.91 | 1.88 | 1.85 | 1.81 | 1.76 | 1.75 | 1.73 | 27 |
| 28 | 1.93 | 1.90 | 1.85 | 1.83 | 1.80 | 1.76 | 1.73 | 1.70 | 1.66 | 1.65 | 28 |
| 29 | 1.85 | 1.81 | 1.78 | 1.76 | 1.73 | 1.70 | 1.66 | 1.63 | 1.60 | 1.59 | 29 |
| 30 | 1.78 | 1.75 | 1.71 | 1.70 | 1.66 | 1.63 | 1.60 | 1.58 | 1.55 | 1.53 | 30 |
| 32 | 1.65 | 1.62 | 1.59 | 1.57 | 1.54 | 1.50 | 1.47 | 1.45 | 1.42 | 1.40 | 32 |
| 34 | 1.53 | 1.49 | 1.47 | 1.44 | 1.42 | 1.39 | 1.35 | 1.35 | 1.32 | 1.30 | 34 |
| 36 | 1.42 | 1.39 | 1.37 | 1.34 | 1.32 | 1.30 | 1.27 | 1.25 | 1.22 | 1.20 | 36 |
| 38 | 1.32 | 1.30 | 1.27 | 1.25 | 1.24 | 1.22 | 1.19 | 1.17 | 1.14 | 1.14 | 38 |
| 40 | 1.22 | 1.21 | 1.18 | 1.16 | 1.14 | 1.13 | 1.09 | 1.08 | 1.06 | 1.04 | 40 |
| 42 | 1.14 | 1.11 | 1.09 | 1.08 | 1.06 | 1.04 | 1.01 | 0.99 | 0.98 | 0.98 | 42 |
| 44 | 1.07 | 1.04 | 1.03 | 1.01 | 0.99 | 0.98 | 0.96 | 0.94 | 0.93 | 0.91 | 44 |
| 46 | 0.99 | 0.98 | 0.97 | 0.95 | 0.93 | 0.92 | 0.90 | 0.88 | 0.87 | 0.85 | 46 |
| 48 | 0.93 | 0.92 | 0.90 | 0.88 | 0.87 | 0.85 | 0.83 | 0.82 | 0.80 | 0.78 | 48 |
| 50 | 0.86 | 0.84 | 0.82 | 0.81 | 0.79 | 0.77 | 0.76 | 0.76 | 0.74 | 0.72 | 50 |
| 55 | 0.72 | 0.71 | 0.69 | 0.69 | 0.67 | 0.66 | 0.66 | 0.64 | 0.62 | 0.61 | 55 |
| 60 | 0.59 | 0.59 | 0.57 | 0.57 | 0.55 | 0.54 | 0.54 | 0.52 | 0.52 | 0.50 | 60 |
| 65 | 0.48 | 0.46 | 0.46 | 0.46 | 0.44 | 0.44 | 0.43 | 0.43 | 0.41 | 0.39 | 65 |
| 70 | 0.37 | 0.37 | 0.37 | 0.35 | 0.35 | 0.35 | 0.33 | 0.33 | 0.33 | 0.32 | 70 |
| 75 | 0.27 | 0.27 | 0.27 | 0.26 | 0.26 | 0.26 | 0.24 | 0.24 | 0.24 | 0.22 | 75 |
| 80 | 0.18 | 0.18 | 0.16 | 0.16 | 0.16 | 0.16 | 0.16 | 0.15 | 0.15 | 0.15 | 80 |
| 85 | 0.08 | 0.08 | 0.08 | 0.08 | 0.08 | 0.08 | 0.08 | 0.06 | 0.06 | 0.06 | 85 |
| 90 | 0.00 | 0.00 | 0.00 | 0.00 | 0.00 | 0.00 | 0.00 | 0.00 | 0.00 | 0.00 | 90 |

## Table III   For Reducing to Elongation of Polaris Observations Made near Elongation

| *Time (m) | 1° 0′ | 1° 10′ | 1° 20′ | 1° 30′ | 1° 40′ | 1° 50′ | 2° 0′ | 2° 10′ | *Time (m) |
|---|---|---|---|---|---|---|---|---|---|
| 0 | 0.0 | 0.0 | 0.0 | 0.0 | 0.0 | 0.0 | 0.0 | 0.0 | 0 |
| 1 | 0.0 | 0.0 | 0.0 | +0.1 | +0.1 | +0.1 | +0.1 | +0.1 | 1 |
| 2 | +0.1 | +0.2 | +0.2 | 0.2 | 0.2 | 0.3 | 0.3 | 0.3 | 2 |
| 3 | 0.3 | 0.4 | 0.4 | 0.5 | 0.5 | 0.6 | 0.6 | 0.7 | 3 |
| 4 | 0.5 | 0.6 | 0.7 | 0.8 | 0.9 | 1.0 | 1.1 | 1.2 | 4 |
| 5 | +0.9 | +1.0 | +1.1 | +1.3 | +1.4 | +1.6 | +1.7 | +1.9 | 5 |
| 6 | 1.2 | 1.4 | 1.6 | 1.8 | 2.1 | 2.3 | 2.5 | 2.7 | 6 |
| 7 | 1.7 | 2.0 | 2.2 | 2.5 | 2.8 | 3.1 | 3.4 | 3.7 | 7 |
| 8 | 2.2 | 2.6 | 2.9 | 3.3 | 3.7 | 4.0 | 4.4 | 4.8 | 8 |
| 9 | 2.8 | 3.2 | 3.7 | 4.2 | 4.6 | 5.1 | 5.6 | 6.0 | 9 |
| 10 | +3.4 | +4.0 | +4.6 | +5.1 | +5.7 | +6.3 | +6.9 | +7.4 | 10 |
| 11 | 4.1 | 4.8 | 5.5 | 6.2 | 6.9 | 7.6 | 8.3 | 9.0 | 11 |
| 12 | 4.9 | 5.8 | 6.6 | 7.4 | 8.2 | 9.0 | 9.9 | 10.7 | 12 |
| 13 | 5.8 | 6.8 | 7.7 | 8.7 | 9.7 | 10.6 | 11.6 | 12.6 | 13 |
| 14 | 6.7 | 7.8 | 9.0 | 10.1 | 11.2 | 12.3 | 13.4 | 14.6 | 14 |
| 15 | +7.7 | +9.0 | +10.3 | +11.6 | +12.8 | +14.1 | +15.4 | +16.7 | 15 |
| 16 | 8.8 | 10.2 | 11.7 | 13.2 | 14.6 | 16.1 | 17.5 | 19.0 | 16 |
| 17 | 9.9 | 11.5 | 13.2 | 14.9 | 16.5 | 18.2 | 19.8 | 21.5 | 17 |
| 18 | 11.1 | 12.9 | 14.8 | 16.7 | 18.5 | 20.4 | 22.2 | 24.1 | 18 |
| 19 | 12.4 | 14.4 | 16.5 | 18.6 | 20.6 | 22.7 | 24.7 | 26.8 | 19 |
| 20 | +13.7 | +16.0 | +18.3 | +20.6 | +22.8 | +25.1 | +27.4 | +29.7 | 20 |
| 21 | 15.1 | 17.6 | 20.1 | 22.7 | 25.2 | 27.7 | 30.2 | 32.7 | 21 |
| 22 | 16.6 | 19.3 | 22.1 | 24.9 | 27.6 | 30.4 | 33.2 | 35.9 | 22 |
| 23 | 18.1 | 21.1 | 24.2 | 27.2 | 30.2 | 33.2 | 36.2 | 39.3 | 23 |
| 24 | 19.7 | 23.0 | 26.3 | 29.6 | 32.9 | 36.2 | 39.5 | 42.8 | 24 |
| 25 | +21.4 | +25.0 | +28.5 | +32.1 | +35.7 | +39.2 | +42.8 | +46.4 | 25 |

| *Time (m) | 2° 10′ | 2° 20′ | 2° 30′ | 2° 40′ | 2° 50′ | 3° 0′ | 3° 10′ | 3° 20′ | *Time (m) |
|---|---|---|---|---|---|---|---|---|---|
| 0 | 0.0 | 0.0 | 0.0 | 0.0 | 0.0 | 0.0 | 0.0 | 0.0 | 0 |
| 1 | +0.1 | +0.1 | +0.1 | +0.1 | +0.1 | +0.1 | +0.1 | +0.1 | 1 |
| 2 | 0.3 | 0.3 | 0.4 | 0.4 | 0.4 | 0.4 | 0.4 | 0.5 | 2 |
| 3 | 0.7 | 0.7 | 0.8 | 0.8 | 0.9 | 0.9 | 1.0 | 1.0 | 3 |
| 4 | 1.2 | 1.3 | 1.4 | 1.5 | 1.6 | 1.6 | 1.7 | 1.8 | 4 |
| 5 | +1.9 | +2.0 | +2.1 | +2.3 | +2.4 | +2.6 | +2.7 | +2.9 | 5 |
| 6 | 2.7 | 2.9 | 3.1 | 3.3 | 3.5 | 3.7 | 3.9 | 4.1 | 6 |
| 7 | 3.7 | 3.9 | 4.2 | 4.5 | 4.8 | 5.0 | 5.3 | 5.6 | 7 |
| 8 | 4.8 | 5.1 | 5.5 | 5.9 | 6.2 | 6.6 | 7.0 | 7.3 | 8 |
| 9 | 6.0 | 6.5 | 7.0 | 7.4 | 7.9 | 8.3 | 8.8 | 9.3 | 9 |
| 10 | +7.4 | +8.0 | +8.6 | +9.2 | +9.7 | +10.3 | +10.9 | +11.4 | 10 |
| 11 | 9.0 | 9.7 | 10.4 | 11.1 | 11.8 | 12.4 | 13.1 | 13.8 | 11 |
| 12 | 10.7 | 11.5 | 12.3 | 13.2 | 14.0 | 14.8 | 15.6 | 16.5 | 12 |
| 13 | 12.6 | 13.5 | 14.5 | 15.4 | 16.4 | 17.4 | 18.4 | 19.3 | 13 |
| 14 | 14.6 | 15.7 | 16.8 | 17.9 | 19.0 | 20.2 | 21.3 | 22.4 | 14 |
| 15 | +16.7 | +18.0 | +19.3 | +20.6 | +21.9 | +23.1 | +24.4 | +25.7 | 15 |
| 16 | 19.0 | 20.5 | 21.9 | 23.4 | 24.9 | 26.3 | 27.8 | 29.3 | 16 |
| 17 | 21.5 | 23.1 | 24.8 | 26.4 | 28.1 | 29.7 | 31.4 | 33.0 | 17 |
| 18 | 24.1 | 25.9 | 27.8 | 29.6 | 31.5 | 33.3 | 35.2 | 37.0 | 18 |
| 19 | 26.8 | 28.9 | 30.9 | 33.0 | 35.1 | 37.1 | 39.2 | 41.3 | 19 |
| 20 | +29.7 | +32.0 | +34.3 | +36.6 | +38.8 | +41.1 | +43.4 | +45.7 | 20 |
| 21 | 32.7 | 35.3 | 37.8 | 40.3 | 42.8 | 45.3 | 47.9 | 50.4 | 21 |
| 22 | 35.9 | 38.7 | 41.5 | 44.2 | 47.0 | 49.8 | 52.5 | 55.3 | 22 |
| 23 | 39.3 | 42.3 | 45.3 | 48.3 | 51.4 | 54.4 | 57.4 | 60.4 | 23 |
| 24 | 42.8 | 46.0 | 49.3 | 52.6 | 55.9 | 59.2 | 62.5 | 65.8 | 24 |
| 25 | +46.4 | +49.9 | +53.5 | +57.1 | +60.7 | +64.2 | +67.8 | +71.4 | 25 |

*Sidereal time from elongation.

## Table IV   Trigonometric Formulas

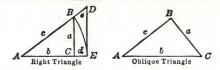

Right Triangle          Oblique Triangle

---

### RIGHT TRIANGLES

$$\sin A = \frac{a}{c} = \cos B \qquad\qquad \sec\ A = \frac{c}{b} = \operatorname{cosec} B$$

$$\cos A = \frac{b}{c} = \sin B \qquad\qquad \operatorname{cosec} A = \frac{c}{a} = \sec B$$

$$\tan A = \frac{a}{b} = \cot B \qquad\qquad \operatorname{vers}\ A = \frac{c-b}{c} = \frac{d}{c}$$

$$\cot A = \frac{b}{a} = \tan B \qquad\qquad \operatorname{exsec} A = \frac{e}{c}$$

$$a = c \sin A = c \cos B = b \tan A = b \cot B = \sqrt{c^2 - b^2}$$

$$b = c \cos A = c \sin B = a \cot A = a \tan B = \sqrt{c^2 - a^2}$$

$$c = \frac{a}{\sin A} = \frac{a}{\cos B} = \frac{b}{\sin B} = \frac{b}{\cos A} = \frac{d}{\operatorname{vers} A} = \frac{e}{\operatorname{exsec} A} = \sqrt{a^2 + b^2}$$

$$d = c \operatorname{vers} A \qquad\qquad\qquad\qquad\qquad\qquad\qquad e = c \operatorname{exsec} A$$

---

## Table IV   Trigonometric Formulas (*continued*)—Oblique triangles

| Given | Sought | Formulas | |
|-------|--------|----------|---|
| $A, B, a$ | $b, c$ | $b = \dfrac{a}{\sin A}\cdot\sin B$ | $c = \dfrac{a}{\sin A}\cdot\sin (A + B)$ |
| $A, a, b$ | $B, c$ | $\sin B = \dfrac{\sin A}{a}\cdot b$ | $c = \dfrac{a}{\sin A}\cdot\sin C$ |
| $C, a, b$ | $\frac{1}{2}(A + B)$ | $\frac{1}{2}(A + B) = 90° - \frac{1}{2}C$ | |
| | $\frac{1}{2}(A - B)$ | $\tan\frac{1}{2}(A - B) = \dfrac{a - b}{a + b}\cdot\tan \frac{1}{2}(A + B)$ | |
| $a, b, c$ | $A$ | If $\ s = \frac{1}{2}(a + b + c)$, $\sin \frac{1}{2}A = \sqrt{\dfrac{(s - b)(s - c)}{bc}}$ | |
| | | $\cos \frac{1}{2}A = \sqrt{\dfrac{s(s - a)}{bc}}$, $\tan \frac{1}{2}A = \sqrt{\dfrac{(s - b)(s - c)}{s(s - a)}}$ | |
| | | $\sin A = 2\dfrac{\sqrt{s(s - a)(s - b)(s - c)}}{bc}$ | |
| | | $\operatorname{vers} A = \dfrac{2(s - b)(s - c)}{bc}$ | |
| | area | $area = \sqrt{s(s - a)(s - b)(s - c)}$ | |
| $C, a, b$ | area | $area = \frac{1}{2}ab \sin C$ | |
| $A, B, C, a$ | area | $area = \dfrac{a^2 \sin B\cdot\sin C}{2 \sin A}$ | |

## Table V   Convergence of Meridians, Six Miles Long and Six Miles Apart, and Differences of Latitude and Longitude
### (U.S. Bureau of Land Management)

| Lat. | Convergency | | Difference of longitude per range | | | Difference of latitude for— | |
|---|---|---|---|---|---|---|---|
| | On the parallel | Angle | In arc | | In time | 1 mi. | 1 Tp. |
| ° | Lks. | ′ ″ | ′ | ″ | Seconds | ′ | ′ |
| 25 | 33.9 | 2 25 | 5 | 44.34 | 22.96 | | |
| 26 | 35.4 | 2 32 | 5 | 47.20 | 23.15 | | |
| 27 | 37.0 | 2 39 | 5 | 50.22 | 23.35 | 0.871 | 5.229 |
| 28 | 38.6 | 2 46 | 5 | 53.40 | 23.56 | | |
| 29 | 40.2 | 2 53 | 5 | 56.74 | 23.78 | | |
| 30 | 41.9 | 3 0 | 6 | 0.26 | 24.02 | | |
| 31 | 43.6 | 3 7 | 6 | 3.97 | 24.26 | | |
| 32 | 45.4 | 3 15 | 6 | 7.87 | 24.52 | 0.871 | 5.225 |
| 33 | 47.2 | 3 23 | 6 | 11.96 | 24.80 | | |
| 34 | 49.1 | 3 30 | 6 | 16.26 | 25.08 | | |
| 35 | 50.9 | 3 38 | 6 | 20.78 | 25.39 | | |
| 36 | 52.7 | 3 46 | 6 | 25.53 | 25.70 | | |
| 37 | 54.7 | 3 55 | 6 | 30.52 | 26.03 | 0.870 | 5.221 |
| 38 | 56.8 | 4 4 | 6 | 35.76 | 26.38 | | |
| 39 | 58.8 | 4 13 | 6 | 41.27 | 26.75 | | |
| 40 | 60.9 | 4 22 | 6 | 47.06 | 27.14 | | |
| 41 | 63.1 | 4 31 | 6 | 53.15 | 27.54 | | |
| 42 | 65.4 | 4 41 | 6 | 59.56 | 27.97 | 0.869 | 5.216 |
| 43 | 67.7 | 4 51 | 7 | 6.29 | 28.42 | | |
| 44 | 70.1 | 5 1 | 7 | 13.39 | 28.89 | | |
| 45 | 72.6 | 5 12 | 7 | 20.86 | 29.39 | | |
| 46 | 75.2 | 5 23 | 7 | 28.74 | 29.92 | | |
| 47 | 77.8 | 5 34 | 7 | 37.04 | 30.47 | 0.869 | 5.211 |
| 48 | 80.6 | 5 46 | 7 | 45.80 | 31.05 | | |
| 49 | 83.5 | 5 59 | 7 | 55.05 | 31.67 | | |
| 50 | 86.4 | 6 12 | 8 | 4.83 | 32.32 | | |
| 51 | 89.6 | 6 25 | 8 | 15.17 | 33.03 | | |
| 52 | 92.8 | 6 39 | 8 | 26.13 | 33.74 | 0.868 | 5.207 |
| 53 | 96.2 | 6 54 | 8 | 37.75 | 34.52 | | |
| 54 | 99.8 | 7 9 | 8 | 50.07 | 35.34 | | |
| 55 | 103.5 | 7 25 | 9 | 3.18 | 36.22 | | |
| 56 | 107.5 | 7 42 | 9 | 17.12 | 37.14 | | |
| 57 | 111.6 | 8 0 | 9 | 31.97 | 38.13 | 0.867 | 5.202 |
| 58 | 116.0 | 8 19 | 9 | 47.83 | 39.19 | | |
| 59 | 120.6 | 8 38 | 10 | 4.78 | 40.32 | | |
| 60 | 125.5 | 8 59 | 10 | 22.94 | 41.52 | | |
| 61 | 130.8 | 9 22 | 10 | 42.42 | 42.83 | | |
| 62 | 136.3 | 9 46 | 11 | 3.38 | 44.22 | 0.866 | 5.198 |
| 63 | 142.2 | 10 11 | 11 | 25.97 | 45.73 | | |
| 64 | 148.6 | 10 38 | 11 | 50.37 | 47.36 | | |
| 65 | 155.0 | 11 8 | 12 | 16.82 | 49.12 | | |
| 66 | 162.8 | 11 39 | 12 | 45.55 | 51.04 | | |
| 67 | 170.7 | 12 13 | 13 | 16.88 | 53.12 | 0.866 | 5.195 |
| 68 | 179.3 | 12 51 | 13 | 51.15 | 55.41 | | |
| 69 | 188.7 | 13 31 | 14 | 28.77 | 57.92 | | |
| 70 | 199.1 | 14 15 | 15 | 10.26 | 60.68 | 0.866 | 5.193 |

## Table VI   Azimuths of the Secant
### (U.S. Bureau of Land Management)

| Lat. | o mi | 1 mi | 2 mi | 3 mi | Deflection angle 6 mi. |
|---|---|---|---|---|---|
| ° | ° ′ | ° ′ | ° ′ | | ′ ″ |
| 25 | 89  58.8 | 89   59.2 | 89   59.6 | 90° | 2   25 |
| 26 | 58.7 | 59.2 | 59.6 | E or W. | 2   32 |
| 27 | 58.7 | 59.1 | 59.6 | " " " | 2   39 |
| 28 | 58.6 | 59.1 | 59.5 | " " " | 2   46 |
| 29 | 58.6 | 59.0 | 59.5 | " " " | 2   53 |
| 30 | 58.5 | 59.0 | 59.5 | " " " | 3    0 |
| 31 | 58.4 | 59.0 | 59.5 | " " " | 3    7 |
| 32 | 58.4 | 58.9 | 59.5 | " " " | 3   15 |
| 33 | 58.3 | 58.9 | 59.4 | " " " | 3   23 |
| 34 | 58.2 | 58.8 | 59.4 | " " " | 3   30 |
| 35 | 58.2 | 58.8 | 59.4 | " " " | 3   38 |
| 36 | 58.1 | 58.7 | 59.4 | " " " | 3   46 |
| 37 | 58.0 | 58.7 | 59.3 | " " " | 3   55 |
| 38 | 58.0 | 58.6 | 59.3 | " " " | 4    4 |
| 39 | 57.9 | 58.6 | 59.3 | " " " | 4   13 |
| 40 | 57.8 | 58.5 | 59.3 | " " " | 4   22 |
| 41 | 57.7 | 58.5 | 59.2 | " " " | 4   31 |
| 42 | 57.7 | 58.4 | 59.2 | " " " | 4   41 |
| 43 | 57.6 | 58.4 | 59.2 | " " " | 4   51 |
| 44 | 57.5 | 58.3 | 59.2 | " " " | 5    1 |
| 45 | 57.4 | 58.3 | 59.1 | " " " | 5   12 |
| 46 | 57.3 | 58.2 | 59.1 | " " " | 5   23 |
| 47 | 57.2 | 58.1 | 59.1 | " " " | 5   34 |
| 48 | 57.1 | 58.1 | 59.0 | " " " | 5   46 |
| 49 | 57.0 | 58.0 | 59.0 | " " " | 5   59 |
| 50 | 56.9 | 57.9 | 59.0 | " " " | 6   12 |
| 51 | 56.8 | 57.9 | 58.9 | " " " | 6   25 |
| 52 | 56.7 | 57.8 | 58.9 | " " " | 6   39 |
| 53 | 56.6 | 57.7 | 58.8 | " " " | 6   54 |
| 54 | 56.4 | 57.6 | 58.8 | " " " | 7    9 |
| 55 | 56.3 | 57.5 | 58.8 | " " " | 7   25 |
| 56 | 56.2 | 57.4 | 58.7 | " " " | 7   42 |
| 57 | 56.0 | 57.3 | 58.7 | " " " | 8    0 |
| 58 | 55.8 | 57.2 | 58.6 | " " " | 8   19 |
| 59 | 55.7 | 57.1 | 58.6 | " " " | 8   38 |
| 60 | 55.5 | 57.0 | 58.5 | " " " | 8   59 |
| 61 | 55.3 | 56.9 | 58.4 | " " " | 9   22 |
| 62 | 55.1 | 56.7 | 58.4 | " " " | 9   46 |
| 63 | 54.9 | 56.6 | 58.3 | " " " | 10   11 |
| 64 | 54.7 | 56.5 | 58.2 | " " " | 10   38 |
| 65 | 54.4 | 56.3 | 58.1 | " " " | 11    8 |
| 66 | 54.2 | 56.1 | 58.1 | " " " | 11   39 |
| 67 | 53.9 | 55.9 | 58.0 | " " " | 12   13 |
| 68 | 53.6 | 55.7 | 57.9 | " " " | 12   51 |
| 69 | 53.2 | 55.5 | 57.8 | " " " | 13   31 |
| 70 | 89° 52′.9 | 89° 55′.3 | 89° 57′.6 | " " " | 14′  15″ |
| | 6 mi | 5 mi | 4 mi | 3 mi | |

**Table VII   Offsets (In Links) from the Secant to the Parallel**
(*U.S. Bureau of Land Management*)

| Lat. | o mi | ½ mi | 1 mi | 1½ mi | 2 mi | 2½ mi | 3 mi |
|------|------|------|------|-------|------|-------|------|
| ° | | | | | | | |
| 25 | 2 N. | 1 N. | 0 | 1 S. | 1 S. | 2 S. | 2 S. |
| 26 | 2 | 1 | 0 | 1 | 1 | 2 | 2 |
| 27 | 3 | 1 | 0 | 1 | 2 | 2 | 2 |
| 28 | 3 | 1 | 0 | 1 | 2 | 2 | 2 |
| 29 | 3 | 1 | 0 | 1 | 2 | 2 | 2 |
| 30 | 3 | 1 | 0 | 1 | 2 | 2 | 2 |
| 31 | 3 | 1 | 0 | 1 | 2 | 2 | 2 |
| 32 | 3 | 1 | 0 | 1 | 2 | 2 | 3 |
| 33 | 3 | 1 | 0 | 1 | 2 | 2 | 3 |
| 34 | 3 | 2 | 0 | 1 | 2 | 3 | 3 |
| 35 | 4 | 2 | 0 | 1 | 2 | 3 | 3 |
| 36 | 4 | 2 | 0 | 1 | 2 | 3 | 3 |
| 37 | 4 | 2 | 0 | 1 | 2 | 3 | 3 |
| 38 | 4 | 2 | 0 | 1 | 2 | 3 | 3 |
| 39 | 4 | 2 | 0 | 1 | 2 | 3 | 3 |
| 40 | 4 | 2 | 0 | 1 | 3 | 3 | 3 |
| 41 | 4 | 2 | 0 | 2 | 3 | 3 | 4 |
| 42 | 5 | 2 | 0 | 2 | 3 | 3 | 4 |
| 43 | 5 | 2 | 0 | 2 | 3 | 4 | 4 |
| 44 | 5 | 2 | 0 | 2 | 3 | 4 | 4 |
| 45 | 5 | 2 | 0 | 2 | 3 | 4 | 4 |
| 46 | 5 | 2 | 0 | 2 | 3 | 4 | 4 |
| 47 | 5 | 2 | 0 | 2 | 3 | 4 | 4 |
| 48 | 6 | 3 | 0 | 2 | 3 | 4 | 4 |
| 49 | 6 | 3 | 0 | 2 | 3 | 4 | 5 |
| 50 | 6 | 3 | 0 | 2 | 4 | 4 | 5 |
| 51 | 6 | 3 | 0 | 2 | 4 | 5 | 5 |
| 52 | 6 | 3 | 0 | 2 | 4 | 5 | 5 |
| 53 | 7 | 3 | 0 | 2 | 4 | 5 | 6 |
| 54 | 7 | 3 | 0 | 2 | 4 | 5 | 6 |
| 55 | 7 | 3 | 0 | 3 | 4 | 5 | 6 |
| 56 | 7 | 3 | 0 | 3 | 4 | 6 | 6 |
| 57 | 8 | 3 | 0 | 3 | 5 | 6 | 6 |
| 58 | 8 | 4 | 0 | 3 | 5 | 6 | 6 |
| 59 | 8 | 4 | 0 | 3 | 5 | 6 | 7 |
| 60 | 9 | 4 | 0 | 3 | 5 | 7 | 7 |
| 61 | 9 | 4 | 0 | 3 | 5 | 7 | 7 |
| 62 | 9 | 4 | 0 | 3 | 6 | 7 | 8 |
| 63 | 10 | 4 | 0 | 3 | 6 | 7 | 8 |
| 64 | 10 | 5 | 0 | 4 | 6 | 8 | 8 |
| 65 | 11 | 5 | 0 | 4 | 6 | 8 | 9 |
| 66 | 11 | 5 | 0 | 4 | 7 | 8 | 9 |
| 67 | 12 | 5 | 0 | 4 | 7 | 9 | 9 |
| 68 | 12 | 6 | 0 | 4 | 7 | 9 | 10 |
| 69 | 13 | 6 | 0 | 5 | 8 | 10 | 10 |
| 70 | 14 N. | 6 N. | 0 | 5 S. | 8 S. | 10 S. | 11 S. |
| | 6 mi | 5½ mi | 5 mi | 4½ mi | 4 mi | 3½ mi | 3 mi |

# Index

Hard surface, heavy duty road, four or more lanes ..........................

Hard surface, heavy duty road, two or three lanes ..........................

Hard surface, medium duty road, four or more lanes ..........................

Hard surface, medium duty road, two or three lanes ..........................

Improved light duty road ..........................................................

Unimproved dirt road—Trail ......................................................

Dual highway, dividing strip 25 feet or less ...............................

Dual highway, dividing strip exceeding 25 feet ...........................

Road under construction ...........................................................

Railroad: single track—multiple track ......................................

Railroads in juxtaposition ........................................................

Narrow gage: single track—multiple track ...............................

Railroad in street—Carline ......................................................

Bridge: road—railroad .............................................................

Drawbridge: road— railroad ....................................................

Footbridge ..............................................................................

Tunnel: road—railroad .............................................................

Overpass—Underpass ..............................................................

Important small masonry or earth dam .....................................

Dam with lock .........................................................................

Dam with road ........................................................................

Canal with lock .......................................................................

**Symbols for roads, railroads, and dams.**

Buildings (dwelling, place of employment, etc.)..........

School—Church—Cemeteries..........

Buildings (barn, warehouse, etc.)..........

Power transmission line..........

Telephone line, pipeline, etc. (labeled as to type)..........

Wells other than water (labeled as to type)..........oOil..........oGas

Tanks; oil, water, etc. (labeled as to type)..........• • ● Water

Located or landmark object—Windmill..........

Open pit, mine, or quarry—Prospect..........x

Shaft—Tunnel entrance..........

Horizontal and vertical control station:

    tablet, spirit level elevation..........BM △ 3899

    other recoverable mark, spirit level elevation..........△ 3938

Horizontal control station: tablet, vertical angle elevation..........VABM △ 2914

    any recoverable mark, vertical angle or checked elevation....△ 5675

Vertical control station: tablet, spirit level elevation..........BM × 945

    other recoverable mark, spirit level elevation..........× 890

Checked spot elevation..........× 5923

Unchecked spot elevation—Water elevation..........× 5657 ..........870

**Symbols for structures and stations.**

| | |
|---|---|
| Perennial streams.......... | Intermittent streams.... |
| Elevated aqueduct.......... | Aqueduct tunnel.......... |
| Water well—Spring.......... | Disappearing stream.......... |
| Small rapids.......... | Small falls.......... |
| Large rapids.......... | Large falls.......... |
| Intermittent lake.......... | Dry lake.......... |
| Foreshore flat.......... | Rock or coral reef.......... |
| Sounding—Depth curve.... 10 | Piling or dolphin.......... |
| Exposed wreck.......... | Sunken wreck.......... |
| Rock, bare or awash—dangerous to navigation.......... | |

| | |
|---|---|
| Marsh (swamp).......... | Submerged marsh.......... |
| Wooded marsh.......... | Mangrove.......... |
| Woods or brushwood.......... | Orchard.......... |
| Vineyard.......... | Scrub.......... |
| Inundation area.......... | House omission area.......... |

**Symbols for hydrography and land classification.**